U.S. AMPHIBIOUS SHIPS AND CRAFT

U.S. AMPHIBIOUS SHIPS AND CRAFT

AN ILLUSTRATED DESIGN HISTORY

By Norman Friedman

Ship Plans by A. D. Baker III and Norman Friedman

Naval Institute Press
Annapolis, Maryland

Naval Institute Press
291 Wood Road
Annapolis, MD 21402

© 2002 by Norman Friedman

All rights reserved. No part of this book may be reproduced or utilized in any form or by any means, electronic or mechanical, including photocopying and recording, or by any information storage and retrieval system, without permission in writing from the publisher.

Library of Congress Cataloging-in-Publication Data
Friedman, Norman, 1946-
    U.S. amphibious ships and craft : an illustrated design history / Norman Friedman.
      p. cm.
    Includes bibliographical references and index.
    ISBN 1-55750-250-1 (alk. paper)
    1. Amphibious assault ships—United States—History. 2. Landing craft—United States—History. 3. Amphibious warfare—History—20th century. I. Title.
V895 .F75 2001
623.8'25—dc21

                                               2001044914

Printed in the United States of America on acid-free paper ∞
09 08 07 06 05 04 03 02 9 8 7 6 5 4 3 2
First printing

In memory of Dr. Reuven Leopold (1938–2000),
a great friend of the U.S. Navy,
who designed the *Tarawa, Spruance,* and FDL classes.

Contents

List of Tables   ix

Acknowledgments   xi

List of Abbreviations   xiii

Part One. Developing Amphibious Warfare

1   Introduction   3

2   Developing the Techniques   17

3   Transports   23

4   Landing Craft   67

5   The British Connection   103

Part Two. World War II Developments

6   Attack Transports and Conversions   151

7   Increasing Demand for Landing Craft   207

8   Fire Support   223

9   Command and Control   261

Part Three. Postwar Developments

10   Perfecting the Technique   287

11   The Fast Submarine and the Fast Amphibious Force   311

12   The Bomb and Vertical Envelopment   347

13   New Landing Craft   383

Part Four. The Sixties and Beyond

14   Fire Support Revisited   403

15   A New-Generation Amphibious Flagship   425

16   Over the Horizon Assault   433

17   Preparing for the Future   457

Appendixes

A   Maritime Prepositioning and Sealift   467

B   Landing Craft/Ship Programs   481

C   Amphibious Ships   495

Notes   621

Bibliography   645

Index   653

Tables

3-1. *Henderson* vs. Standard Troop Transport Designs, 1914–20   26

3-2. Maritime Commission Hull Specifications   33

3-3. Attack Transports Converted Pre–World War II   44

3-4. Attack Cargo Ships Converted Pre–World War II   53

4-1. Small Landing Craft, 1937–42   80

4-2. Tank Landing Craft, 1938–42   92

5-1. British-Inspired Amphibious Ships and Craft   104

6-1. War-Built Large Attack Transports   168

6-2. Wartime Small Attack Transports   175

6-3. Wartime Attack Cargo Ships   176

6-4. Wartime Quasi-Attack Transports   187

6-5. Postwar Plans for Amphibious Ships and Craft   205

7-1. Wartime Amphibian Development   218

8-1. Fire Support Ships and Craft   227

9-1. Amphibious Flagships   273

9-2. Control Landing Craft   278

10-1. Estimates of Amphibious Requirements, 1948   288

10-2. New Landing Craft   291

10-3. Post–World War II LVTs   301

10-4. New LSTs   305

11-1. The 20-kt Amphibious Force   312

12-1. Helicopter Carrier Alternatives, April 1955   351

12-2. LPH Alternatives, 8 August 1956   356

12-3. Ships for Vertical Envelopment   361

12-4. LHA Design Alternatives, 1965–66   372

13-1. Post–World War II LCU/LCAC Development   390

16-1. Reductions in AFOE Amphibious Footprint, 1981–83   447

A-1. Time to Reach Crisis Location, 1965   471

A-2. Prospective MPS Ships, 1980   475

B-1. LVTs Completed during World War II   492

Acknowledgments

No book like this one can be written without extensive help. As in past books in this series, I benefited enormously from the assistance of the old NAVSEA preliminary design branch. In a sense this book and others in the series constitute a history of that branch. As always in this series, A. D. Baker III, editor of *Combat Fleets*, provided valuable advice and assistance with photographs as well as his superb drawings. Steve Roberts generously provided information on the massive U.S. merchant shipbuilding programs of the two world wars, and on World War II auxiliary programs, which included attack transports and attack cargo ships. Christopher C. Wright, editor of *Warship International*, provided valuable input. All three read and commented on early versions of the manuscript. Chuck Haberlein of the photographic section of the Naval Historical Center was extremely helpful. Mark Wertheimer of the Naval Historical Center provided invaluable assistance, particularly with material concerning underwater demolition. I am grateful to the staff of the U.S. Navy Operational Archives, present and past, including Cal Cavalcante, Kathy Lloyd, Ken Johnson, and John Walker. Among other things, I used the U.S. Navy General Board files while they were at the Operational Archives; they are now located at the National Archives in Washington, D.C. I benefited greatly from the assistance of Barry Zerby at the National Archives at College Park, Maryland, for post-1940 records, and Rebecca Livingstone and Rick Peuser at the National Archives in Washington, D.C., for pre-1940 records, including those of the Marine Corps and the Maritime Commission. Dr. Evelyn Cherpak provided vital access to the archives of the Naval War College. Some of the material on small landing craft was provided by the NAVSEA small boat section; it was gathered at the time I wrote the design history of U.S. small combatants. For some more recent material, I am grateful to William B. Darden of the Naval Surface Warfare Center (Norfolk detachment, concerned with small combatant craft). This book required unusually (for this series) heavy use of the files of the ships histories division of the Naval Historical Center, and I am particularly grateful to its long-time head John Reilly, who recently retired, and to Reilly's successor, Ray Mann. The staff of the Marine Corps Historical Center at the Washington Navy Yard provided useful material, particularly on maritime prepositioning and on postwar amphibious vehicles. Christopher C. Cavas, editor of *Navy Times*, provided me with a copy of his database on U.S. warship chronology, including fates. Charles Dragonette provided important merchant ship information. Steve Zaloga provided data on amphibious vehicles. Dr. Thomas Hone was also helpful.

Booklets of plans of modern U.S. amphibious ships were provided by Thomas L. Olsen of the Puget Sound Detachment in Boston and by D. L. Caskey of NAVSEA headquarters. For some plans of older ships, I am indebted to the Cartographic Service of the National Archives at College Park. The National Archives also provided some key photographs. Larry Ferreiro of the American Bureau of Shipping helped unearth some important plans of wartime transports. I am also grateful to Thomas Walkowiak of The Floating Drydock, who supplied some crucial plans on short notice. Paige Eaton of Textron provided valuable information on the LCAC SLEP program and on Textron's LCU(R) proposal. Jennifer Till and Dawn Stitzel of the U.S. Naval Institute staff patiently and effectively ran down several vital but obscure photographs.

Because U.S. amphibious ships and craft are intimately related to those developed by and for the Royal Navy, this book is unusual in this series in incorporating much British material. For access to that material I am grateful to the staff of the Public Record Office at Kew, and also to the staff of the Brass Foundry branch of the National Maritime Museum at Woolwich

Arsenal. Capt. Christopher Page, RN, head of the Royal Naval Historical Branch, arranged access to some crucial World War II material.

I greatly appreciate the considerable assistance I received in the preparation of this book, including the detection of some errors. I alone am responsible for any errors remaining. Except as otherwise noted, photographs in this book are courtesy of the U.S. Navy. Throughout the lengthy development of this book, my loving wife Rhea supported and strongly encouraged my efforts. Without her I could not have written it.

Abbreviations

AAAV	advanced amphibious assault vehicle
AAFSF	amphibious assault fuel supply facility
AALC	advanced assault landing craft
AAV	amphibious assault vehicle
AAW	antiaircraft warfare
ABFS	amphibious bulk fuel system
ABS	American Bureau of Shipping
ABSD	drydock
AC	alternate current
ACE	air component element
ACU	assault craft unit
ACV	air cushion vehicle
AD	destroyer tender
Adm.	Admiral
ADO	advanced development objective
AE	ammunition ship
AE	assault element
AEW	airborne early warning
AF	stores ship
AFOE	assault follow-on element
AFS	combat stores ships
AG	general-purpose auxiliary
AGC	command ship
AGF	fleet flagship
AGP	torpedo boat tender
AGS	survey ship
AH	hospital ship
AK	cargo ship; assault versions redesignated AKA in 1943
AKA	attack cargo ship; 1943–69 designation; redesignated LKA in 1969
AKN	net carrier and minecraft mother ship
AKS	stores issue ship
AKV	aircraft transport
ALC	assault landing craft (British)
ALPS	amphibious landing platform slipway
AMC(U), AMCU	coastal minehunter
AMFIST	amphibious fire support study
AN	netlayer
AOE	one-stop replenishment ship
AOG	gasoline tanker
AP	transport ship; assault versions redesignated APA in 1943
APA	attack transport ship; 1943–69 designation; redesignated LPA in 1969
APB	unpowered artillery lighter; pre–World War II designation
APB	barracks ship
APD	transport destroyer
APF	administrative flagship
APH	casualty evacuation ship
APL	unpowered barracks ship
APL	Applied Physics Laboratory of Johns Hopkins University
APM	mechanized transport; redesignated LSD in July 1942
APP	designation applied to specially built personnel landing craft through 1940
APSRON	afloat prepositioning ship squadron
APSS	transport submarine
APY	special troop transport (short-radius); redesignated LCI(L) in July 1942
AR	repair ship
ARB	battle damage repair ship
ARC	cable layer ship
ARG	amphibious ready group
ARL	landing craft repair ship
ARSD	diving tender
ARST	salvage tender
ARV	aviation repair ship
ARV(A)	aircraft airframe repair ship
ARV(E)	aircraft engine repair ship

ASCAC	carrier-type antisubmarine warfare center	CIC	combat information center
ASDV	tender for swimmer delivery vehicle	CinCLant	Commander in Chief, Atlantic
		CinCPac	Commander in Chief, Pacific
ASROC	antisubmarine rocket	CinCSouth	Commander in Chief, South
ASW	antisubmarine warfare	CIWS	close-in weapon system
ATACMS	army tactical missile system	CM	minelayer
ATF	amphibious task force	CMC	Commandant Marine Corps
ATU	amphibious task unit	CNA	Center for Naval Analyses
AV	seaplane tender	CNO	Chief of Naval Operations
AVB	aviation base ship	CO	commanding officer
avgas	aviation gas	COEA	cost and operational effectiveness analysis
AVP	small seaplane tender	COHQ	Combined Operations Headquarters (British)
AVR	amphibian vehicle requirement		
AVS	aviation supply ship	COIN	counterinsurgency
AW	water tanker	Col.	Colonel
AZ	airship tender	Comdr.	Commander
B&W	Babcock & Wilcox	Cominch	Commander in Chief, U.S. Fleet (also Office of)
BAPM	mechanized transport for Royal Navy (British); redesignated LSD in July 1942		
		CONFORM	concept formulation
		COTS	container offloading and transfer system; note that COTS is currently used to designate commercial off the shelf
BARC	barge, amphibious resupply, cargo		
BAT	naval tug for Royal Navy (British)		
		CP	controllable pitch
BB	battleship	CPO	chief petty officer
BHP	brake horsepower	CPS	collective protective system
BJU	beach jumper unit	CRP	controllable reversible pitch (propeller)
BLT	battalion landing team		
BP	between perpendiculars (refers to waterline at bow to rudder post)	CSNP	causeway section non-powered
		CSP	causeway section powered
		cu ft	cubic foot/feet
BPC	beach protection craft (British)	CURV	cable-controlled underwater recovery vehicle
BPDMS	Basic Point Defense Missile System (the Sea Sparrow)		
		CV	aircraft carrier
Brig. Gen.	Brigadier General	CVE	escort carrier
BuAer	Bureau of Aeronautics, created in 1921	CVE(T)	escort carrier transport
		CVHA	helicopter assault carrier
BuEng	Bureau of Engineering	CVHE	helicopter escort carrier
BuOrd	Bureau of Ordnance	CVL	light carrier
BuShips	Bureau of Ships, created in 1940 when C&R and BuEng merged	CVS	support (ASW) carrier
		DASH	drone antisubmarine helicopter
BuWep	Bureau of Weapons (formerly BuOrd)	DC	direct current
		DCNO	deputy chief of naval operations
C&R	Bureau of Construction and Repair	DD	destroyer
		DDAP	destroyer transport
CA	heavy cruiser (8-in guns)	DDG	missile destroyer
CAB	captured air bubble	DE	destroyer escort
Capt.	Captain	DEC	destroyer escort, control
CC	command ship	DLG	missile frigate (large destroyer)
CE	Combustion Engineering	DLGN	missile cruiser
CEP	circular error probable (50% of rounds fall within)	DNC	director of naval construction (British)
		DP	dual purpose
CF/CD	concept formulation/contract definition	DPM	draft presidential memorandum

DUKW	amphibious (dual-drive, wheels and propellers) version of the army's standard 2½-ton truck	hr	hour(s)
		HSC	heavy support craft (British)
		HTS	high-tensile steel
DWT	deadweight ton(s)		
DX/DXG	new destroyer; became the *Spruance*- and *Kidd*-class destroyers	IAS	interdiction assault ship
		IFF	identification friend or foe
		IFS	inshore fire support
		IHP	indicated horsepower (for reciprocating steam engine)
ECM	electronic countermeasures		
ELCAS	elevated causeway	ILS	infantry landing ship (British)
ELINT	electronic intelligence	in	inch(es)
ENTPS	enhanced near term prepositioning squadron ship	INSURV	Board of Inspection and Survey
		IR	infrared
		ISTDC	Interservice Training and Development Centre (British)
EUSC	Effective U.S. Control Fleet		
		IX	unclassified vessel
FAM	fast aerial mine (British)		
FDL	fast deployment logistics	JCC	joint craft, control
FDT	fighter direction tender	JCS	Joint Chiefs of Staff
FF	frigate	JPTDS	Junior Participating Tactical Data System
FFD	forward floating depot		
FIE	fly-in element		
FLEX	fleet landing exercise	kg	kilogram(s)
flo-flo	float-on, float-off	km	kilometer(s)
FMC	Food Machinery Corporation	kt	knot(s)
FMF	Fleet Marine Force, a permanent USMC force formed in 1933 as an integral part of the U.S. Navy fleet	kW	kilowatt(s)
		LAMPS	light airborne multi-purpose system
FRAM	Fleet Rehabilitation and Modernization; FRAM I and FRAM II were 1960s programs	LantPhibex	Atlantic Fleet amphibious exercise
		LARC	lighter, amphibious resupply, cargo
ft	foot/feet	LASH	lighter aboard ship
FW	Foster Wheeler	LAV	light armored vehicle
FY	fiscal year	lb	pound(s)
		LBP	landing boat, personnel
gal	gallon(s)	LBS	landing boat, support; redesignated LCT(S) in July 1942
GCI	ground-controlled intercept		
GEM	ground effect machine	LBV	landing boat, vehicles
GHz	gigahertz (equal to one billion hertz)	LC(FF), LCFF	landing craft, flotilla flagship
		LC(X)	landing craft, experimental; late 1990s
GL	grenade launcher		
GM	metacentric height	LCA	landing craft, assault (British)
GPS	Global Positioning Satellite	LCA	landing craft, assault
GRC	giant raiding craft (British); became LCI(L)	LCAC	landing craft, air cushion
		LCC	landing craft, control; later used for amphibious flagship (ex-AGC)
GRP	glass-reinforced plastic		
GZ	maximum righting arm (a measure of stability)	LCF	landing craft, flak
		LCF(L)	landing craft, flak, large; redesignation of British BPC
HE	high explosive		
HERO	hazards of electromagnetic radiation to ordnance	LCG	landing craft, gun
		LCG(L)	landing craft, gun, large
HF	high frequency	LCH	landing craft headquarters (British)
HMS	His/Her Majesty's Ship (British)		
HP	horsepower		
HPI	high performance improvement	LCI	landing craft, infantry

LCI(G), LCIG	landing craft, infantry, gunboat	LCVP(T)	landing craft, vehicles and personnel, planing; later T designation used for twin
LCI(L), LCIL	landing craft, infantry, large		
LCI(M), LCIM	landing craft, infantry, mortar		
LCI(R), LCIR	landing craft, infantry, rocket	LF	low frequency (radio or sonar)
LCI(S), LCIS	landing craft, infantry, small	LFS	landing ship, fire support
LCM	landing craft, mechanized; number appearing in parentheses indicates type, e.g., LCM(1), LCM(2), etc.	LFSR	landing ship, fire support, rocket
		LFSW	landing ship fire support, weapon
		LHA	landing ship, helicopter, assault
LCM(K)	landing craft, mechanized, hydrokeel	LHD	landing ship, helicopter, dock
		LKA	cargo ship, attack; 1969 redesignation of AKA
LCM(R)	landing craft, mechanized, rocket		
		LLC	light lift craft
LCP(L), LCPL	landing craft, personnel, large	LOGLAND	logistic support of land forces (study)
LCP(M)	landing craft, personnel, medium		
		LPA	transport ship, attack; 1969 redesignation of APA
LCP(N)	landing craft, personnel, nested		
LCP(R), LCPR	landing craft, personnel, ramped	LPD	landing ship, personnel, dock
		LPH	landing ship, personnel, helicopter
LCP (Sy)	landing craft survey (British)		
LCQ	landing craft administrative flagship (British)	LPR	amphibious transport, personnel, raiding and reconnaissance
LCR(L)	landing craft, rubber, large	LPV	landing ship, personnel, fixed-wing aircraft (as opposed to LPH)
LCR(S)	landing craft, rubber, small		
LCS(L)	landing craft, support, large		
LCS(M)	landing craft, support, medium (British)	LRG	long-range gun
		LRGS	long-range gun system
LCS(S)	landing craft, support, small	LRO	long-range objective
LCSR	landing craft, swimmer reconnaissance	LRR	long-range requirement
		LSC	light support craft (British)
LCSR(K)	landing craft, swimmer reconnaissance, hyrokeel	LSD	landing ship, dock
		LSE	landing ship, engineer (British designation for ARL)
LCSR(L)	landing craft, swimmer reconnaissance, large		
		LSFF	landing ship, flotilla flagship; ex-LCFF
LCT	landing craft, tank; number appearing in parentheses indicates type, e.g., LCT(1), LCT(2), etc.		
		LSFH	landing ship, fire support, heavy
		LSH(L)	landing ship headquarters, large (British)
LCT(A)	landing craft, tank, armor (tank fire support)	LSI	landing ship, infantry (British)
		LSI(L)	landing ship, infantry, large (British)
LCT(R)	landing craft, tank, rocket		
LCT(SP)	landing craft, tank, self-propelled artillery	LSIL	landing ship, infantry, large; 1945 U.S. redesignation of LCI(L)
LCT(U)	landing craft, tank, utility		
LCU	landing craft, utility; previous designations were LCT, then LSU	LSIM	landing ship, infantry, mortar; redesignation of LCI(M) after World War II
LCV	landing craft, vehicles	LSM	landing ship, medium
LCVP	landing craft, vehicles and personnel	LSM(R), LSMR	landing ship, medium, rocket
		LSSL	landing ship, support, large; redesignation of LCS(L)(3) after World War II
LCVP(H)	landing craft, vehicles and personnel, hydrojet; later H designation used for hydrofoil		
		LST	landing ship, tank; number appearing in parentheses indicates type, e.g., LST(1), LST(2), etc.
LCVP(K)	landing craft, vehicles and personnel, hydrokeel		

LST(H)	landing ship, tank, hospital	MCS	mine countermeasures mother ship
LST(M)	landing ship, tank, mother ship	MEB	marine expeditionary brigade
LSTS	landing ship, tank, small; LCT designation used briefly	MEF	marine expeditionary force
LSU	landing ship, utility; ex-LCT, later LCU	MENS	mission need statement
		MEU	marine expeditionary unit
LSV	landing ship, vehicle	MF	medium frequency
Lt. Col.	Lieutenant Colonel	MG	machine gun
Lt. Gen.	Lieutenant General	MHz	megahertz (equal to one million hertz)
LVA	landing vehicle, assault		
LVT	landing vehicle, tracked; commonly referred to as the amtrac; number appearing in parentheses indicates type, e.g., LVT(1), LVT(2), etc.	MIT	Massachusetts Institute of Technology
		Mk	mark (number—ordnance, ship designation)
		MLC	mechanized landing craft (British)
LVT(A)	landing vehicle, tracked, armored	MLRS	multiple long-range rocket system
LVT(U)	landing vehicle, tracked, unarmored	mm	millimeter(s)
		MMROP	Marine mid-range objectives projection
LVTAA	landing vehicle, tracked, antiaircraft	MOB	mobile offshore base
LVTCR	landing vehicle, tracked, command-control	mogas	motor vehicle gasoline
		mph	miles per hour
LVTE	landing vehicle, tracked, combat engineer	MPS	maritime prepositioning squadron
LVTH6	landing vehicle, tracked; close-support version carrying a 105-mm howitzer	MRO	medium range objective
		MSC	Military Sealift Command
		MSTS	Military Sea Transportation Service
LVTP	landing vehicle, tracked, personnel		
LVTR	landing vehicle, tracked, recovery	MT	measurement ton(s) (40 cu ft equals 1 measurement ton)
LVTX	landing vehicle, tracked, experimental	MUTE	multiple unit for transmission elimination
LVW	landing vehicle, wheeled		
LX	landing ship, new or undefined design	NASSCO	National Steel and Shipbuilding Corporation
LXA	assault amphibious ship	NATO	North Atlantic Treaty Organization
LXS	support amphibious ship		
		NAVMAT	Naval Materiel Command
m	meter(s)	NAVORD	Naval Ordnance Systems Command
MAB	marine amphibious brigade		
MAF	marine amphibious force	NAVSEA	Naval Sea Systems Command
MAG	marine air group	NAVSEC	Naval Ship Engineering Center
MAGTF	marine air-ground task force	NAVSES	Naval Ship Engineering Station
Maj.	Major	NAVSHIPS	Naval Ship Systems Command
Maj. Gen.	Major General	NAVWAG	Naval Warfare Analysis Group
MARS	multiple army rocket system	NBC	nuclear-biological-chemical
MAU	marine amphibious unit	NCDU	naval combat demolition unit
MAW	marine air wing	NCO	noncommissioned officer
MCDEC	Marine Corps Development Center	NDRF	National Defense Reserve Fleet
		NGS	naval gunfire support
MCLWG	major caliber lightweight gun	NL	navy lightering
MCMOBE	Marine Corps mobility enhancement	NLFED	Naval Landing Force Equipment Depot

nm	nautical mile(s) (2 kiloyards equal about 1 nautical mile)		
NRDL	Naval Radiation Defense Laboratory		
NSDM	national security decision memorandum		
NSFS	naval surface fire support		
NTDS	naval tactical data system		
NTPF	near-term prepositioning force		
NTPS	near-term prepositioning squadron		
NUGM	naval universal gun mount		
OA	overall		
OEG	operational evaluation group		
OFS	offshore fire support		
OMFTS	operational maneuver from the sea		
ONI	Office of Naval Intelligence		
ONR	Office of Naval Research		
Op-xx, Op-xxx	element of the Office of the Chief of Naval Operations (OpNav). Note: Elements with a single number, such as Op-03, are the main divisions. Additional digits indicate subdivisions, and the meanings of these numbers changed over time. Later all the "Op" codes were replaced by "N" codes that, in theory, match the codes of the other services. The significance of all these codes changed over time. For example, at the end of the 1940s Op-03 was deputy chief of naval operations (DCNO) for operations; Op-04 was DCNO for logistics, which included ship procurement; and Op-05 was DCNO for air. There was no Op-09, but the 90-codes were for a series of assistant chiefs of naval operations (ACNOs). Thus branches of Op-03 were responsible for, among other things, amphibious and mine warfare. Op-04 included the secretariat of the Ship Characteristics Board. About 1971 OpNav was reorganized around DCNOs for warfare areas: Op-02 for undersea warfare (essentially submarines), Op-03 for surface warfare, and Op-05 for air warfare. Op-03 also included a branch for amphibious and mine warfare, but on a very different basis from that of the past. Thus at various times Op-343 and Op-37 (or Op-037) were responsible for amphibious warfare and, sometimes, mine warfare. Op-093 (or Op-93, the two are interchangeable) was responsible for long-range objectives studies through the late 1950s and the 1960s.		
		OPDS	offshore petroleum discharge system
		OpNav	Office of the Chief of Naval Operations
		OpTevFor	Operational Test and Evaluation Force
		OR	operational requirement
		OSD	Office of the Secretary of Defense
		OSS	Office of Strategic Services
		OTH	over the horizon
		P&C	passenger and cargo
		PA&E	program assessment and evaluation
		PAC	parachute and cable; British World War II device used to snag low-flying aircraft
		PC	patrol craft; 173-ft subchaser
		PCC	patrol craft, control; 173-ft subchaser
		PCE	patrol craft, escort; 180-ft subchaser
		PCEC	patrol craft, escort, control
		PCS	patrol craft, small; 136-ft subchaser
		PF	frigate
		PGM	gunboat
		PGM(K)	gunboat, hydrokeel
		PGY	press boat
		PhibLant	Atlantic Fleet's amphibious command
		PhibPac	Pacific Fleet's amphibious command
		phibron	amphibious squadron
		PIP	product improvement program
		PLRS	position location reporting system
		PMS	program manager, ships
		PO	program objective
		POL	petroleum-oil-lubricants
		POM	program objectives memorandum

PPI	plan-position indicator (for radar)	SLEP	service life extension program
PSD	propulsion systems demonstrator	SLF	special landing force
		SLWT	side-loaded warping tug
psi	pounds per square inch	SOFAR	sound fixing and ranging
PT	motor torpedo boat	SOSUS	sound surveillance system
PTA	proposed technical approach	SRBOC	super-rapid blooming offboard chaff
		SRI	Stanford Research Institute
R&D	research and design	SS	Steam Ship; international designation for a merchant ship (sometimes MV, motor vessel, is used for a diesel-powered ship)
RA	7-man rubber boat		
RAF	Royal Air Force		
RAM	rolling airframe missile		
RAP	rocket-assisted projectile	SS	hull designation for submarine
RB	rubber boat	SSBN	nuclear-powered ballistic missile submarine
RCT	regimental combat team		
RDF	rapid deployment force	SSDS	ship self-defense system
RDMF	rapidly deployable medical facility	STOL	short take-off or landing
		STOM	ship-to-objective maneuver
Rear Adm.	Rear Admiral	STRAC	Strategic Command (U.S. Army)
RFP	request for proposal	STS	special treatment steel (armor)
RLT	regimental landing team		
RMHS	remote magnetic heading system	T-	prefix designation for MSTS/MSC ship
RN	Royal Navy		
ro-ro	roll-on roll-off	T-ACS, TACS	MSTS/MSC crane ship
rpm	revolutions per minute	T-AE	MSTS/MSC ammunition ship
RPV	remotely piloted vehicle	T-AGOS	MSTS/MSC sonar surveillance ship
RRF	Ready Reserve Fleet		
		T-AH	MSTS/MSC hospital ship
SACC	supporting arms coordination center	T-AKR	MSTS/MSC cargo ship, ro-ro version
SACEUR	Supreme Allied Commander, Europe	TAC	tank assault carrier; called Winette (British)
SADARM	sense and destroy armor munition	tac-log	tactical logistics
		TAKRX	MSC multi-purpose mobilization ship, SL-7 version
SAIP	Ship Acquisition and Improvement Panel	TAKX	MSC multi-purpose mobilization ship
SAL-GP	semi-active laser-guided projectile		
		TAVB	MSC aviation depot ship
SC	subchaser	TDP	total development program
SCAJAP	Shipping Control Administration Japan	TLC	tank landing craft (British)
		TLC-C	tank landing craft carrier (British)
SCB	Ship Characteristics Board		
SCC	110-ft subchaser, control	TLL	tank lighter, light tank capacity; alternate WL designation
SCIB	Ship Characteristics and Improvement Board		
		TLRO	tentative long-range objective
SDV	swimmer delivery vehicle	T-LSV	MSTS/MSC landing ship, vehicle
Seabee	naval construction battalion	TOR	tentative operational requirement
Seabee	sea barge		
SEAL	Sea-Air-Land Team	T-POM	tentative program objectives memorandum
SEF	sea-based expeditionary force		
SES	surface effect ship	transdiv	transport division
shipalts	ship alterations	TS	Higgins support boat; alternate LBS designation
Ships xx	NAVSHIPS element		
SHP	shaft horsepower	TSOR	tentative specific operational requirement
SLC	support landing craft (British)		

UDT	underwater demolition team	XAPG	fire support ship; pre–World War II conversion designation
UHF	ultra high frequency		
USAF	U.S. Air Force	XAPM	artillery transport, mechanized; pre–World War II conversion designation
USCGR	U.S. Coast Guard Reserves		
USMC	U.S. Marine Corps		
USNS	U.S. Naval Ship (MSTS/MSC prefix)	XAPN	artillery transport, nonmechanized; pre–World War II conversion designation
USS	U.S. Ship (U.S. warship)	XAPP	converted troop barges Type A; pre–World War II conversion designation
UV	ultraviolet		
VHF	very high frequency		
Vice Adm.	Vice Admiral	XAPT	converted troop barges Type B; pre–World War II conversion designation
VLC	vehicle landing craft		
VOPNAV	vice chief of the Office of the Chief of Naval Operations		
		XAPV	aircraft transports; pre–World War II conversion designation
vs.	versus		
VSS	VSTOL support ship	XAS	submarine tender; pre–World War II conversion designation
VSTOL	vertical or short take-off and landing		
		XCA	auxiliary cruiser; pre–World War II conversion designation
VTOL	vertical take-off and landing		
		XCM	minelayer; pre–World War II conversion designation
WL	waterline		
WL	tank lighter, light tank capacity; also sometimes TLL designation used; WLs and WMs were redesignated as LCM in July 1942	XCV	aircraft carrier; pre–World War II conversion designation
		YAGR	ocean radar picket
		yd	yard(s)
WM	tank lighter, medium tank capacity; WLs and WMs were redesignated as LCM in July 1942	YDT	diving tender
		YF	covered lighter
		YFB	ferry
		YFU	harbor utility craft
		YG	garbage lighter
WSA	War Shipping Administration	YMS	small minesweeper
		YRBM	accommodation barge
XAH	hospital ship; pre–World War II conversion designation	YTD	diving tender
		YTL	yard [harbor] tank lighter; redesignated LCT(5) in July 1942
XAP	expeditionary transport ship; pre–World War II conversion designation		
		YTL	small tug
		YV	drone control craft
XAPA	animal transport; pre–World War II conversion designation		
		ZA	Z-boat, armored; redesignated LVT in July 1942
XAPB	converted artillery barge; pre–World War II conversion designation		
XAPF	administrative flagship; pre–World War II conversion designation		

U.S. AMPHIBIOUS SHIPS AND CRAFT

Part One
Developing Amphibious Warfare

What makes modern amphibious forces special is their ability to move both personnel and heavy equipment over beaches anywhere in the world. Doing that requires considerable ingenuity. Here a fast *Newport*-class LST (landing ship, tank) opens her bow before extending a ramp down which tanks can drive. The LPD (landing ship, personnel, dock) in the background carries smaller landing craft.

1
Introduction

This book describes the development of the U.S. amphibious capability, spectacularly demonstrated during World War II and continuously developed since. Unlike other books in this series, this one involves not only the U.S. Navy but also the U.S. Marines and, to a lesser extent, the U.S. Army. Although navies have always been amphibious to some extent, the modern capability demonstrated in World War II is different in two ways. First, it brought ashore mechanized armies, which required a wide range of specialized landing craft. Second, it successfully confronted beach defenses armed with machine guns.

The U.S. Navy and the U.S. Marine Corps are the military arms of the Department of the Navy. Because of the historical U.S. insistence that responsibility ultimately be placed on civilians rather than on professional military staffs, they are unified at the political rather than at the staff level: both answer to the appointed civilian serving as the Secretary of the Navy. Thus the Marines' budget is part of the overall Navy Department budget. Whatever their effective independence, the Marines have always had to live under the thumb of the larger U.S. Navy service. Their forte, the amphibious operation, is but one of many ways in which seapower can be exercised, and by no means always the navy's favorite. The U.S. Army has periodically tried to eliminate the Marines, which it sees as a rival ground force. During the 1947 fight over service unification, the Marines feared that the navy would trade them away to the army in return for its support of their independent air arm. That did not happen, in part because the Marines capitalized on their great World War II amphibious success; the army disingenuously claimed that amphibious operations required no special expertise.

Modern U.S. amphibious development began roughly at the start of the twentieth century. At that time neither the navy nor the Marine Corps had a professional military staff, for example, for war planning, and Congress much preferred it that way. The professional chiefs of the U.S. Navy (the Admiral of the Navy) and of the Marine Corps (the Major General Commandant) both reported to the Secretary of the Navy. Independent of both were the bureaus that provided them with materiel. For this book the important ones were Construction and Repair (C&R), which designed, built, and repaired ships and lesser craft; Steam Engineering (later Engineering), responsible for ships' machinery; Ordnance; and Yards and Docks (responsible not only for permanent installations in the United States but also, it turned out, for advanced bases). In 1921 a new Bureau of Aeronautics (BuAer) was created; it would provide the Marines' new helicopters and other aircraft after World War II. In 1940 the two ship design organizations, C&R and the Bureau of Engineering, merged as the Bureau of Ships (BuShips).

The experience of the Spanish-American War showed that war plans could not easily be extemporized. It was already clear that it was difficult to coordinate the bureaus. In 1900, Secretary of the Navy John D. Long established a General Board of senior officers, headed by Admiral of the Navy George Dewey, the hero of Manila Bay, mainly for war planning. War plans in turn made it obvious just how important advanced bases or amphibious capability might be. Because so many decisions hinged on the way they would contribute to wartime performance, the General Board was often asked to advise the Secretary of the Navy. For example, in 1904 it was asked to devise a long-range building program.

The General Board worked closely with the Naval War College at Newport, Rhode Island. Annual classes at the college tested new tactical and strategic concepts on the gaming floor, and the War College is credited, for example, for developing the massed air attack concepts that shaped the role of U.S. aircraft carriers. The college also tested the island-hopping strategy the United States employed against Japan. Before the end of World War II, the

War College was the navy's closest equivalent to modern "think tanks."

Several prominent younger officers argued that U.S. warship designs were poorly coordinated and reflected the bureaus' innate conservatism rather than the service's real needs. After a conference at Newport in 1908, President Theodore Roosevelt made the General Board responsible for the overall design requirements, called "characteristics," of all new U.S. warships. Typically the board held hearings to determine just what was needed, and it often referred back to the demands of the war plans.

The pre-1914 navy's professional head was the Admiral of the Navy and the Bureau of Navigation was the operating staff. Because the Admiral of the Navy headed the General Board, he could use it as his planning staff, producing proposals that the Secretary of the Navy generally approved. This arrangement proved inadequate, and in 1912 Secretary of the Navy Truman H. Newberry was provided with a naval staff, comprised of two sections that were headed by an Aid for Operations and an Aid for Material. When Admiral Dewey died in 1914, the post of Admiral of the Navy lapsed. The following year the Aid for Operations became the Chief of Naval Operations (CNO), in effect Dewey's statutory successor. He was also deputy to the Secretary of the Navy. The CNO's staff was the Office of the Chief of Naval Operations (OpNav). Although in theory the Aid for Operations advised the Secretary of the Navy on Marine Corps matters, the CNO did not become head of the Marine Corps. The Marines retained their Major General Commandant position, which later was renamed Commandant Marine Corps (CMC).

There was no successor to the Aid for Material; the bureaus remained independent of the CNO, hence of the operating forces of the navy. Some became notorious for their indifference to the views of operating personnel. The failure of the Bureau of Ordnance to solve the torpedo problem of 1941–43 is famous. The prewar and early wartime fight over landing craft may have had a similar bureaucratic origin. During World War II, the CNO, Adm. Ernest J. King, tried to bring the bureaus under his direct control, but that attempt did not survive the end of the war.

Because OpNav ran the navy, it gained enormous authority during World War I, and at the end of the war it seemed that the General Board might be eclipsed altogether. OpNav took over important roles such as fleet training and war planning. By the end of the 1930s OpNav was preeminent. Moreover, because he was the Secretary of the Navy's deputy, the CNO became acting secretary in the secretary's absence, which was necessary when several of President Franklin D. Roosevelt's secretaries became incapacitated by illness. However, the General Board retained advisory power over ship characteristics and many other administrative questions. In 1941, for example, it held hearings on the appropriate size of the Marine Corps.

At the beginning of the twentieth century marines served on board U.S. warships. They manned guns and were available as the core of ad hoc landing parties. Many within the navy saw such roles as anachronistic. Some marines argued against abandoning their shipboard role. Others were intrigued by a new one. The United States lacked overseas bases, yet its basic policy was forward defense: the U.S. Navy was justified largely as a means of fighting wars as far as possible from the United States. A modern fleet could not operate very far from its base. The fleet would often have to create its advanced base—and defend that base. As the U.S. Navy's ground force, the Marines were the obvious candidates for this mission. The difference from past practice would be that substantial Marine units would be organized and maintained, rather than split up among many warships. In 1898 marines seized Guantanamo, a key forward base in the attack against Cuba. After the war Admiral Dewey contended that Manila could have been taken and properly garrisoned if his Marine force had been large enough, thus the long Philippine insurrection could have been avoided altogether.

Advanced bases were needed in two quite different contexts. One was defense of the Monroe Doctrine: keeping the European powers from seizing any new colonies in the New World. Germany ("Black"), the rising power, seemed the most likely enemy. A German fleet bent on seizing a colony would first establish an advanced base in a more or less deserted place in the Caribbean. To fight it, the U.S. fleet needed its own base; the Marines could set up and defend it.

The other likely enemy was Japan ("Orange"), which wanted to control East Asia and the entire Pacific Ocean. After the success against Russia in 1904–5, there was a real fear that the Japanese would next seize the Philippines, which had been recently acquired by the United States, and possibly even Hawaii. The U.S. fleet could not be based in the Philippines, partly because it would not be available to face Black, and partly because the Philippines lacked the necessary infrastructure. Thus to fight Japan, the fleet would have to steam west across the Pacific, a not inconsiderable feat. The Russians found in 1905 that the voyage from Europe to Japanese waters so sapped their fleet that it was destroyed at Tsushima. Among their problems was the lack of any nearby advanced base.

In 1907, with tension between the United States and Japan rising, President Theodore Roosevelt sent the U.S. battle fleet, the "Great White Fleet," to the

Western Pacific. He sought to demonstrate that it could operate successfully far from home. The cruise demonstrated just how badly the fleet needed advanced bases. The solution was to build a fleet train, which could create a base by anchoring in an island harbor. The train naturally included transports for the marines who would seize and defend the base. Out of the advanced base mission grew the modern U.S. Marine Corps' amphibious mission.

In 1910 the Marines established an Advanced Base School at New London; the school moved to Philadelphia in 1911. In 1914 Secretary of the Navy Josephus Daniels approved the formation of a permanent advanced base regiment on each coast. During World War I, however, marines functioned mainly as ground troops, alongside army troops. In 1918, with World War I over and the German threat supposedly neutralized, the Orange problem still had much of its prewar form. The Marines' commandant ordered that the advanced base mission study be resumed, because it was so clearly relevant to any Orange war plan. In March 1920, Maj. Gen. John A. Lejeune, who would soon become commandant, told the House Naval Affairs Committee that the primary role of the Marine Corps was to furnish expeditionary forces to operate with the fleet. In 1921 Maj. Earl ("Pete") Ellis, USMC, who is generally considered the prophet of World War II Central Pacific amphibious operations, produced a classified plan, "Advanced Base Operations in Micronesia" that envisaged the seizure of Japanese-occupied islands.

The Marines already filled a vital expeditionary mission, and there was some debate within the Corps as to whether or not it was more important than seizing advanced bases. As Col. Ellis B. Miller, USMC, pointed out to the U.S. Naval War College in January 1932, international law distinguished between naval interventions, including punitive expeditions, and declared wars fought by armies. "By our status as part of the naval service, Marines may be landed on foreign territory where the local government has failed and proved itself impotent to preserve order, without it being considered an act of war." Marines were thus sometimes called "Presidential Troops" or "State Department Troops" because they could be dispatched without an act of Congress. Thus the Marines fought a series of "small wars" in places like Haiti and Nicaragua. Miller pointed out that the "immediate availability of the Marines permits a display of firmness and decision on the part of our government which would be impossible if the administration had to resort to the Army for this work and wait for Congressional action." One consequence of this peacetime, although warlike, mission was that the Marines, unlike the army, did not think in terms of mobilization; they always had to be ready to fight. Only in China, where the United States had a treaty permitting the deployment of troops, did both marines and soldiers serve abroad between World War I and World War II.

Because the U.S. Marines expected to fight wars in small units, they integrated forces at lower levels than the army. For example, in the 1920s a Marine brigade normally included artillery and aircraft, whereas the army included artillery at the division level and aircraft were generally integrated at an even higher level. Only later would the army form lower-level all-arms task forces. Miller pointed out that at Tientsin in 1928 the 3rd Marine Brigade was the only allied force that had artillery and aircraft; this affected not only the Chinese but also the other allied units, because refusals of U.S. support generally meant that plans had to be aborted. This practice of forming small all-arms units deeply affected transport ship design.

On the other hand, Marine units were smaller than their army counterparts, simply because the Marines planned for much shorter lines of communication, leading directly over the shore to the fleet. Army operations deep inland required much more overhead (in current terms, a much higher tail-to-tooth ratio). The Marines generally thought in terms of the smallest possible units, because of the need for seaborne mobility.

The U.S. strategy for an Orange war was for the fleet to fight a decisive naval battle in the Far East, after which Japan could be starved by blockade; in 1929 bombing was added as a way of pressing the Japanese to surrender. It was entirely possible that the Japanese would seize the Philippines, or at least the existing U.S. base in Manila Bay, well before any U.S. fleet could appear to challenge them. Thus U.S. success might well turn on the ability to create the necessary string of bases. There was never, before World War II, any expectation that the United States would raise an army large enough to invade Japan. The Marines would seize territory because it was valuable for further naval operations (and, after 1929, air operations), not because it was valuable in itself; this concept ran counter to U.S. Army views.

The only viable route to the Western Pacific lay through the island chains of the Central Pacific, which before World War I had been German colonies. As a World War I ally of the British, the Japanese had conquered several of them. The League of Nations mandated them to Japan in 1919, despite furious U.S. protests pressed by the U.S. Navy. To the extent that aircraft based on the islands might attack a U.S. fleet passing through, islands would have to be seized simply to protect the fleet. Similarly, to the extent that the U.S. Navy would rely on seaplanes for its airpower, it needed islands on which to base them. Advanced Base Force Marines would have to fight to gain the bases the fleet would need, or to deny them to the Japanese. This was rather different than what had been envisaged before World War I, simply seizing and defending largely uninhabited islands or bays in the Caribbean.

The U.S. position in the Pacific was greatly affected by the February 1922 Washington Treaty. It set a ratio of capital ship tonnage, which was broadly equivalent to strength, between the major navies: five for the United States, five for Britain, three for Japan. To secure Japanese acceptance of the short end of the ratio, the treaty prohibited fortification of U.S.-held Pacific islands such as Guam and the Philippines. The U.S. Navy protested that inevitably Guam would soon be lost in any Pacific war, hence it would have to be retaken or a substitute taken. The prohibition was justified in part by the fact that Congress had refused to fortify the islands over the more than two decades they had been in U.S. hands; there was little hope that it would do more in the future. If anything, the treaty made Japanese possession of the island chains more valuable, because it placed no limits on land- or sea-based naval aircraft, which could be based on the islands, or on the submarines and light surface craft that could also be based there. Seaplane and submarine tenders also were not limited in any way. Although the Japanese apparently did not fortify their islands in peacetime, they were certainly able to erect effective defenses when necessary. Moreover, U.S. planners had to accept that the Japanese might seize key U.S. possessions (as they did) early in a war. They would have to be reconquered, or alternatives seized.

Given the emerging war plan, the Marine Corps established small provisional east and west coast expeditionary forces and held fleet landing exercises in 1923, 1924, and 1925. In 1925 a joint army-navy maneuver was held on Oahu. It was already accepted that both the U.S. Army and the Marines would be involved in future overseas expeditions, and in 1926 the new Joint Board produced a manual, *Joint Action of the Army and Navy*.

The U.S. Marines' expeditionary role obviously was becoming more important. For example, approximately two-thirds of the 17,363 marines were on duty outside the continental limits in July 1928; 9,187 were at foreign posts and 1,540 were at sea. This proportion, maintained during 1927–29, is probably why major landing exercises were curtailed after 1926. The Marines' role as guardians of American interests in the Caribbean ended only in 1934, when the Roosevelt administration renounced any right of intervention as part of its new "Good Neighbor Policy."

Given the officially sanctioned emphasis, the amphibious portion of the Marine Corps' school course grew from 2 hrs in 1924–25 to 49 hrs in 1926–27 and to over 100 hrs the following year (and to more later). In 1931 the school began to prepare a manual on landing operations. Completed in 1934, it was published on 25 May 1935 as the *Tentative Landing Operations Manual*. That year the revised version of *Joint Action of the Army and the Navy*, the first since 1927, appeared. For the first time it contained a chapter on joint overseas expeditions, defining many amphibious concepts that later became standard, such as the transport area offshore, the beachmaster, and the shore party commander. The U.S. Navy formally adopted a revised version of the Marines' manual titled *Landing Operations Doctrine* (FTP 167) in 1938. In 1933, moreover, the commandant established a Marine Corps Equipment Board, which would evaluate the new equipment required to execute the new doctrine. Among its earliest tasks, the Equipment Board was to lay out a program of landing craft development. It helped test the experimental craft in 1936, and its pressure provided the Marines the amtrac, which later received the designation LVT (landing vehicle, tracked).

Landings would be executed by a new permanent Fleet Marine Force (FMF), formed in 1933 as an integral part of the U.S. fleet. Now there was a portion of the Marine Corps whose specific role was to train and equip for amphibious war. It would not be detached for minor war operations in the Caribbean or in China. Once formed, the FMF could press the U.S. Navy to provide it with the transports and landing craft needed to execute wartime tasks as part of overall U.S. naval strategy. In October 1934 the FMF comprised two regiments, which would later be called battalions, consisting of 3,186 marines. Although few men were added, by June 1936 one of the regiments had been raised to the next higher status (brigade: three regiments in wartime), and by June 1937 both regiments had been raised to brigade status. They approached full strength in April 1940, with a total of 9,837 marines. The brigades in turn formed the nuclei of two Marine divisions. In February 1941, with war clearly imminent, FMF strength was two understrength divisions plus base defense battalions, a total of 22,786 men.[1]

Thus by 1941 the Marines had formed the units with which they would fight World War II. The basic tactical unit was the battalion landing team (BLT) or combat team, of about 1,600 men. Three BLTs, plus additional units, comprised a regimental landing team (RLT). Three RLTs, again with additional units, formed a Marine division, which was sometimes termed a combat force. A BLT was often characterized as one-ninth of a division, although in fact the division had assets not divided among the BLTs. This triangular organization was typical of ground combat

The U.S. Navy built its amphibious force, like the rest of its fleet, to project power far from home, whether or not it had nearby bases. Thus it developed large amphibious transports to carry troops and their materiel overseas, and small landing craft to bridge the water gap between transports and shore. Here the attack transport *McCawley* (APA 4), a converted liner, uses cargo nets to load her troops into LCVPs (landing craft, vehicles and personnel). The two-boat davits shown gave way to triple Welin davits when she was refitted in 1943.

units; it derived from the perception that the "span of command," the number of independent units an officer could handle, was three.

The interwar U.S. Navy never had the resources to buy many amphibious ships as part of its peacetime force. Instead, it planned to mobilize civilian resources. Unfortunately, by the 1930s the U.S. merchant fleet, much of it built at the end of World War I, was aging, and it had few fast ships particularly suited to naval use. In 1936, however, a Maritime Commission was formed to revive the merchant marine by building new ships; its 10-year program called for 50 ships per year, built to new standardized designs. Headed by Vice Adm. Emory S. Land, the former navy chief constructor, the commission planned explicitly to provide necessary naval auxiliaries—including transports—in the event of an emergency, which by 1936 looked increasingly likely. The commission provided wartime attack transports and attack cargo ships.

By the time war broke out the Maritime Commission's most important role was to build enough ships to maintain sea communications in the face of severe enemy attacks. Its most numerous products, the Liberty ships, were designed for mass production but were too slow for amphibious operations. The allocation of hulls between army, navy, and general transport use had to be decided at the Joint Chiefs of Staff (JCS) level. Programs were often tied directly to overall U.S. and Allied war strategy.

Until the fall of 1938 U.S. planners concentrated almost exclusively on the Orange problem, for which small Marine expeditionary forces were generally well suited; the army would contribute forces for any large operation. After the Anglo-French climbdown at Munich in November 1938, however, there was increasing fear of some sort of combination between the Fascist powers (Germany and Italy) and Japan. The threat from Europe would be far deadlier than that from Japan, because the conqueror of Europe would possess so much of the world's production potential. This possibility became real after World War II broke out in September 1939. The U.S. fleet was split between the Pacific and the Atlantic. Also in 1939, Congress approved fortifying Guam (the Washington Treaty having lapsed), on the theory that unless Guam could be defended in wartime it would have to be reconquered, at enormous cost. To support the work, the U.S. Navy obtained its first transport since 1921.

When France fell in June 1940 many in the U.S. government must have suspected that ultimately the country would be drawn into war. Through the fall of 1940 that subject could not be raised officially, because the November 1940 presidential campaign turned on keeping the United States out of the European war. By December, however, the war planners were explicitly emphasizing the need to defeat Germany, which was clearly the more menacing of the Axis powers. That would require a land campaign. The British had been ejected from the continent, thus beginning the campaign that would require one or more massive amphibious operations, on a scale quite different from that imagined in the Pacific. At the least, the United States might have to furnish naval and air assistance to the British, and quite possibly it would provide expeditionary ground forces for service in Africa or Europe.

Now interested in amphibious warfare, the U.S. Army began to buy amphibious craft in competition with the navy. Perhaps the most important consequence of the army's involvement was that the British, rebuffed by the U.S. Navy in 1941 in their attempt to have tank landing ships (designated LST; landing ship, tank) built in the United States, could appeal successfully to the U.S. Army's chief of staff, Gen. George C. Marshall.

Suddenly it seemed necessary to provide a division-size Fleet Marine Force in each ocean. Each division needed its own amphibious lift. In April 1941 the Marines proposed a third division—a reserve if the two existing ones were deployed—and also to provide the sort of reinforcements that might be needed if the United States entered the war. That increased the lift requirement, both in ships and in landing craft.

As war approached, amphibious forces, including army and marine units designated Amphibious Corps, were formed in both oceans. The navy envisaged permanent assignment of specially trained army units. The army preferred to assign divisions on a provisional basis, and thus automatically rejected any suggestion that special amphibious divisions be formed or trained.

As had happened in World War I, OpNav gained power after the United States entered the war in 1941. Appointed CNO in 1942, Adm. Ernest J. King was also made chief of the operating force, as Commander in Chief, U.S. Fleet. The abbreviation of King's title was

The assumption before World War II was always that transports could be improvised in wartime. With the invasion of North Africa imminent, the liner *Santa Clara* was hurriedly converted into the attack transport USS *Susan B. Anthony* (AP 72). After participating in both the North African and the Sicilian invasions, she returned home for a refit. An inclining experiment demonstrated that she had virtually no reserve of stability, so she was stripped of her attack transport fittings, to be employed only as a point-to-point transport. The 20-mm gun tubs around her funnel were a late addition that presumably contributed heavily to her stability problems.

The U.S. Army's view of amphibious assault emphasized the danger of enemy air attack; the army much preferred craft that beached, hence could not be sunk while discharging their personnel or cargo. LCI(L)87 is shown disgorging her troops in pre–D-day maneuvers. The pair of gangways were characteristic of this type of amphibious ship. Note the raised bridge characteristic of U.S. units.

changed to Cominch to avoid the unfortunate pronunciation, Cincus, of the original C-in-C designation. During the war, King ran the two parallel organizations: Cominch for operating forces and OpNav for the navy support organization, including shipbuilding. In practice his vice chief, Adm. Frederick J. Horne, was responsible for the shipbuilding program, and many wartime documents therefore refer to VOPNAV. In theory the General Board retained its power over ship characteristics, but in practice it was largely superseded. It surrendered responsibility for major amphibious ships to an Auxiliary Vessels Board, formed in 1941, and it had already in effect surrendered responsibility for landing craft to the Landing Boat Board created prewar.

In April 1942, against navy objections that operations should combine marine and army forces, the joint planners proposed that the Marines be used in the Pacific and the army in the Atlantic. The navy successfully resisted the army's attempt to set up its own amphibious training center, but only at the end of 1942 did the army finally agree to have the navy train all landing boat crews and operate all the boats it could. In May the navy agreed that continental operations were an army function; island amphibious operations rightly belonged to the navy. Atlantic landings led into continental campaigns: North Africa, Sicily, Salerno, Anzio, Normandy, and Southern France. Thus the Marines were responsible for the thrust through the Central Pacific, beginning with Tarawa in 1943. They seized Iwo Jima and Okinawa, the latter with army units. Despite the agreement under which the navy controlled all Pacific amphibious operations, the army (with Australians and New Zealanders) was responsible for South Pacific landings, as Gen. Douglas MacArthur island hopped along New Guinea toward the Philippines. Thus it was the U.S. Army that landed at Leyte Gulf in the Philippines in October 1944, and then at other Philippine islands, such as Mindanao. The initial South Pacific landing, at Guadalcanal, was carried out by the U.S. Marines because at the time it was the only U.S. service capable of mounting such a landing. Had the war continued, both services would have landed together on the main Japanese islands.

In the aftermath of World War II, interservice rivalry intensified. The U.S. Army promoted service unification. Among its goals was the elimination of the Marines: if the services were all brought under a single executive department, the Marines would no longer have any special status alongside the navy. As a separate service, the Marine Corps survived; the fight actually strengthened its hand. In 1947 the Marine Corps gained statutory authority over amphibious operations. By law, the Marines were mandated to maintain two active divisions and their associated air wings, with a third in reserve. Implications for the navy included a requirement to lift a full Marine division and air wing for an amphibious assault. That had been the goal in 1941, when the Marines first formed two full divisions. In later terms, the division and associated units, including an air wing, formed a marine amphibious force (MAF) or a marine expeditionary force (MEF). As ship and assault technology changed during the post-1945 period, the great question was whether sufficient new amphibious shipping for the required two-MAF lift would be built.

The defense budget was sharply cut in 1949. President Harry Truman demanded a balanced budget, and the Republican Congress would not raise taxes to pay for a larger military program. U.S. national strategy shifted toward heavy reliance on nuclear weapons delivered by bomber. President Truman, a World War I army veteran, was no friend of the U.S. Marines, and he allowed Secretary of Defense Louis Johnson to cut active Marine strength as well as amphibious shipping.[2] Plans developed in 1947–48 were suspended from early 1949 through the outbreak of war in Korea in June 1950.

Amphibious forces, however, were considered integral to the operational fleet, to an extent unimagined before World War II. Thus a 15 November 1948 Fleet Employment Plan for fiscal year (FY) 49–50 showed

In a classic assault maneuver, boats circle the attack transport *Chilton* (APA 38) while they are waiting to load. With the advent of atomic weapons, it seemed that such maneuvers were no longer practical. The boats plus the transport stayed together in one place so long that they became an attractive target for a single nuclear weapon. Modern techniques cut unloading time and place the valuable transport well beyond the horizon.

for each of the Atlantic and Pacific fleets an amphibious force and a Fleet Marine Force alongside its carrier strike force, its submarine force, and its reconnaissance and antisubmarine (ASW) units.

The U.S. Navy found itself deploying the Sixth Fleet continuously in Mediterranean waters beginning in 1946 as evidence of sustained American interest. In the navy's view, its most valuable role in a future war would be to attack the strategic flanks of any Soviet advance. That had an amphibious flavor. To the extent that Western Europe is a broad peninsula projected from Asia, the North Sea and the Mediterranean form the flanks of any advance through it. The small U.S. ground force could best be delivered (by sea) against the exposed Soviet flanks, rather than be chewed up in a frontal Soviet assault. That is, the U.S. Navy argued that amphibious mobility so amplified the power of a small ground force that it could make up for the sheer size of the Soviet steamroller.

On this basis the Sixth Fleet gained an amphibious component early in 1948, explicitly to encourage stability in the area, particularly Greece, to deny the Saudi oilfields to the Soviets in the event of war, and more generally to demonstrate U.S. interest in Europe and in the Mediterranean. Initially the Fleet Marine Force battalion was formed by augmenting units already aboard combatant ships. A few amphibious ships were deployed. Between early 1948 and 1955 the force grew from 1,086 to 1,700 men, and enough shipping was provided for the resulting full-strength battalion landing team. This force was included in the first SACEUR (Supreme Allied Commander, Europe) Order of Battle Report, dated 30 September 1951, with a footnote, "not restricted to the Mediterranean," to indicate that it might be used elsewhere in Europe.

The flanking strategy had been officially rejected by 1949. Before the new air-oriented strategy had gained much ground the Korean War broke out. Korea demonstrated the flanking strategy in practice. After the North Koreans had penned U.S. and South Korean forces into the small Pusan perimeter, the U.S. Marines spearheaded an amphibious assault in their rear, at Inchon in September 1950. The North Koreans were surprised, and their forces collapsed.

The post–World War II Marine Corps made explicit the long-standing connection between ground and air arms, defining a series of marine air-ground task forces (MAGTFs). The BLT plus a composite squadron became the marine expeditionary unit (MEU). An RLT plus a composite marine air group (MAG; helicopters plus attack and fighter aircraft for sustained operations) became a marine expeditionary brigade (MEB). A division plus a marine air wing (MAW) became a marine expeditionary force (MEF). During the Vietnam War the expeditionary forces were renamed amphibious forces because Vietnamese memories of the French expeditionary forces were too raw, thus becoming marine amphibious unit (MAU), marine amphibious brigade (MAB), and marine amphibious force (MAF). By 1975 definitions were more flexible. The MAU was still one-ninth of an MAF; the associated full amphibious unit was the amphibious task unit (ATU). The amphibious task group/MAB was two-ninths to five-ninths of a division/air wing team. The amphibious task force (ATF)/MAF was the landing force elements of five-ninths to two division/air wing teams.

The Sixth Fleet's floating battalion was a BLT, a component of the 2nd Marine Division, rotated half-yearly. In 1958 the Seventh Fleet, in the Western Pacific, acquired a similar unit, drawn from the 3rd Marine Division, most of which was permanently stationed on Okinawa. During the 1960s these floating BLTs were called special landing forces (SLFs). Eventually the SLF became the core of the present amphibious ready group (ARG).

Entering office in 1953, the new Eisenhower administration was concerned, just as Truman's had been, with controlling spending. Like Truman, President Dwight D. Eisenhower clearly saw nuclear forces as a deterrent to general war. However, he accepted that smaller conflicts were possible and indeed likely. He relied on the U.S. Navy–U.S. Marine Corps team to deal with local outbreaks; once more they were considered "Presidential Troops." Eisenhower's reactive concept was exemplified by the landing of marines in Lebanon in 1958. It may have been significant that as a senior general, Eisenhower had strongly espoused a flanking concept of European defense.

Ships are expensive, both to build and to maintain. Through the Eisenhower administration, as the navy modernized, it had faced dramatically escalating shipbuilding costs. Both new amphibious construction and the operational amphibious force were often cut as partial compensation.

There were several alternative measures of adequate amphibious forces. The bare minimum was the shipping to support the two floating BLTs, each about a ninth of a full division. During the 1950s and 1960s it took at least three ships in service to maintain one forward-deployed ship. Thus merely to maintain the two floating battalions required a total of about two-thirds of a divisional lift. In an emergency much larger Marine forces had to be deployed from the United States, so in theory the entire two-division lift had to be maintained in service. Surviving World War II amphibious ships more than sufficed for the two-division lift, but through the 1950s they became obsolescent. At the end of the Eisenhower administration active amphibious assault lift amounted to a full division equivalent in the Pacific and half a division equiv-

Because they need no ports, amphibious forces offer enormous flexibility. The attack transport *Montrose* (APA 212) is shown in Vung Tau harbor, Vietnam, alongside the accommodation ship *Benewah* (APB 35), in May 1967. The latter had been built on an amphibious (LST) hull, and was then serving as a riverine forces' base ship. Among the craft alongside the pontoon are an LCM(6) and a riverine troop carrier adapted from an LCM(6). The existence of numerous shallow-draft amphibious craft made it possible to produce a U.S. riverine fleet on very short notice.

alent in the Atlantic. During the 1961 Berlin crisis enough ships were activated to increase the Atlantic lift to a full division, and in a 6 October 1961 memorandum the secretary of defense recommended that the two-division force be maintained indefinitely. As a measure of the increase involved, 111 amphibious lift ships were active at the end of FY 61 (30 June 1961), compared to 131 at the end of FY 62 (30 June 1962).

By that time several new kinds of amphibious ships had been developed, so the 1961 recommendation amounted to a request for substantial new construction. The General Board, so important during the pre-1941 development of amphibious technology, was gone, having been dissolved in 1950.[3] With much new technology under development, however, it was increasingly important to make long-range plans as a basis for choice. On 20 April 1954 OpNav constituted an Ad Hoc Committee to Study the Long-Range Shipbuilding Program. Its creation coincided with early planning for U.S. Marine helicopter carriers,

and it affected their design. In November 1955 the committee reported plans for the 1960–70 era. OpNav formed a long-range objectives group (Op-93), which wrote its initial report on the same subject. The group survived until about 1970. Although its reports were generally concerned with a period about 5–10 years in the future, in fact they were commentaries on existing programs. The group had its greatest impact under Adm. Arleigh Burke's tenure as CNO from 1956–62.

The existence of such long-range organizations encouraged the materiel bureaus to project what platforms and weapons they could develop. In 1954, BuShips produced and briefed a "dream book" of conceptual ships, including amphibious types. The exercise was repeated by BuShips' successor organization, Naval Sea Systems Command (NAVSEA), in 1973. About 1980 NAVSEA again began to develop a series of concept designs, this time under the rubric of concept formulation (CONFORM); this effort lasted at least through 1986.

Although in theory the services were increasingly unified beginning in 1948, in fact they remained fairly independent until the Kennedy administration took office in 1961. President Eisenhower had greatly increased the potential authority of the secretary of defense. Because he was in effect his own defense chief, the changes had little impact. Secretary of Defense Robert S. McNamara, appointed by President John F. Kennedy, used the new powers to the full.

McNamara turned the Office of the Secretary of Defense (OSD) into a defense staff, thus controlling the services partly by demanding new kinds of analysis to justify their programs. He derailed the U.S. Navy's attempt to buy a new amphibious fire support ship, designated LFS (L, amphibious; FS, fire support), by demanding a mathematical analysis of its role; the analysis was fatally misleading. McNamara's effort had an important and almost certainly unintended consequence. In order to compare programs across service lines, he had to separate out their functions. For example, sea control was clearly distinguished from power projection. Yet the U.S. Marines had often seen their power projection role as the seizure (or defense) of places just to promote sea control.

Meanwhile, in the 1960s OpNav completed its triumph: the bureaus themselves came under OpNav control as Systems Commands, although it is not clear that this actually increased the influence of the operating forces. BuShips became Naval Ship Systems Command (NAVSHIPS) and then merged with the BuOrd successor, Naval Ordnance Systems Command (NAVORD), to become the current Naval Sea Systems Command (NAVSEA). The SCB became less and less relevant, because the bureaus involved were no longer independent. The Office of the Deputy CNO for Surface Warfare (Op-03) took over the task of specifying the characteristics of new surface warships, including new amphibious ships. By the 1980s the office responsible for amphibious development (Op-37) also controlled the mine countermeasures ships, which have the important role of clearing beaches prior to assaults. In the 1980s, the SCB was revived as the Ship Characteristics and Improvement Board (SCIB), but its power was limited. Currently ship characteristics are developed largely by the relevant OpNav offices.

The Naval War College had long lost its role as the U.S. Navy's think tank. During World War II the navy created an operational evaluation group (OEG), which ultimately became the Center for Naval Analyses (CNA). CNA's single greatest contribution to amphibious warfare was its advocacy of what became the LHA (landing ship, helicopter, assault) in the mid-1960s. Much later, in the 1980s, the Naval War College's war gaming role gave it considerable importance in the development of the "maritime strategy," in which amphibious forces were quite important, as described below.

The Kennedy administration was hardly content to limit intervention abroad to small marine forces backed by the navy. Probably heavily influenced by the army, it accepted that quite substantial wars in the Third World could and should be fought before the question of nuclear weapons arose. The army, which Eisenhower had gutted, was rebuilt. Army mobility had to be improved. The new administration would move its troops by airplane rather than ship. However, their equipment still had to go by sea. The new administration built new fast sealift ships. It also created floating stockpiles of army equipment. At the same time the U.S. Marines received more modern amphibious ships. The two programs fit together: the Marines would "kick in the door" by landing across beaches. Once a port and an airfield had been secured, the army would build up. This idea, which reemerged in the 1980s as the "maritime prepositioning squadrons," is described in Appendix A.

By the mid-1960s the earlier requirement for two-divisional lift had translated into two divisions of modern 20-kt ships. They were so expensive, however, that in 1963 Secretary of Defense Robert S. McNamara mandated that modern ships be bought only for the assault echelons of one and a half divisions; the rest could use older slower ships.[4] One rationale was that slower ships could cover the short distances in the Atlantic as quickly as fast ones could cross the Pacific. As funds became tighter, numbers were cut further. In 1969 the Nixon administration called for one and two-thirds divisions lift (fast only); and in May 1970 the U.S. Navy proposed a cut to one and one-third divisions. In the late 1970s President Carter cut lift further, to 1.15 divisions.

During the Vietnam War, the Seventh Fleet maintained a ready amphibious force in the South China Sea. The North Vietnamese never knew whether or not invasion was contemplated. At least until negotiations began in 1968, their great fear was that the United States would tire of the war in the south and attack them directly with ground forces. Given the well-publicized U.S. capacity to invade against opposition, the North Vietnamese had to retain significant forces for local defense, forces that could not be sent to fight in the south. As an indication of how serious the possibility was, for years in the 1960s evaluations of amphibious fire support included the demands of an assault near Vinh.

Marines never invaded North Vietnam, but their amphibious technology allowed them to advance across any suitable beach. That gave them valuable mobility, because their Viet Cong and North Vietnamese opponents tended to consider the sea a safe barrier rather than an open flank. Marines sometimes landed in the rear of enemy units facing American units ashore, outflanking them. The U.S. riverine tactics practiced later in the war, by army troops as well as by marines, were to some extent an extension of this idea, that amphibious capability made the sea a highway as viable as the land.

On the other hand, most marines served on the ground in Vietnam, in effect as an alternative to the U.S. Army. With the war over, the special role of the Marine Corps came into question. How valuable was amphibious capability, given more effective Soviet-supplied coast defenses? Vietnam might be seen as an extension of the earlier expeditionary role. Some argued that the Marines Corps had performed particularly well in Vietnam, precisely because of its experience. Perhaps small wars were its future. It did not help that Secretary of Defense James Schlesinger asked in 1974 and 1975 whether an amphibious assault force was still needed. Moreover, with the Vietnam War over, it seemed unlikely that the United States would engage in any further conflicts of that type.

In 1975 a new commandant, Gen. Louis A. Wilson, appointed a board under Maj. Gen. Fred E. Haynes Jr. to reexamine the role of the U.S. Marine Corps. Haynes concluded that the Corps had a global mission as a strategic ready force, capable of fighting in high-intensity as well as low-intensity wars. Certainly it could be argued that central war, or at least a major non-nuclear war in Europe, had become more possible. The Soviets had a large nuclear arsenal backing a large army. The North Atlantic Treaty Organization (NATO) might well be deterred from using nuclear weapons against an advancing Soviet army. Consequently, NATO amphibious forces might have important roles on the flanks of an advancing Soviet army.

The old idea that particular land areas could help decide naval power reemerged. Because they might be peripheral to the main land campaign, the army might reject defending them. For example, much of the Soviet fleet was based on the Kola Peninsula in northern Russia. NATO planned to erect a barrier to keep Northern Fleet submarines out of the Atlantic shipping lanes in war. If the Soviets could seize Norway, they could break up the barrier by outflanking it. Marines ran their first exercises in Norway in 1975, and they began to stockpile equipment there. By 1978 they were practicing in brigade strength.

In the 1980s, the Reagan administration espoused a new national strategy to be used in the event of a general war, including the navy's maritime strategy, which revived earlier ideas. The navy would seek decisive engagements in which it would destroy Soviet seapower. Winning sea control, it could then mount assaults on the flanks of any Soviet ground advance, forcing the Soviets back.

Amphibious forces figured in this strategy in three distinct ways. First was the existing idea of denying the Soviets control of Norway. U.S. and NATO amphibious forces could threaten the flank of any advancing Soviet land force.

Second, it was known that the Soviets took the U.S. ability to land near places like Severomorsk and Leningrad seriously. Significant ground units had to be withheld to protect these places. This effect is called virtual attrition, because it reduces the strength of the main Soviet attack (attrition), albeit not by actually destroying the forces involved. Anyone studying the World War II Normandy campaign can see virtual attrition in action. The Germans' defense of Normandy was badly weakened because they had to maintain forces along the whole European coast.

Third, amphibious forces could land on the strategic flanks of an advancing Soviet army. With its flanks threatened, the army might well have to halt. Inchon was a perfect example of a flanking attack from the sea that forced a victorious ground force to withdraw.

The Reagan administration invested heavily in new amphibious ships and craft, including the revolutionary LCAC (landing craft, air cushion) vehicles. It called for a buildup back to one and a half divisions.

Virtual attrition continued to be important after the end of the Cold War. During the run-up to the Gulf War, marines staged several big amphibious exercises, which convinced the Iraqis that they might well attack from the sea. The Iraqis in turn retained large formations to defend their seaward flank. The force involved was much larger than the Marine force because defenders had to placed anywhere marines might come ashore. They could not reposition themselves quickly enough to counter any such attack as it appeared. Nor could they turn to meet the actual attack, over land. A relatively small Marine force thus enjoyed considerable leverage.

What happens now? Since World War II, there have been only three important opposed landings: one by U.S. forces at Inchon in 1950, and two by the British, at Suez in 1956 and in the Falklands in 1982. However, amphibious capability is increasingly important to the United States. What matters is not the ability to land against opposition, but rather the ability to land mechanized forces without permission or preparation, that is, when no port is made available. This ability makes it possible for the United States to project its power into a very unstable Third World. Almost as much as the "Where are the carriers?" question, the availability of marines on board amphibious ships has determined where and how the United States has been able

As heavy-lift combatant ships, amphibious ships help support a global U.S. Navy when it operates in littoral waters. USS *Portland* (LSD 37) carried patrol boats (PB Mk IIIs) to the Gulf on the superdeck, normally used to carry vehicles, above her well deck. She is shown at Manama Harbor, Bahrein, on 30 January 1988. Many types of U.S. small combatants, as well as some minecraft, were designed to fit the well decks of U.S. amphibious ships. In this particular case, the well deck apparently carried cargo containers.

to intervene since World War II. Examples include Lebanon in 1958, the Dominican Republic in 1965, Grenada in 1983, and Haiti in 1994.

The following chapters detail the development of U.S. amphibious ships and craft to support amphibious capabilities. U.S. amphibious craft and ships developed during World War II all had hull designations beginning with an "L" prefix, for example, LCVP (landing craft, vehicles and personnel) and LST (landing ship, tank). The equally important transports had an "A" (auxiliary) prefix, initially AP (transport) and AK (cargo ship). In 1943 assault versions were redesignated APA (attack transport) and AKA (attack cargo ship), but that still seemed, quite misleadingly, to suggest that these ships were somehow less amphibious than the others.[5] Only in 1969 were the ships again redesignated as LPAs and LKAs, emphasizing their amphibious roles. Because few LPAs and LKAs were used after 1970, they are referred to in this book mainly as APs or APAs, and AKs or AKAs, respectively.

Measurements, for example, length, are written in the form that appears on the source document. Most measurements are denoted using fractions of feet and/or inches (e.g., 3½ ft), but occasionally decimal specifications are given (e.g., 3.6 in). Also, some metric measurements are used in specifications (e.g., 1.1 m).

As previously noted, maritime prepositioning and sealift are described in Appendix A. Appendix B details the procurement and, to some extent, the fates of landing craft and some numbered landing ships. Appendix C lists important amphibious ships with their dates of construction and their fates. Numbered LSTs, LSMs, LCI(L)s, and LCS(L)s are included. This appendix also includes transports (AP), some of which were in effect attack transports although not designated as such, as well as evacuation transports (APH) that served as attack transports.

The ultimate stage of any amphibious operation is getting troops and their materiel to the beach. During the interwar period, enormous efforts were made to develop the appropriate landing craft. Here marines embark from an APD, a converted World War I destroyer, into ramped personnel landing craft (LCPRs).

2
Developing the Techniques

During the interwar period, the U.S. Marines' model for a modern landing was Gallipoli, the ultimately failed British-led attempt to seize the Dardanelles in 1915. After World War I Gallipoli was widely regarded as proof that modern weapons, particularly machine guns, would preclude amphibious assault. The Marines disagreed. The attackers had gained and maintained their foothold. Their key failure had been to waste precious time. An excessive delay in mounting the expedition gave the Turks time to organize a defense; a few months earlier, the attackers would have prevailed quite easily. Even when they did attack, slightly more initiative would have taken the British through the initially thin Turkish defenses. Ultimately the operation failed because the British were unable to reinforce by taking more forces from the Western Front, not because they were driven into the sea.

One important lesson of Gallipoli was that ships had to be combat loaded: equipment had to be accessible in the order it would be required by the assault troops. Freighters were normally commercially loaded, cargo being stuffed in as tightly as possible. For accessibility, cargo had to be stowed much less densely. Ultimately that required more transports, differently arranged from typical freighters. By the 1930s the Marines distinguished between commercial loading, organization unit loading (organizations and units were kept intact aboard ship, but cargo was still loaded as tightly as possible), and unit combat loading or simply combat loading (in British parlance, tactical loading). For Gallipoli, ships that had been organization unit loaded had to be unloaded before the assault and reloaded tactically, contributing greatly to the fatal delay.

The British were clearly aware that Gallipoli had not been a total failure. They nearly executed a big landing in Flanders in 1917–18, to deny the Germans a vital U-boat base, and they did execute a smaller-scale raid on the U-boat base at Zeebrugge. They also conducted amphibious operations in Mesopotamia. In 1918 they built a landing craft capable of disgorging a light tank. The British view, that amphibious operations could succeed, was almost certainly transmitted to Americans during the latter stage of World War I, when the U.S. Navy operated very closely with the Royal Navy.

British amphibious ideas had little impact on the World War I U.S. Navy, which was deeply affected by other British concepts of the time, such as the aircraft carrier, the fast capital ship, and special types of submarine. One reason may have been that the Royal Marines were not analogous to the U.S. Marine Corps; with a worldwide empire, the British were far less interested in seizing advanced bases. Because the bond with the Royal Navy did not survive long into the postwar period, the Marines were not affected by continuing postwar British interest in landing craft development and in amphibious tactics.

Because their targets were so far from the United States, the Marines expected to use conventional long-haul shipping—transports—to get within range. Small boats would leap the remaining gap between transports and beach. World War II examples of such long-range attacks included Guadalcanal, mounted from Australia, and the North African invasion, mounted from the United States and from the United Kingdom. Clearly the transports would be vulnerable to attack during the approach to the target, but the Marines reasoned that an island could be isolated by a naval campaign. The aircraft based on the island could be destroyed in a preinvasion attack. This sort of isolation, entirely feasible for islands in the Central Pacific, was less practical in Europe, where an enemy would have ready access to enormous inland resources. This difference helps explain the difference between U.S. Marine and U.S. Army amphibious concepts.

The Marines analogized an assault across water onto a defended beach to a World War I assault against a trench system defended by machine guns.

Marines were always an amphibious force. Serving aboard major U.S. warships, they expected to go ashore in ships' boats. The great change after World War I was the assumption that marines would have to land on beaches defended by machine guns, so it became vital that they be able to land quickly in sufficient numbers to overrun enemy defenses. Experiments showed that standard motor launches like those shown here could not deal with surf conditions. This photograph is taken from the collection of Lt. (j.g.), later Admiral, Arleigh A. Burke, who was assigned in 1929–30 to help plan the use of boats in assaults.

As in World War I, the solution would be a vehicle capable of protecting troops while it moved them as close as possible to the trenches. In World War I, that was a tank; for the Marines, it was a landing craft. The main threat to such craft, particularly if they were armored, would be the maritime analog of antitank artillery: small quick-firing antiboat guns near the beach. Guns probably would not reveal themselves until the landing craft approached the shore. The best counters would be support craft capable of closing the beach and opening fire on the enemy guns when they revealed themselves.

In 1930 the Marines therefore convened a Gun Fire Board at Quantico. Its 3 March 1931 report recommended both naval gunfire support experiments and tests of small guns (machine guns and 1-pounders) aboard ships' standard boats carrying troops ashore. For example, Naval Gunfire Exercise "Cast" (C), conducted at the end of 1931, tested the ability of battleships' director-controlled 5-in/25 guns to support a landing against minor opposition. The Marines convinced the navy to develop shore bombardment shells (high capacity high explosive [HE] with the appropriate fuses) and fire control techniques, to make the supporting ships responsive to the needs of those coming ashore. Naval guns would initially concentrate on known enemy strongpoints (preparation fire). As the boats approached the shore, it would keep defenders' heads down (suppression fire). Naval gunfire would also defend the beachhead while those who had landed set up their own organic defenses.

Tanks would help marines overcome enemy defenses. Tests of a Christie amphibious tank in 1924 failed, although the Marines did use six of these craft in China in 1927. When returning to the subject in the 1930s, the Marines had the navy develop self-propelled tank lighters, the forerunners of the LCM (landing craft, mechanized), from which a tank could drive onto a beach.

In the early 1930s the basic U.S. Marines assault

unit was a single battalion, which was called a marine regiment. The Marines noted that during World War I two or three machine guns had often stopped attacks mounted by entire battalions. The best protection would be dispersal; instead of using a few large ships' boats, the Marines would prefer to land in numerous smaller ones. The smallest practical ones would carry a squad, the smallest tactical unit. Riding such boats, Marine assault troops could assume attack formations before hitting the beach, out of range of hostile fire, rather than halting on the beach to re-form. Dispersion would complicate boat control (in World War II considerable effort went into solving this problem), but during the interwar period the Marines thought control would still be simpler than in a dispersed land operation.

Because the FMF was part of the U.S. fleet organization, the annual training cycle had to accommodate it, in the form of new fleet landing exercises (FLEXs). They in turn highlighted developments in both thinking and materiel. The FMF was formed too late in 1933 to arrange a 1934 exercise, but FLEX 1 was held January–February 1935 in the Culebra area. FLEX 1 involved 1,500 FMF troops, two-thirds of them carried aboard two battleships. Landing craft were 18 ships' boats (motor launches and motor whaleboats) and one unpowered artillery lighter.

FLEX 2, held January–February 1936 in the Culebra area, involved nearly 1,800 FMF troops (three-quarters of them aboard two battleships) landed by 23 standard ships' boats. It included advanced gunfire exercises, ships attempting to hit the reverse (invisible) sides of hills and firing together against single targets. There were attempts, not entirely successful, at night landings and at landings through smoke screens. Lessons included the need for specialized transports and specialized landing craft. Twenty army observers were present, the CNO having reminded the army the previous year that under the approved Orange war plan it would furnish most of the troops for amphibious operations. By mid-1936 the army was trying to rotate amphibious training among four divisions. Its FY 38 budget, prepared in 1936, included participation in the 1938 FLEX.

Held January–March 1937 in the San Clemente–San Diego area, FLEX 3 was the first to be held in the fleet's main operating area, the Pacific. It was much larger than its predecessors, with two FMF brigades (about 2,700 troops) and one army brigade (about 800 troops). As before, combatant ships (three battleships and a cruiser) were surrogate transports. The new naval gunfire techniques, including control by shore parties, were tested; ships fired the new shore bombardment shells. These innovations succeeded, but shellfire against underwater obstacles failed. Accidents confirmed the earlier conclusion that standard ships' boats were unsatisfactory, even dangerous, for landings through surf. Most troops landed from standard ships' boats, but three experimental landing craft were also used for the first time. (See Chapter 4 for discussion of the experimental landing craft program.) The exercise included close air support by 83 Marine Corps aircraft.

In 1938 FLEX 4, in the Culebra area, again combined marines (about 1,850) and army troops (about 600). As before, there were no specialized transports, so the troops rode three battleships and a navy auxiliary. FLEX 4 included the transfer of troops to destroyers at sea for surprise landings. Four experimental personnel landing craft operated alongside 28 standard ships' boats. They were joined by an experimental self-propelled tank lighter, the previously mentioned forerunner of the LCM.

FLEX 5, held January–March 1939 in the Culebra area, employed the 1st Marine Brigade (about 2,200 troops). This time there were enough experimental landing craft to permit useful comparisons among them.

In January–March 1940, in the Culebra area, FLEX 6 was limited to the navy and marine personnel; the U.S. Army decided to call a separate amphibious exercise (discussed below). FLEX 6 introduced night landings of patrols from destroyers and submarines, using rubber boats with outboard motors.

In the Culebra-Vieques area in February 1941, FLEX 7 was the largest prewar fleet amphibious exercise, employing the 1st Marine Division (although not in divisional strength) and the army's 1st Division Task Force (two battalion combat teams). By this time the Atlantic Squadron (soon to become the Atlantic Fleet) was quite powerful. The landing was supported by 3 battleships, 3 carriers (with the 1st Marine Air Group aboard), 7 cruisers, and 12 destroyers. There were finally specialized transports: three navy and two army, supplemented by two destroyer transports. They were entirely inadequate. The fleet commander estimated that he needed ten transports, four destroyer transports, and four cargo ships for the Marine division alone. The transports had just been converted and their crews were untrained; they had to struggle to keep their ships in service. None of the ships could service her landing craft and none could produce adequate fresh water. One could not hoist tank lighters and another had unreliable engines. Even using standard ships' boats in addition to the new landing craft, it was barely possible to land two battalion combat teams simultaneously.

The U.S. Army became seriously interested in amphibious operations in 1939. Brig. Gen. Frank A. Keating, commanding the 3rd Division at Fort Lewis,

Pre–World War II transport conversions were rudimentary. As first converted, USS *McCawley* (ex–SS *Santa Barbara*) retained her dummy forefunnel and her single-boat liner davits. The main visible changes were 3-in/50 antiaircraft guns fore and aft and degaussing cable to protect her against magnetic mines. The landing craft are Higgins boats, recognizable by their spoon bows.

Washington, ordered troops of his 15th Infantry to practice amphibious assaults. In these simulated "alfalfa assaults" troops disembarked from trucks and assaulted the opposite end of a field. The army asked for a joint west coast exercise in place of FLEX 6. The division, about 9,000 troops and 1,100 vehicles, would embark at Puget Sound on board five army transports and an army cargo ship. The navy would provide supporting units, from battleships down. Unfortunately the exercise was irrelevant to the opposed landings the Marines envisaged. The fleet commander rejected any landing through surf because the boat crews were insufficiently trained. There were no specialized landing craft; it took all the boats of the battleships and cruisers to land a 1,550-man combat team. Moreover, the transports' commanders were unfamiliar with convoy procedures and lacked necessary means of signaling to the accompanying fleet. Without specialized lighters, it was extremely difficult to land artillery, not to mention vehicles. The exercise also showed that the army transports could not be combat loaded. The exercise degenerated to the transfer of the division from Puget to Monterey under ideal conditions.

In January 1940 the army proposed a 1941 exercise in which an east coast division would land in Puerto Rico from naval vessels. FLEX 7 was already scheduled for January–March 1941, and the navy proposed that the two exercises be combined. The army suddenly discovered that the supply of landing craft was quite limited. As a consequence, the navy asked that army participation be held to the size of the Marine contingent, about 2,000 men, and it rejected an army proposal for a second landing exercise on the force's return to the east coast, on the ground that the Atlantic Squadron could not spare the ships. The army concluded that it might have to obtain its own boats.

When France fell in May–June 1940, the army became more interested in amphibious operations. The 1st and 3rd Divisions were ordered to train for landings, using improvised equipment, such as engineer—river assault—boats instead of landing boats, and pack rather than normal artillery. The army began to buy landing craft.

The situation the U.S. Army envisaged was quite different from that analyzed by the U.S. Marines. Because it sought beachheads into Europe, it thought in terms of large assaults. The Germans would oppose a landing with several divisions, with more in reserve. Nearby air bases would be difficult to neutralize. Quite possibly enemy aircraft would destroy transports offshore. The U.S. Army therefore much preferred to move troops ashore via landing ships that could beach to discharge men and materiel. Once beached, the ships would be quite vulnerable, but by that time their destruction would mean relatively little. The army approach, which produced the LST (landing ship, tank), demanded much larger numbers of amphibious craft. Among its advantages were that many more troops and tanks could be unloaded more quickly onto a beach, because lighters did not have to shuttle back and forth. Because beaching-ship production never matched demands, the army had to settle, at Normandy and in southern France in 1944, for a combination of its favored tactic and a navy-style transport operation. For their part, the U.S. Navy and the U.S. Marines found the LST far more useful than they had imagined.

The Marines Corps and the U.S. Army also differed in their approaches to the element of surprise. The Marines' islands were so small that defenders would probably know which beaches were to be hit. The preinvasion bombardment would certainly announce that an attack was imminent. It would be best, then, to forego surprise and assemble the force offshore in daylight. The army always hoped to forestall the inevitable counterattack. The sheer size of continental targets like Northern France made it difficult for a defender to be sure of exactly where the attack would fall. The army therefore preferred to attack in darkness, accepting all the attendant problems of assembling craft into waves at the right beaches. Normandy, the dawn attack, was the exception.

The army approach to surprise also precluded any substantial preassault bombardment, because that too would give away the position of the landing beach. Thus, for example, there was no softening-up bombardment at Normandy, despite the existence of substantial German fortifications there. The army perception seems to have been that naval gunfire would be effective, if at all, only if it were used for a considerable period. The choice at Normandy was to rely on the only weapon deemed capable of delivering massive firepower in a very short time, daylight bombing. The need to allow bombers at least one pass over the beach defenses determined that the attack would begin at dawn. In the event, bombing was badly oversold, and the beach fortifications were untouched, with serious consequences at Omaha Beach. Delays in landing on the British beaches allowed time for a protracted naval bombardment, which proved extremely valuable. Of all the major landings in Europe, only in Southern France was the preferred naval model, of a daylight landing using heavy bombardment preparation, followed.[1]

For the army, much depended on how quickly forces could be built up, as the enemy tried to mount his own buildup. The buildup, moreover, would be huge, because the army wanted to penetrate far inland after the landing. At Normandy, the fear was that no conventional, or at least usable, port would be captured for the first few weeks or months. Hence, an artificial port, the "Mulberry," was created. This problem did not arise in small Pacific islands, which were worth seizing precisely because their reefs provided natural harbors.

The two services differed in other ways. Aware that follow-up supplies might well be lost to enemy fire, the army had each soldier carry as much as possible, typically about 80 lbs. Both in North Africa and later, heavily loaded troops sometimes fell and drowned as they left their landing craft. To marines concerned mainly with seizing a small area and then digging in, the battle was unlikely to last very long, so there was little point in carrying many days' worth of rations or ammunition. The result was much greater mobility.

The army operated a substantial fleet of transports prior to World War II, but they were intended for point-to-point service, delivering troops to operating ports in friendly hands. Although other wartime army troopships were converted from freighters, *James O'Hara* was built to a special U.S. Army transport design. In April 1943 she was taken over by the U.S. Navy and converted into an attack transport, and designated as APA 90. She is shown newly converted, with triple Welin davits but as yet without her landing craft, on 2 May 1943. In 1946 she reverted to the army, but was reacquired by the navy when sea transport services were amalgamated in 1950.

The greatest single difference between a combat-capable transport and a freighter was the transport's ability to handle large numbers of landing craft. USS *Hamblen* (APA 114) is shown on 13 June 1945. Note the triple Welin davits, carrying one landing craft outboard, another on deck, and another raised above the deck. They were key to the ability to handle the large number of landing craft modern amphibious tactics required.

3
Transports

The U.S. Marines' interest in specialized transports—ultimately in ships adapted to the amphibious role—can be traced back to the period immediately after the Spanish-American War, as the United States acquired both formal and informal empire in the Caribbean and Central America. Beginning in 1903 the Marines experimented with maintaining a battalion afloat, in effect a forerunner of the current amphibious ready group (ARG).

By 1909 the Marines had deployed expeditionary forces nearly every year since the Spanish-American War. Typically such a force was split among battleships or cruisers. Adding 125 or more marines to the complement of a combatant ship caused crowding and interfered with the usual work of the ship. Because marines did not travel in tactical units, they could not coordinate before reaching their objective. Nor could they exercise in formation at sea. Yet marines might have to remain on board their transports for some time before beginning operations. Their situation was very different from army troops, which were always barracked ashore; they spent only the sea passage on board ship. Formation training at sea was irrelevant to them.

The U.S. Navy had five transports, of which two were in reserve. All were badly crowded. The U.S. Marine Corps needed a transport shaped to its special needs. Two were included in the long-range building program proposed late in 1908 by the General Board, which included a fleet train. Although marines were needed to seize and defend the base built around the train, transports came last among the board's priorities.

Two transports were needed for the two different Marine advanced base defense regiments (about 1,300 men each). One regiment was responsible for fixed defenses, the other for mobile defense of the base. Fixed defenses included both guns and controlled minefields, with the necessary signal and observation posts. The mobile force combined two infantry battalions with mobile artillery. These units, formed in 1913, were the first Marine Corps regiments; the mobile defense unit (2nd Marines) fought at Vera Cruz, Mexico, the following year. A provisional 3rd Regiment was formed from fleet detachments. The entire two-regiment Marine Brigade fought in Haiti later in 1914. Within two years the Marines had formed a second pair of regiments for the Pacific.

The mobile defense regiments differed from contemporary army regiments in that they included supporting arms. The army was organized in divisions (single-purpose regiments, such as infantry, plus supporting arms), but the Marines expected their base defense regiments to fight independently. These integrated units proved effective in expeditionary warfare, for example, in the Caribbean. Such a unit had to arrive at its objective as an integrated whole. Thus its transport had to carry not only men and their ground transport (animals and later vehicles), but also field artillery and engineer equipment. This concept, that one ship would carry a whole tactical unit, shaped Marine transport design through World War II and beyond. For the army, the smallest self-supporting unit was the division; regiments were specialized (e.g., infantry). The Marines formed no divisions until 1940, and even then they expected to deploy in much smaller units. The main change over time was that units of various designations grew. By the end of World War I, the battalion was the basic tactical unit, nominally about the size of the pre–World War I regiment.[1]

As the General Board feared, Congress preferred combatants to auxiliaries, and not even the few the board requested were authorized in FY 09, the fiscal year ending 30 July 1909, or in FY 10. By the fall of 1910, however, prospects for auxiliary construction were good enough for the General Board to draw up characteristics. The importance of the fleet train, including the Marine transport, was reflected in the board's demand that its ships have an unusual degree of underwater protection, including torpedo bulkheads, a double bottom, and a watertight deck. Required sea speed was 14 kts (16 kts was desirable) with twin screws. For maximum flexibility, auxiliaries would burn both coal and oil fuel. Endurance was 8,000 nm at 10 kts, similar to contemporary battleships, and stores endurance was 60 days.

Unlike other auxiliaries, the transport needed moderate draft, so it could get close inshore to unload. On the other hand, she might carry the Marines' floating battalion, thus she needed both good habitability and deck space for troops to exercise. The ship would carry the Marines' field artillery, which was 8 guns and 32 horses to pull the guns and ammunition. Her six 5-in guns would be easily dismountable so that they could be set up ashore to defend the advanced base. Other base defenses would include 50 mines, to be laid by the transport's boats.

The transport would be specially fitted to unload her cargo quickly, onto unimproved beaches, via boats and launches, to get the advanced base force ashore before enemy opposition could materialize. She would carry the first specialized U.S. landing craft, a lighter or scow to land guns, mounts, animals, and stores. Her two 50-ft picket launches would be specially designed for towing and "rough work," with a small searchlight that would help them defend the base against enemy boats. They would also lay defensive mines, using a mast and sheer legs to take mines aboard and a trough and rail to lay them.

Congress authorized no transports in FY 11, so the General Board proposed them again in 1911, fruitlessly, for FY 12. In June 1912 no fewer than nine battleships, each carrying about one-third of a battalion, lifted the 2nd Marine Regiment from the United States to Cuba. In July the Marines' commandant reported that the only existing transport, *Prairie*, was grossly inadequate. He hoped that the collier *Vestal* might be adapted to carry one complete Marine regiment. However, *Vestal* was converted instead to a repair ship (AR 4).

The FY 14 request, drawn up in the fall of 1912, included two transports. Congress approved only one, which became *Henderson*, Transport No. 1 (later designated AP 1). She was named after a former Marine commandant. On 25 September 1912 the General Board submitted roughly the earlier characteristics to the secretary of the navy. The ship would carry 64 officers and 1,250 troops. After conducting a feasibility study, which was completed in February 1913, the board particularly demanded clear exercise space. Troop exercises at sea would include marching in columns of four, 8⅔ ft shoulder to shoulder. The General Board therefore wanted at least 10 ft of clear deck space on either side of the main deck.

C&R submitted a preliminary design on 18 August 1913. Rear Adm. Richard M. Watt, the chief constructor, later said that C&R had considered the ship an unusual design opportunity, hence it had been given far more consideration than might have been usual. C&R consulted not only with the Marines (Lt. Col. Eli K. Cole of Marine Corps Headquarters), the General Board, and the Bureau of Ordnance, but also with Lt. Col. Chauncey B. Baker, who was in charge of Army Transport Service, and his naval architect assistant, Mr. Anthony.

Contemporary warship designs were generally weight critical; it was relatively easy for a designer to provide the necessary volumes. The transport, however, was volume critical, even deck area critical. To make matters worse, protection was important. Torpedo bulkheads were placed abeam the machinery spaces. The inner bottom was carried up to the second platform deck throughout the ship's length, from bilge to bilge, and the ship was well subdivided by transverse bulkheads. Late in August 1913, the General Board demanded that ammunition be moved from a vulnerable position above the waterline; it was interchanged with cargo spaces.

The Marines kept enlarging their regiment. In December 1911 they asked for 1,465 troops including officers; in mid-1913 they wanted 75 officers plus 1,600 enlisted men, far beyond what the characteristics required. The designers offered 350 more troops, but no permanent accommodations for the additional officers. In mid-1916, the Marines protested that the 32 draft animals specified were hardly enough to make the defense mobile. They wanted at least 100, and got 60. Before the ship was complete the Marines had switched to tractors. Horse spaces were then used to house Marine warrant officers.

To provide the gangways for exercise, C&R planned a 36 ft wide superstructure on a 62 ft wide main deck (63 ft beam), leaving 13 ft on either side, including 15-in waterways, over the full 300 ft of the main deckhouse. The 36 ft could not easily be reduced, because it was set by the width of the galley needed to feed the set number of marines. Galley facilities could not be submerged into the hull, because the whole of the second deck (below the weather deck), which was specially ventilated, was devoted to troop quarters. Much of the hull not needed for troops or machinery was reserved for the 133,000 cu ft of the advanced base outfit. Given the demands placed on available hull volume, the ship needed an unusually massive superstructure, covering about a third of her length, with an average width of 38 ft. It rose three deck heights above the main deck.

USS *Henderson* (AP 1) was designed specifically to deploy marines. She is shown loading them at their Quantico, Virginia, base. Her 5-in guns were originally intended to be brought ashore to help defend an advanced base, for which purpose special artillery lighters were designed and built. By the time this photograph was taken, however, *Henderson* was being used only as a point-to-point transport, and she carried only standard ships' boats. Because she was needed so badly to support marines abroad, particularly in China, she did not figure in prewar amphibious exercises. During World War II she was converted into a hospital ship.

Unfortunately, detailed calculations showed that the ship would be too stiff. The only remedy was to reduce beam to 61 ft, thus reducing the main deck to 60 ft. That in turn cut the gangways to 10 ft inside the waterways. When contract plans were ready and offered for bids in January 1914, the General Board discovered that even amidships there was only 8 ft clearance from deckhouse to bitts, and only 6 ft abreast the after deckhouse, which could not readily be narrowed.

The General Board's modifications would entail redesign. Watt pointed out that the secretary of the navy had only reluctantly agreed to the project. The navy had already advertised for bids. Any substantial change would require new bids, and the secretary of the navy might well simply cancel the project. Capt. Albert G. Winterhalter, the aid for material (the secretary's advisor on such issues), agreed.

The ship had eight, rather than the six planned, 5-in guns (of the older 5-in/50 type, rather than the current 5-in/51) for base defense. Because the two guns at each end could not be dismounted without special lifting arrangements, only four could be brought ashore. Largely to bring the guns ashore, *Henderson* was designed to carry four artillery lighters (or scows), to beach rather than come alongside a pier. In 1919 the Marines said that the two scows built had "been of the utmost value in unloading and transporting heavy guns and materials." They were the distant ancestors of the World War II LCM (landing craft, mechanized). Standard ships' boats would carry personnel ashore: six 40-ft motor sailing launches, a 36-ft motor barge, two 30-ft whaleboats, and two 21-ft motor dories.

To defend the advanced base, *Henderson* had magazine space for 100 moored contact mines. She did not carry some other defenses advocated by the General Board: torpedo tubes to be set up ashore or a boom defense. The latter was abandoned because of the vast space involved, and because it would be needed only in special cases.

The transport duplicated the underwater forms of other U.S. auxiliaries: the ammunition ship, repair ship, hospital ship, and destroyer and submarine tenders. There was some question as to whether 14 kts was fast enough, but to attain higher speed she would have needed a finer hull, with less internal volume. *Henderson* was rated at 13.5 kts sustained speed. She strained her engines during World War I, and afterwards was not considered good for more than 12 kts. Rated endurance was 8,000 nm at 10 kts on oil and 5,000 nm at 10 kts on coal. Table 3-1 compares *Henderson* with three classes of merchant ships built during and just after World War I. Some of these ships became World War II attack transports.

Henderson was commissioned on 24 May 1917. With the United States at war, she was no longer needed for Marine expeditions, but instead was refitted to transport troops to Europe. The navy now distinguished between expeditionary transports, which would land their troops as combat units, and point-to-point transports to bring troops to assembly areas abroad. Because the latter traveled in convoy (for protection), they were also called convoy transports. During World War II a similar distinction was drawn between attack and convoy transports. As a point-to-point transport, *Henderson* accommodated 56 officers, 20 warrant officers, and 2,100 troops in summer. In winter she was reduced to 1,600 troops, her designed capacity. Compared to a converted liner, her capacity was limited because her superstructure was constricted and considerable cargo space had been reserved for the advanced base outfit.

Table 3-1. *Henderson* **vs. Standard Troop Transport Designs, 1914–20**

	Henderson (AP 1)	535-Footer Design 1029	502-Footer Design 1095	Hog Island Type B Design 1024
Length (ft-in)				
OA	483-10⅛	535-2	522-8	448-0
BP	460-0	517-0	502-0	437-0
Beam (ft-in)	60-10	72-0	62-0	58-3
Draft, mean (ft-in)	22-0.5	30-7	26-2	23-9
Displacement (tons)				
Light	10,000	12,200[a]	10,495[b]	8,400[c]
Full load	11,277[d]	21,325	16,984	11,100
Gross tonnage	7,493	12,500	9,255	7,500
Net tonnage	3,992	NA	NA	NA
DWT	NA	13,000	13,100	8,000
Power (SHP/shafts)	4,000/2	12,500/2	7,000/2	6,000/2
Speed (kt)	13.5	16–17.5	14.0	15.5
Fuel				
Oil (tons)	1,401	4,138	NA	NA
Coal (tons)	989	—	—	—
Radius (nm/kt)	6,480/10[e]	9,800/17	13,000/13	10,440/NA

NOTES: Design numbers are U.S. Shipping Board numbers, equivalent to later Maritime Commission design numbers. Light and full load figures are actual ship weights. Gross tonnage is a measure of internal cargo-carrying volume, 100 cu ft being counted as a ton. Only enclosed spaces are counted, an important non-tonnage space being the upper 'tween deck on a ship with a tonnage well and tonnage openings in bulkheads. Many port and canal charges are based on gross tonnage, since in theory it measures the earning capacity of a break-bulk ship. Dash indicates that specification is irrelevant; NA indicates relevancy but data not available.

[a]Light and full load displacement, draft, and power data are for *Wharton*, 1945, as typical of this type. Gross tonnage was a typical figure. Load displacement when acquired was 19,411 tons. Gross tonnage at that time was 13,788 tons.

[b]Light, full load, and gross tonnage displacement data, as well as draft, power, and radius data, are for *Refuge*, ex-*Kenmore* (AP 62), 1945. Gross tonnage is the prewar figure.

[c]Light and full load displacement, draft, and power data are for *Argonne*, 1945. Gross tonnage is an approximate prewar figure.

[d]As of September 1940, 10,000 tons was considered "normal" rather than light displacement. Data are taken from a BuShips design sheet comparing *Henderson* with the new Maritime Commission transport design, which became *Doyen*.

[e]Endurance was given in September 1940 as 8,400/13 and 10,000/10, burning oil only. At this time ship complement was 23/445, and troop capacity was 86/1,609; cube was 125,928 cu ft.

In 1919 C&R estimated that *Henderson* could have accommodated 3,381 troops in summer and 2,881 in winter if she had been redesigned as a convoy transport. She could have been fitted with a second platform deck forward and it, together with the first platform, could have been devoted entirely to troop quarters. Much of her cargo capacity would have been eliminated or turned into provision spaces for the troops. The difference reflected the fact that *Henderson* was, in effect, an attack transport.

Marines were quite pleased with *Henderson*. Troop quarters on the second deck and the first platform deck were crowded but well ventilated even in bad weather. When the marines came on board, however, the ventilators were out of order and closed, and the air ports of their quarters had to be kept closed at sea. Unfortunately, the distilling capacity of 9,000 gal per day (less than 5 gal per man, not allowing for water used by boilers) was quite insufficient. Also, the marines on board had far less space (95.1 cu ft each) than the ship's crew (260.5 cu ft on average).

The second transport was proposed in 1913 for FY 15. April 1913 characteristics generally followed those of *Henderson*, with draft limited to 23 ft. Mines were cut to 50. Cargo booms would lift 10-ton loads. After the ship was dropped from the FY 15 program, much the same characteristics were approved on 16 May 1914 for FY 16, except that (as in *Henderson*) the ship would accommodate 1,600 troops and 60 animals. Again the ship was dropped, to be revived in 1915 for the FY 17 program. Characteristics approved on 30 August 1915 showed special attention to cargo handling. As before, four unpowered scows would land guns and other heavy material. Of the ship's boats, three 50-ft motor sailing launches and three 40-ft motor sailing launches would land 800 men with their arms. Another pair of motor sailing launches (of both lengths) and one of four motor dories would all be used for urgent minelaying, to protect the base. Stores would be landed by two 40-ft and two 36-ft sailing launches. Two steamers (one 40 ft and one 50 ft) would be provided for the ship's use and general use. The commanding officer or embarked flag officer would have a 36-ft motor boat (barge). There were also two whaleboats, four motor dories, and two punts for general use and for use as the ship's lifeboats. It is not clear which boats would have towed the scows.

No transport, however, was included in the FY 17 building program. FY 18 characteristics, completed 19 October 1916, showed cargo boom capacity increased to 15 tons, a necessary requirement because equipment was becoming heavier. The mine outfit was restored to 100. Once again, Congress rejected the project.

Now involved in World War I, the General Board in 1917 strongly supported construction of an FY 19 transport (for the fiscal year beginning 30 June 1918). This time Congress approved construction. The transport's characteristics illustrate the Marine Corps' thinking. Because *Henderson* was being used as a point-to-point transport, she had accumulated no relevant experience. Ultimately the main improvement in the new transport was higher speed, 16 kts. A debate over her battery brought out the fact that the transport might have to shell landing places and surrounding territory to protect landing parties and troops in force. The Marines recalled the U.S. naval bombardment of the naval school at Vera Cruz, Mexico, on the morning of 22 April 1914.

It was by no means clear to the Marines, in 1920, whether their future lay with expeditionary warfare (as in China or Nicaragua) or with the prewar advanced base role. In the expeditionary role, the ship would carry a full infantry brigade (2 regiments, each regiment having 44 officers and 1,073 enlisted men) plus the brigade headquarters (9 officers and 100 enlisted) and its intelligence unit (1 officer and 10 enlisted), its depot detachment (2 officers and 27 enlisted), and its medical unit (11 officers and 44 enlisted), for a total of 111 officers and 2,327 enlisted men, about 50 percent more than the regiment previously planned.

The ship would carry 189,300 cu ft of cargo, including 6 months' provisions (80,000 cu ft) and advanced base material (65,000 cu ft). In a reversal of an earlier decision, the new transport would be able to carry 32 horses, "which are an absolute necessity in expeditionary service" in countries hardly adapted to motor vehicles. By this time the U.S. Navy had a new base defense gun in the form of a 7-in/45 dismounted from obsolete battleships. To handle it, the Marines needed more boom capacity, because the gun weighed 28,000 lbs and its mount 48,000 lbs.

The new ship never materialized, although the name *Heywood* was officially designated in 1919 and construction was assigned to the Philadelphia Navy Yard. It is not clear whether any official decision not to build her was ever made. Authorization for a second Marine expeditionary transport remained on the books between the two world wars. Beginning around 1934, a new 8,300-ton transport was included in long-range shipbuilding plans, but it was given only a very low priority.

Given their experience with *Henderson* and the characteristics of the slightly improved FY 19 ship, Marine Corps Headquarters laid out characteristics of an ideal transport in May 1919. Although the tactical unit was now the two-regiment infantry

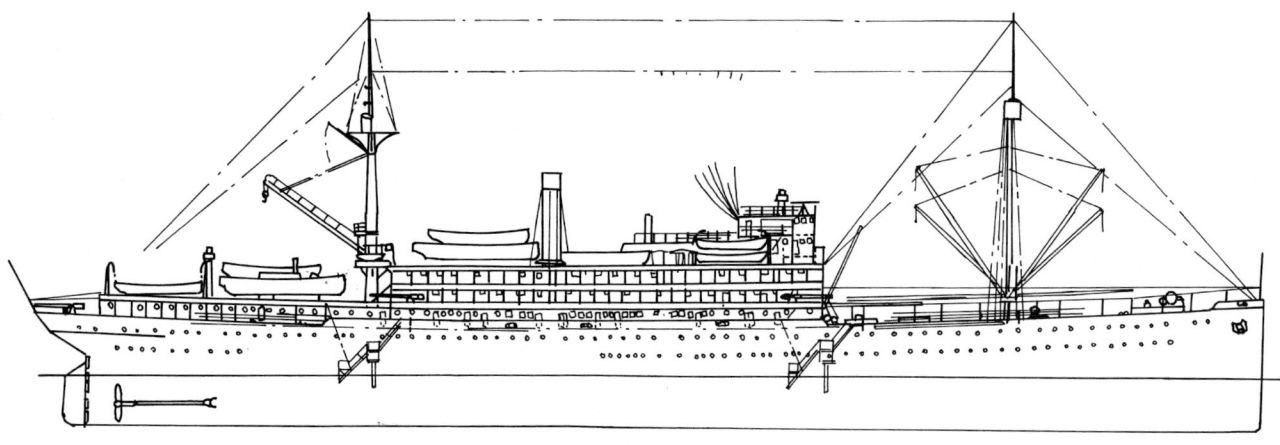

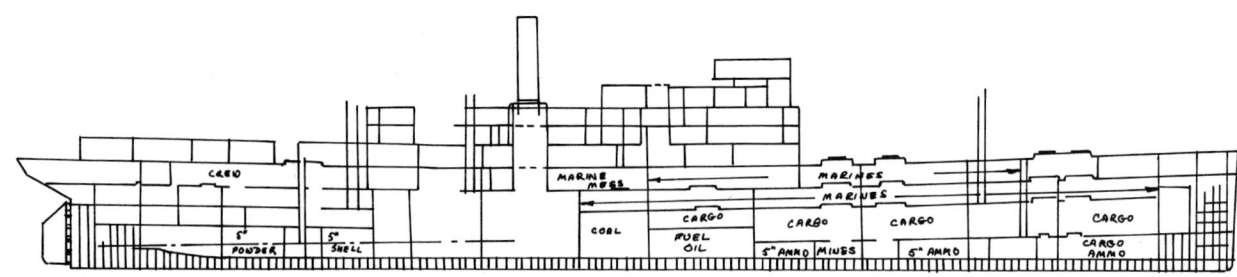

The Marines' abortive transport: the contract design for *Heywood*, approved 14 June 1919. Her proto-attack transport role shows in her unusually large complement of boats, with a heavy crane to handle them, and in her heavy gun battery (six 5-in/51). In the inboard profile, note the mine stowage; mines would have been part of the defense of any advanced base the Marines set up for the fleet.

brigade, the Marines wanted one-regiment ships. A full brigade ship would be too large, her draft excessive. Several smaller ships could be unloaded more quickly, could distribute material more conveniently for landing, and could land stores at the same time at more points along a beach, thus enabling a quicker buildup. The ideal ship would have about the same capacity as *Heywood*: 100 officers and 1,600 enlisted men, with special accommodations for 30 noncommissioned officers (NCOs). Cargo capacity would be reduced to 100,000 cu ft, presumably to release space for accommodations. As before, it would be essential to provide sufficient deck space for drills and instruction. The ship also had to provide sufficient space for unit headquarters offices, again because she would be the floating base for her tactical unit. Equipment was becoming more mechanized, thus the ship needed hatches large enough to pass heavy guns and tractors without dismantling, each served by at least one 15-ton crane. The Marines also emphasized combat loading: "Advanced Base material must be stored [so as] to be readily accessible and [to] permit of rapid handling in transportation ashore." High speed (18 kts) would help her evade submarine attack, a theme that would reemerge many times. High speed would also make it easier for the advanced base force to steam under the protection of the fleet.

The reality was that civilian ships would have to be mobilized in wartime. Plans were drawn to make that mobilization efficient. American-registered ships were assigned serial numbers. C&R produced sketch conversion plans. The first Naval Transportation Service Readiness Plan (WPL-10), dated 29 April 1924, listed civilian ships for specific Orange war roles. About 32 expeditionary transports (XAP) would be needed; the fleet's marine expeditionary force would use 10 to execute the Mobile Base Plan. Each would carry 1,000 or more officers and men, along with their equipment, at a sustained speed of 14 kts or more, with a steaming radius of 6,000 nm. These specifications would roughly match *Henderson*, except for capacity. Armament would also match *Henderson*'s: four to eight 5-in or 6-in single-purpose guns and four 3-in or 5-in antiaircraft guns. The 1925 version of the mobilization plan also referred to space for lighters on board the XAPs.

The most important available hulls had been laid down as troopships during World War I's emergency shipbuilding program. The 535-ft design (Design 1029) was completed in December 1917, and as many as 70 were initially planned. Contracts were let for 46 ships: 8 to Bethlehem Steel Sparrows Point (3 were canceled 24 August 1920); 20 to Bethlehem Steel Alameda, to be built in a new yard (10 were canceled 22 June 1918, and the rest on 15 October 1918; the yard was never built); 16 to New York Shipbuilding at Camden (7 were reordered as 502-footers); and 2 to Newport News. Ten were begun as troopships but completed as "State class" liners. Six more were redesigned as passenger-cargo ships before they were laid down. See Table 3-1 for details.

Depending on details, ships had a cruising radius of 9,800 nm at 17 kts, or 11,700 nm or 14,000 nm (in some cases the No. 5 hold had deep oil tanks). Planned capacity was 3,500 troops, but in point-to-point service actual capacity was variously given as 3,700 or 5,250.

Of the ten laid down as troopships, one became the U.S. Army's troopship *Southern Cross*, later USS *Wharton* (AP 7). Her characteristics influenced specifications the navy adopted for transports in the early 1930s. She was considerably larger than *Henderson*. As an army transport, her light displacement was 12,200 tons, and 21,325 tons fully loaded. Troops were carried on the third and fourth decks, with messes and heads on the second deck (below the main deck), and 167 officers and 18 warrant officers in the deckhouse. Marines were particularly impressed that the ship's fourth deck could be used either for troops or for cargo. When it was used for troops, the ship carried 3,030 enlisted men and 153,800 cu ft of cargo; when used for cargo, she carried 1,910 enlisted men and 276,100 cu ft of cargo (alternatively, only part of the deck could be used for troops). The ship had one 30-ton boom (forward); the others had capacities of 6 or 8 tons. Wartime armament would be six 5-in or 6-in guns and four 3-in antiaircraft guns. Six ships of this type became wartime attack transports.

The seven 502-footers (Design 1095) also were mentioned in interwar mobilization plans. They may have been planned from the beginning as a less

Chateau Thierry, an army transport, was a typical product of the World War I building program. She is shown in naval service, designated AP 31, and equipped with APA-type triple Welin davits, although not with landing craft. She helped carry the floating reserve during the Sicilian invasion, after which she was returned to the army.

expensive alternative to the 535-footer; the contract was signed on 1 July 1918, just after ten of the Alameda ships were canceled. Designed troop capacity was 2,700, but by 1924 rated capacity varied from 3,000 to 4,500. They were unusual among Shipping Board products because they had reciprocating steam engines rather than steam turbines. Ships cruised at 13 kts and were rated at 14 kts, but they had enough power to make 15 kts. Five became U.S. Army transports (two of which became army hospital ships). One became the U.S. Navy convoy transport *Kenmore* (AP 62) and later the hospital ship *Refuge* (AH 11). Six ships (of a contract for thirteen) were changed to the larger 535-ft design and completed as passenger-cargo ships from the keel up.

Also important in pre–World War I thinking were the 15-kt, 448-ft, 8,000 deadweight ton–class (Design 1024, Hog Island Type B) ships built by the new mass production yard at Hog Island, near Philadelphia. A total of 70 were ordered; 35 were canceled on 31 March 1919 and another 23 were canceled on 6 November 1919. Rated troop capacity was 1,140. At the end of World War I the army decided that, instead of operating a mix of 535-ft and Type B troopships, it would use only Type Bs. The U.S. Army was allocated all 12 surviving Type B ships: *Cambrai, Cantigny, Chaumont, St. Mihiel, Argonne, Somme, Aisne, Ourcq, Marne, Chateau Thierry, Tours;* the twelfth was delivered incomplete to the navy and eventually became the airship tender *Wright* (AZ 1).

As the army's troopship requirements declined, it transferred ships to the navy. *Argonne* and *Chaumont* became AP 4 and AP 5, respectively, in 1923; *Argonne* later became a submarine tender. *Argonne* and *Chaumont* served as transports in some pre–World War II exercises. Attempts to arm *Chaumont* foundered because of her notoriously poor stability. The two Type B ships still in army service in 1941, *Chateau Thierry* and *St. Mihiel*, were turned over to the U.S. Navy as transports (AP 31 and 32, respectively), both later becoming army hospital ships (1944 and 1943, respectively). *Chaumont* also became an army hospital ship.

The U.S. World War I building program also took over a class of eight 11,800 deadweight ton freighters being built by the British (with "war" names) at the Union Iron Works. Five, bought by the Baltimore Mail Line in 1931 and converted into passenger ships (with "city" names, for example, *City of Baltimore*), became the *Heywood*-class attack transports during World War II. A sixth, *Courageous*, became a Normandy block ship.

Besides these new ships, early mobilization plans included fast (13–15 kt) liners of prewar design and construction, completed from 1895 (SS *St. Louis*) through 1917. Because it was generally accepted that 20 years was the useful life of a merchant ship, few of these ships survived to be considered for conversion to attack transports when the crisis came in 1940.

The 1931 mobilization plan called for 120 XAPs, including ships for later conversion to other types. The U.S. merchant fleet was not nearly large enough, so 53 XAPs would have to be built for the purpose. Only half the XAPs would be permanently employed as transports, some for point-to-point duty. Other roles for ships suitable as XAPs included 2 administrative flagships (XAPFs), 30 auxiliary cruisers (XCAs), 7 aircraft carriers (XCVs), and 17 minelayers (XCMs), as well as submarine tenders (XASs) and hospital ships (XAHs). Other auxiliaries would include 50 animal transports (XAPAs), 8 fire support ships (XAPGs), 5 artillery transports, mechanized (XAPMs), 8 artillery transports, nonmechanized (XAPNs), and 3 aircraft transports (XAPVs). Of 9,483,200 gross tons of U.S.-registered ships above 1,000 tons, about 46 percent was scheduled for conversion, leaving few to carry essential cargoes. Conversely, any demand for essential cargo or transport service would preclude many essential conversions.

Soon after taking office in March 1933, the new Roosevelt administration ordered major public works intended to revive the U.S. economy. There was serious discussion of new merchant ship construction. President Franklin D. Roosevelt was well aware of the U.S. Navy's need for auxiliaries, having served as assistant secretary of the navy during World War I. He also knew that Congress generally preferred to buy combatants, and that the U.S. fleet was badly under strength. If new merchant ships had the appropriate characteristics, they might be adaptable to wartime conversion. Other major powers already subsidized ships with features that made them particularly adaptable in wartime. Later it would emerge that particular Japanese liners had been designed specifically for conversion into fast tenders and carriers.

The existing U.S. Shipping Board was the obvious vehicle for a peacetime program of new merchant ship production. During 1933 it developed designs for new cargo ships designated C1, C2, and C3, as well as refrigerator ships, a passenger ship (P), and a tanker (T). Because the ships might be taken over by the navy in wartime, their designs had to reflect the navy's needs for speed and capacity. The General Board was therefore asked to develop new characteristics for auxiliaries, including transports. There was no expectation that any such ships would be laid down for the navy in the near future, and the characteristics were not sent to C&R for design work. The Shipping Board circulated characteristics of its designs to the navy in January 1934.

The board's draft transport characteristics, completed at the end of 1933, were based partly on a review of the characteristics developed in 1914–19 for *Henderson* and *Heywood*. Transports might be used either for advanced base work (identified later as attack transports) or for point-to-point work (convoy transports). Attack transports would carry the weapons and equipment of the embarked troops; their extra cargo would take up space that a convoy transport might allocate as troop quarters. Increasing mechanization demanded not only much more cargo space but also stowage for gasoline.

The U.S. Marines would use the same ship for radically different forces, for example, pure infantry (165 officers and 3,000 enlisted) and an all-arms marine base defense force (2,500 enlisted). The all-infantry force would require 150,000 cu ft of cargo, the all-arms force 275,000 cu ft. Such figures were far beyond World War I estimates. The Marines soon pressed for at least 300,000 cu ft, three times what had seemed sufficient in 1919. The U.S. Army's troopship, the 535-ft *Southern Cross*, seemed to provide an answer: the 500-man difference between the two forces could be traded off for 125,000 cu ft of cargo space on the lowest deck. In its draft characteristics the General Board required three or more stowage decks below the main deck, with the third—the lowest—deck suited for either troops or cargo. Thus the lowest deck required the same sort of ventilation as the others.

Equipment was becoming heavier, so the General Board asked for one, but preferably two, 30-ton booms, which was twice the lift *Heywood* would have had. Some large equipment, and some large boats, could be stowed over hatches, to be unloaded first. As before, the ship should be designed for quick unloading; cargo ports in her side would be desirable. She would carry 2 months' provisions and 10 days' supply of gasoline (12,000 gal in drums) for troop vehicles. Her magazines would accommodate ammunition for field artillery: 6,000 rounds of 75 mm and 12,000 rounds of 3-in antiaircraft.

The ship would usually depend on her own boats for unloading. The ideal boat outfit comprised two 45-ft artillery lighters (direct descendants of the artillery scows of the past), six 50-ft motor launches, eight 40-ft motor launches, sixteen 33-ft motor whaleboats (a nonstandard type, these may have been envisaged as landing craft), one 36-ft motor launch, and two 26-ft motor whaleboats. Because the artillery lighters were unpowered, at least some of these craft would have been detailed as tugs. Altogether, these boats could land two infantry battalions and one battery of artillery in a single trip.

As before, the ship needed relatively high speed (15 kts sustained, loaded, using twin screws), mainly to reduce vulnerability to submarine attack. This remained the standard criterion for troopships and World War II front line amphibious ships. Endurance was set at 6,000 nm at 15 kts. There also was some interest in higher speeds. In August 1934, based on comments by the commanding officer of *Henderson*, C&R suggested raising sustained speed to 18 kts. The CNO wanted the ship to be able to steam with a 15-kt convoy (i.e., with a sustained speed greater than 15 kts), but agreed that in an emergency a marine expeditionary force might need a much higher speed. Minimum draft remained important. Later the U.S. Marines said that they wanted to limit draft so that the ship could steam up the Potomac to anchor at Quantico; embarking marines and their material at the marine base would save time, labor, and money. C&R read this as limiting deep draft to 24 ft, or preferably 23 ft. The idea of landing the ship's own guns was abandoned. However, the transport still needed a considerable battery, partly to support her troops ashore: four 6-in/47 single mounts (1,000 rounds per gun), and not fewer than eight antiaircraft machine guns (1.1 in or 0.50 caliber). As it turned out, government-subsidized merchant ship construction was not authorized until a new agency, the U.S. Maritime Commission, was created in 1936.

In September 1934 the commander of the new Fleet Marine Force pointed out that neither of the two existing navy transports, *Chaumont* (a passenger-cargo vessel) nor *Henderson*, could be made available to it. Both were fully occupied in carrying passengers and freight to and from the Orient, which for the Marines meant China. The authorization for *Heywood* was still in force. The navy was growing, and the establishment of the Fleet Marine Force suggested that its leadership understood the need for an assault ship. The commander wanted the ship to carry his full offensive strength: a reinforced three-battalion regiment, including three artillery batteries, other special-arms troops, and his force headquarters, for a total of about 2,500 officers and men, and 2 months' supplies. He would attack with two of his three maneuver units (holding the third in reserve), so he wanted boats to land two infantry battalions and two artillery batteries in a single trip. Well aware of the value of aircraft for close support, he recommended that the ship have a flat flight deck. Existing treaties, which soon changed, limited the number of U.S. carriers, but the deck might still be valuable for exercise and drill. To keep up with the fleet, he wanted a sustained speed of 20 kts. This was hardly the converted merchant ship for which the new characteristics were being written. The CNO agreed with the Fleet Marine Force, but nothing came of these ideas.

The General Board submitted characteristics for an ideal transport on 8 January 1935. It was becoming apparent that the transport might not carry sufficient landing craft. Thus the 1933–35 studies of auxiliary characteristics include a tanker arranged for deck stowage not only of crated aircraft but also of troop and artillery barges, for which she would have a 20-ton boom. For the moment, there was no money for transports; there was not even enough money for major combatant ships, which had a much higher priority.

In 1936 the U.S. Maritime Commission was created. Unlike the Shipping Board, it had money to build new ships; indeed, its program dominated U.S. commercial (non-naval) shipyards before and during World War II. The Maritime Commission designed standard freighters (C1, C2, C3) that could be modified to suit particular customers. The C designation indicated a cargo ship (freighter), the number an approximate size. C2 and C3 were conventional engines-amidships designs; for C1 there were both engines-aft and engines-amidships types. The Maritime Commission designed both steam turbine and diesel versions. Suffixes indicated propulsion and variants. C2 and C3 were faster than most freighters, because they were intended for possible naval use. However, only the C3 made the ideal speed of 16.5 kts. Many C2s and C3s served as assault flagships, transports, and cargo ships. From a naval point of view, the one great defect of the designs was that all machinery was concentrated in one space, driving a single screw.

The fleet of fast freighters built by the Maritime Commission became an important wartime resource. *Dorothea L. Dix* (AP 67, ex–SS *Exemplar*), a C3-E, was hurriedly converted into an attack transport to support Operation Torch, the invasion of North Africa; she is shown off Norfolk on 8 October 1942, with landing craft on deck. Although never redesignated as an attack transport, she served as such through the rest of the war. She and five other "Torch" conversions were called XAPs. Sisters, all built for American Export Lines, were the attack cargo ship *Almaack* (ex-*Executor*) and the naval cargo ship *Hercules* (ex-*Exporter*). The first four ships of this type were the first C3s built. Of five more of this class, two were converted into point-to-point troop ships run by civilian shipping companies.

Table 3-2. Maritime Commission Hull Specifications

	C1-B	C2	C3	Victory (VC2-S-AP2)	Mariner (C4-S-1A)	*American Challenger* (C4-6-57A)
Length (ft-in)						
BP	395-0	435-0	465-0	436-6	528-6	529-0
OA	417-9	459-6	492-0	455-3	560-0	560-6
Beam (ft-in)	60-0	63-0	69-6	62-0	76-0	75-7 (molded)
Draft (ft-in)	27-6	25-9	28-6	28-6	29-10	28-6
Displacement (tons)						
Light, average	3,797	4,373	5,212	4,442	7,684	NA
Light, U.S. Navy	5,775	7,293	8,236	4,420	NA	NA
Full load, U.S. Navy	9,104	10,850	12,349	15,580	21,093	21,053
Gross tonnage	6,750	6,085	7,773	7,609	9,216	11,186
DWT	7,815	8,794	12,595	10,850	13,418	10,714
Power (SHP)	4,000	6,000	8,500	6,000	17,500	16,500
Speed (kt)	14	15.5	16.5	15.5	20	21
Radius (nm)	13,000	12,000	15,600	NA	NA	NA
Capacity (cu ft)	452,420	562,849	730,549	453,210	736,723	625,775

NOTES: All data are for steam versions. Average light displacement tonnage is the average for the class, as merchant ships. The next two specifications (for light and full load) are typical navy tonnages. One difference between navy and Maritime Commission light tonnage is permanent ballast. Navy data are given for C1-B (*Otus*, AS 20), for C2-Cargo (*Arcturus*, AKA 1), for C3-Cargo (*Griffin*, AS 13), and for VC2-S-AP2 (*Boulder Victory*, AK 227 class). Other figures are Maritime Commission design data. Some Victory ships (AP 3 and AP 5) had 8,500 SHP engines. The postwar Mariner (C4-S-1a) and *American Challenger* are included for comparison (discussed in Chapter 12); both were considered for conversion to amphibious ships. Mariner was a Maritime Commission design and *American Challenger* was a private (Gibbs & Cox) design for U.S. Lines. Because construction was subsidized by the then Maritime Administration, it received a standard design designation. For both ships, unlike the earlier Maritime Commission types, engines had special National Defense ratings: 22,000 SHP for the Mariner, 23,500 for *American Challenger* (for 24 kts). The smaller capacity of the later ship, for much the same dimensions, reflects her finer lines, adopted to improve her cruising performance. NA indicates data not available.

In the C2 and C3 designs, the deepest holds (Nos. 1 and 3) were forward, with a shallower No. 2 hold (above deep tanks) between them. Abaft the superstructure were two shallower holds, Nos. 4 and 5. No. 3 (both Nos. 1 and 3 in a C3) had two 'tween decks; the others had one. In the standard freighter configuration, the ship had five masts or sets of kingposts, two at the ends of her superstructure, plus two free-standing forward and one free-standing aft. Details for the basic C2 and C3, as well as the smaller C1-B, which became a British attack transport, and the VC2 (Victory ship), which became a wartime attack transport, are shown in Table 3-2. The postwar Mariner and *American Challenger* are included for comparison.

The U.S. Marines began to experiment with converted destroyers, which would later be called transport destroyers (APDs). In September 1936 the fleet war plans officer suggested them as a way of preventing or reducing transport losses to enemy submarines defending islands. Troops would transfer at sea for a night run (over 200 miles) from the hostile shore. The destroyers would carry surf boats and material. They would force passages into lagoons, if necessary, in darkness, providing covering 4-in and machine gun fire as well as close support from within the lagoons. During the run-in, troops would be carried below decks. Although few historical cases of such operations could be cited, on at least two occasions at Gallipoli destroyers were used to carry troops ashore. The U-boat threat, dramatized by the sinking of two old battleships, had stopped the British from using large transports.

The Marines soon added another mission: subsidiary attacks in support of the main combat team attack. A destroyer could carry an infantry company plus part of the combat team's machine gun company or perhaps some artillery or mortars. Five destroyers could carry a transport's combat team. They could not carry heavy equipment such as tanks, however, and they might be quite vulnerable. Moreover, a converted destroyer would lose most of her military value.

On 16 March 1937 the Fleet Marine Force recommended that five old destroyers be modified as fast transports and tested in a fleet landing exercise. With destroyers in short supply, the Landing Boat Board suggested that none be converted until suitable landing craft were available. The fleet commander suggested that one or two ships, slated for decommissioning, be modified, and one ship was tried out during the 1938 fleet landing exercise. On 19–20 January 1938 the old destroyer *Jacob Jones*

By the late 1930s the Marines were interested in using converted destroyers as short-range transports. USS *Manley,* the prototype, is shown at Staten Island about December 1938, shortly after her conversion. Initially designated a miscellaneous auxiliary (AG 28), she was redesignated a destroyer transport (APD 1). Replacing her torpedo tubes were six sets of davits: four conventional sets and two frames designed to handle the new landing craft. *Ted Stone via U.S. Navy*

(DD 130) successfully carried 100 marines for 19 hrs. Fortunately, she had been fitted with an extra head for a midshipmen's cruise. Her galley proved entirely adequate. The tests included transfers at sea, but admittedly they were conducted under favorable conditions, with high visibility, moonlight, and no surf. A Marine board reviewing the experiment believed that similar results could have been achieved under less favorable conditions, however.

The board estimated that up to 200 men could be carried, without excessive fatigue, for up to 24 hrs; no more men could be carried for a shorter time. Probably 100 men could be carried for 48 hrs. The board envisaged a special conversion, in which the two forward boilers would be surrendered, thus reducing the speed to 27 kts, to provide accommodation on two deck levels, with equipment stowed under the lower level. A new forced ventilation system would supply air. A deckhouse replacing the uptakes would provide six double rooms for troop officers and the platform for a centerline gun. Landing boats would replace the ship's torpedo tubes. Head facilities would have to be enlarged, refrigerating and icebox capacity tripled, and distilling capacity increased by about 3,000 gal per day. A ship thus converted could accommodate her marines for 5–6 days. In fair weather, a largely unconverted ship would probably suffice if her torpedo tubes were landed to provide boat space. The ship probably would need additional crew to operate her landing boats.

The Marines thought that standard Norfolk Navy Yard 18-ft outboard motorboats or skiffs (capacity 10 men) would be adequate for landing, because the heaviest equipment envisaged was an 81-mm mortar. Nested in fours, the ship could carry four nests in place of four sets of triple torpedo tubes. Boats were light, so they could be launched by hand from simple pipe davits. The Marine board suggested tests

Manley is shown in the fall of 1942, fully converted, with her forward boilers replaced by troop spaces, her two waist 4-in guns replaced by a single centerline weapon, and four 20-mm machine cannon added. Barely visible atop her mast is a Canadian-supplied radar using a single Yagi antenna. The boats are 36-ft Higgins boats (LCPLs). The ship in the background is one of the three *Joseph Hewes*–class attack transports, two of which were lost in November 1942.

of the standard life raft (Carley float) or a rubber raft used for aircraft; the latter was tested in FLEX 5 in 1939. Also present for FLEX 5 was the first destroyer converted for transport use, USS *Manley*. She was initially designated AG 28, a general-purpose auxiliary rather than as a DDAP, a destroyer (DD) transport (AP), as the Marines proposed. Later destroyer transports were designated APD, that is, transport (AP) destroyer (D), and *Manley* became APD 1. *Manley* carried four surf boats in davits, which replaced her four sets of torpedo tubes.

The fleet's Landing Boat Board envisaged a much more elaborate conversion, in which not only the torpedo tubes but also all but the forward 4-in gun were landed. Six sets of davits and a ramp at the stern for a tank lighter would be installed. No accommodations were added, because the theory was that troops would not be on board more than 12 or 24 hrs; part of the main deck could be enclosed with heavy side screens and an awning to accommodate troops in bad weather.

Even an unconverted destroyer might carry marines overnight toward the beach. Such ships would use landing craft taken from the transport. To this end the U.S. Marines proposed that the special-purpose transport (discussed below) carry 30 small surf boats or outboard motor squad boats small enough for an unconverted destroyer to take aboard.

Five more destroyers were converted in 1940, as part of a larger transport program (described below). Each was initially assigned six 30-ft surf boats. By March 1941 the Marines were exploring a new mission, presumably independent raiding, and were interested in runs lasting up to 30 days. Despite resistance by the OpNav war plans and fleet training divisions, the elaborate conversion (to accommodate 130 marines), roughly along the lines proposed by the Marines' 1938 board, was approved late in April, with four davits. Although all torpedoes were landed, the ships retained their depth charges and sonars, and thus their antisubmarine warfare (ASW) capability. Their main batteries were three 4-in guns, valued for fire support; later they were replaced by dual-purpose 3-in/50s. A typical APD accommodated 148 marines and 25 tons of deck cargo, with four landing craft. Conversions of old destroyers continued through World War II, followed by destroyer escort conversions. During the war they carried reconnaissance teams and underwater demolition teams, typically a total of three

Wartime conversions of old destroyers to transports were more elaborate than prewar ones, as evidenced by USS *Kane* (APD 18), shown carrying four LCP(R)s in her davits on 11 April 1945. As in other conversions, two of her four boilers were replaced by troop spaces, and her 4-in single-purpose guns were replaced by six 3-in/50s.

ships (one team each) for a division-scale assault. Later they were also used as amphibious flagships.

In June 1938 the OpNav war plans division called for a new Marine transport. It had consistently argued that scarce peacetime funds should be spent on combatants and on the auxiliaries directly supporting them. Yet without suitable transports the Fleet Marine Force could not continue to develop its vital amphibious capability. *Antares,* which was particularly fitted to handle landing force equipment, had just been assigned to the fleet as a stores ship. Alternatives to new construction were the training battleship *Wyoming,* an existing active transport or cargo ship (*Henderson, Chaumont, Vega* [AK 17], or *Sirius* [AK 15]), or recommissioning *Capella* (AK 13) or *Spica* (AK 16). *Wyoming* was rejected because of her deep draft; the necessary changes would reduce her efficiency as a military unit. The transports in commission were so heavily used that they could not be diverted, even for a few exercises. The decommissioned ships were unsatisfactory, and they were so old that they would soon have to be replaced. The authorization for *Heywood* was still on the books.

The commander of the Fleet Marine Force asked his planning section to develop characteristics for an ideal transport. He took into account recent Japanese experience, displayed during a landing at Shanghai in 1937. Maj. Victor H. Krulak had observed a specially designed attack transport, the Japanese army's *Shinshu Maru,* launch small landing craft out of side ports as well as down a stern ramp resembling that of a whale factory ship. His reports undoubtedly influenced the 1938 Marine proposal. It referred to the possibility of developing

Roper (APD 20) is shown off Charleston Navy Yard, 21 November 1943. The object atop her bridge is a loudspeaker. The boats are LCP(R)s.

a means of hoisting boats out through cargo ports as "in a foreign Navy now engaged in hostilities" (i.e., Japan).

Characteristics for an ideal transport were submitted to the naval staff on 14 July 1938. The ship would carry a combat team: an infantry battalion, an artillery battery, and supporting arms. That equated to the same 100 officers and 1,600 enlisted men often required in the past. The Marines soon termed this a battalion landing team (BLT). It became the basis of almost all World War II transport designs. Just as three battalions formed the core of a regiment, three ships could carry a regimental landing team (the term brigade was sometimes substituted for regiment). Each would carry a proportionate share of regimental arms and stores. The Marines did not yet have the next higher formation, a division or, in reinforced form, a combat force.

Much would depend on how quickly essential equipment could be unloaded. It would be better to have more numerous smaller holds rather than a few deeper ones; the Marines chose four. Each would have a capacity of 25,000 cu ft, topped by a 'tween deck space 12 ft high, to accommodate airplanes, artillery, and vehicles. In later terms this was a compromise between cube and square. "Cube" is cubic space for cargo that can be stacked vertically to fill the entire space. "Square" is deck area for items that cannot be stacked vertically, such as vehicles. The difference between cube and square mattered more as the number of vehicles multiplied. Apart from ammunition, they were the most important cargo on board. The need for square would make it impossible to make full use of the height between decks in the deep holds of typical freighters. For example, in 1941 the Marines rejected a proposal by the secretary of the navy to motorize their divisions. It would have taken 12 more transports—twice as many as planned—to provide enough square for a division's troop carriers. At this time a Marine division required 831 vehicles, compared to 2,073 for a U.S. Army's motorized division.

An August 1940 fleet handling exercise at San Clemente, moreover, showed how important topside deck space could be. Ex-battleship *Utah* and transport *Chaumont* acted as transports for troops and equipment. Equipment could be laid out on *Utah*'s broad decks before boats came alongside. *Chaumont* had to transfer equipment directly from her deep hold to the boats alongside, a far more laborious process. Landing boats could not be lowered filled; troops had to assemble and clamber down the ship's side. There was too little deck space aboard *Chaumont* for such assembly, so loading was slow. Moreover, troops

Clemson (APD 31), a late wartime conversion of an old destroyer, is shown off Charleston Navy Yard, 21 April 1944. Note the depth charge throwers aft, which gave her a significant ASW capability. Circles indicate changes during the previous refit, which converted her into an APD.

embarked for long voyages needed deck space for exercise. As Adm. Ernest J. King said at a prewar General Board hearing on transport design, "we don't want [the troops] to be cave dwellers."

The Marines were well aware that ships supporting an assault had to be "combat loaded," a point that had emerged at least as early as Gallipoli when ships had to be unloaded and reloaded before the operation. Whatever was needed first in combat had to be most accessible. That made for inherently inefficient loading, but it was absolutely vital. By 1941, assault transports were called "combat loaders."

Transports had to have numerous landing craft so unloading could occur quickly, thus limiting exposure to air and submarine attack. Depending on other ships for landing craft would risk crippling a transport if the other ships were lost. Like ship's boats, large landing craft would be stowed on deck, atop hatch covers, because no cargo could be unloaded until boats were in the water to receive it. The Marines envisaged three platoon-size surf boats atop No. 1 (10-ton boom), three tank lighters (each carrying two light tanks) atop No. 2 (30-ton boom), three more platoon surf boats atop No. 3 (10-ton boom), and a water/cargo lighter and two 40-ft motor launches atop No. 4 (10-ton boom). The motor launches would land large numbers of men or loads of equipment or stores; they were not merely peacetime ship's boats. The hatch covers could be unloaded in parallel.

The first wave would consist of all the infantry in the platoon-size surf boats plus the six vital tanks. Once they were away, the cargo lighter could bring antiaircraft guns and a squadron of aircraft ashore, followed by the general cargo the force would need to sustain itself for 30 days. Cargo included gasoline and ten units of fire.

The Marines expected the specially built transport to supply boats for more numerous merchant ships, which would be converted in wartime. By March 1939 the big platoon boat had been abandoned and a 30-footer with one-third its capacity adopted (see Chapter 4 for details). As many as 36 craft would be needed. Hoisting them out one by one would have been prohibitively slow. The alternative was davits. At the same time the Marines called for a fifth tank or artillery lighter, and for more ship's boats (seven 50-ft and seven 40-ft motor launches). C&R argued that 14 big motor launches would crowd the ship's decks and be difficult to handle. Surely the 30-ft landing craft could substitute for them.

C&R studies of transport conversions showed a few lighters and eighteen 30-ft boats (using single davits). Not all the boats could be hoisted out immediately. The boats the Marines wanted for the ideal transport totaled 1,160 ft in length, more than twice the likely length of the ship. C&R told the Marines to settle for fewer boats. Even so, boats would have to be nested or stowed in pockets under other boats; they could not all be hoisted out simultaneously, as the Marines badly wanted to minimize unloading time. As written, the characteristics called for four tank lighters and thirty 30-ft (platoon) surf boats; the Marines wanted another six, to serve other ships. Ship's boats were cut to four 50-ft and six 40-ft motor launches. At least 16 of the 30-ft landing boats and as many as possible of the motor launches and lighters would be simultaneously hoistable.

The Marines always maintained that all of the troops' equipment and supplies should be on board the ship carrying the troops. That now included not only combat vehicles but also aircraft. As had been argued so forcefully in 1935, the Marines wanted a sustained speed of 20 kts, with an endurance of 6,000 nm at that speed.

The Marines estimated that the combination of troops, vehicles, and sustaining cargo added up to a 12,000 gross tons ship 600 ft long. C&R made no design study, but estimated the required size by comparing the characteristics of several existing liners. Early in 1939 it pointed out that the 535-ft *Southern Cross* came quite close to the desired troop and cargo capacity. Slightly over a year later two sister ships would be taken up for conversion to become the first U.S. attack transports: *Harris* (AP 8, later APA 2) and *Zeilin* (AP 9, later APA 3).

The characteristics for the ideal transport were formally approved by the Office of the Secretary of the Navy (actually signed by the acting secretary, CNO Adm. Harold R. Stark) on 25 August 1939. Although she was never built, the approved characteristics inevitably affected choices that were made when merchant ships were converted.

In 1939 the U.S. Navy finally received permission to fortify U.S. Pacific possessions. For support it acquired Grace Lines' *Santa Rita* as USS *William Ward Burrows* (AP 6). She was considered too slow

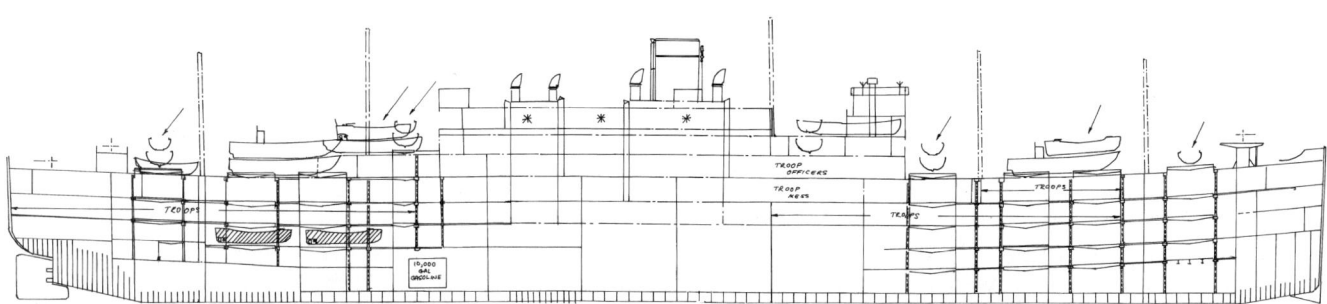

The October 1940 design for the conversion of the attack transport *Zeilin*, a World War I 535-footer, to carry 1,642 marines (119 officers, 63 NCOs, and 1,460 enlisted), plus 207 navy boat crewmen, and a navy crew of 491, including officers. This drawing was dated 16 September, as approved by BuShips letter of 7 October, and altered accordingly. The triple Welin davit had not yet been adopted, and the design shows how difficult it was to fit in sufficient boats without that device. Davits on the boat deck would have accommodated only six 36-ft landing boats, three on each side. Shading indicates 30-ft landing boats (a total of 11) in the bottom of a hold aft. Arrows indicate 36-ft landing boats on deck, some of them carried athwartships. In all, the ship would have carried 14 such boats: 6 on the boat deck, 5 on the main deck, and 3 on the bridge deck. She would also have carried five tank lighters (45-ft LCM(2): two abreast forward, and three aft, two abreast atop a pair of boats, one of them an artillery lighter) and one artillery lighter (port side aft), all on the main deck. Other ship's boats would have been two 26-ft motor whaleboats and nine 40-ft motor launches (two on the main deck, four on the bridge deck, and three on the boat deck). Partly to serve the boats, she would be fitted with new booms: 30-ton booms to handle LCMs on the after side of the kingpost in the bow, on the forward side of the foremast, and on the forward side of the kingpost aft (lengths: 50 ft for the booms forward, 55 ft for that aft). All other new booms would have 10-ton capacity: six on each of the kingpost forward, the foremast, and the after kingpost; eight on the mainmast; and two on the kingpost/vent between funnel and bridge. Guns are indicated by crosses (one 5-in/51 right aft, pairs of 3-in/50 antiaircraft guns abreast in the bow and stern) and Vs (four 0.50 caliber machine guns on the flying bridge). The enclosure at the after end of the bridge deck was a new fireproof movie locker, presumably for the embarked troops. Actual conversion was quite protracted, presumably in part due to the advent of the triple Welin davits and the abandonment of the 30-ft landing craft.

Zeilin (AP 9, then APA 3) was one of the original quartet of attack transports. She had been laid down during World War I as a troop transport, but completed as a liner. These ships could be recognized by their perpendicular bows and sterns. *Zeilin* is shown as initially converted, without radar and with liner-style davits. The big boat forward with the curved bow is a 45-ft LCM(2); nested in it is an LCV. The boats in davits are ramp-less Higgins LCPLs.

to be an attack transport. The navy also acquired a faster ship, which might have been considered quite suitable as a Marine attack transport: the army troopship *Southern Cross*, on which some of the Marines' thinking had been based. She was taken over as USS *Wharton* (AP 7) in November 1939, after war broke out in Europe. In May 1940 the OpNav war plans division, which had been instrumental in obtaining both ships, proposed that *Wharton* be converted into the special transport the Marines badly needed. The idea was rejected because the ship was being converted (to a naval transport) to commercial standards; it would cost $3.5 million to make her a combat loader. However, the OpNav proposal focused interest on the urgent need for an attack transport.

In June 1940 the world crisis suddenly became very immediate when France fell to the Germans. French possessions in the New World might well become German outposts. The United States might have to seize French or even British possessions in the New World. Apart from the slow elderly *Henderson*, no ships could quickly deploy the Fleet Marine Force. Besides being fully employed, *Chaumont* was considered too slow.

In June 1940 President Roosevelt ordered that four ships be acquired from the Maritime Commission: two World War I–built 535-footers and two smaller British-built diesel-powered Grace liners. It was imagined that appropriate merchant ships could be converted into transports in as little as five days. However, the 535-footers practically had to be rebuilt, due to their poor condition. Taken over in July 1940, they were not ready until September–October 1941. This experience led the director of fleet maintenance to urge, as early as August 1940, that the CNO not accept any more elderly merchant ships, as "it is well known that the merchant marine practice is to 'run the legs' off the vessels with the minimum of overhaul and upkeep." Ships could not really be evaluated until their machinery was broken down for examination.

The navy initially planned to fill the 535-footers with 3,478 marines, using all available space on the second, third, and fourth decks. Some would sleep in portable bunks over cargo hatches. This was hardly what the Marines meant by a combat loader. They rejected berthing on the fourth deck, which was reserved for cargo, or over hatches, which would negate quick access to cargo. That cut capacity to roughly a combat team (1,525 marines), and roughly doubled cargo capacity (from 109,613 to 230,258 cu ft). Even with this reduction, planned potable water stowage (655 tons) would suffice for only 20 days (at 4 gal per man per day); the ship needed much more. The navy planned to plate over some of the hatch area, which the Marines rejected on the grounds that it would reduce the speed of loading or unloading, and that they might well want to handle crated aircraft. Similarly, the Marines objected to limited troop ammunition (5,450 cu ft); they might want to stow ten units of fire for a base defense battalion (36,000 cu ft).

Leonard Wood (AP 25, later APA 12) was similar to *Zeilin*. She is shown on 12 September 1942, after a refit, with changes marked. They include a lattice radar mast forward and extra 20-mm guns, visible aft. She also shows triple Welin davits, carrying boats outboard.

Tasker H. Bliss (AP 42) was generally similar to *Zeilin* but she had been rearranged while under construction as a liner; note the small hold abaft her bridge. She was lost on 12 November 1942, before she could be redesignated as an APA. The two boats lying across her foredeck are LCM(3)s; note the skids for a smaller boat to be stowed atop the forward one.

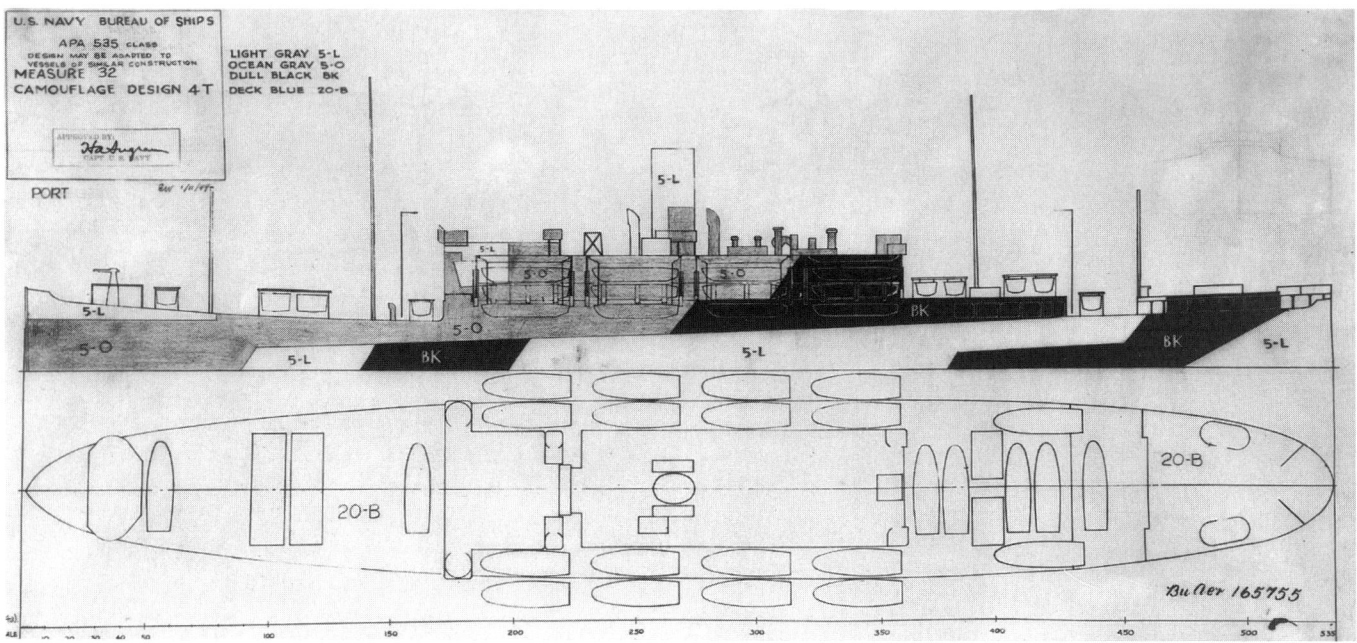

A wartime camouflage scheme shows the basic arrangement of the converted 535-ft liners, with the plan for deck boat stowage. Although such drawings are crude, they reflect the details of the more sophisticated plans, now apparently lost, from which they were drawn.

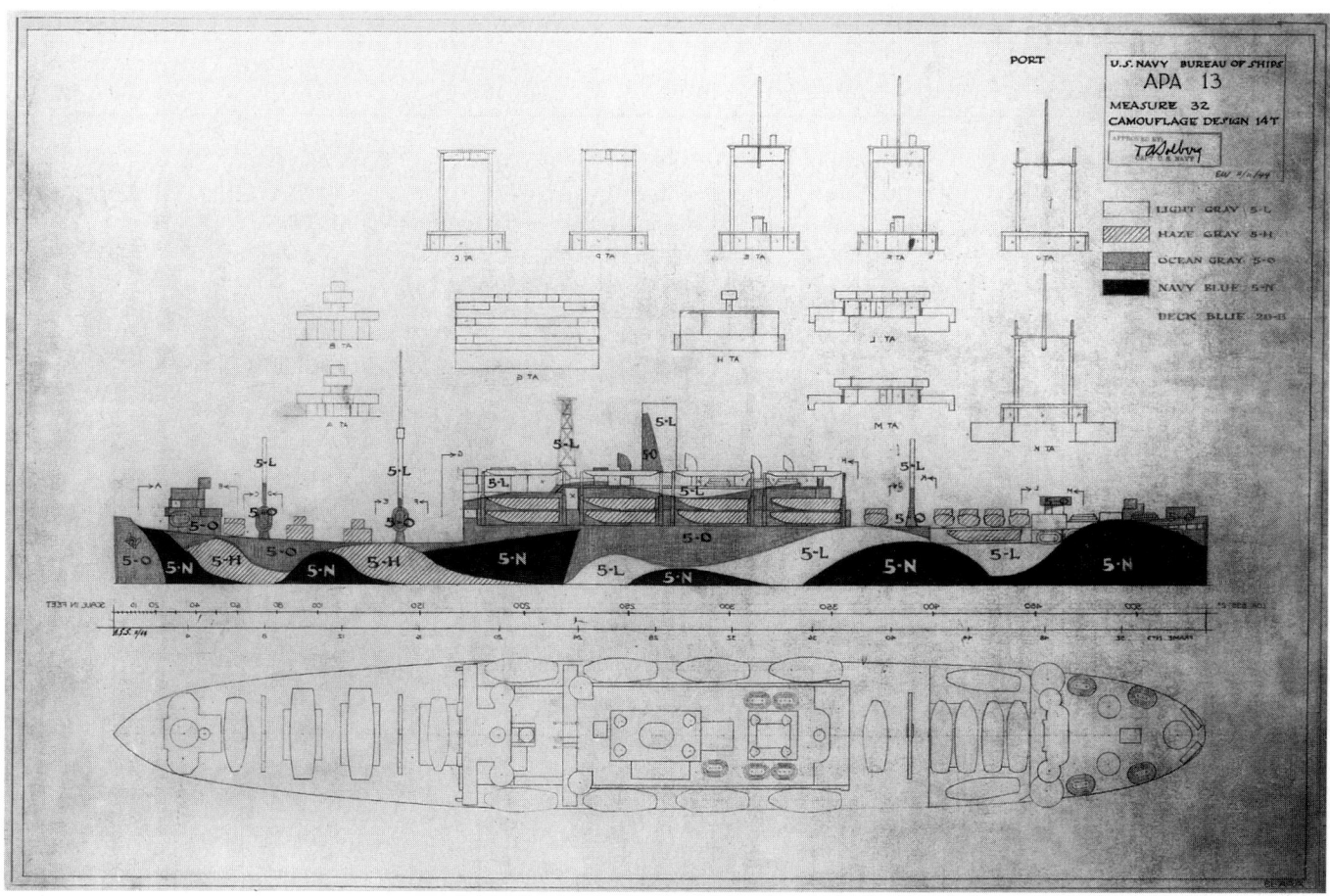

A later wartime camouflage plan for the APA 13 (*Joseph T. Dickman*) class.

The Marines also wanted provision for 12,000 gal of gasoline in drums (3,000 cu ft), with a special hoist, to serve a defense battalion for 30 days. Planned booms were 30 tons fore and aft, supplemented by 6-ton booms. The Marines pointed out that many loads would exceed 6 tons, so some of the 6-ton booms had to be strengthened to 10 tons to avoid overworking the 30-ton booms.

The navy plan allowed for twenty (later twenty-two) 30-ft landing boats, two tank lighters, two artillery lighters (in August 1940, three tank lighters and one artillery lighter), beside ship's boats (nine 40-ft motor launches and two 26-ft motor whale-boats). The Marines considered 26 landing boats the minimum for a combat team, and preferred three tank lighters and one artillery lighter. They thought the additional boats could be stowed on the boat and bridge decks. By March 1941, landing boat allowance was eleven 30-ft and fifteen 36-ft Higgins boats plus five tank lighters and one artillery lighter. Plans also called for carrying about eight amphibious tractors in the hold. Later such ships took ten sets of triple Welin davits each, accommodating a total of 30 Higgins boats.

As armament C&R offered two 6-in/50 (minimum one 5-in on the centerline), four 3-in/50 antiaircraft, and four 0.50 caliber antiaircraft. When completed, *Harris* mounted one 5-in/51, four 3-in/50, and eight 0.50s, which was standard at the time. Plans called for replacing the machine guns with 20-mm cannon. Because the main threat to such ships was air attack, by mid-1943 *Harris* had her 5-in gun replaced with a pair of twin 40-mm, and she also had ten 20-mm. This was typical among attack transports converted prewar. At the end of the war *Harris* had a quad 40-mm gun in addition to the two twins, plus twenty 20-mm. See Table 3-3 for 1945 data comparing *Harris* and other ships that were converted to attack transports during the early years of World War II.

Built in 1928, the two smaller British-built diesel-powered Grace liners, *Santa Barbara* and *Santa Maria*, acquired from the Maritime Commission in 1940 became *McCawley* (AP 10, later APA 4) and *Barnett* (AP 11, later APA 5), respectively. They were in better condition than the 535-footers, but even so it took about two months to convert them. Each could lift 1,350 men. Unfortunately, spares for the foreign-built Sulzer diesels were not available in the

Table 3-3. Attack Transports Converted Pre–World War II

	Harris (APA 2)	*Barnett* (APA 5)	*Heywood* (APA 6)	*Harry Lee* (APA 10)	*President Jackson* (APA 18)
Length (ft-in)					
WL	534-0	465-0	486-0	450-0	465-0
OA	535-0	486-6	507-0	475-4	491-10
Beam (ft-in)	72-0	63-9	56-0	61-6	69-6
Draft, full load (ft-in)	30-6	25-9	25-3	27-0	27-6
Displacement (tons)					
Light	13,529	9,432	8,789	9,989	10,192
Full load	21,300	14,080	13,525	14,564	14,430
DWT	8,600	4,600	6,100	4,500	NA
Boilers	8 Yarrow	—	4 B&W	4 B&W	2 B&W
Conditions (psi/°F)	265/Sat[a]	—	275/510	350/620	450/750
Power (SHP/shafts)	12,000/2	8,000/2[b]	9,500/1	7,200/1	8,500/1
Speed (kt)	16.0	14.5	16.8	16.0	17.5
Generators (qty-kW)					
Main	3-200	4-270, 1-300	3-150	1-300, 3-150	4-300
Auxiliary	NA	NA	1-60	1-100	NA
Fuel oil (gal)	1,245,741	416,016	524,142	503,344	461,275
Landing craft fuel (gal)	19,960	Ship fuel[c]	17,192	21,356	approx. 23,000
Radius (nm/kt)	13,700/15	9,000/12	9,000/12	12,500/12	10,700/15
Complement (officers/CPOs/enlisted)	28/23/519	30/22/464	27/23/442	27/23/407	28/23/486
Capacity					
Troops (officers/enlisted)	90/1410	94/1,288	72/1,203	80/1,048	57/1,322
Tons	2,200	1,700	2,900	1,200	3,500
Sq ft	17,252	6,468	NA	5,392	5,079
Cu ft	200,885	144,064	124,847	166,710	178,762
Water, fresh (gal)	25,741	205,741	232,000	168,050	NA
Distilling rate (gal per day)	40,000	24,000	24,000	24,000	40,000
Booms (qty-capacity)[d]					
30 ton	6	1	2	1	2-40
10 ton	6	2-15, 4-10	4	5	14
5 ton	2-7.5	12	2-7	3	—
Boats					
LCM(6)	4	—	4	—	4
LCM(3)	—	2	—	2	—
LCVP	18	23	21	13	23
LCPL	2	1	1	1	2
LCPR	1	1	1	1	1
Armament					
3-in/50	4	4	4	4	4
40-mm	2 × II[e]	2 × II	2 × II	2 × II	2 × II, 2 × I
20-mm	10	18	16	16	12

NOTES: Data are as of July 1945. Dash indicates that specification is irrelevant; NA indicates relevancy but data not available.

[a]Saturated (steam), i.e., at 212°F.

[b]Diesel-powered, by two Busch-Sulzer diesels; power is given in BHP, not SHP.

[c]Diesel powered ship; ship and her landing craft shared the same fuel supply.

[d]Only quantity is listed unless capacity is different. For example, *President Jackson* had two 40-ton booms instead of 30-ton booms.

[e]Multiple gun mounts are indicated by Roman numbers; therefore 2 × II indicates two twin mounts.

Barnett (APA 5), the former Grace liner *Santa Maria*, was one of the second pair of attack transports converted. She is shown on 19 February 1943.

United States. They were commissioned, respectively, in July and August 1940. Initially they had two funnels (one false), and carried their boats in standard merchant-type davits. Later the false funnel was removed and triple davits were installed. By March 1941 rated boat capacity was 15 Higgins boats (six 30-ft and nine 36-ft in *Barnett*) plus two tank lighters and amphibians in the hold. They were not equipped with artillery lighters. Like the 535-footers, *Barnett* lost her 5-in/51 in 1943 and had two twin 40-mm mounted instead; *McCawley* was lost before that could be done.

None of the merchant ships had enough distilling capacity. It was standard merchant marine practice to carry just enough fresh water, in double bottoms or tanks, to get from port to port; naval vessels had to be self-sustaining. Overall, it would be far better to get some of the new ships being built by the Maritime Commission.

In July 1940 President Roosevelt asked Admiral Land, the maritime commissioner, to design and build two small attack transports, which became the *Doyen* (AP 2) class. They appear to have been inspired by contemporary British ships (see Chapter 5).

In October 1940 the Joint Board ordered the navy to provide sufficient amphibious lift for an entire division of 15,000 men. To meet the new demand, in October through December 1940, five Baltimore Mail liners, which had long figured on mobilization lists, were taken over as the *Heywood* class (AP 12–16). Smaller than the 535-footers, they could lift about 1,000 men each, and they could accommodate only four sets of triple davits. In March 1941 rated boat capacity was four 30-ft and eight 36-ft Higgins boats plus two tank lighters. AP 13 was lost off Guadalcanal in August 1942 before she could be redesignated as an attack transport; the others became APA 6–9. The American Export Lines' *Exochorda*, built in 1930, became USS *Harry Lee* (AP 17, later APA 10); she was one of the company's prewar "four aces" used on the Mediterranean run. Her three sisters were taken over in 1942. Her rated boat

McCawley (AP 10, later APA 4) is shown under second-stage conversion at Norfolk Navy Yard, 10 September 1941. The stump of her dummy forefunnel now supports her 2½-m navigational rangefinder. At this stage the ship still retained her high superstructure, which would be cut down in wartime. The boats shown are just below the level of the former boat deck, and double davits replaced the original liner-type davits. Later they were replaced by triple Welins.

capacity in March 1941 was six 30-ft and eight 36-ft Higgins boats plus two tank lighters. Thus, by the end of 1940, the navy had acquired ten attack transports, although few were yet in service.

The big joint army-navy amphibious exercise in the Caribbean in the spring of 1941 demonstrated how vital it was that a landing force be accompanied by heavy equipment. A conference between representatives of the Atlantic Fleet, the 1st Marine Division, and the Navy Department decided that there should be at least one cargo ship for every three transports. The transport would carry the troops and their vehicles and essentials for the first 10 days (rather than the first 30 days) of combat. The cargo ship would carry more vehicles and their operators, as well as backup cargo. The special cargo ships were later called attack cargo ships (AKA) rather than cargo ships (AK). The four ships together would form a transport division (Transdiv), carrying a regimental combat team. Because a division is far more than 9 infantry battalions, it would require 12–14 personnel transports and 3 or 4 cargo ships.

No transport could accommodate sufficient boats to land her troops and vehicles in a single wave. The 20 assault boats of a typical combat loader, *McCawley,* provided only 340 spaces, against a troop capacity of 1,338. In 1941 plans called for substituting 36-footers for 15 of these boats, to increase capacity to about 540. Nine of the new amphibians, forerunners of the LVT (landing vehicle, tracked), in the ship's holds would add another 180, but that was still far too few. In September 1941 a single combat team required thirty-nine 36-ft personnel boats.

The number of davits was set by the ship's length (boats hung end to end). At first each davit could carry

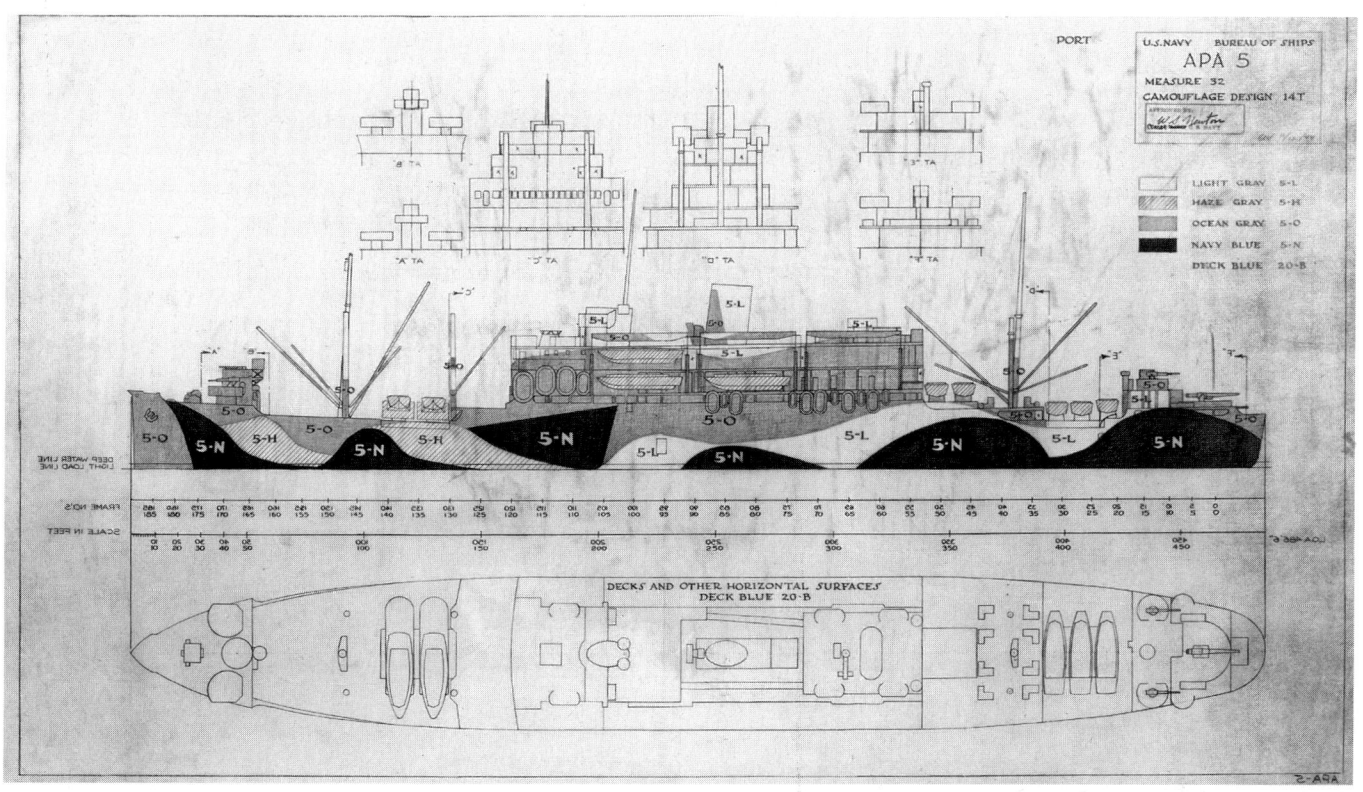

A wartime camouflage scheme provides a simple general arrangement drawing for *Barnett*. Empty gun tubs contain automatic weapons; she has one 5-in and two 3-in guns aft.

William P. Biddle (AP 15, later APA 8) is shown as initially converted, off Mare Island on 21 February 1941. She carries only a very few landing craft, and the main conversion feature evident in the photograph is newly installed gun armament.

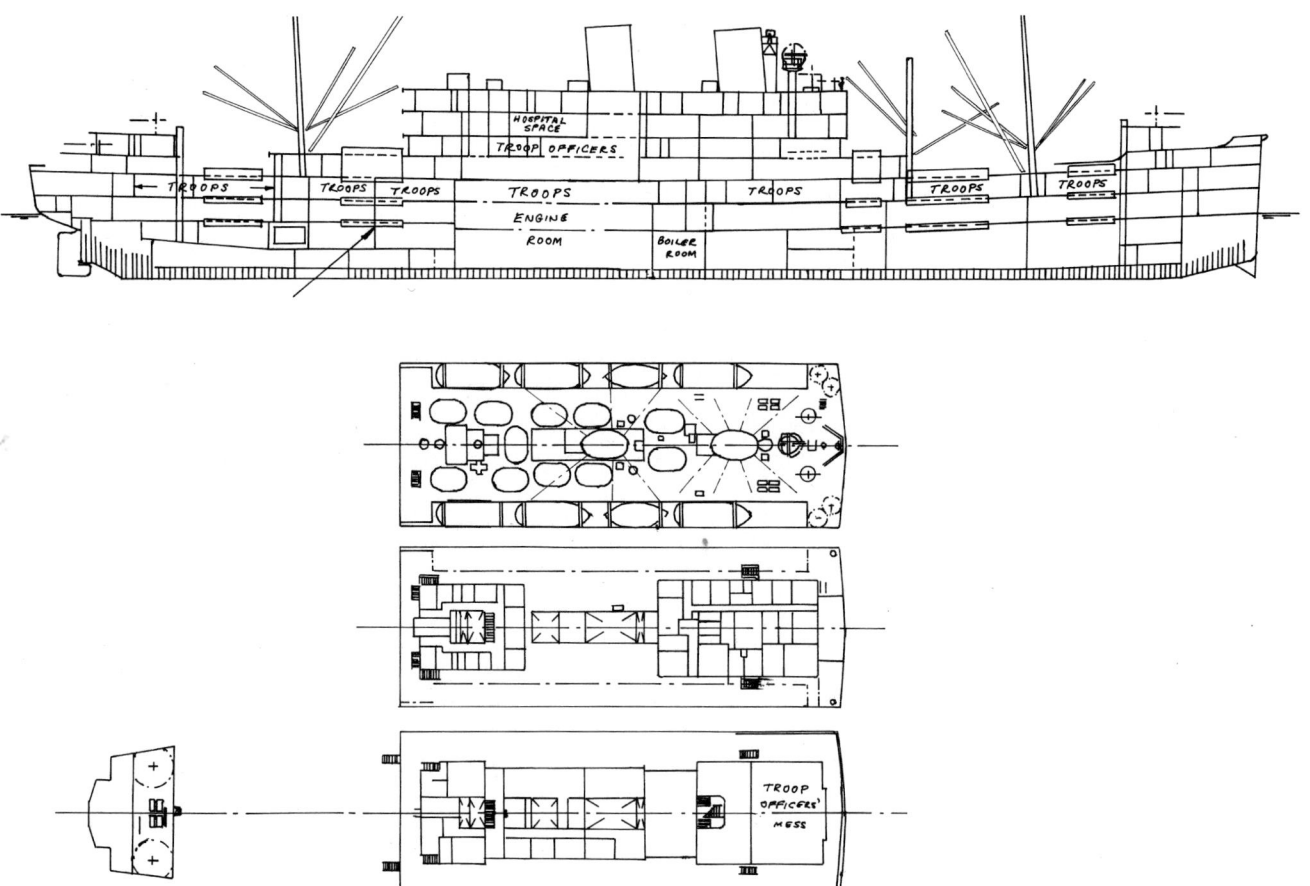

The BuShips sketch plan to convert the liner *Santa Barbara* into the attack transport *McCawley*, 22 July 1940. Liner deck designations were retained: from top to bottom they were sun deck, boat deck, promenade deck (with docking bridge aft), bridge deck, upper deck, main deck, and third deck. The hold deck has not been drawn. She had diesel machinery (the forward stack was a dummy, later removed), but also had a small boiler for ship services, shown forward of the large engine room. A new watertight bulkhead, indicated by an arrow, was installed aft largely to ensure that the ship had two-compartment stability. Near it is the newly installed 10,000 gal tank to fuel the ship's landing craft (diesels had not yet been adopted for these craft). The small landing craft shown are 30-footers. The six on the boat deck would go into existing davits, but according to a note on the original plan, the davit heads would have to be reinforced to take their weight. Between the first and second landing boat on each side of the boat deck is a 26-ft motor whaleboat. The only other ship's boat is a 40-ft motor whaleboat in which a landing boat nests, atop No. 4 hatch cover, aft on the bridge deck. Given the lack of multi-boat davits, it was difficult to provide space for all the necessary landing boats above decks. Thus 30-ft landing boats are shown on two levels of No. 2 hatch, on the main and third decks. Other 30-footers are carried athwartships on the bridge deck (2 boats) and on the upper deck (5 boats), for a total of 15 personnel boats. The two craft forward on the upper deck are tank lighters (LCM[2]). All of the booms drawn in were new. The two long ones, on the centerline fore and aft, were 30-tonners to handle tank lighters. The others were all 5-tonners. Rated troop capacity was 99 officers, 24 NCOs, and 1,368 enlisted. Armament was one 5-in/51 aft, pairs of 3-in/50 antiaircraft guns fore and aft on raised platforms, and four 0.50 caliber machine guns on the bridge. The large range finder above the bridge is an altimeter (an antiaircraft fire control device). Just forward of it are a pair of 2½-m range finders, later replaced by a single unit atop the stump of the dummy funnel. At the apex of the wind break is an Mk II range keeper, a fire control device. According to the drawing, as of mid-1940 the ship displaced 13,100 tons at her normal merchant service draft of 25 ft 1¾ in. Gross tonnage (a measure of volume) was 7,857 tons; deadweight was 6,380 tons, and net tonnage was 3,380. Troop cargo capacity was 185,850 cu ft. Distilling capacity was 12,000 gals per day. The ship carried 1,079 tons of fuel oil, for a steaming radius of 9,400 nm at her rated speed of 16 kts.

only one boat. The British introduced double-banked davits, in effect doubling a ship's boat capacity. The U.S. Navy soon learned that the British filled their boats before launching them, thus getting troops ashore more quickly. The navy was impressed by the speed with which some of the British attack transports could lower their boats by using power davits. In July 1941 the Welin Boat and Davit Corporation, in association with BuShips and the supervisor of shipbuilding at Newport News, developed a triple-banked davit, which became standard on board wartime U.S. transports. Two were stowed one atop the other under the davit trackway, the third was swung outboard from the davit arms. The first boat could be lowered directly into the water. The falls were then retracted and the traveling arms of the davit raised along the trackways

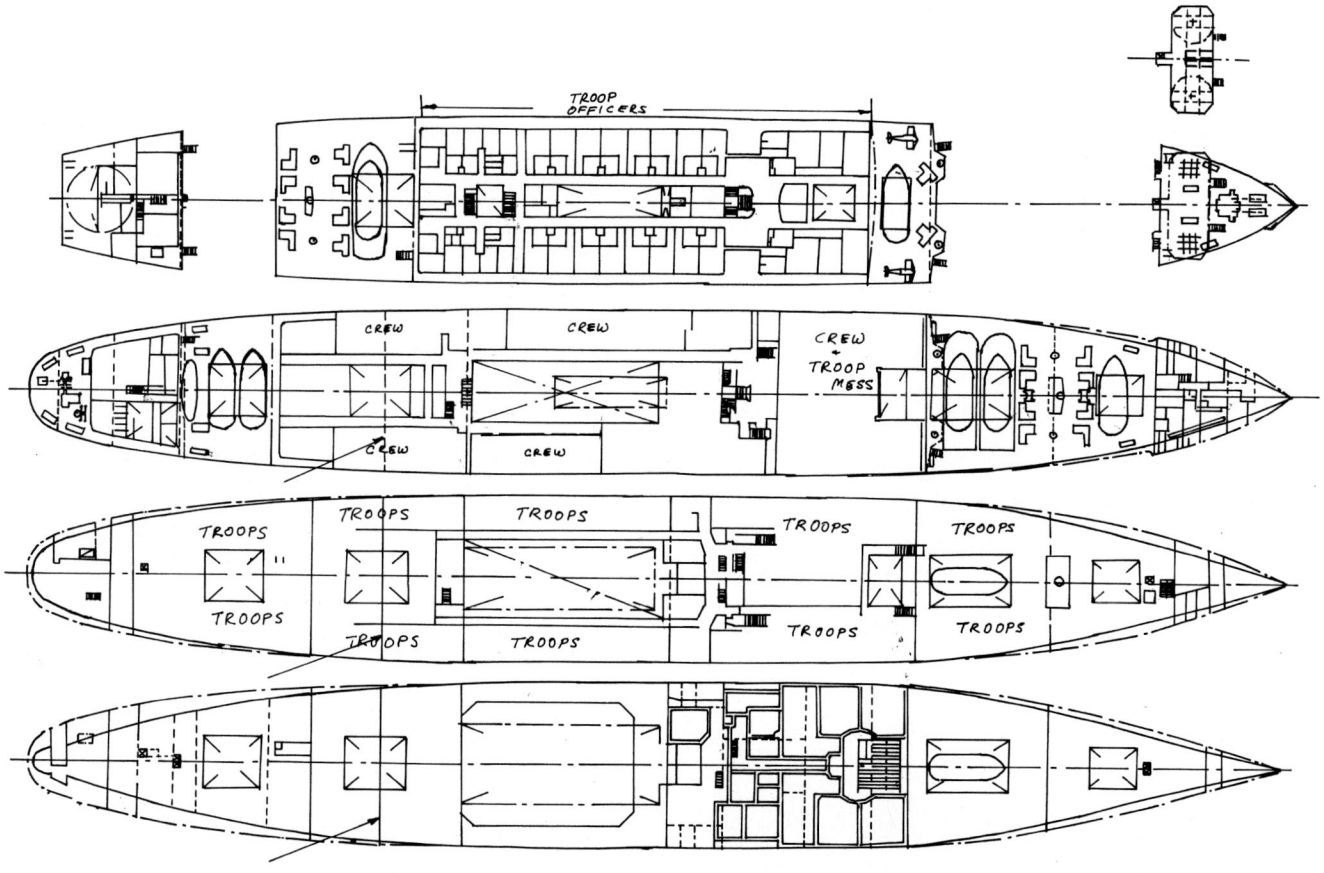

to plumb the upper of the tiered boats. After it was launched, the same arms could lift out the lower of the two. The davit was strong enough to lower a loaded landing boat. Unfortunately, no landing craft could be hoisted out if it was loaded with vehicles. Because no ship could carry enough boats to land all her troops, at least some troops had to go into boats already in the water. These craft ran in circles, each waiting its turn to load. When a group had been loaded, it assembled in line abreast (a line of departure) off the beach, then charged the beach together. This procedure was inherently slow, but even with triple (and, in a few cases, quadruple) davits no transport could carry enough boats. Initial triple davit installations were aboard the four C3-class attack transports transferred by the Maritime Commission in June 1941: *President Jackson* (AP 37/APA 18), *President Adams* (AP 38/APA 19), *President Hayes* (AP 39/APA 20), and *Crescent City* (AP 40/APA 21). The only attack transports that never received these davits were three of the converted "four aces": *Joseph Hewes* (AP 50, ex-*Excalibur*), *John Penn* (AP 51), and *Edward Rutledge* (AP 52, ex-*Exeter*). Their sister, *Harry Lee* (AP 17/APA 10), did receive triple Welin davits.

The total of ten new combat loaders (attack transports) would require three supporting cargo ships. The navy already had some World War I–built ships acquired in 1921, AK 13–17, but they were too slow for attack transport service. The Auxiliary Vessels Board considered the Maritime Commission's C1 and C2 suitable. The U.S. Navy had bought one Maritime Commission diesel C2, *Arcturus* (AK 18, ex-*Mormachawk*), in 1939 to service the new Pacific fortification program. Similar hulls were later acquired as AK 23–26 (later AKA 6–9), AK 28 (later AKA 11), and AF 11 (stores ship). The navy acquired a slightly larger, diesel C2-T (*Procyon*, AK 19, ex-*Sweepstakes*) in 1940. *Arcturus* and *Procyon* were not modified at the time of acquisition, but eventually they became AKA 1 and AKA 2, respectively. In mid-1941 one each was in the Atlantic and the Pacific fleets, together with a recently acquired C2 stores issue ship (AKS), which could also do AK duty (*Castor* [AKS 1] in the Pacific, *Pollux* [AKS 2] in the Atlantic).

In addition to these ships, on 5 August 1940 the secretary of the navy approved acquisition of a third cargo ship, a C1-A, which became *Formalhaut* (AK 22). She was the only such ship to be designated an attack cargo ship (AKA 5); in 1944 she reverted to AK status as an ammunition ship (mine transport), but was not redesignated AE 20 until she was in reserve postwar. *Sangay* (AE 10) was a sister. On 15

William P. Biddle is shown off Norfolk, newly refitted, on 12 September 1942, with changes, such as a new radar mast, marked.

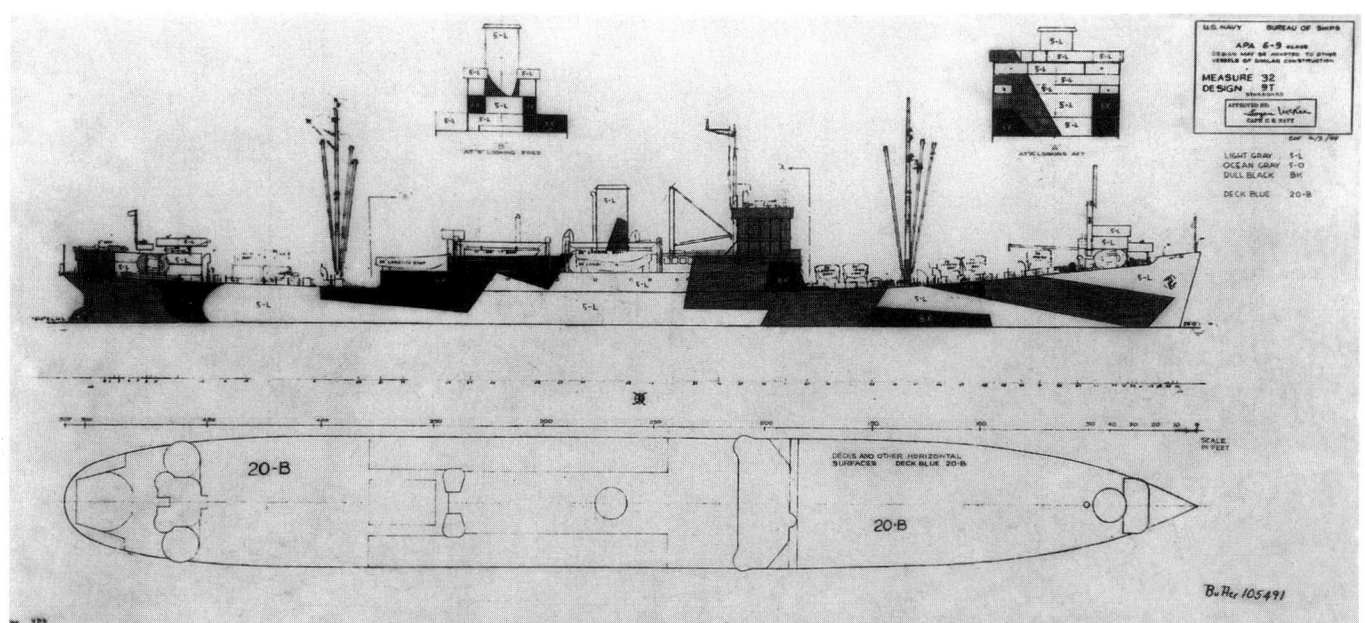

A wartime camouflage design shows the general arrangement of the *Heywood* (APA 6–9) class. Another ship of this class, *George F. Elliott,* was lost as AP 13 before she could be redesignated an APA.

January 1941 the secretary of the navy approved acquisition of five more cargo ships. The first two were diesel-driven C2-Ts, still under construction, similar to *Procyon*. They became *Bellatrix* (AK 20, later AKA 3) and *Electra* (AK 21, later AKA 4). Sisters served as ammunition ships: AE 3–6, 8–9, and 13.

Three more cargo ships had been approved but not authorized. In February 1941 the Auxiliary Vessels Board asked for another four, specifically to accompany attack transports, for a total of seven, but nothing was done.

The main feature of the attack transport was the ability to carry numerous tank lighters and tanks, thus requiring heavy-lift booms. To support the booms, heavy pole masts were stepped between pairs of kingposts, fore and aft, braced to the kingposts. Because commercial heavy-lift booms were considered inadequate, four special 40-ton booms and hooks were made for each of the two ships. Other major new features were extra crew accommodations and troop quarters, and a 12,000 gal per day low-pressure distilling plant. Armament was a 5-in/51, four 3-in/50 antiaircraft guns, and eight 0.50 caliber machine guns. Naval radio equipment was installed. Thus fitted, the ships were expected to handle nine LCM(2) tank lighters and six 36-ft landing boats. In September 1941 the boat allowance for the new cargo ship *Bellatrix* was nine tank lighters, seven Higgins boats, and two motor whaleboats. These craft could usefully supplement those on board the combat loaders, which typically carried only two tank lighters. A regimental combat team, for example, required 21 tank lighters, of which the transports could contribute 6 and the accompanying cargo ship 9. By the end of 1941 the standard cargo ship allowance was eight tank lighters, ten Higgins boats in davits, and four amphibians in the holds. See Table 3-4 for 1945 data comparing *Arcturus* and other ships that were converted to attack cargo ships during the early years of World War II.

As of February 1941, the navy had ten combat loaders either ready or under conversion. A marine division required 12, however, so another 14 were needed to support the two marine divisions. In its initial report in February 1941 the Auxiliary Vessels Board asked for the 14 ships plus 1 convoy transport. For the moment President Roosevelt rejected the request. The Lend-Lease bill was being debated in Congress, and the president probably feared that acquisition of numerous transports would signal his expectation that the United States would soon enter the war, rather than merely supply the British.

A distinction was drawn between ships taken over and ships that would become available upon mobilization (M-day). Army transports would come under navy control on M-day, but for at least three months they would be used as point-to-point troop transports. Ideally, before M-day they would be fitted for quick combat loader conversion. Of 14 army troopships, 7 had been equipped (or would be equipped by 1 July 1941) to carry 12 or more landing boats and to handle heavy equipment: *Hunter Liggett, Leonard Wood, American Legion, Joseph T. Dickman* (ex–*President Roosevelt*), *Henry T. Allen* (ex–*President Jefferson*),

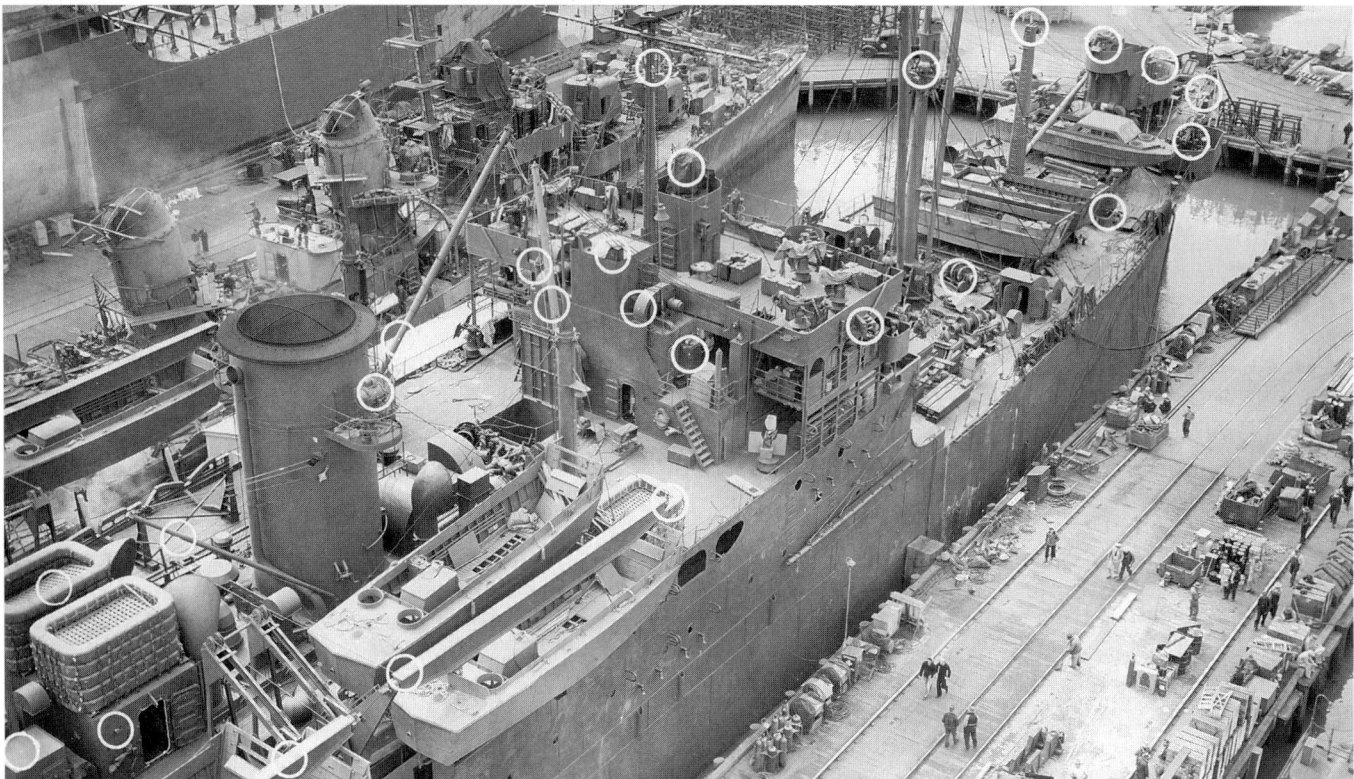

Heywood (APA 6) is shown refitting at San Francisco, 23 May 1945, with alterations indicated. The 5-in gun aft has already been replaced by a twin 40-mm mount, with a Mk 51 director. Boats carried athwartships on deck aft are LCM(3)s with LCVPs nested inside. Other boats are LCVPs, with a single LCPL, provided with a canopy as a ship's boat, forward. OpNav had rejected a 1943 call for this modification, but by 1945 LCPLs were commonly being fitted with such canopies by ships' crews, to act as Captain's gigs. Note the new antiaircraft gun mounts right forward, a symptom of the growing air (largely kamikaze) threat.

Table 3-4. Attack Cargo Ships Converted Pre–World War II

	Arcturus (AKA 1)	Bellatrix (AKA 3)	Alhena (AKA 9)	Almaack (AKA 10)	Formalhaut (AKA 5)
Hull type	C2-Cargo	C2-F	C2-S	C3-E	C1
Length (ft-in)					
WL	435-0	435-0	450-0	450-0	390-0
OA	459-0	459-1	479-6	473-1	412-3
Beam (ft-in)	63-0	64-8	66-5	66-5	61-0
Draft, full load (ft-in)[a]	25-1/26-7	25-10	21-5/27-0	25-4/30-3	23-6/23-5
Displacement (tons)					
Light	7,476	8,045	7,151	7,074	4,036
Full load	10,850	10,760	11,154	14,480	10,630
Boilers	—	—	2 B&W	2 B&W	—
Conditions (psi/°F)	—	—	450/750	450/750	—
Power (SHP/shafts)	6,000/1[b]	6,000/1[c]	6,300/1	8,500/1	4,800/1[d]
Speed (kt)	15.0	16.5	16.5	18.0	16
Generators (qty-kW)					
Main	4-250	2-285	2-300	2-300	3-300
Auxiliary	1-5	1-15	NA	1-100	—
Fuel oil (gal)	374,936	340,000	496,090	528,108	393,000
Landing craft fuel (gal)	Ship fuel[e]	Ship fuel	NA	17,900	Ship fuel
Radius (nm/kt)	19,733/12	22,600/12	11,00/12	19,345/12	NA
Complement (officers/CPOs/enlisted)	24/17/302	24/17/302	25/18/302	24/18/302	19/17/306
Capacity					
Troops (officers/enlisted)	12/100	4/50	12/150	11/156	19/155
Tons	4,410	4,515	4,885	5,175	4,500
Sq ft	23,174	23,098	31,270	33,343	NA
Cu ft	296,743	366,082	439,245	372,913	257,808
Water, fresh (gal)	63,648	54,000	NA	160,646	101,400
Distilling rate (gal per day)	13,000	10,000	16,000	12,000	10,000
Booms (qty-capacity)[f]					
30 ton	5	3-40	4	3	1
10 ton	7	4	6	4	4
5 ton	—	8	—	7	6
Boats					
LCM(6)	6	6	6	6	—
LCM(3)	2	2	2	2	2
LCVP	9	9	10	8	8
LCPL	2	1	1	1	2
Armament					
5-in/38	1	1	1	1	—
3-in/50	—	4	—	4	4
5-in/51	—	—	—	—	1
40-mm	2 × II[g]	—	4 × II	—	2
20-mm	16	18	18	18	12

NOTES: Data as of July 1945. At that time only *Arcturus*, of all the attack cargo ships, was not fitted for fueling at sea. *Arcturus* and *Bellatrix* were diesel-powered. Some data for *Formalhaut* are as postwar redesignation as cargo ship AK 22, although most data are as of late 1944. Full load tonnage given for *Almaack* is for her sister ship *Hercules* (AK 41). *Almaack*'s trial displacement was 8,600 tons; limiting displacement was 15,265 tons. Dash indicates that specification is irrelevant; NA indicates relevancy but data not available.

[a]Draft average if only one figure given; draft fore/aft if two figures given.

[b]Powered by a Sun Doxford diesel.

[c]Powered by a Nordberg diesel.

[d]Powered by a Nordberg diesel; power is also given as 4,000 BHP.

[e]Diesel-powered ship; ship and her landing craft shared the same fuel supply.

[f]Only quantity is listed unless capacity is different. For example, *Bellatrix* had three 40-ton booms instead of 30-ton booms.

[g]Multiple gun mounts are indicated by Roman numbers; therefore 2 × II indicates two twin mounts.

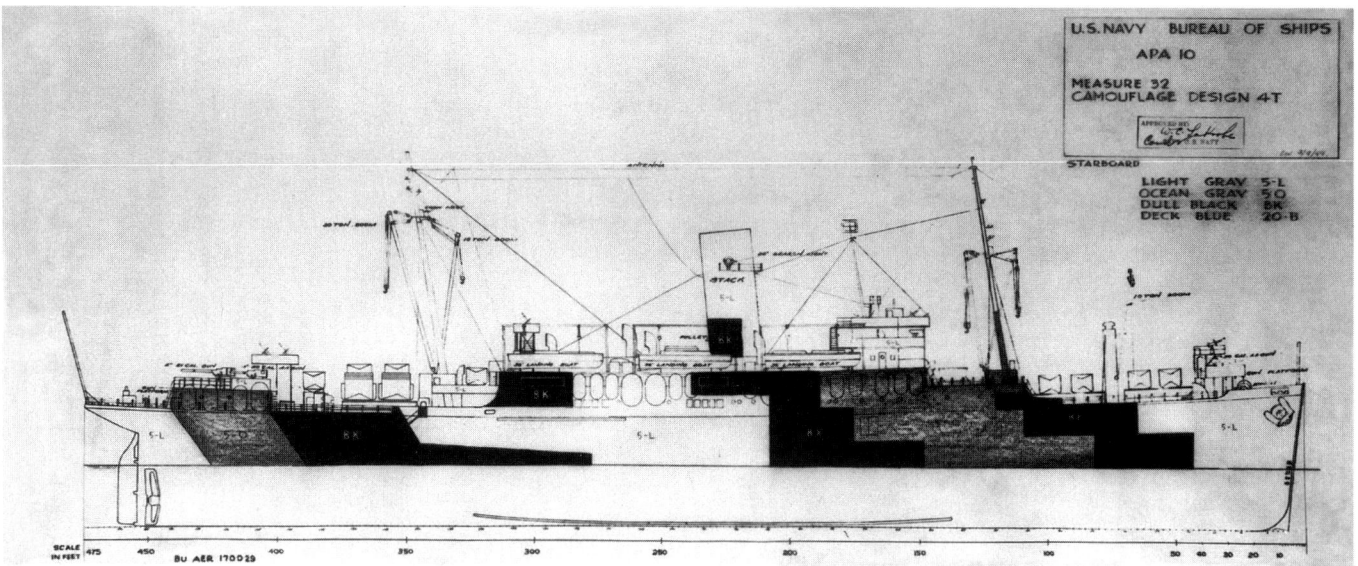

Harry Lee (AP 17, later APA 10) was acquired in October 1940. Her three American Export Lines sisters, the rest of the prewar "Four Aces," were acquired in January 1942.

Republic, and *J. Franklin Bell* (ex–*President McKinley*). All but *Republic* had one 30-ton boom, and thus they could handle tank landing craft. Two more, both World War I 448-footers, were converted for modified combat loading, without landing boats: *Chateau Thierry* and *St. Mihiel*. Their largest booms had 10-ton capacity. Although not yet in navy service, the army troop transports were soon designated AP 24–35, of which AP 25–27, 30, and 33–35 were considered full combat loaders. Of these ships only *Republic* (AP 33) was not later redesignated as an APA.[2]

In April, the secretary of the navy asked the Maritime Commission for, among other ships, a steam-powered C2-S-A1 (*Exceller*) as the third new cargo ship to support attack transports, but she was not allocated, and indeed never saw naval service. The other six cargo ships had been approved by the secretary but neither authorization nor appropriations were forthcoming. Reconsidering the situation, the Board on Auxiliary Vessels in May 1941 asked for a total of ten new cargo ships, all to be manned by navy crews and ready for service not later than 1 August 1941.

By mid-May 1941 the Auxiliary Vessels Board listed 14 ships for conversion to combat loaders (attack transports). Among new ships, the most attractive choice for conversion was the Maritime Commission's C3 passenger and cargo (P&C) type. The navy tried to convert large identical groups, in hopes of limiting the number of different types of attack transports. In some cases commercial considerations made it difficult to acquire an entire class at one time. The board chose six of the *President Jackson* class (C3-P&C) and three of the *Delbrasil* class (C3-Delta). These ships could take

John Penn (AP 51, later APA 23), one of the prewar "Four Aces," is shown off Norfolk, 13 September 1942, with alterations indicated. She and *Joseph Hewes* (AP 50, later APA 22) could be distinguished by their pole radar mast; her sister *Edward Rutledge* (AP 52, later APA 24) had a lattice. These ships retained liner-type davits during their careers; they were never fitted with triple Welins.

six triple-banked davits. It added five older ships: three 535-footers (*President Cleveland, President Pierce, President Taft*); any one of the remaining three of the "four aces" (*Excambion, Excalibur,* and *Exeter*); and Grace Lines' *Santa Clara.*

In May 1941, an assault on the Azores was seriously considered, but there were not enough troopships in the Atlantic. The navy therefore transferred the four Pacific Fleet combat loaders to the Atlantic. To replace them, it received four army combat loaders; all were 535-footers: AP 25–27 and 30. Also earmarked for the expedition were six army transports, including the huge liners *Manhattan* and *Washington,* then on army charter, and the larger commercial *America.* The navy took over all three as AP 21–23, respectively.

In return the army received seven C3-P or equivalent (535-footers): *President Coolidge, President Pierce, President Taft, President Cleveland, Panama, Delargentino,* and *Oriente.* All had been acquired by early November 1941. *President Cleveland, President Pierce,* and *Delargentino* were later transferred to the navy, and became combat loaders (AP 42, AP 43, and AP 64/APA 31, respectively).

In June 1941 the army was ordered to turn over all of its remaining troopships. The remaining 535-footers were tentatively numbered AP 42–44. AP 41 was a redesignated World War I cargo ship, initially AK 45. The designations AP 45–49 were assigned to army transports, not taken over. Overall, agreement between the two services proved elusive. The army argued that, with shipping scarce, it could not withdraw ships for extensive conversions. In September 1941 the Joint Board recommended that the army transports be converted to navy standards as quickly as possible, so long as army transport needs were

The triple Welin davit made it possible for reasonably sized ships to carry enough landing craft. Two boats were stacked under the davit, and the strongback, shown in the raised position, slid down to support a third outboard. Once it had been lowered, the strongback could ride back up to carry the upper and then the lower boat outboard for launching. This pair of davits is aboard *Du Page* (APA 41), 7 May 1945.

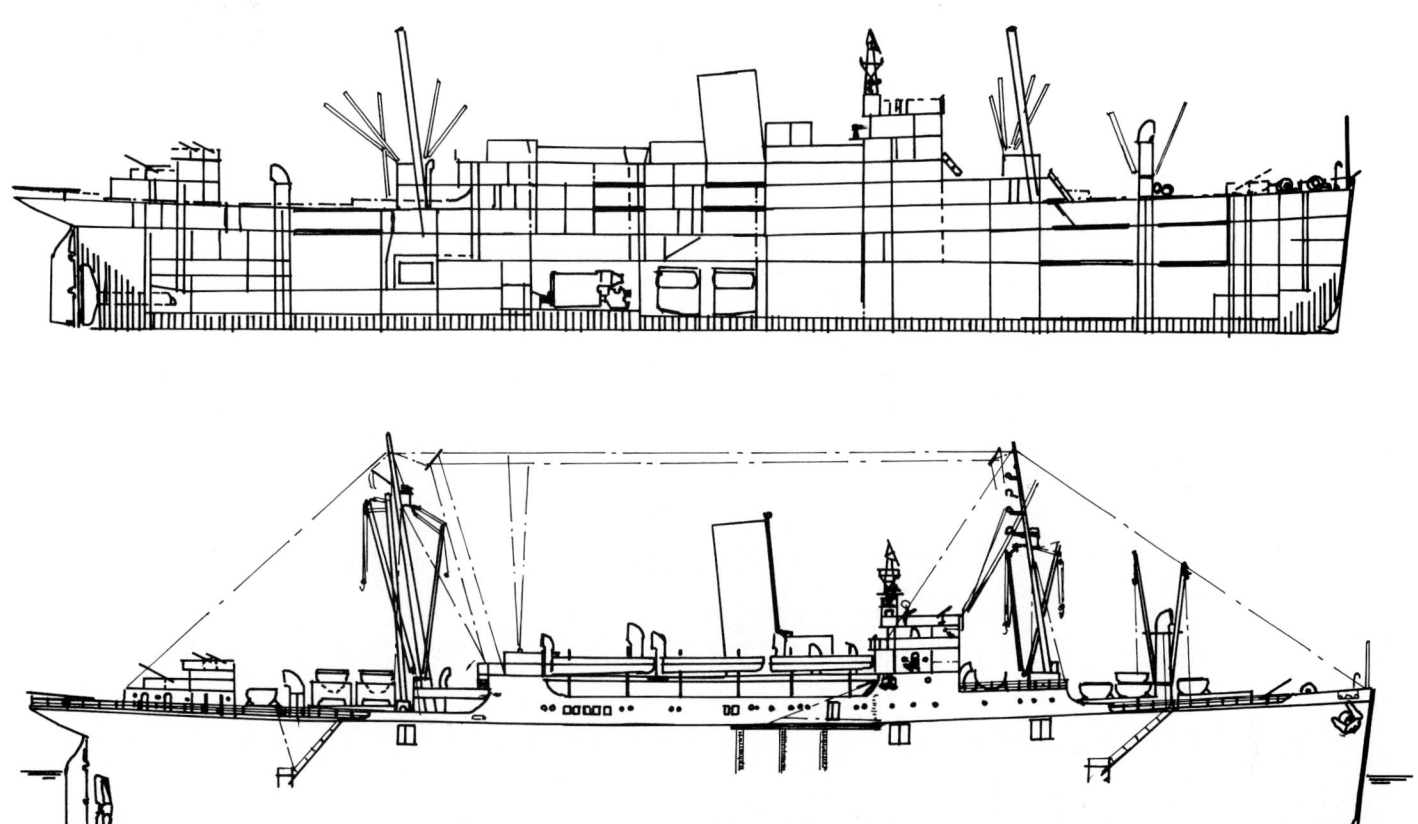

John Penn (APA 23, ex-*Excambion*) as converted to an attack transport, according to plans for herself and her two sisters dated 1 January 1942. They matched her sister ship, *Harry Lee* (APA 10, ex-*Exochorda*). These ships were unusual because they had only single davits; only late in the war did the surviving ship, *Harry Lee*, receive triple davits. Beside the six 36-ft personnel boats in the davits, the ship carried three more athwartship fore and aft, two of them nested in tank lighters (LCMs). Each of the two masts carried a new 30-ton boom to handle the LCMs. The ventilator forward supported two new 10-ton booms, and the foremast retained the original 5-ton boom and added a new 10-ton boom. The mainmast had two new 10-ton booms. Armament was a 5-in/51 right aft, paired 3-in/50 antiaircraft guns fore and aft, and quartets of 0.50 caliber machine guns atop the bridge and on the platform above the 3-in guns aft. Note the lattice tower for an air search radar, fitted to the new 1942 conversions and then retrofitted to *Harry Lee*.

met. In October the army's General Staff agreed, but its transportation branch refused to release ships unless they could be replaced, while out of service, by navy or Maritime Commission ships. The bottleneck was broken only by the outbreak of war in December. Thus, of the other six combat loaders, AP 35 followed in August, but AP 34 was not turned over until December. Neither of the two limited conversions (AP 31 and 32) became an APA. The navy found it difficult to man all of the transports; one ship, *Kent* (AP 28), reverted to army control as early as March 1942.

The Azores operation would also require cargo ships, of which the navy had as yet only two modern ones in the Atlantic, *Arcturus* and the stores issue ship *Pollux*. The Auxiliary Vessels Board wanted at least three more, two to replace the two navy cargo ships for the operation, and one to service the greatly expanded Atlantic Fleet operations. That made a total of 13, plus the 3 already provided, including the C1-A. All were soon authorized. By June, nine cargo ships

had been acquired for limited conversion as attack transports, as AK 23–31. They fell into groups. Five were C2-Cargo, similar to *Arcturus* and *Procyon*, and were designated AK 23–26 and 28 (later AKA 6–9 and 11); *Alhena* (AK 26, later AKA 9) was a modified C2-S. The other four were variants of the C3: *Almaack* (AK 27, later AKA 10) was a C3-E; *Delta* (AK 29, later AR 9) was a modified C3, the Matson version, and *Hamul* (AK 30, later AD 20) and *Markab* (AK 31, later AD 21) were C3-Cargo versions. Sisters to *Almaack* served as a cargo ship (AK 41) and as a point-to-point transport (AP 67). A sister to *Delta* served as a repair ship (*Briareus*, AR 12). Sisters to *Hamul* served as seaplane tenders (AV 8 and 9) and as transports (AP 69–71 and 76). AK 41 and AP 67, 69–71, and 76 also served as quasi-AKAs and APAs (see Chapter 6).

The Auxiliary Vessels Board sought, without success, to obtain two new destroyer tenders. With *Bellatrix* and *Electra* already under conversion, it felt that it could spare two of the nine new cargo ships ear-

USS *Procyon* (AKA 2), a C2-T, was one of the first attack transports. This photograph was probably taken in May 1943, when she was carrying up to seven LCM(3)s stowed athwartships (there was space for another alongside the boat aft), at least five of them with LCVPs nested inside, plus a pair of LCVPs stowed fore-and-aft between her superstructure and her after kingposts, and another four stowed fore-and-aft forward of her forward kingposts. Unlike her sister ships, she was not fitted with a pair of Welin davits (for LCVPs) on her deckhouse. All boats were handled by booms, the kingposts having been reinforced to support 30-ton booms to handle LCMs. By 1945 *Procyon* had quadrupods in No. 2 position forward as well as aft, and her two kingposts forward had been connected by a lattice structure, but she still had no Welin davits for LCVPs. At that time she had four twin 40-mm guns in place of the 3-in/50s originally fitted; she retained the single 5-in/38 aft. *Miller Collection of U.S. Naval Institute; U.S. Navy photo*

Bellatrix (AKA 3) displays the extra kingposts needed to handle LCMs. Like most AKAs, she shows triple Welin davits for LCVPs alongside her deckhouse. This photograph was taken in 1957.

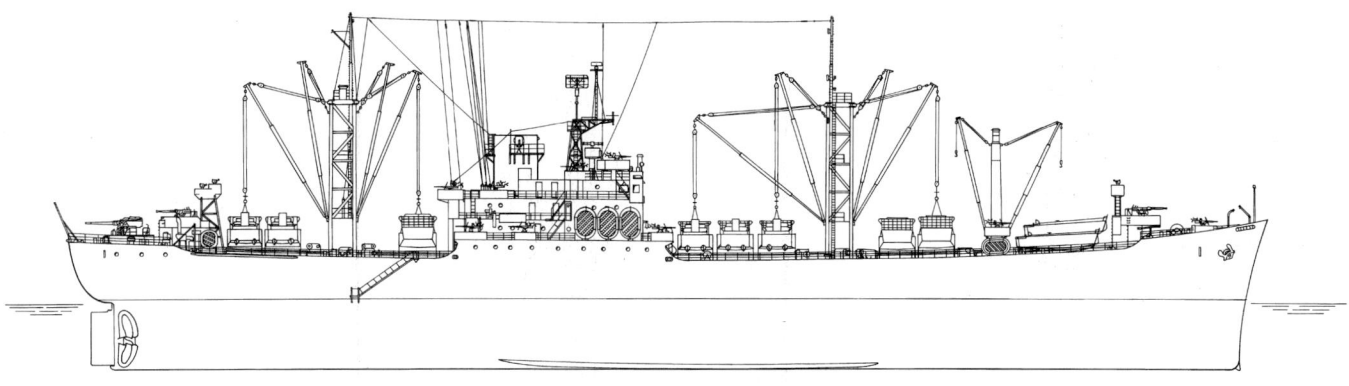
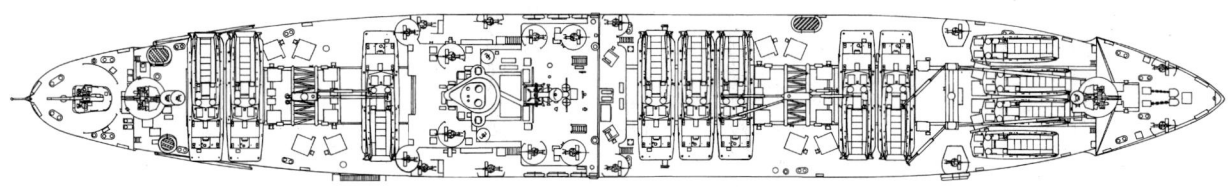

Arcturus (AKA 1) in December 1944, with 8 LCM(3) athwartships on deck, each with an LCVP stowed in it, and four LCVP in Welin davits. Each quadrupod mast supported 30- and 10-ton booms, the former on the centerline, and the latter staggered to allow for LCVP stowage on deck. *A. D. Baker*

marked as attack types; *Hamul* and *Markab* would be withdrawn for conversion as soon as the two new attack cargo ships were ready. Meanwhile they would undergo the most limited possible conversion, to serve as interim attack cargo ships. Improvements would be limited to facilities to hoist, handle, and stow the nine tank lighters and six Higgins boats; the 12,000-gal distilling plant; and appropriate radio facilities. When completion of the cargo ships was delayed, the board in November ordered conversion of *Markab* or *Hamul* even before the two attack cargo ships were ready. Both had already completed their austere conversions, and had carried landing craft. They were converted to destroyer tenders one after the other, *Markab* going first. In November *Delta* was withdrawn from the list of attack cargo ships to be converted into a repair ship.

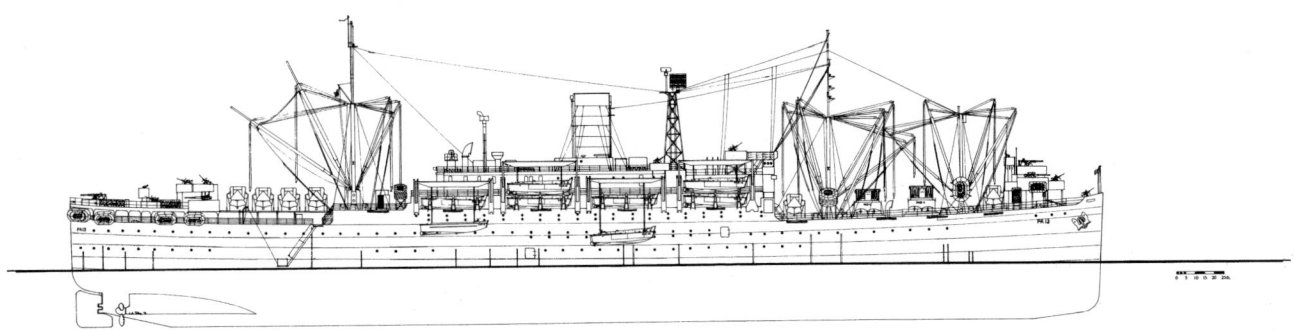

The Coast Guard–manned *Joseph T. Dickman* (APA 13) as drawn by Dr. John A. Tilley. No date is given, but the ship's armament is as in 1945, with four single 3-in/50, two director-controlled twin 40-mm (replacing the original pair of 5-in/51s), and eighteen 20-mm. In August 1945 she was also credited with eight 0.30 caliber machine guns (in November 1945 with sixteen 20-mm and five 0.50 caliber machine guns). The ship shows the usual variety of landing craft: two LCM(3) forward athwartships, at least two LCPR, an LCPL aft on a davit, and numerous LCVPs. Of this type, *Leonard Wood* (APA 12) and *Hunter Liggett* (APA 14) were also manned by the Coast Guard. In all, the wartime Coast guard manned 9 attack transports, 5 attack cargo ships, 76 LSTs, 28 LCIs, 22 transports, 15 cargo ships, 18 cargo tankers, and 33 smaller craft, as well as its own ships and craft and 288 army vessels. *U.S. Coast Guard; drawing by Dr. John A. Tilley*

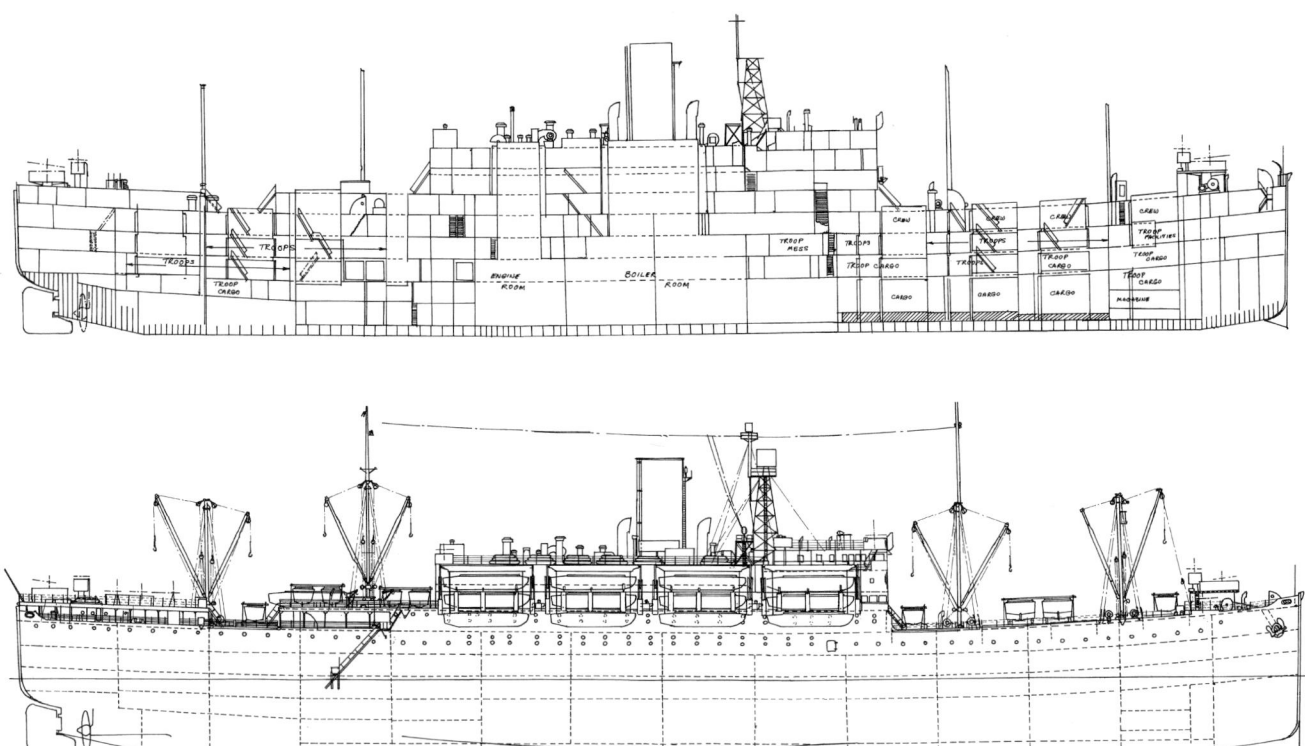

Hunter Liggett (APA 14) as of May 1945. Described as a 535-footer, she was 534 ft long between perpendiculars (actually, on the waterline) and 535 ft 2⅛ in overall, with an extreme beam of 72 ft. She accommodated 122 troop officers and 1,458 enlisted troops, plus 38 of her own officers and 422 of her own enlisted men. She carried 2 LCMs and 33 LCVPs (the booklet of plans still shows the former as 45-footers and the latter as Eurekas (and so depicted them on the original plan) but by 1945 all were modern types. Her plans show four triple Welin davits on each side (24 LCVP) plus four stowed athwartship and two lengthwise at the after end of "B" deck (6 LCVP), plus one athwartships on hatch No. 1 (forward), one athwartship on hatch No. 4 (forward), and one athwartships on hatch No. 8 aft, both on "C" (main) deck . The two LCMs are shown stowed athwartships forward, atop hatches Nos. 2 and 3, forward. Guns are indicated by crosses. The ship had twin 40-mm guns at each end, plus a pair of 3-in/50s on her poop deck and another on deck right forward. She had 20-mm guns at each corner of her flying bridge, two at the after corners of her bridge deck, and two on each side of her poop deck between the twin 40-mm gun and the 3-in/50, for a total of ten mounts. The antennas depicted on the radar mast are an SH surface search set above an SC air search set.

The booklet of plans on which these sketches are based was modified after a major refit, during which the earlier pair of quadruple 1.1.-in machine cannon was replaced by twin 40-mm guns. For unknown reasons this modification did not make its way into the official Bureau of Ordnance *Armament Summary*, which was still showing 1.1-in machine cannon as late as November 1945, with the 40-mm guns as an approved replacement. The ship was credited with sixteen single 20-mm guns, which were to have been replaced by ten twin mounts (as the plans imply was done). Indeed, as of November 1945, *American Legion* (APA 17) was also still listed with quadruple 1.1s. Planned ultimate automatic armament for the class was one twin and one quadruple 40-mm plus ten twin 20-mm. Only *Zeilin* (APA 3) seems to have been fitted with this battery. The plan views of *Hunter Liggett* (APA 14) show her main deck, the poop and boat decks above, and the bridge deck two levels above that. "G" indicates generators installed during conversion: two in a new generator house aft, and an emergency generator space on the bridge deck. "SL" indicates searchlights. Much of the piping on the bridge deck was associated with thermal tanks, presumably for primitive air-conditioning. Note the two paravanes on deck and the two lashed to the forward side of the superstructure as spares. The long rectangles forward represent LCMs.

The others were converted similarly to *Bellatrix*, but they simply used pairs of heavy kingposts to rig their heavy booms. Unlike *Bellatrix*, these and later ships had 30-ton booms and hooks. The production of those items was a bottleneck in the conversion process; initially the C2s had only one such boom. *Alchiba* (AK 23) was typical. As of mid-September 1941, she had one 30-ton, two 10-ton, and sixteen 5-ton booms. The Atlantic Fleet train commander recommended, and OpNav supported, adding a second 30-ton boom, using existing kingposts, and strengthening all 5-ton booms to 10-ton capacity; the C3s got a third 30-ton boom. *Alchiba* could take four tank lighters on deck, served by the boom, and another three in her largest hold, under the boom. She could also accommodate 14 or more Higgins boats. *Almaack*, with a 25-ton boom, could handle Higgins but not BuShips tank lighters. Her hold was too small to accommodate them, so she was limited to three on deck. *Alhena* also had too small a hold,

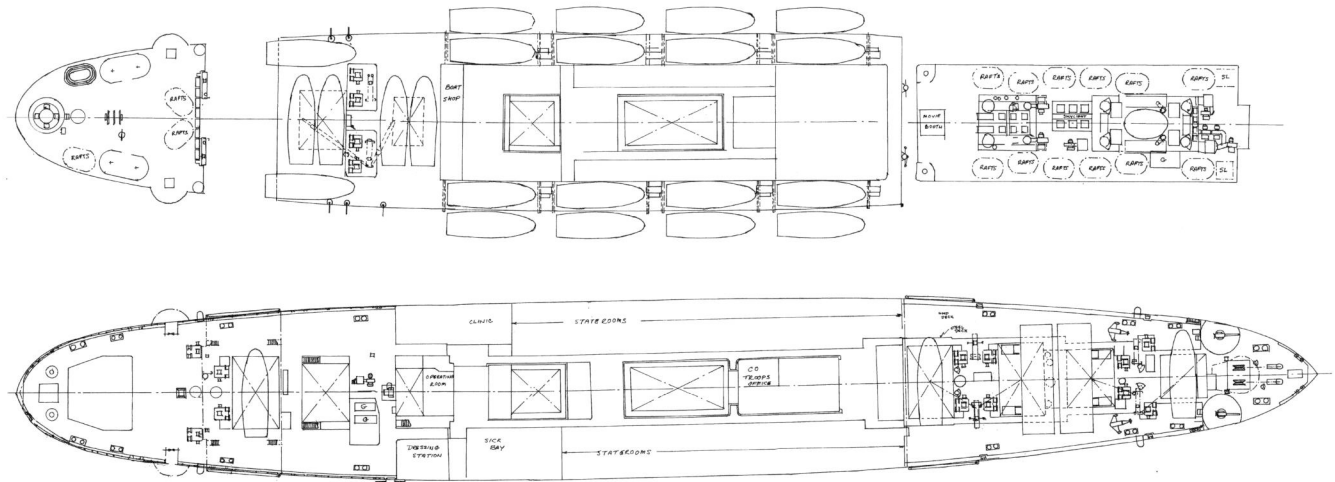

although she could handle three BuShips lighters with her 30-ton boom. Required capacity was nine tank lighters and six Higgins boats. *Formalhaut*, the C1, could not take the weight of two 30-ton booms plus a full boat complement.

Charleston Navy Yard drew up conversion plans, which were approved early in September 1941. Merchant ships did not meet naval survivability requirements, so Charleston planned to close up tonnage openings, which reduced the ship's paper tonnage, to reduce port and canal charges, between the holds above the second deck, and to permanently close the watertight doors from the engine room in the shaft alley. Initially gun foundations would be installed, but the stack would not be raised and no extra quarters would be built. The design called for using the upper 'tween deck of No. 4 hold to berth and mess the enlarged crew, and one of the deep tanks

President Adams (APA 19) was one of the first group of modern Maritime Commission–built liners taken over as attack transports.

beneath for the big new distilling plant. The other tanks would be used for seawater ballast. Other important conversion items were quarters for troops (12 officers, 12 NCOs, 150 enlisted), permanent degaussing, splinter protection, and naval radio equipment.

By September conversion plans included not only AK 23–28 but also AK 18 (*Arcturus*, later AKA 1). *Procyon* (AK 19, later AKA 2) and *Formalhaut* (AP 22) were added by 1 November. These three early ships took the place of the now transferred AK 29–31 (*Delta*, *Hamul*, and *Markab*). As of early September, *Arcturus* had one 30-ton, three 10-ton, and fifteen 5-ton booms. She could accommodate four BuShips tank lighters on deck, and three more in her 50-ft hold; thus she was comparable to the AKs being converted. Given the bottleneck, none of these ships got her full complement of four 30-ton booms until mid-1942. For example, *Procyon* was fully converted at Mare Island in July–August, receiving her forward pair of heavy booms and LCM chocks to match. Ultimately AKAs typically had one 30-ton and one 10-ton boom for each of four hatches, with lighter booms for a fifth. As initially converted, ships also lacked triple Welin davits amidships, although these were later installed. Ultimately ships typically had a pair of triple Welin davits amidships carrying six LCVPs. Hatch covers carried eight LCM(3), with an LCVP, LVT, or DUKW (amphibious truck) nested in each.

Initial armament was one 5-in/51, four 3-in/50, and eight 0.50 caliber machine guns (with foundations to take 20-mm cannon). Several entered service with 3-in/23s in place of their 3-in/50s, but by 1943 all had director-controlled 3-in/50s. Eventually the single-purpose 5-in/51 was replaced by a dual-purpose 5-in/38, and the 3-in/50s by twin 40-mm guns. The machine guns were replaced by 20-mm machine cannon.

An unusual feature of AKA armament was a pair of depth charge throwers (three 300-lb charges

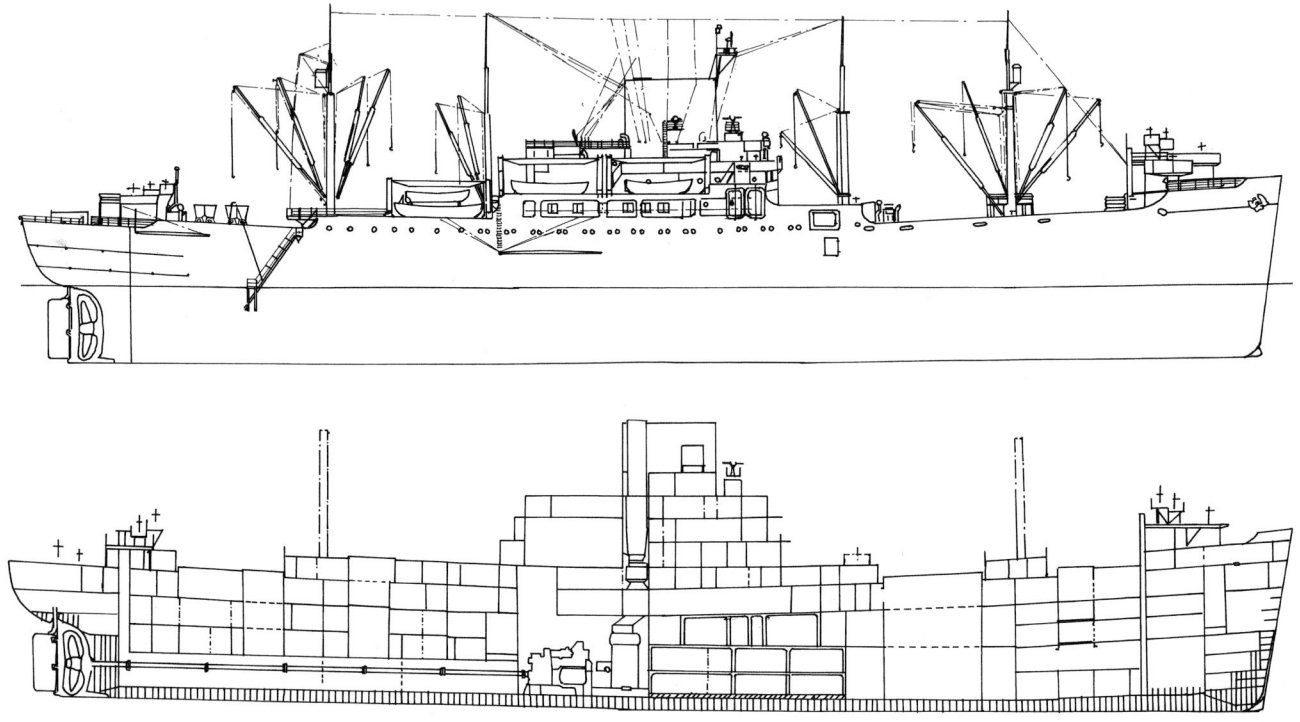

President Adams (APA 19) in April 1949, modified as a dependent transport. Like other Maritime Commission–built ships, she benefited from the advanced steam conditions being adopted by the U.S. Navy. Compare the size of her machinery spaces with those of the World War I–built *Hunter Liggett* (see illustration appearing earlier in this chapter). To the extent that a transport was volume critical, more compact engines made it possible to carry comparable numbers of troops and amounts of materiel on a substantially smaller hull. Modifications for the dependent transport role included elimination of troop accommodation in the forward hold, which was needed to carry dependents' cargo, including cars. Ballast is indicated by shaded lines. The object inside the uptake housing is the emergency radio room. The ship retained her triple Welin davits, but in this incarnation they carried conventional lifeboats. *President Adams* did retain two LCPLs, which are shown on deck aft. Passenger capacity was 88 in staterooms, 18 NCOs or CPOs, and 996 enlisted passengers—not too far short of wartime troop capacity. Enlisted passengers were berthed mainly on the three decks below the main deck, around the trunking of the hatch to No. 1 hold. Armament, shown as crosses in the drawing, amounted to two twin 40-mm, two 3-in/50, and two twin 20-mm guns aft; two twin 20-mm just abaft the hatch for No. 2 hold; and two 3-in/50 and one (centerline) twin 40-mm gun forward. The radar on the platform stepped from her funnel is an SC-2, and below it is the usual navigational range finder fitted to attack transports. The very similar *President Hayes* (APA 20) and *Thomas Jefferson* (APA 30) differed in having a pole radar mast separate from her funnel. *President Jackson* (APA 18) had the radar platform stepped from her funnel.

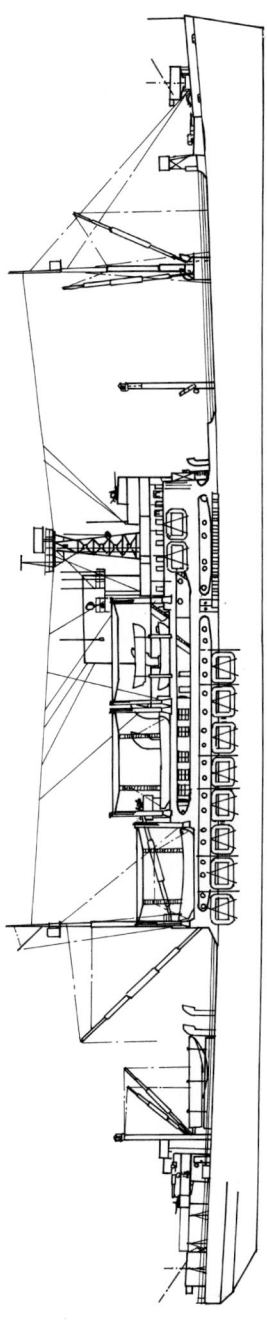
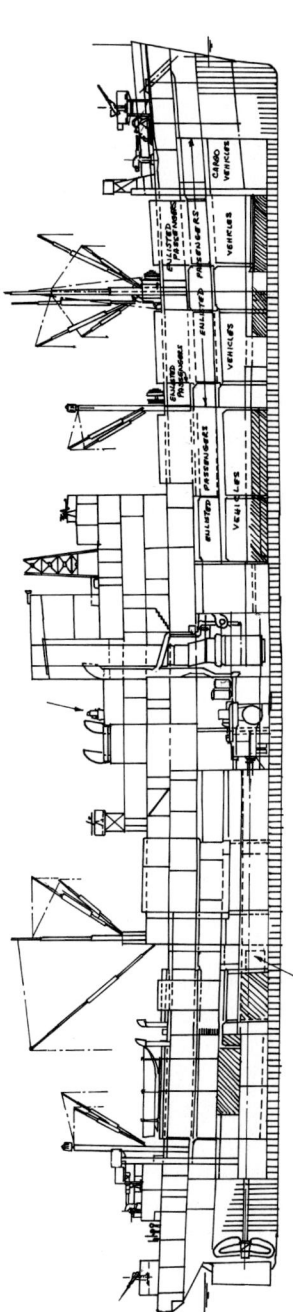

Crescent City (APA 21) as a dependent transport, March 1947. As converted, she could carry 18 NCO and 871 enlisted passengers, whose accommodations approximated the positions of wartime troop quarters. With dependents on board, however, the ship also had to provide facilities such as a nursery. The only landing craft carried were a pair of LCPLs, but the wartime Welin davits survived. In the outboard profile, they accommodate standard merchant-type boats, four 36-ft motor launches and four 26-ft motor whaleboats. Surviving armament comprises pairs of 3-in/50 guns fore and aft, with a twin 40-mm on the centerline fore and aft, and six twin 20-mm (paired atop the conning station, on a platform between the 36-ft motor launches, and on the platform aft just abaft each 3-in/50). The foremast carries two 10-ton booms, the forward kingpost 5- and 10-ton booms, the mainmast a 30-ton boom (for LCMs) and a 10-ton boom, and the kingpost aft a pair of 10-ton booms. The kingposts are also ventilators for the holds. The fog oil (smoke screen) generator and tank are indicated on the inboard profile. This ship was converted into a temporary hospital evacuation ship at Pearl Harbor in March 1945, and then used as an evacuation transport at Okinawa.

Photographed off San Francisco on 18 May 1945, *Tryon* (APH 1) shows the exotic raked lines left over from her liner design. Her small funnel projects from a streamlined structure, also left over from the liner design, used as a solarium. Unlike APAs, she did not carry LCMs.

each), which were removed early in 1944. In mid-1943 these weapons were ordered installed on board Atlantic and Pacific Fleet service force cargo ships and tankers capable of 15 kts or more. They seem to have been intended to provide a limited degree of self-defense for ships fast enough to sail unescorted.

Alhena differed from the other AKs; her heavy masts were free-standing (not flanked by kingposts) fore and aft. After being torpedoed off Guadalcanal on 29 September 1942 she was rebuilt at Sydney, with quadrupod masts (as in later AKAs) fore and aft. *Arcturus*, *Alchiba*, and *Almaack* were all refitted late in the war with new-type quadrupod masts, which replaced the kingposts. *Almaack* was unusual among AK/AKAs because she had a counter, rather than cruiser, stern.

With the situation worsening and with Lend-Lease approved March 1941, President Roosevelt ordered the Maritime Commission to begin turning new ships over to the navy. It began with three C3-A passenger-cargo combinations: *President Jackson* (AP 37), *President Adams* (AP 38), and *President Hayes* (AP 39), later reclassified as APA 18–20, respectively; and the C3-Delta *Crescent City* (AP 40, ex-*Delorleans*; later APA 21). Like the earlier ships, they were initially armed with one 5-in/51 and four 3-in/50. They exchanged their single-purpose 5-in guns for two twin 40-mm, but *Crescent City* exchanged her 5-in/51 for a 5-in/38. Secured by early November 1941, these ships brought the combat loader force to 18, against a requirement for 24. By this time a third divisional lift was being discussed, initially for another amphibious division and later for an army armored division, thus bringing the requirement to 36 combat loaders.

Because the situation was urgent, initial conversion was quite limited. BuShips warned that in the time available it would be impractical to provide gun batteries (other than machine guns), landing craft stowage and handling facilities, and anything more than make-shift berthing and messing facilities. Additional evaporators could be installed, provided they had sufficient priority and ships had sufficient power; otherwise, ships would make do with additional fresh water tankage, for about 30 days' supply.

As in an aircraft carrier, embarked craft added considerably to a ship's complement. For example, when commissioned after preliminary conversion in January 1941, *George F. Elliott* (AP 13) and comparable ships had 15 officers and 225 enlisted men. By September 1941, with boats aboard, it was 28 officers and 334 crewmen. With even more boats and tank lighters, as already contemplated, it would be 375 enlisted men.

One other type of transport was conceived before the United States entered the war. In October 1941 the Bureau of Medicine and Surgery proposed a casualty evacuation ship (APH), which would take casualties from a beach during combat. The Auxiliary Vessels Board recommended converting a 502-footer. At this stage the ship was expected to function as a convoy transport when not used to evacuate casualties. By the

As first converted, *Tryon* had a smaller, more streamlined, funnel. The characteristic solarium structure atop her deckhouse is shown clearly. Photographed on 10 October 1942, she is still carrying merchant-type lifeboats rather than landing craft.

time plans were developed in November 1941, the ship was apparently envisaged as a combat loader that could evacuate casualties after a landing. Thus she was to have carried landing craft (nine 36-footers and two 45-ft tank lighters). As a convoy transport, she would have carried 132 officers and 2,318 troops; alternatively, she could carry 1,014 patients. Armament would have been the usual combat loader battery, a single 5-in/51 aft, four 3-in/50, and eight 0.50s (with foundations to take 20-mm cannon). Unfortunately, shipping was so tight that no ship was available. The 502-footer *President Madison* became available in March 1942, and the board proposed that she be acquired and considered for APH conversion. Instead, *President Madison* became the navy convoy transport *Kenmore* (AP 62).

Meanwhile, three C2-S1-A1s under construction for Alcoa were being completed with APH features as the *Tryon* class (named after a former navy surgeon general); they became the only APHs. These ships could and did function as attack transports, at least on the way into a landing. They could carry roughly the same troop load as an APA, or they could evacuate about 700 casualties afterwards. Shorter than other C2s, they had been designed to carry 102 passengers. As an APH, each had a distinctive solarium surrounding her small funnel. They also had unusual raked masts and an unusually large superstructure overhanging the hull. With C3 machinery, they were also somewhat faster than other transports, at 18 kts. They were completed carrying conventional lifeboats. Before the end of 1942 they had been fitted with six triple Welin davits, but apparently never carried tank lighters (LCMs). Armament, at the end of the war, was one 5-in/38, four twin and four single 40-mm, and six twin 20-mm guns.

A transport and her landing craft: the converted destroyer escort *Blessman* (APD 48) at Mare Island, 7 August 1945, with an LCP(R) visible in her starboard Welin davit. Changes made during her final wartime refit are outlined in white, including a new director for her twin 40-mm guns. *Blessman* was a former *Buckley*-class destroyer escort, converted in 1944.

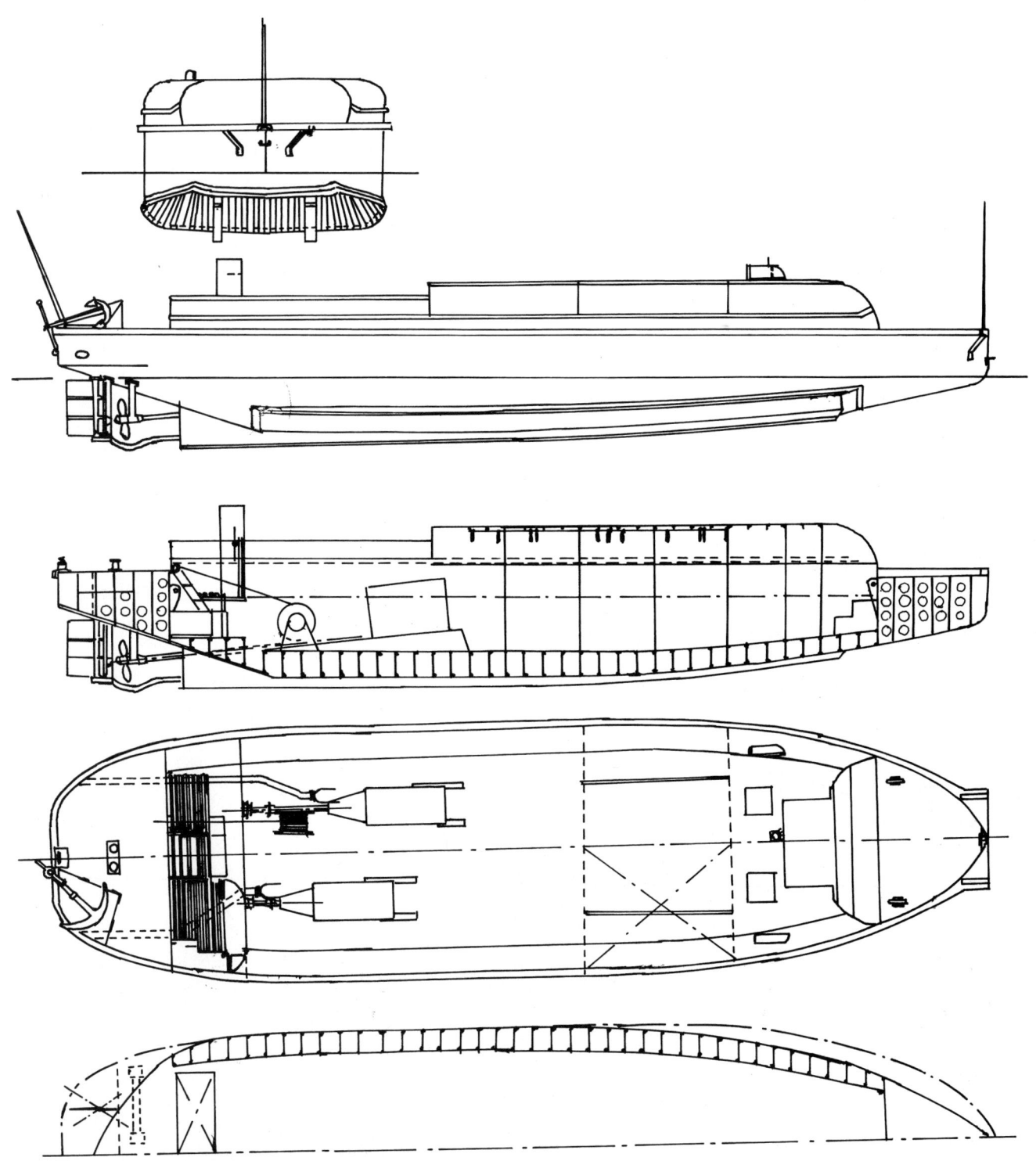

Barge A, the personnel landing craft derived from the Marines' original "beetle boat." This illustration, derived from the C&R design sketch produced about 1926, shows her two V-4 gasoline engines and her pair of 100 gal gasoline tanks aft. Capacity was 120 troops in "expeditionary force heavy marching order." The vertical lines above the troop compartment are subway-style grab straps to help heavily laden troops stand up when the craft beached. Dimensions were 50 ft 4 in (overall) or 50 ft (molded) x 13 ft 4 in (overall) or 13 ft (molded) x 3 ft. A portable roof section made of ⅛-in bullet-proof steel, flanked by hinged sections, is indicated. The original drawing was undated, but was probably drawn about 1925, based on its C&R number (122030).

the beach at Culebra, troops quickly disembarked dry; others who had landed in motor sailing launches were wet to above the waist. According to the formal report of the exercise, the beetle showed a marked tendency to sheer even in smooth water, and it was clearly unsuited to rough water. Shafts and propellers were so high that when the boat rolled it brought one and then the other out of the water, causing the boat to yaw dangerously. The amidships engines and fuel took up too much space.

C&R circulated a sketch design of an improved version, Barge A, in May 1925, and OpNav recommended that one be built for tests. The first of three, ordered in 1926, was completed in 1928. Because the barge was no longer required to carry a 75-mm gun, its canopy was made much lower, for reduced topweight and hence better seakeeping. The engines (4- rather than 6-cylinder) were moved aft and the propellers moved deeper and protected by skegs. These 50 × 13 ft craft were protected against small arms, and could be armed with machine guns or 37-mm cannon. They could carry 120 men at 10 kts. They weighed about 42,500 lbs, placing them beyond normal crane capacity, and hence precluding further production. One underwent surf tests in mid-1930.

The boat's armor could not protect against anything heavier than a light (e.g., 0.30 caliber) machine gun. When one swamped and sank en route to an exercise, it revealed a third design defect: the overhead canopy kept water in, so that any flooding could easily prove fatal. As late as 1938–39, however, when characteristics of their ideal transport were being planned, the Marines thought in terms of platoon-size landing craft clearly descended from the beetle boat.

In reviewing the plans of Barge A in July 1925, the OpNav war plans division argued that only a few could be carried by an expeditionary force. Something smaller, which could be distributed much more widely, was needed. The division proposed a steel boat carrying 50–70 men with light removable

The rationale for Barge A was that wheeled artillery could be winched ashore by an unpowered artillery lighter, shown here being loaded.

Artillery lighters were attractive because they were less expensive than powered landing craft. C-1981 was delivered on 17 December 1940. Although carried by prewar transports, such lighters were not used during World War II.

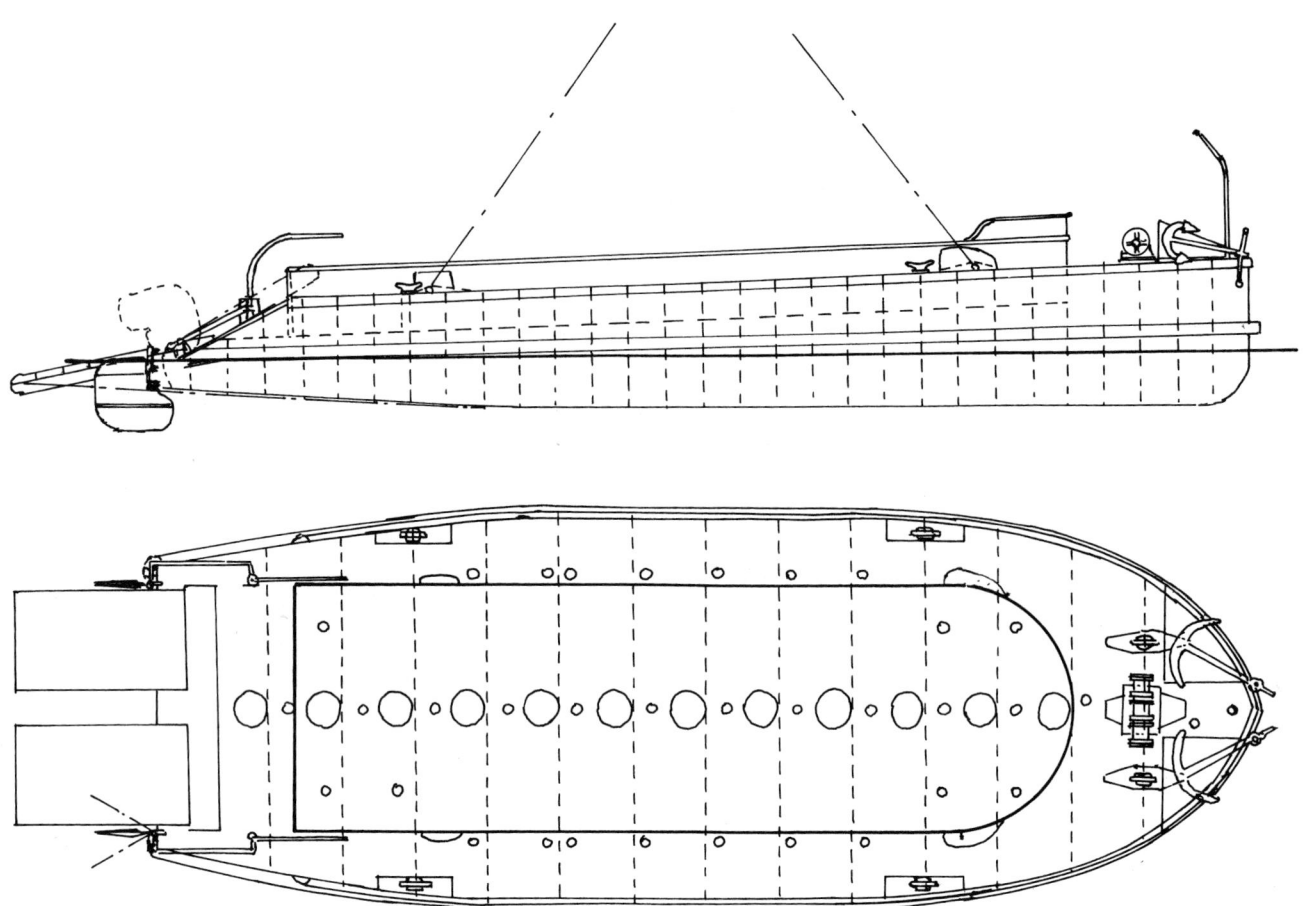

The 45-ft artillery lighter, as designed about 1926. Figuring prominently in interwar U.S. thinking, it remained in production as late as 1940. Carried aboard early transports, the lighter was towed to the beach, then winched in to shore stern first. Note the stern ramp shown in folded and unfolded positions. The rationale for a non–self-propelled lighter was that the troops landed by powered boats would secure the beach; artillery was for follow-up. Hoisting lines are indicated.

overhead protection, capable of towing an artillery lighter. The following year OpNav surveyed merchant ships likely to be used as transports in wartime, and concluded that the barge should not weigh much over 12 tons; the limit was later set at 25,000 lbs. In March 1926 C&R offered a 23,000-lb 40-footer (9-ft beam). OpNav suggested that it be enlarged to 42 × 10 ft (25,000 lbs) to accommodate 15–20 more men. It does not appear that the craft was built.

Earle wanted a self-propelled artillery lighter, but C&R protested; the Marines agreed that the artillery boat could be towed and then winched ashore. Heavy artillery would not be needed until after the marines, landing from their protected boats, had secured the beach. The resulting Lighter B (APB), 45 × 15.5 ft (38,100 lbs; maximum draft 2 ft, aft 1 ft 2 in), was built at Norfolk Navy Yard in 1927. Unfortunately, Lighter B was too heavy for nearly all ships' cranes. Nor could it carry the next major artillery piece, the 6-in gun (35,700 vs. 26,000 lbs for the 155-mm). In 1934 the Marines called for a new craft, which C&R soon produced as Experimental Lighter C (see 30-ft Landing Craft section in Appendix B for production information).

Artillery lighter specifications were incorporated in the designs of early U.S. attack transports, as noted in Chapter 3. They were valued because they were much less expensive than powered lighters, and much easier to build. Several prewar exercises showed that they were of little tactical value. For example, the two that participated in the February 1940 exercises were used mainly to transport potable water rather than equipment (the LCM was soon modified to perform this service). They were difficult for surf boats to tow; many surf boats had clutch trouble as a result. They continued to be built in small numbers, however. None saw World War II combat service.

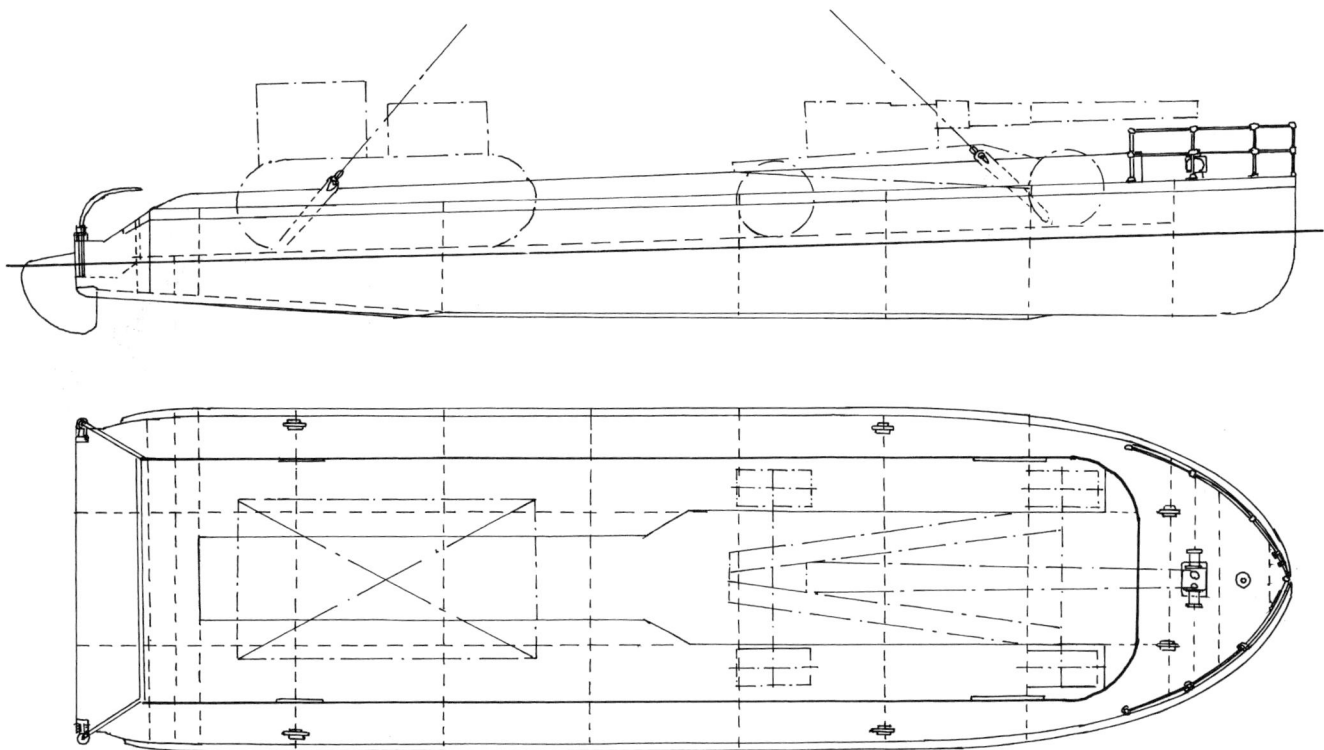

The 50-ft artillery lighter, from a 17 February 1926 Bureau of Construction and Repair sketch (C&R 128784). Because the 45-ft lighter could not accommodate the 155-mm gun the Marines adopted in the 1920s, the enlarged 50-ft lighter was designed to carry both the gun and its tractor (outlined in dashed lines). Hoisting lines are indicated. The 50-footer, however, was never built.

Without specialized craft, the Marines depended on standard ships' boats. Thus, the archives of the U.S. Naval War College contain a 1928 inventory of the fleet's boats, a guide to selecting those to be used for a landing. In 1927, the 50-ft motor launch was rated at 142 marines, with full equipment, for the first wave. Later waves would carry more equipment, so the same boat would carry 114. Abandon-ship capacity was 190. The boat could carry 10 tons of equipment, although for tanks, guns, and caissons figures had to be reduced by about 40 percent because loads had high centers of gravity. Lesser boats were the 40-ft motor launch (68/54 men in first and later waves, 5 tons), 36-ft motor launch (52/42 men, 3¾ tons), 33-ft motor launch (38/30 men, 2½ tons), 30-ft motor launch (30/24 men, 2 tons), and 24-ft motor launch (14/11 men, 1 ton).

None of the ships' boats was particularly suited to landing in surf, but they did offer an interim capability. The Marines developed a canvas surf screen to protect the boats' exposed engines as they ran through surf, but it did not always work. During FLEX 3 in 1937, ships' boats generally tried not to come too far onto the beach, to avoid damage. Troops had to jump over their sides into fairly deep water. Many of the surf covers failed; two 33-ft motor launches were swamped by waves breaking over the screen before troops could disembark. Many rudders and propellers were badly damaged.

The only way a cash-strapped U.S. Navy could plan for massive wartime expansion was to mobilize civilian resources. In May 1928 OpNav issued guidelines for selecting civilian craft. Converted troop barge Type A (XAPP) would be stripped of decks and deckhouses to accommodate at least 40 troops. It could be up to 50-ft long, with a draft of up to 4 ft, reinforced to permit grounding. The barge would transport troops over distances of up to 8 miles, in the face of opposition, with a radius of action of 40 miles. It would be armed with one to three 0.30 caliber machine guns in its bow. These characteristics were not too different from those of World War II landing craft. Mobilization plans assumed that not enough barges would be available, so many troops would ride ships' boats towed by converted troop barges Type B (XAPT), with similar range and limiting draft, armed with two machine guns. Because Type B was mainly a tug, decks and deckhouses would not have to be removed. Both types of barges appeared in the 1929 list of civilian ships and craft to be mobilized in the event of an Orange war. As late as 1940 the APP designation was being applied to specially built personnel landing craft. It was soon apparent, however, that no really suitable craft existed. Civilian craft had unprotected propellers, and they were generally not strengthened for hoisting. Their decks, which would have to be removed, were part of their structural strength, their very varied engines would be a maintenance nightmare, and their odd sizes would make stowage difficult. As in the case of troop carriers, mobilization plans called for a converted artillery barge (XAPB), presumably powered, to carry field artillery guns and tractors ashore in the face of opposition.

By 1932, the Marines no longer wanted large landing boats like the Type A barges. The troops they landed would be bunched too tightly in the face of enemy machine guns. There was a problem, however. Transports could accommodate only so many boats, and during the 1930s the Marines' new landing doctrine called for landing all combat troops in one trip. Moreover, as far as possible transports should carry all the boats and lighters they needed. That was practical with standard ships' boats but not with the smaller surf boats developed in the late 1930s (see discussion below). Thus as imagined in 1938 the ideal transport might carry as many as five lighters (tank or artillery) plus 36 × 30 ft, 7 × 40 ft, and 7 × 50 ft motor launches, giving 1,876 troop spaces. Minimum capacity was set at five lighters, and 18 × 30 ft, 7 × 40 ft, and 7 × 50 ft motor launches, for 1,498 boat spaces. That is why plans for the ideal transport generally included substantial numbers of ships' boats alongside surf boats.

Any specialized amphibious craft seemed rather expensive in the Depression era. Thus the CNO in July 1936 suggested that a towed lighter might do dual work: perform general-purpose cargo lighter duty at naval bases and, in wartime, carry tanks or heavy artillery. The Bureau of Yards and Docks designed a 50,000-lb, 50 × 17.5 ft lighter that could carry a 95,000-lb load. Although the Marines were reluctant to adopt this cargo lighter, in September 1936 the District Craft Board suggested building one as a matter of urgency. This craft seems not to have figured in prewar exercises, although on one occasion the Marines did use an unpowered 500-ton lighter to transport tanks, and they did consider using converted coal barges.

Boat Rig "A" was an extemporized solution to the problem of bringing motorized equipment ashore. The Marines limited themselves to the small Marmon-Harrington tank (shown), specifically to simplify transport ashore.

As of 1936, the Marines were more interested in the pontoon floats that the Bureau of Yards and Docks was developing. Rafts of pontoons might be used to land heavy equipment. For the moment, the Marines concluded that it would be too difficult to assemble such rafts at sea. However, the pontoons matured spectacularly during World War II (see Chapter 6 for details).

To disembark light artillery and combat train equipment, the U.S. Navy and U.S. Marines developed Boat Rig A, a platform atop a 50-ft motor launch leading to a bow ramp. It weighed 4,700 lbs and could be installed in 30 minutes. The rig was tested successfully during Grand Joint Exercise No. 4 at Oahu in 1932. Unfortunately, the load was so high that the boat became unstable. The ramp was carried broken down; once the boat beached it took 8 men 10 minutes to set up. During FLEX 4 in 1938 the Marines tried something more elaborate, an "iron horse," which was built at Quantico. Once the ramp had been erected, its inboard end was hooked onto the iron horse, whose legs held it up in the water. The launch could back away to pick up more vehicles, the ramp remaining at the beach. The horse thus saved the considerable time required to rig and unrig the ramp each time the boat carried a vehicle to the beach. It was pronounced a success. As of August 1940, 11 boat rigs and 6 boat ramps existed, and another 10 boat rigs and 5 ramps were under contract, scheduled for delivery about 1 December. Plans for naval expansion included 40 rigs and 20 ramps.

In March 1934 the Marines proposed a program to develop specialized craft capable of negotiating surf. To minimize losses in the first wave, they wanted each boat, which they called an X-boat, to carry only a squad (12 men) in addition to a crew of 6 (1 officer, 2 gunners, 1 loader, 1 coxswain, 1 deckhand). Tests had shown that the squad would hit the beach in attack position, and that its advance would not be delayed to permit adjoining squads to get into position (as would be the case with larger boats). The X-boats had to close rapidly with the beach, so speed was set at 15 kts. Armament would be four 0.30 caliber machine guns paired on aircraft mounts (ultimately some boats might be armed with a mortar), and the boat would be armored against small-arms fire. Boats in later waves could carry more troops. A companion Y-boat would carry a section (two squads and a section leader). In a battalion landing, the first wave would comprise 13 X-boats, the second 13 Y-boats; later waves would be standard navy boats.

At about the same time Andrew Jackson Higgins visited the Marine Equipment Board to promote his "Eureka," a fast, 20-kt shallow-draft hard-chine boat, which he had designed in 1926 specifically for loggers and fur trappers who had to cruise the bayous. British historians later claimed that it had been designed for bootleggers and moonshiners. With its solid wood spoon-bill bow, it could jump the bars and submerged logs common in Louisiana bayous; it could also easily run up on a shore. A skeg protected its propeller and rudder. Thus it already fit the beaching requirement the Marines had in mind. Although interested, the Marines had no money.

The Marines elaborated their program in the "Estimate of the Situation for [FY] 1937," prepared in 1935, backed by details submitted in October 1935. It called for 60 X-boats, 60 Y-boats (in which X-boats could nest), 15 self-propelled lighters, 6 artillery lighters, 12 boat ramps, and 24 boat rigs.

In their *Tentative Landing Force Manual*, the Marines noted that fast crash boats might be useful for reconnaissance "and other special purposes," which presumably meant feints and deception. During World War II the 63-ft crash boat, which resembled a PT boat, was used by special deception teams called beachjumpers.

C&R and the Bureau of Engineering (BuEng) developed specifications for a 30-ft diesel-powered landing craft that merged the X-boat and Y-boat into a single type capable of carrying 18–21 marines. Diesel engines would considerably reduce the fire hazard faced by transports and landing craft crews. For the moment, only gasoline engines were available in the desired combination of high power and small size. The 30-ft length was set by the typical spacing of ships' davits. Hopes of quickly converting merchant ships into transports depended heavily on using such standard fittings. C&R conversion plans, however, often had boats stowed on deck under cargo booms. To permit stowage two high, C&R wanted all projections above deck to be made portable. On this basis, ships to be converted into transports could carry 16–18 boats, which could land about a quarter of the marines on board in the first wave.

As of May 1935 C&R envisaged a shallow-draft sea-sled type landing boat, a type often considered as a fast attack boat. Plans called for using the following year's appropriations (FY 36) to buy one or two sea sleds for trials in surf. The FY 36 program also included outboard motors for standard ships' boats, which might transport follow-on waves of a landing. This idea had been discussed for some years, but had not been tested because of lack of funds.[1]

The following year the navy advertised for prototype shallow-draft landing craft (BuS&A Schedule 6151). It is not clear just when or why it decided to

The first specially built personnel landing craft of the new mid-1930s program was the 30-ft "Bureau" boat. Unlike the fishing craft, it had a round stern to deflect seas coming from astern. The navy built a wooden version alongside this metal one. Note the gun tub forward, for a 30-caliber machine gun, which was intended to disrupt enemy machine gun nests defending the beach.

go beyond the sea sled. Higgins did not bid, although apparently he received the advertisement. The navy bought five boats: a 32-ft boat ("Red Bank") by Red Bank Yacht Corporation, a 32 ft 7 in boat ("Bay Head") by Hubert S. Johnson, a 33 ft 2 in sea sled by Greenport Basin and Construction Corporation, a 32-ft boat ("Freeport") by Freeport Point Shipyard, Inc., and a modified whaleboat by Welin Boat and Davit Corporation (a major lifeboat maker). The Red Bank, Bay Head, and Freeport boats were all modified traditional fishing designs, which C&R hoped would behave well in shallow water. All were tested in the fall of 1936 at Cape May, New Jersey. Then they were shipped to the Philadelphia Navy Yard, modified to correct some deficiencies, and retested in the spring of 1937 during FLEX 3.

The sea sled, previously favored, pounded too violently when running fast into head seas. The modified fishing boats ("sea skiffs") clearly outperformed standard ships' boats, the best being the Red Bank boat. Its main defect was excessive freeboard. C&R modified the Red Bank design into a 30 ft 8 in boat and ordered it to be built at Philadelphia Navy Yard in February 1937. Unlike the fishing boats, it had a round stern that could deal with seas coming from astern. This was C&R's first landing craft since barge Type A, about a decade earlier. In December, C&R ordered a 30-ft metal version of its design from Welin. It ordered wooden versions from two builders, Luders and Jacobson & Peterson. Bureau surf boats performed well during the FLEX 4 exercise, although vibration damaged propeller shafts and bearings, and the Marines wanted armor. Exposed propellers and rudders tended to dig in when the boats tried to retract from a beach.

These exercises showed again that standard navy boats were inadequate. The transport area was 10,000 yds off the beach. Standard boats made one trip to the beach while the experimental ones made two or three. Due to the low speed of the navy boats, the 1st Battalion of the 5th Marines landed piecemeal, arriving over an hour late at its third objective.

Meanwhile, on 12 January 1937, the secretary of the navy established a continuing board in the Navy Department for the Development of Landing Boats for Training Operations (later it became simply the Landing Boat Board). The board's senior member was the assistant director of the fleet maintenance division. A month after the Landing Boat Board was established, the training squadron set up a board to direct tests of landing craft. The Marines had their own Equipment Board.

Letter designations were assigned to the various boats available for amphibious operations: A for the 50-ft motor launch, B for the 40-ft motor launch, C for the 36-ft motor launch, U for the motor whaleboat, V for the 45-ft artillery lighter, W for the tank lighter, and X for the 30-ft bureau surf boat. During the 1938 FLEX, the A-boat was deemed unsuitable for an initial wave because of its considerable draft, and because it carried too many men (one hit would kill too many). It was also difficult to beach. The B-boat seemed best among standard navy boats. The C-boat was suitable in calm water, but its low freeboard when loaded made it dangerous in rough seas or surf. The U-boat was considered unsuitable. It could founder in a moderate sea when loaded with one squad, with a machine gun in the bow. After several incidents the use of sand bags, to anchor machine guns, in these boats was barred by the commander of the naval attack force.

By this time Higgins was very much interested in selling a landing craft to the U.S. Navy. He thought his Eureka was far superior to the bureau's designs, claiming that it was far better adapted to grounding and then to retracting. In 1937, Comdr. Ralph S. McDowell, responsible for landing craft in C&R, asked Higgins to visit him in Washington, D.C.; they spent a week together redesigning the Eureka. In January 1938 Higgins asked for a test, and C&R asked him to demonstrate a Eureka when the destroyer *Somers*, with one of its constructors aboard, visited New Orleans that March. When the tests were successful, C&R ordered a Eureka (C-1781) in May, alongside five 30 ft 6 in bureau-designed wooden landing boats, equipped with bullet-proof steel side protection, from Philadelphia Navy Yard. In November 1938 C&R ordered four Eurekas (C-1793–1796) from Higgins, two wooden and two steel. Because Higgins was constantly improving his design, these craft were different from the boat tested in May.

By this time it was understood that a landing craft could be no more than 30 ft long (10,000 lbs carrying a 5,000-lb load); should accommodate 18 troops (as many forward as possible) plus a machine gun forward; make at least 10 kts (75-nm radius); needed the shallowest possible draft; should be able to beach through surf and then retract under its own power (assisted by a stern anchor); and needed protection against small-arms bullets for its gasoline tank, engine, and coxswain.

Between 28 January and 4 February 1939 the Atlantic squadron tested experimental personnel and tank landing craft off Culebra. Eighteen boats were available. All of the modified fishing boats had square sterns. They were intended to meet seas bow-on; but landing boats had to be able to take astern seas. The twin-screw Bay Head boat had twin skegs (and screws) and a very flat stern, which was considered too wide and too square for surf work (seas coming from astern could damage it). It was considered too heavy, and its twin screws were objectionable because they could not be protected by a keel skeg. The skeg of the Freeport boat projected about 6 in below the flat keel. When the boat beached, it dug into the beach, making retraction difficult. By way of contrast, the Red Bank boat had a flat bottom section (boxed garboard) blending into a skeg no deeper than its keel, so it could beach without digging in. However, in the 1939 tests it was rated too heavy and too clumsy; it drew too much water. The Welin whaleboat was too slow (8.2 kts). The sea sled had already been rejected, although for a time it appeared to be C&R's favorite. The six Philadelphia boats were too heavy, drew too much water, and were too slow (8.5 kts). The Jacobson & Peterson version of the C&R boat did not handle as well as the Welin version of the boat. The two Higgins steel boats were too heavy (11,500 lbs) and too slow (8.4 kts, despite their high power of 250 HP).

That left a Higgins' wooden Eureka (in two versions) and a Welin bureau boat; a Luders bureau boat was not available for tests. The original Higgins was fast enough and light enough, but it pounded heavily at low speed and had a square stern. The newer version was too heavy (11,000 lbs), due to the addition of a stern sponson and a closed cooling system that gave considerable trouble during tests. It was also criticized for its high engine power, 250 HP (i.e., high fuel consumption), which Higgins claimed was necessary for the boat to retract properly. On the other hand, the second new Higgins boat rode well, even in 3-ft waves, was fairly dry, and had good beaching characteristics; it could retract without using a stern anchor. This particular boat was quite popular, and was used by the Marines' commandant during the exercises. The sponson was designed to split stern seas when beaching. Another innovation was a small rudder installed forward of the propeller to assist in steering when going astern. It turned with the main rudder. However, it proved ineffective when the boat backed into wind and sea. According to a contemporary account, Higgins boats quickly turned and headed to seaward as soon as the incoming tide floated them, but they would not steer when their skegs were grounded but their forebodies were afloat.

The Welin version of the bureau boat made an excellent impression. It was the lightest (8,011 lbs) and the fastest (10.92 kts, even on 160 HP). It handled and behaved well at sea, and had excellent lifeboat characteristics, which might make it an attractive replacement for standard ships' boat. However, it was very wet, and difficult to embark and disembark; the Jacobson & Peterson boat had a similar embarking/disembarking problem. The Welin version did have very good beaching qualities, and it retracted without using a stern anchor. It operated effectively in 5-ft surf. Minor improvements—slightly more bow flare (for dryness), steps leading up to the bow and a bow grab rail, for disembarking—might make it entirely satisfactory.

All the boats were difficult to bring alongside to load troops. A C&R observer of the 1939 tests thought that was inevitable, given the boats' cut-away forefoot. Necessary for beaching, the cut-away also made it easier for the boat's bow to fall off before the wind. Experienced boat handlers could overcome the problem.

Also present at the tests were flat-bottom 18-ft skiffs powered by outboards and rubber landing rafts powered by paddles or outboards. Both had been suggested by the Marines' board evaluating the destroyer transport (see Chapter 3). The skiffs were only moderately successful, but the Marines liked the new rubber rafts. They could accommodate nine men and a machine gun. They were being developed from standard seaplane rafts by the Marines at Quantico, in conjunction with Goodyear Rubber and the Naval Aircraft Factory.

In June 1939 C&R ordered one boat from Welin (C-1817) and one from Higgins (C-1818). Following an August 1939 competition, it ordered eight bureau-type metal boats and five wooden Eurekas, all from Higgins (C-1840–1847 and C-1830–1834), in October.

C&R also bought three experimental boats. One was a metal bureau-designed boat with a Kort nozzle, which allows maneuverability in shallow water (C-1879). The other two (registry 14000 and 14001) were wooden boats of Philadelphia Navy Yard design, generally following the bureau's metal-boat lines. Wood was favored for ship's boats because it was easier to repair, whereas the boats stowed at Quantico and San Diego should be metal because they would not deteriorate if they were kept properly painted. One of the boats had a diesel and the other a gasoline engine. Both had machine guns in ring mounts. Later the ring mount was discarded, on board both bureau-type boats and Eurekas, in favor of a pair of pedestals at the fore end of the troop compartment.

The Eureka proved far superior to the bureau-designed boat. It could run much better through surf, its propeller was better protected against underwater obstructions, and it could easily retract from a beach, even when broached. For example, after a minor landing exercise at San Clemente on 25 May–1 June 1940, the beachmaster considered the Higgins boat far superior to the C&R types. The

The 30-ft Higgins boat was the basis for the small personnel boats with which the U.S. Navy fought World War II.

Table 4-1. Small Landing Craft, 1937–42

	C&R	Eureka	LCPL	LCV	LCPR	LCVP	LCP(N)
Length (ft-in)	30-6	30-6	36-8	36-3	35-10	36-10	32-0
Beam (ft-in)	8-11	10-10½	10-10	10-10	10-9	10-5¼	7-10
Draft (ft-in)[a]							
Light	2-3	2-3	2-6	1-6/2-6	2-6	NA	NA
Loaded	NA	NA	3-6	2-2/2-10	3-6	2-2/3-0	0-10
Displacement							
Light	11,010 lbs	12,500 lbs	13,000 lbs	7 tons	13,000 lbs	9 tons	1,650 lbs[b]
Loaded	NA	NA	18,000 lbs	11 tons	NA	NA	3,400 lbs
Crew	3	3	3	3	3	3	2
Capacity							
Troops	18	18	30–36	36	30–36	36	13
Load (lb)	NA	NA	6,700–8,100	10,000	6,700–8,100	8,100	NA
Armament							
0.30 caliber MG	1	1	2	2[c]	2	2	None
Power (HP)[d]	250	250	225	225	225	225	50
Speed (kt)	NA	NA	8, loaded	9, loaded	11, loaded	9	9.6, loaded
Radius (nm/kt)	NA	130/8	68/9	69/11	102/9	105/11	NA

NOTE: NA indicates data not available.

[a] Draft aft if only one figure given; draft fore/aft if two figures given.

[b] 1,400 lbs without engine.

[c] On bulwarks.

[d] The C&R and 30-ft Higgins boats were powered by 250 HP Hall-Scott gasoline engines. Fuel capacity was, respectively, 80 and 120 gal. Data for wartime types, except LCP(N) are for the versions powered by 225 HP Gray diesels. Actual powerplants varied considerably, engines including 165 HP Gray diesels, 150 HP Superior diesels, 105 HP Buda diesels, 115 HP Chrysler Royal gasoline, and 150 HP Palmer gasoline. Many LCPLs were powered by 250 HP Hall-Scott or 225 HP Kermath gasoline engines.

commander of the naval attack force noted sarcastically that "all Navy boats built hereafter should be designed with the expectation that they will run aground instead of hoping that they will not." Higgins had probably been the only builder to think that way. C&R ordered 62 Eurekas from Higgins. To reduce weight, their fuel capacity was halved (120 vs. 240 gal), for a cruising radius of about 65 land miles at 14.5 mph (not knots).

Higgins later claimed that the bureau (by this time called BuShips, happening after C&R merged into BuShips) tried to protect its designs despite their evident deficiencies. Thus it ordered 45 of its own metal design alongside the 62 Eurekas. Further trials at Virginia Beach in September 1940 showed so clearly that Higgins' Eureka was superior that it was chosen for all future procurement. The Eureka had also beaten off a new challenger, a 30-ft Chris-Craft with twin engines, that was tested in August 1940. Table 4-1 compares the C&R and Eureka designs with later versions of the Eureka—LCPL (landing craft, personnel, large), LCV (landing craft, vehicles), LCPR (landing craft, personnel, ramped), LCVP (landing craft, vehicles and personnel)—and with the wartime privately designed LCP(N) (landing craft, personnel, nested).

Eureka deliveries depended on the supply of their Hall-Scott engines. In July 1940 the British ordered 50 enlarged (36 ft 8 in, generally described as 36-ft) Eurekas. Their order probably saved Higgins financially while BuShips dickered over further orders. The Eureka seemed to fit the urgent requirement for a raiding craft, now that the British had been ejected from continental Europe. The first craft arrived in Portsmouth in November 1940. The British described them as magnificent sea boats. Modifications included provision of a long gangplank stowed atop the boat's canopy, which could be launched over a bow roller. Without such a gangplank, troops had to clamber over the sides into the water. With its ship-like bow, the Eureka was considerably faster than its British equivalent, the assault landing craft (ALC, later LCA), 15 vs. 10 kts. A British 1941 evaluation, made after a week's tests on the Cornwall coast, praised the Eureka's performance but noted that its engines were noisy and unprotected. The British also wanted self-sealing tanks. The British boats were powered by the same 250 HP Hall-Scott engine as U.S. craft, so there was

By the time the United States entered World War II, the original 30-ft "Eureka" had been discarded in favor of the 36-ft LCP(L) shown here. Note the characteristic Higgins spoon bow. The main defect of this craft was that the troops it carried had to leap over its side to get to the beach.

some question as to whether sufficient engines could be found. Many U.S. craft were completed with lower-power gasoline engines or diesels.

In August 1940 Higgins pointed out to the U.S. Navy that the 36-footer, powered with the same engine as the 30-footer, was inherently faster. It also was more seaworthy and had twice the capacity (36 vs. 18 marines), yet it was only slightly more expensive than the 30-footer. The Marines strongly advocated it, despite BuShips' arguments for a 30-ft length. In September 1940 the Landing Boat Board observed that 36-ft boats could readily be accommodated aboard large transports. There were already enough 30-footers to supply small transports (APDs), if indeed they could not accommodate 36-footers (as it turned out, they could). On 20 September 1940 the secretary of the navy approved the Landing Boat Board's proposal to order 335 enlarged Eurekas. In October 1940 a 36-ft Eureka was successfully tested against a 30-ft Eureka at Virginia Beach. The 36-footer was adopted for all future production. It was effectively the Y-boat the Marines had requested in 1935. The 36-footer was soon fitted with a pair of machine gun rings at the bow.

The British 36-footer had deeper deadrise than the U.S. 30-footer, which BuShips thought might improve speed but reduce the boat's ability to retract from a beach. At his expense, Higgins built two 36-ft Eurekas, one with a British and one with an American hull form, for October 1940 tests. Higgins found the low-deadrise version superior, so he offered only this version to the U.S. Navy. November 1940 tests were very successful; the boat had "exceptionally good retracting qualities," even when driven onto the beach for almost its whole length. According to a later evaluation, its main disadvantage was that troops on board often got very wet during the approach, causing problems in cold weather.

When mass production began, boats were ordered from other builders and the design was standardized. For example, the canopy, which had characterized British Eurekas, was no longer fitted. The British much preferred the original version. They also complained that, unlike the home-grown LCA, the Eureka lacked armor. In 1941 the British decided to bolt 10 lbs of ¼-in plating to the outside of hull, at a cost of several knots. The postwar British official history questioned whether this small protection was worth the loss of speed.

The U.S. view was that no boat could be armored very effectively. The best protection was the highest possible speed. A May 1941 conference on transports and floating equipment for the Fleet Marine Force concluded that armor should be compromised in favor of speed, capacity, seaworthiness, and retractability. The armored British LCA was 3 ft longer than a Eureka and 3 tons heavier; it was also 1–2 kts slower, and found it more difficult to retract from a beach. At the conference, a British officer familiar with his navy's amphibious practices remarked that unarmored Eurekas would not be used in the first wave of an assault for two reasons: they were unarmored, and troops leaping over the sides would be too vulnerable. U.S. naval architects argued that armor was not an option. The British applied about 6,500 lbs of ¼-inch armor to the LCA, but the entire payload of the Eureka was only 8,100 lbs. Since any considerable increase in weight would cost performance, armor would have to come out of that 8,100 lbs. As it was, a recent decision to increase Eureka fuel capacity from 120 to 300 gals would cost performance.

The Eurekas were not entirely unarmored. U.S. LCPLs had three transverse 10-lb (¼-inch) armor bulkheads, one in the bows, one forward of the engine, and one in the stern forward of the fuel tank.

The U.S. Navy called the 36-footer a Y-boat, but later changed that to T, presumably to avoid confusion with yard or district craft. It was briefly called an LBP (landing boat, personnel). In July 1942 the 36-footer was designated an LCP(L) (landing craft, personnel, large), and the smaller type was designated LCP(M), the medium equivalent. The LCP(M) designation, however, was more widely used for a very different British 38-footer. The British called the Eurekas "R" (raiding) craft, to distinguish them from armored assault craft. To the initial 50 the British added 112 and then another 100, for a total of 262 boats. The British received many more later under Lend-Lease.

Eurekas and other contemporary U.S. landing craft were powered by gasoline engines, which were the most compact available. Their fuel made them quite dangerous, however. They could also stall when their ignitions became wet. In December 1939 the U.S. Navy installed a prototype lightweight diesel, with aluminum cylinder block and pistons, on board a C&R boat. This engine performed so well that in April 1940 it was recommended that all future landing craft be diesel powered. For the moment, many Higgins boats had gasoline engines because no compact 250 HP diesel was available, thus one virtue of the competing C&R boat was that it could run on a 150 HP diesel.

The General Motors (Gray Marine) 6-71 was adopted. Its normal rating was 225 HP: one powered a Higgins boat, two an LCM (landing craft, mechanized), three an LCT (landing craft, tank). For higher power four were clutched to one bull gear as a "quad"; each of the four could be separately engaged or disengaged. The quad was devised specifically for the 110-ft subchaser; it later powered the LCI(L) (landing craft, infantry, large). Overall, the 18 models of 6-71 diesels accounted for 80 percent of installed power of U.S. amphibious ships and craft. At first engines did not reach their planned 2,000 hrs between overhaul, partly because many operators used them only at full and idle throttle settings; the solution was to add a "normal" setting (the old full setting became the "battle" setting).

General Motors was also producing variable-pitch propellers, and a 36-ft Eureka was tested with one in August 1941. It offered easier operation, simplicity (fewer gears), less weight, and greater efficiency, albeit at a cost in speed (the blades were inefficient). BuShips recommended further tests, looking toward adoption. The war intervened, and the conventional geared transmission already in production survived. However, the wartime LCI(L) used variable-pitch propellers.

Marines exiting a Eureka had to jump over its sides. In China, the Marine observer, Maj. Victor Krulak, had watched Japanese ramped beaching craft in 1937. He proposed a similar solution to C&R. His report was buried at first, probably because the prewar U.S. Navy bureaus often seemed to display contempt for anything "not invented here." Krulak persisted, however. In September 1940 BuShips formally rejected his idea on the ground that production was now so urgent that no changes could be made in the basic Eureka design. Moreover, after the 335 enlarged Eurekas, no further landing craft were planned. Another 188 boats were ordered in February 1941, however.

By this time Krulak had senior Marine allies. When Higgins visited the U.S. Marine base at Quantico in April 1941, Maj. Ernest E. Linsert, secretary to the Equipment Board, showed him one of Krulak's photographs. The board's president, Brig. Gen. Emile P. Moses, asked him to install a ramp on

This Higgins prototype vehicle landing craft was based on his 36-ft "Eureka" or LCP(L). Its success in turn led to development of the LCV.

a Eureka. Higgins built three prototypes at his own expense. These were not, however, Krulak's ramped personnel boats. Rather, they were small vehicle carriers that could also, necessarily, carry troops.

Linsert saw the prototype the next month, watching tests with both a truck and with Higgins employees simulating troops embarking and disembarking. In May 1941 BuShips ordered one of the 188 boats modified with a 6 ft 3 in bow ramp, its engine relocated to the stern. It differed in detail from his prototype, and Higgins protested some of the modifications as impractical. The engine was now too far aft to use a conventional propeller shaft, so it was given a vee-drive. Prompted partly by Higgins, BuShips decided that another 87 boats would be modified if the new version was successful. Because the others were already under construction, modification would have been too expensive. The boat was tested with ten men, a 20 HP tractor, and a 75-mm gun on board, a total of 14,500 lbs. That was definitely too much (the trials board considered 10,000 lbs the most the boat could carry). Heavily loaded, the boat trimmed by the head, and the ramp scooped up water. Because the engine had to be placed aft, the boat's stern dug into the beach after it was unloaded, and the ramp flapped down, making it difficult to close. The dug-in stern made the boat difficult to retract. In 1941 exercises, several boats were swamped on the beach with their ramps open. The vee-drive caused other problems.

By late August 1941, 100 more ramped boats were on order, as part of an order for 262 more Higgins boats. They were held up pending construction of a modified prototype, with its engine moved forward as far as possible to correct trim, yet leaving a 17-ft cargo space. Moving the engine might make it possible to revert to direct drive. The coxswain's platform would be moved to the centerline (rather than on the port side, as on the 88 earlier boats), and a bulkhead (for strength) installed just forward of the engine.

The LCV was conceived as a complement to the personnel-carrying LCP(L), carrying trucks and light vehicles to a beach after the troops from the LCP(L)s had secured it. Unlike the LCP(L), it was unarmed and its steering station (raised to clear the ramp) was not protected.

The bow would be redesigned to do away with the canvas cutwater of the earlier boats, the object being to make the cargo flat as long as possible and to keep the engine as far forward as possible. October 1941 trials proved successful, due partly to a redesign of the bow that increased buoyancy forward. After trials, the engine was moved slightly further aft to improve trim. Late in November 1941, three prototypes were tested. One had a straight-drive engine located far enough aft to provide a 17 ft 9 in cargo space; another had the engine further aft, for a 19 ft 3 in cargo space; the third had a vee-drive, with the engine further forward. The last was rejected; judged as no better than the others, it was more complex and more expensive. The boat with the 19 ft 3 in cargo space performed perfectly, and it was adopted. Of the original 88 ramped boats, 68 had been completed with vee-drives, but 20 were still without engines; they received engines placed far enough forward to use straight drives.

The ramped Higgins boat was essentially unarmored, because it was considered a follow-up craft rather than an initial assault boat. The trucks it carried would not be wanted in the first wave, and it was too small to deliver a tank. Thus it was acceptable not to armor the coxswain's station; set high for visibility, armor would have added too much top-weight. The craft did have a steel ramp, and that was made of ¼-in STS (special treatment steel) to provide some protection "should occasion demand."

The ramped boat was produced in quantity as the TR or LBV (landing boat, vehicles) or the VLC (vehicle landing craft). It was apparently sometimes called a "jeep lighter," because it could transport jeeps and similar vehicles. In July 1942 it became the LCV (landing craft, vehicles). In March 1942 the Atlantic Fleet amphibious force recommended that, instead of providing transports with equal numbers of LCV and LCPL, they have only LCVs. In May the Landing Boat Board rejected the idea, on the

The LCP(R) was a ramped personnel carrier, envisaged as a direct replacement for the rampless Higgins boat. LCP(R)s are shown aboard the fast transport (ex-destroyer) *Crosby* after a Mare Island refit, 19 February 1943. Note the new ladders, which make it easier for troops to get into boats alongside, and the newly installed 3-in/50 antiaircraft guns.

grounds that the LCV was not as seaworthy and could not be lowered when loaded. It was altogether unarmored, although some craft apparently were given athwartships armored bulkheads, and it lacked the firepower of the LCPL. Nor could it be stowed as easily, because its coxswain station projected from its deck. It could not be stowed two high in davits (a boat could be carried swung away from a davit, or on deck). Thus, switching to an all-LCV force would cut numbers. On the other hand, at Operation Torch in November 1942 the LCV was clearly much more useful than the LCPL. The U.S. Navy bought a modified version, the LCVP, to replace all other craft. It is described in Chapter 7.

In January 1942 the British noted that the LCV was just too narrow to take a Bren gun carrier (6 ft 3 in vs. 6 ft 11 in); BuShips agreed to modify 50 of them to suit.

Another ramped version emerged in 1942: the personnel ramp boat, with a 3 ft 4 in ramp and a 500-lb capacity. It seems to have been closer to Krulak's idea, and in effect it was Buships' response to Krulak's initiative. It also seems to have been an attempt to retain such Eureka virtues as the low silhouette and armor with the advantages of a bow ramp. Late in January 1942 the navy asked Eureka builders to shift production to one of two alternative narrow-ramp types: version A (minimum modification to provide a bow ramp) or version B (9 in wider chine, wider quarterdeck, STS armor on sides and quarterdeck). Owens Yacht protested that it could not shift production, but Chris-Craft agreed, and version A was chosen. Chris-

LOA- 35'-11-1/2"   Max Displ- 26,600 LBS   Capacity- 36 MEN OR 8100 LBS
Max Beam- 10'- 9-1/2"   Hoist WT- 18,500 LBS   Speed at Max Displ- 10 KTS
Mach & H P- 1- 225 HP DIESEL   24 VOLTS   Fuel Cap- 180 GAL
Cruising Range at Full Power & Full Ld- 110 NAUTICAL MILES   ARMAMENT- 2-30 CAL MG
Purpose- TO LAND AND RETRIEVE PERSONNEL OR EQUIPMENT DURING AMPHIBIOUS OPERATIONS

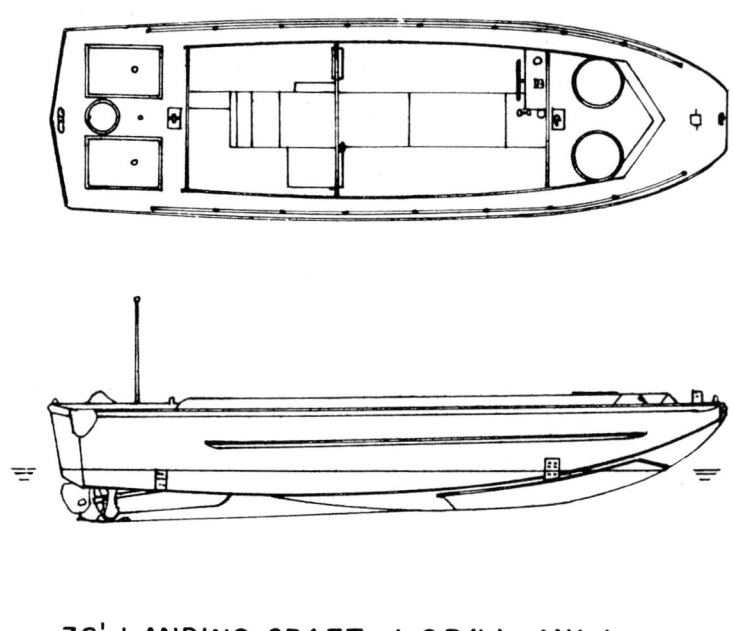

36' LANDING CRAFT LCP(L) MK I

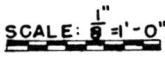

LCPL Mk 1, the famous World War II "Higgins boat." Note the pair of tubs forward for 0.30 caliber machine guns. *U.S. Navy*

Craft became the sole producer of the new type, which was called Eureka "Ramp Type A." Modifications were minimal; the new craft had much the same troop compartment as a Eureka, with two machine gun tubs alongside the ramp and a narrow troop passage between them. The forward armored bulkhead of the LCPL was given a hinged section, which could be dropped to the deck so that troops could run toward the narrow bow ramp. This version was initially designated TP; in July 1942 it became the LCP(R) (landing craft, personnel, ramped). After being superseded by the LCVP, the LCP(R) survived as a support craft for underwater demolition teams (UDTs); its higher speed was valued, and its ramp made it easier to ship equipment. Like the LCPL, it could be lowered loaded from a davit, but hoisted out only when unloaded.

July 1942 tests by the army's 9th Division showed that men could disembark from an LCPR or LCV in 30 seconds, compared to 55 seconds for the LCPL. In heavy weather, both LCPL and LCV were slightly drier and safer going to windward, the LCV ramp offering better protection against green water over the ramp. The LCPL lifted and rode over the sea better than the LCPR.[2] The LCPR achieved its quick disembarkation because it had access doors in its midships armored bulkhead, thus eliminating the bottleneck as men disembarked. If the LCV had a similar feature, it could have been emptied in less than 20 seconds. Thus the LCV seemed the best of the three versions of the Higgins boat; it was no surprise that it was soon developed into the LCVP (described in Chapter 7).

The development of the tank landing craft, ultimately the LCM (landing craft, mechanized), was more tangled. The Marines wanted to include tanks in their initial assaults at least as early as 1930. Because they could not wait for the beach to be secured, the unpowered artillery lighter simply would not do. The Marines envisaged using the same

Like the Higgins boat, the LCP(R) was intended to carry a pair of 0.30-caliber machine guns, as shown here. Sometimes the guns were provided with small flat shields, as on the photograph of an LCP(L) above.

self-propelled landing barges for troops, tanks, and artillery.

In 1934 the new Fleet Marine Force wanted a light tank company of 15 vehicles, each weighing no more than 6,000 lbs fully loaded, yet armed with a 1.1-in machine gun or 37-mm cannon plus a machine gun, and armored against weapons up to 37-mm. To land the tank, the Marines wanted a 10–12-mph self-propelled lighter to carry a 7,000-lb load, with a bow ramp over which the tank could drive under its own power.

Although the army thought such a tank could be built, in fact the Marines had to settle on a Marmon-Herrington 6½-tonner (13,000 lbs); its fixed turret carried a pair of machine guns. Its weight was set by the expected boom capacity of converted transports. By 1940 the Marines had funds for ten such tanks. They expected a developed version to weigh about 9 tons. These vehicles were far lighter than those being developed by the army.

In December 1935 the Marines formally asked C&R to design a self-propelled tank lighter, to carry a 12,000-lb load at 13 kts. In July 1936 C&R offered to design a lighter that could accommodate both the tank and much heavier 155-mm artillery pieces, but the Marines pointed out that very few items of equipment were much heavier than 10,000 lbs, which was the weight of the largest tanks they then planned. Fewer artillery lighters would be needed if smaller ones were available. Therefore they preferred the smaller-powered lighter, tailored to their tanks, to be supplemented by a few unpowered artillery lighters. In the end, what killed the artillery lighter was that tanks grew heavier than artillery pieces, so there was no longer any point in having two kinds of lighter. C&R's prototype tank lighter was tested successfully during the 1938 fleet amphibious exercise, FLEX 4. According to the official report of FLEX 4, the small tank lighter was more valuable than the more capacious artillery lighter because it could shuttle so rapidly between ship and shore.

The boat's configuration was unusual. Two wide side walls, which provided buoyancy, were almost rectangular, with nearly vertical front ends, but the ramp was sharply sloped, leaving a pocket under the bow. The pocket entrapped head seas. The wing tanks maintained buoyancy and stability even when a tank lowered into the craft accidentally sprung open its bow door, as once happened. The broad side walls left only a narrow cargo space, which was difficult to load by crane when the boat bounced about in a seaway. The walls were wide enough to accommodate machine gun turrets, giving them wide arcs of fire against targets ashore. At 46,000 lbs, the boat was considered rather heavy for the load it carried, a 5-ton tank. The lighter had an armored pilothouse but a manually operated ramp, which might be difficult to operate under fire. It had power retraction gear. C&R's 1938 prototype and later tank landing craft are described in Table 4-2.

In March 1938 the Landing Boat Board decided that any follow-on tank lighter should be no heavier than the standard 50-ft boat, so that it could be hoisted aboard battleships. C&R therefore redesigned the craft, drastically reducing its weight by eliminating the wing tanks. The resulting 40-ft version, tested early in 1939, carried two 5-ton tanks. It broadly resembled the con-

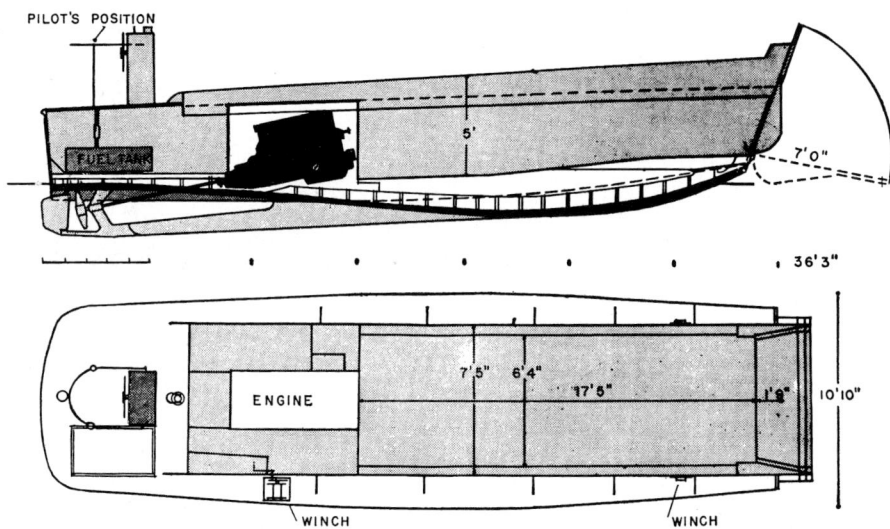

The LCV, from the April 1944 U.S. Navy manual of amphibious craft. The LCV was designed to carry small vehicles. In effect it was a ramped version of the original Higgins boat. On at least one transport's booklet of plans it was labeled a "jeep lighter." *U.S. Navy (ONI 226)*

temporary British tank lighter, then called an MLC (mechanized landing craft), in that its tank deck was above the waterline. Buoyancy was concentrated in the boat's bottom, under the tank deck. Its bow was somewhat rounded in profile, the ramp forming the bow; there were no side wall projections, because the side walls added no buoyancy. The only usable enclosed volume was right aft, under and abaft the pilothouse. Because the side walls were so narrow, plans called for mounting machine guns abaft the pilothouse, a choice criticized because they would have only limited arcs of fire. The 40-footer trimmed down by the stern due to a lack of buoyancy aft and too much buoyancy forward. A contemporary report found it poorly formed for open sea work. Head seas caused it to lift with considerable violence. The conclusion was that the boat's bow should be as narrow as the width of the ramp would permit, with well-rounded instead of square waterlines, and with deadrise to the forward sections. The appropriate bow form could be found, for example, on Cape Charles car floats and on Isle of Wight (England) automobile ferries.

The fleet wanted the new type, but the C&R observer at the January–February 1939 exercises preferred the earlier one. He wanted something much more rugged and seaworthy. The 40-footer's machinery installation was so cramped that it took about 4 hrs to clear a sand trap; on one occasion a boat had to run at one-third speed because her engines overheated when sand clogged the cooling system. By early 1939, moreover, it no longer seemed so important that major combatant ships carry tank lighters; surely auxiliaries, with heavier cranes, would be used.

By 1940 it was clear that the Marines were unlikely to buy enough tanks to interest a manufacturer; the army, however, was about to buy tanks in numbers. Its light tanks, moreover, had a considerable tactical advantage: their machine gun turrets could fire in any direction. That was particularly important for a tank that would be landed in very small numbers. In April 1941 a marine told the General Board that henceforth the Marine Corps would use only army tanks: "We can't get any others, and furthermore they are probably better than anything we can produce." The army was now designing light tanks armed with 37-mm guns capable of dealing with any enemy tanks that might form part of a beach defense. No such weapon could be installed in the Corps' proposed 9-tonner.

Army light tanks already weighed 11½ tons or, with diesel power, 13 tons; the M3 of 1941 weighed about 16 tons. The next step up, the 25-ton M3 medium tank, was protected against 37-mm antitank fire rather than 0.50 caliber machine guns. In the spring of 1941 the navy's war plans division wanted the Marines to adopt it. The Marines were less than enthusiastic.[3] Worse was coming. The standard wartime U.S. medium tank was the 30-ton M4 Sherman.

The existing tank lighters could not take the M3 light tank, let alone the mediums. In 1939 the maximum weight of a next-generation tank lighter was set at 25 tons by boom capacity; shortly the weight would be increased to 30 tons. BuShips designed a

LOA- 35'- 11 -3/4"   Max Displ- 26,600 LBS   Capacity- 36 MEN

Max Beam- 10'- 9-1/2"   Hoist WT- 18,500 LBS   Speed at Max Displ- 10 KNOTS

Mach & HP- 1- 6 CYL 225 DIESEL   12 VOLTS   Fuel Cap- 1?0 GAL

Cruising Range at Full Power & Full Ld- 110 NAUTICAL MILES   ARMAMENT- 2 -30 CAL MG

Purpose- TO LAND AND RETRIEVE PERSONNEL DURING AMPHIBIOUS OPERATIONS

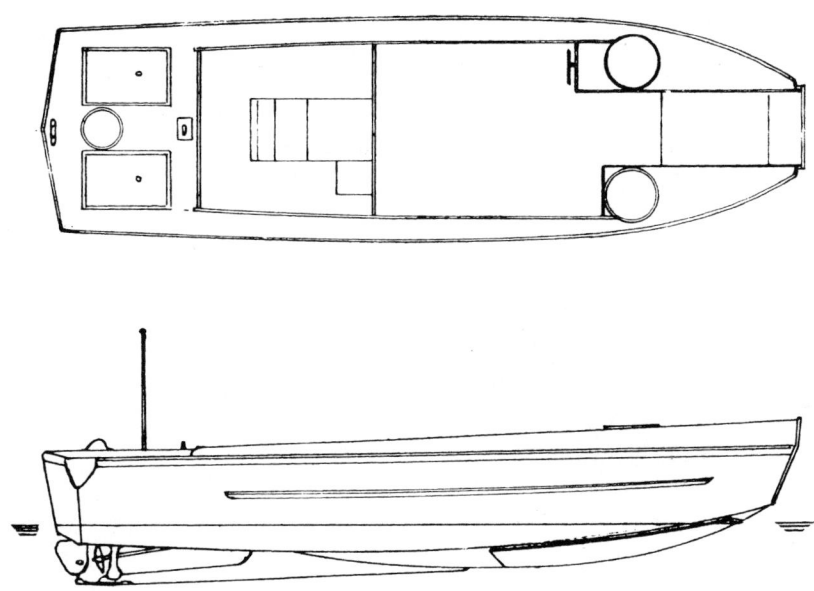

LCPR, the ramped equivalent of the Higgins boat for personnel only. After 1942 it was used only for special purposes. Like the Higgins boat, this craft shows machine gun tubs forward. *U.S. Navy*

23-ton (about 50,000 lbs) lighter capable of carrying two small tanks, one army light tank, or a 155-mm artillery piece. Soon obsolete, the lighter could, however, carry the disassembled parts of the new 7-in coast defense gun.

This first production lighter was essentially a stretched version of the 40-footer. It had a cargo space wider than the widest load contemplated, on the theory that otherwise the tank might jam diagonally while the lighter rolled as it was being loaded in a seaway. When carrying a 26,000-lb load, the tank top (tank deck) was 10 in above the waterline. The craft was therefore made self-bailing, with scuppers at tank deck level. Above-water armor was limited to the pilothouse, on the theory that the armor of the tank being carried would protect the crew. Tanks could not be chocked down, because no crewmen could go out on the exposed tank under the fire deck to undo chocks after the boat beached.

Similarly, the ramp had to be remotely operable (raised and lowered electrically) from inside the protected conning station. A 20-lb (½-in) belt of STS steel extended from about 8 in below the light waterline to the level of the tank deck and across the bottom of the lighter at its forward end; STS also covered the engine space. An inner bottom bulkhead forward (25-lb STS) protected the buoyancy of the craft from machine gun fire. The boat carried two standard pedestal (army-type) mounts for 0.50 caliber machine guns. After completion, boats were provided with double bottoms so that they could carry potable water from ship to shore.

Compared to the earlier lighters, this one had new lines adding displacement in way of the tunnel, to make flow to the propellers less turbulent. According to a May 1941 war plans division report on amphibious materiel, in practice the craft was limited to 6 kts in a calm sea and to 4 kts in a mod-

This 38-footer was the first approach to a tank landing craft. She is shown disgorging one of the Marines' small fixed-turret tanks and being hoisted by the seaplane tender *Curtiss* (*right*), 5 February 1941.

erate sea, because the craft was overweight and the engines were unreliable. To prevent overheating, they had to be run at half speed or less most of the time. Because the craft was so close to maximum boom capacity, it could not be unloaded from transports rolling 3 degrees or more in a rough sea.

Prototype no. 13975, ordered from Norfolk Navy Yard in 1939, was tested alongside the 38- and 40-footers on board USS *Capella* in the early 1940 amphibious exercise. It was successful, so the Landing Boat Board asked for five more in May 1940. By the time bids were advertised the figure had grown to 13 and then to 82 by January 1941. All, except two built at Norfolk, were built by contractors. In February 1941 the Marines asked for 28 more 45-ft tank lighters, and the required number kept growing.

One of the prototypes sank in December 1940, in a foretaste of later problems with the bureau (by this time C&R had merged into BuShips) designs. It had water in the port double bottom tank (the starboard tank was dry), although no water was supposed to be on board when the boat carried anything on her tank deck. The tank was loaded off-center, unfortunately on the side already down due to the water. The free surface of the water undoubtedly helped cause the craft to roll over and sink. BuShips' solutions were to add freeboard, to further subdivide the bottom tanks, and to provide guides that would center the load on the tank deck and limit its movement to the sides.

Tank lighters were originally designated as W-boats; the small tank lighters (45 to 48 ft) became WL (light tank capacity) and the later 50-footers (discussed below) were WM (medium tank capac-

ity). The designation TLL was sometimes substituted for WL. In July 1942 all WLs and WMs received the LCM (landing craft, mechanized) designation. Note that the LCMs were not the same as the LCTs; the LCT (landing craft, tank) designation, also created in July 1942, was used for much larger craft (discussed in Chapter 5).[4]

Production tank lighters were heavier than the 52,000-lb prototypes. Some were as heavy as 59,000 lbs, far too close to the new limit of 30 tons. BuShips observed that the British MLC, later designated the LCM(1), weighed only 18 tons. Even so, it could carry 54,500 lbs with 9 in freeboard, whereas the C&R lighter was limited to 29,100 lbs with only 6 in freeboard to the tank top. C&R considered the British hull form inferior, with too square a bow (the U.S. boat had a better bow rake), and a square stern that would not be particularly desirable for retracting from the beach (the U.S. boat had a rounded stern). On the other hand, the British boat offered better sea protection from aft, and its underwater stern form was considered better. Also, the U.S. tunnel stern offered little if any advantage and was more expensive to build. The British square bilge was considered undesirable. The rounded U.S. bilge was stronger and reduced the underside area touching the bottom. The British design had an exposed rudder, while the U.S. rudder was under the overhanging stern.

BuShips let a contract to a New York engineering consulting firm, Tams Inc., to compare the bureau and British designs, as a guide to a new lightweight bureau lighter. In BuShips' view, Tams offered no real insights. Of the 5-ton difference between the two designs, Tams credited half to structural steel and half to machinery. The company suggested eliminating all electrical auxiliaries, most transverse subdivision (as in the British design), and the tunnel

Table 4-2. Tank Landing Craft, 1938–42

	1938 Prototype	1939 Prototype	LCM(2) Prototype	Higgins Prototype
Length OA (ft-in)	38-0	40-0	45-0[a]	45-0
Beam (ft-in)	13-6	12-6	14-1[b]	12-9
Draft (ft-in)[c]	3-7, loaded	2-0	2-8	2.48 ft
Displacement				
Light	45,000 lbs	25,000 lbs	52,000 lbs	18.75 tons
Loaded	NA	NA	NA	32.14 tons
Tank deck[d]	NA	NA	23 × 7.5 ft	NA
Load capacity	two 6.5-ton or one 9-ton tank	NA	26,000 lbs	NA
Armament				
0.30 caliber MG	2	NA	—	NA
0.50 caliber MG	—	NA	2	NA
Power (HP)	two 75 Ford	NA	two 100[f]	one 330
Speed (kt)	6[g]	8–10	9.5[h]	9
Radius (nm/kt)	NA	NA	75/7.5	NA

NOTES: Dash indicates that specification is irrelevant; NA indicates relevancy but data not available.

[a]Length WL is 40-7.5. Length 47-2.5 over the anchor bracket. Production craft typically weighed 58,000 lbs.

[b]Beam WL is 13-6.

[c]Draft aft if only one figure given; draft fore/aft if two figures given.

[d]Length only if one figure given; width unknown.

[e]The standard wartime handbook claimed that the C&R type could carry 120,000 lbs of cargo. LCM(2) was credited with the capacity for 100 troops; LCM(3) had only a capacity for 60 troops, due to its smaller tank deck.

The first U.S. production landing craft was the 45-ft LCM(2). It was very similar to a prototype 40-footer (Y-312).

Higgins Production	C&R 47-ft Prototype	Higgins 50-ft Design	C&R 50-footer	LCM(3)
48-0	47-0	50-0	49-6	50-0
13-3	13-6	14-0	15-9	14-1
2.95 ft	2.9 ft	3-6	3-4	1-2/2-1.25, light; 3-6/4-6, loaded
19.64 tons	19.8 tons	17.71 tons	23 tons	52,000 lbs
31.69 tons	31.85 tons	4.49 tons	44.49 tons	52 tons
29.75 ft	30 ft	31.5 ft	41 ft	9 ft 5 in x 31 ft 6 in
NA	NA	27 tons	60,000 lbs[e]	60,000 lbs
NA	—	NA	NA	—
NA	2	NA	NA	2
two 175	two 225	330	450	two 165
8.5	10.4	7.6	8.6	8, loaded
NA	NA	60/8	240/8	140/11

[f]Lincoln Zephyr engines.
[g]The 1938 prototype was sometimes credited with 9.5 kts.
[h]Given as 7.5 kts in a standard wartime handbook.

An LCM(2) unloads a truck on Guadalcanal, November 1942. Some of these craft were also used in North Africa and Sicily, and at Salerno. They can be recognized by their round sterns, which are very different from those on Higgins-designed LCVs and LCM(3)s.

This Higgins prototype tank lighter, October 1941, is carrying an M3 (General Grant) medium tank. The depth of the tank in the lighter's hold is evident; notice that only the small turret atop the tank's rather high hull is visible.

stern, the latter replaced by a deep skeg as in the British design. BuShips rejected the Tams report as incompetent. Only a complete redesign would save enough weight. As of March 1941, 14 LCM contracts had not yet been let, and BuShips had hoped that these craft would be built to a new lightweight design. After the Tams fiasco, on 13 June BuShips decided that, in view of the ongoing emergency, it would have to settle for the existing design, which would bring the total of such craft to 110, sufficient for two Marine divisions. Another 56 had just been approved for a third division, but they were not needed until the transports were ready, which might take some time. Thus it seemed unnecessary to continue building the existing unsatisfactory type.

One reason the bureau's 45-ft lighter was so heavy was that it was well-armored, with ¼–⅜-in STS plating. By early 1942 BuShips was more aware that, as in the case of the Eureka, the overall vulnerability of a landing craft depended as much on its agility as on the plating covering its vitals. A lighter craft was also easier to retract from a beach. As a consequence, the later LCMs were essentially unarmored, with ¼-in steel limited to their control positions.

When the Landing Boat Board reviewed Linsert's report on the ramped boat (the proto-LCV), it decided to ask Higgins to develop a 45-ft tank lighter; if it was successful, he would get a 50-boat contract. Linsert may have been inspired by the sight of a lighter that Higgins had built for Colombian customs, to carry a bulldozer that could ride out via a ramp. The analogy to a tank lighter was clear. Higgins' biographer reported that the previous year, 1940, Higgins had been so sure of a Corps of Engineers contract for a combination dredger tender and towboat that he had built one on speculation. Higgins was already developing a tank lighter design, and it was adapted to the existing hull, which was stripped of its deckhouse. To hold length to the specified 45 ft, the ramp tilted inwards. Building shallow-draft tugs for Peru, Higgins offered to convert them into tank lighters. He stripped off the boat's deckhouse and replaced its bow with a ramp. The inside of the double bottom formed the craft's hold, with its bottom well below the waterline. The boat was therefore inherently stable, hence it was much better adapted than the bureau design to rough water. The prototype was ready in a remarkably short time.[5]

BuShips preferred its raised-deck type. Its preliminary designers had already rejected sunken-deck designs of their own, much preferring a raised deck with a long ramp. The raised-deck arrangement offered more deck space, because the engine was in the hull rather than abaft the tank deck. It could better survive underwater damage; without longitudinal subdivision it would not list when flooded. Although both types had about the same freeboard to the tops of their bulwarks, the C&R type would drain freely from its above-water tank deck, whereas the Higgins type needed pumps. BuShips did concede that Higgins' design needed less beam for a given degree of stability.

It turned out that the bureau design, as well as the British MLC, had a more fundamental problem. Because the tank cargo was carried so high, the craft could easily become top heavy. This problem, moreover, would worsen as the craft was loaded with heavier tanks. The boat could support its weight, but the high center of gravity ruined its stability. With the center of gravity of his cargo far lower, Higgins had no such problem.

In June 1941, while it was still developing a lightweight successor to the 45-footer, BuShips ordered 50 Higgins tank lighters as a stopgap. After successfully testing the prototype, the order for the other 49 was confirmed. However, further construction was not expected. BuShips badly wanted to differentiate these craft from its own design, resorting to such devices as calling them "vehicles" rather than boats.

BuShips built only one of its improved small tank lighters, a 47-footer with a 30-ft tank deck. Hoisting weight was cut, partly by using a lighter structure and partly by eliminating the power windlass for ramp and anchor and fresh water tanks. The twin diesels drove through straight shafts rather than vee-drives. The design was later modified with a square- rather than vee-section cargo compartment, thus leading to a wider square-shaped ramp opening; the bottom of the vee determined the width of equipment the earlier craft could carry. BuShips tested the boat with several loads: a pair of 11,300-lb marine tanks, a 24,000-lb army tank, and a 155-mm gun (26,000 lbs).

These designs became obsolete when the Marines decided to adopt the army's medium tank. Also, the army had not shown the Marines' taste for lightweight equipment to ensure mobility. In September, the Landing Boat Board received a list of important army equipment the new tank lighters could not transport: the medium tank; the new 155-mm gun, which BuShips' 45-footer could take; the 7½-ton prime mover; the 4-ton cargo truck; the 4-ton wrecker; the 10-ton wrecker; the 4-ton tractor; and the power shovel. With its large deck, the BuShips craft could be modified to take some of this equipment, but its 9 ft 3 in ramp was too narrow for the medium tank. Yet all of the equipment listed could certainly be carried by existing cargo ships either on deck, in the hold, or in the 'tween decks, handled by 30-ton booms. There had to be some way to bring the equipment ashore.

In October the Auxiliary Vessels Board asked for a lighter to handle the new 30-ton medium tank. Once more, Higgins developed a design and produced a

One of Higgins' stopgap tank lighters is shown en route to Espirito Santo Island in the New Hebrides, November 1943. She is distinguishable from the later and slightly larger LCM(3) by her vertical ramp and by the distinctive grillwork atop it. "K 23" on the bow probably indicates USS *Alchiba*, AK 23 (later AKA 6).

prototype. In December 1941 BuShips felt compelled to order his craft, because it had no completed design of its own. Probably Higgins' greatest advantage was that his craft's structure was much simpler than BuShips' design.

BuShips saw Higgins' craft as no more than another stopgap. A 9 December internal memo strongly recommended further development of the 50-ft bureau design. It offered far more growth potential because it had much more reserve buoyancy, as well as a roomier tank deck. For example, the bureau lighter could carry two typical vehicles to Higgins' one—and there would never be enough lighters. Bureau designers contended that experience showed that it was much easier to load lighters with wider decks; only the bureau configuration offered a deck the full width of the lighter's hull. Because the ramps were part of their above-water structure, they could be wider than what Higgins offered: 10 ft rather than 8.5 ft. With machinery at the same deck level as the tank, Higgins' design offered a considerably shorter tank deck: 290 vs. 500 sq ft for the bureau design. Higgins offered much better stability, however. Metracentric height (GM) in loaded condition was 4.5 ft rather than 2.97 ft. Because the deck of the bureau lighter was 1 ft 4 in above the waterline, its bulwarks were pierced with freeing ports for any water taken on board.

The British were aware of Higgins' design, and on 20 November the Admiralty asked its delegation in Washington to investigate the design, in hopes of supplementing the supply of MLCs (mechanized landing craft, the equivalent British design). They reported that Higgins' boat had better performance and beaching characteristics than the MLC. It could be derrick hoisted or towed, but it could not be car-

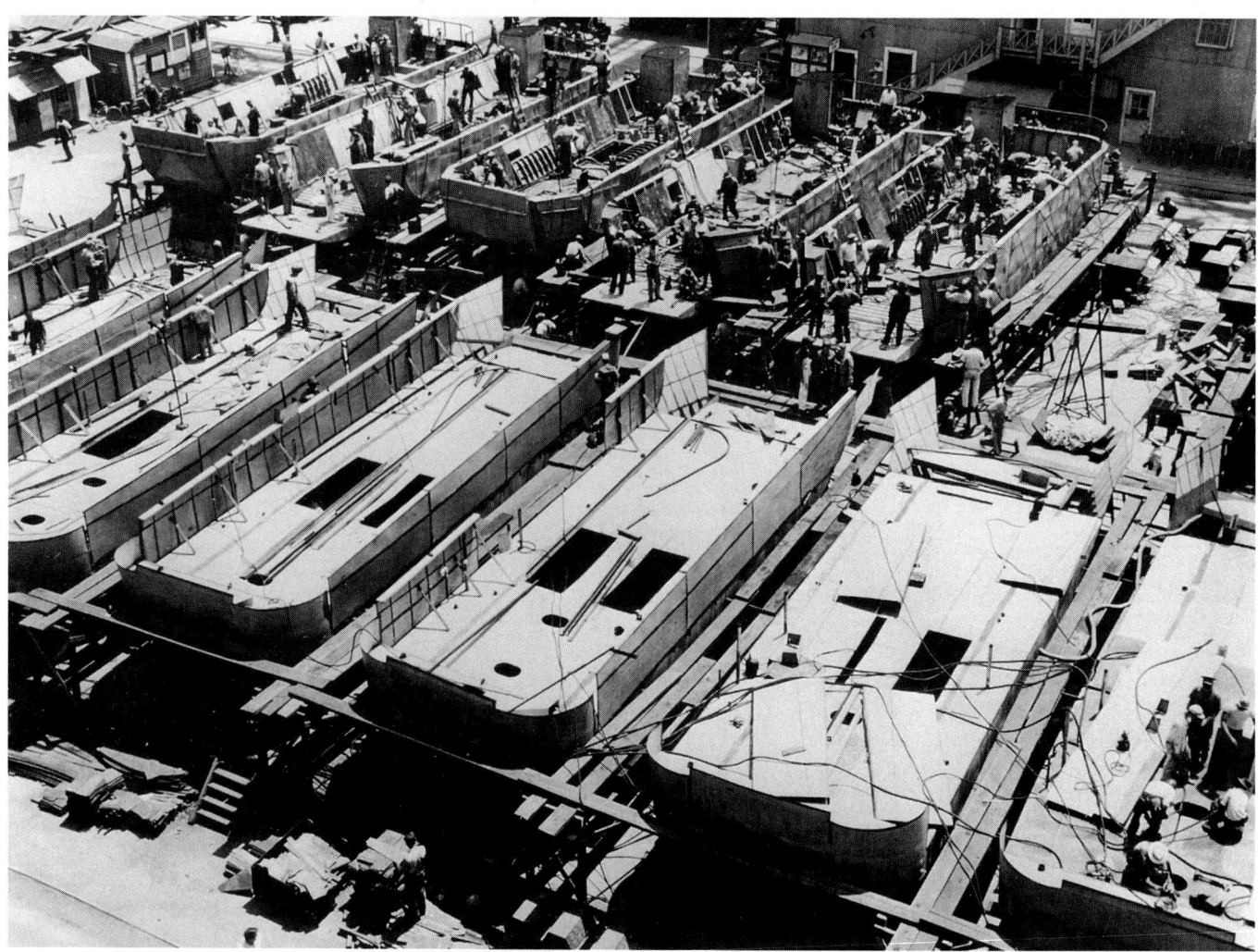

Bureau 50-ft tank landing craft are shown under production at Norfolk Navy Yard, during the summer of 1942. Compared to the more successful Higgins design, they show a very broad tank deck above the waterline, unencumbered by engines. Rectangular holes in the decks of the craft in the foreground are for engines placed below the tank deck. Failing a crucial test, this design never entered combat. For a time, however, official documents referred to it as an alternative type of LCM(3).

ried in British gantry ships. With the highest priority, production would not exceed 25 per month, so the delegation planned to ask for 150. These craft were still in the design stage. On 28 November the British Admiralty's delegation sent in its requisition, asking for some minor modifications: the internal width of the hold had to be at least 9 ft, the boat needed bullet-proof protection, a stern anchor was to be fitted. Higgins was asked to investigate whether the boat could be hoisted with a 10-ton load on board. Boats would all be powered by 165 HP Gray Marine diesels. The previous day Higgins had offered to deliver one lighter per day, starting 3 weeks after an order was let. After seeing a test of the 45-footer in December, the British increased their order to 250. They planned to call the 50-footer the MLC Mk II.

Because BuShips much preferred its own 50-footer, it opposed the order. Through early 1942 the bureau tried to standardize production on its design. Higgins protested that the bureau design was not merely inept but also quite unsafe. A comparative test was ordered.[6] On 25 May 1942 one of each type of lighter, loaded with a 59,000-lb army M3 medium tank, left Norfolk to land the tank in the surf outside Fort Henry. A fair surf was running at nearly low tide, and wind speed was 23 mph. The Higgins lighter beached successfully through 3–4-ft surf. The bureau lighter failed. It clearly displayed the problem inherent in its virtue, its big tank deck: the tank was slightly off center and too far forward. When the lighter took on water and trimmed by the bow, water could not drain out of the after scuppers. It leaked into the engine room in the hull; about a foot accumulated in the bilges. The boat had to slow to about 2 kts to allow the engine room bilge pump to clear enough for the boat to regain way. The boat had to turn away from the beach. In so doing she nearly capsized, perhaps due to free water surface in her hull.

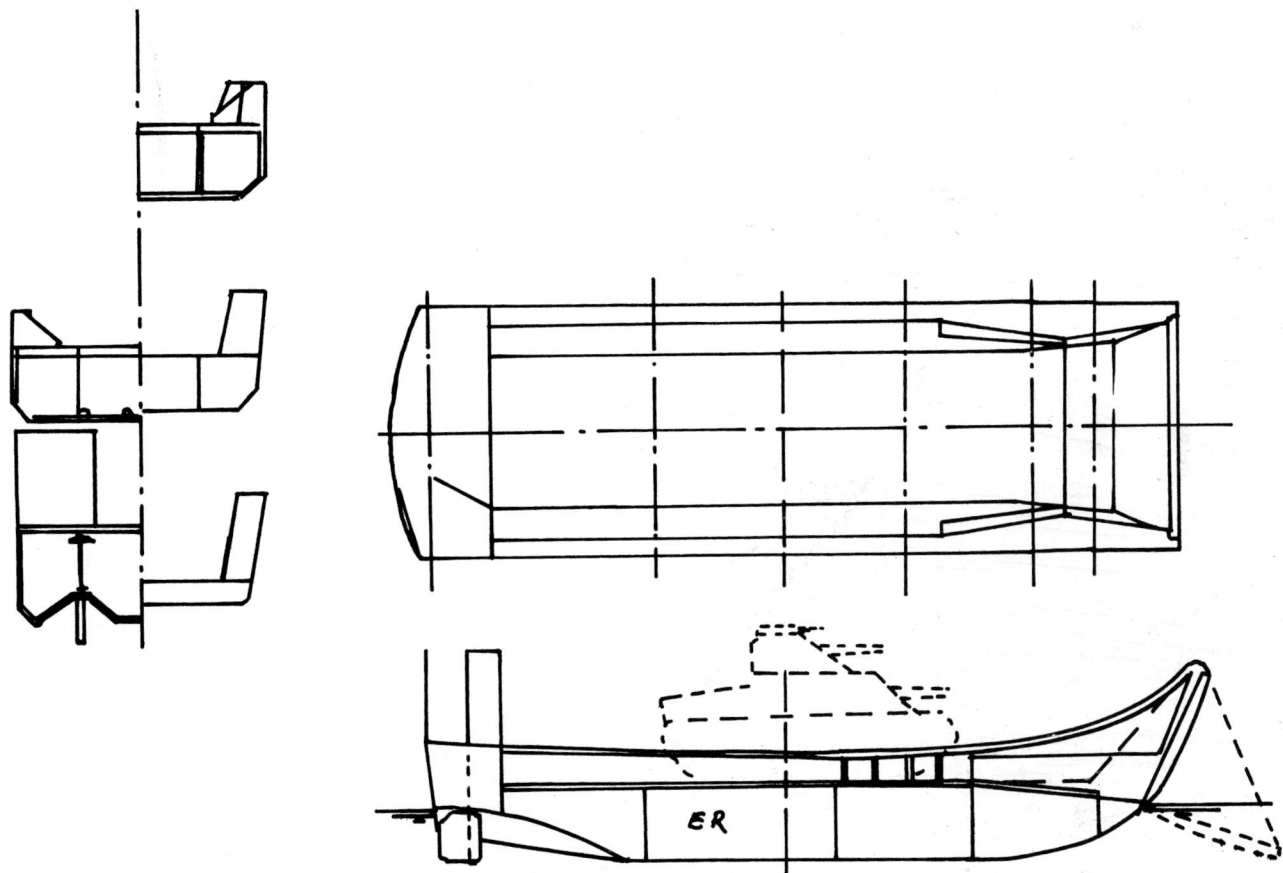

The failed BuShips LCM(3) design, from BuShips preliminary design section, 1941. Detail views are structural sections; note the tunnel for the propeller and rudder on the bottom section. The stern compartment largely above water housed steering gear and ramp hoist machinery. Engines were in the hull compartment second from aft. The other hull compartments were watertight, for buoyancy. Water sloshing in these compartments (to form a free surface) almost capsized an LCM of this type during rough-water trials in 1942. By way of contrast, Higgins opted for housing the engine in the well deck, putting the well deck below water, and minimizing watertight volume below it. The tank indicated by dashed lines is the early-war M3 (Grant). Length was given as 49.5 ft. The tank deck was 1 ft 4 in above the waterline, whereas the much more successful Higgins lighter, which went into mass production, had its tank deck below the waterline. The below waterline tank deck proved decisive for stability in rough water.

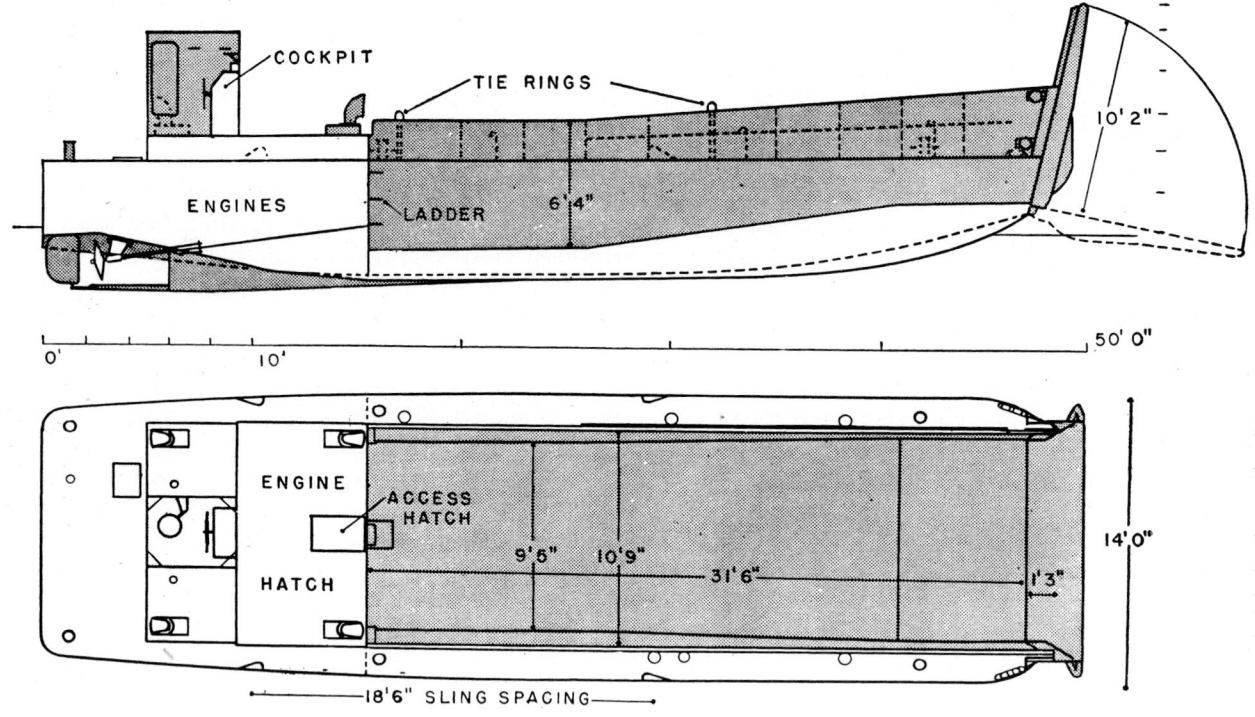

The Higgins version of LCM(3) became the standard tank-carrying landing craft of World War II. When armed, this craft carried a pair of pedestal-mounted 0.50 caliber machine guns atop the engine hatch forward of the conning position. Compared to the later LCM(6) Mod 1, this craft had much wider gunwales, from which lines could easily be handled. *U.S. Navy (ONI 226)*

Capt. Edward Cochrane, chief of preliminary design, was on hand. Despite his association with the bureau design, he was quite objective. He reported that the Higgins lighter not only got through successfully, she would probably have done so under somewhat worse conditions. The bureau lighter failed. She might have done somewhat better had the tank been further aft, but that was beside the point. Cochrane doubted that, however loaded, she would have performed as well as the Higgins boat. Nor was there much point in thinking through possible improvements; there was no time for further development. Cochrane recommended an immediate shift of production to the Higgins 50-footer.[7] Only bureau-designed lighters in advanced stages of construction were completed. All material amassed for others was diverted to the Higgins type.

When standardized designations were chosen for British and American landing craft in July 1942, the British MLC became LCM(1) (i.e., LCM Mk 1). The 45-ft bureau type became LCM(2); the bureau 50-footer never received a formal designation. The Higgins 50-footer became LCM(3). LCM(2) saw combat in 1942 at Guadalcanal and in North Africa, and in 1943 in Sicily and at Salerno.[8]

LCM(3) was little changed during or, indeed, after the war.[9] Early ones had wooden shelters for their coxswains, to avoid affecting the magnetic compasses. Due to a shortage of nonmagnetic steel, later ones had mild steel shelters, hence they had unreliable compasses. Unlike Higgins personnel boats, the LCM had to be hoisted by boom, and it was not strong enough to be hoisted loaded; most ships did not have booms with sufficient capacity to hoist loaded LCMs, even if the boat been strong enough. LCM(3) was first used in North Africa in November 1942. After the invasion of Empress Augusta Bay in November 1943, it was decided that Pacific AKAs and even large cargo ships (AKs) should carry eight of them. Instead of a tank, an LCM(3) could carry 60 troops. It proved a valuable supplement to the standard troop-carrying LCVP at

A standard Higgins LCM(3) is loaded aboard an attack cargo ship at the New York Navy Yard during World War II. Note the skegs protecting the two propellers, each with a rudder at its after end. Other LCMs on deck display the characteristic bow ramp.

Normandy in 1944. A wartime LCM development, the LCM(6), is discussed in Chapter 7.

The Higgins personnel boats and the LCMs satisfied long-understood requirements. The third major pre-war landing craft, an amphibian called Alligator, was unexpected. It was adopted because the Marines saw an experimental commercial craft as the solution to a problem they badly wanted to, but could not, solve.

Boats could bring the Marines ashore, but any post-landing buildup would be slow, because boats would have to be unloaded at the shore before their cargoes could be brought inland to dumps, perhaps aboard small trucks brought in by landing craft. A truly amphibious craft could solve the problem by taking cargoes directly inland. In the late 1930s the Marines were not developing any such craft, because all available money was going into surf boats.

Fortunately, a private inventor was vitally interested in just this problem. In 1933, Donald Roebling, son of the bridge builder John Roebling, began work on what became the LVT (landing vehicle, tracked). He envisaged a vehicle to rescue aviators who crashed into the Florida Everglades. The environment he faced was not too different from what Higgins saw in the Louisiana bayous, and naturally Roebling developed an amphibious vehicle. Roebling used tracks for propulsion both in water and on land. The tracks were not particularly efficient, so his Alligator was not very fast in either environment, but the key point was that it could travel in both using the same powerplant and the same mode of propulsion. Roebling's amphibian came to the Marines' attention when, at a dinner, Rear Adm. Edward C. Kalbfus showed Maj. Gen.

The Marines' amphibious tractors became vital World War II landing craft. Unlike other tractors, they were not derived from any projected requirement. Instead, the Marines bought them because Donald Roebling's invention so clearly fit their projected type of warfare. This LVT is on the beach at Fedala, 4 December 1942.

Louis M. Little, who commanded the Fleet Marine Force, a 4 October 1937 article in the widely read *Life* magazine. Little contacted the commandant, Maj. Gen. Thomas Holcomb, who alerted the Marine Corps Equipment Board. In April 1938 the board asked for an Alligator, to be tested in the 1939 FLEX 5 exercise. The board visited Roebling in September 1938. C&R resisted procurement; after spending money on various kinds of landing boats, the cupboard was bare. Alligators could not be bought until the urgent landing boat program was further along.

For his part, at first Roebling insisted that his Alligator be used only for peaceful purposes. In September 1939, after the outbreak of war in Europe, a visit from Brig. Gen. E. P. Moses, president of the Equipment Board, changed his mind. Early in 1940 he allowed BuShips to study his plans. Roebling completed a redesign of his Alligator in January 1940. Although BuShips reported favorably, at first the U.S. Navy refused to provide money, on the ground that conventional landing craft were more vital. However, within a few months the Marines did get the $20,000 they needed.

Given the BuShips report, in March 1940 the Landing Boat Board recommended that two experimental amphibians be bought. They were tested at Guantanamo Bay in November 1940. This first version could carry 24 Marines or 4,500 lbs of cargo over water at 6 mph (range 50 miles) or over land at 10–15 mph (range 75 miles). It had a raised cab forward for the driver and commander and an open cargo bay. There was no armor. The Marines considered the amphibian underpowered, but they recognized its potential. After the November trials, they ordered 200; the first production unit was delivered in July 1941. By that time the U.S. Army was also buying the craft. LVTs were first used, to carry cargo, at Guadalcanal. The prototype survives at the Marine Corps museum in Quantico.

As early as mid-1941, however, calculations of the number of troops that could be brought ashore by transports' boats included the personnel capacity of the LVTs. This role seems to have been forgotten until mid-1943, when the assault on Tarawa, fringed by a coral reef, was being planned. There, the LVT became a troop carrier. This function was so different from that envisaged prewar that later LVTs are described with wartime craft (see Chapter 7).

Under the prewar system, the other letters of the alphabet had been exhausted, so the Alligator was designated a Z-boat, with the armored version designated ZA. In July 1942 it was christened LVT, landing vehicle (rather than craft), tracked. The initial version became LVT(1).

The navy was also interested in an amphibious truck. Trucks were needed to move materiel from the beachhead inland to dumps. At the New River exercise in 1941, it became apparent that landing craft could not bring enough trucks ashore. By late 1941 the Amphibian Car Corp of Buffalo, New York, was offering a swimming truck. OpNav asked the Landing

Three types of amphibious vehicles, all of them extemporized for World War II, are shown: an LVT in the middle, amphibious Jeeps in the foreground, and DUKWs in the background. All are part of an assault on Perry Island, Eniwetok Atoll, 22 February 1944. *U.S. Coast Guard*

Boat Board to consider it, not as a way of landing cargo, but rather as a means of getting trucks ashore without taking up valuable landing craft space. Nothing came of this initiative, but the army produced an amphibious (dual-drive, wheels and propellers) version of its standard 2½-ton truck, the DUKW.[10] A navy-type designation, LVW (landing vehicle, wheeled), was rarely used. DUKWs carried cargo to the beach and also transported it further inland; indeed at some landings problems occurred because, once they landed, DUKWs were used as conventional trucks and thus were taken away from their vital buildup role. At Sicily in 1943 they were vital in rigging pontoon causeways. At Normandy DUKWs carried cargo all the way from ships inland, the shore party commander having decided against using beaching craft to avoid double handling. In many operations DUKWs equipped with "A" frames were used to unload small beaching craft (LCM, LCVP, LCT) at the beach. On the other hand, DUKWs (and LVTs) were inefficient cargo carriers. They presented the boom operator on a cargo ship with a very small target, and they could not handle a full sling load of cargo. Compared to an LVT, a DUKW was not as efficient a personnel carrier. Compared to an LCVP, it was not nearly as good a seaboat.[11]

The other prewar landing craft, built in vast numbers in wartime, were 7- and 10-man rubber boats, initially designated RA (or RB-7) and RB (or RB-10), respectively. They were redesignated LCR(S) (landing craft, rubber, small) and LCR(L) (landing craft, rubber, large) in July 1942. They were used by raiders, by reconnaissance units, and by underwater demolition teams.

The LST was probably the most important product of British influence on American amphibious development in World War II. Here LST 715 lies at Mare Island alongside ARL 30 (USS *Askari*), herself converted from an LST (LST 1131). Visible just abaft the forward gun tubs is the elevator that transported vehicles carried externally, on the upper deck, to the tank deck so that they could be offloaded over the bow ramp. The structure abaft it is a short ramp leading to the elevator, opened up to reveal stowage between deck and ramp. Landing craft in the davits are LCVPs.

5
The British Connection

In developing its amphibious arm the U.S. Navy was heavily influenced by the Royal Navy. After France fell in 1940, Britain's prime minister, Winston Churchill, was determined to project British power back onto the continent. To do that he needed landing craft and ships. With his own shipbuilding resources badly stretched, he looked to the United States. British orders were directly responsible for some of the most important wartime U.S. classes: the LST (landing ship, tank), the LSD (landing ship, dock), the LCI(L) (landing craft, infantry, large), and the LCT (landing craft, tank). Table 5-1 shows details for the British-inspired amphibious ships and craft. The British were also largely responsible for the invention of the amphibious flagship, the amphibious control craft, and close-in fire support craft, which are discussed in later chapters. British experience and thinking deeply affected U.S. practice, particularly after mid-1941, and especially in the Atlantic theater.

After a long controversy as to whether air attacks made landings impossible, the British in 1938 formed an interservice board to develop the necessary techniques. It ordered prototype assault (personnel) landing craft (ALCs) designed to be carried in ships' davits. The board also wanted a support landing craft to beach with the ALCs, providing covering fire against enemy troops on or near the beach. The British also developed mechanized landing craft (MLCs) capable of carrying single tanks. They could be hoisted by a standard cargo boom.

Unlike the Americans, the British had numerous liners they could use as transports. Their directorate of sea transport and their prewar amphibious developers (ISTDC, Interservice Training and Development Centre) examined British and Dominion ships for possible conversion, just as American war planners had examined the U.S. merchant fleet since the 1920s. Their requirements can be compared with U.S. ideas: high speed (17 kts), capable of being disguised (not a U.S. consideration), potential for adequate antiaircraft armament, capable of carrying an infantry battalion without its transport (embarked for less than 7 days), which would be landed in waves of 15 LCAs (landing craft, assault) each. Minimum range was 1,000 nm. The 2,000-man British brigade consisted of four 500-man battalions, each less than half the size of a U.S. Marine battalion. In 1940 the British chose sufficient tonnage to carry and land a brigade, in the form of four fast (18 kt) "Glen" liners; they actually got only three, and had to take over two smaller Dutch ships to provide the necessary lift. Each Glen could accommodate the landing craft the men would use: 12 ALCs, 2 MLCs, 2 SLC (support landing craft). In service a Glen typically carried slightly more: 34 officers and 627 enlisted men plus boat crews (4 officers and 77 men) plus her own crew. Like the Marines' transport, it was an integrated tactical unit, but on a much smaller scale. The British liners were eventually called ILS (infantry landing ship); under the unified designation system adopted in July 1942, they became LSI (landing ship, infantry). Unlike U.S. attack transports, they had little or no space for vehicles, and thus they had to depend heavily on accompanying fast freighters.

It appears that Glen conversions so impressed President Roosevelt that in July 1940 he personally ordered two ships of similar capacity from the Maritime Commission (headed by Adm. Emory S. Land). These *Doyen*-class ships were the first amphibious transports designed as such for the U.S. Navy. Construction was slow; they did not enter service until 1943. The Maritime Commission sketch design was designated P-X-L (i.e., a passenger ship; the commission was not yet using a separate S series for naval vessels). On 4 September Land sent a sketch to President Roosevelt. The ship would displace 6,300 tons, and she would make 18.5 kts on a pair of C1 powerplants (total 8,000 SHP). She would carry 60 marine officers and 500 marines, with

Table 5-1. British-Inspired Amphibious Ships and Craft

	LSD 1	LST 1[a]	LCT(5)	LCT(6)	LCI(L) 1	LCI(L) 351
Length (ft-in)						
WL	454-0	316-0	105-0	105-0	153-0	153-0
OA	457-9	327-9	114-2	120-4	158-5.5	158-5.5
Beam (ft-in)	72-0	50-1.5	32-0	32-0	23-3	23-3
Draft fore/aft (ft-in)[b]						
Light	8-2.5/10-0.5	2-4/7-6	0-3.375/3-0.875	0-6.25/3-6.5	3-1.5	3-1.5
Beaching[c]	30-9.5/29-9.5	3-10.75/9-9.5	2-10/4-1.675	3-4/4-0	2-8/4-10	3-0/5-0
Loaded	15-5.5/16-2	8-2.75/14-1.25	NA	NA	5-4/5-11	5-8/5-8
Displacement (tons)						
Light	4,032	1,625	133	143	216	246
Beaching	14,078	2,366	283	284	234	250
Loaded	7,930	4,080	NA	NA	380	393
Power (SHP/ shafts)	7,400/2[d]	1,800/2	675/3	675/3	1,600/2	1,600/2
Speed (kts)	17	12.1	8	8	16 max, 14 cont	16 max, 14 cont
Fuel (tons)	1,770	590	11.12	11.12	130	110
Radius (nm/kt)	7,400/15	6,000/9[e] 24,000/9[f]	1,200/7	1,200/7	4,000/12 500/15, 1,500/12[g]	8,000/12
Complement[h]	17/17/309	7/104	1/12	13	3/21	4/24
Capacity						
Troops[i]	22/310	16/147	—	8	6/182	9/200
Tons	1,450	500[j]	150	150	75[k]	75[l]
Cu ft	250,000	118,800[m]	—	—	NA	NA
Water, fresh (tons)	NA	NA	—	—	None	37
Boats	3 LCT or 14 LCM(3)	2 LCVP	—	—	—	—
Armament						
5-in/38	1	—	—	—	—	—
40-mm	2 × IV, 2 × II[n]	2 × II, 4 × I	—	—	—	—
20-mm	16	12	2	2	4	5

NOTES: Dash indicates that specification is irrelevant; NA indicates relevancy but data not available.

[a]Data for LST is as of 1945, instead of as when built.

[b]Draft aft if only one figure given; draft fore/aft if two figures given.

[c]For LSD, beaching is term used to designate that ship is ballasted down to launch small craft. For all others, it is the lightly loaded condition in which the ship beaches.

[d]Data is for LSD 1–8. LSD 9–21 and 25–27 are 7,000 SHP; LSD 22–24 are 9,000 SHP.

[e]When loaded to beaching draft.

[f]In oceangoing condition.

[g]When loaded to beaching draft.

[h]Complement is given as officers/CPOs/enlisted when three figures are listed and officers/enlisted for two figures; when only one figure is given, breakdown is unknown.

[i]Officers/enlisted when two figures are listed; when only one figure is given, breakdown is unknown.

[j]When loaded to beaching draft.

[k]Alternative load, not in addition to troops.

[l]In oceangoing condition.

[m]As described in 1943; later a figure of 90,000 cu ft was used. Square was 7,920 sq ft. Tank deck area was 6,328 sq ft on an LST 542; the upper deck added another 5,000 sq ft. Thus the 1943 figure presumably allows for only a small fraction of the upper deck. LCT(5) square was 2,060 sq ft.

[n]Multiple gun mounts are indicated by Roman numbers; therefore 2 × IV indicates two quadruple mounts.

space for 150 extra. Proposed armament was two 5-in/38 dual-purpose guns and two 1.1-in quadruple machine cannon. Land expected that the first ship could be delivered 14 months from contract date, with additional ships 60 days thereafter.

Before ships could be built, their design had to be reviewed by the U.S. Navy and the U.S. Marines. The results make it clear that the ships bore no relationship to the characteristics already developed. For example, the desired capacity of 500 marines was about a third of a combat team. The Marines made it clear that they did not contemplate splitting up the team into smaller units. The ships did differ from British conversions in that they would carry tanks, in accordance with U.S. practice.

The new ships were generally described as intended for Caribbean operations, their shallow draft justified by local conditions. They were sometimes described as marine battalion transports, which would suggest their British origins (the U.S. Marine battalion was nearly three times as large as the British one). It seems likely that they were intended as prototypes for European operations, on the theory that the Glens were so adapted. By going to Land at the Maritime Commission rather than to the U.S. Navy, President Roosevelt avoided congressional scrutiny. In July 1940 he was arranging to run for an unprecedented third term. That fall any debate about transport construction would surely have raised questions about whether the president believed the United States could stay out of the war. Building transports might have implied shipping U.S. troops overseas.

The president having informally asked for the ships, the secretary of the navy included them in a 5 August 1940 list of ships to be ordered from the Maritime Commission. Design was formally authorized on 12 August; by then it was already under way. The ships were to be as small as practicable, about 5,000 tons standard, with minimum draft, and capable of 18–19 kts (18 kts sustained sea speed). Besides the 500 troops, they would have space for another 150 (which could be used instead for cargo).

For survivability, they would be built to a two-compartment flooding standard, but they would lack naval features such as duplicate controls for each machinery compartment or extensive damage control fittings. Later the formal characteristics added that the inner bottom should be carried up the sides to the deck next above the full load waterline, at least abreast of machinery spaces. That would not provide torpedo protection, but it would help protect them from grounding and even minor collisions.

Late in August 1940 the Maritime Commission sent its sketch design to the navy. Capacity was 60 marine officers and 672 enlisted marines, about half a combat team. Cargo capacity, exclusive of ammunition, would be 89,500 cu ft. Later the Marines accepted a reduction to 75,000 cu ft—a quarter of that in the "ideal" transport—to gain berthing space. Like the ideal transport, this one had four holds, in this case paired fore and aft. Instead of the usual weather deck hatch, the smaller of the after holds was served by a small hatch in the first deck above it and thence by monorail and cargo ports in the ship's side.

Once the idea of the smaller transport had been broached, the Marines liked it. Each ship could carry about half a combat team, and two ships would carry more boats than one large one, thus allowing quicker unloading. Smaller ships, moreover, should be more flexible. For example, they could reinforce particular parts of a beachhead or landing area. A Marine representative told the General Board that smaller ships lent themselves better to combat loading. Their holds were shallower (hence could be worked far more quickly), and compared to a single ship they would have twice as many hatches and booms and more deck space; overall they should be roomier than a single combat loader designed to take a full combat team. They would have more deck space to take boats or other materiel that must be carried either on deck or in 'tween deck spaces, and they would have more deck space on which to exercise troops.

The Marines had always liked shallower draft and better maneuverability, because they expected to operate in constricted waters such as the lagoons of Pacific atolls. They found the design acceptable and even attractive as long as ships were built in pairs, which had been planned in the first place. They did admit that combat teams were not meant to be split, and that it would be essential to reunite halves quickly upon landing.

The Marines' defense of the small transport was so passionate that Rear Adm. Ernest J. King (later the wartime CNO), who was chairing the hearing, suggested that it might be good to build, or convert, more of them. The counterargument was that there were already small transports (APDs) for small jobs and big transports for big ones. Did the U.S. Navy really need an intermediate class? Even so, King asked for a list of merchant ships suited to conversion to small attack transports. It turned out that all would require extensive conversion for combat loading. None was self-supporting in fresh water (an important consideration), and many drew far too much water. Although several had sufficient cargo space, it was of the wrong kind, without deck hatches or booms, because the cargo generally carried was small items that were handled through side ports.

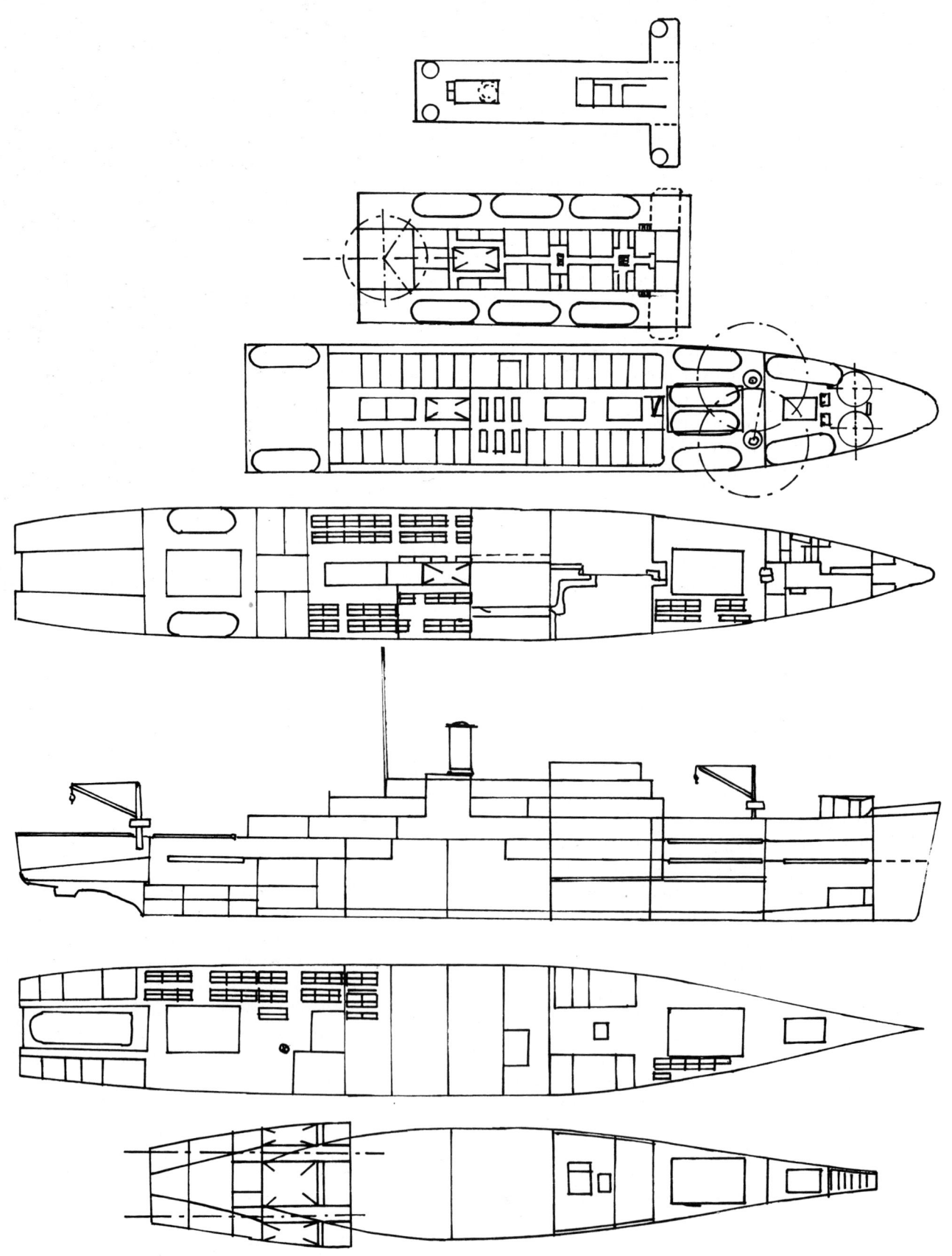

Although not ordered by the British, the two small U.S. attack transports were probably inspired by early wartime British conversions of fast liners. Newly completed, USS *Doyen* is shown on 17 June 1943 off the California coast. Designed and built by the Maritime Commission without direct naval supervision, she was not sufficiently stable, and had to be modified before entering service. Placed in Maritime Commission (rather than U.S. Navy) reserve in 1946, she became the Massachusetts Maritime Academy training ship *Bay State* in 1961. She reverted to Maritime Administration custody in 1973.

The Maritime Commission design had been drawn up so hastily that it did not show positions for landing craft; the commission's designers had taken the 18 boats of the ideal transport as their standard. BuShips doubted that more than 14 could be stowed or expeditiously handled. Surely 12 would suffice. Later it was pointed out that 18 landing boats could accommodate all the troops on board in one trip.

Probably the most innovative feature of the design was a launching ramp or slipway aft, closed by doors, for the single tank lighter. It would make for quick launching and bring the tank aboard quickly into action; the lighter would return to get the second tank on board the ship. BuShips critics pointed out that the ideal transport had four tank lighters, whereas this ship, in theory half an ideal transport, had only one. Artillery, tractors, and other heavy equipment could be stowed only in the very limited deck space or in the holds. The Marines considered one tank lighter inadequate, so an artillery lighter was added, stowed forward. To handle it, BuShips moved one of the two 30-ton booms forward; the Maritime Commission had located both aft, to handle the tank lighter and tanks and other heavy equipment. The approved design also showed twelve 36-ft Eurekas and two 40-ft motor launches, the latter to handle large numbers of men and substantial loads of equipment and stores.

The design originally showed 10- and 30-ton cranes. In October 1940, however, BuShips pointed out that most lifts would be 2 tons or less, and that heavy booms could not handle them very quickly. Also, such cranes might not be obtainable in time. BuShips therefore asked for an alternative arrangement of

The small transport (*left*), which became *Doyen*, from a sketch by the Maritime Commission, 1940. The small boats shown are 30-ft landing boats: six of them in davits on the boat deck, four of them forward on the upper deck, two in davits aft on the upper deck, four aft on the main deck forward of the tank lighter slipway. Note the split machinery spaces. Dimensions were 409 ft (overall) or 400 ft (waterline) x 56 ft x 17 ft 4 in (maximum molded) or 14 ft (standard displacement). Displacement was 4,500 tons standard and 6,350 tons maximum. Power was 8,000 SHP (two shafts) for 19 kts; endurance was 10,000 nm at 16.5 kts. Troop accommodation was 60 officers and 672 enlisted. Ship's complement was 22 officers, 32 CPOs, and 180 enlisted. Armament comprised one 5-in/51 (125 rounds), four 3-in/50 antiaircraft guns (500 rounds per gun), and four 0.50 caliber machine guns, all on the navigating bridge (10,000 rounds each). The drawing shows the two 3-in/50 side by side forward and the 5-in/51 right aft on the boat deck, but the location of the other two 3-in/50 guns is unclear (they may have been intended to occupy a raised platform above the 5-in aft). Boats were one 45-ft tank lighter, 8 landing boats handled by davits, and 10 landing boats handled by cranes.

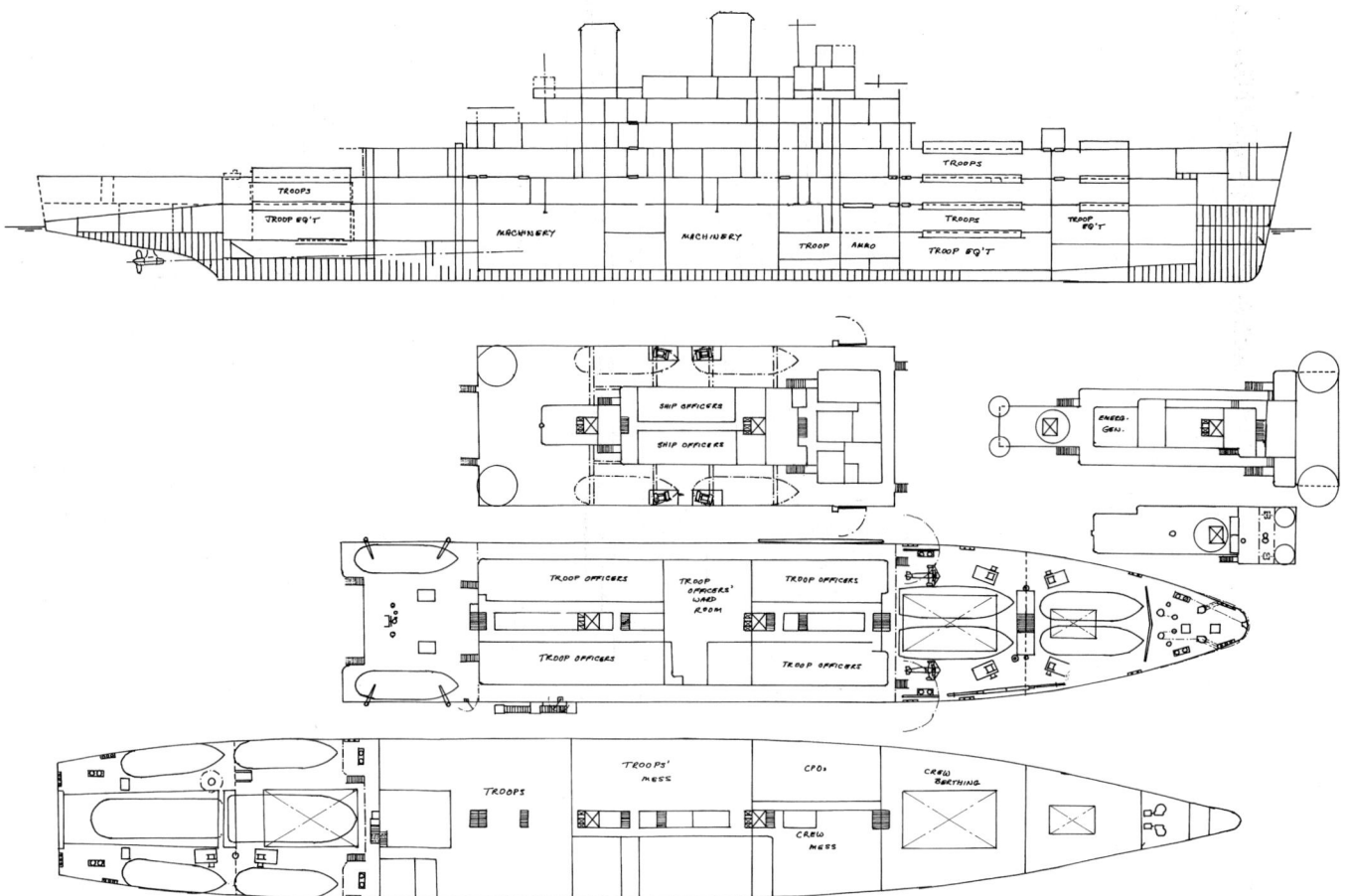

Doyen (P1-S2-L2) plans approved by the Maritime Commission for construction by Consolidated Steel, 25 February 1941. The most obvious change from the first sketch design was to two separate machinery spaces, hence to two funnels. The landing craft slipway aft survived, and was built into the ship, only to be plated in because it could not accommodate the 50-ft LCM(3), which had become standard by the time the ship was built. The craft shown in the slipway is a 45-ft LCM(2), abaft a 45-ft artillery lighter atop the 20 × 32.5 ft hatch forward of it. Note that the ship has no multiple Welin davits for her 36-ft landing boats: four in davits on the superstructure deck, two in davits aft on the upper deck, two on the upper deck forward atop the 10 × 15.75 ft hatch, and four on main deck aft abeam the artillery lighter and the slipway with the tank lighter, for a total of only twelve such craft. The 40-ft motor launches are stowed atop the forward 20 × 32.5 ft hatch. A big 30-ton boom is shown stowed on deck forward. Boat boom (44 ft) and accommodation ladder are shown on one side only, but they would have been provided on both sides of the ship. The large gun tubs on the superstructure deck and on the navigating bridge deck would have accommodated quadruple 1.1-in machine cannon; the smaller tubs were for 0.50 caliber machine guns. The plans did not show directors for the 1.1-in guns or the usual larger-caliber guns.

booms for 10-ton lifts from the large hatches (one forward and one aft) and 2-ton lifts from all four cargo openings simultaneously. The cranes were replaced by conventional kingposts and booms, with one 30-ton crane aft to handle the tank and artillery lighters.

The Maritime Commission design showed no gasoline stowage; in the end the characteristics called for below-decks stowage for about 10,000 gal in drums. The characteristics required a sustained speed of 18 kts (fully loaded, 4 months out of dock) and an endurance of at least 10,000 nm at 16.5 kts, the speed of other auxiliaries. The Maritime Commission planned 8,000 SHP geared turbines in two compartments. This plant was heavier than the 11,000 SHP plant BuShips had just developed for the minelayer *Terror;* the bureau also considered it too vulnerable. It preferred an 8,000 SHP diesel-electric plant it had just designed for large netlayers. The ship's landing craft would be diesel powered, and a diesel plant would not need the big engine casings of the steam plant. Alternating engine and motor rooms would make the plant much more survivable. The approved characteristics, then, showed diesel power. However, by late October BuShips had accepted the Maritime Commission steam plant, because diesels would increase cost and impose delays. BuShips did require that the single stack be split in two to increase area available for living space.

At first it seemed that the new ships, in common with the ideal transport, would be armed with a mixture of 5-in guns and antiaircraft machine guns. It proved impossible, however, to find appropriate positions for four such weapons, and two would not provide all-around fire. The Bureau of Ordnance (BuOrd) reluctantly accepted a battery of four 3-in/50, which it described as useless without fire controls, and two quadruple 1.1-in machine cannon. During the war the 1.1s were replaced by two twin 40-mm, and the ships also carried eight 20-mm machine cannon.

The secretary of the navy approved the formal characteristics for the modified Maritime Commission design on 14 October 1940. *Doyen* (AP 2, later designated APA 1) and *Feland* (AP 18, later designated APA 11) were built to this design. They were the first U.S. attack transports to be designed as such. Neither was completed until 1943, long after numerous standard Maritime Commission ships had been converted to larger attack transports.

The Maritime Commission supervised all aspects of construction. The "peculiar" contractual arrangements made it difficult to order changes and monitor construction. Thus it was an unpleasant surprise to the navy when *Doyen* turned out to be 550 tons overweight, 3 ft over draft, and only marginally stable. The commission argued that the navy's demands for changes during construction were responsible. They included the addition of 390 troops (to the original 710) in April 1942, the replacement of 45-ft tank lighters with 50-ft lighters, and changes to armament (including the installation of a Mk 50 director). The navy replied that the original weight estimates had been erroneous, and that the commission had failed to control construction, allowing the builder to install too much sheathing internally and even to use excessively thick splinter shields. Clearly, drastic changes were in order. By the time the ships were ready, the 45-ft tank lighter had been superseded by the 50-ft LCM(3). To accommodate the larger craft, the slipway required considerable alteration. Due to its hull shape, an LCM(3) required a steel cradle for launching, with buoyancy tanks to refloat it for recovery. Given the gross overweight the navy decided to seal up the slipway, the most innovative element of the design, to form a buoyancy space. The ship clearly could not carry tank lighters (or, for that matter, tanks) at all, so she did not need her 30-ton crane. Her cargo ports had

Seen from aft, *Doyen* shows her transom stern. She was designed with a stern slipway for launching an LCM(2) landing craft. By the time the ship had been completed, LCM(2) had been superseded by the larger LCM(3), and the slip was closed up to form a buoyancy space.

Feland (APA 11) in 1945.

to be sealed up to improve seaworthiness. Four of her planned landing craft would be eliminated, leaving sixteen. On the other hand, the navy wanted to add a pair of twin 40-mm guns, forward and aft. Later the navy removed the 30-ton boom forward, as well as the mainmast; a light tripod was substituted. The starboard kingpost was relocated to the centerline and provided with two 10-ton booms. As completed, the ships had four sets of triple Welin davits, accommodating 12 LCVPs; 4 more LCVPs were stowed on deck.

On trials in October 1943, even though she had been lightened, *Doyen* heeled 13 degrees when turning at full speed (19 kts), with only 2 (rather than the full load of 16) LCVPs on board. The ships became a byword for the problems of small transports, and their example was cited when BuShips rejected a July 1943 JCS proposal to convert some C1-Bs to attack transports.

During the September 1940 General Board discussion of the Maritime Commission design, the question arose of whether it might not be better to adapt the existing design of the minelayer *Terror*. She had considerable internal deck space, intended to house mines. Her design had already been adapted as the basis for a heavy netlayer. Using existing plans would considerably speed production. *Terror* was, moreover, about the same size as the proposed Maritime Commission transport: 4,700 tons light compared to 4,250 tons light, with about 10 percent greater usable volume. She was slightly longer, 440 vs. 400 ft, and slightly beamier. Fully loaded she would displace 8,125 tons at a draft of 18.2 ft, compared to 5,300 tons at 17.5 ft. Estimates made during the netlayer design showed that cargo capacity would be 100,000–125,000 cu ft.

There was a definite feeling that the Maritime Commission design had been put together far too hastily. The *Terror* design was far more mature. Moreover, those designing the netlayer version of the minelayer had been well aware of the need for transports. The netlayer was designed to carry numerous vehicles, if necessary, instead of nets, on her main deck. Vehicles could be brought aboard using the existing 30,000-lb crane, or they could be driven on board. The netlaying or minelaying space on the second deck could easily be adapted to stowing vehicles. As in the new transports, it might be possible to launch tank lighters down the ramp already provided in the stern of the minelayer *Terror*.

The *Terror*/netlayer idea bore fruit in 1943 (discussed in Chapter 6).

The British used the term "combined operations" for amphibious warfare. In July 1940 they formed a Combined Operations Headquarters (COHQ) specifically to promote amphibious assault techniques. A U.S. naval mission to Britain was soon established. Given Prime Minister Churchill's intent to ensure the closest possible cooperation, by the fall of 1940 the U.S. Navy and the Marines Corps were well aware of all British amphibious experiments. Contemporary U.S. Navy files show masses of British information, supplied even before Lend-Lease became law in March 1941. For example, they include a detailed 28 February description of British tank landing craft, including early problems encountered. The final design of these craft had been forwarded to the builders only the previous November, and the report described major changes already being made.

In June 1940 newly appointed Prime Minister Winston Churchill personally asked the Admiralty for a ship capable of carrying a tactical unit of tanks several hundred miles, then unloading them onto a beach. Victory over Germany would ultimately require a cross-channel invasion, including tanks. The initial solution harked back to Gallipoli.[1] Troops had run down ramps hung from the bows of a beached troop-carrying steamer, *River Clyde*. Surely the idea could be revived, this time using tanks and a new 40-ft portable bridge that had been developed by the Royal Engineers. To this end the Royal Navy took over two Southern Railroad train ferries and eight horse ferries being built for Turkey; they could carry, respectively, 14 and 12 tanks. All had open cargo decks and ramps down which tanks could proceed, at least onto a pier. None was satisfactory. They were slow (12 and 9 kts), they had very limited range, and they drew too much water to land their tanks while using the portable bridge.

Rowland Baker, of the British Royal Corps of Naval Constructors, had first worked on landing craft in 1938. Now he quickly designed a very austere shallow-draft tank ferry, which the British called a tank landing craft (TLC). It carried its tanks (six 25-ton or three 40-ton) on its main deck, unloading them over a bow ramp. The TLC could carry tanks across the English Channel, or even to Gibraltar, but not from Britain to the Middle East, where they were wanted. Churchill wanted enough lift for an armored division, which meant 60 tanks.

In September 1940 the British attempted their first major amphibious operation, against Dakar, which was held by the Vichy French. The attack failed; a post mortem convened by Prime Minister Churchill concluded that it could not have succeeded owing to a lack of tanks. In the past, it had been imagined that the troops could capture a port, which would then be used to unload tanks from conventional merchant ships. Now it was evident that tanks would be needed merely to capture the port. The TLC, already on order, could not have survived the sea passage to Dakar. Churchill said that no offensive could be mounted until this problem was solved. He told the Admiralty's director of plans to produce a ship capable of carrying a division's 60 tanks across an ocean (as part of a fast attacking convoy) and land them directly on an enemy shore.

At a 26 September 1940 meeting with the ship designers at Bath, the director of Combined Operations asked for the means to transport and land directly on a beach in not more than 5 ft of water sixty 25-ton tanks as an assault echelon. He also wanted some means of launching 120 swimming tanks, and of transporting and hoisting out 80 small tank landing craft (MLCs). The immediate reaction of the director of naval construction (DNC) was to fit out three small Belgian *Prince Baudouin*–class ferry steamers with doors, a ramp, and a launching pier. For the swimming tanks he proposed larger ships with ballast tanks and bow doors. For the MLCs the best solution would be tankers or cargo ships with special cranes; the conversions were termed "gantry ships." All of these solutions presented major problems. The MLC had been designed within the limits of merchant ship booms, so it could accommodate nothing larger than a 16-ton tank, yet 40-ton tanks were being developed. Building flotation apparatus for such tanks would be difficult at best. It would be possible to cut the bows away from the shallow-draft Belgian steamers and remove their middle decks for head room, thus accommodating 20 tanks. Unfortunately their endurance was only 500 nm. Yet another possibility was an improved TLC with an endurance of 1,300 nm and a speed of 14 kts.

For the sort of operations Churchill hoped to mount in 1941 he really needed a new kind of ship. Building one would have been a considerable challenge in peacetime; in September 1940 Britain was under heavy air attack and her shipbuilding industry was hard-pressed to build cargo ships and urgently needed warships. The DNC thought it would take 18 months merely to build the ship, after considerable time had been taken to design it. He also thought that a 60-tank ship would be too large; it would be better to limit the ship to 20 or 25 tanks.

Fortunately the British had access to a deep fund of knowledge of the world's merchant ships, partly because most of them were insured by Lloyd's of

London. A day or two after the prime minister's meeting, the director of Combined Operations mentioned the tank carrier problem to A. T. Sheffer, a Lloyd's of London surveyor attached temporarily to the DNC department, the equivalent to C&R in the United States. Sheffer had recently worked on shallow-draft tankers specially built by Harland & Wolff of Belfast to cross the bar of the Maracaibo river in Venezuela en route to the refinery on Aruba in the Caribbean. He thought that these tankers could be converted to oceanic tank carriers within two to three months; they would meet all of Churchill's requirements except for speed. The controller of the Admiralty, responsible for ship construction, flew to Belfast to see the drawings of these ships, and conversion of three was agreed upon within 48 hours. There was some question as to whether they could safely cross the Atlantic in winter, but they made the passage, and the first conversion conference was held aboard the tanker *Bachaquero* on 14 March 1941.

While these conversions proceeded, DNC considered three alternatives: a ship ("Winston") capable of carrying all sixty 25-ton tanks, three smaller ships ("Winettes") each carrying twenty tanks, and fifteen oceangoing LCT, each carrying four 40-ton or six 25-ton tanks. The Winette was unanimously approved at a December 1940 Combined Operations conference, and a program of three such ships (one division) was submitted by the director of plans on 14 December 1940. They were so important that on 20 December 1940 the controller approved their construction, with priority for completion by April 1942, at the expense of nine corvettes that were badly needed as convoy escorts.

The Winette or TAC (tank assault carrier) could carry twenty 25-ton or fifteen 40-ton tanks on its lower deck, plus 150 tons of trucks. Beside the strengthened bow with its 124-ft truss bridge, the ship had large side ports through which trucks could be unloaded. It was credited with a maximum speed of 18 kts and with an endurance of 8,000 nm at 14 kts.

The three TACs represented the limit for landing ship construction in Britain. By this time the British hoped to go on the offensive in 1942. On 21 December 1940, it was estimated that apart from replacing casualties for 1942 they would need 7 more Winettes and the equivalent of 330 TLCs. The Winettes would have to be built in Canada or in the United States, and there was no hope of obtaining 330 TLCs. On 24 December, the controller approved building the Winettes abroad, and also construction of some enlarged tank ferries ("Brunettes") that had been proposed by Alexander

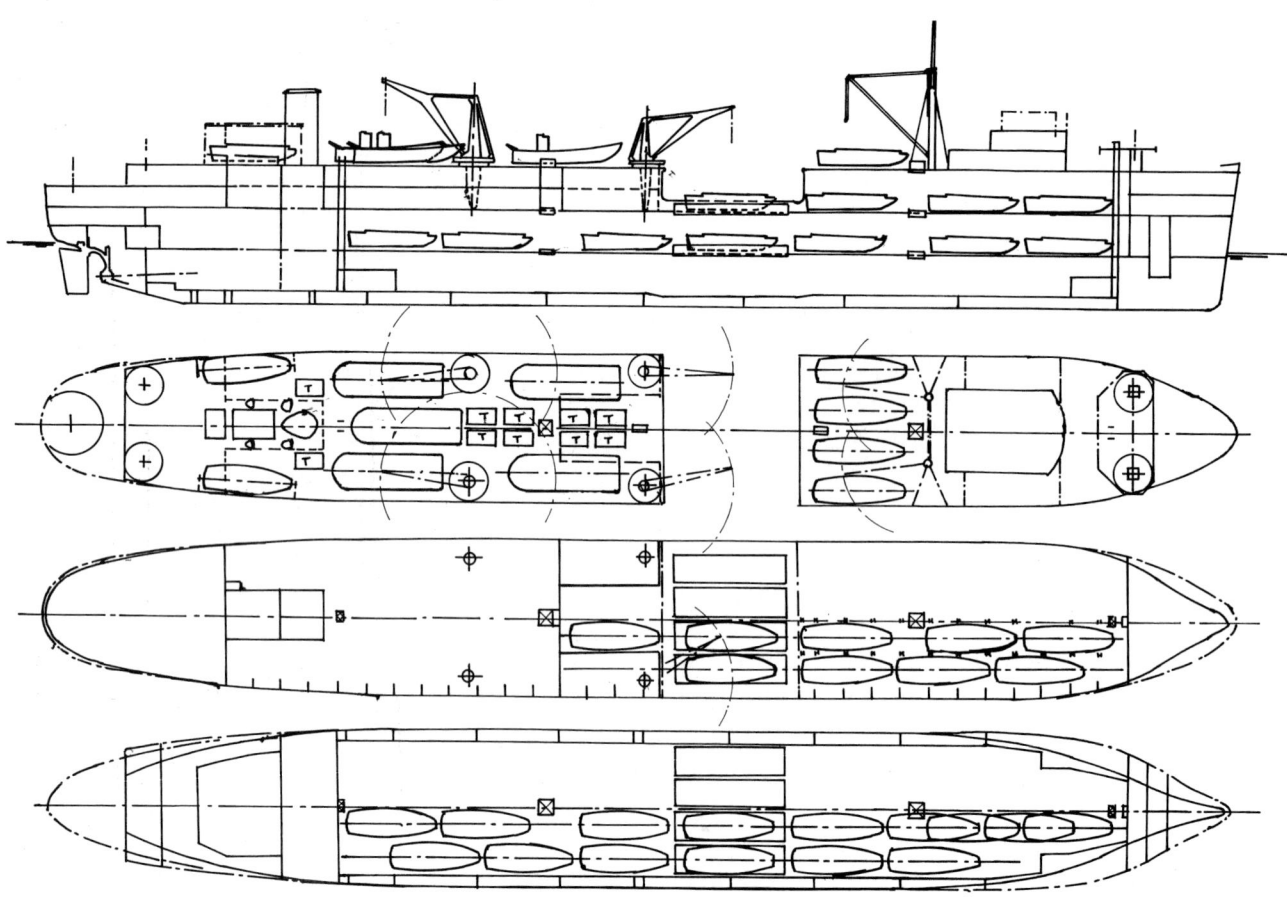

Brown & Company (for a time it appeared that such craft could be built in idle shipyards). Within six weeks the Brunette project had to be abandoned in favor of an "elongated TLC," later called LCT(3). Like earlier TLCs, it was hardly oceangoing.

The Winettes could not be built in Canada, so early in 1941 the British approached the United States. By this time the Lend-Lease Act was clearly in prospect. When the act passed in March, further British development of the Winette began to have direct impact on U.S. amphibious development. The requisition for the Winettes was issued in May 1941. Having selected Gibbs & Cox as design agent for British Winettes, the Admiralty provided preliminary design plans and data on 30 July 1941. The contract to Newport News was let that September. As a cover, the Winettes were described as truck transports. BuShips became aware of the Winette some time during the spring of 1941; U.S. files show sketch plans dated February 1941.

In addition to the large tank carriers, the British developed specialized ships to carry and launch loaded LCMs. Some ships were fitted with special heavy-duty davits, two ships were fitted to launch the craft over stern slipways, and oil tankers were fitted to carry them on deck and launch them using gantry cranes fore and aft. The U.S. Navy was aware of these ideas, and by February 1941 OpNav had asked for the design of a ship that could launch boats and lighters over a stern slipway.

Like the British, the Americans initially thought in terms of train ferries, in their case the big "Seatrains" built to carry freight cars to the Caribbean. Many countries had railroad ferries, but the Seatrains were unusual for their combination of size and steaming range. In April 1941 BuShips sketched plans for two alternative Seatrain conversions, one to carry landing craft and one to carry aircraft. On 16 April it was told to concentrate on the aircraft transport version. At this time mobilization plans called for two of them to be taken over by the navy as aircraft transports (AKVs); they could carry either aircraft or heavy landing force equipment. For example, when SS *Seatrain New Jersey* was acquired by the navy in October 1942, she was designated an AKV even though the army had just converted her to carry 200 tanks. In 1941 the U.S. Army wanted to build 50 Seatrains specifically to carry tanks (200 tanks per ship). By July 1941 BuShips

A BuShips feasibility study sketch of a Seatrain railroad car carrier converted into a landing force tender, in effect, the U.S. alternative to the LSD, 27 March 1941. Seatrains figured prominently in U.S. prewar and early wartime planning, just as big train ferries figured in early British thinking about assault ships. Landing craft were about the same size and weight as the railroad cars they were designed to carry, in this case between the United States and Cuba. When war began, the U.S. Navy headed off a proposed 50 Seatrain program on the ground that the ships could not survive damage. On 1 April 1941 BuShips submitted three alternative schemes for Seatrains as landing force tenders or aircraft transports. On 16 April OpNav decided that the ships would be converted to carry aircraft. This sketch shows Scheme 3, a composite arrangement: as an aircraft transport she would carry assembled aircraft on deck and crated ones below. As a landing force tender she would carry five tank lighters (LCM(2), all of them on the superstructure deck aft) and fifty 36-ft Higgins boats (18 on the main deck, 26 on the 'tween deck). For simplicity, only half the boats on each of the main and 'tween decks are shown. The tank lighters would be handled by the big 30-ton crane aft. The aft superstructure deck also accommodated ten tanks (indicated by small squares), but note that all but four were beyond the reach of the big cranes. The cranes overlooking the break in the upper deck would suffice to handle the rolling cradles shown, used to move boats along the two railroad car–carrying decks. The capacity of the forward cranes, which would handle the landing boats on the forecastle, is shown as only 2 tons, which seems insufficient. Not shown are a pair of 26-ft whaleboats alongside the forward superstructure. Only enough troops to handle the ship's boats would have been carried. Armament would have been one 5-in gun right aft, pairs of single 3-in/50s on platforms fore and aft, and two 0.50 caliber machine guns alongside the bridge. Typical aircraft capacity was 14 crated Helldivers (SB2C) on the superstructure deck, 18 on the main deck, and 26 on the 'tween deck. Alternatively, seven assembled Lockheed light bombers (Venturas) could have been carried on the superstructure deck. The aircraft transport option was selected. Presumably the Seatrain study was the U.S. tank-carrier project (in effect, an alternative to the LST or LSD) to which BuShips referred in the spring of 1941. Seatrains were used to carry tanks, although generally not in combat. The exception was the landing in North Africa, when specialized ships were not available in sufficient numbers, and *Lakehurst* (ex–*Seatrain New Jersey*, at that time designated APM 9), was part of the Southern Attack Transport Group, together with four attack transports (AP 8/APA 2, AP 65/APA 32, AP 67, and AP 71; the last two were quasi-attack transports or XAPs) and an attack cargo ship (*Titania*). APM was a short-lived designation meaning transport, mechanized artillery; the eight LSDs ordered for the Royal Navy were initially designated BAPM (i.e., British APM) 1 through 8. *Lakehurst* herself was briefly designated APM 1, then redesignated APM 9 to allow for the eight ships ordered for the Royal Navy. They soon became BAPM 1 through 7 and the U.S. APM 8, then LSD 9 through 16. Acquired by the navy under bareboat charter in October 1942, presumably as a last-minute addition for the North African operation, *Lakehurst* was transferred to the U.S. Army in 1943. Two other Seatrains were taken over by the U.S. Navy as aircraft transports (APVs, later AKVs, the V indicating heavier-than-air aircraft): *Kitty Hawk* (ex–*Seatrain New York*, APV/AKV 1) and *Hammondsport* (ex–*Seatrain Havana*, APV/AKV 2). *Lakehurst*, which was built in 1940, was 483 ft × 64 ft 1 in × 27 ft 2 in. Her 8,000 SHP turbines drove her at 16 kts, fast enough for amphibious convoys. Armament was one 5-in, four 3-in/50, and eight 20-mm guns. She was initially designated AKV 3, but redesignated as an APM in December 1942. Built in 1932, the other two AKVs displaced 10,900 tons light and 16,480 tons fully loaded; dimensions were 478 ft × 63 ft 8 in × 27 ft 2 in (26 ft 3 in, in navy service); 8,000 SHP turbines drove them at 16 kts. The army bareboat chartered a sister ship to *Lakehurst* as a tank carrier: *Seatrain Texas*. All four ships were built by Sun Shipbuilding of Chester, Pennsylvania. The initial railway car carrier was built as Seatrain in 1928, and later renamed *Seatrain New Orleans*. She was not taken into wartime service.

could report to the CNO that it was already making some preliminary studies of a specialized tank carrier, and that the Seatrain was too limited. She would unload only very slowly, even in a favorable sea. Seatrains were also quite vulnerable. No new ones were built. The army was persuaded to accept 50 C4 troopships instead.[2] Eventually they came under naval control (see Chapter 6), as point-to-point ships.

BuShips informally discussed oceangoing tank carriers with the British during the summer of 1941, and on 5 July asked the CNO whether it should begin design studies of a U.S. tank carrier equivalent to a Winette. The Auxiliary Vessels Board referred the question to the commanding general of the Atlantic Fleet amphibious force, Gen. Holland M. Smith, USMC. At the big New River exercise in August, he found that transports unloaded tanks far too slowly into one-tank landing craft. The Winette was not quite good enough. In mid-September Smith proposed a special tank carrier to carry 18 tanks (13.5 or 28 ton) and 18 tank lighters (LCMs), each of which could rapidly be launched with its tank aboard. This was in addition to the British-style truss bridge, to unload tanks directly onto the beach when the beach slope permitted. The Auxiliary Vessels Board agreed that this proposal should be the basis for a design study. Like the Winette, it should have an 8,000-nm radius. Speed might be reduced to 16 or 17 kts.

This was rather ambitious; it was a cross between the Winette and the landing ship dock the British were then conceiving. It may have been inspired by British projects for heavy LCM carriers using gantries to launch the craft. If the ship could not be built, the general wanted three TACs added to the Atlantic Fleet amphibious force. This appears to have been the first proposal for U.S. LSTs.

Early in October the Auxiliary Vessels Board echoed Smith's complaint that lighters could not possibly bring tanks ashore quickly enough. Worse, there was no lighter capable of carrying a medium tank. To realize the combat power of such tanks, a special tank-carrying ship was needed. The obvious solution was to take over the Winettes for the Atlantic Fleet amphibious force. They were not altogether satisfactory. For example, they were quite vulnerable, because they had been designed to a one-compartment standard. The board suggested that BuShips immediately begin a design to meet the Atlantic Fleet's requirements, three of which were to be built. Meanwhile three of the TACs ordered for the Royal Navy should be diverted to the U.S. Navy, and three more built in their place. If the British were not amenable, three new ships should be built at once. The board also proposed that BuShips study possible adaptations of existing U.S. merchant ships, and a tank lighter to take the 30-ton medium tank (see the Chapter 4).

On 9 October CNO Adm. Harold R. Stark approved these recommendations and forwarded them to the secretary of the navy, who approved them the same day. By that time, the situation had changed dramatically. Even the British no longer liked the Winette. The three tanker conversions had been completed in June. They had flat bows with horizontally hinged doors and 51-ft ramps (with 37-ft extensions, for a total of 68 ft beyond the door hinge) over which tanks could land. Their blunt bows limited them to 10–12 kts, and to much less in a head sea, because there was nothing to divert the sea from the flat bow door. Although the ramp was designed to take a 25-ton tank, it could take a 40-ton Churchill driven slowly enough. Vehicles were generally driven aboard, although the ships had two 25-ton cranes to lift tanks through a hatch. Using the cranes, they could also take aboard two "heavy support craft" or LCMs, although in practice this was not done. There was also some hope, which had to be abandoned, that the ships would retain some oil tankage so that they could be used as oilers when not assaulting a beach. Unfortunately the ships could only be used on very steep beaches, because loaded draft forward was 5 ft 6 in (13 ft aft).

The full Winette was a much more elaborate turbine-driven ship designed to make 17 or 18 kts, thus allowing it to keep up with a fast convoy. For speed and seakeeping, she needed substantial draft. However, she had to land tanks or other vehicles dry-shod on a 1:35 or steeper beach. She was therefore fitted with a power-extended 90-ton 120-ft ramp. It was imagined that using a long causeway would allow tanks and vehicles to be landed on comparatively flat beaches; if the slope was 1:60 or less the ship would ground first by her stern, and the causeway would simply land in the water forward. On a really steep beach the ramp would not be needed at all. Thus the rather vulnerable ramp would be useful only for the limited range of beaches between 1:35 and 1:60 slope.

By August 1941 the British had realized that the French beaches, the ultimate targets of the Winette, were closer to 1:100 or even 1:200. This same problem of flat beaches forced redesign of the smaller tank landing craft. A very flimsy hull was accepted as the price of adapting the type, later designated as the LCT(4), to a 1:150 beach. The Winette's draft would have to be reduced. Bulging would allow her to land on 1:75 beaches and would not reduce speed below 15 kts. It was too late to do anything about the three ships being built in Britain. That left the seven U.S. ships.

In July 1942 the original Winettes were redesignated as landing ships, tank Mk 1 or LST(1); in this sense they were the progenitors of the famous World War II LST. Their design, however, was not related to that of the later mass-produced LST. They were considered expensive failures. For example, it took 13 minutes to extend their ramps, far too long for combat. Ironically, the concept of combining a conventional hull (for good ship performance) with a long power-operated ramp was revived for the fast U.S. LST of the 1970s, the *Newport* class. It can also be seen in the recent Singaporean LSTs. The unpleasant U.S. experience of the late 1940s shows just how difficult it is to design a fast LST with any more conventional arrangement.

The solution was an entirely new type of ship. On 19 July 1941 Maj. R. E. Holloway, Royal Engineers, sent the assistant director of training at the war office a copy of a patent (accepted 30 October 1924) taken out by Otto Popper of Bratislava, secretary of the Danube International Commission in 1920–29. Popper had conceived a barge transporter for the Danube. It was a 350 × 70 ft ship with a flat low deck between forecastle and poop. It would now be called a flo-flo: it flooded down to take Danube River barges on board, then pumped out its tanks to lift them out of the water. The barges were clearly comparable to TLCs. The transporter had been conceived in the early 1920s as part of a broader transmodal scheme, in which Danube barges would be loaded directly onto oceangoing freighters. A model had been built, and calculations carried out by C. N. Triestino, Isherwood & Company of London, Blohm & Voss, and the Vienna Experimental Institute for Shipbuilding. Lloyds Registry had approved the concept. The barge transporter was clearly unsuitable now, because it carried its barges on an open deck. Holloway suggested that it would need side walls, which would restrict the number of barges; the forecastle would have to be removed; and the barge would have to be maneuvered through bow doors to be loaded on board. After about a month, the director of training passed the idea forward to Combined Operations.

Britain's chief constructor, DNC Stanley Goodall, was less than enthusiastic. The idea was quite old, yet it had never been turned into a real ship. The concept "is so extravagant as to be impractical under existing circumstances in the shipbuilding industry." A large and expensive ship would carry only two TLCs (i.e., only twenty 25-ton tanks or ten 40-ton tanks); it would take 50 ships to carry comparatively few tanks. One of his assistants added that the ship was hardly seaworthy, stating that it would need absolutely calm water to load or unload its barges.

Yet Capt. Thomas A. Hussey of COHQ knew that the "Popper" ferry could be turned into the solution to the problem. Late in August he asked whether, after all, the idea was not entirely practical. He countered the DNC's arguments; the ship could be built in the United States. She could carry more tanks by stowing them below decks, hoisting them into landing craft by derrick for a second wave. The ship might carry crated aircraft, munitions, or general cargo. On 5 September Hussey's chief, Admiral of the Fleet Sir Roger Keyes, pointedly asked DNC Goodall about the status of the design.

With all this preparation, at a 19 September 1941 Admiralty meeting to discuss the Winette problem, Hussey displayed the patent sketch of the Popper ferry. The sole remaining qualm was that specially trained personnel would be needed. However, ordinary sailors successfully operated a somewhat similar ship that had been built to dock the submarines of the Brazilian navy. Staff requirements were developed at a 30 September 1941 Admiralty meeting. The ship would carry two LCT(3) or LCT(4), carrying up to 24 tanks, as well as their crews and administrative staff, at a cruising speed of 14 kts (radius 5,000 nm at this speed). Armament would be limited to close-range antiaircraft weapons. Besides her well deck, the ship would have cranes to hoist vehicles weighing up to 10 tons onto the upper deck.

DNC Goodall allowed Rowland Baker, who had designed the TLC, to design the new craft. He transformed the flo-flo into a powered drydock with an enclosed well deck, suitable for ocean passages. The forward two-fifths of the hull was decked over to accommodate her officers and crew as well as the crews for the vehicles and the landing craft in her well deck. The main ballast tanks were in the wing spaces alongside the well deck, with fuel oil, boilers, and engine rooms. Compared to a Winette or LST(1), the new ship could use her embarked craft to land tanks in shallower water, and they could disperse more easily after having landed (in two places rather than one). Because the new ship carried about as many tanks as a Winette, three would form a tactical unit. To avoid delays, it was decided on 2 October 1941 that only the last four of the seven Winettes being built in the United States should be reordered as TLC carriers.

Hussey's revolutionary idea was realized in the TLC carrier or TLC-C (later designated as LSD). In retrospect, although the mass-produced LST was the most important specially designed amphibious ship of World War II, Hussey's ship was far more fruitful postwar, because it proved so much easier to build a fast well-deck ship than a fast LST. Moreover, the well deck could accommodate new kinds of landing craft.

When their original ideas for an LST proved impractical, the British invented an alternate way to move tanks across the ocean: the LSD, which could carry smaller landing craft in a big well deck, shown here on board HMS *Eastway* (LSD 9), under construction at Newport News on 21 May 1943.

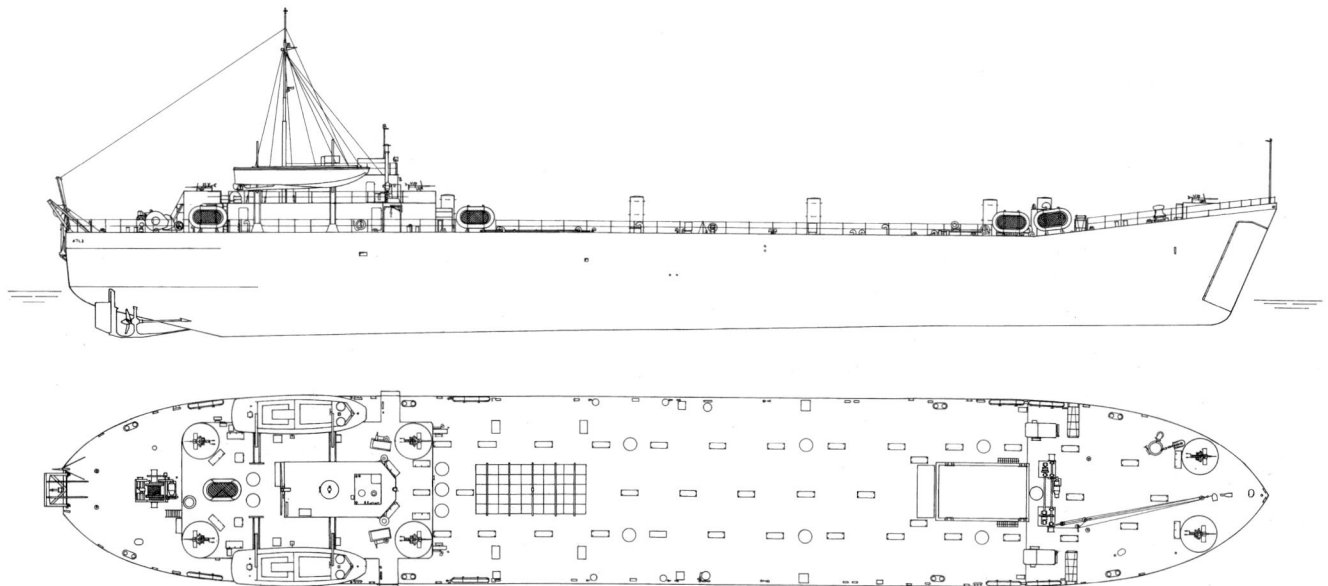

LST 1 (ex–ATL 1) as completed, 1942. Note the extremely light armament, only six single 20-mm guns, and the LCPLs (Higgins boats) in the two davits. The plan view shows the tank elevator forward. The prominent tubes projecting from the upper deck are ventilators for the tank deck, to remove fumes generated by vehicles starting up. As LSTs demonstrated their value during the war, their armament increased dramatically. They also were fitted with radars. *A. D. Baker*

Through the fall of 1941 British estimates of the number of tanks needed for an invasion of France grew. As of October 1941 the British Joint Planning Staff estimated that 2,150 TLCs would be needed. They could not possibly be built in Britain. Nor could they cross the Atlantic on their own power from the United States. Nor could sufficient shipping space be found to carry them disassembled, and Britain did not have enough industrial capacity to reassemble them. The British already imagined an invasion landing heavy mechanized forces far from an established port, as was done at Normandy. The force ashore might have to be maintained for as much as a month, requiring that about 5,000 vehicles per day be landed. The ideal solution was an inexpensive beaching ship that could carry trucks as well as tanks across the Atlantic under its own power: an Atlantic TLC (ATL). The British Combined Operations mission asked that 200 ATLs be built in the United States. When the original Winette became the LST(1) in July 1942, the new design became LST(2). It was an amphibious workhorse, built in enormous numbers, but its gestation was painful.

The British would have to carry many of the invasion tanks aboard conventional cargo ships. Instead of landing them aboard one-tank LCMs, the British proposed a tank ferry. Because it could not possibly cross the Atlantic under its own power, the British suggested that it come (in one piece) as deck cargo on board the Atlantic TLC. When U.S. and British amphibious craft designations were merged, the TLCs became LCTs Mks 1 through 4, and the tank ferry became the LCT(5). The British asked for 200 LCT(5) to go aboard the 200 ATL.

On 22 October 1941 the British Admiralty delegation in Washington received basic specifications for the Atlantic TLC: a speed of 10 kts; endurance for Atlantic passage; 40-ton tanks; minimum 1:150 beach slope, which could be reduced to 1:50 if that were a limiting factor. Capacity was to be in units of three tanks each plus auxiliary motor transport such as tracked carriers for personnel and Roto truck trailers. It forwarded the data to BuShips. Rowland Baker produced a sketch design, presumably as a feasibility study. It was either never forwarded to Washington or misplaced upon arrival; there is no trace of it in U.S. records.

On 4 November 1941 Capt. Cochrane, chief of BuShips' preliminary design section, discussed the Atlantic TLC with John C. Niedermair, his civilian technical director. Cochrane's role was extremely important. When the British delegation arrived in the United States on 18 November, asking for 200 LSTs and 400 LCTs, Rear Adm. S. M. Robinson, BuShips' chief, pronounced the whole project fantastic, and CNO Adm. Stark agreed. According to the British, only Cochrane understood from the start that these craft were absolutely necessary; he insisted on trying to design them. Baker later said that Cochrane was responding in large part to the challenge of designing a new kind of ship, stating that he would have been interested even if he had not considered them essential.

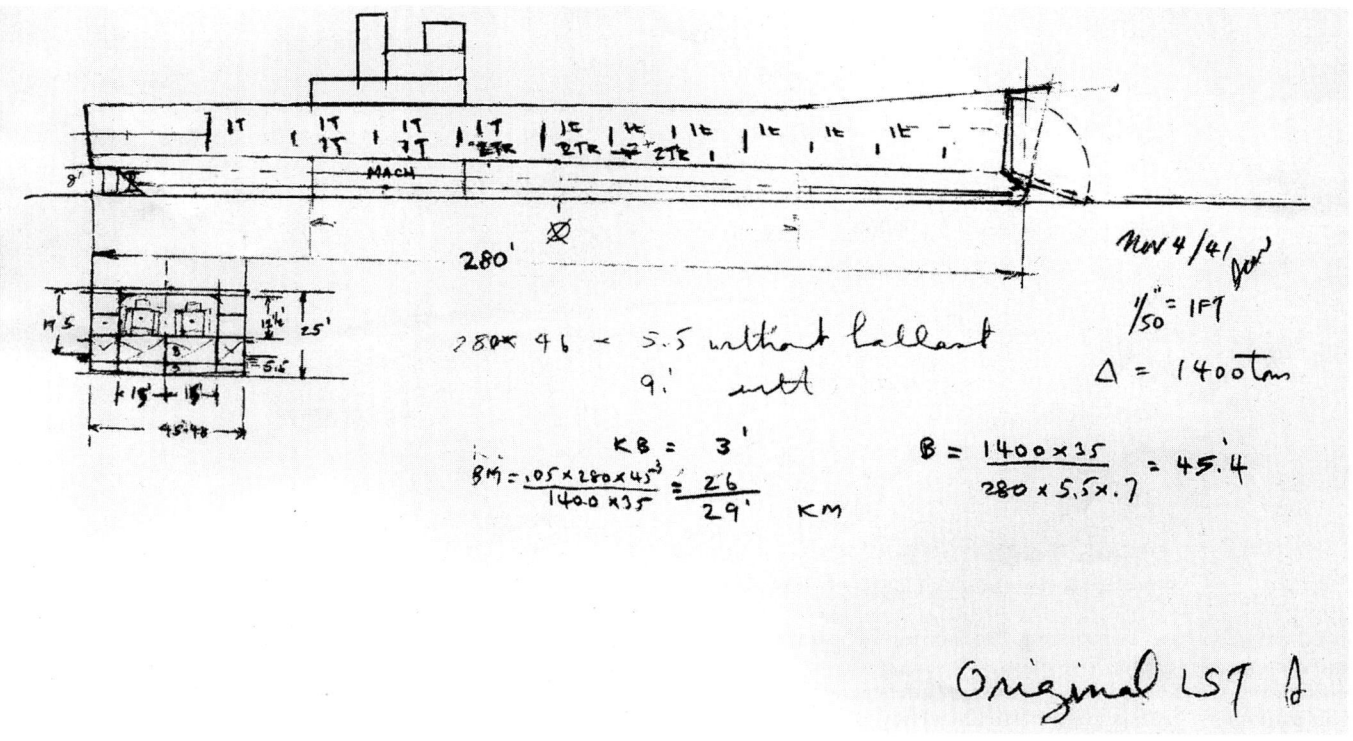

John Niedermair's sketch design for the LST, 4 November 1941, including a cross-section (lower left) showing ballast tanks (B) under the tank deck. Calculations include an estimate of the position of the metacenter above the keel, or base line (BM) and an estimate of the beam (B) based on the ship's length, her required draft, and a typical block coefficient (0.7), at lower right.

Niedermair produced a small pencil sketch of a 900/2,800 ton (light/loaded), 280 (between perpendiculars) × 45 × 10 ft (hull depth 25 ft) Atlantic TLC. It would be powered by two 1,000 BHP geared diesels (total effective power 1,800 BHP), for a maximum speed of about 10 kts. Projected armament was two 3-in/23 antiaircraft guns and six 20-mm, which Cochrane thought too expensive for a mass-produced ship. Niedermair's expanded 1/16-in scale sketch, including cargo deck layout, was forwarded by airplane the next day to the Admiralty.

The TLCs could not cross the Atlantic because they were not seaworthy. Their draft was grossly inadequate, because they had to be able to beach on a gently sloping shore. How could a TLC cross the Atlantic, and beach? Niedermair had long been involved in submarine design, and his solution was quite simple. By using large submarine-style ballast tanks, the ship could have two alternative drafts, a deeper one for ocean passage and a shallower one for beaching. The tanks would be pumped out quickly before the ship approached the landing zone. No such solution had ever previously been applied to surface ships. To provide the necessary hull strength, Niedermair made the hull proportionately deeper than that of the British TLC, and he decked it over; lighter vehicles could be carried on the upper deck. An elevator would bring them down to the tank deck so that they could be landed.

Niedermair's design completed, a special Combined Operations mission (including Baker, acting as the DNC representative) arrived in Washington on 8 November 1941 with formal staff requirements. Like a Winette, the new ship would carry twenty 25-ton tanks, as well as their crews. She would beach on a 1:50 slope with a bow draft of 3 ft 6 in, far less than that of a Winette. With such a shallow draft, she would not need an elaborate bow ramp. On the other hand, there was no demand for high convoy speed. Range was to be 10,000 nm at 10 kts; given their tankage, U.S. LSTs were eventually credited with a maximum range of 39,000 nm. Niedermair's design already met most of these requirements, which largely repeated those of the 22 October dispatch. Having examined Niedermair's drawing, on 21 November—less than a month after it sent the dispatch—the Admiralty delegation gave BuShips the formal staff requirements, titled "Design on the general lines of the outline drawing prepared by the Bureau of Ships."

After further study, the design was expanded to 300 × 50 × 7 forward/13 aft ft (25 ft hull depth); displacement would be 3,350 tons in oceangoing condition (1,100 tons light). By the end of the preliminary design phase it had grown again, to 316 ft (328 ft overall). Design displacement was 1,400 tons light, 3,700 tons seagoing, and 2,140 tons in beaching condition. The larger dimensions made for shallower draft in beaching trim. Also during the design, the bow doors were widened from 12 to 14 ft, as the designers expected mechanized equipment to grow; even 14 ft imposed serious constraints late in World War II. Cargo capacity in beaching condition was 500 tons, a requirement that was followed in all later, and often much larger, LSTs. Capacity was 20 medium or 10 heavy tanks. Endurance was 12,000 nm at 10 kts if some ballast tanks were used for fuel.

BuShips prepared a formal preliminary design. However, initially the U.S. Navy saw little point in the LST. It did not fit the style of amphibious warfare the Marines had developed. U.S. shipbuilding capacity was fully employed in building merchant ships and warships. At a 28 November meeting CNO Admiral Stark told Captain Hussey that there was no U.S. requirement for such craft, and that British expectations were unrealistic. That was a real problem, because in theory all Lend-Lease hardware was U.S. equipment lent to allies; it could not be procured unless the United States could use it. Ship designs came through the BuShips design division, after which the U.S. Navy had to agree that a U.S. use for the article in question could be foreseen. A skeptical Stark referred the British to Robinson, who wanted no part of the project.

Not to be defeated so easily, Hussey turned to Maj. Gen. Sir Henry C. B. Weymss of the British mission to the United States. He turned to the U.S. Army's chief of staff, Gen. George C. Marshall, who agreed to push the project. Meanwhile Rear Adm. J. W. S. Dorling of the British Joint Planning Staff mission got BuShips to admit that, if the order were given, it could build the required ships without delaying any other vital projects. Robinson passed the British on to Vice Adm. Howard L. Vickery of the Maritime Commission. However, Rear Adm.

LST 173 illustrates the original LST configuration, with prominent tank-deck ventilators on the upper deck, on 17 January 1944. The ventilators made it difficult to stow vehicles there, and impossible to stow an LCT. The ship is armed with a 3-in/50 aft and with 20-mm antiaircraft guns.

Herbert S. Howard and Captain Cochrane of BuShips' design division certainly understood well enough to pressure Vickery to accept the job. This arrangement was formalized at a 3 December 1941 meeting between representatives of the British Admiralty delegation, BuShips, and the Maritime Commission. Vickery in turn brought in the Dravo Corporation of Pittsburgh, whose chairman, J. V. Berg, was in Washington looking for projects. Gibbs & Cox was made design agent.

Before much could be done, the U.S. Navy took back control of the project. After Pearl Harbor, landing ships became a U.S. requirement, hence very much the business of BuShips. When Prime Minister Churchill arrived in Washington on 22 December, he took over the cause, and he personally pressed it with President Roosevelt. The president apparently approved the LST program for the British even though he considered such specialized ships a mistake.

It was no longer necessary to work indirectly, so the Maritime Commission withdrew as builder. Instead, ships would be built to U.S. Navy order. As in several other amphibious programs, work was accelerated by beginning construction while building drawings were being completed. The first LST contract, for 300 rather than 200 British LSTs, was placed on 23 January 1942, as part of the larger "1,799" program of ships for the British. The lead yard, Dravo Corporation of Neville Island, Pittsburgh, selected Gibbs & Cox to produce working plans while BuShips produced contract plans. The last of the latter was signed on 10 March 1942. The first keel was laid on 10 June 1942, the first was launched on 7 September, and the first was completed on 28 October. In all, 23 were commissioned before the end of 1942. Ultimately an LST could be built in two months.

Even after the contract was signed, the program had only a low priority, because it corresponded neither to U.S. naval nor to perceived army amphibious requirements. Only in May were LSTs given higher priority (exceeded only by carriers), due to the president's personal intervention, backed by Gen. Marshall and by presidential assistant Harry Hopkins. OpNav ordered another 190 (90 to be built by the Maritime Commission), to be sure that 300 ships were delivered on time. However, 100 of those ships were canceled in June 1942, leaving 390 in the program.

Also in June the Atlantic Fleet amphibious force formally requested both LSTs and TLC carriers (LSDs), and about the same time the Pacific Fleet amphibious force made a similar request. Both argued that tanks were essential to an amphibious assault, but that the usual technique of landing them from LCMs launched by transports was far too slow. OpNav tried to reinstate the canceled units in July. That proved impossible; other programs had taken up the capacity involved. It was not until January 1943 that Admiral King approved a 20-ship continuation program, for 1944 delivery (LST 491–510). By the end of the war, LST 1155 was on order.

LSTs were first used in the spring of 1943 to supply islands in the South Pacific. Their first assault

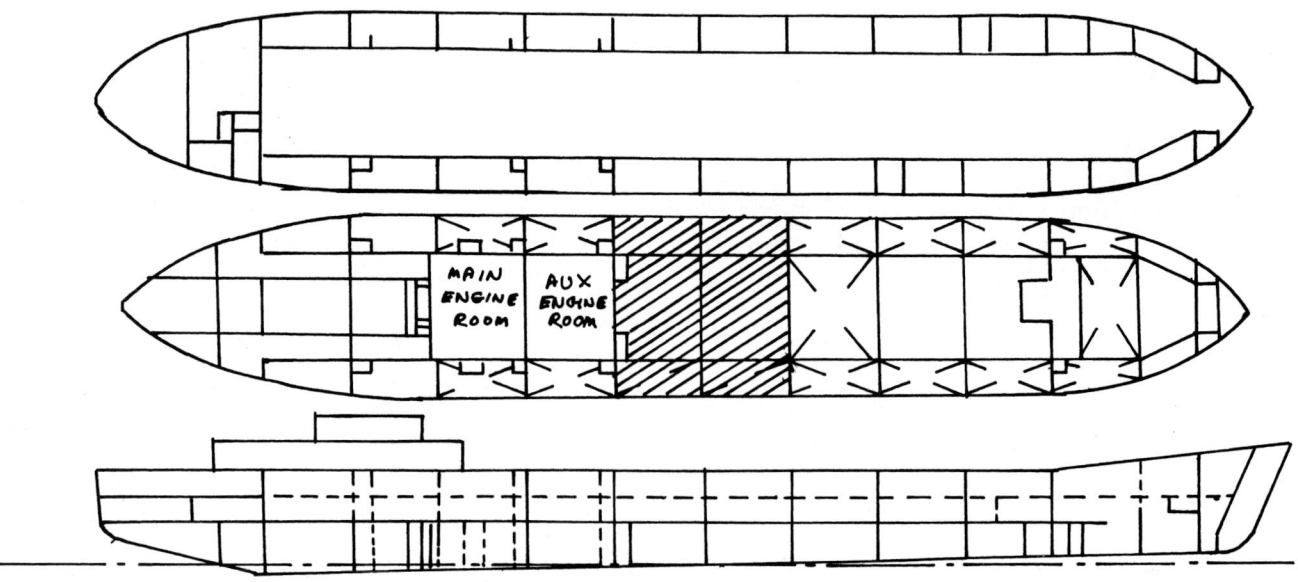

LST compartmentation, from a sketch in the 1945 edition of the *LST Operator's Manual*. The tank deck and the hold are shown. Crew and troops lived on the deck above, both in the wing walls and in the stern area. Ballast tanks are indicated in the plan view of the hold. Shaded tanks could take either diesel oil or ballast.

The LST's bluff bow and nearly square bilges made her slow and a poor seakeeper. Through the postwar period U.S. designers sought a faster alternative. LST 172 is shown in the floating drydock ABSD 1, on 29 September 1944.

was Rendova on 30 June 1943, and they had a major role in Sicily the following month. Afterwards they figured in most Allied amphibious operations.

Much of the detailed shape of the LST was developed at the contract plan stage, in January 1942, in close collaboration with the British, particularly Baker. To protect the rudders and propellers from grounding and ground obstructions, the BuShips designers built a skeg on each side, forming a tunnel similar to that in the new *South Dakota*–class battleships. The rudders and propellers were placed inside the skegs, much closer to the centerline than on the Maritime Commission's plans. This arrangement would keep the stern anchor under control, so that it would not drag into the tunnel.

Vehicles on the tank deck had to start their engines before the bow door opened. Forced ventilation by fans was considered impractical. Instead, at Baker's suggestion, the vehicles' exhausts were connected to ventilation ducts (uptakes), and fresh air was supplied through hatches. The result was the forest of portable fan-fed vertical ducts characteristic of war-built LSTs. Initially they were distributed around the upper deck, four being just forward of the deckhouse; later those distributed along the deck were relocated to leave the upper deck clear for an LCT (see below) or for vehicles. The upper deck also carried air intakes for the tank deck (initially scoops, later flush gratings).

Trucks carried on deck had to be lowered to the tank deck. In January 1942 two alternatives were developed, a ramp and an elevator; the elevator was simpler to arrange but slower to operate. Baker helped develop the designs, and he helped choose the elevator solution. Based on his extensive experience with the British TLCs, Baker helped design the bow doors and ramp.

Ships were to be built in large numbers, so lines were simplified and the structure was designed for welding, with a minimum number of different steel

Beginning with the invasion of Sicily, many LSTs were given a total of six davits for LCVPs, so they could function as semi-attack transports. LST 708 is shown.

thicknesses. The structure was very strong. LSTs proved remarkably survivable. Although the enemy soon realized that they were prime targets, only 37 were sunk during World War II.

By January 1942 a hull model had been made and was undergoing testing, in record time, at the David Taylor Model Basin. It was soon clear that the bluff hull with nearly square bilges would behave poorly at sea and could not be driven very fast, but these limitations had to be accepted. Several alternative sets of lines were tried, but the first turned out best. The combination of light displacement, shallow draft, and broad beam made for an unusually large metacentric height (GM), hence a very quick—and uncomfortable—roll. Cargo might move if it was not properly secured. After the war, several LSTs converted to bulk carriers sank, probably because their loads shifted in storms. On the other hand, the great GM allowed the ship to take an unusual amount of damage without taking a dangerous list. When lightly loaded, with little of the ship immersed, an LST would tend to "sail" with the wind. The shape and shallow draft forward could cause severe pounding in a seaway, apparently causing cracks in the decks of many LSTs. Given such unpleasant characteristics, BuShips was surprised that it received so little criticism.

Apparently LST crews generally had so little previous seagoing experience that they accepted LST behavior as normal.

Shipbuilding capacity was already largely occupied, so new yards were needed. That would take time; initially large numbers of LSTs were allotted to the big east coast navy yards and to major commercial yards. Later the big bridge-building and steel companies built new yards in the Ohio and Mississippi River valleys. When the LST program was expanded again in November 1943 the navy again turned to yards on the coasts.

LSTs often carried LCTs on deck, but visibility over the LCT was poor. Many LSTs therefore had an extra bridge level added. Because boat capacity was inadequate, many LSTs were fitted with two additional sets of davits, thus further constricting the view from the original conn. In some ships, a lattice tower carrying a small conning position was built atop the original bridge. In the run-up to the Sicilian invasion (Operation Husky) in August 1943 it became apparent that not enough attack transports or LCI(L)s would be available. Rear Adm. Richard Conolly ordered six davits installed on 36 LSTs; each LST could thus land the assault element of an infantry company. The extra davits were fitted forward, to keep enough deck space clear to accommo-

LST 125 is shown after a Mare Island refit, 17 August 1945. By late in the war, upper decks had been cleared on LSTs. Bridges were raised a level to clear the top of a stowed LCT. Most 20-mm guns were replaced by 40s; the 3-in gun was replaced with a director-controlled twin 40-mm gun (another is right forward). The ship has a surface-search radar at her masthead.

date an LCT. At Sicily a typical six-davit LST carried five landing craft and one LCS(S) (landing craft, support, small), an assault support craft. Five of these LSTs were allocated to each assaulting battalion. The British were impressed. Using LSTs for the infantry drastically reduced the number of ships involved in the assault, and thus the sea room required. Instead of 36 LSTs and 18 LSIs (landing ships, infantry; the attack transports) to land a division, they could use 36 LSTs and a few much smaller LCI(L)s, the latter for reserves. On this basis the British specified four pairs of 15-ton davits for their next-generation LST(3). Conversions to six davits each were made in wartime, although after the war most or all LSTs reverted to the original two-davit standard.[3]

Armament increased considerably as the value of the LST became clear. To the original six 20-mm guns were added one 3-in/50 aft and one army-type single 40-mm gun forward. Crews improvised, and many LSTs were much more heavily armed than their designers intended. At Bougainville in November 1943, for example, LST 354 was armed with three army-type 37-mm antiaircraft guns, four 40-mm, fourteen 20-mm, ten 0.50 caliber, and a 3-in/50 aft. Most LSTs in the operation had twelve or more 20-mm.[4] The automatic weapons of the vehicles they were carrying were also often mounted temporarily on deck. The LST 542 class was wired to take power-operated director-controlled twin 40-mm guns fore and aft, although few ships seem to have been completed in this form.[5] By 1944 the U.S. Navy had authorized an interim armament of seven single 40-mm and twelve 20-mm guns, and the planned ultimate battery was two twin director-controlled 40-mm guns with Mk 51 directors, four singles, and the twelve 20-mm. The standard British armament was one 12 pounder, six 20-mm, and four parachute and cable (PAC) or fast aerial mine (FAM) parachute-supported barrage weapons.[6]

To keep production up, the U.S. Navy tried to avoid changes. On 9 August 1943, however, some minor changes were approved for LST 491–541. To

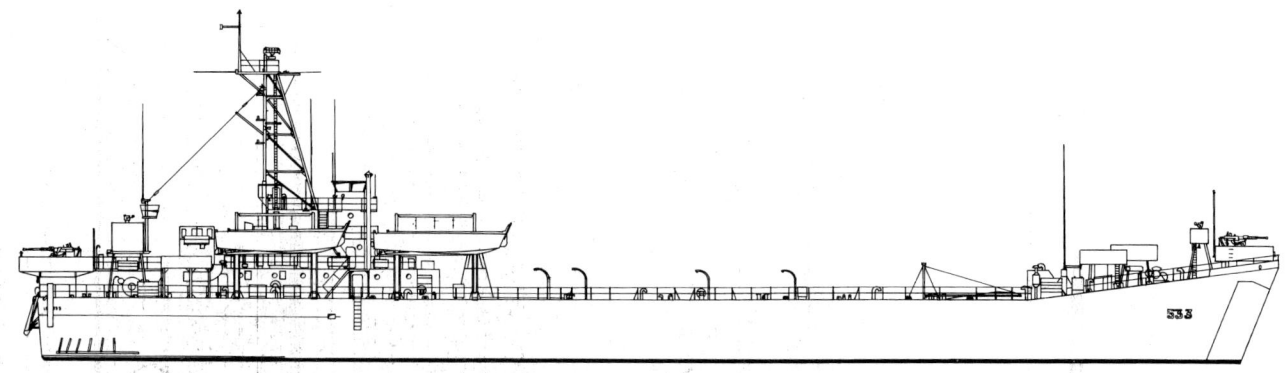

Cheboygan County (LST 533) about 1960, showing the tripod mast added to most active LSTs in the 1950s, with the twin 40-mm guns still aboard. Single 40-mm gun tubs remained, but were empty by this time. A. D. Baker

move stalled vehicles, a snaking winch was installed at the after end of the tank deck. A forklift and a 3½-ton crawler crane were added to the ship's allowance. Lashing staples were installed on the tank deck to handle homogenous cargo. The bow door was stiffened, and the Welin davits were modified to handle LCVPs. The elevator from main to tank deck was too slow, it sometimes jammed, and it could not easily accommodate some important vehicles, such as LVTs. A ramp replaced the elevator from the main to the tank deck, beginning with LST 513 (except for LST 531).

LSTs not too far advanced were modified more extensively to form a new LST 542 class, production of which continued through the war. A distilling plant was added to increase the LST's endurance. Cabling was also improved. It was impossible to solve some basic problems, however. As in the converted amphibious transports, ventilation was inadequate, radio spaces were cramped, and internal communication was insufficient. The austere design made no provision for emergency gear to operate bow doors or ramps.

One problem was not adequately addressed. Vehicles typically had to back into an LST when loading. If a turntable was installed, vehicles could simply be driven on, then turned around. In 1943 it was estimated that would save 20 percent of loading time, thus the British and Americans installed turntables on board their next-generation LSTs. British observation of actual loading in June 1944 suggested that the turntable was overrated; a faster elevator might make more of a difference.

Probably because they were available in such large numbers, and because they were so capacious, LSTs were adapted to numerous roles beyond the planned one of carrying tanks and similar vehicles (including artillery). For a short trip, an LST could carry up to 550 troops. Alternatively, an LST could transport 300–600 tons of combat-loaded cargo for beaching, or 1,400 tons of commercial-loaded cargo to a dock. In Europe, LSTs carried railroad loading stock, including locomotives, on rails laid on their tank decks. This inverted the early war attempts to carry tanks on board train ferries: railroad cars and tanks imposed similar loads on a ship.[7]

In the Pacific, the amphibious LVT (amtrac) was far more useful than the beaching LCVP, at least for the initial wave of an assault. The LST proved ideal to bring LVTs to their launch position off a beach. Initially it seemed that the LVTs would have to be carried on a temporary upper deck, to distribute their weight; a design for such a deck was dated 29 November 1943. Later, simple shoring was used. The LST carried only the tractors and their drivers; troops from attack transports went into LCVPs, from which they were transferred (at sea) to LVTs before heading to the beach. LSTs were organized in "tractor groups," indicating their main function. The LSTs were far slower than other amphibious ships, so they had to travel in special slow convoys that set out before the main amphibious force. Those on board had to stay much longer at sea on board their rather uncomfortable craft.

At Kwajalein in January 1944 LVTs were used to unload LSTs at sea. It proved difficult to reload the craft on board LSTs for the return trip. They could roll up into the ship, but backing out was more difficult. This was one reason for the requirement for a turntable in later LSTs.

LSTs were used to evacuate casualties. Surgical teams and hospital facilities were set up on the tank deck. First orders to outfit LSTs specifically for this purpose, with a crawler crane and one or two forklifts, were issued in February 1944. A more elaborate 1945 conversion produced the LST(H) (landing ship, tank, hospital), which could be distinguished by an extended deckhouse.[8]

Several LSTs were fitted with flight decks for U.S. Army liaison airplanes to support the Mediterranean invasions. This version, on board LST 906, incorporates stowage alongside the deck. *U.S. Army Aviation Digest*

The most sophisticated form of an LST "aircraft carrier" was LST 775, with the "Brodie system" of wires onto which a light airplane could land. The trainable launch platform on deck is reminiscent of World War I practice, when many British and a few American battleships were given flying-off platforms on their turrets. Many British and some Japanese cruisers also had flying-off platforms.

An LST could carry 1,000 tons for transfer to combatant ships; at the end of the war the Pacific Fleet reported that the LST was ideally suited to reammunition gunfire support ships at the scene of an amphibious operation. LSTs carried fog oil and smoke equipment, and served as water ships for small craft.[9]

Surely the most unusual adaptation of the LST was to operate light artillery-spotting aircraft. A light flight deck was built over the ship's upper deck. First tested on Lake Bizerte, it was used by LST 386 in the Sicilian invasion in August 1943.[10] LST 337 was converted for the assault on Salerno; lack of wind made take-off hazardous, and only one aircraft was successfully launched. However, the idea was clearly attractive, so LST 16 was converted specially for the Anzio landing in January 1944. The U.S. Navy modified LST 525 and LST 906 at Palermo to support the invasion of southern France. Unlike the earlier conversions, LST 906 had a narrow 220 × 16 ft take-off deck, leaving space for stowed Piper Cubs alongside.

These quasi-carriers were not used at Normandy.[11] All had one major defect: the ship's bridge blocked any attempt to land back on board. This problem was solved by the "Brodie" system, in which a cable strung between two booms was used to recover the airplane.[12] Because the forward boom blocked the bow, the take-off platform was made turnable. The system was tested on board LST 776 in the Gulf of Mexico on 28 August 1944. She participated in the Iwo Jima and Okinawa invasions. By that time the cable was also being used for take-offs.

LSTs were rebuilt for other purposes. The fleet badly needed forward-area repair ships and tenders, particularly for the mass of new craft powered by internal combustion engines. On 16 December 1942, for example, OpNav ordered three LSTs (two east coast, one west coast) converted to landing craft repair ships (ARLs): LST 25 (replaced by LST 10), 489, and 490. Two days later OpNav ordered three more converted to battle damage repair ships (ARBs), and two to motor torpedo boat tenders

LSTs were valued both during and after the war because their tank decks could easily accommodate shops and storage facilities. Shown on 17 December 1952, LST 1069 served as a mine squadron flagship, and was redesignated MCS 6 in 1959. The new deckhouse built up abaft the bridge housed the necessary spaces. The bridge has been raised. She has twin 40-mm guns on her centerline, fore and aft, plus six singles forward and two singles aft. There were also four twin 20-mm guns. Note the crawler crane on deck. At least three other LSTs also served as mine squadron flagships: LST 735, 799, and 802. These ships operated mine-spotting helicopters from their upper decks.

(AGPs).¹³ Ultimately the program produced 41 landing craft mother ships (ARLs), 12 battle damage repair ships (ARBs), and 10 motor torpedo boat tenders (AGPs). There were also aircraft repair ships (ARV[A]s and ARV[E]s) and salvage tenders (ARSTs).¹⁴ Another 16 LST hulls were designated as powered accommodations (barracks) ships (APBs), of which six were ordered as APBs and not as LSTs. The ex-LSTs were all actually small craft, mainly landing craft, tenders.¹⁵

Perhaps the most astounding feature of the wartime LST was its longevity. Many were still in service 15 years after the end of the war. Some were refitted under the Fleet Rehabilitation and Modernization (FRAM) program of the 1960s.¹⁶ Nearly half a century after they were built, many still service foreign navies.

Fortunately, the British logic for higher speed in their LST(1) was negated by the way the war developed. The slow LST was less vulnerable than might have been expected because it entered service in quantity after the U-boat threat in the Atlantic had declined dramatically, and the Japanese showed little interest in attacking amphibious convoys far from their targets. That could not have been anticipated in 1940–41. Also, despite being slower, LSTs could land vehicles much more rapidly than conventional transports, thus they made an important tactical difference in many landings. Faster ships, available in much smaller numbers, would not have been nearly as effective.

The first British data for the more sophisticated TLC carrier were given by Comdr. R. C. Todhunter, of the British Admiralty delegation, to Captain Cochrane, BuShips' chief of preliminary design, on 10 November 1941. Rowland Baker later brought more detailed data from the British DNC's office. BuShip preliminary designers prepared a complete set of plans, to serve as contract plans for U.S. construction. Compared to the British plans, their lines were redrawn for greater length and beam at the load waterline, to give the greater stability thought necessary. Depth to the third (tank) deck was increased by a foot. BuShips suggested to the British that it would be easiest to reorder all seven Winettes not yet laid down as TLC carriers.

At first the U.S. Navy apparently saw little value in this odd design, but on 26 November 1941 the Auxiliary Vessels Board decided that the U.S. Navy could use it better than the "truck transport" or TAC or LST(1) (Winette). The board recommended that the U.S. Navy build ten, soon cut to eight. Instead of two British TLCs, a U.S. TLC carrier would carry 16 U.S.-type tank lighters (LCMs), each tank lighter car-

USS *Donner* (LSD 20) is shown on 5 January 1951.

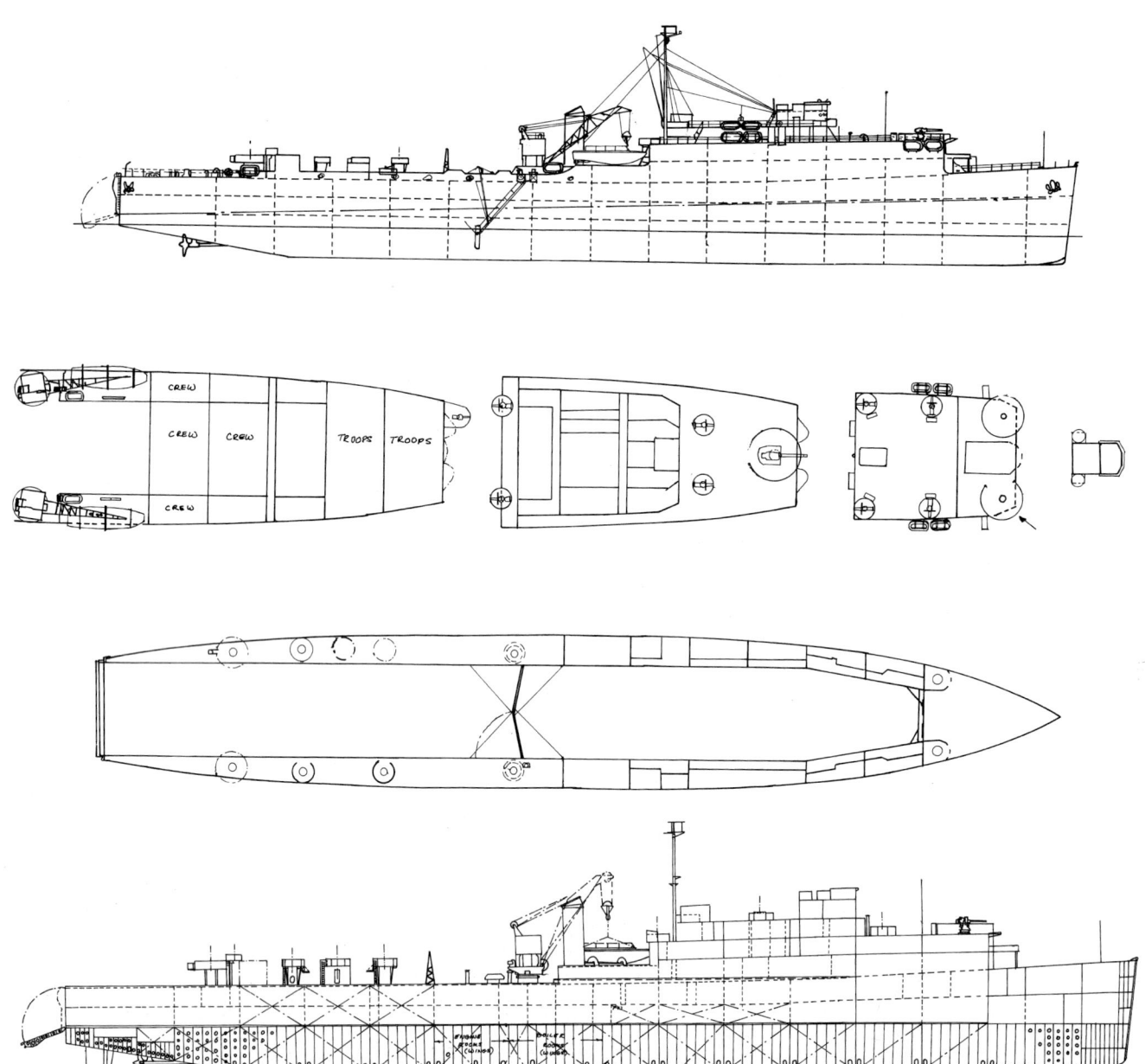

LSD plan, based on a U.S. booklet of plans, probably originally drawn about 1944, for LSD 9–12, which went to the Royal Navy. The LSD was probably the most significant World War II amphibious ship from the point of view of later development, because only the LSD could easily be developed for high speed. In the side view, the dot-and-dash line is the ship's knuckle. Dashed lines indicate internal compartmentation. The armament indicated, a 5-in/38 gun forward and a pair of large gun tubs for 40-mm mounts (indicated by arrows), is actually characteristic of U.S. rather than British ships. In the booklet, the 40-mm mounts are indicated as twins, ultimately to be replaced by quadruple mounts. Ballast tanks are indicated on the inboard profile. The gun shown is a 3-in/50.

rying one tank. At the original planning stage a U.S. medium tank battalion included 48–54 medium tanks and ten 2½-ton supporting trucks. Thus a tank battalion could be accommodated aboard four TLC carriers (up to 64 vehicles). By November 1941, only three TLC carriers were needed for each tank battalion (48 tanks in LCMs, trucks on deck as second loads in the LCMs). The original figure of ten TLC carriers was thus reduced to eight, providing accommodation for two tank battalions, allowing for two spares. The board wanted a common U.S.-British design, to carry 220 troops (vehicle crews), with a sustained speed (fully loaded, with 25 percent reserve power) of 15 kts, to keep up with fast troop convoys. To simplify fuel supply, the ship might be powered by gasoline engines, although this proposal was not adopted.

As finally designed, the LSD could carry two 193-ft British LCTs (side by side), three U.S.-type

LCT(5)s (two side by side, with one forward of them) or fourteen U.S.-type LCMs, or about 1,500 tons of vehicles in its well deck, with 12,000 gal of fuel for vehicles or boats. It would accommodate the crews of the 24 tanks on board the LCTs as well as a headquarters unit. Besides the craft in her well deck, the ship would carry two Higgins Eurekas, which could be hoisted out by crane or by davit. Her cranes and winches would lift 10-ton trucks, so she also might be used as a motor transport with her well deck filled with trucks. The design called for a gate between the two British LCTs to prevent water in the well deck from surging as the ship pitched in a seaway. Light displacement was only 4,500 tons (5,746 tons fully loaded), compared to 7,100 tons at sea carrying TLCs and trucks, or 13,490 tons when flooded down to float out the craft on board. On this basis draft fully loaded would be 14.4 ft, but 29.1 ft when the ship was ballasted down to float out her landing craft.

The planned powerplant was 7,000 SHP twin-screw geared turbines, for a speed of 17 kts (as in the Winette); endurance was 8,000 nm at 15 kts. Initially gear-cutting capacity was badly stretched. LSD 1–8 had Skinner Uniflow reciprocating engines. In June 1943 BuShips proposed a diesel-electric powerplant using six General Motors 16-278A diesel generators, the type used in submarines. It was not adopted; all later units had the geared turbines initially envisaged.

Gibbs & Cox had been design agent for the American Winettes, which were to have been built at Newport News. These companies naturally inherited the new project. They received design plans on 22 December 1941 and contract arrangement plans on 1 January 1942. Preliminary plans went to the BuShips' contract design section on 18 January. As a cover, the ships were called mechanized transports (APM, or BAPM when ordered under Lend-Lease). As part of the general redesignation of amphibious

In December 1968, USS *Donner* (LSD 20), is little changed from her World War II appearance. The most visible modifications are the extension of the deckhouse under her bridge (the 5-in gun forward has been removed) and the addition of another, enclosed, bridge level (the old bridge windows have been plated in). Although built in limited numbers, the fast LSDs were probably the most important of the World War II–designed amphibious ships.

A temporary superdeck over an LSD well deck could carry vehicles. USS *Comstock* is shown off Chu Lai, June 1965, with her stern gate down. She has not yet flooded her ballast tanks to bring the well deck under water. Alongside is LCU 1481, a direct descendant of wartime LCTs designed to fit an LSD well deck.

types ordered in July 1942, they became landing ships, dock (LSD).

According to a U.S. March 1942 assessment, "judged by the number of tanks transported, these ships appear rather less efficient." They were, however, the only means of carrying the heaviest tanks to a beachhead. Unlike the LST, the LSD was well adapted to the style of amphibious warfare the United States intended to execute in the Central Pacific.

There were some problems. LSDs snap-rolled. Their pumps operated too slowly. Ballast valves sometimes stuck and it was impossible to get the tanks completely emptied. Stern gates did not always seal properly, so well decks sometimes flooded. When amphibious vehicles were carried, their exhaust fumes sometimes filled the fore end of the well. U.S. operating personnel tended to stop the LSD before flooding her down, whereas the British recommended running at low speed, to take advantage of weather conditions. In 1944 the design was changed, the stern gate cut down from the original 19 ft at its sides to 10 ft in the middle, to save weight.

The original main armament was one 3-in/50. U.S. units initially had eight 20-mm guns; British units had eight 20-mm and four single 2-pounders. The 20-mm guns took up upper deck space originally designed to accommodate ten 2½-ton trucks. By late 1944 both navies had sixteen 20-mm guns, and the U.S. version had a single 5-in/38, two quadruple and two twin 40-mm, and sixteen 20-mm; LSD 15–27 later had twelve twin 20-mm. Planning for production took up most of the first half of 1942. The first keel was laid on 22 June 1942. The first ship

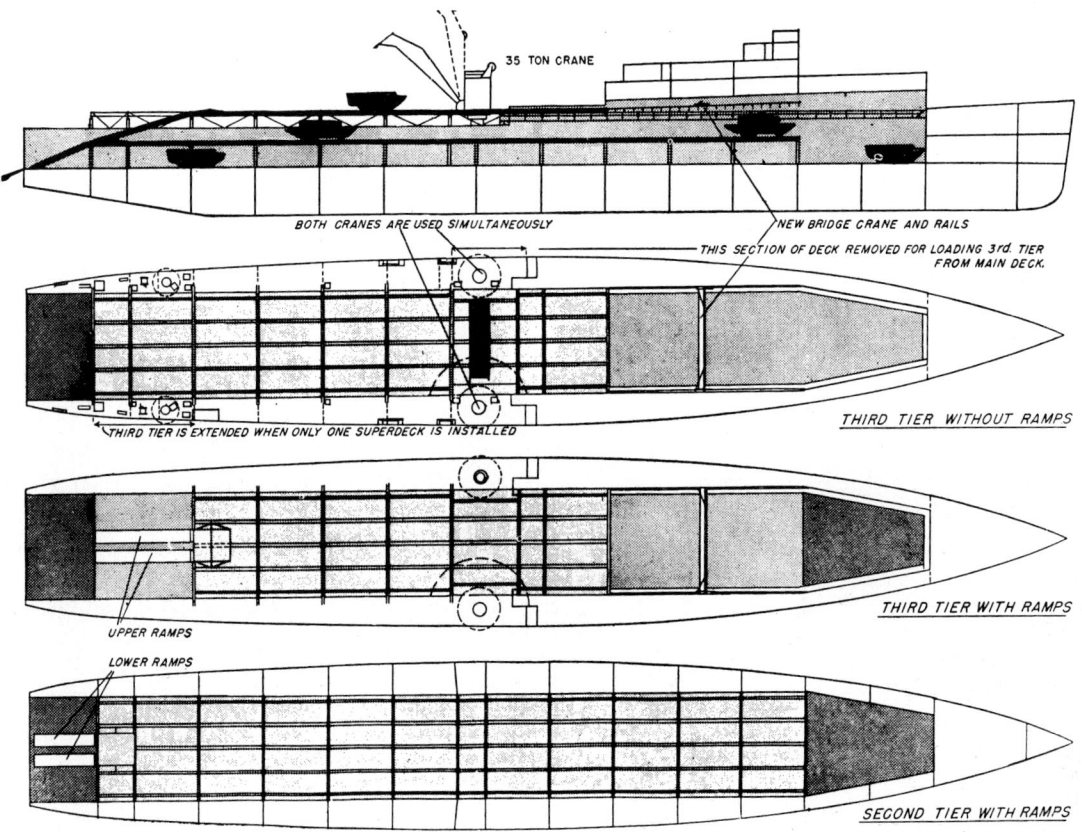

Superdeck LSD plan, 1945. LSD 13–27 were fitted with superdecks to carry vehicles. They had a new bridge crane spanning their well deck that was intended to help assemble the temporary decks in the well. The top temporary deck was supported by the side weather decks. The intermediate or second deck was supported on stanchions, and could be removed without affecting the top or third temporary deck, so that the ship could carry LCMs in her well deck. Ramps at the after end could be used to unload LVTs and DUKWs. Alternatively, the top deck could be extended right aft, and vehicles unloaded using the ship's two 35-ton cranes. With two temporary decks and ramps, an LSD could carry 92 LVTs or 108 DUKWs; with one, 64 LVTs or 74 DUKWs could be carried. Portable decks added 473 tons to the ship's light (Condition 2) displacement. *U.S. Navy (ONI 226)*

commissioned was *Ashland* (LSD 1) on 5 June 1943. Twenty-five were built.[17]

The major wartime modification was a deck that could be erected over the well deck in 48 hrs, doubling carrying capacity for amphibians and vehicles. Vehicle capacity could be further increased by adding a mezzanine or spar deck under the additional deck, the two being connected by a ramp. Vehicles were driven in from the stern and then up the ramp to the deck. A major issue at the end of the war was whether such decks could be made installable in forward areas. Work on the extra deck began at the end of November 1943.

In wartime the LSD was often used to transport small warships such as PT (motor torpedo) boats and minesweepers, and to dock landing craft for repairs. Some complained that LSDs were not available for their nominal primary role. The two big cranes on the weather deck particularly fitted LSDs as repair ships, and to this end they were fitted with machine, carpenter, blacksmith, welding, and shipfitting shops. Not designed for this purpose, the LSD lacked stowage for material and spares, and had insufficient personnel. So important was the secondary wartime role that in 1947 the Marines suggested that the type be split in two: one to carry amphibious vehicles craft, the other a floating drydock to repair landing craft.[18]

The British had always seen the tank ferry, which became the U.S. LCT, as a companion to the LST. The British delegation that brought the final LST staff requirements to Washington on 21 November 1941 also brought a Thornycroft sketch design (100 × 32 × 5½ ft) and staff requirements for the tank ferry, which they called a shallow-draft tank lander. The Thornycroft design showed a ramp forward and an open stern. It was designed to carry three 50-ton tanks or six 10-ton trucks at 10 kts with a 500-nm endurance. There was no accommodation at all. As in the case of the LST, the ferry was to be able to land trucks in 1 ft 6 in of water, but in this case on a slope of 1:150 (with an allowance of 3 in for run-up while grounding). Tanks could land in 3 ft of water under similar conditions. The steering position

LCTs were designed so that they could be shipped overseas in sections and bolted together in the water. This LCT(6) is shown on 24 June 1944.

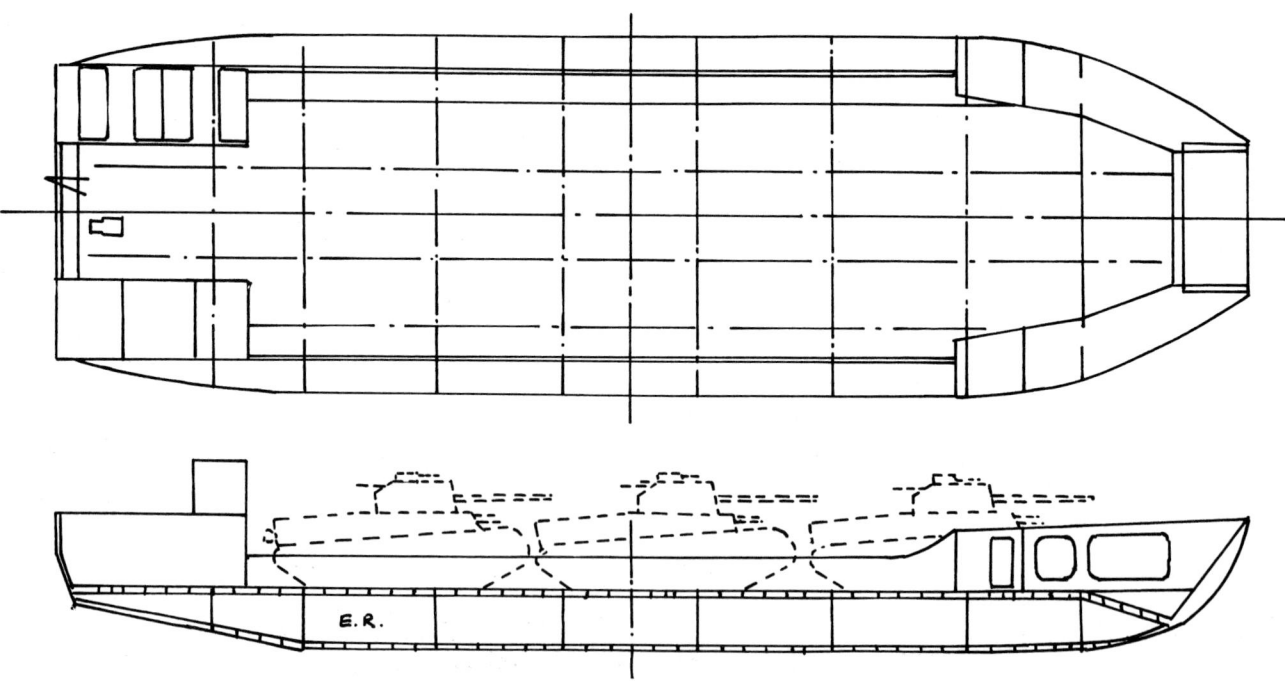

The LCT, from sketch by the BuShips preliminary design branch, 21 January 1942. Three M6 heavy tanks are shown on board. Dimensions were 100 ft × 52 ft × 3 ft 1 in forward and 3 ft 9 in aft; displacement was 112.5 tons empty and 265 tons with tanks on board. Power would have been provided by two 450 BHP twin Gray diesels, for a speed of 10 kts. Endurance would have been 500 nm at 10 kts. At this stage the craft was designed to be transported in two sections, the after one (including machinery) weighing 53 tons, and the forward one 45 tons; LCTs were actually built in three sections.

would be given ¼-in protective plating. If the craft was gasoline powered, it would be given self-sealing tanks. The Thornycroft drawing was considered no more than a guide; it had insufficient deck area for trucks. The Admiralty wanted at the least to have its well deck widened to 18 ft so that vehicles could be carried three abreast. Accommodation (for 1 officer and 10 enlisted men) would have to be provided, and the after end closed in so that only a single centerline capstan and stern anchor would be needed. Overall, the Admiralty wanted the simplest possible design, not only for ease of production but also for ease of maintenance. For example, it would have no radio or gyro-compass, and it would not be degaussed.

BuShips' preliminary design section quickly developed a new design. By August 1941 the bureau already had completed plans for a 100-ft self-propelled steel lighter to carry tanks; mentioned in a 3 August paper on British amphibious craft, it was described as the closest U.S. approach to the British TLC. Details have not, apparently, survived.

The new tank ferry retained the dimensions of the Thornycroft craft, but was welded rather than riveted, with longitudinal framing. It substituted a simpler flush-deck arrangement, which DNC representative Rowland Baker approved, for Thornycroft's well deck. Thus it was, in effect, an enlarged version of the BuShips LCM. It proved quite successful; presumably it was large enough to be sufficiently stable, whereas the LCM was not. Its two 3 ft 6 in screws (one per diesel engine) were protected by tunnels. Beside the specified 10-lb nonmagnetic bullet-proof steel around the wheelhouse, the design showed 10-lb STS for the front plate of the ramp, where enemy bullets might be expected. Although the design did not show self-sealing fuel tanks, the 13.2-ton margin would more than suffice to provide them. To provide for transportation aboard freighters, the hull was broken by an amidships butt joint. As completed on 18 January 1942, the BuShips design showed a craft 100 (107 overall) × 32 × 3⅓–9 ft when loaded with heavy tanks. Hull depth was 5 ft

Shallow-draft landing craft needed protection for their propellers. LCT(6) 840 is shown at the builder, Kansas City Structural Steel Company, 3 June 1944.

6 in, and displacement was 112.3 tons without tanks or 265 tons loaded. With two 450 BHP twin Gray Marine diesels, it was expected to make 10 kts loaded. A fuel capacity of 6.5 tons offered an endurance of 500 nm at 10 kts, as specified.

Because the original weight estimate proved overoptimistic, the craft had to be lengthened to 105 ft during the contract design phase to maintain its required very shallow draft. A completely new design, with the same cross-section amidships but with new bow and stern lines, was prepared. The structure was changed to transverse framing to make construction simpler. Although initially this version showed a single amidships joint, the hull was built in three sections.

The planned pair of twin Gray Marine diesels was soon replaced by three single 225 HP Gray Marine diesels. Following U.S. practice, engine controls were placed in the wheelhouse, the engine room being too congested for watchkeeping; small craft engines were run unattended until they reached expected overhaul time. Each diesel drove a 3 ft 2 in propeller in a tunnel, and the stern lines were redrawn to match. The craft could not back off a beach if the tunnels were out of the water. BuShips installed a skirt plate to force water up into the tunnel.

The design showed no protection or armament. Because STS was in short supply, mild steel had to be substituted on the pilothouse and the ramp gates. The Admiralty asked that plastic armor be provided around the pilothouse. All unnecessary fittings and insulation were eliminated. In August 1942 the craft were ordered armed with two 20-mm guns aft, port and starboard. British craft received an upper bridge upon arrival in the United Kingdom.

Manitowoc Shipbuilding Company, which was also building submarines, was chosen as the lead yard. It was soon clear that the craft would have to be produced by minor inland builders. However, there was also a possibility that a large experienced builder could mass produce them. In April, then, 100 were allocated to New York Shipbuilding, normally a cruiser and capital ship builder. The British wanted 200 of these tank ferries, but by 13 February 1942 plans called for building another 200 for the U.S. Navy. The first contract was let on 2 March 1942, and the first keel was laid 4 May 1942. The first tank ferry was launched on 10 June and delivered on 29 June.

The U.S. Navy initially designated the tank ferry a YTL, indicating a harbor ("yard") tank lighter. In July 1942 the YTL was redesignated LCT(5). As mentioned previously, LCTs Mks 1 through 4 were the earlier much larger British tank lighters designed for substantial ocean passages.

The LCT was not used quite as intended. Conceived as a ship-to-shore ferry, at Normandy it carried troops and vehicles all the way from shore to shore. The LCT was not furnished with towing gear, but the fleet asked for it. In the South Pacific LCTs were towed up to 700 miles, to participate in assaults; nominal range was 1,200 nm at 7 kts, although this was seldom realized. For the invasion of the Marshalls in January–February 1944, 45 LCTs proceeded under their own power and 15 were carried on board LSTs. In other places the LCT was not considered seaworthy enough for long-distance towing. When an LCT was in a forward area, her officers and crew had to live on board for much longer than planned; their quarters proved inadequate.

Originally the LCT was to have been hoisted from an LST, which carried her across the Atlantic, by using a 150-ton crane, but Baker and the British Admiralty delegation in Washington pointed out that Britain had few such cranes. In July 1942 Baker suggested that the LCT might be launched directly from an LST trimmed down on one side. The technique was successfully tested at Newport News in February 1943. It became common in Pacific assault areas. Four tankers were fitted with 160-ton cranes specifically to hoist LCTs onto LSTs for onward movement to an assault area.[19] They also lifted PT boats.

In February 1943 the 3rd Marine Division complained that the LCT hogged and sagged even in a comparatively calm sea. This should not have been a great surprise, given its very shallow hull. Despite its very shallow draft, it often beached too far out when heavily loaded. The leading edge of the ramp was generally about 2 ft above the beach, poorly suited to wheeled vehicles. In April BuShips ordered tests of a portable upper deck on board LCT 233, to increase vehicle capacity, but it was not adopted. The British proposed that the LCT be lengthened; Captain Todhunter, who was in the British delegation to Washington, proposed adding 40 ft amidships. Tank tests showed that the longer version would have been faster with the same engines. This idea apparently led to the much larger LSM (landing ship, medium) described in Chapter 6. By that time it seemed that adding 25–30 ft would not preclude hoisting an LCT onto the deck of an LST for transport overseas. Having in effect invented the LCT(5), the British criticized it at an August 1944 roundtable as having "a terrible shape," too large a crew, and too little speed.[20]

BuShips wanted the LCT modified to help unload LSTs. An LCT broadside to an LST ramp could take vehicles aboard through a portable section cut into one or both of its bulwarks.[21] Alternatively, the LST ramp could mate with the after end of the LCT, the

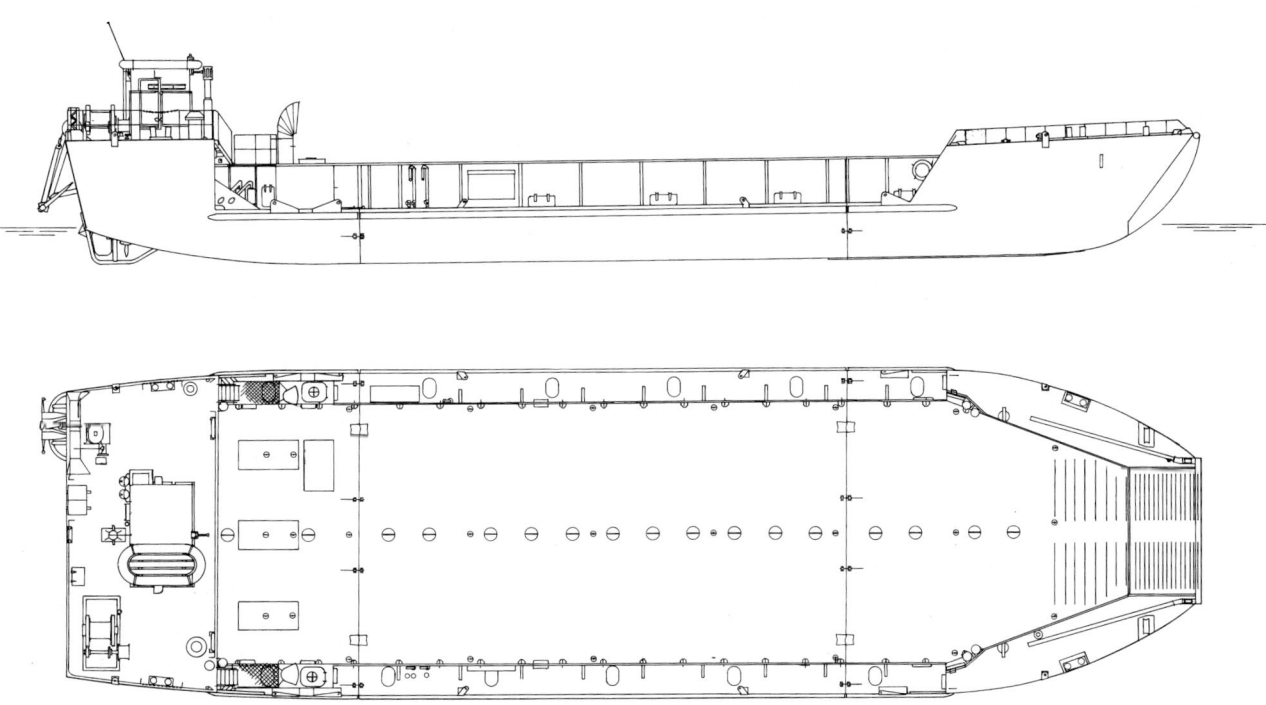

British conceived, and BuShips designed, the LCT to ferry tanks from ships to a beach, or across the English Channel. This LCT(5) carries Grant (M3) medium tanks.

LCT(5)-1 (ex–YTL 1) as completed, June 1942. Two single 20-mm guns were later added. *A. D. Baker*

LCTs were often carried as LST deck cargo, launched by heeling the ship. This is LST 214.

deckhouse there being cut through and a gate or ramp installed. In effect this was a reversion to the original Thornycroft concept. Rejected late in 1942, the drive-through design was adopted as LCT(6) for LCT 501 and later units; production began in April 1943. Living spaces were placed on either side aft, and the conning station relocated to the top of the starboard deck structure. Like the LCT(5), it could be transported in sections, in this case three of them.

Both the U.S. Navy and the Royal Navy favored the LCT(5) over the LCT(6). The British pronounced the broadside-on LCT(5) quite satisfactory, although vehicles lacked space to turn onto its fore-and-aft axis. It could only be used in fairly calm water. Despite the intent to use the craft to marry to an LST, it proved very difficult to moor the LCT(6) end-on to an LST except in the calmest water. The U.S. Navy felt that the higher stern of the LCT(5) permitted it to unload in surf conditions that would swamp an LCT(6). At Anzio in January 1944 LCT(6)s were criticized for their limited deck space and for their high bulwarks; the U.S. Army wanted the bulwarks cut down. Existing LCT(5)s were all disposed of at the end of World War II, perhaps because they had seen harder use than the LCT(6)s. The first postwar LCT design (LCU 1466), however, reverted to the closed-stern configuration of the LCT(5).

To some extent the LCT could be considered an alternative to the LCM. At Sicily the LCM was considered the most useful shipboard landing craft, due to its capacity. At Normandy the same conclusion was reached, but it was also said that if LCT(5)s were available the LCM was a poor second. In parallel with studies of the LCI(L), described below, in September 1945 BuShips studied the feasibility of converting surplus LCT(6) into mail and passenger tenders for fleet bases. Postwar, the Marines wanted the LCT to be made substantially faster, to match other assault landing craft.

When the LCT(5) was first being considered, there was brief interest in an alternative, a modified version of the "Sea Otter" offered by a com-

Pontoons made it possible for LSTs to unload over beaches whose slopes were too shallow for beaching. This LST is show at Leyte, October 1944.

BuShips modified the basic LCT so vehicles could drive through her. LCT(6) 839 is painted in Southwest Pacific dappled island camouflage.

mercial group, Ships Inc. It proposed to solve the shipping crisis by building small freighters powered by car-type gasoline engines grouped around vertical shafts, with right-angled gearing to propellers. Given such an arrangement, any type of engine could be used. The craft would be built out of the standard strip mill plating normally used in cars. On 3 December 1941 Hussey, Baker, and BuShips' Captain Cochrane met with representatives of Ships Inc. Contemporary British accounts describe the idea as absurd, and nothing came of this approach.

While the LST and LCT were being developed to support the invasion of France, the British were raiding German-held Europe. They began with converted Royal Air Force (RAF) crash launches and U.S.-supplied Higgins boats. By early 1942 larger raiding craft, capable of carrying more men, were wanted, but the standard British personnel assault craft, the LCA, lacked the speed and capacity for cross-channel raiding. Early in 1942, the chief of Combined Operations, Lord Louis Mountbatten, requested a "giant raiding craft" capable of carrying a full infantry company (200 troops) at 20 kts for 200 nm. The smaller craft did not carry enough troops, and it seemed dangerous to keep large LSIs waiting off the enemy coast without air superiority. It would be better to develop a small self-propelled craft that could cross the English Channel by itself. This was really not too different from the requirement posed by a cross-channel invasion. The troops would disembark directly onto "moderate" beaches in no more than 2 ft 6 in of water. The craft would be armed with antiaircraft guns and protected against small-arms fire; she would have the lowest possible silhouette consistent with good seakeeping. Initially Mountbatten envisaged merely seating the troops, but soon an endurance of 48 hrs was required.

According to the British official history, these requirements translated into a steel craft 150 ft long using the sort of high-power machinery used by coastal attack craft. It could not be built in Britain except at the expense of the much more important LCT program. The only alternative to U.S. construc-

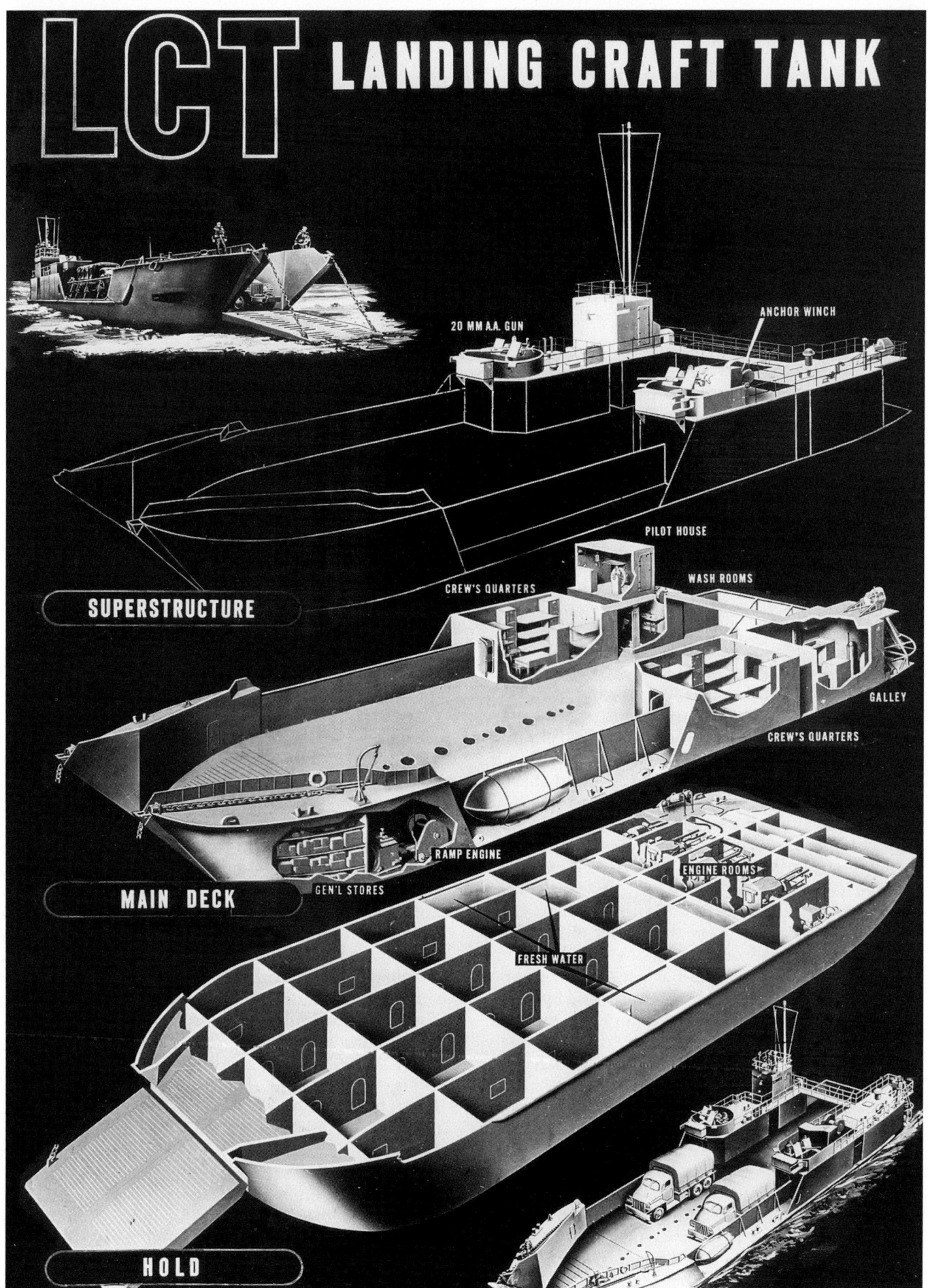

A wartime navy poster shows details of the LCT(6) design. Note the walkway over the vehicle deck.

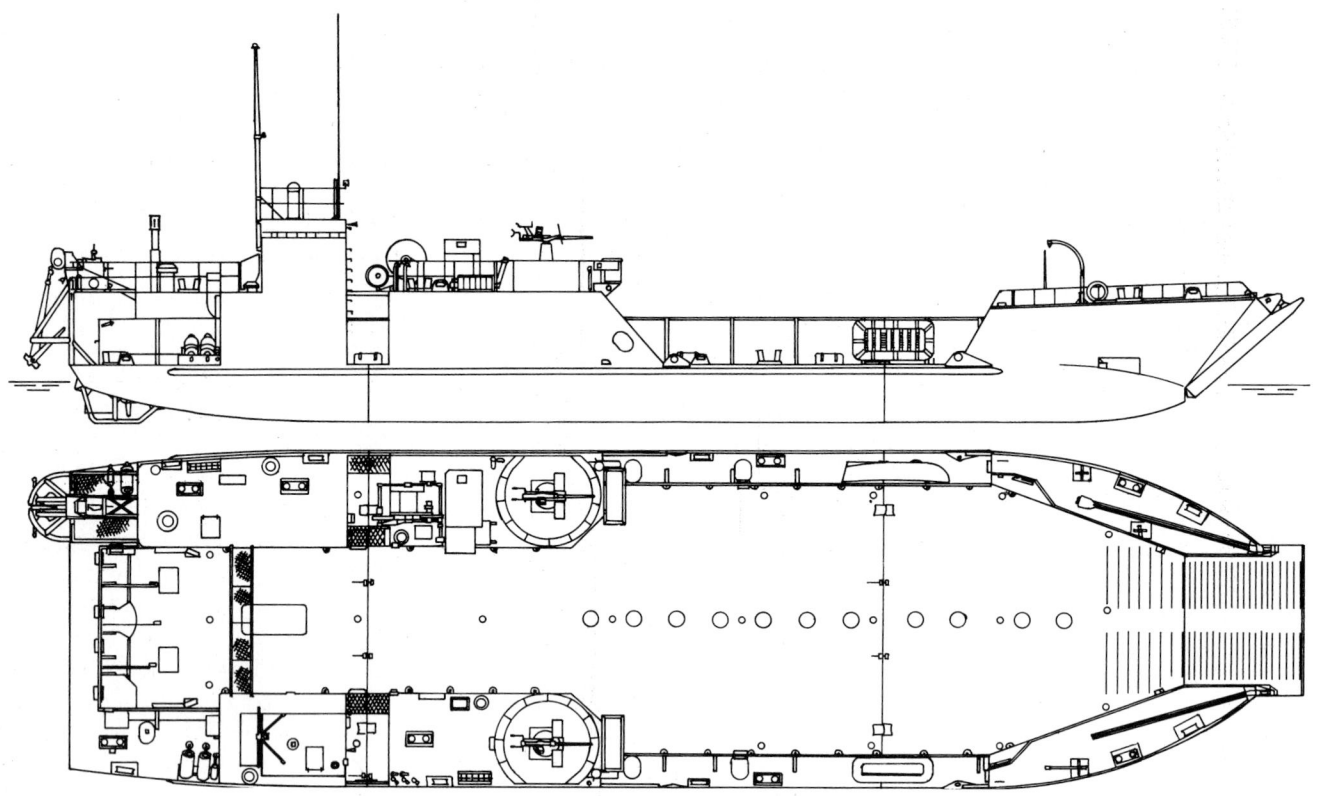

LCT(6)/LCU 501 class. Note the two single 20-mm guns and the drive-through configuration. *A. D. Baker*

tion was to build a less satisfactory wooden version in Britain, carrying only 100 troops at 15 kts. A design by Fairmile, which otherwise built motor launches and coastal attack craft, entered production as the LCI(S) (landing craft, infantry, small). The Americans were approached to build the big steel LCI(L) (landing craft, infantry, large).

The LCI was conceived as British interest in raiding ebbed. Too large for earlier types of raiding, it could move large numbers of troops across the English Channel. According to a 30 April 1942 British Admiralty delegation letter to Rear Adm. Alexander H. Van Keuren, then chief of BuShips, based on telegrams from Britain, the need for it "has been made clear by preliminary studies of the big operation [i.e., the invasion of France] and by recent experience of raids on enemy territory." By this time it seemed that U.S. building capacity was completely occupied. The only way to get enough such craft quickly enough would be "to give more or less carte blanche to a firm of boat builders such as Higgins"—who had already achieved miraculous levels of production. By this time, however, Higgins was behind in deliveries of both PT boats and landing craft. The miracle was that the LCIs were produced, roughly on time, in a much more conventional way.

Despite the note to Van Keuren, the LCI was always characterized as a raiding craft. Probably the British thought that it would have to be sponsored by the U.S. Army, which still imagined that troops could cross the English Channel to France using the smallest beaching craft, LCPLs and LCMs. It would take some months for the absurdity of the idea to sink in. Until then, the army would hardly be willing to buy a larger beaching personnel transport like the LCI, and it still had no use for the navy's transports. Presumably the raiding designation would have made the LCI(L) acceptable while the army changed its mind.

Certainly the LCI(L) fit the army's view that only beaching craft were worth using, because they would not really become vulnerable until they beached—and unloaded. The tie to the French invasion is suggested by the number the British wanted, 300. This would accommodate 60,000 troops, roughly the size of the contemplated assault force. LCI(L)s were used extensively at Normandy.

The program moved remarkably quickly. On 30 April 1942 the Admiralty sent staff requirements (characteristics) to its mission in Washington. Presumably mindful of the LST story, the Combined Operations naval representative gave a copy to Col. Albert C. Wedemeyer of General Marshall's staff at the combined U.S.-British staff talks. The chief of the British Admiralty delegation in Washington, Rear Admiral Dorling, sent the U.S. Navy another copy with a letter explaining the need for the craft. A British design was sent by air, but it apparently had no influence on what followed.

On 2 May, Comdr. E. H. Logsden, RN, a member of the British Admiralty delegation, gave a set of staff requirements to Captain Cochrane, the head of preliminary design at BuShips. Logsden asked for 300, requesting they be available for service in the United Kingdom by 15 April 1943 (i.e., in time for the 1943 landing in France then contemplated). Given the same lack of shipping cited for the LCT, the new craft had to be seaworthy enough to cross the Atlantic on its own power, given extra fuel stowage. Desired maximum speed was 20 kts (not less than 15 kts), and endurance 500 nm at 15 kts, rather more than what Mountbatten had asked. The Admiralty suggested using 1,200 HP Stirling gasoline engines, and asked for silent running at slow speed for the final approach to the beach (as in its earlier LCA). Instead of specifying the number of troops, Logsden set a seating area of 1,500–2,000 sq ft (approximately 100 × 18 ft), thus setting minimum dimensions. Maximum draft forward would be 2 ft, for beaching. Instead of the usual heavy ramp, troops would disembark down three or four light gangways, each 2 ft 6 in wide.

There would be no fixed armament; the troops on board would use their own weapons for self-defense. When no troops were carried, the craft would be armed with four 20-mm cannon. It would have bullet-proof protection against fire from ahead and from as far aft on each side as practical.

When Logsden met Cochrane on 2 May, he said that he hoped to see "a certain builder" (probably Higgins) about the project on the fifth. BuShips' sensitivity about its LCM problems probably spurred Cochrane to ask Logsden to allow BuShips do some preliminary work before involving a private builder. By 5 May he hoped to have some preliminary ideas.

The U.S. summary requirements differed from Logsden's. According to an undated sheet in the design file, the new craft would land assault troops on a beach with a slope of 1:100, with a draft (forward) of not more than 30 in. Orderly but rapid disembarkation would be very important. Speed should be better than 15 kts. Passage might be as long as 24 hrs. The craft would be able to carry cargo broken down into man-size packages; there would be no large hatch or boom. The main design factors would be speed of construction (to meet the Admiralty deadline), technical limitations (among untrained operators), and avoiding the use of critical materials. Armament would be four 20-mm cannon. The British protection requirements translated into 15-lb high-tensile steel (HTS) on the bridge and pilothouse, and 10-lb HTS on the side and bulwark over the forward third of the ship's length.

BuShips' preliminary design section tried a ship based on the 173-ft subchaser then in production,

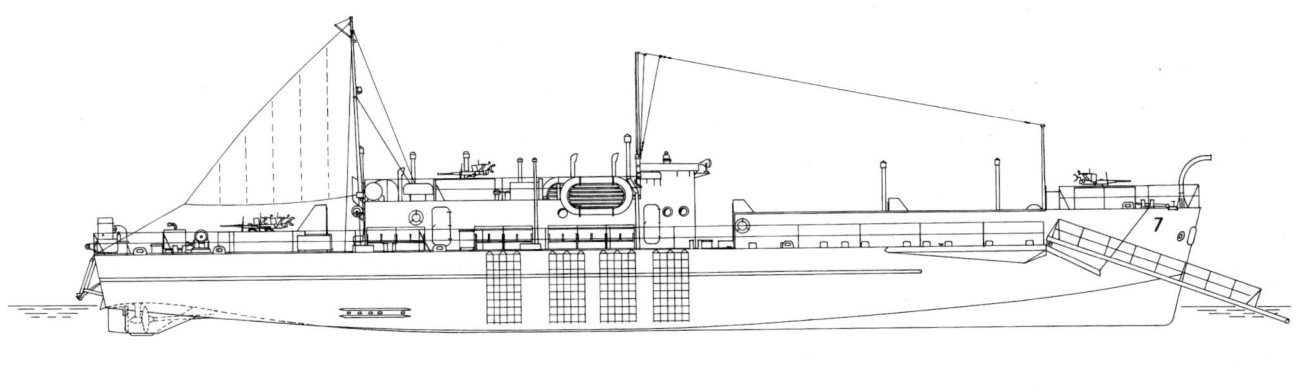

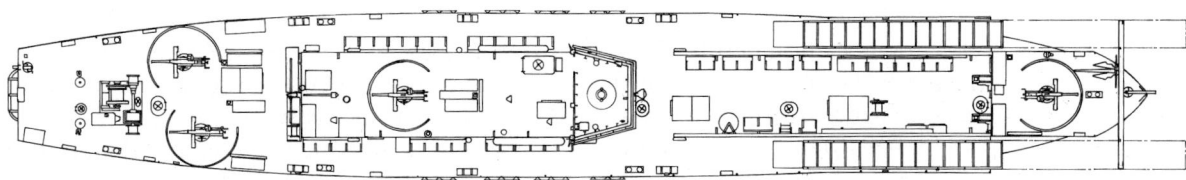

LCI 24, as completed in Royal Navy configuration, with steadying sail rigged and with portable "park benches" in place on main deck. She shows debarkation nets and bow ramps deployed. U.S. units had their bridges raised. *A. D. Baker*

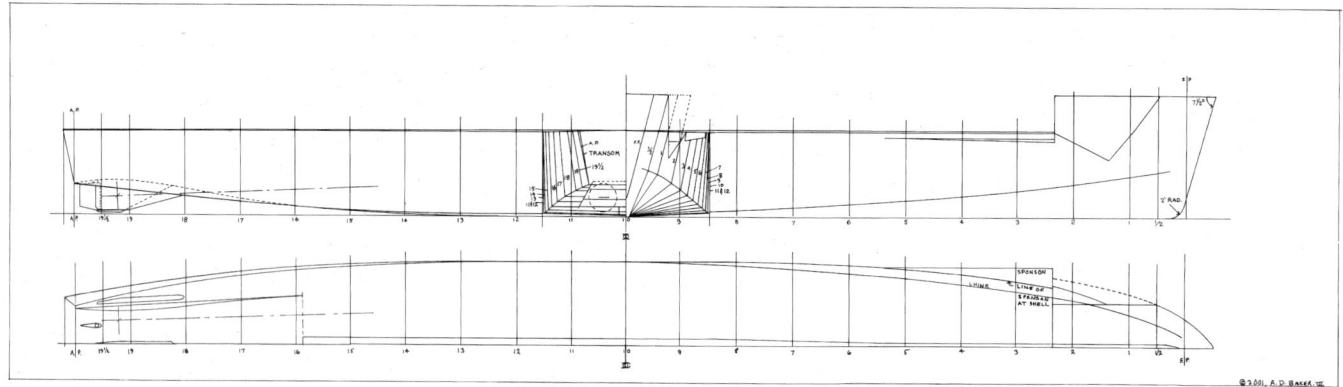

LCI(L) lines, showing how the hull form was simplified to ease production. *A. D. Baker*

with the lines altered to give shallower draft (about 2 ft forward and 4 ft 8 in aft). Rough lines and a general arrangement drawing were prepared. This work convinced Logsden to abandon his plan to visit the private builder, possibly a ploy on his part to spur BuShips' interest.

The British had already interested the U.S. Army in the design, and at a 10 May conference its representatives assured the preliminary designers that construction of a large number of the craft would probably receive a very high priority. Meanwhile it had become clear that the 173-footer was somewhat too large. On 11 May Cochrane ordered a new study of a smaller steel ship. Waterline length (153 ft) was estimated on the basis of troop capacity, and beam (23 ft) was set to give sufficient deck width and shallow enough draft. A small-scale plan was drawn, and at 2:30 on 11 May the BuShips group, including Cochrane, met Royal Navy representatives.

The British were generally satisfied with the design, but they considered its draft excessive. Initially they wanted a mean draft of 3 ft 6 in (2 ft forward, hence 5 ft aft), thus creating a 1:50 beaching slope on a 150-ft ship. Later the Admiralty asked for a maximum of 3 ft 6 in and a beach slope of 1:100 (because French beaches were so flat), even if that would require more draft forward. The designers pointed out that mean draft could not be reduced by much; in the end the British accepted 2 ft 6 in forward and minimum draft aft. The British also wanted a stern anchor.

After an 11 May discussion with army Gen. Brehon Sommervel, Cochrane froze basic details. Draft would be 2 ft 5 in forward and 4 ft 5 in aft in landing condition (225 tons), with a hull depth of 11 ft 6 in. Propellers would be designed for 17 kts, but speed would probably be something over 15 kts. To get this performance, propellers had to be 4 ft in diameter, so they had tunnels to protect them. BuShips initially wanted to use two derated PT boat–type Packard engines (about 1,000 HP each), then shifted to three 500 BHP diesels, one per shaft, and finally to quad 6-71s on each of two shafts. Endurance in landing condition would be 500 nm at 15 kts, but the craft would cross the Atlantic (4,000 nm) at 10–12 kts carrying extra fuel and 3 weeks' stores. Crew would be 24, including officers, and capacity would be over 200 troops.

Considerable effort went into adapting the craft for quick production; she had few curves either in her hull or in her superstructure. That did not rule out a degree of sophistication. The LCI(L) must have been one of the first U.S. applications of controllable-pitch propellers, to provide good efficiency at both seagoing and beaching displacements.

The preliminary design for this special troop transport (short-radius), designated APY, was completed on 21 May 1942. To speed construction, BuShips' preliminary design section, rather than the contract design section, prepared the principal contract plans and specifications. The first APY contract was signed on 3 June, just a month after Logsden's first meeting with Cochrane. The lead yards were George Lawley and New York Shipbuilding. The first keel was laid on 7 July 1942 and the first unit was commissioned on 9 October 1942.

By that time the class had been redesignated LCI(L) (landing craft, infantry, large). Besides the U.S. APY, earlier British designations had been Giant Y (the Eureka was the Y) or giant raiding craft (GRC) Mk II. GRC Mk I was the British LCI(S). After 1945 the LCI(L), which was clearly more ship than craft (it was not designed to ride aboard a conventional ship) was redesignated LSIL (landing ship, infantry, large). By that time the British no longer had what they called LSIs, which were equivalent to U.S. attack transports.

BuShips tried some alternative configurations in hopes of getting the best performance. LCI(L) 209, the first to be completed, had three skegs, one on the centerline between the propellers and one outboard

Although nominally a raiding craft, the LCI(L) seems to have been conceived as a cross-channel transport specifically for the invasion of France. The raiding craft heritage shows in the very low bridge, selected to reduce the craft's silhouette. LCI(L) 213 is shown on trials, 30 October 1942, with her boarding ramps stowed. All four guns are 20-mm: two on the centerline (fore and aft), two on the sides aft.

of each propeller. LCI(L) 1, the other prototype, had no outboard skegs. Both had skirt plates and filler plates as in the LCT, to allow them to back even in very shallow water. Trials—LCI(L) 209 on 1 October and LCI(L) 1 on 25 September—showed that the LCI(L) could beach and retract quite well, although its propellers were quite vulnerable. The boat without skegs, built by New York Shipbuilding, had a smaller tactical diameter (75 rather than 100 yds). Neither boat was able to maneuver while backing. To remedy this defect, there was much interest in redesigning the LCI(L) underwater hull aft during 1943. For the moment, however, BuShips ordered that all LCI(L)s be completed with skegs.

Compared to British LCI(L)s, U.S. units had their conning stations one deck higher, with 2¾-in plastic armor around their wheelhouses, conning stations, and guns. They carried two depth charge racks and two 300-lb depth charges, omitted from LCI(L) 351 and later units.

For the British, the real test was whether LCI(L)s could cross the Atlantic under their own power. The first such group comprised LCI(L) 6, 7, 97, 244, 210, 161, 162, and 163, accompanied by three naval tugs (BAT 6, 9, and 10). They left Norfolk for Bermuda on 20 November 1942, manned by extremely inexperienced but enthusiastic crews. In a typical crew of 23, 12–15 had never been to sea before. The LCI(L)s maintained 11 kts in a heavy swell, first on the beam and then astern, with breaking seas up to 6–8 ft high. Wind was up to Force 4. The craft rolled unpleasantly fast and yawed, making steering difficult. The senior British officer, Comdr. M. B. Sherwood, concluded that they were good sea boats in moderate weather (the boats had not experienced heavy weather). Engine room communication was bad because engine rooms were too noisy. In March 1943 the U.S. Navy reported that in a head sea boats pounded and waves of vibration were seen weaving along their hulls from stem to stern. U.S. craft now

Later units were modified: the new LCI(L) 351 shows her new high round bridge structure, relocated 20-mm guns, and an enlarged deckhouse. She is off Boston, 20 May 1943. Later a fifth 20-mm gun was added atop the forecastle.

LCI 368 is shown on 17 January 1944.

A wartime navy drawing shows the internal arrangement of an LCI(L) 351–class craft.

considerably exceeded their design displacement, typically drawing about 6 ft, and were capable of no more than 13.9 kts at sea.

LCI(L)s became available in the late fall of 1942, too late for the North African landings or for Guadalcanal. They entered combat early in July 1943 in the South Pacific, landing troops at Kiriwina on the New Guinea coast. LCI(L)s were also at the Kiska (but apparently not the Attu) landings in the Aleutians. The unit from the Aleutians later participated in the Central Pacific campaign. Later in July large numbers of LCI(L)s participated in the landing on Sicily. From then on LCI(L)s were engaged in both Pacific thrusts and in all the big Atlantic and Mediterranean landings. At Salerno, according to the Eighth Fleet after-action report, LCI(L)s were considered the most useful craft in the operation. They unloaded British attack transports (LSIs), which had insufficient boats of their own. They were used for traffic control and for salvage work. They also easily handled barrage balloons.

While the LCI(L) 1–350 series was being built it became obvious that the craft could usefully make much longer open-ocean passages with troops aboard. They were rearranged as "sleepers" rather than "day coaches," with bunks for about 190 men instead of seats for 250 troops in the four holds and on the main deck. Amenities, however, were not improved.

Later craft were redesigned in view of early service experience; LCI(L) 351 was completed in May 1943. The new version could be recognized by its rounded (tubular) conning tower. The tower was raised a level, and the former wheelhouse was taken over for officers' quarters. Troops had encountered problems coming topside from exposed hatches, so the deckhouse was extended to the sides and fore and aft to enclose the hatches to troop holds No. 2, 3, and 4 (rather than only 3, as previously). The enlarged deckhouse accommodated staff or troop officers and galley facilities for the troops, a necessary revision based on operational experience. No. 1 troop hatch was led into a slightly raised forecastle. Both the forecastle and the enlarged deckhouse offered more toilets and washing facilities for the troops. Hatches were enlarged to take stretchers. Extending the deckhouse considerably increased the craft's range of stability, making acceptable

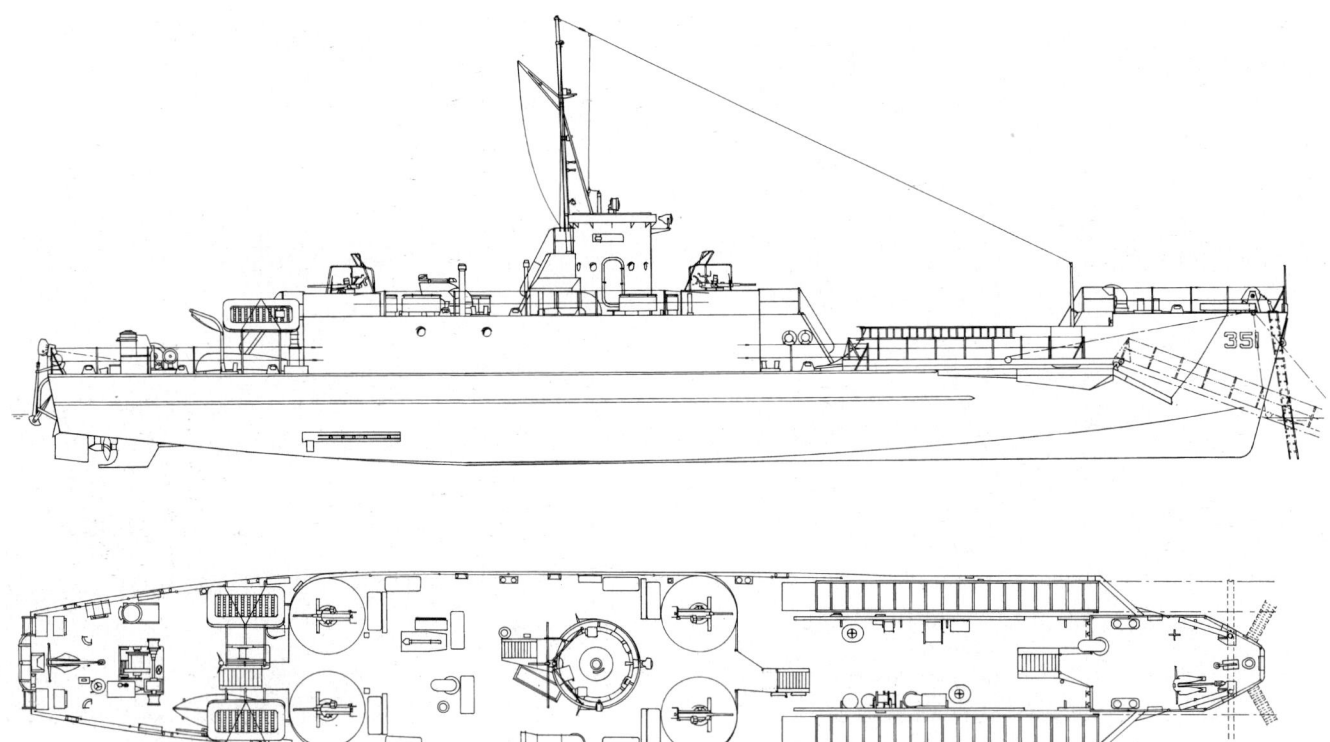

LCI(L)-351, the revised version of the LCI(L), newly completed, May 1943. LCI(L) 351–353 lacked the bow 20-mm gun of later units and had the 12-ft wherry stowed on its side. LCI(L) 354–371 had the wherry stowed vertically. LCI(L) 351–373 carried the disembarkation ladders shown. *A. D. Baker*

increased displacement and reduced metacentric height. Given a new capacity to fuel at sea, the craft could use their seaworthiness to make the much longer passages their troops could now more easily withstand, given their much better habitability. The four 20-mm were relocated to the corners of the enlarged deckhouse.

Troop capacity was 9 officers and 206 enlisted men. Designed as a "sleeper," the new version had increased refrigerator space, potable water, galley space, messing facilities, and toilets. Ventilation was modified to provide for sleeping troops. Steam heat replaced the oil stoves of the earlier craft. The troop gangways were lengthened from 28 to 36 ft. Instead of being manually operated, they were positioned by the bow anchor winch. Their sponsons were faired into the bow, to make the ship drier. The operators wanted them replaced by a centerline ramp, but that could not be done yet.

The new version had solid skegs in way of the propellers but no side skegs. The side skirts were not as deep as in the earlier series. Turning circle was about 80 yds, only slightly worse than that of the earlier type without skegs. At the operators' suggestion, the engines were provided above-water exhausts in addition to underwater ones, to be used when the boat was not in combat. Ballasted to 56 tons to landing displacement, LCI(L) 351 made 15.9 kts, a knot slower than LCI(L) 209, the Lawley prototype of the earlier series; the Lawley prototype had run about 20 tons lighter than landing displacement. An attempt to provide control while backing failed altogether.

BuShips listed design modifications in a 21 January 1943 memorandum, and the modified design was completed that February. After the first craft was inspected in May 1943, additional improvements were ordered. A single 20-mm gun was added on the forecastle. Shore connections were installed, so troops and crew could remain on board while the craft was moored. Air ports were added to the troop officers' stateroom, the junior officers' staterooms, and the galley, all with light-excluding ventilator screens.

Work continued to improve performance. In March 1943 BuShips called for higher speed and better maneuverability. The existing hull form afforded excellent protection to the propellers, but at a high cost in propulsive efficiency. In April, the David Taylor Model Basin tried placing a rudder immediately abaft each propeller. In June it tried adding a cylindrical backing rudder forward of the propeller. In November, model tests of a nozzle rudder (without any conventional rudder) that surrounded the propeller seemed to show that turning

Later LCI(L)s such as LCI(L) 795 had internal ramps down which carts and small vehicles could roll. *U.S. Naval Institute Collection*

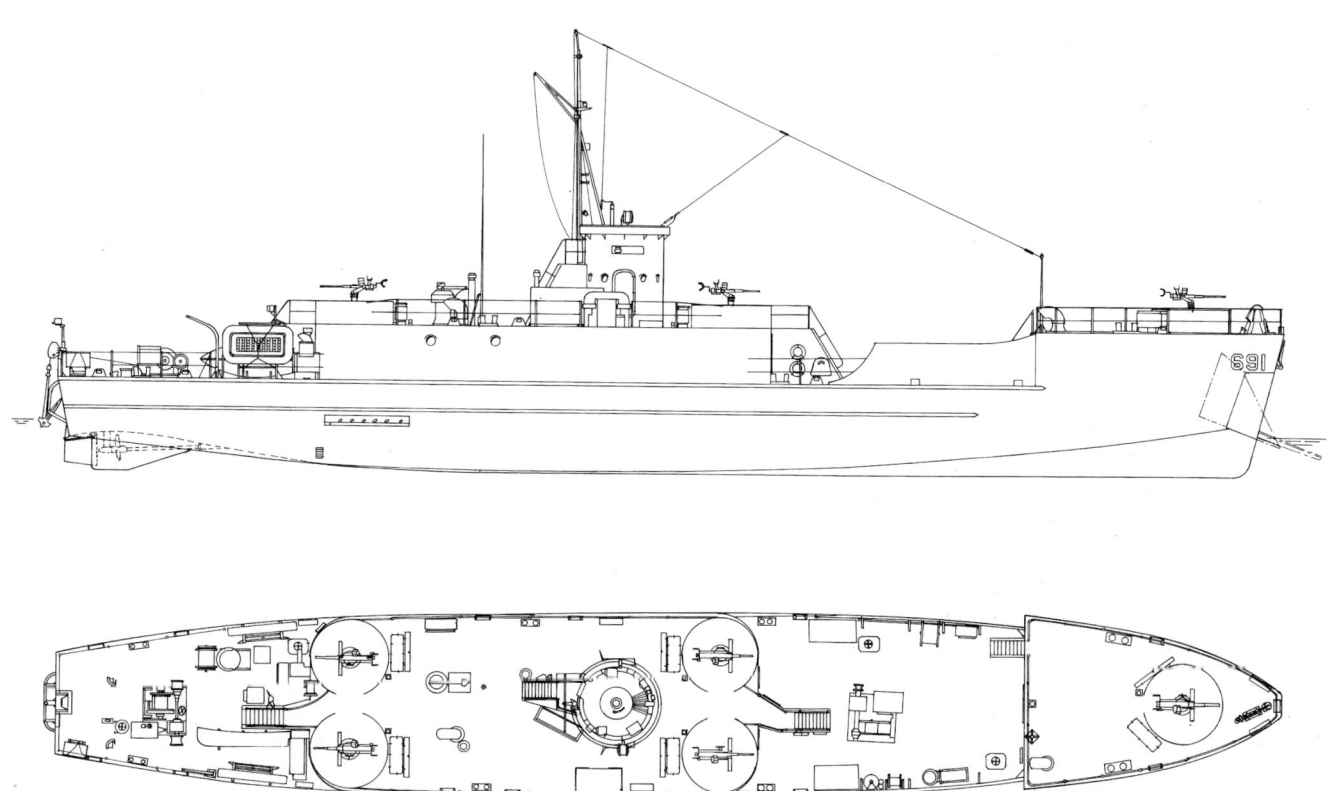

LCI(L) 691, the ultimate version of the LCI(L) with a centerline ramp, in June 1944. Units with the bow ramp were LCI(L) 641–657, 691–716, 762–780, 782–821, 866–884, 1024–1033, and 1068–1098. Numbers are not continuous because different builders introduced the bow ramp into their series at the same time. Units configured as flagships (LCFF) had the whip antenna. Many had gun shields. Most were fitted with radar in 1945. LCI(L) 715–716 had two additional flag containers on their upper decks. Units converted to rocket gunboats in 1944–45 had a single 40-mm army-type mount in place of the bow 20-mm gun, plus rocket launchers. *A. D. Baker*

performance could be improved without any loss of propulsive efficiency. The new arrangement would have been incorporated in LCI(L)s built in 1944–45.

Craft built in 1944 had their side gangways replaced by a centerline ramp led from a bow door: LCI(L) 641–657, 691–716, 762–780, 782–821, 866–884, 1024–1033, 1068–1098. The prototype, LCI(L) 402, had two single bow doors separated by a vertical post. The production version eliminated the post, thereby allowing small vehicles to be driven out of the ship, down the bow ramp. An additional ramp, 20 × 4.5 ft, slid down over the hinged ramp, to make a net extension of 16 ft to the beach. The forecastle was rearranged, with head room between ramp deck and bow overhead increased 4–6 in.

Work on a next-generation LCI(L) design, LCIL 1100, may have begun as early as October 1943. In January 1944 OpNav asked BuShips to increase speed by 2 kts. BuShips protested that the hull form was unsuited to high speed. The most powerful small-ship engines available were the "pancake" diesels on board some subchasers. Nearly all of those on order had already been delivered, and no more were planned. The pancakes were rated at 1,200 BHP, which might have offered more speed, but the continuous rating was only 770 HP, compared to 660 HP for the existing engine. Thus, as of April 1944 the main change seems to have been increased-capacity alternating current (AC) rather than the earlier direct current (DC) power. This series was never built.

Overall, the LCI(L) was far more ship-shaped than contemporaries such as the LST and LCT, because it had to be much faster. It gained its fine lines by dispensing with a bow door. Part of the price was that it could not carry or discharge much cargo. Despite those lines, an LCI(L) rolled and pitched quite easily, causing the troops on board to become seasick. On the other hand, higher speed made the LCI hull adaptable to other roles, particularly close-in fire support (discussed in Chapter 8). Thus in 1945 virtually all remaining LCI(L)s were ordered to be converted to gunboats for the invasion of Japan. Some unconverted LCI(L)s carried searchlights with which to support U.S. troops facing Japanese night attacks. Others supported "frogmen," the underwater demolition teams (UDTs), and were unofficially called LCI(D)s, although they were never designated as such.[22]

At the end of the war, BuShips found itself studying an LCI(L) conversion as a mail and passenger tender for fleet anchorages. Adm. William F. Halsey had already cited the LCI(L) as the best way to take large numbers of men ashore for liberty at fleet anchorages. However, none was converted. Later LCI(L)s were converted to coastal minehunters, designated AMC(U); see Appendix C for details.[23]

By the end of the war, the LCI(L) concept of moving a substantial force of troops directly ashore no longer seemed very attractive. Once beached, any ship or craft became a valid target; it would be better to split troops up among LCVPs or LVTs. Because she had to beach to unload, an LCI(L) could not offload her troops to LVTs and thus could not participate in any assault in which a reef fringed the beach. Thus LCI(L)s were ill-suited to Central Pacific landings. In 1945 the Pacific Fleet recommended against further development.

Part Two
World War II Developments

The LSM exemplified the speed of World War II amphibious development. Conceived in mid-1943, delivery began in May 1944. LSM 256 is shown in the well deck of *Casa Grande* (LSD 13), 30 June 1944. Ships were later fitted with much heavier batteries, including 40-mm guns forward.

6
Attack Transports and Conversions

Initially the declaration of war in December 1941 accelerated the attack transport program. Late that month the secretary of the navy verbally requested that the Auxiliary Vessels Board recommend ships to be acquired to add a division's lift: nine C3 combat loaders and three C2 cargo ships. The board recommended the rest of the prewar "four aces" (American Export Lines' *Excalibur, Excambion,* and *Exeter*) and three groups of C3s: Farrell Lines' *African Comet, African Meteor,* and *African Planet;* Delta Lines' *Del Uruguay;* and American President Lines' *President Van Buren* and *President Monroe* (*President Garfield* was substituted because *President Monroe* was on a mission under army control). The nine were initially designated AP 50–52 and 55–60; AP 53 and 54 were two large liners. Planned 1943 redesignations were APA 22–30, but *Joseph Hewes* (AP 50, ex-*Excalibur*), *Edward Rutledge* (AP 52, ex-*Exeter*), and *Thomas Stone* (AP 59, ex–*President Van Buren*) were all lost before they could be redesignated. These ships and the 7 army combat loaders brought the total to 34 combat loaders; another two C3s would soon be authorized.

The three accompanying attack cargo ships were *Libra* (AK 53, later AKA 12), *Titania* (AK 55, later AKA 13), and *Oberon* (AK 56, later AKA 14). These C2-Fs were near-sisters of the earlier AKs, with the *Bellatrix* (AK 20, later AKA 3) arrangement of heavy-lift pole masts braced by kingposts alongside. Armament was one 5-in/51, four 3-in/50, and eight 0.50 caliber machine guns (with foundations for 20-mm cannon). They would carry nine 45-ft tank lighters and seven Higgins boats. Conversion included fitting paravanes. Complement was 209,

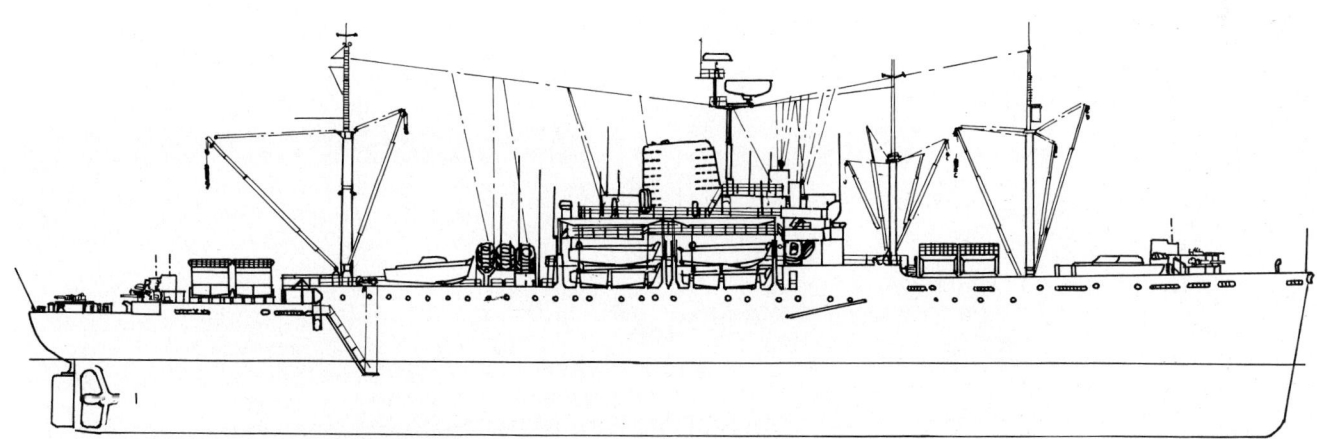

George Clymer (APA 27) as she appeared in the 1950s. Note her unusual armament of two slow-firing 3-in/50 forward and two single rapid-fire mounts aft, in each case with twin 40-mm guns nearby. Radars are SPS-10 (surface search) and SPS-6C (air search). The foremast carries a 30-ton boom on the centerline, facing aft, and 10-ton booms port and starboard. The mast next aft carries two 8-ton booms on its forward side and two 5-ton booms on its after side. The mainmast carries 10-ton booms port and starboard and a 30-ton boom aft on the centerline to handle the two LCMs on deck. The craft on deck abaft the superstructure are an LCVP and an LCPL on the centerline, with an LCPL to port and an LCVP to starboard. As a relief flagship, *George Clymer* had a flag bridge above her navigating bridge, as shown here.

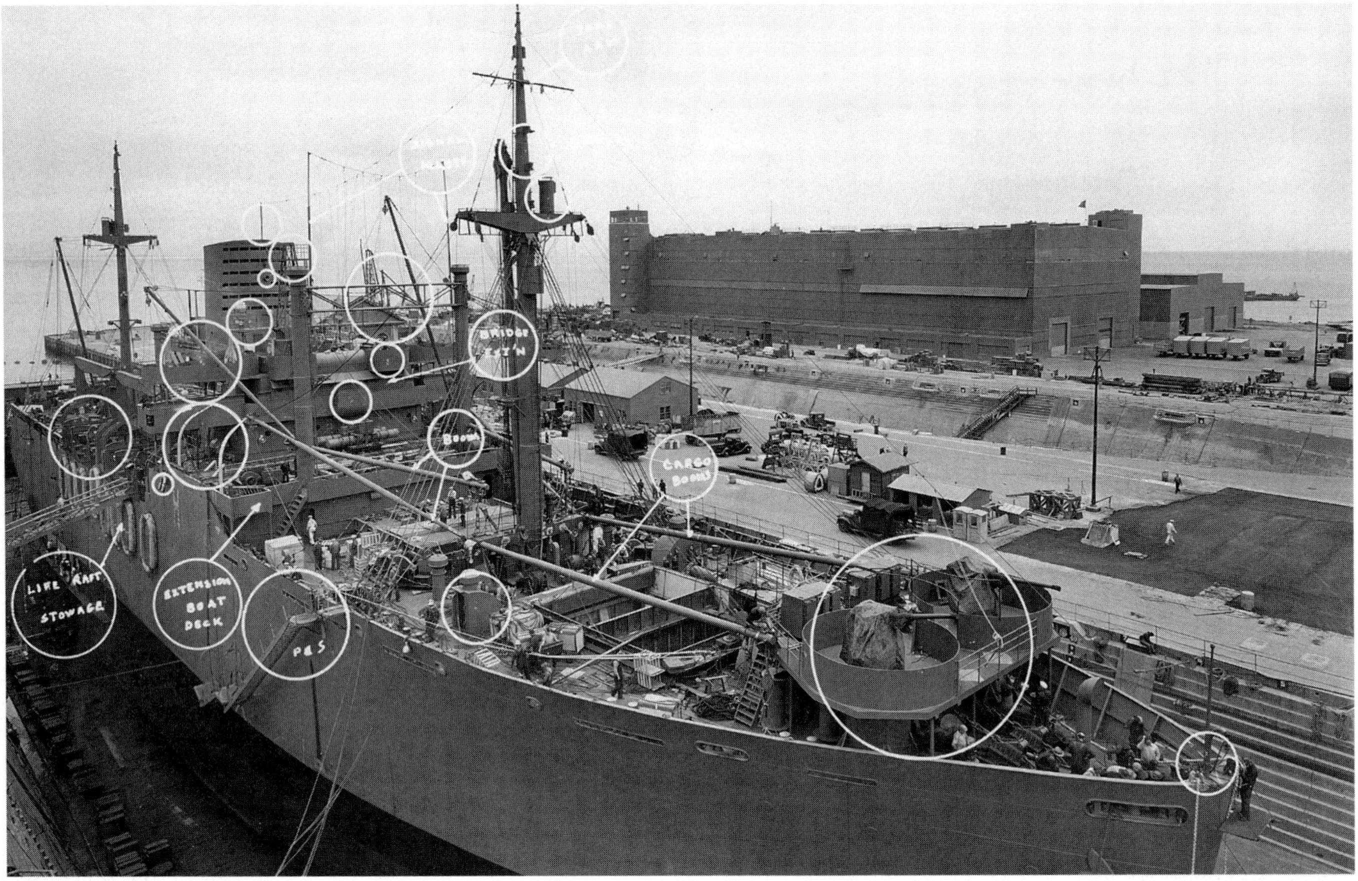
BRIDGE EXT'N
BOOMS
CARGO BOOMS
LIFE RAFT STOWAGE
EXTENSION BOAT DECK
P&S

BOOMS
LIFE RAFT STOWAGE
LADDER & PLATFORM

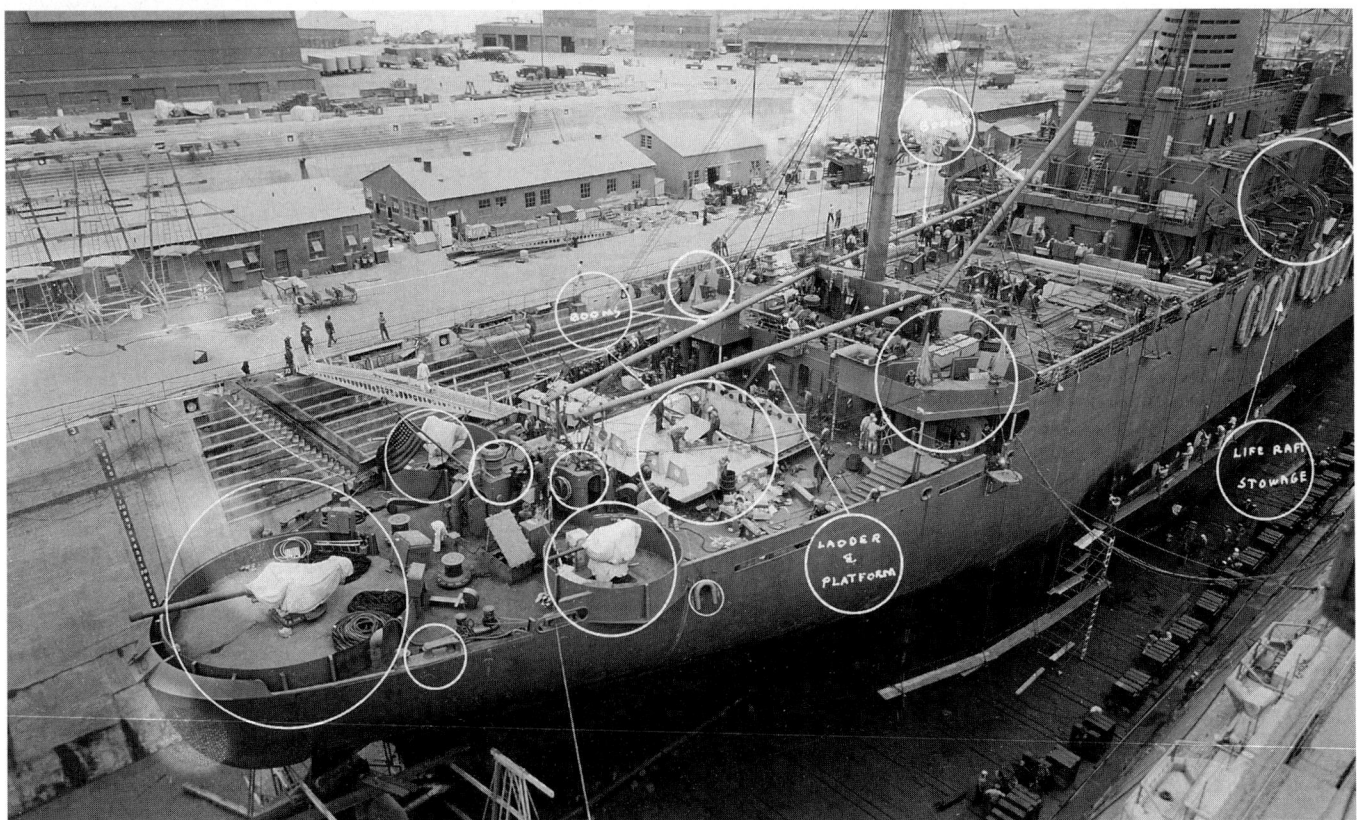

Newly converted to a combat loader (APA) in San Francisco, *Arthur Middleton* (ex–SS *African Comet*; AP 55, later APA 25) shows just how much had to be added to a cargo liner to convert her into an attack transport. She needed heavier booms to handle landing craft (note the boat chocks atop the hatch in the photo of the ship's after end). Note also the catwalk to assist in handling personnel boats launched from the new triple Welin davits, and the boat chocks positioned under the davits. The small lattice mast ultimately carried a radar antenna. *Arthur Middleton* and two sisters, also converted into APAs, were C3-Ps built by Ingalls. A fourth was taken over by the army and converted into the troop transport *George W. Goethals*. In contrast to the APAs, she had all of her kingposts removed because she would not handle cargo. Four heavy double davits (for lifeboats) were fitted on each side. The massive funnel casing, which the APAs retained from the original commercial design, was eliminated, leaving only a narrow smoke pipe. Three similar ships, earmarked for U.S. Lines, were also converted into army transports: *Henry Gibbins*, *David C. Shanks*, and *Fred C. Ainsworth*. All four were transferred to the Military Sea Transportation Service in 1950.

with space for 174 troops, including 12 officers and 12 NCOs.

As in the case of transports, conversion was simple and not altogether satisfactory. In December 1942, after the invasion of North Africa, the commanding officer of *Titania* listed required changes, which the Atlantic Fleet amphibious force circulated in hopes of influencing the next six AKA conversions. He wanted his ship to be capable of maintaining 15 kts (convoy speed) fully loaded, and he wanted her ballasted so that she would be stiff enough to handle her landing craft in all weather. BuShips approved his suggestion that sufficient electric power be provided so all heavy booms could be worked simultaneously, for faster unloading, but rejected a request for hatch covers that could be worked more quickly. BuShips rejected what must have been an important request for better internal communication, in the form of loudspeakers for each boat handling station, troop space, control space, and engine room. Presumably wiring all these spaces together would have been prohibitively complex. There was also insufficient space for the ship's crew, some of whom slept in hammocks rigged in the mess halls. In wartime the mess halls were always in use, so crewmen could never get sufficient rest. Additionally, the masts and booms all interfered with the ship's antiaircraft guns. The CO

Newly refitted, *Arthur Middleton* (APA 25) is shown off San Francisco, 30 August 1945. Improvements visible include a twin radar-controlled 40-mm gun forward, just abaft the two 3-in/50s, and an enclosed control position just forward of the lattice radar mast.

The big C3 conversions were valued postwar, and many remained active. USS *George Clymer* (APA 27, ex AP 57), is shown at San Francisco on 7 May 1965. Her 3-in/50 guns are gone; without director control, they were useless against most aircraft. She did, however, retain her pair of radar-controlled twin 40-mm guns, fore and aft. This ship was a relief AGC or squadron flagship.

Thomas Stone (AP 59, later APA 29) was typical of the five C3-P&C (passenger and cargo) class liners: *President Jackson, President Adams, President Hayes, Thomas Stone,* and *Thomas Jefferson* converted into attack transports (APA 18–20, 29, and 30, respectively). Two more, *President Polk* and *President Monroe,* were converted into point-to-point troopships and quasi-attack transports in 1943, as AP 103 and 104. *Thomas Stone* is shown, newly converted, on 27 May 1942. She was later used to exemplify the potential for C3 conversions, as opposed to a proposed new BuShips transport design. Torpedoed off Spain on 7 November 1942, she was disabled, her stern buckling and her rudder and propeller knocked out. She placed her troops in landing craft for the 155-mile run to Algiers, so that they would not miss out on Operation Torch, but that proved impractical. The landing craft had to be abandoned, and the troops were delivered to Algiers (after the fighting was over) on board the British corvette *Spey*. The ship herself was towed to Algiers, where she was further badly damaged during a 25 November 1942 air raid. She drifted aground, and efforts to salvage her continued through the spring of 1944. While under salvage, she was reclassified as APA 29. However, by 1944 there were enough attack transports in service that salvage was no longer so important; she was declared a total loss and stricken.

wanted better arcs astern, and he wanted the ship's funnel cut down to avoid blanking some 20-mm arcs, but this request was denied.

The army saw little point in such ships. Its main role was to build a sufficient force in Britain to invade France and then overrun Germany. For that it needed many point-to-point transports. The Maritime Commission offered a 3,800-man conversion of its new C4 freighter. The navy protested that these new ships, not scheduled for completion until the end of 1944, would interfere with vital ongoing warship work. There was very little slack in the U.S. shipbuilding industry. Navy demands for fast C2s and C3s for assault work inevitably competed with army demands for the same ships for convoy trooping. Nor did the army think it would need attack transports to cross the English Channel. Its planners apparently considered Britain a sort of unsinkable

Charles Carroll (AP 58, later APA 28) was one of four sisters, all laid down for Delta Lines, converted to attack transports: *Crescent City* (ex-*Delorleans*, AP 40/APA 21), *Charles Carroll* (ex-*Deluruguay*), *Monrovia* (ex-*Delargentino*, AP 64/APA 31), and *Calvert* (ex-*Delorleans*, AP 65/APA 32). A fifth, *Delbrasil*, became the point-to-point transport (actually quasi-attack transport) *George F. Elliott* (AP 105). Another sister, the original *Delargentino*, was scheduled for naval use as AP 47 but was not taken over; she became the army transport *J. W. McAndrew*. *Charles Carroll* is shown off Hampton Roads, 3 September 1942, as newly completed. *Crescent City* differed in having a third set of triple Welin davits on each side, abaft the two on board *Charles Carroll*. *Charles Carroll* stowed boats on deck in this position. *Calvert* had a tall pole radar mast instead of the short lattice evident here. By 1957 she had a short narrow extension to her funnel, but otherwise was little changed. *Monrovia*, which also had a pole radar mast, was similarly modified postwar.

transport ship. In the spring of 1942, then, the U.S. Army expected simply to run thousands of Higgins boats and Higgins tank lighters directly from Britain to France. It took some months for the British to persuade the army of the impossibility of this idea.

Unwilling to risk large transports off the French coast, in April the army proposed to use American excursion vessels, such as those on the Norfolk-Washington run, for the post-assault buildup. The U.S. Navy and the Royal Navy doubted that such craft could make a transatlantic crossing, but the Combined (U.S.-U.K.) Chiefs of Staff approved attempts using British crews. Fourteen set out, but only four arrived undamaged; on arrival they were useless as anything except accommodation ships for the Combined Operations training center. Even so, the shortage of landing ships and craft was so frustrating that in January and August 1943 President Roosevelt revived the idea of sending U.S. river and lake craft across the Atlantic to support the cross-channel invasion.

In March 1942, U.S. planners wanted sufficient lift for an amphibious corps (two divisions) in each ocean, equated to 44 large combat loaders, 8 small combat loaders (the equivalent of 4 large ones), and 12 attack cargo ships. The planners added 3 large transports and one of each other category in each ocean to allow for war losses and accidents: 50 large and 10 small attack transports and 14 attack cargo ships. Against this, 18 attack transports were ready or being converted (plus 18 authorized); 2 small transports were being built (no more were ordered); and 8 attack cargo ships were ready (plus 5 authorized).

Only the Maritime Commission could provide new hulls for transports or attack cargo ships. Hard pressed to make up for enemy attacks on shipping, it shifted production from the fast C2 and C3, suitable for assault work, to the slower but much more easily produced Liberty ship. Although the commission did not generally use the same yards as the navy, it did compete for steel and machinery, particularly the turbines needed for fast (i.e., assault suitable) ships. Moreover, to operate across the Pacific, the navy needed tenders, which often used the same hulls as assault transports. Attack transports and cargo ships had only seventh priority in the U.S. naval shipbuilding program. As the war shifted to the offensive in the Pacific, where they were most important, they rose to fourth place in October 1943 and then to first place in August 1944.

In April, the Maritime Commission formally stated that it was prepared to include combat loader features in two new C3s under construction: *Delargentino* and *Delorleans*. The Auxiliary Vessels Board had recommended purchase and these ships were included in the March 1942 figures. The new *Delargentino* became *Monrovia* (AP 64/APA 31) and the new *Delorleans* became *Calvert* (AP 65/APA 32).

Given the shipping shortage, there was no serious attempt to obtain more attack transports until the summer of 1942. Then plans for the invasion of North Africa (Operation Torch), formally approved on 25 July 1942, required more ships. On 1 August Adm. King directed partial conversions to combat loaders and naval manning of ten ships already in use as transports, to become AP 42, 43, and 66–73: the army's *Hugh L. Scott* (AP 43), *Tasker H. Bliss* (AP

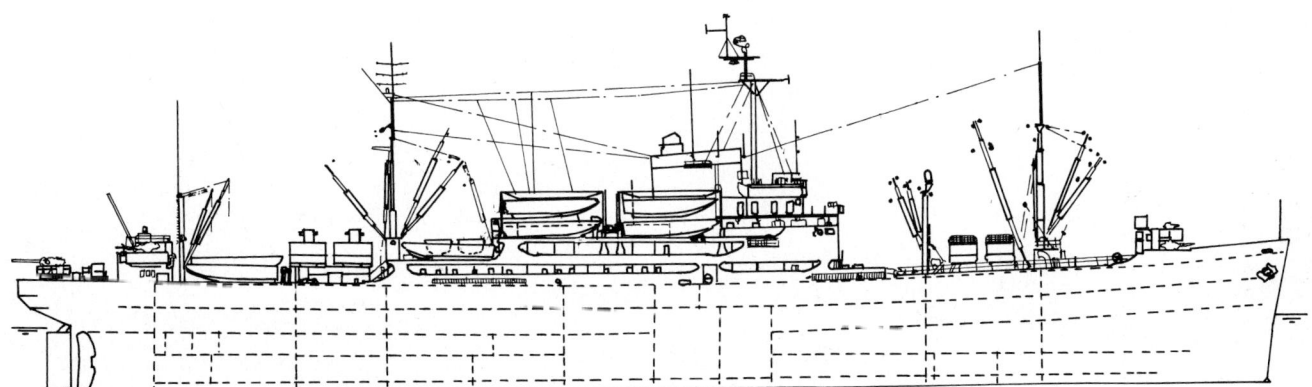

Monrovia (APA 31) postwar, probably in the late 1960s, after the air search antenna had been removed from her foremast. The radar shown is an SPS-10B surface search set. Each of two kingposts forward (port and starboard) is topped by a radome for her WLR-1 ESM set. The mainmast shows the triple antennas used for the postwar UHF ship-to-air radio. Aft the ship has a 5-in/38 on her fantail, with two 3-in/50 in raised gun tubs forward of it. Forward she shows a pair of 3-in/50s flanking a twin 40-mm gun. Dashed lines indicate internal compartmentation. *Monrovia* was a relief flagship. Thus deck levels, counting up from the main deck, were the superstructure deck, navigation bridge, and flag plot. Atop the flag plot, with its windshield, is an Mk 56 director (an Mk 51 is mounted with the guns forward). Although her small landing boats are drawn as LCPLs on the original navy drawing, in fact they were LCVPs. Beside the personnel boats, she has four LCMs on deck, atop hatch covers.

For the invasion of North Africa, Grace Lines' *Santa Lucia* was hurriedly converted into the combat loader *Leedstown* (AP 73). She was one of six attack transports lost during the operation, and this loss rate, following the loss of two transports at Guadalcanal, convinced OpNav that it might be wise to build smaller transports, the loss of which would not be as devastating.

Florence Nightingale (AP 70, ex-*Mormacsun*) was converted to an attack transport for the North African invasion, but was never redesignated as an APA. Instead, she was referred to informally as an XAP, a dual-purpose AP/APA. Unlike a full APA, she had two rather than four sets of triple Welin davits, stowing virtually all of her landing craft on deck. *Florence Nightingale* is shown on 1 October 1942. Her sister ship *Mormacsea* became a civilian-operated transport. Other C3s differed in mast arrangement. They included the XAPs *Elizabeth C. Stanton* (AP 69), *Lyon* (AP 71), and *Anne Arundel* (AP 76). *Elizabeth C. Stanton* lacked the foremost and aftermost pairs of kingposts. She received a funnel extension in a 1944 refit. *Lyon* and *Anne Arundel* were rigged much like *Florence Nightingale*, although *Anne Arundel* ultimately had cross-pieces to all three forward pairs of kingposts. All of these XAPs participated in the major European invasions, but afterwards only *Anne Arundel* functioned as an attack transport in the Pacific.

42), and *Ancon* (AP 66), and the Maritime Commission's *Exemplar* (AP 67), *Monterey* (AP 68), *Mormacstar* (AP 69), *Mormacsun* (AP 70), *Mormactide* (AP 71), *Santa Clara* (AP 72), and *Santa Lucia* (AP 73). Of this group, only *Monterey* (which was to have been renamed *Alameda*) was not acquired. Ships would be converted on a not-to-delay basis, with the maximum number of landing craft and necessary fuel stowage, in no more than 30 days. The army was to pay for and monitor conversion of *Exemplar*, *Mormacstar*, *Mormacsun*, and *Mormactide*; the navy would convert the others.

Hugh L. Scott and *Tasker H. Bliss* were 535-footers. *Ancon* (AP 66) was a modern Panama Railroad Company ship (not a Maritime Commission design), earmarked for APA conversion as early as September 1941. Never redesignated an APA, she served as an attack transport in North Africa in November 1942, then was converted to a command ship (AGC). Two of *Ancon's* sisters remained in army service. None of the ships converted to combat loaders was redesignated an APA. *Hugh L. Scott*, *Tasker H. Bliss*, and *Leedstown* (ex–*Santa Lucia*, AP 73) were lost before the new designations were put into use.

After the North African invasion, the other ships in the group (AP 67, 69, 70, 71, and 72) were not designated as APAs, although they were still fitted as combat loaders.[1] *Susan B. Anthony* (ex–*Santa Clara*, AP 72) was evaluated in 1943 as insufficiently stable, and her landing craft fittings were removed.[2] The others were assigned to the Naval Transportation Service between operations, but they could also function as attack transports. They were classed as XAP rather than as pure AP, to indicate their special characteristics. In official accounts of invasions

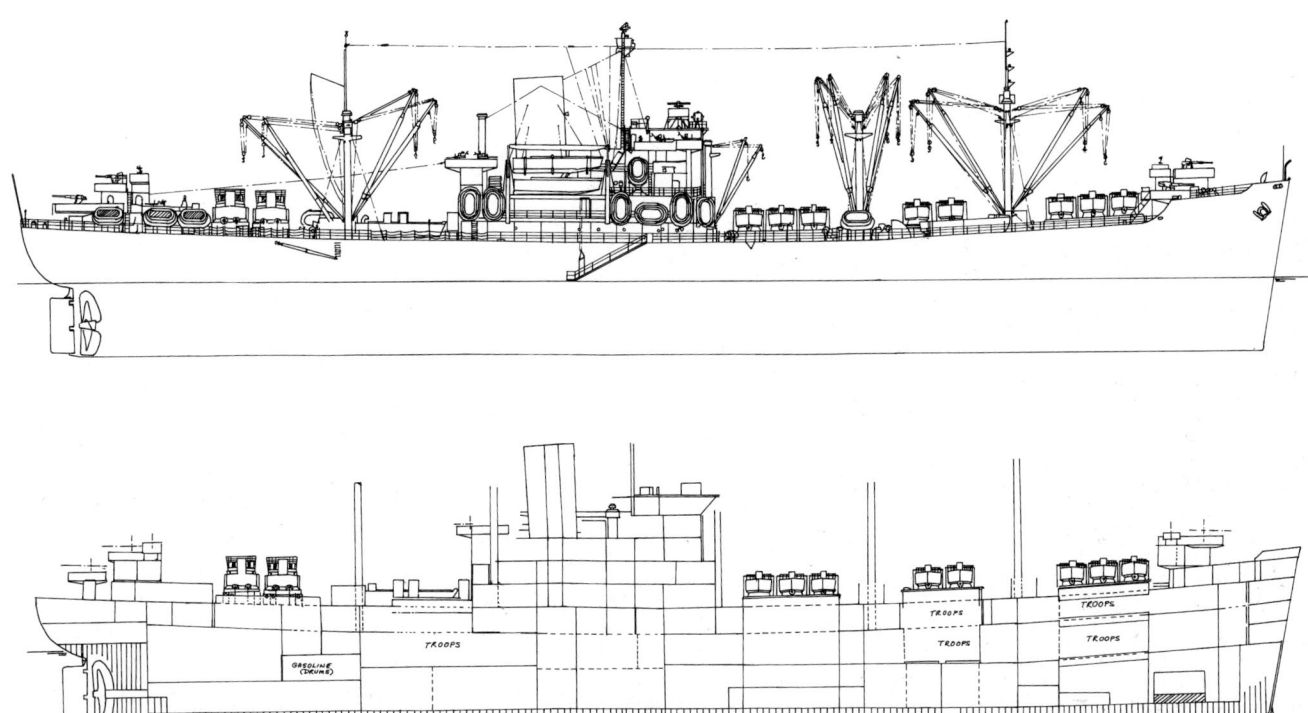

Florence Nightingale (AP 70), August 1944. She was converted into an attack transport specifically for the invasion of North Africa. Called an XAP, she could be used either as a point-to-point transport or in the attack role. Thus she was part of the assault force for North Africa, Sicily, and Italy (Salerno), after which she served as a point-to-point transport between the United States and Europe. She then served in the assault on Southern France. She then reverted to point-to-point service. Compared to a true attack transport, she had only a single set of triple Welin davits and, at least as important, no specialized command spaces such as a CIC or, as one XAP commander protested, a combatant-level radio room and complement. On the other hand, her booms could handle very nearly an AKA load of tank landing craft. She is shown as rigged for North Africa, with a mixture of 36-ft Higgins boats (LCPLs), LCVs (marked "jeep lighters" on the original drawing), and LCM(3)s. In addition to the four or six LCPL in her triple Welin davits, she has two on deck, lengthwise, forward, their bows abeam the foremast and two nested in the LCMs aft. In addition she has eight LCVs athwartships forward, plus two more (lengthwise) aft. Finally, she has two LCMs athwartships aft, with LCPLs nested in them. To handle boats and cargo, she has a single 30-ton boom on the centerline aft (with three 10-ton booms) plus 10-ton booms on the foremast and the big kingposts forward, and 5-ton booms on the small ALFA. Armament comprises pairs of 3-in/50 fore and aft, with twin 40-mm director-controlled guns on the centerline, and twelve single 20-mm (two in the bow, four atop the bridge, four around the kingpost just abaft the amidships superstructure, and two aft).

XAPs were counted as combat loaders. All four were attack transports for the invasion of Sicily in July 1943; *Lyon* (AP 71) and *Elizabeth C. Stanton* (AP 69) were at Salerno in September 1943. Only *Dorothea L. Dix* (AP 67) was at Normandy in June 1944, but all four were at the invasion of Southern France in August 1944.³ All reverted to point-to-point duty, but apparently they were earmarked for attack transport duty in the projected (abortive) invasion of Japan.

On 27 August 1942 Admiral King asked the JCS for another two divisions of lift, to supplement the two-division lift already in existence (or projected) and to make up for expected combat losses (a combat loader had been lost at Guadalcanal). That equated to 24 attack transports and 6 attack cargo ships, to be delivered at the rate of 6 per month. Plans called for a new BuShips preliminary design, to be developed and built by the Maritime Commission. OpNav had already issued characteristics for a new APA on 7 August 1942. However, King's request to the JCS asked that suitable ships be obtained from those built or building for conversion.

Early in September 1942, Admiral King verbally requested two more ships, which were being converted by the army, for completion by 12 September, in the expectation that the navy would take them over: *Mormacyork* (C3) and *Delsantos* (C2). They became *Anne Arundel* (AP 76) and *Thurston* (AP 77), respectively; neither was redesignated an APA. Conversion entailed accommodation for troops, installation of 30-ton booms, the maximum number of landing craft davits, and increased fuel stowage. Conversions were hurried because the North African invasion was scheduled for November. For example, in at least one place on board *Anne Arundel*, wood was substituted for a steel watertight door. Inadequate stability was accepted for Operation Torch, but ships were modified when they returned home. *Anne Arundel* needed 1,000 more tons of ballast, reducing her cargo capacity from 4,000 to 3,000 tons. Although rated as transports (AP), both ships functioned as attack transports for the Sicily, Normandy, and Southern France invasions. Both accompanied *Dorothea L. Dix* (AP 67) to the Pacific in the fall of 1944, and all three ships were included in the official book listing Pacific amphibious ships (as opposed to transports and freighters used to reinforce a beachhead once it had been taken).

BuShips submitted a preliminary APA design early in September, based on the existing *Dixie*-class

Thurston (AP 77) was the last of the XAPs. She is shown during the invasion of Iwo Jima, 20 February 1945. Unlike the others, she was a C2-F, smaller than the C3s usually selected for conversion to attack transports. Three sisters (of seven C2-F built) became attack cargo ships: *Libra* (AKA 12), *Titania* (AKA 13), and *Oberon* (AKA 14); they had heavy pole masts stepped between their second and third pairs of kingposts specifically to lift LCMs. Another sister became the stores issue ship *Pollux* (AKS 4).

(AD 14) destroyer tender hull and machinery. Deck heights and bulkhead locations were changed to suit the new mission. For quick production in yard building civilian ships, the framing was changed from warship-type longitudinal to transverse, and the partly riveted structure was changed to all-welded. The ship's inner bottom was run up her sides to make her more survivable. This ship would be 520 ft (waterline length) × 73 ft × 20 ft 8½ in (depth at side 45 ft), almost as large as the converted World War I ships, displacing 14,200 tons fully loaded, compared to 14,870 tons (465 ft × 69 ft 6 in × 24 ft 11 in, 42 ft 6 in hull depth) for a C3 conversion. Light displacement would be 8,395 vs. 8,374 tons for the C3, the latter including 1,000 tons of ballast. The C3 was more stable, with a higher GM (4.4 ft, with 1,000 tons of ballast, compared to 4.2 ft), a larger righting arm (GZ: 4.7 ft at 48 degrees vs. 4.5 ft at 45 degrees), and a greater range of stability (over 90 vs. 85 degrees). Moreover, the new ship's stability would likely deteriorate during construction, whereas the C3 conversion already existed. On the other hand, the double bottom offered protection against near-misses unmatched by the C3, and the new ship would have 14 main transverse bulkheads rather than 10.

In the new ship, twin-shaft machinery (two units separated by an auxiliary machinery space) would develop 11,000 SHP (12,000 SHP overload) for a speed of 18 kts (10,000-nm endurance), compared to 16.75 kts on 8,500 SHP and 10,500-nm endurance for a converted C3 like *Thomas Stone* (AP 59). The C3 had a better fuel rate, 0.63 lbs/SHP/hr vs. 0.85, thereby needing far less fuel (1,530 vs. 2,423 tons).

The new ship would carry about as many troops as the C3 (91 officers and 1,370 men vs. 98 officers and 1,373 men), but less cargo (140,000 cu ft or 1,472 tons rather than 188,000 cu ft or 2,794 tons of cargo plus 492 tons of ammunition cargo). She also carried less gasoline (22 vs. 32 tons). Part of the disparity was due to the much larger fuel load and greater tonnages of stores to serve her larger crew. She had one 30-ton boom serving four hatches (two 30 × 17.5 ft, two 24 × 17.5 ft); other booms had at least 10-ton capacity. She would have at least four sets of triple Welin davits per side, stowing 36 Higgins boats (LCPR, LCV); at the end of September plans called for 2 LCMs and 31 Higgins boats. Armament was four (later two) single 5-in/38, one quadruple and two twin 40-mm (originally three quadruple 1.1-in), and six single 20-mm. Complement (25 officers and 450 crew) was larger than that of a C3 (24 officers and 328 crew) but OpNav thought it too small for the battery and the boats.

BuShips' conversion section doubted that the 520-footer was worth the effort required to build it in wartime. As of September, it was developing plans for a C2 conversion that could carry 1,400 men at 16 kts, with two-compartment subdivision, and with the standard conversion battery of a single 5-in/38, four 3-in/50, and eight 20-mm. This ship would carry 2 LCMs and 23 Higgins boats. In contrast to the 520-footer, most troop officers would be berthed chief petty officer (CPO) style. A C3 could carry a few more troops, would make about 17 kts, and could carry about 26 Higgins boats. A ship like the 520-footer would make sense only if she were built longer and wider, perhaps with an additional deck level, with deck heights sufficient for troops to be berthed four high, to accommodate two combat teams instead of one. Otherwise it would make better sense to build a larger number of smaller ships. Even then, the proposed subdivision would make cargo handling quite difficult. Placing the 5-in guns on the centerline, as proposed, would also limit boom handling to four holds.

Moreover, given the vast ongoing naval building program, no building slips for 520-footers would be available until 1 August 1944. Because many existing APAs were converted C3s, BuShips suggested converting more, which could be obtained from the Maritime Commission, rather than building a special naval design. This suggestion was included in the report on the new design, but the paragraphs involved did not come from BuShips' designers.

Shipping was very tight. As of September 1942 only 30 fast (15 kts or over) ships to be completed over the next 6 months had not yet been allocated to the army or navy: 14 C2s, 13 C3s, and 3 of a type designed for the Waterman Line. The proposed program would absorb all 30. Although the JCS Joint Military Transportation Committee agreed that the navy needed the ships, it suggested that some be placed in a pool for assignment as needed to the army and the navy as convoy transports. The army feared that ships allocated to the navy would not be available for point-to-point trooping, for example, for the buildup in England. It argued that the navy should not have more than two or three divisions' worth of attack transports. When the civilian War Shipping Administration (WSA) tried to retain control of the ships, however, the army shifted ground: better navy than civilian control. The civilians appealed to President Roosevelt, arguing against creating another pool of dedicated (frozen) shipping that could not be allocated as needed. The Joint Chiefs solved the problem by certifying that the ships would be used continuously for vital war operations. In a 24 September 1942 letter the WSA pronounced itself satisfied by saying that it would

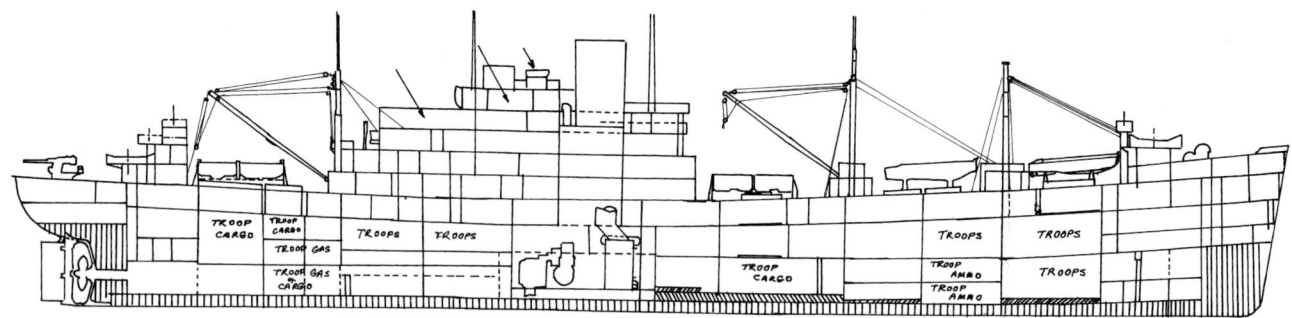

Fremont (APA 44), configured as a relief amphibious flagship, in November 1963. An arrow indicates the flag bridge forward of the signalman's shelter. On the deck below, an arrow indicates the flag plot. Forward of it is the flag operations office. A third arrow indicates the large compartment for a troop radio central. When the booklet containing this drawing was produced, *Fremont* had been converted to UHF radio, and that equipment shared the troop radio central. Forward of this space is the crypto room and, forward of that, the secure teletype room. Below the crypto room is the ship's CIC. Ballast in the forward holds is indicated by diagonal shading. Boats were 4 LCM(6) carried athwartship on deck, fore and aft; 16 LCVP (11 in the triple Welin davits amidships, one nested in each of the four LCMs, and one on deck, lengthwise, alongside two LCPLs, between the foremast and the forward kingposts), and 4 LCPL (two alongside the LCVP forward and two more carried lengthwise forward of that nest). The foremast carried an SPS-40 air search radar, a 30-ton boom to service the two LCMs between it and the superstructure, and two 10-ton booms; the kingpost forward of it, which carried a big discone antenna, carried a 10-ton boom. The mast above the bridge carried the ship's SPS-10 surface search antenna. Abaft it was a new mast carrying UHF antennas, and then the mainmast, carrying a pair of 30-ton booms to service the two LCMs abaft the superstructure. Armament amounted to a 5-in/38 right aft and quadruple 40-mm mounts fore and aft; by this time the ship no longer carried 20-mm guns. The ship accommodated 129 troop officers and 1,209 enlisted troops, the high proportion of officers indicating her flag status. Her crew accommodation amounted to 59 officers, 45 CPOs, and 331 enlisted.

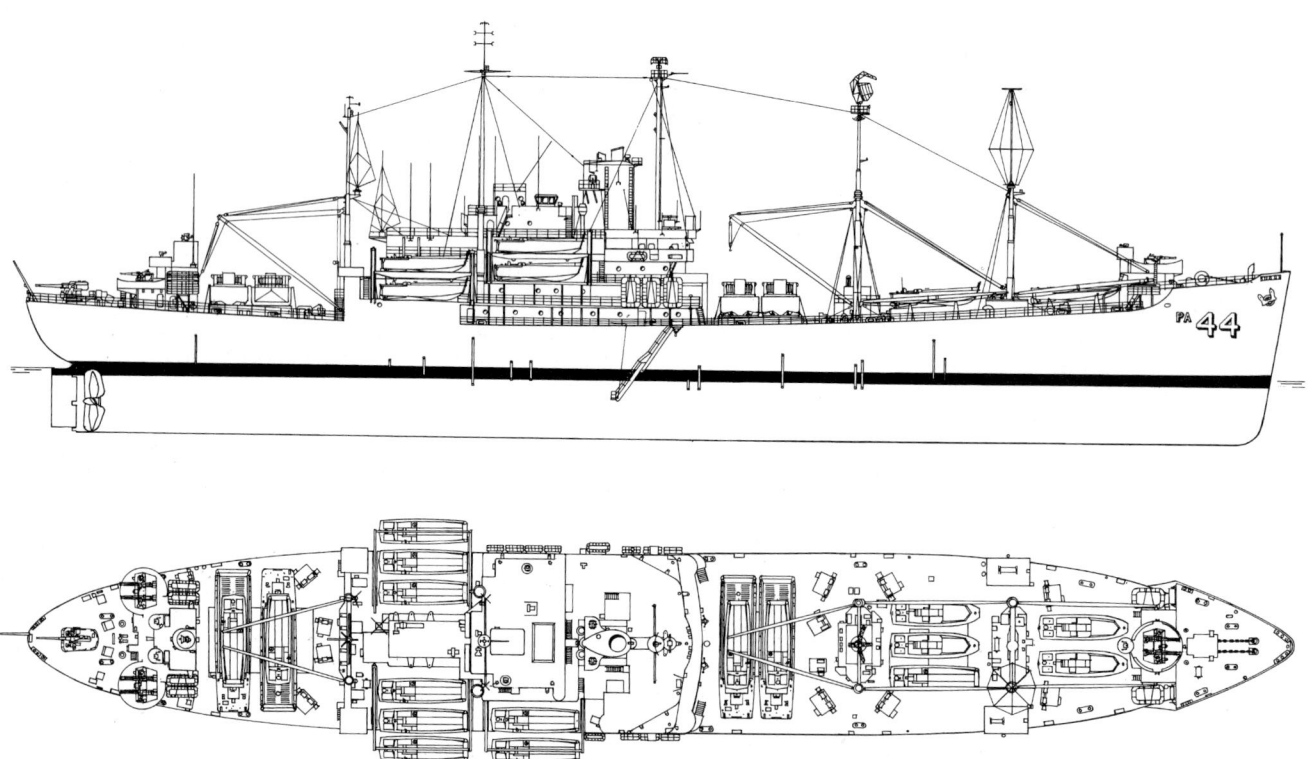

Fremont (APA 44), a relief flagship, in November 1963. Note that her command spaces had been so enlarged that one of her triple Welin davits had to be landed. Special communications equipment included a discage antenna forward. Note, too, the unusual total of four LCPLs on deck forward. She carried four LCM(6) and thirteen LCVP. By this time, all of her 20-mm guns had been landed, but neither she nor other World War II amphibious ships had their 40-mm guns replaced by 3-in/50s (as in destroyers and cruisers). The big air search radar forward is SPS-40. *A. D. Baker*

determine allocation in consultation with the Joint Chiefs from time to time. That having been done, the program for a new series of attack transports began within a month of Admiral King's proposal.

The Maritime Commission was ordered to convert and transfer 24 combat loaders as AP 78–101 (later APA 33–56) and six C2-S-B1s as AK 64–69 (later AKA 15–20), the *Andromeda* class. This program roughly paralleled the conversion of three C2s to AGCs (see Chapter 9).

The transport series began with the *Bayfield*-class C3-S-A2s (AP 78–84, later APA 33–39), laid down with "sea" names beginning with *Sea Bass*. They had three sets of triple Welin davits on each side. The *Custer* (AP 85–93, later APA 40–48) class was distinguished from the *Bayfield* class. Although both were derived from the same Maritime Commission C3-S-A2 design, the *Bayfield*s were built by Western Pipe and the *Custer*s by Ingalls; they therefore differed in detail. Many of these ships were squadron flagships with relief AGC capability (see Chapter 9).

Because C3s were in short supply, three C2-S-B1s were converted into the *Ormsby* class (AP 94–96, which became APA 49–51). Requirements set at the end of 1942 included accommodation for 91 troop officers, 20 NCOs, and 1,452 enlisted men, served by 26 LCVPs (18 in 3 sets of triple-deck Welin davits, 8 on deck) and 2 LCMs. Armament was set at two 5-in/38 (fore and aft), two quadruple 1.1-in machine cannon, and twelve 20-mm. Similarly, three *Sumter* class (AP 97–99, which became APA 52–54) were converted from C2-S-E1s. The series was completed by two *Windsor*-class (AP 100–101, later APA 55–56) ships, converted from C3-S-A3s. Table 6-1 shows details for the various types of large attack transports built during the war.

Barnstable (APA 93), a repeat (non-flagship) *Bayfield*-class attack transport, is shown off San Francisco after a refit, 18 May 1945.

The repeat *Custer*-class attack transport *Hamblen* is shown on 13 June 1945. In effect, the *Custer* class was the Ingalls equivalent to the Western Pipe–built *Bayfield*. Detail differences included a somewhat different bridge structure and a heavier pair of kingposts forward.

Sheridan (APA 51) was an *Ormsby*-class attack transport, among the few converted from C2 hulls. Note the third pair of triple Welin davits forward.

The accompanying *Andromeda*-class AKAs, which were the basis for all later wartime units, carried considerably more boats than their predecessors: eight LCM(3)s and sixteen 36-footers, six of which would ride in Welin davits. Given its experience in Operation Torch, the Atlantic Fleet amphibious force wanted more LCM(3) stowage, but BuShips refused. Troop capacity was the same as in the past, 12 officers, 12 NCOs, and 150 enlisted; capacity was actually 10 officers and 169 troops, plus the crew of 38 officers and 257 enlisted men. The ships had the same 12,000 gal per day distilling plant. Cargo capacity was 350,000 cu ft. Specifications included bulk stowage for 20,000 gals of diesel fuel for the ship's boats and 35,000 gals of gasoline in 5-gal cans. Specified armament was one 5-in/38, four 3-in/50, and twelve 20-mm; by the end of the war twin 40-mm had replaced the 3-in/50s.

During conversion, at an early September 1942 New York Navy Yard conference, G. H. Copenhaven of the yard's design (fittings) section suggested that the big solid masts and flanking kingposts of the *Libra* (AK 53) be replaced by quadrupod masts built of four straight pipes merging at the top. The 30-ton booms could be topped to the center of the structure, the 10-ton booms to its sides. This much-simplified design would be considerably lighter, and it would require no stays, which interfered with antiaircraft arcs. BuShips adopted the idea for all AKAs, beginning with *Andromeda*. Although it planned also to use the quadrupod for APAs, that did not happen. Of earlier ships, it appears that only *Alhena* (AK 26, later AKA 9) was refitted with quadrupods. As completed, the *Andromeda*s had a total of four 35-ton, two 10-ton, and four 5-ton booms.

Given the loss of a combat loader at Guadalcanal, in the fall of 1942 OpNav considered abandoning large transports in favor of smaller ones about the size of the two *Doyen*s under construction (their serious stability problems were as yet unsuspected). On

The *Sumter* class was similar in arrangement to the *Ormsby* but had a higher bridge structure and lacked a shelter deck amidships. Just completed, *Sumter* (APA 52) is shown at Norfolk, 9 September 1943, her davits as yet empty of landing craft. The two empty gun tubs aft were soon filled by a pair of quadruple 1.1-in machine cannon, ultimately replaced by twin 40-mm guns.

Leedstown (APA 56) was a *Windsor*-class attack transport, recognizable by her old-fashioned counter stern. She is shown on 19 August 1943. Circles on the photograph mark alterations, in this case the addition of life rafts.

30 September 1942 it issued new characteristics for a ship that could be completed either as an attack transport or an attack cargo ship: 5,000 tons, 15 kts, with a cruising radius of 8,000 nm. The APA version would accommodate 30 troop officers and 600 men, similar to the still-incomplete *Doyen* class, plus at least 60,000 cu ft of cargo, less than half that of the large transport. She would have 10- and 15-ton booms to serve at least two sets of triple-bank Welin davits, to accommodate at least 15 LCPR and LCV. Maximum loaded draft would be 15 ft. Armament would be two 5-in/38 and six 20-mm. The AKA version would accommodate 15 troop officers and 183 men, with 85,000 cu ft of cargo, and maximum boom capacity of 20 tons. If practical, she would stow 12 LCPR and LCV in triple-bank davits (later the 12 boats were changed to LCVPs, and 2 LCMs were added). Desired speed was soon increased to 18 kts.

According to Cochrane's note on an October 1942 route sheet, "the stability required for the guns, booms, and boats makes the biggest hitch here—two hulls should be alike. AK should have arrangements for rapid discharge of cargo, but I am not sure just what it should be. I suspect that she should have some side doors to facilitate loading boats alongside with chutes. Dredger-type elevators for handling case goods in holds in way of side doors. 360 ft LWL is probably the minimum. Do not pinch too tightly." By the end of October 1942, the main remaining question was machinery. Possibilities included the new geared turbine destroyer escort plant (6,000 HP) and paired submarine diesels, two per shaft (5,000 SHP would give 17 kts).

BuShips submitted a preliminary APA design on 1 November 1942. This ship would be 380 (LWL) × 53 × 15 ft (29 ft depth) and would displace 5,100 tons fully loaded. Like the earlier big transport, she had two separate machinery plants (5,000 SHP, for 17 kts), and she was well subdivided for survivability, having 12 main transverse bulkheads. The main change in the AKA, requested by OpNav, was separation of combat gasoline stowage from the forward

Table 6-1. War-Built Large Attack Transports

	Arthur Middleton (APA 25)	Thomas Jefferson (APA 40)	Monrovia (APA 31)	Bayfield (APA 33)	Custer (APA 40)
Hull type	C3-P	C3-AP&C	C3-Delta	C3-S-A2	C3-S-A2
Length (ft-in)					
WL	465-0	465-0	465-0	465-0	465-0
OA	489-0	492-0	491-0	491-7	492-0
Beam (ft-in)	69-6	69-6	65-8	69-7	69-5
Draft, loaded, fore/aft (ft-in)	24-0/27-3	18-2/27-6	23-6/25-6	18-9/25-5	27-9/27-3
Displacement (tons)					
Light	11,040	10,192[a]	8,646	8,920	7,845
Full load	15,490	14,370	13,590	12,900	11,700[b]
Boilers	2 FW	2 B&W	2 B&W	2 CE	2 FW
Conditions (psi/°F)	465/765	450/750	450/750	465/765	465/765
Power (SHP/shafts)	8,500/1	8,500/1	7,800/1	8,500/1	8,500/1
Speed (kt)	18.9	17.2	17.0	18.0	18.8
Generators (qty-kW)					
Main	3-300	4-300	3-350	3-250	3-200
Auxiliary	1-60	1-75	1-75	NA	1-36.8
Fuel oil (gal)	442,974	454,577	557,300	356,420	356,300
Landing craft fuel (gal)	47,000	14,500	23,984	31,825	33,500
Radius (nm/kt)	11,000/15	12,000/15	15,500/15	10,450/12	8,900/15
Complement (officers/CPOs/enlisted)	27/23/474	27/23/486	27/25/517	45/25/517	28/23/482
Capacity					
Troops (officers/enlisted)	78/1,162	69/1,197	100/1,137	82/1,420	90/1,785
Tons	2,700	NA	2,700	4,700	4,700
Sq ft	3,950	5,038	2,628	10,007	NA
Cu ft	143,352	176,288	118,238	180,536	127,846
Water, fresh (gal)	107,300	158,207	91,334	98,200	98,679
Distilling rate (gal per day)	40,000	40,000	40,000	38,000	38,000
Booms (qty-capacity)[e]					
30 ton	1-50	4	1	2	2
10 ton	8	6	10	6	6
5 ton	2	2	—	—	2
Boats					
LCM(6)	4	4	4	4	4
LCM(3)	—	—	—	—	—
LCVP	18	24	20	18	21
LCPL	3	1	3	3	2
LCPR	2	1	2	2	1
Armament					
5-in/38	—	—	1	2	2
3-in/50	4	4	4	—	—
40-mm	2 × II[f]	2 × II	1 × I, 1 × II	2 × II	4 × II
20-mm	10	18	11 × II	18	18

NOTES: Data are as of July 1945, except as otherwise indicated. However, the 1946 Naval Vessel Register gives light displacements for existing and discarded ships, and those figures have been used here. Dash indicates that specification is irrelevant; NA indicates relevancy but data not available.

[a]As an indication of loads, the somewhat similar *Thomas Stone* (APA 29) displaced 8,374 tons light, including 1,000 tons of ballast; light draft was 15-1.5. Fully loaded she displaced 14,870 tons, including 159 tons for troops (98/1,373) and their effects, 2,794 tons of troop cargo, and 492 tons of ammunition cargo. The ship's naval ammunition amounted to 163 tons, for a single 5-in gun plus four 3-in/50 and eight 20-mm. Cargo capacity was 188,000 cu ft.

[b]Trial displacement; 1945 documents do not give light and full load figures, so for ships discarded soon after the war these data often are not available.

[c]Full load tonnage given for *Ormsby* is for similar C2-S-B1 hulls (e.g., AKA 15–20); Ormsby's trial displacement was 13,910 tons.

[d]The full load displacement (13,577 tons) for sister ship *Burleigh* (APA 95) included 1,304 tons of concrete ballast; she carried 2,000 tons of cargo.

[e]Only quantity is listed unless capacity is different. For example, *Arthur Middleton* had one 50-ton boom instead of any 30-ton booms.

[f]Multiple gun mounts are indicated by Roman numbers; therefore 2 × II indicates two twin mounts.

Ormsby (APA 49)	Sumter (APA 52)	Windsor (APA 55)	Frederick Funston (APA 89)	Baxter (APA 94)	Haskell (APA 117)
C2-S-B1	C2-S-E1	C3-S-A3	C3-S1-A3	C2-S-E1	VC2-S-AP5
435-0	445-0	450-0	465-0	445-0	436-0
459-2	468-8	473-1	492-3	468-7	455-0
69-0	63-0	66-5	69-7	63-0	62-0
22-0/26-0	22-6/24-0	24-0/26-0	22-1/27-0	22-6/24-0	22-9
7,300	8,355	8,276	10,967	8,355	6,873
13,893c	13,893	13,500	14,700	13,910d	10,680
2 FW	2 B&W	2 B&W	2 FW	2 B&W	2 B&W
450/750	450/750	450/742	520/765	450/750	465/750
6,000/1	6,000/1	8,000/1	8,500/1	6,000/1	8,500/1
17.0	14.7	18.5	16.5	15.0	17.0
3-250	2-120	3-120	4-300	3-300	3-300
1-15	NA	NA	1-75	NA	NA
294,233	363,804	517,845	525,029	360,000	327,755
29,274	29,316	29,793	13,844	29,651	40,000
12,000/15.5	13,507/12	16,000/15	13,000/15	12,000/15	7,200/15
42/23/446	28/23/451	28/23/454	28/23/458	28/23/481	28/23/451
91/1,524	80/1,362	82/1,432	138/1,455	94/1,636	86/1,510
2,700	1,300	1,600	1,500	1,400	1,800
NA	NA	NA	NA	NA	NA
147,585	146,451	152,691	171,702	131,959	126,848
90,754	129,820	103,500	176,000	167,000	137,790
34,000	35,000	40,000	80,000	20,000	40,000
2	2	3	1	2	1-35
6	6	5	10	8	8
2	2-3.5	2-3	—	10	4
2	2	4	4	—	2
1	3	—	—	3	—
19	15	16	16	19	22
1	1	1	1	1	1
1	1	1	1	1	1
2	2	2	1	2	1
—	—	—	2	—	—
4 × II	4 × II	2 × II	2 × II	4 × II	1 × IV, 4 × II
12 × II	10 × II	18	16	10 × II	10

magazine and hoist. The AKA design submitted in October showed two main cargo holds forward and one aft, with four 10-ton booms forward and two 20-ton booms aft; she would carry no LCMs, only 12 LCP(R) or LCV in davits. Armament was two 5-in/38 and eight 20-mm. BuShips described her as resembling a three-island ship except with a small deckhouse aft instead of a poop.

The problem was machinery. BuShips favored a twin-screw diesel-electric powerplant (2,500 SHP per shaft), using diesels like those in LSTs and PCEs (but with 16 rather than 12 cylinders), for higher speed. The motors were similar to those used in fleet tugs, and the generators to those used in destroyer escorts. Unfortunately, production of diesels, motors, and generators was fully committed. APAs could be built only at a cost to those other programs. A 3,200 IHP triple-expansion plant (offering barely 15 kts) would be larger and it would it would consume more fuel. In November, OpNav recom-

Shelby (APA 105) was a repeat *Windsor*-class attack transport. Note that she had no set of triple Welin davits forward. She was photographed off Sparrows Point in Chesapeake Bay, 20 January 1945.

USS *Matthews* (AKA 96) displays the quadrupod mast developed by the New York Navy Yard for attack cargo ships. This photograph was taken at Mare Island, 5 March 1952. The boats are LCVPs nested inside LCMs, with an LCPL (with canopy) visible forward of the paired kingposts.

Arneb (AKA 56) was a typical repeat *Andromeda*-class attack cargo ship. She was refitted in 1949 for Arctic operation, which probably explains the gunhouse fitted to her 5-in/38 aft. This photograph was taken by the Philadelphia Naval Shipyard, 22 April 1949.

mended diesel-electric power. Later BuShips would suggest that the cargo ships (but not the attack transports) could get single-screw powerplants, which might be easier to supply. Troopships, however, needed the extra security of twin screws, even though C3s had single screws.

A September 1942 BuShips survey of all private yards building for the navy plus one that wanted navy contracts showed that fifty 380-ft transports could probably be completed before 1 July 1944 (end of FY 44), if machinery was available.

Meanwhile, amphibious needs expanded. Resources were so tight that ships had been withdrawn from the South Pacific to mount the North African invasion. However, Operation Torch opened a new amphibious theater, in the Mediterranean, where operations might have to proceed in parallel with those in the South Pacific, as well as in a new area, the Central Pacific. At the end of 1942 just over two divisions of lift were available (28 attack transports and 14 attack cargo ships), but another 25 transports and 6 cargo ships were under conversion.

The two small transports ordered in 1940 had not yet been completed.

Admiral King told the JCS in a 28 December 1942 letter that he wanted the special small transports BuShips was designing. The large ships in service took too long to unload and too many troops and too much material were lost when one was sunk; five (plus a convoy transport) had been lost during the recent North African operation. The big ships drew too much water. Because they had been designed as civilian ships, they were too vulnerable and they were not reliable enough. Each small attack transport would carry about half a reinforced battalion (half the capacity of a conventional large one), and each small cargo ship would have about a quarter of the capacity of a conventional one.

King estimated that in 1943–44 the navy would have to replace 23 large attack transports and 8 large attack cargo ships. He recommended construction of 32 each of the two new smaller types. The JCS approved the program on 5 January, with caveats to reassure the army that it would receive enough transport tonnage. The JCS asked the Maritime Commission to build the 64 ships or, if that was not practicable, 27 transports and 27 cargo ships of the same basic type as the *Doyen*. It appeared that destroyer escort geared turbine powerplants would become available beginning in the last quarter of 1943 at the rate of about two per month.

Initially the Maritime Commission preferred the *Doyen* design: plans already existed, and its powerplant was in production for C1-type cargo ships. The navy wanted a new design because the *Doyen* draft was excessive (18 ft vs. the desired 15 ft). *Doyen*-type geared turbine machinery might be worked into a new design; the Maritime Commission also suggested Nordberg diesels, which were used in some C1s. BuShips now preferred steam, because diesels required technical expertise that might not be available at forward bases. It suggested a new (as yet untried) geared turbine plant planned for late-production destroyer escorts. However, any surplus might disappear if the escort program expanded, which still seemed possible. The Maritime Commission preferred its own machinery production lines. In mid-January 1943 it agreed to consider a simplified shallower-draft *Doyen*, or installation of its machinery in the hull of the BuShips design.

The commission's studies led it to develop a completely new design, mainly because the only suitable powerplant it could readily obtain was the twin-screw 6,000 SHP turbo-electric, which was too large for the BuShips design. The powerplant lengthened the ship to 400 ft and incidentally relieved some overcrowding in troop spaces. Although draft increased to 15.5 ft, by the time the ship arrived at her target that would have decreased to the desired 15 ft. Cominch found these changes acceptable. The new ship was described as intermediate between the BuShips 380-ft design and the *Doyen*. The Maritime Commission rejected the navy's proposed hull form in favor of one that combined features of a small tanker then being built for the navy (the T1-MT-E2) and the *Doyen* hull. The commission considered the tanker's bow better for shallow draft, especially near a shoreline. The stern would be similar to the *Doyen*'s stern, which the commission argued was superior for propulsion. It also argued that, coupled with the V-shaped bow, it gave maximum stability on the limited displacement. The *Doyen*s, however, encountered problems a few months later. The commission complained that in every other design navy changes had raised the vertical center of gravity 1–2½ feet during construction; this time the commission would try to provide so much stability at the start that such interference would not ruin the ship. Hence she might seem excessively stiff at the early design stage. As of February 1943, the commission hoped to deliver the first ship early in 1944. On 17 February the Maritime Commission formally agreed to build the ships, and on 25 February 1943 the JCS authorized construction to proceed. Major features were developed within the Maritime Commission, with detail design conducted by the design agent, George G. Sharp, Inc.

As of February 1943 the ship was 400 (LWL) × 58 × 15.5 ft, displacing 6,000 tons fully loaded. Sustained speed was 16.5 kts. Capacity had grown to 46 troop officers and 800 men, with 60,000 cu ft of cargo and 15 landing craft (12 in davits, 3 on deck; handled by 10-ton booms). The parallel AKA had 90,000 cu ft capacity (5 troop officers, 200 enlisted men, 12 landing craft). As specified, armament was two 5-in/38 and eight 20-mm guns. A BuShips–Cominch–Maritime Commission conference on 25 February 1943 decided to limit the equipment to be carried, to keep the small APA from approaching the size of large APAs and losing the advantages of small size and quick construction. Therefore, the APA would carry no LCMs at all, the AKA would carry two (plus 12 LCVPs). Thus only the AKA would need a 30-ton boom to handle LCMs. Both the AKA and APA carried four sets of triple Welin davits; the APA carried 3 LCVPs on deck, in addition to the 12 in davits). More detailed AKA characteristics in March 1943 added capacity for 17,500 gals of gasoline in 5-gal cans (for troops), and 10,000 cu ft of troop ammunition. Armament was changed to one 5-in/38, four twin 40-mm, and ten 20-mm.

Visually, the AKAs could be distinguished by their broken decks (APAs had a poop, forming a continuous upper deck) and by their quadrupod mainmasts, which carried two 5-ton booms and a single 30-ton boom (originally they were to have had two 10-ton, two 20-ton, and two 30-ton booms). The pole foremast carried two 5-ton booms. Unlike larger AKAs, the smaller AKAs had only three holds, and bale cubic capacity (110,570 cu ft) was considerably less than that of a C3 or Victory APA (150,000 cu ft). Cubic capacity of the APA version was 89,480. This version had two holds, fore and aft, each served by a pair of booms (one 5-ton and one 15-ton); the ammunition hold forward was served by two 2-ton booms. Maximum safe cargo capacity, in tons, was 980 for the AKA and 650 for the APA, compared to 2,910 for the later Victory APA. The Victory ship carried 24 LCVPs and 2 LCMs, nearly twice as many boats as the AKA. Her ship's complement was also about twice that of the small APA, accommodating 57 officers, 30 CPOs, and 447 enlisted vs. 20 officers, 21 CPOs, and 240 enlisted.

The new small transport *Butte* (APA 68) is shown off San Pedro, 3 December 1944. The SG radar on the foremast is for surface search; on the mainmast is an SC-2 air search set. Because both masts supported booms, she needed the odd mast fastened to her forefunnel for signal flags. The two gun tubs right forward and the two aft house twin director-controlled 40-mm guns; the weapons atop the bridge and at the after end of the superstructure are 20-mm guns. *Butte* served as a target ship for the Bikini nuclear test (Operation Crossroads) and was retained afterwards for structural and radiation studies; she was scuttled off Kwajalein on 12 May 1948. A total of 20 ships of this class participated in the Bikini tests, of which *Gilliam* and *Carlisle* were sunk in the initial ("Able") test and *Brule* and *Fallon* were badly damaged and nearly capsized in the second ("Baker") test.

Polana (AKA 35) exemplifies the AKA version of the S4 design. She is shown off Elizabeth City, North Carolina, 6 March 1945. Laid up in the Maritime Commission (rather than navy) reserve in June 1946, she was scrapped in September 1966. The most visible difference from the APA version was the heavy quadrupod mainmast, but note also the cut-down quarterdeck, the propeller guard, and the pipe aft, from gasoline stowage (for vehicles) in the stern. Some ships of this class had extensive postwar careers: *Devosa* (AKA 27) served during 1946–48 as the U.S. Merchant Marine Academy training ship *Kings Pointer; Hydrus* (AKA 28) served during 1946–58 as New York State maritime training ship *Empire State II; Melena* (AKA 32) was the California state training ship *Golden Bear* during 1946–70; *Pamina* (AKA 34) became the survey ship *Tanner* (AGS 15); *Renate* (AKA 36) became the survey ship *Maury* (AGS 16); *Selinur* (AKA 41) was the Pennsylvania state training ship *Keystone State* during 1946–47; *Sirona* (AKA 43) was Pennsylvania Nautical Academy's training ship *Yankee States* during 1946–47; *Turandot* (AKA 47), after being handed over to the Maritime Commission, was reacquired in 1954 and converted into the cable ship *Aeolus* (ARC 3); *Vanadis* (AKA 49) similarly became the cable ship *Thor* (ARC 4); and *Xenia* (AKA 51) and *Zenobia* (AKA 52) were sold to Chile in 1946 and renamed *Presidente Errazuriz* and *Presidente Pinto*, respectively.

The APAs were built by Consolidated Steel (APA 57–88, *Gilliam* class) and the AKAs by Walsh-Kaiser (AKA 21–52, *Artemis* class). Because it was a new design, progress was slow. Some ships were not delivered until the end of the war. Table 6-2 shows details for the wartime small attack transports and Table 6-3 shows details for wartime attack cargo ships.

Apparently not well liked, the APAs were soon transferred to Maritime Commission custody after the war. Eighteen were used as targets for the 1946 Bikini atomic tests. *Burleson* (APA 67), which had transported test animals back from Bikini, was the only unit retained in navy reserve. She was reclassified IX 67 in 1956. She was a static training hulk at Little Creek, used for disembarkation training. The navy retained four AKAs, which were converted to surveying ships (AGS) and to cable layers (ARC). When the Ship Characteristics Board considered converting a cargo ship as a support ship for radar pickets in the Arctic in October 1947, a C2 was chosen over this type on the grounds that the Kaiser type lacked the other ship's engine reliability, seakeeping ability, and cargo capacity.

Table 6-2. Wartime Small Attack Transports

	Doyen (APA 1)	Catron[a] (APA 71)	Catskill (LSV 1)	LSM
Hull type	P1-S2-L2	S4-SE2-BD1	—	—
Length (ft-in)				
WL	405-0	400-0	440-0	196-6
OA	414-4	426-0	453-10	203-6
Beam (ft-in)	56-0	58-0	60-1	34-6
Draft, loaded (ft-in)[b]	18-6	15-6	19-6/19-10	6-4.5/8-3.5
Displacement (tons)				
Light	4,351	4,247	5,876[c]	520
Full load	6,510	7,081	7,600	1,095[d]
Boilers	2 B&W	2 B&W	4 CE	—[e]
Conditions (psi/°F)	450/765	450/750	400/700	—
Power (SHP/shafts)	8,000/2	6,000/2	11,000/2	2,880/2
Speed (kt)	19.0	17.5	20.5	13.2 (928 tons)
Generators (qty-kW)				
Main	2-500	2-250, 1-150, 2-100	4-500	2-100
Auxiliary	1-75	NA	NA	2-20
Fuel oil (gal)	306,526	406,310	494,659	51,450
Landing craft fuel (gal)	15,434	15,340	32,999	—
Radius (nm/kt)	9,500/15	5,256/15	7,057/15	4,900/12 (928 tons)
Complement[f]	27/23/396	21/19/316	33/23/434	4/54
Capacity				
Troops (officers/enlisted)	65/780	29/957	76/749	2/46
Tons	400	600	NA	165 max
Sq ft	2,387	NA	NA	NA
Cu ft	37,522	117,779	126,765	NA
Water, fresh (gal)	62,600	45,350	71,142	NA
Distilling rate (gal per day)	24,000	24,000	40,000	NA
Booms (qty-capacity)[g]				
30 ton	—	—	—	—
10 ton	2-15, 2-10	2-15	—	—
5 ton	2	1-5, 4 -2	4-4[h]	—
Boats				
LCM(6)	—	—	—	—
LCM(3)	—	—	—	—
LCVP	14	13	14	—
LCPL	1	1	1	—
LCPR	1	1	1	—
Armament				
5-in/38	—	1	2	—
3-in/50	4	—	—	—
40-mm	2 × II[i]	4 × II	4 × II	1 × II
20-mm	10	10	20	4

NOTES: Data are as of July 1945. Dash indicates that specification is irrelevant; NA indicates relevancy but data not available.

[a] Data on Catron, a Gilliam (APA 57)–class ship, are taken from the ship's loading characteristics pamphlet, produced on board and revised to 4 September 1945. Draft (light) fore/aft was 7-6/12-6; maximum draft was 16-0. Designed displacement was 4,100 tons light, 6,800 tons loaded. Designed speed was 18 kts.

[b] Draft mean if only one figure given; draft fore/aft if two figures given.

[c] Displacements are 1945 figures. As inclined 29 June 1944, LSV 1 displaced 5,106.3 tons light, 8,107.5 in Condition V, and 9,030.3 in Condition VI (overload).

[d] LSM displaced 743 tons in landing condition; draft fore/aft was 3-5.75/7-0.5. Data given are for the loaded condition.

[e] Diesel boilers. Two direct-drive Fairbanks-Morse 38D81/8 for LSM 1–145, 201–232, 253–267, 310–319, 354–388, 414–428, 459–478, and 540–552; General Motors 176-278A for the others.

[f] Complement is given as officers/CPOs/enlisted when three figures are listed and officers/enlisted for two figures; when only one figure is given, breakdown is unknown.

[g] Only quantity is listed unless capacity is different. For example, Doyen had two 15-ton booms as well as two 10-ton booms.

[h] LSV 1–4 each had four 4-ton booms, but LSV 5 and 6 had one 15-ton and two 4-ton booms.

[i] Multiple gun mounts are indicated by Roman numbers; therefore 2 × II indicates two twin mounts.

Table 6-3. Wartime Attack Cargo Ships

	Libra (AKA 12)	*Andromeda* (AKA 15)	*Algol* (AKA 54)	*Tolland* (AKA 64)	*Artemis* (AKA 21)
Hull type	C-2F	C-2SB1	C2-SB1	C2-S-AJ3	S4-SE2BD1
Length (ft-in)					
WL	435-0	435-0	435-0	435-0	400-0
OA	459-3	459-2	459-3	459-2	426-0
Beam (ft-in)	63-0	63-0	63-0	63-0	58-0
Draft, full load (ft-in)[a]	24-2/27-4	25-9/26-6	25-9	24-6/24-3	13-7/18-2
Displacement (tons)					
Light	6,944	6,556	6,470	6,318	4,087
Full load	11,600	13,910[b]	11,130	13,050	6,800
Boilers	2 FW	2 CE	2 FW	2 B&W	2 Wickers
Conditions (psi/°F)	465/765	465/765	450/750	450/750	450/750
Power (SHP/shafts)	6,000/1	6,000/1	6,000/1	6,000/1	6,000/2
Speed (kt)	16.4	16.0	16.2	16.0	17.8
Generators (qty-kW)					
Main	2-250	2-300	3-300	3-300	3-150
Auxiliary	—	1-115	NA	NA	2-10
Fuel oil (gal)	345,000	449,301	431,562	438,000	370,000
Landing craft fuel (gal)	23,500	28,179	29,330	32,000	16,340
Radius (nm/kt)	18,180/12	16,046/12	NA	NA	9,500/12
Complement (officers/CPOs/enlisted)	24/18/302	24/18/326	24/18/326	24/18/326	20/19/282
Capacity					
Troops (officers/enlisted)	12/178	12/66	8/66	8/54	5/247
Tons	4,605[c]	NA	4,900	5,275	980
Sq ft	23,212	24,022	NA	NA	9,596
Cu ft	383,047	248,346	302,209	295,410	104,948
Water, fresh (gal)	89,236	93,416	89,510	96,135	45,000
Distilling rate (gal per day)	12,000	12,000	15,600	12,000	20,000
Booms (qty-capacity)[d]					
30 ton	1-40, 4-30	4	4	4-35	2
10 ton	6	2	2	6	—
5 ton	2	4-4	4-3	6	8
Boats					
LCM(6)	6	6	6	6	—
LCM(3)	3	2	2	2	2
LCVP	10	13	13	13	13
LCPL	1	1	1	1	1
Armament					
5-in/38	1	1	1	1	1
40-mm	4 × II[e]	4 × II	4 × II	4 × II	4 × II
20-mm	18	18	18	16	10
0.50 caliber MG	10	—	—	—	—

NOTES: Dash indicates that specification is irrelevant; NA indicates relevancy but data not available.

[a] Draft mean if only one figure given; draft fore/aft if two figures given.

[b] Trials displacement.

[c] Note that a postwar table of AKA capacities gave 4,100 tons for a typical C2 conversion. A sheet comparing wartime ships (C2 conversions) with the Mariner conversion (*Tulare*, AKA 112) gave capacity as 3,551 tons or 390,946 cu ft, on a full load displacement of 10,747 tons.

[d] Only quantity is listed unless capacity is different. For example, *Libra* had one 40-ton boom as well as four 30-ton booms.

[e] Multiple gun mounts are indicated by Roman numbers; therefore 4 × II indicates four twin mounts.

In February 1943 conversion of another 20 AKA of conventional design was approved. For the moment nothing was done, because no more appropriate hulls were available.

When the army released the transports *Frederick Funston* and *James O'Hara* to the navy in April 1943, they were ordered converted to attack transports (APA 89 and 90, respectively).[4] They carried 30 LCVPs (18 in three sets of triple-bank Welin davits) and 2 LCMs. Armament was set at three 3-in/50, two quadruple 1.1-in machine cannon, and twelve 20-mm. Both ships were returned to the army in 1946 to serve again as point-to-point transports. Transferred to MSTS in 1950, they became T-AP 178 and 179.

The navy did obtain a few transports, AP 102–105, on the ground that only navy-manned ships could

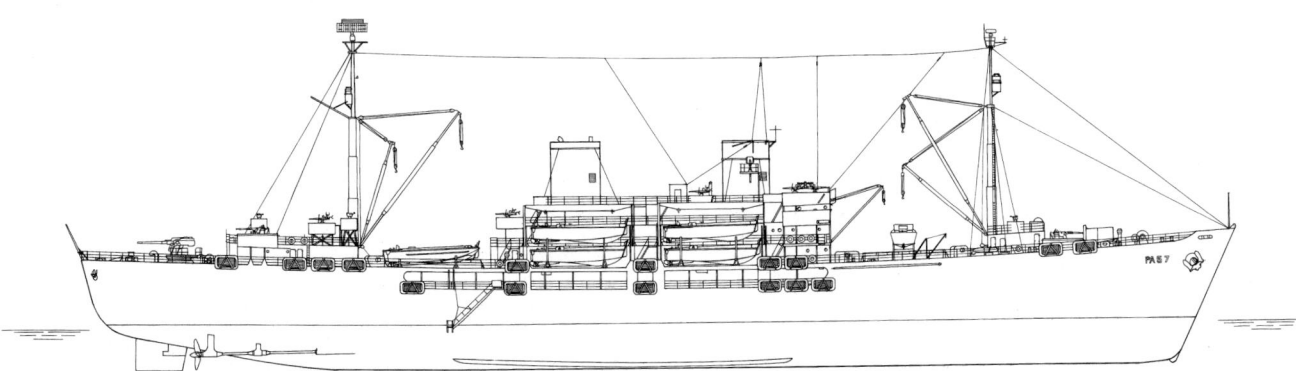

Gilliam (APA 57) as built, about July 1944. The S4 attack transports (S4-SE2-BD1 type) could be distinguished from the corresponding attack cargo ships (S4-SE2-BE1) by their flush upper decks (high poops) and by the absence of a lattice (quadropod) mast aft, for boat handling. *A. D. Baker*

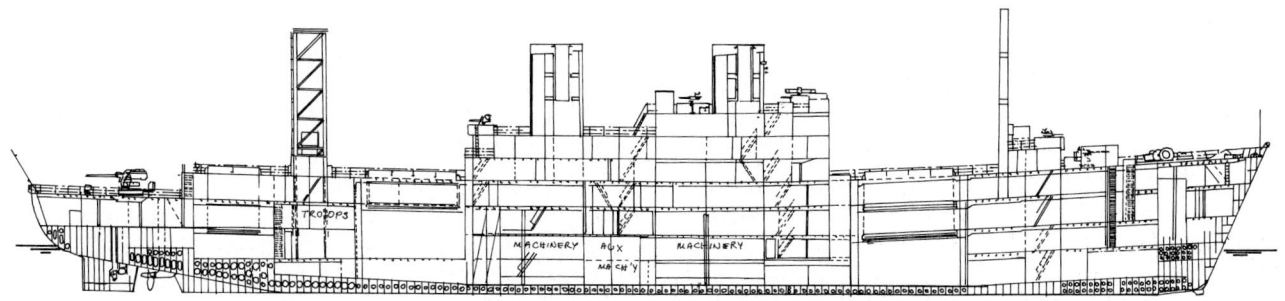

An *Artemis* (AKA 21)–class attack transport (inboard profile), from a Maritime Commission drawing dated 14 January 1944.

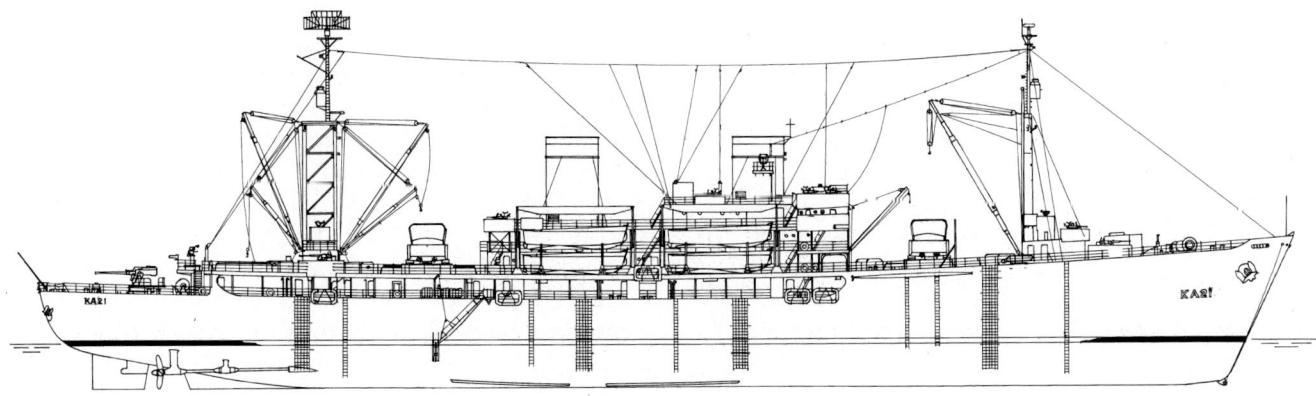

Artemis (AKA 21) in September 1944, showing debarkation nets strung down her sides. Her bow shows paravane cable. Note the two LCMs stowed athwartships, and the LCVPs on triple Welin davits; the balance between LCVPs and LCMs was unusual for an attack cargo ship. *A. D. Baker*

transport troops directly into combat zones. Despite their considerable limitations, they were used as attack transports in the Pacific. These limitations highlight the extent to which an APA was more than a transport. *La Salle* (AP 102) was acquired after the VCNO pointed out, in February 1943, that it was vital to deliver construction units (Seabees) directly to combat zones immediately after beachheads had been secured. Merchant crews were not allowed in such zones, so valuable time was lost transshipping the units. An accompanying C1 cargo ship was acquired at the same time as USS *Auriga* (AK 98). Conversion was hurried, leaving insufficient quarters for the navy crew, which was far larger than the previous civilian crew, and insufficient distilling and electric capacity. Thus in July 1943 the ship was authorized to receive a 300 kW turbo-generator and a 20,000 gal per day distilling plant. The worst problem, however, was poor survivability. She could not survive the flooding of the main holds forward (Nos. 1 and 2) or aft (Nos. 4 and 5). The proposed solution was to fit bulkheads in Nos. 1 and 5 holds. The ship's half-sister, the XAP *Thurston* (AP 77), was even worse, but she could not be spared for a major modification.

An AP was not intended to go into a combat zone, so she had a low priority for installation of a combat information center (CIC), and she lacked a combatant ship's communications team. When such a ship entered a combat zone, there were real consequences. Anchored off Guadalcanal, *La Salle* missed an air raid warning message and did not get under way in time. Her commander complained that without a combatant-size communications team, her signalmen and signal strikers were being exhausted by continuous dawn-to-dusk watches.

By October 1943 the need for assault transports was far more urgent than the need for a specialized Seabee carrier. OpNav authorized temporary stowage (nominally for transportation, not use) of 100 tons of landing craft on *La Salle* and on ten other ships, which were later designated AP 166–175. *La Salle* was expected to carry about 2,000 tons of cargo. It was far too late to install new Welin davits, so all landing craft had to be stowed on deck: two LCM(3) athwartships across No. 3 hatch, and six LCVP (one lying fore and aft on each side outboard of No. 2 hatch with a third athwartships across it, and one lying fore and aft on each side outboard of No. 5 hatch, with another diagonally across it). The other ten ships were to be similarly equipped. In combat the ship carried as many as seven LCVPs plus her two LCMs, launching all using her booms. By late 1944 *La Salle* had fought in five assaults; in her commander's words, she was "operating as an APA with AP facilities."

Late in July 1943 the Auxiliary Vessels Board recommended immediate acquisition of three large C3 liners: *President Polk* and *President Monroe*, sisters to *President Jackson* (APA 18), and *Delbrasil* (renamed *George F. Elliott*), a sister to *Crescent City* (APA 21). They became AP 103–105, respectively.[5] Again, the justification was the need to deliver troops directly to combat zones, which came close to making the ships attack transports. Again, as in the case of *La Salle*, because the ships were never designated as APAs, they lacked key features. The three were already serving as troopships, but navy manning doubled the size of their crews. Yet conversion time was very short: only 21 days for *President Polk*. The BuShips cover sheet for correspondence about deficiencies due to the short conversion time read "another case of an incomplete conversion." The ship received an evaporator plant (for the embarked troops), ballast, and some radar. However, she was never fitted with a combat information center. By this time it was clear that the ship might have to serve as a part-time APA, so in October BuShips ordered her equipped to handle six LCVPs. At Lingayen Gulf in January 1945, however, she carried four LCM(3) (picked up from the army at Finchafen in November 1944) and ten LCVP, all of which were stowed on deck: three LCVP each on Nos. 1 and 5 hatches, and one LCVP nested in each LCM(3) on Nos. 2 and 4 hatches. Among her deficiencies, she had no beachmaster to help her boats unload. Conversions were not systematic, so *President Monroe* had skids for five LCVP on what had been her promenade deck. Unfortunately they faced the wrong way. She carried two LCM and eight LCVP, and formed part of a transport division (with three APAs, including two of her half-sisters, APA 18 and 19, and an AKA). The third liner, *George F. Elliott*, received even less of a conversion. At Lingayen Gulf she had only one LCM(3) and one LCVP, so she had to rely mainly on other ships' boats to unload. They tended to vanish after one or two trips. She had been operating as an APA since Saipan (June 1944).

The new minelayers and fast netlayers were also converted into attack transports, albeit not designated APA. The four fast (20 kts) AN 1 class netlayers (453 × 60 × 21 ft, 5,177 tons), based on the minelayer *Terror*, were nearing completion in the spring of 1943, but they were not needed. They had been conceived to carry and lay a heavy antisubmarine indicator net (100 tons per mile), which would fire a flare when fouled by a submarine. Even such large ships could carry only 6 miles of such nets; the much smaller HMS *Protector*, on which they were based, carried 3 miles. As it turned out, nets were drastically lighter than expected. One weighing 37 tons per mile was tested in 1941, and in March 1943 a 4 ton per mile net was successfully tested. An LCVP could lay a mile of such net in 30 minutes.

As had been evident in 1940, the ships' long net or mine deck could accommodate substantial numbers of trucks or other vehicles. In March 1943 BuShips proposed a transport conversion of the incomplete AN 3 and 4. In April Cominch directed conversion of two similar *Terror*-class minelayers (CM 6 and 7, redesignated AP 106 and 107) and of the netlayers, which became AP 108 and 109. In June 1943 the OpNav division of base maintenance certified that it did not need the first two netlayers; they too could be converted to transports (AP 160 and 161). Plans in March 1943 initially called for 14 LCVPs and 800 troops, 350 tons of cargo on the second deck, and 492 tons in the hold. Armament was set at four 5-in/38, four twin 40-mm, and eight 20-mm guns. By November armament had grown to twenty 20-mm (eighteen in LSV 6), but plans showed only 280 tons

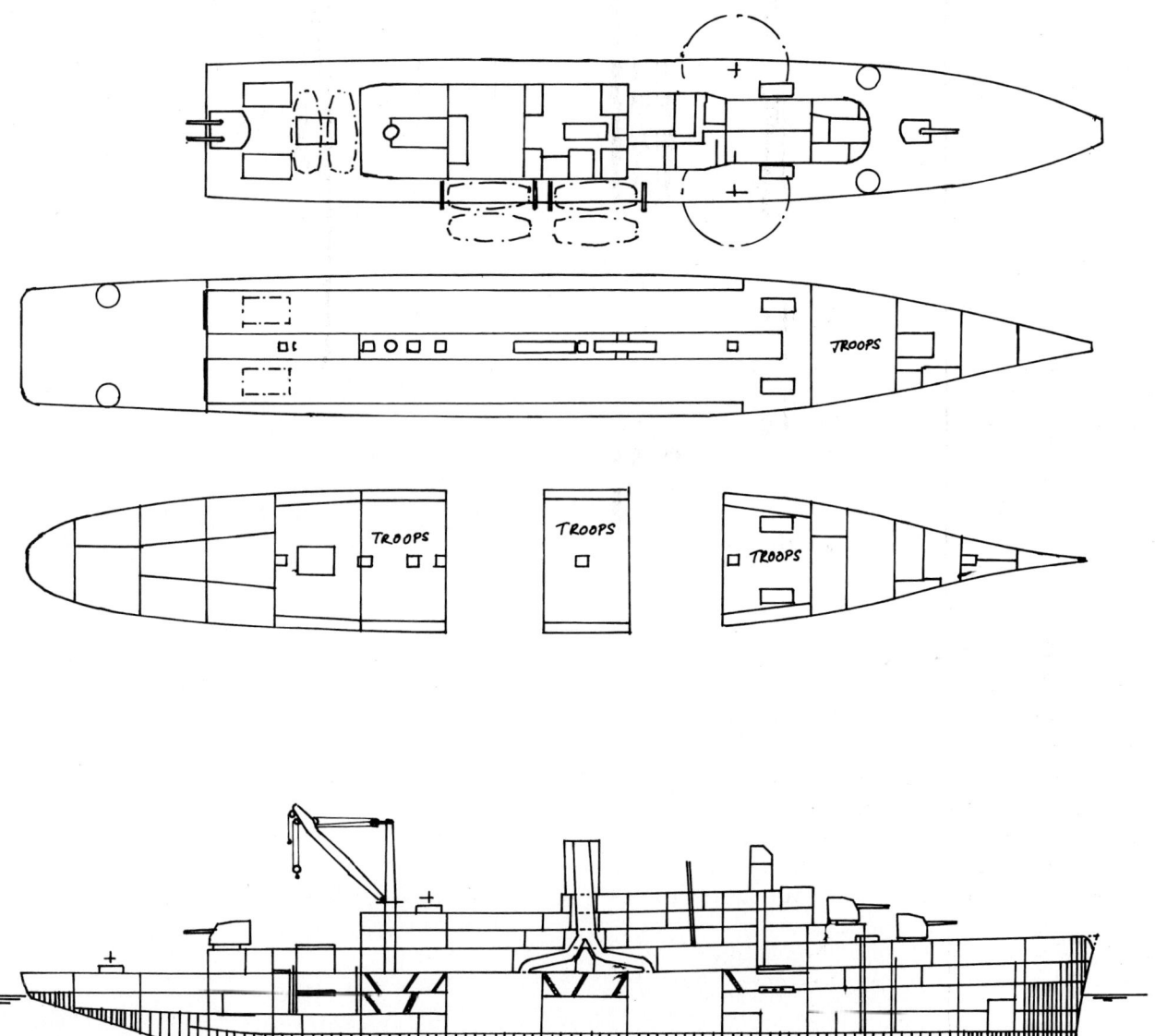

The sketch design for conversion of a net transport (AN) to a transport (AP), 7 January 1943. The ship was completed as an assault transport (LSV). Clearly showing the extensive net stowage, well suited to carrying vehicles, the ship would have unloaded vehicles by crane rather than via a ramp at the stern. At this stage capacity was given as 40 vehicles (each 20 × 8 × 8.25 ft, 3 tons) and 50,000 cu ft. Troop capacity was 37 officers and 775 enlisted men (crew was 13 officers and 290 enlisted). Armament was four 5-in/38, two twin 40-mm, and eight single 20-mm guns. The ship was expected to displace 8,575 tons fully loaded and to carry 1,860 tons of fuel, for a steaming radius of 9,000 nm at 16.5 kts. Rated power was 11,000 SHP, for a sustained speed of 18 kts. The design shows two sets of triple Welin davits on each side and two LCVPs stowed athwartships on deck abaft the superstructure on the upper deck. The ship could not have carried LCMs because she had no heavy cranes, only the two 4-ton units forward and the unit aft with 5,000-lb and 27,000-lb hooks.

The new transport (ex-minelayer) *Catskill* is shown at Willamette, Oregon, 12 July 1944. Her main asset was the ramp, visible in both photographs, leading from her former mine deck to the water, down which amphibians could be launched. The ship was laid down as a sister to the large minelayer *Terror* (CM 5). In 1955 all six LSVs were earmarked for future conversion as minecraft mother ships, and were redesignated MCS (mine warfare command and support). However, only the ex-minelayers were actually converted. They were reacquired from Maritime Commission reserve in 1963–64 and rebuilt to carry helicopters and to support minesweeping boats and launches. *Catskill* served in this role during 1967–70.

of cargo on the former mine or net deck and 668 in the hold. Some fuel oil tanks would be used for potable water.

By July 1943, the ships were described as combined amphibious vehicle carriers and transports with assault features, to carry LVTs and army DUKWs. They were redesignated LSV (landing ship, vehicle), a new category. After hatches were enlarged amphibians on the second (mine or net) deck could be launched over a stern ramp. LSV 1 and 2 (the ex-minelayers) were each fitted to carry 44 DUKW; LSV 3 and 4 each carried 19 LVT and 29 DUKW; and LSV 5 and 6 each carried 21 LVT and 31 DUKW. Ex-netlayers could be distinguished by their cut-off bows and single stacks.

To BuShips' surprise, although the ships had a large metacentric height, they had only a small righting arm (GZ) if damaged, because the watertight deck was relatively low in the ship. By this time the number of LCVPs in later ships had been cut to 10 to simplify conversion, but that had to be increased to 16, which in turn meant adding two heavy triple Welin davits (LSV 1 had 16 boats). In July 1944 the CNO approved landing two 5-in guns as weight compensation. The ex-netlayers were completed with two superfiring single 5-in/38 forward and one twin mount aft. After they entered service, the superfiring 5-in gun was replaced by a pair of twin 40-mm mounts, and the twin aft by a single 5-in/38, leaving only two 5-in/38s. The ex-minelayers retained their four single 5-in/38.

All could carry sixty ¾-ton weapons carriers and 30,000 gal of gasoline in drums below decks. Exclusive of troop ammunition, cargo capacity was 40,000 cu ft (the preliminary design showed 50,000). Normal troop accommodation (LSV 4) was 81 officers and 799 enlisted men, but another 1,072 enlisted men could be carried in place of vehicles (27 DUKWs or 108 jeeps). The ships had one 12-ton and four 4-ton cranes. LSV 3 and 4 had more troop facilities than the others.

The LSVs were not particularly successful in service; they were converted mainly because hulls intended for other purposes became surplus. When first employed, they were criticized for their limited vehicle capacity and lack of cargo space. The vehicle

Monitor (LSV 5) was laid down as a netlayer, based on the minelayer design, but was released for conversion to an amphibious transport when new nets proved far lighter than those inspiring her construction. Newly completed, she is shown at the New York Navy Yard, 14 June 1944. The ex-netlayers could be distinguished from the ex-minelayers by their cut-off bows (presumably adopted to handle nets), single funnels, and twin 5-in gun mount aft (instead of two singles in the ex-minelayer). The stern has a hard knuckle, rather than the curve of the ex-minelayers.

The ex-netlayers had insufficient stability, and they had to be stripped of some armament as compensation. The final ship of the series, USS *Montauk* (LSV 6), is shown off New York, 12 October 1944, with a pair of twin 40-mm guns in place of her No. 2 5-in gun, and a single 5-in mount aft in place of the original twin. After the war she was partially converted to a minecraft mother ship as *Galilea* (AKN 6), but the conversion was canceled and she was laid up.

deck was cramped and low, and compartments on either side of the ramp on the first platform deck limited movement of vehicles onto the ramp. Because the vehicle deck was narrow, vehicles could not redistribute themselves once on board. The ships were used, however, in the Philippines and Okinawa invasions. All were retained in reserve postwar, in the expectation that their hulls would prove valuable for other purposes. *Montauk* (LSV 6) was converted to a net carrier and minecraft mother ship (*Galilea*, AKN 6) but was laid up without entering service as such. In 1951 another was proposed for conversion to an administrative flagship, but again nothing was done. LSV 1 and 2, the former minelayers, became mine countermeasures support ships in the 1960s. The LSV designation was revived in the 1950s for roll-on roll-off ships built by the Maritime Administration for the Military Sealift Command (see Appendix A).

In April 1943, the navy was assigned 20 Maritime Commission P2s and 30 C4s, to be operated to army schedules as point-to-point troopships. Evidently there was some interest in using the engines-aft C4s as attack transports. Hurried BuShips calculations showed that they might accommodate about 10 LCVPs, but would have neither the space nor the deadweight needed. Given their deficient stability, they certainly could not carry LCMs.

When the Allies invaded Sicily in July 1943, assault transports were in such short supply that many LSTs had to be fitted with extra davits as surrogate APAs. Two ex-army convoy transports, neither of them a combat loader, had to be used to help lift the floating reserve at Sicily: *Orizaba* (AP 24) and *St. Mihiel* (AP 32). The pre-Sicily shortage of attack transports also explains the special six-davit (infantry assault) modifications to many LSTs (described in Chapter 5). Given this problem, on 9

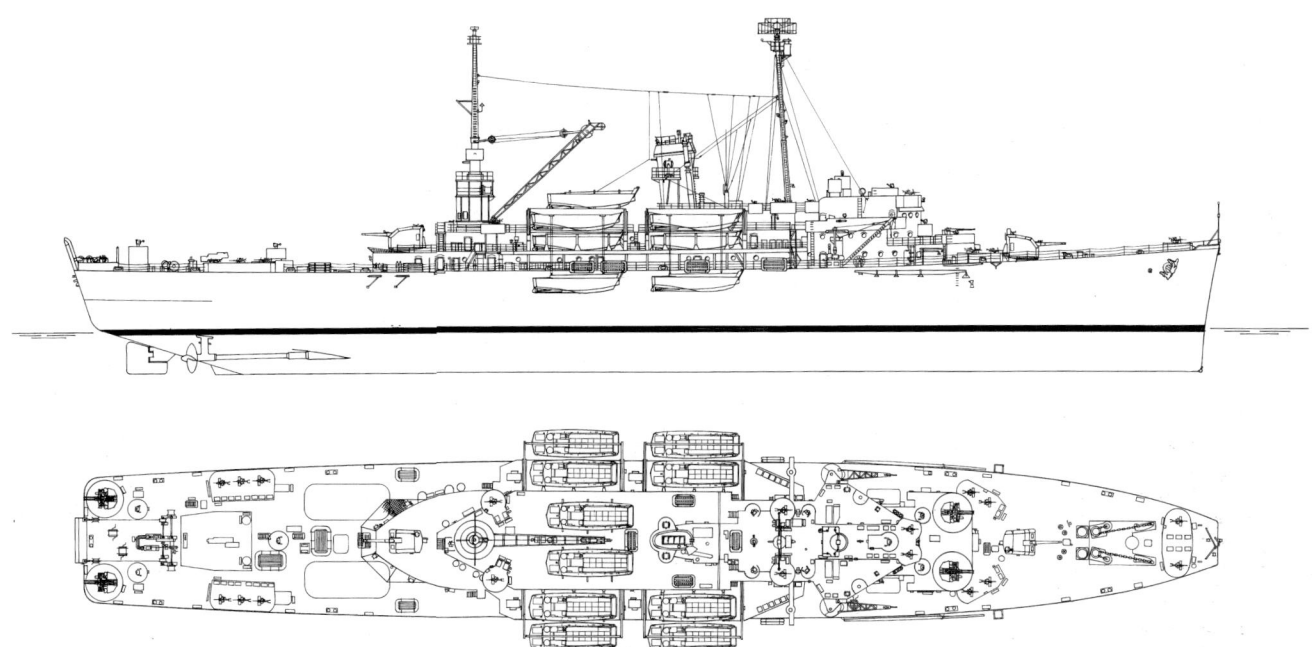

Montauk (LSV 6) in March 1945, with 5-in armament reduced for stability. *A. D. Baker*

July 1943 Admiral King requested immediate conversion of eight APA and six AKA, to be followed by one APA per month and one AKA every two months. The Auxiliary Vessels Board earmarked six C3s and two C2s for immediate APA conversion, and approved a continuing program of six C3s. It also listed six C2s for AKA conversion, plus five for a continuing program. Another C3, being built at Ingalls, was available, so the program was extended to 15 APA and 11 AKA, for delivery during 1944.

The APAs in this program were APA 91 (*Windsor* class: Bethlehem–Sparrows Point design, C3-S-A3), APA 92 and 93 (*Bayfield* class: Ingalls and Western Pipe and Steel design), APA 94 (*Baxter; Sumter* class: Gulf Shipbuilding C2-S-E1 design), APA 95 and 96 (*Bayfield* class), APA 97 and 98 (*Windsor* class), APA 99–102 (*Bayfield* class), APA 103 (*Windsor* class), APA 104 (*Bayfield* class), and APA 105 (*Windsor* class). In September 1943 the VCNO ordered that the ships be modified versions of the APA 40 conversion. They would carry 26 LCVP and 2 LCM(3), 16 of the LCVP to be handled by two sets of Welin davits on each side (i.e., one stowed on deck next to each davit). Armament was set at two 5-in/38 (fore and aft), two twin 40-mm, and eighteen 20-mm. Cargo would include 35,000 gal of gasoline (in 5-gal cans) for embarked vehicles, plus 20,000 gal of diesel fuel for the ship's boats. Plans for *Baxter* showed differences from the *Sumter* class: kingposts, booms, and winches were added aft, hatch area was increased, and boat stowage was rearranged to eliminate the strongback across the hatch opening, to ease cargo handling. In November 1943, all these ships were ordered equipped with three rather than two LCMs, for quicker cargo handling across the beach (see below). The cargo ships were AKA 53–63, repeat *Andromeda*s (AKA 15 class).

Given the lack of suitable hulls, a British request for new assault transports, which they called LSI(L)s, was particularly difficult to fill. Given losses of large transports during Operation Torch, the Ministry of War Transport wanted to withdraw the largest of the British LSI(L)s (15,000–23,000 tons) for point-to-point trooping. Troopships were badly needed, and there were no further suitable 15-kt ships. LSI(L)s were retained in the United Kingdom for use in specific operations. Moreover, the big ex-liners were too valuable to risk in assaults. Yet by mid-1943 it was clear that attack transports would be needed for the invasion of France. The ideal alternatives were the Dutch "Jays" and U.S. combat loaders, the former operated by the U.S. Army and the latter by the U.S. Navy; but they were not available. In July 1943 the JCS released nine C1-Bs to the British for conversion in the United Kingdom, to replace the large liners. The British asked for four more to cover expected 1943 losses. This *Empire Weapons* class was converted specially for the Normandy invasion, with 18 LCA, the maximum number of troops, and the best possible close-range antiaircraft armament. Because they would not make long voyages, they could carry troops in particularly austere conditions.

Given the U.S. Navy's urgent need for more attack transports, the JCS asked whether the last four might not be suitable for U.S. use. BuShips' strong rejection explained why only the scarce C3 was really worthwhile. The C1-B was far too small; she could accommodate no more than 1,000 troops (about half a combat team plus beach parties), and even then they would be jammed in. The ship would be about the size of the unsuccessful *Doyen*, but much slower. Deckhouses would have to be added for officers and troop officers, and ballasting would be needed to restore stability. Ballast would also cost troop capacity. Despite her short length, the C1-B had large holds, so when loaded she was a one-compartment ship. Splitting the loads would make her a two-compartment ship, but that would be an illusion, because the compartments would be so short that any mine or torpedo would open at least two of them. Compared to the new small attack transports, a C1-B was shorter and much slower. Moreover, design capacity was badly stretched, so it would be difficult to prepare plans for a completely new conversion. The ships were being built on the west coast by Consolidated Steel, but no west coast yards could take the conversion job.

The British got all 13 ships. In July 1944 the U.S. Navy asked them to provide some assault transports for special Pacific employment, and the British provided Force X: a command ship (*Lothian*), two squadron flagship LSI(L), and four *Empire Weapons*. In effect they confirmed BuShips' fears. According to the official British history of amphibious ships, the LSI(L)s "entirely failed to satisfy the Americans."

They were poorly adapted for work in the tropics, and they did not meet American habitability standards. None of the LSI(L)s was used operationally. Overall, the attempted deployment of Force X highlighted the difference between an LSI(L) and a U.S. APA. The U.S. ships carried fewer landing craft and fewer tanks, but they did carry the troops' motor transport and immediate supplies. This difference made it impossible to replace an APA with an LSI(L), as had been hoped. Later LSI(L)s, including some C1-Bs, were tropicalized to work with the British Pacific Fleet.

In a sense, though, the U.S. Navy did find itself operating a pair of C1-B attack transports. In November 1943 the Auxiliary Vessels Board proposed that the navy acquire ten C-type cargo vessels as navy-manned transports (AP 166–175).[6] These were the ten ships (besides *La Salle*) fitted so they could transport landing craft, which soon meant fitted as semi-APAs. Because hulls were scarce, they were a mix of types. Six (AP 166–170 and 173) were C2-S-B1s built by Moore, laid down as sisters to *La Salle*. They lacked triple Welin davits, stowing their landing craft on deck. Two (AP 171 and 175) were C2-S-AJ1s, built by North Carolina Shipbuilding. They could be distinguished by their pair of triple Welin davits, one on each side, just abaft the superstructure. The other two (AP 172 and 174) were C1s, similar to 25 army troopships. Unlike the army troopships and the British *Empire Weapons*–class attack transports, they did not have deckhouses

In 1943 the U.S. Navy managed to obtain ten transports. Nominally converted for point-to-point use, they were actually semi-attack transports, and they served as such. *Winged Arrow* (AP 170) was a C2-S-B1. Of this group (AP 166–170 and 173), *War Hawk* (AP 168) had landing craft davits (for single LCVPs) abaft her superstructure. These ships were effectively sisters to *La Salle* (AP 102), which also served as a quasi-APA.

Starlight (AP 175) was a converted C2-S-AJ1 comparable to the XAP *Thurston,* with a single triple Welin davit abaft her superstructure on each side. Chocks for LCMs are visible on her forward hatch coamings. This photograph was taken on 6 May 1944.

Arlington (AP 174) was one of only two C1-Bs that the U.S. Navy converted into troop transports—and quasi-APAs. Unlike C1-Bs converted into British *Empire Weapon*–class attack transports or into U.S. Army troopships, she did not have a long deckhouse extended over the holds fore and aft of the original superstructure. Sister ships in U.S. Navy service were the cargo ship *Auriga* (AK 98), the submarine tender/engine repair ship *Otus* (AS 20/AG 20), and the hospital ships *Comfort* (AH 6), *Hope* (AH 7), and *Mercy* (AH 8). *Arlington* served as pre-commissioning training ship for west coast APA crews.

extended fore and aft over their No. 3 and No. 4 holds.

The program could not be entirely successful. Ships were converted by the War Shipping Administration (WSA) without naval review. WSA clearly thought they were point-to-point transports like the many others it had already converted. As a result, some ships lacked facilities for boat handling.[7] For its part BuShips refused to recognize that the ships were quasi-APAs. Yet except for *Arlington* (AP 174), which was kept on the west coast to train APA crews, the ships served as APAs in the Pacific.[8] Table 6-4 shows details for six wartime quasi-attack transports. As an indication of their combatant status, AP 166–175 were among Naval Transportation Service transports assigned to the Commander in Chief, Pacific (CinCPac); the others were AP 76, 77, and 102–105. Assignments of transports to amphibious roles caused a severe shortage of point-to-point transports, and 18 Liberty ship freighters had to be modified as short-haul Pacific transports.[9]

The Atlantic Fleet's amphibious training commander, Adm. Alan G. Kirk, considered existing C3-type APAs deficient, and in October 1943 he called a conference with representatives of OpNav and of BuShips. The APAs and AKAs had to be able to unload all combat equipment and cargo within 12–24 hrs, to minimize exposure to enemy attack. Similarly, to get the cargo off the beach as quickly as possible, ships had to provide the maximum possible number of 2½-ton trucks. Given these concerns, the LCM(3) was particularly important, and more had to be carried.

The command proposed that a quadrupod mast be stepped between Nos. 2 and 3 hatches, fitted on either side with one 30-ton and two 10-ton booms to plumb both hatches. If the quadrupod were installed, the three LCVPs normally carried on Nos. 2 and 3 hatches should be replaced by two LCM(3)s on each hatch, increasing the number of LCM by four and reducing the number of LCVP by two (one would nest in each LCM). All hatches except those for troop ammunition or gasoline should have 10-ton booms; the others could have 3-ton booms. As many of the cargo hatches as possible should be arranged to take the largest piece of combat equipment, the 2½-ton truck. All hatches should load directly into the holds, and all hatch closures should be openable at sea, without disturbing boats (e.g., using roller doors instead of sliding hatch covers, the boats being supported at each end outside the hatch cover). Nos. 2, 3, and 5 hatches should be 14 × 7 ft, the others at least 8 × 8 ft. An existing practice of fitting heavy horizontal beams across hatches to strengthen the ship should be abandoned, as it took time to remove and stow the beams, and they took up deck space. Thus, discharging cargo would entail only putting craft in the water and opening hatches. Once heavy cargo had been unloaded, it was normal practice to shift to yard and stay rig (the boom would be fixed). Booms had to be fitted so that they could quickly shift cargo handling method. The large holds (Nos. 2, 3, and 5) should have special ventilation to remove gasoline fumes from fueled vehicles, with overhead clearance in the lower holds of 12 ft (8 ft 6 in otherwise, to conserve space). Gasoline stowage should be 12,500 gal, stored in 5-gal containers. Troop and ship officers' messes should be separated, so troop officers could hold preattack conferences. Potable water supply should be 15 gal per man per day, to deal with the demands of the tropics.

A January 1944 review of the latest APA designs (APA 91, 92, 94, and 117) showed that all greatly exceeded the desired 12,000-gal gasoline stowage. Although quadrupod masts were not installed, all got the proposed boat stowage; in the smaller APA 94 class, it was necessary to use single davits in way of No. 2 hatch, carrying one LCVP port and starboard. LCM(3) stowage, however, had not been increased. Except for APA 92, which had two, each ship had only one 30-ton boom. Cargo hatches were larger than in previous designs, and vehicle holds were provided with special ventilation. Cargo spaces were being given extra deck area by installing flats in holds, which reduced the number of deep holds. The review noted that desired overhead clearances were seldom met. Because the ships were designed by the Maritime Commission, it was too late for the navy to modify hold depth. It was sometimes impossible to provide a separate troop officers' mess. Total evaporator capacity was 46,000 gal per day, more than enough by the standards set in October.

In November 1943 OpNav asked BuShips to try to stow at least four LCMs on board each APA. What would be the cost in LCVPs and ballast? BuShips considered four factors: space, handling facilities, stability, and cargo capacity. To gain space, some of the LCVP normally stowed on hatches would have to be removed. More heavy booms would be needed, so some kingposts might have to be substantially modified. Stability was already a problem, as armament, landing craft numbers, and communications equipment steadily grew. Some older and smaller ships were already close to inadequate stability. Taking into account the need for more ballast, adding two LCMs would cost about 200–300 tons of cargo capacity. In most cases ballast would leave insufficient headroom for vehicle stowage in lower holds.

Table 6-4. Wartime Quasi-Attack Transports

	Dorothea L. Dix (AP 67)	*Anne Arundel* (AP 76)	*Thurston* (AP 77)	*Comet* (AP 66)	*Arlington* (AP 174)	*Starlight* (AP 175)
Hull type	C3-E	C3-Cargo	C2-F	C2-S-B1	C1-B	C2-S-AJ1
Length (ft-in)						
WL	450-0	465-0	438-5	435-0	395-0	435-0
OA	473-1	492-0	459-3	459-2.5	417-9	459-1
Beam (ft-in)	66-5	69-6	63-0	63-0	60-0	63-0
Draft, full load (ft-in)	24-0	24-0	23-0	25-9	27-8	27-8
Displacement (tons)						
Light	6,888	7,980	6,932	7,440	5,668	6,363
Full load	14,480	14,907	13,898	13,893	9,104	10,700
Boilers	2 B&W	2 FW	2 FW	2 FW	2 B&W	2 CE
Conditions (psi/°F)	495/750	465/765	465/750	450/750	450/750	450/750
Power (SHP/shafts)	8,000/1	8,500/1	6,000/1	6,000/1	4,000/1	6,000/1
Speed (kt)	17.5	18.4	16.5	15.5	14.7	16.5
Generators (qty-kW)						
Main	1-150, 3-300	3-300	3-300	3-250	3-300	3-300
Auxiliary	1-60	NA	NA	NA	NA	NA
Fuel oil (gal)	528,441	1,096,200	394,800	450,000	440,000	357,670
Landing craft fuel (gal)	18,500	13,230	NA	46,264	NA	28,395
Radius (nm/kt)	18,000/15	NA	NA	NA	NA	NA
Complement[a]	22/19/389	22/19/391	48/408, accom	20/16/185	26/ 323, accom	35/325, accom
Capacity						
Troops (officers/enlisted)	74/1,608	77/1,204	77/1,229	74/1,469	59/1,337	64/1,479
Sq ft	NA	NA	NA	2,094	NA	2,685
Cu ft	164,417	181,815	NA	106,438	90,546	112,460
Water, fresh (gal)	161,991	NA	NA	110,565	NA	470,033
Distilling rate (gal per day)	20,000	NA	NA	20,000	NA	20,000
Booms (qty-capacity)[b]						
30 ton	4	1	2	2	1	2
10 ton	2	10	11	6	—	4
5 ton	—	—	—	6-5, 2-3.5	4	6
Boats						
LCM(3)	2	2	2	2	2	4
LCVP	18	18	14	6	6	10
LCPL	—	1	1	3	2	—
LCPR	—	1	1	—	—	—
Armament						
5-in/38	—	—	—	1	1	1
3-in/50	4	4	4	4	4	4
40-mm	2 × II[c]	2 × II	2 × II	—	—	—
20-mm	10	10	14	12	12	12

NOTES: AP 67, 76, and 77 data are for mid-1945; other data are for late 1944. Boat list for *Thurston* is as at Iwo Jima. Light tonnages are from the 1946 Naval Vessel Register. Neither it nor any other contemporary source gives full load displacements. Full load data therefore are for similar ships. Thus *Dorothea L. Dix* (AP 67) data are for *Hercules* (AK 41); *Thurston* (AP 77) data are for *Pollux* (AKS 2, lost 1942); *Arlington* (AP 174) data are for *Otus* (AS 20); and *Starlight* (AP 175) data are for *Mount Hood* (AE 11). Trials displacements: *Dorothea L. Dix*, 12,580 (limiting displacement); *Anne Arundel*, 11,700; *Thurston*, 13,910; *Comet*, 13,910; *Arlington*, 12,930; *Starlight*, 13,910 tons. Dash indicates that specification is irrelevant; NA indicates relevancy but data not available.

[a]Complement is given as officers/CPOs/enlisted when three figures are listed and officers/enlisted for two figures.

[b]Only quantity is listed unless capacity is different. For example, *Comet* had six 5-ton booms as well as two 3.5-ton booms.

[c]Multiple gun mounts are indicated by Roman numerals; therefore 2 × II indicates two twin mounts.

Changes were ordered in March 1944. Older APAs (APA 2, 3, 5–10, and 12–17), converted C2s (APA 49–54), and some C3s (APA 55, 56, and 82–90) could not take additional solid ballast. They should land four LCVPs stowed under booms for every LCM added. Some of these ships were already carrying four LCMs. *American Legion* (APA 17), which carried 33 LCVPs, would continue to carry only 2 LCMs.

Barnett (APA 5), which carried 25, would also continue to carry 2 LCMs. The *Sumter*s (APA 52–55) would receive 5 LCMs and would carry 19 LCVPs. Most converted C3s (APA 18–21 and 25–48) would encounter no problems, because they already carried so much cargo (3,000 tons deadweight). They would land only two or four LCVPs (one, in a very few cases). At least some ships already carried four LCMs, and BuShips found itself demanding urgent weight compensation. Ships' commanders were often apparently unaware of just how marginal their ships' stability was, and boats 25 ft above water contributed greatly to topweight. The small APAs (*Doyen* and *Gilliam* classes) continued to carry no LCMs.

In the fall of 1944, as modified, a typical large APA carried four LCM(3) on hatch covers, with either an LCVP or a DUKW nested in it; six DUKW could be carried in place of the LCM. The four Welin davits carried 12 LCVP. DUKWs could be carried in place of LCVPs. Small craft capacity was a serious problem. According to the Eighth Fleet after-action report for the invasion of Southern France, ships could and did carry additional LCVPs at their rails, supported only by their cargo booms, in good weather.

By this time attack transports were being more heavily armed. Early in 1944 BuOrd called for at least one (preferably two) 5-in/38 on the centerline, the maximum possible number of twin and quadruple 40-mm guns (at least four mounts, to cover all four quadrants), and the maximum number (at least twelve) of 20-mm guns sited to distribute fire equally on the bows and the quarters, with maximum arcs. Unfortunately, ships converted from liners, particularly the 535-footers, had large superstructures, leaving little space for guns. Thus in August 1944 the commanding officer of *Harris* (APA 2) argued for at least one 5-in gun for his 21,300-ton (full load) ship, but admitted that there was not enough space. *Heywood* (APA 6) had enough space but insufficient stability. To mount the battery recommended in February 1944 (one 5-in/38, two twin 40-mm, twelve 20-mm in place of the planned ultimate battery of four 3-in/50, two twin 40-mm, and ten 20-mm), she had to take on board 400 tons of ballast. Any further increase required weight compensation, as she would be at her limiting displacement.

Survivability was a constant concern. The merchant ships from which transports were converted were built with capacious holds; these holds were menaces to any ship suffering underwater damage. Yet breaking up those holds with bulkheads would make a ship less efficient as she unloaded off a beachhead, and thus would keep her longer in a dangerous area. Moreover, to get transports quickly enough, the navy tried to minimize conversion time, often to only a few weeks. There was little enough time to install essentials such as distilling plants for the troops' drinking water, sufficient quarters for the ship's expanded crews, and wiring for radars and radios. Thus wartime BuShips files are littered with orders to install necessary bulkheads to block cargo ports (in the case of the *Harry Lee* [APA 10] class) or, more generally, to install bulkheads to break up holds (as ordered by the Pacific Fleet amphibious force for all new attack transports and quasi-attack transports, as late as February 1944).

At the "Quadrant" conference in Quebec in August 1943, the U.S. and British governments agreed that a major Pacific offensive would begin in 1944. The U.S. Navy pointed out that operations would be limited by a lack of combat loaders, and that "experience to date has been that actual requirements exceed early estimates." Point-to-point transports (APs) were not attractive alternatives, partly because the loss of one such ship would kill so many troops, and thus might jeopardize an operation. Admiral King therefore urgently requested a program to convert, by July 1945, 40 APA and 20 AKA using C2, C3, or Victory hulls, with the maximum possible number to be fitted into the 1944 program. He also wanted three more AGCs (see Chapter 9), in addition to the five already approved, partly to make up for expected losses.

King's request may have been the first formal proposal to use Victory hulls, a new class the Maritime Commission had announced to the navy in March 1943. The fast Victory was to replace the slow Liberties in mass production, hence it might be produced in sufficient numbers to satisfy the navy. Indeed, its initial designations, EC2-S-AP1 through EC2-S-AP4, suggested that it was a modified Liberty (EC2). As in a Liberty ship, its design was simplified for mass production. Many of its features, however, approximated those of the more sophisticated and far less numerous C-series ships. As of March 1943, the AP3 version was scheduled to receive C3-type machinery, and thus to make 16.5–17 kts, just what the navy wanted for a transport. It was probably less attractive to the navy that a Victory was designed to a one-compartment standard (i.e., she would succumb to a single torpedo hit). The merchant cargo design included foundations for two 5-in/38 (fore and aft) and eight 20-mm guns.

The JCS went much further than King had requested. In November 1943 it approved a "must" program of 130 APA (APA 106–235) and 30 AKA (AKA 64–93) against an ultimate requirement, set by the joint staff planners, for 133 APA, 53 AKA, and 13

AGC, to be completed in 1944 and early 1945. As part of this program, ten C2s were converted to AGCs (see Chapter 9).

The only viable candidate for most of the 130 APAs was the Victory ship (VC2), the mass production design that was just beginning to replace the slow Liberty. Because it was simpler than a C2 or C3, it could be built by the emergency yards set up to build Liberties. It was fast enough, however, to be a viable attack transport. To produce enough ships in time, the Maritime Commission had to use mass production methods, which in effect froze the design quite early in its development. For example, the commission agreed to install combat information centers (CICs) only in ships laid down after 1 August 1944; some ships laid down as late as 28 August still had to have CICs installed by the navy after completion. The sheer urgency of the "must" program also caused problems. Essential onboard spares were often delivered late, so ships had to be accepted on the assurance that spares would be received within ten days. Ships had to be rushed into service. For example, *Talladega* (APA 208) had only a one-day shakedown, and thus reported for sea in unsatisfactory condition. There was much confusion due to changing requirements, set after the "must" program had been frozen.

The Maritime Commission conversion plan was based on the larger APA 91's conversion plan. It was reviewed by navy representatives at a 5 November 1943 VCNO conference. As might be imagined from its Maritime Commission designation, a Victory was C2 rather than C3 sized, yet she had to accommodate as many troops (officer accommodations were cut to 79) and nearly as much troop cargo as a C3 conversion. Troop cargo was cut to 134,000 cu ft. Cargo ammunition and gasoline space were cut. Some facilities were below the standard realized in C3 conversions: ship's stores, ship's magazine (all ammunition was stowed, but floor space was reduced, so handling was more difficult), supply department stores, refrigerated stores (full cubic space, but with some crowding), and laundry facilities. The ship lacked the usual barber, tailor, and cobbler shops. Steaming radius was cut to 12,000 nm at 15 kts. The ship had only a single 35-ton boom. Fore to aft, the ship had four masts with the

Maritime Commission mass production merchant ship capacity made it inevitable that the final, very numerous, class of wartime attack transports would be a version of the "Victory" ship. USS *Sherburne* (APA 205) is shown off the southern California coast, 4 October 1944. Designated a VC2, the Victory was C2 sized, and thus compared poorly with C3-size attack transports. Even for an APA, the design was unusually cramped, with landing craft stacked three high and handled by the ships' booms. Mass production provided the desired numbers, but only at the cost of rejecting changes during production. Thus some Victory APAs were delivered without combat information centers.

Despite their limitations, many Victory APAs remained in service postwar. Laid up at the end of World War II, *Talledega* (APA 208) was activated for the Korean War, and served until 1969. The ship's electronics dates this photograph to the early 1950s: the SR air search radar on the main mast was replaced with SPS-6 in the late 1950s. The curved extensions to the yardarm carry the new UHF ship-to-air radio antennas. Note the drastic reduction in the number of landing craft, compared to the 1944 photograph of *Sherburne*.

following booms: two 10 ton forward/two 4 ton aft; two 10 ton; two 10 ton forward/two 4 ton aft; and one 35 ton and two 10 ton, serving hatches Nos. 2 through 5. Ships had two sets of triple Welin davits on each side. Armament was cut to one 5-in/38, one quad 40-mm (centerline forward), four twin 40-mm, and ten 20-mm. Planned boats were 1 LCS(S) (landing craft, support, small; see Chapter 8), 2 LCPL, 23 LCVP, and 2 LCM(3).

The 119 Victory APAs of the "must" program were assigned to only three companies (at four emergency yards): Oregon Shipbuilding (36), California Shipbuilding (34), and Kaiser (49, at two yards). In May 1944, six Victory APA (APA 181–186) under construction by Oregon Shipbuilding were ordered to be built as hospital ships instead. Then C4 hulls were substituted, and the six ships were completed as merchant ships. (See below for late Victory APA orders.)

The Victory APAs (*Haskell* class) became by far the most numerous attack transport class. They long survived the war. By about 1963 they accounted for 55 percent of the active attack transports and 100 percent of those in reserve, thus suggesting that the other classes were considered more valuable. Four (APA 212, 215, 222, and 237) were modernized under the FY 64 FRAM II program.

The Maritime Commission decided to have conventional yards build 11 ships (4 by Western Pipe and Steel, the other 7 by Ingalls). They were APA 106–116 (all *Bayfield* class), and they followed the plans prepared in September. Boat complement was set at 1 LCS(S), 2 LCP(L), 23 LCVP, and 2 LCM(3), of which 16 LCVP would be handled by two sets of triple Welin davits on each side. They would accommodate 91 officers, 20 warrant officers, and 1,450 troops. In December, the LCS(S) and LCPL were

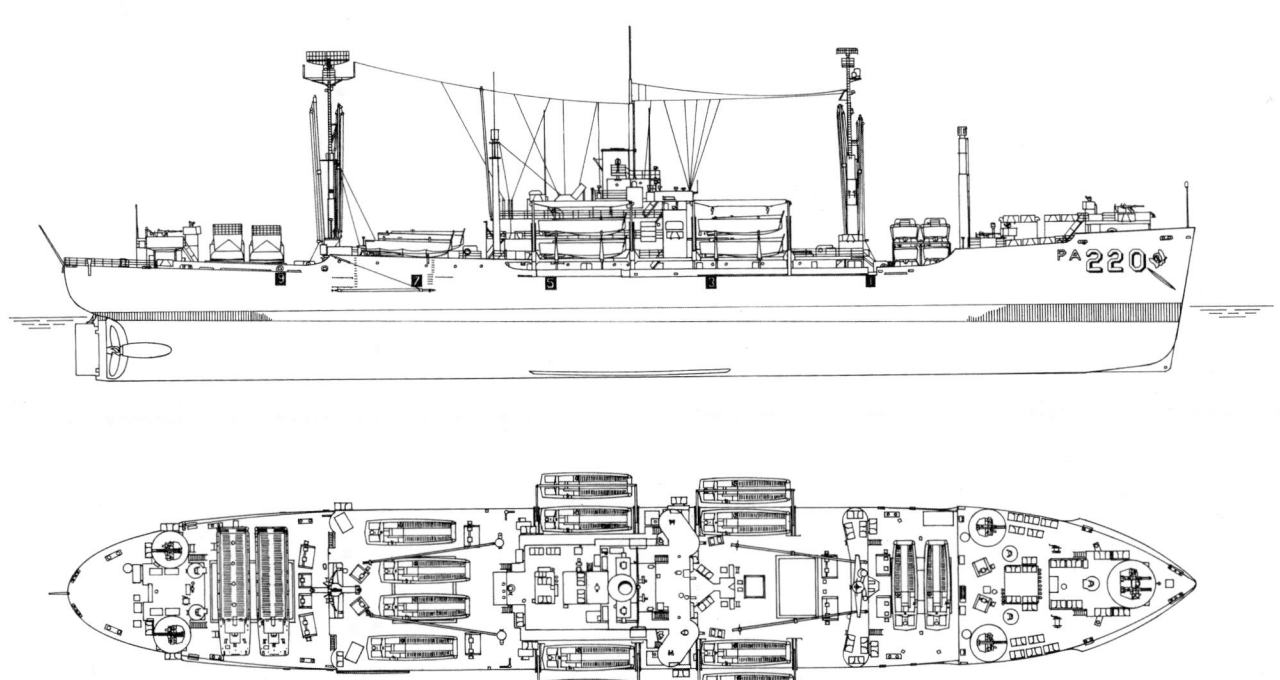

Okanogan (APA/LPA 220) in 1967, with a Vietnam War loadout of two LCM(6) and 18 LCVP (and no LCPL/LCPR). By this time, all 20-mm guns had been landed. The air search radar aft was SPS-6C. *A. D. Baker*

replaced by three more LCVP, and the number of troops cut to 1,400.

The accompanying 30 AKAs were of two types: a new-build version of the C2-S-AJ3 (*Tolland* class: AKA 64–87) and six repeat versions of the AKA 53–63 class (AKA 88–93). Of the former, half were converted by the Maritime Commission and half under navy auspices in Baltimore.

In July 1944, 15 more AKAs were ordered: 7 C2-S-B1 (AKA 94–100, to be similar to AKA 53–63, by Federal Shipbuilding, converted by the Maritime Commission) and 8 C2-S-AJ3 (AKA 101–108, by North Carolina Shipbuilding). Of the latter, four would be completed similar to AKA 64–75, and four delivered incomplete for conversion by the navy similar to AKA 76–87.[10] The C2-S-B1s would be armed with one 5-in/38 aft, four twin 40-mm (two forward, two aft), and eighteen 20-mm guns; they would carry eight LCM(3) and sixteen 36-ft landing craft (six of them in triple Welin davits). Masting and other features would match those of the earlier AKA 15 class.

As in the case of the transports, the navy operated three cargo ships that functioned as AKAs: *Hercules* (AK 41), *Mercury* (AK 42), and *Jupiter* (AK 43). Unlike the transports, they were not part of any concerted program. Taken over in 1941, they were initially placed in service as civilian-manned naval cargo ships. In 1942 they were converted to

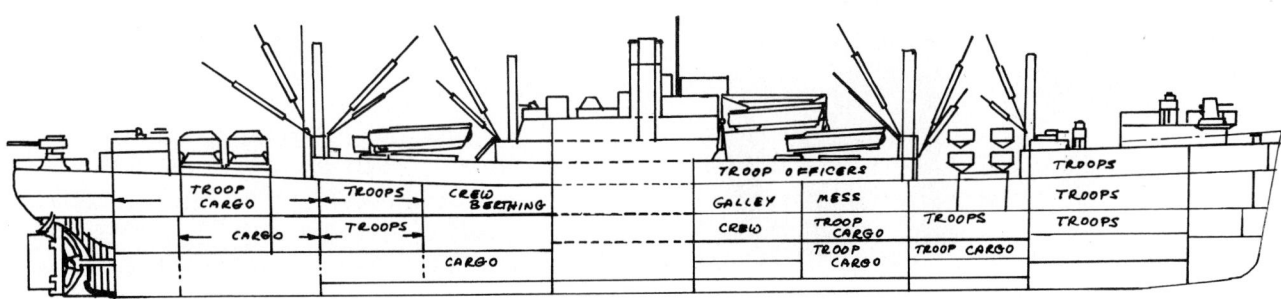

Haskell (APA 117) class inboard profile, based on drawings in a 1944 ship's information book. The copy used was for *Rawlins* (APA 226), but was clearly standard for the entire class.

be navy manned, so that they could be sent into combat areas, but they were still used as point-to-point cargo ships. In 1944, however, they were shifted to an AKA role, replacing AKAs within transport squadrons during major operations. By the end of the war all three had been shifted to other roles.[11]

Another 12 Victory-type APA were ordered in March 1945, more than making up for the cancellations (and exceeding the 133 planned AKA). In addition, 3 AKA (AKA 109–111: C2-S-AJ3 type, similar to AKA 100) were ordered, bringing the total to 48, short of the 53 envisaged late in 1943. When the war ended in August, 8 attack transports and all three attack cargo ships were canceled, leaving 117 Victory-type APAs.

By January 1944 the Pacific Fleet amphibious force was complaining that the early assault transport conversions based on the old Shipping Board hulls had inadequate antiaircraft armament; inadequate radar and radio; insufficient main generator capacity, which made the first two deficiencies difficult to remedy; poor ventilation; insufficient potable water; insufficient refrigerated space; insufficient compartmentation to survive damage; and insufficient boom capacity for quick unloading, particularly of heavy equipment.

C3 and Victory hulls were far more satisfactory, but they still suffered from inadequate ventilation and insufficient distilling capacity. An October 1945 Pacific Fleet evaluation suggested that Victory APAs be provided with an additional 300 kW generator

Matthews (AKA 96) was a repeat *Andromeda*-class attack cargo ship. She is shown at Mare Island after a refit, 5 March 1952, little changed from her World War II appearance. Laid up after World War II, she was reactivated for the Korean War, serving until 1968.

USS *Seminole* (AKA 104) was a repeat *Tolland* (AKA 64)–class attack cargo ship. She is shown after a refit at San Francisco, 11 February 1954, hoisting one LCM and preparing to hoist a second. Numbers painted on her side indicate boat stations for loading. Never laid up after World War II, she was decommissioned and stricken in 1970.

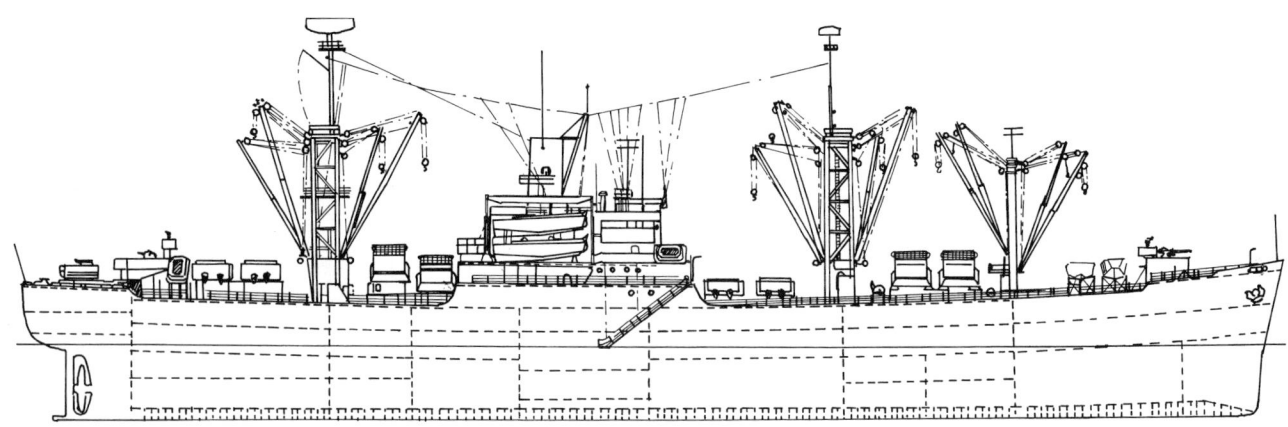

Union (AKA 106) postwar, probably in the 1960s. The big antenna atop her kingposts forward is for television; a similar antenna is visible just abaft her bridge. Attack cargo ships differed in rig. *Union* was a unit of the *Tolland* class (AKA 64–87 and 101–108), which had radio masts stepped from their funnels and air search radars atop their forward quadrupods. The externally similar *Andromeda* class (AKA 15–20, 53–63, and 88–100) had a radar mast forward of the funnel. At least the first six (AKA 15–20) were completed without any radar mast at all. Some (AKA 15, 16, 20, 53–63, and 94–100) had a pole radar mast; the others (AKA 17–19 and 88–93) had a lattice. *Union* has a pair of LCPLs (one Mk 4, one Mk 2) forward, athwartships, plus eight LCM athwartships and four LCVPs in Welin davits. The kingpost forward shows two 5-ton and two 10-ton booms. Each quadrupod supports two 35-ton booms (fore and aft) as well as two 5-ton booms. The radar forward is SPS-4, a surface search set; aft is an SPS-12 air search antenna. Armament is pairs of twin 40-mm guns fore and aft, with a 5-in/38 right aft.

and air conditioning, particularly in the sick bay. Some attack transports should be modified as hospital evacuation ships.

After the war, the Marines pointed out that, although able to carry a coherent combat unit, an APA had little provision for its commander, and none for the radios he needed to exercise command when part of his unit was ashore but he and his staff were still on board ship. They wanted space permanently assigned for the commander and staff, large enough to accommodate the next higher headquarters staff that would be required for a regimental combat team (RCT). For preassault briefings, a space large enough for 200 men was needed. As for exercising command, in wartime the Marines had resorted to makeshifts, using portable radios on deck. Too often they interfered with the ship's radios, and equipment and personnel were exposed to weather and also to enemy attack. Each AKA or LSV should have RCT radios, similar to requirements for a BLT. Some ships would carry RCT commanders. Some would serve as divisional headquarters.

Transports unloaded too slowly. The Marines wanted wider ladders into the troop compartments (in the holds) and debarkation stations wide enough for four men to go over the side at one time. They wanted davits that were strong enough so boats could be loaded at the rail. The Marines also revived earlier proposals to use side ports rather than the usual deck hatches. Ideally the ship should carry enough boats (or equivalents) to land all 1,300 troops at one time: 1,300 boat spaces, about 2 LCM and 30 LCVP. Future ships would probably have to accommodate more troops, hence more cargo: 75 officers and 1,500 enlisted men, with 200,000 cu ft for cargo.

As vehicles grew, holds and hatches had to match. In 1947 that meant at least one vehicle hold (minimum height of at least one deck would be 10 ft) with a hatch at least 25 × 15 ft, served by a pair of 55-ton booms. Future ships should have two vehicle-capable holds. All cargo spaces should be fitted with the same fire-fighting systems, so that vehicles, ammunition, and other inflammables could be loaded as desired in any hold. The Marines suggested that any new ship should have her engines aft, like a tanker.

Early attack cargo ships were criticized, in mid-war, for much the same faults as the attack transports: limited compartmentation, limited potable water, insufficient generator capacity, insufficient ventilation, cramped living quarters, insufficient radar and radio, and poor internal communications. Conversion from cargo ships was not as simple as had been imagined. By the end of the war, however, the C2 conversions were considered entirely satisfactory.

In 1947 the Marines expected larger vehicles, thus requiring more gasoline. For existing ships, one forward and one aft boom should be strengthened to 55-ton capacity. A future AKA should have at least one boom with the capacity to lift the heaviest combat loads anticipated. The gasoline hold opening should be enlarged so that four 55-gal drums could be loaded simultaneously.

Another wartime lesson was the need for standardization, both within classes and between APAs and AKAs. Because so many Victory ships were in reserve, in the 1950s the Ship Characteristics Board (SCB) issued characteristics (SCB 93) for a VC2 conversion to an AKA that could operate with VC2 APAs. By this time the navy was shifting toward a 20-kt amphibious force, but clearly VC2s would dominate the APA force for many years to come. Nothing came of the project.

The Marines' ideal was a fast universal (transport/cargo) assault ship using either amphibians or landing craft as required to land troops very quickly. It might combine APD and LSV characteristics: high speed, a powerful armament, and the ability to launch amphibians, yet small enough for adequate dispersal in the face of nuclear weapons.

The other wartime transports were converted destroyers (APDs); they proved quite useful. At the end of the war, 11 of the old flush-deckers still served as APDs (APD 6, 9, 11, 13, 17–20, 22, 23, and 36). In 1943, with the Battle of the Atlantic winding down, the decision was made to convert 100 new destroyer escorts (DEs) to replacement APDs: 50 *Buckley* class (APD 37–86) and 50 *Rudderow* class (APD 87–136). Of the *Buckley*s, six (DE 668–673) were converted on the slip. Two ships were lost before they could be converted to APD 41 and 58. Five ships planned for conversion were completed instead as radar pickets. To make up for the radar picket conversions, three *Rudderow*s were earmarked as APD 137–139 (of which the first two conversions were canceled at the end of the war), and an additional *Buckley*-class ship (DE 637) became APD 40. Ships could be distinguished by their high (*Buckley*) or low (*Rudderow*) bridges. Conversion entailed installing a pair of double Welin davits and building up an enlarged deckhouse to accommodate 12 officers and 150 enlisted men, slightly more than an ex–flush-decker. As fast underwater demolition team (UDT) transports these APDs carried 14 officers plus 100 enlisted men. Ships were given a lattice mast aft, from which a cargo boom that was capable of loading a small vehicle into an LCVP was stepped; the flush-deckers had no such capacity. Cargo capacity was 50 tons on deck, plus 35 tons of troop ammunition and 3,350 cu ft (50 tons) of other cargo. Postwar, APDs were used as amphibious control craft for RCTs; they normally carried a 15-man taclog (tactical logistics) party and 2 control craft (LCPLs) and their crews. Endurance as a transport was 30 days, but as a control craft it was reduced to 7 days. Postwar, a converted destroyer escort typically carried two LCPR and two LCVP. Typical armament was one 5-in/38, three twin 40-mm, and six 20-mm guns plus one or two depth charge tracks aft. Units used for UDT support landed their cranes and had a quad 40-mm mount on their fantails.

Burke (APD 65) was one of 50 *Buckley*-class destroyer escorts ordered converted to fast transports. The distinctive jungle camouflage scheme was applied to many amphibious units at the end of World War II. The main structural change was the enlarged deckhouse, which provided troop quarters. Theoretically the two Welin davits could have supported three landing craft each; in practice each carried only two. The booms aft could handle light vehicles. Some APDs had them replaced by a quadruple 40-mm Bofors. As APDs, these ships retained their destroyer escort sonar and depth charge tracks, although neither Hedgehogs nor depth charge throwers, so they had a limited antisubmarine capability. *Burke* is shown on 24 April 1945.

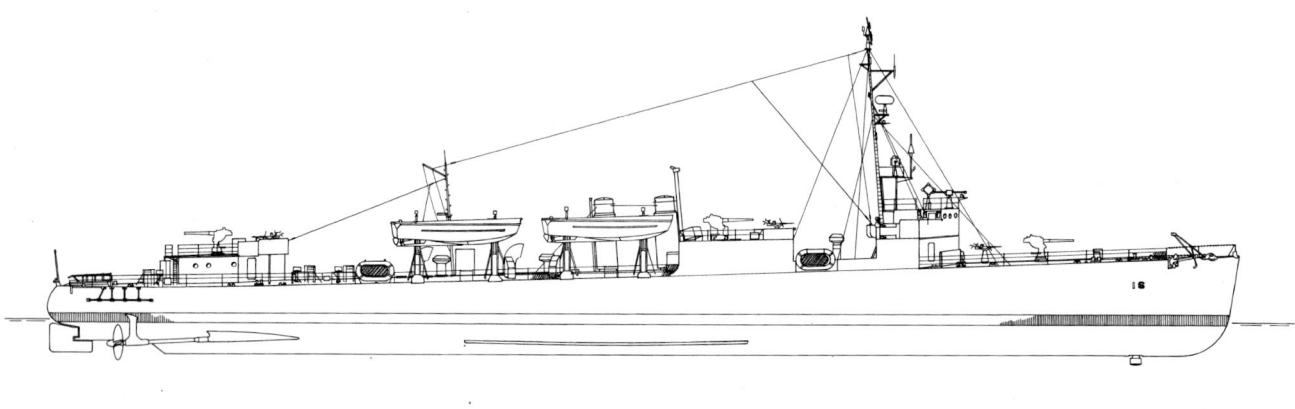

Kinzer (APD 91) was a fast transport conversion of a *Rudderow*-class destroyer escort. She is shown, newly converted, off Charleston on 16 November 1944.

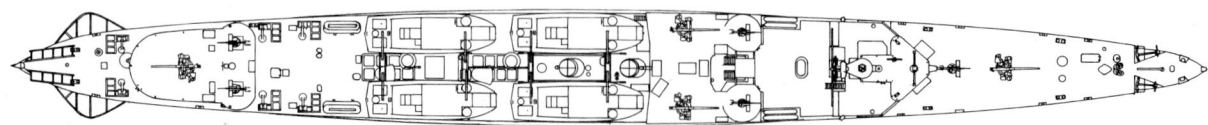

Ward (APD 16), as of 1944, shortly before her loss, with four LCPR aboard. She still shows the old Mk 4 conical-base 20-mm mountings. The surface search radar in the cylindrical housing is SE. *A. D. Baker*

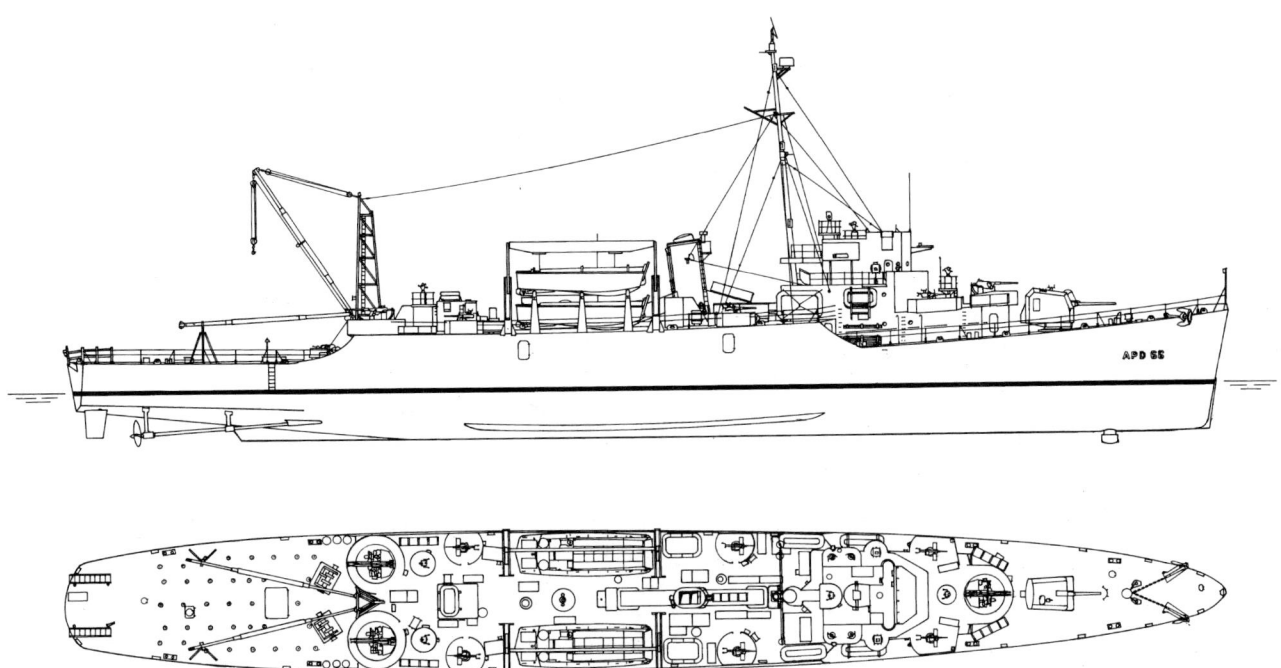

Enright (APD 66) in September 1945, as converted from a *Buckley*-class destroyer escort. Some units had a quadruple 40-mm gun aft in place of the crane. The boats in davits are LCVPs. A. D. Baker

It is not clear why so many ships were converted. Possibly, with the decline of the U-boat threat in the Atlantic, numerous hulls became available in 1944. Although many destroyer escorts were canceled, there was pressure to complete at least some surplus hulls to avoid destroying shipyard morale. It was feared that work cancellation would discourage workers at yards about to shift to building amphibious craft for the Pacific.

At the end of the war the APD was used to support reconnaissance or raids by small units (as had been envisaged prewar) and to support UDTs. Beach reconnaissance required special navigational equipment, and the equipment in turn might be used to guide landing craft to a beach. Ultimately, the APD replaced the existing large control craft converted from subchasers.

As for the LST, it turned out that beaches in Europe often sloped too gently for them to beach as designed. A folding ramp extension was considered, first in November 1942 and again in January 1943, but it was not adopted. Fortunately, for the Sicilian invasion in July 1943 a solution was at hand in the form of pontoons developed (for entirely different purposes) by the U.S. Navy's Bureau of Yards and Docks. They were built out of standard buoyant steel boxes, which could be made into piers and even landing barges ("Rhino" ferries). Comdr. John W. Laycock, head of the bureau, had conceived these building blocks as early as 1935, but the idea had languished. In 1940, however, the navy was urgently interested in advanced bases. Laycock could not interest any company in his idea. In December 1940 he began building a model out of cigar boxes; it was ready in February 1941.

Box size (5 × 7 × 5 ft) was set by typical steel plate size and by the need to keep a string of boxes stable afloat. Limiting weight was 1 ton, and the upper surface would support a 10-ton load from each of two dual tires. Each box could be connected to others by punched steel angles welded to plates cutting off its corners. Any number could be connected end to end or side to side. Three strings side by side were about 20 ft wide, and draft unloaded was about 16 in. A box with a curved end could form the end of a pontoon barge. On 18 February 1941 the Pittsburgh–Des Moines Steel Company received a fabrication contract; tests on the Ohio River below Pittsburgh were very successful. The boxes were adopted as navy lightering (NL) equipment, Pontoon T-6 and Pontoon T-7 (the barge bow). The first Seabee (naval construction battalion) units took pontoons with them on their first deployment in January 1942.

At British request, early in 1942 the Americans began to develop rigid floating causeways, 175 ft long (made up of two parallel strings of pontoons,

The converted destroyer escorts were considered so valuable that seven of them were modernized under the FRAM II program: APD 60, 89, 90, 119, 123, 130, and 135. *Ruchampkin* (APD 89), modified between November 1963 and June 1964, is shown. Changes included extending the 02 level forward to enclose an enlarged combat information center. She also received more modern electronics (antennas which were supported by a tripod mast), and a pair of Mk 32 triple antisubmarine warfare torpedo tubes (on the 01 level, just abaft the bridge). By the time of this photograph, the tubes had been landed. The radomes on the tripod are for a WLR-1 radar intercept system; the ship also had a surface search radar, but no air search set. The boat is an LCPL with a canopy.

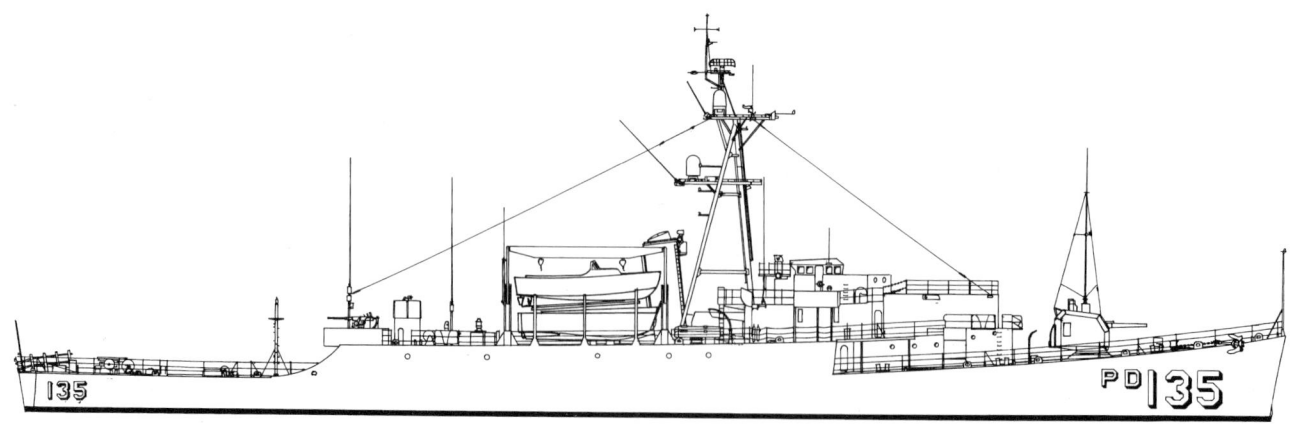

Weiss (APD 135) in 1964 as modified under the FRAM II program. The new deckhouse was an amphibious warfare planning center, the amphibious equivalent to a CIC. *A. D. Baker*

each 30 units long; 2 × 30 pontoons) to be used in pairs to bridge a 300-ft water gap between an LST offshore and shallow water. Initially they were to have been launched from the upper deck of an LST. In practice they were hung on shelves on the ship's side. Released as the ship approached the shore, the pontoons were carried on by their own momentum. They were brought into line with the ship by guy lines. The pontoons were first used in the invasion of Sicily, and later in the Pacific. They survived postwar. The British equivalent was the "Mexafloat."

Pontoons were also assembled into Rhino ferries (6 × 30 pontoons, 400 tons) powered by Murray-Tregurtha outboard motors. They were first used at Normandy, where their great capacity and their ability to land vehicles on a beach of almost any gradient were most impressive. Two Rhinos, which were normally towed to the operating area, could empty a normally loaded LST. They were widely used, particularly in the French landings. There were also smaller pontoon barges (e.g., 3 × 7 pontoons, 50 tons; and 6 × 18 pontoons, 250 tons). The postwar Pacific Fleet evaluation, however, described Rhinos as somewhat difficult to handle, with unreliable engines.

Instead of pontoons an elaborate artificial harbor, the "Mulberry," was built at Normandy to permit conventional ships to unload despite the great rise and fall of the tide. The alternative, for LSTs, was to moor as close to the beach as possible, then allow the tide to run out, leaving the LST high and dry. Had they not been able to dry out on the beach, the LST would have been limited to a brief period when there was just enough water to allow the ship to lower her ramp onto the beach while still afloat.[12]

By mid-war there was interest in designing a new fast LST; the "large slow target" was too vulnerable. Unfortunately, in order to be able to land on a gently sloping beach the LST needed a very broad full hull form, to provide very shallow draft forward. The wide ramp demanded very blunt bow lines. It proved impossible to develop a more efficient hull, so BuShips resorted to brute force. Although 1,300 SHP sufficed for the 10 kts of the existing LST, a May 1944 powering estimate showed that even 6,000 SHP would increase speed to only 13.75 kts. The speed really wanted, 15 kts or more, was out of reach as long as most of the space in the LST was used for cargo. Even 6,000 SHP meant steam. Late in World War II, both the Americans and the British developed steam LSTs.

A geared turbine machinery arrangement drawing, dated 29 November 1943, indicates that work probably began earlier in the fall of 1943. The machinery itself weighed little more than the original diesels, but it was much less efficient. Thus a May 1944 estimate showed a net increase of less than 50 tons, including 40 tons of fuel to maintain the same endurance (250 miles round trip) in landing condition. Such small increases would add only about 3 in of draft in landing condition. BuShips wanted to redesign the LST altogether, but it had to admit that it could squeeze a steam plant into the existing hull. It pointed out that delay due to designing and manufacturing the new steam plant would exceed any delay due to a new hull design. OpNav agreed. On 21 December 1944 it approved the BuShips proposal that the last three LSTs (LST 1153–1155) be built to the new design, formally proposed on 27 November.

BuShips wanted three experimental 6,000 SHP steam plants, each consisting of two boiler-turbine units separated by an auxiliary machinery space. Besides greater speed, the new designs offered greater backing power, to help the ship retract after beaching. Two of the plants had conventional separate astern turbines. The third, with a clutch and reversing gear, was being bought for its prototype turbine reversing gear. As at Normandy, the ship might dry out on a beach, when the steam plant could not run because no water could pass through its condenser. The steam LST therefore had a 100 kW diesel generator, which could be cooled by water from the ship's ballast tanks. She also had two steam turbo-generators.

BuShips circulated a preliminary contract design on 2 March 1945. The new LST was lengthened to restore the original landing draft and keel slope: 3 ft 4 in forward, 10 ft 7 in aft, compared to 3 ft 11 in forward and 9 ft 9 in aft in a fully loaded wartime LST. Waterline length grew from 316 to 368 ft (beam was 54 rather than 50 ft). The larger ship could carry more than twice as much cargo in ocean voyage condition (3,010 vs. 1,400 tons); limiting displacement increased from 4,080 to 6,000 tons. Beaching cargo capacity, however, was still 500 tons. The larger hull could accommodate more troops (180 rather than 149, plus 251 rather than 103 on a temporary basis). Troops would be berthed on the port side and the crew on the starboard side, the troop area being arranged so that it was easily convertible to casualty evacuation. Postwar, the Marines would press for accommodations sized to a tactical unit, in this case a reinforced rifle company (12 officers and 300 enlisted men).

Very long vehicles sometimes hung up on the steep ramps of existing LSTs. Because it was larger, the new LST's ramp (from main to tank deck) had a gentler angle, 19 rather than 25 degrees. Unlike the ramp on previous LSTs, this ramp could be used whether or not the bow doors were open, and it

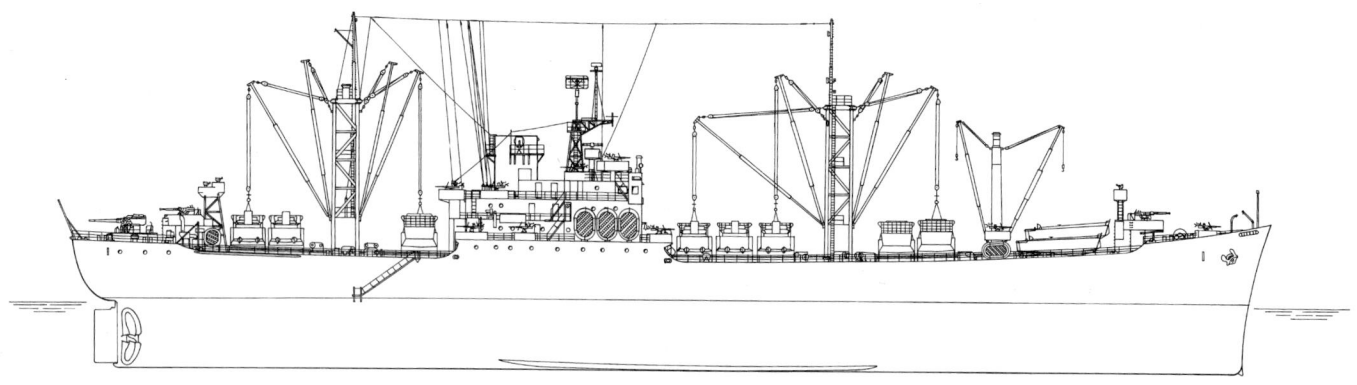

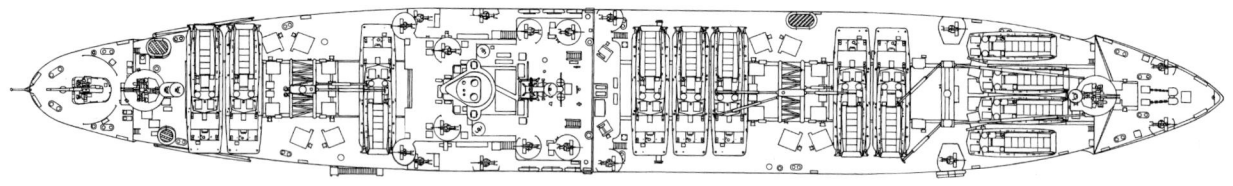

Arcturus (AKA 1) in December 1944, with eight LCM(3) athwartships on deck, each with an LCVP stowed in it, and four LCVP in Welin davits. Each quadrupod mast supported 30- and 10-ton booms, the former on the centerline, and the latter staggered to allow for LCVP stowage on deck. *A. D. Baker*

could handle a 50,000-lb LVT. The bow ramp would carry an 80,000-lb load unsupported at its end, or 160,000 lbs if supported by a beach. The forest of natural-circulation ventilators on the main deck of the existing LST had obstructed vehicle stowage. Instead, the new design used forced ventilation, air being drawn in along the tank deck and ejected aft. Unlike the earlier LST, this one could carry 50,000-lb LVTs on her main deck without any special shoring (i.e., about four times the weight loading of the earlier class), and her tank deck was designed for "the heaviest tanks now under development," presumably the super-heavies the army then had in mind.

War experience had shown that cargo carrying was quite important. Therefore, the cargo hatch was enlarged. Two 5-ton booms could lift loads about 12 ft over the side. Two double-banked davits could take four LCVPs, which were considered valuable for pontoon handling and cargo unloading. Postwar, the Marines pressed for 10-ton booms.

The longer hull demanded a heavier structure, and the ship had a double bottom in way of the machinery spaces, the ballast control room, and the forward magazine; LSTs had been damaged while beaching over rocks and coral heads. The forward 50 ft of the bottom was also reinforced to 40 lbs of plating to reduce bottom damage while beaching.

The existing LST was armed only for self-defense. The new LST would also provide fire support, designed to be equipped with two 5-in/38 (each controlled by an Mk 57 director; one replaced a planned quadruple 40-mm mount). The old one had no such large-caliber gun. To save weight, the six 40-mm mounts (two twins, four singles) of the earlier LST were replaced by two twin 40-mm. The twelve single 20-mm guns of the earlier design were cut to six twins. The steam plant, a heavier armament, and more boats all demanded a larger crew: 178 rather than 106 enlisted men. Of the three U.S. ships, LST 1155 was canceled in January 1946; the other two became the basis for postwar designs.

The British equivalent, the LST(3), makes an interesting comparison. The British badly wanted LSTs of their own specifically for a planned campaign in Southeast Asia. By the time the LST(3) was designed in 1943, the British had been badly frustrated by U.S. unwillingness to release scarce amphibious shipping for their attempt to reconquer their prewar Southeast Asian empire. They wanted higher speed, 15 kts, because Southeast Asian distances were so great, but, like BuShips, they soon

LST 1153, the prototype steam-powered fast LST, is shown off Boston, 25 August 1947. Her main battery is a pair of 5-in/38s, fore and aft, backed by a pair of twin 40-mm guns forward, and by 20-mm guns. This ship was the basis for all postwar LSTs until the radically different *Newport* class. LST 1153 was named *Talbot County* in 1955.

realized that it was out of the question. Even to get 13 kts, they had to accept deeper bow draft, 4 ft 6 in, which seemed excessive. They were particularly interested in holding down draft so that the ships could use small ports. The ship was powered by a pair of frigate reciprocating engines, because the British lacked turbine production capacity. The British had already rejected BuShips' offer of 50 sets of LST diesels, and it was not revived because they could not produce the necessary auxiliary diesels; they would have had to provide a boiler simply to run auxiliaries. Several frigates were canceled late in 1943 specifically to release engines for this program. With far less power, LST(3) made nearly the same speed as the new U.S. design, about 13 kts. BuShips received details of the British design (completed in December 1943) in April 1944, before its own letter to OpNav proposing a new LST had been sent. A note on the BuShips route slip (enclosing details of the British ship) commented that BuShips' engines would be about half the weight, twice the efficiency, and a quarter of the upkeep of the British engines. To speed construction in their own yards, the British had to adopt riveted construction, which increased hull weight about 12 percent. In the end, LST(3) demonstrated the sort of limits imposed by the basic LST concept, which affected not only LST 1153 but also all postwar U.S. LSTs.

A new type of amphibious ship, the LSM (landing ship, mechanized), emerged out of the LCT. In the spring of 1943, the U.S. Army planned to put larger tanks into production. The LCT(6) would have to be lengthened by 24 ft to gain enough buoyancy to accommodate them. There was some question as to whether the longer craft could be carried by an LST; it might have to travel on its own bottom, probably over considerable ocean distances. The shallow LCT structure could not bear the stress of the longer hull. Even if stresses had been tolerable, the LCT had too little freeboard for real seakeeping. The British had encountered much the same problem with their big but shallow-draft TLCs.

BuShips decided that the fast LCI(L) had created a new amphibious niche. The enlarged seagoing LCT could be made fast enough to run in convoy with LCI(L)s. Like the LST, it would be ballasted for beaching or for ocean voyages. By mid-1943, BuShips was working on a 150-ft, 15-kt tank carrier, with space for tank crews (as in an LST). By late June, Cominch was very interested. The new craft was, in effect, a smaller, faster LST. Quite possibly Cominch liked the idea of more smaller beaching ships because the Guadalcanal and North African landings had demonstrated just how vulnerable amphibious ships could be; this was when Cominch was pressing for small attack transports. Production was approved, and it was ordered expedited. BuShips was ready to circulate a preliminary design by late August.

The new LCT design kept growing to provide the combination of freeboard and strength needed for ocean passages. In preliminary form, with a blunt LCT-style ramped bow, it was 196 ft 6 in long on the waterline (the LCT was 105 ft). It had nearly three times the landing displacement of an LCT(6), but cargo weight was only 165 tons, compared to 150 tons for the LCT. On the other hand, deck area was about twice that of an LCT(6), so it could carry about twice as many lighter vehicles (like the LCT; unlike the LST, however, it had no upper deck). In beaching condition an LST carried only about three times as much cargo as the new craft, but for ocean voyages it carried about twelve times as much. The new craft could carry 75 troops (vehicle crews) and either five M4 tanks, six LVT, or nine DUKW. Like an LST, it would beach trimmed down by the stern; required keel slope was 1:75 (the design showed 1:78), compared to 1:250 for an LCT.

The LSM, an entirely new type of landing ship, was created remarkably fast during World War II, despite numerous demands on ship design and ship construction resources. Newly completed, LSM 152 is shown off the Charleston Navy Yard during the summer of 1944, painted in Pacific jungle camouflage.

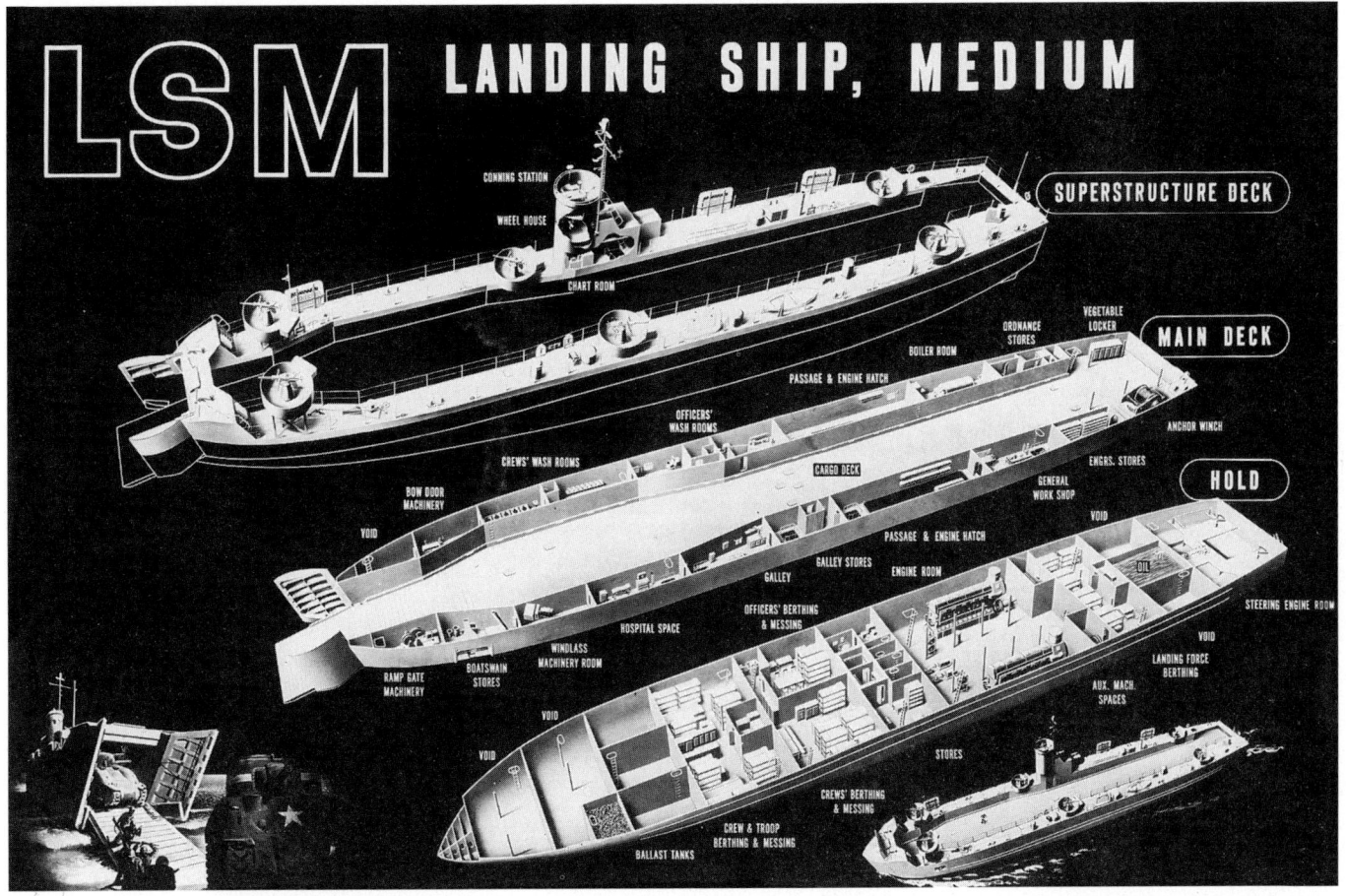

A wartime official drawing shows the internal arrangement of an LSM. As built, LSMs generally had a single 40-mm gun in the bow instead of the two 20-mm shown. Some had both the 20s and the 40-mm, and ultimately the single gun was replaced by a twin director-controlled mount.

The blunt bow was hardly an asset for a ship that might encounter rough weather, so BuShips placed LST-style bow doors forward of the ramp. Plans originally called for quad Gray Marine diesels, as in LCI(L)s. In September 1943 BuShips chose a pair of 10-cylinder Fairbanks-Morse diesels (2,700 BHP each), as in submarines and destroyer escorts, which would drive the craft at 15 kts (12–13 kts sustained at oceangoing displacement). LST diesels were rejected because they were slightly too large and required propellers with excessive diameter.

Details of the new LCT design were fixed in September 1943. Unlike the LCT(6), it would have a fixed solid transom; it would not be used to unload LSTs. On the other hand, the width of its deckhouse was minimized so that it could help unload attack cargo ships. Like the later LSTs, it would have a crawler crane to help unload cargo. These decisions suggest an intent ultimately to replace the LST with the new LCT. Armament was initially set at six 20-mm guns. By November 1944, however, two of the 20-mm had been replaced by a twin 40-mm mount; the interim battery was a single 40-mm.

Initially the craft was called an LCT(7), but it was soon clear that it deserved a completely new designation: landing ship, medium (LSM).[13] On 13 September 1943 the VCNO ordered BuShips to build as many as possible in 1944. Orders continued through April 1944. Production began in December 1943, and the first completed ship was delivered 14 April 1944. Others were converted to rocket-firing ships (see Chapter 8) and to auxiliaries (see Appendix B).

The LSM hull was criticized for its flimsy exterior plating, which could be damaged when going alongside major units to help them unload. A rubbing strake (a split 10-in pipe) was sometimes installed to help. The skeg was not rigid enough, and the bow doors were often damaged and misaligned. There was not enough main capacity simultaneously to fight a fire and to cool the engines. Cargo capacity seemed small in proportion to the size of the ship, although the ratio was

Like LSTs, LSMs were attractive candidates for alternative roles, given their large open deck space that could be filled with equipment. LST 445 was converted into a radar test ship (later a radar training ship) to support the new operational development force. She is shown, newly converted, off Boston Navy Yard on 27 September 1945. The foremast carries the antenna of an SP height finder. The main mast carries the antenna of an SR air search set with, below it, TDY jammers (microwave in the radome, metric-wave on the platform abaft it). The original LSM mast carries an SU surface search antenna. The ship was armed with a twin 40-mm director-controlled gun forward. The mainmast was later removed, and the ship converted to a launcher for target drones, designated YV 1 (named *Catapult* in June 1957).

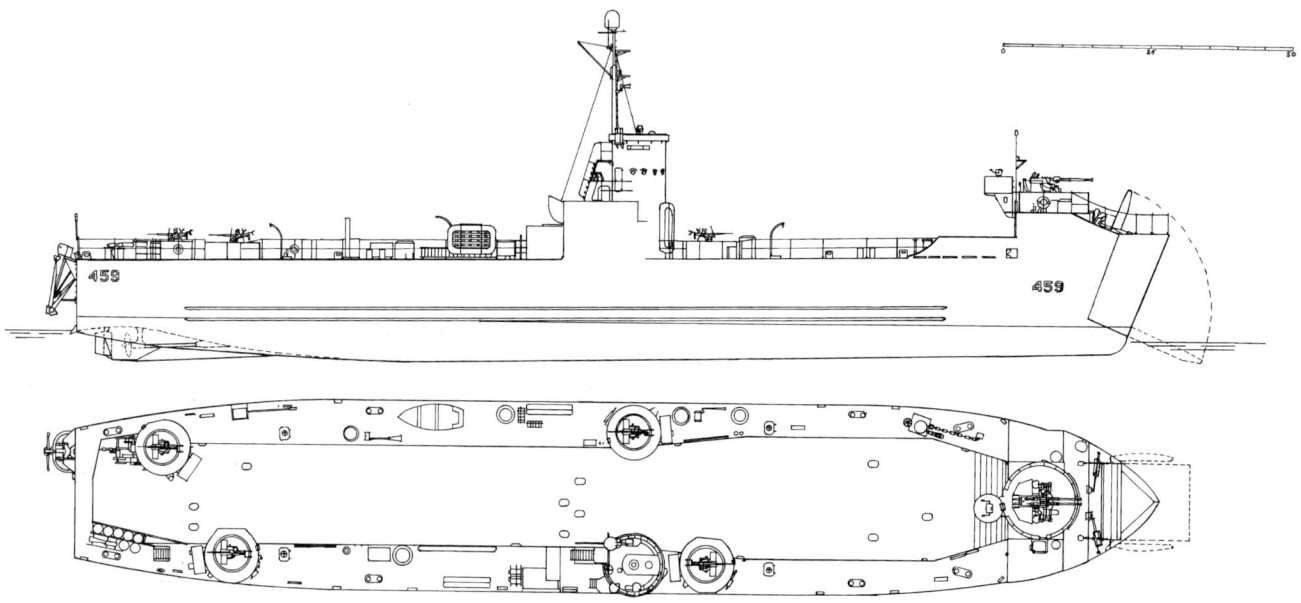

LSM 459 as completed, with the planned ultimate LSM armament, including a twin 40-mm director-controlled gun mount in the bow. Many LSMs were completed with single 40-mm guns forward. She also has the hull stiffeners applied to many LSMs in view of their thin hull plating. *A. D. Baker*

Table 6-5. Postwar Plans for Amphibious Ships and Craft

	Available	Plan 1[a] (2 May 1945)	Plan 1A (14 December 1945)	Plan 2 (31 March 1946)
AKA	107	15/66	15/66	12/24
APA	227	43/39	29/49	26/52
APD	93	—	18/74	12/80
LSD	21	9/12	5/16	4/17
LSV	6	3/3	1/5	1[b]/5
LST	817	30/124	30/114	24/120
LSM	488	30/204	30/204	24/210
LSMR[c]	57	30/27	30/18	24/24
LCI(L)	343	30/90	30/126	24/108
LC(FF)	49	12/16	11/17	0/20
LCI(M)	60	0	0	0
LCI(R)	52	0	0	0
LCI(G)	140	0	0	0
LCS(L)	124	78/46	30/94	0/99
LCT(5)	257	0	0	0
LCT(6)	917	30/444	30/444	24/146

[a]Figures are Active/Reserve. In Plan 1, reserve includes low-manned units assigned to service forces, and inactive units. The number on hand includes ships built and authorized.
[b]Slated for conversion to ASW netlayer and mine countermeasures mother ship (AKN).
[c]LSMR, LCI(M), LCI(R), and LCI(G) are fire support craft, described in Chapter 8. LC(FF) was a headquarters craft, described in Chapter 9.

better than in the LST. Postwar evaluators could see no particular logic to the speed of the LSM. It was too slow to keep up with amphibious convoys (15 kts), although faster than the LST. At an August 1944 Anglo-American conference on wartime amphibious development, the Americans characterized the British equivalent, the LCT(8), as better and handier, but with much inferior machinery: 460 HP Paxman-Ricardo diesels rather than 1,600 HP Fairbanks-Morse engines.

The LSM was a remarkable achievement in a very short time. By the time it was ready, however, the shortcomings of a quick design were far too evident. It had little role in a postwar amphibious force, but was widely distributed to foreign fleets (see Appendix B).

Postwar demobilization plans drawn up in 1945–46 (summarized in Table 6-5) give some idea of the relative value accorded the different types of amphibious ships. The 1946 plan provided a full division of amphibious lift in each of the two active fleets (Atlantic and Pacific); there was not yet a planned structure of forward-deployed fleets. These figures were cut, particularly after 1949, as the naval budget contracted dramatically.

The LCVP was probably the most important U.S. landing craft of World War II. Beginning in the fall of 1942, it carried most American, and many Allied, troops to landing beaches. Classic photographs of troops leaving a landing boat on D-day show LCVPs. Two LCVPs are moored among LSTs at a North African port in 1943.

7
Increasing Demand for Landing Craft

Perhaps the most important wartime development of beaching craft was that the personnel and vehicle versions of the Higgins boat were combined into what became the standard wartime small U.S. landing craft, designated the LCVP (landing craft, vehicles and personnel) in the fall of 1942. Compared to an LCPL (landing craft, personnel, large) or LCPR (landing craft, personnel, ramped), it had a much blunter bow, hence it was about a third slower although it used the same engine. Compared to an LCV (landing craft, vehicles), the main changes were the addition of about 1,500 lbs of ¼-inch STS armor to the sides and the ramp, heavier lifting gear (so that it could be lowered with an 8,100-lb load), and relocation of the coxswain's station into the boat (alongside the engine) so that it could be triple nested. Two gunners' cockpits, with foundations for 0.30 caliber machine guns, were sunk into the deck abaft the coxswain's station; when armed, LCVs used stanchion mounts on their bulwarks. The well deck was slightly shorter (17 ft 3 in vs. 17 ft 6 in) but wider (7 ft 5 in vs. 6 ft 4 in at its bottom) than that of the LCV. Compared to its predecessors, the LCVP was considerably heavier (about 17,800 lbs) due to the armor, increased fuel, and heavier lifting gear. The increase of 3,500 lbs over the LCV cost some freeboard, but that was considered acceptable. Like the earlier LCPL and LCPR, it could be lowered when loaded only from davits; it had to be unloaded to be hoisted out by a boom.

The new boat, essentially a modified LCV, was first tested on 5 October 1942 by the Atlantic Fleet amphibious force. LCVPs soon replaced all other types of boats, except for special purposes. As Appendix B shows, orders for LCVs and LCPRs were rewritten for LCVPs. Given the numbers produced, the LCVP was ubiquitous; it literally made large-scale U.S. amphibious operations possible. LCVPs were first used in the North African invasion in November 1942.

In a wartime assault, eight LCVPs in two waves of four normally carried an assault company of a battalion landing team. Later LCVPs brought light vehicles ashore. By 1945, the Marines generally brought their first-wave troops inshore on board LVTs rather than to the surf line aboard beaching craft; the troops boarded their LVTs at sea, outside the range of enemy guns, from LCVPs. Thus there had to be a relationship between LVT and LCVP troop capacities.

Neither the LCVP nor its LCPL predecessor was designed to carry any particular tactical formation; it just happened to accommodate 36 troops. Normally, then, LCVPs were not fully loaded. For Normandy, the decision was taken to reorganize formations specifically so that the minimum number of boats would be used. Boats were in short supply due largely to the expected demands of the invasion of Southern France, timed for shortly after Normandy (and designed to pull German central reserves out of Northern France, so that they could not resist a breakout from the Normandy beachhead). Thus each assault company (6 officers, 193 enlisted men) was organized into a command boat plus five assault sections and a support section (each section would have 1 officer and 29 enlisted men), each on board one LCVP, rather than the usual three rifle platoons (41 men each) and a supporting weapons platoon (60-mm mortar and 0.30 caliber machine guns).[1]

LCVPs were also adapted to other roles. Some became drone explosive boats (Apex). Others were modified as shallow-water minesweepers.[2]

Some wartime U.S. users considered the LCVP too fragile. Presumably the original Eureka was enormously strong partly because its bow consisted of a single solid block of wood. With the block replaced by a ramp, its bracing effect was gone. The boat's wooden skeg proved quite vulnerable in shallow water. According to a British report on U.S. Pacific operations in November 1944, LCVPs were

A wartime navy poster shows the internal arrangement of an LCVP. It was similar in arrangement to the LCV, but its cox station was recessed into the hull so it could fit a triple Welin davit.

not robust enough for post-assault use, whereas metal LCMs stood up well. The early postwar Pacific Fleet report on ships and craft suggested that future LCVPs should be made of light metal alloy with high tensile strength, for ruggedness. At an August 1944 Anglo-American round table on amphibious craft, Rowland Baker, the British amphibious expert, admitted that the LCVP had better surf performance than the British LCA. The manuscript of the British official report on amphibious craft added that the LCVP was faster and more seaworthy. Baker pointed out that the LCA was better protected and had greater carrying capacity. It lacked astern power, but its twin screws offered better maneuverability to negotiate beach obstacles.[3] Probably in hopes of convincing the U.S. Navy to build ships that could carry the LCA, the British lobbied the Americans to change their standard 20 ft 6 in davit spacing to the British standard of 27 ft 3 in, but that was impractical because too many American transports were too far along by that time. As it developed, the LCVP long outlasted the British LCA, even in British service. The U.S. Navy continued to build LCVPs of essentially Higgins' hull form, both in wood and later in plastic, and many other navies built U.S.-type LCVPs of their own postwar (see Appendix B).

More generally, landing craft were subject to very rough treatment, and some had to be converted specifically to salvage others. For example, in the Marshalls invasion during January 1944 an LVT was assigned specifically for salvage; 136 others were used as troop carriers. At Normandy there was a specially equipped LCM, carrying a bulldozer, motor-driven pump, tow lines, repair materials, and a specially trained crew. For Salerno, four LCI(L)s and two LCTs were used as inshore salvage craft. Each APA, AKA, and XAP had one LCM(3) fitted with a special pump and salvage gear, to augment beach repair parties. For Normandy, 18 LCM(3) were used in craft recovery units. It was expected that, because the beaches were so flat, more than the usual number of craft would broach or be damaged. The LCMs were equipped with fire pumps and special towing bitts and towing gear. Two of each three carried a bulldozer to assist in retrieving stranded craft. According to the after-action report, they proved quite useful, particularly after the storm that struck the invasion beaches a few weeks after the landing.

Troops from an attack transport board an LCVP by climbing down cargo nets. Due to bad weather at Normandy, U.S. attack transports there had to load their boats at the rail rather than in the water. That could not be done with the outboard boats in triple Welin davits, because troops had no way to step directly into them. The British official naval staff history of the Normandy landing asked whether the large numbers of craft APAs carried were worth the sacrifices involved in carrying them, given the inability to lower all of the craft loaded in rough weather.

Damaged craft were carried back to England for repairs aboard the LSD HMS *Oceanway;* on one trip she carried 17 LCM(3) from Omaha Beach back to Plymouth. During the invasion of Southern France each of three special inshore salvage units on average comprised three LCI(L), one LCT(5), two LCM(3) (in addition to those on the APAs and AKAs, one per ship being equipped and assigned to assist in salvage on the ship's assigned beach), one warping barge, and two small tugs (YTL). The LCI(L) were modified by installing heavy bitts, stern chocks, heavier stern anchor and anchor wire, and miscellaneous gear. Six LCI(L) were fitted for salvage and fire-fighting, and three for salvage only. Four LCT(5) were fitted with sheer legs for shallow-water salvage. Inshore salvage work included helping stranded or broached craft retract, and even fighting grass fires ashore.

In 1946–47 the Marines proposed characteristics for a new LCVP that BuShips was about to design. They wanted a larger cargo space, lighter loaded draft—and higher speed. Perhaps their most important requirement was overhead splinter protection against proximity-fused bombs and shells. In April 1946 the Marine commandant called for a 50-ft LCVP, which might carry vehicles of up to 15 tons. By November 1946 the Marines were considering a 40-footer capable of carrying 48 troops, slightly more than an entire infantry platoon. Six, rather than eight wartime LCVPs, could carry the assault company of the battalion landing team. Beach defense weapons, however, were becoming more accurate, so they were more likely to sink individual landing craft before they could beach. Bigger boats would be just as easy to sink as smaller ones. Also, the existing LCVP carried as many troops as an LVT,

LOA- 35'-9" Max Displ- 26,600 LBS Capacity- 36 MEN OR 8100 LBS
Max Beam- 10'-6-1/4" Hoist WT- 18,500 LBS Speed at Max Displ- 9 KTS
Mach & H P-1- 225 HP DIESEL 24 VOLT Fuel Cap- 180 GAL
Cruising Range at Full Power & Full Ld- 110 NAUTICAL MILES ARMAMENT- 2- 30 CAL MG
Purpose- TO LAND AND RETRIEVE PERSONNEL OR EQUIPMENT DURING AMPHIBIOUS OPERATIONS

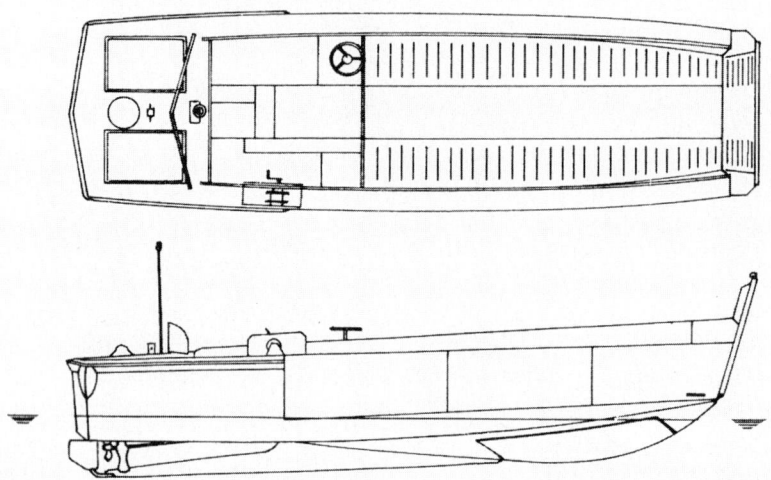

LCVP, the standard World War II infantry landing craft, descended from the Higgins boat and the LCV. Unlike the LCV, it had a steering position sunk into the hull, so that boats could easily be nested in triple Welin davits. In this postwar LCVP illustration, the craft lacks the two gunners' cockpits (for 0.30 caliber machine guns) set into the hull just forward of the prominent breakwater visible aft. *U.S. Navy*

An LCVP runs toward a beach at high speed. Although the maximum speed (9 kts) may seem unimpressive, it was 50 percent faster than the craft's natural speed of 6 kts. That is why postwar attempts to build a really fast LCVP proved so difficult.

hence it could transfer its whole load to the amphibian. A 48-man LCVP would carry too many for one LVT and too few for two. The Marines abandoned the 40-footer, although they continued to support the commandant's recommendation for the 50-footer. However, in the end BuShips tried for an improved 36-footer, and had to settle for a plastic version of the wartime boat (see Chapter 10).

In 1943 the U.S. Navy bought an electric surfboard-type landing craft, the "Flying Mattress," specifically for underwater demolition teams. This silent craft was designed to carry one or two swimmers toward a coast from a craft offshore. It was devised by the wartime Office of Strategic Services (OSS). Records of naval use are sparse. However, it was reported in use by UDTs at Saipan in June 1944, where the commander of a UDT team rode one to within 300 yards of the beach. It was also used by beach jumpers (deception operators) in the Mediterranean, for example, to land commandos for a diversionary operation in the South of France.[4]

Three abortive personnel landing craft were developed. In 1942, looking toward the demands of an invasion of France, a 28-ft craft was conceived. Its length was set by standard liner davit dimensions, the idea being to use ships that had not been refitted as attack transports. It does not appear that any of these boats was built.

There was also a 32-ft plastic "Chemold" LCP(N) (landing craft, personnel, nested), proposed in October 1941 by Western Plastics of Los Angeles. It was marketed to the Pacific Fleet Marine Force, which in turn asked BuShips to buy it. The bureau was interested in plastic hulls but less so in a new landing craft. It wanted test craft of various types. That would have been very expensive, because each design required its own mold. Western Plastics persisted, and BuShips bought prototypes. Having no internal stiffeners, the boat could be nested on deck, greatly increasing a ship's boat capacity. Nesting made it impossible to provide the boat with an inboard (enclosed) engine, so the boat was powered by a 50 HP Evinrude 2-cycle 4-cylinder outboard. In trials, with a slight overload (3,500 lbs) the boat maneuvered well in a foot of water and retracted through 5-ft surf. It could make 13.7 kts unladen and 9.6 kts laden. The boat could be hoisted by a derrick or an LCVP davit. On the other hand, troops on board would be cramped, the boat was not sturdy, and outboard engines were unreliable. The boat could not, therefore, replace the LCVP. Production for underwater demolition teams (UDTs) was canceled at the end of the war.

There were also 26-ft Outboard Landing Craft Mks 1 and 2. These shallow-draft skiffs were bought in 1942 from Miami Shipbuilding specifically to run over reefs into lagoons. The two versions had alternative lines, and both were tested in July 1944. Like the Chemold boat, they used outboard motors. However, the motor could not be turned around to

The "Chemold" LCP(N) under test off the California coast.

help the boat retract. Although test results were encouraging, the boat was apparently too fragile to be useful. It could not, for example, survive surf. Only four were made.

The Higgins LCM(3) could only just transport the existing version of the M5 Sherman tank, the army's standard type. By 1943 the army wanted a larger landing craft; a heavier version of the Sherman was in prospect. Its Engineer Amphibious Command developed characteristics, and Higgins built a 62 ft 9 in boat, which the navy called LCM(X), powered by four Tucker-Higgins gasoline engines. In this form the boat could carry a 20,000-lb load at 13.5 kts or a 70,000-lb load at 10.7 kts. It had ½-in armor around its fuel tanks, control station, and machine gun turret, and ¼-in armor on its hull. It was armed with three 0.50 caliber machine guns, one on the control station and two in an Mk 17 turret. Compared to an LCM(3), it could carry a heavier load (equivalent to the improved Sherman tank then under development), it had more cargo space, it was a slightly better seakeeper (due to its greater freeboard and increased deadrise), and it had better armament. It also had a greater radius of action, 250 nm. The prototype was tested in the fall of 1943 by the army's 4th Engineer Special Brigade at Camp Gordon Johnston, in Carrabelle, Florida. The army was enthusiastic.

The alternative was LCM(6), a standard LCM(3) with a 6-ft section inserted amidships.[5] LCM(3)s converted to LCM(6)s could be recognized by a broken deck line; new-build LCM(6)s had a continuous deck line. LCM(6) could always be distinguished by the number of vertical reinforcing strakes on the bulwark: 10 on each side for LCM(3), 12 for LCM(6). An LCM(6) could just lift the 32-ton improved version of the Sherman. The LCM(6) was apparently developed at the request of forces in the Southwest Pacific. It required more steel than an LCM(3), so in October 1943 the navy pointed out that production would have to be cut from 150 to 135 per month in the first quarter of 1944. The army accepted that in December 1943, and promised to transfer enough steel from its allocation to the navy to produce enough LCM(6)s. At that time the army wanted a total of 1,559 LCMs of all types, a figure later increased to 2,109. No such production was possible, of course, if the army chose LCM(X) instead. The first 48 LCM(6) were produced late in 1943.

Because the navy was in charge of landing craft production, it had to choose which landing craft the army would get. The engine was a key problem. The Tucker-Higgins gasoline engine was not in production. Two Gray Marine twins (four diesels) would offer nearly the same speed as the Tucker-Higgins engines, but such engines were in short supply because they needed elaborate gearing. Powered by the usual pair of Gray Marine 6-71s, LCM(X) would make only 10.5 kts carrying a light load (20,000 lbs), and only 9.0 kts carrying 70,000 lbs, the improved tank. An LCM(6) powered by two Gray Marine diesels would be just as fast, and would require less steel. As of early 1944, a total of 1,141 LCM(6)s were programmed for 1944, against the army's requirement for 1,400. BuShips estimated that it could produce 150 LCM(X) per month once the design had been fixed, which it thought could be done by May. By that time 541 LCM(6) would have been built; to reach the desired 1,400 craft, 859 LCM(X) would be needed. To get them during 1944, the production rate would have to rise to 175 per month, which would be possible only if they had an overriding priority. BuShips recommended against the new LCM(X).

In January 1945 it was discovered just before the attack on Iwo Jima that the new Shermans exceeded the safe load limit of the LCM(3), and loading had to be rearranged quite suddenly. LSDs were to have transported LCTs and LCM(3)s loaded with the tanks, but that was no longer possible. Yet the LSDs were still wanted for small craft docking and repair; their 36 LCM(3)s were loaded with naval resupply ammunition instead of tanks. The tanks were embarked aboard 6 LSMs, which supplemented the planned 25.

Presumably this experience demonstrated that the navy needed the LCM(6), which had previously been used only by the army. A few were shipped on tankers, but most went overseas in sectionalized form. For Leyte Gulf in October 1944, however, a few went on board AKAs and LSDs, in effect proving that they could replace LCM(3)s aboard navy transports. In January 1945 experiments on board APA 14, AKA 3, and APA 228, as well as studies of ships' plans, showed that the LCM(6) could replace nearly all LCM(3)s aboard attack transports and attack cargo ships. AKA 3, for example, could replace six of her eight LCM(3) with LCM(6); there was insufficient space around the other two. Small transports would have some difficulty in a heavy sea due to restricted working space and close clearances. Contracts were issued to convert 200 (and then 500 more) LCM(3) to LCM(6) for naval use. Afterwards a typical attack transport carried two of each type of LCM. By early 1945 the LCM(6) was credited with the ability to carry 80,000 lbs through heavy surf, thus exceeding the desirable load of 70,000 lbs.

The LCM was not without detractors. After the war the Marines complained that it drew too much: with an M4 on board, 4 ft 4 in forward and 3 ft 4 in aft. They wanted the craft rebalanced, to draw less forward (preferably 1 ft forward and 3 ft aft, although they would accept 3 ft forward and 4 ft 6 in aft). They also wanted a larger cargo space with a wider (14-ft) ramp. These desires helped define the modified LCM(6) that became the postwar standard type (see Chapter 10).

There was also a tactical issue. LCMs could bring tanks to a beach only so quickly. To get sufficient numbers, larger tank carriers, such as LCTs, would have to be used, but they presented concentrated, lucrative, targets. At Normandy, where the Germans had fortified the beaches extensively, an alternative was tried, the dual-drive or swimming tank. A canvas shield erected around a tank gave it sufficient freeboard to float, if only barely, with most of the tank underwater. A propeller geared to its engine propelled it. Launched in deep water, then, the tank could propel itself to the beach, where it would drop the canvas and go immediately into action. The advantage of this approach was that tanks could be widely dispersed. Separate tanks presented much smaller targets than LCTs. Experiments in launching the special dual-drive tanks began with LSTs, but the LCT was selected because the larger number of such craft would disperse the tanks more widely. It was necessary to set the right angle between tank deck and ramp, and also to support the ramp, which would be in deep water and thus not supported by a beach. U.S. LCT(6)s were specially modified, each carrying four tanks; they could be launched in less than eight minutes. Launching was generally successful, although many of the special Shermans swamped afterwards in rough seas, particularly off the U.S. beaches.

The LCM and the LCVP were often used as cargo carriers and, as has been noted, by 1944 there was a strong preference for craft that could go all the way inland, to avoid double labor. At Saipan and Guam in mid-1944 a new technique, palletization, was tried. Bulk cargo was split into small waterproof packs, roughly the size of the smallest landing craft that could handle it. Because it was a single unit, a pallet could easily be handled. In a sense this was the forerunner of the container concept, and many postwar amphibious ships and other auxiliaries were designed to handle palletized loads.

By 1944 the army planned to shift to the 51-ton M26 Pershing tank. A 70-ft LCM would be required. It could not have been carried by attack transports and cargo ships, and although one was designed, OpNav decided in May 1944 not to build it. Such tanks would have traveled on board LCTs (themselves transported by LSDs) or else on board LSTs and LSMs. The idea was revived in June 1945 when OpNav asked for a design study. U.S. forces had just crossed the Rhine into Germany, so design requirements included the ability to transport the craft over land (either whole or knocked down) to support river crossings. In December 1945 BuShips reported two alternative versions, both 71 ft 7 in (overall) × 21 ft. Mk 2 incorporated a new W-shaped (or double-V) bottom, which the designers claimed would save at least 8 percent on power; without full-scale tests, however, they could not say whether it would beach and retract well. They also thought it would show considerably easier motion in a seaway, although they needed trials to be sure. On the other hand, the W-form had a deeper draft, 5 ft 6 in vs. the 3 ft 10 in of the conventional hull. Both versions displaced 240,000 lbs fully loaded (100,000 lbs light), and both could carry 140,000 lbs of cargo at 10½–11 kts using three standard Gray Marine diesels. One unusual feature was that the bow ramp folded down so that a tank on board could fire forward as the boat approached the beach. Work continued on both versions; a major design problem was the need to control all three engines simultaneously. The big LCM became the postwar LCM(8) (see Chapter 10). The LCM(6) was also redesigned.

The LVT became an assault craft. Even before the war, the Marines knew that LVTs could the cross coral reefs that stopped boats. LVTs could penetrate 8-ft surf and could move freely once ashore, over sand, mud, gravel, rocks, snow, and ice. An LVT could also climb a 60 percent grade. By 1945 the Marines sent their initial waves ashore on board LVTs rather than LCVPs, because the LVTs did not have to stop at the surf line, where an enemy could zero weapons. In the flexibility it offered an amphibious assault, and in its ability to move inland from the beach, the LVT was the direct ancestor of today's LCAC (landing craft, air cushion). LVTs were used mainly in the Pacific, where they made it possible to land on islands, such as Saipan, fringed by reefs. They landed troops in North Africa in 1942, and they fought in Italy (Po River), at the Rhine Crossing, in Belgium, and at the Scheldt Estuary. A few participated in the Normandy invasion; the army in Europe became much more interested in LVTs after troops emerging from landing craft suffered badly on the Normandy beaches.

By 1941 the Food Machinery Corporation (FMC) had been brought into the LVT program. It redesigned the craft as the LVT(2) Water Buffalo, powering it with a rear-mounted 250 HP Continental radial engine (as in contemporary M3 light tanks), freeing more space for cargo (albeit with a drive shaft running through). LVT(2) could carry 6,500 lbs of cargo or 24 troops rather than the

During World War II, marines came to prefer tracked amphibians to landing boats, because the amphibians could negotiate coral reefs and they did not have to stop at the water's edge to disgorge their occupants. Compared to an LCVP, an LVT's main drawback was its low speed, about half that of the boat. The LVT(2) Water Buffalo was the first wartime model. Although nominally an unarmed cargo carrier, it was equipped with track rails that could take one 0.30 and one 0.50 caliber machine gun. The open cargo hold lent itself to further improvisation. At Roi-Namur early in 1944 twelve LVT(2)s in each of two Amtrac battalions were equipped with pairs of 4.5-in rocket launchers that became disabled by surf as the vehicles reached the shore.

initial 4,500 lbs or 20 troops. Range increased to 75 miles in water or 150 miles on land, compared to 50 or 75 in LVT(1). Speed in the water increased, due in part to more efficient grousers on the tracks, from the 4 mph of LVT(1) to 5.4 mph. Production began in early 1943, the LVT(1) design having been frozen in the interest of maximizing production. LVT(2) entered combat at Bougainville in November 1943.

The army bought a limited number of armored LVT(A)2 cargo carriers; first tested in January 1943, it entered production in April. Some were used for fire support. The army's 2nd Engineers Special Brigade Support Battery was an LVT(A)2 carrying four 4.5-in Mk 7 barrage projectors, three 0.50 caliber machine guns, and a pedestal-mounted 37-mm Mk 4 cannon (as in P-39 fighters; this gun was also mounted on PT boats). The Marines used some of their unarmored cargo-carrying amtracs for similar support missions.

The earliest indication that a prewar landing craft, an amphibian called Alligator, might be more than a cargo or troop carrier was a 27 June 1941 request by the Marine commandant for a version armed with a 37-mm gun, for close support of landing troops; the CNO endorsed the idea in July. There was already interest in close-support boats, and the CMC envisaged a boat capable of crawling ashore to keep supporting those emerging from the boats it accompanied. He wanted a 0.50 caliber machine gun or a 37-mm cannon, three 0.30 caliber machine guns, and armor to resist 0.30 caliber machine guns (0.50s if possible). The boat aspect of the design showed in the planned crew: a coxswain, an engineer, five gunners, and two assistant gunners and ammunition handlers. The CMC soon produced additional details. The single 37-mm gun would be turreted, firing over the operator's cab. The 0.30s would provide abeam and astern fire, and one gun would fire ahead from the cab. The diesel-powered vehicle would make 10 kts in the water (to keep up with landing craft) and 25 mph ashore. Armor would be ½- and ¾-in STS; the turret would be of ½-in STS. Weight might be up to 40,000 lbs.

Amphibious tractors were specialized beasts; BuShips was happy to turn development over to the expert, Donald Roebling, sending him a formal request on 9 August. He in turn rejected the 40,000-lb behemoth; he felt that 20,000 lbs was a limit, and even that was too large. Armament was cut to one 0.50 caliber machine gun in a small turret plus two 0.30s, and armor to ¼- and ½-in STS. In January 1942 Roebling offered a design using the 0.50 caliber turret of the Marines' small Marmon-Herrington tank (CTL3M) plus two 0.30s in fixed sponson mounts and a flexible 0.30. There was already a standing requirement for such vehicles, but the design did not yet exist.

The turret seems to have been the problem. Marmon-Herrington was no longer making ultra-light tanks. It also appears that the Marines were impressed by the 37-mm turret that Borg-Warner had mounted in its abortive amtrac (discussed below). Roebling adopted the tiny turret of the T9E1 (later M22) Locust airborne tank, which carried a single 37-mm gun, but in September 1942 the army pointed out that it was not available in sufficient quantity. The only alternative, the one selected, was the turret of the M5 light tank, carrying a 37-mm antitank gun and a coaxial 0.50 caliber machine gun. The vehicle also carried a 0.30 caliber machine gun

Marines ride an LVT(A)1 into combat in a photograph released in March 1944. The turret cannon is a 37-mm antitank gun; the shielded machine guns are 0.30s. Another 0.30, not visible, is mounted in the turret alongside the cannon. This type was first used in the invasion of Kwajalein, beginning 31 January 1944.

in a ball mount. A pilot model, presumably using the T9E1 turret, had already been completed in June. The new vehicle was designated LVT(A)1. Because the gun was gyro-stabilized, the vehicle could fire accurately even while afloat. Despite its designation, this vehicle was based on the improved LVT(2), with its greater internal space. The first production version appeared in August 1943.

In 1941 the Morse Chain Company, a division of the Borg-Warner Corporation, received a navy contract for an improved LVT track and suspension. Late in September 1941 it decided to spend its own money to build a prototype of an alternative LVT, which it called Model A. The company offered a bolt-on module carrying a 37-mm turret in its cargo section. The navy encouraged further work but had little interest in buying this vehicle. It was not sufficiently better to justify diverting resources.

Late in 1942 the Marines planned an assault on reef-fringed Tarawa. Craft at Guadalcanal were tested in surf, and they were assigned to carry assault troops. This choice was not universally popular. The general commanding the 1st Marine Division in the Solomons considered the LVT far too valuable a cargo carrier to be diverted to troop carrying. Rear Adm. Richmond Kelly Turner, commanding the amphibious task force, argued that the craft were too lightly protected, too slow, too difficult to control, and unseaworthy. However, because they alone could get over the reef, Maj. Gen. Holland M. Smith, USMC, insisted on their use for the first three waves; there were too few to land the whole force. The LVT was now much more than a follow-up cargo carrier.

Marines ride LVT(4)s towards Iheya Jima, a small island near Okinawa, 3 June 1945. Like contemporary landing boats, these craft were ramped (at their rear ends). Each was provided with four socket mounts capable of taking a 0.30 or 0.50 caliber machine gun. Two gun shields are visible here, but only one is occupied by a machine gun (a 0.50 caliber M2).

The LVTs saved the day; marines got ashore in good order, whereas those riding landing craft in later waves were stopped at the reefs, and they suffered badly. There were problems, however. The waves mixed tired craft, which had been at Guadalcanal, with new LVT(2)s. Overloading, wind, sea, and an ebb tide reduced the LVT(1)'s sea speed so badly that the waves were seriously delayed. Lacking radios, the ships offshore lifted their fire at the designated time, but this was well before the LVTs arrived. Many of the unarmored craft were badly damaged (71 of the 125 used were lost), and their gunwales were too high for disembarking marines. The turreted LVT(A)1 was not available for the attack, but it was now clear that such craft should precede troop-carrying LVTs, to suppress beach fire. It was even suggested that these craft should replace light tanks, to save shipping space.

By this time all landing craft were ramped. The Navy-Marine Continuing Board for Tracked Landing Vehicle Development called for a ramped LVT. FMC moved its engine forward to create LVT(4), which was tested in August 1943. The first large order was placed in November, and the craft entered production in December 1943, to become the most numerous type. This version was more efficient than a DUKW for carrying vehicles and for unloading cargo ships. The army considered but rejected an armored version, which would have been LVT(A)3. Optional armor kits were produced, however.

Meanwhile Borg-Warner persevered. To meet the new demand for an open space at the stern, it replaced the single engine with a pair of the 110 HP

A mixed group of LVT(A)1s and LVT(A)4s attacks at Iwo Jima, 19 February 1945. The LVT(A)4s are armed with single 0.50 caliber machine guns atop their turrets, rather than the pair of 0.30s originally provided. The ship in the background is an LCI(G).

Cadillac engines normally installed in M5 light tanks, driving through a differential. This Model B was tested in August 1943, but it was not accepted for production. However, a modified Model D was accepted for production in April 1944 as the LVT(3) Bushmaster. Moving the engines to the side walls opened up a larger cargo bay, which could carry 8,000 lbs of cargo (but still only 24 troops). Because it had an automatic rather than a manual transmission, LVT(3) was far more successful than LVT(4) in the mud of Okinawa, where it first saw combat. It was the main early postwar type.

It was soon clear that the 37-mm gun of the LVT(A)1 could not deal with the sort of beach defenses the Japanese erected at Tarawa. Col. William S. Triplet, commanding the 18th Armored Group that trained army amtrac units at Monterey, California, had ordnance experience, and knew that army light tanks could be armed with short 75-mm howitzers in place of their 37-mm guns. On his initiative, FMC placed a howitzer turret on an LVT(A)1. Triplet successfully demonstrated this improvement to the navy, and got it approved. The production version of this modification in March 1944 was the LVT(A)4. The vehicle was criticized for its lack of overhead protection (snipers could fire into it) and for its lack of machine guns (to kill troops approaching to attack it); it had only a 0.50 caliber gun in a ring atop its turret, rather than the rear machine guns and coaxial gun of the LVT(A)1. An improved "Marianas" model had two 0.30s atop the turret and a gun in a ball mount on the bow (some vehicles already had this gun). Unlike the 37-mm gun, the 75-mm howitzer was not gyro-stabilized, so LVT(A)4 could not fire accurately while afloat. Gyro-stabilization was added in LVT(A)5, which appeared in April 1945, too late for action in the war. The (A)5 designation had previously been applied to an unsatisfactory revised version of (A)4 with its turret moved forward and two 0.30s in scarf rings abaft it. It trimmed by the bow and was not stable enough.

In June 1944 a version armed with the 75-mm turret of the new M24 light tank was proposed. An LVT(A)1 with an enlarged turret ring proved too unstable to fire while afloat, and at the end of the war there was interest in a redesigned version with a somewhat lower hull and the turret modified and

Table 7-1. Wartime Amphibian Development

	LVT(1)	LVT(2)	LVT(3)	LVT(4)	LVT(A)1	LVT(A)4	DUKW
Length (ft-in)	21-6	26-1	24-1.5	26-1	26-1	26-1	31-0
Width (ft-in)	9-10	10-8	10-10	10-8	10-8	10-8	8-2.875
Height w/top down (ft-in)	8-2	8-1	8-5.5	8-2.5	10-1	10-2.5	7-1.375
Weight (lb)							
Empty	17,300	24,400	30,600	27,400	29,050	39,460	13,000
Loaded	21,800	30,900	38,000	33,350	32,800	41,000	18,600
Crew	3	3	3	3	6	5	1
Capacity							
Cargo (tons)	4,500	6,500	8,000	6,500	1,000	5,000	5,000
Troops	20	24	24	24	—	—	12
Power (HP)	146	200	440	200	200	250	94
Fuel (gal)	80	110	150	NA	106	110	40
Speed							
Water (kt)	4	5.4	5.2	5.5	5.4	5.2	5.5
Land (mph)	15	25	25	15	25	15	50
Endurance							
Water (nm)	50	75	75	75	75	75	50
Land (miles)	75	150	150	150	150	150–175	400 at 35 mph
Armament							
Cannon	—	—	—	—	one 37-mm	one 75-mm	—
0.50 caliber MG	1	1	1	4	—	1	—[a]
0.30 caliber MG	1	1	1	4[b]	3	3	—

NOTES: LVT(A)2 was an armored LVT(2), produced only for the army. The abortive LVT(A)3 was an LVT(4) with built-in armor instead of the standard armor kit. DUKW data are provided for comparison. Dash indicates that specification is irrelevant; NA indicates relevancy but data not available.

[a]One DUKW in four was fitted with a ring mount capable of taking a 0.50 caliber machine gun.

[b]Alternative machine gun armament.

moved forward. The war ended before this vehicle could be built, but a 76-mm gun LVT figured in the postwar development program (described in Chapter 10).[6] Table 7-1 shows details of seven wartime amphibians.

In June 1945 both Borg-Warner and FMC produced lightweight (mainly aluminum) versions of their LVT(3) and LVT(4), respectively, but these craft never entered production. By this time the Marines wanted to replace the existing gasoline engines with lightweight diesels; they expected the army to resist, because its vehicles generally burned gasoline. An adapted General Motors 6-71 diesel (as in the LCVP) was proposed for the LVT(4), and lighter-weight 3-71 (90 BHP) or 4-71 (100 BHP) engines for the LVT(3).

The LVT was so successful that by 1947 the Marines wanted to replace the LCVP with a big tracked amphibian. Marine Corps Headquarters laid out requirements. It would have to be faster than an LCVP (11 kts), and as seaworthy (unfortunately, until the late 1990s the best the Marines could get was 7 kts); endurance would be 150 miles. It would have much the same payload (9,000 lbs of cargo or 36 personnel in a 24 × 8 ft hold with an 8-ft ramp), with armor to stop small-arms fire at point-blank range. The best BuShips could offer, through the 1950s, was a much faster LCVP (see Chapter 13).

The great failing of the LVT was its slow speed in the water, because its tracks were so inefficient. In April 1944 BuShips' preliminary design section proposed a 50-ft 15-kt hard-chine boat to carry an LVT close to the shore. In May it sketched an LVT carrier based on an LCM(3), using a more ship-shaped bow, but the same powerplant. Any such arrangement was far too cumbersome. In August 1944 the preliminary design section began work on a faster LVT with better lines (30 ft long, with finer ends). The grousers were inefficient in water, so the designers suggested using waterjets, which were rejected as inefficient, or propellers, which could be housed in wings emerging from the vehicle's sides. The army's DUKW already used propellers in the water. The project was dropped about January 1945 because of other priorities.

Throughout the war, shortages of landing craft was a constant theme in strategic decision making. Landing craft requirements exploded when the U.S. Army was assigned to prepare for the invasion of France, a project called Operation Sledgehammer, in 1942. Small landing craft were crucial because, as noted above, the army expected to use them to carry the assault force—two infantry divisions and two tank regiments—all the way across the English Channel. Because the trip was so long, no craft could return for a second run and still contribute much to the impact of the landing. The army wanted enough craft to lift the whole force in a single trip. None would be needed for the 28 follow-up divisions, which would land at a port seized by the assault force. Allied strategy for 1942 literally depended on whether a huge force of landing craft could be produced and assembled quickly enough in Britain.

Although a March 1942 plan called for 1,500 landing craft, soon the army wanted 8,100: 1,400 LCPL, 3,200 LCV, and 3,500 LCM. The army's deputy chief of staff thought 20,000 craft a more reasonable estimate. These figures included British craft, many of them already on hand. The U.S. contribution was set at 503 LCPL, 1,200 LCV, and 600 LCM, but this did not include U.S.-built craft earmarked for the British, which made up much of the total. British production was already stretched to the full. Early in April 1942 the War Department asked the U.S. Navy to provide the necessary craft in the United Kingdom no later than 15 September 1942.

President Roosevelt backed Operation Sledgehammer, so he ordered the landing craft program accelerated. The army's chief of staff, Gen. George C. Marshall, went to London to urge the project. Without any studies in hand, on 4 April 1942 President Roosevelt ordered 600 LCM produced by 1 September 1942. To get that many in time, on 15 April BuShips ordered 1,100. It also ordered 600 LCP(L), 1,200 LCV, and 100 LCT(5), the latter for 8 LSDs then being built for the U.S. Navy. These orders completely filled U.S. capacity, the bottleneck being a lack of engines. The navy had only enough facilities to train operating and maintenance personnel for its own transports; the army set up schools for its cross-channel craft.

The production goals could not be met. The navy could promise only about 250 craft by August 1942. It thought that perhaps a third of the army's numbers might be met by September. In April 1942 a JCS study pointed out that lack of landing craft would limit any 1942 operation to about a fifth of the required scale. If the navy's own requirements for transports were met, nothing would be available for Operation Sledgehammer. That pushed any invasion of France back to 1943 at the earliest; unless landing craft production was considerably accelerated, even a 1943 operation (on the same scale as that projected for 1942) might be difficult.

President Roosevelt chaired another White House meeting on landing craft on 5 May. If not enough craft could be produced by the spring of 1943, the invasion of France would have to be deferred. Yet it was essential that the Allies take the offensive against the Germans in 1942. Thus the landing craft shortage led directly to the decision to invade French North Africa, code-named Operation Torch, in November 1942.

Too much depended on huge numbers of small craft, which the British pointed out would be difficult to handle cohesively. Also, troops probably would not be fit to fight after long trips (estimated at 8 hrs, 80 nm) in them. With only 22 LSTs likely to be available as of the spring of 1943, the necessary large numbers of tanks could not land over a beach.

It was estimated that by early September the following craft would be available in Britain (with British contributions in parentheses): 4 (4) LST, 196 (151) LCT, 495 (95) LCM, 602 (202) LCV, and 595 (195) LCPL. They could lift about 21,000 troops, 3,000 vehicles, and 300 tanks; the assault force of a 1943 cross-channel operation, however, was 76,500 troops, 18,380 vehicles, and 2,250 tanks. Note the large number of vehicles and tanks. An 8 May 1942 Combined (U.S.-U.K.) Chiefs of Staff paper estimated a shortfall of 15,000 vehicles and 58,000 men. The solution was larger beaching ships: LSTs for vehicles and the new LCI(L)s for troops. The revised assault force for the spring of 1943 was set at 200 LST, 300 LCI(L), 570 LCT, 300 LCM, 500 LCPL, and 80 LCV (including British equivalents). Despite the drastic cut in LCVs (down by 800), the U.S. Navy was reluctant to cancel these craft because it might need them in the Pacific.

Despite adding larger ships the army still planned to send some small landing craft all the way across the English Channel. Moreover, the cuts in small craft depended on LST and LCI(L) production. The combined staff planners pointed out that landing craft numbers might be cut to practical proportions if a few attack transports were used, but both the U.S. Army and the British rejected the idea.

U.S. planners estimated that to obtain sufficient craft they would have to order 172 LST, 300 LCI(L), and 295 LCT(5), and give them priority over all other naval construction. This was impossible, given the raging Battle of the Atlantic and the demands of the Pacific. All of this was aside from the requirements of attack transports. The BuShips priority of landing craft construction, which was tenth at the outbreak of war, rose to highest priority on 4 July 1942. It seemed that all the required small amphibious craft could be in Britain by 1 April 1943, together with 160 LCT(5).

Requirements for the cross-channel attack kept increasing, however. A September 1942 estimate showed, in addition to the LSTs and LCI(L)s previously listed, much larger numbers of other types were required: 35 British LCI (with about half the capacity of a U.S. LCI), 340 LCT(5), 286 British LCT, 750 LCM(3), 212 British LCM(1), 707 LCP(L), 425 British LCA (ramped personnel boats), 400 LCV, and 400 of the new 28-ft boats, plus British close-in support craft. Close-in support craft needed were 12 beach protection craft (BPC), later redesignated as LCF(L) (landing craft, flak, large), for antiaircraft; 47 light support craft (LSC); and 5 heavy support craft (HSC), which the British already had.

In July 1942 the European amphibious effort was redirected to Operation Torch, the November 1942 invasion of French North Africa. Because it was mounted so far from bases, Operation Torch relied mainly on large ships (14 attack transports, 9 partially converted transports, and 9 attack cargo ships), and thus needed fewer small specialized landing craft (105 LCM and 707 personnel craft). Even so, it had to be delayed from early fall to November 1942 to allow for delivery of many craft and for crew training. There was insufficient shipping, including landing craft, to permit the desired landing east of Algiers. Shortages of landing craft would haunt all later operations.

At the same time as Operation Torch, craft were needed both to replace losses at Guadalcanal (58 LCM, 160 personnel craft) and to support troops ashore there (37 LCM and 160 personnel craft), as well as to train and equip new ships (229 LCM and 646 personnel craft) and to replace craft already in Britain. In 1942 the Chiefs of Staff estimated that in 1943 the United States would need, for all purposes, 1,625 LCM(3), 4,836 Higgins boats (including LCV), 415 LCT(5), 294 LCI(L), and 330 LST. Expected U.S. production offered thin margins in most cases—347 LCM(3), 55 LCT(5), 60 LST, no LCI(L)—but a deficit of 3,467 Higgins boats.

The U.S. Chiefs of Staff felt that Operation Torch would preclude any 1943 invasion of France, but the British disagreed: Germany might be so weakened by 1943 that the invasion would be practical. As a result, it was necessary, late in 1942, to produce craft both for Torch and for the projected 1943 operation. The accelerated landing craft/ship program cut into other construction, particularly of ASW ships. Bottlenecks were an increasing problem; the JCS even had to decide on the allocation of a few hundred diesel engines between landing craft and tanks, hence the fall 1942 cancellations of LCI(L)s and LCT(5)s.

In the fall of 1942 the shipbuilding situation began to relax. The U.S. Navy began asking for more transports, and it ordered more landing craft. There were never enough landing craft, partly because LCVPs quickly wore out under harsh operational conditions. Shortages of landing craft and other amphibious shipping notoriously haunted operations in Italy and complicated planning for Normandy. They made it impossible to mount the Normandy and Southern France landings simultaneously, as had been planned. In the Pacific, operations were slowed because it was impossible to form several groups or forces that could strike different targets in quick succession. Plans for assaults in Southeast Asia were long delayed by a lack of amphibious craft that could not be made up by local production in India.

Wartime operations demanded far more fire support than had been imagined prewar. Rockets were part of the solution. Because they imparted no recoil to a ship's deck, they could be fired by lightly built craft such as converted amphibious ships. They offered enormous, if inaccurate and short-lived, firepower. LCI 1078 has just been converted into a rocket-firing support craft in this 29 December 1944 photograph, taken in San Francisco. The Mk 7 rocket launchers, which fired 4.5-in rockets, are the frames visible at the break of her forecastle and abaft her forward 20-mm mounts. A fifth is visible on her port side, just forward of the after 20-mm gun tub. These launchers could elevate between 25 and 40 degrees; maximum range was 1,100 yds. The frame above the launching rail held rounds ready to fire; it was fed by hand.

8
Fire Support

Both the U.S. Marines and the British recognized that ships offshore could not supply enough firepower to overcome local resistance on the beach. Machine guns or light cannon, therefore, must arrive with the first wave of boats. The Marines experimented with guns, including 1-pounders, aboard ship's boats beginning in 1931. The fire support requirement explains why even the early U.S. Higgins boats (LCPLs) were provided with a pair of machine gun mounts in their bows. There was also U.S. interest in adding shore bombardment firepower. The June 1929 Naval Transportation Service Readiness Plan (WPL-10) included eight fire support ships (XAPG), each armed with six 155-mm army Schneider howitzers (with navy-supplied mounts) on her centerline, in two separately controlled groups of three in her well decks. These ships, which had not figured in earlier plans, would have been converted from 250-ft 10-kt cargo ships. Eventually the World War II amphibious fleet included equivalents of both the boat guns and the XAPGs.

The prewar British solution was to convert 40-ft assault landing craft into support landing craft (SLC). Later they tried a larger 47-footer. These became the light support craft (LSC) and the heavy support craft (HSC), respectively. As of August 1941, the British wanted one SLC for each group of ten ALC (assault landing craft).

In June 1941 the Landing Boat Board ordered development of a U.S. equivalent to the LSC, and by September 1941 it was in hand. The obvious candidate was the U.S. standard landing craft, the 37-ft Higgins boat. It could occupy the same davits as the personnel boats, so it could accompany those boats from transport to beach. In fact Higgins modified the boat's lines, giving it deeper deadrise. By January 1942 a pilot model was running tests. Because the Higgins boat was the T-boat, the support version became the TS. Then it became the LBS (landing boat, support). In July 1942, when designations were standardized, it became LCS(S) (landing craft, support, small). The two British types became the LCS(M) (landing craft, support, medium) and the LCS(L) (landing craft, support, large).

The U.S. Navy tried to develop a larger armored support craft, armed with a 37-mm light tank turret, that could fit the stowage space otherwise used for an LCM(3). As of August 1942, 100 were planned. The design proved difficult, however. The boat could be only 48–50 ft long, and it had to be quite beamy to preserve stability, despite the top weight of the turret and 5,000 lbs of deck armor. There was also 4,100 lbs of side armor, an 1,800-lb transverse bulkhead, and 900 lbs on the coxswain's station. The only available powerplant was two 225 HP Gray Marine diesels. After trying five alternative sets of hull lines, BuShips reported in October 1942 that its boat could make only 13.5 kts, rather than the desired 16 kts. Using Hall-Scott gasoline engines would bring speed up to slightly less than 15 kts. To make 16 kts, some drastic weight reduction would be needed: 9,500 lbs for Gray Marine diesels, 5,000 lbs for Hall-Scott engines. The former amounted to eliminating the 37-mm turret and ammunition, all the deck armor, and half the side protection. The latter equated to eliminating the turret and 2,400 lbs of the deck armor. The boat might make the required 16 kts if the turret and armor were given up and a new single 600 HP Hall-Scott Defender gasoline engine was installed. The Defender engine, however, was needed for 45- and 63-ft crash boats and for a planned bomb target boat; also the amphibious force much preferred diesels. BuShips concluded that the concept was impractical. Fortunately, as it was being dropped, a new requirement, for an amphibious control craft, eventually designated LCC (landing craft, control), was arising. On 30 October the Atlantic Fleet amphibious force issued characteristics for a 20-kt control boat and navigation leader with a maximum length of 60 ft. BuShips

The first specialized U.S. fire support craft were LCS, adapted from the LCPL design, which a transport could carry. That is why this LCS(S) Mk 1 is marked P25; she is assigned to AP 25 (USS *Leonard Wood*, later designated APA 12). The Mk 2 version could be recognized by its twin 0.50 caliber machine gun; this Mk 1 is armed with three 0.30 caliber machine guns and two 12-round Mk 1 launchers for 4.5-in rockets.

commented that it could meet most of the desired characteristics, except speed, using a stretched version powered by a pair of Gray Marine diesels and removing the armor and the heavy guns. This craft could exceed 16 kts and it became the LCC (described in Chapter 9).

The LCS(S) program seems to have been deferred initially to permit production of the Higgins personnel boats for the abortive 1942 landing in France. In August 1942 the Small Boat Development Board conference decided to order the first 100 craft. Plans and specifications were at the preliminary stage by 15 August. Compared to a Higgins personnel boat, the LCS(S) had a covered deck and a cockpit amidships. Tracks on the sides of the cockpit could take two single 0.50 caliber machine guns and three 0.30 caliber on skate mounts. The Atlantic Fleet removed one 0.50 and two 0.30 caliber machine guns to provide weight for two rocket launchers (56 rockets) and eight smoke pots at the stern. Any additional loading was unacceptable, partly because the working load of the existing Welin davit was 20,600 lbs; later this limit was relaxed.

The U.S. Navy used LCS(S) for the first time during the North African landings in November 1942. At Salerno these support boats proved quite useful. Initially they laid smoke to cover the landing craft from enemy fire. The Eighth Fleet after-action report mentioned a support boat from USS *Dickman*, which was covering part of 6th Corps Green Beach. Enemy machine gun fire pinned troops of the 142nd Infantry down on the beach. The boat closed to within 80 yds on the left flank of the beach and fired salvoes of 3–4 rockets each from one flank to the other, forming a sweeping barrage of 34 rockets at a range of 750 yds. All enemy fire ceased

Loading 4.5-in rockets on board an LCS(S) Mk 1, March 1944. The electrically fired Mk 1 launcher could elevate from 0 to 47 degrees; at 45 degree elevation its 4.5-in rocket could travel 1,100 yds. The launcher was not trainable; it was aimed by pointing the boat. This weapon armed many types of U.S. fire support craft, up to converted LCI(L)s. The other weapons are two 0.50 caliber machine guns forward and two 0.30s at the after end of the cockpit.

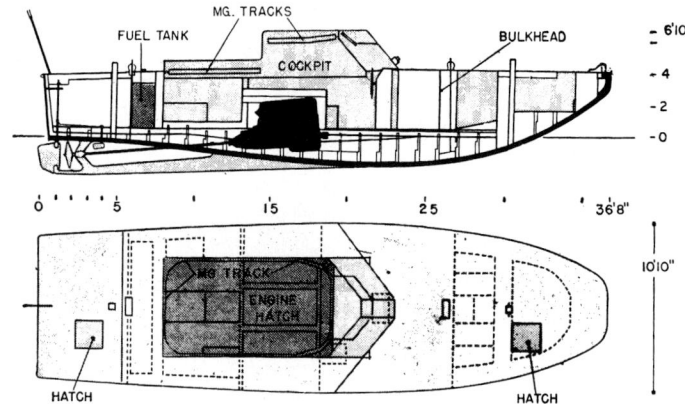

LCS(S) Mk 1, 1944. Note the tracks installed to take single 0.50 and 0.30 caliber machine guns (two 0.50 or three 0.30 or one 0.50 and two 0.30). A pair of box-type barrage rocket launchers, not shown, could be carried alongside the cockpit. The Mk 2 version differed in having one twin 0.50 caliber machine gun in the cockpit, together with two single 0.30s; it also had the usual pair of barrage rocket launchers. Unlike Mk 1, it had a 225 HP Gray Marine diesel instead of the earlier 250 HP Hall-Scott gasoline engine. *U.S. Navy (ONI 226)*

during this attack, and when it resumed it was far weaker—and directed at the LCS(S), not at the troops on the beach. Later a German prisoner said that his machine gun had been destroyed altogether by rocket hits, and that his crew had been badly demoralized. The after-action report concluded that such craft had made up to some extent for the lack of pre–D-day bombardment. The LCS(S) were transported to the beach on the davits of LSTs and attack transports. They were not well liked by those ships, because they were heavy, cumbersome, and used gasoline rather than the diesel fuel of the other landing craft. However, Eighth Fleet concluded that they had been so useful that they would be well worth retaining. At Normandy, LCS(S) were used mainly to make smoke to cover the landing craft. After the first waves had landed, they were used for local traffic control. LCPLs were used similarly. Table 8-1 compares LCS(S) with four subsequent classes of fire support ships and craft, built during and after World War II.

British support craft were more heavily armed. Late versions of LCS(M) carried a pair of 0.50 caliber machine guns in a power turret, two 0.303 caliber Lewis machine guns, and a 4-in smoke mortar. The lengthened LCS(L)1 added the turret of a Daimler armored car, carrying a 2-pounder cannon (45-mm) and a machine gun. Later the British adapted their wooden LCI(S) as LCS(L)2, with a 6-pounder (57-mm) tank gun in a turret, two 20-mm guns, a twin 0.50, a 4-in mortar, and two 0.303 Lewis guns. This version seems not to have been in service when the U.S. Navy developed its second-generation assault craft, in the spring and summer of 1943. As for LCS(S), the British tested one, but preferred their LCS(M).

There was always some question as to whether the LCS(S) offered enough firepower. Thus, in mid-August 1942 BuShips was studying possible conversion of the obsolescent 70-ft PT boat to a heavy support boat. Nothing came of this project, but later torpedo boats were converted into "gunboats," primarily to attack Japanese barges in the Solomons.

At Dieppe in the fall of 1942 the British found that destroyers offshore were unable to deal with German artillery and counterattacking tanks; their solution was to mount medium-caliber (4.7-in) guns on board LCTs. Converted craft were designated LCG (landing craft, gun). They also became interested in a landing craft conversion specifically for air defense: the LCF (landing craft, flak). A version derived from LCT(2) was armed with two twin 4-in guns and three 20-mm; others based on LCT(3) and LCT(4) were armed with 2-pounders and 20-mm guns.

With the LCS(S) not yet available and the North African landing looming, the Atlantic Fleet amphibious force asked for a stopgap. Early in October BuShips offered a fire support version of the obsolescent LCM(2), mounting a 37-mm antitank gun and two 20-mm machine cannon. It cautioned that the boat could not be hoisted out when carrying this extra weight (6,500 lbs); the weapons would have to be lowered into the lighter once it was in the water. However, the foundation for the 37-mm gun would be built in. The project highlighted the problems of designing any sort of support craft that transports could carry to a landing. Nothing seems to have been done at the time.

Table 8-1. Fire Support Ships and Craft

	LCS(S) Mk 1	LCS(L)3	LSM(R) 188	LSM(R) 501	*Carronade* (IFS 1)
Length (ft-in)					
WL	NA	153-0	196-6	204-6	237-0
OA	36-8	158-0	203-6	206-3	245-0
Beam (ft-in)	10-10	23-3	34-6	34-6	39-0
Draft (ft-in)[a]					
Light	3-6	4-0.25	4-5	NA	NA
Full load	NA	4-9/6-6	6-6	7-2	10-0
Displacement[b]					
Light	20,000 lbs	250	605	850	1,040
Standard	22,000 lbs	312[c]	783[d]	994[e]	NA
Full load	N/A	387	968	1,280	1,500
Power (SHP/shafts)	250/1	1,600/2	2,880/2	2,880/2	3,100/2
Speed (kt)	12	15.5	13.2	12.6	15
Radius (nm/kt)	115/12	5,500/12	4,900/12	NA	NA
Complement[f]	3 or 4	5/65	5/76	138	12/150
Main generators (qty-kW)	NA	2-60	2-100, 2-20	2-100, 2-20	2-200
Armament					
5-in/38	—	—	1	1	1
3-in/50	—	1[g]	—	4[h]	—
40-mm	—	2 × II[i]	2	2 × II	2 × II
20-mm	—	4	3	8	—
0.50 caliber MG	2	—	—	—	—
Rocket launchers	2 barrage	10 Mk 7	75 Mk 36, 30 Mk 30	10 Mk 105	8 Mk 105

NOTES: Dash indicates that specification is irrelevant; NA indicates relevancy but data not available.

[a] Draft aft if only one figure given; draft fore/aft if two figures given.

[b] Displacement is given in tons, unless otherwise specified.

[c] Light service condition; draft 3-9.5/5-8.

[d] Attack displacement; 826 tons for LSM(R) 196 series.

[e] Attack displacement; 994 tons for LSM(R) 401 (197-3 length WL, 203-6 length OA).

[f] Complement is given as officers/enlisted when two figures are listed; when only one figure is given, breakdown is unknown.

[g] LCS(L)3 main battery was one 3-in/50 and two twin (2 × II) 40-mm on initial series (LCS(L) 1–10, 26–30, 41–50, 58–60, 79, 80); one 40-mm and 2 × II 40-mm on second series (LCS(L) 11–25, 31–40, 51–57, 61–66, 81–89, 109–124); and 3 × II 40-mm on third series (LCS(L) 67–78, 92–108, 125–130).

[h] 4.2-in mortars.

[i] Multiple gun mounts are indicated by Roman numerals; therefore 2 × II indicates two twin mounts.

Perhaps the strongest lesson of Dieppe was that enemy defenders had to be stunned as the assault reached the beach. The British army was already experimenting with rockets intended to provide smoke cover; a 4.5-in round carried a 60-lb smoke head 3,500 yds. In March 1942 the British Combined Operations coordinator had suggested using rockets from large landing craft, and an explosive warhead was an obvious next step. An LCT(2) was taken over for conversion at the end of 1942. Given the shortage of such craft, the rocket launchers had to be arranged so that the ship could be converted back into a landing craft within 48 hrs. An LCT(2) could accommodate 840 rockets. The project became urgent when Operation Husky, the invasion of Sicily, was planned. In early March 1943, before trials had been conducted, the conversion of six LCT to LCT(R) (landing craft, tank, rocket) was approved. Fortunately the 11 April 1943 trials were quite successful. An LCT(R) typically fired all of her rockets in a single salvo, so there was no opportunity to fire and then correct aim. To solve the problem of precise location, the craft were fitted with adapted bomber radars (H2S).

The Royal Navy provided the inshore fire support at Salerno: nine LCG (4.7-in guns), seven LCF (flak, antiaircraft), and three LCT(R). In addition it converted two small landing craft (LCA) to mortar-firing Hedgerows (four rows of six spigot mortars each, firing 60-lb bombs; in effect they were oversized ver-

Given standard rocket launchers, standard landing craft could be converted into bombardment craft. This LCVP was armed with eight Mk 7 launchers. Note that this was an operational boat, attached to APA 3 (USS *Zeilin*).

sions of the Hedgehog ASW weapon) intended to explode minefields in the path of a landing force. Eighth Fleet considered none of these craft really satisfactory. The British LCTs were slow and not very maneuverable. The LCG could not fire near its bow, hence it could not engage mobile artillery ashore. Hedgerow was a hurried conversion carried out at Algiers. One sank on firing trials because the deck reaction from firing blew out its bottom; the other was used in the assault. This experience fed back into the U.S. discussions of large fire support craft, which bore fruit only in the Pacific.

Before these craft had a chance to prove themselves in Sicily, the Admiralty began planning their use at Normandy; in July it decided to convert 30 LCT(3) by April 1944. They used a new naval launcher, which increased rocket capacity to 1,080 rounds. These craft supported U.S. troops at Normandy and in Southern France, and their success helped encourage U.S. adoption of rockets for bombardment. Specialist fire support landing craft at Normandy amounted to 36 LCT(R), 29 LCF, 25 LCG, 10 LCS(L)2, 4 LCS(L)1, 26 LCS(M), 36 LCS(S), and 45 LCA(HR). LCA(HR) or Hedgerows were mortar craft intended to blast through underwater obstructions; nine were assigned to each British (but not American) assault force. Except for the LCS(S), all of these craft were British. However, the crews for British-built craft operating off the U.S. beaches were American.[1] There were also 144 LCPL smokers, 48 LCT(A), 5 LCT(CB), and 16 LCT(HE), the latter three types being U.S. LCT(5)s carrying vehicles capable of firing while afloat. For the invasion of Southern France the Royal Navy provided 30 LCT(R), 6 LCF, and 5 LCG. In place of the LCA(HR), the U.S. Navy provided "Woofus" boats (discussed below). As at Normandy, there were also U.S. LCS(S).

On 1 May 1943 Cominch (Admiral King's office) called for further developments. There was no U.S. equivalent to the British LCT conversions, partly

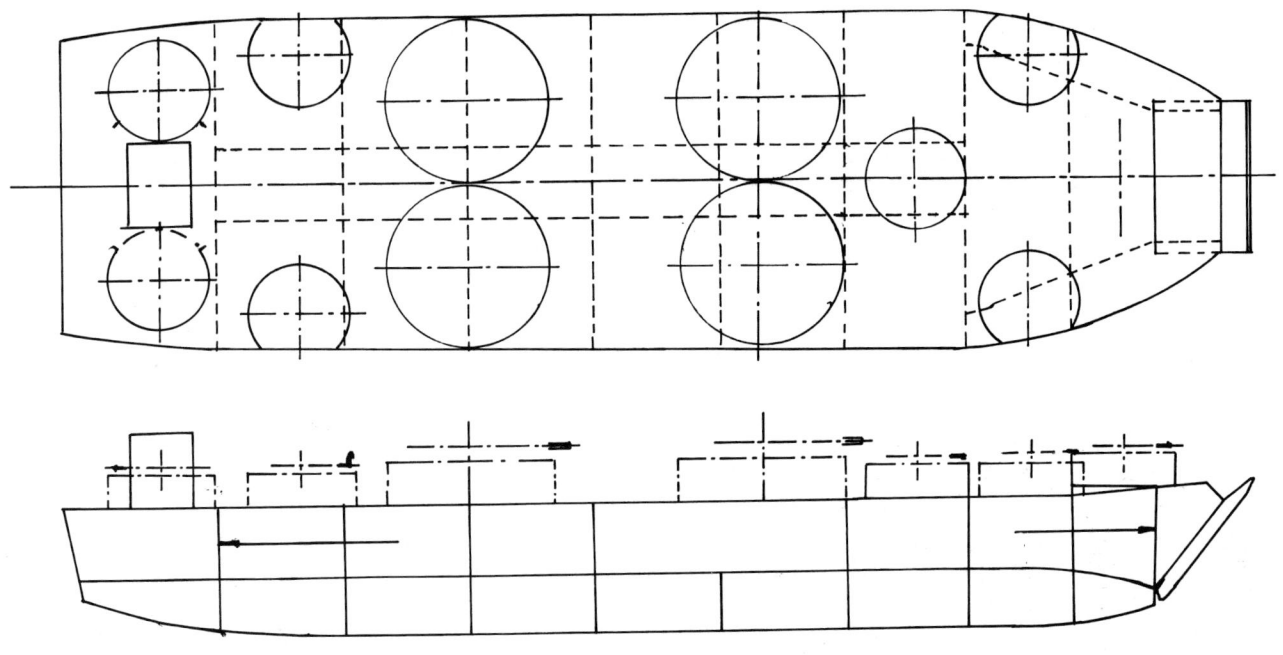

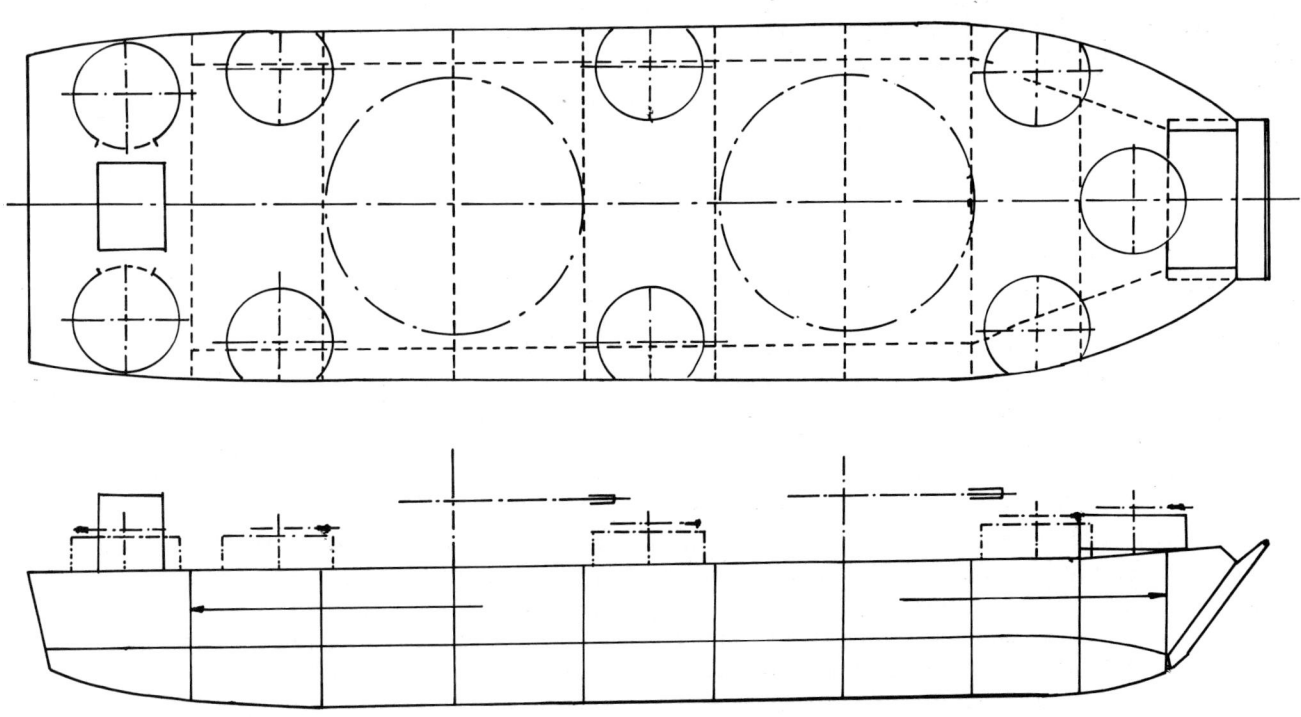

Sketch designs to convert an LCT for fire support, June 1943. Arrows indicate the extent of planned side plating over crews' quarters and magazines. Crosses indicate guns. Main armament would have been either two 5-in/38 or four 3-in/50, in each case supplemented by single 20-mm guns (nine for the 5-in version, eight for the 3-in version).

because the LCT(5) was far less seaworthy than the British LCT. As before, Cominch envisaged an LCS(L) comparable to the British version, armed with an army light tank turret (37-mm gun) forward and a 20-mm or twin 0.50 caliber gun aft. To be stowable aboard attack cargo ships, the LCS(L) could not be any longer than 50 ft, the length of an LCM. The existing LCS(S) should probably mount 20-mm rather than 0.50 caliber machine guns. BuShips' utter failure to produce such a craft the previous year had been forgotten. Adm. John L. Hall, the amphibious commander in Northwest African waters, considered that both Operation Torch and training exercises showed that such craft were essential. Like the LCS(S), they should be carried by an attack transport. They could be as large as an LCM, with similar beaching characteristics. Each attack transport should carry at least two; others might go on board cargo ships. They should neutralize pillboxes and personnel at a range of about 2,000 yds, thus they would need rockets and smoke mortars, as well as armor against enemy light machine guns on shore. Support craft should be at least 3 kts faster than the landing craft they accompanied, to maneuver as required. The Atlantic Fleet amphibious force commander, Adm. Alan G. Kirk, considered the LCS(S) too slow and unstable. He wanted a diesel craft, preferably based on the hull of the 36-ft picket boat rather than that of the Higgins boat, armed with two 20-mm guns and two multirail rocket launchers. Kirk reminded OpNav that neither the U.S. nor the British had succeeded in developing an LCM-size LCS(L).

The next step up, the LCG, should have at least two medium-caliber guns forward; it would have to be at least 100 ft long. A U.S. LCF would be armed with at least eight antiaircraft cannon; perhaps the LCG and LCF roles could be combined. Speed would have to be at least 12 kts, so the craft could accompany waves of personnel craft to the beach, and also so the LCF and LCG could escort LSTs on ocean voyages. The oceangoing LCF and LCG would need an endurance of at least 1,000 nm, with provision to fuel at sea. These craft would be able to beach to provide a stable gun platform. Hall wanted something larger, more like the British LCG/LCF, which could actually escort transports. Kirk preferred to focus on an LCT(5) conversion, which might be transported on board an LST or in an LSD. The Atlantic force suggested, entirely unrealistically, that it be armed with two to four shielded 5-in/38 (destroyer) guns or four M10 tank destroyer turrets (armed with 3-in guns), plus at least four 20-mm antiaircraft guns. Other army gun mounts worth considering were the two tank weapons, the 75-mm gun and the 105-mm howitzer turrets. The Atlantic Fleet was also interested in the British rocket craft, a version of which it wanted based on the LCT(5). Such craft, called LCT(U) (landing craft, tank, utility—i.e., multipurpose), could be extemporized for particular operations. The accompanying U.S.-type LCF would be armed with six to eight twin 40-mm director-controlled guns, plus at least four 20-mm.

Seventh Fleet (Southwest Pacific) amphibious force pointed out that, with the advent of the LST and the LCT(5), the tank-carrying LCM was no longer so important. All LCMs aboard attack transports should therefore be replaced by LCS(S)s; LCMs would still be carried by attack cargo ships. It would certainly be a good idea to replace the two 0.50s with a 20-mm gun, but it would be best to retain the 0.30s and the rockets. It would be desirable to add a smoke mortar and smoke projectors, as in British craft. The 37-mm gun seemed too small for the LCS(L).

In July, the Pacific Fleet amphibious force proposed that each attack transport carry at least four LCS(S); none would go aboard attack cargo ships. Like the other forces, it wanted more firepower: one 20-mm and two 0.50 caliber machine guns in the LCS(S), with two rocket launchers. For the LCG, which should be considerably faster than an LCS(S) (15–17 kts vs. 12 kts), it wanted a 37-mm or 40-mm gun in a dual-purpose turret forward, with two rocket launchers and two 20-mm. This was not too different from what others called an LCS(L), but the fleet wanted it to be much more seaworthy. For movements of 500 miles or less, which would not involve attack transports, LCGs would provide support (two per BLT beach). A destroyer would be more effective than an LCG against shore and air targets; only where a reef precluded a close approach to the shore would the LCG be needed. Existing landing craft could be fitted out with rockets as needed on a temporary basis.

In mid-May the BuShips designers proposed adapting the existing LCI(L) hull. Their first sketch showed two 3-in/50, two director-controlled twin 40-mm guns, and four 20-mm. The twin 40-mm was heavier than the 3-in/50. When the designers were asked for improved antiaircraft fire forward, they moved one of the twin 40s to fire over a 3-in/50 forward, and eliminated the other 3-in/50 to save topweight. Rockets, sound gear, and depth charges could be accommodated by sacrificing two of the 20-mm guns. The craft could be armored with 10-lb STS on gun bulwarks and pilothouse, and with 10-lb HTS on the forward hull above the waterline.

An LCG(L) version would be armed with four 3-in/50 (two forward and two aft) and with four 20-mm; an LCF version would have three twin 40-mm

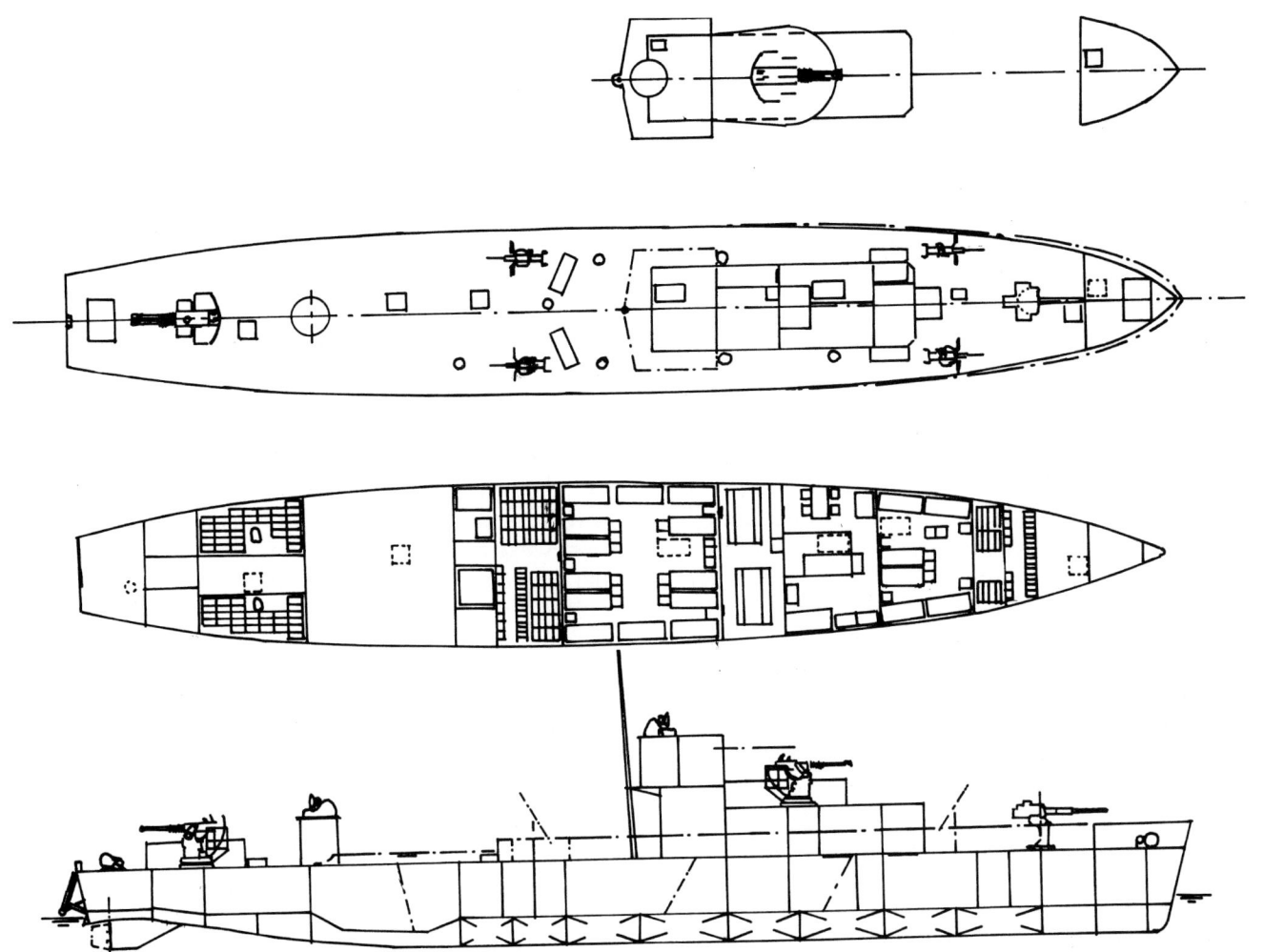

Sketch design for what became the LCS(L) Mk 3, 5 June 1943. The most obvious difference from the final design is the much lower bridge, in this case topped by an Mk 49 40-mm director. Note that the projected armament was one 3-in/50, two twin 40-mm (as fitted), and four single 20-mm guns; at this stage there were no rockets at all. The small boxes on the platform deck plan view are magazine stowage. The ship was expected to make 15 kts, and complement was 4 officers and 57 enlisted. Protection amounted to 10-lb STS bulwarks around the 40-mm guns and on the hull side forward abeam the 3-in gun and the two forward 20-mm guns, and 7.65-lb mild steel bulwarks further aft, back to the after 40-mm director. Bulwarks on the hull are shown as dot-and-dash lines.

(two forward, one aft) and four 20-mm. This craft would be at least as fast as desired (over 15 kts), and would have a longer endurance than required (4,000 nm at 14 kts). It could, for example, escort an LST convoy, given sound gear and depth charges. BuShips proposed that a single type, combining 3-in and 40-mm armament, be built to simplify production and assignment. While OpNav was considering BuShips' proposed LCI conversion, much the same idea occurred to Lt. (jg) James E. Hollis, USCGR, commanding LCI(L) 321 in the Eighth Fleet, which was assaulting Sicily. His letter was passed along via the Eighth Fleet amphibious force.

The alternatives were much less attractive. An LCT(5) armed with either 3-in/50 or 5-in/38 guns could no longer be transported on board an LST, nor could it be knocked down for assembly in a forward area. It would probably be too slow and too unseaworthy. The steel 173-ft subchaser (PC) was already overweight. Ultimately it was adapted as a gunboat (PGM), mainly for barge busting, but it could not offer the kind of firepower wanted for an amphibious operation. Nor could it beach. Smaller craft—PCS, SC, and PT—were wooden, hence ill-adapted to fire support, although the SC and PT could be (and were) armed as small gunboats for barge busting.

For the LCS(L), BuShips offered the LCC hull, which had evolved from an earlier failed attempt at designing an LCS(L); this time it would be armed with one 20-mm fore and aft, one twin 0.50 caliber machine gun, and two rocket launchers (72 rockets), with 10-lb STS armor. The extra weight, 2,500 lbs, would cut speed to about 13 kts, which was too slow.

As for the LCS(S), the designers replaced the single 0.50 and two 0.30s on skate tracks with a twin lightweight 0.50 caliber mount. It was impossible to install a lightweight 20-mm gun instead of the twin 0.50, because its weight would be too high in the boat. This version retained the rockets of the Atlantic Fleet boats (two 12-round rocket launchers, which could fire in six salvoes). Preliminary layout studies for additional armament were dated 24 May and 1 and 14 June 1943. On 21 June BuShips was asked for plans and specifications for 120 (possibly 135) more craft, for which armament and machinery had not yet been chosen. On 8 July plans and specifications were requested for 255 more boats, using 225 HP Gray Marine diesels in place of the original gasoline engine (a 250 HP Hall-Scott Invader). This was the Mk 2 version, LCS(S)2.

The LCS(S) proved successful, if too slow, in Sicily; a loaded Mk 1 had difficulty maintaining 10 kts. Typically a six-davit LST carried one in addition to her five LCVP. The commander of landing craft and bases in Northwest African waters proposed that all LSTs have six davits, so that each could carry either an LCS or an LCVP with special equipment (e.g., radio-equipped jeep, 81-mm mortar to throw a grapnel, or a 4.2-in smoke mortar). At the next landing, the LCS(S) silenced beach defenses at Salerno, and it also laid smoke to cover the advancing waves of landing craft. It proved unpopular with transport commanders, however, because it was heavy, cumbersome, and it required a special supply of gasoline (other craft used diesel fuel). Most commanders wanted it replaced by a support version of the LCVP. The task force commander, however, wanted the LCS(S) or some equivalent retained as an integral part of attack transport equipment. By this time the major problem of the gasoline engine had already been solved, although the fleet was probably unaware of that.

For both Sicily and the follow-on against Salerno, the U.S. fleet had to rely on British LCGs. Adm. Richard Conolly considered them too slow, with inadequate fire control. He called for a combination LCG/LCF, and supported Hollis's proposed LCI(L) conversion. Moreover, an attempt to use subchasers (SCs) as control craft had not been entirely successful. An LCI could combine all three functions (support, flak, control) in one hull by adding a single 40-mm gun, an additional 20-mm gun, an 81-mm mortar, a reliable gyro-compass, SC radar, smoke equipment, and an improved bridge and chart room with better navigational facilities.

Finally, representatives of OpNav, Cominch, BuShips, BuOrd, the Marines, and the Sixth Fleet amphibious force attended a 27 September 1943 conference on future support craft. The forces afloat wanted two types, a seagoing craft capable of beaching and a small ship-borne type to accompany assault waves. For the larger craft, a converted LCI(L) would clearly be superior to a converted LCT(5). If some LCI(L) already authorized were converted, deliveries could begin in about 6 months (i.e., late March 1944) and reach 15 per month the following month. If new hulls were ordered, deliveries could begin in about 9 months (late June 1944), reaching 15 per month the following month. The conference chose the latter. The only bottleneck would be production of the 6-71 diesels needed to power the craft. By this time (discussed below), the Southwest Pacific service force was already extemporizing fire support craft from LCI(L)s, and the Pacific Fleet was about to follow.

The conference endorsed the existing LCS(S)2; a version of the 36-ft picket boat was considered and rejected. An LCS(S) could already be carried in the lower and outboard tiers of the Welin davits on attack transports and cargo ships as a jury rig, and a service rig was being developed. LST davits would be modified to take its greater weight as soon as possible. The conference decided to increase LCS(S)2 production.

By the fall of 1944, the LCS(S) was no longer wanted, however, presumably because it could not deal with the fortifications the Japanese were erecting. In September assignment of LCS(S) to Pacific AKAs and APAs was canceled. LCS(S) under construction were ordered completed as LCPL, although they did not fit many davits because they were a foot too long (they were to be used as Captain's gigs). In November all existing ones in the United States were ordered converted, to alternative plans developed by Higgins and by the Naval Landing Force Equipment Depot (NLFED) at Albany, California. The Higgins version was more comprehensive; the Albany version was easier to accomplish on a boat completed as an LCS(S). As it happened, depots were busy with more urgent work, so at the end of the war many LCS(S) were still in their original form. After 154 had been declared surplus in 1946, 50 were taken back for transfer to China, as part of the U.S. aid package. Ten of them were intended for Chinese Maritime Customs in lieu of small tugs (YTLs) previously assigned. Some survived (in storage) as late as 1955, when BuShips proposed converting them to replace worn-out LCPLs. That proposal failed, but as late as 1962 an LCS(S), C-60522, was proposed for conversion to replace

LCPL C-60535 at the acoustic range at Carr Inlet, Washington.

Two other shore bombardment craft were suggested, a modified destroyer escort (DE) and a modified escort (PCE). Some of the destroyer escorts already had a formidable battery of two 5-in/38 guns. If the torpedo tubes were deleted, a third could be added and the antiaircraft battery increased to three twin director-controlled 40-mm guns and four 20-mm. The smaller PCE could be armed with one 3-in/50, three twin director-controlled 40-mm, five 20-mm, and ASW weapons. The conference agreed to pursue both possibilities, but the destroyer escort conversion never materialized; PCEs were built or modified with the armament listed above.

There was an urgent need in the Solomons for small gunboats to deal with heavily armed and armored Japanese barges. The best counter seemed to be a U.S. equivalent, a converted landing craft. Such gunboats were not intended for amphibious fire support, but inevitably they offered that capability.[2] LCI(L) 22, 24, 68, and 69 had been converted at Nouméa by August 1943, with one 3-in/50 in a prominent gun tub forward of the bridge, one 40-mm antiaircraft gun in a smaller tub abaft the bridge, four 20-mm, and six 0.50 caliber machine guns. They had limited armor. The first two (LCIL 24 and 68) were used very successfully for the first time in the Treasure Island operation of October 1943. They led the landing craft to the beach, turning aside only to avoid beaching. They protected the flanks of the assault waves by attacking gun positions on a nearby island, and also silenced machine guns on the beach itself. A second series of conversions—LCI(L) 21, 24, 61, and 64–67, 70—were similarly armed, but had the 40-mm gun in the bows. These craft were eventually redesignated LCI(G) (landing craft, infantry, gunboat). In addition, one LCM(3) was converted, her ramp removed and a more ship-like bow welded on, giving her a speed of 14 kts. She had a single 3-in/23 gun and two 37-mm guns, with light armor.

The South Pacific force also converted LCIs into rocket craft. In April 1944 it reported that LCI(L) 31, 34, and 73 had been armed with 24 Mk 1 barrage rocket launchers, offset up to 2½ degrees on each

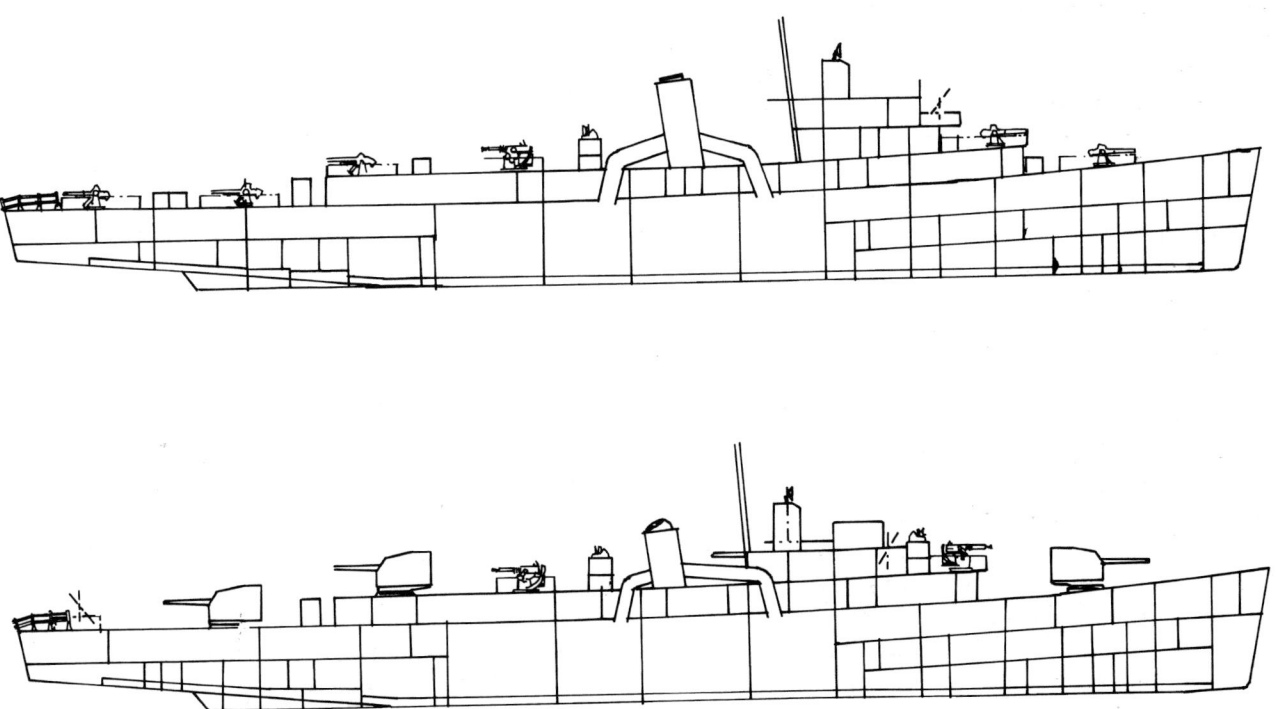

"Spring style" sketches of proposed conversions of steam destroyer escorts for shore bombardment (with five 5-in/25 guns) and antiaircraft support (with three 5-in/38), 7 September 1943. In the bombardment version, light armament would have been two twin 40-mm and two single 20-mm, and the ship had a single depth charge track aft; all the depth charge throwers and the Hedgehog would have been eliminated. In the antiaircraft version, armament was three 5-in/38, three twin 40-mm, and two 20-mm, plus the single depth charge track right aft. Both versions show Mk 49 directors for their 40-mm guns, devices soon abandoned in favor of the simpler Mk 51. The bridge director was presumably to have been an Mk 37. The only boat shown on the original drawings was a 26-ft motor whaleboat just abaft the bridge.

LCI(G) 67 was converted by the Southwest Pacific amphibious force. She is shown high and dry after having been blown ashore on Okinawa by the 15 October 1945 typhoon. Note the artwork on her bridge face. She shows a single 40-mm gun and a single 3-in/50 (in the gun tub) forward, and a 20-mm gun right aft; she was credited with four 20-mm and six 0.50 caliber machine guns, plus ten Mk 7 rocket launchers. The earliest conversions, including LCI(G) 22 and 68, had their single 40-mm guns abaft the bridge, the bow being clear. Gun positions were protected by 2½-in plastic armor. Not present are the Mk 7 barrage rocket launchers normally carried on the main deck outboard of the ship's bulwarks.

side from the centerline, and 12–14-in inboard from the gunwale. Rockets were all carried in No. 2 troop compartment. LCI(L) 31 and 34 had six control panels on bulkheads in the well deck; LCI(L) 73 used two panels on the bridge. Rack capacity was 600 rounds, but boats had stowed 1,000 for short periods. Between the racks of rockets, a 40-mm gun was mounted atop No. 2 troop compartment.

When BuShips received details of the LCI(G) early in December 1943, it feared that the weights added would make the craft topheavy. As a guide for conversion it developed two parallel designs, Scheme B retaining some troop capacity (its deckhouse would be eliminated in favor of additional guns) and Scheme C, Lt. Hollis' proposal, which retained all its troop capacity (the troops would land by rubber boat rather than over the bow ramps).

Scheme A was the new-construction gunboat. Scheme B had one 3-in/50, one single 40-mm, four single 20-mm, and six 0.50 caliber machine guns. Scheme C had three single 40-mm, five single 20-mm, and four 0.50 caliber machine guns. Neither Scheme B nor Scheme C seems to have been adopted by the Pacific Fleet service force.

The Scheme A preliminary design was completed late in January 1944. It had the LCI(L) hull form and dimensions, except that the recess for the bow ramps was eliminated. The forecastle deck was carried aft and lowered; the bulwarks were moved out to the edge of the main deck and extended from the forecastle to about three-quarters length. The entire deckhouse was replaced with a smaller one somewhat further forward. The circular pilothouse (10-lb STS) was modified. The open conning station was

LCI(G) 23, another early South Pacific conversion, in San Francisco harbor after the war, in the fall of 1945.

LCI(R) 73, an early South Pacific conversion, is shown on 28 May 1944. She differed from LCI(G)s in having more Mk 7 rocket launchers, visible along her main deck, and no 3-in/50. However, both conversions carried and used rockets.

A rocket-firing LCI(R) attacks at Leyte Gulf, October 1944.

removed from the top and an open platform provided at the navigating bridge level. The battery was one manually operated 3-in/50 DP (350 rounds, 48 of them ready use), two director-controlled power twin 40-mm mounts (fore and aft: 1,200 rounds per gun, 372 ready service per gun on the shield), and four twin 20-mm (4,000 rounds per gun). By March 1944 it was assumed that the craft would also carry rockets, although they were not yet shown on the plans. To maintain shallow draft and high speed, no more than 3 tons could be accommodated, equivalent to two Type 8 projectors with four complete loads (88 rockets). The approved armament later included ten Mk 7 rocket launchers.

Estimated speed was 14.5 kts clean at 350 tons at 1,300 SHP continuous rating (maximum rating was 1,600 SHP), with a cruising radius of 3,200 nm at 13.5 kts (allowing 25 percent extra fuel for adverse weather). The new battery required an additional diesel generator (30 kW) and switchboard. To accommodate additional radio equipment, the chart and radio room was enlarged by about 50 percent. Planned complement was 4 officers and 57 enlisted men (in service, 5 officers and 65 enlisted men).

The new craft was designated LCS(L)3, the two earlier versions (Mk 1 and 2) were British conversions of their wooden 105-ft LCI(S). Authorized production was 130 craft. The LCS(L)3 was far too late for Normandy, so there, as in the Mediterranean, the U.S. Navy had to rely on borrowed British support craft.

In the Central Pacific in 1944, the U.S. Navy used converted LCI(L)s apparently unrelated to those used in the South Pacific the previous year. None had the 3-in gun of BuShips' Scheme B, a choice made, according to a postwar Pacific Fleet report, after considerable experimentation. All had 2½-in plastic armor. There were four versions, all eventually reclassified as LCI(G). Design work began in the fall of 1943; on 3 December 1943 the Fifth Fleet (Central Pacific) amphibious force asked the Pearl Harbor Navy Yard to convert ten LCI(L) (LCI[L] 77–82 and 345–348) to gunboats. They would retain their ramps and troop capacity. Armament would comprise two 40-mm guns on the centerline, one on the forecastle and one on a platform extending across the bulwarks, with a splinter shield low enough to allow fire at targets 100 yds away on the water. Two Mousetrap projectors, firing 7.2-in rockets, would be fitted inside the bulwarks at Frame 21, with blast shields abaft them. A blast test would be conducted to see whether Mousetraps could in fact be mounted this way. Ten Mk 1 barrage rocket launchers (4.5-in rockets) would be fitted, five on a side, abreast the deckhouse outboard of the ship's side, 5 ft apart, parallel to the centerline. Six 0.50 caliber machine guns would be added, their positions depending on whether the Mousetraps could in fact be used. The ships would retain three 20-mm guns (one on top of the deckhouse, two on the fantail). The blast test soon showed that the Mousetrap installation was impractical, and weight calculations caused elimination of two of the 0.50s. Thus the Pearl Harbor units had two 40-mm guns, ten rocket launchers (Frames 45–78), and three 20-mm guns (one atop the deckhouse at Frame 36, two on the main deck, on either side at Frame 44). By the end of the war they also had six 0.50 caliber machine guns, valued for dealing with Japanese swimmers and suicide boats, which could get so close that 40-mm guns could not depress sufficiently to hit them. By late in the war some had thirteen Mk 1 rocket launchers. Early BuShips correspondence files describe these craft as Group 1, but in the 1945 handbook of U.S. and British amphibious craft they are Group A.[3] Each box launcher carried twelve rockets, firing to a fixed range. Ammunition (720 rockets, 3,000 rounds of 40 mm, in former troop holds 1 and 4) sufficed for a 3-minute salvo, after which the ship withdrew to reload. By April 1945 the rocket battery had been replaced by 10 lightweight Mk 7, whose range could be adjusted between 500 and 1,100 yds.

These first units were all of the initial LCI design. Pearl Harbor then converted a group of the later boats, LCI(L) 365, 366, and 437–442, which were initially designated Group 2, and later Group B. The LCI(L) 351 series already had a pair of gun tubs at the fore end of the deckhouse, which were easily converted to take single 40-mm guns. Other weapons were eight box-type (Mk 1 Mod 1) rocket launchers, two 20-mm guns (the original after mounts), and five 0.50 caliber machine guns. These units later had 10 Mk 7 rocket launchers atop their deckhouses. Later most were modified to carry a total of 42 Mk 7 launchers (16 on each ramp sponson, with the ramps removed).[4] Ammunition amounted to 754 rockets and 4,500 rounds of 40 mm.

On 17 December the Fifth Fleet amphibious force asked the Pacific Fleet service force to have another seven craft converted in San Diego (LCI[L] 449–453, 455, and 457). San Diego Naval Repair Base converted its seven units to the original design, adapted to the LCI(L) 351 hull. They became Group 3 (Type C). Wartime correspondence suggested that they would have the same armament as Group 2, but in fact they were armed like Group 1.

Then more conversions were ordered at private and naval facilities on the west coast. They were classed as Groups 4 and 5 or Type D: LCI(L) 372, 373, 454, 456, 458–460, 464, 465, 467–475. Later orders added LCI(L) 396, 397, 404–407, 422, 438, 441, 449–453, 455, 457, 461–463, 466, 470, 558–561, 565–580, 725–730, 751, and 752. Again, they were intended to repeat Type B, but they were more refined. Given lighter-weight rocket launchers (10 Mk 7), the two forward waist 20-mm guns could be retained; they were moved forward to just abaft the bow 40-mm gun on the forecastle. The 10 Mk 7 launchers were on top of the deckhouse. Ammunition was stowed in Nos. 1 and 4 troop magazines (720 rockets and 4,000 rounds of 40 mm). LCI(G) 405 had a 60-mm mortar in addition to her other weapons. As with Group B, some of these units were later modified to carry a total of 42 rocket launchers.

The program proceeded quickly; the service force was responsible for design work and calculating safe liquid loadings to maintain stability despite considerable added topweight. Not until April 1944 did BuShips ask the service force for details of both the conversion and of safe loadings. Pearl Harbor sent plans of the initial version.

The 12 used at the Marshalls in January 1944 were LCI(L) 77–80, 365, 366, and 437–442. They slightly preceded the first wave of landing craft, firing rockets at 800 and 1,100 yd range. Thirty of these craft supported the Marianas invasion in June 1944. Late in the operation, these gunboats supported troops ashore by firing into Japanese-held caves. One boat fired 5,000 rounds of 40-mm, 20,000 rounds of 20-mm, and 1,000 rockets. They supported destroyers maintaining nightly "Flycatcher" patrols against Japanese suicide boats and barges attempting to end-run American lines on the islands. Unmodified LCIs carried loudspeakers, through which interpreters called on the Japanese to surrender. Another served as an artillery observation post. In other operations LCI(G)s protected large moored units by making smoke, supported underwater demolition teams, and supported larger picket or fighter direction ships. Later these tasks were also performed by LCS(L)s. The boats' rockets were very useful, but early wartime installations were unsatisfactory. They took too long to reload, they carried insufficient numbers of rockets, and they suffered from a very high misfire rate of 15–20 percent.

The later LCI(M) (landing craft, infantry, mortar), which were first used at Pelelieu in September 1944,

The LCS(L) Mk 3 was successor to the extemporized LCI(G). The newly completed LCS(L) 50 is shown on 19 September 1944. This class retained the LCI(L) hull but was completed redesigned. Early production units like this one had a 3-in/50 forward. A second series replaced it with a single 40-mm Bofors that, like the 3-in gun, was not director controlled. Late-production units had a director-controlled twin 40-mm gun forward. The hooded objects just abaft the 3-in gun are Mk 7 rocket launchers.

was armed with three 4.2-in "chemical" mortars developed by the Army Chemical Corps to fire smoke rounds on standard army mounts, one on the centerline abaft the forecastle, and one either side abaft the dismantled troop ramps.[5] Elevation was fixed, the charge being varied to change range. To absorb recoil, each mortar had its base plate embedded in a box half-filled with sand and half with dry sawdust. The LCI(M) carried 1,200 rounds in her No. 2 troop compartment, under the well deck. To serve the mortars, the ship's complement was nearly doubled; 20–25 men were added. Some LCI(L) were modified to carry 1,200 replacement 4.2-in rounds. Compared to rockets, mortars were more flexible, because their range could be varied between 600 and 3,200 yds. They were fixed in train, however, so the entire ship had to be aimed at a target. It was difficult to keep a ship from swinging even in a very calm sea. The mortars could not be fired to maximum range because the ship's deck could not take the resulting stress. Some units were initially designated LCI(G). They amounted to LCI(M) 351–356, 359, 362, 431, 582, 588, 594–596, 630–633, 638, 658–660, 664, 669, 670, 673, 674, 739–742, 754–757, 760, 801–810, 951, 952, 975, 1010–1012, 1023, 1055–1059, 1088, and 1089.[6]

The LCI(M) experience showed that mortars could really be useful on board a fire support ship: they had a unique ability to reach the reverse slopes of hills. By the end of the war BuOrd had developed a new mortar (Mk 1-0) on a 3-in/50 gun mount, which could be trained and also had a recoil and cooling system. Maximum range increased to 4,400 yds, and the rate of sustained fire more than doubled to 21 rounds per minute. The mortar was so valuable that in October 1945 the Pacific Fleet recommended that it be installed on board any LCS(L) successor, even at a cost in size.

The Pacific Fleet developed an LCI(R) with one 40-mm gun, four 20-mm, and six Mk 30 rocket launchers for 4.5-in finned rockets. LCI(R)s equipped with Mk 30s (at least LCI[R] 704, 705, 1068–70) could fire 36 rockets in 4.5 seconds, after which they needed 30 minutes to reload. The rocket launchers were paired just abaft the bow 40-mm gun and atop the deckhouse between the two 20-mm guns on each side, at Frames 56 and 66 (the 20-mm were at Frames 45 and 77). Ammunition stowage was 1,000 rounds of 40 mm in No. 1 troop hold and 1,100 rockets in No. 2 troop hold, the crew occupying No. 3 hold. As of February 1945, this group amounted to 18 units: LCI(L) 647–649, 704–706, 762–767, 785, 1024, 1026, 1068–1070. Units were rearmed with (or initially armed with) Mk 51 spin-stabilized rocket launchers; the original Mk 30 was criticized as too flimsy because rockets dispersed when it warped. Mk 51 fired 5-in spin-stabilized rockets out to a range of 5,250 yds, thus allowing rockets to be laid down in a thousand-yard pattern. A craft armed with Mk 51 could ripple-fire 72 rockets from 6 launchers in 4–5 seconds, then required 10–15 minutes to reload. As of February 1945, 18 boats were armed with Mk 51s (1,700 rockets each): LCI(L) 642–646, 650, 651, 707, 708, 769–772, 1028–1030, 1077, 1078. LCI(R) 773–777 were added by June 1945. By August there were another 16, including reclassified South Pacific conversions: LCI(R) 31, 34, 71–74, 224–226, 230, 331, 337, 338, 340–342.[7]

The LCI(R) appeared late in the war, and according to a postwar Pacific Fleet evaluation it saw too little action to prove the concept. Many LCI(R)s were fitted with electronic countermeasures equipment, so that the unit fighting at Okinawa was RCM (radar countermeasures) and Rocket Division Two. In its after-action report, it complained that after adding its weight to the assault day bombardment, it was used mainly to counter kamikazes and for "flycatching," fending off Japanese swimmers and suicide boats.[8] Except for the first day (L-day), the division fired only 19 support missions in March and April. Then its commander began to sell the idea of rocket bombardment to the army, which was presumably

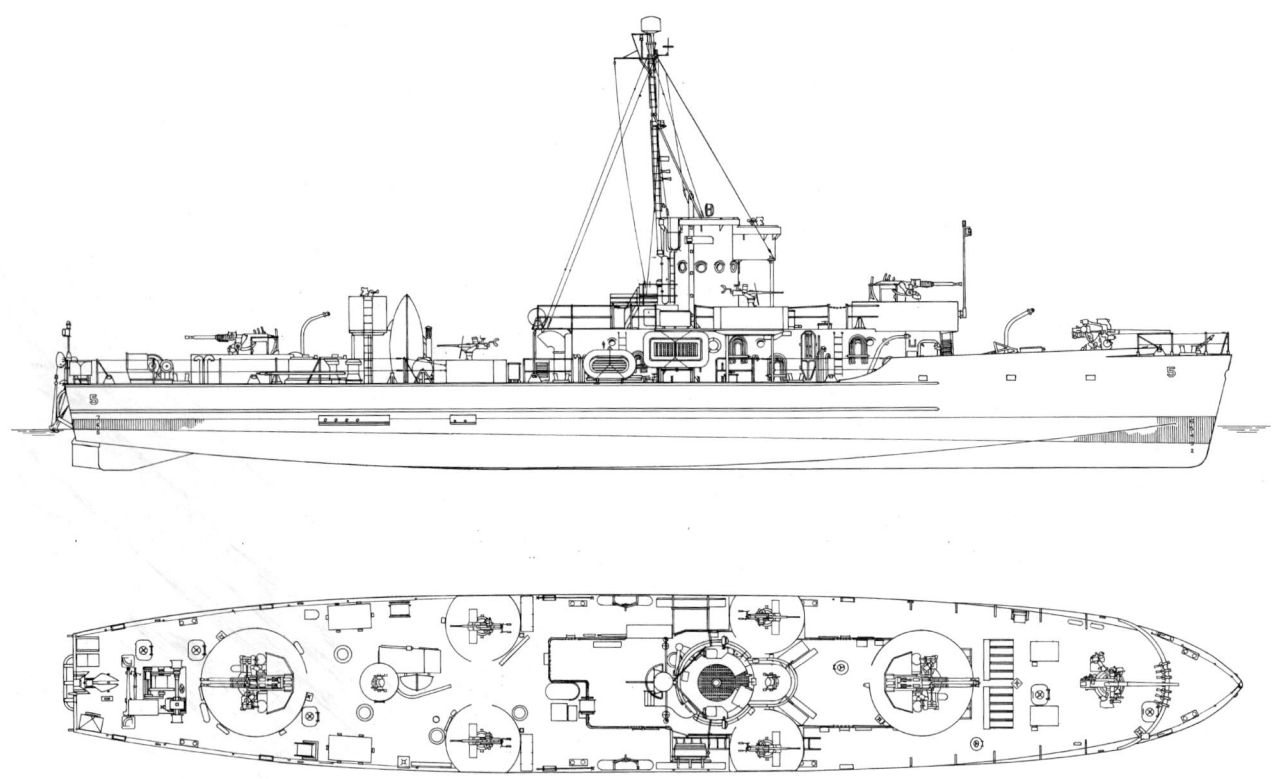

LCS(L)-5 is shown as completed, August 1944. She was one of the original group of LCS(L), armed with a single 3-in/50 forward. A second series had a single army-type 40-mm gun in this position; the final version had a twin 40-mm. All units were wired for the director-controlled twin 40-mm gun, but they could not be modified in forward areas. Abaft the 3-in gun are Mk 7 rocket launchers, stowed flat on deck. Units differed in detail. Thus LCS(L) 1–4 had their 12-ft wherries stowed horizontally, on their starboard side. From LCS(L) 10 onward, the signal flag bag was farther aft and to starboard. *A. D. Baker*

more interested due to the heavy resistance it was encountering ashore. In May, mainly in the second half of the month, the division carried out 75 fire support missions; it carried out 145 during the first 21 days of June 1945. Thanks to their shallow draft, boats could approach as close as 200 yds to the shore. They used SO-8 radars with visual position reflectiscopes (VPRs) to position themselves, and this equipment was found accurate enough for rocket fire. Typically the units of six boats operated together, with only the flagship in a division plus some standby flagships being equipped with a radar; the Okinawa after-action report called for fitting radar to one in three craft rather than one in six (the flagship). Radar countermeasures equipment was ordered used only twice, and each time there were several failures.

LCI(R) effectiveness depended not only on weaponry but also on communications. By Okinawa, amphibious task force command and control was quite sophisticated, but that meant that each unit had to connect to numerous different command nets. LCI(R) fire support missions were assigned on a gunboat control circuit. A separate circuit, at a fixed frequency, connected all the LCI(R)s. Unfortunately, many other types of small combatants, such as LCI(M)s and LCTs, lacked their own fixed frequency circuits. Their flagships sometimes usurped it, and the LCI(R)s could no longer hear their radio net controllers. Given an effective radio range of about 20 miles, nets too often overlapped, and it became clear that different frequencies (channels) should have been assigned to the different types of amphibious craft. There was also a local command net. The LCI(R) had enough radio sets for all three, but only one operator to listen for them. There was also a technical problem. One of the radio sets used a loudspeaker, so the ship's commander could hear directly what was being received. Another used earphones, so only the operator received and transcribed traffic. The gunboats also had to listen to guard force–wide radio nets. A heavily used fleet common net operated at much the same frequencies as the LCI(R) and gunboat nets, and it was reserved mainly for larger ships. However, there was also a local harbor frequency, traffic on which was so heavy that craft often had to wait 24 hrs to get the chance to send their messages.

LCI(G) 346 and 348 were initial Central Pacific conversions, recognizable by the square gun tub forward of the bridge. Classed as Type A, they had two single 40-mm forward plus three 20-mm (in the usual positions atop the superstructure and right aft), six 0.50 caliber machine guns, and ten Mk 7 rocket launchers. LCI(L) 346 was a press boat (PGY) for the Iwo Jima operation, rushing copy from war correspondents aboard ships to a communications ship that radioed it home.

LCI(G) 80 (Type A) lies alongside LCI(G) 449 (Type C) in an advanced base drydock (ABSD 6). In effect Type C was the Type A conversion applied to an LCI 351–class hull.

LCI(G) 455 was a Type C conversion. Armament was two 40-mm guns and six 0.50 caliber machine guns, plus ten Mk 7 and two Mk 22 rocket launchers.

Each craft had to guard an expeditionary force fox (one-way) circuit, listening for messages addressed to it from the force commander. In order to handle the heavy traffic for an entire massive force, signalers on the flagship had to send their Morse code messages very fast; junior radiomen on the LCI(R)s and similar craft could not always cope. An LCI(R) or other LCI conversion had only a single receiver at the appropriate frequency, but from time to time higher-level commands asked that all units listen to other broadcasts, in the same frequency band. All these problems applied not only to LCI(R)s but also to LCFF (flagship) conversions used to lead divisions of LCI gunboats.

The LCS(L)3 was clearly successful, to the point that in October 1945 the Pacific Fleet recommended retaining it but disposing of all the converted LCIs. Its only unsatisfactory feature was its limited rocket fire. Postwar, these craft were redesignated LSSL (landing ship, support, large). All were laid up, and they were soon considered surplus. Many were transferred to postwar allies, particularly Japan and Italy, as small gunboats (see Appendix B).

By late 1943 BuShips was not altogether impressed with the British LCG and LCF. During February–April 1944, BuShips developed plans for a fire support conversion of the LCT(6). In August the bureau reported a study of an LST mounting 155-mm howitzers, as a shore bombardment ship; it also studied a parallel conversion of a tanker. A BuShips officer suggested that an LST with more 40-mm and 20-mm guns might be quite as good as an LCF. It might also be worthwhile to adapt the LST as a rocket-firing ship. The existing British LCT(R) typically fired all of her rockets in a single salvo, but the LST might accommodate so many that she could fire several salvoes.

These projects were never realized, but in March 1944 1st U.S. Army Headquarters circulated a memo describing LCT(SP) (landing craft, tank, self-propelled [gun]) and LCT(A) (landing craft, tank,

LCI(G) 580 and 726 were Type D conversions of LCI(L) 351–class craft, with single 40-mm guns replacing three of their 20-mm guns (the guns were relocated as shown). Rocket launchers were carried on the former gangways; note that LCI(G) 726 retained the gangways themselves and the cross-bar used to support them when they were deployed. The frames of two Mk 7 launchers are visible on the starboard gangway of LCI(G) 726. All five port side launchers are visible on board LCI(G) 580. Craft of this group had ten Mk 7 rocket launchers. An earlier but similar Type B had eight Mk 1 launchers. Because they were heavier (a loaded Mk 1 weighed 675 lbs, a loaded Mk 7 weighed 515 lbs), only the two after 20-mm guns were retained; these craft also had five 0.50 caliber machine guns.

LCI(G) 725 and 726 (Type D) attack at Saipan, 15 June 1944, where ships of this type were credited with a major role in overcoming Japanese resistance.

LCI(G) 456 (Type D) prepares to fire 4.5-in rockets from her Mk 7 launchers at Pelelieu, 19 September 1944. The cruiser *Portland* is in the background. Note the pipe guard to prevent the bow 40-mm gun from firing into the ship; it had no director system to do so.

LCI(R) 644 is shown at a 1945 Pacific landing, rockets filling her Mk 7 launchers abaft the forward 40-mm gun, abaft the forward 20-mm guns at the fore end of her deckhouse, and just forward of the after 20-mm guns. The battleship in the background is USS *Nevada*, recognizable by her funnel extension.

LCI(R) 1078, not yet even redesignated, is shown off San Francisco, newly converted, 29 December 1944. Her Mk 7 launchers are barely visible. Note that the one just abaft the forward 20-mm gun has been raised to fire over it. The ship has no radar at all, just a "ski-pole" IFF transponder at her masthead.

armor—i.e., tank-armed). LCT(SP) carried 155-mm self-propelled guns, and were expected to provide support fire while afloat, from about 4,000 yds behind the first wave of landing craft. LCT(A) were armored and would carry two or three tanks each. The two forward ones would fire their guns; another would be equipped with a T34 or T40 rocket launcher. As used at Normandy, the LCT(A) had STS armor over fuel tanks, engine rooms, ramp winch rooms, crew spaces, and pilothouse. That added 75 tons, and the craft trimmed by the stern. She carried three tanks, two firing and one in reserve abaft them. A total of 48 LCT(A) were involved, 16 in the western and 32 in the eastern task forces.[9] Related conversions were the LCT(CB) and LCT(HE). LCT(CB) or "concrete buster" carried three Shermans armed with 17-pounder high-velocity guns specifically to destroy concrete fortifications. LCT(HE) was LCT(SP), carrying self-propelled guns, typically 105-mm M7 howitzers. At Normandy one unit of such craft began shelling the beach at a range of 9,000 yds, ceasing fire at 1,000 yds because the guns could not depress any further without hitting the LCT bow ramps carrying LCT. At Normandy were 5 LCT(CB) (2 in the western, or U.S., task force and 3 in the eastern, British, task force) and 16 LCT(HE) (8 in each task force).

The army also armed LCMs. Its 2nd Engineers Special Brigade fitted one as an antiaircraft craft, with one 37-mm gun forward, two 20-mm (one starboard side on the engine deck forward of the wheelhouse, one abaft the wheelhouse), two single 0.50 caliber forward, and four 0.50 caliber in two twin aircraft turret, on the decked-over half of the cargo space. This craft also had two T-45 (twelve 4.5-in rockets each) automatic rocket launchers, one on each side. An LCM(R) (landing craft, mechanized, rocket) had 32 T-45s, for a total of 384 rockets, plus four single and two twin 0.50 caliber machine guns.

Craft were also armed specifically to breach obstacles on beaches. In 1944 the Eighth Fleet produced "Woofus," an LCM to clear a lane through a minefield on a beach, with 120 7.2-in rockets in Mk 24 launchers. It was used in the invasion of Southern France.[10]

By mid-1944 it seems to have been realized that the 3-in/50 planned for the LCS(L) was not powerful enough. In August 1944 BuOrd proposed mounting a power-operated 5-in/38 on board an LCI(L). In September BuShips pointed out that the LCI hull was far too light and lacked sufficient stability. However, the LSM would be an ideal platform. It was very stable, its hull was strong, and it was almost as fast as an LCI(L). It offered three 5-in guns in boxes on the tank deck, the bow 40-mm being removed and the four 20-mm being relocated. Supporting structures for the 5-in guns could be prefabricated for installation and assembly at a forward base. There was also some interest in a 155-mm–armed LSM, presumably parallel to the LST version.

By this time rockets had proven their value. An LSM rocket installation was first sketched in September 1944. In October the Pacific Fleet asked for 12 rocket support ships, presumably based on BuShips' work. This project was combined with the BuOrd proposal to produce the LSM(R) (landing ship, medium, rocket), the ultimate wartime fire support craft. The first 12, LSM(R) 188–199, were converted on an interim basis. Each was decked over to form three handling rooms, each with two magazines (fore and aft). Rocket heads and motors were stored separately, assembled in the handling rooms, and passed up by hand. Each ship mounted a single enclosed 5-in/38, two single 40-mm guns, and three 20-mm, as well as 75 four-rail Mk 36 rocket launchers and 30 six-rail Mk 30 rocket launchers, for a total of 480 rails. The 5-in finned rockets were considered equivalent to destroyer shells. All 480 could be fired in 30 seconds, but it took 2½ hrs to reload all the launchers. Rocket range was 4,000 yds. The last four ships (LSM[R] 196–199) were completed with 85 Mk 51 automatic rocket launchers firing longer-range spin-stabilized rockets (range 5,250 yds). Each launcher held 12 rockets (for a total of 1,020), and could be stepped between 30- and 45-degree elevation. Each launcher had its own blast-deflecting scoop. These ships could fire all their rockets in 1 minute. The first reload took 45 minutes, later ones required 1½–2 hrs.

These "interim" ships were used very successfully at Okinawa. Prior to the preliminary landing at Ie Shima, LSM(R)s pinned down the Japanese troops to prevent them from mining the invasion beaches. During the consolidation phase at Okinawa, they patrolled against suicide boats, destroying three one night. They provided call fire as enemy troops retreated. Their high-trajectory fire cleared protective ridges just inshore of the beaches. The only failure was an attempt to use LSM(R)s to support destroyers on radar picket stations, the hope being that barrages of rockets could deal with incoming kamikazes. On this duty three were sunk and another damaged.

The great limitations were that the launchers were fixed and reloading was so slow. BuOrd developed a double-barrel power-loaded launcher, Mk 102, based on the twin 40-mm mounting. Because it was based on the 40-mm mounting, the launcher was power trainable and power elevatable. Hence it could shift

"Woofus" was an LCM(3) armed with rockets to blast a path through obstacles blocking a beach. This prototype is armed with rocket launcher Mk 24, which was designed specifically for such installation, to clear a beach of barbed wire, mines, and light obstacles. It comprised 120 rocket guides (for 7.2-in rockets, the type used in Mousetrap antisubmarine launchers) in eight fore-and-aft rows of 15 guides (launchers) each: two rows on the starboard side of the well, two on the port side, and the other four in the well itself. Guides were at fixed elevation angles between 25 and 45 degrees. At 45 degree elevation, a 7.2-in rocket had a range of 293 yds. In the invasion of Southern France, "Woofus" LCM(R)s followed drone minesweeping boats directed by LCCs; they preceded the troops.

its aim point even though the ship did not move, and the rocket launchers could be used like guns. The launchers were remotely controlled by a version of the Mk 51 director used to control 40-mm fire, with two directors per ship. A dead-reckoning tracer kept track of ship and target positions. Unfortunately, without an appropriate computer, fire could be controlled precisely only when the ship lay-to or was at anchor. The directors, designed to control fire at visible targets, merely ensured that all the launchers pointed together. The settings for train and elevation had to be calculated by hand. Hitting moving targets was virtually impossible. Similarly, spotting corrections had to be hand calculated. With the ship lying-to, in 1950 LSM(R) 515 managed to get 200 yds right and 100 yds over when opening fire at 8,250-yd range.

Accurate reverse slope fire turned out to be important when LSM(R)s were used in Korea, at Hungnam in 1950. The new ability to split the battery, either to each side or using two range bands, also turned out to be important in Korea. Splitting an LSM(R) battery would have been important in a planned operation in which rocket ships would have fired from a narrow channel into a small city on each side. Dual-range bands were used at Hungnam. The ability to fire at several targets from one position proved important at Inchon in 1950.

Each Mk 102 mounting could fire 30 rounds per minute. The mounting's hoist, hung from it, fed the barrels when they were vertical. Then they depressed hydraulically and fired. A few such launchers could replace the forest of manually loaded units on board the earlier LSM(R), and they could sustain fire much longer. An improved lighter version of the launcher, Mk 105, became available in 1945.

The Pacific Fleet laid out characteristics for an "ultimate" LSM(R), based on the new rocket launcher, in November 1944. She would support troops out to about 4,000 yds beyond the beach, conduct call fire (interdiction, harassment, destruction, illumination), and fire at high trajectories to destroy reverse slope targets. There was some hope, too, that rockets could be used against aircraft and against

The LSM(R) was the ultimate wartime fire support ship. LSM(R) 194 is shown, newly completed, off Charleston Navy Yard on 2 December 1944. Her decked-over well deck is covered by 75 Mk 36 four-rail rocket launchers, with 30 Mk 30 launchers outboard.

LSM(R) 193 displays her rocket launchers. Outboard are six-rocket Mk 30s (rockets were loaded above and below each of three rails). They swung inboard for loading, outboard for firing, and could be elevated manually at 5 degree intervals. At 17 degree elevation the 5-in finned rocket had a range of 5,550 yds. Mk 36, inboard, consisted of four modified aircraft rocket launchers (Mk 4) at a fixed 45 degree angle of elevation. It also fired the 5-in finned aircraft rocket. Reloading so many launchers was a laborious process; the LSM(R) offered terrific, but not sustained, firepower.

the high-speed suicide boats the Japanese were beginning to use.

By mid-December 1944 BuShips had produced plans and weight estimates to convert an LSM to a continuous-loading rocket ship. Because the bow would be sealed, plans for improved bow lines (for better seakeeping) were developed by the end of the month. Like their predecessors, the ships were armed with one enclosed 5-in/38; they also had two twin 40-mm mounts, four twin 20-mm, and four 4.2-in mortars. The rocket battery was ten Mk 102s controlled by two Mk 51 directors, so a ship could split her battery (e.g., for simultaneous direct and indirect fire).

In all, 48 ships were converted to this "ultimate" configuration. At the end of the war 24 of them were working up for the invasion of Japan. Although they saw no action, they were clearly quite valuable. In October 1945 the Pacific Fleet recommended that one division be retained postwar and that development continue. These "ultimate" LSM(R)s were used very effectively five years later, in Korea, and three (LSM[R] 409, 525, and 536) were reactivated in 1965 for Vietnam.

A postwar Pacific Fleet evaluation criticized them for inadequate living quarters and limited fresh water supply. Their mortars lacked continuous ammunition supply and suffered from 5-in/38 blast interference. The report suggested that two rocket launchers be removed, and the ammunition compartments for one of them turned into additional living space. Later the Marines pointed out that the arrangement of the rocket launchers limited arcs for long-range fire.

The mortars, which were potentially far more accurate than the rockets, were neither stabilized nor

LSM(R) 196, shown off Charleston on 16 December 1944, had a simplified battery of 85 gravity-fed Mk 51 rocket launchers firing spin-stabilized rockets with a maximum range of 10,000 yds. The launch rail could elevate between 30 and 45 degrees. Each launcher could accommodate 12 rockets.

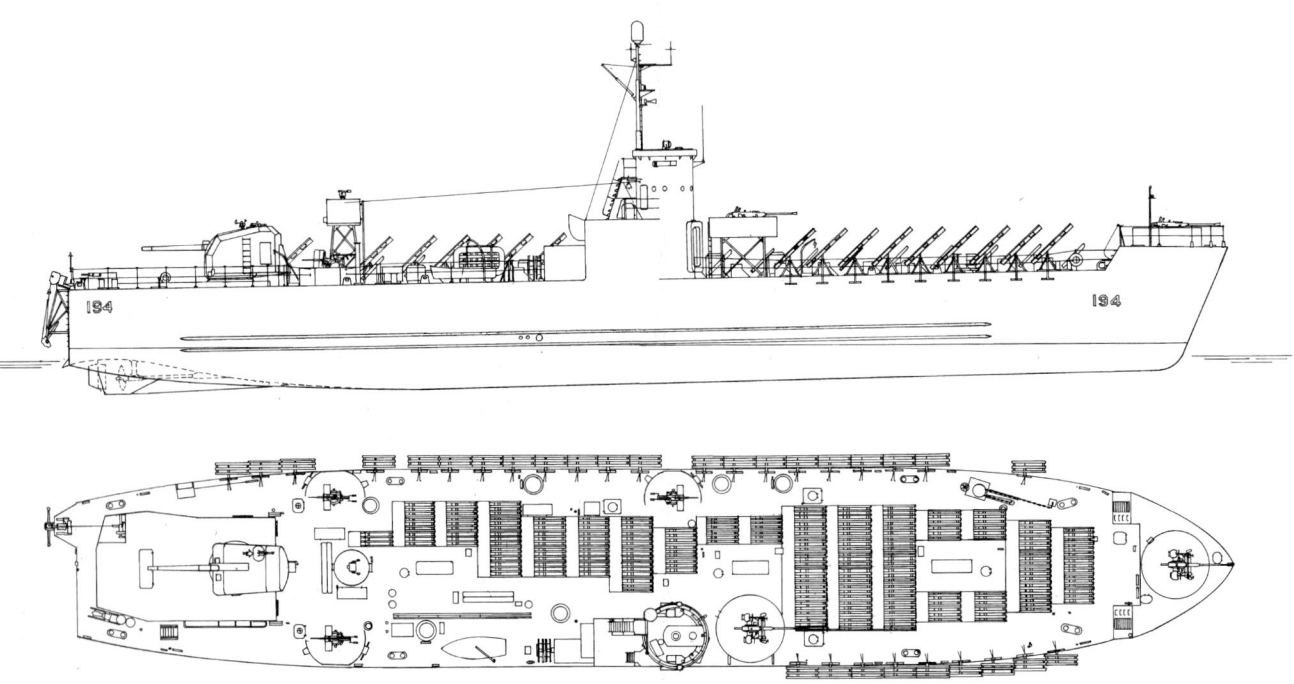

LSM(R) 194 was typical of the improvised LSM(R)s that saw action at Okinawa, with their banks of hand-loaded rocket launchers. She was a member of the LSM(R) 188–195 group. Note the hull stiffeners. All of the launchers had to be hand-reloaded, which took several hours. *A. D. Baker*

centrally controlled, hence they could not be fired precisely enough for close troop support. Removal was approved in February 1950. By February 1954 it had been decided to land two of the ten rocket launchers. Ultimate armament was set at one 5-in/38, two twin 40-mm, four twin 20-mm, and eight Mk 102 rocket launchers. Ships so modified were given forecastles; the units given to Germany retained their ten rocket launchers, and never had forecastles.

In June 1945, presumably with the kamikaze threat in mind, the BuShips' hull design section studied an LSM with its well decked over and its bow door closed, armed with a Mk 37–controlled twin 5-in/38 gun, and equipped with search radar. This project may have been a preliminary to converting LSM 446 into a radar evaluation ship for the new Operational Test and Evaluation Force (OpTevFor). Contract plans for the 5-in conversion were completed on 31 August 1945.

In May 1945, with the invasion of Japan in sight, it was clear that many more support ships would be needed. CinCPac asked for more LCI(L) to LCI(G) to supplement the existing 207 ships (125 LCS[L]3 and 82 LCI[G]), bringing the total to 400. Conversions would come as close as possible to the capabilities of an LCS(L)3. To support this fleet, LSMs would be converted to ammunition ships. Cominch approved the project at the end of May, and at least 20 were to have been converted each month. On the basis of a recent LCI(L) 351–class inclining experiment, planned ultimate armament was two twin 40-mm (each with a Mk 51 director), three twin 20-mm, and two single 0.50 caliber machine guns. In addition, the usual two 20 kW diesel generators would be replaced by three 30 kW units. A 1,000 gal per day evaporator, a 200 gal per minute motor-driven fire pump in addition to existing pumps, two 125 cu ft refrigerators, and SO-4 radar would be installed. This applied to both LCI(L) 1–350 and LCI(L) 351 classes. In the latter, about 40 percent of the deckhouse would be cut away. Cominch hoped that conversions could be completed at the rate of at least 20 per month. East coast yards would complete ships to the ultimate configuration. West coast yards would convert LCI(L)s to a less ambitious interim configuration corresponding roughly to Type A and B LCI(G)s for LCI(L) 1 and LCI(L) 351–class ships, except that they would carry no rockets. Thus the LCI(L) 1–350 class would have two single 40-mm guns, three 20-mm, and four single 0.50 caliber machine guns. The LCI(L) 351 class would have three single 40-mm and four single 20-mm guns (later two 0.50 caliber machine guns were added on the deckhouse abaft the 40-mm guns). These craft would be flycatchers (anti–swimmer/suicide boat patrollers) and "smokers" (anti-kamikaze smoke screen generators), so two of their fuel tanks would be converted to take fog oil. Later two 1,000 watt, 12-in searchlights were added to both versions, specifically to help crews spot swimmers and small boats at night at ranges out to 700 yds. They had to be capable of switching quickly on and off, to spot the swimmer without making the boat a viable target for Japanese coastal guns. The searchlight modification applied to all the new LCI(G)s.

In June the first units were ordered to east coast yards: LCI(L) 401, 403, 408, 412–415, and 428 to Boston and LCI(L) 506, 538–541, and 556 to Charleston. Actual conversion would be done by local private yards. The Boston yard was Lawley, which was the lead LCI(L) yard; Charleston subcontracted work to two Jacksonville, Florida, yards. Another 22 LCI(L)s were due back from the Mediterranean about 15 July 1945 (three 351 class, the rest 1–350 class), which were to be distributed between Boston (5) and Charleston (17). Charleston units would be converted in Texas yards.[11] The full 36-boat LCI(L) Flotilla 8 would follow, so that as of July 1945 a total of 72 ultimate conversions was imminent. Another 5 Atlantic LCI(L)s would become available when they completed their current training duty, for a potential total of 77 ultimate gunboats. Contract plans and specifications were ready only on 25 June 1945. In mid-July it was decided that one in every six LCI(G)s would be fitted as a flagship, the initial ones being LCI(G) 408, 428, 506, 530, 576, 953, and 954. Because east coast capacity was limited, it was decided in mid-July to convert Flotilla 8 to ultimate configuration on the west coast.

Meanwhile, 27 more ships were scheduled for interim conversions on the west coast: LCI(L) 2, 17, 19, 41–43, 46, 190, 192, 195, 196, 220, 233, 234, 398, 417–421, 514, 516, 517, 528, 542, 563, 948. The 36 ultimate conversions on the east and Gulf coasts, the 36 ultimate conversions on the west coast, and the 27 interim conversions on the west coast filled only about half of the 200-gunboat requirement. Any more conversions had to come from LCI(L)s active in the Western Pacific, but they could not easily be withdrawn from current and prospective operations. Thus by the end of July 1945 it was clear to Cominch and BuShips that only half the program could be completed.

Work on the first conversions was in progress when the war ended and Cominch stopped work on 16 August 1945. At that time the Lawley units were 5–50 percent complete; they were ordered sold as is. On many the engines had been removed as part of the conversion process. Of the Charleston units, LCI(G) was furthest advanced, and could be com-

Thirty-six LSMs were rebuilt completely as LSM(R)s; the prototype, LSM(R) 401, is shown. Her main armament was ten continuously loading rocket launchers. Abaft the launchers on each side were two 4.2-in mortars, with a twin 20-mm gun abaft them.

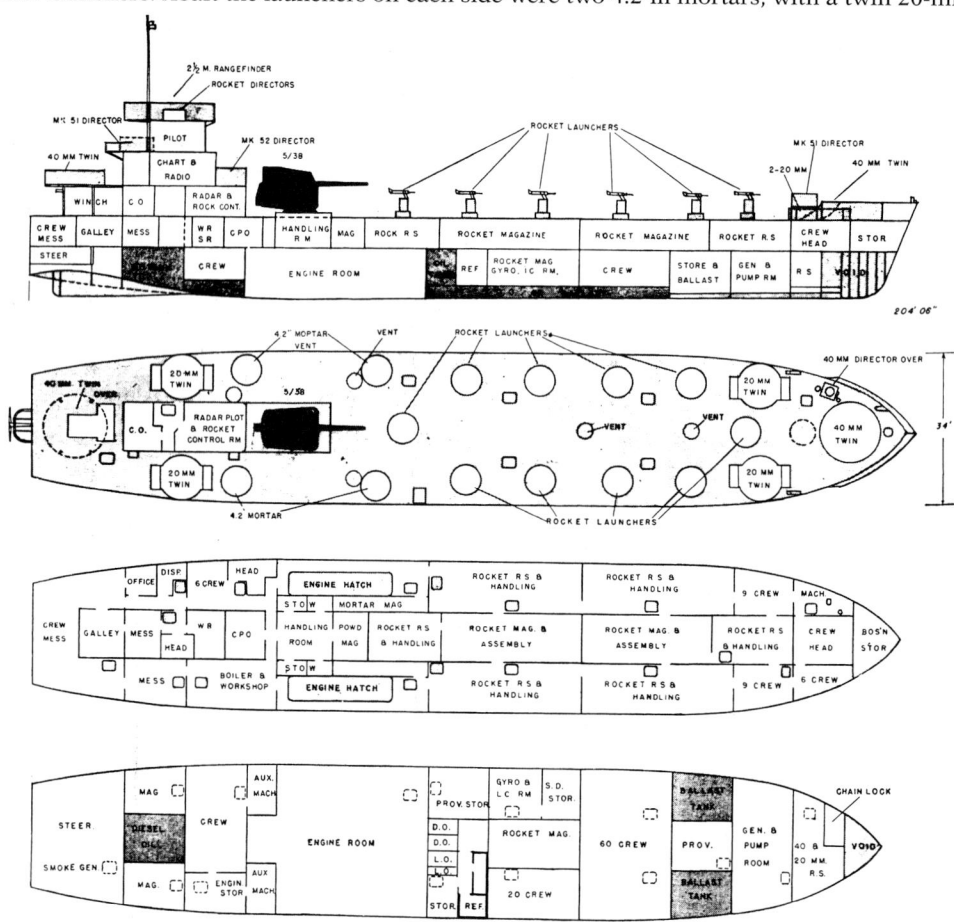

Wartime drawings of an LSM(R) 401–class ship. Showing the extent of her rocket magazines, many of them were above water just below her main deck. Placing rocket magazines below the waterline, hence protected to some extent, was a driver in the later *Carronade* design. *U.S. Navy (ONI 226)*

pleted by 7 September. She seems to have been the only "ultimate" LCI(G) completed.[12] CinCLant considered completing some units at Jacksonville (if that could be done inexpensively) so that they could be displayed on Navy Day in October 1945, but it is not clear if that was done.

By 1948 the Marines wanted a new kind of fire support craft. Their wartime experience, from the Marianas onwards, was that regardless of the weight or duration of preinvasion fire, some beach defenses survived. Thus the attackers needed a support craft that could direct short-range fire as they closed the

The specially rebuilt LSM(R)s were retained postwar. Two rocket launchers and all the mortars were removed, as in this view of LSM(R) 412. The long tubes on the 20-mm gun tubs forward were for spare 40-mm barrels.

beach. During the war that had failed due to poor observation or inadequate communication by shore fire control parties, and also because the supporting ships were unable to determine just where friendly troops were. The Marines considered neither the LSM(R) nor the LCS(L)3 a solution. Instead they wanted a heavily armored craft able to withstand 3-in hits, with a low silhouette to provide beaching capability and with one or more high-velocity weapons capable of penetrating 3 ft of Class A–reinforced concrete at 1,000-yd range. The weapons would be controlled from heavily protected stations with high-power optics. The craft would be maneuverable enough to evade enemy return fire.

New ships might also embody a new development, ship-mounted mortar-locating radar; mortars had imposed many of the Marines' wartime casualties. Existing naval radars had been tested in August and September 1945, and the cruisers' Mk 8 and Mk 13 sets proved adaptable. By 1948 BuShips was working on a new KPQ-1 shore-based set. The Marines wanted it aboard all combatants, such as the LSM(R), which might be used for naval gunfire support.

By 1948 the Ship Characteristics Board had on its books two future fire support craft, SCB 36 and SCB 37, which were in effect the LCS(L) and LSM(R) successors. There was a growing belief that both projects should merge into a single 20-kt hull. The key seemed to be the new Mk 105 launcher, with its high volume of fire. Admittedly there was a conflict between the demand for shallow draft and the desire to meet the new 20-kt standard (see below for discussion of the origin of the high speed desired for the amphibious task force).

When the General Board asked various commands whether it was worthwhile to retain 18-kt destroyer escorts (in the face of future fast submarines), in March 1948 BuOrd answered that they might usefully be converted to either offshore radar pickets or to inshore support ships. By April, how-

At the end of the war numerous LCI(L)s were ordered converted to gunboats at U.S. yards, in preparation for the invasion of Japan. LCI(G) 538 is shown, newly converted, with a pair of twin director-controlled twin 40-mm guns fore and aft, two more 20-mm guns at the break of the forecastle, a single 20-mm gun on the centerline aft, and a pair of 0.50 caliber machine guns in the former 20-mm positions at the forward end of the superstructure. She represented the closest possible approach to an LCS(L). Because the program was stopped with the end of the war, she was probably the only one of her type completed. Like other LCI(G)s, she was clearly a makeshift, and she was not retained after the end of the war.

ever, it was not clear that such a DE hull would meet the Marines' requirements. The ship had to be both fast and small enough to maneuver in shallow water, where she might well ground. Thus she had to be compartmented enough, and strong enough, to continue operating after such a grounding.

On 19 May 1948 the SCB issued draft characteristics for SCB 37 (SCB 36 having been amalgamated with it), which was planned as an FY 50 design study. The Marines' hopes for a high-powered gun battery were dashed; SCB 37 was in effect a 20-kt LSM(R), armed with a single 5-in/38, with 4.2-in mortars, and with enough rocket launchers to provide 300 rounds per minute (as in the LSM[R]).

Interest in this, as in other amphibious ships, lapsed as finances tightened in 1949. However, with the outbreak of war in Korea in 1950, interest revived. The 20-kt speed criterion could not easily be met. In February 1951 BuShips compared a 15-kt design (14 kts sustained) it had developed with faster (20-kt) and slower (13.5-kt) designs, the latter comparable to an LSM(R). The 238-ft, 15-kt hull required 3,100 BHP. To make 20 kts (17.5 kts sustained), the ship would need about twice the power (6,080 BHP), and she would grow to 281 ft, with a foot more draft (9.8 ft); displacement would be 1,595 tons fully loaded, compared to 1,340 tons. On the other hand, because length was set largely by the arrangement of the armament, cutting speed to 13.5 kts would reduce overall length only to 231 ft, and full load displacement only to 1,340 tons; the required power would be cut to 1,570 BHP.

BuShips reported its feasibility study to the SCB on 1 June 1951. To reduce damage due to inadvertent grounding, her inner bottom would extend over 80 percent of her length. To make refloating easier after grounding, her diesel oil tanks would be fitted for ballasting. For example, if she grounded with half the fuel tanks ballasted, she could reduce her draft by about half a foot by pumping about 100 tons of water overboard.

The new Mk 105 rocket launcher fired 48, rather than 30, rounds per minute, so the ship could reach the desired volume of fire with eight rather than ten launchers. Adding two more would make the ship about 18 ft longer and 40 tons heavier. By this time plans already called for cutting LSM(R) batteries to eight launchers. The ship would carry 6,060 rockets. Her single 5-in gun would fire forward rather than aft (as in an LSM[R]) because it seemed that the ship would need protection on the way in, before she fired her rockets, rather than on the way out, after her rocket fire had disrupted enemy gun emplacements. The LSM(R) mortars were eliminated.

As in an LSM(R), BuShips wanted to stow the rockets on the platform deck, slightly above the waterline, where they would be accessible to the launchers. This big magazine was entirely unprotected. One consequence was that the emergency diesel generator had to be placed in the hold rather than above the waterline, as the draft characteristics required. Moving the generator above the waterline would add 14 ft to the ship's length.

To the Board of Inspection and Survey, this was little more than a slightly improved LSM(R). Surely it could be something more. Fire support ships would likely be needed only briefly during an assault. They should have some alternative role (e.g., as ASW escorts), but that would require speed comparable to that of a destroyer escort. If BuShips' low speed was accepted, the ships could not operate with any fast task force, or with an advance force. Moreover, their rockets might have alternative roles, such as dealing with the fleets of sampans then threatening Formosa (Taiwan). A faster ship could use them effectively. The SCB replied that size had been held down to limit the ship's cost and to maintain shallow draft and maneuverability. Higher speed might not be so valuable; the ship could still operate with the advanced force of a joint amphibious task force. There was also some question as to whether barrage rockets could deal effectively with sampans, given their limited range and poor accuracy at medium and low elevation.

In the preliminary design, length was increased to 245 ft; full load displacement was estimated as 1,475 tons (1,200 tons standard). The hull form was based on one developed for an abortive new 220-ft aluminum minesweeper. However, the forebody was made fuller to accommodate the rockets, and the stern was kept narrow to reduce the ship's tendency to yaw and broach when approaching a shore in a following sea. Model tests showed that the ship would turn easily but that she would not be directionally stable, so 7 ft was added to the after end of the skegs and rudder area was increased. Tank tests suggested that lengthening the ship would increase her tactical diameter at 15 kts from 2.1 to 2.7 ship lengths, which was acceptable.

Estimated trial speed (two 1,550 BHP diesels) was 15.25 kts. The ship had controllable-pitch (CP) propellers to avoid any need for reversing gears or reversible engines, both of which would add complication. Unlike the LSM(R), the ship used a dry muffler (in a stack); in the LSM(R), exhaust discharge near the waterline at the fore end of the deckhouse created a salt spray, which blew through open ports.

Probably the greatest design issue was the vulnerability due to stowing all rocket ammunition above the waterline. The preliminary design showed 30-lb (¾-in) STS on the deck above the magazine and on the sides along it, down to 18 in below the full load waterline. Moving the ammunition to the hold would require longer hoists, but it would make much better use of the volume of the hold. Magazine length would be reduced by 17 ft, and exposed side area by 59 percent (only a strip 3½ ft wide would be exposed above the waterline amidships). Because the magazines were no longer under the exposed weather deck, they no longer had to be refrigerated. As a side benefit, moving the crew above water would improve habitability. The only real drawback was more complicated replenishment. This possibility was first considered in May 1951, and at first it was rejected for fear that it would delay completion of the ship. In February 1952, however, the rearrangement was approved. Because the hoists were lengthened late in the design, 'tween-deck heights were not adjusted; the second deck was not lowered as it could have been.

Overall, the ship was designed for quick wartime production. For example, instead of cambering, the main deck had flat plates sloped to the deck edge. The ship was powered by stock (World War II–made) engines, which, it was pointed out in 1955, might not be available for later ships in the class.

One ship, *Carronade*, was built under the FY 52 program. Although she was essentially an upgraded LSM(R), she was given a new classification, IFS (inshore fire support). In April 1955 the two naval officers supervising construction of *Carronade*, the supervisor of shipbuilding at the Puget Sound Bridge and Dredging Company and the naval inspector of ordnance at Seattle, wrote a joint letter criticizing the design. Little effort had, it seemed, gone into providing for ammunition replenishment; it would take 24 hrs to fill the ship's rocket magazine. It was incredible that the ship was not air conditioned, given the new emphasis on habitability. BuShips said that this had been done deliberately to save space and weight. Part of the excessive overhead space could have been used for air conditioning machinery. Because the second deck, over the magazine, was a ballistic (protective) deck, it could not be pierced. That complicated such trivia as installing shower drains. The ship service generator

The postwar USS *Carronade* was, in effect, a modernized LSM(R). Newly completed, she is shown off Seattle on 4 May 1955.

was too small (200 rather than 250 kW), and its diesel too loud.

A new-design IFS was tentatively included in the FY 56 program. However, it was deleted in March 1955. Adm. William K. Mendenhall Jr., DCNO (deputy chief for naval operations) for logistics, decided not to include it in the FY 57 program, although the Atlantic and Pacific fleet commanders did show some interest. That ended fire support craft development before the Vietnam War. *Carronade* herself was decommissioned in 1960. She was recommissioned for service in Vietnam. Some of the slower LSM(R)s survived in reserve; three were recommissioned for Vietnam. Two were transferred to West Germany, one to South Korea.

While *Carronade* was being built, there was revived interest in small rocket craft. In February 1951 OpNav asked for studies of rocket armament

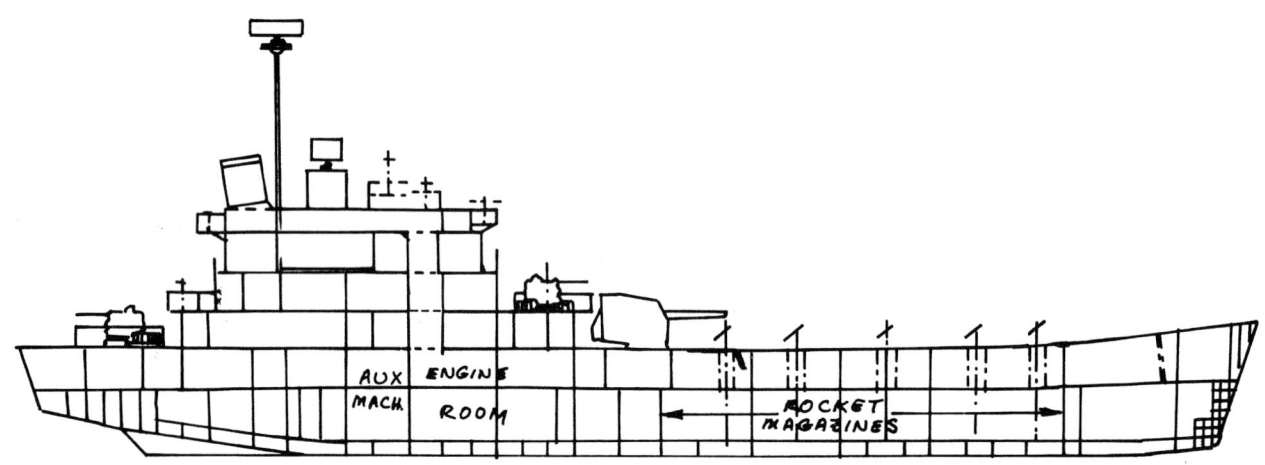

Carronade (IFS 1) inboard profile. Ammunition capacity was 350 rounds of 5-in, 8,000 rounds of 40-mm, and 6,000 5-in rockets.

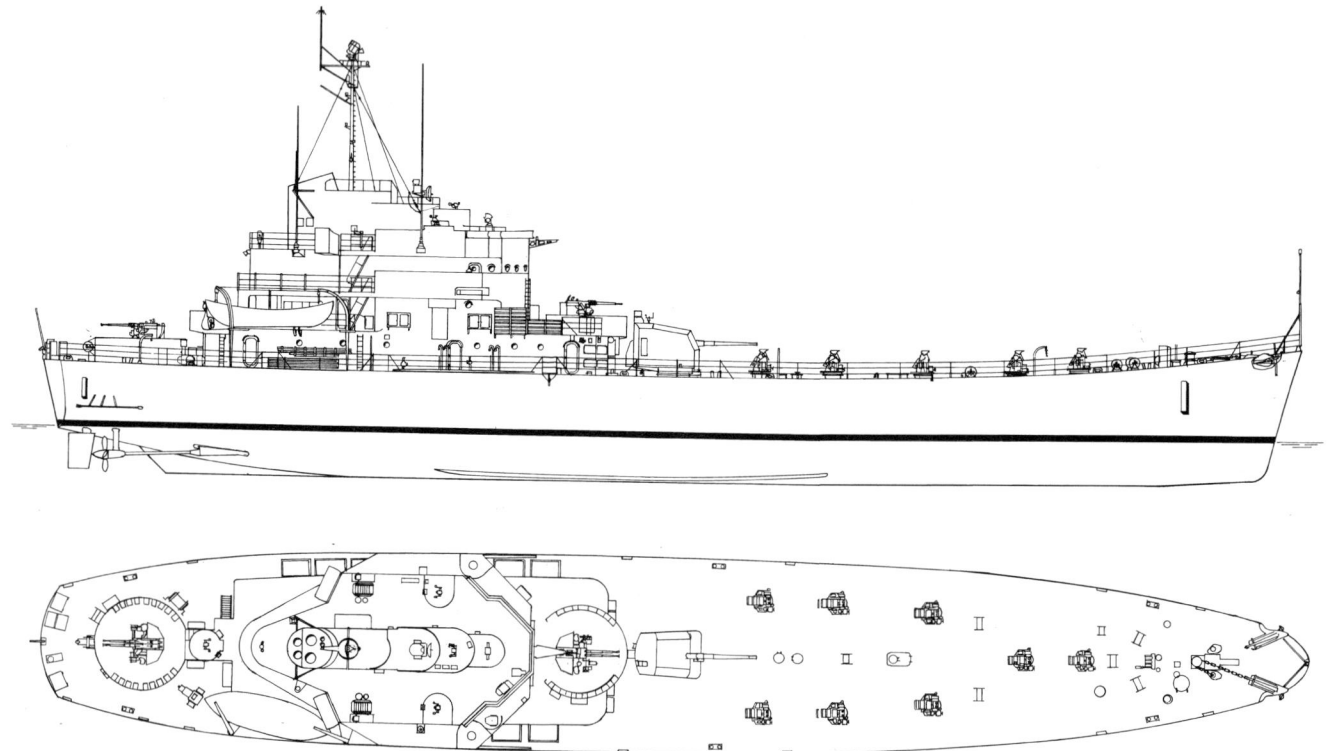

Carronade (IFS 1) as completed. Although designated in a new series, *Carronade* was in effect the final development of the wartime LSM(R). *A. D. Baker*

FIRE SUPPORT **259**

White River (LSM[R]/LFR 536) about 1968, in service in Vietnam. Visible changes from her original configuration include the elimination of two rocket launchers, the big radio antenna atop her 5-in gun house, and the enclosure of her bridge. The scrap view shows the longer bow of the earlier group of LSM(R)s, typified here by *Clarion River* (LSM[R] 409). *A. D. Baker*

for 20 LCM(3) and 36 LCVP, apparently for transfer to France for combat in Indo-China. BuOrd had considerable supplies of wartime rockets and launchers: 4.5-in Mk 7 (12 rounds each) and 5-in spin-stabilized Mk 51. Without any change to her structure, an LCM(3) could carry six Mk 7 on her sides, four on her stern, and four over her engine compartment. A portable raised platform could carry 40 in her cargo space, for a total of 648 rockets. An LCVP could carry 4 outboard, 2 on her stern, and 12 in her cargo space, for a total of 216 rockets.

War experience showed that specialized ships with extensive electronics were needed to control landing operations. *Mount McKinley* (AGC 7) is shown at Mare Island on 24 August 1945. The big dish on her lattice mainmast is a long-range SK-2 air search radar; the smaller dish forward is an SP for height finding and fighter control. Changes during this refit are indicated by circles; note the HF whip antenna newly installed abaft her deckhouse.

9
Command and Control

Although the need for fire support was understood before World War II, the need for special command ships, at many levels, was not. At one end of the spectrum were headquarters ships or amphibious flagships, which were needed to coordinate an entire amphibious operation. At the other were small control craft needed to guide waves of boats to the appropriate beaches.

The British were the first to appreciate both needs. Before the war, they had assumed that any operation would be commanded from a major unit, such as a cruiser or a battleship. At Dakar in 1940 they found that because such units also had important naval responsibilities, they might well have to leave the area of the landing, for example, to deal with enemy naval forces. A dedicated flagship was needed. Because the British wanted to emphasize the joint nature of the operation, and because they understood that army commanders had to make their headquarters aboard until they could be established ashore, they called their ships headquarters ships. Just as a flagship maintained a plot of the naval situation, a headquarters ship would maintain a current picture of the situation ashore.

In 1940 there were no surplus ships large enough. However, the idea recurred, and in a 14 January 1942 report, the Inter-Service Committee on Communications, working for Combined Operations, proposed a headquarters ship. Advantages cited included availability for training prior to a major operation, because no major surface combatant could be sidelined long enough. The idea was approved by the Chiefs of Staff committee on 28 January, but the shipping shortage was such that the first ship, *Bulolo*, was taken up only in March (to be ready in June 1942). The British converted another ship, as HMS *Largs*, to serve the American landing at Oran; she turned out to be too small for the purpose, with only 200 sq ft vs. 1,600 sq ft in *Bulolo*. *Bulolo* later served as communications ship for the Casablanca Conference in January 1943.

About mid-1942 the British coordinator of Combined Operations, through his Washington representatives, pressed the U.S. Navy to develop headquarters ships. Conversion plans and staff requirements (characteristics) for *Bulolo* and *Largs* were provided. It was probably more significant that Rear Adm. Kent Hewitt, during his visit to Britain in June 1942, was much impressed by *Bulolo*. He probably sold the idea in the U.S. Navy, to an extent no British representatives could have done. The assistant chief of staff for readiness in the Cominch's office urgently requested three ships for flagship (AGC) conversion as early as 28 July 1942; the flagships were called combined operation and communication headquarters ships. Three C2s, under construction for delivery beginning January 1943, were earmarked in a 19 September 1942 Auxiliary Vessels Board report. The secretary of the navy formally requested ships for conversion, from the War Shipping Administration, on 30 September 1942. They were explicitly described as similar to ships the British were using to conduct amphibious operations. The Joint Chiefs approved the conversions on 15 October. Initially designated administrative flagships (APFs), the ships became the *Appalachian* class, AGC 1–3 (C2-S-B1 conversions).

Thus the U.S. decision to build special amphibious flagships predated all U.S. operational experience, both at Guadalcanal (August 1942) and in North Africa (November 1942). The need for such ships was certainly dramatized at Casablanca in November. Like the British flagship at Dakar in 1940, the U.S. flagship, the cruiser *Augusta*, had to leave the area during the landing to deal with a naval emergency.

The new AGCs were built around three main spaces: a joint operations room (support control room), which maintained the air and surface picture; a war command room in which senior officers of all the services involved sat; and a flag plot. The ship also had a combat information center (CIC) capable of fighter control, and a voice filter room, which received, filtered, evaluated, and passed on information received by voice radio. Other input passed through an intelligence office, and there was also a photo interpretation room. There were offices to accommodate army and navy officers. A print shop and a map reproduction room produced the documents needed for an assault. By November 1944 CICs and some other spaces, such as photo interpretation rooms, had been ordered air conditioned. Overall, however, the ships were also considered inadequately ventilated for the tropics. Ships' crews did somewhat improve the situation, but the problem was never really solved. Thus, in 1947 the Marines complained of insufficient insulation between boiler uptakes and officers' staterooms. To accommodate the additional spaces, the ships were built up by a deck height between their pairs of kingposts, covering all but their former foremost and aftermost holds. They retained one hold, forward.

When these ships were completed in 1943, they had the most extensive radio communications afloat, including receivers to monitor enemy radio channels for intelligence. It turned out that voice radio was dominant in amphibious operations, so in November 1944 six operator positions in the central receiving room were ordered converted to control radio-telephone circuits.

Ships were initially fitted with long-range SK air search radars on a lattice mainmast; they retained single pairs of kingposts, with booms, fore and aft. Ships also had the necessary air navigation beacons (YE plus the supplemental YG), and by the end of the war most also had TDY jammers. By the end of the war, the standard was SK or SK-2 plus an SP height finder on the after port kingpost (the carriers' SM radar antenna was too heavy). The booms on the fore sides of the after kingposts were eliminated, because the after holds had been filled in. There were antenna interference problems. Short-range TBS antennas had to be relocated to outriggers, one forward and one aft of the radar tower so that they would not interfere with the ship's long-range SK air search set. Vertical wire receiving antennas would be replaced by whips; the number of horizontal ("flat-top") antennas aft would be reduced (and relocated by removing the ship's booms there), and inverted-L low frequency (LF) antennas fitted forward. More generally, the numerous radio transmitters, all operating in very close proximity, interfered with each other. Suddenly there was a real need to develop some alternative type of antenna. The interference problem led to the postwar development of broadband sleeve antennas, which are now com-

Blue Ridge (AGC 2), shown on 6 October 1943, was one of the initial class of three specially built AGCs. To gain the necessary internal space, her hull was built up one deck between her two goalpost masts. Note the two whaleboats in davits; in service such ships carried only standard landing craft. Note also the freighter booms retained even though the hatches they served (fore and aft the forward kingposts and forward of the after kingposts) had been plated over. Some proved useful to serve the ship's boats.

mon. The radar and radio equipment also filled considerable space; by November 1944 radar and radio transmitters were ordered raised a level to free space on the flag bridge.

Postwar, SK/SK-2 was replaced by SPS-6B and then by successor air search radars: SPS-17 and SPS-37. When heavy height finders (SPS-8A, then SPS-30) appeared, they typically replaced SK/SK-2 on the lattice mainmast, with the much lighter two-dimensional antenna being mounted forward or aft. Airborne early warning (AEW) was a less visible development. In effect, an AEW airplane could be an additional search radar, if the AGC had the necessary terminal. These devices were installed on board both carriers and AGCs, although initially they were so heavy that they could be installed only in an emergency. The lightweight version, for permanent installation, became available in 1948.

A fourth ship, the ex-army transport *Ancon* (AP 66), earmarked for conversion to an attack transport, was selected soon after the three C2s; she became AGC 4. Because she had already been completed as a transport, *Ancon* was ready before the C2 conversions, and thus was the first operational U.S. AGC. She was first used in the invasion of Sicily, along with several improvised ships.[1] The three initial C2 conversions were soon supplemented by a fourth, *Catoctin* (AGC 5). *Catoctin* was considered a

Ancon was converted into an attack transport specifically to support the invasion of North Africa, then converted into the first U.S. amphibious flagship. She was one of three fast passenger-cargo liners ordered in 1938 by the Panama Railroad Steam Ship Company, some of the very few U.S. merchant ships of the period *not* built under Maritime Commission auspices. Because the company was owned by the U.S. government, these ships were immediately available for military use. The first, *Panama*, was taken over in June 1941 and converted into the army transport *James Parker*. She figured in early naval planning for combat loaders. *Ancon* underwent a similar conversion, having been taken over on 11 January 1942, but between June and July 1942 she was converted into a combat loader by the Moore Drydock Company of San Francisco. She served as flagship of Transport Division 9 in North Africa, after which she was rebuilt at Norfolk as a command ship. She served as the training ship for the Maine Maritime Academy in 1962–72 under the name *State of Maine*. The third ship, *Cristobal*, was also taken over as an army troopship. In this 15 October 1942 photograph of *Ancon*, taken off Norfolk, the evident signs of conversion are the guns forward and aft and the air search radar antenna placed, unusually, at the fore end of the ship's funnel.

Although not the first ship ordered as an AGC, *Ancon* (AGC 4) was the first completed as such. She is shown, newly completed, on 24 April 1943. The big radar antenna on her lattice mast is for an SK, a long-range air search set. Note that she retained her four attack transport davits. *Ancon* was unusually well adapted to AGC service because, having been built as a liner, with space for 202 first-class passengers, she had a voluminous superstructure.

different class because, although she was also a C2-S-B1, she came from a different builder (Moore Drydock rather than Federal Shipbuilding), hence she differed in detail from the others.

Although the first conversions were approved in the fall of 1942, no command ship was available for the North African invasion. *Ancon*, which would be the first, served there as an attack transport. She continued to transport troops after the invasion, and did not enter the conversion yard until 16 February 1943. However, she was ready for the next major Atlantic operation, the invasion of Sicily, in July 1943. In that operation she accommodated the commander of the Central Task Force. There was no intent to add fighter control to her flagship role. Instead, an LST was selected, presumably because her tank deck provided sufficient volume for the plot used to vector fighters. The radars themselves could be set up on the ship's large flat upper deck.

For the Sicilian operation, then, LST 355 was fitted with a special ramp so that she could load army-type fighter-control (GCI, ground-controlled intercept) equipment onto her upper deck. Unfortunately she was unable to set up VHF voice radio communication with either the fighters or with the flagship *Ancon*. Even so, the experiment was considered encouraging, and it was repeated at Salerno the following month with two LSTs (the U.S. LST 385 and the British LST 305). One of the LSTs was hit during the assault, and was too badly damaged to function as a GCI ship. Again, it was impossible to set up VHF communication. *Ancon*, which was not completely fitted for fighter control, had to take over that task. Because Morse code had to be used for reporting, the plot aboard *Ancon* lagged the reports. The Germans took advantage of the situation to create false plots and tracks on the reporting channels. Also, when the battle began and

Ancon was well out to sea, her radar operated quite effectively. As she approached the shore, it detected too many land targets and was saturated. One lesson was that no single radar could suffice in such a situation; radars throughout the area had to be linked together. That in turn raised the communications load on the AGC.

Eighth Fleet concluded that it was unwise to combine the fighter director and flagship roles. A flagship had to be large enough to accommodate staffs, and to be less conspicuous she would anchor in the transport area. Due to her value, she would leave the beach area as soon as possible. That in turn would influence the air commander to go ashore as soon as possible, which might be before air control was fully set up ashore. Even so, the U.S. Navy concluded that it would have to combine the AGC and GCI functions. At this time the three initial C2 conversions were being designed. They had radar and CIC features equivalent to those of contemporary fleet carriers. *Appalachian* (AGC 1) and *Rocky Mount* (AGC 3) entered combat as control ships at Kwajalein in January 1944.

Special fighter control ships were used both for Normandy and for the invasion of Southern France. For Normandy, the British had three more elaborately converted LSTs, which they designated fighter direction tenders (FDTs). One such ship participated in the invasion of Southern France.[2] For the invasion of Southern France three U.S. LSTs were fitted with special ramps by the ship repair yard at Naval Operating Base Palermo, so that they could take GCI equipment on board. The United States also used GCI LSTs in the Pacific; for example, three were at Iwo Jima.

Ancon's team was the only successful interception team, although several were present at Sicily. Lessons learned included the need for antennas that could stand up to sea conditions, for automatic orientation of the PPI scope fed by the radar, for reduced side- and back-lobes (to limit false echoes), for identification friend or foe (IFF) and radar beacons, and for elimination of mutual interference between radar and radio antennas. The placement of the ship needed the same consideration as the placement of land radars, and the ship needed full liaison with other units (including a power boat) and full "Y" service (ELINT).

Another important point was the number of amphibious flagships needed for an operation. Initially plans called simply for one flagship for the entire operation, whatever its magnitude. By the time of Normandy, the British wanted one for each division going ashore; they had to convert a ship specifically for the operation. The six surviving "Treasury"-class Coast Guard cutters were all converted into amphibious flagships, their places as convoy flagships being taken by large prewar-built destroyers. However, only *Duane* was redesignated; she briefly became AGC 6. The overall flagship of an

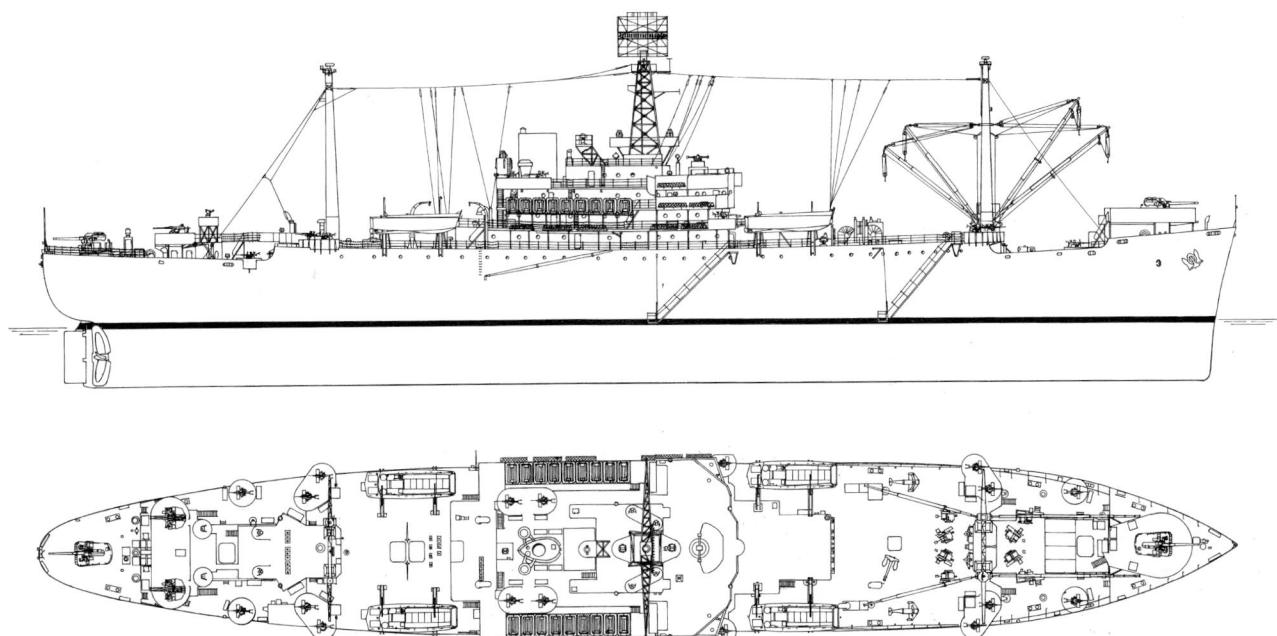

Rocky Mount (AGC 3) in August 1944, with an SK long-range air search radar on her lattice mast. She does not yet have a height finder. Note her extended 01 level, to gain volume for quarters and offices, and the reduction to four boats. Only her forward hold remained in use, largely for vehicles the staff might need when ashore. *A. D. Baker*

operation did not need elaborate facilities, and modified attack transports were satisfactory in this role. They were not redesignated; they were generally referred to as relief AGCs (see below).

When they approved the "must" program, the JCS Planners also asked for 13 more AGCs. In December the Auxiliary Vessels Board advised Admiral King that there was no substitute for C2 hulls, so he asked the Maritime Commission for eight C2-S-AJ1s, then under construction, for conversion to AGC 7–14. The first three would be similar to AGC 5. In the other five, instruments and equipment not needed by a task group commander would be omitted. Any saving in space would be used for troop and cargo lift, a provision that suggests just how badly lift was needed. This provision was soon reversed. *Auburn* (AGC 10) and *Eldorado* (AGC 11) became the standard for AGC joint operations' support aircraft command centers. In *Eldorado*, the naval gunfire communications center, operations office, tactical plot, strategical plot, and communications office were placed just abaft the flag bridge, on the same level. This change was planned for all AGCs, but not effected on all of them. In October 1945 the Pacific Fleet lessons learned report suggested that instead of being placed abaft the flag bridge (i.e., accessible to the amphibious commander), it should have been in the joint operations room, so that it could coordinate artillery and air support. Overall, the AGCs were criticized on the ground that information was not sufficiently concentrated and accessible to the force or group commander. This problem probably could not have been solved prior to the advent of very powerful computers, because it was impossible to display in any one place all of the many plots carried in all the different spaces.

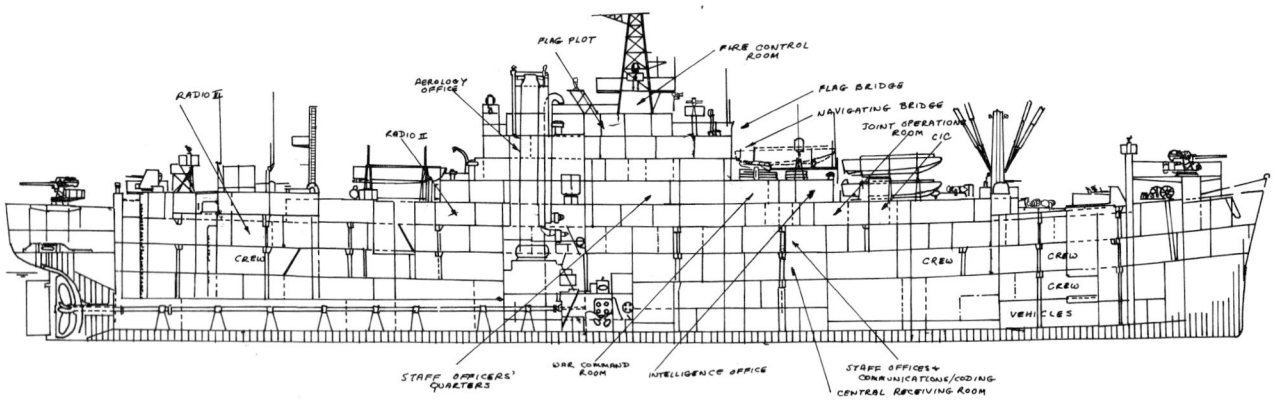

Panamint (AGC 13), April 1945. The object on the 01 level forward of the superstructure is a 5-in loading machine flanked by a pair of 36-ft personnel boats similar to those abaft the deckhouse. The ship's foremast shows four 10-ton booms. Note that very little hold area remains. What is left of No. 1 hold is devoted to vehicles, with crew berthing above around the trunked hatchway. The deckhouse immediately abaft the two boats is devoted to hospital space. Guns shown here are single 5-in/38s fore and aft, with a pair of 20-mm guns forward and a twin 40-mm aft.

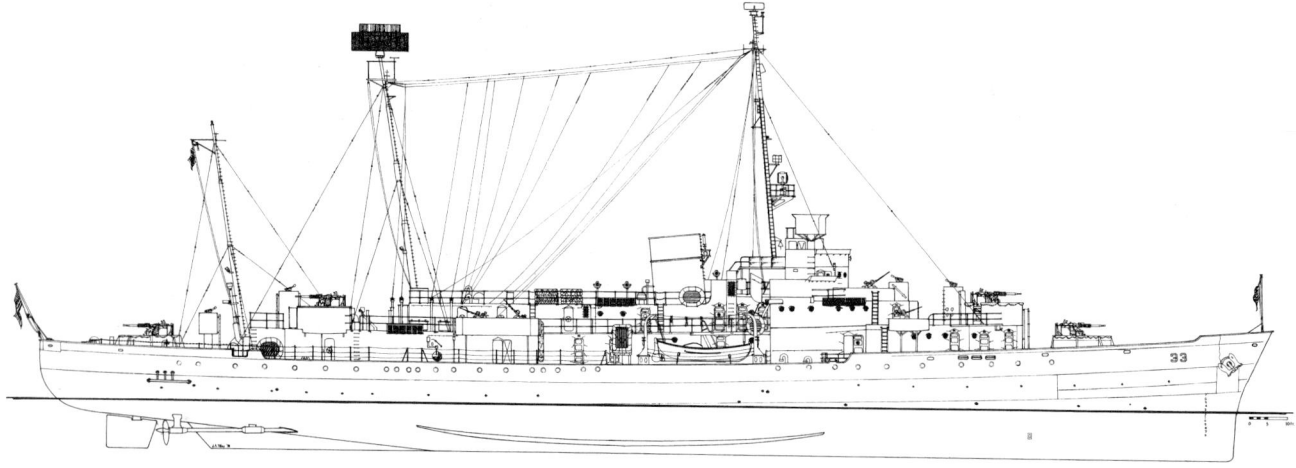

The Coast Guard cutter *Duane* as converted to an amphibious flagship in 1944 with all of her 5-in guns having been landed in favor of 40-mm weapons. Note her forest of long wire antennas and the apparent absence of whips.
U.S. Coast Guard; drawing by Dr. John A. Tilley

Eldorado (AGC 11) emerges from Mare Island, 22 November 1944. Improvements included addition of a height-finding SP radar (with a surface search SG alongside it) on the forward pair of goalposts. The framework atop the after goalpost is an aircraft homing beacon.

As of January 1944 the Maritime Commission was unwilling to arrange for the five more hulls required, although Admiral King had asked the Joint Chiefs to request them from the Maritime Commission. In July 1944, however, another three C2-S-AJ1s were ordered converted to AGC 15–17. These ships became the *Adirondack* class, leaving a requirement for two more; this was met, if at all, by using Coast Guard cutters. The *Adirondack*s were similar to AGC 10–14, being only slightly modified as a result of a July 1944 conference. They could be distinguished by their pole masts (in place of paired kingposts) aft; without any after holds, there was no point in retaining kingposts and booms there. Accommodation was increased to 175 officers and 950 enlisted men. As BuShips had suggested, vehicle stowage and the boat repair shop were turned into crew berthing. The boat complement was four LCPLs and a 26-ft

Mount McKinley (AGC 7) emerges from a Mare Island refit, 25 August 1945. The two small radomes on the crossyard of her after goalposts are radar direction finders, which were used both as sensors and to support jamming.

plane personnel boat (ships actually carried four boats in davits and two more on deck). Armament was set at two 5-in/38, four twin 40-mm, and twelve 20-mm; the first three AGCs initially had only two twin 40-mm guns, both aft.

Biscayne (AVP 11), a small seaplane tender, became AGC 18. She was converted at Mers-el-Kebir by the repair ship *Delta* between 2 and 31 May 1943, but was not redesignated until 10 October 1944. Table 9-1 shows details for seven ships that were either designated or used as amphibious flagships. In March 1944 BuShips proposed that several projected large seaplane tenders be completed as AGCs, but this was not done.[3] Amphibious flagship conversions of both destroyer escorts (DEs) and frigates (PFs) were also considered and plans drawn, but they were not pursued.[4]

By April 1944 the Pacific forces afloat were criticizing the new AGCs; they wanted more facilities and many more personnel. Unfortunately the C2 hull had been squeezed to its limit. Accommodations grew from 130 officers and 738 crew in *Appalachian* (AGC 1) to 156 officers and 877 crew in *Mount McKinley* (AGC 7); and the forces were asking for 185 officers and 1,000 crew. No deckhouses could be added without violating draft (i.e., flooding survivability) and stability limits. In June 1944 BuShips offered two alternatives: the ship's functions could be split or a larger hull could be adopted. Neither was acceptable. The whole point of the AGC was that it unified command functions. As for a larger hull, the AGC was already being criticized as too large and too easily identifiable. Moreover, no larger hulls were available. Perhaps something could be cut. BuShips noted that the forces afloat were deemphasizing aircraft control, and suggested that ships in company could provide boats and vehicles.

These three detail views of *Mount McKinley* (AGC 7), at Mare Island on 24 August 1945, give some idea of the extent of the electronic outfit of such ships. AGCs were the first U.S. warships to encounter catastrophic mutual electronic interference from their radios. In all the photographs, what look like dashes are insulators on antenna wiring. Changes are circled. In the view from aft, note the new HF whip antennas. The object atop the starboard kingpost is a TDY radar jammer; a radome on deck forward houses a complementary microwave jammer. The large white object is a movie screen. Atop the bridge is an existing navigational range finder. In the view of the area just abaft the bow, note the radome for the jammer and the new SP height-finding radar. The surface search antenna alongside it has been raised to clear it. Note the added whip antennas. The boats with deckhouses are modified LCPLs. Note also the added 40-mm guns and the radar director for the forward 5-in gun.

Photographed on 18 March 1946, *Pocono* (AGC 16) eliminated some of the merchant ship features of the earlier AGCs; she had only a pole mast aft, and her two forward kingposts were made considerably lighter. The port kingpost carries radar direction finders. Compared to wartime AGCs, she has a much-reduced boat outfit.

One complication was that the CIC was not well suited to the army air controllers (Army Air Force Tactical Control) who would operate from the ship while the beachhead was being established. For Salerno, *Ancon* was fitted out with just such a center, and in February 1944 a U.S. Army Air Corps major commanding the control center aboard the new command ship *Catoctin* (AGC 5) suggested using the existing flag plot to build a similar center because the CIC was not large enough. The army required a large horizontal filter board to handle the large volume of air traffic because the usual naval vertical board could not suffice. Also, the army used plastic markers indicating time and position to help separate out large numbers of unrelated plots. These, too, could not be used with a vertical board. The use of a horizontal board, 5 × 6 ft, in turn forced up the size of the air control center. In addition, the army used a 5 × 6 ft horizontal operations table at which two army and one navy (or RAF) controllers sat, shoulder to shoulder, to coordinate up to 12 squadrons operating in adjacent areas, at various altitudes, carrying out various missions. Operations in the Mediterranean showed that the three intercept officers and the senior controller had to be able to observe the plot of the unfolding situation. To make matters more complex, the ship's radars all reported in coordinates centered on the moving ship, whereas operations over land were controlled in terms of a grid set on shore. Thus a gridded operations board was needed, and ship's data had to be translated into its terms.

The British later commented that because the U.S. Navy designed and built its whole series of amphibious flagships to essentially a single design, developed at an early stage of development, the ships were not as well adapted as the more gradually evolved British headquarters ships. For example, according to the British, a later British ship, HMS *Lothian*, much impressed the U.S. Navy when she

Table 9-1. Amphibious Flagships

	Appalachian (AGC 1)	Ancon (AGC 4)	Mt. McKinley (AGC 7)	Adirondack (AGC 15)	Biscayne (AGC 18)	Blue Ridge (LCC 19)	Duane (W 33)
Hull type	C2-S-B1	NA	C2-S-AJ1	C2-S-AJ1	NA	NA	NA
Length (ft-in)							
WL	435-0	471-6	435-0	435-0	300-0	579-11	308-0
OA	459-2	493-6	459-0	459-2	311-4	636-5	327-0
Beam (ft-in)	63-0	64-0	63-0	63-0	41-0	82-0	327-0
Draft, loaded (ft-in)[a]	19-9/23-6[b]	26-3	22-1/23-10	24-0	11-7/14-2	28-10	12-5/15-0
Displacement (tons)							
Light	7,397	9,946	7,201	6,849	NA	16,790	1,903[c]
Loaded	12,691	13,144	13,040	12,270	2,563[d]	18,646	2,842
Boilers	2 CE	2 Yarrow	2 B&W	2 CE	—[e]	2 FW	2 B&W
Conditions (psi/°F)	465/765	465/765	450/750	450/750	—	600/873	400/412
Power (SHP/shafts)	6,000/1	6,000/1	6,000/1	6,000/1	6,400/2	22,000/1	6,200/1
Speed (kt)	16.9	19.0	16.9	16.0	17.5	22	19.3
Generators (qty-kW)							
Main	3-300	3-300	3-300	3-500, 1-100	2-200, 2-100	NA	1-300
Auxiliary	NA	1-150	1-100	NA	NA	NA	NA
Fuel oil (gal)	1,137,675	408,177	1,126,848	852,600	82,514	NA	154,141
Landing craft fuel (gal)	30,724	11,590	28,140	29,820	Ship fuel[f]	NA	—
Radius (nm/kt)	48,460/15	10,852/15	43,948/12	NA	4,500/16.3	13,500/16	7,012/12
Complement[g]	54/31/611	54/31/553	54/31/611	54/579[h]	17/13/212	42,742	21/15/199
Staff[i]	101/267	NA	103/338	NA[j]	30/148	250	NA
Capacity							
Troops	55/110	195	61/275	31/286	—	1/15	—
Water, fresh (gal)	104,800	255,059	103,891	NA	12,442	NA	30,620
Distilling rate (gal per day)	24,000	12,000	33,000	30,000	10,000	NA	16,400
Boats							
LCVP	2	6	2	—	—	2	—[k]
LCPL	4	4	4	4	—	3	—
LCPR	2	2	2	1	—	—	—
LCC	2	2	2	—	—	—	—
Armament							
5-in/38	2	2	2	2	2	—	—
40-mm	4 × II[l]	4 × II	4 × II	4 × II	5 × II	2 × II[m]	2 × IV, 3 × II[n]
20-mm	14	14	14	12	6	—	8

NOTES: Dash indicates that specification is irrelevant; NA indicates relevancy but data not available.

[a]Draft mean if only one figure given; draft fore/aft if two figures given.

[b]Draft in light condition: 10-8/15-8.

[c]Displacement figures are for *Duane*'s sister ship, *Taney* (W-37), inclined 8 January 1945.

[d]Trial displacement.

[e]Two General Motors 12-288 diesels, with geared drive.

[f]Diesel-powered ship; ship and her landing craft shared the same fuel supply.

[g]Complement is given as officers/CPOs/enlisted when three figures are listed and officers/enlisted for two figures.

[h]Includes staff.

[i]Complement is given as officers/enlisted when three figures are listed; when only one figure is given, breakdown is unknown.

[j]Staff breakdown unknown; included with complement figures.

[k]Boats for *Duane* are two 26-ft motor whaleboats and two 26-ft Monomoy surf boats (typical Coast Guard complement, without any special AGC provision).

[l]Multiple gun mounts are indicated by Roman numbers; therefore 4 × II indicates four twin mounts.

[m]As built, two Sea Sparrow missile launchers were carried 1974–92; as of 2000 the ships were armed with two 25-mm Bushmaster, two 20-mm Phalanx, and four 0.50 caliber machine guns.

[n]All other Coast Guard 327-foot cutters converted to AGCs were armed with two 5-in/38 (fore and aft) instead of the two quadruple 40-mm mounts on board *Duane*.

was sent to the Pacific as flagship of Force X in July 1944. On the other hand, the U.S. ships were much superior in such facilities as photography, printing, and the visual display of signal messages.

Postwar, the Marines wanted spaces permanently allocated for parallel use by the naval attack force and the landing force staff, such as a joint operations room, joint communications room, and joint message center. The Marines pointed out that the landing force staff embarked on board the AGC would come ashore in the course of the operation. To do so it needed both its own boats and vehicles. Thus in 1947 they asked that the AGC have at least one hold and one boom for their staff vehicles and mobile signal equipment. They wanted AGCs equipped with LCMs as well as LCVPs, to carry headquarters equipment ashore, but they were never provided. The Marines also wanted two small fast boats to act as dispatch boats, to carry documents (and officers) around the amphibious force. To some extent LCPLs filled this role. Ultimately each ship would need a helicopter pad, the helicopter acting as messenger. Once the landing force commander was ashore, he would still need a link with the ships offshore, which would be commanded from the AGC. Thus the Marines wanted sufficient equipment and personnel for both ends of a radio teletype and telephone link between the attack force commander afloat and his ground force counterpart.

The AGCs were also criticized as too slow; in October 1945 the Pacific Fleet wanted 2 more kts (adapting a C3 rather than the C2 hull would have provided the higher speed). Later the Marines called for speed higher than that of accompanying transports, so that the AGC could maneuver among them. Faster AGCs figured in early postwar projected building programs, but they did not materialize until the 1970s (see Chapters 11 and 15).

For the future, the Marines wanted enough AGCs for all landing force commands down to divisional size. A multi-division (corps) landing might well employ AGCs on two levels, one for the command and one for each of up to three divisions. An army-level (multi-corps) landing, as at Okinawa, would add a higher level. To some extent, too, functions could be split up. At Iwo Jima in February 1945, *Eldorado* (AGC 11) was principal AGC, but *Estes* (AGC 12) controlled underwater demolition and pre-assault bombardment, and *Biscayne* (AGC 18) controlled the destroyers screening the invasion area.

The existence of AGCs was a major wartime secret. Apparently the Japanese were entirely unaware of their significance. None was lost in combat. When their significance was revealed at the end of the war, it seemed likely they would be attacked in any future conflict. Thus the Marines called for a new type with significant passive protection and armament.

Attack transports were modified into secondary AGCs, sometimes called "relief AGCs." Such ships were occasionally used as group flagships. They were not quite AGCs, but they did have important roles. For example, at Normandy, *Bayfield*, a relief AGC, was flagship of the Utah Beach assault force (*Ancon* was flagship at Omaha Beach). In the invasion of Southern France, *Bayfield* was flagship of "Camel" force. This description applied to *Monrovia* and *Samuel Chase*, which were fitted out in North Africa for the invasion of Sicily. In April 1943 *Leonard Wood* (APA 12) was ordered partially converted to a headquarters communication ship as a "second *Monrovia*." The 535-footers *Harris* and *Zeilin* became squadron flagships. These conversions were made partly in response to a February 1943 request by the Atlantic Fleet amphibious commander for one APA to be converted into a transport squadron flagship. Important features included a flag office, an operations office, and accommodations for flag personnel. BuShips protested that ships would have to be completely rearranged, because space was so tight. A sketch of a redesigned C3 showed new deckhouses and reduced troop capacity (78 rather than 91 officers, 1,309 rather than 1,465 enlisted men). The desired flag bridge could not be fitted. Ballast would have to be added. Even then some troops would have to sleep in five-high tiers of bunks, some hatches had to be partly blanked off, and some officers had to sleep on the second platform. A June 1943 request for better radars and fighter direction equipment, plus a flag radio space adjacent to flag plot, was rejected out of hand.

In June 1943 six new *Bayfield*- and *Custer*-class transports were ordered fitted out as squadron flagships and relief AGCs during construction, to carry the general commanding a division and his staff. The first to be fitted were *Cambria* (APA 36), *Chilton* (APA 38), *Elmore* (APA 42), and *Henrico* (APA 45). APA 36 and APA 42 were further converted for temporary duty as group flagships. *Bayfield* (APA 33) and *Cavalier* (APA 37) were later fitted. These ships could be recognized by the additional superstructure block built abaft their funnels. It accommodated a flag plot and other spaces, with a flag bridge atop the flag plot. A combined flag operations and intelligence office was fitted abaft the flag plot. A further but lower deckhouse abaft the normal superstructure, between the after Welin davits, had the CIC, flag office, and joint operations room on the boat deck level, with a general's cabin on the main

Several attack transports were completed as, or modified as, relief AGCs, usable as overall flagships for an amphibious operation (AGCs were needed as divisional flagships). USS *Burleigh* (APA 95) is shown on 5 November 1944. Her special function is indicated by the two big deckhouses erected abaft her funnel. Otherwise, unlike an AGC, she functioned as an attack transport.

deck. It also contained the staff officer bunk room, the radio transmitter room, and two radio receiver rooms (Radio II and Radio IV). As weight compensation, the forward set of triple Welin davits was removed.

In January 1944 *Burleigh* (APA 95), *Cecil* (APA 96), *Dade* (APA 99), *Mendocino* (APA 100), *Montour* (APA 101), *Riverside* (APA 102), and *Westmoreland* (APA 104) were ordered converted as transport squadron flagships. In addition, *Callaway* (APA 35), *Clay* (APA 39), *Custer* (APA 40), *Du Page* (APA 41), *Elmore* (APA 42), *Fayette* (APA 43), *Knox* (APA 46), *Lamar* (APA 47), *Leon* (APA 48), and *Barnstable* (APA 93) all had the necessary communications but were not designated squadron flagships. In April 1945 *George Clymer* (APA 27) and four other ships, to be designated, were ordered converted. The latter were *Arthur Middleton* (APA 25), *Charles Carroll* (APA 28), *Monrovia* (APA 31), and *Calvert* (APA 32). Of these ships, *Monrovia* had already been partially converted. Plans were drawn to modify AKA 53–class attack cargo ships, but none was altered.

At the end of the war AGCs faced another kind of threat. With their extensive communication and command facilities, they might well be called upon to control nonamphibious operations. How could they be held to their primary function?

New AGCs were not built until the late 1960s. Nine ships were laid up soon after the war and never recommissioned. *Adirondack* (AGC 15) and *Mount Olympus* (AGC 8) followed in 1955–56. The survivors, *Mount McKinley* (AGC 7), *Eldorado* (AGC 11), *Estes* (AGC 12), *Pocono* (AGC 16), and *Taconic* (AGC 17), all remained in service until 1969–71. By the mid-1950s all ships had helicopter platforms aft. Heavy postwar height-finding radars were fitted to the lattice mainmasts in the late 1950s or early 1960s, at about which time forward pairs of davits

Photographed on 1 December 1945, Henrico (APA 45) shows the extent of relief AGC modifications: the massive additional deckhouses and also the plated-in portion of the main deck.

The relief flagship role continued postwar. *Fremont*, a *Custer*-class attack transport, is shown after a Charleston refit, August 1956. Her flag facilities had just been enlarged, replacing one of her two port side triple Welin davits.

Postwar AGCs were fitted with helicopter pads. *Estes* (AGC 12) shows typical late-1950s radars: SPS-12 forward for long-range air search, SPS-8 for height finding atop the lattice mast, with a topmast carrying SPS-10 (surface search). By the end of the decade, the after set of kingposts typically carried a Tacan aircraft beacon.

Photographed on 6 May 1964, *Estes* shows the effects of modernization. AGCs were always short of enclosed volume, so the deckhouse abaft the bridge was built up. The height finder was replaced by a more massive SPS-30, and the air search set by SPS-37. The radome on the forward side of the lattice mast is part of the WLR-1 radar intercept set.

were removed; boats were carried on deck, handled by booms on the forward kingposts. Two ships, *Mount McKinley* (AGC 7) and *Eldorado* (AGC 11), were refitted in 1963–64 under the FRAM II program. The most prominent sign of conversion was a considerable forward extension of the superstructure (02 level) forward.

The relief AGCs also remained in service. Probably because of their flagship roles, in the 1950s *Fremont* (APA 44) and *Henrico* (APA 45) were unusually heavily armed, with three quadruple 40-mm guns as well as the usual 5-in gun. *Henrico* lost one set of Welin davits so that her superstructure could be enlarged; in the 1960s she was fitted with a ULQ-6 jammer, normally reserved for important combatant ships. *Cambria* (APA 36) and *Fremont* were refitted under the FY 64 FRAM II program, and a helicopter platform installed.

Both the Royal Navy and the U.S. Navy found that waves of boats often missed their beaches. In January 1942 the British decided to convert some Higgins boats into what they called landing craft survey, LCP (Sy). One craft per flotilla was given an additional compass, a chart table, and microwave radar.

The considerably later U.S. equivalent was the 56-ft LCC (landing craft, control). Table 9–2 compares the LCC Mk 1 with the postwar LCPL Mk 4. The Atlantic Fleet amphibious force first proposed an LCC on 31 October 1942, and had reiterated its support after the North African landings, in a 16 December 1942 message. Operation Torch graphically demonstrated how easily landing craft could lose their way en route to a beach. The idea was endorsed in the 31 December 1942 report (report 6-42) of the Landing Boat Board. The board consid-

Table 9-2. Control Landing Craft

	LCC Mk 1	LCPL (postwar) Mk 4
Length (ft-in)	56-0	35-9.5
Beam (ft-in)	13-7	11-2.5
Draft (ft-in)	3-11.5 max	3-6, loaded
Displacement	30 tons	19,320 lbs
Power (SHP/shafts)	450/2	300/1
Speed (kt)	13.5	19, loaded
Radius (nm/kt)	500/10	140/19
Armament		
0.50 caliber MG	3 twin mount	—
Complement	14	8

Notes: LCC Mk 2 was beamier (15 ft) and drew half an inch less; it was armed with two rather than three twin 0.50 caliber machine guns. Dash indicates that specification is irrelevant.

Effective control of amphibious operations required ships and craft of many sizes. An LCC(1) is shown on trials. Her design heritage as a heavy support craft shows in her trio of twin 0.50 caliber machine guns. The long whip antennas extend above her radio room. Abaft them is radio direction finder mast. The radome covers the antenna of an SO series surface search radar, a type also installed on board PT boats.

ered a guide craft more vital than the LCS(L) that it had recommended the previous August (see Chapter 8). The board therefore recommended that work on the LCS(L) be abandoned in favor of an LCC. BuShips had already stated that it could not provide the desired LCS(L) characteristics.

The board asked for a boat 50–60 ft long, with a maximum beam of 15 ft and a maximum weight of 50,000 lbs (i.e., within the capacity of AKA cranes). It wanted at least 17 kts using two standard 225 HP Gray Marine diesels and a cruising radius of 500 nm. Armament was set at four twin 0.50 caliber machine guns, and armor protection (over the bridge only) of ¼-in STS. The boat would have to loiter at low speed (5 kts) off a beach, so she needed sufficient keel area to keep her controllable at that speed. She would have a small-boat gyro-compass, which was being developed, three standard TCS radios, a special ZBX radio receiver, a QBG passive sonar, an echo depth finder, a dead-reckoning tracer, and two 8-in signal searchlights. She would also have a smoke generator.

On 4 January 1943 both the VCNO and the secretary of the navy approved the board's report. Development was quick because BuShips had already designed the necessary hull for its abortive LCS(L). By 17 January a conference had produced a preliminary design for a boat capable of about 15¼ kts. Of the four twin machine guns planned, the forward one was omitted to save weight, improve boat arrangement, and simplify ammunition handling and stowage. Plans and specifications were produced in mid-March 1943. The major change was a redesign of the boat's operations room in June. As fitted out, LCCs had SO-series short-range radars on a folding mast, which allowed shipboard stowage.[5]

The Mk 2 version was completely rearranged. It had one fewer twin 0.50; its mounts were fore and aft rather than two abaft the bridge plus a stern mount. In place of two or three identical tall whips for TCS radios these craft typically had one for TCS and two shorter ones for SCR-610 voice radios for beach communication. They also had a fixed loudspeaker forward of the bridge, and a prominent IFF antenna, so that they could easily be identified and tracked. This version also had a microfilm chart projector (NPM). Many Mk 1s were apparently completed to, or modified to, Mk 2 configuration; the only important difference ultimately was that Mk 2 was slightly beamier, 14 ft 6 in rather than 13 ft 7 in.

In addition, the LCS(S) was adopted as a scout boat, to mark beaches, by the U.S. Navy's wartime scout and raider units. The scout version was specially silenced, and it had its armor cut down to reduce its silhouette. The scout LCS(S) was first used at Sicily; in North Africa LCPRs had been used (one unit at Sicily used LCVPs). Beach marking was particularly crucial because the landing was done at night. Plans called for scout boats to be launched by LSTs and APAs about 90 minutes before the landing began. Each was equipped with navy (TBY) and army (SCR-536) radios, as well as night binoculars, twin IR lights on a rotating base, a pair of battle lanterns with colored lenses on swivels, and a signal light with which to indicate the beach designation. Also on board were two army scout and raider rubber-boat swimmer teams. Approaching the beach, the scout boat would use reference vessels offshore to locate the appropriate beach.[6] The swimmers would go ashore, locating each flank of the designated beach so the scout boat could be positioned midway between them. The scout would then anchor offshore, turning on its IR lights at H-30 and then using its signal light beginning at H-15. Scouts retained full LCS(S) armament to deal with enemy resistance on the beach. For example, during the Sicilian landing, a boat silenced enemy fire with her 4.5-in rockets. They also helped guide boats to the beach. After daybreak the scouts acted as traffic control boats. At Normandy scouts and raiders used both LCS(S) and LCCs to mark beaches and conduct rapid hydrographic surveys under fire. The LCS(S) was apparently not used as a scout boat in the Pacific.

Experience at Tarawa demonstrated the need for the kind of coordination a small command boat could offer. When the LVTs were delayed, there was no way that delay could be communicated back to the fire support ships offshore. To extend the Marines' prewar analogy between World War I assaults and amphibious assaults, this was not too different from the problems World War I troops encountered when they did not meet the schedule on which a "walking" artillery barrage was predicated. If they lagged too badly, the barrage lifted too long before they reached the enemy's positions. Unsuppressed, the enemy could wipe out the attackers.

In later landings, the initial waves rode LVTs, and a transfer zone (LCVP to LVT) was set up about 1,500 yds from the beach. Without radios, the assault craft (LCVP or LVT) could be controlled only visually or by loud-hailer or loudspeaker, from craft nearby. The boats' wave commander, riding one of the troop-carrying craft, relied on signs and flags and flashlights for communication with the rest of the boats.

Inevitably many levels of command were required, from command of each wave of boats arriving together at a beach through successive waves as well as maintaining contact with higher echelons. Control could not be exerted directly from

This LCC Mk 1 acted as a survey boat off the D-day beaches; note the "Survey" flag. Her radar mast has been folded down, and she lacks several of the usual radio whip antennas forward.

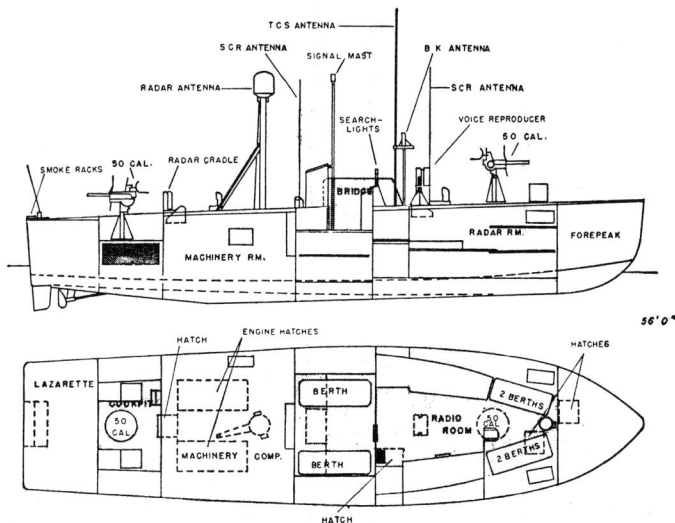

LCC Mk 2, 1945. The LCC was conceived as an inshore landing craft control boat. Mk 1 had a narrower beam, 13 ft 9 in, rather than 14 ft 6 in. Mk 2 had single or twin 0.50 caliber machine guns fore and aft, whereas Mk 1 had a pair in ring mounts side by side abaft the cockpit and a third twin mount in a ring mount right aft (some boats apparently lacked this mount). At least some Mk 1s had only two tall radio whip antennas (for SCR) forward, and no radar at all; others had the folding radar mast shown, carrying SO (the same radar as on PT boats). Plans called for a third whip abaft the first two, for the navy's TCS radio. The Mk 2 shown was much more completely equipped. BK was an IFF antenna; SCRs were army radios (SCR-610, SCR-608, and SCR-508 FM radio), and TCS was a standard navy radio. *U.S. Navy (ONI 226)*

the transports because they could not get close enough to the beach. Maneuvers were complex. For example, no single transport could carry the entire boat group (typically 6 LCM and 40 LCVP) that would land the battalion landing team (BLT) on board, so the first task of control was to form up the full group, using boats from AKAs and from APAs carrying divisional reserves.

At Eniwetok in January 1944, for example, the boats formed up into waves in an area 800 yds deep, inshore of the transports. Having formed up, they ran another 100 yds to the line of departure, from which they were dispatched to the beach. Timing was tight. Off one beach, the waves of LVTs (eight each) followed each other at two- or three-minute intervals, with LCMs (carrying tanks) as the fourth of seven initial waves.

In the earliest form of the control organization in the Pacific, at the Marshalls landings in January 1944, three subchasers (SCs) were the primary control craft. The chief control officer (who was also chief beachmaster) was embarked on one SC, which was normally at the line of departure, but allowed to roam. Also on board were a representative of the landing force commander, empowered to make decisions on his behalf; the commander of the LVT battalion landing the initial waves; a representative of the G-4 (logistics) division of the landing force commander, empowered to determine the flow of supplies to the beach; and a medical officer concerned with evacuating the wounded. A second SC was stationed permanently at the line of departure, carrying the officer responsible for executing the landing. He dispatched the waves of craft, taking account of the schedule, directions from the control officer, and requests by the regimental commander. A third SC was assigned to the regimental commander for each beach as a temporary command post in direct communication with the landing force commander, the troops in the boats, and the battalion commanders already ashore. Small ASW craft were used as primary controls because they were inconspicuous targets, with shallow enough draft to allow them to approach the beach, but with sufficient internal space for the control parties and their troop advisor parties.

In the Marshalls, two LCCs served as flank guides for each of the leading waves of the assault, under the direction of the control officer, controlling the speed of advance and the direction of the LVTs. Using LVTs complicated the situation. The initial waves simply loaded from the attack transports, but follow-up waves loaded at sea, from shuttling LCVPs. Thus the two LCCs acted as flanking guides for the four waves loaded directly from the APAs, controlling both their direction and their speed of advance. They commanded the LVTs via specially installed loudspeakers. After the first waves had landed, the LCCs took station about 2,000 yds from the beach to supervise the transfer of troops from shuttling landing craft to the LVTs. In at least one case, the BLT assault by *Neville* (APA 9), pairs of LCVPs acted as guides for each wave. Control boats also regulated the flow of equipment and supplies as the tactical situation developed ashore. Cominch regarded the close control of boat waves, and particularly the control of the flow of supplies after the initial assault, in the Marshalls as exemplary.

This system failed at Normandy, where PCs (rather than SCs) were used as the primary control craft, and SCs filled the LCC role. Neither had been adequately trained, because they did not arrive in the theater soon enough. The PCs had only a few days' instructions, took part in one large-scale exercise, and were taken out several times for special drills. The SCs had not received any instruction or training. LCCs were available, but they were not used.[7] The situation was particularly complicated because underwater obstacles prevented many landing craft from beaching and retracting promptly. The planned schedule collapsed altogether. The naval commander, Adm. John L. Hall, reported that only the intervention of the deputy assault group commanders untangled the situation. He concluded that the control craft were inadequate, and he recommended a modified LCI(L) as the ideal command craft. Quite possibly the decision to assemble the assault force at night for a dawn landing was partly responsible for the difficulties. Darkness would have complicated any attempt at controlling small radioless boats. In at least some cases LCS(S) were used to lead waves of boats to the beach.

For the invasion of Southern France, LCCs and British steam gunboats led shallow-draft drone minesweepers to the beach. LCCs controlled the drones. Astern of them came LCMs armed with rockets specifically to clear a path. Each formation of LCVPs following was led by an LCC. Those LCCs not required for immediate control at the beaches went back to the line of departure to help guide subsequent waves. In later waves LCCs were assisted by subchasers (SC). The amphibians used in this assault were DUKWs. They were guided by LCCs and LCVPs, and released about 1,000 yds from the beach.

In the Pacific, the system first used in the Marshalls was extended for later, larger assaults. At Saipan in June 1944 an entire corps (two divisions plus supporting arms) was landed; there were commanders at corps, division, regiment, and battalion level. The force control officer rode a PCS, the two group control officers (each for a Marine division) riding two more PCS. The latter could move freely, but generally stayed near the centers of their lines of departure. These ships used their CICs to keep track of the boats moving toward the beach. They were linked by radio both with higher echelons and also with other control ships, with supporting units, with boat groups and amphibian waves, and with the beachmaster, upon whose instructions they regulated the flow toward the beach, to avoid a pile-up but also to keep the beach steadily supplied. The advent of lightweight two-way voice radios (the army's SCR series) made this sort of coordination practical.

Under each group control officer were two transport division control officers (regimental control) on board PCs, which anchored in the center of the regimental line of departure. The force control officer called the leading waves to the line of departure and dispatched them using flag hoists and voice radio. Subsequent timed waves were dispatched by the transport division control officers on the basis of the landing plan. After all the scheduled waves had beached, each transport division control officer could call up reserve waves as advised by his regimental commander.

Boat group commanders from the APAs carrying the initial assault BLTs were embarked on LCCs, which guided the first wave of each regiment from the line of departure (as in the Marshalls). Using their radars, the LCCs checked their positions every 500 yds and maintained a uniform advance to the beach. The boat group commanders were also responsible for inshore or secondary traffic control at each battalion beach, operating under the transport division control officers. Through wave commanders (riding troop-carrying landing craft), the boat group commander controlled the boats in their assembly areas and directed them to their loading stations in response to calls from the ship. He led the boat group to the line of departure, and led the first wave to the line of breakers. The assistant boat commander rode an LCPR equipped with towing and salvage gear, fire-fighting equipment, first aid kit, and necessary communications equipment. After the last wave, he became salvage officer for his beach.

Each scheduled wave was guided in by LCVP flankers. The after-action report noted some flaws, due in part to divided control. LVT groups got their orders from their army or Marine commanders, but the guide boats were controlled by the naval landing force. That made for inherent confusion; the LVT was not quite a boat, but also not quite a land vehicle. For their part, guide boat officers were unaware of the details of LVT performance, and those aboard the LVTs were unaware of the orders the guides had. Moreover, without radios guide boats at the transfer line could cause considerable complication. They could inadvertently split platoons and confuse the arrangements made by the control officer and the LVT battalion commander. These considerations later helped lead the Marines to develop a command/control version of the LVT.

In the case of Saipan, there was a further complication, because a channel led to some of the assault beaches. An SC and two LCC were stationed in the channel as soon as the outer beach had been secured. They controlled traffic through the channel. The two LCC and an SC had earlier executed a feint (simulated) landing.

After the Marianas operation, permanent control organizations were formed within the amphibious forces. They were tested for the first time at Iwo Jima in February 1945. As before, there were control officers at each level: central (group or force), transport squadron (ground division), and transport division (regiment). Each officer was on board a specially equipped control vessel (PCE, PCS, PC, SC) with a control communications team and troop advisors. As before, the two transport squadron control officers were responsible for dispatching the five initial waves (in this case, to achieve simultaneous landings on all beaches), after which they took control of dispatching reserve waves and furnishing supplies as requested by troops on the beach. In this operation rough seas badly damaged the landing craft; after the first assault the control organization had to concentrate on salvage and on assistance to craft in difficulties.

Later-war organization was exemplified by the control organization of the northern attack force (3rd Amphibious Corps, comprised of two Marine divisions) at Okinawa in April 1945. It employed one PCE, five PCS, four PC, and nine SC, all specially outfitted. One PCS was allocated to the 3rd Amphibious Corps commander and one each to the Marine division commanders. The PCE was for the senior (overall) control officer, two PCS were for transport squadron control officers, and four PC for transport division control officers (with two SC under each as secondary controls). The remaining SC was used by the senior beachmaster while he remained afloat. Thus the SCs took over much of the earlier LCC role (boat group command). The LCC was criticized because it could not remain at sea for the several days over which control might have to be exercised. LCC were still used as flank guides, but that role could be filled by less complex craft; postwar it was taken over by the LCPL.

Some craft were redesignated for amphibious control, as PCC, PCEC, SCC, and DEC, of which all but the SCC survived the war. PCECs, in particular, could be recognized by their extra deckhouse, which contained a radio shack. Early postwar planners called for 13 PCC or their equivalent to support two infantry divisions attacking over 8 BLT beaches. Presumably APDs would replace the PCE and the two PCSs of the Okinawa operation, with PCCs replacing the four PCs and the nine SCs. LCPLs would be wave guides and flankers. In 1953 APDs replaced all the surviving earlier specialized control craft (DEC, PCEC, and PCC). Unlike the others, the APD could be used for more than control once the beach had been secured. It also offered sufficient space for the tac-log parties who represented the troops ashore in the naval control organization, acting as advisors to the control officers. This use of the APDs helps account for their retention in the active fleet through the 1960s and early 1970s, for the decision to modernize these ships under the FRAM program, and for the periodic inclusion, in long-range construction programs, of new APDs. Besides their amphibious flagship role, APDs were valued for their ability to deliver special forces, as in the Korean War. When all of the amphibious ships were given "L" designations instead of "A" designations, the APDs could not be redesignated LPDs; they instead became LPRs.[8]

Other ships were used for amphibious command from time to time. Thus in the early 1950s a small seaplane tender, AVP 52, was used to accommodate the commander of the landing ship component(s) and his staff.

LCI(L)s were also used as flagships. In January 1943 it was decided that some LCI(L)s would be fitted as flotilla flagships, with special facilities for the flotilla commander, doctor, paymaster, communications, and engineers. However, craft were not redesignated.[9] For the Sicilian invasion, eight LCI(L) were converted by the repair ship *Delta* into emergency regimental flagships. They were again used at Salerno. In addition, because the planned control vessels (PC and SC) were not available in time, LCI(L)s were used for this purpose in preinvasion training. They proved successful despite their lack of navigational equipment. According to the Eighth Fleet after-action report, they were ideal for traffic control because of their shallow draft. At Normandy, LCI(L)s were used as ferry control units (flagships of landing craft convoys) and also as local headquarters ships. For Southern France, eleven LCI(L) were fitted with additional communications as LCI(L)(C), or command ships.

By the fall of 1944 there was a perceived need for flagships for landing ship and landing craft flotillas. For example, LST Flotilla 5, in the Southwest Pacific, wanted to convert one of its ships to a flagship. That would have cost valuable carrying capacity. A flagship LST would not have retained full LST capability. Instead, the decision was taken to modify

LCI(L)s. Sacrificing one of them would have little impact on overall fleet carrying capacity. CinCPac formally proposed the conversion on 11 November 1944, although the program was under way at least a month earlier. For example, on 27 October the Flotilla 5 proposal was formally rejected with a reference to the ongoing LCI(L) flagship program.

The LC(FF), an LCI(L) converted into a flagship, was a more elaborate affair. It was proposed by Seventh Fleet (Southwest Pacific) in September 1944, on the ground that flotillas of various kinds of landing craft needed to have their commanders and staffs on the same unit. The troop officers' compartment could be converted to the flotilla commander's cabin, his 8 staff officers could share No. 3 troop compartment, and the supporting 21 enlisted men could bunk in No. 1 troop compartment. No. 4 troop compartment would hold provisions. No. 2 troop compartment was used as an office. The usual pair of 20 kW diesel generators were replaced by 30 kW units. A 2,000 gal per day evaporator was installed. On 19 October CinCPac approved; the loss of troop lift was acceptable. On 11 November 1944 BuShips submitted a feasibility study. Its designers had already worked out the consequences of fitting flagship accommodations. Five LCI(L) were immediately available for conversion, and the 24 LCI(L)s used to train Atlantic amphibious forces could also be released. The first units were LCFF 629 and 993 for LSMs (staff: 7 officers, 11 enlisted), LCFF 994 and 995 for LSTs (staff: 7 officers, 14 enlisted), LCFF 997 and 998 for LCTs (staff: 4 officers, 12 enlisted), LCFF 985 for LCS(L) (staff: 9 officers, 20 enlisted), and LCFF 627 and 628 for LCILs (staff: 9 officers, 21 enlisted). All were done at Pearl Harbor. Ultimately 49 were converted: LCFF 367–370, 399, 423–427, 484–486, 503, 504, 531–533, 535, 536, 569, 571, 572, 575, 627, 628, 656, 657, 679, 775, 782, 783, 786, 788–793, 988, 994, 995, 998, 1031, 1079–1083.[10] LC(FF) were first used at Okinawa, and were still in the trial stage at the end of the war. They were not entirely suitable. They were slow and had inadequate space and antiaircraft armament. Inadequate space made it necessary to combine their radar and navigational plotting rooms, and there was also insufficient space for quarters. In October 1945 the Pacific Fleet recommended that the LC(FF) be superseded by a converted LSM with a decked-over well. It never appeared.

Part Three

Postwar Developments

Enormous postwar effort went into developing plastic landing craft, which were expected to be far more durable than their wartime wooden predecessors. This experimental Mk 5 Zenith plastic LCVP is shown on board USS *Seminole* (AKA 104), October 1956.

10
Perfecting the Technique

By the end of World War II the U.S. Navy had evolved an elaborate and sophisticated technique of amphibious warfare, which formed the basis for postwar developments. The immediate postwar problem was the growing footprint of amphibious forces, as ground forces became more mechanized. A wartime Marine division required about 1.3 tons of lift per man; a postwar division, with reduced manpower, required about 1.4 tons (i.e., equipment did not decrease proportionately to manpower). The wartime army ratio was 1.5; with reduced manpower army requirements would also grow. Clearly matters were getting worse. For example, tanks were becoming much heavier. More mechanized units required more support personnel. Thus a 1948 army analysis of a task force built around one infantry division (19,000 men) for an amphibious assault showed a total of about 35,000 men, the lift for which would be far larger than that needed at Okinawa.

Such a division would require 18 APA and 10 AKA, compared to the 9 APA and 3 AKA of a prewar division. In addition it would have 5 LSD (to carry 15 LCT or an equivalent quantity of LCM), 18 LST, and 9 LCT on board the LSTs; all types that had not even been imagined before 1941. The LSTs would carry 12 pontoon causeway sections and 12 pontoon barges. Control would require two AGCs (one attack force flagship, one transport group flagship), one control ship (a DE), and nine control vessels (PCC or equivalent). Fire support would be provided by 12 LSMR and 12 LCS(L), and the force would be supported by 6 APD.

In 1948–49 navy planners sought to determine which forces, including amphibious forces, they might have to support in the future. Contemporary general war plans envisaged massive assaults in Europe, particularly after U.S. forces had been mobilized. The United States did not yet have very powerful nuclear strike forces, and no one except, perhaps, the U.S. Air Force, had yet imagined that a general war could be decided in a few days or weeks.

Thus, when the General Board estimated future naval requirements in 1948–49, it included the shipping associated with a composite group of three divisions and three air wings. The General Board asked both the Atlantic and Pacific fleets to estimate requirements to lift a composite three division/air wing group. The results, compared to ships on hand, are shown in Table 10-1. They were quite consistent with the one-division figure. The operation was expected to require one LCFF per assault battalion. Some or all of the air wing AKA could have been conventional cargo ships, because the air wing did not land during the assault. The figures were a composite; the two fleets disagreed on details, but gave similar totals. They excluded ships required for training. The LCI(L), which still existed, was no longer wanted.

In Table 10-1 the active figures include naval reserve training ships (14 LST, 19 LCI[L]). Nine LCS(L) were up for sale. Additional requirements (e.g., for training) were 2 APA, 2 AKA, 14 APD, 2 LSD, and 3 AGC (for submarine forces). Reserve figures do not include the 81 APA and 33 AKA of the Maritime Commission reserve fleet, which was separate from the navy reserve fleet. Excess APA were to be retained for replacements; AKAs would come from the Maritime Commission reserve fleet. Excess AGCs would be advanced base command ships, and later would serve expanded amphibious forces.

In 1948 there were 28 active and 86 reserve LCT(6), plus 10 used as district craft, of which 33 were required for the planned force. There were 2,583 LCM, against 341 required; 6,959 LCVP (and LCPL and LCPR) against 2,375 required; 1,760 LVT, against 1,012 required; and 121 LVT(A), against 310 required. By 1949 plans called for using 20 APD as utility transports, the remainder as coastal ASW escorts. Mobilization plans called for building about

Table 10-1. Estimates of Amphibious Requirements, 1948

	Requirements			Ships on Hand	
	3 Divisions	3 Wings	Total	Active	Reserve
APA	59	9	68	20	54
AKA	22	45	67	26	3
APD	6	0	6	12	80
LSD	6	5	11	6	15
LST	111	15	126	50	91
LSM	27	130	157	15	145
CVE(T)[a]	0	15	15	0[b]	0
LSV	0	5	5	0	5
AGC	5	0	5	7	8
LCFF	12	0	12	4	21
PCC	12	0	12	0	0
LSMR	16	0	16	8	40
LCS(L)	12	0	12	0	97
LCI(L)	0	0	0	30	52

[a] CVE(T) was an escort carrier used to transport aircraft that operated from a shore base.
[b] CVEs were available in numbers, but not in transport configuration.

50 APA, 30 AKA, 12 AGC, 45 LSD, 1,000 LST, 200 LSMR, and at least 1,000 LCT.

The wartime programs had produced a flood of landing ships and craft. Amphibious forces were a relatively low priority, and so were replacements. In 1948 OpNav estimated that the reserve fleet would suffice to maintain the active fleets' amphibious strength for the next decade or more. It did caution that wartime types had been lightly built, were of questionable workmanship, and had relatively nondurable machinery. At some point they would all have to be replaced, and sheer numbers would be difficult to provide under tight peacetime budgets. BuShips reported increasing numbers of LST structural distress or failure, but it was deemed neither necessary nor desirable to begin a general program of structural modification. Indeed, more than 30 years later, many LSTs were still operational in foreign navies.

The obvious new requirement was for ships to be able to operate in the Arctic, because the Soviet Union was the only likely enemy. Puget Sound converted LSD 5 and 6 for Arctic operations, together with 6 LCT (1273, 1330, 1363, 1375, 1462, 1463) and 16 LCM for each of the 2 LSDs. This work was completed in March 1949.

Of the smaller types, the LSM was not wanted for future construction, although the LSM(R) was a different proposition. Certainly an LSM could land tanks and other equipment when an LST could not, and it could beach under surf conditions that would swamp lesser craft. However, the same equipment could go aboard LCTs carried by an LSD, and there was no hope of increasing LSM speed to the 20 kts demanded postwar (see Chapter 11). LSMs were retained because they were so useful in forward areas where LCTs would find it difficult to operate. They were released for conversion as cable layers, to support the sound fixing and ranging (SOFAR) undersea sound system project and then sound surveillance system (SOSUS), and later for other tasks. By 1955 they were being replaced by LSTs on a one-for-three basis: an LST carried three times the cargo with one and one-half times the crew.

The LCI(L), redesignated LSIL (landing ship, infantry, large), was dismissed as small and relatively unseaworthy. Existing ones were retained for use in forward areas during the consolidation and garrison phases of a campaign. By January 1950, as funds tightened, an Ad Hoc Committee on Reserve Fleet Requirements recommended immediate disposal of 25 LSFF (landing ship, flotilla flagship; formerly LCFF), 50 LSIL, and all 104 LSSL. The Korean War slowed disposals, so that in 1955 there were still 64 LSIL on the navy list.

Landing craft stocks would probably last about five years. In November 1948 there were 673 active and 304 reserve LCM, with another 1,606 in stock. There were 2,297 active and 2,338 reserve LCVP, LCPL, and LCPR. In stock were 3,828 LCVP, 330 LCPL, and 504 LCPR, all of them wooden. Mobilization plans called for building 13,200 LCM, 32,000 LCVP, and 2,500 LCPR, but no LCPL. Maintenance money was quite limited. It seemed unlikely that wooden boats, which were generally stored in the open, would last much more than five years.

In 1947 BuShips set up a program to develop a postwar LCVP. The new plastics were much more durable and they could be cast in virtually any shape. They were also easier to fabricate. Molds were becoming much less expensive, and it seemed that plastic fabrication required little skilled labor. In July 1947 BuShips estimated that it would take only 10 manhours to make a plastic hull (later increased to under 100 manhours), compared to 500 manhours for the existing type.

A plastic hull could easily be cast to test any desired LCVP hull form. BuShips argued that very little was known of the ideal shape of a future LCVP; the only types tried had been the box-garboard Jersey Sea Skiff, which became the prewar bureau type, and the Higgins form. In designing the 70-ft LCM(8) in 1945 BuShips had discovered that a W-form was more efficient than the standard V-bottom. It would be well adapted to the main desired improvement, twin screws (and rudders). Given twin-screw maneuverability, fewer boats would broach and be left stranded on the beach, as demonstrated by the much lower percentage of LCMs than LCVPs that had stranded. The twin propellers would have much smaller diameter. Each would be driven by a 105 HP diesel, then under development. Other improvements were reduced draft (by 6 in) and higher speed (½ kt faster loaded, but ½ kt slower empty). The new craft would have the same payload as the wartime craft, and it could be stowed in the same davits, or nested in an LCM. The new Mk 2 design would have a wider ramp opening into an enlarged cargo space so it could accommodate the Marines' new ¾-ton Carryall (ambulance).

BuShips planned two parallel programs. In one, Norfolk Naval Shipyard would build two plywood prototypes with improved hull forms. Compared to plastic, plywood was tougher and better developed, but it was clearly an interim material.

The other program, approved on 25 June 1947, called for building a series of hull test vehicles, each resembling an LCVP but with a molded-in ramp rather than an operating ramp. Like the parallel plywood prototypes, all had twin engines. The new lightweight diesel not yet developed, the prototypes were powered by Chrysler Crown gasoline engines. They were designated A-1 through E-1. A-1 had a rounded V-bottom. B-1 had a moderately sharp W-bottom. C-1 had a rounded W-bottom and was tested with both twin and single screws. D-1 had a sharp V-bottom. E-1 had a sharp W-bottom. F-1 was a new plywood LCVP, G-1 was a standard LCVP (for comparison), and H-1 was a standard LCVP powered by two Chrysler Crowns, for comparison with the others. Tests began in October 1949.

B-1 was satisfactory until it broached. C-1 proved more stable than a standard LCVP, but slower to retract. It rode well, but had trouble loading alongside a transport. It was the only boat in the initial series that did not suffer serious engine trouble. D-1 had considerable trouble retracting, and lacked sufficient buoyancy fore and aft. E-1 broached in two of four landings. Its sharp W-bottom apparently tended to dig into sand, making retraction difficult. F-1, the plywood modified LCVP, suffered continuous engine trouble. H-1, a standard LCVP modified with twin screws, broached once and then suffered from engine trouble.

Alongside these craft the navy tried a hydrojet, which it called LCVP(H) (landing craft, vehicles and personnel, hydrojet), or J-1. It was rejected as too slow, but because the jet could be rotated it was extremely maneuverable and did not broach as readily as a standard LCVP; it also retracted more easily. On the other hand, it could not steer a steady path.

Although the W-hulled C-1 proved best, it was nonetheless dropped. In November 1951 the CNO decided that future LCVPs would have one engine (hence one screw), so that, as in the past, LCMs could use two LCVP engines. Many other craft could also use a diesel of about 225 HP, whereas there was no point in buying a separate small diesel specifically for LCVPs. Thus all of the LCVPs built postwar have essentially duplicated the wartime type, albeit in many cases in plastic rather than wood.

Meanwhile Puget Sound Navy Yard built a series of ten plastic LCVPs of conventional design, designated Mk 3. Their main new feature was that powerplant, rudder, and propeller were all shipped in one unit into a stern recess. Once a satisfactory hull form had been chosen, it could be mass produced using the plastic techniques perfected by Puget Sound. Production of 100 was planned, but an August 1952 evaluation report pointed out serious structural problems. Boats on chocks were seen to sag noticeably. The resin used was not being cured properly, and the strength of the laminates was low and variable. (Readers who built plastic kits at this time may remember how, every so often, they would degenerate into powder.) The plastic industry was not yet quite mature. After its first failures, Puget Sound built a series of five vacuum-injected boats with double skins and integral flotation material.

With the failure of the Puget Sound project, BuShips had to buy 425 wooden LCVPs. In order to buy even a few plastic boats in FY 55, it had to have a successful prototype by 1 June 1954. That proved impossible. The bureau financed several parallel plastic LCVP projects, but none was ready in time. Palmer Scott & Company of New Bedford built Mk 4, with a single skin and integrally molded frames (flotation material was added after completion). W. R. Chance and Associates produced Mk 5 (1954), using a single plastic sandwich shell (plastic laminate faces with a phenolic impregnated paper honeycomb core 3 inches thick). Mk 7 was a fiberglass laminated hull (1955). Mk 8 (1955) was Puget Sound's Styrofoam-core hull. Puget Sound was also working on a double-skin hull with a honeycomb core.

There were still serious problems. In January 1954 the honeycomb sandwich was rejected because water was seeping in noticeably, the boat was very stern heavy, it could not be cradled in an LCVP cradle, and it was considerably slower than a plywood boat.

Plastics did eventually work. The U.S. Navy built its first 28-ft fiberglass personnel boat in 1947, and by 1966 it had built over 3,000 plastic boats, including 199 LCVP and 138 LCPL. In mid-1956 a first run of 60 plastic LCVPs was being delivered. Mk 7 became the standard plastic version.

In September 1953 BuShips reported that LCVP stocks were critical, and that 500 new LCVP would be needed each year for the next three years (FY 55–57); the first postwar production was 350 LCVP under the FY 53 program. Then the FY 55 program grew to 1,000 LCVP, 100 of which were replaced by LCPLs (eventually 871 were built). Large numbers of LCVP were included in later programs, but few were built (see Appendix B for details).

In 1948 Norfolk Naval Shipyard began developing an LCM(6) replacement (Mod 1). Table 10-2 compares the LCM(6) with the later LCM(8). Length was limited to 56 ft so the craft could go on board existing attack transports and attack cargo ships, and width was limited to 14 ft 1 in so it could be transported by rail. To get maximum ramp and cargo deck width, the designers had to make the side walls

Purpose	Wave guide and control
Capacity	8 crew and equipment
Length overall	36′ ½″
Beam	11′ 3¼″ maximum
Draft	3′ 8″ loaded
Full load displacement	22,500 lbs
Hoisting weight	18,500 lbs
Hoisted by	Sling or davits
Construction	Round bottom, fiberglass reinforced plastic
Speed	15 knots at full load displacement
Fuel capacity	160 gallons
Range	100 nautical miles at full power and full load
NavShips Drawing No	LCP(L) 36-1428729
Stock No	SNSN S9-L-15182-1305
Engine details	1—6-cylinder diesel, 300 hp. at 2,300 r.p.m. emergency; 250 hp. at 2,100 r.p.m. continuous, GM6-71 turbocharged, V-drive, 24 volt electrical system. Stock No. S75-E-53737-4695
Propeller details	1—22″ D by 23″ P by 2″ bore, rh. rotation
Note: Experimental boat, only one built.	

LCP(L) Mk 3, an experimental plastic boat. She was part of the BuShips attempt to replace wooden craft with more durable plastic ones. Only one was built. *U.S. Navy*

Table 10-2. New Landing Craft

	LCM(6) Mod 1	LCM(6) Mod 2	LCM(8)	LCM(8) Aluminum
Length OA (ft-in)	56-1.5	NA	73-7.75	74-3
Beam (ft-in)	14-0.25	14-4	21-0.125	21-0.5
Draft, loaded (ft-in)	3-10	3-6	5-2	NA
Hoisting weight (lb)	56,000	53,500	134,000	115,000
Displacement, full load (lb)	124,000	121,500	254,000	230,000
Capacity (lb)	68,000	68,000	120,000	115,000
Power (SHP/shafts)	450/2	450/2	650/2	590/2
Speed (kt)	9	9	9	9
Radius (nm/kt)	130/9	130/9	190/9	150/9.2 max
Complement	5	5	5	5

NOTE: NA indicates data not available.

LOA- 56'-1-1/2"   **Max Displ-** 124,000 LBS   **Capacity-** 60,000 LBS

Max Beam- 14'-0-1/4"   **Hoist WT-** 56,000 LBS   **Speed at Max Displ-** 9.5 KNOTS

Mach & H P- 2-6 CYL 225 HP DIESEL   24 VOLTS   **Fuel Cap-** 450 GAL

Cruising Range at Full Power & Full Ld- 130 NAUTICAL MILES

Purpose- TO LAND GENERAL VEHICULAR CARGO AND PERSONNEL DURING AMPHIBIOUS OPERATIONS

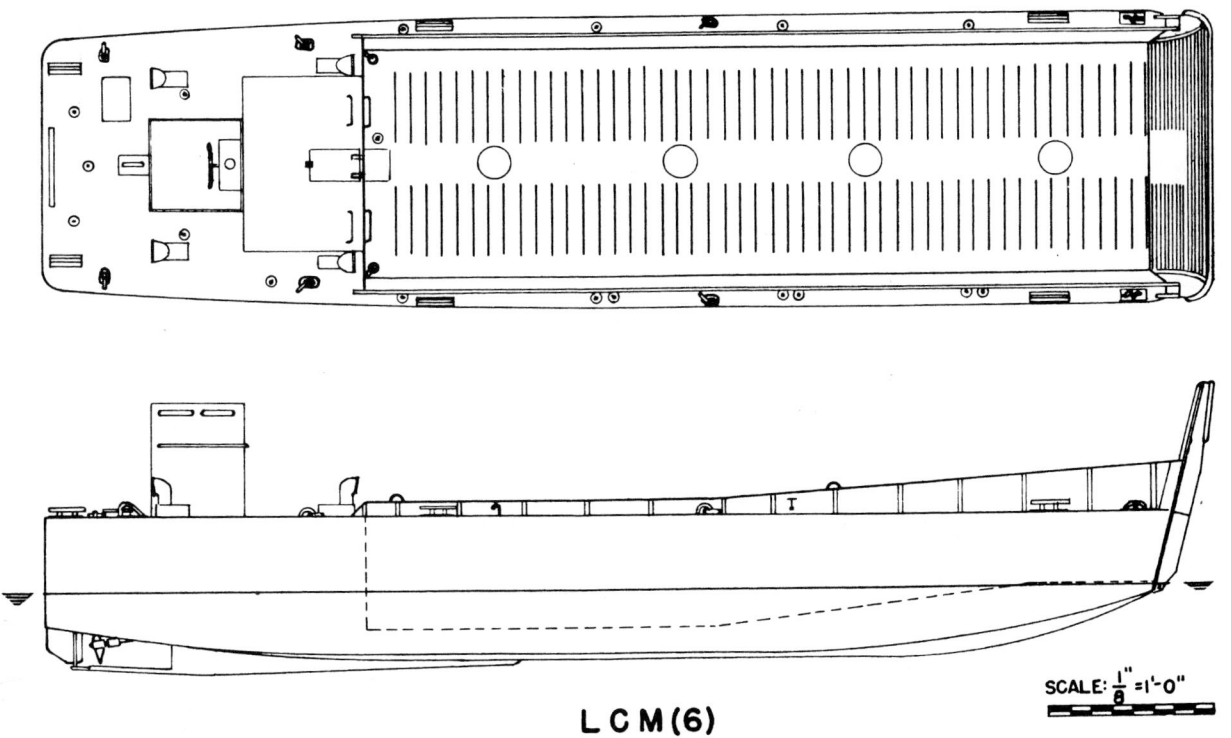

LCM(6)

LCM(6) Mod 1, a postwar redesign of the original LCM(6) with much narrower gunwales that allowed for more cargo space. *U.S. Navy*

thinner, even though in that case the craft would lack the deck width for line handling. As in contemporary LCVP and LCM(8) designs, BuShips tried alternative V- and W-bottoms, prototypes being completed in late 1949 and early 1950, respectively. In this form the craft could lift 70,000 lbs but in August 1950 OpNav asked that the design be reviewed to take the new 50-ton T42 medium tank. It turned out that the LCM could be redesigned without affecting its external dimensions, using very narrow side walls (11 in wide rather than 1 ft 8½ in). In 1951 tests, both the V- and W-bottomed prototypes outperformed the World War II LCM(6). When light, the V-bottom LCM retracted fastest (but slowest when heavy). The W-bottom version was criticized because it could not be "walked" sideways as easily as the others. Both craft offered wider well decks, and their sides were compartmented up to the gunwale. Tests of new hydraulic and pneumatic controls were less successful, and the somewhat flattened V-bottom hull with wartime-type mechanical controls was recommended for further production.

There was also interest in an alternative Snadecki chain drive, brought to BuShips' attention in September 1948 by retired Comdr. Forrest H. Wells, who had commanded a drydock during World War II. Unlike a conventional shaft directly connecting engine and propeller, Snadecki's engines drove a shaft protruding through the transom; it drove the propeller shaft, parallel to it, by a chain. Unlike those in a conventional LCM, the Snadecki engines were horizontal. Propellers and rudders could easily be raised, as units, for repairs. The Snadecki LCM was noticeably faster in free route, and it was more maneuverable than a conventional LCM. It drew less water aft, and it was easier to hold in position at the beach. Because the axis of the propeller was parallel to the baseline of the boat, the stream of water entered the propellers at a slight up angle. That raised the stern slightly and made retraction easier. However, the propellers and rudders were entirely unprotected from aft or below, hence they were subject to damage when backing down. In a conventional LCM, the skeg extended aft of the rudder, and the rudder post fit into it. When the boat grounded, the shock was absorbed by the hull, whereas a grounded Snadecki boat would suffer propeller and rudder damage. In mid-1951 BuShips recommended against buying the Snadecki drive, but it was not finally rejected until January 1953. At that time it was criticized for poor maneuverability (due to the short rudders) and for the loss of stern buoyancy due to the cut-away stern (to accommodate the drive units). The weight of the units tended to dig them into the sand. The evaluation report also noted that whereas a beached boat was normally lifted by wave action, the cut-away under the stern tended to trap waves, the bulk of which would break over the stern sheets to drench the coxswain and the engines.

In November 1951 the Army Transportation Corps requested 328 LCM(6)s. Because the 9 ft 2 in ramp of World War II craft was insufficient, only the new Mod 1, with its 11-ft ramp, would do. Given this initial requirement, contract drawings were prepared, and production of postwar LCMs began. This new LCM(6) was designated SCB 94 for production purposes. A 1954 project for a new LCM(3) was designated SCB 141. Like LCM(6) Mod 1, it would provide an improved ramp and ramp hoist engine, and consideration would be given to reducing engine noise. This craft was not built, however; the only new LCM(3)s were cut-down LCM(6)s.

These LCMs were no faster than their World War II predecessors. By December 1951 there was interest in a 100,000-lb 15-kt LCM, 65 ft long, for which a model was tested at David Taylor Model Basin (in a parallel program a model equivalent to a 210,000-lb craft at 15 kts was also tested). Nothing came of this effort.

Given the advent of the M26 Pershing tank, in 1946 the Marines wanted priority given to the development of the big LCM sketched the previous year. They pointed out that the trend was toward ever-heavier vehicles. The FY 49 program included two prototypes (V- and W-bottom), now designated LCM(8) (SCB 95); LCM(7) was a wartime British design. Remembering that Higgins had created such successful landing craft in wartime, in August 1948 it was proposed within BuShips that he be asked to develop a parallel design. At this time the BuShips landing craft program consisted of two LCM(6) Mod 1, two LCVP, two or more 70-footers, and the plastic LCVP hull form test vehicles. The LCVP and LCM(6) were already Higgins types; it was natural to ask him to develop an alternative to the 70-footer. In a 6 October letter, he was given only the basic requirements: maximum beam of 21 ft, speed at least 10.5 kts on 675 SHP, and at least 500 nm at 8 kts. A tank in the well deck should be able to fire while the craft approached a beach.

Higgins' approach was characteristically original. During the design, it occurred to him that landing craft often could not negotiate reefs, as happened at Tarawa. He therefore placed two sets of standard tank treads on each side of the bottom of his craft, driven by a hydraulic take-off from the two 330 HP diesels. They were housed when not in use. The craft was considerably longer than the BuShips version (90 ft), displacing 98 tons light and 165 tons fully loaded. Its cargo flat (56 ft × 12 ft 6 in) was, ironi-

The postwar LCM(6) Mod 1 replaced the wartime type. It could be recognized by the grating above its bow (the wartime type had "ears") and by its very narrow bulwarks. Mod 1s are shown exercising at Camp Pendleton, 23 August 1961, and in the well deck of USS *Tarawa*, 7 July 1976.

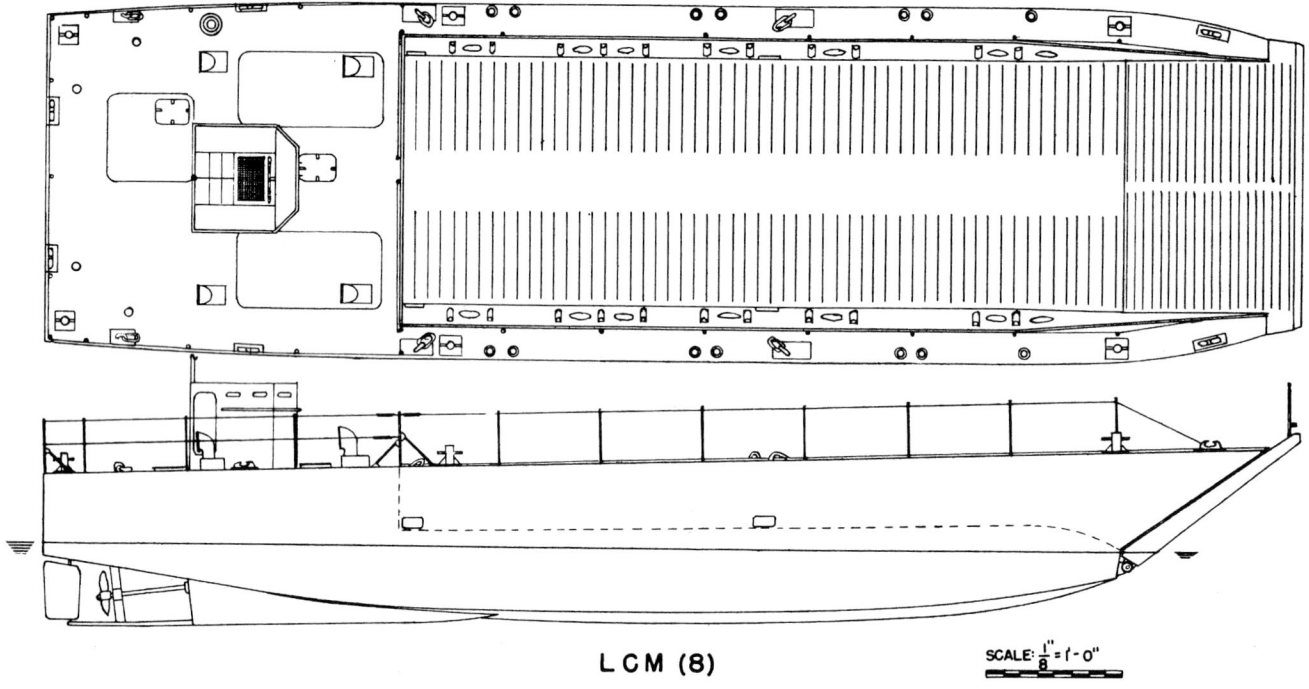

```
LOA- 73'- 7·3/4"        Max Displ- 254,000 LBS        Capacity- 120,000 LBS
Max Beam- 21'- 0·1/8"   Hoist WT- 134,000 LBS         Speed at Max Displ- 9 KNOTS
Mach & H P- 2- 12 CYL 370 HP DIESEL   24 VOLTS        Fuel Cap- 1146 GAL
Cruising Range at Full Power & Full Ld- 190 NAUTICAL MILES
Purpose- TO LAND A HEAVY TANK OR ONE OF THE LARGER TYPES OF VEHICLES DURING
         AMPHIBIOUS OPERATIONS
```

LCM (8)

SCALE: 1/8" = 1'-0"

LCM(8), conceived to carry the new generation of large tanks. Far too large for most navy transports and cargo ships, it could be carried only in well decks or on board the *Charleston* (AKA 113)–class attack cargo ships. *U.S. Navy*

cally in view of Higgins' earlier history, placed 6 in above the waterline, and was therefore self-bailing. Eight of these LCMs could fit into an LSD well deck, compared to nine of the 70-footers. On trials, the treads failed, first from hydraulic problems and then because they imposed excessive ground pressure. Even so, the BuShips evaluator thought the idea might be worth whatever it would take to make it work. It failed in the soft soil near New Orleans, but it would have succeeded on hard coral like that at Tarawa. As it was, Higgins' prototype was rejected outright.

Meanwhile, work proceeded on the two BuShips-designed prototypes. They were sectionalized so they could be carried on board existing amphibious ships, although it could best be transported on board an LSD, which could take nine, compared to three LCTs. As a measure of capacity, an LCM(8) could carry 200 troops, compared to 80 for an LCM(6). It could carry 65 rather than 34 short tons of cargo; cargo sq ft was 719 rather than 423. It was powered by three Gray Marine diesels, rather than the two of a smaller LCM(6), and it was slightly faster (9.2 rather than 9 kts). Greater length brought more fuel tankage, so endurance increased from 130 to 150 nm. On the other hand, LCM(8) was considerably more massive (130 tons rather than 68 loaded).

As of 1949, plans called for Higgins to build 2 more prototypes, and for the first 15 LCM(8) to be built in 1950. Money was tight, however. No production craft would be ordered before 1952, and even then it was ordered for the army, not the Marines. There were reports that the army was developing its own 87-ft LCM(8) equivalent. In FY 52 it was estimated that an LCM(8) would cost about $80,000, compared to $49,500 for an LCM(6).

Initial tests of the two BuShips versions in 1952 were successful, although there was a worrying episode, in which the V-bottom craft, carrying a tank, an armored car, and a jeep swamped in the Potomac River at 10 kts. It went 100 yds, with water building up on its tank deck, then stopped, its bow reared up, and the tank deck drained. It became clear that such craft would probably nose under in

LOA- 36'-0½
Max Beam- 11'-3¼"
Mach & H P- 1- 6 CYL- 225 H P DIESEL 24 VOLTS
Cruising Range at Full Power & Full Ld- 100 NAUTICAL MILES
Purpose- WAVE GUIDE AND CONTROL

Max Displ- 22,500 LBS
Hoist WT- 18,500 LBS

Capacity- 8 CREW + EQUIPMENT
Speed at Max Displ- 15 KNOTS
Fuel Cap- 160 GAL

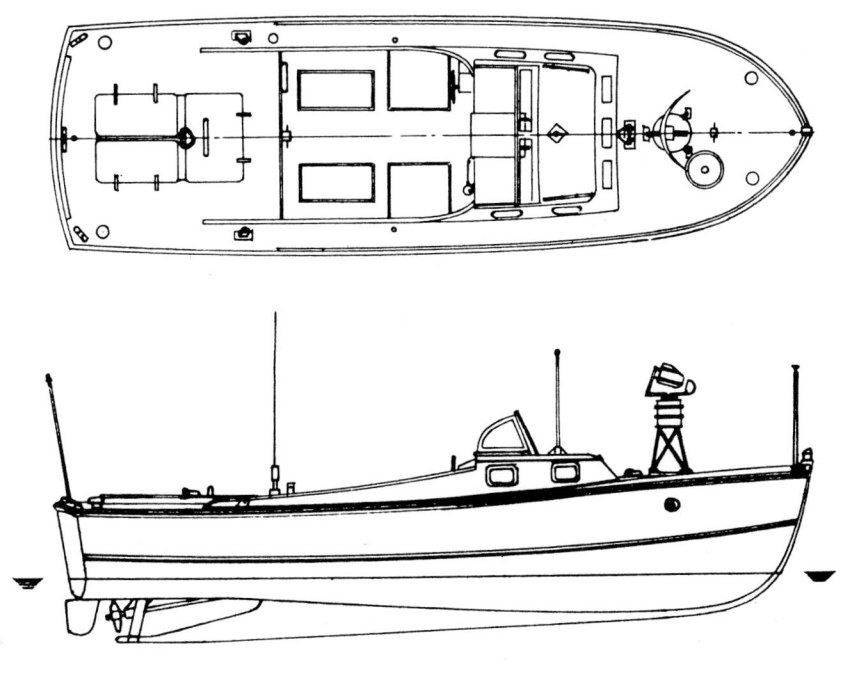

36' LANDING CRAFT LCP(L) MK 2

LCP(L) Mk 2 was the new LCPL, conceived as an amphibious control craft and fitted with a radar. *U.S. Navy*

heavy seas, and therefore the bow ramp would have to be raised.

Initially the LCM(8) was to be used only by the army; the decision was made not to redesign it for transport aboard existing AKA and APA. The army became interested in it as a cargo carrier. The swamping in the Potomac had been tolerable because the heaviest weight in the boat was far enough aft to tip the water back, so that it could drain. The army, however, tested the boat with a uniformly distributed load (i.e., with its center of gravity further forward), and this time the craft shipped far too much green water. The bow had to be changed, and three army LCMs were modified with alternatives. Clearly the hull had to be lengthened. However, the craft had been limited to a 70-ft length specifically so that the craft could be carried athwartships on board existing attack cargo ships; the new AKA would carry them lengthwise, thereby posing less of a problem. The 70 ft in turn had been set by the needs of both buoyancy and trim. The initial solution, to lengthen the craft by 6 ft, was unacceptable. Eventually a length of 73 ft 6 in was adopted. The bow ramp angle was reduced from 55 to 35 degrees so that the craft rode up over waves instead of taking them aboard. The folding ramp was replaced by a one-piece ramp, thus dropping the requirement that a tank be able to fire ahead. Three alternative improved bows were tested. Higgins' bow proved best and was adopted.

As it turned out, the new APA and AKA then being developed did not carry LCM(8)s. When the last class of attack cargo ships (*Charleston* class) was given sufficient boom capacity for it in the 1960s, a new lightweight aluminum version (Mod 2) was developed. Further steel versions were Mods 3 and 5; Mod 4 was a further aluminum version for AKA deck stowage.

The LCC disappeared at the end of World War II.[1] Its effective successor was an adapted LCPL, which was considerably faster than the LCVP or LCPR. Immediately after the war these craft were used by boat commanders during the ship-to-shore phase, carrying charts. By 1954 there was a compact enough short-range radar (SPN-11) to go aboard an LCPL, which could then function as a boat wave guide. At this time LCPLs were also used as personnel boats (as ships' barges, gigs, officers' motor boats, and to carry liberty parties). Their beaching capability, although remaining, was now secondary; it was increasingly important that they be enclosed to protect charts and equipment from spray. LCPLs used as gigs already had fixed canopies over the forward half of their cockpits.

The first postwar LCPLs were built under the FY 53 program, to equip postwar amphibious ships. The inclusion of four in the FY 54 program gave BuShips the opportunity to redesign the type for its new role, under SCB project 128. The boat would be equipped with short-range navigational radar, a radar beacon, and voice radio. Each APA or AKA would carry two, which would serve as primary control when boats had to make a long approach through shallow water, over the horizon from the controlling ships. Otherwise it would assist the primary control officer in controlling waves of boats from the transport area to assembly area and on to the beach. This Mk 4 steel version had a conventional bow and a cabin, and its new hull form was intended to offer a better ride as well as higher speed, using a diesel of the same family as that of the LCVP. A plastic Mk 3 existed only as a prototype. So that the Mk 4 could be lowered from davits fully loaded, the LCPL would have a combat weight (without personnel) of 19,000 lbs. The SCB specified a combat speed of 17 kts (11 kts fully loaded) and a radius of 100 nm at 17 kts.

Using the existing 225 HP Gray Marine diesel, a Mk 4 could make only 15 kts. Design studies showed that it would take more power to gain the desired additional speed. Adding another Gray Marine diesel would increase speed (loaded) from 9.5 to 13

An LCM(8) approaches the beach; another is visible in the background. Nearby are an LCPL used as a control craft and an LCM(6) (marked "NV"). The LST in the background, *Grant County*, is carrying pontoons.

A plastic LCPL Mk 11 is shown at the Elizabeth City Shipyard, 19 April 1962. In effect a successor to the wartime LCC, she had three radios (indicated by whip antennas) and a radar installed. The foredeck was recessed to make line handling easier in rough weather. Metal LCPLs were used in Vietnam as river and harbor patrol boats.

LCPL MK 4

Purpose	Landing craft control boats (Primarily) Officer personnel boats (secondarily)
Capacity	LCCB 8 persons (crew) OPB 20 persons including crew of 3
Length overall	35′ 9½″
Beam	11′ 2½″ maximum
Draft	3′ 6″ loaded
Full load displacement	19,320 lbs (combat loaded)
Hoisting weight	18,000 lbs
Hoisted by	Slings or davits
Construction	Steel, V-bottom
Speed	19 knots combat loaded
Fuel capacity	160 gallons
Range	140 nautical miles at full power and full load
NavShips Drawing No.	LCP(L)36-MK4-145-1758730
Stock No.	S1905-627-6401
Engine details	Diesel, 6121T-General Motors, 300 hp. at 2,300 r.p.m., hull cooled 24-volt electrical system. Stock No. S2815-554-1925
Propeller details	24″ D by 22″ P by 2¼″ bore, rh. rotation

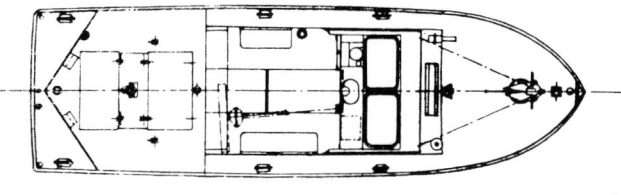

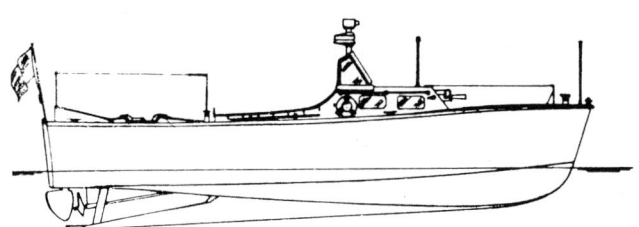

LCP(L) Mk 4, a steel equivalent to Mk 3, conceived as a command boat. It was used in riverine combat in Vietnam. *U.S. Navy*

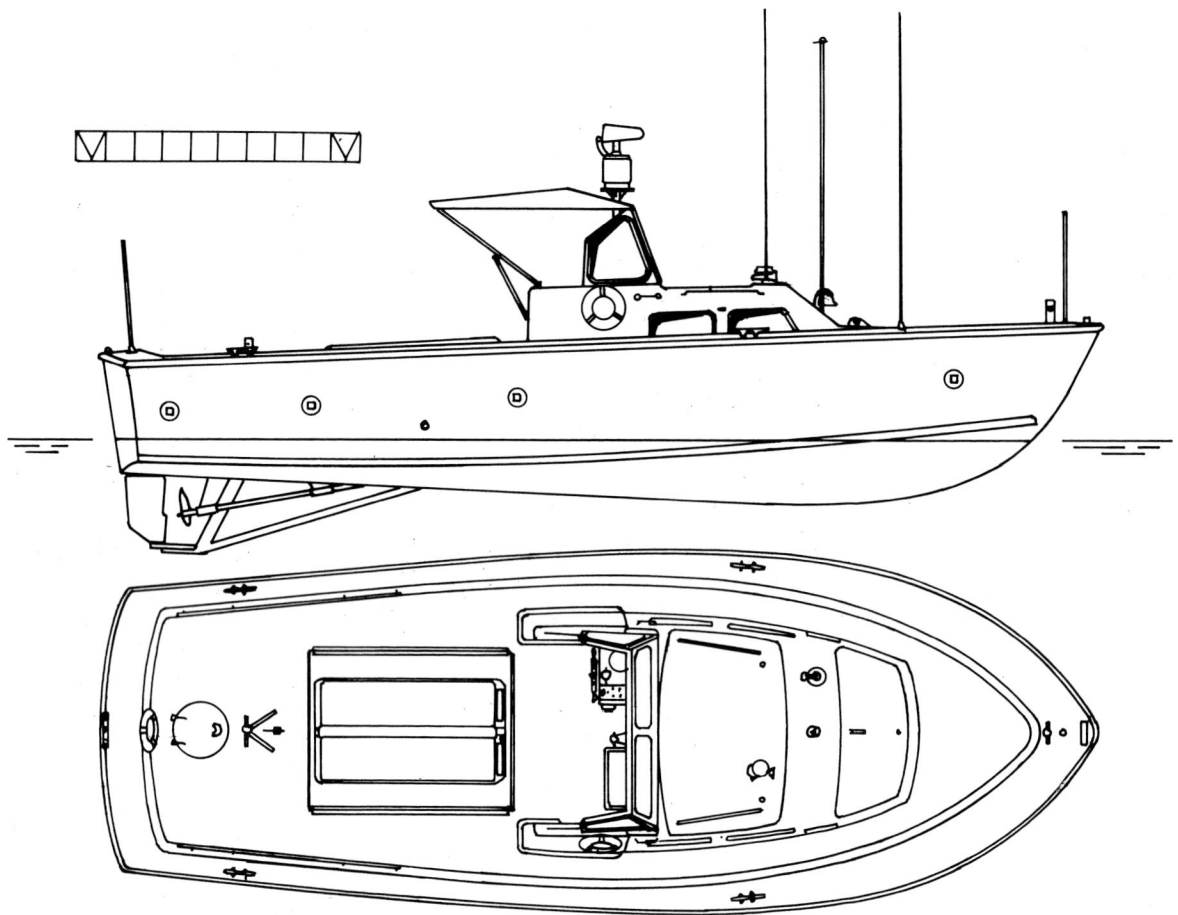

LCPL Mk 11, a fiberglass version of Mk 4, used as a command boat but not in riverine combat due to its vulnerability. *A. D. Baker*

kts, but that would bring the boat's weight above the hoisting limit. The alternative was a Packard lightweight diesel, which offered 300 HP and probably 450 HP (18 kts) in a developed version; there was also a lightweight 600 HP V-12 diesel (21 kts sustained). However, all were very expensive. A Gray Marine diesel (64HN9) cost about $4,000, a Packard 300 HP engine without spares cost about $30,000, and the 600 HP V-12 engine was about twice as expensive as the Packard. As of December 1954 the solution seemed to be a projected (FY 56) turbocharged version of the Gray Marine diesel (64HN10), but it, too, was not bought.

Another approach to higher speed, tried in the late 1950s, was a lengthened LCPL. However, the next production version was the 36-ft fiberglass Mk 11. The steel Mk 4 was used as a riverine gunboat in Vietnam. In the 1990s the navy's small-boat designers, the combatant craft detachment in Norfolk, modified the LCPL and converted it to metric measurement as the 11-m LCPL (no Mk number).

The LVT was so valuable that in 1948 it was subject to a major new development program. Mobilization plans called for building 33,000 LVTs. BuShips was the design and procurement agency for LVTs for all the services, so it sponsored both new designs and major changes to existing craft.

Characteristics for three categories of next-generation LVTs had been laid down in August 1945: personnel carrier, armored turreted 105-mm howitzer craft (Prototype A), and armored turreted 76-mm gun craft (using the turret of the M24 light tank). The Marines wanted safer diesel powerplants, but recognized that only gasoline engines might provide sufficient power. Thus, in each category they wanted two parallel prototypes: one gasoline, one diesel. In September 1945 the secretary of the navy approved construction of nine pilot models (four personnel carriers, two each of the gun carriers, and a cargo carrier). After the war, the program added an LVT command vehicle. The Marines became interested in more exotic forms of assault, so they added a small lightweight cargo carrier (8,000 lbs) suitable for a new assault seaplane (discussed below), a large unarmored cargo carrier designated LVT(U) (landing vehicle, tracked, unarmored), and a submersible personnel carrier. To hold the total to nine vehicles, half of the personnel carriers and the gun carriers were eliminated; the 105-mm craft would have a diesel engine, and the 76-mm would use gasoline. Prototypes built under this program included Baldwin-Lima-Hamilton's LVTPX1 (1949 contract, completed 1952: 29.2 × 11.6 × 8.3 ft, with a 7,000-lb capacity; 23 mph on land, 6.3 in water; terminated due to transmission problems), Food Machinery Corporation's LVTPX3 (contract early 1946, completed December 1950: 30 ft × 10 ft 11 in × 10 ft 2.5 in, 52,290 lbs with 10,000-lb capacity and integral armor, driven at 30 mph on land or 8 mph in water by an 810 HP Continental engine), Continental Aviation's lightweight cargo carrier (contract 1949, delivered 1950: 26,000 lbs, 27.9 × 10.8 ft; it exceeded specified weight by 8,000 lbs; water speed was 8 mph), Marmon-Herrington's 76-mm gun carrier (1951; never tested), and Pacific Car and Foundry's 105-mm howitzer carrier (contract 1946, delivered 1949, with an 810 HP Continental engine). None of these craft entered production.

By 1948 the Marines looked ahead to a very fast amphibious command vehicle, capable of 25 mph in the water, but also capable of operating over rough muddy terrain and over snow. It would carry six men, three of whom would be radio operators. This project was outside the LVT development plan as it was then understood, and the vehicle was never realized.

Perhaps the most interesting feature of the LVT program was a plan to use Boeing 502 gas turbine powerplants to power the lightweight LVT and to boost the turreted vehicles. Gas turbines were attractive because they were so compact, with a continuous output of 700 SHP (400-hr lifetime) or 1,600 HP in an emergency (1-hr lifetime). That compared to 500 HP for standard diesels or gasoline engines. In the turreted vehicles, they could quickly be started up to drive auxiliaries such as the turrets, when the main powerplant was shut down. They could provide emergency propulsion when the main piston engine was shut down. Even their exhaust heat might be valued in cold weather. None of the prototypes used this engine, but the Boeing 502 did figure in other contemporary navy programs.

The most exotic of the new craft was a lightweight LVT that could be carried by a big seaplane. The requirement was established by the Marine Corps' commandant on 23 December 1947. Weight was not to exceed 10,000 lbs, yet the LVT would carry a 3,500-lb payload, including 15 fully equipped troops,

The massive LVTP5 became the Marines' standard amphibian for the 1950s.

The LVTH6 was the fire support companion to the LVTP5, in effect the successor to the wartime LVT(A)5.

with the same performance as a conventional LVT. In 1948 BuShips had a sketch design of a 9,000-lb vehicle, with a stowed height and width of 8 ft, with better seagoing characteristics than an LVT(3). It would make 10 mph in water.

As of late 1948 the companion submersible LVT was in the preliminary model test stage. It would operate like a submarine, with positive buoyancy, submerging dynamically. Controls might be waterjets or planes. Underwater, it would have to use propellers rather than the usual LVT tracks. No prototype was ever built.

The Marines also wanted a very large LVT, in effect a tracked LCM, which they called a 60-ton cargo carrier. It had been formally proposed by the Marine Corps' commandant on 17 April 1947; in February 1948 the BuShips amphibian vehicle development section was informally approached on this issue by Op-34. The big LVT would transport heavy vehicles such as the M26 tank equipped with a bulldozer blade. By the fall of 1948 BuShips had a sketch design of a 60 ft × 20 ft × 9 ft 5 in craft, 120,000 lbs light, with a cargo capacity of 120,000 lbs or 1,400 cu ft (hold area 263 sq ft, ramp width 12 ft 6 in). To reduce ground pressure, it had belly tracks as well as the usual side tracks, and it could increase its freeboard by inflating a bellows under its stern ramp. A lightweight 1,000 BHP diesel would drive it at 10 kts in the water, loaded. A 1949 design showed 100,000 lbs of cargo in a 42 ft 6 in × 14 ft hold, with a speed of 11 kts and an endurance of 150 nm at that speed. Two such LVT(U)s were included in the FY 50 budget. An alternative proposed at the time was a big (68 × 28 ft, 150,000 lbs unloaded) but very slow "walking barge" capable of carrying 100,000 lbs at 1 mph on land or at 5 mph on water. A prototype was tested, but apparently it was never particularly satisfactory.

With the Soviet Union the only likely enemy, the Marines would have to fight under cold conditions. Personnel carriers would have to be enclosed and heated. Enclosure was also important if nuclear weapons were used. Beginning in 1948, Long Beach Navy Yard modernized the Marines' existing LVTs under SCB project 60A for the LVT(A)5 and SCB project 60B for the LVT(3). At this time 191 LVT and 15 LVT(A) were active in Marine service, with another 1,569 LVT and 106 LVT(A) in reserve; the army had another 3,500 LVTs. To meet the new requirements, LVT(3) received aluminum cargo doors over its open compartment and a machine gun turret; it was redesignated LVT3C (covered). LVT(A)5 was provided with a turret cover and a false bow, which would allow greater buoyancy.

The LVT program contracted dramatically when the defense budget was deeply cut in 1949. It was revived when war broke out in Korea in June 1950, and particularly after the Marines demonstrated just how effective an amphibious assault could be, at Inchon in September. The emphasis shifted from an ideal future LVT to something more immediately producible, a trend also visible in other amphibious craft such as the LST and the LCT. In December 1950 the Marines let a contract to the Ingersoll Kalamazoo Division of Borg-Warner, which had produced the wartime LVT(3), for a new series of vehicles. They were developed in 1950–51, and production beginning in 1952. All vehicles were for the Marines; the army had given up on amtracs.

The LVTP5 was in effect the amphibian LCVP the Marines wanted, except that it could not carry a small vehicle.[2] It could carry 34 marines or 6 tons of cargo at a maximum water speed of 6–7 kts; in 1961 it was described as about a quarter as efficient as a boat in water. The close-support version, LVTH6, carried a 105-mm howitzer in a gyro-stabilized turret. It was numbered Mk 6 to avoid confusion with the earlier (and surviving) LVT(A)5. In fact it was the first prototype to appear (August 1951). The other members of the family were command-control (LVTCR-1), air defense (LVTAA-1, with a twin 40-mm gun), recovery (LVTR-1), and combat engineer (LVTE-1). There was no direct equivalent to the earlier cargo-carrying LVTs; instead there was to have been the much larger LVT(U). Of the various types, only the antiaircraft version did not enter production.

These new LVTs entered service in 1955–56. Credited with a 15-year lifetime, they would have to be replaced about 1970. Thus, they equipped the Marines in Vietnam. They had been designed for amphibious assault, sometimes described as an 80 percent water/20 percent land environment. Indeed, in one way the LVTP5 was superior to open landing craft; it could handle 15-ft surf, while open landing craft would swamp in surf over 6–8 ft. In Vietnam, the Marines operated mainly ashore, and the LVTP5 became, in effect, an armored personnel carrier. Its bulk, necessary to ensure buoyancy at sea, became a major disadvantage, and its tracks, designed largely for propulsion in the water, were quite inefficient. Ground pressure was excessive. It also seemed to require far too much maintenance. The replacement became the LVTP7 family and later AAV7, an AAV (amphibious assault vehicle). Table 10-3 compares the LVTP5 with the LVTP7/AAV7 (described in Chapter 13) and the advanced amphibious assault vehicle (AAAV; described in Chapter 17).

In 1951 Food Machinery Corporation (FMC) proposed a new design, which became LVTPX2, with roughly the characteristics of the existing LVT3C, mainly the power train arrangement. It would use components of the M59 armored personnel carrier FMC was building for the army, such as the engine, transmission, and suspension. The company argued that, because it used components already in produc-

Table 10-3. Post–World War II LVTs

	LVTP5	LVTP7/AAV7	AAAV
Length (ft-in)	29-8	26-0.75	30-5[a]
Width (ft-in)	11-8.5	10-8.72	11-11
Height (ft-in)	10-0.5[b]	10-8.5	10-6
Weight (lbs)			
Empty	64,200	38,450	63,300
Loaded	81,780	50,350	76,000
Crew	3	3	3
Capacity			
Cargo (lbs)	12,000[c]	10,000	9,800
Troops	34	25	18
Power (HP)	810	400	800[d]
Fuel (gal)	456	180	391
Speed			
Water (kt)	6.8	8.4	20–25[e]
Land (mph)	30	40	45
Endurance			
Water (nm)	57	55 at 8 mph	65
Land (miles)	190	300 at 25 mph	400
Armament			
40-mm GL	—	1	—
30-mm	—	—	1
7.62-mm	—	—	1
0.50 caliber MG	—	1	—
0.30 caliber MG	1	—	—

NOTES: Dash indicates that specification is irrelevant.
[a]On land; on water a bow plate deploys, and total length is 34-8. On water, skirts deploy to cover the tracks, and overall width is 14-7.5.
[b]To top of machine gun cupola; 8-7.5 to top of hull.
[c]On water; could carry 18,000 lbs on land.
[d]On land; 2,600 HP on water.
[e]Speed in transition mode between high water speed and land configuration is 9 kts.

The LCA was conceived as a tracked alternative to the LCM. This prototype LCA(X)1, built by Food Machinery Corporation, is shown under test about 1968. Powered by Solar Saturn gas turbines, it had steerable and retractable propellers that could run it at 12 kts in the water. Payload was 60,000 lbs.

tion, this new LVT could easily be mass produced in an emergency. It was considerably smaller than the LVTP5, and some suggested using it to complement the larger vehicle. Prototypes were built in 1952, and FMC's new LVT was accepted as LVTP6 in 1956, but it was never placed in production. There was also a fire support version, LVTHX4, in 1953. The next FMC LVTP was designated LVTPX12, which seems to have indicated a "double" or super LVTP6.

Work proceeded on the LVT(U). Baldwin-Lima-Hamilton's LVT(U)X1 was delivered in 1951. LCVP-sized (36.2 × 12.2 × 8.7 ft, 54,000 lbs), it could make 5 mph in water (30 mph on land) using a single 500 HP gasoline engine. Cargo capacity was 30,000 lbs, loaded via a stern ramp. The vehicle was never completed; its contract terminated because of the number of changes required during construction. Pacific Car and Foundry's LVT(U)X2 ("Goliath") appeared in 1958; the contract was let in 1951, and construction had begun in September 1952. Like a big LCM, it could carry 60 tons, equivalent to any tank then in service. It was nearly as large as an LCM, 44.6 × 21 ft, powered by two 500 HP engines, and could achieve 7 mph in the water using a pair of retractable steerable propellers with Kort nozzles. Like the BuShips design, it had four tracks. This vehicle had integral sheet steel and ⅜-in armor. It maintained 20 percent reserve buoyancy by making the machinery space and other compartments watertight, and by filling voids with plastic foam.

The next version of the big amphibian was the LCA (landing craft, assault). The tracked LCAX1 (1962) was powered by 2,280 HP gas turbines, and could make 12 kts in the water, using propellers. It was of roughly LCM size (56 × 21 × 5 ft maximum draft), with bow and stern ramps, and weighed 127,000 lbs including 60,000 lbs of cargo. It had bow and stern ramps. FMC's tracked LCAX2 (1968), which also carried 60,000 lbs of cargo (total weight was 90,000 lbs), used two 725 HP diesels driving waterjets to reach 10 mph in water (land speed was 20 mph). It was somewhat smaller than LCAX1, 52 × 21 ft, and it too had a stern ramp and bow gate. In 1970 the LCA was the only craft available to move substantial cargoes from ships to beyond the beach. One prototype existed and had been service approved; the navy wanted a total of 58 as interim replacements for the existing LCM(6), beginning with 16 in FY 71 (at a unit cost of $625,000). They were not built.

There was also a wheeled amphibian, LARC (lighter, amphibious resupply, cargo), which figured in proposals for well-deck ships in the mid-1960s. These were apparently Marine Corps projects, although the army tested and used the craft. Certainly the Marines' LVW (landing vehicle, wheeled; described in Chapter 13) seemed to resemble LARC-5, and contemporary army equivalents were developments of the wartime DUKW. Borg-Warner began work on LARC-5, with a 5-ton load (or 15–20 troops), in 1958, the first production contract following on 6 June 1961; in all, 950 were built in 1962–68 (plus 7 prototypes). The vehicle was used in Vietnam. Dimensions were 35 × 10.2 × 10.2 ft, and the powerplant was a 300 HP Cummins diesel. Like the wartime DUKW, LARC used a propeller in the water; maximum water speed was about 8.2 kts. The larger LARC-15 and -60 had their cabs at the rear, with a bow ramp. About 100 LARC-15s were made. LARC-60, originally designated BARC (barge, amphibious resupply, cargo), was in effect a wheeled equivalent to the abortive LVT(U)X. It was developed by Pacific Car and Foundry under a 1951 contract, and thus preceded LARC-5 and -15. Whether or not it was conceived by the Marines, BARC was ultimately an army project. None of these vehicles matched the capacity desired in the LCA, and all were considered resupply rather than initial assault vehicles. The over-the-beach assault problem was solved only with the development of the LCAC (landing craft, air cushion) in the 1980s (see Chapter 16).

Paralleling the revival of LVT production was revived LST construction. A project for a 20-kt LST had been proceeding since 1947 (see Chapter 11 for details). It was temporarily abandoned in favor of a 15-kt ship, which at least would be able to steam in company with existing transports. The designers hoped to eliminate the heavy pounding the most recent design (LST 1153) endured in a seaway. They hoped to use diesel rather than steam power, with twin or triple screws; steam plants risked clogged condensers in shallow water, needed more personnel, took more time to get under way, and had less cruising radius than diesel ships. It turned out that to get the required speed the ship needed 12,000 SHP, twice the power output of the 13.5-kt LST 1153.

The SCB already had a project for a 15-kt LST, SCB 9. As of July 1946 two had been included in the proposed FY 48 program, but they were never funded. The most important requirement was a speed of not less than 15 kts in cruising condition. Draft would be not more than 5 ft forward in beaching condition; presumably it was hoped that accepting the increased draft would make a better hull form possible. Unlike the late-war LST 1153, this one would have a main battery comprising the new twin 3-in/50 antiaircraft gun and a planned advanced close-in antiaircraft gun. Ventilation and living facilities would be improved, and SCB 9 would carry 500 troops (vehicle crews and attached troops). The between-deck ramp would have a capacity of 25 tons. It appears that no design work had been done on SCB 9; available effort was concentrated on the more difficult 20-kt project.

Like its predecessors, the new LST conceived in 1950 would carry 500 tons of cargo in beaching condition (draft 3 ft 6 in forward on a 1:50 beach), and it would accommodate an LCT, now called an LSU (landing ship, utility) and later redesignated as LCU (landing craft, utility), on its deck. It would transport and land heavy equipment up to and including a heavy tank. It would accommodate 200 troops, with mess and sanitary facilities for up to 400. Armament was set at two (later three) twin 3-in/50 guns (600 rounds per barrel) and six twin 20-mm (3,000 rounds per barrel).

A sketch design was ready by early October 1950. It showed a 410 × 60 ft hull drawing 3 ft 6 in forward in landing condition (3,800 tons). Overall length later increased to 430 ft. Compared to earlier LSTs, the bow was made longer and finer to reduce resistance. To achieve that, the bow ramp had to be lengthened from 29 to 36.5 ft. It had to be folded to fit inside the ship.

The tank deck would be at least 300 ft long, with a clear width of 30 ft and a clear height of 14 ft; the bow door would be at least 17 ft high and 16 ft wide between curbs. Full load would be about 6,500 tons including 1,400 tons of dry cargo and 500 tons of liquid cargo; maximum displacement would be 7,500 tons. In landing condition radius would be 1,000 nm at 15 kts; full load radius would be 8,000–10,000 nm at 15 kts. Improvements over LST 1153 included better deck cargo handling, a better stern anchor arrangement (more like that in the preferred LST 542 class), and better boat davits, to protect boats from being swamped while being lowered in a seaway.

By this time BuShips was badly overburdened, so it let a contract to Gibbs & Cox to develop the design. That company chose a spoon-shaped stern with skegs supporting twin shafts; BuShips had used triple shafts, in which case a third shaft would have run in the tunnel formed by the skegs. Gibbs & Cox increased beam to 62 ft and added freeboard forward, in the form of a forecastle instead of BuShips' steep sheer, so that the bow ramp would not have to fold when brought inboard.

On 9 January 1951 the CNO stopped the project; this LST was far too big and expensive for mass production. He asked instead for a diesel version of LST 1153 with the new armament (three twin 3-in/50 and six twin 20-mm) and two new features: a turntable at the forward end of the tank deck for LVTs (the tank deck had to be widened to 38 ft to allow for it) and provision to berth and mess 383 embarked troops so that an LST could carry all the personnel to man all the vehicles it could transport to the beach. The new design was designated SCB 9A.

Again BuShips contracted with Gibbs & Cox for preliminary and contract design. For mass production, Gibbs & Cox adopted commercial rather than navy specifications, at a cost of 25 tons. Other new features added weight: 15 tons for the turntable, 20 for the troop accommodation, 25 for steel rather than aluminum furniture. To gain the necessary

Wahkiakum County (LST 1162) was one of 15 first-generation postwar LSTs. Newly completed, she is shown off her builder's yard on 6 August 1953. Note the pair of kingposts and the crawler crane on deck. Unlike the late-war steam LSTs, she is armed only for self-defense, with three of the standard postwar twin 3-in/50s plus twin 20-mm cannon. Her boats are an LCPL and three LCVPs.

buoyancy (to avoid increasing landing draft), the hull was enlarged, using lines similar to those of LST 1153. The stern was much the same as the company had used for the abortive 15-kt LST. Overall length, as of late March 1951, was 384 ft, well short of that planned for the 15-kt LST, and beam was 55 ft 6 in. Estimated landing displacement was 3,168 tons (draft 3 ft 4 ¾ in forward). It was expected that 6,000 BHP diesels would drive the ship at 13.7 kts, and endurance was set at 10,000 nm at 12.5 kts.

As in LST 1153, the crew was separated from the troops; the crew were berthed in the starboard wing wall alongside the tank deck, the troops were to port. A 12-ft space was taken from the after end of the tank deck (whose length was reduced to 295 ft) for extra troop accommodations, with messing and laundry spaces above. To accommodate troops, some stores had to be relocated from the wing walls to below the tank deck, an area used in previous LSTs only for liquid cargo or ballast. Compared to LST 1153, the new design had a slightly larger bow door. The characteristics called for the ramp to support a 60-ton tank when on the beach, or to allow the launch of a 35-ton amphibian. To accommodate future tanks, the designers provided for a 75-ton tank or a 50-ton amphibian. The between-deck ramp could take a 25-ton vehicle, and it could be used with the bow door shut. Major improvements on past practice were the motor-driven turntable and hydraulically operated bow doors. Troop accommodations were better than in previous ships. Plans called for using four diesel engines with twin controllable reversible-pitch propellers. They eliminated any need for reversing gear and gave greater backing power. Engines were controlled from a pilothouse console.

Fifteen of this LST 1156 class were built under the FY 52 program. The first was later renamed *Terrebonne Parish,* so that became the class name. Table 10-4 shows the details for three LSTs: LST 1153, LST 1156, and the later LST 1171 (discussed later).

At least the first few of the LST 1156 class (LST 1156–1165) suffered badly from hull vibration. BuShips' preliminary design section blamed the Gibbs & Cox stern, which was later described as cheap and easy to build, but prone to vibration. The engineering officer of LST 1156 was enthusiastic

Table 10-4. New LSTs

	Talbot County (LST 1153)	*Terrebonne Parish* (LST 1156)	*De Soto County* (LST 1171)
Length (ft-in)			
WL	316-0	368-0	442-0
OA	382-0	384-0	442-0
Beam (ft-in)	54-0	55-6	62-0
Draft, fore/aft (ft-in)			
Beaching	3-5/11-4	3-7/11-0	4-0/13-0
Full load	9-6/16-3	8-2/16-1	9-11/16-8
Displacement (tons)			
Light	2,250	2,586	3,828
Beaching	3,203	3,330	4,636
Full load	6,000	5,777	7,804
Power (SHP/shafts)	6,000/2	6,000/2	13,700/2
Speed at trial (kt)	14.1	14.2	16.9
Generator capacity (kW)	600	600	900
Radius (nm/kt)	10,000/10	10,000/10	9,000/15
Complement (officers/enlisted)	10/171	10/147	10/162
Capacity			
Troops (officers/enlisted)	15/160	15/377	30/604
Tons[a]	500	500	500
Cu ft	NA	700	1,400[b]
Sq ft	13,607	13,607	17,245
Armament			
5-in/38	2	—	—
3-in/50	—	3 × II[c]	3 × II

NOTES: Dash indicates that specification is irrelevant; NA indicates relevancy but data not available.
[a]At landing displacement. Full load cargo is 1,395 tons for LST 1156 and 1,825 tons for LST 1171, compared to 1,212 tons for LST 542.
[b]This is combat cargo cube, not the cubic capacity of the tank deck. A World War II LST had no combat cargo cube at all. For comparison, a typical World War II ship (LST 491) had a vehicle parking area of 11,328 sq ft.
[c]Multiple gun mounts are indicated by Roman numbers; therefore 3 × II indicates three twin mounts.

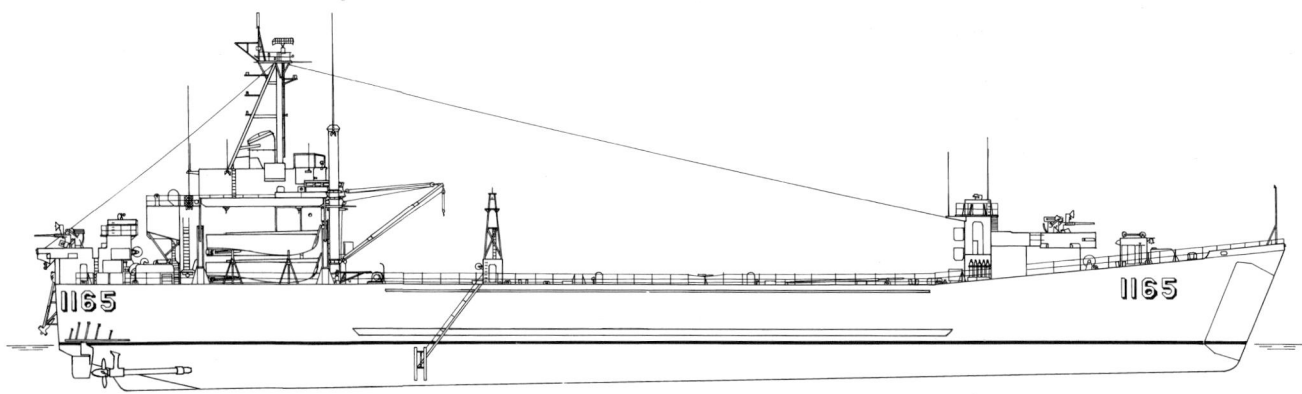

Washoe County (LST 1165) about 1966, little changed externally since completion except for the enclosed bridge and tripod mast (which most units lacked). *A. D. Baker*

about his controllable-pitch propellers and considered his plant very reliable. The ship's commanding officer acknowledged the slamming problem, which was similar to that of earlier LSTs.

With the Korean War over, in September 1953 OpNav asked BuShips whether a faster LST could be built. Requirements were 18 kts in landing condition and 16.5 kts on trial (fully loaded), both well beyond the 15 kts planned in 1950. BuShips had already begun a new design based on the October 1950 characteristics updated to incorporate some LST 1156 features. It would land the usual 500 tons of cargo on the usual 1:50 beach and it would avoid the serious vibration problem already being experienced with the LST 1156 class. Initial studies used the Gibbs & Cox forebody with a modified afterbody to improve flow into the propellers.

To overcome vibration, the preliminary design section tried two alternative sterns: twin skegs with the propellers between, and a large single skeg with propellers raised for better flow. Model tests suggested that both techniques would be successful. The trials of LST 1173 showed that the new stern was successful. The deckhouse was moved 32 ft forward so that its sides lined up with the longitudinal bulkheads forming the after end of the tank deck (these bulkheads were extended aft). The after 3-in mount and its director were moved forward.

The SCB issued initial characteristics for this SCB 119 design on 3 November 1953. Generally they matched those of LST 1156, except for speed and power. There were some new features: a helicopter landing area (for a 31,000 lb HR2S), air conditioning for ship control and crew living spaces, additional crew berthing, and a heavier bow ramp that could take a 75-ton amphibian.

Initial weight estimates suggested that forward draft would be excessive, so the preliminary design section increased beam to 65 ft. As in the past two classes, Gibbs & Cox received the contract for contract design. Its review showed that machinery weight had been underestimated, and it suggested that greater length would permit several alternative diesel combinations on the same shafts. At the company's suggestion, the extra buoyancy was provided by lengthening the ship by 16 ft rather than by adding beam (and, therefore, resistance). At first it seemed that the ship would need triple screws (11,400 BHP), but by September 1953 a simpler two-screw 13,500 BHP arrangement had been adopted. A steam plant was considered but rejected.

The preliminary designers considered this new design very close to the performance limit imposed by a traditional LST hull form, with its shallow draft, its hard bilge, its spoon-shaped bow, and its requirement to land 500 tons on a 1:50 (2 percent) beach with a bow draft of 3 ft 6 in. The higher speed still desired, 20 kts, would require something closer to a conventional hull. The new LST also seemed to violate the spirit of the original LST; it was rather large for its 500-ton beaching load. Its cost had risen to the point where it could no longer be considered expendable.

At the end of March 1954 the SCB relaxed the beaching requirements from a 1:50 to a 1:47 beach, but kept the forward draft of 3 ft 6 in, which was set by the requirement that vehicles and men disembark without too much difficulty. To retain ability to beach on a 2 percent slope would have required a bow draft of 3 ft 10 in, which was considered unacceptable. On the other hand, the bow draft requirement was often violated. A typical World War II LST was loaded to 6 ft 6 in forward draft, then run so far up a beach (at flank speed) that it could discharge its tanks on dry land. It took about 45 minutes to retract. Postwar, LSTs often ran up to half their length up a shore on beaching; that was often necessary to keep the ship in position. Thus an LST could

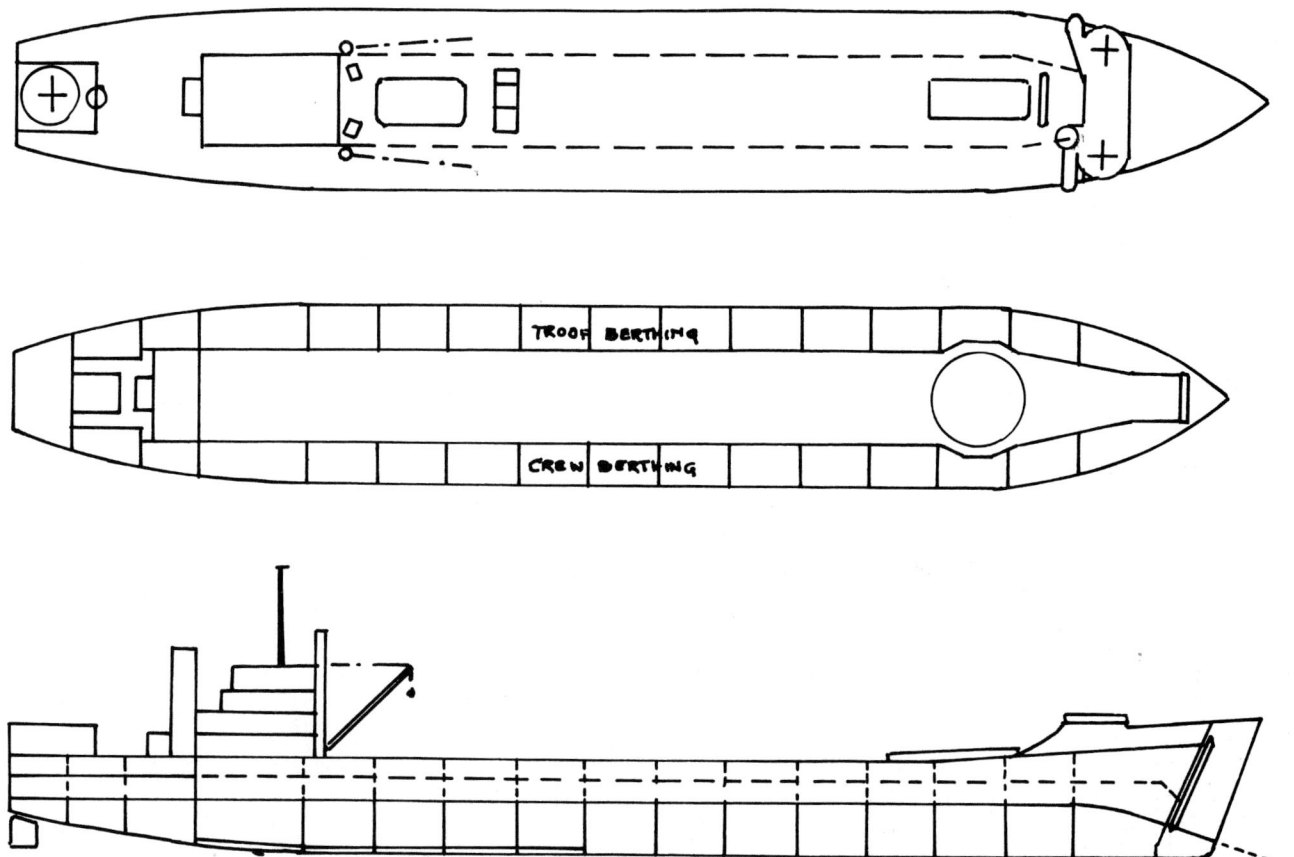

A preliminary design sketch of the SCB 119 LST, produced about 1952. Note the turntable at the fore end of the tank deck. The tank deck occupies only 32 ft of the total beam of 62 ft, with troops and crew occupying much of the wing wall space.

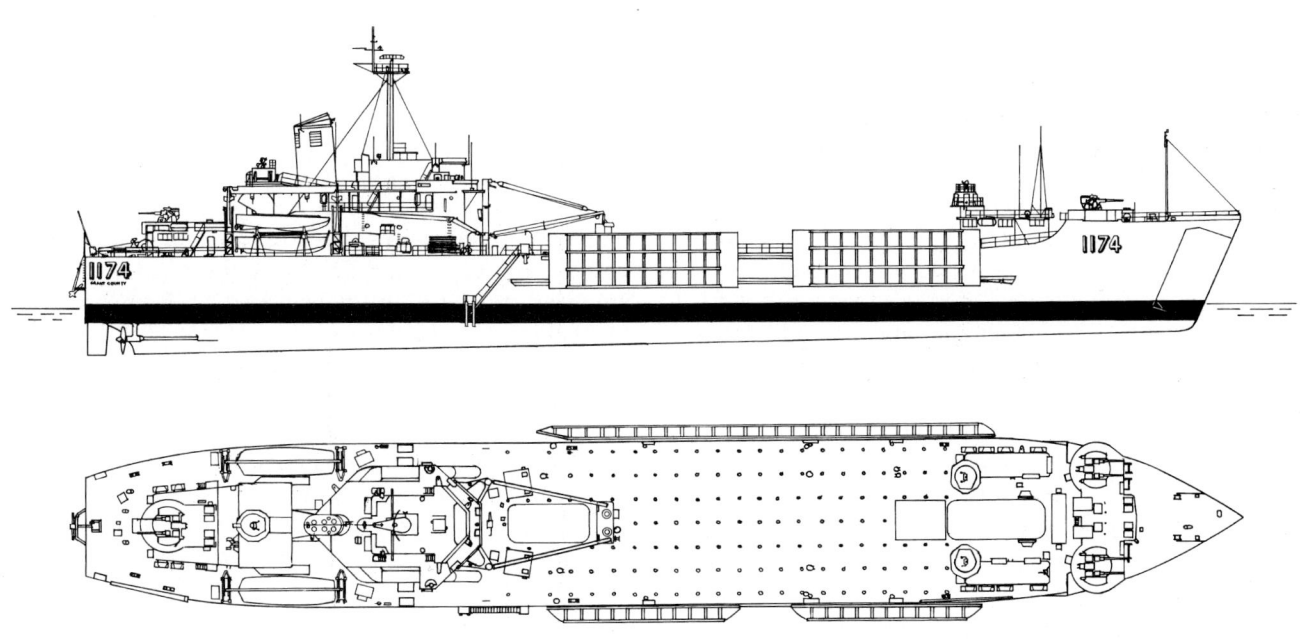

Grant County (LST 1174), showing pontoons rigged along her sides. She represents the ultimate development of the original World War II LST concept. *A. D. Baker*

York County (LST 1175) was the ultimate development of the World War II LST concept. She is shown on 30 October 1957. The higher speeds being demanded by 1957 required a radically different type of ship, realized in the form of the *Newport* (LST 1179) class.

be loaded to 7 ft draft forward, hitting a 2 percent beach at a speed that would carry it 200 ft up.

Changes in 1956, reflecting experience with the LST 1156 class, substituted 7,000 gal of bulk gasoline stowage for the earlier 2,500 gal in drums. LST 1175 and 1176 had 250,000 gal of aviation gas capacity. That made it possible to refuel helicopters with aviation gasoline and/or vehicles with motor gasoline. At the same time 194 troops were added to the 410 already specified. These LSTs carried four LCVP in new link-type davits. Reflecting the increasing importance of helicopters, they had ultraviolet (UV)–lit landing areas to accommodate five HRS helicopters. These LSTs were fully air conditioned. All had antinuclear washdown and vent isolation.

The prototype, LST 1171 (*De Soto County*), was included in the FY 54 program, and six more were built. All but LST 1176 had six reversing diesels. In service, the electric clutches on the diesel engines were criticized as unreliable. Slipping made for awkward control at minimum speed. Close propeller spacing limited maneuverability. LST 1176 had four larger engines driving controllable-pitch reversible propellers, and as a consequence could turn inside a destroyer. The ships were also criticized for their slow oil transfer rate at sea.

Among other things, LSTs were valued because they carried causeways on their sides. As in World War II, the causeways were needed if materiel was to be landed on flat beaches. By 1955, however, it was clear that an amphibious force might need much more than its LSTs could carry, and BuShips studied a possible merchant ship conversion specifically to carry and deploy causeway sections.

The new threat posed by fast submarines led to a demand that all ships of the amphibious force be capable of maintaining a speed of 20 kts. *Fort Snelling* (LSD 30) exemplifies this program. She is shown off her builder's yard, 12 January 1955. *Ingalls Shipbuilding Corporation*

11

The Fast Submarine and the Fast Amphibious Force

The vast wartime amphibious force split into a fast components (including LSDs) capable of 14–17 kts, and the much slower LSTs, with craft such as LSMs and LCI(L)s intermediate. Speed was an important protection against submarine attack: wartime convoys were grouped by speed because speed determined how vulnerable they were. Prewar and wartime submarines had underwater speeds below 10 kts, so a 15-kt ship was nearly immune. Late in World War II, the Germans introduced a new generation of fast submarines, particularly the Type XXI, which was capable of about 16.5 kts submerged. It seemed obvious that the Soviets, who had captured German submarine technology in 1945, and who had possessed the world's largest submarine fleet before World War II, would adopt the new German designs. On the horizon were even faster submarines, such as the Germans' hydrogen peroxide–fueled Type XXVI (fortunately only a design concept in 1945), intended to sustain 25 kts submerged for 10 hrs.

For the moment, with Type XXI a reality, 20 kts rather than 15 kts seemed to be a minimum speed for new U.S. amphibious ships. Higher speed was also prized for quicker dispersion in the face of air attacks, and for an ability to work with fleet units. In 1946 the SCB listed a series of projects for 20-kt amphibious ships: SCB 14, a fast APA; SCB 15, a fast AKA; SCB 16, a fast amphibious flagship (AGC); SCB 17, a fast LSD; and SCB 32, a fast LST. The 20-kt speed objective was ratified by a 13–16 January 1947 Amphibious Type Conference, whose report was later approved by the VCNO. Higher speed was no great problem for transports and even for LSDs, but the LST was a very different proposition. Table 11-1 shows details for APA, AKA, LSD, and LST classes (discussed in this chapter) that achieved the 20-kt requirement for the amphibious force.

There was a parallel demand for faster unloading, so ships would not have to remain in their vulnerable offshore positions. Hence, there were references to palletized cargo in the 1947 AKA study (described below). There was also a need for higher speed to get cargo to and over the beach, which translated into faster craft shuttling to the beach and also to amphibians capable of moving cargo beyond the beach.

The SCB expected the APA and AKA (and a future fast AGC) to share a common hull and powerplant, adapted to quick production. Apparently the AKA requirement was the more demanding, so BuShips' preliminary design section started with it (see below); the APA would be derived from the AKA. The secretary of the navy approved the AKA design study for inclusion in the FY 48 program on 19 July 1946. Estimated cost of building a ship, as of 1948, was $13.5 million, including $2.525 million for equipment. In May 1948 the same hull was also proposed as the basis for a new escort carrier (SCB 43) and a new one-stop replenishment ship, which later became the AOE.

The Amphibious Type Conference ultimately wanted a single class to replace both APA and AKA. It would carry about 500 troops and associated cargo, it would use amphibians instead of boats, and it would probably have a stern gate similar to that of the LSV. In retrospect it is surprising that the conference was willing to go below the battalion landing team (BLT) level in troop capacity in one ship.

BuShips wanted to encourage commercial production of suitable ships that could be taken over in an emergency. The Amphibious Type Conference called for some special survivability features: two-compartment flooding standard, bulkheads unpierced up to the second deck, and twin-screw machinery suited to split-plant operation (later the Atlantic Fleet amphibious force would advocate a combination of twin screws and twin rudders for better maneuverability in the face of air attacks). Some of these features, and the requirement that the ship

Table 11-1. The 20-kt Amphibious Force

	Paul Revere (APA 248)	Tulare (AKA 112)	Charleston (AKA 113)	Thomaston (LSD 28)
Length (ft-in)				
WL	528-0	528-0	550-0	NA
OA	564-0	564-0	575-6	510-0
Beam (ft-in)	76-0	76-0	82-0	84-0
Draft, loaded (ft-in)[a]	27-0	28-0	25.44 ft	19-0
Displacement (tons)				
Light	10,709	9,050	13,727	6,880
Loaded	16,838	15,970	18,648	11,270
Boilers	2 CE	2	2 CE	2 B&W
Conditions (psi/°F)	NA	NA	600/850	600/850
Power (SHP/shafts)	22,000/1	22,000/1	22,000/1	24,000/2
Speed (kt)	21	21	22	23
Radius (nm/kt)	10,000/20	10,000/20	9,600/16	13,000/10
Complement (officers/enlisted)	45/484	31/362	25/311	21/384
Capacity				
Troops[d]	98/1,980	18/301	15/211	29/312
Tons	1,900	4,476	5,280	2,400
Sq ft	NA	NA	32,900	10,200
Cu ft	137,678	450,000	66,100	3,500
Booms (qty-capacity)[g]				
70 ton	2-60	3-60	2	—
40 ton	1-30	1	2	2-50
15 ton	3-10, 2-8	6-10	8	—
5 ton	1	—	—	—
Boats				
LCM	7	9	9[h]	18[i]
LCVP/LCPL	15[m]	11	2	4
Armament				
3-in/50	4 × II[o]	3 × II	4 × II	8 × II
20-mm	—	—	—	6 × II

NOTES: Dash indicates that specification is irrelevant; NA indicates relevancy but data not available.

[a]Draft mean if only one figure given; draft fore/aft if two figures given. Measurements are given in ft-in unless otherwise specified.

[b]Diesel power: four Colt-Pielstick 16PC2.5V400 in LSD 41 and LSD 49 classes.

[c]Diesel power: six Alco 16-251 engines.

[d]Officers/enlisted when two figures are listed; when only one figure is given, breakdown is unknown.

[e]Full load: 2,000 tons; deck areas: tank deck 12,704 sq ft, main deck 11,391 sq ft. Presumably the figure given is total usable deck area.

[f]Combat cargo cube was given as 4,200 cu ft in a 1969 summary of the capabilities of the new LST.

[g]Only quantity is listed unless capacity is different. For example, Paul Revere had two 60-ton booms instead of any 70-ton booms.

[h]Four LCM(8) and five LCM(6).

[i]LCM(6); alternatives are three LCU or nine LCM(8).

not roll excessively in a roadstead while unloading, might be difficult to meet in a converted merchant ship. A 6 November 1947 design report warned that mass production might be difficult, because only 20 building ways suited to so large a ship were available; the specialist wartime yards, which had built the mass of new shipping, had all shut down. The ships would be armed with the new twin 3-in/70 antiaircraft cannon, which featured in many contemporary design studies. Not appearing in service for a decade, they then proved unsuccessful.

The 20-kt transports would be accompanied by a 20-kt LSD, which was a Priority 4 design study for FY 48. Capacity was set at three LCT or two LSM; the ship could also carry PT boats. The secretary of the navy approved a design study for a fast LSD at the same time as he approved the fast AKA. As of 1948, estimated construction cost was $13.5 mil-

Anchorage (LSD 36)	Whidbey Island (LSD 41)	Harpers Ferry (LSD 49)	Newport (LST 1179)
534-0	580-0	579-11	522-2
562-0	609-5	609-5	561-2
84-0	84-0	84-0	69-6
20-0	19-7	19-9	5-11/17-2
8,200	11,471	11,894	4,975
13,680	15,745	16,695	8,576
2 CE	—[b]	—	—[c]
600/850	—	—	—
24,000/2	41,600/2	41,600/2	16,500/2
22	22	22	22
14,000/12	8,000/20	8,000/20	14,250/14
18/304	21/289	21/312	15/247
336	560	400	430
NA	NA	NA	500[e]
15,200	13,500	16,600	17,300
1,400	5,100	50,700	3,400[f]
—	1-60	—	—
2-50	—	1-30	—
—	1-20	—	2-10
—	—	—	—
9[j]	10[k]	4[l]	—
4	3	2	4[n]
4 × II	—	—	2 × II
—	2 RAM[p]	2 RAM[q]	—

[j]LCM(8) without mezzanine deck, or six with mezzanine deck; alternatives are 2/3 LCAC (with/without mezzanine) or 1/3 LCU (with/without mezzanine) or 50 LVT.

[k]LCM(8); alternatives are 4 LCAC, 3 LCU, or 64 AAV.

[l]LCM(8); or two LCAC or one LCU.

[m]Ten LCVP and five LCPL.

[n]Typically three LCVP and one LCPL or Seafox.

[o]Multiple gun mounts are indicated by Roman numbers; therefore 4 × II indicates four twin mounts.

[p]Plus two 25-mm, two Phalanx, and eight 0.50 caliber machine guns.

[q]Plus two 25-mm, two Phalanx, and eight 0.50 caliber machine guns.

lion, including $3,137,500 in equipment. There seemed to be little point in building a prototype fast LSD, because the design did not present any really novel problems. Moreover, there would be no need for a fast LSD until the fast AKA, APA, and AGC had materialized.

In 1946 no one knew how to design a 20-kt LST, and the SCB wanted a design study included in the FY 48 program. Improvements, compared to earlier LSTs, would be a tank deck strong enough for 75-ton tanks, capacity for at least 19 LVT, and a main deck strong enough for 25-ton vehicles.

Characteristics for a fast AKA (SCB 15) were drawn up in 1947. Sustained speed was set at 20 kts, and endurance at 10,000 nm at 20 kts. The ship would carry at least 425,000 cu ft of cargo, with a high ratio of deck area to hold volume. Cargo would include four units of fire for one-third of divisional artillery (26,000 cu ft) and two and a half units of fire for one-third of all organic divisional weapons (7,500 cu ft) (i.e., for the third of a division carried by the transport division of which the AKA was a unit). Ammunition would be distributed about equally fore and aft. She would also carry 4,000 drums of motor fuel (160,000 gal), also distributed fore and aft. In addition to a crew of 495, the ship would carry troops: 16 officers and 300 enlisted men (ship unloading platoons and vehicle crews).

In an extension of the combat loading concept, the AKA would be arranged so that cargo could be stowed selectively and unloaded as needed. To do that the holds would have maximum deck area but minimum height. If cargo were not stacked vertically, and if it were spread sufficiently horizontally, virtually all of it would be accessible at all times. Half the cargo spaces, including those for troop ammunition, would be suitable for palletized cargo (1½-ton lifts), which would be handled by forklifts, hydraulic jacks, conveyer belts, chutes, rollers, and similar gear. Special stowage for troop communications and optical gear would be accessible without opening a hatch. To accommodate vehicles, at least three cargo compartments in the holds or lower 'tween decks would have 10 ft clearance under the hatch girders (later reduced to 8 ft). Maximum vehicle size was set at 11 ft wide × 11 ft high × 35 ft long.

The combination of high speed and less dense cargo stowage would make the AKA a very large ship; in the fall of 1947 estimated full load displacement was 17,500 tons (550 × 75 × 25 ft), and capacity was about 3,000 short tons. The existing converted C2s generally carried 4,000 long tons (4,480 short tons) of cargo on 11,000 tons full load displacement (459 × 63 × 26 ft), so the new ship might seem less than a great bargain. To some extent the reduced cargo capacity might be attributed to a larger crew and to a split engine room, for better survivability.

By May 1948 U.S. defense spending was up due to a strong sense that the Cold War was heating up. Recent events included the Czech coup in February 1948, and the Soviets were tightening access to Berlin. The FY 48 Maritime Commission appropriation was $84 million, and President Truman had just asked that the FY 49 figure be increased from $29.5 to $90 million to buy badly needed passenger ships and tankers. It appeared that Congress would add even more. In this climate, the Maritime Commission formally approached the U.S. Navy to ask about speed as a national defense feature in new designs. It was also quite interested in building ships that might meet new operational requirements, such as a 20-kt AKA.

Studies to estimate the cost of higher speeds were based on C3 and P2 hulls. BuShips told the SCB in May 1948 that to increase the sustained speed of a C3 from 16½ kts (8,500 SHP) to 20 kts (20,000 SHP) would cost about $1.5 million; the basic C3 would cost $5 million. For a 590-ft P2, it would take 16,000 SHP to increase trial speed to 20 kts and 27,000 SHP for 22 kts, which the navy considered equivalent to 20-kt service speed. The cost differential was about $1.65 million. The new AKA would have to be a modified commercial ship, so the question was how much speed a commercial operator would want. Because it was unlikely that shipping lines would want speeds much above 16.5 kts, the navy would have to subsidize them for the higher power, and also for the somewhat larger hull needed to carry the same amount of cargo, with a finer form suited to higher speed. At this time the navy was considering subsidizing a pair of new American Export Line passenger ships, which became *Independence* and *Constitution*. The line was willing to pay for 22 kts sustained speed (40,000 SHP), but it offered to buy 24-kt ships (55,000 SHP) if it were subsidized for the difference in powerplant cost, about $2 million. An operational evaluation group study showed that really high speed, about 33 kts, would buy some protection against submarines, but nothing close to that figure was contemplated for amphibious ships.

Another major issue was boom capacity. In December 1947 OpNav asked the Marines how many items of equipment exceeded current cargo limits (25 × 10 × 7.5 ft, 30 short tons). It turned out that a Marine Corps division had 398 pieces of oversize equipment, including its tanks (35 short tons for the current M4A3) and a truck-mounted revolving crane (34 ft 8 in long). The new M26 tank would weigh 50 short tons (24 ft 2 in × 11 ft 6 in × 9 ft 2 in). Since the Marines used standard army equipment, these figures also sufficed for the army. The Marines advised that 8 percent of the cubic capacity of an AKA should have more than 8 ft clearance. Earlier AKAs had one 30-ton and one 10-ton boom for each of their four hatches, so initial draft characteristics called for one 60-ton and a 10-ton boom for each of four hatches. Because the new AKA would be so much larger than its predecessors, it was expected to have two more hatches, which could be served by two 10-ton booms each. The SCB asked, in October 1947, whether there was any point in having more than two very heavy booms, because the existing LCM(6) was limited to 34 short tons. Heavy booms were expensive in terms of winches and electric power—and topweight. Booms were cut back to 60-tonners serving two hatches and 40-tonners over another three, with additional high-speed 5- or 10-ton booms. It is not clear whether there was an expectation that the new ship would carry the heavier LCM(8) then being developed, or whether it would unload heavy vehicles into LCTs from accompanying LSDs.

Originally the SCB had called for four 50-ton booms over two large hatches. A tabulation showed that a few navy cargo ships already had 50-ton booms: two Victories (AK 231 and 227) and at least five Liberties (AK 71, 90, 110, 127, 225). Neither class could make anything close to the required speed, but these ships demonstrated that it would not take a new kind of mast to support the desired heavy booms.

In May 1948 the SCB proposed that all active AKA and APA and all new construction ships have 60-ton capacity on all hatches capable of taking the new M26 tank (larger than 12 × 25 ft, with overhead clearance of 9 ft 2 in or more). On active ships and conversions that could not provide this much lift, the minimum would be 45 tons. All other jumbo booms should be at least 35-ton capacity, and 45-ton capacity on new ships. All of these capacities were in long (2,240 lb) rather than short (2,000 lb) tons. In November 1948 Op-343, in charge of amphibious ships, added that any ships converted to AKAs should have at least 45-ton booms. Existing transports being upgraded should have their jumbo boom capacities increased to 45 tons. In October 1948 BuShips offered to install a 60-ton boom to serve hold No. 3 on board all active AKA. On AKA 13–100 it would replace an existing 30-ton boom (on AKA 104–108, a 35-ton boom). On AKA 13 and 14 a quadrupod would replace the existing kingpost, and on other AKAs the quadrupod would have to be reinforced. On AKA 13–100, 35-ton booms could replace the existing 30-ton booms at hatches 2, 4, and 5 (AKA 104–108 already had 35-ton booms there). *Arneb* (AKA 56) was already undergoing conversion to the new standard.

On APA 18–27 a 60-ton centerline boom could replace the pair of 30-ton booms serving hold No. 2. On the active Victory APAs (APA 217–237) a 60-ton boom would replace the single 35-ton boom serving hold No. 5. If some reduction in hatch length was acceptable, 60-ton booms could replace 30-ton centerline booms on APA 33–45 at hold No. 3 and port 30-ton booms on APA 44 and 45. APA 30 had no hatches large enough to be worth plumbing with a 60-ton crane. As for other jumbo booms on APAs, 35-ton booms could replace existing 30-ton booms at hatch No. 5 on APA 18–27, 30, and 33–45. APA 30 had a 26-ton boom instead of 30-ton, and the 26-ton boom at hatch No. 2 would remain unchanged, because this was a split hatch. Victory APAs had only a single jumbo boom, which (as noted above) would be replaced by a 60-ton boom. To make the speed of the new boom equal to that of existing 30-ton booms, 50 HP winches would be replaced by 100 HP winches.

On 14 September 1948 BuShips submitted a pair of AKA sketch designs, with engines aft or amidships; it had already discarded one sketch design, for a larger ship. Both had 525 (waterline) or 564 (overall) × 76 × 24 ft hulls (9,000 tons light, 16,000 tons full load) with bale capacity of 425,000 cu ft. The design allowed for 4,000 short tons of cargo. Quadrupod masts fore and aft supported two 60-ton, two 40-ton, and four 10-ton booms. Boats were 2 LCP(L), 16 LCVP, and 10 LCM(6) (rather than the 12 requested). The ship's complement was 40 officers and 450 enlisted men, and troop capacity was 16 officers and 300 enlisted men, for the embarked vehicles.

The designers chose single-screw rather than the specified twin-screw powerplant (20,000 SHP: 22 kts trial, 20 kts sustained speed) because it corresponded to contemporary commercial practice. The engines-aft powerplant was considered somewhat less vulnerable because torpedoes generally hit amidships, and there was about 10,000 more cu ft of hold volume. Ships would probably have to be built for commercial operators, however, then taken over in an emergency. Commercial operators clearly preferred the engines-aft arrangement. BuShips estimated that a merchant ship version would displace 19,700 tons and would make 21.5 kts. In August 1948 the design was discussed with a commercial yard, Bethlehem Quincy. Its representatives pointed out that operators were interested in capacity, which would be reduced by the fine ends needed for high speed. Fine ends also made for holds difficult to load. Commercial operators might accept the finer hull form if they could be given a cost differential to cover it.

Armament was four twin 3-in/70 (fore and aft and amidships, with two Mk 56 and two on-mount Gunar directors), 4 twin 20-mm, and four NAE torpedo decoy projectors. The 3-in/70 was a new gun specified in many contemporary U.S. designs, and the NAE launchers were standard in U.S. designs of this period. Arrangement was complicated by the two 3-in mounts amidships, and the designers asked OpNav whether they could be omitted.

The APA version would embark 105 officers and 1,750 enlisted troops, plus a crew of 90 officers and 575 enlisted men. Cargo capacity would be 900 bale tons (90,000 cu ft) of assault cargo, 11,000 sq ft of vehicles, and a 20-day supply of fuel for the vehicles. Boat teams would form below decks, near embarkation stations; troop compartments would be clearable in 60 seconds. To handle the new heavier vehicles, the ship would have one long heavy lift boom (60 tons) forward and one aft. For improved habitability, the ship would be air conditioned (underlined in the Amphibious Type Conference report) and would have a more powerful distilling plant (50,000 gal of potable water per day). As in the

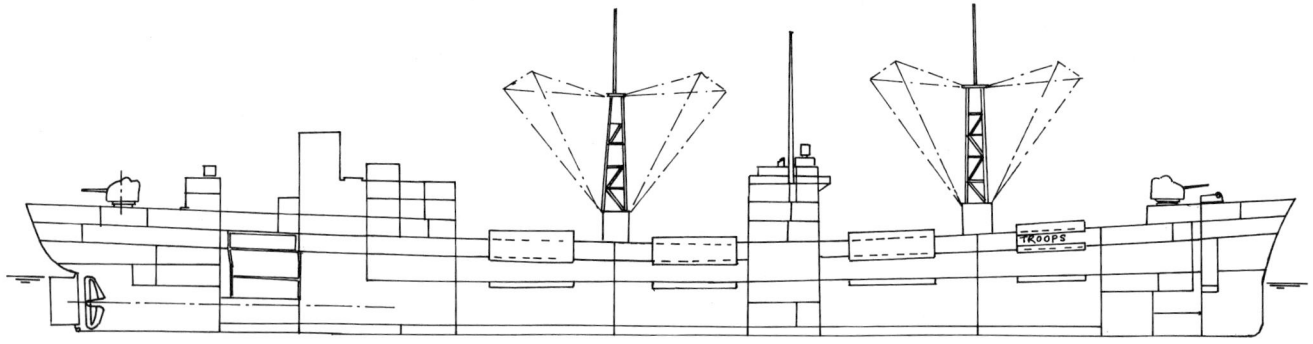

The projected engines-aft version of the SCB 15 AKA (Scheme B), from a sketch dated 1 September 1948. Dimensions were 525 (waterline) × 76 × 23.9 ft (16,200 tons fully loaded); power was 20,000 SHP for a trial speed of 22 kts (20 kts sustained). Armament was 4 twin 3-in/70 (1,100 rounds per barrel, 180 rounds per mount ready service) and 4 twin 20-mm. The ship would also have had four NAE antitorpedo decoy projectors, a common feature of U.S. Navy sketch designs of this period. The 3-in guns would have been mounted fore, aft, and in the waist. The fore and aft mounts would have had Mk 56 directors; the two waist mounts would have had on-mount Gunar directors. Complement would have been 40 officers and 450 enlisted, with a capacity for 16 troop officers and 300 enlisted troops. Boats would have been 10 LCM, 16 LCVP, and 2 LCPL, served by two 60-ton, two 40-ton, and four 10-ton booms. Capacity would have been 425,000 cu ft, with 1,000 tons of ballast in the holds. Refrigerated capacity would have been 30,000 cu ft. A similar engines-amidships design (Scheme A) would have been 525 (waterline) or 564 (overall) × 76 × 23.7 ft (16,100 tons fully loaded), with similar characteristics except for a capacity of 435,000 cu ft.

case of attack cargo ships, Op-343 wanted 45-ton booms in all conversions and in all upgraded ships.

The associated AGC would have a sustained speed of 22 kts, because she would displace less than the AKA/APA. To a crew of 54 officers and 601 enlisted men would be added navy (96 officers and 537 enlisted) and army (59 officers and 644 enlisted men) detachments. The ship would have a helicopter hangar and pad.

The Maritime Commission was the great hope for early construction of such ships, just as it had been before World War II. Unfortunately, the Maritime Commission was interested mainly in two types that it felt the United States lacked: fast liners to compete with foreign ships, and larger tankers. As the Cold War began to accelerate, however, the commission readily added two war cargo ships to its program: an 18.5-kt mobilization design (in effect a successor to the wartime Liberty ship or Victory ship: she became the *Schuyler Otis Bland*) and the Navy's 20-kt APA/AKA (in cargo ship form). On 22 July 1948 President Truman approved $10 million to design an APA/AKA. The ship was tentatively designated S-X-DY, because it was considered a special naval type (S series). It became the "Mariner" class.

BuShips' preliminary design section developed a fourth (525-ft) sketch design, with engines amidships, which it hoped would be suitable for commercial operators. Because BuShips had other design priorities, it could not continue work. It asked the Maritime Commission to develop the design as a cargo ship or AKA. The commission's chief technical designer pointed out that this was a real choice; a ship designed for commercial operation and then converted would differ substantially from one conceived as an AKA and then converted for civilian work. The heavy gun battery planned for the AKA severely limited the design of the ship. Placing guns amidships forced the pilothouse so far aft that the boilers had to be placed at the after end of the machinery space (to put the stack far enough abaft the bridge). Commercial freighters, however, generally had their boilers forward of the engines, to minimize the length of the propeller shaft. The BuShips designers agreed that the battery set by the characteristics was probably unrealistically heavy. Similarly, commercial operators generally did not want 60-ton booms or the accompanying quadrupod masts. The BuShips position was that sufficient structure should be built into the hull to support such masts when they were fitted, but that masting above deck could conform to merchant marine practice. Working deck space should be provided for the necessary large winches, which again need not be installed when ships were built.

Because commercial operators might not find the ship attractive, considerable subsidies might be needed. Many modern cargo ships had been built just before and during the war. In October 1948, 3,194 of the total of 3,883 U.S.-controlled merchant ships were dry cargo freighters. Only 689 were privately owned, compared to 2,428 government owned, and 1,820 freighters (mostly slow Liberty ships) were laid up. The only merchant ships being built in the United States were 55 tankers. Although a huge fleet of war-built tankers had survived, there was already pressure to build much larger ones, and existing ones could not meet that requirement. This market would soon collapse.

By March 1949 the Maritime Commission had a sketch design of an AKA, from which a freighter had been derived. By this time BuShips had agreed that the navy would reconsider the heavy gun battery, which had caused problems the previous fall. It might be cut to two mounts aft (one superfiring) and one forward (by December, and the 3-in/70 had been replaced by the much lighter 3-in/50, but the navy wanted two forward and four aft). By June 1949, the boom arrangement of the AKA version had been set. Space would be provided for ten LCM(6), with the expectation that the ship might well carry some LCM(8) instead. The DCNO for logistics, Adm. Robert Carney, approved a preliminary version of the ship's plans on 24 January 1950. Meanwhile the Maritime Commission developed a new design (ED) based on S-X-DY, but without some navy features such as the quadrupod masts. It hoped ED would be more attractive to commercial operators. It would have a very similar hull, with the same seven holds, hence would offer a similar potential for conversion.

Plans were completed during the summer of 1950. By that time U.S. commercial shipbuilding was at so low an ebb that the Maritime Commission pressed for an immediate program simply to keep shipbuilders in business. The Executive Office and Congress agreed. Contracts for the first 25 ships were let early in February 1951, and seven weeks later ships were being fabricated. The $300 million program amounted to 35 ships, all of which were

The Maritime Commission designed its new "Mariner" class freighter specifically for wartime conversion as a fast amphibious ship. Of 35 Mariners built, only 3 were ever converted. Two others served as test platforms for the Polaris and Poseidon programs. SS *Diamond Mariner* is shown at Todd Shipyard in San Pedro prior to conversion to the attack transport *Paul Revere* (APA 248). The decommissioned LSM in the foreground gives an indication of scale. The other craft is an unpowered accommodation barge (APL).

given Mariner names. Five eventually served the navy, three of them in amphibious roles.

The Maritime Commission designed its new ship, which it designated a C4-S-1A, as an attack cargo ship generally meeting the requirements of SCB 14/15, then converted it (on paper) to a freighter for construction. Dimensions generally matched those estimated by BuShips: 560 ft 10 in (528 ft on the waterline, 525 ft between perpendiculars) × 76 ft × 29 ft 6 in (scantling draft 31 ft 6 in). As an AKA, she would be limited to 25 ft draft. Dimensions were on the large side for a merchant ship, and there was some question as to whether there would be many commercial takers. As a merchant ship, she would displace 20,800 tons fully loaded (8,500 tons light). As an AKA, she would displace 16,850 tons fully loaded (9,500 tons light). In merchant ship terms, the ship was 9,700 tons gross, and over 12,500 tons deadweight.

For two-compartment survivability, the ship had four transverse bulkheads watertight to the main deck, except for necessary openings to hold down her rated tonnage, which affected port fees. There were seven holds (four forward, three aft), each with two 10-ton and two 5-ton cranes; Nos. 4 and 6 had 60-ton cranes for heavy vehicles. Special hatches in the second deck were provided with coamings less than 9 in high to take vehicles above them.

Because she had to be commercially viable, the ship had only a single screw, driven by two double-reduction turbines and two boilers in a single compartment. On the other hand, the machinery spaces were protected by cofferdams on each side, a very unusual merchant ship feature. Vital propelling machinery and control equipment were made shock resistant. The machinery spaces were enlarged to take the extra evaporators and condensers that might be needed to serve troops. Continuous power output (80 percent) was expected to drive the ship at 17.4 kts. However, using the rated continuous overload power, 22,000 SHP, the ship would make a trial speed of 22 kts, which corresponds to a continuous sea speed of at least 20 kts, as an AKA. The higher power would be obtained by reducing main turbine bleeding and by accepting reduced plant efficiency and design factors higher than normal commercial practice (but within naval limits). Fuel would suffice for 10,000 nm at 20 kts. The freighter would be built with the generating capacity needed for naval service, but without the two 300 kW emergency diesel generators planned for the AKA/APA, for which space was reserved.

A horn frame on the stern was expected to reduce the propeller noise, hence the ship's detectability, usually associated with a skeg and a propeller with an even number of blades. Vital propelling machinery and control equipment was shock hardened. To provide sufficient water for troops, the engine room could accommodate extra evaporators and condensers. The ship was provided with an oversize rudder, for a turning circle of about 650 yds.

The ship's deckhouse was laid out with the AKA conversion in mind. Parts were splinter-proofed, and the entire deckhouse was gas-tight. The stern was strengthened for installation of a helicopter deck. A compartment in the forepeak was provided for listening equipment, presumably for torpedo detection; the transport would have antitorpedo beacon launchers. Special hatches in the second deck had low coamings so the 'tween decks above them could be used for vehicles.

The SCB issued approved characteristics for the new AKA, its project 77, on 26 July 1951, in time for the FY 53 program. Cargo space would be 10,625 measurement tons (4,250 long tons in weight). On a weight basis, she carried about as much as a World War II AKA, but on a volume basis she carried far more (450,000 vs. 364,000 cu ft). Clear deck heights in vehicle stowage spaces were 8 ft 4 in, 12 ft, and 14 ft. Ammunition would be slightly more than previously planned (27,000 cu ft of artillery, 8,000 of organic ammunition). The design also allowed for a liquid cargo of 2,000 tons. Whereas a Mariner had seven holds, only five of them would be used by the AKA version, which operated at a much lighter displacement. At merchant ship draft, the Mariner was a one-compartment ship, but at AKA draft she was a two-compartment ship, as the navy required. The ship would have two quadrupod masts, the forward one with two 60-ton and two 10-ton booms, the after with one 60-ton, one 40-ton, and two 10-ton booms. The 60-ton booms were for tanks, the 40-ton boom for landing craft. The forward kingpost would carry two 10-ton and two 5-ton booms. Troop accommodation was set at 20 officers and 300 enlisted (plus 39 officers and 396 enlisted men of the ship's company). The ship would carry 28 landing craft: 2 LCPL, 16 LCVP, and 10 LCM(6). Six LCVP would be carried in two triple davits with LCPLs outboard, so they would be launched first, as guide boats; the other ten would be nested in the ten LCMs. If the new LCM(8) were modified to permit hoisting it aboard, four would be stowed, and total craft stowage could be reduced accordingly. This requirement was canceled in October 1952. Armament would be six twin 3-in/50 (two abeam forward, and pairs of mounts superfiring over each other aft) plus six twin 20-mm (the NAE launchers had long been abandoned). The various additions for AKA service required more electric power, so the ship would receive an additional ship's service generator and

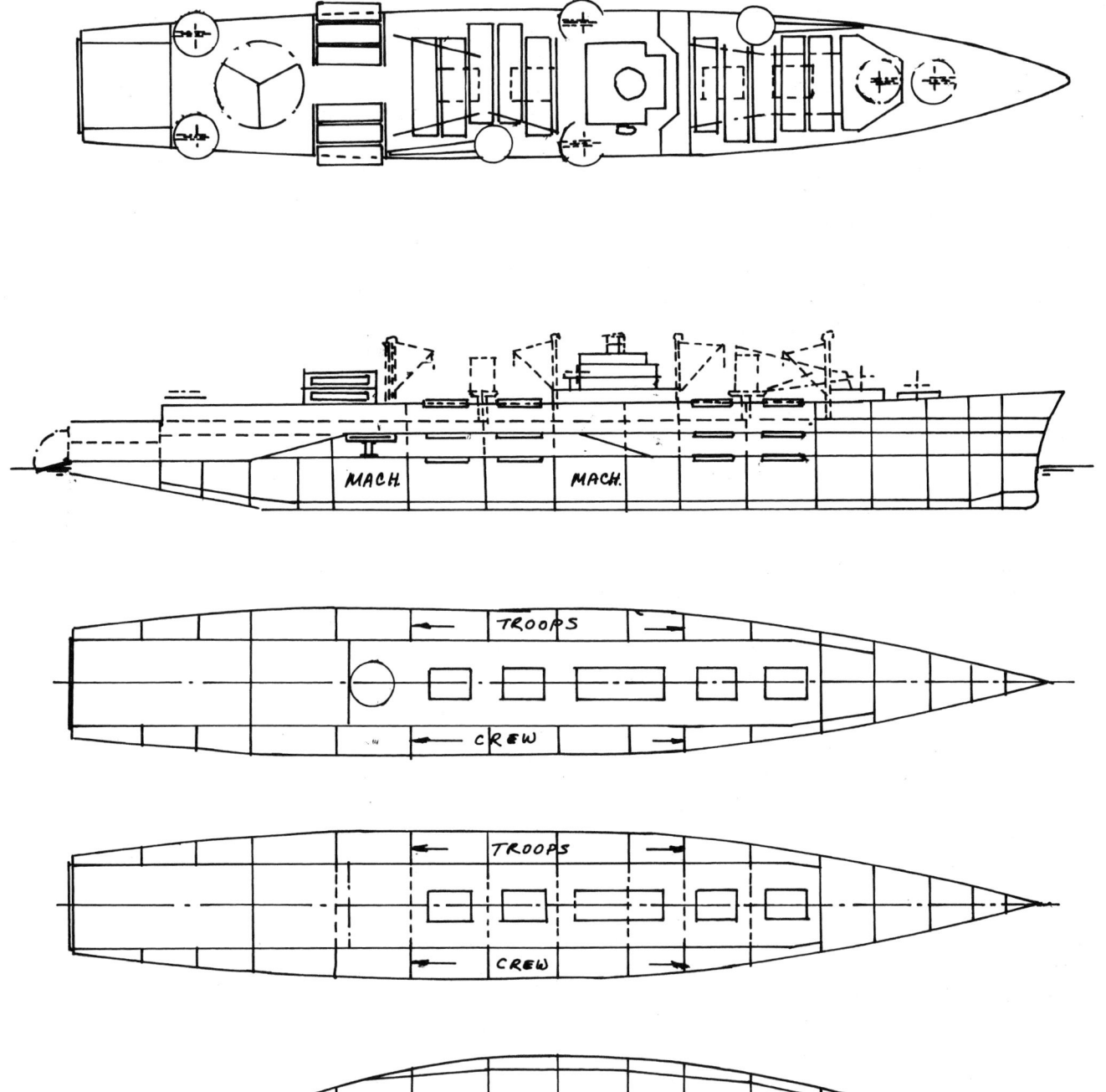

The SCB 77A sketch design, developed for the 1953 shipbuilding program. Note the short well deck aft, onto which vehicles could roll down a ramp, as well as the turntable just forward of the ramp. The ship was basically an LSD, with a portable mezzanine and superdeck to carry vehicles. The ship would have unloaded by driving the vehicles directly onto an LCU or LCMs grounded on the ramp at the forward end of the well. Roll-on/roll-off efficiency would have been limited by the presence of hatches, booms, and winches. The ship would also have had a pair of quadruple Welin davits aft, abaft which was a helicopter pad. The davits would each have accommodated three LCVP and one LCPL. The five oblongs forward represent LCM(6)s with LCVPs nested in them, as onboard contemporary attack transports. The two cranes shown had capacities of 50 tons. In the side view, note that the two machinery spaces are separated by a pair of holds, for survivability. Planned dimensions were 575 overall × 85 × 23 ft (fully loaded); displacement would have been 9,200 tons light and 18,100 tons fully loaded. The ship was intended to make 22 kts (trial) on 24,000 SHP, as in a contemporary LSD; endurance would have been 10,000 nm at 20 kts. The weapons shown are twin 3-in/50 guns. This illustration was reproduced from a sketch produced in June 1955 for comparison with the newer AKR concepts.

two 300 kW emergency diesel generators would replace the single 75 kW unit.

Before SCB 77 could be built, interest arose in an alternative, SCB 77A, based on the new commercial ro-ro (roll-on roll-off) concept. Instead of moving cargo on board by crane, then offloading into warehouses from which trucks would take it inland, a ro-ro ship was arranged so that trailers could be driven aboard and lashed down. At the pier, they could be driven off, down ramps. In volume terms, ro-ro was far less efficient than conventional loading, but it made for much faster turn-around, which was exactly what amphibious forces needed. BuShips became interested in ro-ro both for assault transports and for point-to-point ships.

This idea had actually been tried during World War II. For the attack on Empress Augusta Bay in the South Pacific, bulk cargo was loaded onto LSTs on board trucks. Upon arrival, they simply drove off, unloading the ship in about 2 hrs. By way of contrast, a conventionally loaded LST at Salerno unloaded about 100 tons per day, so she took 4–6 days to unload completely. The ro-ro technique used in the South Pacific did reduce the ship's payload by about a quarter. Postwar, commercially operated LSTs were the first ro-ro ships.

In August 1952, Lt. Gen. Lyman L. Lemnitzer, who was the U.S. Army's deputy chief of staff for plans and research, passed the CNO a Munitions Board recommendation to the Defense Production Administration that the Military Sea Transportation Service (MSTS) build ro-ros for quick movement of heavy vehicles. In the last six months of 1951, 19 percent of all army troop support and military aid cargo had been wheeled and tracked vehicles such as tanks and trucks; between March and May 1952 there were about 88 cargo loads for which ro-ros would have been desirable. At the least, Lemnitzer wanted enough ro-ros to move the heavy equipment of an infantry division. In September, Capt. Norman S. Short, the BuShips member of the Navy Transportation Committee (concerned with MSTS), passed a copy of Lemnitzer's memo to BuShips. At this time the committee planned to report to the CNO that the navy should design a ro-ro, similar to a Sea Train. Capt. Russell Snyder, a senior BuShips officer, told the preliminary design section to consider a ro-ro AKA. The alternatives were an LSD with some ballast traded for cargo, an LSV type with a stern ramp, or a modified LST without any beaching requirement. Preliminary designers began work on a new AKA to be designed around efficient cargo handling, rather than an existing hull.

It was well known that existing AKAs could not easily unload heavy vehicles in rough weather. In July 1952 the commander of *White Marsh* (LSD 8), C. B. Bright, reported that he had been able to unload LCUs (landing craft, utility; ex-LCTs) in rough weather. With only 2 ft of water in the well deck, running at 10 kts, he backed at full power. The resulting surge sufficed to launch an LCU. Similarly, an LCU could embark, then ground in the well deck when engines were switched to full ahead. Based on his AKA and LSD experience, Bright proposed a large "triphibious" ship based on the LSD hull form. Triphibious meant a combination of surface and vertical (see Chapter 12) attack, and the ship would be large enough to carry a third of a full Marine regiment (13,000 men) rather than a BLT (as in an APA). The ship would be 850 × 90 × 24 ft (25,000 tons fully loaded). At this size she could be driven at really high speed, probably 30 kts sustained on 150,000 SHP. The size of a fleet carrier, she could accommodate 100 helicopters (each about the size of a conventional airplane) on flight and hangar decks, plus 86 LCM (50 of them in the well deck, the rest nested on the weather decks) or 150 LVTs (on the mezzanine deck and in the well deck). She might also carry hydrofoil boats, which were then of considerable interest.[1] The Marines liked the idea, and late in October 1952 BuShips' preliminary design section began work on a ro-ro AKA it inspired. Its characteristics were based on the Mariner conversion, the Marines' concept, and LSD 28. For their part, the Marines formally proposed the triphibious ship in November 1952.

To the preliminary designers, the ro-ro concept made little sense unless the ship could quickly discharge the vehicles; the World War II practice of lifting them into craft alongside was far too slow. Nor could it be effective in the rough northern waters in which the navy then expected to fight. Bright's idea solved the problem: the vehicles could be driven into an LCU, which could then run to the beach. The well deck was just large enough for an LCU, three LCU(6), two LCM(8), or two army BARCs to enter and load at one time. Because the well deck was small, the ship could ballast down very quickly to fill it. The rest of the well deck was used for two levels of vehicle parking, with a ramp from the upper down into the well. Vehicles from the lower deck could run up to the upper via a portable ramp, and there was a 24-ft turntable on the upper vehicle deck near the ramp into the well. As in an LSD, craft in the well deck would be stowed loaded. Once they were launched, other craft could dock; vehicles would enter them via the ramp from the upper vehicle deck.

The ship had three cargo holds, to be used only for palletized combat loaded cargo (e.g., ammunition, gasoline in cans). All were forward of the machinery. Three 3-ton traveling cranes would transfer cargo to the upper deck, where it could be loaded into vehicles. Two 35-ton topside cranes would handle landing craft stowed on deck. They also served two hatches, large enough for a vehicle. The deckhouse was set forward, as in an LSD.

The ship's stern form precluded a single-screw plant, so two machinery spaces were provided, separated by the ballast tanks. In the final version of the design, embarked landing craft were 1 LCU, 10 LCM(6), and 10 LCVP, with the LCVP nested in the LCM on deck.

BuShips did not initially sell this idea to the SCB. Thus, in mid-November 1952 the SCB issued first preliminary SCB 77A characteristics modeled on those of the converted Mariner, but made the explicit point that although the ship might be of about Mariner size, the size and arrangement of spaces on board would not be governed by the Maritime Commission design. At this point the main difference from SCB 77 was that, the requirement for handling LCM(8)s having been abandoned, the ship's two quadrupod masts would each carry two 45-ton and two 10-ton at each quadrupod; four 10-ton booms would be stepped from a forward kingpost. The ship also had sufficient open deck space for an HO3S-1 helicopter. This was not yet a ro-ro ship. She would stow 10 LCM(6), 20 LCVP, and 2 LCPL. Preliminary designers considered placing a portable ramp at the stern to launch the embarked landing craft via a marine railway, but this proved impractical. It also tried an elevator similar to a carrier's deck-edge unit, to launch an LCM, carrying it from the boat level (01 deck) into the flooded part of the well. It, too, had to be rejected as too complex.

These failures convinced BuShips' preliminary design section to modify the entirely new approach described above, to carry an LCU, 10 LCM(6), 16 LCVP, and 2 LCPL. This version had three holds, one between the machinery spaces (instead of ballast tanks) and three forward; the turntable was eliminated and the travel of the 3-ton cranes extended all the way to the wet well. Late in 1952 the army asked that all hatches be large enough to handle tanks, and that cargo booms be provided to facilitate pierside unloading when the ship was used for point-to-point supply. To allow for the large new hatches (18 × 24 ft) the deckhouse had to be moved back to near amidships and the two machinery spaces made adjacent between Nos. 2 and 3 holds (immediately below the deckhouse). A small deckhouse was left forward to support a superfiring twin 3-in/50, and another was placed aft in way of the davits for the LCVPs and LCPLs. LCMs on deck were placed athwartship rather than fore and aft as previously. Kingposts carrying 5-ton (later 10-ton) booms could work the hatches and also provided vehicle deck ventilation and exhaust, as in LST 1156. Because tanks might have to be handled, the 35-ton boat cranes were replaced by 50-ton cranes (later the 35-ton capacity was restored because it was deemed impossible to handle heavier loads by crane). The traveling cranes were eliminated as unnecessary. Some hold volume was lost because propeller shafts now passed through holds 3 and 4. In a later version, one machinery compartment was moved abaft holds 3 and 4, so the machinery was separated properly for survivability. This split-machinery version was selected for further development.

By this time the ship was 575 ft overall (560 ft on the waterline) × 85 ft × 23 ft, with two 272 × 50 ft vehicle decks (heights 18 and 13 ft). A preliminary design comparison sheet, completed before the design was complete, showed SCB 77A with substantially more capacity than SCB 77 (543,000 vs. 425,000 cu ft, 8,500 deadweight tons vs. 8,100), and the heaviest boom had been cut to 35 tons. The most striking comparison was in vehicle deck area with 8–12 ft headroom, 43,536 sq ft vs. 26,606 in the Mariner, which also had 6,576 sq ft of deck with less than 8 ft headroom. A C2 had about the same vehicle deck area as the Mariner, 25,746 sq ft with headroom 8–12 ft, and 6,248 sq ft with headroom up to 8 ft. More developed designs prepared in March 1953 showed vehicle loads of 752 tons on the upper deck and 361 tons on the lower deck, which compared to 158 tons on the upper deck and 141 tons on the mezzanine deck of the new LSD, although the LSD could carry another 2,101 tons in her well deck. Total cargo weight was 4,500 tons, compared to 2,400 for the LSD.

In January 1953 it was estimated that the ship would need 32,000 SHP to make the desired sustained speed (a sketch design showed 33,000 SHP). Possibilities included twin-screw versions of the steam plant then being installed in the new *Neosho*-class (AO 143) tanker and the diesel planned for an abortive new stores ship. The diesel was lighter but only one company made it, and it could not be stretched (e.g., to make up for added displacement) beyond the 16,000 SHP per shaft for which it was rated. In March the speed requirement was relaxed slightly, so the ship would have the AO 143 steam powerplant, 24,000 SHP.

In March 1953 the preliminary design section asked for cost estimates, both for a new SCB 77A and for a Mariner built as an AKA from the keel up. It also wanted the cost to convert an existing Mariner to a ro-ro. Estimated cost for FY 54 was

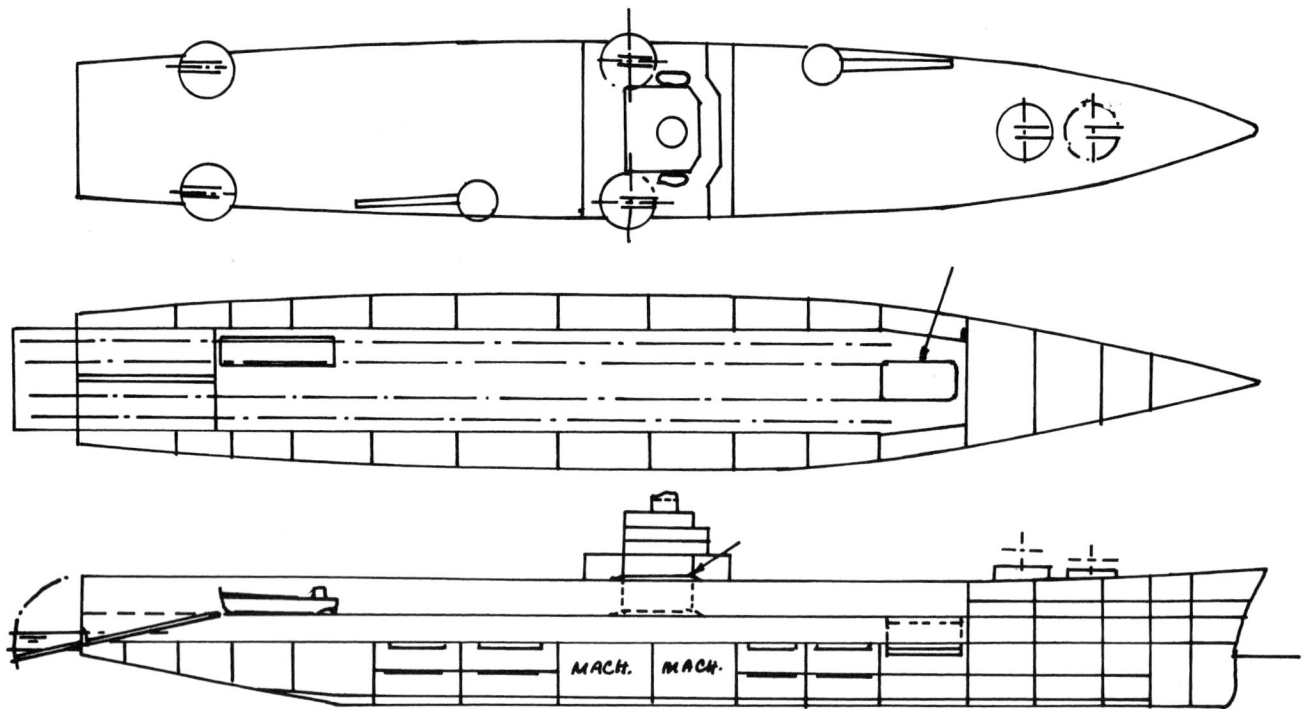

AKR Scheme A, July 1955. The AKR (R for roll-on/roll-off) was conceived by the BuShips Committee for Advanced Design Concepts; Schemes A and B were submitted as part of a 15 July 1955 memorandum. As sketched, the AKR was made the same size as the 1953 SCB 77A study to simplify comparisons. The idea was to speed delivery to the beach by launching fast combat-loaded boats, which were imagined as hydrofoils, from within a transport. This type of operation would eliminate the need for boats to load alongside. In Scheme A, a pair of marine railways (lines of rollers) would carry combat-loaded boats. Conventional craft, like the LCM(6)s shown, would ride in special cradles, but a boat could be designed to ride the railway without a cradle. Boats would be launched from a depressible platform that mated with the hydraulically operated stern ramp. Cradles could be jettisoned after launch, or they could be retrieved for re-use. The preliminary design branch estimated that 14 combat-loaded LCM(6)s carrying 400 tons of cargo could be launched in less than an hour. After they were launched, the ship would steam out to sea to unload other cargo. Wheeled vehicles would be driven down the stern ramp into LCMs or LCUs. Topside space was available for 10 more LCMs (total 24) and 16 LCVPs. They would be hoisted into the water in conventional fashion. Vehicles stowed on the main and fourth decks would move to the unloading deck by elevators (see arrows) connecting the main to the third (boat stowage) deck amidships and the fourth and third decks forward. For quick unloading, cargo would all have been palletized, to be handled by forklift and elevator for loading into vehicles and boats. The deck forward of the bridge is cleared for boat or vehicle stowage; that abaft the bridge, for boat or vehicle stowage and as a helicopter landing area. The two compartments amidships would have been machinery spaces, with the holds on either side showing hatches at two deck levels. The two cranes would have had 35-ton capacity, and the guns shown are twin 3-in/50s. Dimensions were 575 overall × 85 × 23 ft; displacement was 9,700 tons light and 18,000 tons fully loaded. The ship would have made 20 kts sustained on 24,000 SHP (as in an LSD), and endurance would have been 10,000 nm at 20 kts. Total cargo capacity would be 3,800 tons, 400 tons of it assault loaded. Cube was 500,000 cu ft; vehicle deck area (square) was 28,000 sq ft. By way of comparison, a Mariner AKA offered 425,000 cu ft. SCB 77A offered 500,000 cu ft and 23,000 sq ft. Estimated cost, in July 1955, was $33.6 million for the first ship and $26.3 million each for follow-ons, compared to $32.8 million for the lead SCB 77A and $25.5 million each for follow-ons.

$34.75 million (cut by May to $29.2 million), compared to $28.4 million for a keel-up Mariner. The difference was small, and in May 1953 OpNav planned to include an SCB 77A in the FY 54 program. Work was ordered stopped on 29 June 1953, however, and a Mariner conversion was substituted. Even so, SCB 77A was a step toward ships such as the later LHA (landing ship, helicopter, assault). This well-deck ship was attractive partly because it was becoming impossible to load modern tanks into boats alongside. Without a new type of cargo ship, they would have to go onboard LSTs and LSDs, although they were by far the least efficient amphibious weight lifters, devoting only about 10 percent of their displacements to their payloads.

The Maritime Administration, the successor to the Maritime Commission, was asked to handle the Mariner conversion, on the theory that upon mobilization it would have to convert other Mariners. Bethlehem Steel prepared contract plans, and Gibbs & Cox prepared working drawings. By the end of 1953, the Mariner program had achieved little commercial success. Of 35 ships, 20 were operating under MSTS charter; some would soon be

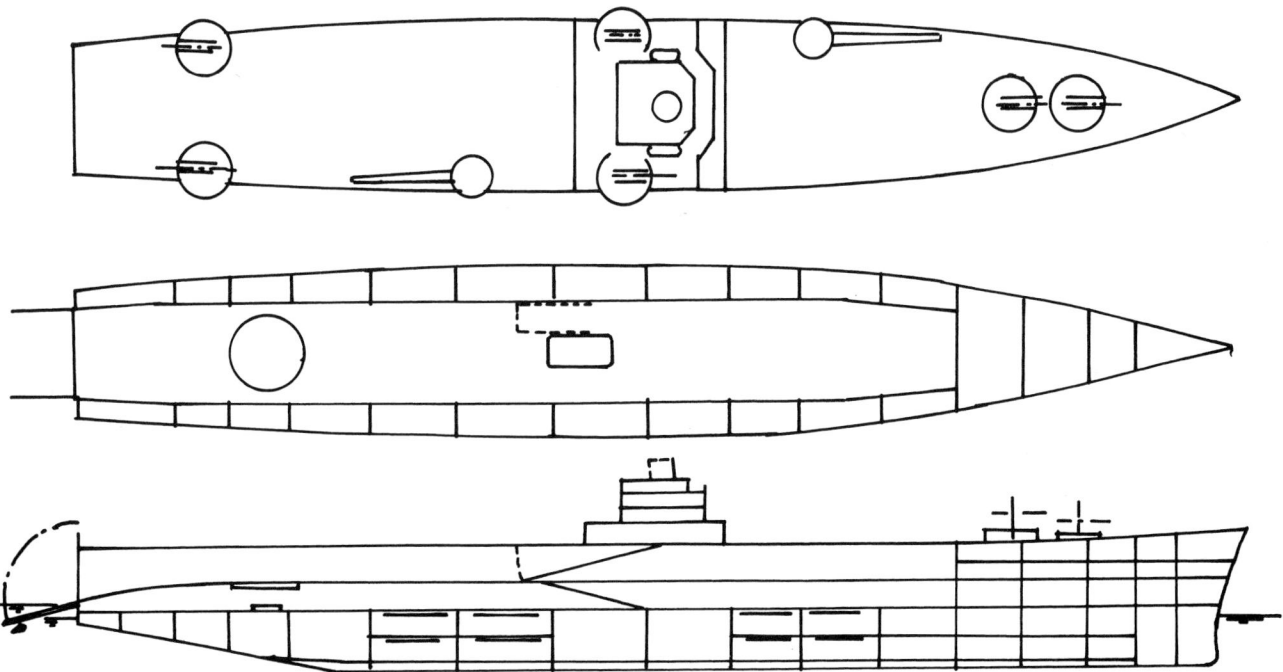

AKR Scheme B, July 1955. Scheme B differed from Scheme A in having ramps rather than elevators to connect her upper and vehicle stowage decks. She would have carried mainly LVTs. A portable portside ramp connected the third deck with the upper deck. A fixed centerline ramp connected the fourth deck (vehicles) with the third (LVT vehicles) deck. The boat complement was reduced to 10 LCM(6) and 16 LCVP. Scheme B had a turntable aft on the third deck. BuShips preliminary designers estimated that this type of ship could launch 35 LVTs from her third deck in about 30 minutes; because such vehicles were themselves cargo, in effect she would be offloading 1,200 tons of cargo in that short time. However, the LVTs would be carrying only 175 tons of cargo. Other vehicles would be driven down the ramp into the LCMs carried topside. Eliminating the rollers of Scheme A would simplify cargo handling. Dimensions for Scheme B matched those of Scheme A, but displacement was less: 8,800 tons light and 18,800 tons fully loaded. Performance would have matched that of Scheme A. Total cargo capacity would be 4,800 tons, of which 1,200 would be assault loaded. Cube would have been 525,000 cu ft, and vehicle parking area would have been 30,000 sq ft. The Scheme B design was also considered well suited to pure ro-ro operation, similar to Scheme C. In Scheme C the ship would carry no boats at all; it could unload 2,500 tons of vehicles onto a pier in less than 3 hrs. Total cargo capacity, for an unopposed landing (onto a pier) would be 5,100 tons, compared to 4,800 tons for SCB 77A (unopposed) or 4,550 tons for a Mariner (unopposed). On an opposed basis SCB 77A would carry 4,500 tons, and the Mariner 4,250 tons. Cube would not have changed from that of Scheme B, but vehicle parking area would have increased to 38,000 sq ft. Estimated cost of Scheme B in June 1955 was $31.8 million for the lead ship and $24.5 million each for follow-ons. The cost of Scheme C, with only four LCVP, was $30.9 million for the lead ship and $23.6 million each for follow-ons. BuShips preliminary designers much preferred the Scheme C design, with its unobstructed vehicle deck and its wing walls for buoyancy and accommodations, to the new MSTS ro-ros, which had side ports and numerous between-decks ramps, and thus sacrificed watertight integrity. Moreover, the MSTS ro-ros stowed their vehicles in holds, hence unloading involved much turning and backing. One way to use a ro-ro in an assault was to land a stable pier; in 1955 the U.S. Navy was experimenting with the deLong pier, a floating barge that could jack itself clear of the water on piles.

dropped due to the truce in Korea. Three were under charter to Pacific Far East Lines, one had been lost, and only three had been sold. That left eight incomplete, hence they were the only ones available for conversion. Plans for the AKA originally called for using the last ship under construction at Newport News, which would be delivered upon launching, with her machinery laid up. Later the second ship at Bethlehem San Francisco, *Evergreen Mariner*, was selected. Converted to *Tulare* (AKA 112) under the FY 54 program, she entered service in March 1956. The FY 56 program included a second ship, but the project had to be dropped due to lack of funds. Further Mariner AKA conversion was planned in FY 57 and again in FY 58, both times abortively.

Meanwhile, in December 1953 the existence of a large fleet of underused Mariners inspired the DCNO for logistics, Adm. Roscoe F. Good, to propose converting Mariners into combination AKA/APA for a full division of 20-kt lift, the required heavy equipment being transported by LSDs. He thought that these conversions should be conducted over the next two or three years. BuShips estimated, however, that it would take 28 combination AKA-APA plus 15 LSDs to lift a single reinforced division (30,248 men plus a 2,123-man naval beach group) with its equipment. There was some interest in

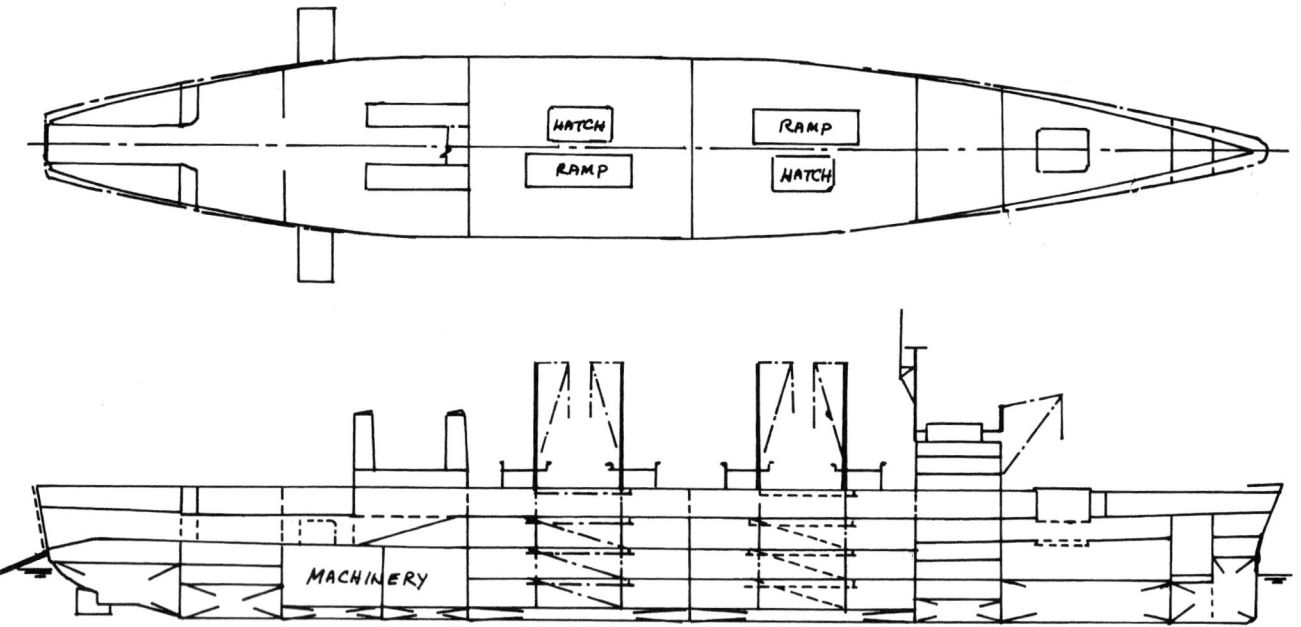

A BuShips preliminary design sketch of a pure ro-ro military cargo ship, 1955. Probably intended to contrast with the AKRs, she shows the massive internal ramps that would have made for poor survivability. Dashed lines aft indicate a pair of side cargo doors, the ramps for which are shown in the plan view.

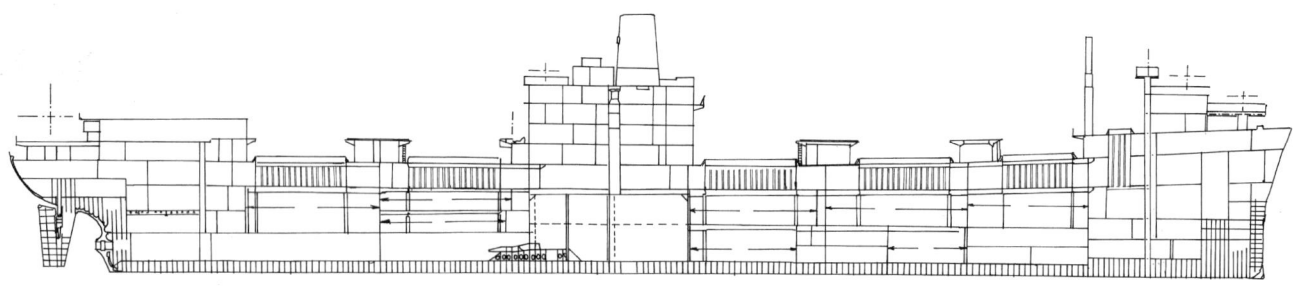

Tulare (AKA 112) inboard profile, October 1963. Arrows in the holds indicate stowage for fueled vehicles. She carried 9 LCM(6) athwartship on deck, 14 LCVP (6 in her two triple Welin davits, 7 nested in LCMs, and 3 athwartships forward in double-banked stowage), and 3 LCPL (1 nested in an LCM aft, 1 nested in an LCM just forward of the superstructure, and 1 in the double-banked stowage forward). The forward quadrupod supported two 60-ton and two 10-ton booms, the after one 60-ton, one 40-ton, and one 10-ton boom. The kingpost forward supported two 5-ton and two 10-ton booms. Armament amounted to six twin 3-in/50, paired side by side fore and aft, and on the after part of the superstructure.

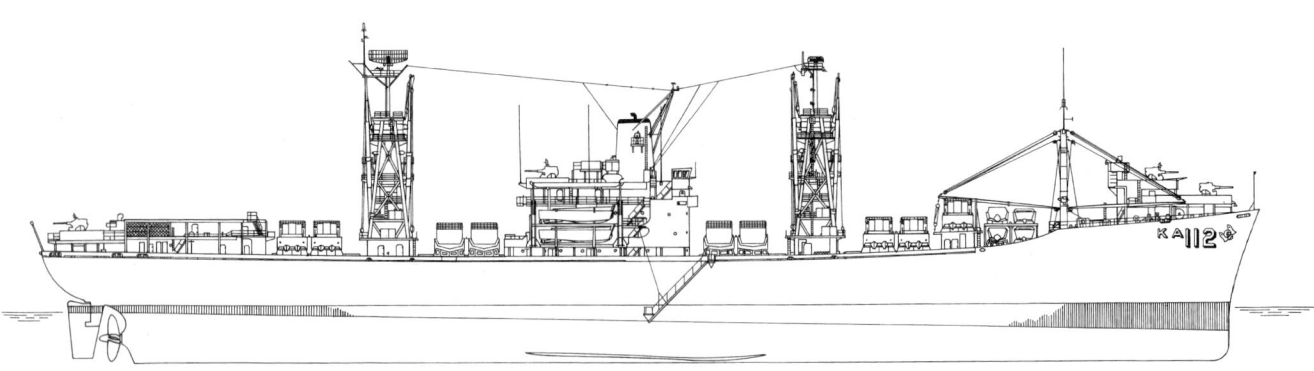

Tulare (AKA 112) in 1967. The big air search radar is SPS-12. *A. D. Baker*

The converted Mariner attack cargo ship *Tulare* (AKA 112) is shown en route to trials, 7 December 1955. She was the first U.S. attack transport with a helicopter platform. Her quadrupod masts carry standard contemporary radars: SPS-10 forward for surface search, SPS-6 aft for air search.

changing the AKA conversion to an AKA-APA, but nothing was done. Two ships were converted into APAs, however (discussed below).

Throughout the early postwar period, the argument against building new attack cargo ships was not only that there were so many in reserve, but also that others could be retrieved from the Maritime Administration. When the Korean War broke out, six ships (AKA 3, 4, 96, 97, 99, and 103) were reacquired from the Maritime Administration, under the FY 52 program. Another six (AKA 53, 54, 88, 92–94) were reacquired under the FY 62 program.

Two Mariners (C4-S-1As) were converted into the *Paul Revere* (APA 248) and *Francis Marion* (APA 249) under the FY 57 and FY 59 programs, respectively; the first of the type had been dropped in March 1955 from the FY 56 program, and the second from the FY 58 program. Like the Mariner AKA, they had a helicopter platform aft, served by a cargo elevator. Troop capacity was slightly greater than that of their World War II predecessors (91 officers, 1,561 enlisted); cargo capacity was 2,050 tons or 138,000 cu ft. Both were completed with amphibious squadron flag facilities. Boats were 7 LCM(6), 12 LCVP, and 3 LCPL.

These were the last World War II–style attack transports. All subsequent ships were helicopter ships or hybrid helicopter/well-deck ships. Traditional APAs remained important, however. Two *Bayfield*-class ships (APA 36 and 44) and four *Haskell* (Victory)–class ships (APA 212, 215, 222, and 237) were modernized under the FRAM II program. The last wartime ship was *Chilton* (later designated LPA 38), decommissioned in 1972. The Marines' triphibious ship, however, survived the SCB 77A dis-

The converted Mariner attack transport *Paul Revere* (APA 248) is shown, newly completed, at Todd Shipyard, San Pedro, 30 August 1958. This class was unique among U.S. attack transports because it had AKA-style quadrupod masts for handling LCM(6)s. Note the enclosed control station around the funnel.

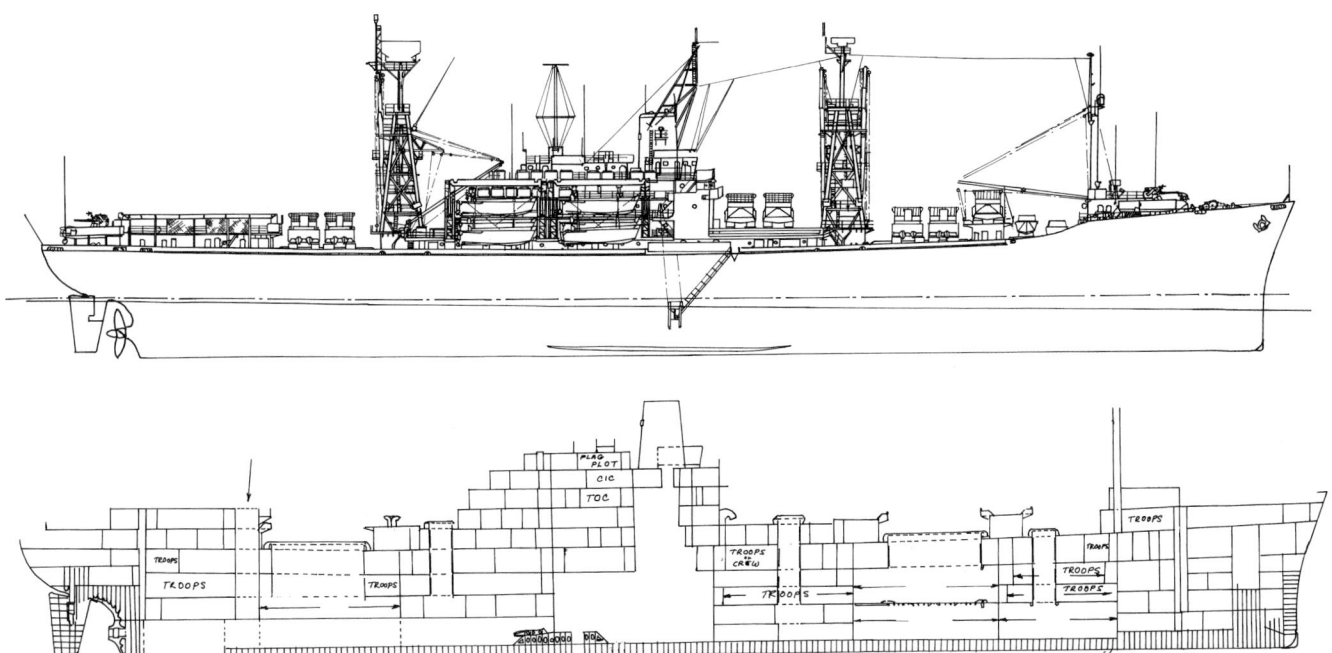

Paul Revere (APA 248), February 1967. She was a squadron flagship; the object at the foot of the funnel is a flag shelter. In the inboard profile, TOC on the level below the CIC indicates the troop operations center. The arrow indicates an elevator running up to the helicopter deck aft. Arrows in the holds indicate spaces for fueled vehicles. Comparison with the *Tulare* illustration shows just how much of the original Mariner hold space was devoted to troop berthing. Radars shown are SPS-10 and SPS-12 (aft). The forward quadrupod shows 60-ton booms on its forward side, a 30-ton boom on its after side, and 10-ton booms fore and aft. The kingpost forward of the mast shows an 8-ton boom; atop it are radomes of the WLR-1 ESM system. The quadrupod aft shows a 5-ton boom on its fore side and 10- and 30-ton booms on its after side. This mast also carried a 60-ton, 70 ft boom. The small radome atop the mast is an HF/DF (URD-4). Probably the most striking feature is the large number of communications antennas, not only the usual whips and wires but also a big broad-band discage atop the superstructure and short fat monopoles (URC) on the foremast and atop the funnel. Ship's boats amount to 12 LCVPs in four triple Welin davits (but only 10 according to the Booklet of General Plans) plus seven LCM(6) stowed athwartships and two LCPL forward, also athwartships. Capacity was 96 troop officers and 1557 enlisted troops.

aster, under the designation APA-M; APA-M is described in Chapter 12.

For about a decade AKA construction was abandoned in favor of well-deck ships. In the 1960s a much larger conventional AKA, the *Charleston* class (AKA 113–117), was developed. The first four were built under the FY 65 program. Construction beyond the one in FY 65 was dropped in favor of LHAs, which carried heavy cargo (see Chapter 12). These were the first AKAs specially designed for the role. Their holds had cargo elevators, which could reach any level, and cargo was moved by forklift. To handle the heavier new tanks, they had a pair of 78.4-ton booms (using the German Stuelken system), rigged from kingposts distinctively splayed out instead of the earlier quadrupods. There were also two 40-ton and eight 15-ton booms. They carried four LCM(8) lengthwise on skids at their sides, as well as five LCM(6) on hatches and on the main deck. Booms were arranged so that landing craft could be loaded simultaneously on both sides. The powerplant was an automated version of the earlier Mariner plant, with two top-fired boilers controlled by an enclosed central control station in the main machinery room. Similar automated propulsion was installed in some combat stores ships (AFS) and ammunition ships (AE).

The fast LSD, SCB 17, was larger than wartime LSDs, not only because of its higher speed and more powerful gun battery (planned as 3-in/70s, like those on board the abortive AKA/APA), but also because it had to accommodate 21 LCM(6)s. Above that, the designers provided 10 percent for expansion to match envisaged growth in landing craft and in transport helicopters. The LSD had a portable superdeck, parts of which could be transported aboard LCTs, to accommodate the maximum possible number of LVT amphibians, which would have to be moved to and from the well deck. Like the fast AKA/APA/AGC, the 20-kt LSD hull could also be used for another purpose, in this case as a tender for the very large seaplanes under development after World War II (some of which had important amphibious roles).

USS *Charleston* (AKA 113, then LKA 113) was lead ship of the final class of attack transports. The big craft carried on her sides are LCM(8)s; an LCVP nests in an LCM(6) carried athwartships. The big splayed kingposts are for the German Stuelken heavy-lift boom system. Note that only the bow guns have weather shields. This photograph was taken on 18 November 1968.
Newport News Shipbuilding and Dry Dock Company

USS *El Paso* (LKA 117), newly completed, is shown on 25 November 1969. *Newport News Shipbuilding and Dry Dock Company*

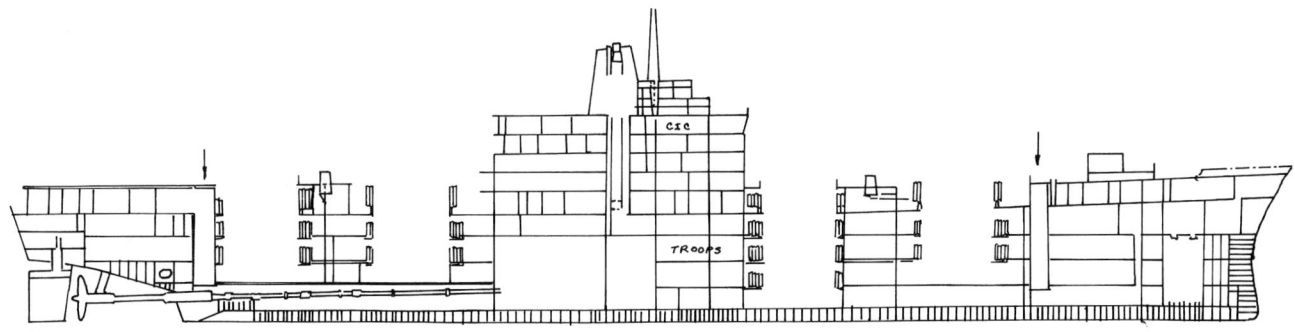

Charleston (AKA 113), May 1990. She was the lead ship of the only class of attack cargo ships to be built from the keel up as such. Each of four holds was served by 15-ton booms, but each pair was also served by 70-ton Stuelken booms. Nos. 1 and 3 holds, where landing craft were stowed, were also served by 40-ton booms. Rigging was arranged for easier set up, and hook speeds (i.e., unloading rates) were much higher than in the past. The ship could be unloaded simultaneously from port and starboard. Arrows indicate the cargo elevators, a new feature of this design. They were intended for breaking out palletized cargo, troop ammunition, and drummed fuel, which could be handled by forklift to main deck stations or to the helicopter platform without having to be broken out in sequence from deep holds. Fore and aft access was limited to the main deck, and sufficient space outboard of the deckhouse and of the hatches was provided for the forklifts. In past designs, access would have been difficult until stowed boats were launched. Vehicle holds accounted for 78 percent of the ship's cargo space; holds for vehicles have special covers that can be partially opened. The boat allowance was four LCM(8), five LCM(6), seven LCVP, and two LCPL.

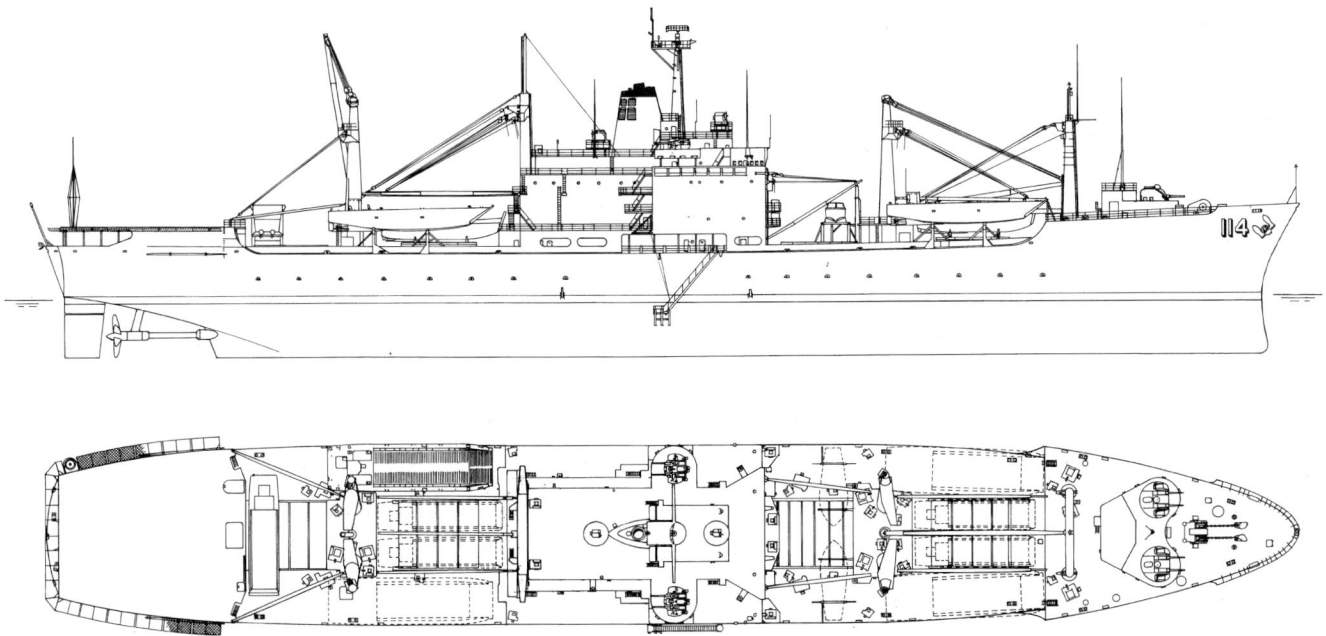

Durham (LKA 114) in July 1969 with all four 3-in twin AA mounts aboard. In the plan view, all but one of the LCM(8) landing craft are shown as dotted lines so that the deck details can be seen. *A. D. Baker*

Estimated dimensions, in November 1947, were 550 × 80 × 20 ft loaded (12,000 tons full load). Estimated complement, including a repair party for landing craft, was 25 officers and 350 enlisted men, to which were added 5 officers and 105 enlisted men to operate the embarked landing craft and vehicle crews (25 officers and 325 enlisted men). Troop berths were separated from crew berths, and located so they were not traversed when not in use. The ship would carry bulk fuel for her embarked craft and amphibians: 25,000 gal of diesel oil and 10,000 gal of gasoline (for LVTs). Boats were two LCVP and two LCPL or LCPR in davits, plus two lifeboats.

As of 1948, a class of eight was planned, the first to be authorized in FY 50. Like the fast AKA/APA, the LSD project stalled because money was short. It was revived as SCB 75 after the Korean War broke out. In November 1950 the SCB requested studies of the impacts of various features. In December, BuShips reported three alternatives:

1. A 19-kt ship (490 × 80 × 17¼ ft, 7,150 tons standard or 10,650 tons fully loaded, compared to 5,675

tons standard or 9,200 fully loaded for a wartime LSD) built around a slightly wider well deck than that of the existing LSD, but with heavier armament (two 5-in/38 and four twin 3-in/50), greater endurance, and a better bow form. It could carry 17 LCM(6) (vs. 12 for a wartime LSD) or, like a wartime LSD, 1 LCT and 37 LVT or 3 LCT and 14 LVT. This LSD could accommodate 9 LCM(8), compared to 5 for a wartime LSD. To achieve a trial speed of 21.5 kts (19 kts sustained), this design would need 20,000 SHP (as in the AKA/APA), compared to 7,000 SHP in a wartime LSD. Endurance would be 9,000 nm at 16.5 kts, compared to 9,000 nm at 12.5 kts in a wartime LSD.

2. A 20-kt ship (500 × 80 × 17½ ft, 7,350 tons standard and 10,850 tons fully loaded) with similar characteristics. An output of 24,000 SHP would give a trial speed of 22.5 kts. Endurance would match that of the smaller ship. This ship would cost 6 percent more than the smaller ship, and about 56 percent more than a wartime LSD.

3. An updated SCB 17 (550 × 86 × 17¼ ft, 8,425 tons standard or 12,650 tons fully loaded), which could accommodate 21 LCM(6), 11 LCM(8), 1 LCT and 47 LVT, or 4 LCT and 18 LVT. It would be more survivable than the others and better able to maintain its speed in a seaway, using the same powerplant as the 20-kt ship, for the same trial speed; it was also 11 percent more expensive.

BuShips favored the smallest satisfactory ship, because it would be far easier to produce in quantity. In January 1951 the SCB chose the compromise 20-kt ship. Minimum acceptable capacity was three loaded LSU (landing ship, utility; ex-LCT), later redesignated LCU (landing craft, utility), with a gate across the well deck located so an LCU could be docked and undocked while equipment remained dry in the forward part of the well. A portable mezzanine deck would add as much vehicle stowage (16 trucks) as the superdeck. Clear height from the well-deck bottom to the superdeck, initially set at 29 ft, was reduced to 26 ft to hold down the size of the ship. The ship would be used as an emergency drydock for amphibious craft, so the design had to anticipate likely growth in the LSU and in the LCM. Given the potential for Arctic operations, the ship was to be able to operate in cold temperatures down to 0°F; ships' bows were strengthened for ice work.

The machinery was placed under the well deck rather than in the wing walls for better protection. BuShips adopted a unit machinery arrangement (one boiler and one geared turbine per shaft), each boiler being capable of supplying 150 percent of the required steam per shaft. The machinery spaces were separated by 32 ft of ballast tankage. Compared to the earlier wartime LSD, this one had a smoother hull form.

Gibbs & Cox, designers for the Korean War LSTs, received a contract for the design. Their ship was slightly beamier than what BuShips had proposed (500 [WL] × 82 [WL] × 17.1 ft fully loaded, 10,800 tons) and was expected to make 21.3 kts on trial on 24,000 SHP. At full load it could lift 1,750 tons of cargo, including temporary decks; there was also a maximum load condition, with a 2,700-ton cargo lift.

The well deck had to be floodable to 10-ft depth to bring an LSM on board. That took 12,000 tons of ballast. More ballast required more pumping capacity, in this case four pumps, each with more than three times the capacity (12,500 vs. 4,000 gal per minute) of wartime pumps. They could deballast the ship in about 60 minutes, which was much faster than for the earlier LSD. On the other hand, available space had to be split between ballast and pumps, so there was a limit to pump capacity. Ballast so filled the ship's underwater volume that in the initial design 5-in ammunition was stowed above the normal waterline. Overall, compared to that of the earlier LSD, the cargo well had a greater depth of water and an improved gate.

The big open well cut away what in a conventional ship would have been much of the strength deck. Portable decks could not contribute to the ship's strength. In the new LSD, the strength deck forward was carried up from the weather deck to the 01 level in the deckhouse structure, which was over the forward end of the well deck. The deeper the structure at that point, the stronger.

Compared to the wartime LSD, the new one could carry about as many boats, and it could carry more LVTs (48 vs. 35). It carried about twice as many marines (29 officers and 312 enlisted vs. 18 officers and 150 enlisted) and somewhat more cargo (2,400 vs. 2,000 tons, 400,000 vs. 250,000 cu ft). Complement was comparable, 21 officers and 303 enlisted vs. 18 officers and 300 enlisted (in the 1960s). Endurance was similar, 13,000 nm at 10 kts vs. 13,800 nm at 9 kts.

The ship was more versatile than her predecessors. She could take a portable superdeck and a mezzanine deck over her 50 × 396 ft well deck; loads were 158 and 140 tons, respectively, the superdeck load being equivalent to 18 combat loaded 2½-ton 6 × 4 trucks. A helicopter platform on deck presaged the later shift to the LPD (landing ship, personnel, dock) concept (described in Chapter 12). Cargo was handled by two 50-ton cranes (60-ton cranes were initially rejected in favor of 35-tonners on the

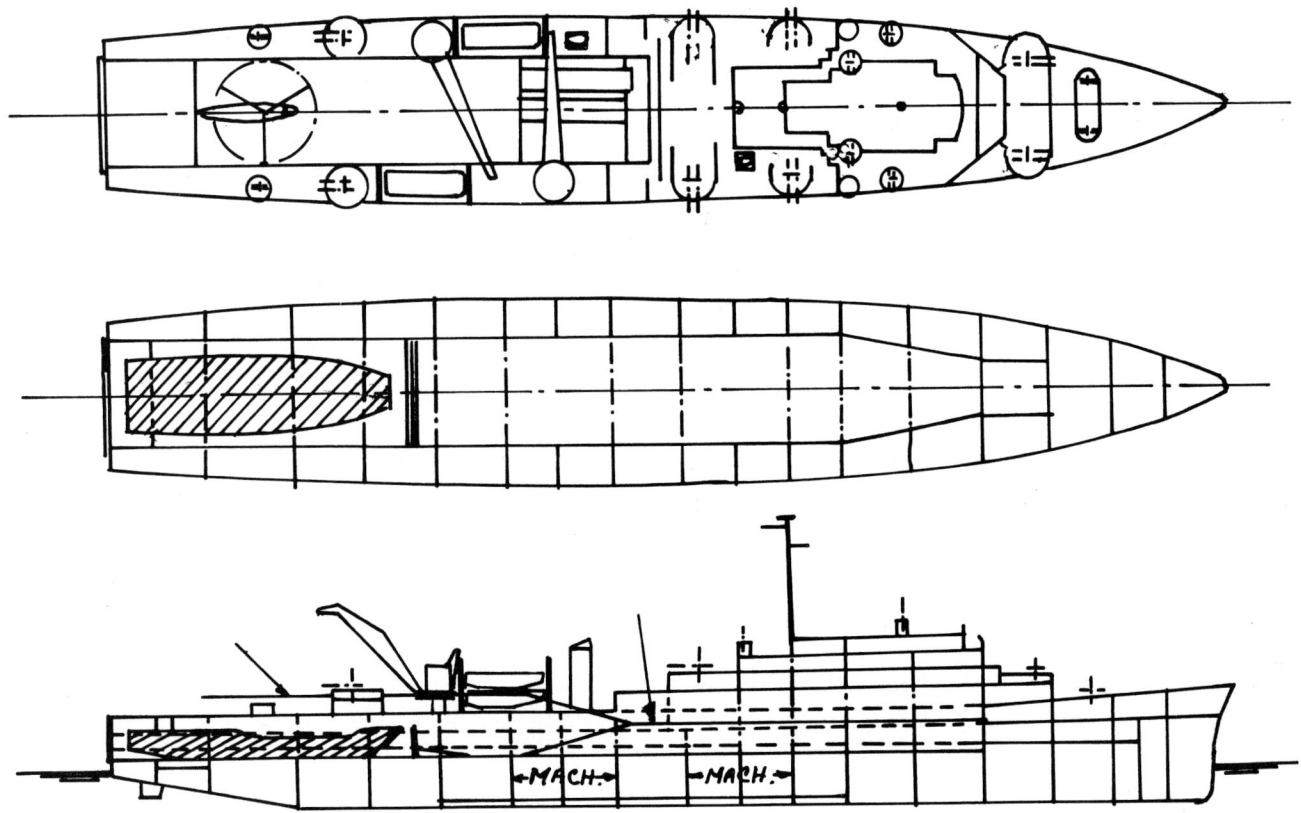

A preliminary design sketch of the *Thomaston* (LSD 28)–class design, about 1953. The arrow shows the temporary mezzanine deck installed. The helicopter deck was also portable; when it was removed, the ship could carry a variety of service craft up to harbor tugs. The water barrier made it possible to carry more vehicles while retaining some landing craft capacity.

ground that they were slower and more expensive). A 7½-ton bridge crane served the covered part of the well deck. As a cost saver, air conditioning was omitted; it was restored in the last two ships of the class, however.

Eight ships of this *Thomaston* (LSD 28) class were built: LSD 28–31 under the FY 52 program, LSD 32 and 33 under the FY 54 program, and LSD 34 and 35 under the FY 55 program. Three ships were dropped from the FY 53 program (in the planning stage) to help finance the carrier *Saratoga* (CV 60). LSDs were built before AKAs or APAs because they were so versatile; in November 1948 the amphibious arm of OpNav, Op-343, pointed out that there had never been enough during World War II. During the Korean War, all available LSDs were activated. By 1952, LSDs were being justified for the FY 55 program not only for their amphibious capabilities but also for their ability to tend and transport small minesweeping boats.

Conversely, as the fast new LSDs entered service, earlier ones became available for conversion to other purposes. For a time it seemed that they would become tenders for the new jet seaplane bomber, the P6M Seamaster; conversion of LSD 1 into AV 21 began but was stopped when the bomber was canceled. Another conversion was canceled before it began. LSD 4 was reclassified as a mine countermeasures craft mother ship (MCS 7) in 1962.

In FY 59 the navy began buying a modified welldeck ship, the LPD. No more LSDs were programmed until the mid-1960s, even though the 22 wartime-built units were wearing out. The first pair reached overage as of 30 June 1957. Long-range planners in 1955 wanted two new LSDs in the FY 57 program. In the fall of 1958 two were earmarked for transfer to Taiwan, so planners considered adding two new LSDs to the FY 61 program (in fact only one was lent). Funds were insufficient.

To stave off replacement, LSDs were modernized under the FRAM II program. A 19 December 1958 program summary showed 15 ships: 3 in FY 60, 5 in FY 61, 4 in FY 62, and 3 in FY 73.

Ultimately the World War II LSDs did have to be replaced. Five improved *Thomaston*s were built as the *Anchorage* (LSD 36) class: one in FY 65, three in FY 66, and one in FY 67. The new LSDs were built under SCB projects 404.65 (LSD 36) and SCB 404.66 (the other four). Compared to the earlier class, they were considerably lengthened, from 510 to 553 ft.

The new 20-kt *Monticello* (LSD 35) is shown off her builder's yard, 13 March 1957. Compared to wartime LSDs, she is more heavily armed, with eight twin 3-in/50 guns as well as a pair of twin 20-mm in the bows and another atop the bridge. Such weapons required considerable manpower, and the trend of the 1960s and beyond was to eliminate them. The battery was cut to six twin mounts during the 1960s and to only three during the 1970s (two forward of the bridge, one abaft the bridge). The ships of this class were run hard, and only two were deemed worth transferring to a foreign navy, Brazil. *Ingalls Shipbuilding Corporation*

The improved *Thomaston*-class *Pensacola* (LSD 38) is shown off Quincy, Massachusetts, where she was built, 25 February 1971. She has only half the armament of the *Thomaston* class, designed two decades earlier. Note the large opening forward of the superdeck, allowing the ship's cranes access to her well deck. This ship was decommissioned and transferred to Taiwan on 22 September 1999, after nearly three decades of service.

Mount Vernon (LSD 39) is shown on 5 April 1972.

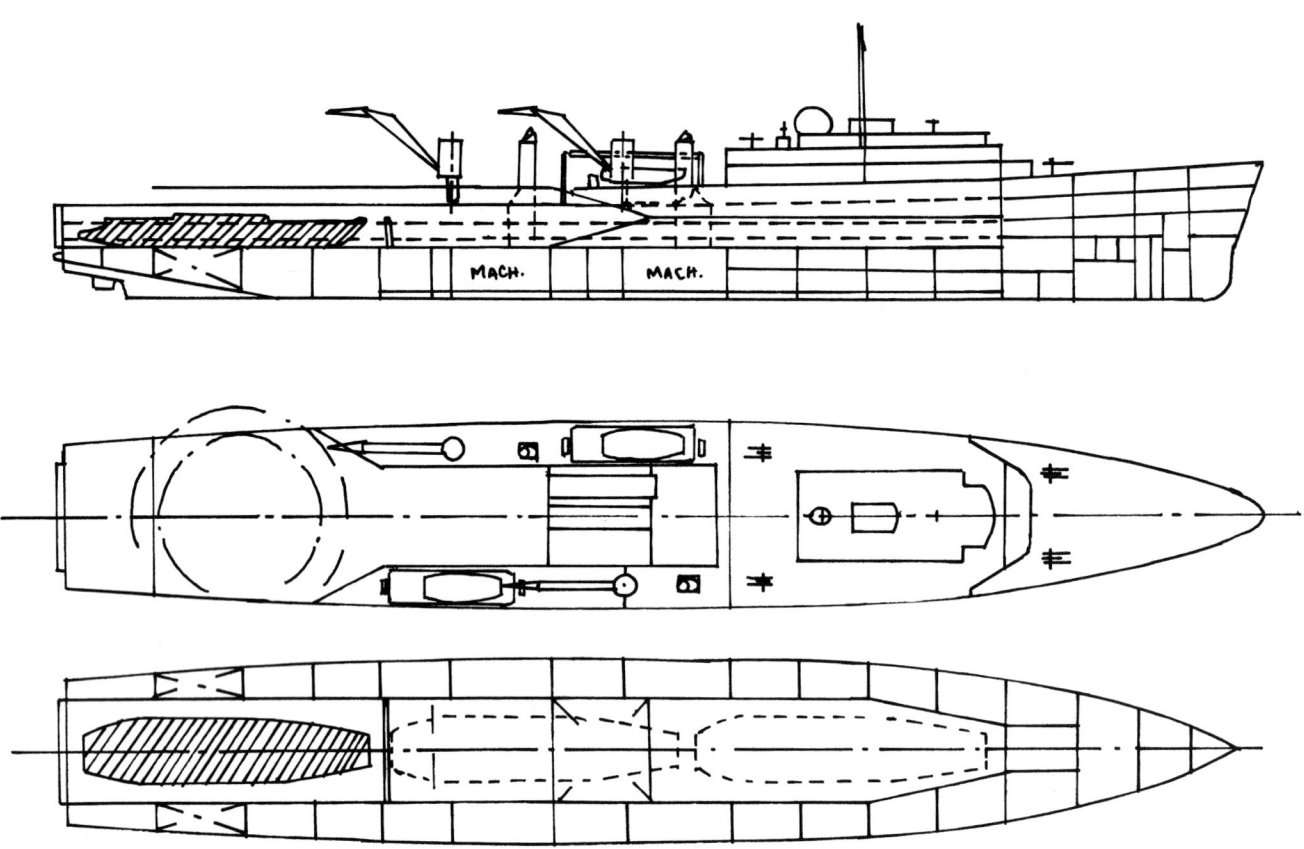

A preliminary design sketch of the SCB 404.65 (*Anchorage*, LSD 36, class) LSD, about 1964. Note how the well deck could accommodate three LCU 1610s. This ship was lengthened 40 ft largely because the previous *Thomaston* class could not easily accommodate the new fast LCU; it had to be angled in the well deck to fit. Added length also added troop and cargo capacity. The 01 level was extended fore and aft for better habitability. Vertical conveyors improved cargo handling. The pilothouse was raised half a deck to improve visibility and also to improve ventilation of the forward part of the well deck; a plenum chamber occupied the half-deck void under the pilothouse. The two LCUs outlined in dashes are forward of the erectable water barrier. Boats on deck are LCM(6)s with LCPLs nested in them. The circles aft are the safety circles for a CH-53 helicopter on the ship's flight deck. The dome on the bridge is Miser (a microwave satellite relay), a projected satellite communications dish, not ultimately installed on board these ships. Crane capacity shown on the sketch was 50 tons. Dimensions of the preliminary design were 555 ft (overall) or 540 ft (waterline) × 84 ft (extreme) × 18 ft 6 in (fully loaded); full load displacement was 13,650 tons. Speed was 20 kts (24,000 SHP) as in LSD 28, and complement was 51 officers, 33 CPOs, and 709 enlisted. Armament shown is four twin 3-in/50.

The well deck was lengthened from 391 to 430 ft and it was widened by 2 ft. Light displacement increased from 6,880 to 8,600 tons in LSD 36 (8,100 tons in the others). Other changes included a tripod mast to support the new SPS-40 air search radar. The next class of LSDs, associated with the new LCAC (landing craft, air cushion), is described in Chapter 16.

The new series of fast amphibious ships inevitably included a 20-kt LST. Work began in August 1947. Such a fast LST was considered so valuable that by 1948 plans called for substituting it, if it succeeded, for six of the eight planned fast LSDs. Fast LSDs carrying landing craft were the obvious alternative to fast LSTs, if the fast LST design proved impractical. Requirements were the usual 20 kts sustained speed and 10,000 nm at 20 kts steaming radius. The ship would carry 75-ton tanks on a 300-ft tank deck (30 ft wide, with 14 ft clear overhead), with 25-ton vehicles above on the main deck, plus troops (20 officers, 410 enlisted men). Compared to the other fast amphibious ships, armament would be quite limited (two single 3-in/50 plus pedestal-mounted machine guns).

The preliminary designers soon realized that it was hopeless to try to drive a blunt bow at high speed. Much of the usual resistance could be attributed to the very full hull, with its high prismatic coefficient (0.820). Bow doors could be a very small part of a much longer, beamier hull with more rounded bilges (beyond the doors the bilges could sweep up more steeply). This more conventional approach was rejected. Instead, the designers chose a radical alternative: the ship would have a conventional bow, with the usual blunt end with doors placed at the stern. Approaching the beach, she would turn around and, in effect, back in. The propellers could not be in their normal position at the

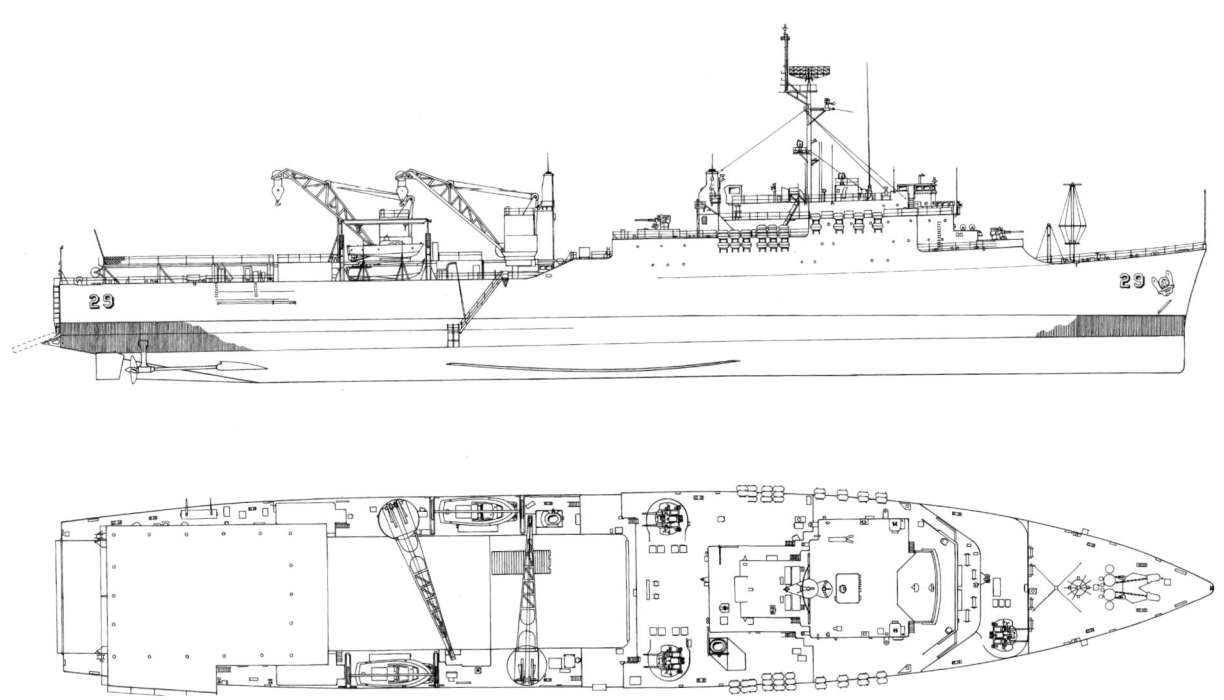

Plymouth Rock (LSD 2) in November 1982, with her portable deck sections installed and her 3-in battery reduced (note that only one of two forward mounts remains). Ships were completed with a total of eight twin 3-in/50, including two abaft the heavy cranes and two forward of the two after mounts, which survived. The ship's bridge has been enclosed and a discage radio antenna added right forward. The big air search radar is SPS-6C. *A. D. Baker*

stern because they would have buried themselves in the beach. Instead, they were placed at the bow, as tractors rather than the usual pushers. That still left the problem of a rudder. In the final design the rudder was placed near the doors, and retracted for landing.

The designers tried various combinations of dimensions at the design (trial) speed of 22 kts, in each case maintaining the usual 2 percent keel slope and a landing draft of 3 ft. Lengths over 500 ft required less power but made for an unwieldy ship. Beams over 70 ft offered little advantage in reducing required power. By August 1948, two sets of dimensions had been chosen for further study: 500 × 65 × 14 ft (7,610 tons, 22,500 SHP) and 470 × 70 × 13.7 ft (7,670 tons, 26,700 SHP). The former was chosen. With 3 ft draft at the stern and 13 ft at the bow, the ship would land at 4,500 tons. Steaming radius would be 7,000 nm at 15 kts, far short of that demanded of the other fast amphibious ships. The bow-propelled ship would be a much better seakeeper than existing LSTs, but it was likely to have steering problems, and the propellers would be relatively vulnerable.

Design work on the fast LST stopped in 1948, and during the Korean War (as described above) much slower LSTs were designed and built. The desire for a really fast LST survived, however, and in October 1953 the stern-driven 500-footer was shown to the SCB as an alternative to the 15-kt LST, which became LST 1171. The SCB showed little interest. However, the stern-first solution was applied, albeit in drastically altered form, to the smaller LCU design (see Chapter 13).

About this time the naval architectural firm of George G. Sharp offered what it called a "carport" LST, essentially a barge with a slot aft for a specially designed towboat. Sharp claimed that its broader hull form (with a bow door) could be driven at the required high speed. It also offered a conventional single-body version. BuShips analyzed and rejected both. Sharp had improved the ship's hull form partly by raising the hinge line of the ramp, but that might be unacceptable for a beached LST. Model tests suggested, moreover, that the advantages of the broad-beam (low prismatic) hull had been overstated. BuShips preferred the 17-kt design it was then pursuing. Another possibility, which was not pursued very far, was simply to moor the ship parallel to the beach, far enough out to solve the beaching draft problem, disgorging vehicles via a side opening onto a long causeway.

In 1954 the commander in chief of the Atlantic Fleet revived the fast LST concept, recommending that enough be bought to transport an entire Marine division in a single lift. His amphibious command (PhibLant) disagreed. It thought the future belonged to high speed forces; the minimum acceptable should be 25 kts. Although a very fast LST (20–25 kts or even 25–27 kts) could certainly be designed, it would not be cheap, nor could it easily be mass produced. For example, if it had to be steam powered for sufficient power, there were limits on the size of gears that could be mass produced during mobilization. PhibLant revived a July 1950 proposal to abandon LSTs in favor of LSDs. It preferred the well-deck SCB 77A AKA (described above), sometimes called an AKA-LSD, to LST 1171, which cost so much for only a 2-kt advantage over LST 1156. An APA-LSD could be designed on the same lines.

LSD-type ships, however, would not be able to put nearly as much cargo over the beach early in an assault. For the moment, the LST was not quite dead. In 1954 a single fast LST was included in the tentative FY 57 program. On this basis, in July 1955 the DCNO for logistics included two (later cut to one) in the FY 58 program; eight slow ones originally planned had just been dropped in favor of helicopter carriers.

The BuShips designers looked again at more conventional hulls. If they accepted greater draft for-

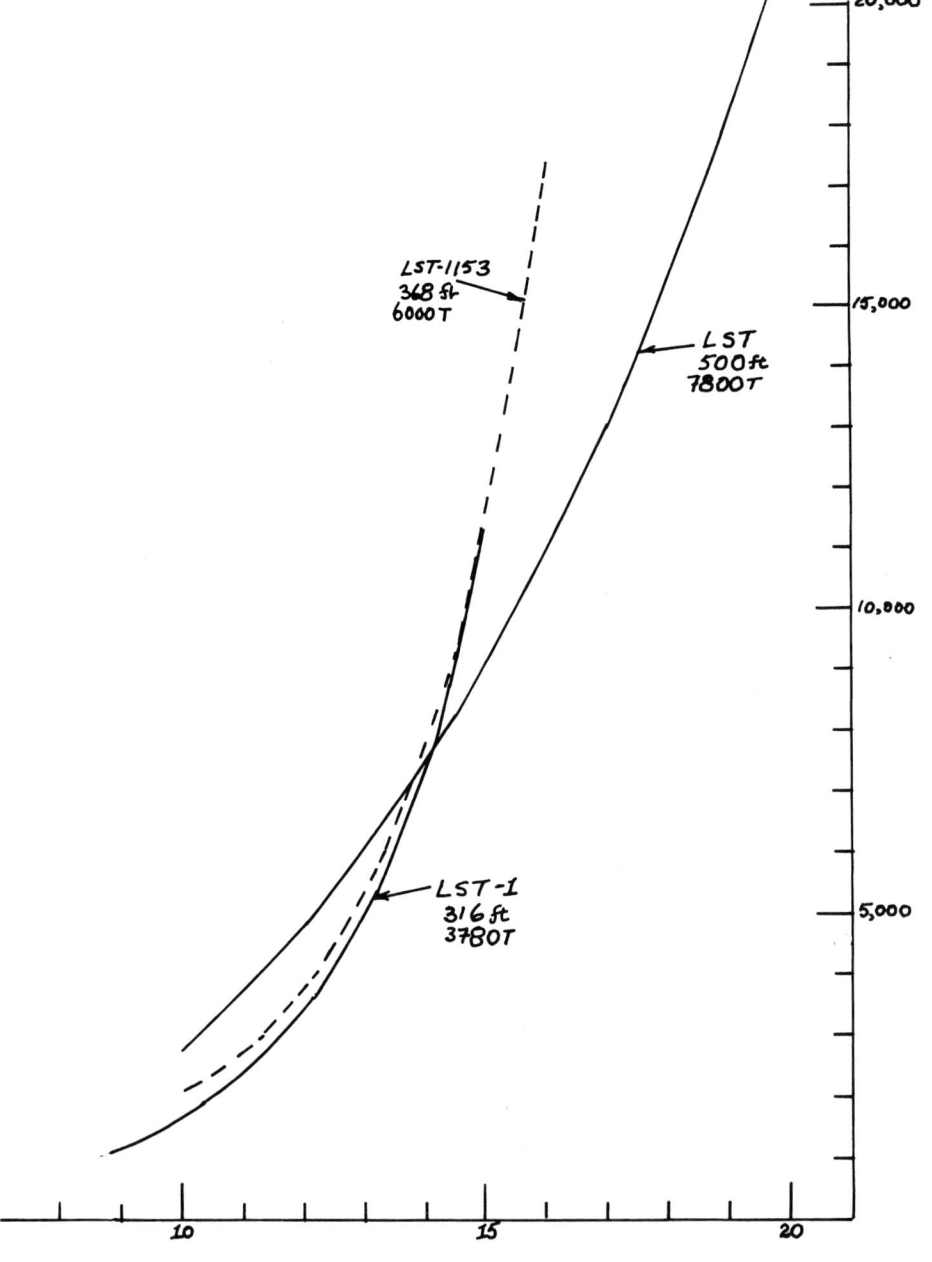

ward they could have a better hull form. In that case the LST would beach further out, so it would need a longer ramp. That would complicate the design, but it might be far better than the bow-propelled ship conceived in 1947–48. Thus the U.S. designers reverted to much the same solution the British had adopted in their unfortunate Winette 15 years earlier. There is no indication that the BuShips designers were aware of that history, however.

By early 1955 the designers thought that a 446 × 62 × 6 ft ship with a landing displacement of about 4,410 tons could be driven at 20 kts by diesels totaling 16,000 BHP. Once beached, the ship would launch a 350-ft causeway, which would float out and sink to support vehicles driving ashore. While on board the ship, the causeway could be filled with fuel oil. This ship could accommodate the usual 500 tons of cargo in beaching condition, or 1,000 tons when fully loaded, and it could also accommodate the 400 troops desired in new LSTs. Endurance, 10,000 nm at 20 kts, would match that desired in other fast amphibious ships.

On 22 July 1955 the SCB issued preliminary characteristics (SCB 152) for such a ship. As developed through 1956, it grew slightly, to 477 × 62 × 6 ft (15 ft fully loaded), displacing 4,705 tons in landing condition (7,200 tons fully loaded). This hull could be driven at 21 kts by 17,000 SHP machinery. Like contemporary LSTs, this one would carry 75-ton tanks and 25-ton vehicles, using a 270 × 32 ft tank deck (with a turntable at its forward end). For dry cargo, she would have a 15 × 30 ft hatch on the upper deck. Like other LSTs, she would carry pontoons on her sides. Armament would be three twin 3-in/50 and six twin 20-mm, as in other major amphibious ships.

Trial speed would be 21.5 kts (20 kts sustained). All control, living, berthing, and messing spaces would be air conditioned.

Between 1956 and 1959 BuShips briefed this causeway, or causeway-regurgitating, LST alongside a submarine LST, as part of a package of advanced ship concepts. There was also some interest in a false-bow LST, which would raise its seagoing bow to expose a blunt LST bow before beaching. It seems not to have been developed in any detail; although many current short-sea ferries use much this configuration, they do not have to beach.

There was also a catamaran or twin-hull LST. It offered better seakeeping and better behavior in surf. As drawn in 1960, it was 550 × 106 × 16 ft (mean draft), with a landing displacement of 12,500 tons (14,000 tons fully loaded). It could be driven at 20 kts on 28,000 SHP (endurance 6,000 nm at 20 kts). It could carry 3,000 tons of cargo in beaching condition, compared to 500 tons for a monohull. It seems to have been rejected as too exotic; no one had yet built a catamaran that large, so no one knew what stresses it might encounter at sea.

Thus the only acceptable choice was a more or less conventional hull with some kind of long ramp, to make up for excessive bow draft. BuShips presented a feasibility study on 2 April 1958. A 450 × 66 × 15.5 ft ship (4,600 tons in landing condition, with 6 ft bow draft; 7,540 tons fully loaded) would achieve 22 kts on 20,000 SHP; endurance would be 10,000 nm at 20 kts. Landing cargo capacity would be the usual 500 tons (2,850 tons in ocean condition).

In March 1960, however, the CNO suggested a less expensive LST combining the best features of the 1156 and 1171 classes. He set 16 kts as a mini-

The logic of the fast LST: BuShips' preliminary design curves of horsepower required to reach high speeds, 1958. A 500-ft 7,800-tonner is compared with the World War II LST and its 1945 successor. Note that the added length in LST 1153 balances off her considerably greater displacement, so that her speed-power curve very nearly matches that of the original LST. To reach 16 kts, LST 1153 would have required 17,200 SHP, a prohibitive figure. At that power the 500-footer would have made nearly 19 kts. The graph shows designed trial SHP required to make various sustained sea speeds. The curves are for machinery operating at 80 percent of trial power, allowing 25 percent for fouling and weather, factors that increase the rated power required for a given speed by 56 percent. The 500-ft LST would have had a 22,500 SHP steam plant, slightly less powerful than that used later for the 20-kt LSD. According to a 4 February 1949 memo, model tests showed that at 7,800 tons it would take 30,100 SHP to drive the ship at 24 kts, and 34,400 SHP to drive her at 25 kts; at the landing displacement of 4,500 tons the ship would make 24 kts on 19,100 SHP and 25 kts on 22,100 SHP, both below her rated power. On the other hand, according to a table in the design file, a 500-ft LST of conventional form (8,900 tons) might need as much as 47,000 SHP to reach 22 kts. A 400 × 80 ft LST (8,000 tons) would need 60,000 SHP. Another table showed that the new LST would cost four and a half times as much as a World War II LST per ton of landing load. Another graph showed just how rapidly the cost per ton of landing lift rose with sea speed. Moreover, the big steam LST had much less seagoing range than her smaller World War II predecessor, 3,700 nm at 20 kts (the later standard for amphibious ships was 10,000 nm at that speed). According to a 3 March 1950 table in preliminary design files, the fast LST would have carried 2,300 tons of cargo in seagoing condition, compared to 1,850 tons for a World War II LST. Seagoing displacement would have been 7,800 tons, but landing displacement would have been only 4,500 tons, so that endurance in landing condition would have been only 90 nm at 20 kts (compared to 222 nm at 10 kts for the World War II LST). Endurance in seagoing condition would have matched that of the World War II ship, 10,000 nm at 10 kts (but only 4,200 nm at 20 kts, up from 1948 figures). Compared to the existing steam LST 1153, the new ship would have had almost four times as much power, for a dry machinery weight of 670 rather than 267 tons. Total oil fuel capacity would have been 1,100 rather than 922 tons.

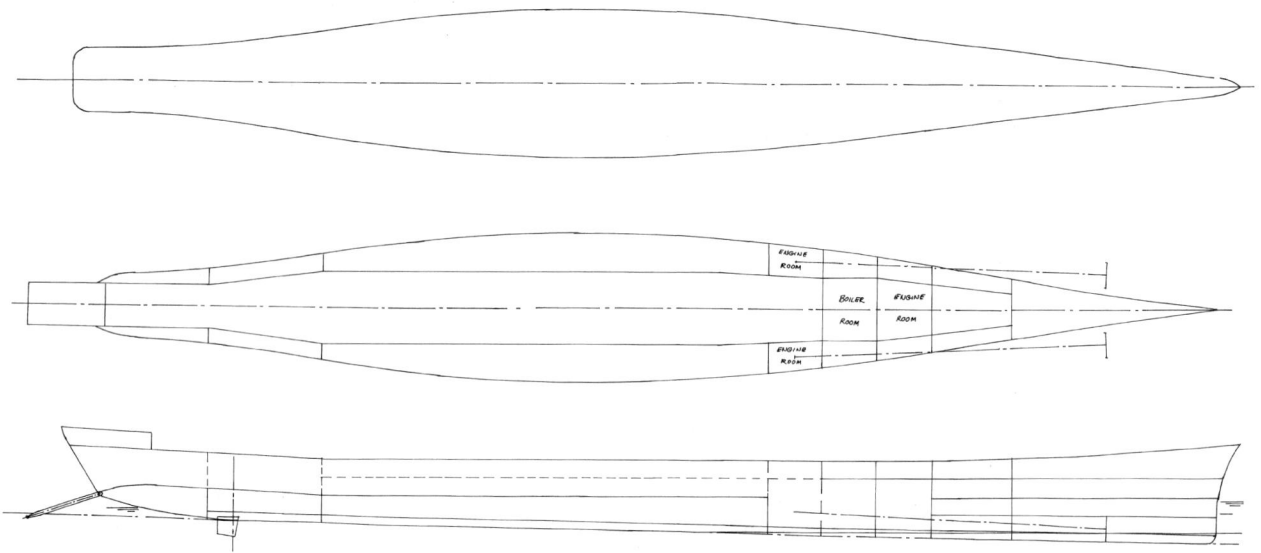

A BuShips preliminary design branch sketch of a fast LST with propellers in her bow, August 1948. No full spring–style drawing was prepared, only this hull sketch and an accompanying structural cross-section (dated 4 August 1948). The bow, with its propellers, is at right, the stern, with its ramp, at left. The stern waterline is the beaching waterline. Waterlines at right are the seagoing 14-ft waterline and the beaching waterline (5 ft at the bow, which became the stern when the ship beached). Requirements, as formulated in May 1948, included a tank deck 300 × 30 ft with 14 ft clear height, strong enough to take a 75-ton tank; the main deck and the 'tween deck ramp had to be able to take 25 ton vehicles. Armament would be two 3-in/50 guns. Complement would be 7 officers and 175 enlisted, and troop capacity would be 20 officers and 410 enlisted. Endurance was set at 10,000 nm at 10 kts, as in a World War II LST. Maximum length was 500 ft. This concept was renewed several times. For example, in May 1955 the BuShips preliminary design branch (then designated Code 420) produced a bow-propeller LST sketch design, to accompany an alternative design for an LST with a false (i.e., streamlined) bow. A typical final stern-first design was 500 (waterline) × 65 ft, with beaching drafts of 3 ft forward and 13 ft aft; displacement was 4,500 tons for beaching but 7,800 tons fully loaded. The ship could beach with the usual World War II LST load of 500 tons. Sustained speed was 20 kts (22,500 SHP), and endurance was 7,000 nm at 15 kts.

mum acceptable speed. BuShips' preliminary design section developed a simplified 1171 hull, estimating a cost of $12.8 million (vs. $14.5 million for an 1171) for a repeat ship.

The main innovation was to accept 6 ft beaching draft at the bow so that lines would be closer to those of the recent 20-kt studies. The hope was that, with less resistance, the ship could carry much more beaching cargo on a reasonable displacement. For example, a ship with the same lines as the 20-kt LST, but carrying 1,500 tons of cargo in beaching condition, and with power reduced to 12,000 SHP, would make about 18.5 kts.

In October 1960 PhibLant representatives gladly accepted the 6 ft design landing draft. PhibLant was already loading its LSTs to 6 ft draft, and it claimed that about half the time a ship so loaded could be driven onto a beach. Ships typically beached at 8 kts, at which speed their draft forward was reduced by about a foot.

Moreover, operations now almost always involved pontoons, typically the 3 × 15 type (21 × 90 ft), of which an LST carried up to four. Because they projected about 7 ft above the ship's main deck, an LST could not operate helicopters when pontoons were aboard. Typically six pontoons were set end to end by warping tugs carried on board an LSD in an operation taking about 6 hrs. To marry the LST up with a pontoon causeway, she had to be controllable at speeds as low as 3 kts. To a BuShips analyst, the LSTs were now almost ro-ros. Perhaps it was time to abandon the type. Instead, assault vehicles could be ferried from well-deck ships. The Marines held out for the "traditional" LST capability; the PhibLant staff was much less firm.

PhibLant was more interested in high speed (at least 20 kts) and better maneuverability than in a larger payload, although desiring about 900 tons rather than the usual 500 tons. It would accept a smaller payload to get larger numbers of ships. The BuShips analyst visiting PhibLant wondered whether it might be acceptable to halve the number of tanks aboard any one LST. Certainly troop capacity could be cut from the 600 on an LST 1171 to about 450. Similarly, to get numbers PhibLant would sacrifice habitability, including air conditioning.

The interviews at PhibLant led back to the 20-kt LST. A feasibility study was reported to the CNO on

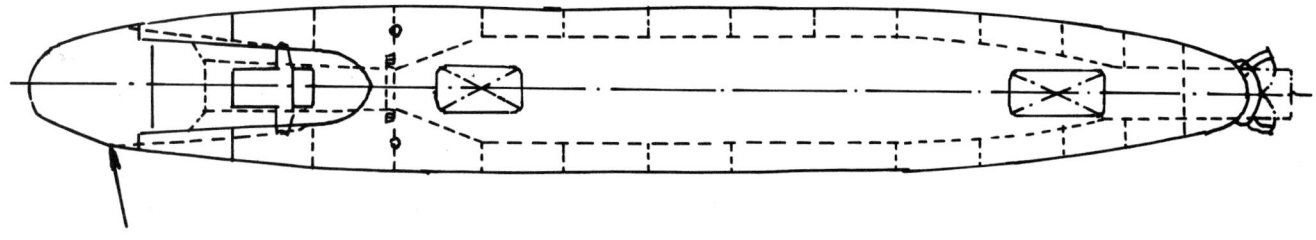
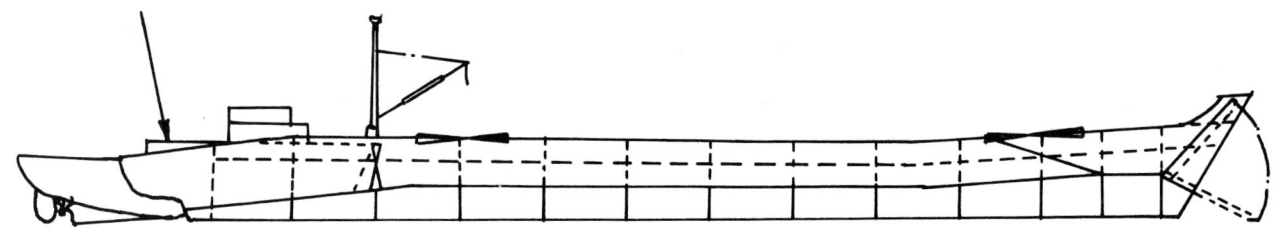

The "carport" LST, from a sketch of the concept produced by the BuShips preliminary design branch that was based on a proposal by the naval architectural firm of George G. Sharp in June 1953. Arrows indicate the towboat nestled in the carport at the after end of the unpowered barge section. Under way, hydrodynamic forces would have locked the towboat in place. The virtue of the concept was that, because few towboats would be needed to power a large number of LST barges, they could have been made very powerful, hence capable of propelling the LST barge at higher speed. Moreover, once beached, the barge would form a long causeway from which deeper-draft (hence better-shaped) LSTs could have unloaded. The towboat would have been powered by 8,000 BHP geared diesels, offering a trial speed (for the combination) of 18 kts and a service speed of 17 kts; endurance would have been 3,000 nm at 17 kts. Dimensions were 420 (overall) × 75 × 3 ft landing draft forward, with 11.2 ft landing draft aft. Displacement would have been 3,470 tons light and 4,175 tons in landing condition, with 550 tons of cargo. According to a 19 March 1954 BuShips preliminary design analysis, Sharp overestimated the ship's speed by about 2 kts (to 20 kts) and its design benefited considerably from a new bow form that was better for speed, but was much worse for beaching. For example, when extended, the bottom of the ramp just reached the ship's keel line. LST ramps extended lower than this to allow for beach washout, which often occurred under an LST's bow. Although the carport concept was rejected, the causeway idea remained attractive, and it explains why later LCUs were built in drive-through form.

31 January 1961, at a cost of $24.8 million for the lead ship and $23 million for a follow-on, based on the 1958 studies. A larger ship was offered in mid-1962 ($29 million in FY 64). On this basis single-sheet (simplified) characteristics were issued in June, and a feasibility study begun in November 1962. Once again alternatives such as a stern door and a catamaran were examined. Bow beaching was the only acceptable technique. The Marines insisted on no more than 4 ft of bow draft, and the Office of Naval Intelligence (ONI) coast and landing beach section revived the requirement for a 1:50 (2 percent) slope. An initial study produced a 500 × 66 ft ship (8,400 tons fully loaded).

Detailed characteristics required substantial increases in endurance, accommodations, stores, and electronics; soon the ship was 525 ft long. That was too much, so in April 1963 the working level section of the SCB tried to prune the ship back. Required endurance was cut to an earlier level, 6,000 nm, and the requirement that the ship be able to land helicopters while she had causeways aboard was dropped.

The most important difference from earlier LSTs, indeed the key to the design, was a longer 100-ft ramp, which would be extended from the main deck forward. With the ship on a 2 percent beach, with 6 ft at the bow, the longer ramp would allow vehicles to land in 4 ft of water, at a 20 degree slope. This ramp was tested on board *Mahnomen County* (LST 912).

A short stern ramp was provided specifically to launch LVTs when the ship was appropriately trimmed down, because the long bow ramp was unsuitable. Now the ship required turntables at both ends, to handle vehicles forward and LVTs aft. The stern gate also made it possible for the LST to marry up with an LCU, so vehicles not in the scheduled assault wave could drive through the LST over its ramp to a causeway or beach.

To simplify topside arrangement, the deckhouse was moved forward of amidships. That made for a better stack arrangement, a better position for the helicopter landing area, and a less exposed position for the pontoons carried on the ship's sides. Because vehicles had to be able to occupy the upper

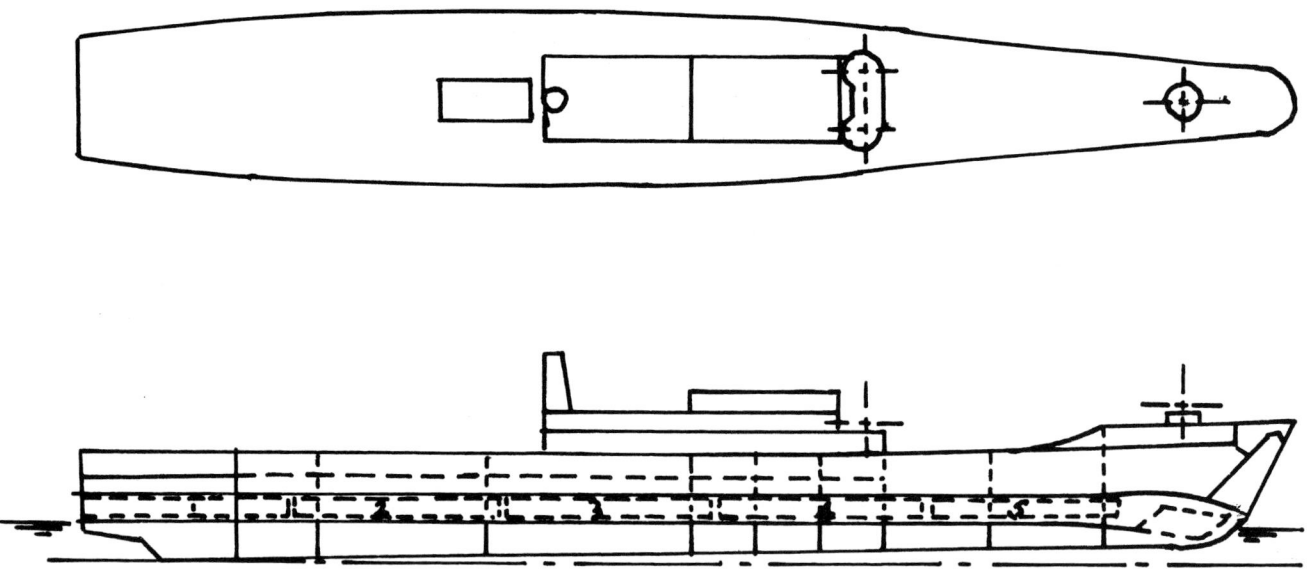

The causeway-regurgitating LST, from a preliminary design sketch produced about 1958. This was one of two fast LST designs. She carried her causeway internally (sections are numbered 1 through 5) on a trackway. An alternative design used a different method of deploying a causeway. In each case, the causeway allowed the ship to stand off down the beach, and thus permitted her to beach at a deeper draft, hence with better hydrodynamics. Regurgitating the causeway made possible a landing draft of 6 ft 6 in, considerably greater than that of a conventional LST. Dimensions were 468 ft 6 in × 66 ft × 15 ft 6 in (fully loaded); displacement would have been 7,540 tons fully loaded. On 20,000 SHP (steam or diesel) the ship would have made 22 kts on trial. As in earlier LSTs of about half this displacement, landing cargo was only 500 tons. Armament would have been three twin 3-in/50. Troop capacity would have been 430, and the ship's crew would have numbered 191. The alternative scheme, presumably that adopted for LST 1179, would have limited landing draft to 6 ft. BuShips also considered an LST with side ports and extended ramps.

deck, a vehicle tunnel would be cut through the deckhouse.

The ship was cut to 500 (517.75 overall) × 68 (69.9 maximum) × 15 (full load) or 11 (landing) ft (5,573 tons landing displacement, including the skeg; 8,482 tons fully loaded). Power would be provided by six 3,400 BHP Fairbanks-Morse diesels; the first three had General Motors 16-645-E5s, the others Alco 16-251 diesels. Armament was two twin 3-in/50, with space and weight reserved for a point defense missile (the abortive Sea Mauler). Model tests indicated that the new LST would behave much better than earlier ones at sea. She was given antiroll tanks; because she had to beach, she was given no bilge keels. Overall, the design was driven by the need to provide sufficient volume. Had bow draft been set at less than 6 ft, the governing factor would have been buoyancy, and the ship would have been much larger. The ratios of length to beam and depth were set by arrangement, structure (strength), and the need for sufficient stability.

BuShips considered this result disastrous, and in April 1963 it recommended simply repeating the earlier LST 1171 design. The new LST was no longer a simple small expendable mass production ship. In October 1964 Vice Adm. John B. Colwell, the Pacific Fleet amphibious commander, remarked that it was longer and beamier than a World War II APA. Its vehicle square footage was 19,000 sq ft, compared to 14,600 for its immediate predecessor, LST 1173, and to only 11,000 for LST 1156. Combat cube (cargo cubic footage) was 4,200 cu ft, compared to 1,400 for LST 1173, 700 for LST 1156, and none for LST 542. The ship could normally carry 24 LVTs, compared to 20 for LST 1173, 17 for LST 1156, and 14 for a World War II LST. Clearly size was the price of the 20-kt requirement. Admiral Colwell felt "very strongly that we must not create the impression that the new LST must have a 20-knot speed to be useful." The SCB, however, stood firm for a 20-kt ship.

A preliminary design (SCB 247) was completed late in August 1964. A bow thruster simplified mooring to a causeway. Engines were normally controlled from the bridge, and controllable reversible-pitch (CRP) propellers offered quick speed changes. The wing walls on the second and third decks were continued down to divide the lower decks into three transverse watertight sections. Off-center flooding might have endangered the ship, so six cross-connecting ducts were provided, to limit list after damage to 15 degrees. The two large ballast tanks aft ran the whole width of the ship. To keep sand from contaminating cooling water during beaching or in shal-

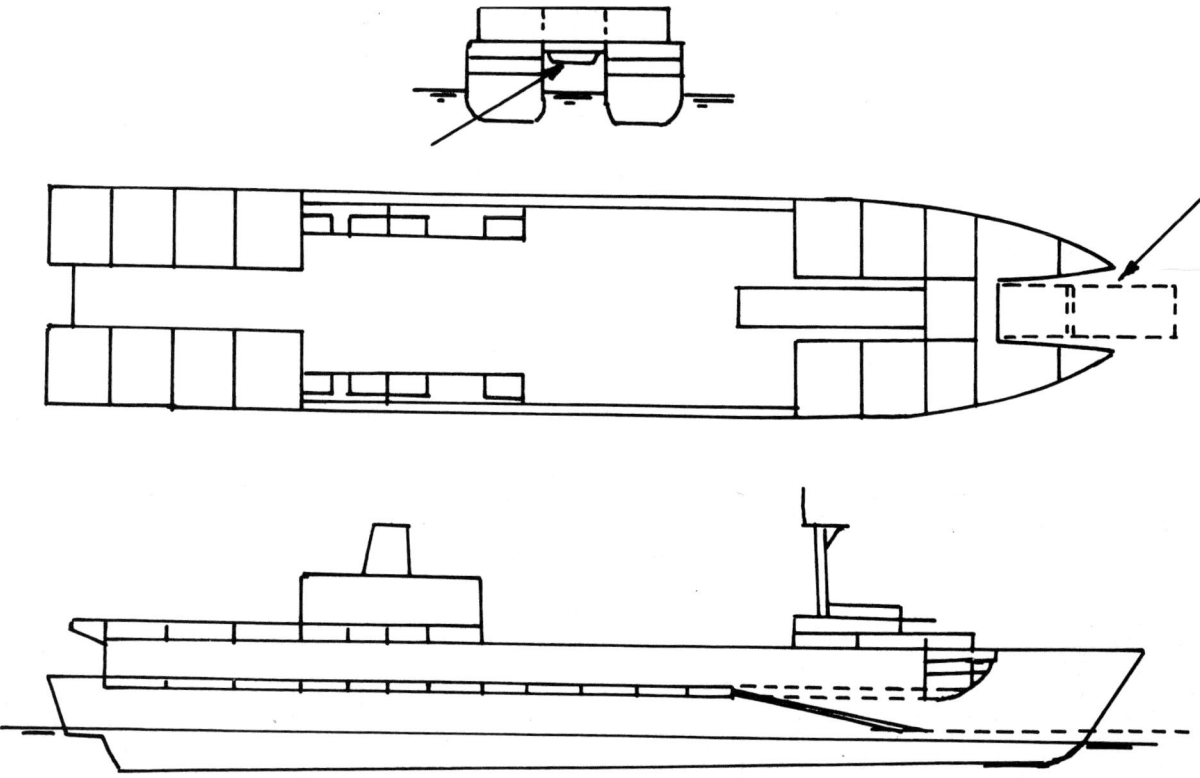

The catamaran LST, from a sketch produced about 1960. For a ship requiring substantial deck area, the advantage of a catamaran is that the two supporting hulls can be quite narrow, hence suited to high speed. For an LST, then, a catamaran configuration eliminates the bluff bow problem altogether. Recently the U.S. military has shown considerable interest in high-speed catamarans for point-to-point transport of troops and vehicles, on exactly this basis; they offer a very high ratio of deck area to displacement (hence resistance). Unlike the sketch shown, they are not intended to beach. The design shown was 550 × 106 × 16 ft in landing condition, displacing 12,500 tons in landing condition (14,000 tons fully loaded), carrying 3,000 tons in landing cargo. It offered 20 kts sustained speed on 28,000 SHP, with an endurance of 6,000 nm at 20 kts. The ramp led down onto a causeway that was normally stowed between the two hulls. The 16 ft landing draft was acceptable only if the ship could extend her causeway a long way up the beach.

low water, the ship was provided with eight sea chest sand traps.

As completed, the ship was larger than had been envisaged; it was 563 ft long with a 190 ft 6 in bow ramp. Designed full load displacement (with 2,000 tons of cargo) was 8,400 tons; beaching displacement (500 tons of cargo) was 5,264 tons. The ship was designed to beach on the usual 2 percent slope with a bow draft of 6 ft and a stern draft of 16 ft. In tests, a ship beached and retracted on a 2.5 percent (1:40) beach at 7,200 tons, and on a 10 percent beach at 7,300 tons. Limiting displacements were 8,500 tons to maintain speed, 8,700 to maintain structural strength, and 8,775 to maintain sufficient subdivision (i.e., to keep the lowest watertight deck high enough out of the water).

The lead ship, *Newport* (LST 1179), was dropped from the FY 64 program, but it was included in FY 65. By 1980 the class had lost much of its ability to beach because of undocumented weight growth. For example, between refits LST 1182 grew 345 tons, only 48.5 of them due to approved ship alterations (shipalts). The problem was out of hand, as authorized alterations were typically about twice the weight reductions Naval Sea Systems Command (NAVSEA; the former BuShips) was desperately seeking. All now exceeded their designed beaching displacement by about 400 tons. Early ships of the class (LST 1179–1181) now exceeded limiting displacement, based on recent inclining experiments. Later ships were at or near the subdivision limit, that is, the limit at which decks would be sufficiently low in the water that subdivision would no longer offer the required margin of safety. Above the 8,500-ton design displacement, damage would put the ship's stern under water unless liquid loading instructions were very precisely followed, which might be impossible in service. For example, the ships could no longer carry their 2,000 tons of cargo.

The LST survived through the 1980s largely because it offered a unique, if unexpected, capacity:

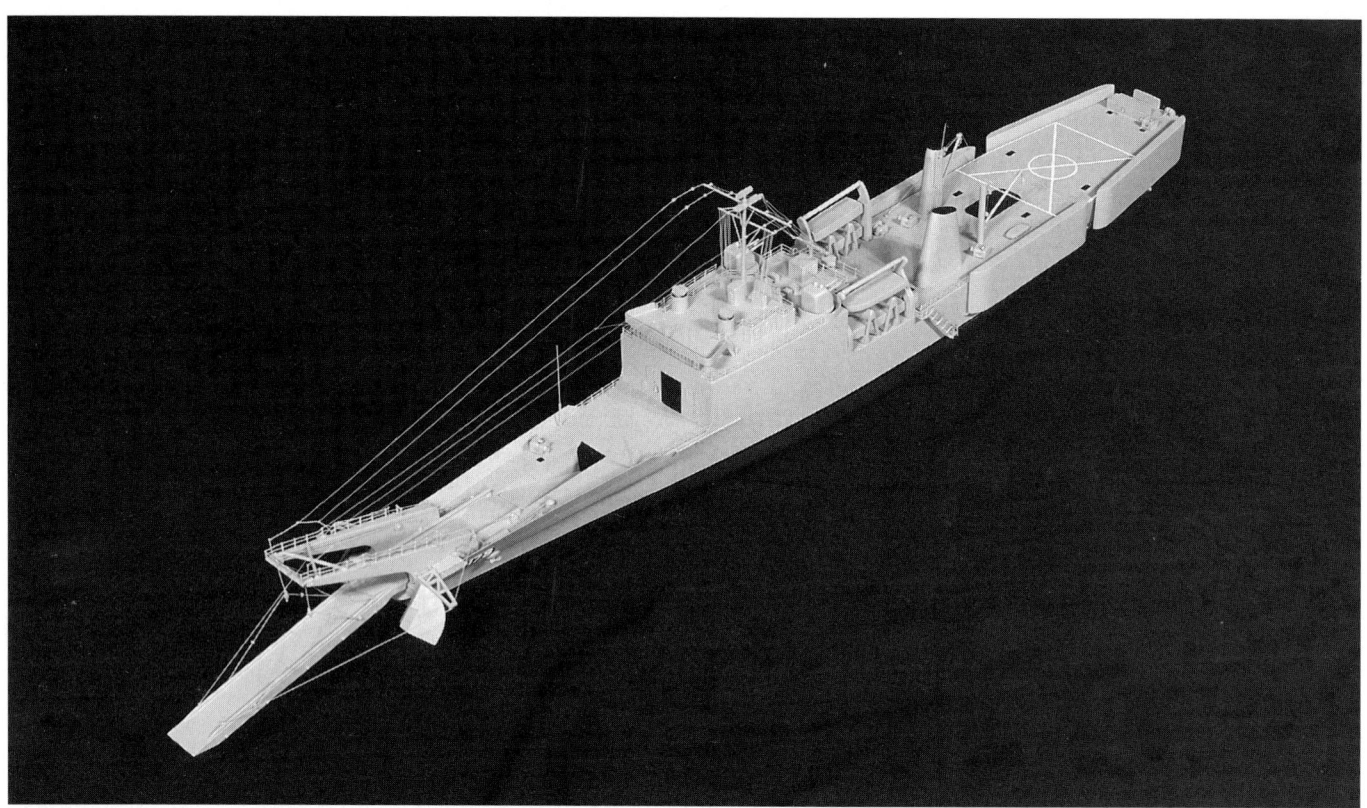
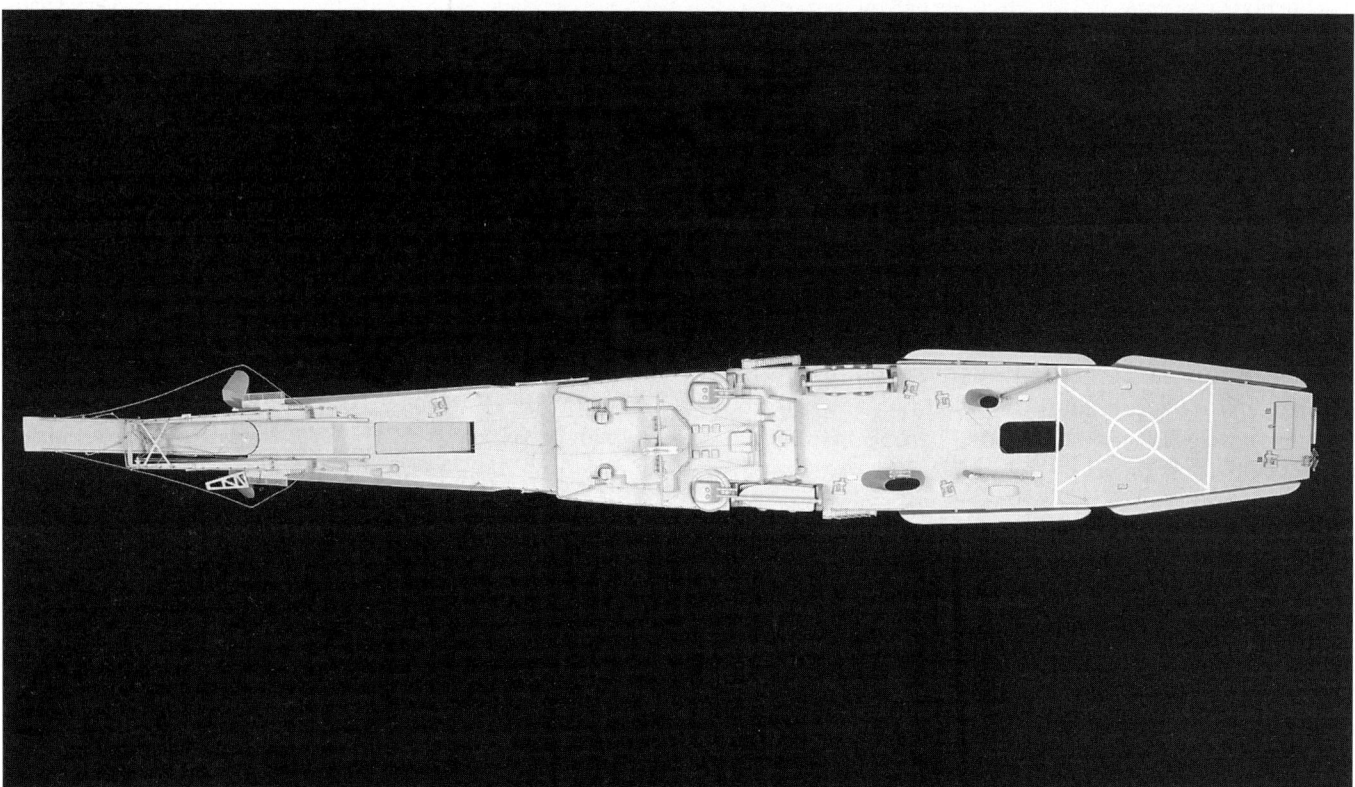

An official model shows how the bow ramp on the *Newport*-class 20-kt LST was expected to work. The oblong shapes aft are pontoons to connect the far end of the ramp with the beach in shallow water. This solution to the LST problem was essentially that chosen by the British in 1940 with their *Boxer* class: to build a more or less conventional hull and concentrate on the problem of extending a long supported ramp from its bow. Singapore has adopted much the same solution in its new *Endurance* class.

Sumter displays some characteristic features of the *Newport* class, such as the tunnel through the superstructure (in this case, with a truck emerging), permitting the entire upper deck to be used for vehicles. When this photograph was taken, the ship still had two enclosed twin 3-in/50s, but their directors had been landed (note the empty tubs forward of the mounts). Also visible is the bow ramp folded back on deck abaft the gallows used to deploy it. *Sumter* was leased to Taiwan on 1 July 1995.

Racine (LST 1191) is shown, newly completed, at her builders, 14 June 1971. Note the raised stern ramp and the kedge anchor, with its winch, used to help the ship retract after beaching—a common LST feature. Barely visible near the port uptakes is the after opening of the tunnel through her superstructure.

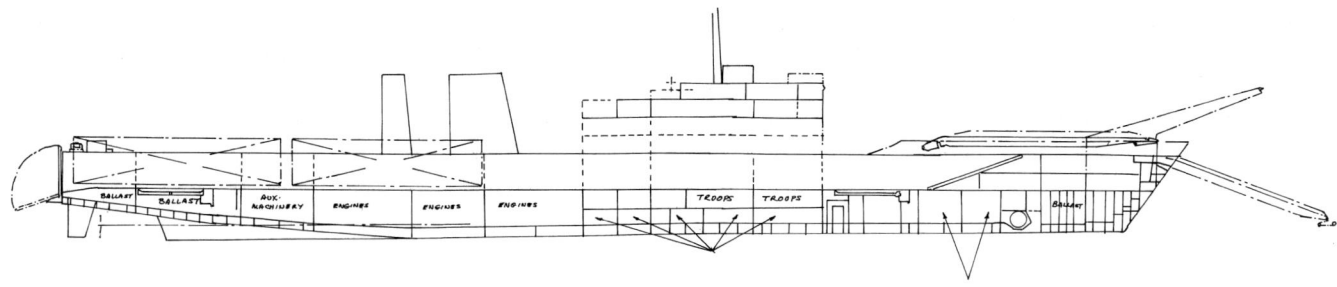

Newport (LST 1179), showing her long aluminum bow ramp in stowed position, May 1991. The ramp could pivot at the ship's bow, allowing her some freedom of movement once at the beach. The ship had a separate stern ramp to launch amphibians because the new bow ramp, unlike previous LST bow ramps, was ill-suited to this purpose. The superstructure was placed amidships to simplify stack arrangement, improve visibility forward, and provide a less exposed position for the pontoon stowage. Because the whole upper deck was used for vehicle stowage (as in previous LSTs), a tunnel had to be cut through the superstructure. The two oblongs aft, outlined in dots and dashes, are pontoons. The small circle forward is a side thruster for better maneuverability in tight situations. Arrows indicate two forward tanks for JP-5 jet fuel or ballast, with a JP-5 pump room abaft them; aft, they indicate tanks for diesel oil or ballast.

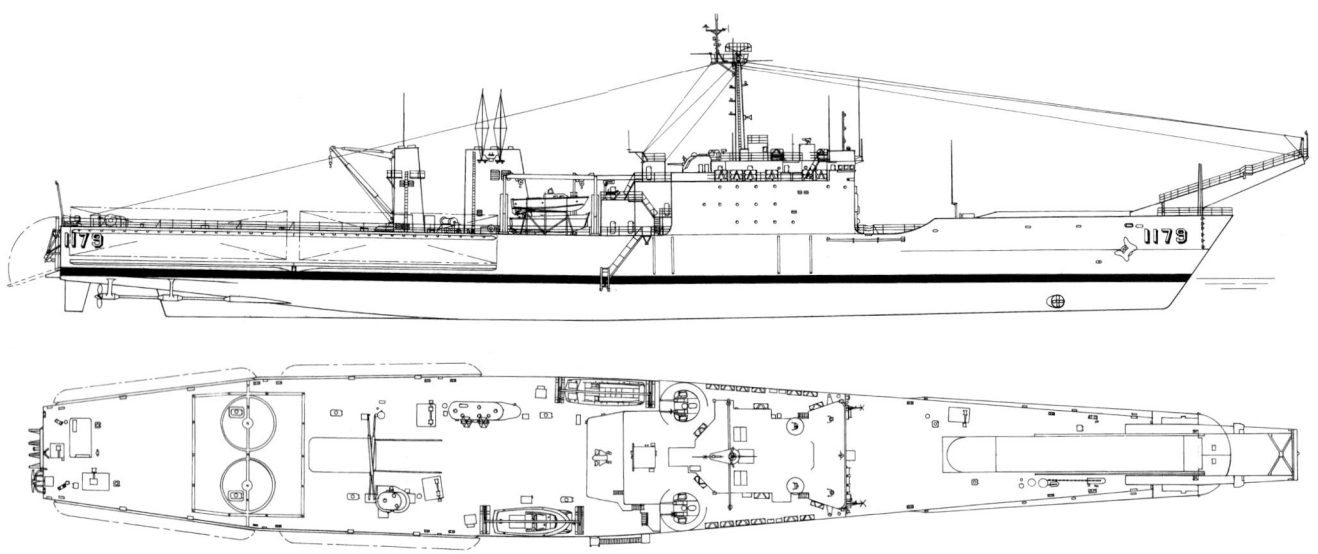

Newport (LST 1179) as completed. Note the side-thruster forward, for precise manueuvering. Dot-and-dash lines aft indicate the positions of the quartet of pontoons these ships generally carry. *A. D. Baker*

bulk tankage for 200,000 gal of fuel (JP-5 and motor vehicle gasoline [mogas]).[2] One marine amphibious force (MAF) alone making a landing would need 19 LSTs for this purpose, 95 percent of the LSTs—and normal availability was only 85 percent. When the LSTs were retired (only two remained active in 2000), expedients, such as special bladders for well-deck ships, had to be developed.

The LST died because it was inefficient as a cargo transporter. BuShips was right; an LST 1179 really was a rather expensive way to deliver 500 tons onto a beach. As part of the 1960 study of less expensive LSTs, OpNav calculated the relative rates (weight per hr) at which equal-cost fleets of various ships could deliver cargo, taking a fleet of World War II LSTs as 1.0. Because they were so much more expensive, later LSTs delivered far less: 0.73 for a fleet of LST 1153s, 0.56 for a fleet of LST 1171s, 0.47 for a fleet of 500-ft 20-knotters (as then understood). However, a fleet of conventional cargo ships (Mariners) could deliver at a rate of 5.0.

There was one other important surviving type, the APD. Because it already met the 20-kt criterion, replacement was not an issue through the 1950s. By the 1960s, however, the ships were aging.[3] To replace the APD, in August 1965 the Pacific Fleet commander proposed a new amphibious raid and reconnaissance ship. In September 1966, in reviewing the planned FY 69–75 program, he remarked that nothing had reduced the requirements for beach surveys (by UDTs) or for support of larger groups of reconnaissance and raiding personnel.

Indeed, "operations along the coast of Vietnam confirm the need for these ships almost daily." By this time the APD was no longer mainly a boat control ship, but she was valued as a coordinating center for small area interdiction, blockade, surveillance, and/or destruction. The new ship would also conduct electronic reconnaissance and would support deception operations. A design study was carried out, but the ship was not built. The APD capability was lost when the ships were decommissioned.

The advent of the atomic bomb forced the Marines to adopt helicopter assault techniques, and the navy, therefore, had to build large-deck amphibious ships. *Saipan* (LHA 2) is shown in 1977. *Ingalls Shipbuilding Corporation*

12
The Bomb and Vertical Envelopment

Besides the fast submarine, the great military fact of late 1945 was the atomic bomb. In July 1946 the U.S. Navy tested atomic bombs against an array of ships, including attack transports, at Bikini. Shortly after the test, the Marines formed a special board, headed by Maj. Gen. Lemuel C. Shepherd Jr., the assistant commandant, to study the future of amphibious warfare. In the 19 December 1946 report the board noted that the ships had been anchored much closer together than in a normal wartime anchorage, yet few had been very badly damaged. Their resistance to air blast and underwater shock seemed remarkable. The Bikini test seemed to show that the lethal radius of a bomb was about 2,000 yds, so ships would have to disperse fairly widely. They would lose some of their ability to be mutually supporting, particularly in antiaircraft fire, but that would be a small price to pay for survival.

On the other hand, a landing force or a conventional ground force had to concentrate to develop its power. It seemed to follow that nuclear weapons were likely to be far more effective against landing and ground forces. An enemy, however, would be unlikely to use nuclear weapons against such units when they were in close contact with its own forces, for fear of self-destruction.

Surely, then, the landing force would be most vulnerable during the usual ship-to-shore phase, when it was concentrated for the landing but far enough from its target that an air or underwater blast would not endanger the defenders. Troops aboard the small slow craft involved would be exposed to flash, blast, and radiation. The Marines calculated that a single bomb dropped on the boat lanes at Iwo Jima would have destroyed the combat effectiveness of the two divisions landing on the island.

The assault troops would have to disperse, but spreading the force in time and space would make it easy for defenders to defeat the landing force in detail. Dispersion in depth would stretch the landing of an assault battalion from 20–27 to at least 100–140 minutes. Because the troops would also have to disperse laterally (by "checkerboarding," i.e., staggering the landing of adjacent waves) the landing period would stretch further to 200–280 minutes, ten times the wartime optimum. Dispersal of the transports would complicate matters further, because landing craft would have to make longer trips.

This Marine analysis typified the dilemma of ground forces facing a combination of nuclear and non-nuclear threats. The obvious solution to the nuclear threat, dispersion, left troops weak because their own weapons depended on concentration for firepower. Thus, they would be vulnerable to concentrated non-nuclear fire. One solution, to arm small units with nuclear weapons, was not conceivable in 1947, because nuclear weapons were far too large and too expensive. When the army tried it about a decade later, it left open the real possibility that enemy nuclear weapons could destroy the dispersed units. Moreover, the operation would be intended to seize vital territory (e.g., for bases); Marine nuclear weapons might contaminate it.

A more interesting possibility was to concentrate only when in close touch with the enemy. To the Marines, carrier-based transport helicopters might replace existing small landing craft. Because they were so fast (120 kts, hence 2 nm per minute was expected), they could disperse in flight, yet concentrate to land. A mass of helicopters might spread out over 54 miles in depth, yet it could deliver all its troops in less than half an hour, the optimum time. Waves flying 6 nm apart could land at 3-minute intervals. Moreover, the helicopters would avoid altogether the nuclear hazards demonstrated at Bikini: tidal waves and radioactive water. The helicopter's speed would easily make up for dispersion of the assault shipping. Unlike paratroops or glider troops, helicopter-borne marines would arrive as coherent combat units, delivered exactly where intended.

The Marines' commandant made the case in a 19 December 1946 letter to the CNO. The existing Piasecki HRP-1 could lift a ton (8–10 men); the navy planned to buy 48 for delivery in 1948. This was clearly too little, but it could be used to develop tactics. Piasecki was developing the much larger R-16 (later H-16) under joint army-navy sponsorship to carry 30–35 men, or a squad plus a ton of accompanying equipment and supplies; the Marines saw it as a flying LCVP. As of 1946, it appeared that the R-16 could be in production within five years. A complete regimental combat team could be lifted in one flight using 262 such helicopters. By 1948 the H-16 was expected to carry only 20 troops, but this did not really affect the promise of the helicopter.

Given helicopters, the Marines could abandon their earlier frontal tactics. At the least, helicopters could land troops where beaches were clearly unsuitable, greatly increasing the Marines' flexibility and making it far more difficult to erect a defense. In this sense the helicopter might be considered analogous to the LVT. Helicopter-borne marines could outflank an enemy, or land in his rear to seize a beachhead. Naval gunfire would not reach deep enough to cover landing zones. Instead, aircraft from the carriers launching helicopters would provide preparatory and covering fire.

Helicopter pads could easily be added to existing amphibious ships, as indeed they were during the 1950s. The 1946 board, however, called for a specialized helicopter carrier, with a continuous flight deck, accommodating the usual APA load (a BLT) and a hangar. Small elevators would lift palletized cargo directly to the helicopters. The board considered the existing light fleet carrier (CVL) an attractive conversion prospect, on the basis of its speed, size, and likely troop capacity.

The Marines recognized that helicopters probably would never be able to move heavy equipment or supplies. The Marines imagined using seaplanes to bring in the first heavy equipment, followed by conventional landing craft. Thus the helicopter would not entirely displace conventional craft, but it would carry the great majority of personnel ashore.

Given the rather primitive state of helicopter development in 1946, this was a very bold initiative. The Marines formed an experimental squadron, HMX-1, and in May 1948 five of its aircraft flew from the escort carrier *Palau* (CVE 122) during Operation Packard II, a command post amphibious operation involving a simulated landing. That August HMX-1 received its first large helicopters, Piasecki HRP-1 "flying bananas," capable of carrying eight troops (plus the crew of two). The Marines published their tentative helicopter manual (Phib-31) in November.

Raids would still be valuable, if not more so. For example, the 1946 board imagined clandestine destruction of an enemy long-range missile base or establishment of a friendly one. This would be the APD mission, but at a much greater range than in World War II. The Marine board offered two possible replacements for the existing APD.

One was a large transport seaplane, which the Marines called a "flying LST." Unlike a land-based airplane, it would not require prior seizure of a large airfield. It could be used both in long-range unopposed operations and to supplement helicopters by bringing in heavy equipment (i.e., as an alternative to an LST). The Marines argued that during World War II only such seaplanes could have delivered reinforcements to besieged islands such as Wake, Guam, and the Philippines or, for that matter, have evacuated them. Future U.S. forces would surely be scattered thinly around the world, and only seaplanes could react quickly enough to emergencies. They might also be the best way of beating an enemy to strategic places.

In 1947 the Marines drew up requirements for a huge 225-kt attack transport seaplane with a cargo capacity of over 100,000 lbs (350 men), and a radius of 1,500 nm. It would have a 12 × 12 ft bow door. Troops would go ashore in inflatable rafts; cargo would use flotation gear or small amphibian craft carried on board. About 55 assault seaplanes would lift a reinforced Marine regiment. Estimated cost in November 1948 was $10 million ($45 million for the prototype). Although at first the idea had seemed fantastic, by November 1948 two design studies (Martin and Convair) had been completed and were being evaluated. The navy was already interested in a large turboprop patrol seaplane. It selected a Convair design, which became the P5Y. A transport version was built as the R3Y Tradewind.

An alternative was a transport submarine, which the Marine commandant formally proposed in February 1946. Submarines were clearly too expensive to replace the APA, but they offered important tactical advantages. Ideally, one would carry a complete tactical unit, such as a reinforced company, about 300 men. Existing submarines were not nearly large enough. The Marines convinced OpNav to have a pair of fleet submarines (SS 313 and SS 315) converted to transports. Designated APSS, each could accommodate 5 officers and 115 enlisted men, plus 31 tons of cargo in the hold and an LVT with a jeep aboard it plus 21 tons of deck cargo. Another submarine (SS 317; a second conversion was not carried through) was converted to carry 300 tons of cargo. Forty such submarines could deliver a reinforced marine regiment; twelve could lift a BLT. The submarines' guns could cover landing and withdrawal of the unit. No further submarines were converted at the time, however. The existing troop carriers were used for small-scale raiding. In that role they were so useful that a few later missile submarines—ultimately, ballistic missile submarines—were converted to carry Special Forces. The projected submersible LVT, intended for use with troop-carrying submarines, was never built.

These projects lapsed as the military budget tightened and as it became common wisdom that amphibious assault was obsolete. The Inchon assault in September 1950 changed perceptions, however, and in Korea helicopters proved valuable in combat. In July 1951 the commandant of the Marine Corps revived the helicopter assault project in a letter to the CNO. Industry would soon produce a helicopter that could carry 36 men (i.e., a potential airborne LCVP). It was both practical and timely to provide within the fleet a helicopter capacity to land the full assault elements of a Marine division in a continuous wave: 144 helicopters could land the two regimental combat teams, the artillery group, and a command group. That would amount to 10,000–12,000 men and 4,000–5,000 short tons. It was time to obtain a helicopter carrier that could carry and operate at least 18 of the new helicopters (10 of which could sit on the flight deck with rotors unfolded). She would embark 1,500 troops (i.e., a BLT) and at least 600 short tons of cargo. This was the ship the Marines' 1946 development board had envisaged. Eight such ships would launch the full assault echelon of a Marine division.

The CNO doubted that helicopter operations had been sufficiently developed. The Marines' ship would be expensive. Was it really necessary? Would helicopters be the main means of assault? In October the Marines asked that a war-built *Casablanca*-class escort carrier be reactivated to operationally evaluate the vertical assault concept. It would begin with an assault by units of the 2nd Marine Division, lifted by a Marine helicopter squadron (HMR-261) during an Atlantic Fleet exercise, FLEX 52. The CNO vetoed carrier reactivation. Instead, during FY 52 the Atlantic Fleet would do what it could on an intermittent basis, using its escort carriers. BuShips, however, completed a feasibility study of a *Casablanca*-class conversion in the fall of 1952. Fleet commanders supported the project as an excellent use for Reserve Fleet escort carriers, and in November 1952 BuShips' preliminary design section proposed that it be included in the FY 55 program.

The Marines looked beyond a converted carrier to what their Equipment Board called an APA-M, in effect a combination AKA, APA, and helicopter carrier. She would accommodate 20 helicopters (on a full flight deck and in a hangar) for her BLT, which was increased to 2,000 troops. Because heavy cargo

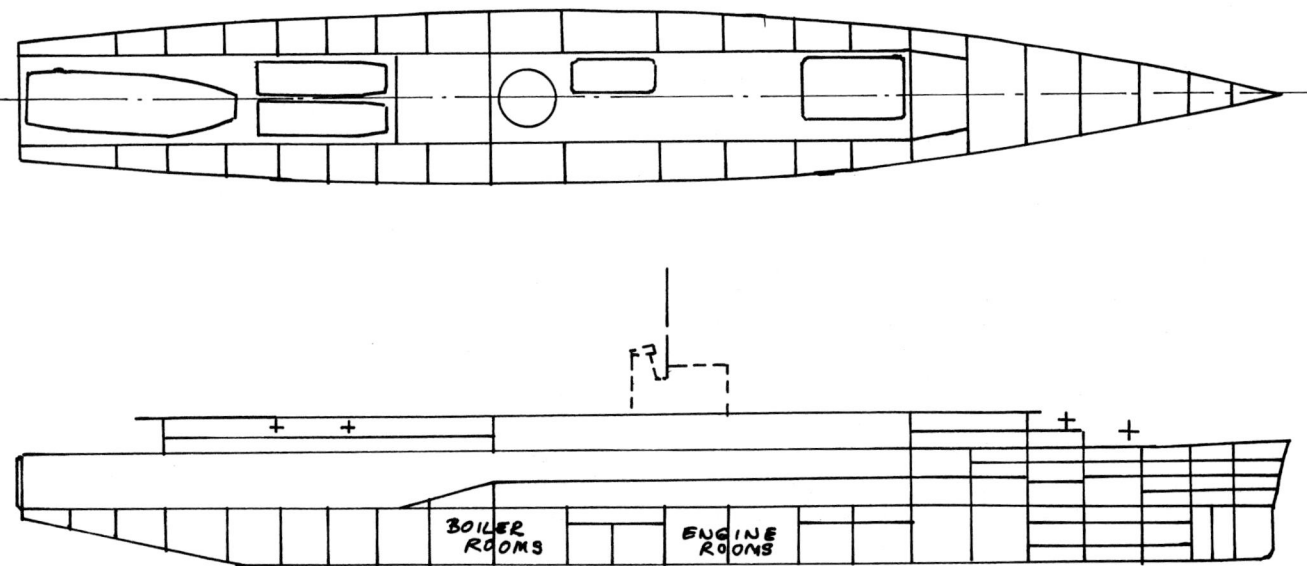

The Marines' APA-M. She would have carried 21 HR2S (CH-37) helicopters, handled by a pair of 60-ton elevators. The well deck shows an LCU and a pair of LCMs. Cargo capacity was 3,500 tons. Dimensions were 700 × 96 ft (25,000 tons fully loaded); power was 70,000 SHP (25 kts on trial), and rated endurance was given as 10,000 nm at 20 kts, the standard postwar amphibious figure. Planned armament was one twin 5-in/38 and seven twin 3-in/50, few of which are shown in this drawing. A vehicle-deck turntable is indicated.

still had to be discharged over a beach, she also had a well deck for an LCU (landing craft, utility; ex-LCT) and 15 LCM. Cargo capacity was 3,500 tons. This ship could carry a BLT and all associated equipment and cargo in a single hull (700 × 96 × 27 ft, 15,000 tons light, 25,000 full load). The U.S. Marines initially wanted 30 kts but that required too much power (100,000–120,000 SHP), so it settled for 25 kts (70,000 SHP). BuShips produced a sketch design in November 1952, and suggested that such a ship be included in the FY 55 program to supplement converted Mariners. Because they involved no really new ideas, conversions could be done quickly; APA-M required a prototype for development and test.

In August 1954 PhibLant recommended building AKA-Ms instead of the fast LSTs that Commander in Chief, Atlantic (CinCLant) wanted. Three such ships, plus two LSD, could carry a full regimental combat team. The conventional alternative would be nine LST, three APA, three AKA, and one LSD. The 5-ship group would cost about 17 percent more than the equivalent 16-ship group. However, if the LSTs were the projected 20-kt type, the comparison reversed, and the non-helicopter group would cost about 23 percent more than the five ships. Two years later the AKA-M concept led to the design of the less ambitious LPD (landing ship, personnel, dock; discussed below).

To test the helicopter assault concept, the CNO now approved converting a small war-built *Casablanca*-class escort carrier (CVE) into an austere prototype helicopter carrier. Characteristics for this SCB 122 conversion, which was designated a CVHE, were issued on 12 May 1954. The designation was soon changed to CVHA, which suggested an assault ship rather than some sort of escort carrier.

Because the conversion was only to test the vertical assault concept, it was quite austere. The ship would accommodate something more than half a battalion team, 38 officers and about 900 enlisted men, berthed at the fore end of the main (hangar) deck. Particular attention would be paid to the requirements of rapid loading and debarkation of troops and assault equipment via helicopters. The flight deck was strengthened to handle 32,000-lb helicopters like the Marines' HR2S troop carrier. They were served by a new 36,000-lb (32 × 45 ft) elevator built into the after end of the flight deck. It could handle one HR2S or two HRS-1 helicopters. The ship's 15 troop-carrying HR2S helicopters required 60 officers and 117 enlisted men, comparable to a crew of 40 officers and 483 enlisted men. Aviation fuel capacity, 120,000 gal, was comparable to the ship's World War II load. The main change to the ship's powerplant was increased electrical generating capacity. All armament would be replaced by eight twin 400-mm guns. This ship, *Thetis Bay*, was recommissioned on 20 July 1956 as CVHA 1 (ex–CVE 90). Later she was briefly redesignated as an LPH (landing ship, personnel, helicopter), the LPH 6, in the new helicopter assault carrier series, before being stricken in 1964.

The climate was changing. On 8 September 1954 the Ad Hoc Committee to Study the Long-Range Shipbuilding Program asked BuShips to compare alternative helicopter carriers. The committee accepted that over the next 10–15 years most existing attack transports and cargo ships would be replaced by helicopter carriers. The core lift for a future Marine division/air wing would be 12 helicopter carriers, each carrying 20 heavy helicopters (HR2Ss). Because plans called for maintaining a complete division/air wing lift on each coast, the fleet would need at least 24 helicopter carriers, each with 20 heavy helicopters. Alternatively, smaller helicopter carriers might be adopted in larger numbers. The committee's figures emphasized the need for an inexpensive helicopter carrier. Any conversion had to use a hull available in large numbers, and the ships involved would have to be surplus yet have considerable remaining lifetime.

BuShips evaluated light and heavy cruisers and even *North Carolina*–class battleships. All had to be rejected. Light carriers (CVLs) might be acceptable, but they were not numerous enough. That left two types of escort carriers, both available in quantity (*Casablanca* [CVE 55] and *Commencement Bay* [CVE 105]), and a Mariner cargo hull. None was entirely satisfactory. The CVEs could not sustain the desired 20-kt speed. Neither CVE, CVL, nor Mariner could accommodate all 20 helicopters, or the desired 1,800 troops and 200 helicopter squadron personnel. Only the Mariner could accommodate the troop cargo load requirement of 2,000 tons. CVLs were deemed to lack remaining hull life.

In April 1955 it appeared that two new-construction LPH would be included in the FY 57 program. BuShips was asked to evaluate both a full-size ship and a reduced version, which it called Ship A and Ship B, respectively. The full ship would carry 1,800 troops and 2,000 measurement tons (at 40 cu ft to the measurement ton, actual weight would probably be 1,000 tons) of cargo at a sustained speed of 20 kts, with 20 HR2S helicopters (200 aviation personnel). All but two of the helicopters would have their rotors folded when on deck. The assumed armament was six twin 3-in/50, and the ship complement was 500. Apparently the ship would carry neither vehicles nor boats to land them; no requirements for boats and vehicles were stipulated. Assumed endurance, as in

Table 12-1. Helicopter Carrier Alternatives, April 1955

	CVE 55	Mariner	CVE 105	Ship A	Ship B
Length OA (ft-in)	512-0	563-8	557-1	650-0	510-0
Beam (ft)	65	76	75	90	84
Draft, loaded (ft-in)	21-0	23-10	31-4	24-0	20-0
Displacement (tons)					
Light	7,200	9,300	10,440	11,000	7,500
Loaded	11,000	16,340	24,410	19,000	12,500
Speed (kt)	18	21	19	20	20
Capacity					
Troops	940	1,880	940	1,880	1,000
Troop cargo (tons)	265	530	265	2,000 MT[a]	2,000 MT
Fuel (gal)	130,000	100,000	150,000	150,000	NA
Helicopters	15	10	18	20	10
Armament					
3-in/50	—	8	8	6	6
40-mm	8	—	—	—	—
GM, fully loaded	3.58	5.55	4.63	NA	NA
Freeboard, loaded (ft-in)	41-0	40-2	40-3	NA	NA
Freeboard % of length	8.4	7.6	7.67	NA	NA
Add blister	No	No	No	NA	NA
Add ballast	No	Yes	Yes	NA	NA
Cost ($ million)	10.3	20.286	15.87	47	31
Expected life (yr)	10–15	15–20	13–18	25–30	25–30
Maintenance/yr ($ thousand)	300	225	240	275	240

NOTES: Dash indicates that specification is irrelevant; NA indicates relevancy but data not available.

[a] Measurement tons; 40 cu ft equals 1 measurement ton.

other recent amphibious ships, was 10,000 nm at 20 kts. Ship B, the half-size ship, had half the troop complement and ten helicopters. Except for speed, the CVE could meet the reduced-ship requirements. Table 12-1 compares some results of the BuShips studies.

An Amphibious Warfare Conference in Washington on 4–6 May 1955 supported the new vertical envelopment concept, to the extent that later in the month the Standing Committee on the Long Range Shipbuilding Program proposed dropping the planned eight LSTs of the FY 57 program in favor of two helicopter assault ships and a 20-kt LST. The eight slow LSTs would have cost $99.6 million, compared to $117 million for the three new ships ($47 million for the lead helicopter carrier, $40 million for the first follow-on ship, and at least $30 million for the fast LST). This change was approved on 14 June, but the LST was later dropped. In July, the DCNO of logistics, the OpNav officer in charge of shipbuilding, proposed that instead of building two new LPH in FY 58, two light carriers (CVL) be converted. That might leave money for two fast LSTs (repeating a design then planned for FY 57), as well as two Mariner conversions (AKA and APA). Although still being pressed at the end of August 1955, it is not clear just how quickly the CVL proposal died.

In mid-July the Marines offered preliminary requirements for their LPH. They confirmed the design assumptions: 2,000 troops including aviation personnel, 20 HR2Ss (plus 3 utility helicopters and 2 cargo helicopters), 2,000 measurement tons of cargo (at least 1,000 short tons). As might have been expected, square footage of cargo stowage was more important than cubic footage; the maximum stowage height would be 12 ft. Vehicle compartments would need adequate ventilation, tie-downs, cable outlets to charge batteries, and air hoses to fill tires. Helicopter fuel capacity would be at least 300,000 gal.

The BuShips preliminary designers offered two alternatives, both about the same size, and both simplified by offering minimum maintenance facilities, even though an HR2S needed its engines, transmissions, and rotor assemblies replaced every 120 hrs. Ship A was broken decked, with take-off access on two levels and no elevators. Ship B resembled a small carrier. Both were 580 (overall) × 85 × 22 ft, smaller than the full carrier of the earlier study. Ship A would displace about 9,500 tons light or 15,800 fully loaded, compared to 10,000 and 16,400 for Ship B. Both would have four twin 3-in/50 and 22,500 SHP engines. Ship A was simpler but her structure would generate areas of turbulence that the aviators disliked. It would have stanchions in its hangar, and cargo handling would be somewhat restricted. It was abandoned. Scheme B offered a clear flight deck and hangar and better distribution of cargo.

The SCB staff tried to hold down the size and cost of the ship by halving the aviation fuel capacity (to 150,000 gal); if the Marines wanted more, it should come out of their 2,000 measurement tons of cargo. The cargo specification was a problem from the designers' point of view, because they worked in terms of weight. Past loadings had run at about 2½ measurement tons to the short (weight) ton, so that 2,000 measurement tons might equate to only 720 short tons' weight. BuOrd wanted eight twin 3-in/50 and four twin 20-mm.

By this time something else had changed. Nuclear weapons were now light and relatively inexpensive. Given organic nuclear weapons, even a lightweight helicopter-borne unit gained considerable firepower. Thus the helicopter carrier had to incorporate a "special" (i.e., nuclear) weapons magazine. Fortunately nuclear weapons no longer had to be assembled on board ship, so such arrangements were relatively simple. The idea of arming small units with nuclear weapons was hardly unique; at the same time the U.S. Army was also toying with dispersed nuclear-armed tactics. The Marines had the important advantage that their organic aircraft might well detect enemy forces far enough away to make nuclear counterattacks viable.

Newly completed, the helicopter carrier *Okinawa* is shown off the Philadelphia Naval Shipyard, 7 June 1962. At this stage her only air navigation aid was a Tacan beacon at the masthead. The big air search antenna was for SPS-12. The radomes on the cross-trees of her mast are for the WLR-1 radar intercept system.

The ship needed more command facilities than an attack transport, at the least to direct her helicopters. That meant ready rooms and briefing rooms for the pilots, a separate troop operations center, and arms and air coordination centers. Each ship would accommodate Marine battalion and helicopter squadron staffs. When operating as flagship of a division, she would also accommodate regimental and helicopter group staffs, as well as transport division staff. She also might serve as a transport squadron flagship.

In fact the techniques the ship would employ had not yet been developed. Although the SCB hoped to issue first preliminary characteristics by 1 September 1955, in August the BuShips designers considered that most unlikely. Meanwhile they sketched a flush-decked (carrier-like) Scheme C: 550 (between perpendiculars) × 88 × 22.5 ft, 10,300 tons light (16,600 fully loaded), with 25,000 SHP machinery. Estimated lead ship cost was $47.5 million ($39.2 million for a follow-on ship). Scheme C became the basis for the SCB characteristics written later in the year. In September 1955, moreover, the Bureau of Aeronautics (BuAer) provided details of the expected 1965 version of the troop-carrying HR2S helicopter: rotor diameter would be 80 ft, vertical height 20 ft, and weight 40,000 lbs. To provide a growth margin, a 50,000 lb weight was adopted for the carrier.

As the designers had supposed, the SCB did not meet to approve first preliminary characteristics for the LPH, now designated SCB 157, until mid-November 1955. By this time it was clear that the ship would have to provide substantial helicopter maintenance. Spares and fuel would support five days of shipboard operations plus the assault landing. Because future helicopters might well use gas turbine powerplants, the SCB accepted a need for more helicopter fuel—300,000 gal, as the Marines had asked. Helicopters were developing very quickly, so the ship had to allow for considerable growth. Helicopter weight was set at 50,000 lbs (40,000-lb elevator load), but the weight limit on the flight deck was set at 60,000 lbs. Hangar clear height was 20 ft (compared to 17 ft 6 in for World War II carriers), and maximum dimensions on the elevator were 34 × 50 ft. Cargo elevators would unload 150 tons per hour onto the flight deck.

The helicopter carrier was extremely important, but until many had been built, there would not be enough to provide a full division's lift on each coast. The ship therefore had to be able to pass through the Panama Canal, for use in either ocean. The ship's value was also reflected in OpNav's strong pressure to arm it with Tartar antiaircraft missiles. BuShips argued that the missile would greatly increase the size and cost of the ship, and BuAer pointed out that missile blast would sweep the flight deck. Even so, the SCB working level voted for Tartar, the only dissent coming from the BuShips representative. Then the board reconsidered, and the final vote was unanimously against the missile. The ship's armament was set at four twin 3-in/50 and four twin 20-mm.

Aircraft carriers used deck-edge elevators. Should the helicopter assault ship? BuShips pointed out that, when down, the elevator would dip into the water when the ship rolled 15 degrees. However, operations would probably stop before the ship rolled 10 degrees. The elevators would, moreover, have to fold up so that the ship could pass through the Panama Canal. The working level voted unanimously to air condition all living and control spaces, at an estimated cost of $1 million.

In November 1955, OpNav's Ad Hoc Committee on the Long-Range Shipbuilding Program, which was trying to estimate the size and shape of a 1970 fleet, reviewed amphibious requirements. To provide the 24 LPHs, the committee planned a combination of new construction and the conversion of 8 heavy cruisers in FY 58–67. Plans were based on building a division/air wing of fast lift on each coast. The Marines equated 2 AGC, 12 LPH, 4 APA, and 3 AKA to a helicopter-borne division/air wing, the cargo ships following up with personnel and materiel to be brought over the beach. The committee added 12 LSD and 12 LST to move heavy equipment over the beach (a September 1954 estimate had included only 9 LSD). The LPH, AKA, and APA provided sufficient troops (31,500) but OpNav experts suspected that they would not carry sufficient equipment. Because the LSD and the LST also carried 9,600 personnel, the resulting force carried too many personnel.

The group proposed a solution. Instead of landing all troops by helicopter, the Marines would settle for a one and one-third division/air wing helicopter lift, 16 LPH instead of 24, so the 8 heavy cruiser conversions could be abandoned. The Marines were satisfied because their primary objective was a full division of helicopter lift as soon as possible. Their ultimate goal, enough lift for three divisions, could be deferred. Excess capacity would also allow a cut from 24 to 16 fast LSTs. The result was a total of 4 AGC, 16 LPH, 8 fast APA, 6 fast AKA, 24 fast LSD, and 16 fast LST. It is difficult to say how well the committee's requirements were met, because quite soon hybrids such as the LPD were being built. Of the types listed, however, 2 AGC, 12 LPH, 1 fast APA, 7 fast AKA, and 20 fast LST were eventually built or converted. Follow-on elements of the assault divisions would be lifted by the 24 slow postwar LSTs and by surviving World War II amphibious shipping.

The committee omitted fire support. No ships could be provided specifically for this purpose, because priority had to go to nuclear strike forces intended to deter nuclear attack. The committee did hope that regular provision of attack carriers and missile cruisers for strike forces would automatically release older carriers (providing sufficient aircraft were provided) and older cruisers for amphibious warfare. Moreover, for the time being enough older gun ships would survive in reserve to be reactivated for fire support.

The committee tried to integrate the new amphibious ships with other U.S. naval requirements. Among those difficult to meet was air support for convoys. The group therefore proposed that the LPH have an alternative ASW capability, so that it could function as an escort carrier. That idea was adopted in the design of the LPH. In practice these ships were never used for ASW, although in the 1970s *Guadalcanal* (LPH 7) was tested in that role as a prototype "sea control ship."

The great question was whether helicopters would take over altogether from surface assault. Opinion within the Marine Corps was divided. In 1954 the Marines' advanced research group proposed an all-helicopter landing concept to fight in nuclear conditions. That is, because the Marines would be lightly equipped, they would have to use nuclear weapons to deal with enemy troop concentrations. By this time it was becoming clear that deterrence might preclude any such tactics, so in 1955 the advanced research group revised its all-helicopter assault. In December 1955 the Marines' Landing Bulletin No. 17 described an all-helicopter assault in which marines would land up to 100 miles from the ships, over a divisional frontage as wide as 50 miles. New aircraft-laid antitank mines might deal with massed enemy armor. They were impor-

The LPH was conceived as an amphibious ship, but like any other carrier it was very adaptable. USS *Tripoli* (LPH 10) is shown entering Subic Bay on 20 July 1972, having supported Operation End Sweep, the clearance of North Vietnamese waters. The helicopters on her flight deck are troop-carrying CH-46 Sea Knights and light UH-1s ("Hueys"). The big radome abaft her island houses a SPN-35 radar for carrier-controlled landings. The frame protruding from the port side of her flight deck supports antennas for a ULQ-6 jammer used for self-defense against Soviet missiles. By this time two of the ship's four twin 3-in guns had been replaced by box launchers for Sea Sparrow defensive missiles, visible on her aft quarter and forward of her bridge. Almost two decades later, *Tripoli* was serving as mine countermeasures flagship in the northern Gulf when she detonated an Iraqi mine.

tant because modern jet aircraft could not loiter over a battlefield, as their World War II forebears or postwar piston aircraft could, hence they could not be effective tank killers. In this idea can be seen the germ of current work on deep-attack antiarmor missiles. The all-helicopter idea was attractive because it eliminated any beach restrictions.

A more conservative view was that helicopter-borne marines would land to the rear of the beach and on its flanks, taking it from the rear. Once the beach was taken, its approaches could be cleared of obstacles and heavy equipment brought over it. On 4 June 1956 the Marines convened a board, led by Maj. Gen. Robert E. Hogaboom, to lay out their future structure. The board ratified the conservative view: while helicopters would in effect replace the assault element, and while helicopter-borne marines would seize critical terrain, surface craft were needed for the follow-up. Later the Marines would always seek the ability to mount either an air-heavy or a surface-heavy assault, depending on conditions. That is, they wanted the ability to put two-thirds of their force either in helicopters or in surface assault craft, the other third using either surface craft or helicopters, respectively. This requirement ultimately doomed the helicopter-only carrier, and led to the creation of well-deck helicopter carriers such as the LHA and LHD. Overall, Hogaboom tended to agree with the growing view within the navy and the Marine Corps, but not within the army and the air force, that the nuclear deterrent ruled out war in Europe, but made war in the Third World more likely. The implication, that mobility was more important than armored firepower, was significant for new-generation surface assault ships. The Marines offered mobility, and to capitalize on that asset Hogaboom advocated a lighter-weight organization, better adapted to quick air-led assaults—and, incidentally, less dependent on over-the-beach shipping.

While the SCB worked out characteristics for a new-construction helicopter carrier, it also looked at a less expensive alternative, a converted *Commencement Bay*–class escort carrier (SCB 159). Unlike the small *Casablanca*, she could support a full battalion landing team, the same number of troops as in the new-construction ship. She was, after all, about as large as a new ship, 561 rather than 591 ft overall, and 2,500 tons heavier. At about the end of 1955 the new-construction LPH was dropped from the FY 57 program in favor of a less expensive CVE conversion. Work on the LPH design continued because it was expected that she would be included in the FY 58 program, as indeed she was.

Characteristics for the CVE conversion were issued on 14 March 1956. Although she had been built from the keel up as a carrier, the *Commencement Bay*–class CVE was an adapted oil tanker design, using the upper (strength) deck of the tanker as her hangar deck. Its marked sheer cost a deck level fore and aft of the hangar itself. Thus the ship had to devote 92 ft of her former 306-ft hangar deck to vehicle parking and offices, with officer berthing above it. Even so, she had only about two-thirds the vehicle space of the new-design ship. Some vehicles would be stowed lower in the ship. Troops had to be squeezed into former wing tanks and voids. Some of them were two levels from their sanitary facilities. Two deck-edge elevators and four cargo booms (one 20 ton and three 4 ton) would be installed.

Because she was essentially a decked-over tanker, the CVE offered 20 percent more aviation gas (avgas) stowage than the new-construction LPH and far more endurance; the endurance would be wasted, however, if the requirement really was only 10,000 nm. Lacking overhead hangar space to accommodate a larger next-generation helicopter, her hangar and flight decks could not be strengthened to take the additional weight. Her larger hull would be more expensive to maintain. The CVE would be far slower in a seaway; a sea that would reduce the LPH to 13 kts would reduce the CVE to 8 kts.

In the new-construction helicopter carrier, not only was the hangar overhead higher, but above it was a continuous gallery deck (as in new attack carriers), which added over 11,000 sq ft of usable space. This space was available for command and control and for ship and troop officer accommodation. Troops themselves were higher in the ship, closer to their staging areas. Also, because the ship was designed from the beginning as a carrier, she carried her avgas in two separate areas fore and aft, thus storage was far safer than in the single tank of the CVE.

BuShips argued further that the low cost of the CVE, $29 million, was illusory, because she would have to be replaced well before the new-construction ship, which would last 20–25 years. Table 12-2 shows comparative characteristics of the two CVEs and the LPH, as they were understood on 8 August 1956.

One converted CVE, *Block Island* (CVE 106), was included in the FY 57 program as LPH 1. Unfortunately costs, particularly for the new missile ships, were rising rapidly. By 1958 it was clear that existing programs would have to be cut to pay for later overruns. The LPH and two escorts of the FY 57 program were canceled. By fall 1956 it was understood that two new-construction LPH would be requested in the FY 58 program. Overall, the LPH

Table 12-2. LPH Alternatives, 8 August 1956

	SCB 122 (CVE 55)	SCB 159 (CVE 105)	SCB 157 (LPH)
Length (ft)			
OA	512	561	590
BP	490	525	550
Beam WL (ft)	65	75	88
Draft, loaded (ft-in)	21-0	31-5	24-6
Hull depth (ft-in)[a]	62-0	71-7	73-0
Displacement (tons)			
Light	7,200	10,440	10,540
Loaded	11,000	24,560	17,500
Power (SHP)	11,200	16,000	25,000
Speed (kt)	18	19.1	21.5
Fuel (gal)	120,000	250,000	300,000
Radius (nm/kt)	7,000/15	35,000/18	10,000/20
Capacity			
Troops (officers/enlisted)	38/900	200/1,765	200/1,765
Tons	265	900	900
Sq ft	1,000	6,000	6,000
Helicopters	15	20	20
Flight deck (HR2S qty)	2 open[b] + 8	2 open + 13	4 open + 6
Hangar (HR2S qty)	5	5	10
Cargo discharge to flight deck (tons per hr)	50	150	150
Hangar deck clearance (ft-in)	17-6	20-0	20-0
Unit aircraft weight (gross tons)			
Flight deck	36,000	50,000	60,000
Elevator	36,000	40,000	40,000
Armament			
3-in/50	—	4	4
40-mm	8	—	—
20-mm	—	4	4
Cost ($ million)	10.6	20.4	49.1

NOTES: Dash indicates that specification is irrelevant.
[a] Height from keel to upper deck.
[b] Rotors open when ready to take off.

resembled the abortive SCB 43 escort carrier then in the design stage. Special efforts were made to provide against the structural problems to be expected in the low temperatures of the Arctic.

Final characteristics were issued 6 November 1956. Unlike the converted CVE, this ship had a secondary mission: ASW using embarked helicopters. The characteristics generally matched those developed earlier; overall length was 590 ft (550 ft on the waterline) and extreme beam was 105 ft, just within Panama Canal limits. A twin-screw powerplant would be capable of split operation. Speed was set at 21.5 kts on trial (20 kts sustained), and endurance at 10,000 nm at 20 kts. Boiler conditions were typical of World War II (600°F, 850 psi) rather than those being adopted for new carriers and destroyers. The ship was originally to have been roll stabilized (optimized for 15 kts), but BuShips argued that she was stiffer (with 8.8 ft metacentric height) than any other ship for which roll stabilization was required.

Unhappy with the 20-kt speed, the Marines asked for 25 kts. That would require about twice the power (50,000 SHP, four boilers). Even at higher boiler pressure, machinery weight would rise by more than 50 percent. The ship would grow to 595 ft and 18,500 rather than 17,450 tons fully loaded. To make endurance 10,000 nm at 25 kts rather than 20 kts would have required a 630-ft ship (21,750 tons fully loaded). Follow-on ship cost would have risen from $41 million for the 20-kt ship to $44.5 or $47 million for the larger, faster, ones.

BuAer pointed out that in 1961, when the ship entered service, helicopters might well be burning jet fuel (JP-5) rather than avgas. It might therefore be possible to eliminate special avgas tanks altogether, particularly if the ship's boilers also burned JP-5. At this time the U.S. Navy was developing a pressure-fired boiler, which used a fuel close to JP-5. Such boilers were actually only installed in a few frigates, but they figured in numerous ship design projects of this period. In this case the decision went against the special boiler. Given the very serious problems pressure-fired boilers encountered, that was fortunate.

Each ship would carry a BLT and its supporting helicopter squadron. Designated ships would accommodate an amphibious squadron (phibron) commander and regimental landing team commander with their staffs. One cargo elevator (14,000-lb capacity) would serve each of at least two cargo spaces on the ship. Estimated total cargo load was 900 tons, equating to 20,000 cu ft of ammunition, 41,000 cu ft (6,000 sq ft) of vehicles, and 20,000 cu ft of cargo, for a total of 81,000 cu ft, which is equivalent to slightly more than the Marines' 2,000 measurement tons. To service the vehicles, the ship would stow 5,000 gal of motor gasoline (mogas) together with its 300,000 gal of avgas.

Because she was operating aircraft, the ship had a more elaborate CIC than the usual transport. As envisaged, the LPH would be much more than an APA with a flight deck. She had flag facilities, including a joint intelligence office near the helicopter direction center. It would not take too much more to provide her with a capacity to control defending fighters; the required SPS-8 height finder would replace the Mk 56 gun director originally planned. At this time the future of VTOL (vertical take-off and landing) fighters seemed quite bright, and interest in the height finder may have reflected failed expectations that the ship would eventually operate fighters. In fact the ships never carried height finders; their air search radars (SPS-12, then SPS-40) occupied the space originally planned for the Mk 56. The characteristics also included combatant-level electronic countermeasures, which in 1956 meant a ULR-5 ESM set (0.1–100 MHz), a WLR-1 ESM set, a pulse analyzer (SLA-2), and an

Guadalcanal (LPH 7) is shown in May 1978. She has big CH-53 Sea Stallions, the helicopter equivalents of LCMs, on her flight deck and on her port deck-edge elevator (in the down position). The smaller CH-46 Sea Knights were roughly equivalent to LCVPs. The framework projecting from her island carries the antennas of a ULQ-6 defensive jammer. The dark radar antenna is for her SPS-40 air search set; the larger lighter-colored antenna is for a SPN-6 marshalling radar to control her aircraft. This combination was not too far short of what a carrier would have.

SLQ-5 jammer (8.8–10.5 GHz). Weather might preclude helicopter operations, so the ship was to have six embarkation stations, from which marines could scramble down nets into boats alongside, as was done in many APAs. The boats would be provided by other ships in the amphibious group.

On 20 May 1957 the SCB voted 6–4 in favor of the new-construction LPH. However, shipbuilding costs were rising. On 29 June the SCB met to review the characteristics for the FY 59 LPH. Adm. Arleigh Burke, the CNO, wanted any changes applied to the FY 58 prototype and also to the FY 57 lead CVE conversion (SCB 159). A study showed that $1 million could be saved by eliminating the stabilizer, another $300,000 by shortening the LPH by 25 ft, and another $300,000 by using single-shaft (Mariner-type) propulsion. The SCB decided to eliminate the stabilizer, to cut flight deck strength to 50,000 lbs, and to reduce the number of helicopter fueling stations. Troop accommodations were slightly reduced, from 200 officers, 35 master sergeants, and 1,765 enlisted men to 171 officers, 35 master sergeants, and 1,709 enlisted men. A proposal to reduce hangar deck height to 17 ft 9 in was rejected, and the board voted 5–4 to retain twin screws. The board reluctantly approved Burke's proposal to limit air conditioning, which was expected to save $1.5 million.

By the end of July the LPH had been redesigned, lengthened but with less beam (600 overall/570 between perpendiculars × 80/105 overall × 26 ft, displacement 11,000 standard and 18,000 tons fully loaded vs. 590/550 × 90/105 × 25 ft, 10,500/17,500 tons). Follow-on ship cost was down from $40 to $38.2 million; prototype cost was cut from $48 to $46.2 million. In August, flight deck strength was cut again, to 35,000 lbs. At this time the highest estimated gross weight of the HR2S was 31,000 lbs. The propulsion issue was revived; all active attack transports and cargo ships had single screws, and adopting single-screw propulsion would save 48 men. Characteristics issued in September 1957 showed the single screw. It was too late to reduce overall fuel oil stowage, which would have thrown off the ship's stability, so the ship had space for 400 tons more oil than needed. The new powerplant, identical to that in a Mariner freighter, produced 22,000 rather than the earlier 25,000 SHP.

Some minor changes were still to come. To improve stability, the ship's beam had to be increased. If the wider beam had been carried up to the flight deck, however, projections at that level would have made the ship too wide overall to pass through the locks of the Panama Canal. The designers therefore placed maximum beam below the waterline, tapering in the ship's sides above that point. Thus the waterline beam increased from 82 to 84 ft. Second deck height was reduced by 6 in. The superstructure would be built of aluminum. The designers wanted to provide the same degree of nuclear blast protection that they were building into the carrier *Enterprise*, but that proved impossible.

As the Marines envisaged the ship, she would operate like an attack transport, launching her troop carriers, having them wait, and then loading them; in this case, they would land to load. There was enough room on the flight deck for four helicopters to load at once. Because only 9 could be stowed on the hangar deck, however, 11 of the 20 would normally ride the flight deck. In that condition the ship could launch only three (two forward, one aft) at one time. The ship's cargo space could accommodate 36 crated Lacrosse support missiles, which would provide the helicopter-borne marines with nuclear firepower. Alternatively, this space could accommodate ASW torpedoes and sonobuoys.

Meanwhile yet another candidate for LPH conversions appeared. *Essex*-class carriers remaining in service immediately after World War II had never been modernized. Now it was too late to do so; in March 1957 they were proposed as austere helicopter carriers. They certainly had sufficient internal volume, and they were much faster than either the CVE or the new-construction ship. BuShips made a limited design study based on USS *Bunker Hill*, which had been laid up after being repaired in 1945. New cargo flats would replace the boilers in two of the four fire rooms, and two main propulsion units would be deactivated, thus speed would fall to 25 kts. Catapults and arresting gear would be removed and the after flight deck and deck-edge elevator strengthened to take heavy helicopters. The forward centerline elevators would be blocked by added accommodations for troops. The larger hull could accommodate 30 helicopters, and on the scale adopted for the new LPH that would require 450,000 gal of avgas, which in turn would need new tanks. At 17 ft 6 in, hangar height would be lower than that of a new-construction ship.

The complement would be only about half that of an operational carrier, but even so new accommodations would have to be installed for troops. Two new decks would fill the forward 340 ft of the hangar deck, up to the gallery deck level. Accommodation would leave much to be desired. Cargo arrangements would be somewhat more complex than in a new-construction ship, partly because of the subdivision and armor built into the older carrier.

BuShips cautioned that the carriers had been heavily used; maintenance costs of all *Essex*-class carriers were rising. Conversion would cost about

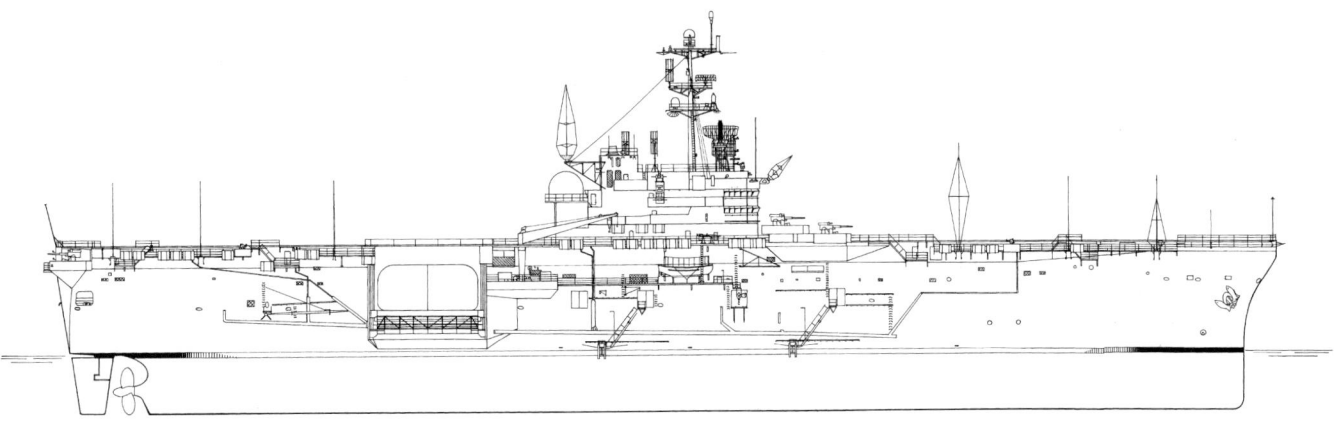

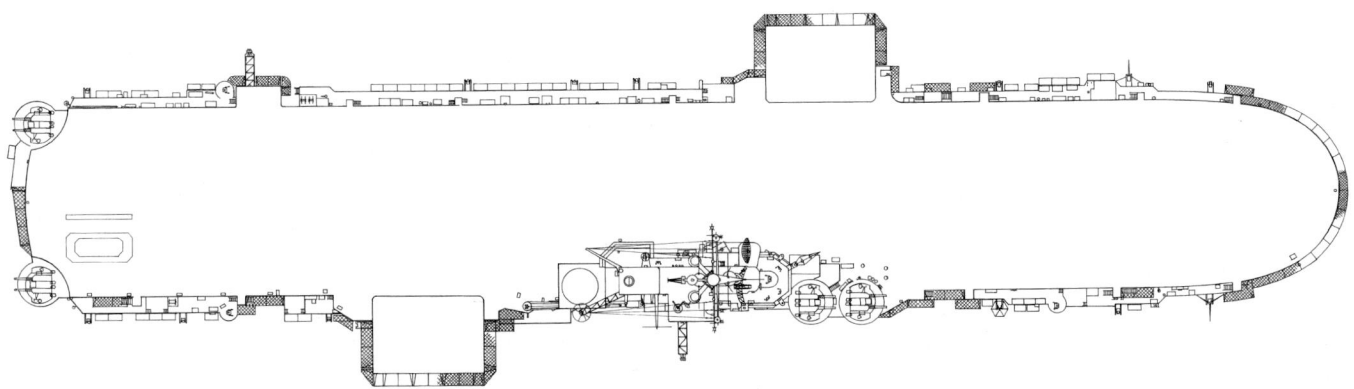

Tripoli (LPH 10), 1972, with SPS-40 air search radar atop her island and a SPN-35 air traffic control radar in the big radome abaft the bridge. She lacked the accompanying SPN-6 marshalling radar for air traffic control. The lattices extending outboard from the ship's island and from her port side aft carry ULQ-6 jammers to counter missile homing radars. At this stage *Tripoli* still had all her 3-in/50 guns; Sea Sparrow point defense missiles were not yet installed. These amphibious ships had a secondary role supporting mine countermeasures helicopters. *Tripoli* was mined in the Persian Gulf on 18 February 1991 while acting as mine countermeasures flagship during the Gulf War. *A. D. Baker*

$25 million, more than half that of a new helicopter carrier, and the ship would need about twice the crew. On the other hand, she would support about 50 percent more helicopters, troops, and cargo, and she would be ready about seven months sooner than a new carrier.

BuShips preferred the new-construction ship. However, in July 1957 Op-05 (DCNO for air) revived the conversion idea before the Standing Committee on the Shipbuilding and Conversion program. The conversion might be much more austere than had been envisaged. For example, instead of being removed, which was expensive, half the boilers might merely be mothballed. BuShips reiterated its opposition.

The conversion concept was tested in Atlantic Fleet amphibious exercise (LantPhibex) 1-58 in early 1958, when the carriers *Tarawa*, *Valley Forge*, and *Forrestal* delivered a regimental landing team. This was also the first major test of the Marines' helicopter assault (vertical envelopment) tactic. Table 12-3 shows details of ensuing representative LPH, LHA, LHD, and LPD ships that were used for vertical envelopment.

By 1958 it had been decided to convert two ships, *Boxer* and *Princeton*, into austere LPH, supplementing seven new-construction LPH. A third, *Valley Forge*, was later added. A fourth, *Lake Champlain*, was dropped from the program due to manning problems. Conversion was so austere that it did not figure in formal shipbuilding and conversion programs; it was done using operations and maintenance funds.

As converted, an *Essex*-class LPH could accommodate slightly fewer troops than a new-construction ship (180 officers and 1,770 enlisted men vs. 215 officers and 1,994 enlisted men) but she could carry more than twice as much cargo (400,000 cu ft or

The *Essex*-class carrier *Princeton* is shown as a helicopter carrier, with Marine SH-34s (originally designated HR2S) on deck. She retained two of her original four twin 5-in gunhouses, as well as two port side deck-edge 5-in guns, for fire support. The radars on her single heavy pole mast were those she used as a conventional carrier: an SPS-12 for air search and SPS-30 for height finding. The most visible sign of modification was the removal of her arresting gear and her catapults.

2,000 tons vs. 81,000 cu feet or 900 tons). Maximum speed was 27 rather than 21 kts, and endurance was 17,600 nm at 12 kts or 9,000 at 27 kts compared to 16,600 nm at 11.5 kts or 10,000 at 21 kts. The carrier was also far more expensive to run. Her ship complement was 93 officers and 1,239 enlisted men, compared to 71 officers and 530 enlisted men on board the new-construction ship. Ships retained their four twin 5-in/38 flight deck guns for shore bombardment, but not their single deck-edge 5-in/38 or their lighter weapons. These were elderly ships; all three were subject to FRAM II modernization. As of 19 December 1958 plans called for FRAM II modernization in FY 61 and FY 62.

Valley Forge (LPH 8), the third conversion, was more austere than the other two. She lacked their added 02 level compartments, which accommodated 316 marines, hence she had a capacity of only 1,500 troops. She was also about 25 percent short of the required square footage for vehicles, and her commanding officer proposed that some might be stowed on the flight deck alongside the island. The ship lacked a public address system for troops on the flight deck, thus making the boarding of helicopters hazardous.

The ship's CO also complained, in 1963, that she lacked evaporator capacity. She could make 60,000 gal of potable water per day, but troops needed 90,000 gal. Her boilers needed 67 gal of fresh water per mile. Capacity was 157,591 gal of potable water and 135,374 gal of fresh water. If her inboard voids were converted to take potable water, capacity

Table 12-3. Ships for Vertical Envelopment

	Guadalcanal (LPH 2)	Boxer (LPH 4)	Tarawa (LHA 1)	Wasp (LHD 1)	Raleigh (LPD 1)	Cleveland (LPD 7)	San Antonio (LPD 17)
Length (ft-in)							
WL	556-0	820-0	777-8	777-10	500-0	NA	NA
OA	592-0	888-0	833-9	844-0	513-1.5	569-9	684-0
Beam (ft-in)							
WL	83-8	93-0	106-0	106-0	84-0	84-0	105-0
Extended	104-0	147-6	132-0	140-0	NA	NA	NA
Draft, loaded (ft-in)	26-7	28-3	26-0	26-8	23-0	23-0	23-0
Displacement (tons)							
Light	10,989	25,800	33,536	28,233	7,962	9,128	NA
Loaded	18,340	36,780	39,967	40,535	13,745	16,550	24,900
Boilers	2 CE	4	2 CE	2 CE	2 B&W	2 FW	—[a]
Conditions (psi/°F)	600/850	565/850	700/900	700/900	600/873	600/873	—
Power (SHP/shafts)	22,000/1	75,000/2	77,000/2	77,000/2	24,000/2	24,000/2	40,000/2
Speed (kt)	21.3	27	24	24	21	21	25
Radius (nm/kt)	10,000/20	17,600/12	10,000/20	9,500/20	9,600/16	7,700/20	NA
Complement[b]	87/507	99/1,239	58/1,005	62/1,084	27/30/444	27/466	24/396
Capacity							
Troops[c]	193/1,806	180/1,770	1,903	1,700	73/589	75/781	720
Tons	900	2,000	NA	NA	2,500	3,900	NA
Sq ft	3,400	8,240	33,700	20,900	12,500	12,000	25,000
Cu ft	49,500[d]	400,000	116,900	109,000	175,000	40,000	25,000
Boats							
LCAC	None	None	1[e]	3[f]	1[g]	1[h]	2[i]
Helicopters[j]	26	NA	41	46	4	4	6
Armament							
5-in/38	—	4 × II[k,l]	—	—	—	—	—
3-in/50	4 × II	—	—	—	4 × II	4 × II	—
3-in/54	—	—	3[m]	—	—	—	—
Sea Sparrow	—	—	2[n]	2 plus 2 RAM[o]	—	—	2 RAM

NOTES: Dash indicates that specification is irrelevant; NA indicates relevancy but data not available.

[a]Diesel: four Colt-Pielstick PC2.5 engines.

[b]Complement is given as officers/CPOs/enlisted when three figures are listed and officers/enlisted for two figures.

[c]Officers/enlisted when two figures are listed; when only one figure is given, breakdown is unknown.

[d]Design documents initially specified 80,000 cu ft, then 40,000 cu ft.

[e]Alternately, 4 LCU or 7 LCM(8).

[f]Alternately, 2 LCU or 6 LCM(8); ships also carry 4 LCPL.

[g]Alternately, 1 LCU and 3 LCM(6), or 4 LCM(8), or 20 AAV.

[h]Alternately, 1 LCU plus 4 LCM(8), or 28 AAV.

[i]Alternately, 4 LCM(8) or 20 AAV.

[j]Helicopter spots, in CH-46E equivalents, on hangar and flight decks.

[k]Multiple gun mounts are indicated by Roman numbers; therefore 4 × II indicates four twin mounts.

[l] At least initially, *Boxer* retained her four single deck-edge 5-in/38s as well. Other LPHs had only two twin 5-in plus two or four deck-edge single mounts. All smaller weapons were landed.

[m]One gun removed early 1990s, the other two about 1997; as of 2001 gun armament is two 20-mm Phalanx and eight 0.50 caliber machine guns.

[n]Replaced by two RAM launchers from 1992 on.

[o]Plus two Phalanx and eight 0.50 caliber machine guns.

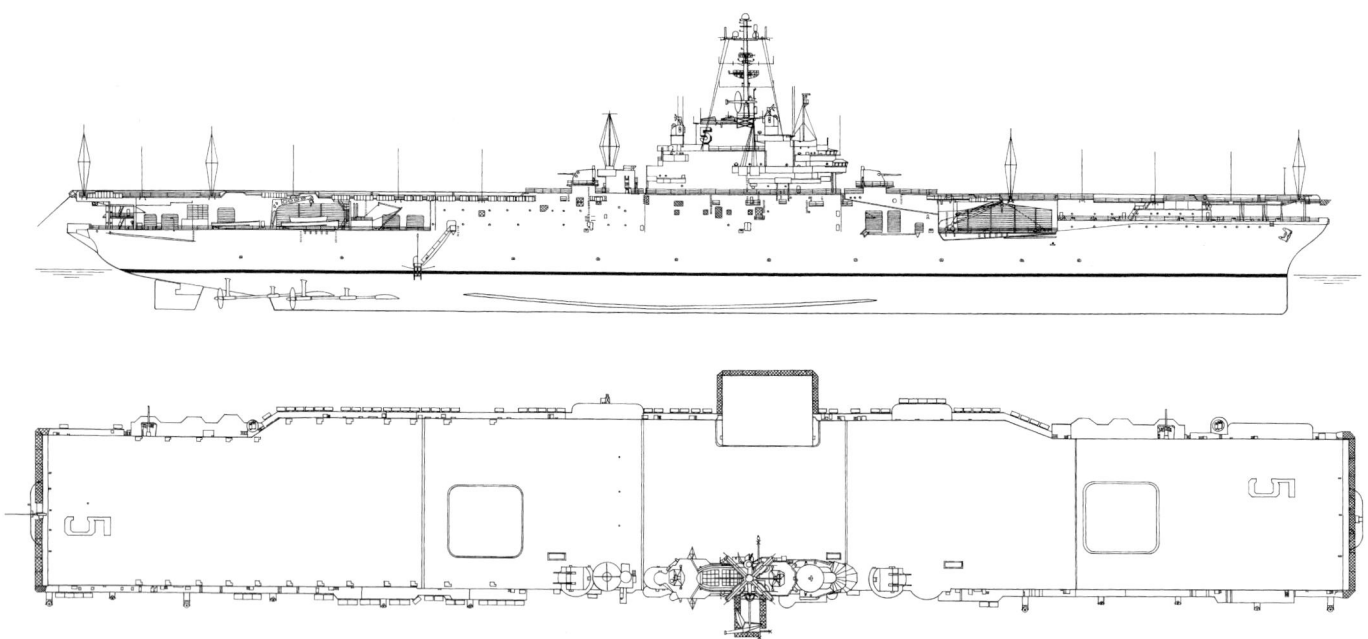

Princeton (LPH 5), an *Essex*-class fleet carrier modified for amphibious assault, in 1965. She retained two of her four twin 5-in/38 mounts and two of her four single 5-in (along the port side of the flight deck). Search radars were SPS-12 for air search, SPS-10 for surface search, and SPS-30 for three-dimensional air search (the big dish on the island). Antennas for the ULQ-6 ECM system were mounted on lattice outriggers. The antennas of the accompanying WLR-1 ESM system were mounted on her mast, which also carried a URN-20 Tacan aircraft homing/navigation beacon at its head. Note the twin funnel caps and the big discage antenna replacing her No. 3 5-in twin mount. *A. D. Baker*

would increase by 80,000 gal. She lacked the permanent air conditioning needed for Southeast Asian operations. Instead, she and other amphibious ships used pooled (cross-decked) air conditioners. They could generally keep internal compartments 7 degrees cooler than the outside air, which might well mean 93 degrees on a 100-degree day.

Existing radars were inadequate. The main air search set, SPS-6C, lacked the moving target indicator needed to operate effectively near land. The CO wanted the ship's SPS-8A height finder replaced by the new SPS-30 then on board full carriers. The ship's old air traffic control radar (SPN-6) would not effectively control the new helicopters (CH-46 and CH-53) about to enter service. The existing generator plant was inadequate. Power could not be supplied to all electronics. Lighting on the damage control deck (second deck, below the hangar deck) was poor. The ship's CO also pointed out that the ASW mission, counted as secondary in the characteristics, might actually be quite important, because the ship would usually launch her helicopters from outside any screen. It is not clear to what extent these problems, manifest on board *Valley Forge*, were typical of the two other converted ships.

On the other hand, the converted *Essex*-class carriers were far better seakeepers than the *Iwo Jimas*; a February 1965 memo from Op-05 stated that "people who have served in *Iwo Jima*–class LPHs assert that they are 'dogs' in moderate seas—everyone seasick." The *Iwo Jimas* were also extremely crowded, and reportedly much disliked by their crews as well as the troops they carried.

To make matters more complex, in the early 1960s helicopters changed. The LPH was designed around a single type of helicopter, the Sikorsky HR2S (S-56). It could carry 26 troops or a small vehicle, and thus corresponded broadly with an LCVP. It was superseded by a pair of helicopters: the Boeing-Vertol CH-46, which won a Marine Corps competition in February 1961 (first flown 16 October 1962, first delivered early 1965), and the Sikorsky CH-53 (first flown 14 October 1964, first delivered September 1966).

The CH-46 Sea Knight was likened to an LCVP. The CH-53 was compared to an LCM, although it could not lift a tank. In its initial version it could lift 38 marines or a 105-mm howitzer. Work on CH-53E, an enlarged version with three (rather than two) uprated engines, began in 1971. It could carry 56 marines or 32,000 lbs of cargo to a 50 nm radius (or 16,000 lbs to 500 nm). It was introduced into the fleet in 1981. The CH-46 was about the size of the earlier HR2S, but the CH-53 was far larger. Thus an *Iwo Jima*–class LPH could accommodate 20 CH-46 but only 11 CH-53.

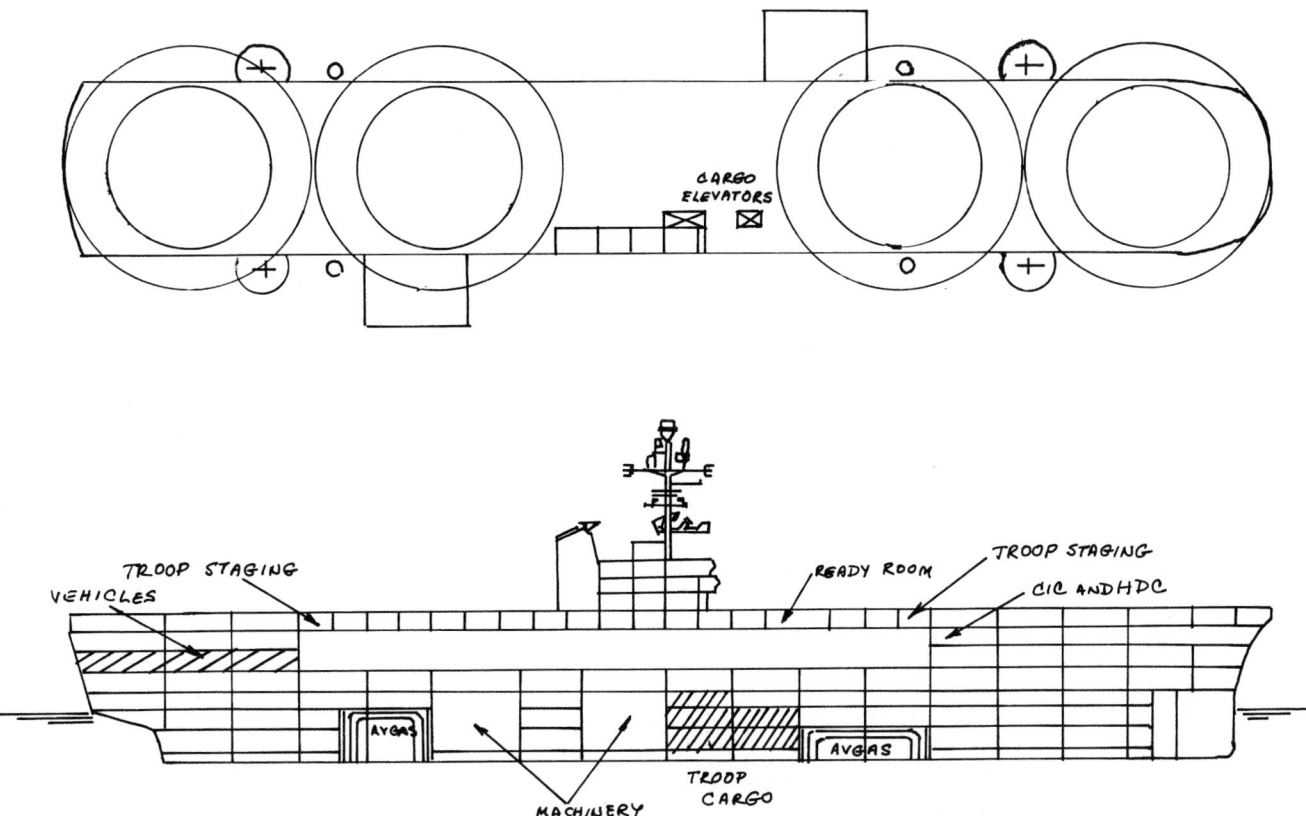

Iwo Jima (LPH 2), from a 28 January 1957 preliminary design sketch. In contrast to the later LHA and LHD, she was in effect a small carrier with a conventional hangar deck amidships; her design was derived in part from that of an abortive escort carrier. The circles on deck are helicopter rotor clearance and safety circles and they show that she could fly only four large helicopters simultaneously. Her 3-in/50 gun mounts are indicated by crosses. The small rectangles marked by crossed lines are her cargo elevators. Her superstructure shows a flag bridge below the pilothouse level. The single mast carries a navigational beacon (URN-3 Tacan) at its head, with SPS-10 surface search and SPS-12 two-dimensional air search radars below it. Ships of this type were ultimately equipped with the full carrier-controlled approach radar suite, comprising a secondary air search radar (SPN-6) and a three-dimensional radar (SPN-35, in a big radome).

By 1960 long-range planners looked forward to the retirement of *Essex*-class ships with steam catapults and angled decks. The tentative long-range objectives paper (TLRO-60) of 22 August 1960, looking ahead to 1972, envisaged replacement of the three *Essex* conversions by modernized ships of the same class, which would be designated LPV.[1] They would carry a mixed load: half a battalion landing team (with helicopters) and about 30 Marine fighter or attack aircraft, to provide integral air superiority and air support for the expected small-scale operations. Such ships could also deliver nuclear weapons in support of allies. A force built around an LPV might provide an economical Cold War presence in the Indian Ocean, augment this deployment as required, or, in an emergency, be deployed to the South Atlantic. Because the ship would retain her conventional carrier capability, she could work with an attack carrier.

In October 1958 the long-range objectives group (Op-93) proposed building 2 LPH annually for FY 60–64, then 2 each in FY 65 and FY 66, for the desired total of 16 (including 2 in FY 58–59). This proved impossible, although later projections continued to provide for two LPH each year. Only seven new-construction ships were built, beginning with *Iwo Jima* (LPH 2) in the FY 58 budget. She was followed by two more in FY 59–60, then two more in FY 62–63 and two more in FY 65–66. The three *Essex* conversions helped fill gaps in capability. They were decommissioned in 1969–70.

The LPV idea was not taken up, but it did not quite die. In April 1965 Op-93 issued MRO-76 (medium-range objectives for the period through 1976). It advocated using the navy's standard C2 carrier onboard delivery airplane to carry paratroops from second-line attack carriers; 40 such aircraft could lift a 1,500-man BLT. The new counterinsurgency (COIN) aircraft could fly from the same ships, providing a degree of fire support. Nothing came of the idea.

Raleigh was the first of a new class of personnel-carrying dock landing ships with helicopter decks. She is shown in Chesapeake Bay late in 1975. The deck aft could take helicopters, but the ship had no hangar.

Parallel to the LPH, another new type of amphibious ship was conceived in 1956, first appearing in the January 1957 report of the newly formed Op-93. It was identified as LPD (landing ship, personnel, dock) to differentiate it from the LPH (landing ship, personnel, helicopter). The Marines preferred a ship that could launch both helicopters and heavy beaching craft such as LCUs (ex-LCTs), but that was not yet deemed affordable.

As designed, the LPD (SCB 187) was a compromise. Instead of the very large ship initially imagined, the existing LSD 28 was modified. It was seen as a complement to the LPH, just as the World War II AKA had provided heavy cargo support and supplemental boats for the APA. The main conceptual difference was that the LPD had a substantial troop capacity, about half that of the LPH.

Space provided by shortening the well deck of the LSD was used for cargo (about 2,500 long tons) and troops (980 marines, including a helicopter detachment). The well deck incorporated an overhead monorail from which palletized cargo could be loaded into embarked craft. There was also a 30-ton deck crane. The well would accommodate nine LCM(6), or three LCM(6) and an LCU, although the LCU might not always be carried. A permanent flight deck built above the well provided two take-off spots. In contrast to an LPH, an LPD would offer only limited helicopter maintenance facilities; her helicopters would always be based on another ship. Her aviation facilities—Tacan (the tactical air navigation beacon system), flight control station, fuel systems, and deck lighting—were provided only for landings to load or offload troops and cargo and for refueling. Like the new LSD, the LPD had a twin-screw steam plant located below the well deck. Living and office spaces were air conditioned.

The LPD was far less expensive than an LPH. In January 1957 it was estimated that, in FY 59 dollars, follow-on LPDs would cost $25 million, compared to $40 million for follow-on new-construction LPHs and $20 million (raised to $29 million in March) for a follow-on CVE conversion. At this time an LSD would cost $20 million, and it was thought that a 20-kt LST would cost $14 million. Prototype costs were $48 million for the LPH and $29 million for the LPD. The LPH and LPD established an important trend toward concentrating amphibious lift in fewer, yet larger, ships of specialized design.

In terms of the 1955 recommendations, 16 LPD could replace the remaining 16 fast LSD plus the fast APA and AKA. However, the amphibious force also used numerous slow World War II–built LSDs, which were wearing out; thus long-range planners generally also wanted new LSDs (discussed in Chapter 11).

The first LPD was included in the FY 59 program. From then on, programs paired LPDs with LPHs. The FY 59 program included one of each type, plus a Mariner AKA conversion. In October 1957 it was suggested that this combination be repeated annually (with an APA instead of the AKA) beginning with FY 60. In fact no more Mariner conversions were authorized. Moreover, this rate of construction would not have produced the 20-kt amphibious force quickly enough. Nor did it provide enough well-deck space for LCUs. More LSDs, which offered more well-deck space, were needed.

Not all commands wanted the LPD. In April 1957 the FY 59 program was being formulated. PhibLant wanted 20-kt lift for one regimental combat team: four APA, one AKA, one LPH, and one LSD. The APA may have been envisaged as APA-M, in which case it was not too far from being an LPD. The Atlantic Fleet asked for new-design attack transport and cargo ships (one each) plus a single converted APA, and a converted LPH plus two LSD. The Pacific Fleet wanted two LPD, two new-build LPH, two LSD, and one LST. The Standing Committee on the Long Range Shipbuilding Program wanted one LPD, three LPH conversions, and one APA conversion for FY 59. Later in April the three conversions were replaced by two new-build LPH, because the estimated cost of conversion was rising; later the two new-build LPH were cut to one.

Three ships were built to the original *Raleigh* (LPD 1, SCB 187) design, with *La Salle* (LPD 3) being the flagship version (SCB 187A), under the FY 59, FY 60, and FY 61 programs. The FY 60 program showed one LPD and one LPH. In 1958 Op-93 had called for two LPD in FY 60, and none in FY 61; in fact one was built in each year. As formulated in 1959, policy called for a quarter of the LPDs to have amphibious squadron (phibron) flag facilities, and another quarter to have transport division (transdiv) flag facilities, a second SCB 187A being planned for FY 62. The only flagship of the *Raleigh* class, *La Salle*, was converted into a fleet flagship (AGF 3) in 1972, serving the commander of the Middle East Force until 1993, and as flagship of the Sixth Fleet from November 1994.

The *Raleigh*s were followed by 12 ships of a lengthened *Austin* (LPD 4, SCB 187B) design, beginning with 3 in FY 62. In the fall of 1958, plans had called for 4 LPD in FY 62, 2 in FY 63, 4 in FY 64, 2 in FY 65, and 6 in FY 66, for a total of 20 for FY 60–66, with 7 more planned for FY 67–69. In fact three were bought in FY 63 (LPD 7–10), three in FY 64 (LPD 11–13), and two in FY 65 (LPD 14 and 15). The single FY 66 ship, which would have been LPD 16, was deferred in favor of the LHA and then canceled in February 1969. Plans for FY 66 were disrupted by the advent of the LHA (described below), and the vast fleet of LPDs envisaged in 1958 never appeared. LPD construction did not resume until the end of the century, with the new *San Antonio* (LPD 17) class (see Chapter 17).

Compared to the *Raleigh* class, the *Austin*s have greater cargo capacity (3,900 tons or 260,000 cu ft vs. 2,500 tons or 175,000 cu ft). LPD 7–13 (SCB 187C) were fitted as phibron flagships, with an extra bridge level. They have a flag plot, ship signals exploitation space, and a supporting arms coordination center (SACC). However, these spaces have not been used for some time. The *Austin*-class well deck was slightly larger, accommodating two LCM(6) as well as four LCM(8); *Raleigh* could take nine LCM(6) or one LCU plus three LCM(6). One ship, *Coronado* (LPD 11), was converted into a fleet flagship (AGF 11) in January 1980. She served as Sixth Fleet flagship in 1985–86, and became Third Fleet flagship on 26 November 1986.

There were soon proposals for additional modifications. In December 1961 the Atlantic Fleet asked for below-decks helicopter stowage, to permit extended independent helicopter operations. Given a shortage of LPHs, this was necessary to provide a minimum helicopter assault capability in the Caribbean on a continuous basis, the need for which was demonstrated during 1962. The necessary capability included not only stowage and maintenance, but also helicopter control during the ship-to-shore movement. In effect the LPD would gain some of the attributes of an LPH, albeit on a much smaller scale.

Both Atlantic and Pacific amphibious forces proposed modifications to make this possible. In October 1962 OpNav authorized BuShips to make this change in LPD 4 and later units. In May 1963 the Atlantic Fleet commander ordered the Atlantic amphibious force to provide an interim helicopter capability in the LPD 1 class.

The *Austin*-class LPD *Trenton* is shown on 18 January 1971. Note the small collapsible hangar extended onto her flight deck. *Lockheed Shipbuilding and Construction Company*

Coronado was the flagship version of the *Austin* class, distinguishable by her extra bridge level. Shown on 28 April 1970, she was converted into a fleet flagship in October 1980.

At a November 1963 conference BuShips was asked for covered main deck stowage for three CH-46s. The best it could offer was stowage for two. Costs would include the 35-ton crane and two LCM(6). Nothing was done. The flight deck could already accommodate up to six CH-46s, with two fly-off spots.

In December 1964 the Marines attacked the idea of an LPD capable of operating truly independently. The minimum unit was the BLT, which the LPD did not carry, supported by a full helicopter squadron. Sustained air operations, moreover, demanded intermediate-level maintenance, which the LPD certainly could not provide. As a "commando," or self-contained unit, the LPD could support company-level operations, which were no longer very useful. Thus, to be effective, the LPD really did have to operate with an LPH. Deploying eight helicopters independently on board a commando LPD, moreover, would leave the squadron with no useful capability at all. The Atlantic Fleet, which was sponsoring the idea, replied that sometimes ships had to operate independently; from the beginning, the LPD should have had helicopter stowage. The existing LPD was too inflexible. For the present, if nothing could be done for the large helicopters, at least a utility helicopter (UH-1E) should be stowable below decks.

This was possible, albeit in a modified way. Most of the ships were fitted with a telescoping hangar, a type originally tested by the Coast Guard icebreaker *Northwind*. Initially there were hopes that it could be large enough to accommodate four CH-46As, but the one actually fitted could take only a single utility helicopter.

Given this interest, in October 1964 OpNav asked BuShips for a cost and feasibility study of an LPD that could hangar and maintain eight CH-46s or four CH-53s. The new design was 24 ft longer and 5 ft wider, with 6 in more draft. Greater length provided less cramped machinery spaces; each space was lengthened by 12 ft and its overhead raised. The new design needed 10,000 more SHP to maintain its speed, thus a new boiler had to be tested. Cost would be $53.7 million for the lead ship and $47 million for each follow-on ship, compared to $39 million for a follow-on version of the existing design. As of December 1964, the lead ship

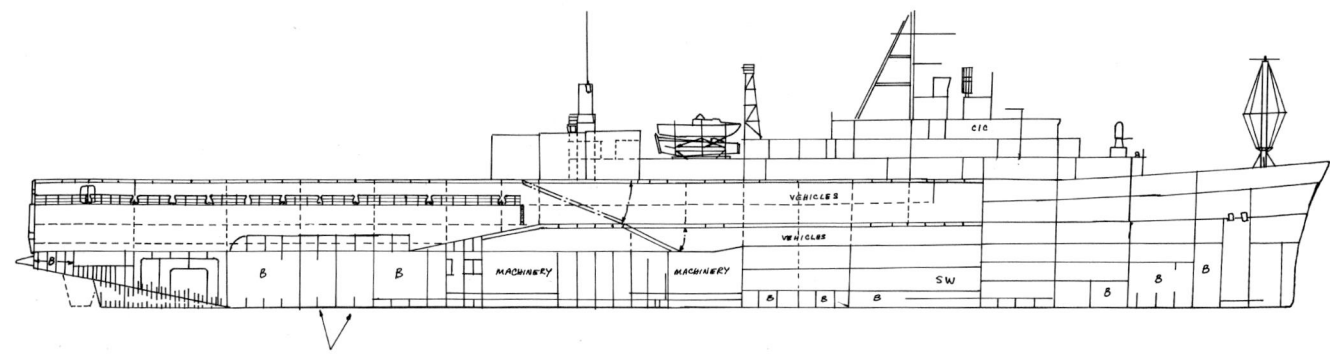

Ogden (LPD 5) inboard profile, February 1998. Tanks marked B are for ballast. The arrowed tank between two ballast tanks is for JP-5 jet fuel, with a JP-5 pump room forward of it. The double tank above the skeg right aft is for aviation gasoline (avgas). The object in the well deck aft is a transverse water barrier. Forward of it is the boat ramp. Forward of the boat ramp are upper (to flight deck) and lower (to lower vehicle stowage) vehicle ramps and hatches. Note the collapsible helicopter hangar at the flight deck level. SW indicates a special weapons (i.e., nuclear weapons) magazine. Compared to the original LPD, this class had a 48-ft plug for added troop cargo and vehicle-carrying capacity, and to enlarge the helicopter landing area. This ship had cargo elevators running from the hold to the upper vehicle stowage and to the main deck, plus conveyors for palletized cargo. Note the 'tween deck ramps for vehicle access. An overhead monorail was installed above the well deck to move cargo into boats docked in the well. In the design, special attention was paid to wave and water surge reduction in the well deck.

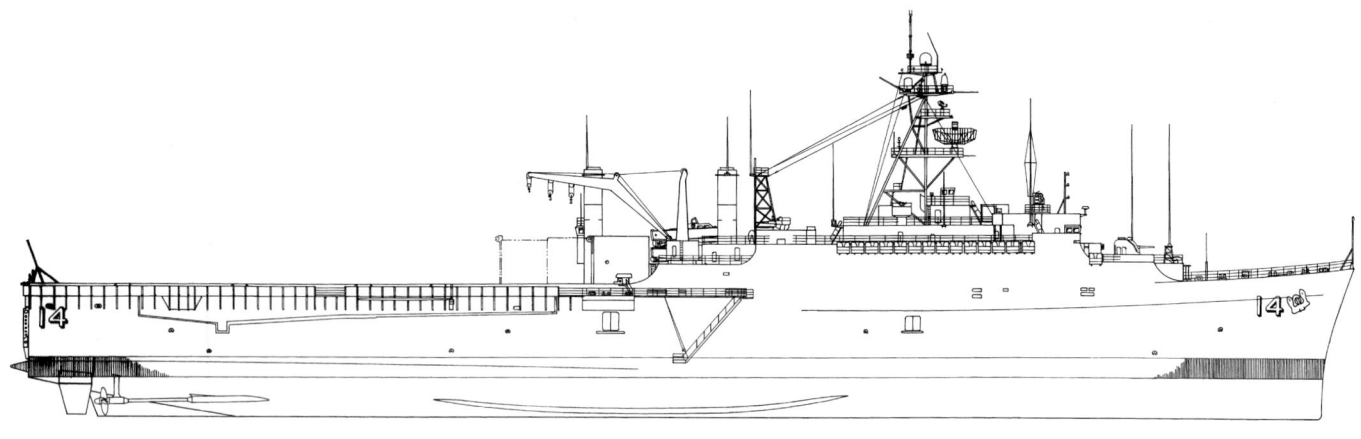

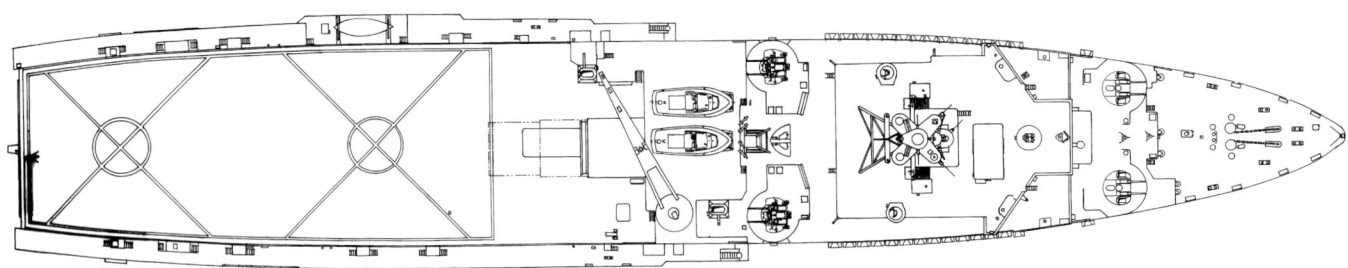

Trenton (LPD 14) in May 1971 with all four twin 3-in mounts aboard, much as completed. The telescoping helo hangar is shown retracted, with the extending portion in dotted lines. *A. D. Baker*

for the new class was tentatively scheduled for FY 67, because the design was already too late for FY 66. In February 1965, however, OpNav rejected the new design as too expensive. It asked for an LPD with a telescopic hangar for two CH-53s, requiring it to cost no more than the current design plus the cost of the hangar. No FY 67 LPD, however, was ever ordered.

By 1960 the Marines were beginning to achieve the 20-kt vertical assault force they wanted. The goal, as expressed in the 1960 long-range requirements document (LRR-60) for the period through

1972, was one full division/air wing of 20-kt lift for each of the two fleets, Atlantic and Pacific. However, the modernization program for LPDs and LPHs could not buy lift for the support echelon (typically 4 APA and 25 AKA or equivalent). Support would have to come from the Military Sea Transportation Service (MSTS; see Appendix A) or from commercial shipping. Unfortunately, the U.S. merchant marine was continuing its steady decline in numbers. Thus in 1960 it seemed that even dock-to-dock lift would be limited largely to war-built Victory ships. The problem of providing sufficient modern backup to the amphibious assault force was never really resolved (see Appendix A for attempts).

Force calculations were being revised. Instead of a full divisional assault at one time and place the 1960 long-range planners used the more likely case of a smaller unit (regiment or battalion) maintained continuously in a forward area. That required enough lift for three such units: one forward, one returning from deployment, and one refitting. This was much the same arithmetic used for continuously deployed carriers. For the Pacific Fleet, the small unit was a regimental landing team (RLT) and its associated marine air group (MAG). Because an RLT was about a third of a division or marine expeditionary force (MEF), the continuous requirement was not too different from a one-shot full-division requirement. Each of three teams required three LPH, five LPD, three LSD, and two fast LST. Backup (20 percent) and reserves would amount to one LPH, two LPD, one LSD, and two LST, plus one each Mariner AKA and APA. Some of the totals involved, 10 LPH, 17 LPD, 10 LSD, and 8 fast LST, exceeded the numbers actually built. These figures help explain why forward deployments were typically limited to a single BLT, about a third of an RLT.

For the Atlantic, it seemed that in many cases a BLT would suffice, although sometimes an RLT/MAG would be required. Because the Atlantic theater was so varied, two BLTs might be needed at any one time. On the other hand, distances were much shorter, so the Atlantic Fleet might be able to rely on slower ships for about 50 percent of its needs. The set requirement, then, was for both fast and slow lift for one RLT/MAG, which might operate either as continuous forward deployment of one BLT (a third of the RLT) or as deployment of the full RLT from the United States. Fast lift for two BLTs would provide backup for deployed forces.

At this time there was interest in the Indian Ocean, so the planners added lift for one light BLT with integral air for routine Indian Ocean deployment, supported by an AKA carrying heavy equipment and stores, to provide staying power. This AKA was presumably the forerunner of later attempts at prepositioning. The Indian Ocean force would be built around an LPV (as described above) and an LPD; two fast BLTs would be built around the other two LPV, four LPD, and two fast LSD. The fast RLT would require three LPH, six LPD, and three fast LST, and another LPH would be needed for backup. The slow RLT would make do with three APA, one fast LSD, five slow LSD, three AKA, and nine slow LST.

Apart from the Indian Ocean and backups, then, as envisaged about 1960, the 1972 fleet would need 12 LPH, 21 LPD, 10 fast LSD, 5 slow LSD, 9 fast LST, and 9 slow LST. The number of LPH, however, was significantly lower than previously adopted. None of these figures would be met, although the advent of the LHA at the end of the 1970s helped.

There was increasing interest in how quickly U.S. forces could get to overseas trouble spots. In October 1960, an internal BuShips memo proposed, in place of the 20-kt assault force, a 30-kt force based on converting obsolescent war-built carriers (*Essex* and *Independence* classes). At about the same time there was interest in converting *Iowa*-class battleships into commando ships, single-package ships capable of carrying troops, helicopters, and heavy fire support (see Chapter 14).

The last pre-Vietnam thinking on amphibious assault tactics was reflected in a February 1963 long-range objectives group report (LRO-74) for the period through 1974. As before, the problem was how to limit the number of expensive new amphibious ships to something affordable. This time the key perception was that the Marines typically planned to bring in two-thirds of their force by helicopter, the remaining third over the beach. Moreover, typically only two of the three units of a triangular formation were committed to an attack, the third would be held in reserve. Thus, of the nine BLTs of a division, only six had to be counted as initial assault elements, and only four of them (5,500 men) would land by helicopter; the other two, 2,700 men, would land by amphibian or landing craft. In theory that would cut the number of LPH per division to four. In fact the Marines also wanted the option to land two-thirds of the assault over a beach, so they needed enough craft to land four rather than two BLTs; that did not affect the really expensive LPH. The three remaining BLTs, the follow-on echelon, plus combat support units (about 16,300 troops) would come over the beach aboard landing craft. They would arrive on the fifth day after an operation began (D+5).

Unfortunately, to produce the desired shock effect, an assault force had to land in a very short time, 90 minutes for 5,500 men and 425 tons of supplies, 50 nm beyond the beach. Plans called for medium (2-ton load) and heavy (4-ton load) VTOL transports. To land the force would require 300 trips by medium VTOLs or 105 trips by heavy ones. Within 90 minutes a medium VTOL could make three trips, a heavy one two, so 100 medium or 53 heavy VTOLs (or an appropriate combination) were needed. An LPH could operate only 7 medium VTOLs and an LPD 2 heavy ones, so planners wanted 8 LPH (56 medium VTOL) and 9 LPD (18 heavy VTOL), although 7 LPH might suffice. This was much like the earlier situation with boats, where no AKA could carry the full boat group needed to land her BLT. Using existing helicopters, a vertical assault required 180 CH-46As for personnel (9 LPH loads) and 36 CH-53As for cargo. Further LPDs and other ships were needed for the assault over the beach.

The full marine expeditionary force (MEF), that is, the division/air wing assault echelon, was set at 3 AGC (overall commander, vertical, and over the beach), 8 LPH, 5 AKA, 13 LSD, 13 LPD, 20 LST, and 2 tankers. The follow-on force, arriving on D+5, would be about 35 MSTS or chartered ships (e.g., 4 transports, 28 cargo ships, and 3 tankers). It would carry about two-thirds of the cargo for the operation. Thus a single MEF landing would have required nearly all the fast amphibious lift ships built over the next decade.

According to another analysis, completed in December for Secretary of Defense Robert S. McNamara, concentration on LPHs had left a serious shortfall in combat vehicle lift. Effective assault capability was only about 60 percent of the desired two division/air wings. This perception, embodied in the draft presidential memorandum (DPM) on FY 65–69 amphibious assault forces, was probably responsible for construction of the *Charleston*-class AKAs. APAs had been abandoned in favor of LPHs and LPDs.

At this time the navy's force goal for FY 76 was two 20-kt MEF-level forces: 6 AGC, 16 LPH, 10 AKA, 26 LSD, 26 LPD, 40 LST, and 4 tankers, plus ancillary ships such as 6 APD and 2 APSS. McNamara supported both the 20-kt capability and the ability to mount a full divisional attack. With limited rather than general war in view, 20 kts no longer seemed so important as protection against submarines. In a world full of crises, however, it offered valuable flexibility. A 20-kt force in the Eastern Pacific could deploy to many areas in the Atlantic about as quickly as a 13-kt force based in the Atlantic.

Certainly smaller units could handle many contingencies, but the Marines maintained (and McNamara accepted) that they could not handle determined opposition. Only a division could seize and hold an area so large that the beachhead itself would not be swept by artillery outside.

It was not as clear that two-division capacity was needed. The classic justification was that two divisional assaults might have to be mounted so close together in time that ships used for one could not possibly be used for the other. McNamara agreed, but his reasoning was faulty. The Atlantic Fleet was concerned largely with the Soviets, the Pacific with the Chinese, and by 1963 the two were close to war. Only if they decided to act together would the old requirement hold. McNamara did name two contingencies, Europe-Mediterranean and Cuba, in which full two-division assaults would be needed.

By 1965 both fleets, having used LPHs, were complaining that it was, at the least, awkward for troops to disembark from an LPH into landing craft supplied by another ship. OpNav ordered a study (BuShips OpNav Study 24-65) of a new LPH carrying over-the-beach infantry assault craft (12 LCVP or air cushion vehicles then under consideration, plus 1 or 2 LCPL). She would have the LPH flight deck and performance. Of her four twin 3-in/50, two might later be replaced by the new Point Defense Missile System (later Sea Sparrow).

In this context, in May the SCB asked whether LPH 11 (*New Orleans*), then under construction, could be modified to carry small landing craft (2 LCPL, 12 LCVP, and 97 more personnel to operate and maintain them). A 52-ft section would be added amidships, increasing displacement by 1,800 tons, but it would not affect the ship's speed. However, the plug would induce a 10 ft trim by the stern, and unless the ship was redesigned her survivability would become marginal. The cost would be $8 million and 10 months' delay; and the builder, Philadelphia Naval Shipyard, was already too heavily committed to a new amphibious flagship and a new fast LST. Nothing could be done, but the last ship, *Inchon* (LPH 12), was given davits for two LCVP landing craft. That had little impact, particularly since the Marines' preferred surface assault vehicle was the LVTP, although, unlike its World War II predecessor, it could not be boarded afloat.

Meanwhile, in November 1964 the CNO had asked the Center for Naval Analyses (CNA) for a study of amphibious force requirements through 1980, which resulted in "Amphibious Assault Shipping in the Mid-Range Period" (NAVWAG 44). In January 1965 CNA asked BuShips for feasibility studies of an LPD with an enlarged hangar and ALPS (amphibious landing platform slipway), an LPD with a marine railway to launch craft. ALPS was a compromise between LSD and LPD, with greater helicopter operating capacity because its

As an interim step toward an all-weather assault ship, capable of landing a force over a beach as well as by helicopter, *Inchon* was modified while under construction and fitted with a pair of davits for LCVPs. She is shown on 11 May 1970. At this time she did not yet have any carrier-controlled landing radars. She had not yet been fitted with defensive missiles, but she did have the lattice support for the ULQ-6 defensive jammer. Between 6 March 1995 and 24 May 1996 *Inchon* was modified to serve as a mine countermeasures flagship and helicopter mother ship; she is the sole survivor of the LPH series. *Ingalls Shipbuilding Corporation*

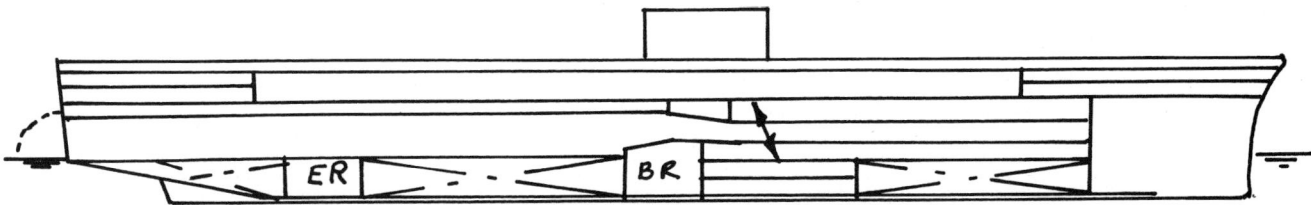

The large general-purpose amphibious assault ship (later the LHA), as offered in March 1965. In this sketch all vehicles are stowed forward of the boat well, with access (indicated by arrows) to the helicopter hangar. The arrows also indicate access to and from the ammunition and troop cargo spaces below the vehicles. Abaft the ammunition and troop cargo is the boiler room, about amidships; the engine room is the small compartment further aft. The big compartments are for ballast. Crew and troops are accommodated in the gallery deck under the flight deck and in the deck above the boat well. This ship would have been 800 ft overall (770 ft waterline), with a beam of 107 ft (to pass through the Panama Canal) and a fully loaded draft of 29 ft; full load displacement would have been 49,200 tons. She would have sustained 20 kts on 65,000 SHP, and endurance would have been the usual 10,000 nm at 20 kts for an amphibious ship. Accommodations would have been 895 ship's company, 2,500 troops, and a margin of 56, for a total of 3,451. The sheer size (hence cost) of this ship led to the series of lesser versions illustrated below.

Table 12-4. LHA Design Alternatives, 1965–66

	Study 6A-65	LARC	Scheme C
Length OA/WL (ft)	800/770	828/788	720/690
Beam (ft)	107	107	107
Draft (ft)	29	30	29
Displacement (tons)			
Light	NA	NA	NA
Full load	49,500	53,440	39,400
Power (SHP)	65,000	70,300	50,000
Hangar deck (ft)	NA	NA	NA
Well (ft)	NA	NA	NA
Accommodations[a]	895/2,300	895/2,300/56	743/1,900/45
Cost, lead/follow-on ($ million)	155/135	160/140	126/109

NOTES: All designs are for 20 kts sustained speed, 10,000 nm at 20 kts endurance. LARC was based on the original design. Schemes G and H were lower-cost versions requested on 23 March 1966. Characteristics were based on Scheme G. Ballasted down, G would draw 40 ft. The Litton design won the competition. Ballasted draft was 23.5/26.5 ft. Sustained speed was 22.4 kts (full speed was 23.9 kts). Cargo: 25,572 sq ft of vehicles, 7,140 sq ft preloaded, 123,205 cu ft; 419,919 gal of JP-5, 10,000 gal of motor gas. NA indicates data not available.

[a]Accommodations are ship/troop/officers when three figures are listed and ship/troop when only two figures.

[b]Includes 400 convertible troop berths.

flight deck could extend over the slipway. At this stage CNA suggested merging the LPH and LPD.

On 23 February 1965 CNA asked BuShips to sketch a large amphibious ship or general-purpose amphibious ship (BuShips OpNav Study 6A-65). Reported on 31 March, the design showed a flat deck, with a hangar (slightly larger than that of an *Iwo Jima*) above an LPD-style well deck, with vehicles stowed forward of the well and living spaces in a gallery deck. Like an LPH, she would carry about a BLT worth of marines. She would also carry a considerable load of vehicles. Vehicle stowage (in effect, in a hangar) required 12 ft overhead clearance quite low in the ship. This ship combined LPH, AKA, and LSD, even LST, features. She could only just transit the Panama Canal.

The Marines objected that the ship seemed better adapted to administrative landings than to amphibious combat. She would be relatively inflexible and quite vulnerable to minor operational casualties. For example, she depended very heavily on her well deck; any problem with its stern gate would prevent her from unloading over a beach. Ballasted down, the ship would draw 40 ft of water (the SCB had hoped for 32 ft), hence she could not unload LVTs anywhere near a beach. LSTs could unload more quickly over a beach, and an AKA had four independent holds.

OpNav thought otherwise. The big amphibious ship offered real economies. By combining several kinds of ship, it could drastically reduce the number of personnel required for a given level of amphibious capability. Between April and June the SCB asked BuShips to develop some design alternatives (see Table 12-4).

As for the Marines' objections, the ship offered a larger hangar than an LPH, a wet well large enough for two LCU, and at least four alongside offloading stations. She also offered improved (i.e., faster) cargo handling, which the Marines badly wanted. BuShips cited a recent comparison between the lat-

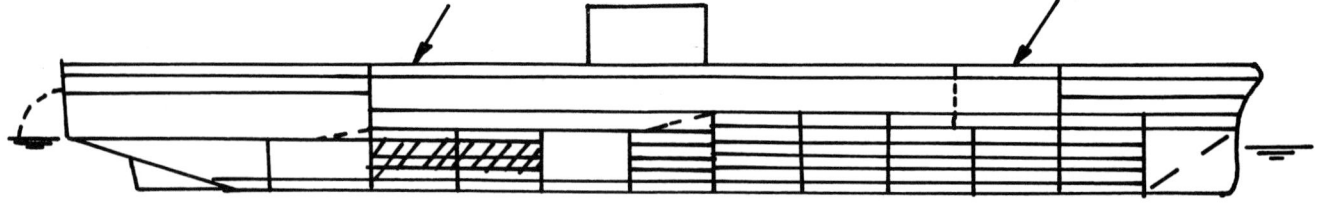

LHA (then called LPH), Scheme B-3 for OpNav Study 24-65, June 1965. She was designed to carry 100-ton surface effect craft in a dry well at waterline level, shown aft. Although not indicated, the island would have been offset to starboard. Two 30 × 60 ft elevators are indicated by arrows, the after one to port. A vehicle deck with 10 ft clearance was shown under the helicopter hangar running much of the length of the ship. The large space under it is the fire room, with an engine room aft, under the fore end of the dry well. Dimensions were 714 (overall) or 684 (waterline) × 95 × 27.6 (full load) ft; displacement was 27,100 tons fully loaded. The ship would have made 20 kts (sustained) on 36,000 SHP (twin-screw gear turbines). Endurance was 10,000 nm at 20 kts. Accommodation was 622 ship's company and 2,152 others (marines) plus a margin of 51, for a total of 2,825.

Scheme D	Scheme G	Scheme H	Litton Design
775/725	770/740	725/680	796/768
107	107	107	106
29	29	29	27
NA	25,400	23,525	24,185
37,700	42,300	37,700	38,000
44,400	52,000	52,000	NA
NA	60 × 360	60 × 420	NA
NA	77 × 377	60 × 280	NA
743/1,900/45	745/2,348/52[b]	745/1,948/52	976/1,825
122/105	113/116	120/103	NA

est conventional American freighter, U.S. Lines' fast *American Challenger* (over the side, 6.5 tons per minute), and *Comet* (ro-ro, 15.5 tons per minute); an AKA would be comparable to *American Challenger*. An LPH was rated at 2.5 tons per minute (cargo elevator capacity). The new ship offered a roll-on stern (into landing craft) like *Comet*, an over-the-side capacity about a quarter of *American Challenger*, and vertical capacity comparable with that of an LPH.[2]

The original design showed only inboard elevators, so the ship could easily pass through the Panama Canal. They limited hangar space; a longer hangar was needed. In Scheme C, the ship was given a port deck-edge elevator (30 × 60 ft), with a similar elevator on the centerline forward. In Scheme D, the helicopter hangar was raised atop the boat well aft. Helicopters would be rolled out of its fore end, so no elevators were needed, but the flight deck was much shorter. Both schemes showed a single boiler room, but all versions had twin screws. Unit machinery would make the ship more survivable, but at a cost in weight and money. A version with four boilers was 10 ft longer on the waterline (780 ft) but cost $1 million less.

In an alternative design, the ship carried 24 LARC-60 amphibians instead of landing craft, with a stern ramp instead of a wet well. The necessary ramp was already being developed for the fast deployment logistics (FDL) ship (see Appendix A). This version had 60 × 50 ft elevators fore and aft. The LARCs would have been stowed on two levels below the hangar, flanked by troop spaces.

Completed in April 1966, NAVWAG 44 strongly supported the general-purpose amphibious ship as the ideal building block for future amphibious forces. By this time the lift goal had been cut from two 20-kt MEF lifts to one and a half, based on the theory that the Atlantic Fleet could make do with half an MEF lift of slower ships because it covered shorter distances. As envisaged in February 1966, the two MEF FY 67–71 building plan included 2 AKA (FY 69), 10 LPD (4 in FY 69, 3 each FY 70 and

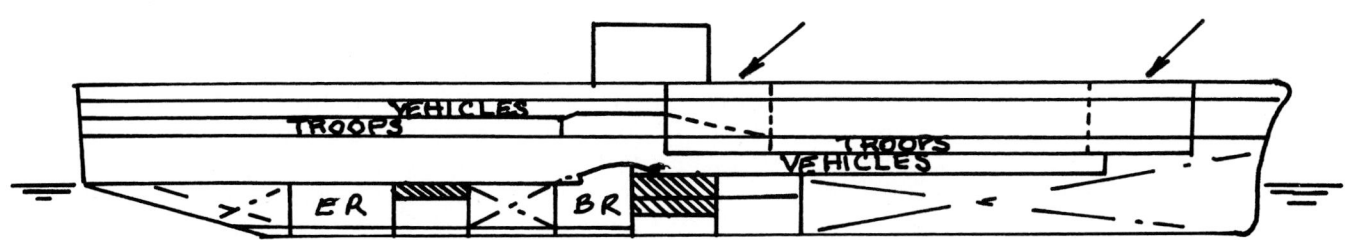

Scheme C, a large general-purpose amphibious ship with a wet well and reduced capacities, June 1965. Shaded spaces are for ammunition and troop cargo. Two 30 × 60 ft elevators, the after one to port, are indicated by arrows. The island would have been set to starboard. The helicopter hangar is the deep space between the two elevator wells. A ramp leads from it up to a vehicle stowage space (10 ft clearance). A deeper vehicle stowage space (12 ft clearance) connects directly with the boat well aft. Below it is an additional 12 ft of vehicle space, just forward of the ammunition/troop cargo space. The crew is housed in the gallery deck below the flight deck. Below them, aft, are vehicles; below the vehicles are troops, with shop space forward of their berths. More troops are housed under the helicopter hangar and forward of the deep vehicle stowage. Presumably this complex arrangement was necessary in order to limit the ship's overall size. Dimensions were 720 ft overall (690 ft waterline) × 107 ft × 29 ft (full load); full load displacement was 39,400 tons. The ship would make 20 kts (sustained) on 50,000 SHP, and endurance would have been 10,000 nm at 20 kts. Accommodation was 743 ship's company and 1,900 troops, plus a margin of 45, for a total of 2,688.

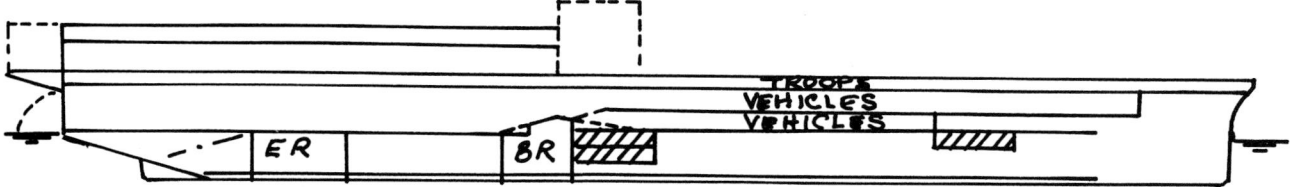

Scheme D, a large general-purpose amphibious assault ship with a stern elevator and reduced capacities, June 1965. Although the flight deck would not have run the length of the ship, length was maximized by offsetting the island to starboard, as in a conventional aircraft carrier. Shaded spaces were for ammunition and troop cargo. Forward of the boat well was a 12 ft vehicle stowage space, with a 10 ft vehicle stowage space under it. Troops were berthed in the gallery deck under the flight deck. At the after end of the flight deck was a large hangar, with shop space to port, and crew living spaces in its gallery deck. This design would have been 775 ft overall (725 ft waterline), with a beam of 107 ft and a full load draft of 29 ft; full load displacement would have been 37,700 tons. The ship would have made 20 kts sustained on 44,400 SHP, with the usual amphibious ship endurance of 10,000 nm at 20 kts. Accommodation was 743 ship's company and 1,900 marines plus a margin of 45, for a total of 2,688.

71), 6 LPH (2 each in FY 69–71), 14 LSD (1 in FY 67, 7 in FY 68, 3 each in FY 70 and 71), and 31 LST (11 in FY 67, 10 in FY 68, 5 each in FY 70 and 71). Cutting to a one and a half 20-kt MEF cut the total requirement to 12 LPH (2 required), 7 AKA (none required), 20 LSD (8 required), 20 LPD (4 required), and 30 LST (10 required).

Too many AKAs were needed simply to provide the vehicles LPHs did not carry. LPDs did not carry enough helicopters. The new ship could greatly simplify the situation. The new ship would replace both LPH and LPD, as well as some AKAs. Building fewer larger ships would cut both construction and manning costs.

The new ship was designated LHA—an assault (A) helicopter (H) ship. It would land a Marine amphibious unit (MAU), successor to the old BLT, supported by at least two more ships, preferably an LST and an LPD (both of which existed in sufficient numbers). The navy now decided that the LHA would be the centerpiece of the 20-kt amphibious force; procurement of LPH and LPD would cease. The navy and the JCS estimated that 4 LHA, 3 LSD, and 9 LST, for a total of 16 ships, could replace 39 older ones: 5 APA, 2 AKA, 10 LSD, and 22 LST. It appeared that 6 LHA and 7 LST would complete the one and a half MEF objective, because there were already 6 LPH and 20 fast LST. Secretary of the Navy Paul Nitze became a strong supporter; on 23 June he ordered procurement of LPH and LPD stopped in favor of LHA, which would become the primary element of the planned one and a half MEF lift.

The NAVWAG 44 conclusions were incorporated into a 22 August 1966 draft presidential memorandum on the amphibious assault ship issued by the secretary of defense. To reach one and a half MEF capacity by FY 72, all necessary ships would be funded by FY 69; the decision on the last half MEF lift would be deferred. Three LHA would be built in each year for FY 68 and FY 69, together with seven LST and the third AGC. An LHA project was formally established in July 1966.

The LHA was about as large as a World War II *Essex*-class fleet carrier, and by May 1966 there was interest in limited facilities for fixed-wing aircraft, especially for the OV-10A light observation/attack aircraft and for future VSTOLs (vertical or short take-offs and landings) that would be used when more sophisticated air support was not be available. The aircraft would not require catapults. Also, the OV-10 had an exceptionally short roll-out on landing, yet some limited arresting gear was suggested. The CNO ordered that no such steps be taken without his personal approval, for fear that the LHA would soon be fitted with arresting gear, barriers, and catapults, to become a small—and expensive—aircraft carrier. Later it was claimed that the island had been made unusually wide to preclude use of an LHA as a carrier; that had the unintentional consequence of making it impossible, in the 1980s, to adopt a transport helicopter larger than the CH-53. This limit on CH-53E rotor diameter may explain its use of an unusual seven-blade rotor.

Meanwhile, the SCB produced draft characteristics for what was now called SCB 409.68, based mainly on Scheme G of NAVWAG 44. The NAVSHIPS designers were asked to shrink it, but that seemed unlikely as long as the well area, square, and cube were held firm. Indeed, initial detailed estimates showed a larger ship, even when it had a smaller well deck. A requirement for eight helicopter operating spots was based on the 720-ft Scheme C; the 770-ft Scheme G would accommodate nine spots in three sections of three each. By way of comparison, an LPH had only four CH-53 spots, and an LPD only two. Hangar capacity was the full group for a BLT: 6 CH-53A, 18 CH-46A, and 2 UH-1E. This was twice as many CH-46 as an LPH could accommodate.

Scheme G showed net cargo stowage of 31,200 sq ft and 86,000 cu ft, which compared to 50,000 and

110,000, respectively, in an AKA. The characteristics showed gross figures of 39,000 sq ft and 112,000 cu ft, equivalent to the net figures in the study. Fuel stowage was 400,000 gal of JP-5 (the characteristics added 25,000 gal of motor gasoline). The provision in Scheme G for cargo space convertible to 400 troop berths was dropped because such spaces would require very expensive "hotel" features, such as galley, heads, and air conditioning.

The Scheme G well was sized to accommodate four LCU and ten LCM(6) equivalents. It was 77 ft wide, to accommodate two LCU and an LCM(6) abreast. The requirement that the ship be able to transit the Panama Canal limited overall beam, so sidewalls were narrow. There was some concern that the natural period of the water in the well and the ship's roll period might be close enough to make for resonance, which had to be avoided; LSDs typically had narrower wells. NAVWAG recognized that it might be necessary to make the well narrower and longer, or to provide a damping system. The SCB noted that LSD wells were often 85 percent of the ship's length, whereas the LHA well in Scheme G was far shorter, at 377 ft. Armament was set at four twin 3-in/50, as in an LPH, with sufficient space, weight, and power for two of them to be replaced by Sea Sparrow point defense missiles.

At this stage there was no interest in providing amphibious flagship facilities. Shipboard command system automation had progressed enough so a small-ship tactical data system (JPTDS), then being planned for destroyers, was specified.[3] Radars included SPS-43 for long-range air search and SPN-6 for air traffic control; there was no three-dimensional radar. Nor was there any provision for a large hospital; the ship would have beds for 2 percent of her complement, thus accommodating about 60 patients.

Tarawa was the first of a new class of all-weather assault ships. She is shown, newly completed, in March 1976. Note the two LCMs on deck; they were easily handled by the ship's large crane. She is armed with NATO Sea Sparrows for self-defense and 5-in guns for fire support, just as the *Essex*-class LPHs retained their 5-in guns.

Peleliu (LHA 5) was the last of the *Tarawa* class. The big radome on her foremast houses the antenna of a SPQ-9 track-while-search radar used to control her 5-in guns; below it is the dish of a SPG-60 radar for Sea Sparrow missile control. The flat plate antenna aft is for an SPS-52 three-dimensional set, used for air control. *Ingalls Shipbuilding Corporation*

In the normal course of events, the BuShips feasibility studies would have developed into a preliminary design. However, Secretary of Defense McNamara believed that private industry could produce more efficient ships if it began at the earliest design stage, much as private aircraft builders produced all phases of an aircraft design. By mid-July 1966 he had decided to follow the same contract procedure used for the new (but ultimately abortive) fast deployment logistics ships (see Appendix A). A request for proposal (RFP) replaced the usual characteristics. Much the same procedure was followed with the *Spruance*-class destroyers. Further study, leading to a total development program (TDP), approved February 1967, added numerous features, including a three-dimensional radar and a full naval tactical data system. Estimated cost rose from the $102 million of NAVWAG 44 to $153 million for the lead ship and $122 million for follow-ons by December 1967. By August 1968, with a design in hand, the navy estimated unit cost at $147 million. As built, the average price was $229 million in FY 74 terms.

Litton, with a design team led by Dr. Reuven Leopold, won the LHA competition against two competitors in July 1968. The same team won the FDL and *Spruance* competitions. All the competitors offered updated ordnance, tactical data systems (with links compatible with those of the new AGC), automated machinery plants, and innovative assault systems for faster and more efficient unloading. Litton's design was described as offering the most efficient and unique assault system. It had a two-computer tactical data system, which in effect offered the same sort of automated performance as the AGC. As required, it had a three-dimensional SPS-52 radar that was valuable for fighter control. It offered all-weather helicopter operation, with a three-dimensional SPN-35 landing control radar. In addition, it offered automated external and internal communications and a wave action dampener in the well deck. Its updated armament consisted of three 5-in/54 lightweight guns and two Basic Point Defense Missile System (BPDMS; Sea Sparrow) launchers. To cut cost, in October 1968 the navy planned to substitute twin 3-in/50s from stock for the three 5-in guns, but that was never done.

Litton's original proposal was described as extremely well integrated and innovative, but it was criticized for some structural weaknesses, specifically in the shell plating, in the after part of the ship (transverse strength), in the stern under the well, and in the island. Design solutions added 977 tons, eliminated stanchions in the well and the hangar, substituted twin for single rudder, and reduced hangar height from 23 ft 6 in to 20 ft (23.6 ft aft). Compared to the notional ship, the most important differences in the LHA were its flagship and significant (300-bed) hospital facilities.

The ship was much larger than an LPH, so she could accommodate more helicopters: 38 rather than 20 CH-46, with 9 CH-46 spots on the flight deck. A typical load was 16 CH-46, 8 CH-53, 4 UH-1, and 6 AH-1 attack helicopters. Hangar deck clearance was 20 ft forward and 23.6 ft aft. The well could accommodate 4 LCU or 20 LCM(6) or 52 LVT, far more than an LPD. Because so large a free water surface could cause stability problems, the well had a center island and water barriers on each side. A fixed ramp from the third deck down to the well deck absorbed wave energy from the water in the well deck. Vehicle stowage was deliberately oversized, to allow for larger future vehicles. Marine capacity matched that of the LPH, 1,903 men plus 119 for the amphibious staff. The LHA could offload the assault echelon of an MAU in 2.9 hrs, which meant 511 measurement tons of cargo per hour. Other capacities included 107,000 cu ft of unitized cargo and 24,416 sq ft of vehicles; 400,000 gal of JP-5 (vs. 300,000 gal in an LPH); and 10,000 (vs. 5,000) gal of motor gas.

This class had two powerplants, each comprising a single unique 600 psi, 900°F combustion engineering boiler and a 35,000 SHP double-reduction geared turbine, plus two turbo-generators (2,500 kW each). Typical output was 60,400 SHP, about 40 percent of a similar-size *Essex*. On trials, the first ship, *Tarawa*, made 25.3 kts. Sustained speed is 22 kts. *Tarawa* made roughly the projected 10,000 nm endurance at 20 kts.

Speed could be important, because fully loaded helicopters found it difficult to take off under hot no-wind conditions. Off Vietnam in 1966, LPHs and LPDs had to steam into the wind, sometimes at 20 kts, to get their aircraft off. Large size meant deep draft when ballasted down. In 1966 the Pacific Fleet commander commented that the ship would be unable to come close inshore in the Gulf of Tonkin or in the South China Sea.

The 9 February 1968 draft presidential memorandum (DPM) on amphibious forces deferred the purchase of the one and a half fast MEF force to FY 71. It approved building one LHA ($153 million) and

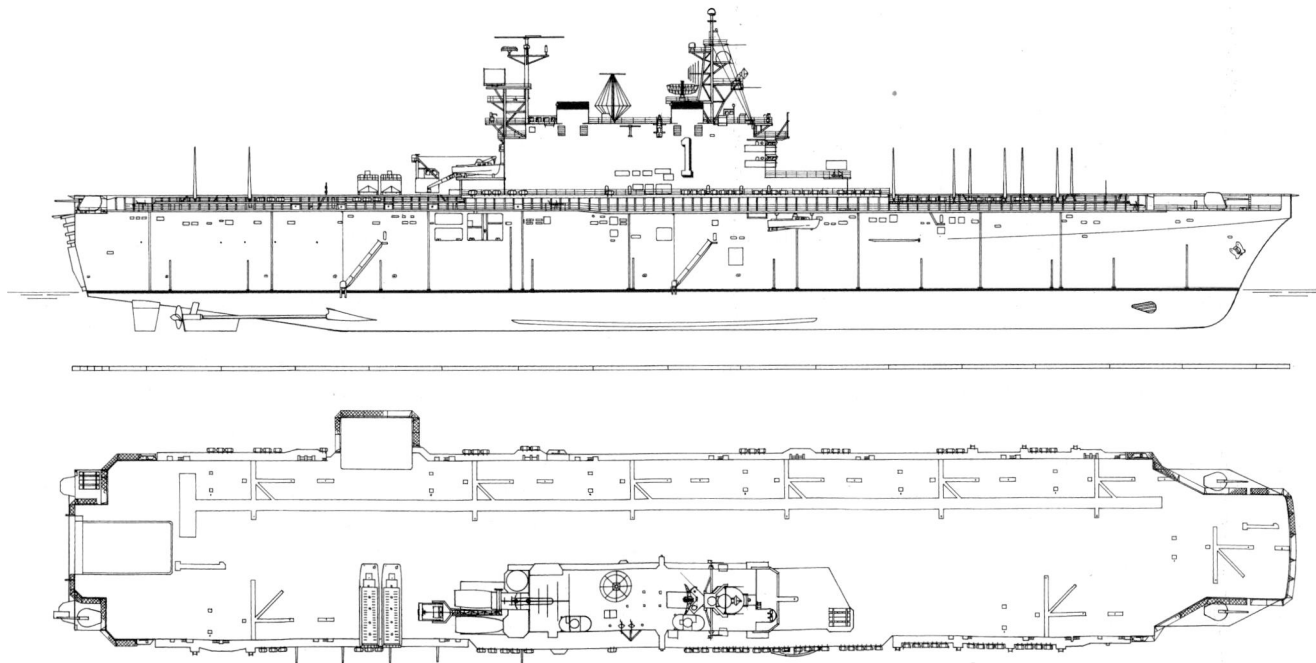

Tarawa (LHA 1) as completed, July 1976. *A. D. Baker*

approved long-lead items for three more, as recommended by the navy. It also deferred the accompanying seven LST from FY 69 to FY 70. By April 1968, with the winning design not yet selected, plans called for a lead ship in FY 69, three in FY 70, and two in FY 71. The contract would be on a multiyear basis; soon one of the FY 70 ships was deferred to FY 71. These 6 LHA would replace 18 previously approved ships: 2 LKA, 2 LPH, 4 LPD, 7 LSD, and 3 LST. The six-ship plan would actually bring navy forces beyond one and a half MEF, as a step toward the desired two MEF (one per ocean) force.

Extending the production run to ten ships would reduce unit cost by about 5 percent, and it would provide enough for the ultimate two MEF goal. OpNav therefore asked for the extension. By FY 76–77 it would be necessary to replace the three *Boxer*-class (converted *Essex*) LPHs. These ships were already very expensive to maintain ($20–22 million per year for *Boxer*s vs. $11–12 million for LHAs); replacing them with three more LHAs would save about 5,000 personnel. The secretary of defense rejected the ten-ship program, but did agree to buy three ships to replace the *Boxer*s, for a total of nine LHA. The resulting total of 16 large decks would support the assault echelons of one and two-thirds MAF (actually, 16 rather than the required 15 MAU). This program was embodied in a national security decision memorandum (NSDM-27), which required the capacity to handle one major and two minor contingencies (respectively, one MAF and one-third MAF).

In 1970, however, the navy programmers were suddenly told that LHAs were costing much more. They had to find more money in the out-years—beginning with $108 million in FY 72, and rising to $236 million in FY 76—or else cut the shipbuilding budget. The FY 72 budget was already bare bones, with six nuclear-powered ballistic missile submarine (SSBN) modernizations, a carrier, three attack submarines, a missile cruiser (DLGN), a missile cruiser modernization, seven destroyers, five minesweepers, and the sixth LHA. To solve the fiscal problem, the March 1970 tentative program objectives memorandum (T-POM) for FY 72 envisaged ending LHA production at five ships, eliminating the $174.5 million projected for the sixth in FY 72; the maximum contract adjustment charge would be $110 million. Op-03 agreed: the sixth ship would merely replace three 20-kt LSDs already in the fleet, and on a ten-year basis five LHAs would cost $110 million less to run.

Moreover, the amphibious force was in better condition than other parts of the fleet, with 34 ships building or authorized, including 2 LCC, 5 LPD, 1 LPH, 4 LSD, 18 LST, and 3 LHA. The program objective memorandum (POM) for FY 71–75 envisaged retiring some relatively modern 20-kt ships: the two Mariner LPAs and eight LSD 28s.

The nine LHA had been needed to maintain the one and two-thirds MEF of lift (total of 16 helicopter ships), which in turn would provide the Atlantic Fleet with at least the assault echelon of one MEF force. At this time the major Pacific contingency commitments were one MEF either in Northeast Asia (Korea) or in Southeast Asia at D+30 to D+35. The United States had also agreed to provide the capacity for a one MEF assault in NATO operations at M+30 (i.e., 30 days after mobilization), as specified in the formal NATO commitment document, DPQ (69). No ships were specifically assigned to NATO. It was reasonable to imagine moving one MEB of lift from the Eastern Pacific to the Atlantic within the 30 days allowed by the NATO timetable. In peacetime, the JCS wanted two amphibious ready groups (ARGs) in the Western Pacific plus one in the Mediterranean, one in the Caribbean, and one LST assigned to Commander in Chief, South (CinCSouth).

The planners justified the cut to five LHAs on the ground that there would be no need to mount an MEF-level assault in NATO. Conditions would change slowly enough for forces to be shifted from the Pacific on a semipermanent basis. As for peacetime, each MEB in the Atlantic could support an amphibious ready group (ARG). For the Western Pacific, it would take a bit more, so that the three MEB of the one MEF lift would support only two, rather than three, ARGs. Thus the one and one-third MEF lift envisaged with five LHAs would support the two required Western Pacific ARGs but only one of the Atlantic ones. There was some question, moreover, about recent drawdowns of forward NATO forces because the withdrawal of the French from NATO required larger amphibious forces in the Atlantic.

Unfortunately, anything more was unaffordable. To reach the approved one and two-thirds MEB would require 9 LHA; to reach the earlier objective of two MEF would require 13 LHA. Four LHA were canceled on 20 January 1971, leaving five to supplement the seven new-construction LPH. The decline in amphibious decks paralleled the decline in numbers of active large-deck carriers.

Without sufficient amphibious carriers or helicopters, in the mid-1970s one of the Western Pacific ARGs and some of the Caribbean ARGs had to be surface configured (BLTs instead of MAUs). Units had no organic helicopters; they had to be cross-decked. The helicopter shortage was worsened by the use of some CH-53s for airborne mine counter-

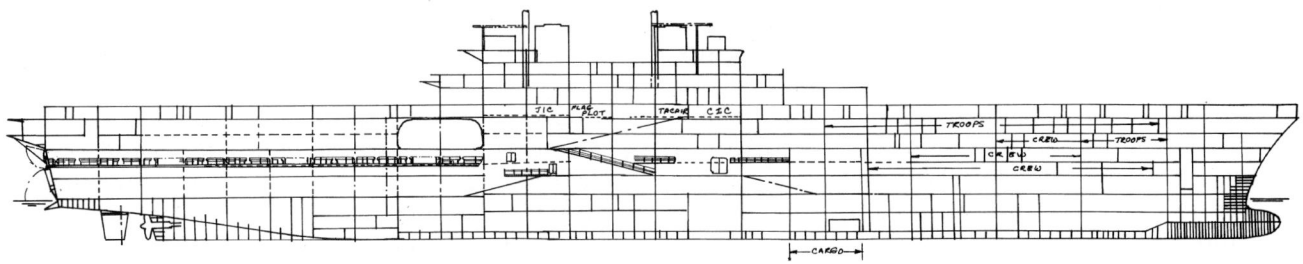

Nassau (LHA 4) inboard profile, January 1998. Note how little internal volume is devoted to the hangar, compared to living spaces and vehicle stowage. Despite her carrier external arrangement, this ship is clearly a large amphibious warship.

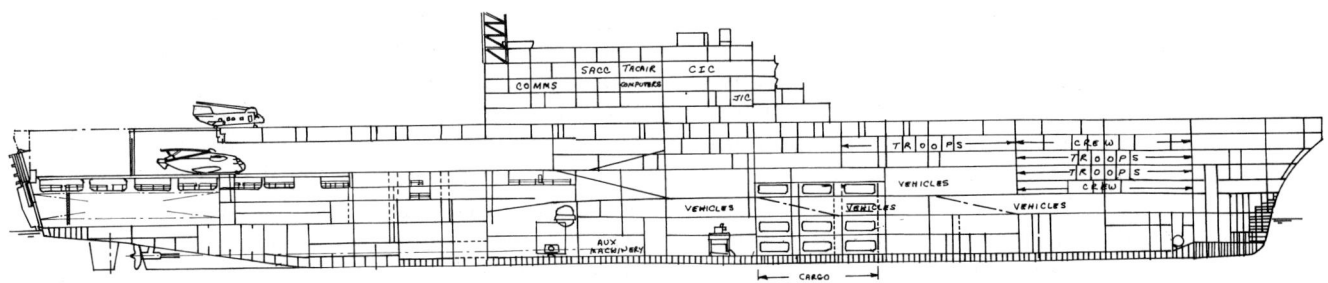

Boxer (LHD 4) inboard profile for comparison with the LHA, 1999. Note that several key command/control spaces have been moved down from the island into the gallery deck. One problem in LHA modernization was limited command/control and computer space.

measures, a force eventually equivalent to two to three LHA loads.

Unlike the LPH, the LHA had no secondary ASW role. As it entered service, however, airborne mine countermeasures technology matured, using adapted CH-53 helicopters towing sleds, which were attached in well decks with the helicopter sitting on the flight deck. Thus at an early stage airborne mine countermeasures became a secondary LHA role. However, LHAs were never used for this role, although LPHs were, and one LPH, *Inchon* (LPH 12), was eventually converted into a mine countermeasures ship (MCS 12).

By the time LHAs entered service, the Marines were buying AV-8A Harriers, six to a MAU. They planned to deploy six Harriers on board the first two LHAs prior to their first overhauls, despite an inadequate landing control system; the existing shipboard radar system, adequate for helicopters, did not give the rate of descent, thus it could not guide a jet VSTOL pilot. The ships would also carry attack helicopters (AH-1J Sea Cobras), and they would need ordnance stowage for both types. Maintenance would be done from vans tied down in the hangar deck, although there was some question as to whether adequate power could be provided.

By 1975 there was also considerable interest in a VSTOL aircraft carrier, tentatively designated VSS.[4] The LHA was not quite a VSS; her amphibious features, such as her well deck, ballast tanks and pumps, and vehicle ramps, were all quite unsuited to any such role. Even so, under a fleet modernization plan the ship was to have been fitted for VSS conversion, with modified magazines and provision for intermediate-level airplane maintenance. An estimate, made about 1975, showed that it would cost $5–8 million to refit the ship for close air support using AV-8s or $4–6 million for ASW, including fitting a carrier-type ASW center (ASCAC). By way of comparison, it would cost over $770 million to build a new LHA with VSS features.

Because shipbuilding costs were escalating, the navy had to ask Congress for about $213 million to complete LHA 4 and LHA 5 by 1976. Cancellation was seriously considered. However, it would take all five ships to maintain the agreed one and one-third MEF lift. Canceled LHAs would have been replaced with pairs of new-generation ships then under consideration, an LX to replace the LSD and an LPH(X), at estimated unit costs of $450 million and $210 million, respectively.[5] At this time an LHA cost about $650 million, so the price seemed quite reasonable. The fourth and fifth ships were completed.

By 1975 a new amphibious ship, the LX (an LSD replacement; see Chapter 16), was being considered. A draft FY 77 budget showed one LX plus one VSS. Because money was tight, the VSS in effect competed with the LX at the secretary of the navy and CNO levels, even though the VSS was not considered an amphibious ship. To get the necessary Marine support for VSS, the navy advertised it as a super-LPH. The Marines initially asked for 1,000 to 1,200 troops, and the VCNO called for design options for up to 1,200. The Marines later softened their position to a combination of 500 ground troops (2 rifle companies) and 304 aviation personnel, plus 21 CH-46 equivalents, somewhat short of an LPH. The resulting VSS would displace about 23,000 tons. Requiring 936 men, 3 companies plus additional command/control and combat support, would have driven displacement up to about 25,000 tons, because the ship would need another deck. The ship could support her marines for up to 6 months, with 15 days' supplies for use once ashore. The VSS design showed a contingency capacity of 500 marines with an amphibious air group comprising seven CH-46, four CH-53, three UH-1N, and six AH-1J.

Alternatively, the Marines were willing to support the VSS as long as its troop capacity was not counted against the required amphibious lift. If helicopters really could replace landing craft, VSS was attractive; eight VSS would more than complete the desired one and a half MAF lift—except for boats. But that battle had already been fought, resulting in the LHA. LX, moreover, more than matched VSS in some ways: it carried more personnel (1,029, as estimated in October 1975), almost as many helicopters (15), and it alone offered the same 20 LCM(6) boat spots as did an LHA.

Ultimately the Marines saw VSS as anything but an amphibious ship, since it had no surface assault capability and probably would never be used for amphibious operations. It would be extremely dangerous, then, to countenance inclusion of VSS in the one and a half MAF goal, since that made VSS look like another LHA. Through the late 1970s, the VSS idea survived, partly on the strength of hopes that a very high performance VSTOL fighter could be designed. There was much talk of abandoning construction of large-deck carriers. Thus, in March 1978 the undersecretary of the navy suggested that the navy consider using an LHA as a VSS "for a year or two." USS *Guam*, an LPH, had already conducted analogous sea control ship trials in 1972–74.

The main objection to a dual role was that the LHAs were so badly needed in their amphibious role. The LPHs were overdeployed; something over 11 large-deck amphibious ships were needed to provide each of the three and one-third deployed MAU with one deck each; there would be one spare deck after the last LHA delivery.

By the late 1970s the navy was planning a new-generation landing craft, the LCAC, to replace the existing LCU. As it turned out, the LHA was ill-suited to the LCAC, and in the 1980s a new generation of follow-on ships, the LHDs (see Chapter 16), was built. Inadequacy to support LCACs presumably helps explain why plans to rebuild the LHAs were abandoned.

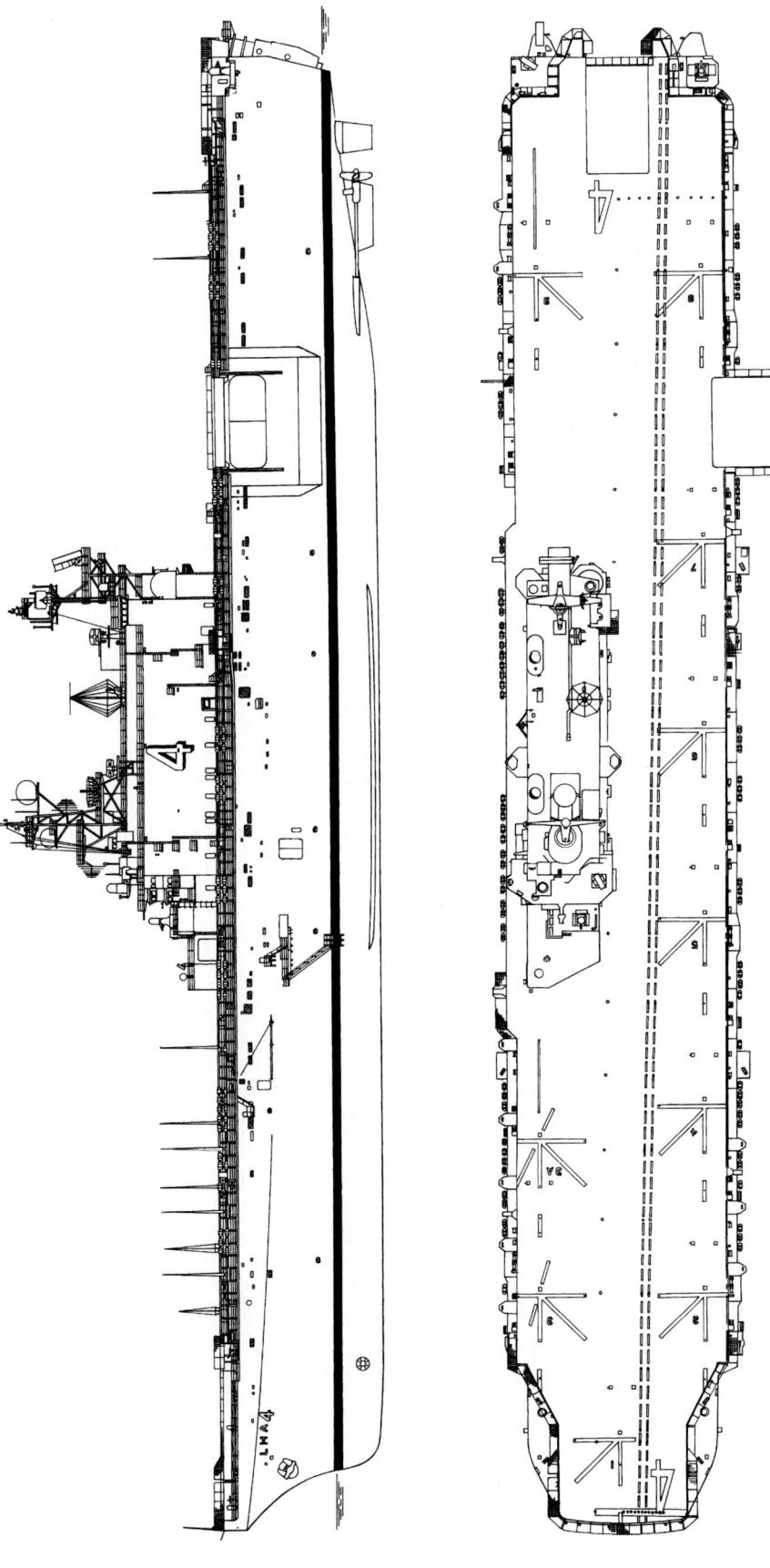

Nassau (LHA 4) as of 2000, showing modifications. Her 5-in guns and her Sea Sparrow missiles have all been landed; her armament is limited to two RAM launchers and two Phalanx close-in defensive guns. Remarkably, she retains the dome of her SPQ-9 fire control radar, atop her lattice foremast. The two objects at the forward end of the island are GBS and, abaft it, the WSC-1 satellite dish. A USC-38 satellite antenna occupies the small radome just forward of the RAM launcher. The larger of the two radomes on the lattice foremast is for a WSC-6 satellite antenna. The lower lattice structure carries an SPS-40 air search antenna. Just below and abaft it is the box antenna of the SLQ-38 countermeasures set. A Tacan antenna (URN-25) is visible at the fore masthead. At its head, the mainmast carries the antenna of the Mk 23 Target Acquisition System usually associated with NATO Sea Sparrow. Below it is the SPN-43 carrier controlled approach radar; the radome of the associated SPN-35A system is visible abaft the island. The big flat-faced radar antenna is for the SPS-48E three-dimensional set. Below it is the radome of a second USC-38. The antenna just forward of the mainmast is the SMQ-11A for receiving meteorological data. *A. D. Baker*

Through the 1950s, the great question was whether a new generation of fast landing craft could be built within dimensions defined by existing major amphibious ships, such as USS *Monticello*, seen here soon after completion, 13 March 1957. The connection between landing craft size and major ship design was very close, evidenced in the need to build new classes of ships to accommodate the current fast landing craft, the LCAC. *Ingalls Shipbuilding Corporation*

13
New Landing Craft

LCVP (landing craft, vehicles and personnel) and LCM (landing craft, mechanized) were capable of about 9 kts loaded during World War II. LVT (landing vehicle, tracked) and LCU (landing craft, utility; ex-LCT, landing craft, tracked) were considerably slower. If the fleet of transports were to disperse to reduce its vulnerability to nuclear attack, landing craft after the war would have to run at higher speeds in order to reach the beach together. Higher speed would not solve the entire problem, because boats concentrating at the surf line could still be destroyed en masse. Ultimately the U.S. Navy needed a fast amphibian that could carry its own load inland.

Until such a craft could be devised, the best the navy could do was to develop faster landing boats. A promising new technology, the hydrofoil, was nearing maturity. For a conventional hull, high speed is determined by the speed-length ratio (speed in knots divided by the square root of length in feet). Anything above about 1.0 is fast, so the natural speed of a 36-ft LCVP is 6 kts. However, once a hydrofoil is "flying" on her foils, which are in effect underwater wings, all that matters is the friction generated by water passing over them; hull length is irrelevant. Even hull shape does not matter, except that the hull must be able to reach take-off speed. The smaller, hence less draggy, the foils are, the higher take-off speed must be. Thus foils promised to make even an LVT fast when foilborne. A planing hull, such as that of a PT boat, is an intermediate case: the boat "flies" on the planing surface of her hull. There must be enough planing surface to support the weight of the hull; a heavier boat needs a larger planing surface. Although such craft were envisaged during the 1940s, the technology did not mature until the mid-1950s.

Thus, the SCB's first landing craft project, SCB 25, was quite conventional. It was an improved LCT with enough space for personnel (probably 2 officers and 18 enlisted men) engaged in protracted operation. Construction became possible when the Korean War broke out. Characteristics were approved by the CNO on 22 November 1950. Performance was slightly better than that of a wartime LCT(6); speed in smooth water was 8 kts (rather than the 10 kts desired), range was 1,200 nm at 6 kts (compared to 700 nm at 7 kts), and capacity was 150 tons (300 troops) on a hoisting weight of 200 tons, compared to 150 tons (250 troops) on a hoisting weight of 150 tons. The new craft had an improved forward ramp gate, and minimum entrance width and minimum width of the vehicle space was set at 14 ft, to accommodate the new heavy tanks. Like the wartime LCT(5), the new LCU had a closed stern. She was sized, like her predecessor, for transport aboard LSTs and LSDs. Length was minimized so that three could fit the standard LSD well deck. Unlike her predecessors, the craft also had to fit under an LSD mezzanine deck. The hull was essentially a simple rectangular form for simplicity of construction. The craft could carry three M48 medium tanks or six M41 light tanks. These craft had the highest cargo-to-ship weight ratio of any landing craft, about 0.90 vs. 0.15–0.35 for an LST or LSM. Perhaps the main innovation in the design was the use of a Kort nozzle for better propulsive efficiency.

LCTs were redesignated LSTS (landing ship, tank, small) on 10 April 1949, then LSU (landing ship, utility) in 1949, and then LCU (landing craft, utility) on 15 April 1952. Thus, by the time the first SCB 25 was built, she was LCU 1466. See Appendix B for program details. The two FY 55 craft (LCU 1608 and 1609, SCB 25A design) were slightly modified to reduce building costs, and they were rated at 9 rather than 7 kts. Plans initially called for building ten of them in FY 56 while a completely new design (SCB 149) was developed for FY 57. However, in May 1956 it was decided that the ten FY 56 craft,

The first postwar LCT reverted to LCT(5) configuration, with a closed stern. LCU 1466, the postwar prototype, is shown at Island Dock Company of Kingston, New York, her builders, 16 December 1953.

LCU 1610–1619, would also be built to the new design, which was scheduled for completion early in August.

SCB 149 was the faster LCU envisaged nearly a decade earlier. By July 1954 BuShips was well aware that fleet operators wanted a faster (12-kt) craft with the same cargo capacity but with more austere accommodations. Design studies began in August 1954. Although 12 kts might not seem very fast today, it was 50 percent above the loaded speed of the earlier LCU 1466, and it considerably exceeded the natural speed of a 100-ft boat, which was about 10 kts. Merely adding power, already limited by available volume, would not suffice. The hull form itself would have to change. The design was complicated by operators' demand that the LCU be usable as a causeway, as in LCT(6) but not LCU 1466; therefore, she had to be open at both ends.

One possibility was the approach being tried for the projected 20-kt LST: the craft would have a more conventional hull, and would approach the beach stern-on. Compared to the earlier LCU, the hull was lengthened (to 135 from 115 ft), beam was reduced from 34 to 29 ft, and the depth of the hull increased to 7 ft amidships. Instead of the bow propellers of the LST, the LCU would use vertical-axis (cycloidal) propellers, which offered fast turning, but would be protected because they lay above the keel line.

This propeller was developed by Professor Frederick Kirsten of the University of Washington during 1921–24; the U.S. Navy tested his first model in 1923. It was not developed further in the United States, but the Germans used the similar Voith-Schneider propeller on board 114 wartime light minesweepers (R-boote). Its blades hang vertically from a horizontal disk. Gearing rotates the blades at half the disk's speed, so they feather as it turns. The propeller can thrust in any desired direction, thus it both propels and steers. It is clearly suited to beaching craft. For example, broaching in surf is no longer a serious problem, because full power is available to maintain the craft's position. Rather than use a stern anchor to pull her off the beach, a craft can alternate thrust to port and starboard. Given these advantages, the U.S. Navy became interested in the Kirsten propeller in 1943, and two were installed on board LSM 458 for tests, completed early in 1946. She could turn within her own length, and also move sideways and stop (from full speed) within half her length. The U.S. Army used Kirsten propellers on

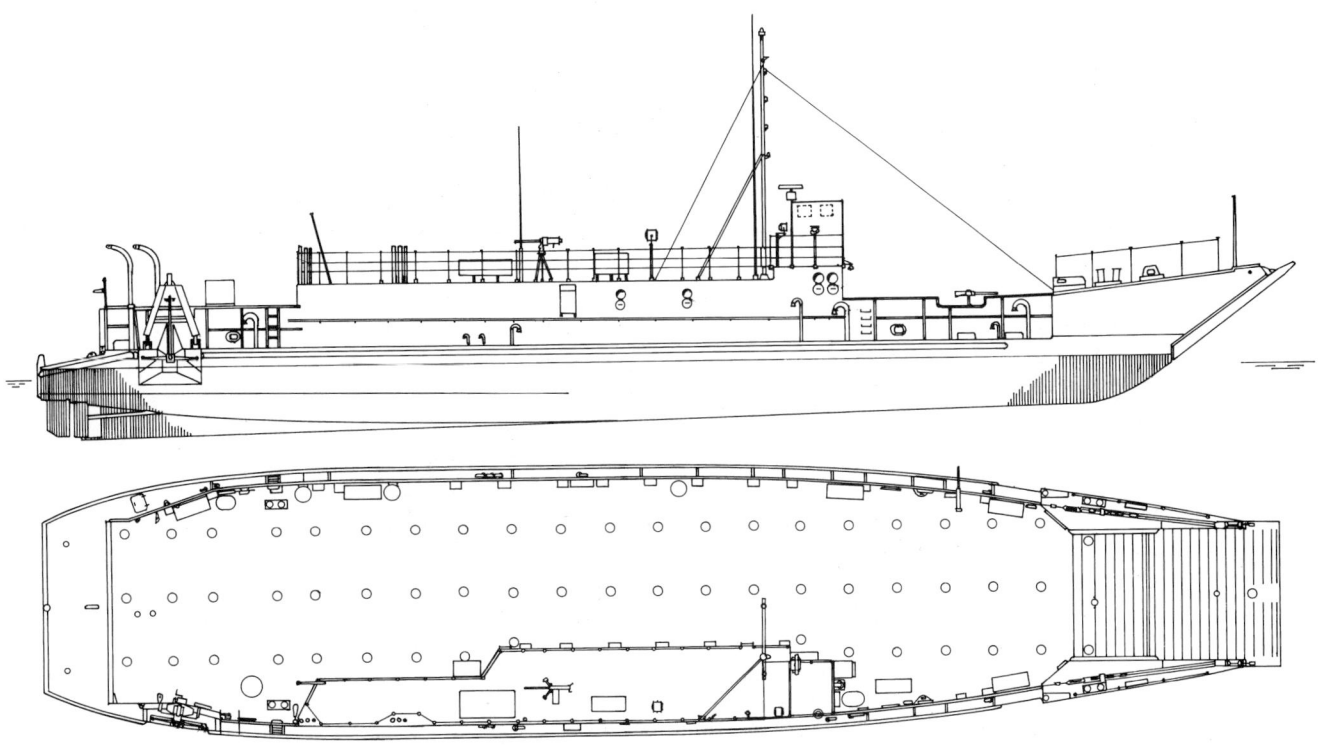

LCU 1491 leads an LCM(8) in a practice landing at Vieques, Puerto Rico, 6 October 1964.

LSU 1626 class about 1968. Note the drive-through configuration and the enlarged superstructure. Armament is two 0.50-caliber machine guns. *A. D. Baker*

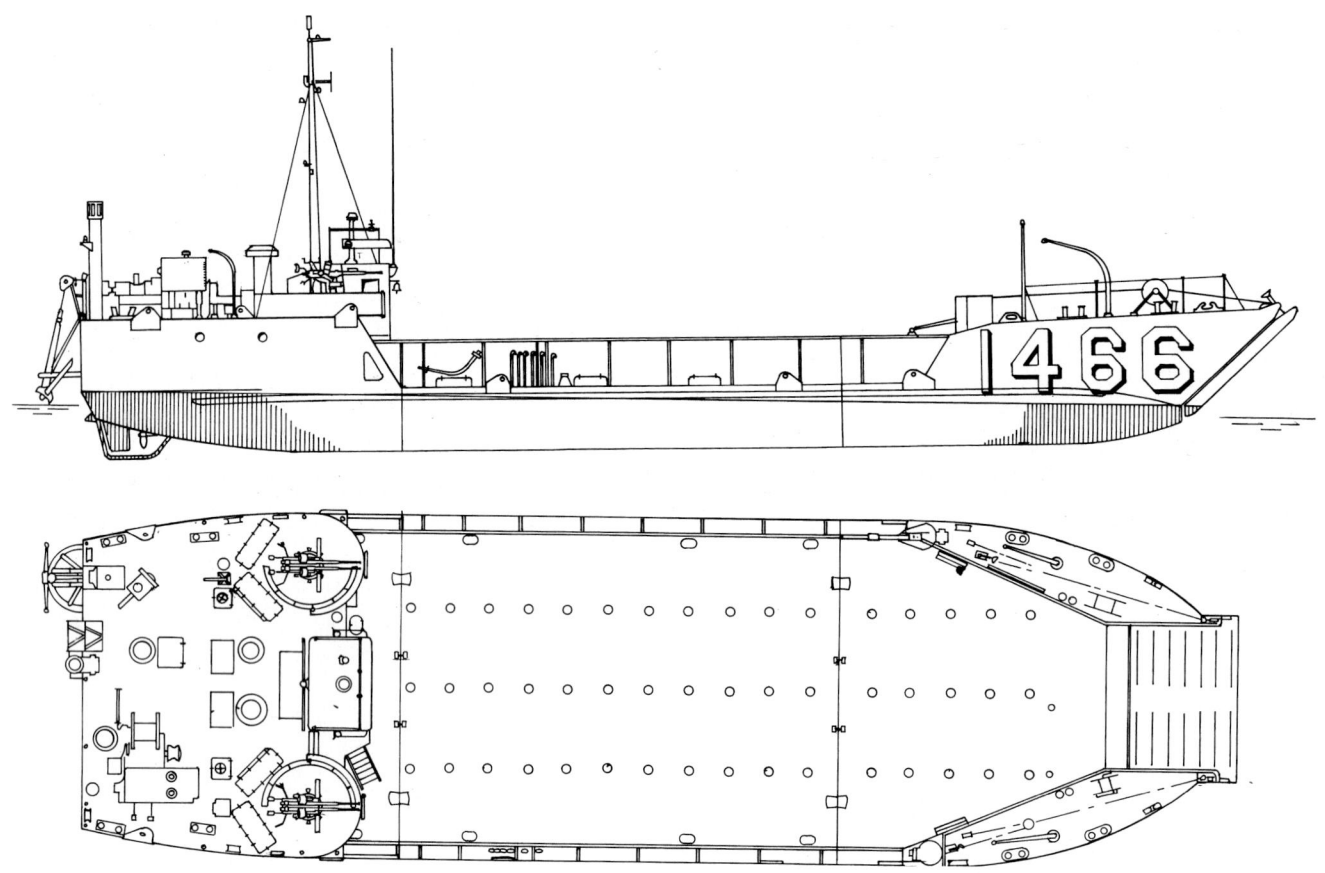

LCU 1466 about December 1953. Note the reversion to the LCT(5) configuration and the pair of twin 20-mm guns. *A. D. Baker*

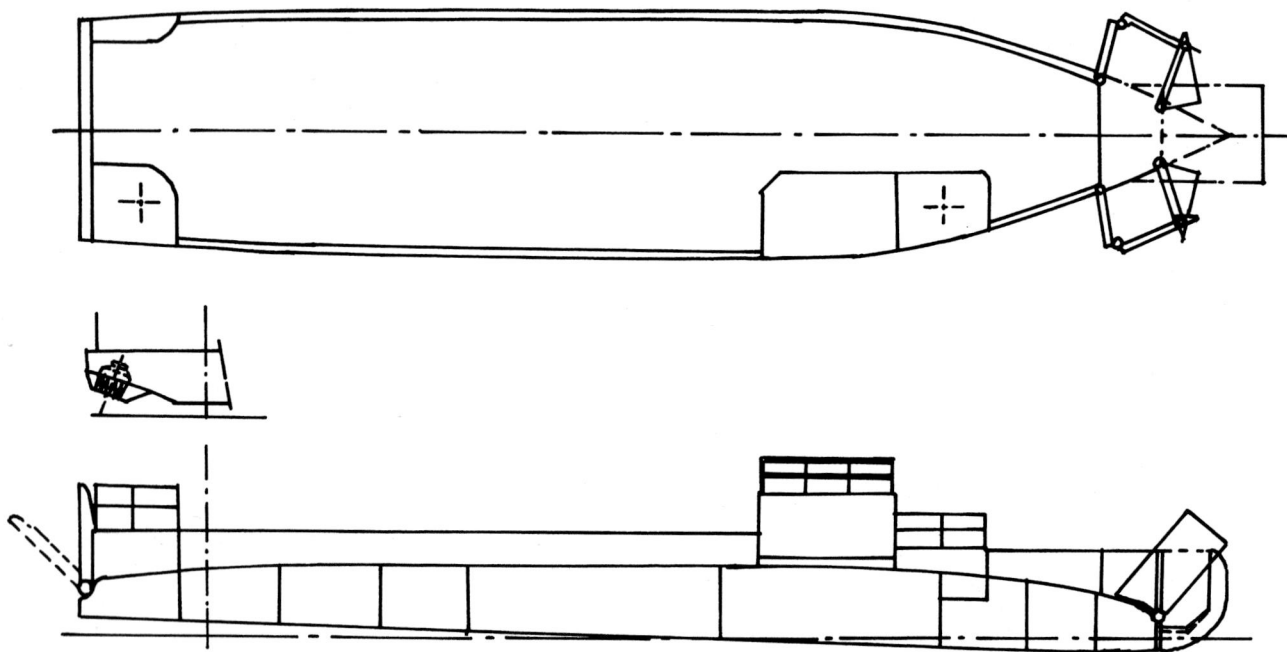

Sketch design of the fast (11 kt) LCU, SCB 149, about 1955. Dimensions were 135 ft 4 in overall × 29 ft × 3 ft 6 in landing draft; displacement was 185 tons light and 375 tons fully loaded. Although 11 kts might not seem very fast, it was an enormous advance on the 8 kts of more conventional LCUs. Endurance was 1,200 nm at 8 kts. The cycloidal propeller planned is shown in the detail view, in a cross-section of the hull. Ramps are shown at both ends; at the fore end the bow hinges up to allow the ramp to deploy. The two crosses indicated 0.50 caliber machine guns. The living space is below the pilothouse, with machinery just abaft it. Capacity was three heavy tanks (M48s) or six light tanks (M41s).

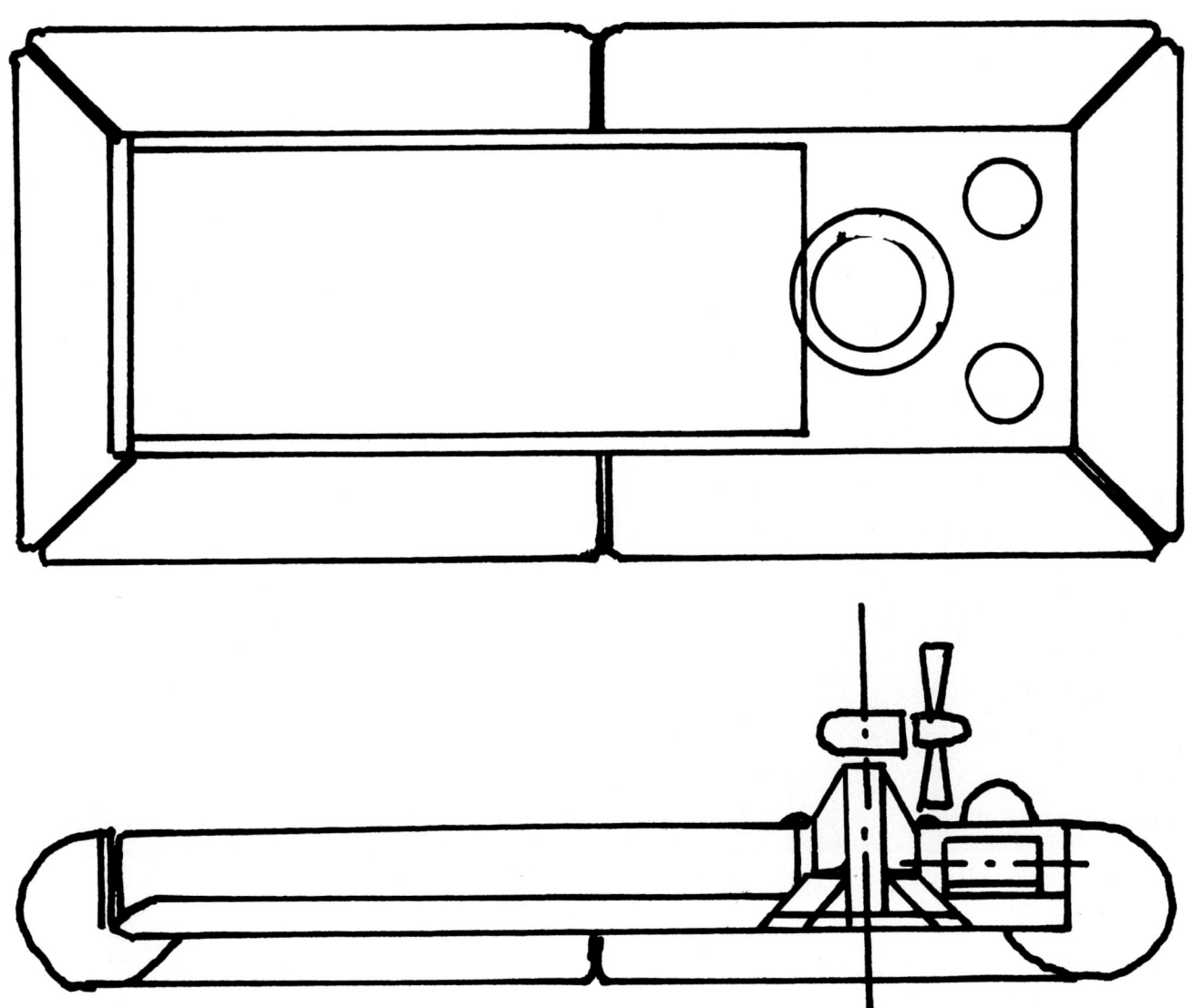

A BuShips concept design for an air cushion LCU, undated, but probably prepared in late 1964 or early 1965. It would have been powered by a single 4,000 HP gas turbine, for a speed of 40 kts and a range of 400 nm at maximum speed without cargo. Dimensions were 86 ft overall (71 ft hull structure inside the skirt) × 40 ft overall (24 ft hull structure) × 25 ft depth (8 ft hull structure). Draft, floating, would have been 1.5 ft, and on the air cushion the craft would have floated 0.1 ft above the water surface. Displacement would have been 45 tons light and 100 tons fully loaded.

board a Mississippi River towboat (LT1-2194), and found that they could operate with a blade bent 90 degrees, and even with stubs of blades.

The stern-first approach had to be dropped because of the drive-through requirement. Neither end could be sharp like a conventional bow; both had to be open. Also, a craft intended as part of a causeway would best approach the beach bow-on. That led back to something much closer to a conventional-looking LCU, with a causeway bow ramp over the finer bow of the new design, and a removable bulwark aft. The underwater hull was slightly modified. With cycloidal propellers the craft could make 11 kts on 800 BHP, which was less than twice the power of an LCU 1466.

An alternative, a twin-hulled catamaran, also failed. In theory, a catamaran offered less resistance because each of its hulls could be quite slim; the boat would still be stable because they were spaced well apart. Length could easily be held to 115 ft so that three could be loaded on board an LSD. However, the two hulls could not be brought close enough together to hold beam below 40 ft.

The SCB based its characteristics, first issued in March 1955, on the modified stern-first design. Pressure for austerity was reflected in limiting accommodation to six; under ordinary circumstances the crew would berth and mess on board a parent ship. Mess equipment was limited to an electric hot plate and a household refrigerator. As in

LSM 458 demonstrates her Kirsten cycloidal propeller, turning within her own length (as indicated by her wake) in Puget Sound, 25 June 1946.

wartime LCUs, these craft would be built in three sections of approximately equal weight, each floatable for waterborne assembly. Later construction cost was cut by eliminating the bolted joints between the three hull sections, on the ground that there was no record of any earlier LCU/LCT having been assembled afloat operationally.

Because the craft no longer had to spin around to beach, the case for cycloidal propellers was weak; the basic design, therefore, had a pair of conventional propellers. Cycloidal propellers offered considerable potential for the future, so one experimental LCU in the class could be built with them. Hull depth was increased to move all accommodations below the main deck; the engine room was set amidships. The bow ramp, which was folded, was handled by hydraulic jacks from two kingposts. The stern also used a ramp, again with two kingposts.

The hull had a sharp bow, a full midships section, and a deadrise (rather than flat) stern. Deadrise minimized resistance and made for some seakindliness. A flat stern, although easier to produce, was rejected because it cost 0.45 kts. A very full spoon bow would have cost 0.75 kts. Small skegs protected against grounding. The first ten had Kort nozzles around their propellers, for greater efficiency. Flanking rudders ahead of the propellers, as in river towboats, offered high maneuverability in surf.

Compared to the earlier LCU 1466 class, these craft were considerably faster. They had the same endurance at 8 kts that LCU 1466 had at 6 kts. Cargo capacity grew to 190 tons (350 troops) on an unloaded weight of 185 tons. Although the new LCU was longer than its predecessor, three could still be squeezed into an LSD well deck.

Three experimental versions, LCU 1620–22, were included in the FY 57 program, but LCU 1622 was canceled. LCU 1620 had a Kirsten cycloidal propeller, LCU 1621 had two right-angle drive units that could rotate through 360 degrees, and the canceled LCU 1622 would have had a vertical-shaft Kort nozzle and a pair of 500 HP gas turbines. The superiority of the Kirsten propeller was demonstrated during Operation Pot Shot in April 1962. All seven Kort nozzle craft were damaged (some became inoperable)

LCU 1653 typified the final series of LCUs, with the drive-through topsides introduced in LCU 1620, but with a conventional hull. She is shown at Defoe Shipbuilding Company, her builder, 5 May 1971.

LCU 1664 is shown carrying a pair of self-propelled howitzers during a May 1987 exercise. Note that her radar mast has been folded down so that she can fit into a covered well deck.

Table 13-1. Post–World War II LCU/LCAC Development

	LCU 1466	LCU 1610	LCU 1626	LCAC
Length (ft-in)				
WL	105-0	127-5	106-3	NA
OA	115-1	135-3	116-6	81-0[a]
Beam (ft-in)	34-0	29-0	34-0	43-8
Draft (ft-in)[b]				
Landing	2-9/5-3	3-6/6-0	3-6/6-6	3-0
Full	6-0	4-10	6-9	NA
Displacement (tons)				
Light	180	172	190	102.2
Landing	347	341	390	169
Full	NA	353	405	184, overload
Power (SHP/shafts)	675/3	1,000/2	680/2	15,820/2
Speed (kt)	8	11	8	50
Radius (nm/kt)	1,200/6	1,200/8	1,200/6	200/40
Capacity				
Troops[c]	None	None	None	24
Tons	167	180	170	NA
Sq ft	NA	2,180	NA	NA

NOTES: NA indicates data not available.
[a]Length over skirts is 87-11; beam over skirts is 47-0.
[b]Draft aft if only one figure given; draft fore/aft if two figures given.
[c]Troop capacity as alternative to vehicles: 300 in LCU 1466, 400 in LCU 1610, 350 in LCU 1625.

because they were not controllable enough either on the beach or alongside an unloading ship. The cycloidal LCU 1620 was undamaged.

By 1959 there was interest in a hydrofoil LCU, but it was abandoned before any design could be developed. Presumably the power requirement would have been prohibitive. The really fast LCU had to wait for the advent of a new technology, the air cushion. Because the new craft, the LCAC (landing craft, air cushion), had so profound an impact on amphibious ship design, it is described in a later chapter (see Chapter 16).

A repeat LCU 1610 series was planned for FY 63, but Pacific Fleet's amphibious command (PhibPac) and LCU Squadron 1 both disliked the design, and wanted to return to the simpler LCU 1466 type. A 6–7 March 1962 Washington conference decided that the LCU 1608 design (itself a modified 1466) would be modified, the main change being a drive-through deck configuration. The preliminary designers favored the cycloidal propeller. The new SCB 149B design differed from LCU 1608 (SCB 25A) in that it had cycloidal propellers (two powerplants in separate compartments) and a deeper hull (7 rather than 6 ft) to improve compartmentation. The hull was slightly reshaped to suit the cycloidal propellers. As in the LCU 1610 class, bolted joints were eliminated. Length increased by 1 ft 6 in. As in the LCU 1466 class, speed was 8 kts. Compared to LCU 1466, this ship had a slightly different complement (2 officers and 12 enlisted vs. no officers and 14 enlisted). Cargo capacity was 170 tons (350 troops) on a hoisting weight of 215 tons, somewhat worse than in LCU 1610.

This SCB 149B design was built in numbers, beginning with LCU 1625 as part of the FY 63 program. She was described as an LCU 1466 hull with cycloidal propellers. However, all later units of the class had Kort nozzles. LCU 1637 was an all-aluminum prototype of an alternative design. Table 13-1 shows the characteristics for three LCUs as well as the LCAC (discussed in chapter 16).

The smaller landing craft could be recast in higher-speed versions. In mid-1948 the Office of Naval Research (ONR) sponsored a design study of hydrofoil landing craft by Ventnor Boat Corporation. The Baker Manufacturing Company of Evansville, Wisconsin, which had gained considerable experience developing hydrofoil pleasure craft, received an ONR contract for two 24-ft boats, one with surface-piercing and one with submerged foils. The surface-piercing "High Pockets," completed in January 1952, generally outperformed its escort, a 63-ft aircraft rescue boat, on its trials. It "flew" in 3–4-ft seas.

By November 1952, BuShips expected that a 25-ton hydrofoil could be included in the FY 55 budget. Possible applications included air-sea rescue, a small gunboat, or a fast landing craft. The choice fell on fast landing craft. BuShips held a design competition for a 40-kt hydrofoil equivalent to an LCVP, powered by a 600 HP engine. The foils had to be

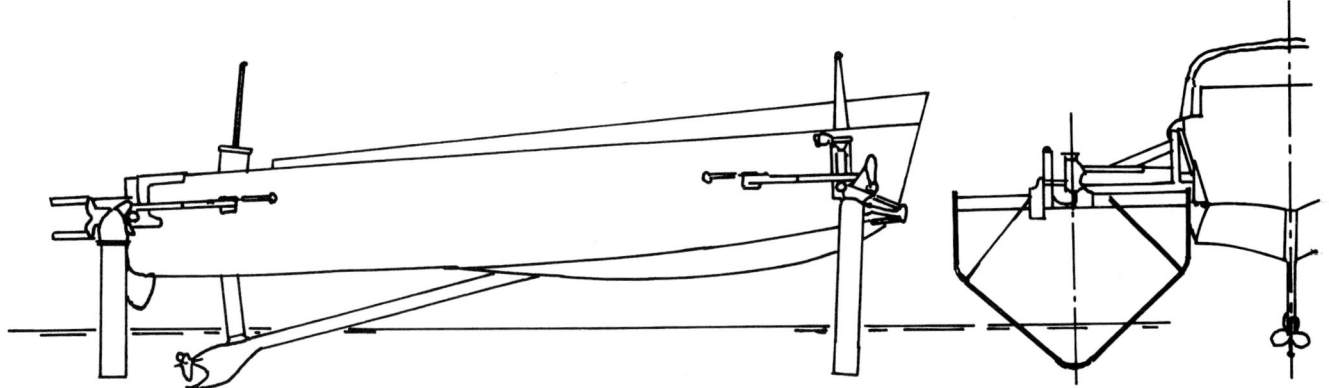

The Baker Manufacturing Company design for a surface-piercing hydrofoil LCVP, about 1956. A 600 HP Hall-Scott gasoline engine would have driven the craft at 40 kts; endurance would have been 135 nm. Overall length would have been 38 ft, and displacement would have been 29,450 lbs (cargo 8,000 lbs). Cost, including the engine, was estimated as $200,000 in 1955.

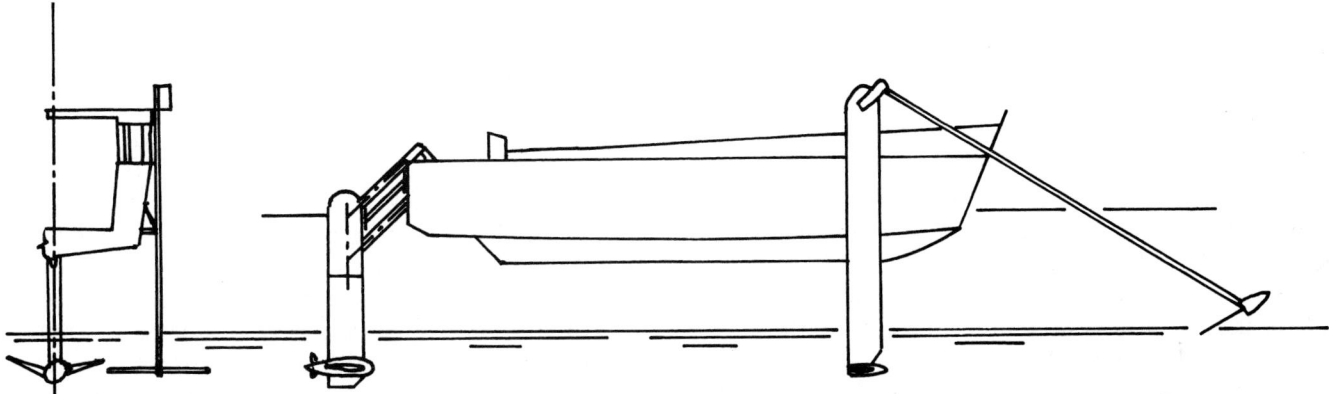

Miami Boatbuilding design for a submerged-foil hydrofoil LCVP, about 1956. She was expected to make 40 kts using a 600 HP Hall-Scott gasoline engine (no diesel engine offered sufficient power in a compact enough package). The hull was that of a standard LCVP, but a new over-the-stern (Cattaneo) drive was installed. Displacement was 31,000 lbs, and cargo capacity was the usual 8,000 lbs of an LCVP. Estimated cost, in 1955, was $243,000. Note the bow probe. A submerged-foil design requires constant control, but it can operate in seas that defeat a surface-piercing hydrofoil. The necessary technology was applied, not to landing craft but to a series of gunboats culminating in the big PHM of the 1970s.

Miami Shipbuilding's hydrofoil LCVP prototype, "Halobates," is shown on trial. Powered by a gas turbine (note the uptake), she had surface-piercing foils that could be retracted as she approached a beach.

retractable so the boat could beach. Nine proposals were received, and contracts were awarded to Baker, Gibbs & Cox Inc. (Grunberg-type boat), and to the Miami Shipbuilding Corporation, the latter for a full-scale test model. This single prototype was built in 1955. At the same time BuShips designed a 30-ton hydrofoil to replace the LCM(6).

In April 1955 the CNO proposed that BuShips and ONR follow the prototype with nine experimental hydrofoil LCVPs of three different types. BuShips proposed cutting to three competing prototypes, to be added to the FY 56 LCVP program, at an estimated unit cost of $200,000 (production cost of a conventional LCVP was about $26,000). They could be paid with savings realized from the delayed delivery schedule of conventional FY 56 landing craft. Further craft could be bought, probably at a unit price of $175,000, in FY 57. As of October 1955, however, nothing had been done in either the FY 56 or FY 57 programs.

BuShips rejected catamarans and a related type, the venturi hull. The reduced wave-making resistance of each of the hulls was more than offset by higher frictional resistance (wetted area was 30–40 percent greater than in a conventional hull) and by interference between the wave systems produced by the two hulls. For an equal load, the catamaran had a heavier hull structure. BuShips was more interested in another alternative, a planing hull (as in a PT boat or a fast motorboat). Although it did not offer the potential of a hydrofoil, a planing hull would be easier to build and handle at sea.

Design studies showed that a 40-ft planing LCVP could make 22 kts on 600 HP or 25 kts on 750 HP. A 70-ft planing equivalent to the LCM(6) could make 20 kts on 3,000 HP, 26 on 4,800 HP, or 29 kts on 6,000 HP. An 80-ft equivalent to the new LCM(8) could make 20 kts on 6,000 HP.

The only compact high powerplant was the new gas turbine. By the fall of 1956 BuShips was experimenting with adapted aircraft engines. A 40-ft personnel boat and a crash boat were powered by Solar T-522J-1 gas turbines. Solar's bigger T-555 offered 600 HP, and Lycoming claimed that its XT-53 (a turboshaft for helicopters) could be developed to produce 800 HP. New fast landing craft needed 1,000 or 3,000 HP engines. It seemed that within three years the T-522 or the General Electric XT-58 could produce 1,000 HP. Aircraft engines already produced as much, but they had not been adapted to marine use. Within about two years a 2,800 HP Bristol Proteus Marine engine and a projected 3,500 HP Packard engine were expected. In October 1956 BuShips' small boat section recommended that the FY 58 program include prototype gas turbine LCVP and LCM(6), to test the suitability of the new powerplants to amphibious craft. For smaller craft, turbo-charged diesels were still competitive, and they could be obtained more quickly. Thus, in February 1956 a 300 HP version of the standard 6-71 diesel was suggested as an alternative to the Solar gas turbine in the LCPL.

In October 1956 BuShips suggested revising the FY 58 small craft program by adding some prototypes. The projected 59 LCVP would be cut to 56. Three prototypes were added as separate line items: one 36-ft planing, one 40-ft planing, and one hydrofoil.

BuShips already realized that more power for higher speed required a longer machinery space and more fuel space, hence longer craft. For a planing craft, the displacement/length ratio had to be kept down, which again implied greater length. A longer craft could not be stowed in davits, however, so it was worthwhile to try the shorter hull. In place of 1 of 19 standard LCM(6), a 70-ft planing craft would be requested. The 97 standard LCPL would be redesigned as 40-footers for higher speed. As a separate item, a hydrofoil beachjumper craft would be requested. Meanwhile the characteristics of the LCVP, LCPL, and LCM(6) would all be reviewed in hopes of improving their speeds by substituting gas turbines for diesels. That was not done, however, and the proposed prototypes were not built.

Higher speed would be expensive. In 1957 BuShips estimated that a production planing LCVP would cost 5.75 times as much as a conventional one, and a hydrofoil LCVP would cost 8 times as much as a conventional one. The divergence would be even greater for larger craft. The existing LCM(6) cost 2.5 times as much as an LCVP, but a planing LCM would cost about 24 times as much as an LCVP (i.e., almost 10 times as much as a conventional LCM[6]), and a hydrofoil LCM would cost 45 times as much as a conventional LCVP. Prototypes would cost two to three times as much as production units, even without taking into account the cost of machinery development.

By this time the LCVP concept seemed obsolete. For example, a BuShips participant at the May 1957 Amphibious Warfare Conference reported that the two amphibious forces and the two Fleet Marine Forces agreed on only one point, that no new LCVPs should be developed, and that the number in the fleet should be cut. LCM(6) was the minimum useful landing craft. Limited by its davit, the LCVP was too small, and it lacked maneuverability. The latest modification merely returned to the unsatisfactory deck-mounted ramp winch, and it added an expensive but useless round counter stern. The BuShips participant called it "an abomination to a seafaring man." Considerable LCVP work was still proceeding, but it was dismissed as a boondoggle: a mass production (Kettenburg) type, plastic versions (Mks 5, 7, and 8), gas turbine power, a hydrofoil Mk 2. The Kettenburg design was presumably the twin-engine plywood type that had been rejected some years earlier. Hydrofoils were rejected because they could suddenly pitch over in rough weather, possibly causing injury to the troops on board.

Moreover, rather than concentrate on making landing craft faster, the designers should focus on the ability to load them rapidly in bad weather and in rough seas. The run-in was the shortest period of the operation; much more time was spent loading and reloading. The BuShips participant pointed out that the new LPD would solve exactly this problem.

At this time the BuShips preliminary designers were working on a hydrofoil LCM (SCB 162), the first for which BuShips was responsible since the 1942 design disaster. Characteristics had been approved in April 1956. The hydrofoils, which had to be retractable for beaching and for coming alongside, were supported by three struts, two at the sides forward and one aft. The designers chose dimensions between the LCM(6) and LCM(8) sizes, and added sponsons to house retracted foils. Compared to a conventional LCM, the cargo deck was raised considerably, partly so that the hydrofoil struts could be attached above the boat's waterline, and partly to get finer lines forward (the bow door was kept wide enough, 12 ft, to load a standard tractor). This figure in turn set the boat's beam. A double bottom would limit the effect of bottom damage. As in any other hydrofoil, the boat had to accelerate to "flying" speed, so she needed a high-speed (planing hull), relatively flat aft. Maximum deadrise forward, to the extent possible given door height and draft, would reduce impact loads as she landed. Raking the bow and giving it a slight V shape on its outside would reduce impact loads when the boat slammed into waves. Bow rake would also make it easier to load long trucks. The pilothouse was placed aft and to one side to improve visibility. As reported in October 1956, the preliminary design showed a 75 ft 9 in (overall) × 16 ft 1 in (24 ft with foils retracted) × 4 ft 1 in (12 ft 5 in with foils down) hull, displacing 50 tons. It could carry only 18 tons of cargo, well short of LCM(3) capacity. Presumably that was partly because it needed sufficient volume for a 3,000 SHP powerplant, which would drive it at 40 kts and achieve 150-nm endurance.

BuShips estimated that a hydrofoil LCM(6) using gas turbines could make 46 kts. By way of comparison, a 76-ft planing LCM(6) (to carry 60,000 lbs on a landing displacement of 170,000 lbs) would make 24 kts using four 1,000 HP gas turbines. Range would be 100 nm at maximum speed. An 80-ft planing version of the larger LCM(8) (120,000 lb load on a landing displacement of 300,000 lbs) would make 20 kts using six 1,000 HP gas turbines. Range at maximum speed would be 100 nm.

BuShips estimated in 1957 that a 40-ft planing LCVP would make 22 kts on 500 HP, and 27 kts on 1,000 HP. Range at maximum speed would be 100 nm. On hydrofoils an LCVP of conventional length might make 42 kts. BuShips built and tested several prototype hydrofoil LCVPs, but they never entered production, partly because by the late 1950s LCVPs of any type were no longer wanted. In March 1959, a table of landing craft characteristics showed a hydrofoil LCM (but not a hydrofoil LCVP) capable of 30 kts foilborne, with a range of 150 nm and 30 kts. A hydrofoil LCU project had just been abandoned.

Borg-Warner's planing-hull amphibian is shown planing. *Borg-Warner Corporation*

Lycoming's hydrofoil LVHX1 is shown "flying" on her submerged foils. The great advantage of hydrofoils is that the craft's hull shape is irrelevant to her high-speed performance. However, submerged foils require an active control system, which was difficult to build prior to the advent of cheap reliable computers. *Lycoming*

Up on blocks, Lycoming's LVHX1 displays her rear strut, which carries both her propeller and a 17 ft 6 in span foil below it. The lower part of the strut rotated to steer the craft. As the craft neared the beach, the foils would be brought out of the water and the wheels lowered. This sort of system made sense in a second-wave support craft, but it was difficult to apply to a first-wave amphibian. Such a craft might well be caught by enemy fire as it stopped to change modes of operation. The current AAAV has a planing hull and retractable tracks, and much effort has gone into very quick transformation from a high-speed watercraft to a land vehicle.

In March 1961 two alternative fast amphibious support vehicles, in effect successors to cargo-carrying LVTs, were on order: a planing-hull LVW and a hydrofoil LVH. Two LVW were delivered in 1963. Powered by a Lycoming T-55 gas turbine (1,500 HP), Ingersoll's LVW had retractable wheels, so it could continue ashore after reaching the beach. It offered a 5-ton payload (19 tons gross) and a speed of 25 kts in the water. Dimensions were 36 × 11 ft 8 in × 10 ft 10¼ in.

BuShips had two alternative LVH under contract, one with water-piercing foils (LVHX1) and one with submerged foils (LVHX2). Two examples of each were completed in 1963. Like the LVW, the LVH had retractable wheels and a 5-ton payload; it was considered more complex, hence less attractive, than the LVW. Lycoming's LVHX1, 36.11 × 10.10 × 11.3 ft (32,830 lbs), was credited with a speed of 45 kts when foilborne in the water, using a 1,225 HP TF1460 gas turbine. FMC's LVHX2 was 37.5 × 10.7 ft × 8 ft (19 tons), with a 15 × 9.7 ft side-loaded cargo bay. Its 1,040 HP Solar gas turbine would drive it at 35 kts in water or at 5–30 kts over land. Surf performance was expected to be poor. Both LVHX1 and LVHX2 were tested extensively, but neither entered production.

In the late 1950s the surface effect ship—sometimes called air cushion vehicle (ACV), ground effect machine (GEM), or hovercraft—was perfected in England. It rode on a cushion of air, thus largely eliminating friction. In this sense it was something like a planing craft. The cushion did more than eliminate friction, however. It lifted the device above the surface, so that it had, in effect, no draft—it could pass over considerable obstacles. Thus an air cushion craft was both very fast on the water and could move on to land, beyond the beach, at high speed. It was what the Marines had always wanted, a fast amphibian that could ignore the barrier at the beach.

A hovercraft could negotiate a much wider variety of beaches than a conventional landing craft. A defender facing conventional landing craft could often predict where they would come ashore simply by excluding beaches that were too flat. The wider the range of beaches a craft could negotiate, the better the chance of preserving the element of surprise.

ONR began funding research into ground effect machines (GEMs) in 1958. It had a predesign for one by July 1960. In November 1961 ONR asked the Marines to fund a design study. It sized its conceptual GEMs to match existing landing craft: 3½-ton load for a DUKW or LCVP, 30 for an LCM(6), 45 for an LCM(8), and 90 for an LCU; the last two were considerably less than the corresponding landing craft carried. A 5-ton GEM (a 105-mm howitzer, half a crew, and a day's ammunition) seemed the most versatile. A 15-ton type was rejected because it could not get beyond the beach and was too large for troops in vulnerable situations. It would be more

For a time the hydrokeel seemed to promise very high waterborne performance. A hydrokeel LCVP is shown. *Bell Aerosystems*

LCVP(H) was the hydrofoil alternative to the hydrokeel LCVP. Two 275 HP Chrysler engines drove her at more than 35 kts when foilborne. She was 40 ft long overall, and 29 ft wide over her foils. The craft is shown on the Potomac, March 1962.

efficient to use 45-tonners; there also might be some point in having a few 90-tonners.

A GEM with a 5-ton payload would be 35 × 11–13.5 ft (13 tons gross), with a 24 × 7.6 ft cargo bay loading via an end ramp. A 1,200–2,800 HP powerplant would drive it at 40–60 kts in the water and at 5–30 kts on land. A tank carrier (45-ton capacity) would be 67 × 40 ft (76 tons gross), making 60 kts on 7,800 HP. The equivalent hydrofoil would be 95 × 20.5 ft (95 tons gross), with 9 ft draft, making 60 kts on 8,000 HP.

By this time it was assumed that, except for LSTs and LPDs launching LVT amphibians, the entire amphibious task force would remain well offshore. The GEM would have a much shorter cycle time at a range of 75 nm (2.9 vs. 3.9 hrs), because the hydrofoil had to retract her foils well before beaching, approach the beach at low speed, and retract at low speed before becoming foilborne once more. Much would also depend on how quickly cargo could be moved from the ships to dumps about 5–10 miles inland. In a conventional operation, 172 5-ton trucks were used simply to move supplies from the beach to the dumps, and 2½-ton trucks took over for runs further inland. A force using GEMs would not need the trucks at all. It took three LSTs simply to carry the trucks.

Standard practice was for 12 rifle companies on board LVTs to hit the beach at H-hour, 6 more companies being held in a floating reserve, and brought ashore at H+6 hrs. If fast GEMs could replace the slow LVTs, the LPDs could stay further offshore, out of range of enemy defenses. Because they were fast, only two-thirds as many GEMs as conventional landing craft would be needed, and fewer than a quarter as many GEMs as LCM(6)s. The well deck of an LPD could accommodate 18 small 5-ton GEMs or 20 hydrofoils (LVHX2).

Unfortunately, small GEMs proved inefficient. The key ratio was cushion area (which determined how much weight the cushion could support) to cushion circumference (which determined how quickly air bled out of the cushion, hence had to be replaced by the vehicle's fans). The larger the vehicle, the smaller the ratio, hence the less power that had to be devoted to replacing cushion air. Large GEMs were not shaped anything like conventional landing craft; their very broad hulls could not efficiently fit existing well decks. Existing and projected amphibious ships could not carry enough of them. Once GEMs had been adopted, in the form of LCACs, new ships would have to be built to accommodate them. In 1962 that was still far in the future.

One other approach was tried at this time, the "hydrokeel," in which an air cushion was injected under a planing hull, drastically reducing friction. The concept was patented in 1959 by the Anti-Friction Hull Corporation; by the early 1960s it was supported by Bell Aerospace. The boat had two sidewall keels and a centerline keel, forming twin tunnels for air. The air was injected behind hinged flaps at the bow. The combination of keels and bow flaps helped keep air from escaping, thus limiting the power required to maintain the cushion or bubble. Later this type of craft (on a larger scale) would be called a surface effect ship (SES). In 1962 it was estimated that a wooden hydrokeel LCVP (not yet built) could make 26 kts on two 300 HP diesels; endurance would be 115 nm at 26 kts. Cargo capacity would be 6,200 lbs. The LCVP version was designated LCVP(K), the hydrofoil version was LCVP(H), and the planing version was LCVP(T). During the early stages of the Vietnam War the hydrokeel LCVP was of interest because it seemed that it could use its air bubble to negotiate marshes and very shallow water. There were also proposals for gas turbine–powered LCSR(K) (landing craft, swimmer reconnaissance, hydrokeel) and LCM(K) (landing craft, mechanized, hydrokeel), and for a PGM(K) (gunboat, hydrokeel).

A hydrokeel prototype was powered by a pair of 300 HP Chrysler M413s plus a 175 HP M318 blower engine. For comparison with it, a conventional LCVP had its single diesel replaced by two diesels using vee-drives; it was redesignated LCVP(T), for twin. LCVP(K) tests began in 1963. By 1965 the David Taylor Model Basin was decidedly skeptical. With 720 HP, LCVP(K) and LCVP(T) performed similarly. For example, they made 31.5 and 32 kts, respectively, at medium displacement.

Another fast new landing craft was the LCSR (landing craft, swimmer reconnaissance), a replacement for the LCPR that had been used to support underwater demolition teams (UDTs) during World War II. They, also, would have to get to their operating areas off a beach much more quickly. There was interest in such a craft, to fit LCVP davits, as early as 1950. In March 1952 BuShips proposed a prototype, which

The reconnaissance boat Mk 2 was based on the LCPL. One is shown launching frogmen; the rubber raft alongside is an intermediate step between boat and water.

was approved in June. Tested in 1956, it had an innovative inverted-V bottom, and it failed altogether, partly due to chronic engine trouble and propeller cavitation. BuShips then used the new LCPL as the basis for an interim design, the Mk 2 36-ft reconnaissance boat. A contract was let in September 1958, and the first boat was delivered late in 1960. By that time it was already clear that such craft were too small for the required load (a UDT platoon) and speed.

After informal discussions in June and July 1960, the SCB was formally asked in August to develop a new set of characteristics. The boat would support a fully equipped UDT platoon (22 men plus 2,500 lbs of equipment), using a Fulton swimmer recovery system. The boat would have to maintain 30 kts at full load in waves up to 5 ft for 6–8 hrs, with fuel for 200 nm at 30 kts. The boat would have to rest upright in an LSD well deck without any special cradle. Maximum size was 56 × 15 ft (roughly the size of an LCM), so BuShips developed a 52 × 15 ft fiberglass hull. After considering and rejecting a 900 HP Packard high-speed diesel, the designers chose the new 1,100 HP Solar gas turbine. Armament was two 0.50 caliber machine guns; later provision was made for a Redeye antiaircraft missile launcher as an alternative.

Ten of these 52-footers (SCB 221) were built in 1961–62, the first seven under the FY 61 program. Four more followed later. Powered by two Saturn T-1000 gas turbines, they were rated at 35 kts (30 kts continuous), with an endurance of 200 nm at 35 kts (1,300 gal of fuel). Dimensions were 52 ft 4.5 in × 14 ft 9.25 in × 5 ft 6 in (loaded); displacement was 36,410 lbs light and 53,630 lbs loaded. Crew was 8, with 22 UDT personnel and 2 tons of cargo. By 1967 these craft had been withdrawn from service due to maintenance problems with their gas turbines. An LCSR(K), proposed in February 1962, was abandoned when the entire hydrokeel program was dropped.

A new 36-ft LCSR(L) (landing craft, swimmer reconnaissance, large) was designed to go in a standard Welin davit. Two were included in the FY 64 program. Competitive Naval Ship Engineering Center (NAVSEC) designs built by Grebe and commercial designs built by Harbor Boat were evaluated about 1967, but no production followed. Like the 52-footers, they were powered by gas turbines. Powered by two ST6-J70 gas turbines, they were rated at 25 kts (200 nm endurance) and had two controllable reversible-pitch propellers. Dimensions were 36 ft 3 in × 11 ft × 3 ft (23,962 lbs). In effect, these craft were superseded by the SEAL transport craft.[1]

By 1961 the LVTP5 amphibious personnel carrier was nearing the end of its useful life, and a new model was planned for 1966 or 1967. The Marines had to decide whether to invest in a really fast vehicle, to attack from beyond the horizon, where transports would be safe from improving enemy shore defenses. The Marine Corps Development Center conducted an amphibian vehicle requirement (AVR) study, which the commandant approved. The study's report recommended the fast air cushion vehicle, disgorging personnel carriers well beyond the beach, as the ultimate solution to long-distance amphibious assault in high-intensity war. The Marines remembered that the main virtue of the LVT was its ability to run deep inland from the sea; the air cushion vehicle could do much the same thing, to project a conventional personnel carrier. Given an increasingly hostile environment inland, perhaps it would be best to concentrate on the land capabilities of the craft. An FMC design showed that any fast-swimming LVT would be quite expensive; the study implied that there was little point in that sort of investment. High water speed required a craft too large to perform well ashore. Moreover, a fast LVT launched at long range would pose major problems of control and navigation, which navy ships had already solved.

The AVR study proposed an interim solution. The Marines estimated that air cushion craft would not be in service in time to replace the LVTP5 series, but that they would be in service by the time the successor vehicle required replacement. Thus, in their next LVT, which they thought would be their last, they chose to emphasize performance ashore. They would minimize danger to the ships by having them launch their LVTs at high speed. Thus an LSD or LPD would run in at high speed, launch, and leave. In 1965 the navy and the Marines began work on this new high-speed underway launch technique. Atlantic Fleet tests were successful, and in March 1968 off Vietnam during Operation Bear Chain *Ogden* (LPD 5) launched her LVTs at full speed while launching helicopters and serving as primary control ship. High-speed launch became official doctrine. As it happened, it depended on very careful trimming of the launching ship, and on keeping the stern gate level. If the ship were trimmed down by her stern, water in the well deck would slosh back and forth, pushing the stern up and down. Also, problems with the *Newport*-class LSTs, the main LVT carriers of the 1970s, limited their launch speed for years to 5 kts. Thus, although high-speed launch remained on the books, by the early 1970s there was a real question of its value.

By 1961 BuShips was working on four alternative LVTs, which would be lighter and smaller, with improved characteristics. The formal requirement for an LVTP5 successor was announced in 1964.

The LVTP7 was adopted to replace the more massive LVTP5. It was about 25 percent faster than its predecessor, but that was still far slower than an LCVP. It would take massive power and a new design approach to produce a really fast amphibious assault vehicle for use in the twenty-first century.

FMC won over Chrysler. The first of 19 LVTPX12 prototypes appeared in 1967. Compared to LVTP5, it had a much better hull shape with a long sloping bow. For the first time in the series this LVT used separate land (tracks) and water propulsion systems (twin waterjets for water), and a diesel rather than a gasoline engine. Because the tracks were no longer compromised by the requirement for water propulsion, speed on land increased considerably, to 40 mph (20–30 mph cruise), to match the Marines' next-generation M60 main battle tank. The tracks were based on those of the T91E1 experimental light tank. The suspension was based on that of the M113 armored personnel carrier. The Marines wanted a water speed of 10 mph to space out waves of craft (to limit nuclear damage) without accepting excessive time between waves. Unfortunately, the waterjets increased water speed from 6.8 to 8.4 mph, which was probably the best the modestly powered 400 HP Detroit Diesels could do. The Marines accepted the figure as a trade-off against weight, fuel consumption, and development cost. These improvements were despite a power to weight ratio slightly worse than that of LVTP5. Thrust deflectors made for much better maneuverability in the water. Although they were inefficient, the craft could use tracks in the water (4.5 mph) so that she could crawl right up onto a beach without waiting to shift power from waterjets to tracks. Later the Marines claimed that this craft was the only amphibian in the world capable of operating in rough seas and in 10-ft surf. The hull was welded aluminum armor, based on that of FMC's M113. At least as important, the new craft needed only a third as many maintenance manhours as its predecessor.

The new craft was considerably lighter and smaller than its predecessor, closer to wartime LVT size (about 50,000 vs. 74,000 lbs loaded), partly because its load was cut to 25 marines or 5 short tons of cargo. The 25 marines were a reinforced rifle squad (with space for crew-served weapons), so three carried a reinforced rifle platoon. Outside dimensions were set by the size of the troop compartment, so the craft could not carry the 105-mm howitzer an LVTP5 could, but seldom did, accommodate. Maximum width was based on well-deck dimensions rather than the traditional criterion of rail transportation (limit 11 ft 8 in), so an LSD could carry four rather than three abreast. Perhaps the only important disappointment was in the armament. The Marines had wanted a 20-mm gun, but problems with it forced them to accept a 0.50 caliber machine gun.

The basic vehicle entered production as LVTP7, with initial deliveries in January 1972. It was credited with a ten-year lifetime, so a replacement (or retirement in favor of air cushion vehicles) was due in 1982. The family included a command craft, LVTC7, and a recovery craft, LVTR7, but not a fire support vehicle. By 1984, 985 LVTP7s were in operation, entirely replacing the earlier LVTP5.

Part Four

The Sixties and Beyond

By the 1960s most of the vast World War II fleet of amphibious fire support craft had gone to allied navies or the scrapyard. The great question through the 1960s was whether carrier aircraft and the longer-range weapons wielded by LSM(R)s, cruisers, and battleships could replace them. In this 1952 photo, two war-built fire support craft, LSSL 77 and 91, await transfer to the Republic of Korea Navy, 1952. Their wartime designation was LCS(L) Mk 3. Note that LSSL 91 has a single 40-mm gun in No. 1 position forward, whereas her sister has a twin mount.

14
Fire Support Revisited

After World War II the fleet's inventory of heavy naval guns that had supported wartime amphibious assaults declined. Aircraft became the strike weapons of choice. The guns' traditional antiship role became less important. Guns seemed less and less capable of dealing with the new jets. Ships were converted to take guided missiles. It proved impossible to justify retaining guns simply to support amphibious operations; in the past, that capability had been one among several uses of a multipurpose weapon.

Briefly, in the mid-1950s, it seemed that nuclear weapons could make up for the decline. The U.S. Navy developed nuclear shells for its battleships and heavy cruisers. Battleship shells (26, 12, and 3.5 kilotons) were stockpiled during 1956, and the *Iowa*-class battleships were modified to carry them. The three *Des Moines*–class cruisers were modified to carry nuclear shells, which entered service in 1957. For a time, it was claimed that nuclear shore bombardment was viable, that armored vehicles could safely roam a beach a few hours after a burst. Under such assumptions, even the nuclear-tipped versions of the new antiaircraft missiles offered significant troop support firepower. The Eisenhower administration said that nuclear weapons should be considered no more significant than conventional ones, but in retrospect it is not certain that such statements were seriously meant. The Kennedy administration, which entered office in January 1961, was certainly determined to erect a clear barrier between conventional and nuclear combat, so the tactical nuclear weapons it inherited were less and less usable.

Beginning in 1959, surviving World War II surface combatants in reserve were discarded. Within a few years, for example, only the four *Iowa*-class battleships were left. Cruisers were disappearing, either to have their guns replaced by missiles or to retirement and scrapping.

The navy and the Marines tried two approaches to this problem. One was a major caliber lightweight gun (MCLWG) that could be mounted on board destroyers, the surface combatants that almost certainly would survive in numbers. The initial impetus, in 1960, was the appearance of a Soviet 130-mm coast-defense gun that could outrange existing destroyer 5-in/54 and 5-in/38 weapons. In 1962 BuShips proposed a new 8-in gun. In 1964 BuShips recommended that the navy adopt the army's 175-mm gun instead, mainly to use new army ammunition. The Defense Department agreed. The 175-mm could reach 40,000 yds, compared to 32,000 yds for a cruiser's 8-in/55, and an automated naval version would fire 10 rounds per minute. On 26 February 1965 the navy issued a specific operational requirement for a 175-mm gun with a rocket-assisted projectile (RAP), to be produced in 1971. Range would be 36,000 yds (50,000 yds with RAP). The gun would fire 147-lb RAP shells at an accuracy of 2.7 mils at 80 percent of range.

Initially the new gun was made part of a larger family of lightweight naval guns, including both single and twin 5-in/54; the single was the Mk 45 that was later widely used on board U.S. destroyers and missile cruisers. A research and design (R&D) contract was issued on 1 March 1966, and at about the same time (FY 66) the gun was separated from the 5-in lightweight program.

The Naval Ordnance Systems Command (NAVORD) issued a ship integration study of the gun mount in June 1968. The existing *Knox*-class (FF 1052) frigate hull could mount two or three guns, a *Forrest Sherman*–class (ASW variant) destroyer could mount two, and a *Dewey*-class (DLG 9–class, later DDG 37–class) missile destroyer could mount one. A *Newport*-class LST (LST 1179 class) could mount three, one forward and two aft.

Unfortunately the army decided to phase out its gun. Reevaluation showed that new ammunition would be needed in any case, so using the army gun would save no money. Meanwhile the project gained urgency as more and more of the fleet's heavy guns were retired, and as the Vietnam War revealed a tactical need for longer ranges. By 1969 the project envisaged a guided projectile. Compared to the 175-mm, the 8-in now offered 20 percent more range, which equated to 44 percent more area coverage. A boosted GP round, with a 25-ft CEP (circular error probable), offered 186 percent more range for the same 125-lb payload (i.e., 720 percent more area coverage). Moreover, new shells offered far greater lethal areas.

In 1969, it appeared that lightweight 8-in gun mounts could be delivered in June 1975 if prototyping was foregone as a low risk. If a prototype was needed, the gun could be delivered in July 1977. Range with a conventional shell would be 43,000 yds; with an extended-range shell (125-lb warhead) it would be 130,000 yds. The 8-in mount (Mk 71) was tested on board the destroyer *Hull*, firing 11.7–12.6 rounds per minute. Its blast was slightly less severe than that of an 8-in bag gun aboard one of the older cruisers. A May 1970 study showed that 8-in long-range terminally guided projectiles, if they could be developed, would be far more effective than a Harpoon-size missile at a range of 60 nm. By this time, moreover, there was a simple solution to terminal guidance: semiactive laser homing. Marines in forward positions could simply illuminate their targets for incoming rounds.

In the early 1970s there was considerable interest in installing the gun on board *Spruance*-class destroyers; indeed, such installation seemed to be a reasonable alternative to building a specialized fire support ship. By the late 1970s the Marines had documented a requirement for 14 MCLWG platforms. The U.S. Navy was operating six amphibious groups, and presumably the Marines wanted two such ships per group. Nothing came of this idea, however. Similarly, the MCLWG figured prominently in studies of new fire support ships.

By 1969 a long-range gun system (LRGS), approximately a 12-in/70, was in the advanced development stage. According to a 1 July 1969 Op-36 point paper, this 300,000-lb weapon would throw a 100-lb projectile 70 nm. Work was already proceeding on longer-range saboted shells (Project Gunfighter) and on laser-guided shells. A saboted 8-in fin-stabilized shell offering the lethality of a 6-in shell could reach 75,000 yds. A full-bore liquid-supported rocket-boosted round might reach 100 miles inland. Another weapon figured prominently in studies of new fire support ships was the lightweight twin 5-in/54 gun (Mk 66), which was expected to fire 96 rounds per minute, and thus rival bombardment rockets as an area neutralization weapon.

Alternatively, it could be accepted that missiles were superseding guns. New-generation ships could fire a shore bombardment missile. The Marines' first attempt was Lacrosse, a surface-to-surface weapon fired into a "basket" and then command-guided by a forward observer. It was conceived by the navy's missile experts, Johns Hopkins University's Applied Physics Laboratory (APL), in the early 1950s as a supplement to artillery and naval gunfire. Technology was hardly mature, and Lacrosse was never deployed in any numbers. Moreover, it was not yet very necessary.

A few years later, the Marines' new vertical envelopment tactics placed helicopter landing zones well beyond normal gunfire ranges. It would no longer be enough to shell the beach and its approaches. Missiles were the obvious solution. Beginning in mid-1960 APL proposed an inertially guided Taurus missile (ZRGM-59) to be fired from Terrier missile launchers. The great advantage of using existing launchers was the ability to equip ships that would have to be present at an amphibious operation in any case, to offer antiair protection. A formal development objective was issued on 4 March 1960, and an operational requirement on 7 August 1961.

The missile's solid-fuel booster would place it in the appropriate ballistic trajectory; range would be set by elevating the launcher, just as a gun would be elevated. Desired minimum and maximum ranges were set at 15 and 50 nm, respectively. The minimum was for a beach assault; the maximum, extendible eventually to 100 nm, was to deal with enemy forces threatening helicopter landing zones and for interdiction. Long range would also allow ships to hit beach targets while they were still well offshore. CEP was 200 yds. Taurus would be limited to destroying soft targets such as massed troops or vehicles. The 1,000-lb warhead could be unitary HE or an area weapon (bomblets or flechettes, to attack vehicles or personnel). Hopefully such area attack could make up for the very small number of shots a missile magazine could accommodate. By August 1964 APL hoped that a project definition program could begin in FY 66 (i.e., some time after 1 July 1965); the technical key to the program was a low-cost inertial platform.

The Taurus study showed that the navy's idea of an LFSW (landing ship fire support weapon) was feasible. In June 1965, Vought proposed a shipboard version of its army Lance missile. It claimed that the weapon could be deployed on board the two oldest missile ships, *Boston* and *Canberra*, as soon as 1968, after tests beginning late in 1965. The 3,250-lb Lance, about the size of a Terrier but with greater diameter (22 in), had a 1,000-lb bomblet warhead

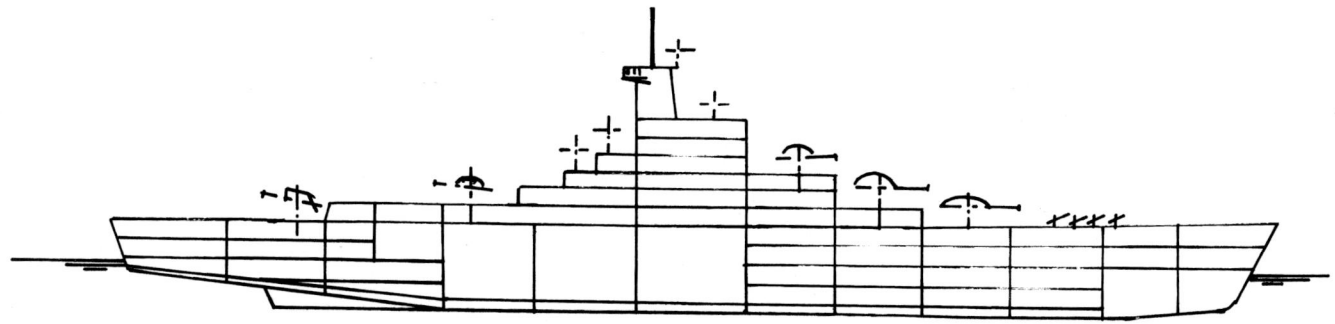

A high-end LFS sketch design, showing LFS armed with the landing force fire support weapon (LFSW), about 1967. LFSW is a guided missile, fired from the launcher right aft. Superfiring over it was a lightweight 5-in gun; the two after directors were for LFSW. Forward the ship had a lightweight 5-in mount superfiring over two lightweight 8-in guns. Forward of the guns are eight Mk 105 rocket launchers. An Mk 86 gun director would have been mounted atop the mast, with a rocket fire controller atop the ship's bridge. Dimensions would have been 530 ft overall (510 ft waterline) × 53 ft 6 in × 20 ft (fully loaded), for a full load displacement of 8,000 tons. The powerplant would have been twin-screw geared diesels (19,400 SHP), for a sustained speed of 20 kts. Endurance would have been 10,000 nm at 20 kts. Accommodation would have been provided for 15 officers and 282 enlisted men. Various LFSW projects were considered, the most viable being a seaborne version of the army's Lance missile.

(carrying BLU-26 or BLU-7 bomblets), and it could be fired at the rate of 15 per hr. Maximum range, 40 nm, would have fallen somewhat short of the navy's requirement, but an extended-range version (100 nm) was in development. System CEP, using automatic radio control (with inertial under development), was 300 yds. Lance may have been less than attractive for shipboard use because it used storable liquid propellant. Nothing was done, but a shipboard Lance figured in some of the fire support ship studies described below.

In the spring of 1967 the Naval Weapons Center proposed a heavy interdiction weapon system suited, like Taurus and Lance, for both direct fire support and interdiction. It could also destroy inland air bases and antiaircraft systems, cutting the attrition of U.S. aircraft that was then becoming quite a serious cost of the Vietnam War. A tentative specific operational requirement (TSOR) was issued on 17 March 1967. The weapon would be launchable by missile destroyers and cruisers, and also by fleet ballistic missile submarines, with a 200-yd CEP achieved by techniques such as terrain reference. A companion TSOR (TSOR 12-12T) in May 1967 called for a 500-nm strike weapon specifically to cut aircraft attrition by destroying enemy aircraft on the ground, as well as radar sites, antiaircraft weapons, and antiaircraft missiles. It had a secondary quick reaction role of destroying massed troops, supplies, and light armor. Nothing came of any of these studies, presumably because the studies were too far from fruition to be relevant to the expanding Vietnam War.

On a more prosaic scale, work began in July 1966 on a new generation of shore bombardment rockets, at about the same time inshore fire support (IFS) *Carronade* and three LSMR (landing ship, medium, rocket) were reactivated to fight in Vietnam. A specific operational requirement, issued 27 June 1967, showed two new rockets under development: an advanced 10,000-yd weapon (C1) and an advanced 20,000-yd weapon (the 61-lb Unit D). Ships were already armed with 2,500-, 5,000-, and 10,000-yd rockets; actual maximum ranges were 2,200, 4,600, and 10,000 yds, respectively. All these weapons could be fired from the Mk 102 and Mk 105 launchers of the existing rocket ships. Like the new gun and Sea Lance, they figured in new studies of specialized fire support ships.

Op-093 raised the problem of declining gun fire support for a 1970-era amphibious operation in an August 1961 study. It was probably not coincidental that this study came at the beginning of the Kennedy administration, which emphasized what it saw as a Soviet threat in the Third World. The navy's program objective for FY 64 (PO-64), released in September 1962, was probably the first official mention of a new amphibious fire support ship, the LFS (landing force, support). It would be armed with a tactical surface-to-surface missile (at least 300 rounds, capable of hitting two targets simultaneously), the 8-in lightweight gun, short-range bombardment rockets, and a self-defense missile (the abortive Sea Mauler). This tentative program also included a non-nuclear support carrier (CVS), which was conceived for amphibious support as well as for ASW.

On 28 November 1962 the Center for Naval Analyses (CNA) released "Capabilities and Requirements for the Fire Support of Amphibious Assault" (NAVWAG 4). Its conclusions were presumably reflected in the 1963 long-range objectives report (LRO-74) for the period ending 1974.

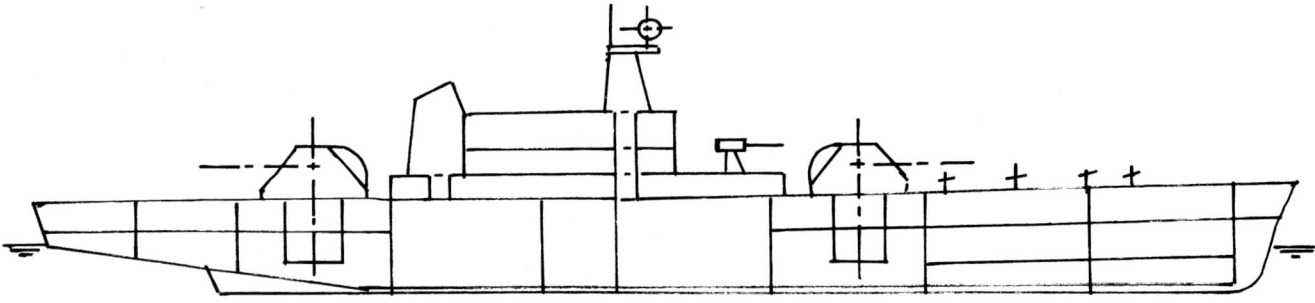

A low-end LFS, from an undated Naval Ship Engineering Center sketch, probably made about 1969. Armament comprised two 175-mm lightweight guns, with a twin 3-in/50 superfiring above the forward mount. Right forward were for Mk 105 rocket launchers, their magazine two decks down below them. The mast carried an Mk 86 director. The original sketch is numbered after that of the LFSW ship, but it is marked to indicate 175-mm rather than 8-in lightweight guns, which suggests an earlier date. Dimensions would have been 320 × 45 × 12 ft (2,900 tons fully loaded); twin-screw geared diesels (10,000 SHP) would have driven the ship at 20 kts. Note the two separate engine rooms, with an auxiliary machinery space between them. Endurance would have been 5,000 nm at 20 kts. Accommodations would have been provided for 11 officers, 8 CPOs, and 158 other enlisted men.

Fire support needs were set by the need to provide each level of command with one or more dedicated ships that could reply quickly enough to calls for fire. Each battalion beach was assigned a destroyer and two rocket ships (neutralization fire). Because the battalion spotter communicated directly with the destroyer assigned to his sector, the time lag between a request and fire was only about 2½ minutes. Given her greater range, a cruiser could cover a pair of battalions (the assault force of a regiment). Her fire had to be requested through the regimental fire support coordination center, which had a processing time of 9–10 minutes. A battleship or another cruiser was assigned to the division as a whole (total processing time, 15–16 minutes). Thus for a division/air wing assault the Marines wanted a battleship, two heavy (8-in gun) cruisers (two regiments), four destroyers, and four rocket ships; in some accounts they wanted another two destroyers as reinforcements where needed. Cruisers and battleships were being phased out of service, and in 1960 all the rocket ships were in reserve.

The Marines admitted that substituting 8-in guns for the battleship's 16-in, to attack hard targets, would be only a moderate risk. The wartime 5-in/38 gun could be replaced by smaller numbers of faster-firing 5-in/54. It was not yet realized that the newer gun jammed when fired at maximum rate. LRO-74 concluded that at least four heavy-gun ships (battleships and/or heavy cruisers) should be retained in reserve. In fact, the U.S. Navy retained all four *Iowa*-class battleships and two decommissioned *Des Moines*–class cruisers through the 1970s.[1]

In the longer term the three heavy-gun ships should be replaced by three LFS. Each would be armed with one or two lightweight 8-in guns plus rocket launchers and, if developed, bombardment missiles. Later papers on the landing fire support ship cited a memorandum released by the OpNav ship characteristics office (Op-36) on 4 November 1963. At this time estimated unit cost was $40 million; by July 1969, it was $142 million for a lead ship and $80 million for a follow-on. Inflation alone would have raised the 1963 figure to $59.1 million for an FY 73 ship, but the weapons had also become more sophisticated.

Similar conclusions, that a new fire support ship was badly needed, were voiced at a 1963 Inter Fleet Amphibious Type Commanders Conference (and repeated at an October 1965 conference of the same type). The need for such ships could be reduced if future amphibious ships as well as future destroyers were armed with one or two lightweight 5-in/54 guns.

In May 1964 the Marines produced MCDEC (Marine Corps Development Center) project 30-64-01, "Naval Gunfire Requirements for the Long Range Period Through 1975 for a Division/Wing Team (MEF)," a study of naval gunfire support (NGS) requirements for the period through 1975. The CNO and CMC (Commandant Marine Corps) convened an ad hoc working group, the Hooper Committee, on NGS. At least three 8-in ships would be needed to support an opposed MEF landing. Without active rocket ships, there was little neutralization fire. Four such ships would be needed in an MEF landing. The committee therefore recommended commissioning two battleships and two heavy cruisers while developing an LFS. Four LSMR/IFS should be commissioned at once, and preparation made to activate four more on an accelerated basis. The navy decided to retain two active heavy cruisers in commission and to activate four rocket ships (*Carronade* and three LSMR), which were assigned to the Pacific Fleet. The Marines wanted one heavy cruiser, four LFS, and seven destroyers to support an MEF landing; the two MEF requirement thus equated to eight LFS. LFS characteristics were not identified.

The Center for Naval Analyses published its own study of naval gunfire support (NAVWAG 36) on 25 February 1965. In support, on 9 October 1964 OpNav asked NAVSEA (BuShips' successor) for cost/feasibility estimates of a minimum ship (Ship A) and an optimum ship (Ship B). Ship A had two 175-mm guns, six advanced Mk 105 rocket launchers, and a twin 3-in/50 for self-defense against aircraft. Ship B had three 175-mm guns and eight Mk 105 launchers. In each case, the twin 3-in/50 guns might eventually be replaced by a Sea Sparrow point defense missile. Like other modern amphibious ships, these would make a sustained speed of 20 kts (using geared diesels); endurance would be 5,000 nm at 20 kts. Neither ship would be particularly large. Ship A would be 320 (between perpendiculars) × 45 × 12 ft, 2,900 tons fully loaded; it would cost $30 million for the prototype and $26.2 million for each follow-on. Ship B would be 370 × 48 × 13.5 ft, 3,840 tons, $37.4 million for the prototype, and $33.0 million for each follow-on.

The CNA study showed that it was vital to reduce time lost in communication, coordination, and weapons assignment. Given better coordination and longer range, a single platform could handle assignments from several commands. Air attack was not a viable alternative because the only precision weapon then available, the Bullpup missile, was unacceptable because rounds too often went "berserk." CNA could not identify any strong requirement for a long-range landing fire support missile, but admitted that a scenario favoring the missile could readily be imagined.

If previous practice had been followed, an LFS would have been designed, probably by some time in 1966, and it would have appeared in the FY 67 or FY 68 program. The logic of the design would simply have been the need to replace earlier fire support weapons. In the past, evolutionary improvements over past ships had been perfectly acceptable in ships with well-defined missions. Missions did not have to be made too explicit because they were widely understood. The LFS, however, was proposed just as Secretary of Defense Robert S. McNamara decreed a radical change in the way that new weapon systems were bought. Called concept formulation/contract definition (CF/CD), its objective is to identify the best approach to a given military problem. The choice might not be so much between different sketch ship designs, as between, for example, ships and carrier-based aircraft. With its emphasis on finding an optimum solution, CF/CD allowed for endless studies before any decision could be made. It also had an apparently unappreciated flaw. In order for CF to proceed, measures of effectiveness had to be computed for alternative approaches. If the mission was relatively simple, such as lobbing a nuclear warhead at a Soviet city, that was not too difficult.

Unfortunately, some military problems, such as fire support, proved elusive. The obvious measure of effectiveness was the percentage of shore bombardment targets a system could destroy. Historical experience suggested that the reality was dramatically different. Shore bombardment often worked mainly by attacking enemy morale, or by forcing the enemy to change tactics. For example, neutralization fire usually kept enemy heads down during a particularly difficult phase of an assault. Often the sudden shock administered by heavy fire proved decisive, as when a cruiser offshore turned back a German armored attack that endangered the beachhead at Anzio in 1944. Unfortunately, in the 1960s the services were unable to argue that their apparently subjective analyses, based on historical analogies or on military judgment summarizing the way in which forces actually reacted to given situations were not as objective as the mathematical models. Nor were the services able to explain why the models did not capture the reality.

Because the navy was unfamiliar with this type of analysis, it had to rely on external think tanks. At one point the navy captain in charge of the LFS effort commented that the model automatically favored aircraft over ships, particularly because it omitted antiaircraft weapons. He knew that naval gunfire had special advantages, but he could not see how to express them in terms of the standard model. This problem ultimately killed not only the LFS of the 1960s and early 1970s, but also an attempt to revive the concept in the 1980s.

The nominal justification for CF/CD, increased efficiency, seems to have concealed a deeper purpose. From the start, Secretary of Defense McNamara fought to gain control over the services and their procurement programs. In CF/CD and its cousin, system analysis, he had analytic tools that the services could not match. Thus, at least at the outset, they found they could not effectively challenge McNamara's decisions. By 1965, moreover, McNamara was fighting a more and more expensive war in Vietnam, yet he could not expand the defense budget to cover it. Because it specified, in effect, a clean start, CF/CD automatically slowed down any program to which it was applied. Thus CF/CD seems to have become a means of deferring expensive new programs. It seems likely that McNamara saw CF/CD as a way of killing off the LFS. McNamara's new style of defense management is evident in navy archives, in an explosion of paperwork connected with new ship projects. Not focused on the ship or the operational requirement, most of the paperwork was an apparently endless elaboration of the way in which the project would be conducted, containing massive and largely content-free flowcharts, thus amounting to CF paperwork.

In February 1965 McNamara announced that CF/CD would be applied to all new ships. Pilot projects were the LFS, the new amphibious carrier (LHA), the mine countermeasures mother ship (MCS), the fast deployment logistics ship (FDL), and the new destroyer (DX/DXG, which became the *Spruance*). McNamara badly wanted the FDL, so its CF phase was abbreviated. DX/DXG would be subject to parallel 18-month abbreviated CF phases so they could be funded in FY 69. McNamara made the LFS the pilot case for full CF/CD. It was vulnerable because there were alternative ways of doing its mission.

To meet immediate fire support needs, there was interest in reactivating a heavy cruiser. To overcome the manning problems that had laid up such ships in the first place, she would be maintained in a state of 30–60 day readiness and brought into full service only as needed. This idea would later be proposed for the LFS. Secretary McNamara rejected this plan on 18 September 1965, leaving the way open for an LFS.

In accord with the new procurement process, the CNO issued an advanced development objective (ADO 46-27X) in June 1965, and exploratory studies began. Ships were sketched with various combinations of 175-mm and lightweight 5-in/54 guns and six or eight Mk 105 rocket launchers. The designers evaluated the impact of increasing speed to 25–30 kts. Ultimately a baseline emerged, rather larger than the Ship B developed a year earlier (452 × 51.75 × 11.16 ft, 4900 tons). Her three heavy guns (fore, aft, and amidships) were supplemented by two superfiring 5-in/54s, yet she had no rocket launchers. One implication was that the heavy gun LFS or LSFH (landing ship, fire support, heavy), which might also carry the Lance missile, should be split from a shallower-draft area-neutralization rocket ship, the LFSR (landing ship, fire support, rocket), which could operate close inshore or up rivers. The latter would replace the existing rocket ships. The LSFH would replace the aging cruisers.

A ship development objective based on the ADO was issued on 10 December 1965. To test possible ship characteristics, CNA naval analyses tried the LFS (two 175-mm, two Lance launchers) in two scenarios: an invasion of North Vietnam (Vinh A), and an attack in Northern Europe (Jutland B, which assumed that the Soviets had seized Denmark and had to be ejected). In each case an LFS was compared to three variants of the projected destroyer: DX(A), with two 5-in guns; DX(B), with two 175-mm guns; and DX(C), with one 175-mm and one 5-in gun. Ships were compared with carrier-based aircraft, with existing cruisers and destroyers, and with Marine organic artillery. Calculations were complex, so the study had to be drastically simplified. It omitted vulnerability, weapon reliability, or the force levels of fire support systems. Nor did it take into account the problems of target acquisition or target saturation; saturation refers to the ability of a ship to engage targets quickly.

The study was limited to single ships supporting single regiments. Thus it did not address the potential offered by several ships firing in direct support, or of one ship in general support of a whole division. The study showed that timeliness was terribly important. Improvements in assigning targets might have dwarfed the effects of particular weapons.

CNA concluded that an LFS armed with two 175-mm and four 5-in guns was more effective in both scenarios than the reactivated *Baltimore*-class heavy cruiser, any of three new destroyers, an LFS armed with missiles instead of 5-in guns, or eight aircraft. The ideal LFS was armed with two 175-mm and four to six 5-in guns. If DX were built instead of an LFS, the unmodified DX(A) was best in the Vinh scenario, and the all–175-mm DX(B) was best in the Jutland scenario. In general, naval gunfire was more effective than carrier close air support against D-day call-fire targets (i.e., against targets suddenly threatening troops as they landed and advanced ashore). The Marines' organic artillery was more effective in the Jutland than in the Vinh scenario, perhaps due to differences in terrain.

Fire support included preparatory fire, to keep enemy heads down during the actual landing at H-hour minus 30 minutes (H–30). As in World War II, mass rocket fire was deemed both most effective and least costly. If rockets could not be used, then a new generation of brittle steel fragmenting shells might be quite effective. Missiles would be useful against some targets, but there were not enough such targets to justify a missile-armed LFS.

Overall, however, aircraft and DX were the least expensive options, presumably partly because, unlike LFS, they were multipurpose systems adaptable to other roles. By omitting vulnerability, the study sidestepped some major questions, such as the high cost of likely aircraft losses.

A parallel gamed study of a Vinh landing was conducted by the Naval Radiation Defense Laboratory

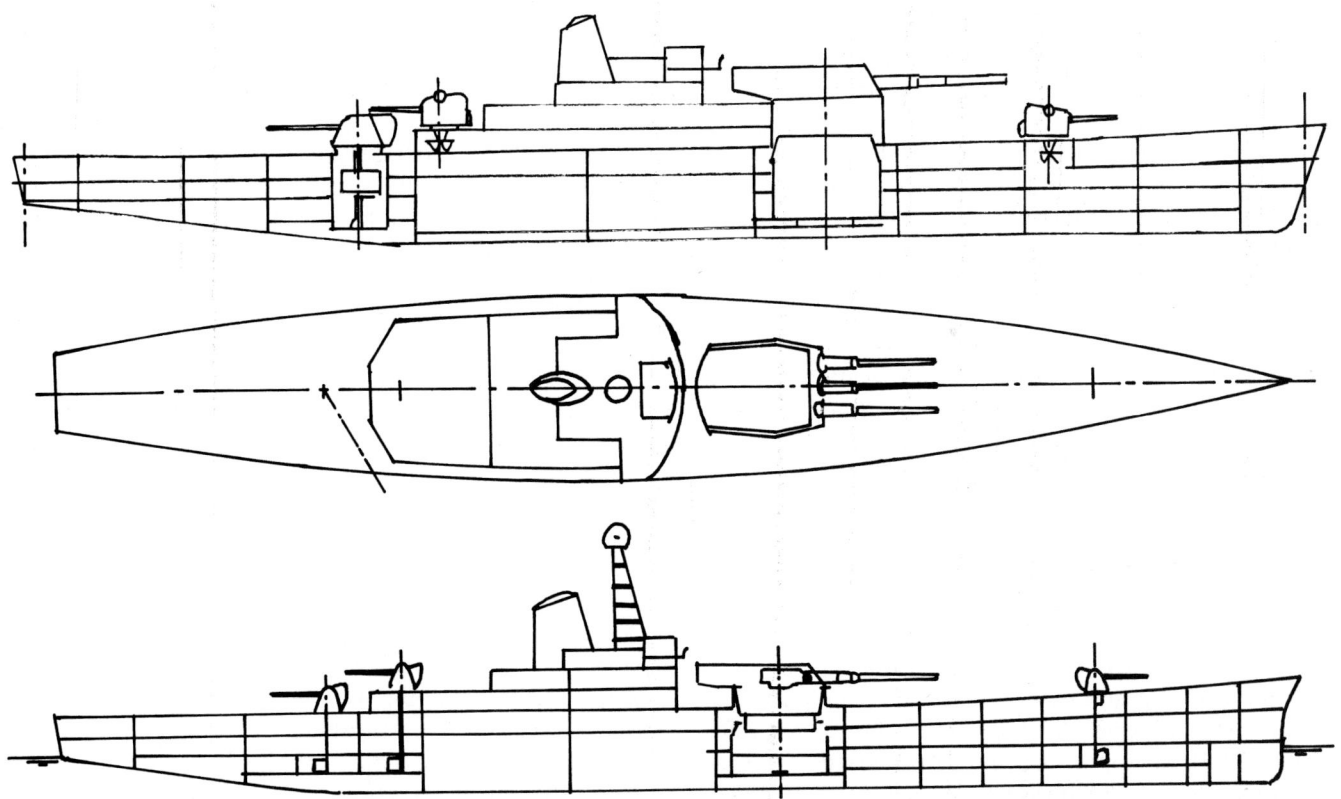

Fire support ships armed with 16-in guns, from sketch by the Naval Ship Engineering Center, late 1967. The version shown only in profile is armed with a triple 16-in/50 turret, with two 5-in/54 Mk 42s, and with a major caliber lightweight gun, at this time a 175-mm/60 (ultimately it was 8-in caliber). Waterline length is 440 ft. According to an undated list of design characteristics, the 16-in LFS was armed with one triple 16-in turret (227 rounds), one 175-mm lightweight gun (400 rounds), and one 5-in/54 (800 rounds), all controlled by an Mk 86 digital (hence adaptable to all calibers) fire control system. Dimensions were 480 ft (waterline) × 72 ft × 16 ft 11 in (9,000 tons); power (diesel) was to be sufficient for 20 kts on two shafts, with an endurance of 10,000 nm. Other than fire control, electronic systems were an SPS-53 surface radar and a surface combatant ECM suite (including WLR-1, SLR-8, SLQ-12, and ULQ-6). The ship would accommodate 14 officers and 288 enlisted men. She would have a landing area aft sufficient for an HU2K (SH-2) utility helicopter. She would not be protected. The single-sheet description notes that according to NAVWAG 36, the 16-in rifle is the desired weapon, but reactivation of a battleship was rejected due to the huge 3,000 manning requirement.

(NRDL), using an LFS armed with two 175-mm and four twin 5-in/54 lightweight guns; the latter were then in development. For support in the objective area, the study compared six LFS, four existing destroyers armed with 5-in/38s, and a combination of LFS and destroyers. This study took enemy defenses into account: missiles (Styxes) and torpedoes fired by light craft and enemy aircraft. The main counter to the surface threat would be the ships' 175-mm guns. It seemed unlikely that ships armed with 5-in/38s would long survive in the face of current, let alone future, Soviet coast defense weapons. NRDL estimated that three LFS would be lost in the approach to the target area and three in the objective area. Therefore, 12 would have to sortie to ensure the presence of 6 at H–30.

In line with the CF idea, a widely varied series of alternatives were considered alongside the baseline. In a 1966 study, the Naval Ship Engineering Center (NAVSEC) considered reactivating a battleship, activating and modernizing both a *Baltimore*-class cruiser and the IFS *Carronade*, modifying an LST 1156 hull, and building a new fire support ship. The new ship might use existing weapons (one version would have had two, three, or four 16-in guns in a destroyer hull) or the 175-mm gun and possibly Lance. NAVSEC also considered other alternatives. The most radical choices were an air cushion vehicle (hovercraft), possibly as a supplement to a conventional surface ship carrying heavy weapons, and a modular ship with removable weapons.

The 16-in gun ship was rejected because this weapon was not really needed to deal with targets, and because there was no navy facility capable of making 16-in guns. New ammunition would be needed. The LST was too slow, despite its considerable capacity (three 175-mm guns, one lightweight 5-in gun, one Lance launcher, and six Mk 105 rocket launchers). Moreover, not enough hulls were available for conversion. A mixed gun and rocket ship was rejected because heavy guns (for destructive fire) and rockets (for neutralization) demanded quite different tactics. The rocket ship, if it were built, should have shallow draft so that it could get closer to its target, and it would need more frequent replenishment. Thus the study favored building two separate ships.

Cost was clearly a major issue. Adding 3-in armor around magazines and machinery would increase ship cost by about 20 percent, and draft about 50 percent. Adding an inner bottom would add about 4 in to the ship's beam. Because they were available from stock, 5-in/38 guns would have little impact on overall ship cost. However, 5-in/54 guns would have to be bought new, and they might amount to as much as half the overall cost of installation.

The Naval Ordnance Systems Command completed its own study in May 1966. It favored the 8-in/55 rapid-fire gun (as on the *Des Moines* class), the 5-in/54 rapid-fire gun (Mk 42), and the 5-in spin-stabilized rocket. None was quite suitable for an LFS, so NAVORD in September recommended the 175-mm gun, the 5-in/54 lightweight gun for low manning, the Lance missile, and the new Basic Point Defense Missile System (BPDMS; Sea Sparrow) for self defense as LFS weapons.

Given these efforts, the LFS was included in the 28 July 1967 draft presidential memorandum on amphibious assault ships, fire support ships, and mine countermeasures forces, which laid out planned programs. Its CD phase would begin in FY 69, leading to construction beginning in FY 70. Actual funding, however, declined steadily from the first year of the CF phase. In November 1967, Op-36, the surface ships characteristics office, expressed concern over the lack of progress.

LFS certainly had a low priority, behind DX/DXG, MCS, and the new attack submarine. However, it was probably more important that its LFS mission and capability had not yet been defined well enough to justify any particular design approach. Despite several recommendations, it had not yet been decided whether rockets and heavy guns would share the same hull. Much more information was needed about new types of shells and about missiles. There had been little attempt to identify alternative methods of supplying fire support. For example, one of the ground rules of LFS studies was that four carriers would always be available for a division-level assault. How many of their aircraft would be available to support the assault? How many would be needed to protect the amphibious task force, to protect it from enemy weapons? Moreover, ideally the LFS would be capable of other missions, to make it cost-effective. What other missions could it carry out?

The 27 September 1967 draft guidance memorandum on shipbuilding practices deferred CF pending a decision as to the final weapon suit for the new destroyer, the DX. It was possible that the new destroyer, armed with the lightweight 175-mm gun, would take over the LFS role altogether. At the same time CNA was arguing that the LFS should be a single-purpose ship, armed only with guns, and intended only to support amphibious operations. Attempts to add other roles would merely complicate and confuse its design. Late in November 1967 Op-36 was still quite undecided. It did not yet have the measures of effectiveness that were central to McNamara's decision-making process.

Late in 1967 the NAVSEC designers revived the 16-in gun idea. CNA's NAVWAG 36 identified the 16-in/50 rifle as the best fire support weapon, although it rejected reactivating a battleship because so large a crew—3,000 men—was required. Rear Adm. Lloyd M. Mustin of the Atlantic Fleet's amphibious command (PhibLant) rejected the 8-in gun as a "pop gun" that could not match the destructive effect of 14- or 16-in guns. He also argued that the emphasis on a divisional (MEF) landing was wrong; the future lay with BLT-scale landings, as was demonstrated in the Dominican Republic and Lebanon during the 1950s and 1960s. In December 1967 Capt. George W. Folta Jr. of Op-36 asked whether there were 16-in gun studies, and asked that they be forwarded to the CNO. Rear Admiral King, chief of Naval Ship Systems Command (NAVSHIPS), ordained that "there will NOT be 16in guns on LFS, period." Some estimates were made, but nothing approaching a formal study; a 26 January 1968 note on the study folder, which ordered the study be stopped urgently, stated "we never STARTED."

In the short time they worked, the designers developed a sketch showing one triple 16-in/50 (without armor) and four 5-in/54 single lightweight guns (an alternative had two 5-in/54 and one 175-mm). In support of CNA studies, they also looked at single and lightweight single 16-in/50 guns, without barbettes (ships could mount one or two such weapons). A ship mounting one single 16-in/50, one 175-mm lightweight gun, and two single 5-in/54 lightweight guns would displace 9,000 tons (480 ft × 72 ft × 16 ft 11 in); two-shaft diesels could drive it at 20 kts.

Because no basic decisions had been taken, by the beginning of 1968 it seemed to the designers that CD would inevitably slip to FY 71. Thus in February 1968 the Naval Ship Systems Command offered plans for construction to start in either FY 70 or FY 71, but considered FY 70 highly problematic; CNA endorsed a start in FY 71. Unfortunately, the logic of the CF/CD process was quite rigid. Once CD funds had been budgeted for FY 69 (beginning 1 July 1968), CD had to begin that year—and some sort of feasibility study had to be completed in time for that to happen. Three years had elapsed since the original proposal for an LFS, yet no agreed ship concept had emerged. A January 1968 internal design memo asked whether the Office of the Secretary of Defense realized just how little had been done on the CF phase of the program. One of the designers wrote on a route slip that "if the Navy really tried to prove how screwed up it could get, it couldn't have surpassed its LFS CF effort."

As the beginning of a greatly accelerated CF phase, OpNav requested a feasibility study of a modified version of the *Knox*-class (FF 1052) hull, which would still be in production at the end of LFS contract definition. Characteristics for this conversion, issued 25 January 1968, showed two or three 175-mm guns (using an Mk 86 fire control system), one point defense missile system, and electronics similar to that of a FRAM I destroyer (but without sonar). As in other versions of LFS, sustained speed was 20 kts, and endurance was 5,000 nm at 20 kts. Alternatives were two 175-mm guns (in place of 5-in/54s) in a *Forrest Sherman*/ASW hull, or rockets on board an LST 1179.

The frigate could not mount the required two 175-mm guns. Their weight would have wiped out her margin for weight growth. To gain back some weight, the existing steam powerplant would have been replaced by a 12,000 SHP (20-kt) steam or diesel plant. Even with this sacrifice, the ship still would not have mounted most of the desired weapons. The *Forrest Sherman* version was not stable enough and thus had to be rejected.

The LST could mount three 175-mm (the third superfiring aft), with a Sea Sparrow launcher atop her bridge. She would require 1,900 tons of liquid ballast, and her three 750 kW generators would be replaced by 1,000 kW units. The LST study showed that a 20-kt LFS could be built on a reasonable displacement, although the enormous liquid ballast requirement ensured that she would not be adapted as an LFS. An LST could also be armed with two 175-mm, four 5-in, and also 5-in rocket launchers.

To complement the heavy-gun ships, the designers tried LSTs converted to inshore fire support (IFS) ships armed with two or three 5-in/54 guns and 8–12 rocket launchers. An early alternative in February 1968, the LFSR, was intended to replace the IFS/LSMR and also the APD, which was then valued mainly for its ability to support underwater demolition teams. On a hull similar to that of an LST 1179, it would mount three twin 3-in/50 guns (for direct fire, illumination, etc.) and 8–12 rocket

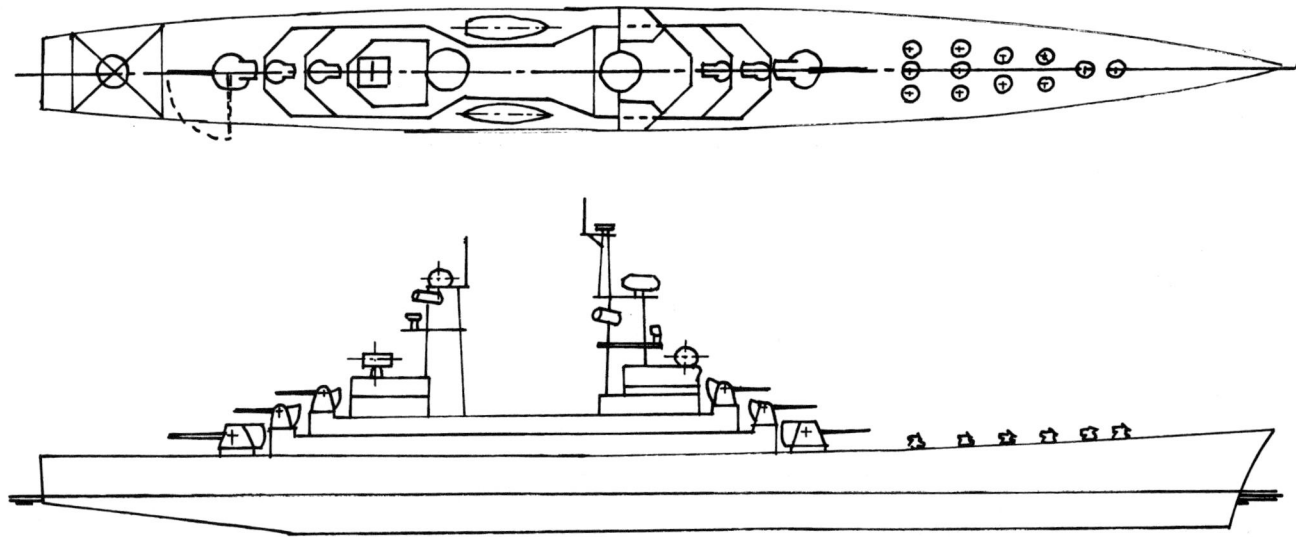

The landing fire support ship armed with both guns and rockets, from a sketch by the preliminary design group of the Naval Ship Engineering Center (NAVSEC 6110), 29 November 1968. The objects on the foredeck are 5-in rocket launchers. The two large guns are major caliber lightweight guns (175-mm); superfiring above them are 5-in lightweight guns (Mk 45s). Above them forward are an Mk 86 gun director and, above that, a rocket director. Atop the aft "mack" is an Mk 86 director; below it is the director for the Basic Point Defense Missile System (BPDMS, Sea Sparrow), whose missiles are carried in the box launcher. On the fantail is a helicopter platform. Dimensions were 550 × 57.8 × 17.2 ft; displacement was 5,487 tons light (7,825 tons fully loaded). Machinery was twin-screw diesels (15,600 BHP) for a speed of 20 kts (endurance would have been 10,000 nm at 20 kts). Accommodation would have been provided for 18 officers and 291 enlisted men. NAVSEC called this ship an LFSR, because it was armed with both guns and rockets. Similar designs showing only guns were also developed.

launchers (10,000 rounds); it would accommodate 15 officers and 200 troops (UDTs). The LST tank deck would accommodate at least two 23-ft UDT swimmer support boats, and it would have a helicopter platform. Compared to an LST 1179, it would retain the stern but not the bow ramp, using a marine railway or comparable boat launch/recovery system. It would have about half the stowage (both cu ft and sq ft) of the LST.

In a later scheme, the IFS would be complemented by an offshore fire support (OFS) ship armed with three triple 8-in turrets and with eight 5-in guns. A variation on this theme showed three forward-firing turrets and two twin 3-in/50 on a 9,500-ton, 450-ft hull. The ship would face the beach, presenting the smallest possible profile, and she would be armored against threats up to 60 degrees off the beam. She would have a helicopter platform, and would use air cushion vehicles, which were then called ground effect machines (GEMs), for ammunition resupply.

NAVSHIPS, the BuShips part of Naval Sea Systems Command (NAVSEA), also proposed a modular "train ship," combining powered drivers (armed with two triple 8-in turrets) with barges (5-in guns and rocket launchers). A related concept was a "universal ship" that could be built either to attack hard targets from long range (armed with three 8-in/55 rapid-fire guns and four 5-in/54) or to attack area targets at short range (two 5-in/54, eight rocket launchers), all in a 9,000-ton, 500-ft hull. This design incorporated a wet well to be used by LCMs supplying ammunition.

There was still some interest in larger guns. The major caliber lightweight gun project was considering a 12-in/70, which never got beyond the proposal stage, so NAVSEA's preliminary designers sketched a 480-ft, 9,000-ton ship armed with a twin 12-in gun forward, a 5-in/54 in the bow, and a 175-mm gun aft of the superstructure, as well as Sea Sparrow. It concluded that preliminary feasibility had not been established. Much the same was said of an alternative scheme employing a triple 16-in turret in place of the 12-in.

With policy undefined, the designers felt free to offer radical alternatives such as a "piggy back" LFS, designated LFSC, to be transported aboard a modified LSD. It was the minimum self-propelled craft (76 × 36 × 9 ft, 700 tons fully loaded, including 150 tons of seawater ballast, i.e., nearly brick-shaped) that could accommodate a 5-in/54 gun and four spin-stabilized rocket launchers. Ammunition was all in modules, and there was no berthing. An entirely different possibility was an LPH (10,500 tons) armed with helicopter gunships.

In line with the CD schedule, at the end of January 1968 completion of a specific operational requirement was set for 3 June, which meant that supporting studies had to be ready by 1 March. They would be fed into NAVWAG 59, the fire support phase of a larger amphibious warfare study that, it was hoped, would help define next-generation amphibious ships. Completed in May 1968, the study was reoriented to concentrate on major caliber guns for the new DX destroyer, which became the *Spruance* class. It recommended that the ship be armed with two 8-in/39, abortive versions of the standard army 8-in howitzer. The main conclusion was that major caliber gunfire could be dramatically effective against soft targets.

Probably it is fairest to say that by early 1968 OpNav sensed that its opportunity to get any sort of LFS was slipping away. The program would die unless it was pressed ahead. Thus on 16 April 1968 the CNO asked for a program leading to a contract award early in FY 71, which meant inclusion in the FY 70 plan. The designers began another round of studies in spring 1968. A 2 May 1968 preliminary design memo pointed out that OpNav had yet to focus its requirements "within a finite envelope of ship system alternatives." Only the existing 5-in/38 and 5-in rocket launcher and the new 5-in/54 lightweight gun were currently available. The 175-mm gun existed in pre-prototype form, scheduled to appear in 1973; if accelerated, however, it could be ready three years from the go-ahead. An alternative single 8-in gun could be ready four years from go-ahead. Heavier guns could be salvaged from the surviving reserve fleet units: a triple 16-in, although it was claimed that its ammunition would be very expensive, and 8-in rapid- or slow-firing guns, for which only high-capacity ammunition was in production. The Lance missile would be available in 1971–72 if it proved feasible; tests were planned for late 1968. By mid-1968, however, it seemed that OpNav would reject Lance altogether.

To justify LFS construction, the CF phase had to prove that the need for fire support justified reactivating a heavy cruiser and/or building a new ship; that the new ship would have a unique capability even if a bombardment missile system (Lance) were available; and that the LFS would be much more cost-effective than a reactivated battleship or heavy cruiser. The battleship was hardly an academic option; USS *New Jersey* (BB 62) was recommissioned in 1967. Pressure for reactivation had come from both the Marines and from outside the navy. Col. Charles Myers, USAF (Retired), saw the battleship as an alternative to tactical aircraft, which by 1966 were being lost at an alarming rate. He called

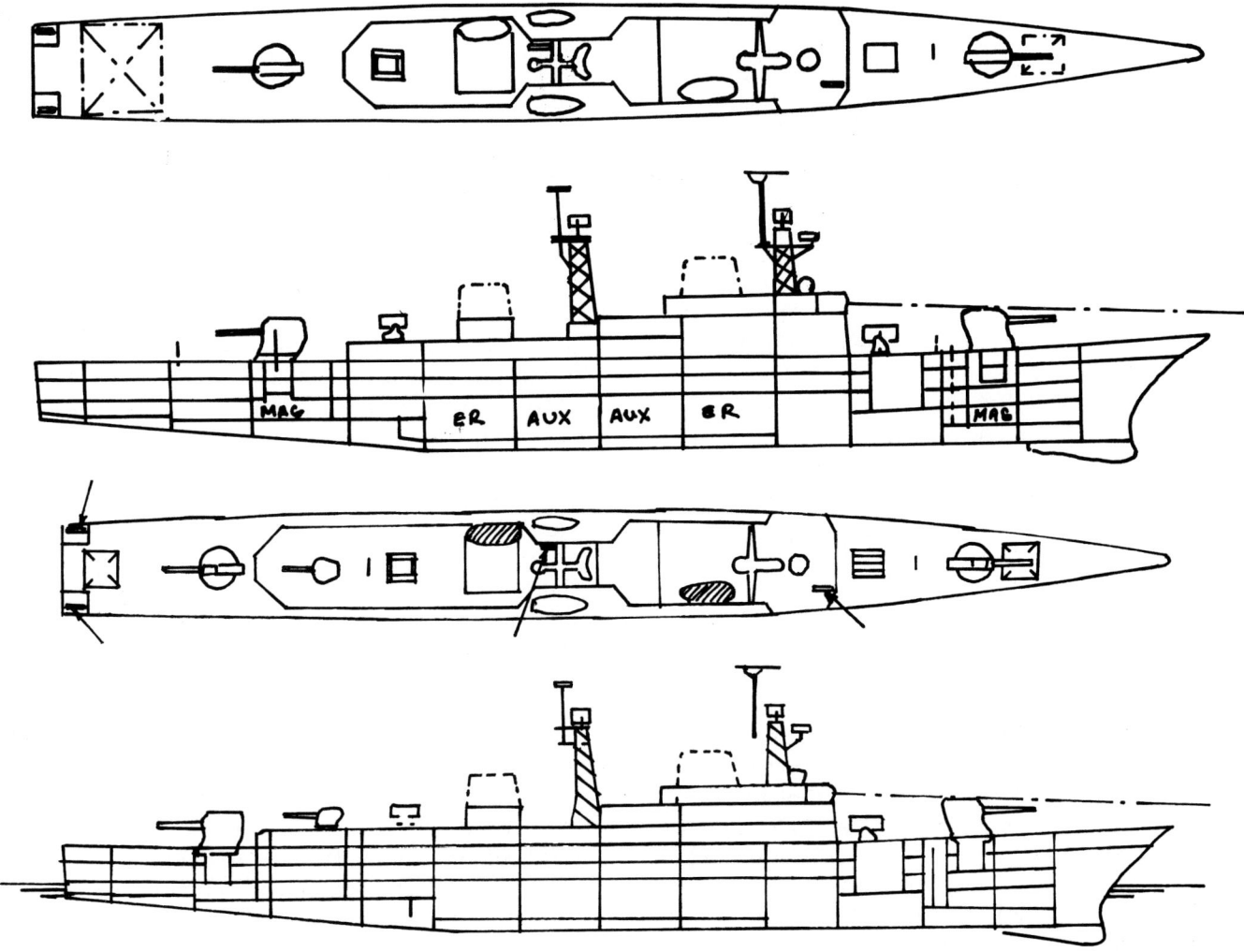

LFS versions of the *Spruance*-class destroyer, 30 September 1970. Arrows indicate Chaffroc (decoy) launchers. The large guns are major caliber lightweight guns (MCLWGs), the standard LFS main armament. Power was reduced to 20,000 SHP. Dimensions were 555 ft 11 in overall (525 ft on the waterline) × 54 ft; the two-MCLWG version would have displaced 7461 tons fully loaded. In addition to the two MCLWGs, it had the Basic Point Defense Missile System (BPDMS, the first version of Sea Sparrow) aft, together with a 5-in lightweight gun (Mk 45). The box launcher forward was for ASROC; the ship retained its big bow sonar.

the battleship an interdiction assault ship (IAS). Myers claimed that, given the peculiar geography of Vietnam, 80 percent of potential targets were within battleship gun range. Moreover, by 1969 the navy was working on a saboted 280-mm round that could reach about 50 nm range. At the very least, Myers' concept would shape future LFS thinking, because he was providing an alternative role for a heavy gun ship. Reportedly the naval aviation community, headed by CNO Adm. David L. McDonald, bitterly opposed reactivation. The battleship option was supported by Senator Richard B. Russell. Reactivation of the *New Jersey* was announced the day after Admiral McDonald retired in 1966. During her brief Vietnam service the battleship fired 5,688 16-in rounds, compared to 771 during World War II. Contrary to views expressed during the LFS design development, ammunition proved to be in plentiful supply. The ship operated with only one turret and half of her 5-in/38 battery continuously manned, as well as only half of her boilers and a quarter of her turbo-generators. That cut the crew to 50 officers, 12 petty officers, and 1,556 enlisted men, about half of her World War II complement. The Marines saw *New Jersey* as an interim step toward reviving gunfire support, but she was returned to reserve soon after the United States agreed to stop using her against North Vietnamese targets when peace talks began in 1968.

By this time navy fire support ships, both destroyers and LSMRs, had been heavily involved both in combat in South Vietnam (e.g., to repel the February 1968 Tet offensive) and in attacking North Vietnam. The LFS was clearly more than an amphibious support ship. CNO guidance dated 3 November 1968 established a secondary mission of independent bombardment, which was linked to close-in defense of other naval forces (advance forces, defensive mine forces) and a tertiary mission of supporting underwater demolition teams, amphibious reconnaissance groups, and beachjumpers. Thus, the LFS was finally a multimission ship. Its main weapon would be the major caliber lightweight gun. The landing force support weapon was explicitly rejected because it had not yet been funded and because new guided rocket-propelled projectiles seemed to offer so much potential.

Experience during Tet, moreover, showed that the ship might need much more ammunition than had been imagined. According to a November 1968 report describing the experiences of USS *Lynde McCormick* (DDG 8), which influenced the LFS project, she had emptied both of her 600-round 5-in magazines several times, firing about 9 rounds per minute. Both of her Mk 42 mounts had failed on occasion after a few rounds, one being repaired quickly with a pencil and twine to replace a pin. Overall, the 5-in mounts were the least reliable part of the weapon system, averaging 20 rounds (30 seconds) before failure when firing in automatic mode. Failures generally took 2–30 minutes to fix, but replacement parts took 2–3 weeks to arrive, so a good machinist was vital. Normal procedure was to use one mount, then transfer to the other when the first failed. The ship wore out a set of barrels in one 5-day period, and during Tet used 200 percent of her rated gun life.

Magazine capacity was insufficient; the ship required fresh ammunition every two days. Replenishment took 3 hrs; refueling, to maintain 60 percent capacity, took the ship off the gun line for 6 hrs at a time. Although rounds were on pallets for quick transfer, they had to be hand carried down passageways to the magazine. A single vertical replenishment helicopter could supply ammunition much more quickly than it could be struck below.

For area neutralization, one LSMR was equivalent to about six destroyers, firing 500–600 rounds per minute. The only important problem was insufficient rocket range. To operate with other amphibious ships, the LFS would need a sustained speed of 20 kts and an endurance of 6,000 nm. However, OpNav wanted size and cost estimates for higher speeds. Later it was pointed out that higher burst speed might help the ship evade shore-based weapons. Thus the ship might need boost gas turbines added to her diesel plant.

Given the CNO's instructions, on 4 November 1968 NAVSHIPS formally asked its preliminary designers for a feasibility study of an LFS, looking toward construction in FY 71–75. OpNav envisaged a ship of 8,000 tons or less, armed with three to five major caliber lightweight guns, costing $60 million or less. Minimum manning and magazine automation were now requirements. NAVSHIPS asked for something no larger than the contemporary LST 1179, mounting two major caliber lightweight guns and four lightweight 5-in; a twin mount, Mk 66, described as an area neutralization weapon, was now in prospect. Alternatives would be armed with and without Sea Sparrow (16 rounds), and with eight, ten, or twelve Mk 105 twin rocket launchers (750 rounds per barrel). The ship would have two Mk 86 fire control systems. Electronics would be on the scale of a destroyer escort (frigate), but without any sonar. Like other amphibious ships, this one would make 20 kts (endurance was set at 10,000 nm at 20 kts) on two diesel shafts. Complement was initially set at 14 officers and 204 enlisted men. The LFS would have a helicopter platform. Late in the study an air search radar (SPS-40B) and destroyer-type naval tactical data system (NTDS) were added.

A draft mission description dated 7 January 1969 shows what the LFS was expected to do. She would attack point targets at ranges of up to 40,000 yds, and area targets at 50 nm (100,000 yds); ultimately she would attack area targets as much as 100 nm (200,000 yds) away. Ranges were predicated on a 20 × 30 mile initial beachhead, and on the ability to mount vertical assaults at ranges of up to 25 miles from the beachhead. The area attack goals are not too different from those demanded of new fire support weapons in 2000.

Accurate fire against hard point targets such as fortifications required both precision navigation systems and a good fire control system; the digital Mk 86 fire control system had the additional advantage that it could also be used to control Sea Sparrow antiaircraft missiles. OpNav wanted the ship to operate and control an observation helicopter. NAVSEC recognized that helicopter stowage would be very expensive in terms of ship size, and decided to ignore this requirement.

The ship would attack troops and armor massing to attack or occupy a helicopter landing zone. Helicopter support might also entail suppressing enemy air defenses, particularly those near the helicopter landing zone. These deep targets were clearly area targets, and at the ranges involved accuracy did not have to be so great. In order to protect mine countermeasures forces and the amphibious task force itself, the ship would have to destroy coast defense weapons, presumably AS-1 missile launchers, from 20 nm away.

The result, submitted 29 January 1969, was 550 × 57.8 × 17.2 ft (5,487 tons light, 7,825 tons fully loaded). She required 15,600 SHP to make design speed. There were twelve Mk 105 rocket launchers. Guns were twin rapid-fire 5-in guns superfiring over 175-mm guns on the weather deck, fore and aft, with a Sea Sparrow launcher superfiring over the after twin 5-in gun and a helicopter deck right aft. There were 750 rounds per 175-mm or 5-in gun. The forward of two "macks" (mast-stacks) was topped by an SPS-40 radar (with the Mk 86 fire control system on its forward side), the after one by the second Mk 86. Complement was 18 officers and 291 enlisted men, a painful increase over what had been envisaged. The study showed once more that rocket launchers on board a big-gun hull carried a disproportionate price; it would be better to build separate gun and rocket ships.

The ship would have to survive in an area possibly swept by enemy fire. As in earlier surface combatants, it would help to place the magazines below the waterline. That was practical because, unlike missiles, shells and their powder charges would be stowed quite densely. Analysis of likely damage showed that the minimum acceptable level of protection was 30.6-lb HY100 (¾-in steel) for bulkheads and overheads of 5-in and 175-mm magazines and projectile handling rooms. This armor would protect stowed ordnance from about 35 percent of the fragments produced by nearby explosions of intermediate 4–6-in shells. Somewhat heavier steel (40 or 60 lbs) would stop 60–90 percent of fragments, and thus would effectively protect magazines from near misses. On the other hand, a direct hit by a coast defense gun would disable or sink the LFS. World War II experience showed a 50 percent probability that a hit on or in a magazine would detonate it. It was considered advisable to protect the machinery spaces, with 30 lbs on transverse bulkheads and decks. A proposed aluminum hull was rejected as far too vulnerable.

NAVSEC produced a new round of feasibility studies in April 1969. By this time the army had given up on the 175-mm gun, so the heavy gun was the new 8-in MCLWG. The baseline was a 20-kt, 6,000-nm diesel ship armed with two 8-in guns (750 rounds per gun) and two twin 5-in lightweight guns (Mk 66, 750 rounds per barrel). Its magazines and other vital spaces would be armored. As in the previous round, the ship had a helicopter platform and refueling facilities, but no helicopter hangar. NAVSEC also considered a ship with three 8-in and three twin 5-in, and it considered adding a point defense missile system to both the baseline and the larger ship.

Eliminating rockets and Sea Sparrow cut the ship to 6,216 tons fully loaded or 4,459 tons light (465 × 58 × 15.3 ft), with a complement of 15 officers and 267 enlisted men. She would need only 12,654 SHP, which could be provided by a nine-cylinder Fairbanks-Morse 38A20 diesel and a 240 HP electric motor on each shaft. Armor (HY100) amounted to about 138 tons in such a ship. Displacement rose slightly because of the shift to 8-in guns. Adding Sea Sparrow added an officer and three enlisted men, and increased displacement to 4,525 tons light and 6,297 tons fully loaded. The ship with three 8-in and three twin 5-in would require a larger crew (15 officers and 278 enlisted men) and would be larger, at 5,044 tons light and 7,118 tons fully loaded (490 × 60 × 16 ft). To cover her larger magazines would take 188 tons of armor. Power would match that of the smaller ship. As in the smaller ship, Sea Sparrow would cost an officer and three enlisted men and a slight increase in size (5,125 tons light and 7,216 tons fully loaded, 495 × 60.3 × 16 ft).

Self-defense and passive protection were both important, because an enemy would certainly shoot back. Among the important threats would be enemy torpedo boats, which would have to come quite close to ensure a hit. A September 1969 Dahlgren study showed that to defeat such craft, rate of fire was more important than caliber, so 5-in guns would do better than 8-in. For targets that had to close to 2,000 yds to make their attacks, two 5-in/54 with a combined rate of fire of 100–180 rounds per minute would, on average, kill three targets attacking together.

NAVSEC also considered converting an engines-aft cargo ship, mounting reserve fleet gun mounts forward. In June 1969 there were six triple 8-in rapid-fire (*Des Moines* class), thirty-three triple 8-in slow-fire (*Baltimore* and *Oregon City* classes), and twelve 6-in rapid-fire (*Worcester* class) mounts in reserve. A C3-S-46a hull could accommodate three triple 8-in and four twin 5-in/38 on a light ship displacement of 8,600 tons, including 1,830 tons of water ballast. It would have been difficult, however, to provide space for the 500 additional men required to serve those guns, for two 1,500 kW diesel generator sets, or for other LFS features.

Meanwhile work proceeded on DX/DXG, which became the *Spruance*- and *Kidd*-class destroyers, respectively. DX was being designed to take one major caliber lightweight gun forward and a 5-in lightweight gun aft. The DX was not too different from an LFS, and it offered important ASW capability. It would be built in quantity whether or not the LFS role was worth funding. If the best LFS was not too different from a pair of DX, it might be best to drop LFS altogether.

By this time the preliminary designers in NAVSEC had a design computer, so it was relatively easy to develop large numbers of alternative rough designs. Thus on 9 April 1969 the center reported a 20-ship parametric study, estimating the consequences of various choices of weapons and other characteristics, followed on 15 April by the 4-ship feasibility study, on 20 May by a single-ship minimum ship study, on 27 August by a 15-ship gun matrix study of armament alternatives, and on 15 October by a 4-ship magazine study. During this period an additional study was made of a possible merchant ship conversion. In October 1969 NAVSEC was working on a 13-ship study to explore variations in powerplant and speed and on a study of a ship with one triple 16-in turret. Further studies of variations in electronic equipment were planned. There probably was little expectation of actually building a ship; it seemed far easier to study the LFS to death.

Given the logic of CF/CD, NAVSEC needed a proposed technical approach (PTA). By October 1969 it planned to offer two designs: minimum and maximum. Because such wide variations were still being explored, it was far too early to offer anything so specific. Indeed, on 21 November 1969 NAVSHIPS asked the project's manager (PMS 378; an office within NAVSHIPS) not to forward the PTA to OpNav because it was too incomplete and too expensive. The logic of the amphibious scenario had not yet been developed adequately. In a departure from past thinking, NAVSHIPS management (Ships 03) commented that the presence of heavy guns seemed to reduce landing force casualties by only about 6 percent, which would strengthen the case to abandon the LFS altogether.

The NAVSHIPS LFS studies inspired academic studies at Massachusetts Institute of Technology (MIT), where many of the navy's naval architects were trained by naval officers and NAVSHIPS civilians. There was some hope that, in this less hidebound atmosphere, students would come to unconventional conclusions. One MIT study, which was quoted in official LFS files, pointed out that much of the cost of an LFS went to portions of the ship that did not contribute directly to her mission. Moreover, the state of the art was changing so rapidly and circumstances could be so varied that it would be best to concentrate investment on "universal" or modular ships. In this case, that meant a helicopter carrier and an LSD. In each case, the ship was a very flexible platform supporting specialized vehicles. Given the expected imminent success of a VTOL or STOL (short take-off and landing) airplane to supplement helicopters, the carrier could handle a wide range of missions: antiair warfare, antisubmarine warfare, amphibious assault, electronic warfare, and mine warfare. The LSD would support complementary small craft, for antisubmarine warfare, Market Time (the blockade of the Vietnamese coast against gun runners), electronic warfare, amphibious assault, and mine countermeasures. Although the modular LFS never materialized, the broader conclusions drawn by the MIT team were certainly valid, and to some extent they are reflected in the creation of the LHA and LHD.

In January 1970 the LFS was due for briefing to Op-03 (in charge of surface warships) so that it could be entered into the long-range T-POM for FY 72. It would be included in the Five Year Defense Plan, funded in FY 72 for construction of a lead ship beginning in 1973. It was expected that seven to ten LFS would be built. This was rather late for exploratory studies, yet at this time Rear Adm. Thomas R. Weschler, director of the OpNav ship characteristics division, mentioned that there was a "good possibility" that the navy would begin development of a new larger caliber 10–13-in RAP round with, presumably, an LFS gun to match. The army's new area bombardment rocket, soon to become the 9-in MLRS (multiple long-range rocket system), had been rejected because it lacked the supposed secondary antiair capability of the standard 5-in rocket.

Some important issues had not yet been resolved. Gun and missile arrangement had not been fixed, and at the ranges involved the flexing of the lightweight hull might well affect gun accuracy. Attempts to automate the magazines (i.e., palletized replenishment) conflicted with attempts to protect them by placing them underwater. The ship was always described as austere, which usually meant abandoning shock protection, yet that might be very important, particularly in a ship with massive magazines. To what extent would allowance be made to upgrade the ship's weapons? For example, of the family of lightweight 5-in guns only the slow-firing single Mk 45 existed, yet calculations envisaged the fast-firing twin Mk 66. Should the design allow for the latter's substitution after completion? Should MCLWG gun mounts be designed for eventual replacement by the projected long-range gun (LRG)? Its projected much larger clear radius requirement and larger diameter mounting would considerably enlarge the LFS. At the least, the ship's beam would have to be wider. If length and displacement were held constant, the ship might have to draw less water; magazines would then come above the waterline.

In February 1970, NAVSHIPS submitted a PTA for a ship slightly smaller than the large ones developed the previous April. She would be armed with three 8-in guns and two twin 5-in (Mk 66), as well as Sea Sparrow. This design emphasized gunpower at the expense of ammunition capacity and handling, and at the expense of fire control (it had only a single Mk 86) and possibly shock. She would have receive-only Link 11, but no computerized combat system. Endurance increased to 10,000 nm at 20 kts, probably because it was relatively easy to accommodate more oil fuel. This 20-kt LFS would have displaced 4,989 tons light and 7,856 tons fully loaded (505 × 58.78 × 17.49 ft), and it would have required 15 officers and 273 enlisted men. NAVSHIPS estimated that in FY 71 dollars eight such ships would cost an average of $80 million ($135 million for the lead ship); 20-year operating cost would be $100 million.

This recommended design was a compromise between maximum- and minimum-cost ships. The maximum ship (9,200 tons, 570 ft, with NATO Sea Sparrow instead of BPDMS [the original Sea Sparrow] for self-defense) would cost an average of $105 million to build and $115 million to operate. She would have two Mk 86 fire control sets, the new Phalanx CIWS (close-in weapons system) self-defense gun, and a computerized combat system (NTDS with three UYK-7 computers). This ship also differed from the others in having a defensive jammer (ULQ-6/SLQ-12); all versions had electronic threat warning systems (WLR-1Cs). As important, for fire support, only the maximum ship would have had a precision navigation system (Omega radio navigational aid and Doppler sonar to measure speed over the bottom), and only she would have had palletized strikedown for quick underway replenishment. A minimum-cost ship (6,600 tons, 475 ft, two 8-in, one 5-in Mk 45) would cost an average of $65 million to build and $100 million to operate. Like the recommended ship, she would have receive-only Link 11.

Now the LFS began to die. In May 1970 the LFS was dropped from the program objective memorandum (POM), although it was included in projected force levels for 1980–81. On this basis CD would have to be completed during FY 75. However, Op-343, responsible for amphibious ships, feared that efforts would be abandoned at the end of FY 70, which meant 30 June 1970. Perhaps most telling was the decision that future meetings of the LFS working group would be quarterly rather than monthly.

The idea of an LFS was not quite dead yet. In 1970 the U.S. Navy was buying a big new destroyer, which became the *Spruance* (formerly DX). Could it be

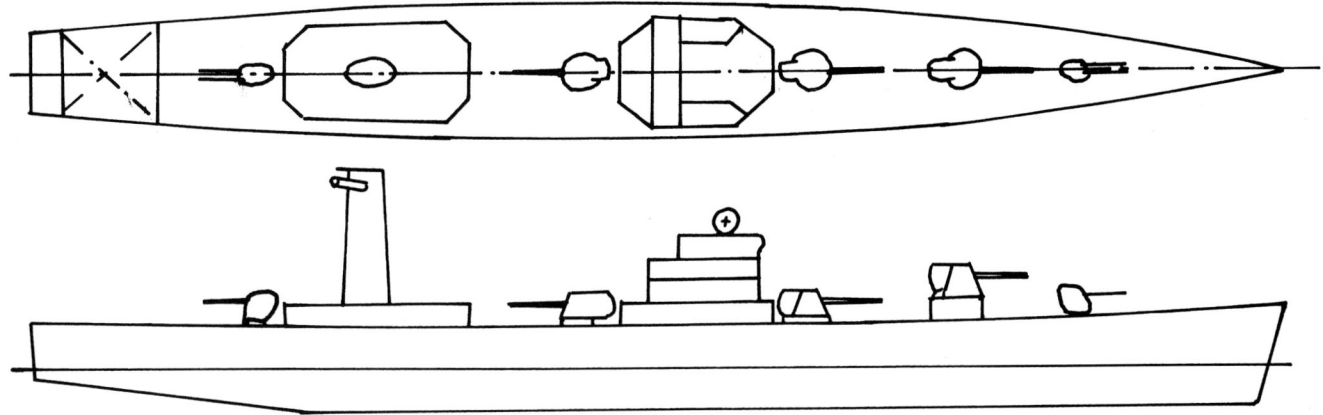

The next few pages show arrangement sketches submitted to the working group meeting for the LFS, 4 February 1970. Completed after the working group's 27th meeting on 15 January 1970, they illustrate design trade-offs. Issues included a desire for uniform gun coverage around the ship, and for minimum ship bending between guns and fire control systems. It seemed likely that requirements for gun separations, clear deck areas, superstructures (including antennas), and helicopter pads would dictate the ship's length. Mounting guns side by side would reduce length but also gun arcs; placing guns at one end of the ship would give good arcs but limit coverage aft. Separating the guns would improve their protection but force vitals into the midship area, where they might be too concentrated. Hopes of minimizing manning made it desirable to automate magazines and to palletize ammunition (for quick replenishment), but that increased magazine volume and made it more difficult to keep all magazines below water. These sketches all show the 8-in lightweight gun and the abortive twin 5-in Mk 66 lightweight gun. The memorandum accompanying the original sketches asked whether the ship should be designed for the existing 5-in Mk 45 single-barrel gun, with provision for later installation of the Mk 66, and whether provision should be made for replacing the 8-in gun with the projected 12 in long range gun (LRG), which would require a large clear circle and larger ship beam. At this time it was assumed that the 8-in production contract would be awarded in June 1971, with first delivery in August 1975, and that the Mk 66 would be delivered at about the same time. No schedule for the 12-in gun was available. Other LFS systems included the "close infighter weapon system (CIWS)," the Mk 86 fire control system, tactical satellite communications, a new version of the NTDS combat direction system ("jeep," or JPTDS), a special version of the Link 14 data link (Link 14W/TIDY), IR surveillance and tracking systems, the SPS-49 radar, and a Doppler sonar.

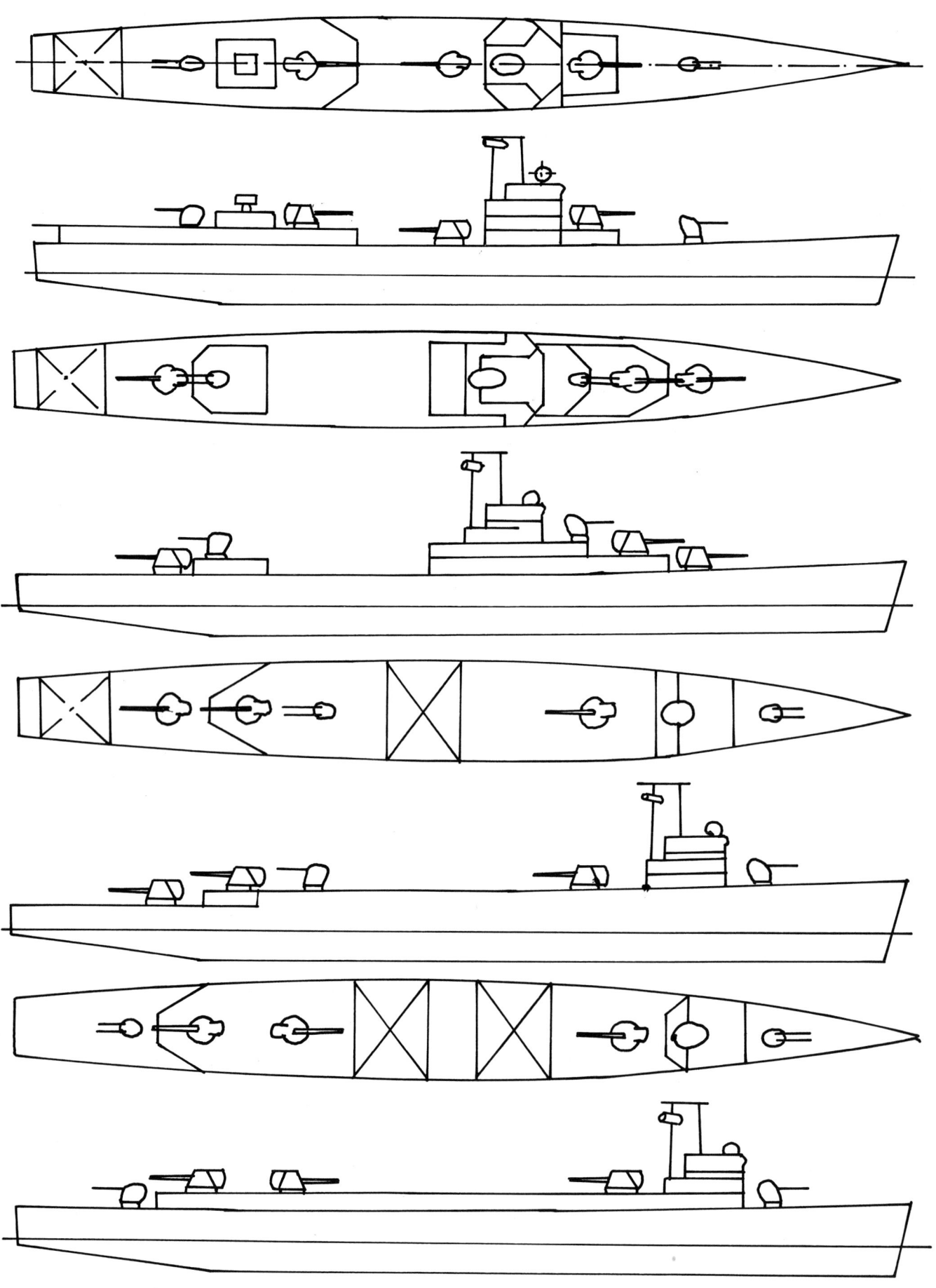

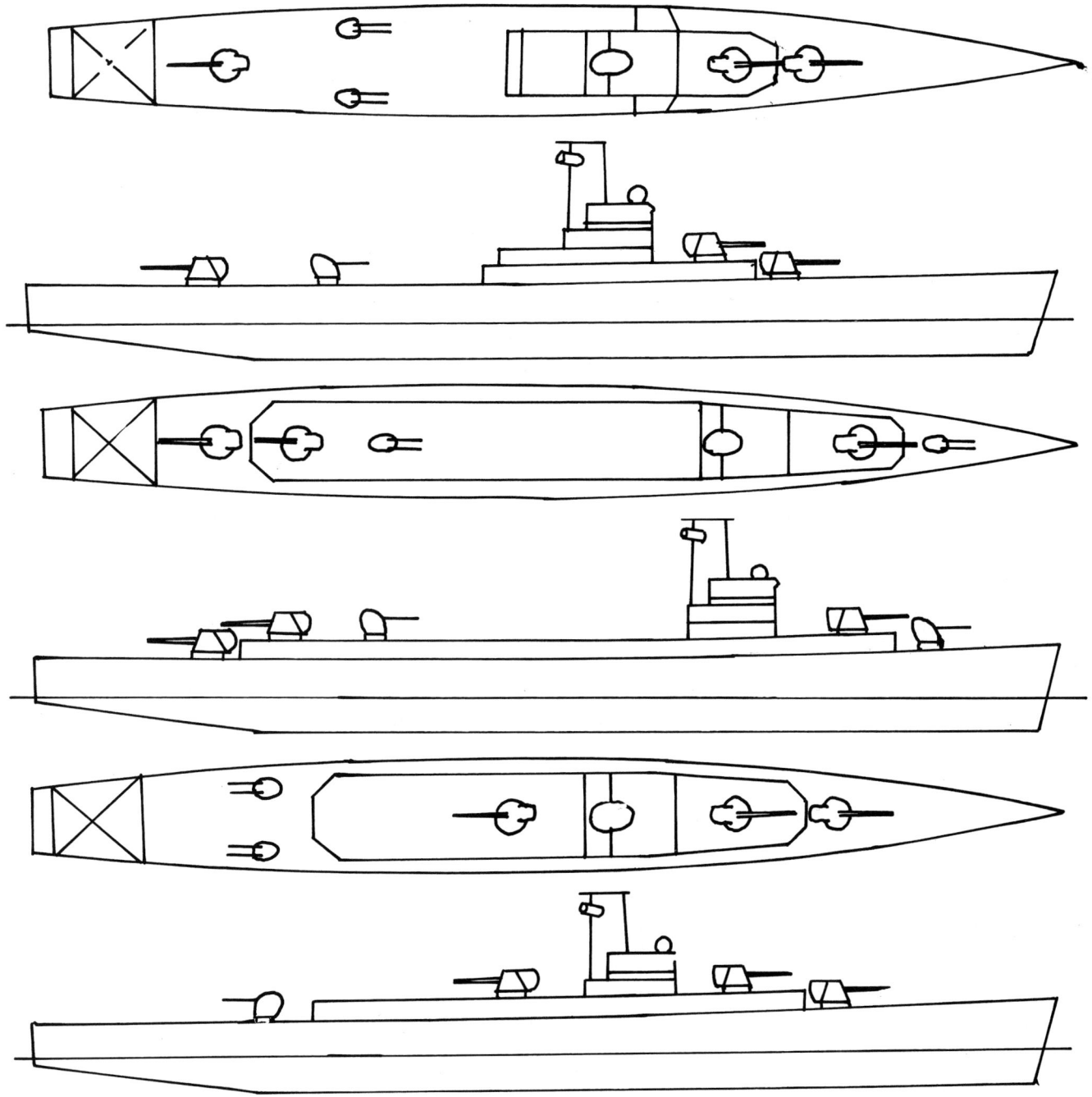

used as an LFS? The baseline design for the destroyer already envisaged possible replacement of the forward 5-in gun by a major caliber lightweight gun (MCLWG). On 14 July 1970 Op-36 asked for a feasibility study of a more specialized LFS built on the same hull, which NAVSEC completed at the end of September. As in the baseline, a single MCLWG replaced the forward 5-in gun. ASROC (antisubmarine rocket) reloads were cut from 24 to 12–16 to add volume for the larger gun and its magazine. Elimination of two of the four gas turbines would provide space for a second heavy gun. The second 5-in gun might or might not be retained; it could be traded off for additional 8-in magazine space, for up to 1,000 rounds per gun. The SCB stopped the project before it was complete, on 30 September. There was no longer much interest in building an LFS, and *Spruance*s were too valuable to use. In June 1971 CNA did calculate characteristics for a rocket ship based on the baseline LFS, with 1 fewer major caliber gun but 12 rocket launchers forward, but there seems to have been little real interest in such a ship.

As for the baseline *Spruance*, the 8-in major caliber lightweight gun (Mk 71) never entered service.

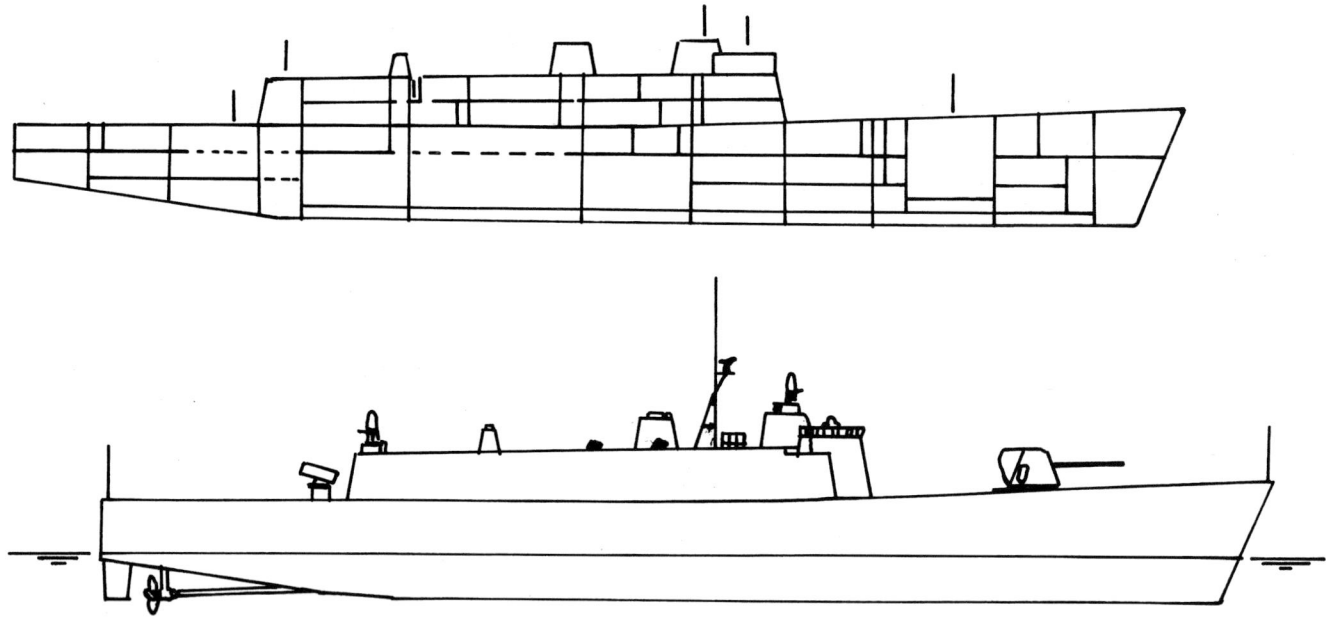

LFS Variant 3, July 1986. At congressional behest, in the fall of 1985 NAVSEA began a new round of LFS studies for an OpNav LFS flag advisory group. A study showed that the two most suitable weapons were an improved 8-in/60 lightweight gun and a marinized version of the army's 9-in multiple rocket launcher (MLRS). Innovations in the designs included the marine integrated fire and air support system (MIFASS) and the TPQ-37 counter-battery radar. LFS Variant 3 was one of seven possibilities. This one was armed with an 8-in lightweight gun forward (250 conventional and 250 extended-range rounds) and an army-type multiple rocket launcher for MLRS rockets aft (900 rockets). The object abaft the forward Phalanx mount is an SLQ-32 countermeasures antenna. Diesel propulsion was chosen because the ship did not have to exceed amphibious task group speed. Compared to *Carronade*, the new LFS designs were far larger because payload weight doubled, volume quadrupled, and speed increased 50 percent. In addition, standards for accommodation, sustainability, and electrical system, and even for pollution, changed dramatically. Dimensions of Variant 3 were 413 × 64 × 15.4 ft (7,191 tons); two PC 2.6 diesels (26,500 BHP) would have driven the ship at a sustained speed of 20.6 kts (21.5 maximum). Accommodation was 219. Estimated average acquisition cost in FY 86 dollars was $251 million, based on a ten-ship program.

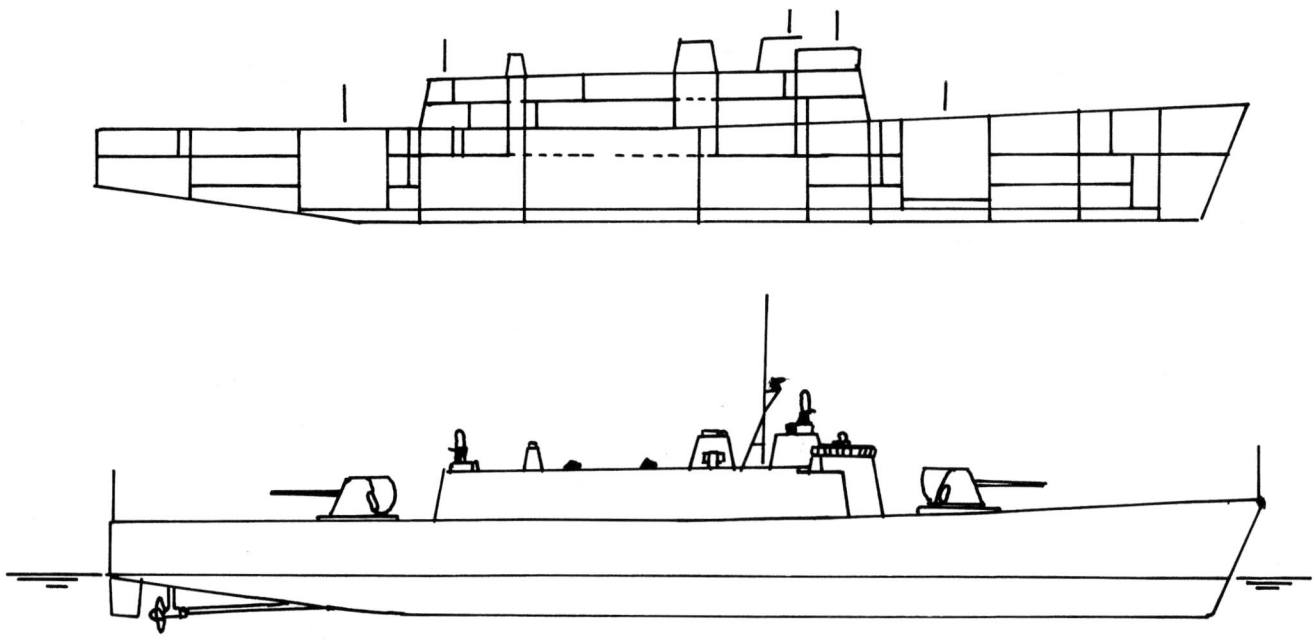

LFS Variant 4, July 1986. She shows 8-in lightweight guns fore and aft and two Phalanx close-in weapons. The object abeam the uptake is an SLQ-32 ECM antenna. Dimensions were 410 × 63.5 × 14.4 ft (6,622 tons); with the same powerplant as Variant 3, the ship would have made 20.8 kts sustained (21.8 maximum). Accommodation was 222. Estimated average acquisition cost in FY 86 dollars was $268 million, based on a ten-ship program.

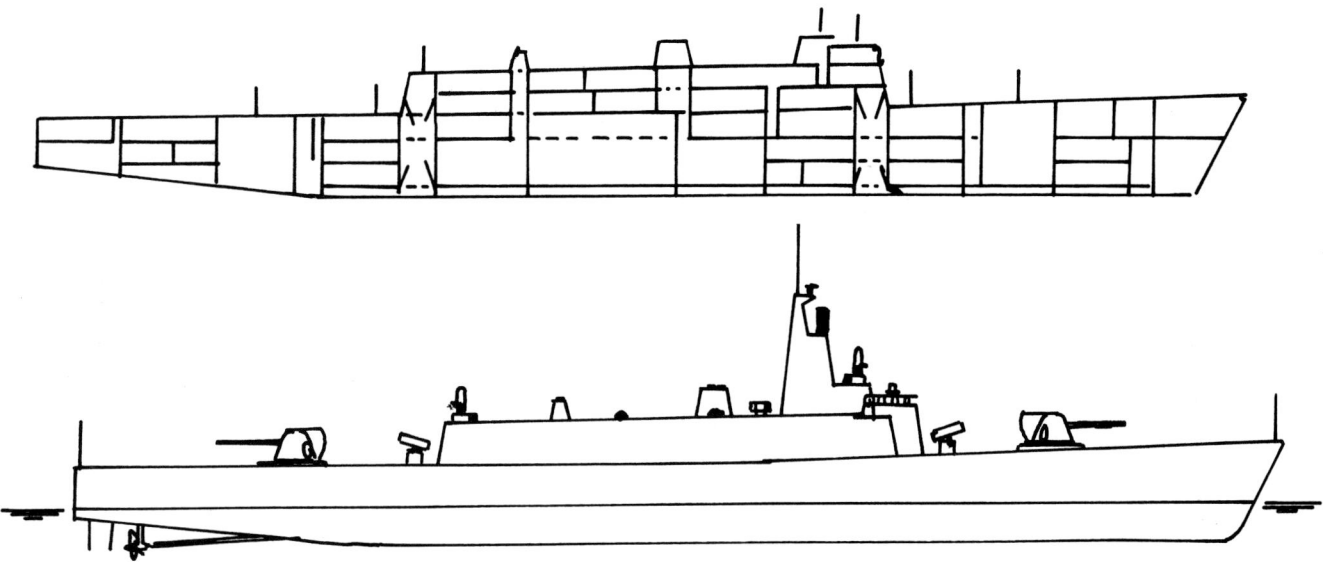

LFS Variant 7, July 1986. The most heavily armed of all seven possibilities, Variant 7 shows both 8-in guns and rocket launchers fore and aft. Dimensions were 503 × 65.5 × 16.4 ft (9,563 tons); the same machinery as Variant 3 would have driven the ship at a sustained speed of 21.0 kts (maximum 21.9). Accommodation was 249. Another design alternative, Variant 5, showed rocket launchers fore and aft, and no heavy guns at all. The Variant 7 estimated average acquisition cost in FY 86 dollars was $352 million, based on a ten-ship program.

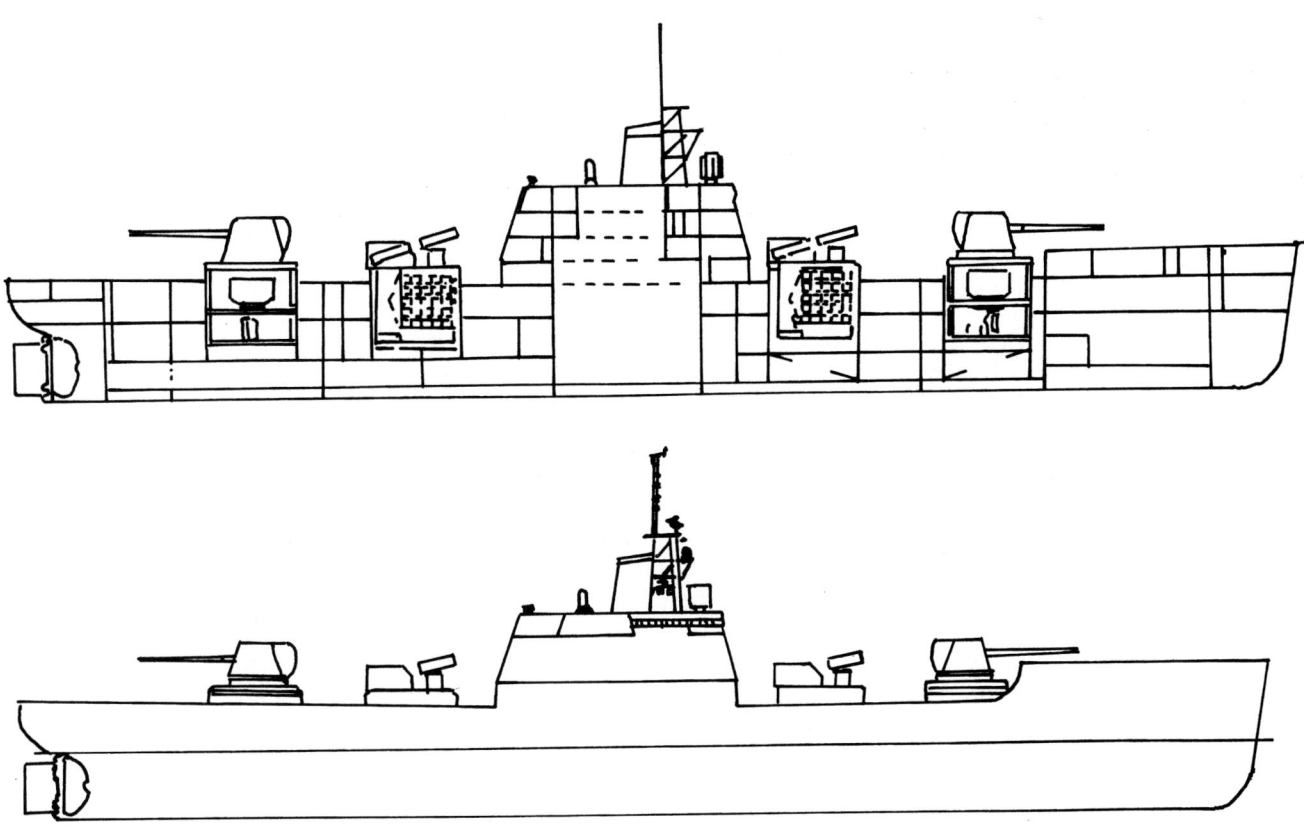

An LFS converted from a World War II–built Victory ship, from a 1985 NAVSEA study. The hull was large enough to carry two 8-in guns and two rocket launchers, the LFS Variant 7 payload, but did not meet damaged stability criteria because both engine and magazine spaces were too large. Because volume was tight, it was impossible to add bulkheads to solve the problem. Additions would have included a new deck house, accommodation for a crew of 242, added electrical generators, deck reinforcement, an augmented firemain system, a new collective protection system, and solid ballast both to improve stability and to immerse the ship's propeller (the weapons constituted a very light load). As converted, the ship would have displaced 12,873 tons, with a draft of 26.5 ft. Estimated average acquisition cost in FY 86 dollars was $152 million, based on a ten-ship program.

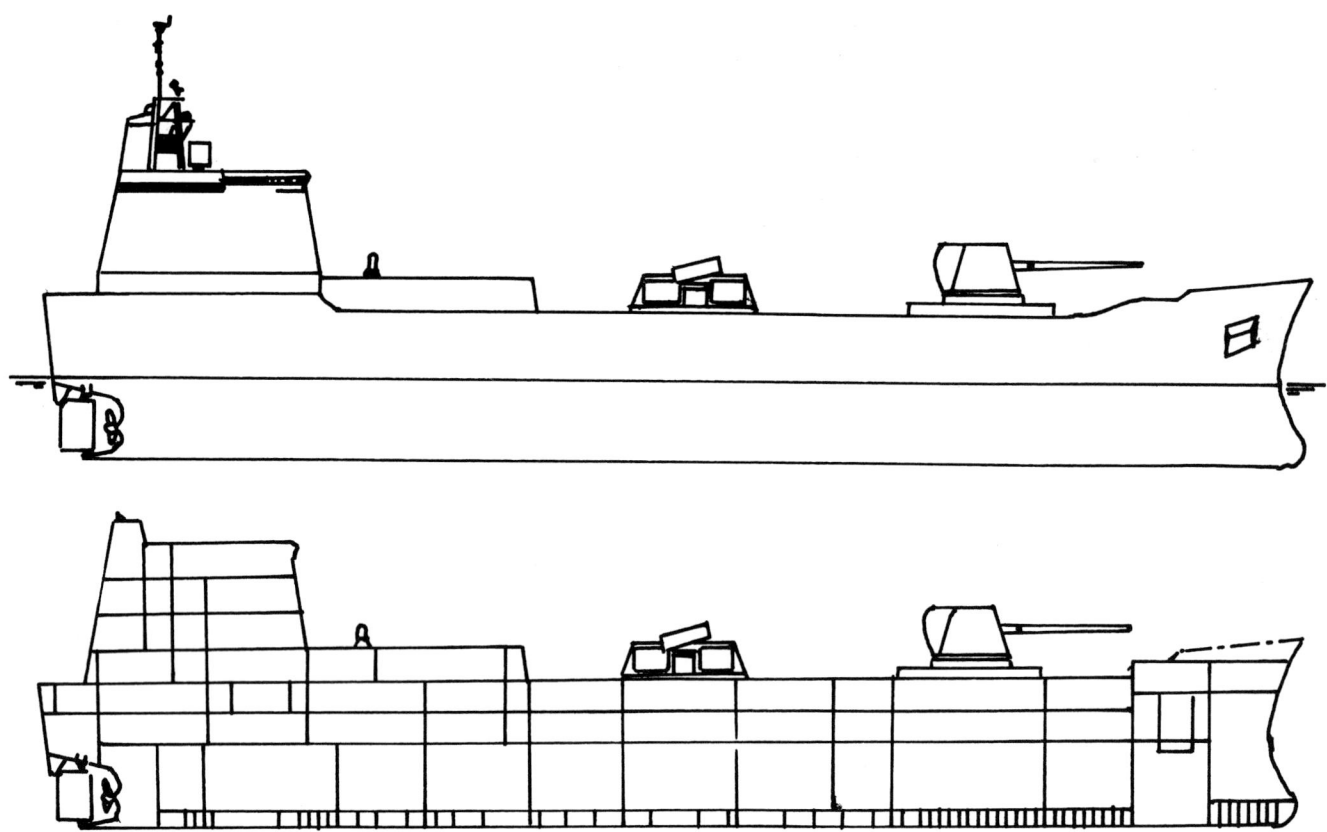

An LFS converted from the Dutch coastal freighter *Schippersgracht*, from a 1985 NAVSEA study. *Schippersgracht* was chosen as representative of modern merchant ships. Dimensions were 80.2 (overall) × 16.0 × 5.97 m. As converted, she would have displaced 5,770 tons (maximum full load was 7,900 tons), with a draft of 20 ft, and a speed of 13 kts. Such ships cost about $3.5 million as of April 1984. This hull proved just large enough to accommodate an 8-in gun and a rocket launcher, the LFS Variant 3 payload. Several watertight bulkheads would have been fitted to provide sufficient damage stability. Even so, the ship could not approach warship standards. For example, her commercial machinery could not meet naval shock requirements. Estimated average acquisition cost in FY 86 dollars was $91 million, based on a ten-ship program.

Reportedly the tests on board *Hull* showed what seemed to be excessive dispersion, due in part to the ship's lightweight and flexible construction. The ship had retained her Mk 68 fire control system, which did not fully incorporate 8-in parameters. She was already due for disposal, and she had already cracked her hull while firing her 5-in guns; the 8-in gun tests worsened the problem. Plans called for solving the accuracy problem by using laser-guided shells, but they would be expensive. By omitting laser guidance from the initial 8-in tests, system cost was considerably understated, but then performance was grossly unsatisfactory. Moreover, it was increasingly clear that the area just offshore an amphibious assault would be quite dangerous. *Spruance*s and their successors, the Aegis-armed *Ticonderoga*s, were just too valuable to risk there.

By the late 1970s there was considerable interest in reactivating the battleships, both for naval gunfire support and as the centerpieces of surface action groups that would supplement carriers to deal with the many trouble spots around the world. That reactivation, occurring under the Reagan administration, demonstrated that the navy would buy a multipurpose ship but not a specialized amphibious assault fire support unit. Conversely, with the four battleships coming into service, it was difficult to justify the construction of an additional class of specialist fire support ships.

Since 1972, engineers at the Naval Surface Weapons Center at White Oak, Maryland, had been working on an 8-in lightweight mount for the Marines. They thought they could build a 70,000-lb naval universal gun mount (NUGM), which could accommodate a 5-in, 155-mm, or 8-in gun; hopefully it would replace the existing 3-in/50, to provide the fleet with more fire support. At the least, the 5-in version could replace any 3-in/50. In an army-navy commonality study conducted in 1979, a new generation of fire support guns was proposed; it was based directly on standard army and Marine guns, using standard ammunition and a new low-recoil non–deck-penetrating mounting. Whether an 8-in howitzer could be accommodated would depend on a ship's strength and stability. It turned out that the weapon could be mounted in three places on board an LST 1179 (Frame 67 to starboard, 132 to port, and 175 on the centerline, all on the main deck). Navy ship design files included a May 1980 attempt to place three 8-in/36 howitzers on board an LST 1179. It turned out that the LST had already run out of space and weight margins, so the concept was impractical. Further fire support issues, revived under the Reagan administration, are discussed in Chapter 16.

By the 1960s, replacement of World War II–built amphibious flagships was urgent. They could not accommodate new technologies, such as automated combat information systems, because they lacked both volume and electric power. USS *Eldorado* (AGC 11) is shown in 1969.

15

A New-Generation Amphibious Flagship

The 20-kt amphibious fleet naturally included fast flagships, initially envisaged as converted Mariner cargo ship hulls. For example, a list of projects for the FY 59 program, dated 4 January 1957, included a new AGC at a cost of $53 million and $46 million each for follow-on ships. However, as long as no 20-kt LST existed, and there were few fast AKAs and APAs, there was little point in demanding fast new AGCs. The situation changed as the number of fast amphibious ships grew and a practical 20-kt LST was designed.

The minimum requirement was one AGC for each ocean, to control a division/air wing assault. However, ideally one would serve the task force commander and another the amphibious group commander. In some formulations, two were needed for the amphibious force, one for the air assault, and one for the surface assault, for a total of three flagships per assault. Thus at least four (in some formulations, six) fast AGCs were wanted. During the fall of 1958, the projected long-range program showed a new AGC each year in FY 61, FY 63, FY 65, and FY 67. Then the AGC was dropped from the FY 61 program; by February 1960 it had been moved to FY 62, at an estimated cost of $54 million. The program continued to slip, so that in the fall of 1962 the secretary of defense approved one new AGC each for FY 64, FY 65, and FY 66, at a cost of $72 million and $65 million each for follow-on ships. Tentative plans called for three more ships (FY 66, FY 67, and FY 68), at a follow-on cost of $60 million each.

The AGCs were not the only flagships of interest. The fleet badly wanted to replace the cruiser flagships. Could an AGC perform both functions? The debate over this question killed plans to include an AGC in the FY 64 program. The solution, to build a single-purpose amphibious warfare AGC, was chosen for FY 65.

Besides higher speed and new radars, technology offered a computerized command system. By this time major combatants were being fitted with NTDS, which was needed to deal with fast jets. It could also keep track of surface craft. Installed on board an AGC, NTDS could help control landing craft and helicopters as well as defending fighters. Computers could also keep track of logistics and assist in planning an assault. Quite aside from the need for computers to help handle larger numbers of faster aircraft, computerization might help hold down the sheer number of amphibious staff members. In 1963 plans called for a double system: a carrier-type three-computer tactical NTDS and a strategic planning system (adding six displays to the usual carrier system), plus two Link 11 radios.

The project began in earnest when OpNav outlined requirements for a new AGC at the request of the Atlantic and Pacific amphibious commanders on 25 April 1960. The fleets commented at length, and BuShips' preliminary designers made a quick study of a Mariner conversion. In December 1960 they compared a new-construction ship to a converted cruiser. The conversion was rejected, but the idea was revived in June 1962, when the Pacific Fleet amphibious commander proposed converting the two remaining *Baltimore*-class cruisers, *Helena* (CA 75) and *Los Angeles* (CA 135), to AGCs. They were scheduled for deactivation in FY 63 and FY 64. Why not convert them instead? The following month the amphibious commander suggested converting three heavy cruisers, beginning in FY 65. A staff study rejected a new-construction AGC as unaffordable, and argued that the recently proposed *Iowa* commando ship[1] was very expensive in personnel and operating cost. The Pacific Fleet staff estimated that a converted cruiser, which it called an LCC (landing craft, control), could retain six of her nine 8-in guns and six of her twelve 5-in. She could be converted in two steps, the first of which would cost only $2.5 million; ultimate cost would be no more than $15 million. To the staff, the ship's main deficiency as an

AGC was her lack of communications, so conversion would entail adding about 42 transmitters, 83 receivers, 28 teletype terminals, and 12 on-line crypto sets. For air control the ship would need a height-finding radar, such as SPS-8A. The fleet would gain the fast AGC it needed, and it would retain some badly needed amphibious support firepower.

This must have seemed too good to be true. The Pacific Fleet commander endorsed the idea, and in November OpNav asked BuShips for a feasibility study. By this time requirements were growing. The ship needed NTDS, the associated communications system (presumably Links 11 and 14), and the big log-periodic radio antenna that had just been developed. Instead of SPS-8, she would have the newer SPS-30 height finder. Later she would be fitted for antiaircraft warfare (AAW) and ASW self-defense, which would entail a Tartar missile launcher in place of her after 8-in barbette and an SQS-23 sonar, which was generally associated with the ASROC ASW rocket.

BuShips looked not only at the two *Baltimore*s but also at the *Des Moines* (CA 134) and *Worcester* (CL 144) classes. Because 8-in firepower was so important for amphibious operations, the heavy cruiser studies had priority; the light cruiser study was conducted only on a preliminary basis. Preliminary results were reported orally on 8 January 1963. All three classes could be converted, but a *Worcester* would lose all but her three forward 6-in mounts. Because that would leave so little firepower, the *Worcester* conversion was dropped. The heavy cruisers could retain their two forward 8-in turrets and half their 5-in; conversion would remove their 3-in guns and all after superstructure.

A flagship needed internal volume and topside space for antennas. Cruisers were weight critical and thus had limited internal volume. A new superstructure aft would house flag spaces, troop commander facilities, a combat information center (CIC) and NTDS, a joint intelligence room, a supporting arms coordination center (SACC), a war command and staff conference room, a meteorological office, other special facilities, added radar and communications equipment, and added personnel. It would not be big enough, so a joint communications center would replace the existing radio central and CIC on the first platform deck. Flag facilities would include a helicopter hangar and pad aft, for two HRB-1s. The same facilities could support two DASH ASW drones for self-defense (36 torpedoes). The ASW sensor would be an SQS-23 sonar in the forefoot. Given the sonar, the ship would also be equipped with an ASROC launcher (with 16 reloads).

For self-defense against air attacks, the initial choice was a projected point defense missile, Sea Mauler. The Bureau of Weapons (BuWeps), the successor to BuOrd, preferred the more massive Tartar (it wanted Sea Mauler as a secondary battery) on the theory that the Tartar launcher would also be able to fire the projected bombardment missile. The Mk 13 Tartar launcher would replace the after 8-in barbette, and the system also required two SPG-51 illuminators and an SPS-39B search radar. For air control, the ship would have a big SPS-30. Davits would be installed for three LCVP plus one lifeboat port and starboard.

The new electronics would need more electrical power: each turbo-generator would be upgraded to 1,000 kW. A *Baltimore*-class ship would need an additional 750 kW diesel generator. Ships would also have to be air conditioned; only *Des Moines*–class cruisers were already partly air conditioned. A *Baltimore* might need a blister to restore her stability, given the considerable added topweight.

None of this would be nearly as inexpensive as the Pacific Fleet staff had imagined: BuShips estimated $82 million for a lead *Baltimore* ($75 million each for follow-ons) or $78 million for a lead *Des Moines* ($71 million each for follow-ons). Conversion would probably require 6 months for contract design and preparation, then 48 months for the actual work. Deleting Tartar could save 12 months. Suddenly new construction looked quite attractive, particularly because a new ship would considerably outlast a cruiser that had already seen a decade or more of active service.

Work on a new AGC design had begun with a single sheet of characteristics issued by the SCB in November 1962; preliminary characteristics were circulated for comment in April 1963. Besides her amphibious role, the ship had a contingent function as flagship for designated major commanders, that is, as a replacement for the cruisers then being used by commanders of the numbered fleets (e.g., Sixth Fleet, Seventh Fleet). For this role she needed nuclear strike planning facilities. It could be argued, too, that she needed higher speed, because she might have to transit between widely separated operating areas within the area of responsibility of a numbered fleet; in fact the LCCs were limited to 20 kts.[2]

The ship needed considerable internal space, thus suggesting that she could best be based on a merchant ship (or naval auxiliary) or a carrier. She also needed a clear upper deck large enough to accommodate several massive radio antennas with minimal interference. Massive broadband high-frequency (HF) discage antennas would dominate topside design, because the AGC would rely on HF radio to reach over the horizon to ships and to commands ashore. The more widely antennas could be separated, the less they would interfere with each other. In July 1963 the BuShips radio experts (Code 450) decided that they wanted a flat deck to act as a ground plane for the antennas, greatly improving their efficiency. There was also another kind of HF antenna, working at the upper end of that frequency range: a directional log-periodic antenna, normally mounted atop a mast. A new technology, satellite relay, was just being introduced for longer-haul communication, back to the United States. The version planned for the AGC was Miser, a microwave satellite relay.

Not including NTDS, the ship had to be able to transmit simultaneously on 40 separate HF channels: 14 at 2–6 MHz, 23 at 7–9 MHz, and 3 at 10–30 MHz. It was impossible to provide a separate antenna for each transmitter, not to mention each receiver. Hence broadband antennas were important.

The preliminary designers chose to modify the design of a new combat stores ship, AFS 1 (*Mars*), with an enlarged superstructure and better transverse subdivision. Like a Mariner, she had a single-screw 22,000 SHP powerplant for a speed of 20 kts; dimensions were 581 (overall) × 79 × 21 ft, for a full load displacement of 13,500 tons. PhibLant wanted higher speed (25 kts) in light of the 1962 Cuban experience. PhibPac agreed, because speed might be particularly important given the sheer size of its operating area; the Pacific Fleet command vetoed the idea. That would have required a different merchant hull (e.g., an *American Challenger*) for conversion. In 1963–64 no such hull was available. Estimated cost was $79 million for a lead ship, $70 million for a follow-on, rather more than had been expected. The Pacific Fleet, which badly wanted a new AGC by 1968, asked whether it might be possible to convert a light fleet carrier (like *Wright*, converted to a national command center as CC 2 in 1962–63).

An initial sketch showed two lightweight 5-in/54 at the ends of the ship, the after mount on the starboard quarter, plus a pair of point defense missile launchers (the abortive Sea Mauler). Unlike the cruiser, the new AGC showed no ASW defensive system. She did have a helicopter deck and a hangar. Forward to aft, the ship showed a discage mast for HF transmission (with a log-periodic antenna on top), another big HF discage amidships, the satellite antenna (in a spherical radome) abaft amidships, and a discage HF receiver aft. The arrangement sketched at this stage did not yet include provision for the search and height-finding radars needed for the AGC air control function.

Code 450 objected to the lack of a satisfactory ground plane; it wanted a flat-topped ship about 630 ft long with a centerline island amidships and a "mack" (mast-stack), as in many contemporary U.S. designs. There was some question as to whether a ship of any length would provide sufficient separation between masts. Early in July 1963, Code 450 asked for four months for tests using brass models, to determine just how badly nearby receiver and transmitter discages would interfere with each other. The preliminary designers considered putting aside the AGC project in favor of other pending auxiliaries, such as a new submarine tender or a Polaris missile cargo ship.

To meet Code 450's objections, the preliminary designers on paper stripped AFS 1 to the main deck and built up a three-deck stem-to-stern superstructure to provide sufficient spaces. Meanwhile the SCB decided that all office spaces would be air conditioned, and that it was no longer necessary to distribute guns or point defense missiles fore and aft.

Code 450 now decided to forego the model tests. The modified AFS with a flat deck would be satisfactory if the island, placed to starboard, were far enough forward, and if the stacks were bent over, as in a light carrier. The preliminary designers were skeptical. The island would be poorly placed and stack gas would be a problem. To solve severe stability problems, she would need more beam. That in turn would reduce her speed. A longer hull might be better, not only for speed but also for antenna separation. When Code 450 suggested that armament should be in sponsons, the preliminary designers realized that the AGC would look like a carrier, perhaps much like an LPH.

Acting on this hunch, the preliminary designers considered the LPH. It was attractive because it was larger, even unmodified it had a flat deck, it had greater variable internal volume for growth potential, and it offered the required speed and endurance. Code 450 was enthusiastic, although it wanted the island moved to the centerline. A sketch was prepared, and the modified AFS was compared with an AGC based on the LPH.

By this time more radios, with different types of antennas, were being added. The sketch showed three masts, with a wire for low frequency (LF) radio transmission strung between them. The foremast carried ultra high frequency (UHF) antennas, a Tacan for aircraft homing, a surface-search radar antenna, and a height-finding radar. The mainmast served simply to support the LF wire antenna. A lattice mast aft carried a log-periodic antenna. A discage and a medium frequency (MF)/HF antenna occupied the deck abaft the foremast, and a big electronically scanned air search antenna was fixed atop a deckhouse abaft the mainmast, with a Miser atop it. Abaft this deckhouse were another MF/HF dome and another discage. Code 450 seems to have been willing to forego the discages in favor of placing a directional HF antenna and Miser atop an island amidships.

Another possibility, which the preliminary designers investigated in August, was to convert a ship in reserve. The choice was fixed by the need for sufficient length to accommodate the antennas, and by the need for a speed of about 20 kts. That left two possibilities: one of three *Independence* (CVL 22)–class light carriers (the two bigger *Saipan*s were already earmarked for conversion to national command centers afloat) or one of seven P2-SE2-R1 class troopships. Another seven Military Sea Transportation Service (MSTS) P2-SE2-R1s were earmarked for APA conversion.

The carrier offered the simplest conversion, because she already had a flat deck that could be turned into a ground plane. On the other hand, equipment would fill her hangar so completely that a new helicopter hangar would have to be built atop the flight deck, aft. A sketch showed a big transmitter discage forward with a log-periodic antenna atop it, then a deckhouse carrying the radars, then another transmitting discage, then the helicopter hangar aft with Miser and the receiver discage atop it. At 623 ft overall, the carrier was the longest ship being considered, hence it offered the greatest separation between antennas. The P2 was 609 ft overall (18,500 tons) and could not make the required speed (rated speed was 19 kts). Plans called for arming her with a lightweight 5-in/54 right forward. Conversion would entail adding a hangar and helicopter platform and trunking her smokepipes; she would not have the flat upper deck Code 450 wanted. At 581 ft overall, the LPH was the shortest ship being considered as a basis for a new AGC. All three ships offered much the same accommodations: 245 officers and 1,300 enlisted in the carrier, 240 and 1,200 in the P2, and 236 and 1,143 in the LPH. Estimated conversion costs (prototype/follow-on) were $75.5/$69 million for both the carrier and P2, and $79/$70 for the LPH.

Blue Ridge clearly shows the aircraft carrier origin of her design in this November 1974 photograph. The U.S. Navy did not yet rely heavily on satellite communications, and at this time her main long-haul antenna was the high-frequency directional Yagi on the lattice mast forward of her bridge. The tower aft was reserved for a big satellite dish, not yet fitted.

Photographed in Freemantle, Australia, on 20 October 1986, *Blue Ridge* shows satellite dishes atop her after tower mast. At this time she retained the long-range high-frequency antenna forward. The broad-band discage forward of it was installed on many U.S. warships during the 1970s. *LSPH E. Pitman, RAN*

By January 1964 the preliminary designers had completed a study of an AGC based on the LPH hull. The bow was altered, the knuckle in the gallery deck removed, and the hull made beamier lower down. The preliminary designers recommended that the stern be modified to reduce propeller cavitation and vibration. Overall length was increased to 601 ft 6 in; displacement was 18,075 tons fully loaded. Hull and machinery amounted to only 20 percent of the ship's cost. Hotel (accommodation and ship's service) spaces accounted for two-thirds of the ship's usable volume of 1.5 million cu ft. Command and electronics accounted for about 290,000 cu ft out of a total of 2.3 million cu ft.

Probably the most important advance incorporated in the design was automated data processing, both tactical (NTDS and intelligence data) and logistics. The ships had three levels of communications: long range (ship to ship, ship to shore), short range tactical (ship to ship, ship to air), and short range for the assault proper. The ships were designated LCC (landing craft, control) rather than AGC, but the earlier AGC numbering was continued, so they became the *Blue Ridge* (LCC 19) class. Two were bought, *Blue Ridge* (LCC 19) in FY 65 and *Mount Whitney* (LCC 20) in FY 66. By this time the SCB had changed its design numbering system. Now the first digit indicated the ship's role, and the fiscal year of initial procurement was given after a decimal point. Thus the AGC, initially SCB 248, became SCB 400.65, the FY 65 flagship.

A 1966 draft presidential memorandum (DPM) on amphibious assault ships, fire support ships, and mine countermeasures led to a new study of amphibious flagship alternatives from February 1967 through April 1968. For FY 70 the secretary of defense approved a total of six flagships (two Atlantic, two Pacific, two backup): three new-construction LCCs and three retained World War II AGCs that the navy wanted to replace with new ships in FY 70–72. The third new ship would have been built under the FY 69 program. A further DPM, dated 9 February 1968, approved a navy proposal to defer the planned FY 69 ship to FY 70, and rejected the navy proposal to replace the three old AGCs. However, as of early 1969 construction of the third new LCC had not yet been approved for FY 70. Funds were withheld pending a study of alternatives.

By this time the navy also badly wanted six new major fleet flagships (two for each forward-deployed numbered fleet, two as backup), to replace aging cruisers. The 1968 Defense Department Posture

By the late 1990s, the U.S. Navy had virtually abandoned high-frequency radio in favor of satellites, and the ship had radomes in place of both the lattice mast and the discage. She also had two Phalanx close-in defensive guns, one right forward and one right aft. Both the 3-in/50s and the Sea Sparrow launchers evident in earlier photographs had been landed. *Blue Ridge* is shown on 10 December 2000. *U.S. Navy photo by Photographer's Mate 3rd Class Daniel Mennuto*

Statement suggested that a modified LCC 19 be used. A fleet staff was much smaller than an amphibious staff; 80 officers and 300 enlisted rather than 200 officers and 400 enlisted, respectively. On the other hand, the numbered fleets wanted 30-kt ships, to run with carriers. Moreover, replacement of both the aging AGCs and the cruisers would require 12 ships. One alternative would be to build only four special-purpose LCCs plus six dual-purpose (amphibious and fleet) flagships. If all the ships were dual purpose, then the four back-ups could be eliminated and the total program cut to eight.

In FY 70 dollars, an LCC 19 was expected to cost $96 million. A 30-kt fleet flagship would cost $106 million, and a dual-purpose 30-kt ship would cost $124–135 million. Adding defensive armament would roughly double the cost of the fleet flagship. An alternative, to create a virtual flagship by dividing staff among flag spaces on board the ships of the amphibious task force, was rejected. One version had been to incorporate additional facilities in a long-hull version of the new LHA (landing ship, helicopter, assault), called LHA-X, and to use two such ships plus an LPD (landing ship, amphibious personnel, dock), which would provide intelligence facilities.

A variety of conversions was considered as alternatives to new construction. Surviving World War II AGCs might be modernized. As in the 1964 study, a reserve troop transport or light carrier might be converted. Other candidates were an LPH already in commission, a Mariner APA, a cargo ship (C4), an LHA, or even a modified destroyer.

The wartime AGCs were now seen as substandard. They were slow, with poor structures for their electronics, and with too little electrical power (three 300 kW turbo-generators and one 100 kW diesel). They would have to be stretched amidships and given new sterns, new propulsion, new superstructures, new auxiliaries, and upgraded accommodation and electronics. Little of the original structure would survive, and the result would probably be more expensive than a new ship. The Reserve Fleet troopship and light carrier conversions were still unattractive. The carrier in particular would have to have her machinery rehabilitated

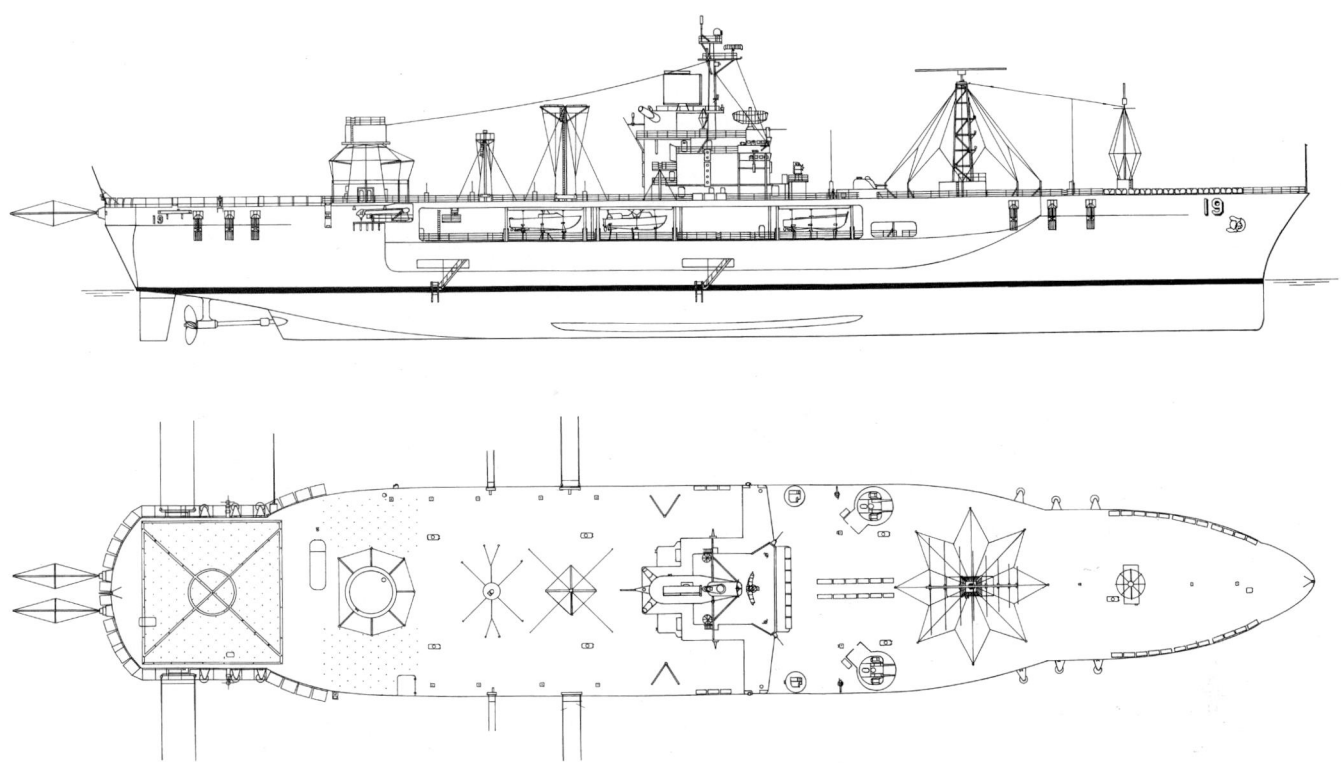

Blue Ridge (LCC 19) in 1971, showing her big troposcatter (Yagi) antenna forward. At this stage, satellite communications were not yet very important. As in World War II, the amphibious flagship was expected to control aircraft until suitable facilities had been set up ashore, so *Blue Ridge* was provided with the best available three-dimensional radar, SPS-48, as well as the usual search set (SPS-40) and a masthead Tacan for aircraft navigation. Sea Sparrow defensive missiles had not yet been fitted.

and her accommodations upgraded. The LPH would be less expensive than building a new LCC 19 because it actually had more generating capacity, but withdrawing a modern ship from the amphibious force would carry an excessive opportunity cost. The Mariner APA lacked generating capacity, and total conversion cost would exceed the cost of a new ship. The C4 cargo ship would need vastly more structure; her steel weight was about 25 percent less than that of an LCC. The LHA was larger than an LCC, hence she could be configured as an AGC, but the cost would be about 25 percent more. The destroyer was rejected because its usable volume (about 320,000 cu ft with boilers removed to cut speed to 20 kts) was only about 21 percent of that of an LCC 19.

The implication was that more LCC 19s had to be built, but that was not done. The hope seems to have been that flag facilities built into the LHA would often suffice, given the degree of computerization that was now possible. The aging cruisers were never replaced. Ultimately the two LCCs served as numbered fleet flagships, because they offered the necessary facilities, apart from speed, and because MEF-level assaults seemed less and less likely. Two LPDs, *La Salle* (LPD 3) and *Coronado* (LPD 11), were also converted into flagships, redesignated AGF 3 and AGF 11, in 1972 and 1980, respectively. Unlike an LCC, they lacked NTDS and air control facilities. As of 2000, two joint flagships, sometimes styled JCC (by contrast to LCC), to replace the worn-out *Blue Ridge* class, are scheduled to be bought in FY 04 and FY 05; they are conceived as modified versions of the new LPD 17.[3]

Whidbey Island was one of a series of new well-deck ships adapted to the new over the horizon amphibian, the air-cushion landing craft or LCAC.

16
Over the Horizon Assault

In 1964 the U.S. Marines issued "USMC in 1985: A Long-Range Study." To the old threat of tactical nuclear weapons was being added the new threat of antiship missiles, particularly surface-to-surface missiles. Even though they envisaged development of new VTOL aircraft, the Marines expected to continue to bring materiel and even troops over the beach. Ships unloading just offshore would become more vulnerable. Valuable amphibious ships would survive better if they could remain beyond the horizon (i.e., about 25 nm out to sea)—beyond, for example, the range of shore defenses—through the early phases of an operation. Simply reaching the beach in much the same interval as in a conventional assault (from 5–10 nm to sea), landing craft would have to be much faster. Amphibious ships could be more widely dispersed against threats such as nuclear attack. Speed reduced exposure to enemy counterattack while the assault force was en route from the transports, with very little self-defense capability. OpNav agreed with this over the horizon (OTH) concept.

The OTH craft ultimately developed, the LCAC (landing craft, air cushion), could traverse a very wide variety of beaches. An enemy would find it more and more difficult to predict just where the assault would come. For example, in the 1980s the Marines pointed out that a force 180 nm out to sea could attack anywhere over 900 nm of shore within 24 hrs. As an example, an assault could occur anywhere on the Atlantic coast from Newport, Rhode Island, to Charleston, South Carolina. Once launched, LCACs could arrive anywhere over a 100 nm frontage.

An enemy who could not predict where the landing would fall could not concentrate troops to oppose it. Nor could he lay mines off the favored beach, because his inventory of mines would inevitably be limited. Thus the Marines might well bypass both heavy coast defenses, including dug-in troops, and mines. The new technique, then, might make up for the drastic decline of both naval gunfire support and mine countermeasures.

On 16 February 1965 the chief of naval material assigned BuShips the task of developing a new generation of fast assault craft. The necessary new technologies seemed to be ripe for development: planing, hydrofoil, ground effect machine (GEM), air-lubricated hydro-ski, captured air bubble (CAB), and hydrokeel. In addition to speed, some of them offered much better seakeeping. There were also new handling techniques: using a ram-tensional davit for boat handling and a probe for quick refueling.

Tentative specific operational requirements (TSORs) for new-generation craft were issued: TSOR 14-17T on 18 January 1965 for an LCVP replacement, TSOR 14-18T on 13 August 1965 for an LCM(6) replacement, and TSOR 14-19T on 13 August 1965 for an LCM(8) replacement. By 1966 these had been characterized as the small, medium, and large landing craft, respectively. Because increased speed generally meant increased complexity and size, it was necessary to estimate the overall increase in assault power due to the improvement.

For the LCVP, objectives included higher speed (at least 35 kts, endurance at least 200 nm) and better surf performance (Sea State 4 rather than 3). The existing LCVP could broach and flood when operating near its 5-ft surf limit, and it could be lost in severe surf. It was also considered vulnerable to underwater obstacles, over which an air cushion vehicle would skate. The desired load was 9,000 lbs or 36 troops, with a ramp at least 7 ft 5 in wide. Minimum interval between overhauls should be 1,000 hrs; minimum inspection interval should be 200 hrs. These reliability figures also applied to the LCM replacements.

For the LCM, but not the LCVP replacement, BuShips was to offer at least one fully amphibious option. The LCMs were to be capable of 40 kts, and to have an 11-ft ramp. Loads were 70,000 lbs or 120 troops for the LCM(6) replacement, and 130,000 lbs or 200 troops for the LCM(8) replacement. Desired endurance was 150 nm for the LCM(6) and 200 nm for the LCM(8) replacement. The designers were also to consider a 380,000-lb payload, which would be an LCU replacement.

To further the program, a Landing Craft Coordinating Group was formed in March 1966; the new landing craft program was expected to begin in FY 67. An outside think tank, SRI, was retained to help winnow the number of design alternatives.

By February 1968 the TSORs had been turned into advanced development objectives (ADOs). The most dramatic change was in the limit on weight and dimensions. In the past, the limit had been boom capacity. Now it would be mainly well deck dimensions because weight limits were abandoned. Vehicles would be driven directly on board the new craft, greatly speeding cargo handling; the old AKA "yard and stay" techniques would be abandoned. This change can be attributed mainly to the decision to build LHAs instead of combinations of well-deck ships and AKAs. All craft, moreover, were to have some over-the-beach (amphibious) capability, and to be able to get onto the beach through 6-ft surf. Craft were to be able to make their design speed in Sea State 2, to proceed and maneuver in Sea State 3, and to survive in Sea State 4, with the larger craft able to survive at the lower end of Sea State 5.

The effect of vertical assault was also being felt. Surface assault waves would be made up almost entirely of LVTs; there was no longer much point in the LCVP. With helicopters carrying not only Marines but also light vehicles, landing craft cargo would be limited mainly to heavy vehicles such as tanks, special equipment, and resupply cargo.

The new craft would be armed with two single 0.50 caliber machine guns on pedestals. They would have provision for some armor protection against small-arms fire. It had to be accepted that an enemy might oppose them with antitank rockets and grenade launchers, which were being used in Vietnam, but there was little hope of armoring landing craft against such threats. The main defense would be dispersion, which again demanded high speed.

In all, 57 candidate craft concepts were evaluated. In 1970 three were identified for further development: Joe, a 160-ton, 35-kt planing craft; Jeff, a 60-ton 50-kt air cushion vehicle; and Jim, a 15-ton air cushion vehicle or planing craft. Jeff was selected for initial development, because it offered true amphibious capacity, hence a fourfold increase in suitable beaches. It could, moreover, operate from a dry well deck, so a ship did not have to ballast down to launch. That made for faster turnaround. A drive-through configuration would make for faster reloading.

Joe became LCM(9), a fast planing alternative to LCM(8). An LCM(9) operational requirement was issued on 24 November 1976. At that time concep-

An LCM(9), from a BuShips sketch made in the late 1970s. The vehicle on board is an AAV-7 amphibian. Plans for 120-ft and 130-ft versions were drawn in 1977 and in 1979, respectively. Capacity was two M-1 Abrams tanks or 132 tons.

tual designs showed dimensions of about 120 × 22 ft, so that four could dock in an LHA, and two could dock abreast in an LSD or LPD well (totals would be six or two, respectively). Length could not exceed 135 ft (the maximum for an LCU); beam could not exceed the 21 ft of an LCM(8). By 1979, LCM(9) was a 20-kt (also given as 30+ kts) aluminum planing craft with a 120-ton payload and a crew of five. It would probably have a drive-through deck, and it would be powered by the same gas turbines as the LCAC (see below). The last LCM(9) sketches showed a 140-ft craft powered by waterjets, capable of carrying a pair of XM-1 tanks (a total of 132 tons). In effect it was a fast LCU replacement. Development was to have begun in FY 81, with delivery scheduled for FY 84, but in mid-1979 it was suspended. The craft was never built. Given the high cost of the LCAC, however, in the late 1990s the Marines became interested in a fast LC(X), nominally an LCU replacement, which may resemble the stillborn LCM(9).[1]

Jim was envisaged as a deck-loaded craft carrying outsized loads too light to be carried efficiently by the larger craft. It was not pursued.

Development contracts were let in 1970 for completing Jeff A (96 × 48 ft) and Jeff B (86.75 × 47 ft). They were called AALC (advanced assault landing craft). Objectives were a speed of 50 kts in Sea State 2, a range of 200 nm, and a payload of 60 tons. They were to be able to negotiate 8-ft plunging surf, sand, swamp, ice, and grassland, and to climb an 11.5 percent (13 percent for the shorter Jeff B) slope. The 60-ton payload was adopted so that the craft could move all types of equipment. Ideally, it would carry three columns of vehicles. The most desirable length to beam ratio for such a craft was 2:1 (90 × 48 ft, which could be accommodated in the 50-ft well deck of an LSD or LPD). Given the length, an LSD could carry three or four craft, an LPD could carry two. LHA stowage was a problem; the 76 ft 6 in wide well could take only one craft, and the full length of 245 ft could not be used because the forward part was divided by a center island. Any change in dimensions, moreover, would cut the Jeff payload, because it was already designed to the limits of the state of the art. To get two Jeffs on board an LHA well would require that the center island be removed, with a moderate to high structural risk. To get three, the well would also have to be extended.

Preliminary designs for these 150-ton craft were completed in October 1970, and detail design contracts were let the following March, with tests expected to begin early in 1976. On trials, Jeff B made 75 kts compared to 58 for Jeff A, with a range of 210 (vs. 140) nm. Jeff B carried the full 60-ton load, while Jeff A carried only 40 tons. Jeff B made the required 13 percent grade; Jeff A made only 9. Both versions of Jeff negotiated 12-ft sand dunes. Jebb B was chosen. Given successful trials, an operational requirement for an LCAC (landing craft, air cushion) was issued on 16 November 1976.

Neither the LCAC nor LCM(9) quite solved the OTH problem. The Marines still planned to come ashore on board slow LVTs, because they offered invaluable mobility ashore. LCACs would land armor, not men. Helicopters offered surprise, like cavalry, but they would do poorly against opposition equipped with hand-held missiles. LVTs had to be launched close offshore, either from large amphibious ships (LSTs or LPDs) or from larger fast landing craft launched at a distance. For example, an LCAC could carry three LVT or two LVT plus landing support equipment. Moreover, an LCAC could carry only a handful of troops, because the noise generated on the tank deck was devastating; troops had to ride in the cockpit or in a special cabin. The gas turbine exhausts generated considerable heat. Thus, even if LVTs were carried, eventually there was a real question as to whether they could have troops aboard. About 1981 NAVSEA's concept formulation (CONFORM) branch proposed a surface effect LVT carrier, which could be carried in a standard well deck; LVT Carrier 2000 was included in the organization's FY 82 program.

The U.S. Navy's program objective memorandum for FY 80 (POM-80) planning document, written in 1977, analyzed alternative landing craft forces. Conventional landing craft were expected to reach the end of their lives in FY 87–91. New landing craft would have to be bought beginning in FY 86. If LCAC was adopted, more lift ships would be needed to provide sufficient well-deck space, and maintenance cost would rise. LCM(9) might be less expensive, but it suffered from all the limits of conventional landing craft, and it would perform poorly in high sea states. Any attempt to make it as fast as an LCAC would raise its cost to LCAC levels. Because it could not run ashore, LCM(9) might be most useful as an LVT(X) carrier.

One possibility would be to combine 69 LCACs with 63 LCM(9)s carrying LVTs to a line of departure. The fleet would have to add 11 new-design LSDs (LSD 41 class) to existing well decks to transport the landing craft. Eliminating standoff would save 26 LCM(9) and 3 LSD, but then the LSTs carrying amphibians would have to come quite close to the beach. Another possibility was an all-LCAC force without standoff (79 LCAC, 6 LSD). Alternatively, LCACs could be eliminated, in which case a no-standoff force would comprise 109 LCM(9) and 5

The LCAC was the key to over-the-horizon assault, because it could get from beyond the horizon to the shore as quickly as a conventional LCM(8) could get from the traditional transport holding position to the beach. An LCAC is designed to carry only vehicle crews (24 troops), in port and starboard cabins. They cannot ride the vehicle deck because of the heat of the gas turbine exhausts. However, an air-conditioned personnel transport module (PTM) can be erected on the vehicle deck to carry 145 fully-equipped troops or 180 evacuees or up to 54 litters to evacuate wounded troops. The first four of nine PTMs acquired in 1994–95 were deployed to the Mediterranean in early 1996, craft so equipped being called MCAC (multi-purpose air cushion craft). Because the air bubble which supports it protects the LCAC to some extent from the shock of an underwater explosion, its builder, Textron, proposed and demonstrated a mine countermeasures adaptation; the necessary sweep kits have been manufactured. *Textron Marine Systems Division of Textron, Inc.*

LSD; using LCMs to transport LVTs to the starting line would require an additional 25 LCM(9) and 4 LSD. LHAs were not expected to carry LCACs, because that would be inefficient.

For the moment, the choice was to develop both the LCAC and the LCM(9). The LCAC project was transferred to PMS 377 (Ship Acquisition) in March 1979. As the LCAC was conceived in 1977, an LCAC would typically be launched at 8 kts, loiter for 20 minutes at 10 kts (below hump speed) to form up, then accelerate over the hump in deep water, transiting 24 nm in Sea State 2. She would carry enough fuel for two such cycles (out and back). Hullborne, the craft had to be able to make 5 kts in Sea State 1 and to make some headway in Sea State 3. The speed goal was 40 kts, with an acceptable threshold for the craft set at 35 kts or greater. By 1982 the LCAC could make 46 kts. It also could exceed the desired mission distance (27 nm vs. 24 nm; maximum range was 223 nm). It could meet the payload goal of 150,000 lbs (threshold 120,000). In 1982 the desired level of availability was 0.80 (threshold 0.70, 0.83 had been demonstrated); the desired reliability was 0.94 (threshold 0.90, 0.95 demonstrated); and desired mean time between failure was 24 hrs (threshold 14, demonstrated 33.7). Mean time to repair was 3.4 hrs (threshold 5). Manning per LCAC was 87, against a goal of 76 (threshold 94). Estimated operational lifetime was 20 years.

LCACs would ride to the amphibious objective area in ships' well decks, so they had to be sized to fit. Generally the mouths of the decks were wide enough, but internal obstructions often limited the number of LCACs a ship could carry. Here an LCAC approaches USS *Pensacola* (LSD 38) during initial underway mating trials in the Gulf of Mexico, February 1985. Note the guide plates on the sides of the ship's well deck. The LCAC is carrying an LVTP7 amphibian; it was often suggested that the solution to low amphibian speed would be to carry them close to the beach on board LCACs. That recalled a 1944 proposal for a fast LVT-carrying craft.

The craft was sized so that three would fit an LSD without a mezzanine (four in the new LSD 41, see below), two in all types of LPD, and one in an LHA. Fitting required 8 in clearance when the craft was off-cushion, and 4 ft between the lowest hard point in the craft and the bottom of the skirt when it was on-cushion.

It was believed that the LCAC would be more resistant to mines, presumably because its air cushion would absorb the shock. Studies showed that only about 3 percent of LCACs would be lost in an OTH assault. However, that assumed that there were no guns or missiles defending the beach, because LCACs were built to aircraft structural standards. The key was that the LCAC could traverse 70 percent of the world's beaches. Given its high speed and long range, an LCAC launched from a ship 20 nm offshore could beach anywhere along 100 nm of shoreline. It was most unlikely that an enemy could so predict the point of LCAC landing as to mass troops or antitank missiles there. Tactics could change drastically. For example, helicopters and LCACs could link up on an enemy's flanks while ships moved to launch LVTs and conventional landing craft directly at the enemy.

Overall, the Marines considered the LCAC the most important advance in amphibious warfare since the helicopter. The design was so successful that by late 1982 there was interest in an air cushion LCU successor; it never materialized, however. The LCAC would replace the LCM(6) and LCM(8) in the initial wave of an assault. Typically LCACs would account for two-thirds of tonnage brought ashore, with LCUs offloading after a beach had been secured.[2]

The LCAC is shown inside the mouth of *Pensacola*'s well. *Textron Marine Systems Division of Textron, Inc.*

The program received provisional approval in February 1980, and the mission needs statement was approved that October. It then had to pass a series of "milestones," beginning with Milestone II in November 1981. Desired LCAC operational numbers (65 for the MAF, 25 for the MAB) were set by the minimum needed for a primary beach assault (two-thirds of the assault force). The Marines actually deployed MAUs rather than MABs. Each incorporated an assault craft unit (ACU) of six LCACs on board two or more ships, somewhat fewer than the proportion associated with an MAB. In November 1982 the navy decided that it wanted 107 LCACs, with the expected mature production rate being about 12 per year.

To conserve LCACs for further assaults, LCUs and LCM(8)s would be used for about 70 percent of unloading from well-deck ships, that is, unloading after the initial assault; one LCU was about equivalent to three LCM(8). Some items could not be brought ashore in any other way. LCM(8) requirements were estimated at 70 percent of troop and vehicle square footage; the LCU requirement was set by required cubic footage. It appeared that an MAF would need 5–8 LCU, an MAB 3–4; an MAF would need 30–38 LCM(8), an MAB, 12–16. Each LKA carried four LCM(8). On the other hand, a 1982 study recommended the MAF use 69 LCAC and 30 LCU to replace 35 LCU, 76 LCM(8), and 90 LCM(6), and the MAB use 31 LCAC and 11 LCU. Thus the total requirement (MAF plus MAB) was 100 LCAC and 41 LCU. Presumably these figures included margins for wastage and for craft out of service.

By this time the Reagan administration was in office. The navy's new "maritime strategy" emphasized the value of holding Soviet and Soviet-held territory at risk, which included amphibious assault. The LCAC, with its ability to clamber over most of the world's beaches, was one key to that capability, so its procurement gained priority. In June 1983 Milestone IIIB, the last before production, was moved up from the third to the first quarter of 1985.

LCAC procurement was combined with a new program for a VTOL troop carrier, the JVX, to replace the existing CH-46.[3] The aircraft chosen for the JVX program was the MV-22 Osprey. With much higher speed, fewer Ospreys were needed to replace CH-46s. It could also strike much deeper, with implications for fire support. The requirement set in 1983 was 180 for the MAF, and 45 for the MAB. With limited cargo capacity (payload 5 vs. 15.2 tons), JVX could not replace the heavy cargo carrier, the CH-53E, which could not match its speed. To the extent that JVX was an essential element of future assaults, it might justify a new generation of large-deck amphibious ships to almost the same extent as an LCAC. At the least, it took something like an LHA, rather than an LPH, to support it. This effect was retarded only because the JVX program took much longer than expected. In 1983 the aircraft was expected to become operational in 1991, whereas in 2000 it was only completing its tests, and indeed the beginning of full production was further delayed by a series of crashes.

In the late 1970s there were also attempts to improve the performance of existing landing craft. A high performance improvement (HPI) program was intended to increase LCM(6)'s speed to 13 kts; 99 craft were bought under the FY 85 program. The HPI version, designated LCM(6) Mod 2, gained its speed by increasing power; it had two 600 BHP diesels rather than the 225 BHP units of earlier craft. Cargo capacity was roughly unchanged, at 68,800 lbs; hoisting weight is 69,600 lbs.[4] An LCM(8) service life extension program (SLEP) included in the 1978 five year plan was expected to extend the craft's usual 15-year life by 10 years.[5]

By the early 1970s the Marines were no longer so sure that they could do without a fast LVT. In 1973 they announced a requirement for a vehicle they called an LVA (landing vehicle, assault). They wanted to be able to launch it from beyond the horizon. These craft would carry 18–22 troops or 6,000 lbs of cargo, at up to 35 kts in water or 55 mph on land. They would be armed with a new 25-mm gun, and would have better fighting vehicle characteristics than the LVTP7. Engineering assessment and small-scale experiments began in 1974, and the program was presented in detail to the commandant in April 1975. Three alternatives were considered: an air cushion vehicle, a hydrofoil, or a planing hull. Contracts were let in 1976, to Bell Aerospace Textron for the air cushion vehicle (with inflatable/deflatable air bag and nonretracting tracks) and to FMC and Pacific Car and Foundry for planing hulls with retractable treads. Expected weight was 55,000–60,000 lbs. Clearly either would need far more power in the water (an estimated 3,000 HP) than on land (500 HP). Thus a successful powerplant had to provide not only enormous power in a small package, it also had to run efficiently at much lower output. The Marines invested in a new technology, Curtiss-Wright's spark-ignition stratified-charge rotary (Wankel) engine, which the manufacturer claimed had the specific output (power to weight ratio) of a gas turbine yet had only about 10 percent more fuel consumption of a very economical diesel. A contract for a 1,500 HP engine was let in 1977; the LVA would use two of them. The new engine was never completed, however, and claims for high swimming speed were never tested.

By 1977 there were plywood mock-ups of the competing designs. A 1978 program review showed serious concerns about the size (30–33 × 11 × 11 ft) required for high water speed, which might make the vehicle less survivable ashore, as well as the complexity (the need, for example, to retract and cover the tracks in the water). The cost, $330 million for development and $1.4 billion for procurement, was also a concern. These problems proved virtually unsurmountable. In January 1979 the commandant raised them with the CNO. Instead of building a long-range fast-swimming vehicle, he proposed extending the existing high-speed launch concept. Now the standoff distance was to be tailored to the onshore threat. The navy would dash its launch ships in toward the beach under the umbrella of task force defenses. Although not ideal, this was acceptable to the navy. On this basis the commandant announced cancellation of the LVA on 2 February 1979.

An alternative was already under consideration. While LVA was evolving, the process of weapon system approval changed; now a formal mission need statement (MENS) was required for each new program. It had to include alternatives. The LVA MENS for amphibious war surface assault was written retroactively and approved by the deputy secretary of defense in October 1978. The alternatives to LVA were an LVT(X), with low water speed, to be launched at high speed; and the army's new infantry fighting vehicle, which became the M2 Bradley. The Marines rejected the M2 on the grounds that it had no ship-to-shore capability whatever, and that it carried about a third as many troops as they wanted, thus they would need three times as many such vehicles as LVTs.

The Marines issued an LVT(X) request for proposal (RFP) in November 1981, with design contracts expected in the third quarter of FY 82. They were already extending the service life of the existing LVTP7, because without some such program these vehicles would reach the end of their ten-year lifetime in 1981–84. To be worthwhile, LVT(X) had to show a substantial improvement over a life-extended LVTP7, yet it could not be too much more expensive. Conversely, the product-improved LVTP7 inevitably became an alternative to LVT(X). Another possibility was a new-production LVTP7 mounting a 25-mm gun.

Requirements set in December 1982 included a main armament at least equivalent to a 25-mm Bushmaster (M242) cannon, with 300–600 rounds; a turret configured so that an antitank missile could be installed; turret stabilization so that the vehicle could shoot on the move; protection against 7.62-mm AP rounds at 300 m; minimum capacity 17 troops; sustained land speed equivalent to a main battle tank (45 mph on roads, 200 mile range at 25 mph); and a water speed of at least 6 mph in 2.2-ft seas (reduced in Sea State 3, capable of surviving in Sea State 4). The heavy seaworthiness requirements would make it possible to launch LVT(X) 10 nm rather than about 2 nm from a beach. If LCACs could bring LVTs so close to the beach, however, it was not clear that any improvement in seaworthiness was worth its cost; only later would it be apparent that noise precluded using LCACs as LVT carriers. It was soon apparent that the 25-mm gun and turret might account for up to 30 percent of vehicle cost. By December 1984 the Marines were considering reducing their requirements to make the vehicle affordable. They considered both a 13-man version (LVTX 13) and a 21-man version (LVTX 21).

LVTP7 lifetime was running out. As an interim step, the Marines chose a life-extension program. In 1977 FMC received a contract to improve the existing LVTP7, producing 14 prototypes of a new LVTP7A1 (later AAV7A1) version beginning in 1979. The most important change was replacement of the 400 HP diesel with a new Cummins engine identical to that in the army's new M2 and M3 infantry fighting vehicles. Other improvements were an automatic transmission, a new fuel tank, an automatic fire suppression system, a thermal sight, and an all-electric turret. The program began in 1983, with upgraded vehicles beginning to enter service in FY 84. It was completed in August 1986 to add another ten years, thus pushing the decision point to the 1990s. In 1987 the Marines began a series of rotating product improvement programs (PIPs) creating a series of improved blocks to the AAV7A1s. Block 1 added an upgunned turret (carrying a 40-mm Mk 19 grenade launcher and a 0.50 caliber machine gun) plus P-900 two-layer applique armor and a fold-out bow vane for better mobility in water, adding buoyancy to keep the nose of the vehicle from plunging. Block 2 added another applique armor kit, this time resembling corrugated plating, plus an automatic fire sensing and suppression system. Block 3 increased power from 400 HP to 750 HP and provided a nuclear/biological/chemical warning/filtering system. A day/night thermal sight was added, and the turret modified so that the 0.50 caliber machine gun could be replaced by a 30-mm weapon. The fire suppression system was improved so that it could deal with a second fire while extinguishing the first. Overall, the effect of the PIP was to permit replacement to be deferred to the early 2000s and thus to kill off LVT(X) entirely.

The new assault craft, the LCAC, engendered a new generation of amphibious ships. A May 1974 amphibious warfare review, the first major one since October 1966 (NAVWAG 44), laid out requirements for the next-generation amphibious ship, LX, to replace the LSD 28 class built in the 1950s. The LSD 28's 30-year life would run out in the 1980s. It would also be necessary to replace the first new-construction LPH, with a type tentatively designated LPH(X). The Ship Acquisition and Improvement Panel (SAIP), successor to the old committee on the shipbuilding and conversion program, issued basic LX requirements on 8 May 1975. The panel specified the capability the class as a whole had to provide: 2,600–4,400 troops and 250,000 sq ft of vehicles and other cargo. A tentative program called for eight ships, to replace the eight LSD 28s: the first in FY 80, three each in FY 81 and FY 82, and the last two in FY 83, at a unit cost of $233.5 million. By July 1975, LX was designated LSD 41.

The Marines envisaged a well-deck ship roughly the size of an LSD, to accommodate 12 landing craft or 4 LCAC, plus 550 troops and some modest cargo (30,000 sq ft and 2,000 cu ft). It would not support helicopters. By July, capacity had been changed to 550–620 marines, with better habitability for extended time at sea. Cargo capacity was 28,000–30,000 sq ft and 5,000–21,000 cu ft. Well-deck capacity had been enlarged to 12–16 boats, but still only 4 LCAC. The ship would be able to launch amphibians while at moderate to high speed. Endurance would be at least 8,000 nm at 22 kts.

The long-term building program was based on the accepted amphibious lift requirement, one and one-third MAF. As of early 1976 the preferred alternative was three LHA, eight LX, and one LPH(X). If no more LHA were forthcoming, more LX and LPH(X) would be needed; an alternative was five LPH(X) and ten LX.

Meanwhile the earlier 20-kt LSDs aged. Similar to the FRAM II programs during the 1960s, a service life extension program (SLEP) was proposed. In 1976 it was estimated that spending $24–31 million could extend a ship's life by five years, until LX was in service. In 1976 the navy planners' program objective memorandum for FY 78 (POM-78) proposed that the first ship be done in FY 84. In the early 1980s fleet commanders objected that work might be far more extensive than imagined, based on much earlier experience with the conversion and restoration of DDG 31. Hull plating and structural members, including keel sections, had been replaced. Surely the 22-year old LSDs would require similarly expensive work. Board of Inspection and Survey (INSURV) reports showed that routine overhauls would cost as much as the 1976 estimate of a full SLEP. Moreover, the ships had some important inherent defects, such as the use of ferrous (corrodable) metals in their firemains. They could not fight fires while maintaining their packaged air conditioning. Habitability was poor and noise levels hazardous. Medical facilities were inadequate. Even without rewiring or any other upgrade a SLEP might cost as much as $110 million. The upgrade would entail extensive ventilation, heat dissipation, and maintenance facilities. NAVSEA wanted 2½ years merely to plan a 24-month SLEP. Without any SLEP, the ships needed a very thorough overhaul just to reach their nominal 30-year lives.

Moreover, whereas an LX could accommodate four LCACs, an LSD 28 could not, no matter how extensively modernized. The SLEP would delay bloc obsolescence, but ultimately it would aggravate that problem. By the early 1980s the shipbuilding program was much larger, and there was interest in accelerating the retirement of these ships rather than modernizing them. The Marines protested; replacements were not yet ready. In the end, the ships were retired as new LSD 41s entered service. Two were transferred to Brazil.

Because funds were tight, plans by 1976 called for stretching out the LX (LSD 41) program, building one each in FY 79, FY 81, and FY 82, then two each in FY 83–85. The first ship would enter service early in FY 84, as LSD 28 SLEPs began. Then the program slipped further. When the FY 79 ship was given up, the CNO accepted a cut to 65 amphibious ships in FY 78, in hopes that a force level of 66 could be maintained in subsequent years. LSD 28s would be replaced by LX on a one-for-one basis: one in FY 81, three in FY 82, two in FY 83, and one in FY 84. As of FY 78 lift would be only slightly greater than that required for a single MAF; it would not reach the desired one and one-third MAF until FY 81, thanks to LHA deliveries. LX would solve the lack of helicopter spots, and by FY 87 this program would provide one and one-half MAF.

An operational requirement (OR) for the new LSD was issued in November 1976; preliminary design began in January 1977. LSD 41 would be an upgraded LSD 36 rather than a completely new design. Marine numbers were set at 300–400, rather than the 325–550 considered for LX. Cargo capacity was set at 12,000 sq ft and 5,000–10,000 cu ft, well below figures set for LX; LX figures, in the end, called for a total of 31,250 sq ft. The new LSD would operate at least one helicopter (actually, the goal of two was met), but she would not stow any; she would also be able to land and fuel a VSTOL aircraft. LX had incorporated helicopter or VSTOL stowage. Speed was set at 22 kts for the new LSD; LX had envisaged a speed as high as 30 kts. Draft was set at 18 (rather than 27) ft.

LCAC facilities dominated the design. To provide sufficient overhead clearance, the mezzanine deck was deleted, with the well deck permanently covered to provide a helicopter deck. In earlier LSDs, vehicles had been stored mainly on the mezzanine deck, so additional deck space had to be found. Vehicles could be stowed in a sheltered area on deck that would otherwise be used as a second helicopter spot. The enclosed well decks of earlier LSDs, moreover, provided insufficient ventilation for LCAC gas turbines. The new LSD therefore had a redesigned well deck. The well incorporated an overhaul crane, as well as a redesigned well-deck entrance (a "ferry chute") to ease launch and recovery of LCACs. An LCAC capture system, employing guidance lights, was successfully tested using Jeff B. As in earlier LSDs, the well deck had a water barrier to permit LCAC operation at the after end while dry cargo was carried forward. To support LCACs, the ship had intermediate maintenance shops. Using the same acquisition manager (NAVSEA's PMS 377) as the LCACs would ensure compatibility.

A striking change was the abandonment of 3-in guns. LSD 36 had four twin mounts; the new LSD would be limited to two Phalanx close-in weapons and two single 20-mm. The CNO had just formally decided that in the future the 3-in/50 would be considered only as a surface weapon, and the LSD had no gunfire support role. Other amphibious ships deactivated their antiaircraft fire control systems. Providing the new LSD with 5-in or even 8-in guns for fire support was briefly considered.

In December 1976 it was estimated that the requirements for the new LSD could be met by a 540 (between perpendiculars) × 84 × 18 ft (8,100 tons light) ship, about the size of LSD 36. The new ship was characterized as a modified repeat version of the earlier ship. The 436-ft well of the earlier LSD 36 was 4 ft too short for the fourth LCAC; moreover, its forward part necked down to 24 ft. Squaring the well and extending it added 16 ft. In earlier LSDs, berthing spaces were in the wing walls. LCACs in the well deck would generate such noise and heat that this was no longer acceptable. The new LSD was therefore rearranged, with troop berthing forward of the well, thus lengthening the ship again, to at least 560 ft. In peacetime, Marines on board would live as well as the crew.

In November 1977 the ship grew another 20 ft, to 580 × 84 × 18 ft (11,006 tons). The other 20 ft of growth provided a second helicopter spot that could be used for the required storm-protected vehicle square footage, and a fixed drive-through ramp from the upper deck to the craft in the well deck. There was also a turntable, so that vehicles could be driven on board and then turned around to go onto the LCACs. Overall, the new LSD was expected to cost $73 million more than LSD 36; displacement was 2,542 tons greater at that stage.

Cargo capacity grew from 11,983 sq ft and 2,000 cu ft to 12,800 sq ft and 5,000 cu ft. The ship could accommodate 338 marines (vs. 300 in LSD 36); under surge conditions 440 could be carried. LSD 36 had no such accommodation margin. Capacity would be 64 amphibians, compared to 39–44 in LSD 36. The ship was given 60-ton cargo cranes specifically so that she could hoist main battle tanks into and out of LCACs. She was also given hard points and space and weight provision to fit a temporary mezzanine deck.

Deletions, to save cost, included flag facilities in some ships, a helicopter hangar, provision to transfer jet fuel to helicopters, provision for airborne mine countermeasures as a secondary function, and MUTE (multiple unit for transmission elimination), a means of controlling all the ship's electronic emissions. LSD 41 did retain her air search radar and her 22-kt speed.

In July 1977 the Ship Acquisition and Improvement Panel (SAIP), in effect equivalent to the earlier SCB, approved a switch to diesel propulsion, which it reaffirmed in October 1978. Diesels would use only two-thirds as much fuel as steam or gas turbines. Power was upgraded to 33,000 (later 36,000) BHP; two medium-speed diesels on each shaft would drive controllable reversible-pitch propellers. The diesel chosen, the French-designed, U.S.-made Pielstick 16PC2.5V400, was described as a world leader, the manufacturer having accumulated over 40 million operating hours. Because the machinery was new, a land-based test site had to be built at the Naval Ship Engineering Station (NAVSES) in Philadelphia. Work began in March 1979.

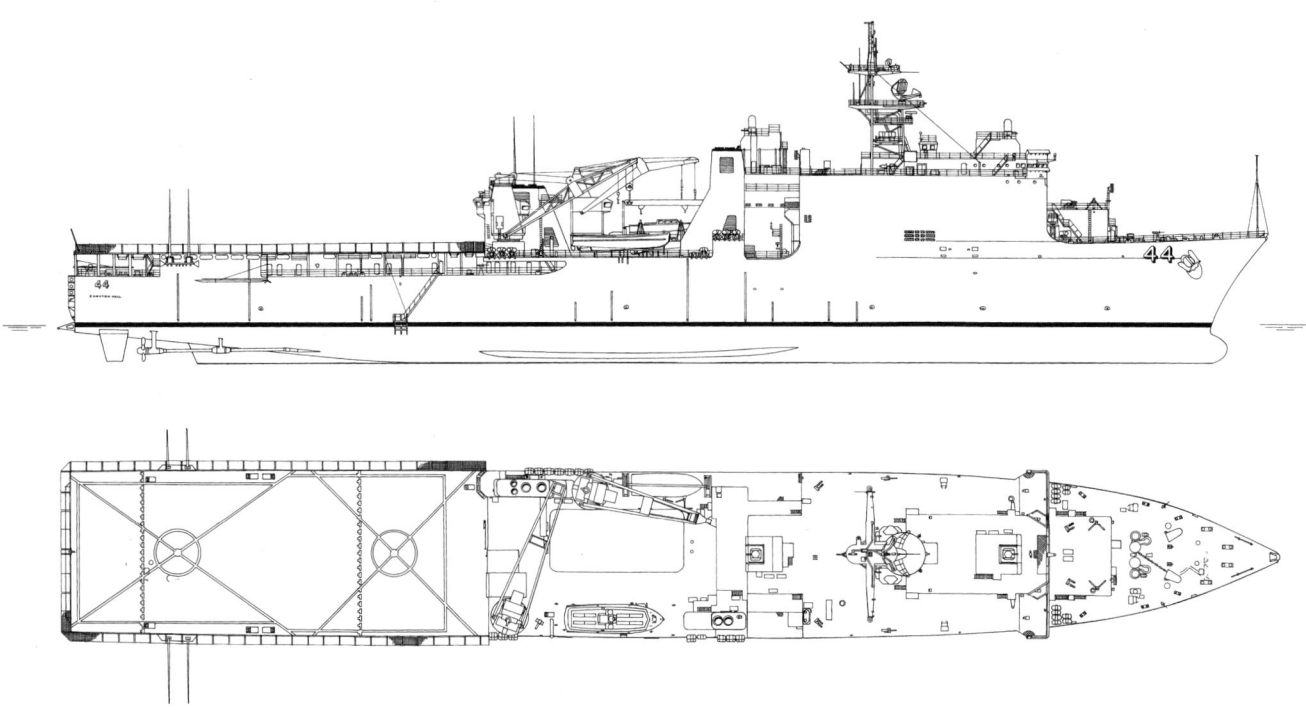

Gunston Hall in September 1991, after the Maintenance Enclosure modification to the Phalanx CIWS installations had been made. *A. D. Baker*

The preliminary design was completed in March 1978. In October 1978 Lockheed Shipbuilding was chosen as the likely lead yard, and in November the SAIP approved characteristics reflecting the existing design. Contract design was completed in July 1979.

Money was tight, so for FY 80 (January 1979) the Defense Department eliminated the ship from its Five Year Defense Plan. Secretary of Defense Harold Brown stated that he needed time to study compatibility with the LCAC, LCM(9), and future amphibious tracked vehicles. LSD 41 was then the only amphibious ship in the five-year program. Critics within the navy pointed out that the ship had been designed specifically for such compatibility. Without LSD 41, the fleet would fall below one MAF lift (counting LCAC spots). The first six ships would replace the lift (in terms of LCACs) of the retiring LSD 28s.

Brown may have been referring to a new CNA study, AMFOR (Amphibious Forces). AMFOR examined the type of fleet that might have been built around the LCAC. Work on this project began in 1978. Like NAVWAG 44, it suggested new kinds of amphibious ships, although in this case none was adopted. However, the CNA proposals are interesting because they illuminate the issues the amphibious force faced at the end of the 1970s. CNA asked whether it was really logical to keep developing the existing classes, particularly the aging LSD, when the amphibious force was shifting to a combination of helicopters and LCACs. The analysts sought to reduce the number of ships required to lift a given marine amphibious unit. To that end they proposed two notional classes: LXA and LXS, assault and support amphibious ships. To some extent the proposed design characteristics reflected criticism of existing ships and designs. As in the earlier study, NAVSEA provided sketch designs to test the feasibility of CNA's ideas.

LXA would carry a fixed fraction of each part of the Marine assault fingerprint (command/control, cube, square, helicopter, boat, and LCAC spots). Thus it would make for standardization within, for example, an amphibious ready group. An LXA would launch and retrieve advanced landing craft (such as LCACs) while under way. LCACs could operate from a dry well, but the ship could ballast down to launch conventional landing craft; ballasted down, the LXA would have 10 ft of water over the sill at her stern, and at least 6 ft at the fore end of the well for LCMs. Stabilization would simplify retrieval of assault craft and reloading in a seaway. The ship would have flight deck level aircraft stowage for helicopters or VSTOLs.

Making the well deck slightly longer than required for the assault craft would provide an adequate vehicle and cargo staging area for loading. As in the LSD, vehicles would load via ramps. The ship would carry POL (petroleum-oil-lubricants) in her tanks (with risers to the well deck, to fill portable tanks) and in drums. To the extent possible, an overhead monorail would be used to place palleted cargo on board the craft. Like other fast amphibious ships, LXA would have a sustained speed of 20 kts (using a twin-screw split plant) and an endurance of 10,000 nm. She would be able to transit the Panama Canal. Armament would be limited to self-defense.

The ship was sized to carry a quarter of an MAU, thus an MAB would require 12 ships and an MAF would require 36 ships. NAVSEA offered a 725 (waterline) × 106 × 21.5 ft hull (27,400 tons). It could be driven at 20 kts by 40,000 SHP diesels, the same type as planned for LSD 41. A sketch showed a forecastle incorporating a hangar, with a flight deck aft above a dry/wet well. Alternatively, the superstructure could be moved to one side to clear a 450-ft flight deck, or a 300-ft ski-jump could be angled forward of the superstructure to hangar six AV-8B VSTOLs. In the baseline design, atop the forecastle were two 5-in/54 lightweight guns, side by side; other armament was a pair of Phalanx and a pair of Sea Sparrow systems. The design also showed a well deck accommodating two LCAC side by side (hence the wide beam, because that amounted to 101 ft width) and a third in a narrower section forward of them. The hangar was sized for ten CH-46 helicopters or their equivalent. Marine squadrons were normally 12 aircraft, so it might be necessary to enlarge the ship. Other elements of the quarter MAU were 1,000 marines, 21,000 sq ft of cargo, 41,000 cu ft, and 190,000 gal of fuel (including 124,000 gal of JP-5 and 20,600 gallons of mogas).

Just as the World War II APA could not carry enough boats, LXA could not carry enough LCACs. CNA proposed a supplementary LCAC carrier, the LXS, carrying four in a well deck and six on a weather deck, all preloaded with armored vehicles. LXA would carry their crews, a total of 600 men. CNA noted that if LCAC cushion pressure could be increased, the craft might be made small enough so that more would fit the LXA well deck, in which case LXS would not be needed.

An August 1978 sketch of an LXS showed an island superstructure with an LCAC elevator abeam it (on the centerline) serving a well deck aft (the elevator being at its forward end). Above the well, whose forward end was open to release exhaust fumes, was a raised helicopter platform. Uptakes were in the island and on the other side of the flat deck. A 60-ton crane, as in the new LSD, was placed on deck abaft the island. Most hotel functions were in the hull. The well deck would accommodate four LCAC (one on the elevator); another six would ride the deck forward of the superstructure (three abreast, two pairs being on sponsons). The concept exploited the fact that LCACs could move about a ship's decks under their own power, albeit with only limited control because wind could blow them sideways. In an emergency, the ship could trim by the head to launch LCACs directly. The elevator could be replaced with a 400-ft ramp.

NAVSEA offered a 770 × 94 × 21 ft hull (24,100 tons) that an LSD 41 powerplant could drive at 20 kts. The ship would carry a limited amount of cargo: 5,000 cu feet of general cargo, 10,000 cu ft of aeronautical materiel, 33,000 gal of JP-5, 1,000 gal of mogas, and 36,000 gal of potable water. If the well were flooded, it could accommodate 18 LCM(6).

CNA also asked whether existing commercial ships might be used in an emergency to augment specialty built amphibious types. Commercial barge carriers, distant relatives of the Popper ferry conceived in the 1920s, were in service, under the designations LASH (lighter aboard ship) and Seabee (sea barge). LASH stacks her barges using a large traveling gantry crane. Seabee has decks, the uppermost well clear of the water. Barges are placed on a large elevator, which lifts them to the decks, onto which they are winched into place. Both had been proposed as early as 1974, in a study called "Plan 2000," as cheaper alternatives to LSDs and LHDs. These merchant ships, nearly new in 1980, were later adapted to the needs of the prepositioning force (see Appendix A).

Just as the British director of naval construction (DNC) had argued in 1940, NAVSEA argued that the barge decks of such ships were open to the sea; the barges normally carried completely filled it, so there was no real flow of sea water. LCACs being

Germantown (LSD 42) was one of the new *Whidbey Island*–class LSDs, designed specifically to accommodate LCACs. *Lockheed Shipbuilding and Construction Company*

Whidbey Island, the prototype of the new LSD series, is seen from aft. *Lockheed Shipbuilding and Construction Company*

carried instead of barges might suffer sea damage. It might be possible to modify such ships quickly using vans built to standard container size; the Maritime Commission had been developing a portable 8 × 8 × 20 ft accommodation container. A similar container might be used for naval communications. NAVSEA considered an adaptation in which the port side of the well deck carried three

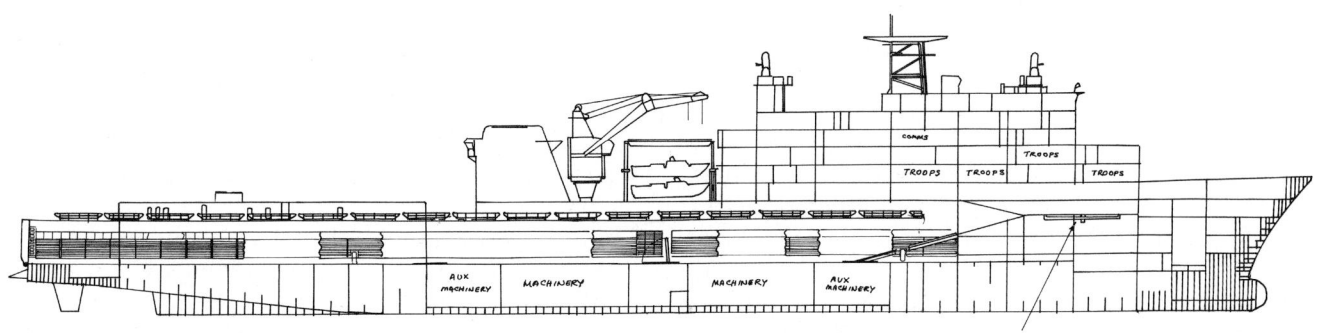

Whidbey Island (LSD 41), June 1998. Her long wet well, in which are indicated wooden battens, is shown. Note the turntable at its forward end, just forward of the ramp, and the erectable water barrier built into the wet well.

LCACs, with two accommodation barges (YRBM 1s) welded down forward of them. The starboard side (two decks high) would be decked over and the enclosed space used for vehicles, ammunition, etc. The forward portion could be made watertight to accommodate auxiliary machinery. Two CH-46s could park on top, with one take-off spot. The communications van would be set down at the fore end of the flight deck.

When the Carter administration decided to rearm in 1980, it included 6 LSD 41 in its FY 82 program, then another 6 in FY 85 and 12 in FY 86. At this time six ships were considered an MAU lift, so the program provided lift for four MAU, a one and one-third MAB. After entering office in 1981, the Reagan administration ordered a major fleet expansion, including amphibious lift. Indeed, expanded lift was central to its maritime strategy program. The goal rose from one MAF to one MAF plus one MAB, which in the past would have been characterized as one and one-third MAB. The proposed Carter program could not be enacted, however. President Carter's rearmament budget had omitted many other essential items, such as new aircraft carriers. Because it was buying carriers, the new administration could not buy nearly as many LSD 41s.

The new administration bought the first ship, *Whidbey Island*, in an amended FY 81 budget, against a planned total of 12. A presidential decision memorandum cut the FY 82 program to two ships, which would not suffice to open the desired production line. An alternative was to buy three each in FY 82 and FY 83; early in 1981 Congress seemed willing to provide long-lead items for as many as six. The administration actually bought one ship in FY 82. It hoped for two per year through FY 87, but had to cut that to single ships in FY 82–84. In December 1982 it was decided to buy five ships on a multiyear basis: one in FY 84, and two each in FY 85 and FY 86. The expectation was that the multiyear winner would be able to invest sufficiently to drive down the cost per ship. Estimated savings, as of April 1983, was $246 million, against a five-ship cost of $1.76 billion.

In October 1981, with LSD 41 procurement in question, the Defense Department's program assessment and evaluation (PA&E) office proposed a 38,000-ton alternative, which it presented as a modified version of the multi-purpose mobilization ship (TAKX) then being developed for the maritime prepositioning squadron (MPS; see Appendix A). PA&E's LKD-X seems to have emerged from the "Assault 90" study then being conducted by CNA.[6] It would be built largely to commercial standards. This approach was adopted in the mid-1990s by the Royal Navy in a new generation of large amphibious ships, beginning with the helicopter carrier HMS *Ocean*. In the U.S. proposal, some military standards would be met: materials would be fire retardant, bulkheads and decks would have increased tensile strength, watertight standards would exceed those of merchant ships, wiring would be of higher quality, and there would be more water pressure for fire fighting.

PA&E argued that the ship would be useful both in amphibious operations and for the new MPS. It offered a well deck comparable to LSD 41, a helicopter deck comparable to that of the LPH (20 spots), and almost as many troops as an LHA (1,580), yet it would cost less than an LSD 41. It offered the cubic footage (341,000 cu ft) then badly needed in U.S. amphibious shipping, plus 66,000 sq ft. Estimated average cost in FY 82 dollars was $286 million. On this basis PA&E suggested a program of five ships.

Neither the U.S. Marines nor the U.S. Navy were impressed. Commercial ships did not normally offer the sorts of facilities envisaged, so it was not clear how commercial shipbuilding standards would apply. The Royal Navy found out much the same thing when it built HMS *Ocean*. LKD-X would stall a perfectly good program for LSD 41s and it might also abort a projected program for an LHA follow-on, the LHD. As for cubic footage, the Marines could always charter ships to provide them with more cargo capacity. Because each of five LKD-X would have no more than the well-deck capacity of an LSD 41, the Marines would find themselves seriously short of LCACs, which they greatly valued. The PA&E proposal died in 1983, when multiyear procurement of LSD 41s was approved.

The question in 1983 was what to build after FY 87, the end of the approved multiyear LSD 41 program. In May 1983 the navy completed a new long-term lift requirement and optimum ship mix study. Helicopter and LCAC numbers were set by a requirement for timeliness. Neither should have to make more than two sorties. For helicopters, that meant 90 minutes of flight time over a 50-nm range, a standing requirement.

The study raised an interesting question. For purposes of analysis, it assumed that all the LVTs would be launched from LCACs, each of which could carry three of them. Conversely, if the Marines did buy a fast-swimming LVT, was it still so vital to maintain the 90 LCAC force required to bring LVTs close to the beach? Given the high cost of new amphibious ships and LCACs, and the desire for the new V-22 VTOL troop carrier, it was not at all clear that the Marines should invest in a very expensive new LVT.

By the early 1980s the main role of the amphibious forces was to place Soviet-occupied territory at risk in the event of a major war. An amphibious task force would have to overrun a Soviet motorized rifle division conducting a mobile defense over a 200-km zone, with one of its regiments at the most likely landing spot. The other two would be held inland as a reserve. The key to U.S. success would be the delay in their deployment to the landing area. OTH tactics would make it difficult for the Soviets to guess the landing site, but it would be essential to build up rapidly once the assault had begun, that is, once the site had been revealed. Helicopters could deliver marines to blocking positions before the first enemy regiment could arrive. Once the enemy appeared, however, the marines would need heavy vehicles and firepower that could only come over a beach, which they would have to seize. They could not, therefore, abandon the ability to make a surface assault, or at the least to land heavy equipment over an unimproved beach. That meant LCACs and amphibians were needed.

In 1981–82 the navy and Marines cut the requirement for specialized amphibious shipping, particularly for cubic footage, by splitting the assault echelon into an assault element (AE) and two other elements: a fly-in element (FIE) and an assault follow-on element (AFOE). The AFOE would rely on strategic sealift ships, some of which would normally be used for prepositioning (see Appendix A). With new technology, conventional sealift ships could unload at an unimproved beach.

The idea of a follow-on was not new, but this time the assault echelon was split. The assault element accounted for 20 percent of troops, 15 percent of the vehicle square footage, and 1 percent of the cargo cubic footage. It would be delivered by LCAC, AAV, and helicopter. It was expected to survive for the first 15 days; the entire assault echelon had to be ashore by D+15. Between May 1981 and 1983 the Marines cut 7 percent in troops, 18 percent in square footage, and 35 percent in cubic footage (see Table 16-1).

However, by 1983 it was unpleasantly clear that Marine modernization was increasing the amount of equipment that had to be embarked, as well as the unit size of that equipment. Because the number of amphibious ships was not growing, they had to be loaded more efficiently, thus they would be less tactically efficient. A 1983 navy internal memorandum commented that such "compromises to amphibious loading doctrine . . . could place landing force at a disadvantage, should it be placed in an unanticipated combat situation."

Other issues also had to be resolved. Existing VHF and UHF radio systems were line-of-sight, which by definition could not work effectively beyond the horizon. Airborne relays were needed, which meant additional helicopters at that time. For LCACs operating beyond the horizon, the Marines adapted the army's position location reporting system (PLRS) and coupled it to the new Global Positioning Satellite (GPS) system. Overall, it seemed likely that the command load would increase as the area over which the operation was conducted expanded. Yet the two LCCs were the only command ships available for operations beyond MAU size; even by the early 1980s, both had been diverted to fleet flagship duties. As before, fire support was a growing problem. Only the laid-up battleships offered anything close to the necessary range.

In 1983 plans for near-term amphibious construction through 1994 (the end of the planning period) called for a total of 14 well-deck ships, of which eight LSD 41s were already under construction. With all eight in service in 1992, there would be a shortfall of 75,000 sq ft, 237,000 cu ft of cargo capacity, and 12 LCAC spots because the 8 LSDs would only provide a total of 90,000 sq ft, 40,000 cu ft, and 32 LCAC spots. These figures took account of the five LHD (discussed below) then planned. Compared to the LSD 41s, the next six well-deck ships would need considerably more cargo capacity, but each could

Table 16-1. Reductions in AFOE Amphibious Footprint, 1981–83

	Troops	Square Footage (thousands)	Cubic Footage (thousands)	Helicopters
1981 MMROP[a]	53,700	1,272	3,850	749
4 November 1982 (revision)	48,400	1,234	3,474	637
1983 (revision)	46,800	1,040	2,440	637
1990 (1983 projection)	50,000	1,040	2,490	633

[a]MMROP was the Marines' mid-range objectives projection.

accommodate two fewer LCACs. At this time the six remaining ships were sometimes described as LPDX, and there was some question as to whether ten LSD and four LPDX, or eight LSD and six LPDX, would be built.

Earlier amphibious ships would retire at age 35, but the LPD 4 class was to be refitted under a SLEP, beginning in FY 87, to last a total of 45 years. The 12–15 month SLEP would have provided new generators and new deballast air compressors as well as Phalanx, SRBOC (super-rapid blooming offboard chaff) launchers, and the new SLQ-32 electronic countermeasures (ECM) system for self-defense. It doubled the number of helicopter spots (from two to four), but it could not, as hoped, double the number of LCAC the ship could accommodate. Certainly it could not solve the cargo capacity problem. In programmatic terms, the LPD SLEP deferred the follow-on well-deck ship. It in turn was canceled due to the budget problems of the early 1990s, although the Phalanx and electronic warfare systems were updated as planned. The designated replacements for these ships were the 12 *San Antonio*–class LPDs (LPD 17 class) described in Chapter 17.

As for the new LSD, plans called for somewhat stiffer standards than its predecessors. Now that Soviet chemical warfare capability was being taken very seriously, she would have a collective protective system (CPS). The tentative operational requirement (TOR) showed 20,000–40,000 sq ft (minimum acceptable 14,200) and 60,000 cu ft (minimum 40,000). There was some interest in increasing endurance from 8,000 to 10,000 nm. Troop capacity might be increased from 504 to 600, and POL capacity might be slightly increased. Added cargo capacity would be obtained by cutting off the well deck, limiting capacity to two LCACs.

Three alternatives were considered: minimum modification to LSD 41, using the fore end of the well deck for vehicles and cargo; modifying the superstructure, machinery, and interior of LSD 41; or a new design to meet the high end of the TOR requirement. NAVSEA claimed that minimum modification, the simplest solution, was not feasible because the ship as delivered exceeded the minimum stability standard. However, she might be viable if she was made 3 ft beamier or if her load was reduced to 14,200 sq ft and 60,000 cu ft. In May 1984 the Ship Characteristics and Improvement Board (SCIB) approved the second alternative with the reduced load, but allowed any change to the superstructure, machinery arrangement, or interior of the hull as long as the external hull form was maintained. Cost savers included a reduction in LCAC maintenance and in the level of CPS protection. As completed, the ships had greater air conditioning capacity than the LSD 41s.

The class was initially called the "LSD 41 cargo variant." Twelve were planned, but the program was terminated in 1992 as part of the post–Cold War wind-down. Four were built (one each in FY 88–92), beginning with LSD 49 (*Harpers Ferry*) under the FY 88 program. The Bush administration rescinded money for LSD 52, but Congress appropriated it again under the FY 93 program, and the ship was built.

While the LSD saga was unfolding, the LHA was redesigned as the LCAC-adapted LHD. The new designation, with D (for dock) replacing A (for assault), seems to have been chosen to avoid any implication that the navy was buying more LHAs; LHD was clearly an entirely new kind of ship. It is not clear why a new designation was needed. An LHDX, capable of accommodating three LCAC, appeared in a planning document as early as May 1979. By June 1981 LHDX was defined as capable of launching

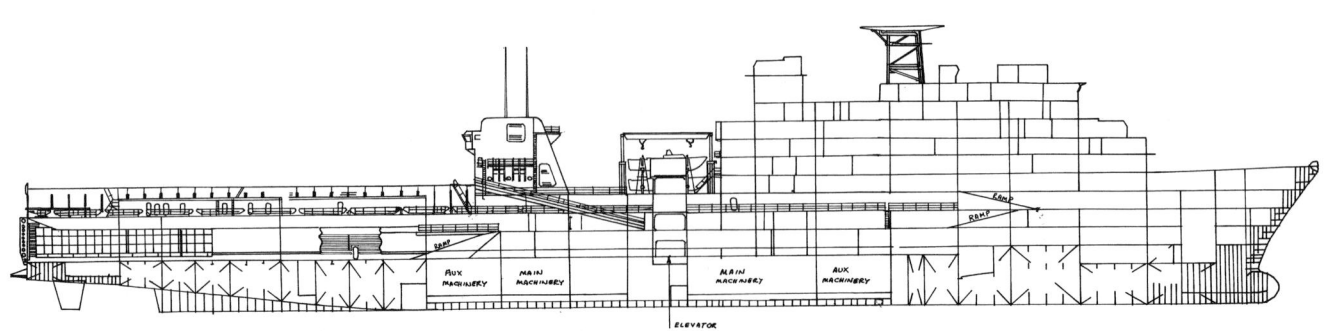

Harpers Ferry (LSD 49), September 1999. She was the modified version of LSD 41, with more cargo space forward of an abbreviated well deck.

preloaded assault craft, helicopters, and VSTOLs, with magazines to support AV-8Bs (whose ordnance capacity was an order of magnitude greater than that of the AV-8A). Compared to an LHA, it would trade cargo cubic footage (60,000 vs. 116,700) for enough well deck to accommodate three rather than one LCAC. It would also carry slightly fewer marines (1,810 vs. 1,903 as planned, but actual numbers are 1,893 vs. 2,075).

Hangar and flight deck facilities would be roughly the same, for 38 CH-46 spots (as built, the LHD had 42 CH-46 spots compared to 43 for an LHA). The aircraft mix, however, would be quite different. The LHD would accommodate 20 AV-8B and 14 CH-46 or 5 CH-53E in carrier mode, but 30–32 CH-46 and 6–8 AV-8B in assault mode. The LHA can accommodate 6 AV-8B, 4 CH-53E, 4 UH-1N, and 16 CH-46. In the LHD, vans would be suspended in the hangar bay for AV-8B maintenance. As modified, the LHAs could also operate AV-8Bs. USS *Nassau* (LHA 4) became the first to operate fixed-wing aircraft in combat, flying off AV-8Bs to attack Iraqi targets during the Gulf War.

The LHD was conceived while some congressional sentiment still favored the VSS (VSTOL support ship). The new Reagan administration had no interest in such ships; it much preferred large-deck carriers. However, the LHD could certainly be designed to be adaptable to a sea control (i.e., VSS) mission, a very reasonable compromise that effectively defused the issue. Operating as a sea control ship, the LHD would carry 20 AV-8B and 4–6 SH-60B LAMPS (light airborne multi-purpose system) helicopters, with van support. Because the ship lacked the specialized LAMPS data link, however, the helicopters would have to transmit their data to other ships. Mission conversion was expected to take 15–30 days. Additional vans and spares would be placed in the upper vehicle stowage space. Ammunition would replace Marine cargo. If the hangar deck were tightly packed, the ship could accommodate up to 28 AV-8Bs.

There was some question as to whether the ship carried sufficient aviation fuel (400,000 gal of JP-5). That would last the AV-8Bs no more than two days. Ship fuel tanks could not be used as "swing tanks" because they had no risers to the flight deck via JP-5 pump room filtering stations.

The British were demonstrating ski-jumps as a way of improving the performance of their Sea Harriers, which are equivalent to the U.S. AV-8s. A 65,000-lb aluminum ski-jump could be built and installed by a U.S. shipyard. Op-05 (DCNO for air) suggested an angled deck, but that was rejected for fear that the ship would look too much like a small carrier, and thus would divert attention from its primary role of amphibious assault.

The Marines' goal was to lift the assault elements of two MAF and one MAB, for which purpose, about 1984, they wanted at least eight LHDs. According to some accounts, 11 were planned. The LHDs and LHAs would form the cores of amphibious ready groups, which would parallel the 16 carrier groups.

By about 1981, plans called for 5 LHDs to work with the other 11 amphibious carriers (LPH and LHA). Advanced ship development funds were provided under the FY 81 program, and ship development funds and long-lead items in FY 82, along with plans calling for the first ship to be programmed in FY 83. That proved impossible. Ingalls, which had built the very similar LHAs, was clearly the favored builder and, indeed, the one chosen. However, it was already building high-priority Aegis cruisers. The yard could not build LHDs at all if it remained the sole source of Aegis cruisers. If a second yard began building the cruisers, it might be possible to order a prototype under the FY 83 program, then two in FY 86, and one each in FY 87 and FY 88. At this point the design included a ski-jump for the AV-8Bs and a package of survivability features. Soon the planned program start was pushed back to FY 87. By early 1984, however, it had been accelerated back to FY 84, with the first contract award scheduled for 17 January 1984. The lead ship, *Wasp* (LHD 1), was included in the FY 84 program, followed by single ships in FY 86, FY 88, FY 89, FY 91, FY 93, and FY 95. Some of these ships replaced LPHs. As of mid-2000, plans call for building LHD 8 as a transitional design powered by gas turbines instead of steam, and then for the introduction of an entirely new design using the LHD 8 plant.

At the end of the 1970s the Pacific Fleet was particularly interested in amphibious assault. Its commander, Adm. Thomas Hayward, saw his wartime role as tying down Soviet forces to keep them from being shifted to Europe: as virtual attrition. To do that he needed the ability to make, or at least to threaten, opposed assaults. In mid-1980 the Pacific Fleet stated an operational requirement for the associated fire support ship. Probably as a result, in August 1980 NAVSEA requested proposals for shore bombardment weapons capable of striking targets 35 km inland on a day/night, all-weather basis, using a ship's own sensors or third-party targeting. A fire support ship would be able to engage two separate targets simultaneously, in seas up to Sea State 3. Among the proposals selected for further study was a naval version of Vought's army MLRS 9-inch

The LHD complemented the new LSD and the projected new LPD; it was an LCAC-adapted evolution of the LHA of the 1970s. The first unit, *Wasp,* is shown on the Ingalls transporter at Pascagoula prior to launch, 30 July 1987. *Ingalls Shipbuilding Corporation*

rocket launcher; the rockets would carry bomblets. MLRS, at that time called MARS (multiple army rocket system) had been rejected for the naval fire support role in 1970. Until this time the Marines had stated their fire support requirements in terms of ships assigned to various sectors of an assault. With far too few bombardment ships, in November 1980 the Marines agreed to substitute an inventory of targets and required levels of destruction.

When Hayward came to Washington as CNO, he brought the new emphasis on assault with him. In July 1981 CNA was ordered to conduct an amphibious fire support study (AMFIST). This time the scenarios were Jutland (in the face of a Soviet motorized rifle division), North Korea (without Soviet intervention), and Southwest Asia, an area of increasing importance. Other standard Marine Corps' scenarios of this period were an East Mediterranean landing (Sinai) and a Pacific choke point landing (Sunda Straits). The last major studies of this type, for both the MAF and the MAB, had been conducted by the Naval Weapons Laboratory in July 1972.

The postulated LFS weapons were White Oak's 8-in/36 NUGM (see Chapter 14), a vertically loaded 155-mm gun deemed attractive because it could use new army ammunition, such as the laser-guided Copperhead shell and bomblet carriers, and the Vought 9-in rocket that could be fired from an ASROC launcher. To these CNA added the new

USS *Essex* (LHD 2) is shown, newly completed, in October 1992. *Ingalls Shipbuilding Corporation*

Tomahawk cruise missile and a bomblet-carrying rocket missile, Martin Marietta's Beachcomber, which was based on a Patriot surface-to-air missile body developed for the army's assault breaker program. By this time, early in the 1980s, remotely piloted vehicles promised effective targeting support; *New Jersey* had used a variation on this theme, the "Snoopy" version of the DASH ASW drone, off Vietnam. The LFS had to be compared in effectiveness with strategic bombers and tactical aircraft, each of which would have been armed with precision-guided as well as unguided bombs.

Fire had to reach as far as possible inland, but as yet gun range was limited. In order to strike 10–12 miles inland, then, CNA assumed that the LFS would stand only two miles off a beach. There she would attract enemy fire. Her magazines would be protected by HY80 steel, because the HY100 mentioned in earlier studies had never reached the producibility stage. Her CIC and other vitals would be buried in her hull. She would be all-steel, hardened against 7 psi nuclear overpressure. The standard of 10 psi for combatants was not completely met.

In support of the study, NAVSEA once again compared a new-construction LFS with conversions of the LST 1179 and *Knox* classes. An LST conversion was attractive because the LST itself was becoming obsolete as the Marines adopted a combination of vertical envelopment and well deck–launched LCACs to move heavy vehicles. Moreover, the LST had to come far too close to a dangerous shore in order to unload. Her elaborate bow ramp experienced continuing maintenance problems. Up to 20 LST 1179s were, then, almost instantly available. The simplest LST conversion maintained most of the ship's amphibious capability, providing one 8-in/36 (1,000 rounds) and one 9-in rocket launcher (600 rounds), with Mk 86 or Mk 92 fire control. Alternatively, in the second conversion option the bow ramp could be eliminated and the between-decks ramp moved to the cargo hatch. The area forward of the deckhouse was freed for a second 8-in gun and a CIWS for self-defense. Additional ammunition might be stowed in tank deck containers. This ship could still deliver pontoon causeways and unload LVTs and other traffic over her after ramp. Eliminating all amphibious capability, the third conversion option, added 500 rounds per 8-in gun.

In the simplest *Knox*-class LFS conversion, the 5-in/54 gun was replaced by a 155-mm vertically loaded gun with 600 rounds, and a digital (hence more adaptable) Mk 86 or Mk 92 fire control system replaced the earlier analog Mk 68. Alternatively, in the second conversion option an 8-in/36 with 500 rounds could replace the 5-in gun; in the third option it could be replaced by a 9-in rocket launcher with 350 rounds. Nothing more could be done, because it was important that the *Knox* class retain the ASW capability represented by ASROC, which occupied the only other available launcher position.

By the spring of 1982, the two congressional military committees had earmarked FY 83 funds for the navy's moribund 5-in semi-active laser-guided shell (SAL-GP), a shore bombardment weapon. The navy was asking for money to develop a vertically loaded 155-mm gun, although it remained a paper study.

In June 1982, presumably in the expectation that the new amphibious fire support study (AMFIST) would soon be completed, the Office of the Secretary of Defense formally asked the navy to determine its naval gunfire support requirements, and to report alternatives, by that September. That required an evaluation of frigate-size (4,000–5,000-ton) ships armed with lightweight 8-in and 155-mm guns and with high-volume rocket systems. New construction seemed likely. Completed in November 1982, AMFIST concluded that guns would have to reach targets at ranges of 40–50 km, beyond what shipboard 5-in guns could reach. CNA liked both a 16-in gun firing improved ammunition and a new medium-range land attack missile. Recommissioned battleships would be more useful than small new bombardment ships carrying shorter-range guns.

Through early 1983 it seemed likely that the vertically loaded 155-mm gun would be developed. Compared to a 5-in gun, it would have fired a much heavier shell. Thus it was estimated that it would take half as many such shells to kill prone troops or a bunker, and many fewer to kill a typical Soviet destroyer—although numbers to kill an air defense battery (SA-4) were about the same. There was some expectation, too, that the 155-mm gun could be adapted to other roles, such as ASW (e.g., to fire sonobuoys and hydrostatically fused shells, which are, in effect, depth bombs). Because of its unusual loading method, the new gun could fire long guided shells as quickly as conventional ones, thus the already low rate of fire of the navy's standard 5-in gun would be halved. Greater shell volume would translate into greater range, 42 km compared to 30 km for the 5-in guided shell, but about the same range, 27 km, for an unguided shell. The new gun might be installed on board existing destroyers, but more likely it, like the earlier MCLWG, would justify construction or conversion of a special-purpose

LFS. Both congressional military committees clearly favored the gun, presumably at the Marines' behest. A Naval Materiel Command (NAVMAT) study recommended that it replace the standard 5-in gun from FY 85 onward, and the gun figured in the POM-85 appraisal of amphibious requirements.

On the other hand, any new start would be expensive. In the early 1980s the navy budget was badly stretched to accommodate many more urgent items, such as improved air defense. Arming different ships in the same destroyer or cruiser class with different guns would cause logistic problems. There was some fear that army ammunition would be unsuitable for shipboard use, because its electronic fusing might be vulnerable to shipboard electromagnetic fields (HERO; hazards of electromagnetic radiation to ordnance). Moreover, the rate of fire was somewhat lower (10 rounds per minute vs. 15 for a conventional 5-in shell). The army's existing Copperhead guided antitank shell would not be very useful to the navy, so further money would have to be spent to develop a new guided shell. Nor could the new gun use the army's standard 155-mm barrel, because the navy needed greater range.

In December 1983 the navy formally rejected the 155-mm vertically loaded gun on the grounds that it offered little real benefit for a life cycle cost of $130 million (in FY 81 terms) greater than that of the standard 5-in/54 aboard destroyers. Without the new larger-caliber gun, there was no point in a new LFS. It would be enough to reactivate the four battleships and to provide the 5-in destroyer gun with a laser-guided shell. Considerable work was done on new 16-in shells, including several carrying bomblets for area neutralization. For the future, the Marines wanted to consider rockets and a new large-caliber gun.

In February 1983 the commandant of the Marine Corps asked the Defense Science Board to study the fire support issue. A report completed in December 1983 advocated reactivating the six ASW *Forrest Sherman*–class destroyers for fire support; and the Vought MLRS could be placed aboard reserve LST 1179s. The report also advocated a fire support role for the SM-2 antiaircraft missile, which is being realized more than a decade later, and using submunitions in Tomahawk cruise missiles, as is currently done. The study also advocated revival of the 8-in major caliber lightweight and soft recoil gun projects. Apparently this study led to a February 1984 navy study of the reactivation of *Forrest Sherman*–class destroyers as gunfire support platforms (without new guns). Unfortunately, the ships had been laid up without having been mothballed, and they had deteriorated so badly that they were no longer deemed viable as mobilization assets.

There was also congressional pressure to build a new naval gunfire support ship. The Senate provided $500,000 for a report, due 1 April 1984, on the potential of such a ship. The study was finally submitted on 4 April 1985 as the "Naval Surface Fire Support [NSFS] Improvement Study," and it was pointedly different from a study of naval GUN fire support. It emphasized that support would have to be provided well inland; an enemy would probably defend the beach itself quite lightly, in the knowledge that most gunfire support would not reach too far inland. Moreover, ships would now operate well offshore. Thus to support helicopters landing 15 nm inland, ships operating from beyond the horizon of the beach (more than 25 nm offshore) should be able to fire effectively as much as 60 nm, to deal with enemy forces before they could engage the marines coming to the landing zone. Effective bombardment at such ranges depended not only on weapons but also on new means of target acquisition, designation, and damage assessment.

Perhaps for security reasons, this study substituted three generic cases for the geographical scenarios of the past: over the horizon assaults against mountainous terrain, which would restrict enemy maneuver; a coastal plain (unrestricted enemy maneuver); and a conventional assault.

The future weapons considered by the study were a bomblet-carrying missile; an 8-in/60 gun, lengthened to improve range compared to the earlier 8-in/55, firing conventional, rocket assisted, and extended-range (ramjet) ammunition; and the army missile system. Due to its long time of flight, the rocket-assisted projectile (RAP) round required a terminal seeker. A faster ramjet (with 70 km range using solid fuel), then under navy development, did not contain a seeker. There was also interest in an antiradar shell for the 8-in gun. The gun might be an upgraded version of the earlier Mk 71 or a vertical loader. Two new rockets were being developed to supplement the army's standard 9-in weapon: a 15.5-in rocket with boost-phase control and an inertially guided 18.5-in rocket. The latter became the army's ATACMS (army tactical missile system), and it was later proposed for naval use. New self-guiding bomblets such as SADARM (sense and destroy armor munition) would enable missiles or shells to attack deployed tanks and other vehicles.

The 8-in/60 was assessed as the least expensive improvement, but also the most marginal. The bomblet-carrying missile offered the greatest improvement. Rockets would be a significant improvement over existing capabilities. Overall, the study showed just how vital tactical airpower was, hence how important it was to suppress enemy air defenses.

Several more exotic weapons were under development, including 5-in and 155-mm rail guns (electric guns) and a pulsed-power 5-in/84 (hybrid solid-propellant gun). They did not figure significantly in the NSFS study, however.

The study included notional fire support combat systems for an *Iowa*-class battleship, a *Spruance*-class destroyer, a *Knox*-class frigate, and a fire support ship (FSS), which might be called a reborn LFS. All would be equipped with counterbattery radar and with a variety of remotely piloted vehicles (RPVs). Forty surveillance RPVs and six antiradar ones would be placed in the FSS, smaller numbers in the others. The envisaged FSS was a converted LPD 4. A sketch showed an 8-in gun forward, with an RPV ramp forward of it, and rocket launchers port and starboard abaft the bridge. Another 8-in gun and two B-module vertical launchers (61 missiles each, for bomblet missiles) were in the well deck aft. The well deck simplified ammunition handling for underway replenishment. The rocket launchers could fire 9-, 15.5-, and 18.5-in rockets, with capacities of 1,200, 300, and 300 rounds, respectively. Each 8-in/60 would have 400 extended-range rounds (100 RAP, 300 ramjet).

Other converted hulls were far less impressive. A *Spruance* could accommodate a 29-round vertical launcher in place of her ASROC, and the 8-in gun (with 336 rounds) could replace her forward 5-in/54. She could mount a rocket launcher aft in place of her other 5-in/54 and perhaps her Sea Sparrow launcher. Capacity would be limited to 400 of the smallest type of rocket. The *Knox* could accommodate one 8-in in place of the 5-in/54 and one rocket launcher aft (in place of the helicopter hangar and pad), but no bomblet-carrying missiles. An RPV launch ramp replaced her ASROC launcher.

Nothing came of this study; the FSS might be attractive, but there were many far more urgent requirements. For the moment the four recommissioned battleships were a far better proposition. The 5-in laser-guided shell, which had seemed so attractive, had to be abandoned because it was too expensive. Then, in the 1990s, with the battleships laid up once again, probably never to be recommissioned, the question of fire support once again became urgent.

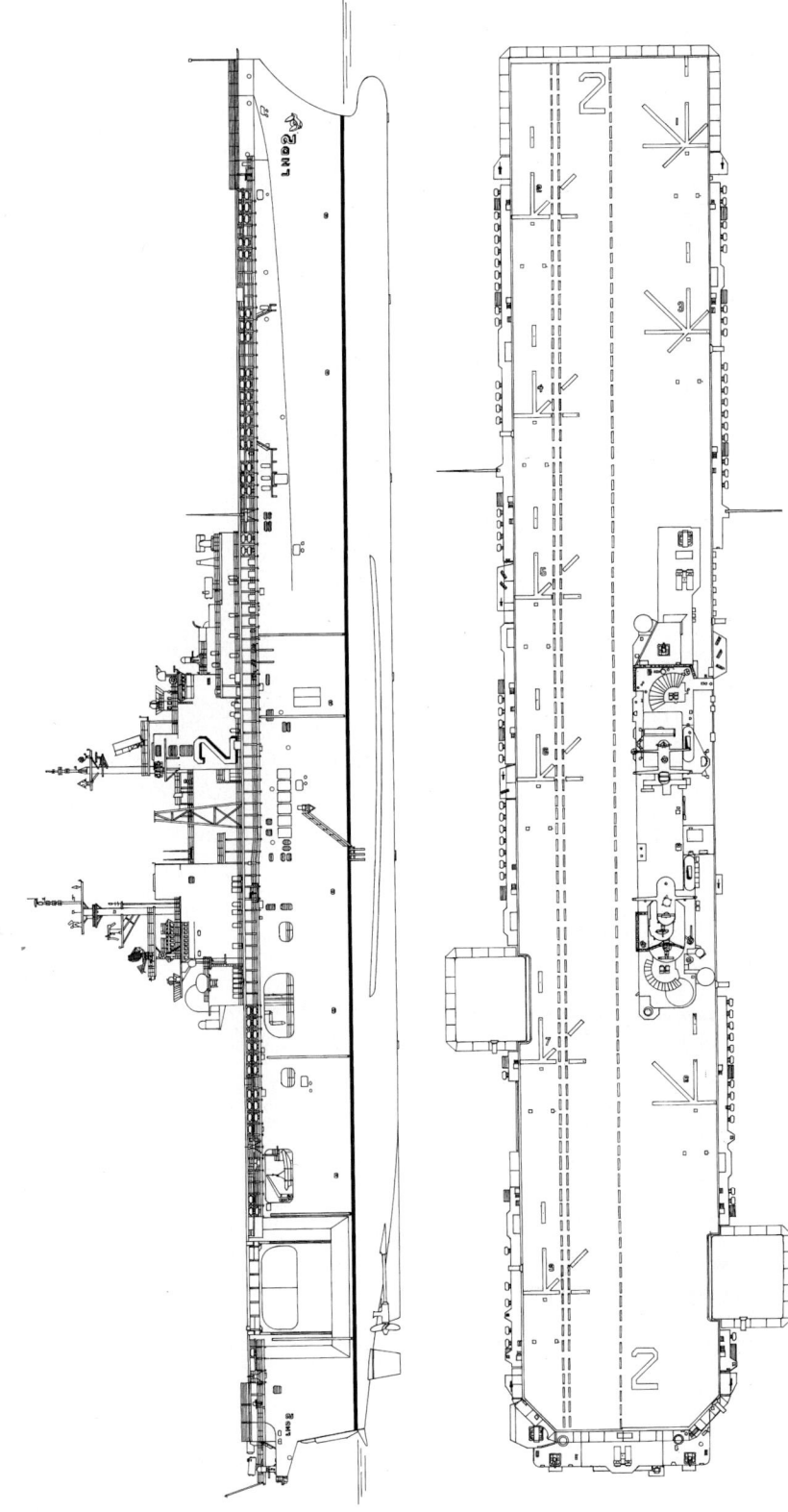

Essex (LHD 2) as completed, armed with a combination of RAM, NATO Sea Sparrow, and Vulcan Phalanx at the fore end of the island and right aft. The object abaft the RAM launcher at the forward end of the island is a blast shield. Above and abaft the Phalanx mount is a USC-38 satellite communication antenna in a small radome, and above that is the Sea Sparrow director. The big flat-faced radar forward is SPS-48E. Above it the foremast carries a small navigational radar antenna; above it a platform carries a surface/low altitude air search antenna, and, abaft it, SMQ-11 to receive meteorological data. The masthead antennas are for the Combat DF system (SRS-1). At the base of the foremast, the box of SLQ-32 ECM antennas is visible. The mainmast upper platform carries the antenna for the Mk 23 Target Acquisition System. Below it is the ring antenna of an IFF receiver. The platform below that carries the antenna of a SPN-43 air traffic control set (the associated SPN-35 approach radar is in the big radome farther aft). Below it is an SPS-49 two-dimensional radar. Alongside the SPN-35 radome is the radome of a WSC-6 satellite antenna, and abaft it is the small radome of a second USC-38 satellite set. Visible only on the plan view is the antenna of a second WSC-6, near the forward end of the island, on the starboard side. *A. D. Baker*

The big 155-mm advanced gun system helps make the new Marine Corps infiltration tactics practicable. This full-scale mock-up was displayed at the 2001 Navy League show. The angled gun house sides and the square housing around the barrel are intended to reduce the weapon's radar cross section. *Norman Friedman*

17
Preparing for the Future

Besides its emphasis on over the horizon (OTH) assault, the Marine Corps' "USMC in 1985: A Long-Range Study," conducted in 1964, explored the potential offered by long-range VTOL aircraft. To some, the experience of Vietnam showed that for limited wars U.S. troops should be based offshore wherever possible. In 1969 the long-range objectives group (Op-93) proposed a new basing concept: Marines should be airlifted directly from sea bases. They would operate entirely without fixed land bases. In the ongoing war in Vietnam there was no classic front line, but the sea was clearly a U.S. sanctuary. Moreover, fighting was always within the littoral area, hence accessible to sea-based forces. The Center for Naval Analyses (CNA) was ordered to assist in conducting a study, "Sea-Based Expeditionary Force–1980" (SEF-80). The concept capitalized on recent and prospective developments in ship design; automation, particularly in handling material; new heavy-lift helicopters, to bring material ashore; and air mobility in general. The army conducted a somewhat similar study, "Project Flattop." Nothing came of these studies, but in them can be seen the germ of the current Marine concept of operational maneuver from the sea (OMFTS) or ship-to-objective maneuver (STOM).

Another major factor in amphibious evolution after Vietnam was the failure to replace diminishing fire support assets. Without fire support, the Marines could not hope to make classical-style frontal assaults against prepared positions. Vertical envelopment and OTH assault, however, still made it possible for attacking forces to outflank enemy positions, and to get into blocking positions before strong enemy forces could arrive on the scene of an assault.

The ultimate objective of Marine amphibious attacks was generally inland from the beach where they landed. The beach mattered because the Marines had to concentrate a mountain of supplies before regrouping to move inland; it was an essential logistical dump. Could the force move directly inland from the sea?

In the parlance adopted in the 1980s, flanking maneuvers were tactical: they were part of a battle. By way of contrast, an operational maneuver was conducted out of contact with an enemy, for example, between battles. Once amphibious forces had to be sent beyond the horizon, it was not too much of a step to imagine them coming ashore far from enemy troops. Instead of maneuvering mainly on land, the Marines would do much of their maneuvering at sea, where they were most mobile.

The Marines became interested in what the army would call infiltration tactics. If they could not blast their way in, they would exploit gaps in defenses. They would not concentrate until they had approached their real objective, generally some distance inland. The new idea recalled the analogy with World War I, which had guided the Marines between the world wars. Most of the time, World War I generals tried simply to smash through enemy defenses. They could not try flanking attacks, because the front line had no flanks; it extended too far. The Germans tried an alternative, infiltration attacks, in which small mobile units found their way through gaps in the defenses too narrow to admit the usual large attacking formations. Indeed, infiltration almost won World War I for the Germans in April 1918.

Infiltration on the scale the Marines envisaged was by no means simple. The supply dump and other centralized functions formerly concentrated at the beachhead would remain offshore. Each small infiltrating unit would have to be supplied, as it moved. To be mobile, it would have to be stripped of much of its normal weight, including the usual artillery. Artillery support would have to come from the sea. It would have to reach deep inland, and it would have to be on call at all times. This was very different from the usual naval gunfire support, which was required only until the beachhead was secured and organic Marine artillery was brought ashore.

At the beginning of the twenty-first century, the Marines have a real hope of realizing this new concept. Satellite navigation, using the Global Positioning Satellite (GPS) system, makes it possible for the infiltrating formations to navigate. Small amounts of supplies can be dropped in places designated by GPS coordinates, and the infiltrators can guide themselves to them.

The fast MV-22A Osprey VTOL troop carrier is the ideal means of distributing marines deep in enemy territory. There is, however, some question as to whether existing amphibious ships can carry enough of them. Will the MV-22A engender new classes just as the LCAC did? To the extent that infiltrating forces will arrive by sea, they will be distributed over a very wide area, thus they will have to arrive on board very fast landing craft. LCACs already exist, and the new tactics make it even more urgent to replace the AAV7 with something much faster.

Current plans call for fire support to be supplied by a new *Zumwalt*-class (DD 21) land attack destroyers. These ships are to be armed with a new 155-mm gun firing GPS-guided shells. At least in theory, each of the two guns on board a *Zumwalt* can fire at the same rate, with the same effect, as the standard six-gun 155-mm howitzer battery that a Marine battalion brings ashore. Because the Marines ashore can use GPS to find target coordinates, and because they have a satellite link back to the ships offshore, they can use the destroyer as a howitzer surrogate. The shells of the 155-mm gun are relatively easy to transfer at sea, whereas missiles are a very difficult proposition. If the destroyer is to provide the sort of sustained support the Marines clearly envisage, then her magazines must be easily refillable. *Zumwalt* is also to carry large numbers of vertically launched missiles that can be used against fixed targets. At least in theory, bomblet-carrying missiles can break up armored formations massing over the horizons of the deployed Marine units ashore. The Marines' own aircraft can help.

Offshore, the Marines envisage ships operating as logistics dumps. Size matters because virtually everything on board the ship must be easily accessible. This is similar to the distinction between an attack cargo ship, which is tactically loaded, and a point-to-point cargo ship, which is loaded to maximum efficiency. As of 2000, the Marines doubted that a new class of enormous attack cargo ships will be built. However, there is an increasing convergence between the sea bases built for maritime prepositioning (see Appendix A) and the needs of sea bases for OMFTS. The main difference is that current prepositioning ships unload by crane, whereas most of the resupply of the infiltrating force would necessarily be by helicopter or VTOL (MV-22A). Thus the future offshore base ship would have to be flat-topped. The Marines see the replacement for the current maritime prepositioning squadron (MPS) ships, which would be needed beginning about 2014, as a natural candidate. It would be affordable because it would not be built to military standards.

Alternatively, the United States may build one or more mobile offshore bases (MOBs). Like the Marines' expanded MPS ship, a MOB would operate as an offshore supply dump, and it would certainly have helicopter/VTOL facilities. Because it would be more like a floating pier, it might be used to replenish ships' missile batteries, and thus might simplify the sort of sustained support the Marines envisage. Like the future MPS ship, in 2000 the MOB is a concept rather than a programmed plan.

Currently, the Marines are often the favored U.S. quick-reaction force, forward deployed on board the amphibious ships of, at present, 12 amphibious ready groups (ARGs). The corresponding units are Marine expeditionary units (MEUs), incorporating battalion landing teams (BLTs). Although in theory amphibious ship numbers are calculated in terms of Marine expeditionary forces (MEFs; the division/air wing teams), the reality is that MEUs and ARGs do the day-to-day work of the Marine Corps. The trend has been to reduce the number of ships in an ARG, hence the number of sailors needed to support each individually deployed marine. Plans call for reducing the ARG to three ships: an LHA/LHD, an LPD, and an LSD 41/LSD 49. There are to be at least 12 of each of the 3 types. With the decline of the surface force, amphibious ships may often find themselves without much defensive support. They may, therefore, need much more sophisticated defensive systems.

In this environment an LPD successor, LPD 17 (*San Antonio* class), is being built. She reflects a revived interest in self-defense. She will have a self-defense combat direction system (Akcita, formerly SSDS [ship self-defense system]), to control her defensive rolling airframe missiles (RAMs). There is also space and weight for future installation of a 16-cell Mk 41 vertical launcher, which can accommodate 64 Evolved Sea Sparrows. The main sensor of the self-defense system is a three-dimensional radar, a recycled SPS-48E. The ship also has substantial passive protection, and a CAE damage control system. At least from LPD 20 on, ships will have reduced radar cross section. All will probably have

The LCAC remains an essential element of any over-the-horizon assault. A service life extension (SLEP) program is under way. LCAC 91, the SLEP prototype, is shown. Because she was used for this new program, she was delivered late, on 27 March 2000. The first rebuilt LCAC was delivered in March 2001. Rebuilt craft will be called LCAC Mk II. As of mid-2001 plans call for two production SLEPs in FY 02, followed by 6 per year, for a total of 74. Textron has sold similar LCACs to the Japanese and South Korean navies. *Carl Gustafson for Textron Marine Systems*

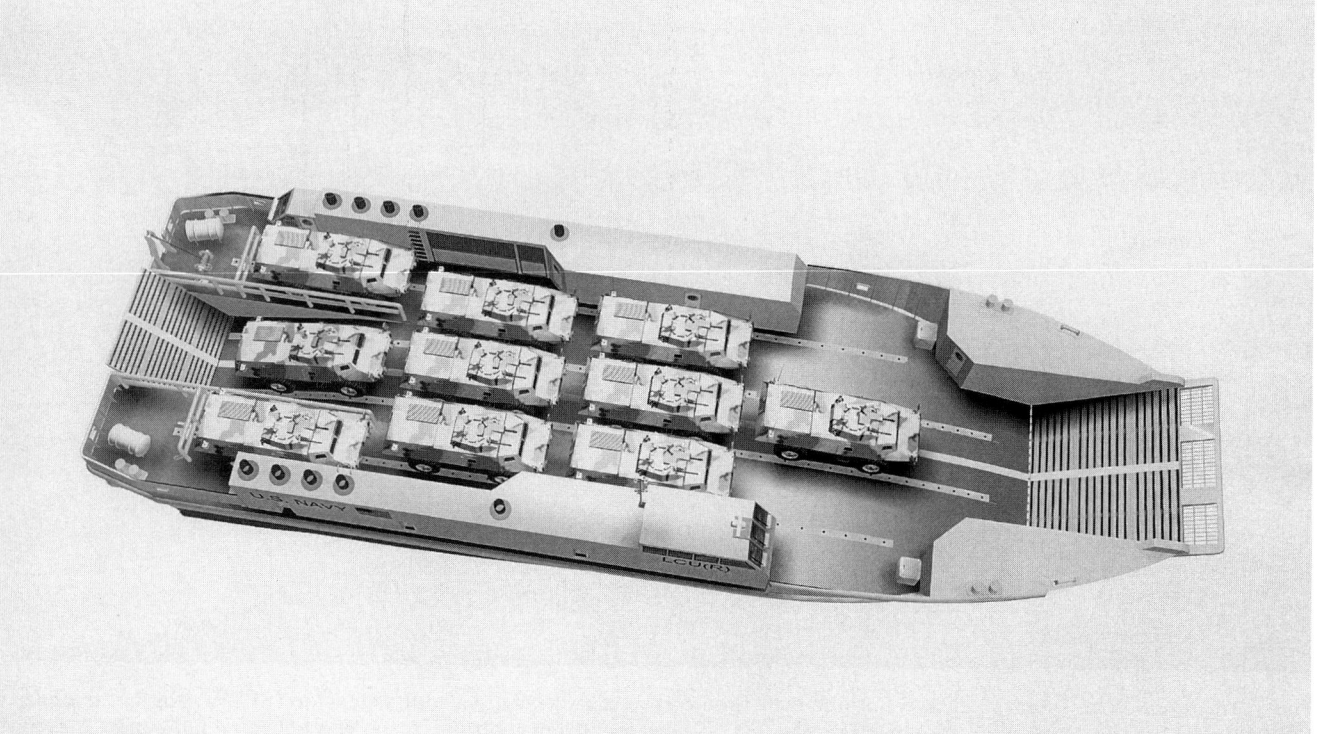

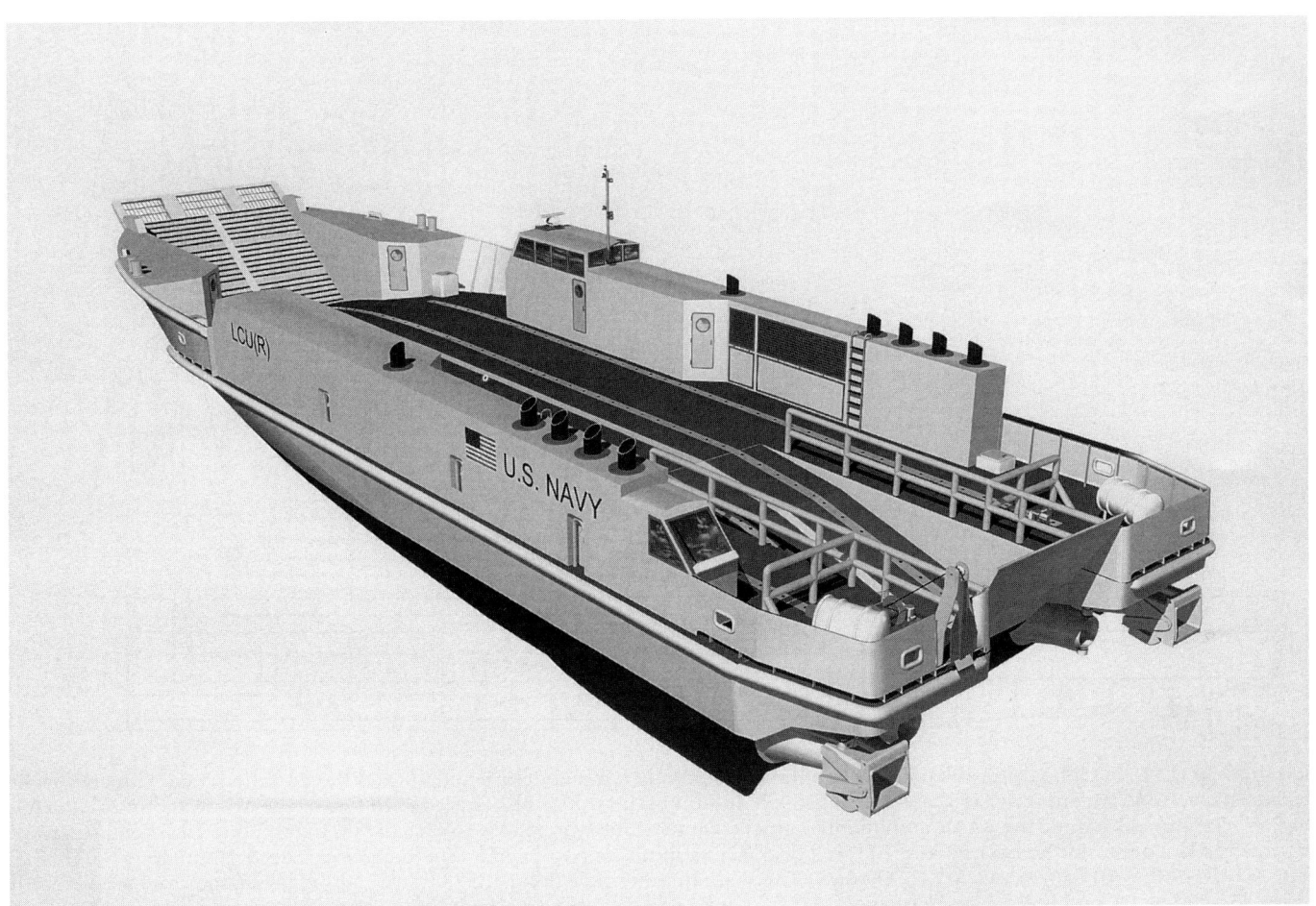

LCACs are expensive, and the U.S. Navy hopes to replace the existing LCUs with something faster yet much less expensive than an LCAC. That recalls the LCM(9) program, for a planing LCM. These drawings show Textron Marine Systems' approach to designing a fast planing LCU(R). The powerplant would be high-powered diesels. Note the waterjets. *Textron Marine Systems*

enclosed masts (for reduced radar cross section), the antennas rotating inside. The well deck, the same size as that of earlier LPDs, will accommodate 2 LCAC, 4 LCM(8), or 20 LVT. Unlike earlier LPDs, this one will have an enclosed hangar for two CH-46E helicopters, three AH-1W, one CH-53E, or one MV-22A. The flight deck will have two spots for helicopters up to CH-53E size. As was predicted in the 1960s, an enclosed hangar requires a much larger hull, in this case displacing almost 50 percent more than previous LPDs (about 25,000 tons vs. about 17,000 for earlier classes, fully loaded). Even so, the ship will carry fewer marines, 708 vs. 835 in LPD 4. Machinery matches that of recent LSDs. The greater length of the new LPD makes up for her greater displacement, so speed is about the same.

Unlike most earlier amphibious combatants, this one was privately designed, on a competitive basis. The request for proposals was issued 8 April 1996, and the program was won by a consortium of Avondale Shipyards, Bath Iron Works, Hughes, Loral, Sperry Marine, CAE, AT&T, and Intergraph. The first two units were included in the FY 96 and FY 99 programs, with two more in FY 00, and two more planned for each year through FY 04. Many features had to be eliminated as costs escalated. They included the Sea Sparrows noted above, an infrared (IR) search and track sensor, and the cooperative engagement capability that would allow the ship to share targeting data from other units. Because space and weight for these features remain, presumably they will be installed after completion. Akcita will first be fitted to LPD 22, then back-fitted to earlier units.

The fast LVT successor is finally being built, under the designation AAAV (advanced amphibious assault vehicle). By 1990 the earlier AAV7 was long overdue for replacement. The Marines wanted a successor that could swim at high speed, from beyond the horizon, but they knew that would be very expensive. The alternatives were a slow swimmer (AAAV without high water speed), a product-

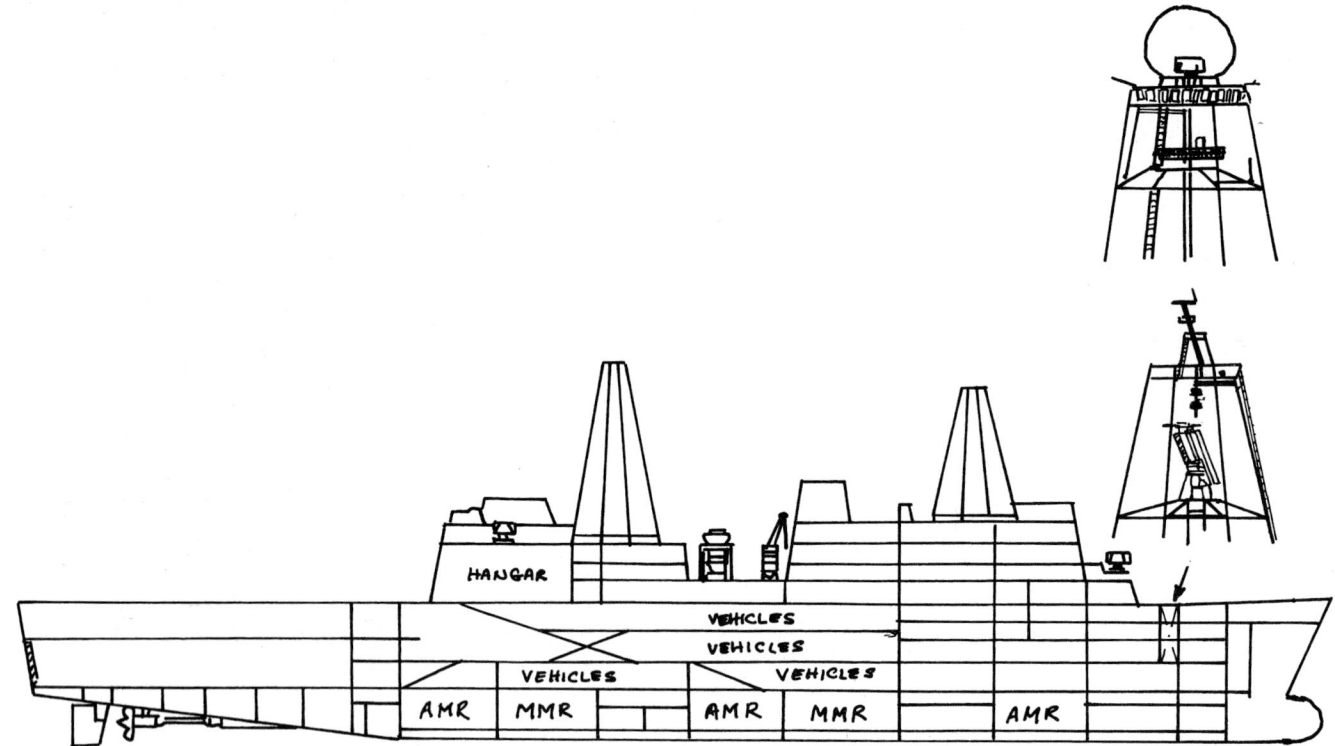

San Antonio (LPD 17) inboard profile, 2000. Note the RAM missile launchers fore and aft and the 16-cell vertical launcher right forward (space and weight provision, not to be installed initially). The ship is also armed with two of the Mk 46 mounts (carrying 30-mm cannon) chosen for the AAAV amphibious vehicle. Dimensions were 684 × 105 ft (25,000 tons). Four diesel engines were expected to drive the ship at over 22 kts. The well deck accommodates two LCACs. The hangar can accommodate two CH-53E, four AH/UH-1, four CH-46, or two MV-22 Ospreys. There are three vehicle decks (25,000 sq ft), three magazines for cargo ammunition (34,000 cu ft), and bulk stowage for JP-5 (315,000 gals) and vehicle fuel (10,000 gals). The ship accommodates 699 troops (with surge capacity for 800) and a crew of 361. The inset shows the insides of the two AEM/S (advanced enclosed mast/sensor) masts. The foremast carries an SPQ-9B antenna to detect sea-skimming missiles, with a circular IFF array (UPX-29) below it; the mainmast carries an SPS-48E 3-D air search radar with Tacan on a topmast. The main ship combat direction system is SSDS Mk 2 (ship self-defense system). The ship will also be equipped with the USG-2 cooperative engagement capability (CEC) system and with the KSQ-1 LCAC over-the-horizon direction system. The ship also has the naval tactical command support system (NTCSS), which connects her to the global command and control system via satellite links, and the NAVSSI integrated navigational system.

improved AAV7A1, the army's M2 Bradley, the older M113 personnel carrier, and the new lightweight LAV 25. Another possibility was to rely entirely on LCACs to land vehicles, and forego any amphibious personnel carrier. A cost and operational effectiveness analysis (COEA) conducted by CNA showed that, based on a requirement to mount MEF- and MEB-size assaults in the Third World, the fast swimmer would be most cost effective. On this basis the Marines created a AAAV program office, and the David Taylor Model Basin built a propulsion systems demonstrator (PSD), a full-scale amphibian carrying a crew of 3 and 14 troops at better than 20 kts in Sea State 3 (and at over 30 mph on land).

At a Marine Corps program decision meeting in April 1991 questions were raised about the viability of the fast-swimming AAAV, so the Defense Department decision to proceed to the demonstration/validation phase (the Milestone I decision) was deferred from May 1991 to February 1992. Issues included power, weight, and engine compartment configuration. However, on 19 July 1991 the AAAV program office released a statement of work for the pending demonstration/validation phase, and contracts eventually went to General Dynamics and to FMC. It was estimated that the demonstration/validation program would take 55 months, so operational prototypes could be tested in FY 96, and the AAAV might become operational in 2006. The Marines wanted to buy 951 of them (personnel carriers and C3 vehicles) at a unit cost of $3–3.5 million. Performance on land would approximate that of the M1A1 main battle tank, swimming speed would be about three times that of the AA7V1 (i.e., better than 20 kts), the vehicle would be armored against 14.5-mm armor-piercing bullets at 300-m range, and its weapon would have a range better than 1,500 m.

The most impressive new amphibious ship is *San Antonio* (LPD 17), the lead ship of a new class. Because she may operate far from supporting units, she is more heavily armed than her predecessors, with two short-range RAM missile launchers and a bank of vertical cells for Enhanced Sea Sparrow missiles, visible on the model forward of the superstructure. Much attention went into reducing her radar cross-section. The knuckle on her hull indicates upper sides angled back to reflect radar beams away from their transmitting antennas. The two tall stack-like structures are masts enclosing the usual radar antennas. Opaque at frequencies other than those used by the ship's radars, they also reflect enemy radar signals away from their transmitting antennas. Solid bulwarks help conceal the usual topside clutter from radars. Raytheon, the LPD 17 system integrator, displayed this model of the ship at the 2001 Navy League show. *Norman Friedman, 2001 Navy League photo; Raytheon, LPD 17 photo*

The Marines' hopes for over-the-horizon assault depend heavily on the success of a new amphibious vehicle, the advanced amphibious assault vehicle (AAAV) shown here. It is much more heavily armed than previous troop-carriers, with a 30-mm Bushmaster gun in its turret. Also unlike earlier amphibians, it is armored against 14.5-mm armor-piercing ammunition and against 155-mm shell fragments. *General Dynamics Land Systems*

General Dynamics won the competition with a planing craft, which retracted its tracks and folded chine plates down from bow, stern, and sides. The craft uses a 2,600 HP engine driving a pair of 23-in waterjets, to achieve well over 29 mph in the water. Speed on land exceeds 45 mph. The turret carries a 30-mm gun. The first prototype was delivered in June 1999, the first production unit is to enter service in 2005, and the last of a planned total of 1,013 (including 78 command vehicles) is to be delivered in 2012.

Appendixes

U.S. strategic military mobility is largely based on strategic sealift because it is still far easier to move heavy weights over water than through the air. Although sealift ships are designed primarily to deliver U.S. military support in unopposed operations, they would also be the follow-up shipping to any future amphibious assault. USNS *Gilliland* (T-AKR 298) is shown on sea trials, 5 May 1997. *Gilliland* was built by Burmeister and Wain of Denmark for merchant service and launched in 1973. She was leased in July 1993 on a long-term basis and then converted at Newport News for military service. Conversion included fitting a heavy slewing stern ramp for vehicles, side doors, internal vehicle ramps, and a helicopter deck forward of the bridge. *Newport News Shipbuilding and Dry Dock Company photo by Stu Gilman*

Appendix A: Maritime Prepositioning and Sealift

Sealift and prepositioning are related to, but not really part of, amphibious warfare. Sealift or point-to-point shipping is distinguishable from amphibious shipping because it uses conventional port facilities. However, using extemporized facilities, it is also the follow-up to an amphibious assault. Because the U.S.-flag merchant marine declined precipitously beginning in the 1960s, sealift is now a major limitation on U.S. strategic mobility.

There are several distinct categories of sealift assets. One is the active U.S.-flag merchant fleet, little of which currently survives. A second is U.S.-owned ships under foreign flags of convenience, a larger entity, sometimes called the U.S.-control fleet. A third is merchant or merchant-type ships operated by the U.S. government. In addition, the Maritime Administration maintains a large reserve fleet, the National Defense Reserve Fleet (NDRF) of laid-up ships. Within this fleet, in 1976 a Ready Reserve Fleet (RRF) was created, comprising ships that could be activated in 4, 5, 10, or 20 days. Initially the RRF comprised the best of the reserve Maritime Administration ships; there was no particular effort to build it up as a sealift asset. A major Reagan administration initiative, described below, enlarged the RRF by buying up modern foreign-built ships, specifically to fill gaps in U.S. sealift. At present selected RRF ships are exercised periodically, and significant numbers are currently operational. Currently, ships taken into the RRF are upgraded, including installation of satellite communications and sealift features.

Scarcity of sealift is not a new problem. World War II strategy was often constrained by a lack of sealift, which could be ascribed partly to the poor state of the pre-1941 U.S. merchant marine. The vast wartime building program created the largest merchant fleet in the world; there were a total of 3,883 ships in November 1948. Over 90 percent had been built between 1939 and 1946, although 5 percent of dry cargo ships and 14 percent of tankers were over 20 years old. However, 1,600 of the dry cargo ships were 11-kt Liberty ships. About half the fleet was already laid up: 1,616 dry cargo ships, 36 tankers, and 194 merchant-type auxiliaries.

Even with this large fleet, in 1948 mobilization requirements for point-to-point transportation and for naval auxiliaries far exceeded the available supply. That was particularly true of passenger ships suitable for transport conversion, because the ocean passenger trade was dominated by foreign fleets. The Maritime Commission, with military backing, began a passenger ship program. The ships had "national defense" features that specially fitted them for wartime service. As of late 1948 five contracts had been awarded, and ten more were expected in FY 50. By 1960 there would be sufficient additional lift for 150,000 troops. This ambitious program collided, first, with the Korean War, when three ships were siphoned off to be completed as transports.[1] It did produce the massive fast *United States,* but a projected sister ship was never ordered, due to the advent of the jet airliner and the consequent collapse of the traditional liner market. One further consequence was that, in the 1980s, the U.S. Navy was unable to identify modern ships convertible to troopships, for use in cases in which no modern airfield was available where U.S. troops were needed; there were, however, still numerous ex–Military Sealift Command (MSC) transports laid up in the National Defense Reserve Fleet.

Merchant ships were generally credited with a 20-year lifetime, so ships completed, for example, in 1943 reached replacement age in 1963; by this time there was absolutely no interest in mass replacement or revival. The Liberties were gradually discarded, but faster war-built ships, principally Victories, and retiring postwar ships were retained in the Maritime Administration's National Defense Reserve Fleet. Some Victory ships were reactivated from Maritime Commission reserve to carry supplies to Vietnam; a few foundered due to their age. Victory ships were still a large part of the much-reduced National Defense Reserve Fleet in the 1990s.

As a complicating factor, merchant ships changed drastically after World War II, often due to U.S. commercial initiatives. The merchant ships of the past, from which transports and attack cargo ships were adapted, were break-bulk ships. Cargo was broken down into packages that filled holds as completely as possible. The same packages were small enough to be handled easily by the ship's cranes, which therefore needed only limited capacity. A few specialized ships could carry and offload heavy items such as tanks and locomotives. Perhaps most important, break-bulk ships could unload using minimal port facilities. In many places they unloaded onto lighters, which then unloaded at piers in water too shallow for the freighters. This was not too different from unloading into landing craft alongside.

The first important postwar development was ro-ro: roll-on roll-off. Trailers could be driven onto a ship, and they drove off at the destination. The LST was in effect the first ro-ro ship, and some LSTs were used commercially in this role postwar. Ro-ro might be considered a logical extension of earlier train ferries. Like them, it generally required specialized port facilities. Since cargo was packed far more loosely than in a conventional ship, a ro-ro had to be far larger to accommodate much the same tonnage. Ideally, a ro-ro was much more like a warehouse than like its merchant ship predecessors.

The next step was the standardized cargo container, in effect the trailer without its wheels, hence it was stackable. A container could be packed onto a ship, or offloaded onto a trailer or a railroad flatbed car. Because even the largest ship carried only a few thousand containers, she could be unloaded quite quickly if the necessary facilities were available. A container port could be infinitely more efficient than its break-bulk predecessor, but it required massive pierside cranes and an adjacent road or rail infrastructure. By 1980, the United States could choose between building new break-bulk ships specifically for military service (and nearly useless for civilian purposes) or finding some way to unload container ships without special port facilities.

Other new ideas included carrying loaded lighters or barges aboard a ship (LASH and Seabee are described in Chapter 16). Such ships were not built in numbers, but they turned out to be quite useful, and some were acquired in the 1980s for U.S. military logistics use. Another type, which the U.S. Navy did not acquire, was the flo-flo (float-on float-off), a semisubmersible that could carry other ships. It has been used to transport mine countermeasures ships, and even to transport badly damaged combatants back to the United States.

Another change, the drastic decline in the liner trade, was mirrored in the number of transports the navy operated. In FY 62, the navy operated 3 armed transports (AP) plus 12 MSTS (civilian-operated, unarmed) transports. The APs were transferred to civilian operation in FY 66, and one ship was reactivated that year, to make a total of 16 in MSTS that year. That declined to 11 in FY 69, and then to 3 in FY 70; none at all was still active by FY 73. Military personnel were all moved by air.

Prior to World War II the U.S. Army operated most American troop transports. During World War II the navy took over many of them and gave them AP numbers, but the army continued to charter others. Indeed, the JCS Joint Military Transportation Committee sometimes had to decide which service got which ships. There were, for example, considerable differences between army and navy practices in controlling shipping. As part of the postwar push for service unification, in May 1946 the JCS asked for a study of what operating procedures would be used if the navy were given responsibility for all sea transport. The main problem, allocating costs, was resolved early in July 1949. On 2 August 1949 the Military Sea Transportation Service (MSTS) was established under the Navy Department. It took over army and navy ocean transport shipping, but not the army's harbor and inland waterways craft. MSTS ship designations were prefixed by the letter "T," and the ships were designated USNS (U.S. Naval Ship) rather than USS. They were operated by civil service mariners (later, largely by contractor personnel). MSTS became the Military Sealift Command (MSC) on 1 August 1970. In the 1980s, as part of another attempt at service unification, MSC was folded into a Joint Transportation Command. The navy resisted the change, because in addition to its sealift function MSC operated some important purely naval auxiliaries, such as T-AGOSs (sonar surveillance ships) and major underway replenishment ships; it was far less expensive to operate many of these vessels on a quasi-civilian basis.

MSTS continued wartime efforts to develop cargo ships well suited to military-type cargoes, particularly heavy vehicles. MSTS adopted ro-ro in T-LSV 7, USNS *Comet* (SCB 236). *Comet* was described as the result of a three-year study by Buships, MSTS, and the U.S. shipping industry. Her merchant ship status was indicated by her Maritime Administration designation, C3-ST-14a. Her designation was

Sea Lift (T-LSV 9, later T-AK 278) was an early ro-ro ship, a modified version of the original *Comet*. She retained standard breakbulk freighter booms, but vehicles could drive aboard via the stern ramp and side doors.

changed from T-AK to T-LSV in 1963. *Comet* met a long-standing army requirement to transport large quantities of all types of army vehicles. Beside her conventional holds (served by booms, with capacities up to 60 tons) she had two holds into which vehicles could roll via either a stern ramp (60-ton capacity) or cargo ports (with portable ramps, with 60-ton capacity). The vehicles could be driven off the same way. Total capacity was more than 700 military vehicles.

Draft was an important consideration in the design. For example, the northern 600 miles of the western coast of Europe included 66 ports leading into highly developed inland transportation systems. The ship could enter 39 of them at her vehicle cargo full load draft of 22 ft, although not at her conventional cargo full load draft of 27 ft.

MSTS bought *Comet* and three T-AKs in FY 55 as part of its nucleus fleet (it also had numerous war-built ships); at the same time it bought an AKD, *Point Barrow* (T-AKD 1), roughly equivalent to an LSD, for Arctic operations. The Maritime Administration designated the design S2-ST-23a. She could carry ro-ro cargo on superdecks.

An LSD, canceled at the end of World War II, became T-LSV 8, USNS *Taurus*. Having languished for a decade, she was converted in 1956 into a passenger and car ferry for Jacksonville–San Juan service. That proved unsuccessful, and she was chartered to MSTS for one trip before her owners went bankrupt and the Maritime Commission foreclosed her mortgage and laid her up in 1958. Then MSTS acquired her. Carrying 125 trailers, she alternated with *Point Barrow* hauling priority ro-ro cargo.

After FY 55 MSTS purchases lapsed. The program was revived as part of the Kennedy administration's attempt to improve U.S. strategic mobility. The FY 63 program included the ro-ro USNS *Meteor* (T-LSV 9, originally *Sea Lift*). This improved *Comet*

(Maritime Administration designation C4-ST-67a) was intended to operate on a shuttle service with the earlier ship, replacing chartered ro-ros. Congress rejected a second unit, planned for FY 64, pending results of operational evaluation between *Comet* and a fast freighter using conventional offloading. Reportedly, commercial interests opposed construction. The second T-LSV was requested again in FY 65. She was bought under an innovative build/charter scheme, which later was used very successfully by the Reagan administration. This ship, *Admiral William B. Callaghan*, pioneered large-ship gas turbine propulsion in the United States. She was completed in December 1967. Later the T-LSVs and similar ships were reclassified as T-AKR, to emphasize their cargo (AK) role (R stands for ro-ro). MSC, the MSTS successor, bought and operated the maritime prepositioning fleet described below.

Sealift issues are as old as amphibious warfare, but prepositioning is a creature of the jet age. Entering office in 1961, the Kennedy administration was much concerned with the threat of numerous Communist insurgencies spread around the world. U.S. forces were limited, however. How could they respond? The earlier Eisenhower administration had considered marine amphibious groups sufficient, but the Kennedy administration wanted the weight that the army offered. Yet troopships could not possibly arrive quickly enough in distant ports. Nor could the United States maintain garrisons around the world.

Initially the Kennedy administration seems to have considered sealift old-fashioned; surely the modern way to deploy forces was to depend entirely on aircraft.[2] There was some interest in prepositioning. In 1962–63 Victory ships were converted into forward floating depots (FFDs) and attached to the Seventh Fleet.[3] From their base at Subic Bay they could quickly move their materiel anywhere in Southeast Asia, which was the major concern of administration strategists. Exercise "Quick Release" in 1964 tested the concept of marrying up airlifted troops with materiel from FFD ships. The Defense Department bought more transport aircraft. Troops could be flown abroad to mate with their equipment. A similar solution was adopted for NATO. Equipment was prepositioned in Germany specifically for the use of troops to be flown there in an emergency. In effect much of the U.S. Army became a "swing" force that could be used in either a NATO emergency or in a Third World insurgency.

The merchant ship solution was abandoned as the Vietnam War absorbed U.S. energies. The ships already converted delivered their loads to Vietnam and then served as point-to-point freighters.

Prepositioning had one last fling, however, during the Johnson administration, which inherited most of Kennedy's advisors, in the form of the fast deployment logistics (FDL) ship. The FDL idea originated in the navy's logistic support of land forces (LOGLAND) study during February 1964 and April 1965. The LOGLAND study pointed out that the fraction of army equipment that could not be transported by air was growing. It coincided with key JCS studies of strategic mobility, which the Kennedy and Johnson administrations saw as key to dealing with Communist insurgencies, such as that in Vietnam. Because the insurgents were locally raised, the Communists could be expected to use them to attack the West anywhere; they did not have to deploy an army in only one place. Hence there was a perceived need for strategic mobility, so that a single U.S. force could deal with many widely separated crises.[4] The argument for sea-based prepositioning, in 1965, was that it would be impossible to preposition equipment on land in areas like Latin America, Africa, and the Sino-Soviet periphery, yet exactly these areas might well erupt. A 25-kt ship near the Philippines could reach a crisis area in Northeast or Southeast Asia in about three days, and Abadan in Iran within ten days. Table A-1, prepared in 1965, showed how quickly FDLs of various speeds could reach likely crisis spots. There was often no strategic warning of emerging crises, so time scales were very attractive. Clearly they could be improved if the ship were forward deployed, in which case she would also provide a degree of presence.

Equipment could be stored aboard an FDL, ready for use, for up to three years. The ship could unload either at a port or over a beach (but not as part of an assault), using its own helicopters and craft. Normally about one-third would be forward deployed, the rest were maintained in U.S. ports. The ships would be paid for as part of the overall Defense Department rapid deployment program, together with airlift and prepositioning. The FDL could stow vehicles for long periods while maintaining, activating, and fueling them in place. Similarly, it could stow numerous helicopters, including the army's big CH-54s, ready to fly. Vehicles and helicopters would be maintained by a special 78-man detachment aboard each ship. The FDL would carry 15 days of supplies, including 1,995 tons of petroleum-oil-lubricants (POL) for its embarked vehicles and helicopters. The FDL could also be converted into an austere troop/cargo carrier (4,000 troops and 4,500 tons of cargo), making up for the lack of liners.

As a measure of capacity, the 12 FDLs needed for the infantry division force (135,000 short tons) equated to 87 Victory ships; the 8 for the armored

Table A-1. Time to Reach Crisis Location, 1965

Destination	From West Coast (days)		From East Coast (days)	
	20-kt Speed	25-kt Speed	20-kt Speed	25-kt Speed
Bangkok	15.0	12.1	27.2	20.7
Pusan, Korea	10.8	8.7	23.4	19.0
Abadan	24.4	19.6	21.4	17.4
Europe (Biscay)	19.1	15.5	7.7	6.2
Caribbean (Dominican Republic)	8.5	6.3	2.9	2.0

division (78,000 tons) equated to 52 Victories. Compared to the Victories, FDLs were much faster, could maintain their equipment on board, and could offload much more quickly. Only FDLs could maintain and operate helicopters. Aging, steam-propelled Victories required larger crews (50 vs. 34) and, overall, would be more expensive to maintain. Unlike MSTS ro-ros, FDLs would always be available for a crisis; they would be kept in readiness because they would not be used for point-to-point shipping. They could unload over a beach, and they could store vehicles in ready condition for up to three years in their dehumidified spaces. The FDL's flight deck and well deck made for some confusion with the LHA, even though clearly the FDL would be limited to administrative landings.

The JCS and the army joined in supporting a 30-ship program, which became the first instance of Secretary of Defense Robert S. McNamara's new total package procurement technique. The program was established in October 1965, and Congress approved the first two in the FY 66 budget, at a total cost of $67.6 million. The navy completed concept formulation in April 1966. On 20 July 1967 McNamara announced that Litton had submitted the best technical proposal, beating out General Dynamics and Lockheed. As design agent Litton selected George G. Sharp, Inc., which prepared the biddable package for industry-wide competitive bids for production.

Litton's winning design was a 40,500-ton (loaded) ship (855.5 overall × 104 × 28 ft) driven at 25 kts by a two-shaft 60,000 SHP steam powerplant. Endurance would be 8,800 nm at 25 kts. Similar to amphibious ships, cargo density was low. Carrying infantry equipment, payload was 12,400 tons; for MSTS resupply, it would be 18,100 tons at a deeper draft of 30.7 ft. Twelve FDLs could carry the equipment of an army infantry division force. The division itself would require four and the armored division would require eight. Capacity was also given as 3.8 million cu ft. The ship would take 48 hrs to load at a pier, 7.5 hrs to offload at a pier, or 17 hrs to offload over a beach (from 3,500 yds offshore). To help unload, the ship would carry three CH-54 helicopters ("flying cranes"), three LCM(8), nine LARC-15, and six LARC-60 (BARC). Aviation features would be a helicopter hangar and a 500-ft flight deck. The normal crew would be 34, but the ship could also carry a 78-man army maintenance detachment and a 270-man army unloading detachment. She would normally be unarmed, but she had space and weight for a Basic Point Defense Missile System. JCS plans, as stated in January 1967, were to maintain 8 ships on the Gulf or east coast, carrying materiel for an infantry, armored, or mechanized division); 5 on the west coast near Puget Sound, carrying materiel for an infantry division; 5 in the Pacific at Hawaii, carrying materiel for one division; and 12 in the Western Pacific, either in the Philippines or near Japan, with materiel for an infantry division force.

No FDL was included in the FY 67 budget. For FY 68 the House approved two of the five requested FDLs. The Senate rejected all five and wanted the FY 66 ships rescinded, on the ground that the FDL would merely encourage the United States to act as a global policeman, that is, become involved in more Vietnam-type conflicts. There was also some skepticism as to how well the ships would be protected. A congressional budget conference confirmed the Senate action. However, the military requirement remained, so the program was resubmitted for FY 69. In April 1968 the Senate approved four ships, but the House deleted them. A conference committee upheld the House action without prejudice to the program, on the ground that money was tight, and the need was not urgent. It was clear, however, that congressional reluctance was tied to the "world policeman" issue. Thinking that Congress might approve a smaller program, the army and navy joined in approving a 15-ship plan beginning in FY 70. Officially the cut from 30 to 15 was justified on the ground that 10 new MSTS-build charter ships could replace the other 15 FDLs, but in fact it was hoped that another 15 could be requested in FY 74. Plans called for three ships in FY 70, then four each in FY 71–73. The navy also preferred a slower building rate, because shipbuilding money was so scarce and so many combatants had to be replaced. As it happened, Congress never approved the program.

Prepositioning seems to have been forgotten for about a decade. Certainly the October 1973 Middle East War demonstrated poor U.S. strategic mobility. Between February 1977, when defense analysis began, and February 1979 the Carter administration seems to have realized an urgent need for strategic mobility for situations outside NATO. One driver was the deteriorating situation in Southwest Asia, where the Soviets were becoming more involved in Afghanistan, and where the Shah of Iran, a major U.S. ally, was overthrown early in 1979. There was a real fear that the Soviets might overrun Iran. In March 1979 the president issued a directive to improve mobility in such cases. U.S. troops could not be stationed in the area; no local government would welcome them, except in an emergency. The Carter administration conceived an alternative, the rapid deployment force (RDF), whose troops could be flown in. Their equipment would be prepositioned on board ships stationed at the nearest U.S. base, Diego Garcia in the Indian Ocean. In August the JCS began to define requirements for the RDF, and to identify which forces could be assigned to it. One MAF was assigned to the RDF for contingencies, as were army ground units.

As part of the new defense guidance, on 2 August 1979 the secretary of defense announced a new Marine Corps mobility enhancement (MCMOBE) concept. Forces would have to be inserted quickly into a crisis situation. It was assumed they would be entering a welcoming country, hence they would not meet opposition on the beach. The building block was an MAB with 30 days' supplies, its personnel airlifted by C-5A transports and its equipment moved by sea. One-third of the equipment for an MAF, with its 30 days of supplies, would be prepositioned on board commercial ships in overseas areas of potential crisis. The other two-thirds would be prepositioned in FY 81–86. MCMOBE was conceived as a global capability.

In addition to the Middle East crises, a NATO exercise, Nifty Nugget, highlighted logistics problems. In September 1979 the newly appointed CNO, Adm. Thomas Hayward, remarked that far too little attention was being paid to the sealift needed to put land forces in position; attention was concentrated almost solely on what the forces would do once they arrived.

The Afghan crisis of December 1979 concentrated attention on the problem, and a near-term prepositioning squadron (NTPS) specifically for Southwest Asia was planned. It was not connected to MCMOBE. Under the RDF concept a mechanized brigade would fly in and mate up with prepositioned equipment. It would hold initial positions until the core of the force, an earmarked army mechanized division, arrived. The division's heavy equipment would come on board fast sealift ships, but its ammunition and its water would be prepositioned. In this way the minimum amount of scarce heavy equipment would be permanently prepositioned; the army could load its divisional equipment as necessary. Ships were also prepositioned with the U.S. Air Force's ammunition and spares, to serve aircraft to be flown into the area.

In January 1980 plans were made to acquire the ships, which would probably amount to two ro-ros, two C4 break-bulk, and two tankers (one for POL and one for water). A 19 January port tour by senior officers made it clear that two of Lykes Brothers' big *Maine*-class (C7-S-95A) ro-ros (*Illinois* and *Lipscomb Lykes*), built in the 1970s, would be key. Each had 175,000 sq ft of vehicle parking. The two break-bulk ships were *American Champion* and *American Courier*, both built in the early 1960s. Both had 70-ton booms, which were sufficient for tanks.

Plans did not specify which service would provide the prepositioned brigade. The Marines began developing a proposal, on a crash basis, on 11 January 1980. Once the six ships had been chosen, their headquarters asked what sort of force they would support. The initial answer was two battalions, but soon the Marines proposed three with heavy combat equipment including three tank companies and supporting artillery. The commandant approved the brief on 6 February, and the JCS soon heard it. The effect of the Marines' brief was to change the emphasis from which service would suffer least by prepositioning to which service could offer maximum prepositioned combat power. About this time the air force asked that some material be prepositioned for it. On 8 February the JCS selected the Marines for the prepositioned brigade, and President Carter approved the plan in March. The Marines formed the 7th MAB as the command element of the prepositioned brigade. The additional equipment required a third ro-ro freighter, USNS *Meteor*, with half the vehicle space of a *Maine*.

In August the seven Marine ships were carrying 15 days of supplies toward the Western Pacific. There were also one each carrying army and air force supplies. The FY 81 and FY 82 supplemental programs bought six enhanced NTPS ships (ENTPSs), adding another 15 days of Marine supplies and more army and air force materiel. The combination of NTPS and ENTPS formed the near-term prepositioning force (NTPF). Eventually there were 13 ships: the 3 ro-ros, 3 break-bulk freighters, a

Part of the initial maritime prepositioning force is shown on station, August 1980. In the foreground is a Military Sealift Command ro-ro ship with her stern vehicle ramp partly extended. Another ro-ro and a tanker lie in the background. The two ro-ros are presumably the *Maine*-class *Mercury* (T-AKR 10, ex–SS *Illinois*) and *Jupiter* (T-AKR 11, ex–SS *Lipscomb Lykes*, ex–SS *Arizona*).

water tanker, 4 fuel ships, and 2 LASH (lighter aboard ship). Of this number, five dry cargo ships served the MAB. To protect the NTPS against a preemptive strike, a third of the ships were underway at all times.

MCMOBE led to formation of exclusively navy-marine maritime prepositioning squadrons (MPSs), to carry three rather than one MAB. Unlike NTPS, it was conceived both as an initial force, to be landed as requested, and as a follow-up to an assault. Moreover, the MAB of the MPS was 16,600 marines (including an air combat element)—three ARGs worth—compared to 12,000 in the NTPS. The NTPS was for Southwest Asia only, and its ships would unload only at piers. Ships were not, moreover, combat loaded in any sense; they were loaded with

The point of prepositioning is to have heavy equipment easily available in forward areas to mate up with its operators, yet is not subject to local political interference. USNS *Mercury* loads a Marine Corps M60A1 tank before taking position in the Indian Ocean, July 1980.

Even the initial prepositioning ships were huge. *Mercury* is shown alongside the tanker *Ashtabula* at Subic Bay, March 1981, taking on containerized cargo.

particular types of cargo, such as bulk and ammunition. By way of contrast, MPS ships were intended to unload in the stream, using their own lighterage. They carried POL that they unloaded via their own pumps. Most important, the MPS was conceived from the outset as a global fast-reaction force. As a measure of capacity, the 22,700 tons aboard an MPS ship equated to 220 C-5 or 504 C-141 sorties.

In March 1980 the Defense Department formally directed the navy to activate enough prepositioning ships for three brigades (one full MAF) in FY 83, FY 85, and FY 87: one for the Persian Gulf/Indian Ocean, one for the Pacific (Korea), and one for the Atlantic and the Mediterranean (NATO). An August 1980 CNA follow-up MCMOBE study calculated requirements. In this study ships were based at Diego Garcia; Sasebo, Japan; and Portsmouth, England. Cases considered were Iceland and the Netherlands in NATO, Kuwait and Yemen in the Gulf, and Korea. Assuming they were preloaded, at 20 kts ships from Diego Garcia could reach Kuwait in 5.5 days; from Sasebo ships could reach Korea in 8 hrs. That was faster than airlift. It took two days for troops and aircraft to assemble, and time had to be allowed between airplanes. Total airlift time after assembly would be 2–5 days. One interesting point was that, apart from Yemen, all the places considered had modern container ports.

CNA compared the existing *Maine*-class combination ro-ro/container ship with the PD-214 "Security class" (C8-M-MA134), a conceptual mobilization ship the Maritime Administration had begun designing in 1974. Preliminary design had been completed in 1978, and in 1980 the Maritime Administration was completing contract design. Because the navy had no MPS design in hand, it had to modify PD-214 to meet the new T-AK(X) requirement.

PD-214 was a break-bulk ship, because it was assumed that it might have to deliver cargo where there were no elaborate container facilities. Moreover, some important military cargo, such as tanks, could not fit standard containers. The designers also produced a jumbo version, a container version, and an austere version. Table A-2 compares the two PD-214s with two existing combination ships, *Maine* and the Waterman Line's modified *Maine*, then scheduled for delivery in 1981.

By August 1980, NAVSEA had modified the jumbo PD-214 (C8-M-MA134J) into a multi-purpose mobilization ship (TAKX). Length increased to 831 ft 6 in overall; beam increased to 105 ft 6 in, about the maximum for a ship able to pass through the Panama Canal. Displacement would be 20,860 tons light or 48,860 tons loaded. Capacity was 170,000 sq ft and 400,000 cu ft, plus 1.9 million gal of liquid load; all these figures exceeded anything offered by a

Table A-2. Prospective MPS Ships, 1980

	PD-214	Jumbo PD-214	*Maine* (C7-S-95A)	Waterman (C7-S-133A)
Length (ft)	560/609[a]	670/719	640	640
Beam (ft)	97	97	102	105.5
Draft (ft)	30	30	32	33
Displacement (tons)	28,870	37,800	33,765	38,500
Power (SHP)	22,500	22,500	37,000	32,000
Speed (kt)	21.5	21.5	23	20.9
Capacity				
Sq ft	NA	NA	144,770	62,025
Cu ft (dry)	NA	NA	158,800[b]	1,039,550
Cu ft (liquid)	NA	NA	28,330	13,229

NOTES: NA indicates data not available.
[a] For the PD-214s, lengths are between perpendiculars and overall.
[b] Dry cargo can be substituted for vehicles, thereby increasing cu ft to 153,360.

Maine. Defense features included decks strengthened for 60-ton tanks, a stern ramp, and heavy cranes to permit the ship to unload herself. Lighterage would amount to four powered causeways, six unpowered ones, and two LCM(8), plus a side-warping tug to install the necessary bulk fuel distribution system. The ship would have a CH-53–size helicopter pad and she could offload LVTs down her stern ramp. By August 1980 it seemed likely that TAKX would be included in the FY 81 program, and requests for proposals to build it were expected shortly.

It would take 12 *Maine*s or 15 PD-214s to support three MABs; estimated operating and acquisition costs for FY 81–87 were $4.1 and $4.3 billion, respectively. The necessary squadrons could be placed in service in 1983, 1985, and 1987. Four *Maine*s existed (all are now in the Ready Reserve Fleet). The quickest solution would have been to buy or charter them and build eight more, but that was not done.

By this time the U.S. Ready Reserve Fleet was down to 13 ships: 1 Victory, 5 18-kt C3-S-33A, 4 Mariners, and 3 old tankers (T2-SE-A2-C) partly converted to container ships. It was backed by the 219-ship National Defense Reserve Fleet: 70 ex–U.S. Navy auxiliaries, 130 Victories, 5 more C3-S-33A, and 5 Mariners. Only the Mariners could exceed 20 kts; only the C3-S-33As could make 18 kts. Thus it would take longer to mobilize. Also, none was large enough.

By 1980, the five-year defense plan called for converting four existing ships (*Maine*s) and buying eight PD-214s. The plan had been to buy 15 TAKX, at a unit price of $103.5 million, but numbers were cut when the unit price rose to $174 million. As submitted, the FY 81 budget called for a lead ship plus $33 million in long-lead items for a second. Congress authorized two TAKX, but then the program stalled. Several shipping lines saw the program as a potential solution to the problem of surplus ships. Waterman Line had ordered its three enlarged *Maine*s for trade with the Soviet Union, but that trade was collapsing in the aftermath of the Soviet invasion of Afghanistan. Sea Land of New Jersey saw its eight fast SL-7s as fuel-eating white elephants. Reportedly it found particularly ardent supporters in Congress, who managed to confuse the MPS and fast sealift requirements. The navy budget was cut, and it was told to choose between prepositioning ships and SL-7s. Reportedly the SL-7s were not in the original FY 81 program, and they were not authorized by the House, but they were pushed hard by the Senate, and they survived into the final defense program. The navy reportedly considered the ships too expensive, and it preferred to buy TAKXs. The SL-7s had been built in Europe, the first having been delivered in December 1973. They could make 33 kts on twin screws, and their full load displacement was 51,815 tons (27,915 tons light).

The incoming Reagan administration issued new guidance in May 1981. Given its many priorities, it could not afford to buy the new TAKX. It decided instead to convert existing ships; presumably that helped deflect the political problems that had nearly killed the MPS program the previous year. The Marines feared that the whole MPS idea was being scrapped, because it was associated with the previous administration, but they were wrong. The administration was determined to gain strategic mobility, but it took a different approach. To pay for the ships, it devised a novel convert-and-charter plan, in which ships were leased from their civilian owners under long-term charters that ensured price stability. The owners gained tax advantages, and the cost of the program was spread out to make it more affordable. Ships remained under the U.S. flag, hence they were available even after the charters lapsed. In 1983 an independent study showed an average saving of over $35 million per ship; this saving was reduced to over $28 million per ship when taking into account taxes lost to the government. The idea of chartering was not new. In 1982 the navy was operating nine chartered sealift tankers. However, conversion-and-charter was novel. The Internal Revenue Service had to rule that the charter really was a service, in which case the investment in the conversion merited an investment tax credit. When that was done, the program proceeded. Currently, leasing is commonplace in military programs.

Moreover, the Reagan administration observed that the NTPS offered much the same capability planned for one of the MPS. Thus it could delay the first full MPS from FY 83 to FY 84. Thus in FY 84 there would be the NTPS plus one MPS, in FY 85 two MPS and the NTPS, and in FY 87 three full MPS. The defense guidance was modified in March 1982 to require initial forces to arrive in Southwest Asia within a week rather than within ten days, as had been previously accepted. The FY 82–86 defense guidance required that by FY 86 one division equivalent of marines (an MEF) be restructured to replace the army mechanized division assigned to the rapid deployment force. It would rely on MPS ships following up the initial marine brigade.

By 1982, plans called for providing enough aircraft and support for two full air component elements (ACEs) for two of the MPS, with a smaller air-liftable element for the NTPS. Two C5-S-78A ro-ro/container ships were converted to aviation base ships (*Wright* [T-AVB 3] in FY 85 and *Curtiss* [T-AVB-4] in FY 86) to support the ACEs. The ACE comprised 24 fighters, 20 light attack aircraft (A-4M or AV-8B), 10 heavy attack aircraft (A-6E), 4 jammers (EA-6B), 6 tankers (KC-130), 4 reconnaissance fighters, 12 utility helicopters (UH-1N), 24 attack helicopters (AH-1T), and 16 heavy-lift helicopters (CH-53E). By way of contrast, the NTPS air element was limited to 12 fighters, 10 heavy attack aircraft, 6 utility helicopters, 8 attack helicopters, 10 heavy-lift helicopters (CH-53D), and 12 troop-carrying helicopters (CH-46).

Under the Reagan administration, a strategic sealift program united the MPS and ships to carry vital equipment from U.S. ports under two categories: mobility enhancement and surge enhancement. Under mobility enhancement were the NTPS and the 13 ships of the MPS, as well as an auxiliary lighterage ship to help unload container ships in the absence of port facilities. Under surge enhancement were the 8 SL-7s (T-AKR 287–294), 6 auxiliary crane ships, 2 hospital ships, and 48 more ships for the Ready Reserve Fleet. The administration enlarged the Ready Reserve Fleet by taking advantage of a slump in world shipping to buy modern merchant ships.

The administration's FY 82 program envisaged acquiring 30 ships: 13 for the MPS (designated TAKX), the 8 SL-7s (designated TAKRX), and 9 T-5 tankers. Eight MPS ships were conversions: three from the Waterman Line (by National Steel and Shipbuilding Corporation [NASSCO] San Diego) and five (one-fifth MAB each) from the Maersk Line (by Bethlehem Steel). General Dynamics Quincy received a contract for five new ships (one-fourth MAB each). Contracts were awarded by the fall of 1982. All were chartered, to enter service in FY 85–86; delivery was up to two years ahead of schedule. The navy estimated that it saved over $500 million by using commercial rather than military standards. The last ships of the group would replace only those ships of the NTPS associated with the 7th MAB, not those used by the army and the air force.

With the NTPS already in place, the first MPS squadron deployed to the Eastern Atlantic in FY 85. Beginning that year, the five dry cargo ships serving the MAB in the Gulf were replaced by ships of the second MPS squadron at Diego Garcia. The third squadron was initially retained in U.S. ports at ten-day readiness under MSC operational control for sealift. As new sealift ships entered service, it went to the Western Pacific in FY 88.

By 1984 the MPS ships were being supplemented by 12 depot ships: 2 break-bulk ships for general cargo and ammunition, 1 ship with a rapidly deployable medical facility (RDMF), 4 LASH, 4 POL tankers, and 1 heavy-lift ship (at Diego Garcia). One of the ammunition ships was in the Mediterranean.

The eight SL-7s had already been bought. They were partially converted (as TAKRXs) to ro-ros at NASSCO, Avondale, and Penn Ship; contracts for the first four were awarded on 10 September 1982. The eight ships sufficed for two army divisions, less nondivisional support units. They could load in 2–4 days. Given their very high speed, they could reach Europe from the continental United States in 3–5 days, and Southwest Asia (via Suez) in 13 days (20 days if they could not use the canal). Ships were maintained on both coasts, beginning with pairs in Jacksonville and New Orleans. The SL-7s were also used for normal Defense Department shipping, replacing the old ro-ros *Meteor* and *Comet*, running about a month each year. Otherwise they were maintained at four days' readiness.

The T-ACS, a crane ship, was particularly important. Properly ballasted, it could unload container ships. A T-ACS could offload an alongside container ship up to Panamax (maximum to transit the Panama Canal) dimensions. She also gained Sea Shed capability (see below). Decks were outfitted to accommodate lighterage sections (see below). The invitation to bid was released in August 1982. The first ship, *Keystone State* (T-ACS 1, ex–*President Harrison*), was delivered 7 May 1984. Eleven ships were planned, later increased to twelve, but the last two were canceled in 1990. Ships converted under the FY 85 program were given vertical motion compensation, and this feature was back-fitted to earlier ships. Plans called for retaining two to four ships to train personnel; in practice only one training ship is normally operational. The rest are in the Maritime Administration's Ready Reserve Fleet. Other surge ships were the Marines' aviation depot ships (TAVBs) and two hospital ships (T-AHs) converted from tankers, *Mercy* (T-AH 19) and *Comfort* (T-AH 20).

Other dedicated sealift comprised the remainder of the MSC fleet plus the Ready Reserve Fleet of merchant ships. In FY 84, MSC had 67 ships, and 32 were in the Ready Reserve Fleet. The review conducted to write the program objective memorandum for 1984 (POM-84) showed that 77 Ready Reserve Fleet ships would be needed by FY 86: 50 dry cargo (in C4 equivalents) and 16 tankers, plus 11 crane ships (TACS). Plans called for increasing the MSC fleet to 72 and the Ready Reserve Fleet to 77 by FY 90.

Because the Reagan administration took sealift seriously, it went beyond the stated goals. In FY 84 the navy obtained 19 break-bulk ships and 2 LASHs from the Maritime Administration and upgraded another 5 ships from the NDRF, thus reaching 66 Ready Reserve Fleet ships, all ready to load within 4 days of mobilization. All the TACS and 16 of the cargo ships could be mobilized in 5 days, the rest needed 10 days. Ultimately the Maritime Administration wanted to obtain all available LASHs while they remained on the market.

In 2000, the Ready Reserve Fleet amounted to 91 ships. Congress had diverted money for seven more in December 1994. Due to underfunding, 16 RRF ships had to be shifted to the NDRF, another 29 reduced to the lowest level of maintenance, and 26 previously kept at 5-day readiness reduced to 10–20-day readiness. As of mid-2000, another 14 break-bulk ships were to be reduced to the NDRF by October 2001. Many current RRF ships have steam plants, which are increasingly difficult to operate. As of 2000, the RRF comprised 31 ro-ros, 10 crane ships, 3 Seabees, 4 LASHs, 7 combination cargo ships (break-bulk, containers, and some ro/ro), 21 break-bulk ships, 7 tankers, 3 small coastal tankers, and 3 troopships; the 2 T-AVBs were also considered part of the RRF.

There was also the U.S.-flag merchant fleet. In FY 84, there were 363 suitable active U.S.-flag merchant ships: 206 general cargo ships, 155 tankers (clean-coated, under 80,000 deadweight tons), and 2 liners (convertible to troopships). Projections showed that the U.S. flag fleet would probably have fallen to 231 ships by FY 90, more than making up for the planned MSC and Ready Reserve Fleet expansion. The FY 93 JCS goal was 107 MSC ships and 77 Ready Reserve Fleet ships.

Some active U.S.-flag merchant ships were included in a sealift readiness program: 160 general cargo ships and 24 tankers (under 80,000 deadweight tons). From 1979, membership had been mandatory for all ships built or operated under U.S. government subsidies. Owners carrying Defense Department cargo had to place half their ships under the readiness program. Other U.S.-flag ships would be chartered. For mobilization, no voluntary plan would be responsive enough, and all militarily useful U.S.-owned ships would be requisitioned. In FY 84 the Maritime Administration's NDRF amounted to 149 general cargo ships, 15 militarily useful tankers, and 18 troop transports (including 2 former state Maritime Academy ships).

Next in potential usefulness was the Effective U.S. Control Fleet (EUSC), the ships registered in Liberia, Panama, Honduras, and the Bahamas that were owned by Americans and covered by intergovernmental agreements. Ships earmarked for mobilization amounted to 23 general cargo ships and 57 tankers suited to military use (i.e., of less than 80,000 deadweight tons), out of a total of over 350 ships.

Finally there were the merchant fleets of NATO and other allied countries. In 1984 the NATO countries were expected to contribute 400 cargo ships. Korea, under a March 1981 memorandum of understanding would contribute 27 cargo ships, plus 5 tankers. NATO ships, however, are for NATO emergencies, so they would probably not be available in the event of war in Korea or the Gulf.

Counting all ships except the MSC or the Ready Reserve Fleet, the available total was impressive: 1,057 ships, of which 805 were general cargo types, 232 were tankers, and 20 were troop transports. However, sealift might be most important outside of NATO, in which case NATO countries might well refuse to participate (as in Vietnam), and the Koreans might not help except in a Far Eastern crisis. Thus, in some very important cases, 400 or more of the cargo ships would not be available. No one could really say how many of the elderly NDRF ships would be usable. The Ready Reserve Fleet ships would really be the core of any surge capability.

In January 1983 the General Accounting Office described mobilization plans for the rapid deployment force. Marshaling of merchant ships would begin seven days before formal mobilization. Within 30 days after mobilization, the navy hoped to have acquired 204 dry cargo ships: 119 from the active U.S. merchant fleet, 37 from the active MSC fleet, 38 from the Ready Reserve Fleet, and 10 from the National Defense Reserve Fleet. Of these ships, 62 would be available on the day of mobilization; without premobilization marshaling, only 18 would be available at mobilization. Of the Ready Reserve Fleet ships, round-the-clock shipyard work would make 30 ready to receive cargo in 5–10 days. Ultimately the navy hoped for 77 Ready Reserve Fleet ships within the first ten days, but the Maritime Administration suspected that shipyard capacity was insufficient, and it has since declined.

Tankers might be a problem. Of 228 U.S.-flag tankers, two-thirds carried crude oil, hence they would have to be cleaned before carrying refined products like jet and motor fuel. The U.S. control fleet amounted to 500 tankers, mostly crude carriers, but only about 50 were suitable for military use.

There was a major gap in troopships, which would be needed if there were no airfield to receive U.S. troops. Only two U.S.-flag passenger ships were in service, in the Pacific, with a capacity of about 4,000 troops each. The others earmarked for wartime service were the laid-up *United States* (capacity 4,000 troops) and two of the state Merchant Marine Academy schoolships (capacity about 1,000 each). About 1984 there were plans to upgrade one World War II troopship (T-AP) in the NDRF each in FY88–90, to go into the Ready Reserve Fleet, but this was not done, and whatever interest there had been in troopships evaporated. The RRF currently includes three ships classed as troopships: an ex-surveying ship and two active state Merchant Marine Academy ships.[5] None of the three was specially converted to this role, however, and total capacity is quite small.

There was some fear that Ready Reserve Fleet ships might not be ready enough. Some had broken down in exercises. For example, in Reforger 82, in which supplies were brought across the Atlantic, the 38-yr-old *Seatrain Ohio* broke down. Such experience certainly seemed to justify the Reagan administration's program of buying up modern foreign merchant ships for the Ready Reserve Fleet.

Under a U.S. Navy–Maritime Administration memorandum of understanding, beginning in FY 84 the navy paid for sealift enhancement features in existing commercial ships. To support the program, a Strategic Sealift Program Office was established in NAVSEA. The Maritime Administration paid for features in new or overhauled ships.

There were two categories: general enhancements and platform improvements. General enhancements included nuclear-biological-chemical (NBC) washdown, the provision of hard points and tiedowns so that offload systems could be carried on deck, and additional lighterage carrying capacity. Communications, navigational aids, and habitability were brought into line with current U.S. standards, especially in foreign-owned ships bought for the RRF; navigational equipment in particular was sometimes superior to that on board U.S. combatants.

Platform improvements made it possible for container ships to carry large military vehicles and other outsize loads. Flatracks were topless and sideless containers (40 × 8 × 12 ft wide) with structural members to adapt container ships to oversize containers. Sea Sheds were oversize (40 × 25 × 12.5 ft) open-top containers with hinged work-through floors for oversize or break-bulk cargo or vehicles. As of 1984 the navy's goal was 1,700 Sea Sheds and 6,800 Flatracks. The ultimate goal was a capacity to lift 75,000 short tons of equipment.

Other enhancements included portable underway solid replenishment stations and helicopter decks. Given those additions, they could transship cargo to ammunition ships (AE, now T-AE) at sea. Tankers were provided with offshore petroleum discharge system (OPDS) units, so that they could discharge their fuel without using terminals. The amphibious bulk fuel system (ABFS) used 5,000 ft of 6-in floating hose to bring fuel ashore. It in turn would be fed by 135,000-gal bladders shuttled from tankers designated amphibious assault fuel supply facility (AAFSF). Under more benign conditions, a commercial system could be installed, using four miles of 6-in piping leading out to an offshore buoy to which a tanker might connect.

An army program developed special equipment for ships in the assault follow-on echelon under a container offloading and transfer system (COTS). An elevated causeway (ELCAS) consisted of 40 self-jacking causeway sections (each 90 × 21 × 5 ft; 31 in the roadway, 9 in the pierhead), to form a platform at which lighters could offload seaward of the surf line onto flatbeds. It included two 150-ton cranes and two vehicle turntables. ELCAS was tested in FY 77 and entered service in FY 83 in the Atlantic and in

FY 84 in the Pacific. As with the World War II Rhinos, causeway sections could be combined into lighters and tugs either for ro-ro discharge or to ferry cargo ashore. A ferry could be assembled out of four causeway sections. Designations were CSP/CSNP/SLWT (causeway section powered/causeway section non-powered/side-loaded warping tug). Special buoys were built so that ships could be moored fore and aft in a confined area offshore.

Overall, sealift was a major triumph of the Reagan administration. The U.S. fast sealift capability was amply demonstrated during the Gulf War, when the MPS did indeed provide battle-ready marines almost instantly to block any further Iraqi advance into Saudi Arabia. U.S. and allied strategic sealift delivered the huge force that eventually invaded Iraq and liberated Kuwait. Without that sealift, the operation could not have been carried out.

By the 1990s, the three MPS squadrons were fixtures of U.S. planning. As MABs evolved, their footprints grew, so that an additional ship had to be added to each squadron under an enhanced maritime prepositioning force program (FY 95, FY 97, FY 99). The extra ships added equipment for an expeditionary airfield, a naval mobile construction battalion, and a 500-bed fleet hospital. NAVSEA awarded five engineering contracts in April 1996, and the first conversion (Phase II) and five-year operating (Phase III) contracts were awarded in February 1997 to Tarago Shipbuilding of Bethesda, Maryland. *Tarago* was reflagged and renamed USNS *1st Lt Harry L. Martin* in FY 90.

In FY 90 Congress added an LSMR (large sealift ship, medium speed, ro-ro) program to make up for insufficient lift. At first plans called for 36-kt ships, but that would have been prohibitively expensive. Plans call for at least 14 ships in 2 classes (TAKR 300–306 and 310–317); hull numbers 307–309 are reserved for further ships.

In FY 92, presumably as a result of Gulf War lessons learned, Congress mandated an army warfare readiness (AWR-3) or army brigade afloat program, to preposition equipment and 15 days' supplies for a heavy brigade and for its 6,000-personnel support element (38 days' supplies). Its ships are positioned to reach the Far East or the Middle East within 14 days. The program began with Afloat Prepositioning Ship Squadron Four (APSRON 4), established in the Persian Gulf in November 1996, consisting of seven activated Ready Reserve Fleet ships. It included five converted ships (TAKR 295–299), each to carry the equipment of an army armor task force built around 58 tanks, 48 other tracked vehicles, and 900 other vehicles. APSRON 4 was disestablished late in 1998 and its ships transferred to the Ready Reserve Force. Some ships, however, were attached to MPS squadrons.

As of 1999–2000, the Marines' plans for 2010's next-generation MPS ships called for a capability to dock LCACs and to handle at least one MV-22 Osprey. The Marines planned to fly troops out to the ships via Ospreys while they were en route to a crisis area. Once there, the ships would be the sea bases of the Marines' new OMFTS strategy, so they have to be large enough to allow full access to all equipment on a random, rather than a combat loaded, basis. The result would be similar in configuration to an LHD, but far larger. It certainly would not be a converted merchant ship, but it might be made less expensively than an LHD if it is built to commercial standards.

By modern standards, the World War II amphibious programs were almost unimaginably huge. LSMs and support landing craft (LSMRs and LCSLs) are shown laid up in the Columbia River at Tongue Point, Oregon, May 1946. Many of the LCSLs would eventually be transferred to friendly navies.

Appendix B: Landing Craft/Ship Programs

Hull designations of uncommissioned (i.e., small) amphibious craft are given wherever possible. Before FY 64 (the fiscal year beginning 1 July 1963), U.S. Navy–built craft were simply given numbers. Contractor-built craft were given numbers prefixed by "C." Most numbers in both series were sequential, but blocks of low C numbers were reserved for experimental craft; some of these low numbers were subsequently assigned in the 1950s. Similar out-of-sequence blocks are evident in U.S. Navy numbers. After World War II authorization for amphibious ships and auxiliaries often included attached landing craft, so bulk craft authorizations and production do not always match.

Beginning with FY 64, craft were given designators. The first digits indicate the length, followed by a two-letter designator; for example, 56CM denotes a 56-ft LCM(6). The first two figures after the letters denote the fiscal year of order, followed by a sequence number; for example, 56CM6514 was the fourteenth LCM(6) of the FY 65 program. Each type was separately numbered.

Given space limitations, builders generally are not listed. Wartime lists are by directive. Wartime transfers were by Lend-Lease; the 1946–47 transfers to China were a continuation of this program. Major units were disposed of either to the Maritime Commission (for merchant service) or to the State Department (for sale to other governments). In at least some cases, such as the Netherlands just after World War II and Israel later, governments bought landing ships and craft through merchant channels.

Major Units

LCI(L) (landing craft, infantry, large)

1942: 300 ordered in February (LCIL 1–300), 50 ordered in June (LCIL 301–350); 48 were canceled in November (LCIL 49–60, 137–160, 197–208). The 300 was the original British request, 50 were added to ensure that 300 would be ready in time.

1943: 72 ordered on 18 January (LCIL 351–422), but 10 per month were added 26 January (LCIL 351–542 assigned 17 February: a full year at this rate); 182 ordered in June (LCIL 543–724); 57 ordered in September (LCIL 725–781), 328 ordered in December (LCIL 782–1109).

1944: 90 ordered in March (LCIL 1110–1199, but only LCIL 1110–1139 were assigned); 100 were canceled on 9 August (LCIL 717–724, 822–837, 885–901, 911–928, 1099–1139). Later another 52 were canceled (LCIL 781, 838–865, 902–910, 929–942) to bring total cancellations to 198 out of 1,139. Total completed: 941.

LCS(L)3 (landing craft, support, large, Mk 3)

LCS(L)1 and LCS(L)2 were British craft. The LCS(L)3 was the U.S. version developed.

1944: In April production authorized for 130 craft, LCS(L) 1–130.

LCT (landing craft, tank) and LCU (landing craft, utility)

1942: 400 LCT ordered in February (LCT 1–400), 100 ordered in April (LCT 401–500); on 30 July, however, LCT 89–118 (30 craft) were canceled.

1943: 120 LCT ordered in January (LCT 501–620, all LCT[6]), 380 ordered in June (LCT 621–1000), 360 ordered in November (LCT 1001–1360).

1944: 105 LCT ordered in March (LCT 1361–1465). Total LCT wartime production: 1,435.

Lend-Lease transfers: 176 LCT(5) to Royal Navy (LCT 2, 4–6, 8–14, 37–57, 73–79, 119–124, 130, 131, 135, 138, 150, 186–194, 225–236, 238–240, 243, 246,

261–267, 269, 270, 272, 273, 275, 281–287, 289, 291, 292, 295–297, 301–307, 309, 310, 312, 313, 331, 334–339, 341, 343–345, 361, 363, 398, 399, 402, 420–30, 432, 433, 435–442, 444, 445, 450, 453–461, 477–480, 483–485, 487–491, 498–500), 2 LCT(6) (LCT 627, 628) to Royal Navy, 17 to the Soviet Union (LCT 599, 561, 563, 744, 745, 1015, 1046, 1047, 1163, 1176, 1434–1438, 1442, 1445).

War losses: LCT 19 (14 July 43), 21 (14 Jan 43, accident), 23 (3 May 43, accident), 25 (6 Jun 44), 26 (25 Feb 44, accident), 27 (6 Jun 44), 28 (30 May 43), 30 (15 Feb 44), 35 (15 Feb 44), 36 (26 Feb 44, accident), 66 (12 Apr 45, accident), 71 (11 Sep 43, accident), 128 (Nov 44, accident), 147 (Jun 44, accident), 154 (31 Aug 43), 175 (21 Feb 45, accident), 182 (7 Aug 44, accident), 185 (24 Jan 44, accident), 196 (27 Sep 43, accident), 197 (6 Jun 44), 200 (Jun 44), 208 (20 Jun 43, accident), 209 (10 Jun 44, accident), 220 (13 Feb 44, accident), 241 (15 Sep 43), 242 (24 Sep 43), 244 (Jun 44), 253 (21 Jan 45, accident), 293 (11 Oct 43, accident), 294 (6 Jun 44), 299 (13 Nov 43, accident), 305 (6 Jun 44), 311 (9 Aug 43, accident), 315 (23 Mar 44), 319 (27 Aug 43, accident), 332 (6 Jun 44), 340 (20 Feb 44, accident), 342 (29 Sep 43, accident), 352 (12 Apr 45, accident), 362 (6 Jun 44, accident), 364 (6 Jun 44), 413 (Jun 44), 458 (7 Jun 44), 459 (9 Oct 44, accident), 496 (Nov 43), 548 (Nov 44, accident), 555 (6 Jun 44), 579 (4 Oct 44), 582 (22 Jan 44, accident), 593 (6 Jun 44), 597 (6 Jun 44), 612 (6 Jun 44), 703 (6 Jun 44), 713 (Jun 44), 714 (Jun 44), 961 (21 May 44, accident), 963 (21 May 44, accident), 983 (21 May 44), 984 (15 May 44, accident), 988 (15 May 44, accident), 995 (Apr 45, accident), 1029 (24 Mar 45, accident), 1050 (4 Aug 45), 1075 (12 Dec 44), 1090 (26 Mar 45), 1151 (26 Jan 45), 1191 (Sep 45, accident), 1358 (4 May 44, accident). Many LCTs were destroyed at the end of the war instead of being brought home.

Conversions: Minehunters: LCT 843, 844, 887–890 became AMC(U) 1–6 in March 1945. Garbage lighter (YG): LCT 353 became YG 68 in September 1946. Ferries (YFB): LCU 610 and 639 became YFB 65 and 66 in June 1951. Diving tender (YDT): LCU 854 became YDT 9.

Postwar transfers: 2 to Cambodia (LCU 622 in 1956, LCU 1421 in 1962), 3 to Chile (LCU 1273 in 1970, LCU 1396, 1458 in 1970), 8 to China in 1946–47 (LCT 512, 515, 849, 892, 1143, 1145, 1171, 1213, of which the Communists got 515 and 1171), 12 to Taiwan in 1959 (LCU 638, 700, 1212, 1218, 1225, 1244, 1271, 1278, 1367, 1397, 1429, 1452); 9 to Denmark in 1961–63 (LCU 715, 765, 810, 1042, 1230, 1294, 1373, 1383, 1422), 3 to Dominican Republic, 4 to France (LCT 799 and 834 in 1946, LCT 3 and 476 [possible source document error] in 1949), 18 to Greece (1946–48: LCT 594, 607, 619, 620, 625, 1227, 1293, 1297, 1300, 1301; 1959–60: LCU 763, 766, 827, 842, 971, 1229, 1379, 1382), 1 to Iran in 1964 (LCU 1431), 2 to Israel (LCU 640, 673), 3 to Peru (LCU 501, 855, 1161), 1 to the Philippines in 1948 (LCT 1117), 1 to South Korea in 1960 (LCU 531), 6 to Thailand in 1946–47 (LCT 753, 800, 861, 904, 1089, 1260), 4 to Turkey in 1967 (LCU 588, 608, 666, 667), 2 to Vietnam (LCU 1051 in 1956, 1221 in 1960), 1 to West Germany in 1958 (LCU 779).

FY 52: 38 LCU (LCU 1466–1503), of which 6 were for transfer. LCU 1478 transferred to Norway. LCU 1479, 1480, 1501–1503 transferred to Indo-China, but LCU 1503 was lost overboard on passage. LCU 1504–1593 (90 units) were built under navy contract for the U.S. Army. Fourteen more were built in Japan for transfer to Taiwan (LCU 1594–1601) and for Japan (LCU 1602–1607). Other transfers: LCU 1471 and 1491 to Spain; LCU 1474 to Lebanon; LCU 1475, 1479, 1480, 1481, 1484, 1485, 1493, 1494, 1498, 1501, 1502, 1594, 1595 to Vietnam; LCU 1596–1601 to Taiwan; LCU 1602–1607 to Japan; LCU 1627 to Burma.

FY 55: LCU 1608 and 1609 (2 units).
FY 56: LCU 1610–1619 (10 units).
FY 57: LCU 1620–1622 (3 units). LCU 1622 was later canceled.
FY 59: LCU 1623 and 1624 (2 units).
FY 63: LCU 1625 and 1626. LCU 1626 ordered for Burma.
FY 65: LCU 1627 and 1628 (2 units).
FY 67: LCU 1628–1637 (10 units).
FY 68: LCU 1638–1645 (8 units).
FY 69: LCU 1646–1666 (21 units).
FY 73: LCU 1667–1670 (4 units) for U.S. Army; an additional 2 unnumbered units ordered for Saudi Arabia are not included in FY 73 budget.
FY 74: LCU 1671–1679 (9 units) ordered for U.S. Army.
FY 85: LCU 1680 and 1681 (2 units, also designated 135CU8501 and 135CU8502) ordered for Naval Reserve training. Total postwar production: 115.

In January 2000, 37 were active in the U.S. Navy; another, the ex-army LCU 1590, has a hydraulic crane permanently mounted in her well deck. Of the 37, 2 are workboats and 1 is an exercise minelayer.

Conversions: Ferries (YFB): LCU 1636–1638 and 1640 became Guantanamo ferry boats, YFB 88–91. Harbor clearance craft, designated as LLC (light lift craft): LCU 1348 became LLC 1; LCU 1459 became LLC 4. Harbor utility craft (YFU): LCU 509 (YFU 54), LCU 524 (YFU 1), LCU 529 (YFU 2, later LLC 5), LCU 539 (YFU 48), LCU 550 (YFU 3), LCU 562 (YFU 4), LCU 592 (YFU 5), LCU 600 (YFU 6), LCU 629 (YFU 7), LCU 637 (YFU 55), LCU 646 (YFU 56), LCU

649 (YFU 84), LCU 664 (YFU 8), LCU 668 (YFU 10), LCU 677 (YFU 11), LCU 686 (YFU 12), LCU 709 (YFU 57), LCU 715 (YFU 85), LCU 716 (YFU 49; later YFU 58, salvage lift craft role), LCU 742 (YFU 13), LCU 743 (YFU 52), LCU 764 (YFU 14), LCU 776 (YFU 15, later YFU 59), LCU 780 (YFU 87), LCU 788 (YFU 16, later LLC 2), LCU 840 (YFU 17), LCU 851 (YFU 60), LCU 869 (YFU 18), LCU 877 (YFU 19), LCU 960 (YFU 20), LCU 973 (YFU 21; later YFU 62, salvage lift craft role), LCU 974 (YFU 22), LCU 979 (YFU 23), LCU 980 (YFU 24), LCU 986 (YFU 63), LCU 1056 (YFU 25), LCU 1082 (YFU 26), LCU 1086 (YFU 27), LCU 1124 (YFU 28), LCU 1126 (YFU 64), LCU 1136 (YFU 29), LCU 1156 (YFU 30), LCU 1159 (YFU 31), LCU 1162 (YFU 32), LCU 1165 (YFU 65), LCU 1195 (YFU 33, later LLC 3), LCU 1203 (YFU 66), LCU 1224 (YFU 34), LCU 1232 (YFU 67), LCU 1236 (YFU 35), LCU 1250 (YFU 36), LCU 1283 (YFU 37), LCU 1286 (YFU 38), LCU 1330 (YFU 47), LCU 1363 (YFU 39), LCU 1373 (YFU 86), LCU 1376 (YFU 40), LCU 1378 (YFU 41), LCU 1384 (YFU 42), LCU 1385 (YFU 68), LCU 1386 (YFU 43), LCU 1388 (YFU 69), LCU 1398 (YFU 44), LCU 1411 (YFU 45), LCU 1430 (YFU 46), LCU 1486 (YFU 50), LCU 1488 (YFU 94), LCU 1608 (YFU 91), LCU 1610 (YFU 100), LCU 1611 (YFU 97), LCU 1612 (YFU 101), LCU 1615 (YFU 98), LCU 1620 (YFU 92), LCU 1622 (YFU 99), LCU 1625 (YFU 93), LCU 1642 (YFU 102). Range support craft: LCU 1618 was converted in 1977 to support cable-controlled underwater recovery vehicle (CURV) and became IX 508 in 1979. Swimmer delivery vehicle (SDV) retrievers, designated ASDV, with right-angle drive: LCU 1621 (ASDV 1), LCU 1623 (ASDV 2), LCU 1628 (ASDV 3). LCU 1641, not reclassified, serves as an exercise minelayer.

Other transfers: Eight ex-army craft were transferred to Colombia in 1990 and 1992, including LCU 1543 and 1583. One unit was transferred to Cambodia in 1969, probably by the U.S. Army (it does not appear in U.S. Navy records). South Korea received ex-army LCU 1542 in 1992, but it was not activated.

Foreign construction: Brazil built 4 LCU 1610 type (*Guarapari* class) in 1974–78. South Korea built 6 in 1979–81 (*Mulkae* 72–78 class). Taiwan built 2 based on the LCU 1610, completed in 1979. The Indonesian *Kupang* class is based on the LCU 1610 design. The Thai *Kaeo* class is based on the LCU 1626 class. The Turkish C 205 class (12 units) is based on the LCU.

LSM (landing ship, medium) and LSMR (landing ship, medium, rocket)

1943: 220 ordered in September (LSM 1–220, originally LCT(7) 1501–1720), 30 were added in October (LSM 221–250; program may have run to LSM 252), 100 ordered in November (LSM 253–352).

1944: 105 ordered in March (LSM 353–457), 100 more ordered in April (LSM 459–558). In December 1944 LSM 401–412 were reordered as rocket ships (LSMR), followed by 501–524 in February and 525–536 in April 1945, for a total of 48 LSMR; another 14 (186–199) were converted after completion.

LST (landing ship, tank)

1942: 300 ordered in January (LST 1–300), 190 ordered in May (LST 301–490); 100 canceled in June 1942 (LST 85–116, 142–156, 182–196, 232–236, 248–260, 296–300, 431–445).

1943: 20 ordered in January (LST 491–510), 219 ordered in June (LST 511–729), 24 ordered in September (LST 730–753), 275 ordered in November (LST 754–1028).

1944: 102 ordered in March (LST 1029–1130), 25 ordered in June (LST 1131–1155, of which LST 1155 was canceled). Total wartime production: 1,054.

FY 52: 15 ships (LST 1156–1170).

FY 53: 20 ships dropped from draft program to help finance the carrier *Saratoga* (CV 60).

FY 54: 1 ship (LST 1171).

FY 55: 10 ships in program, but 2 were dropped during 1954 to pay for 4 YAGR ocean radar picket conversions, and the prototype was put back into the FY 54 program, leaving 7 (LST 1172–1178). Later LST 1172 was canceled to make up for shipbuilding budget overruns, leaving 6.

FY 58: 8 ships planned but dropped in June 1955 in favor of 2 helicopter carriers and a fast LST; the latter was subsequently dropped.

FY 65: 1 ship (LST 1179).

FY 66: 8 ships (LST 1180–1187).

FY 67: 11 ships (LST 1188–1198).

FY 71: 7 ships, deferred from program to program, were dropped. Total postwar production: 43.

Minor Units

Troop Barges

One prototype (No. 9135) and 3 production Barge A (Nos. 6744–6746) were built. The prototype, ordered 16 October 1923 from Norfolk Navy Yard, was completed 3 January 1924. No. 6745 capsized and sank on preliminary trials. Salvaged, it was transferred to the War Department on 4 September 1930, probably becoming the army's first landing craft. It seems unlikely that the follow-on 42-footer was built.

Artillery Lighters

Two were built before World War I for *Henderson*. The prototype U.S. Navy–built 45-footer was No. 11422 (Norfolk, ordered October 1928, and shipped October 1930). FY 36: Nos. 13165 and 13166 (2 boats) were built to a new design, ordered May 1936; the first was received by FMF in San Diego in February 1937. FY 37: No. 13556 built. FY 39: Nos. 13976 and 13977 (2 boats) built; FY 41: C-1979–1983 (5 boats), C-2183 and 2184 (2 boats), and C-3823–3834 (12 boats) built. A total of 19 were on hand by the end of November 1941, and 2 more were built during 1942.

Boat Rigs

The first 2 production rigs were bought under the FY 36 program, followed by 3 in FY 37 and 5 in FY 40. Ramps (for emplacement at the beach, to connect to boat rigs): 2 in FY 36, 3 in FY 37.

LCAC (landing craft, air cushion)

91 craft (LCAC 1–91) authorized. Program plans: 3 in FY 82, 3 in FY 83, 6 in FY 84, 9 in FY 85, 12 in FY 86, 15 in FY 89, 12 in FY 90, 24 in FY 91, 12 in FY 92; not all were authorized or ordered. Of the 91 authorized, 74 will enter the service life extension program (SLEP); the others will be placed in storage. As of 2000, LCAC 91 still had not been delivered. Foreign sales: 6 to Japan, 1 for South Korea (plus 1 locally built, of modified design).

LCC (landing craft, control)

The first production directive was issued 14 June 1943, for 27 (probably C-39044–39058 plus 12 U.S. Navy–built hulls, Nos. 21432–21443). Later series were C-25470–25511 (42) plus 12 navy (Nos. 22666–22677). The last 18 were authorized in December 1943 (C-60140–60157, Mk 2), for a total of 99: 36 LCC(1) and 45 LCC(2) delivered in 1943, and 18 LCC(2) delivered in 1944. A 9 May 1944 directive added another 36 (C-97859–97894), but they were canceled on 3 October 1944. Transfers: 6 were transferred to the Nationalist Chinese in 1946–47.

30-ft Landing Craft

Programs: 3 experimental commercial boats in FY 36, 1 C&R boat in FY 37 (No. 13527), 5 C&R boats in FY 38 (Nos. 13624–13628, ordered February 1938), 11 boats in FY 39 (navy nos. 14000 and 14001 plus 5 C&R type [C-1752, C-1753, C-1755, C-1779, the last built by Welin] and 5 Higgins type [C-1781, C-1793–1796]), 18 boats in FY 40 (12 C&R type [C-1807–1809 and C-1818 plus 8 C&R type built by Higgins: C-1840–1847] and 6 Higgins type [C-1817, C-1830–1834]. A November 1939 description of the FY 40 program showed 15 rather than 18 boats, so some of the units listed were presumably added at the end of FY 40. As of November 1939, plans called for 14 boats in FY 41. Actual FY 41 orders were far larger. They amounted to 48 C&R boats (C-1879, C-1915–1922, C-2111–2142, and 7 between C-1922 and C-2111, numbers unknown) plus 62 Higgins boats (C-1947–1978, C-2020–2049). Another 5 Higgins boats (C-3438–3442) seem to have come from a separate order unrelated to the landing craft program. Totals: 3 experimental craft (fishing boats), 68 C&R boats, 72 Higgins "Eurekas." At the end of November 1941, 116 30-footers were on hand.

Of the C&R boats, the navy's No. 14001 was a diesel version of the wooden type. At least C-1779, C-1807–1809, C-1818, C-1840–1847, C-1915–1922, and C-2111–2142 were steel. U.S. Navy sea sleds were never included among the experimental boats, although in 1935 C&R expected to use them as landing craft. Known 32-ft sea sleds: C-233, C-273, C-275–283, C-1131–1140, for a total of 21 boats. There was also a presumably experimental 30-ft cargo surf boat, U.S. Navy–built No. 13693.

LCP(L) (landing craft, personnel, large), LCP(R) (landing craft, personnel, ramped), LCV (landing craft, vehicle), and LCVP (landing craft, vehicles and personnel)

Directives for all types of 36-ft Higgins boats designated LCP(L), LCP(R), LCV, or LCVP follow.

1940: 335 ordered in September 1940 (C-1923–1946, C-2143–2182, C-2284–2554). The British ordered 262 boats of a slightly different design. The 1940 directive does not include army boats; in May–June 1940 the U.S. Army decided to buy 4 tank lighters, 10 motor launches (command boats), and 40 landing craft for each of two divisions (1st and 3rd) being trained in amphibious operations. The initial army Higgins boats were its Design 185: CL 1–80. They were delivered between December 1940 and February 1941. Only 10 command launches (AC 1–10) were delivered, all built by Chris-Craft.

1941: 188 ordered in May (first requested in February: C-3513–3700; C-3598–3612 and C-3628–3700, 88 boats, were ramped LCVs), 629 ordered in June (C-3800, C-3877–3985, C-3987–4138, and C-4816–5182; C-4039–4138, 100 boats, and C-4987–5182, 196 boats, were ramped LCVs; of the C-4816 group, 38 were completed as LCPL, 133 as LCPR), 332 ordered in November (C-6089–6420, all LCV). The June directive comprised 262 boats for the 3rd Marine Division and 367 boats for boat pools at Norfolk and San Diego. By mid-August the Marines wanted 384 (one-third) of their 1,152 Higgins boats to have ramps. The 262-boat order was changed to 162 Eurekas without ramps and 100 with ramps (originally 89 without and 173 with ramps). That left 196 ramped boats, procurement of which was held up pending the report of a board testing the improved type. The November order was probably for boat pools (Norfolk and San Diego; the figure originally given was 334). In all, 564 LCPL were available at the end of 1941. LCV production in 1941: 110.

1942: 442 ordered built in January (including C-6498–6644, C-6674–6849, C-7560–7588; all LCPR), 500 ordered built in February (C-8460–8959; C-8460–8684 [225 boats] were LCPR, and C-8685–8959 [274 boats, with one hull number used for a cabin cruiser] were LCVs), 1,100 ordered built in March (C-9505–10104; first 600 were LCPR, and the other 500 were possibly army orders), 1,800 ordered in April–May (600 were LCPR [C-10155–10754], 1,200 were LCV [C-10755–11954, of which 10755–11354, 600 boats, were reordered as LCVP]), 600 ordered as LCV in June (C-11355–11954), 74 ordered built in August (ex-army craft turned over to the navy as C-25142–25215), 120 ordered built in September (possibly army orders), 3,300 ordered as LCVP in October (C-17603–17902, C-19971–20970, and C-22161–24160). The 500 of the March directive and the 120 of September were presumably army orders. Known army deliveries for this period were 50 Design 185A boats (CL 81–130). Another 74 incomplete hulls were turned over to the U.S. Navy in August (as noted above). Presumably the others on army contract were canceled and reordered under navy numbers. CL 132 (Design 197) was an experimental Higgins landing craft, 36 ft × 10 ft 5 in × 5 ft, powered by a 165 HP gasoline engine, that was tested in January 1943. British Lend-Lease LCPL, not included in the directives, were C-213–262. Total 1942 production: LCPL, 307; LCV, 1,870 (including 110 begun in 1941); LCPR, 1,563; LCVP, 215.

1943: Presumably orders were for the big European landings and for the limited Pacific offensive. All were initially to have been LCVPs. Directives: 4,000 in January and 1,000 in February (together, C-30782–35781, all LCVPs), 7,259 in June (C-39071–46029, 6,959 LCVP; C-46030–46229, 200 LCPL; C-46230–46329, 100 LCPR), 1,000 in September (C-54047–55046, all LCVP). C-38541–38940 (400 LCPL) were presumably for Lend-Lease. In November 1943, 300 of the September LCVPs were switched to LCPRs (C-54047–54346) and 300 were switched to LCPLs (C-54347–54646). In addition, 174 LCPL were approved: C-69388–69438 (99 canceled) and C-69889–69963 (75 not used). Total 1943 production: LCPL, 282; LCV, 306; LCPR, 24; LCVP, 8,027. LCV production ended in 1943; a total of 2,286 were built.

In January 1943, when it seemed that no more LCP(R) were wanted by U.S. services, the British still wanted 230 (100 for delivery by April 1943, 130 more by August). Note the September 1943 revisions.

1944: When the big Pacific offensive was ordered late in 1943, personnel craft authorizations again rose. 1,905 ordered in March 1944 (1,650 LCVP [C-69964–71613] plus 255 LCPR [C-71614–71868]), 9,600 LCVP ordered in May (C-77639–87238), 225 LCPL ordered in August (C-87239–87688); and 300 LCPR ordered in September (C-87839–88138). In addition, construction of 250 LCPL for the British was approved in April 1944 (C-74121–74370). C-88000–88005 were converted to ramped utility boats. Total 1944 production: LCPL, 547; LCPR, 705; LCVP, 9,290.

1945: January orders were for 225 LCPL (C-87464–87688) and 120 LCPL (C-88139–88258, of which 27 were canceled in August 1945). In August 1945, 3,494 LCVP and 99 LCPL were canceled. Total 1945 production: LCVP, 5,865; LCPL, 550, LCPR, 343. Totals for 1941–45: LCPL, 2,250; LCPR, 2,635; LCVP, 23,397 (one source gives 23,051). These figures may exclude some craft built for the U.S. Army or Lend-Lease.

Lend-Lease transfers: 599 LCPL, 413 LCPR, and 321 LCVP to the Royal Navy; 2 LCVP to Soviet Northern Fleet in 1944.

Postwar LCPL Programs
FY 53: 5 LCPL (C-7259–7263, by Higgins) ordered, presumably for amphibious ships.

FY 54: 4 LCPL ordered for LSD 28–31 (C-27099–27102).

FY 55: 100 LCPL authorized, of which 65 were built. Boats for new amphibious ships were not separately authorized. Total production amounted to 101 wooden Mk 1 (C-3244–3344), including those for amphibious ships. The prototype Mk 4 was Marinette's C-104043, built under this program. This program also probably included the U.S. Navy–built prototype plastic Mk 3 (No. 26986), the only U.S.

Navy–built LCPL. Mk 3 was also unusual in having a rounded rather than a V-bottom.

FY 57: 110 LCPL authorized, then deferred and later dropped. However, 4 prototype 40-ft LCPLs were built: 1 wood Mk 5 (C-2990), 1 plastic Mk 6 (C-2996), 2 steel Mk 7 (C-2993, C-2994). C-2994 had gas turbine power.

FY 59: 37 LCPL Mk 4 authorized (C-4372–4408). Landing craft authorized but deferred in FY 55–57 were canceled in FY 59.

FY 60: 2 LCPL authorized (Mk 9 plastic prototypes, C-4418 and C-4419). The Mk 9s had GM turbo-charged 6121T engines, and were capable of 19.8 kts with modified struts and new propellers. Also authorized were 5 Mk 4 (C-4461–4465) for amphibious ships: 3 for APA 249, 2 for LPD 1.

FY 62: 10 LCPL Mk 11 (C-1369–1378); 2 for stock and 8 for amphibious ships LPH 3 and 7, and LPD 2 and 3.

FY 63: 21 LCPL authorized but 35 Mk 11 were built (C-5693–5727, with side exhausts). Additional units authorized were 10 (2 each) for LPD 4–8, 2 for LPH 9, and 2 for Germany. As many as 20 may have been canceled.

FY 64: 67 LCPL Mk 11 authorized but 69 were apparently built (36PL641–6469). Probably 2 were for transfer. Of the 67, 26 had radar and 41 did not.

FY 65: 19 LCPL Mk 11 ordered (36PL651–6519).

FY 68: 104 LCPL Mk 4 ordered (36PL681–68104).

FY 70: 25 LCPL Mk 4 ordered (36PL701–7025). The FY 68 and FY 70 craft were probably specifically for riverine combat in Vietnam.

FY 80: 75 LCPL Mk 12 ordered (36PL801–8075).

FY 83: 98 LCPL Mk 12 ordered (36PL831–8398, including 20 for LSTs).

FY 85: 4 LCPL Mk 12 ordered (36PL851–854).

FY 90: 8 LCPL Mk 13 ordered (36PL901–908).

FY 92: 12 LCPL Mk 13 ordered (11MPL921–9212). Details: 10.95 × 3.67 × 1.1 m, hoisting weight 9,000 kg (9,500 kg fully loaded), powered by one 455 BHP Detroit Diesel 6V92TA; crew of 3 and capacity of 20 personnel. The hull is made of single-skin GRP (glass-reinforced plastic).

As of mid-1964, the U.S. Navy had 391 LCPL in service and 254 in storage. An April 1966 survey of potential riverine warfare assets showed 114 LCPL Mk 1. There were 256 Mk 4 and Mk 11, of which 4 were currently in South Vietnam. The U.S. Navy had 62 LCPR in service (none in storage). As of January 2000, the U.S. Navy had 2 steel and 108 plastic LCPL in service.

Foreign LCPL construction: Spain built Mk 11 (as Mk 11E) LCPL.

Postwar LCPL transfers: Greece (1 in 1973), Italy (3 in 1953), Philippines (1 undated), Spain (5 in 1958), Taiwan (4 in 1960), Thailand (1 in 1968), Vietnam (18 in 1970, 9 in 1971). Taiwan currently operates LCPR (designated LCVP), and may have built some. The only listed transfers of this type were 1 to Taiwan (undated) and 3 to the Philippines (1 in 1961, 2 in 1964).

Postwar LCVP Production

Records show 100 LCVP (C-200945–201044) were ordered for China: Yangtze. They were part of the pre-1949 package of aid for the Nationalists.

FY 49: 2 U.S. Navy–built LCVP prototypes (Mk 2), Nos. 26704 and 26705, were authorized. In September 1950 installation of a gas turbine (Boeing 502) was authorized for C-34024, an existing boat. When her hull had deteriorated badly (by February 1953) her engine was installed in another existing boat, C-78889, for further tests. C-85474 tested a waterjet (Harley Hydrojet Inc.) under a March 1948 contract.

FY 53: 350 LCVP authorized (C-26703–27052), plus 33 (C-27053–27085) for Indo-China. This program also included C-27086–27098 (13 boats for LSD 28–31). This program probably included 45 boats (C-201045–201089) for Japan. Total FY 53 production: 441.

FY 55: 871 LCVP authorized plus 29 previously deferred; of 1,000 planned for FY 55 LCVP, 100 were replaced by LCPL. Hulls: C-102061–102076 (for AKA 112: 16 boats), C-103194–103651 (458 boats; 4 for AE 21–22 [C-103190–103193] were canceled, C-103651 was terminated; others were for LSD 32–35, LST 1171, 1173–1178), C-103652–104075 (424 boats; due to duplication, C-1040389/44 became 104076/82). There was also a 25-boat contract to Lunn Laminates on 23 May 1955 for 25 Mk 5 plastic LCVP (C-2928–2952), the first being the prototype. They had cored fiberglass-reinforced plastic skins. Navy No. 30433 was an experimental single fiber–reinforced plastic skin Mk 7 LCVP built at Puget Sound, selected for later production. Navy No. 30414 was a Styrofoam LCVP Mk 8 ("Black Box"). Total FY 55 production: 920.

FY 56: 2 experimental LCVP authorized (one plastic [C-2989, with right-angle drive], one hydrofoil [C-2998]). C-2998 ("High Lander") was built on a plastic Mk 5 hull, contract awarded 31 May 1956, and it had two Chrysler gasoline engines (250 HP each, to make 45 mph). Related craft, under the same contract as C-2989, were C-104039 LCVP plastic prototype jig and C-104044 (prototype, NObs 3669). 10 LCVP were built for Greece (C-201145–201154). 9 wooden LCVP were built specifically for transfer to Vietnam (C-6660–6668). 848 LCVP were deferred and never built.

FY 57: 100 LCVP were deferred, and then later dropped.

FY 59 (probably): C-104040 (hydrofoil "Halobates," with a Lycoming T53 gas turbine, capable of more than 30 kts) authorized, C-104041 and C-1040402 (waterjets under NObs 3784) authorized.

FY 61 C-2999 (hydrokeel, LCVPK) contract awarded 15 March 1961 to Anti-Friction Corp, Laurel, Maryland. With 2 Chrysler Marine M413D 275 HP, 1 Chrysler M318D 100 HP blower, it made 30 kts on trial at 27,000 lbs fully loaded.

FY 64: 46 LCVP Mk 7 (36VP641–6446) authorized.

FY 65: 112 LCVP Mk 7 (36VP651–65112) authorized.

FY 66: 144 LCVP Mk 7 (36VP661–66144) authorized.

FY 67: 118 LCVP (36VP671–67118, including 3 for the army) authorized.

FY 68: 32 LCVP (36VP681–6832) authorized.

In mid-1964, the U.S. Navy had 1,022 LCVP in service and 254 in storage. As of January 2000, only 3 remained in service.

LCVP transfers: Argentina (6 in 1970; total 37 in 1946–70), Brazil (2 in 1960, 9 in 1962), Cambodia (21 undated), Chile (1 undated, 1 in 1967, 6 in 1970), Colombia (2 in 1959, 25 in 1970), Ecuador (2 in 1967), Ethiopia (2 in 1962), France (8 in 1957), Greece (6 in 1952, 10 in 1956, 4 in 1958, 10 in 1962, 4 in 1964, 3 in 1969, 2 in 1971), Haiti (1 in 1961), Indonesia (1 in 1958, 1 in 1960), Iran (2 in 1967, 3 in 1968), Italy (1 undated, 23 in 1951, 22 in 1953, 30 later), Japan (2 in 1951, 9 in 1955, 1 in 1956), Korea (17 in 1955, 3 in 1958, 4 in 1963, 1 in 1966), Norway (20 in 1950), Pakistan (1 in 1957, 6 in 1960), Philippines (1 in 1952, 1 in 1955, 1 in 1956, 1 in 1965, 2 in 1971, 2 in 1973), Spain (20 undated), Taiwan (about 95, undated), Thailand (8 in 1963, 24 in 1964–69), Venezuela (12 in 1960, 1 in 1967), Vietnam (43 undated, 5 in 1962, 16 in 1963, 1 in 1968, 2 in 1969, 2 in 1970, 1 in 1971). The official U.S. transfer list does not provide an accurate accounting of craft transfers for either Argentina or Taiwan, probably because some craft disposed of as war surplus were not counted. Reportedly the postwar aid package to China included 25 LCVP.

Foreign LCVP construction postwar: Argentina (8), Brazil (30), France (at least 20, GRP type), Greece (14), Japan (22 plus 28 for export to Brazil), South Korea (unknown number, including 18 for Indonesia), Spain (4 Mk 7E), Taiwan (25–30 as Type 272, in 1970s). Craft were based on U.S. design.

LCP(N) (landing craft, personnel, nested) and 26-ft Outboard Landing Craft

C-13173–13184 (12 boats: 2 delivered in 1942, 10 in 1943), were built by Western Plastics. C-103190–103489 (300 boats, LCP(N) Mk 2), were authorized in April 1945 and ordered 20 July 1945 for delivery by 1 February 1946. They were, however, canceled. Outboard landing craft (26 ft) (C-13520 Mk 1, C-13524–13525, C-13527 Mk 2) were built by Miami Boat about 1944.

LCM (landing craft, mechanized)

LCM(1) was the U.S. designation for British mechanized land craft (MLC). Mks 2, 3, 6, and 8 were U.S. LCMs. Mks 4 and 5 were Australian versions. Mk 7 was a wartime British design.

LCM(2) and Prototypes

The single 38-ft tank lighter (FY 37) was U.S. Navy–built No. 13563. The 40-footer (FY 38) was No. 13694. The 45-ft C&R tank lighter was designated LCM(2) in 1942.

FY 39: 1 tank lighter (No. 13975) ordered 25 May 1939 from Norfolk.

FY 40: 2 U.S. Navy–built tank lighters (Nos. 13997 and 13998) ordered 10 August 1940, plus 8 for the U.S. Army (which became TKL 1–8 [Design 193]).

FY 41: No tank lighters originally planned in November 1939 for FY 41, but program was considerably expanded during 1940. 27 authorized in September 1940 (C-1909–1913, C-2084–2091, C-2254–2257, C-3715–3724) and 38 authorized in December (C-2258–2283, C-2910–2921).

1941 (second half of FY 41): 28 tank lighters ordered in February (including C-3701–3710, C-3791–3794, and 10 army tank lighters, nos. 10–19; C-3701 was the 47-ft C&R prototype built by Higgins, and C-3702–3710 were C&R 47-footers reordered as prototype 45-ft Higgins lighters); 180 ordered in June (117 for the 3rd Marine Division plus boat pools [C-4573–4628, 56 for the Marine division, and C-4754–4814, the other 61] plus 50 Higgins 48-footers [C-4629–4678] as stopgaps plus C-3711–3714 and 9 others, probably army tank lighters later reordered as 50-footers, part of the army's 839–952 series); 1 ordered in October (experimental Higgins tank lighter for the army); 152 ordered in November (authorized in June: 32 C&R and 20 Higgins 47/48-footers plus 25 C&R 50-footers and 75 Higgins 50-footers [C-7350–7424]; the others were later reordered as Higgins type lighters, C-8197–8273); 150 ordered in December (Higgins boats: follow-on order [C-7960–7999, C-8350–8459]). The June order paralleled June 1941 procurement of Higgins 36-ft boats. Most of these craft were eventually reordered as 50-ft LCM(3)s. Thus the final contract for Higgins'

C-3711–3714 and C-4573–4588 (as 50-footers) was let only on 11 December 1941. In January 1945, 24 LCM(2) were on hand, all in the United States.

Transfers: 3 LCM(2), presumably ex-army (Nos. 402–404), to Turkey in 1943.

LCM(3) and LCM(6)
Most of the 1941 LCM(2) tank lighters ordered were eventually reordered as 50-ft LCM(3), with at least orders for 350 changed to be Higgins LCM(3)s. LCM(6) was the designation for the army version, beginning in September 1943.

1942: 1,100 ordered in April (C-11956–12680 [725], plus 150 at Boston Navy Yard [Nos. 18186–18335], 150 at Norfolk [Nos. 18336–18385 plus Nos. 18386–18485 canceled], and 75 at Charleston [Nos. 18486–18560]); 350 ordered in May (C-14270–14369 [100], C-14402–14441 [40], C-14844–14918 [75], and C-18486–18560 [75, canceled], plus 60 army [398–437, 455–474]); 40 ordered in June (army Nos. 873–912); 457 ordered in October (C-21061–21309, C-24174 [250] and ex-army C-24435–24641 [207]). The April orders were intended to ensure delivery of 1,100 by 30 September 1942. Plans initially called for 300 to be built by Bethlehem Wilmington, 25 by Robinson (C-11956–11980), and 775 by navy yards (300 by Philadelphia, 150 by Boston [Nos. 18186–18335], 150 by Norfolk [Nos. 18336–18485], 100 by New York, and 75 by Charleston [Nos. 18486–18560]; Philadelphia was the lead yard). By the end of April 1942, however, Philadelphia and New York Navy Yards had dropped out in favor of private yards (Walsh-Steers [250: C-12131–12280, C-12581–12680] and Ship Associates [150: C-11981– 12130], later reassigned to two other yards). Then the decision was made to shift production to Higgins lighters. At a 29 May 1942 conference at Philadelphia it developed that Boston had already begun its 150 C&R-type lighters (of which it would complete 20, the other 130 were dropped in favor of the Higgins type), Norfolk had begun its 150 (it would complete 50), and Bethlehem had begun 150 (of which it would complete 30). Norfolk could not switch to the Higgins design without delaying LST construction, so its remaining 100 LCMs were reallocated to Bethlehem Steel (they are the first May order listed). Bethlehem was allowed to put back the completion of these craft to 31 October 1942. The craft assigned to Ship Associates were reassigned to two other yards, Dreier Structural Steel and Brewer Dry Dock. Thus 100 C&R-type 50-footers were built by the navy yards and by Bethlehem Steel (in the C-12281–12580 series). Some of the October transfers apparently include the earlier army orders. Other army LCM(3): 546, 547, 810–834 (25), 839–852 (14), 944–1004 (61), plus 27 not numbered.

1943: 830 ordered in January and 2,000 ordered in February (together, C-27992–30781 and C-35952–35991), 2,775 ordered in June (C-46330–49104), 609 ordered in August (C-51721–52329), 1,697 ordered in September (C-52350–54046; LCM(6) included C-52498–52797, C-52850–52987, C-53538–53677).

1944: 900 ordered in March (C-68989–69888), 3,000 ordered in May (74389–77388); however, 1,500 were canceled in June and 1,117 were canceled in July. Half of the March and May programs were LCM(3), the other half were LCM(6).

1945: 2,617 ordered in March (reorder of those canceled in 1944); 1,929 were canceled in August. Conversions to LCM(6) were ordered in January 1945 (contracts for 200) and late in April (500).

Total ordered by directives: 12,464 craft. Official records indicate total wartime Higgins type LCM production was 11,383: 1,255 in 1942; 4,019, including the first 52 LCM(6), in 1943; 5,210, including 1,850 LCM(6), in 1944; 899, including 828 LCM(6), in 1945. The official record for 1945 may be low; another accounting for completions through mid-1945 is considerably higher.

C&R-type 50-footers not completed were simply reordered as Higgins boats, hence they were not assigned numbers. When the army turned over its LCMs to the navy, they, too, did not get navy numbers. The army equivalent was Design 289. A total of 207 (C-24435–24641) were turned over to the navy in exchange for lower-powered LCMs, presumably the C&R type, to be used for cargo handling. Army Design 299 (TKL9) was an experimental Higgins 50-ft tank lighter powered by two 165 HP gasoline engines (50 × 20 ft). An experimental Higgins 59-ft lighter was turned over to the navy as C-25244 in 1943. Specifications were 59 ft 5 in × 15 ft 10½ in × 3 ft 10½ in; four 22 HP Higgins-Tucker engines, for a speed of 11.2 kts; 54,000 lbs with equipment except for tank, guns, and fuel; about 120,000 lbs fully loaded. It was a V-bottom flush-deck type with a closed, mechanically ventilated engine room. Armament was one 0.50 caliber in the power turret and one 0.50 caliber on the scarf ring vehicle-type mount.

Lend-Lease transfers: 671 LCM(3) and LCM(2) to the Royal Navy; 2 LCM(3) to the Soviet Northern Fleet in 1944; 54 LCM(3) to the Soviet Pacific Fleet in 1945 (3 lost during the attack on Manchuria, 17 lost postwar).

Other transfers: 3 LCM(3) were transferred to Turkey in 1943.

1949: 2 LCM(6) Mod 1 prototypes (Nos. 26702 and 26703) ordered.

FY 52: 519 LCM(6) authorized for the army (C-200000–200518, army Nos. 6000–6514, of which 61 were transferred to the navy in May 1960 [23 west coast, 38 east coast]). The highest army number is also given as C-200517. This year's program probably included C-36056–36059 for Indo-China.

FY 53: 100 LCM(6) were dropped from the program in the planning stage, to help finance the carrier *Saratoga* (CV 60). Craft built for foreign transfer, probably under this year's program, were C-201090–201144 (55 boats): C-201090–201095 (6 for Indo-China), C-201096–201114 (19 for Japan), C-201115–201124 (10 for Indo-China), C-201125–201134 (10 for Japan), C-201135–201144 (10 for Indo-China).

FY 54: 150 authorized, plus 10 as part of the outfit for AKA 112 (*Tulare*): C-124514–124673. Probably the program included 16 for Japan: C-124400–124415.

FY 55: 150 included in the draft program (September 1953) but not in the final one. They were SCB 94s, but it is not clear to what extent this represented a redesign.

FY 56: 128 LCM authorized, but they were deferred and never built. They would have been split evenly between LCM(3) (SCB 141) and LCM(6).

FY 57: 62 LCM(6) and one LCM(H) authorized; all were deferred. The LCM(H) was replaced by an LCA; a contract was let under the FY 60 program. The other deferred landing craft were later dropped. FY 56 and FY 57 contracts were to have been awarded together because of procurement delays.

FY 59: 9 LCM(6) Mod 2 (C-4452–4460) ordered for APA 249 (7) and LPD 1 (2). Mod 2 was developed by Lukens Steel for easier mass production.

FY 61: 20 LCM(6) authorized, yet canceled to pay for 3 LCSR. This year 96 LCM were acquired from the army.

FY 62: 23 LCM(6) Mod 2 (C-5819–5839, C-3502, C-3503) ordered; 11 were for transfer abroad (C-3502, C-5822, C-5825, C-5827, and C-5828 went to Thailand in 1965).

FY 64: 65 LCM(6) Mod 2 (C-36773–36837) ordered.

FY 65: 52 LCM(6) ordered; 2 were for transfer abroad (56CM6551 and 56CM6552). 56CM6542 and 56CM6543 were converted into LCM(3); apparently 61 were originally authorized.

FY 66: 48 LCM (56CM661–6648) ordered.

FY 67: 74 LCM(6) Mod 2 (56CM671–6774) ordered, of which 56CM6765–6774 were built as LCM(3) under a March 1968 change order, becoming 50CM671–6710. Conversions to LCM(3) were probably for riverine warfare in Vietnam.

FY 68: 55 LCM(6) (56CM681–6855) ordered, of which 5 were for Thailand; they had hot exhausts.

FY 69: 65 LCM(6) (56CM691–6965) ordered.

FY 75: 14 LCM(6) (56CM751–7514) ordered, of which 56CM751 had a Kort nozzle.

FY 77: 25 LCM(6) ordered; 4 were for Iran (56CM771–7725). The Iranians rejected the boats after the Shah fell. They were offered instead to the Saudis.

FY 78: 50 LCM(6) Mod 3 (56CM781–7850) ordered. This was presumably the HPI version; one account claims 99 HPIs.

FY 86: 2 LCM(6) (56CM861 and 56CM862) ordered.

In mid-1964 the U.S. Navy had in service 175 LCM(3) (plus 105 in storage) and 396 LCM(6) (plus 194 in storage). An early 1966 survey of inventory craft in depots, which meant available for conversion to riverine craft, showed that the army had 72 LCM(6) in depots. The navy had a total of 75 LCM(3), 280 LCM(6), and 12 LCM(8) in depots. These figures do not include operational craft. Another survey of potential riverine craft in April 1966 showed a total of 50 LCM(3) in the fleet, with 100 in South Vietnam, plus 523 LCM(6). As of January 2000, none of the 8 surviving Mod 2 and Mod 3 remained in service as an amphibious craft.

Postwar transfers: Argentina (4 in 1970), Brazil (1 in 1962, 1 in 1965, 2 in 1968), Burma (10 undated), Cambodia (13 undated, 3 in 1955, 3 in 1962, 17 in 1971, 9 in 1972, 7 in 1973), Chile (3 in 1970), Ethiopia (2 in 1960, 2 in 1971), France (2 undated, 25 in 1958), Greece (3 in 1956, 6 in 1958), Guatemala (2 in 1966), Lebanon (1 in 1956), Indonesia (1 in 1960), Iran (2 in 1962, 3 in 1967, 2 in 1968), Italy (5 in 1951, 19 in 1953, 28 later), Japan (35 in 1955–56, 13 in 1961), Korea (10 in 1955, 1 in 1957, 6 in 1962-63, 21 in 1967, 2 in 1971), Nicaragua (1 in 1970), Pakistan (1 in 1958), Philippines (1 undated, 7 in 1955, 1 in 1971, 2 LCM(3) in 1972, 4 in 1973, 36 in 1973–75), Saudi Arabia (8 in 1977), Senegal (4 in 1968), Spain (5 in 1971), Taiwan (2 in 1956, 1 in 1957, 16 in 1958, 1 in 1959), Thailand (12 in 1955, 1 in 1964, 7 in 1965, 2 in 1967, 2 in 1968, 5 in 1969, 17 in 1968–69), Turkey (10 in 1958), Uruguay (2 in 1972), Vietnam (46 undated; also 1 in 1962, 25 in 1963, 4 in 1964, 8 in 1965, 25 in 1966, 2 in 1967, 5 in 1968, 6 in 1969, 16 in 1970, 36 in 1971, 42 in 1972), West Germany (1 in 1958). The figures for Taiwan are probably far too low because many craft not included in these totals were provided to China in 1946–49; 25 LCM were included in the aid package.

Foreign LCM construction postwar: Brazil (6), France (2), Italy (11, related GRP design), Japan

(15), Portugal (LDM type, based on LCM; 2 to Guinea-Bissau), Spain (16), Taiwan (about 250).

LCM(8)

FY 49: 2 prototypes built at Norfolk Navy Yard (Nos. 26715 [V-bottom] and 26716 [W-bottom]).

FY 52: C-200518–200944 (426 boats ordered for the army; army designators were Nos. 8000–8335 for C-200514–200835, and 8421–8430 for C-200935 series. In the early 1960s 57 LCM(8) were transferred by the army to the navy.

FY 53: 100 dropped from the FY 53 program to help finance the carrier *Saratoga* (CV 60). At about the same time the requirement to carry LCM(8) aboard the new AKA *Tulare* was canceled.

FY 66: 28 aluminum boats (74CM661–6628) ordered.

FY 67: 40 Mod 1 steel (74CM671–6740) ordered for the army (army numbers 8500–8539); 43 Mod 2 aluminum with a stern wedge (74CM6741–6743) ordered.

FY 68: 26 Mod 2 (74CM681–6826) ordered.

FY 69: 6 authorized, but deferred pending decision on material.

FY 70: 61 steel ordered for the army (army numbers 8540–8560 and 8580–8619).

FY 78: 6 boats (74CM781–786) ordered for Spain.

FY 83: 26 Mod 5 steel type ordered.

FY 84: 18 Mk 5 (74CM841– 8418) ordered.

FY 90: 12 boats (74CM901–9012) ordered, but another account says 17 aluminum Mod 6 were bought with surplus funds.

In mid-1964, the U.S. Navy had 58 LCM(8) in service and 24 in storage. As of January 2000, the navy had 85 in use as amphibious craft (13 aluminum, 72 steel), and the Army Transportation Corps had 96.

Transfers: Cambodia (2 in 1972), Colombia (1 in 1993), Philippines (2 in 1973), South Korea (10 in 1978), Spain (6 in 1975), Vietnam (6 in 1964, 6 in 1969, 8 in 1970, 20 in 1971, 15 in 1972), Venezuela (1 in 1999).

Foreign construction: Brazil (5), France (31 CTM type; transfers to Djibouti, Ivory Coast, Morocco, Senegal), Spain (2), Turkey (31), West Germany (28 Klasse 521, based on LCM(8); 11 later transferred to Greece).

LCR(L) (landing craft, rubber, large) and LCR(S) (landing craft, rubber, small)

1938: 2 bought under FY 38 program.

1939: 14 (C-1848–1861) bought in August.

1941: 837 ordered in June, including 486 LCR(L) (C-5371–5856); 156 LCR(L) produced. The June order included 830 for the 3rd Marine Division (334) and for boat pools.

1942: 406 ordered in January, 1,645 ordered in June, 300 ordered in July, 600 ordered in August, 1,118 ordered in September, 355 ordered in November (probably including 300 LCR(L): C-15087–15386), 500 ordered in December (LCR(L): C-24642–25141). Total 1942 production: 1,259 LCR(L), 1,301 LCR(S). The January order may have included 11 LCR(L): C-7633–7642 (10 boats) and C-7729.

1943: 835 ordered in March, 2,119 ordered in June (including 466 LCR(S) and 1,153 LCR(L): C-37891–39043), 1,000 ordered in October (LCR(L): C-55251–56250), 400 ordered in December. Total 1943 production: 2,258 LCR(L), 3,194 LCR(S).

1944: 6,700 ordered in January (LCR(L) and LCR(S): A-5448–8147 [2,700]), 6,000 ordered in May (LCR(L): C-64184–64283 [100], C-64318–66417 [2,100], C-66518–68617 [100] plus LCR(S): A-5448–8147 [2,700]), C-97895–101894 [4,000] plus LCR(S): A-8151–10150 [2,000]); 6,000 canceled in July, 100 canceled in August. Total 1944 production: 6,450 LCR(L), 3,460 LCR(S).

Total 1941–45 production: 10,125 LCR(L), 8,150 LCR(S) (including 195 in 1945). Total of directives: 18,287 craft. The army bought 4,403 LCR(L), some of which were presumably included in navy orders (probably included in the 1942 figures).

Plans and specifications for the 16-ft 10-man LCR(L) were half complete in November 1942. Plans and specifications for 1,153 manually inflated boats were issued on 21 June 1943. Then the design was simplified as a lightly built combat type boat. By this time natural rubber shortages governed construction, so by August 1943 the design was being revised so that boats could be built out of synthetic rubber. The new version incorporated the lightweight construction previously ascribed to the combat boat. The 7-man LCR(S) was similarly redesigned.

LCSR (landing craft, swimmer reconnaissance)

The LCSR replaced LCPL or LCPR used in the swimmer reconnaissance role. Hull numbers for the 52-footer were C-1310–1319 (10 boats) and C-5842–5845 (4 boats). The C-1310 series were built under a November 1961 contract. Beginning in May 1975, C-1310–1312, C-1314, C-1315, and C-1317 were transferred to Israel. Others were sold. In 1960 an LCPL Mk 4 was fitted out as 36-ft reconnaissance boat Mk 2.

FY 63: 15-ft plastic swimmer support craft (C-1320–1349, 30 boats) authorized.

FY 64: 36-foot type (36SR681 and 36SR682) authorized.

LCS(S) (landing craft, support, small)

All LCS(S) were built by Higgins. The prototype was C-3986. Mk 1, first batch (C-7475–7499, C-7643–7717 [100 boats], and August 1942 order (C-16610–16659 [50 boats]), were delivered in 1942. A directive for 120 more LCS(S), for delivery in 1944, was issued 15 June 1943, and a contract for the first 255 Mk 2 was let 24 July 1943 (C-49105–49224, C-51386–51520). A 2 February 1944 contract covered 200 boats (C-60483–60682), but was reduced to 153; the others (C-60510, C-60607, C-60616, C-60621, C-60623, C-60634, C-60635, C-60639, C-60640, C-60644, C-60646–60682) were converted to LCPLs under a 10 February 1944 order. An 11 March 1944 directive called for 75 more, at the rate of 25 per month (C-69889–69963). Another 250 were approved on 9 May 1944 (C-77389–77638), but they were never ordered, and all 325 were canceled on 21 August 1944, along with 100 LCI(L). The first 145 LCS(S)(2) were delivered in 1943, the remaining 263 in 1944. There was also a 6-boat Lend-Lease contract. Total production: 565.

Transfers: 2 (C-7653 and C-51393) to the Soviet Northern Fleet in 1944; 1 (C-7479) to the Royal Navy in April 1943; 50 to the Nationalist Chinese in 1946–47.

Electric Surfboat/Surfboard

Production amounted to 256: 200 in 1944, 56 in 1945. Hull numbers were C-68618–68677 (60 boats, obtained January and February 1944 from the Office of Strategic Services), C-68678–68727 (50 boats, contract 3 March 1944 with Ohio Rubber Company), C-74040–74089 (50 boats, contract 25 November 1944 with Ohio Rubber Company), C-74371–74376 (6 boats), C-101895–101914 (20 boats, from the Maritime Commission), and C-101915–102014 (100 boats; contract 3 July 1944 from Ohio Rubber Company). Looking toward the invasion of Japan, on 13 February 1945 the secretary of the navy approved production of 200 more (C-102061–102260), followed on 13 April by 600 more (C-103490–104089, contract 30 June 1945 with Goodyear, for inflatable surfboard Mk 2). Projected deliveries, at Naval Supply Depot Oakland, were 200 by 1 September 1945, 200 by 1 December 1945, and 200 by 1 March 1946. However, 542 craft were canceled 14 August 1945.

Army Storm Boats (10-ft M2)

Navy hull numbers assigned were C-21765–21940 (176 boats), C-64287–64292 (6 boats), C-74377–74382 (6 boats), C-102017–102060 (44 craft).

LVT (landing vehicle, tracked) and AAV (amphibious assault vehicle)

The U.S. Navy classed LVTs as boats, and assigned them boat numbers (C-numbers). The prototype was C-1880.

1940: 200 ordered in November (C-2644–2843, built by Roebling); 180 were for the navy, 20 for the army.

1941: 564 ordered in January (army order), 288 ordered in June (LVT(1): C-4139–4238 and C-5183–5370, all by Food Machinery Corporation, to parallel the Higgins boat orders: 100 for the division, 188 for the pool), 20 ordered in August, 452 ordered in December (including C-6085 and C-6086, prototype FMC LVT(A)1s. The later orders were probably mainly army, plus initial orders from three new builders: Borg-Warner, C-7425–7474 (50 craft); St. Louis Car, C-7500–7529 (30 craft); and Graham-Paige, C-7530–7559 (30 craft). C-11955 was a Roebling LVT(A)(1). Total 1941 directives: 1,324.

1942: 40 ordered in February (presumably add-ons to the 1941 series, using numbers given above), 600 ordered in June (C-16735–17033 [299 LVT(A)2], C-17034 [prototype LVT(2)], C-17035–17334 [300 LVT(A)1]), 1,900 ordered in September (C-18370–18383 [14 LVT(2)], C-18384–19169 [786, including C-18619–18919 LVT(4) and C-18920–19169 LVT(2)], C-19170–19619 [450 LVT(2)]), C-25524–25723 [200 LVT(A)2], C-25724–26173 [450 LVT(2)]), 107 ordered in October (all LVT(1): C-24177–24183 [7], C-24379–24428 [50], C-26524–26558 [35], and 15 others), 551 ordered in November (LVT(3) prototype [C-18361], C-21011–21060 [50 LVT(2)], C-22011–22110 [100 LVT(4)], C-22111–22160 [50], C-26174–26523 [350 LVT(4)]). Total 1942 directives: 3,198.

1943: 2,120 ordered in June (including C-49225–49434, LVT(4); C-49435–49629, LVT(A)1; 49630–49689, LVT(4); 50047–50104, LVT(2); 50105–50549, LVT(3); 50565–50875, LVT(3); and 51153–51354, LVT(4)), 3,936 ordered in November (C-56204–57333, LVT(A)1; 57334–58230, LVT(4);

58231–58461, LVT(2); 58462–59351, LVT(4), 59352–59740, LVT(2), and 59741–60139, LVT(4); one, presumably a prototype, is missing from these numbers), 3,500 ordered in December (C-60683–64182, of which C-62718–62974 and C-63235–62373 were LVT(2), and C-62374–63498 were LVT(3), C-60683–62161 and C-62499–64082 were LVT(4), and C-62162–62717 were LVT(A)4). In January LVTs were specifically omitted from the continuing landing craft program, but in May production for 1944 was specified as 1,725 cargo carriers (later specified as LVT(2)), 195 LVT(A)1, and 200 LVT(A)2. Total 1943 directives: 9,556.

1944: 1,800 ordered in March (C-71869–73668; C-71869–72868 [1,000] were LVT(3), C-72869–73293 [425] were LVT(4), C-73294–73668 [375] were LVT(A)4), 9,600 ordered in May (C-88209–97858). Plans originally called for 1,800 of the May craft to be LVT(A)4: C-88259–89258, C-89759–90258, C90259–90858. However, C-88259–88618 were LVT(A)5 (360 craft); C-88869–90833 were LVT(A)4 (1,965 craft); C-91255–92207 (953 craft) and C-96230–96231 (2 craft) were LVT(3); and C-92340–92473 (134 craft), C-92856–93097 (242 craft), and C-93633–94112 (480 craft) were LVT(4). Experimental lightweight versions were LVT(3) (C-96230 and C-96231) and LVT(4) (C-96234, C-96235 and C-96240). Total 1944 directives: 11,400.

1945: 2,502 canceled in August 1945: 412 LVT(3), 2 experimental LVT(3) (presumably the lightweight version, transferred to another account), 1,838 LVT(4), and 250 LVT(A)5.

Total authorized by directives: 25,678. The discrepancy between this figure derived from the listing of directives (less cancellations) and the number actually built (18,620) is probably due to cancellations and reorders (that is, in effect, double-counting). Table B-1 shows wartime LVT completions by year and by type, as well as breakdown by service users. LVT(3) and (4) figures each include 2 lightweight 1945 prototypes. The 1 December 1945 official history produced by the Continuing Board for the Development of LVTs shows 2,963 LVT(2) but 509 LVT(A)1. It is also possible that 8,351 LVT(4) were made.

In the first postwar series, C-105400–107399 (2,000 craft) were reserved for LVTP5 and LVTH6. Contracts covered C-105400–105469 (70 craft). C-107400–124399 (17,000 craft) were reserved for LVTP5 and LVTH6, of which contracts were initially let for C-107400–107695 LVT(5) (296 craft). Initial postwar production was 70 LVT(H) and the first 4 LVTP(5), under the FY 51 program. Another 39 LVTP(5) were deferred at that time. Production

Table B-1. LVTs Completed during World War II

	Breakdown by Type					
	1941	1942	1943	1944	1945	Total
LVT(1)	72	851	302	—	—	1,225
LVT(2)	—	—	1,540	1,422	—	2,962
LVT(3)	—	—	1	733	2,230	2,964
LVT(4)	—	—	11	4,980	3,359	8,350
LVT(A)1	—	3	288	219	—	510
LVT(A)2	—	—	200	250	—	450
LVT(A)4	—	—	—	1,489	401	1,890
LVT(A)5	—	—	—	—	269	269
Total	72	854	2,342	9,093	6,259	18,620

	Breakdown by Service Users			
	Marine Corps	U.S. Army	Lend-Lease	Total
LVT(1)	540	485	200	1,225
LVT(2)	1,355	1,507	100	2,962
LVT(3)	2,962	2	—	2,964
LVT(4)	1,623	6,224	503	8,350
LVT(A)1	182	328	—	510
LVT(A)2	—	450	—	450
LVT(A)4	533	1,307	50	1,890
LVT(A)5	269	—	—	269
Total	7,464	10,303	853	18,620

through 1957: 1,124 LVTP5 and 210 LVTH6, plus 65 LVTR1 and a few LVTE1. Only a prototype LVTAA1 was built. In addition, 58 LVTP5 were converted to LVTP5(Cmd), for command/control. All vehicles were for the Marines, because the army by this time had quit using amtracs.

In FY 69 2 FMC LVTPX prototypes were built, followed by others in FY 70. In 1970 (presumably under FY 71) the Marines ordered 942 production machines for delivery during 1971–74. In 1983 the Marines had 984 LVTP7 (later redesignated AAV7); another 329 were later bought to equip MPS ships (see Appendix A).

A DUKW runs down the ramp of an LST on D-Day at Iwo Jima. *U.S. Coast Guard*

Appendix C: Amphibious Ships

The following list of amphibious ships is categorized by ship designation:

AGC/LCC	amphibious flagship
AKA/LKA	attack cargo ship
APA/LPA	attack transport
APH	casualty evacuation (H) transport
AP	transport (for personnel, P)
IFS/LFR	inshore fire support ship
LHA	helicopter landing ship, assault
LHD	helicopter landing ship, dock
LPD	landing ship, personnel, dock
LPH	assault helicopter carrier
LSD	landing ship, dock
LSM	landing ship, medium
LSMR	LSM converted to fire support ship
LST	landing ship, tank
LSV	vehicle landing ship
LCI(L)/LSIL	infantry landing craft/ship
LCFF	flotilla flagship converted from LCI(L) (LSFF postwar)
LCIG	gunboat converted from LCI(L) (LSIG postwar)
LCIM	gunboat (mortar) converted from LCI(L) (LSIM postwar)
LCIR	gunboat (rocket) converted from LCI(L) (LSIR postwar)
LCS(L)/LSSL	support landing craft/ship

The headings give both the original World War II designations and the new ones adopted in 1969 (e.g., AGC and LCC for amphibious flagships). Transports (AP) are listed because many of them were actually attack transports, although not so designated. Evacuation transports (APH) served partly as attack transports. For APDs, see Norman Friedman's companion volume on U.S. destroyers (*U.S. Destroyers: An Illustrated Design History*, Annapolis, Md.: Naval Institute Press, 1982).

For each entry, the first column gives the ship's U.S. Navy number. The second column gives the ship's U.S. Navy name (when named) on the first line and the builder's name on the second line. Refer to the following key for the builder's name:

Albina	Albina Engine and Machine Works, Portland, Oregon
Amer Bridge	American Bridge Co., Ambridge, Pennsylvania
Amer Intl	American International Shipbuilding Corp. (Hog Island, World War I yard), Philadelphia, Pennsylvania
Amer Ship	American Shipbuilding Co., Lorain, Ohio
Avondale	Avondale, New Orleans, Louisiana
B&V	Blohm & Voss, Hamburg, Germany
B&W	Burmeister and Wain, Copenhagen, Denmark
Bath	Bath Iron Works, Bath, Maine
Beardmore	William Beardmore & Co., Glasgow, Scotland
Beth-A	Bethlehem Alameda yard, Alameda, California
Beth-F	Bethlehem Fairfield yard, Fairfield, California
Beth-H	Bethlehem Steel Hingham yard, Hingham, Massachusetts
Beth-Q	Bethlehem Steel Quincy yard, Quincy, Massachusetts
Beth-S	Bethlehem Sparrows Point yard, Sparrows Point, Maryland
Beth-SF	Bethlehem Pacific Coast Steel Corp., San Francisco, California
BOSNY	Boston Navy Yard
Brown	Brown Shipbuilding Co., Houston, Texas

Cal Ship	California Shipbuilding Corp., Los Angeles, California
CHARNY	Charleston Navy Yard
Chicago	Chicago Bridge & Iron, Seneca, Illinois
Christy	Christy Shipbuilding Corp., Sturgeon Bay, Wisconsin
Commercial	Commercial Iron Works, Portland, Oregon
Consolidated-O	Consolidated Steel Corp., Orange, Texas
Consolidated-W	Consolidated Steel Corp., Wilmington, California
Cramp	Cramp Shipbuilding Corp., Camden, New Jersey
Defoe	Defoe Shipbuilding Co., Bay City, Michigan
Dravo-Phil	Dravo Corp., Philadelphia, Pennsylvania
Dravo-Pitts	Dravo Corp., Pittsburgh, Pennsylvania
Dravo-Wil	Dravo Corp., Wilmington, Delaware
Federal-K	Federal Shipbuilding & Drydock, Kearny, New Jersey
Federal-N	Federal Shipbuilding & Drydock, Newark, New Jersey
Furness	Furness Shipbuilding Co., Haverton-Hill-on-Tees, England
GD	General Dynamics, Quincy, Massachusetts
Glasgow	yard unknown; this refers to unnamed builder in Glasgow, Scotland
Gulf	Gulf Shipbuilding Co., Chickasaw, Alabama
H&W	Harlan & Wolf, Belfast, United Kingdom
Ingalls	Ingalls, Pascagoula, Mississippi
Jeffersonville	Jeffersonville Boat & Machine Corp., Jeffersonville, Indiana
Kaiser-R	Kaiser Co., Richmond, California
Kaiser-V	Kaiser Co., Vancouver, Washington
Lawley	George Lawley & Sons, Neponset, Massachusetts
Litton	Litton (Litton-Ingalls), Pascagoula, Mississippi
Lockheed	Lockheed Shipbuilding and Construction Co., Seattle, Washington
Manitowoc	Manitowoc S.B. Corp., Manitowoc, Wisconsin
Missouri	Missouri Valley Bridge & Iron Co., Evansville, Indiana
Moore	Moore Drydock Co., Oakland, California
NASSCO	National Steel & Shipbuilding Corp., San Diego, California
NC Ship	North Carolina Shipbuilding Corp., Wilmington, North Carolina
NJ Ship	New Jersey Shipbuilding Corp., Barber, New Jersey
NN Ship	Newport News Shipbuilding and Drydock Co., Newport News, Virginia
NORNY	Norfolk Navy Yard
NYNY	New York Navy Yard
NY Ship	New York Shipbuilding Corp., Camden, New Jersey
Oregon	Oregon Shipbuilding Corp., Portland, Oregon
Penhoet	Chantiers de l'Atlantique (Penhoet), St. Nazaire, France
Pennsylvania	Pennsylvania Shipyards, Inc., Beaumont, Texas
Permanente	Permanente Metals, Richmond, California
PHNY	Philadelphia Navy Yard
PSNY	Puget Sound Navy Yard
Puget Bridge	Puget Sound Bridge & Dredging Corp., Seattle, Washington
Pullman	Pullman Std. Car Manufacturing Co., Chicago, Illinois
Sea-Tac	Seattle-Tacoma Shipbuilding Corp., Tacoma, Washington
SPC	Societe Provence de Construction, La Ciotat, France
Stab Tecnico	Stabilimento Tecnico, Trieste, Italy
Sun	Sun Shipbuilding & Dry Dock Co., Chester, Pennsylvania
Tampa	Tampa Shipbuilding Co., Tampa, Florida
Vulkan	A. G. Vulkan, Stettin, Germany
Walsh-Kaiser	Walsh-Kaiser Co., Providence, Rhode Island
Western-SF	Western Pipe & Steel Corp., San Francisco, California
Western-SP	Western Pipe & Steel Corp., San Pedro, California
Willamette	Willamette Iron & Steel, Portland, Oregon

The third column (the first column of dates) of each entry gives the laying-down (LD) date on the first line and the launch date on the second line. The next column (the second column of dates) gives commissioning (comm) and applicable recommissioning dates. The next column (the last column of dates) gives the decommissioning (decomm) date.

The last column gives the pre-acquisition names of converted merchant ships and the fate of ships. Craft given to the U.S. Military Government of Korea were transferred to the new Republic of Korea (South Korea) navy when the republic was constituted in 1948. These transfers are not otherwise indicated. Some ships given to the U.S. Army in Korea were also transferred. Refer to the following key for the fate of each ship:

BU	broken up (scrapped)
CTL	constructive total loss (write-off)
MarAd	MarAd (sales agent and reserve fleet operator)
MarComm	MarComm (sales agent and reserve fleet operator)
MDAP	Mutual Defense Assistance Program (foreign aid)
merch	merchant ship (conversion of warship)
mil gov	military government
MSC	Military Sealift Command, successor to MSTS
MSTS	Military Sea Transportation System; ships in MSTS service had their hull numbers prefixed by T (e.g., T-LST or T-AP)
N/A	information not available
NDRF	National Defense Reserve Fleet
NEI	Netherlands East Indies (later Indonesia)
NRT	Naval Reserve training ship
rem	removed
ret	returned
ROC	Republic of China (Nationalist China, later Taiwan)
SCAJAP	Shipping Control Authority Japan
str	stricken from U.S. Navy List
SU	Soviet Union (transfers under Lend-Lease)
tgt	target
UK	United Kingdom
UNRRA	United Nations Refugee and Relief Agency (early postwar organization)
WL	war loss
WL(A)	accident
WL(B)	bombed
WL(K)	kamikaze
WL(M)	mined
WL(T)	torpedoed

ACG/LCC

No.	Name/Builder	LD/Launch	Comm	Decomm	Notes and Fate
1	*Appalachian* Federal-K	4 Nov 42 29 Jan 43	2 Oct 43	21 May 47	Str 1 Mar 59, to ROC 64
2	*Blue Ridge* Federal-K	4 Dec 42 7 Mar 43	27 Sep 43	14 Mar 47	Str 1 Jan 60
3	*Rocky Mount* Federal-K	4 Dec 42 7 Mar 43	15 Oct 43	22 Mar 47	Ex–SS *Rocky Mount*; str 1 Jul 60, to MarAd, sold 1 Mar 73
4	*Ancon* Beth-Q	N/A 24 Sep 38	12 Aug 42	25 Feb 46	Ex–SS *Ancon*, ex–AP 66; str 25 Feb 46, ret to army, merch ship 46–63, Maine Maritime Academy 29 Jun 62 as *State of Maine*, replaced 73, BU May 73
5	*Catoctin* Moore	14 Nov 42 23 Jan 43	31 Aug 43	26 Feb 47	Ex–SS *Mary Whitridge*; str 1 Mar 59
6	*Duane*				Number not used; she retained her Coast Guard number (WAGC 33)
7	*Mount McKinley* NC Ship	31 Jul 43 27 Sep 43	1 May 44	26 Mar 70	Ex–SS *Cyclone*; str 30 Jul 76
8	*Mount Olympus* NC Ship	3 Aug 43 3 Oct 43	22 Dec 43	4 Apr 56	Ex–SS *Eclipse*; str 1 Jun 61, to MarAd Jun 66, sold 22 Jan 73
9	*Wasatch* NC Ship	7 Aug 43 8 Oct 43	20 May 44	30 Aug 46	Ex–SS *Fleetwing*; str 1 Jan 60
10	*Auburn* NC Ship	14 Aug 43 19 Oct 43	20 Jul 44	7 May 47	Ex–SS *Kathay*; str 1 Jul 60

No.	Name/Builder	LD/Launch	Comm	Decomm	Notes and Fate
11	*Eldorado* NC Ship	20 Aug 43 26 Oct 43	25 Aug 44	16 Nov 72	Ex–SS *Monsoon;* str 16 Nov 72
12	*Estes* NC Ship	25 Aug 43 1 Nov 43	9 Oct 44 31 Jan 51	30 Jun 49 70	Ex–SS *Morning Star;* str 30 Jul 76
13	*Panamint* NC Ship	1 Sep 43 9 Nov 43	14 Oct 44	Jan 47	Ex–SS *Northern Light;* str 1 Jul 60
14	*Teton* NC Ship	9 Nov 43 5 Feb 44	16 Mar 44	30 Aug 46	Ex–SS *Witch of the Wave;* str 1 Jun 61
15	*Adirondack* NC Ship	18 Nov 44 13 Jan 45	2 Sep 45 4 Apr 51	1 Feb 50 9 Nov 55	Str 1 Jan 61, to MarAd, sold 7 Nov 72
16	*Pocono* NC Ship	30 Nov 44 25 Jan 45	29 Dec 45 18 Aug 51	19 Jun 49 16 Sep 71	Str 1 Dec 76
17	*Taconic* NC Ship	19 Dec 44 10 Feb 45	16 Jan 46	17 Dec 69	Str 1 Dec 76
18	*Biscayne* PSNY	27 Oct 39 23 May 41	3 Jul 41		Ex–AVP 11; to AGC 10 Oct 44, to Coast Guard as *Dexter* (WAVP 385) 19 Jul 46
19	*Blue Ridge* PHNY	27 Feb 67 4 Jan 69	14 Nov 70		
20	*Mount Whitney* NN Ship	8 Jan 69 8 Jan 70	16 Jan 71		

AKA/LKA

This list includes AK that were reclassified as AKA on 1 February 1943.

No.	Name/Builder	LD/Launch	Comm	Decomm	Notes and Fate
1	*Arcturus* Sun	26 Jul 38 18 May 39	26 Oct 40	3 Apr 46	Ex-*Mormachawk,* ex–AK 18; to MarComm 2 Jul 46
2	*Procyon* Tampa	15 Jan 40 14 Nov 40	8 Aug 41	23 Mar 46	Ex-*Sweepstakes,* ex–AK 19; str 12 Apr 46
3	*Bellatrix* Tampa	20 Nov 40 15 Aug 41	17 Feb 42 15 May 52	1 Apr 46 3 Jun 55	Ex-*Raven,* ex–AK 20; str Jul 60, to Peru 20 Jul 63
4	*Electra* Tampa	6 Mar 41 18 Nov 41	17 Mar 42 3 May 52	19 Mar 46 13 May 55	Ex-*Meteor,* ex–AK 21; str 1 Jul 46, reacquired 16 Oct 51, str 1 Jul 61
5	*Formalhaut* Pennsylvania	28 Mar 40 25 Jan 41	2 Mar 42		Ex–*Cape Lookout,* ex–AK 22; reclassified as AK 22, 25 Aug 44
6	*Alchiba* Sun	15 Aug 38 6 Jul 39	15 Jun 41	14 Jan 46	Ex-*Mormacdove,* ex–AK 23; to WSA for sale 19 Jul 46
7	*Alcyone* Sun	10 Aug 38 28 Aug 39	15 Jun 41	23 Jul 46	Ex-*Mormacgull,* ex–AK 24; to WSA for sale 24 Jul 46
8	*Algorab* Sun	10 Aug 38 15 Jun 39	15 Jun 41	3 Dec 45	Ex-*Mormacwren,* ex–AK 25; to MarComm 30 Jun 46
9	*Alhena* Beth-S	19 Jun 40 18 Jan 41	15 Jun 41	22 May 46 owner Sep 46	Ex-*Robin Kettering,* ex–AK 26; ret to
10	*Almaack* Beth-Q	14 Mar 40 21 Sep 40	15 Jun 41	10 May 46	Ex-*Executor,* ex–AK 27; to MarComm 12 Sep 46
11	*Betelgeuse* Sun	9 Mar 39 18 Sep 39	14 Jun 41	15 Mar 46	Ex-*Mormaclark,* ex–AK 28; sold 27 Jun 46
12	*Libra* Federal-K	5 Jun 41 12 Nov 41	13 May 42 28 Aug 50	19 Apr 48 6 Oct 55	Ex–*Jean Lykes,* ex–AK 53; str 1 Jan 77

No.	Name/Builder	LD/Launch	Comm	Decomm	Notes and Fate
13	*Titania* Federal-K	25 Oct 41 28 Feb 42	27 May 42	31 May 55	Ex–*Harry Culbreath*, ex–AK 55; str 1 Jul 61
14	*Oberon* Federal-K	17 Nov 41 18 Mar 42	15 Jun 42	27 Jun 55	Ex–*Delalba*, ex–AK 56; str 1 Jul 60
15	*Andromeda* Federal-K	22 Sep 42 22 Dec 42	2 Apr 43	1 May 56	Ex–AK 64; str 1 Jul 60
16	*Aquarius* Federal-K	28 Apr 43 23 Jul 43	21 Aug 43	23 May 46	Ex–AK 65; to WSA for sale 12 Sep 46
17	*Centaurus* Federal-K	7 Jun 43 3 Sep 43	21 Oct 43	30 Apr 46	Ex–AK 66; to MarComm 11 Sep 46
18	*Cepheus* Federal-K	26 Jul 43 27 Oct 43	16 Dec 43	22 May 46	Ex–AK 67; to MarComm 22 May 46
19	*Thuban* Federal-K	2 Feb 43 26 Apr 43	10 Jun 43	31 Oct 68	Ex–AK 68; str 1 Jan 77
20	*Virgo* Federal-K	9 Mar 43 4 Jun 43	16 Jul 43 19 Aug 66	3 Apr 58 18 Feb 71	Ex–AK 69; str 1 Jul 61, reinstated Sep 65, to AE 30 on 1 Nov 65, str on decomm
21	*Artemis* Walsh-Kaiser	23 Nov 43 20 May 44	28 Aug 44	10 Jan 47	To MarComm 1 Apr 48
22	*Athene* Walsh-Kaiser	20 Jan 44 18 Jun 44	29 Sep 44	17 Jun 46	To MarComm 23 Sep 47
23	*Aurelia* Walsh-Kaiser	5 Feb 44 18 Jun 44	14 Oct 44	14 Mar 46	Str 19 Jun 46, to Maritime Commission 1 Jul 46
24	*Birgit* Walsh-Kaiser	22 Feb 44 18 Jul 44	28 Oct 44	15 Mar 46	Str 21 May 46, to Maritime Commission 1 Jul 46
25	*Circe* Walsh-Kaiser	6 Mar 44 4 Aug 44	10 Nov 44	20 May 46	To WSA 26 Jun 46
26	*Corvus* Walsh-Kaiser	4 Apr 44 24 Sep 44	20 Nov 44	29 Mar 46	To MarComm 31 Oct 46
27	*Devosa* Walsh-Kaiser	22 May 44 12 Oct 44	30 Nov 44	2 Apr 46	To MarComm 2 Apr 46
28	*Hydrus* Walsh-Kaiser	19 Jun 44 28 Oct 44	9 Dec 44	26 Mar 46	To New York State Maritime Academy as *Empire State II* on 26 Mar 46, to MarComm Jun 56
29	*Lacerta* Walsh-Kaiser	19 Jul 44 10 Nov 44	19 Dec 44	25 Mar 46	To MarComm 25 Mar 46
30	*Lumen* Walsh-Kaiser	19 Jul 44 20 Nov 44	29 Dec 44	23 Mar 46	To WSA 11 Jun 46
31	*Medea* Walsh-Kaiser	5 Aug 44 30 Nov 44	10 Jan 45	24 Apr 46	Str 15 Oct 46, to WSA 29 Oct 46
32	*Melena* Walsh-Kaiser	25 Sep 44 11 Dec 44	20 Jan 45	11 Jun 46	To WSA 11 Jun 46, str 3 Jul 46, to California Maritime Academy as *Golden Bear*
33	*Ostaria* Walsh-Kaiser	13 Oct 44 21 Dec 44	31 Jan 45	1 Mar 46	Str 17 Apr 46, to WSA 26 Jun 46
34	*Pamina* Walsh-Kaiser	29 Oct 44 5 Jan 45	10 Feb 45		To *Tanner* (AGS 15) 15 May 46
35	*Polana* Walsh-Kaiser	11 Nov 44 17 Jan 45	21 Feb 45	21 Mar 46	Str 1 May 46, to Maritime Commission 26 Jun 46
36	*Renate* Walsh-Kaiser	21 Nov 44 31 Jan 45	21 Feb 45		To *Maury* (AGS 16) 12 Jul 46
37	*Roxane* Walsh-Kaiser	1 Dec 44 14 Feb 45	12 Mar 45	N/A	Str 3 Jul 46
38	*Sappho* Walsh-Kaiser	12 Dec 44 3 Mar 45	24 Apr 45	23 May 46	Str 15 Oct 46, to MarComm 13 Jan 47
39	*Sarita* Walsh-Kaiser	22 Dec 44 23 Feb 45	22 Mar 45	29 Jan 47	Str 25 Feb 47, to MarComm
40	*Scania* Walsh-Kaiser	6 Jan 45 17 Mar 45	16 Apr 45	2 Sep 47	Str 16 Sep 47

No.	Name/Builder	LD/Launch	Comm	Decomm	Notes and Fate
41	Selinur Walsh-Kaiser	18 Jan 45 28 Mar 45	21 Apr 45	30 Apr 46	Str 8 May 46, to MarComm
42	Sidonia Walsh-Kaiser	1 Feb 45 7 Apr 45	30 Apr 45	25 Feb 46	Str 25 Feb 46, to WSA 29 Jun 46
43	Sirona Walsh-Kaiser	15 Feb 45 17 Apr 45	10 May 45	12 Jun 46	Str 3 Jul 46
44	Sylvania Walsh-Kaiser	24 Feb 45 25 Apr 45	19 May 45	17 Dec 46	To MarComm 30 Apr 47
45	Tabora Walsh-Kaiser	4 Mar 45 3 May 45	29 May 45	31 May 46	To MarComm 1 Jul 46
46	Troilus Walsh-Kaiser	18 Mar 45 11 May 45	8 Jun 45	14 Jun 46	Str 8 Jul 46, to WSA 28 Aug 46
47	Turandot Walsh-Kaiser	29 Mar 45 20 May 45	18 Jun 45	21 Mar 46	Str 17 Apr 47, reacquired 4 Nov 54 for conversion to cable ship *Aeolus* (ARC 3)
48	Valeria Walsh-Kaiser	24 Feb 45 29 May 45	28 Jun 45	18 Mar 46	Str 12 Apr 46, to MarComm
49	Vanadis Walsh-Kaiser	18 Apr 45 8 Jun 45	9 Jul 45	27 Mar 46	Str 5 Jun 46, reacquired 14 Apr 55 for conversion to cable ship *Thor* (ARC 4)
50	Veritas Walsh-Kaiser	26 Apr 45 16 Jun 45	19 Jul 45	21 Feb 46	Str 12 Apr 46, to MarComm
51	Xenia Walsh-Kaiser	4 May 45 27 Jun 45	28 Jul 45	13 May 46	To Chile 30 Nov 46
52	Zenobia Walsh-Kaiser	12 May 45 6 Jul 45	6 Aug 45	7 May 46	To Chile 30 Nov 46
53	Achernar Federal-K	6 Sep 43 3 Dec 43	31 Jan 44 1 Sep 61	18 Feb 56 1 Jul 63	To Spain 2 Feb 65
54	Algol Moore	10 Dec 42 17 Feb 43	21 Jul 44 17 Nov 61	2 Feb 58 20 Feb 70	Ex–*James Baines;* str 1 Jan 77
55	Alshain Federal-K	29 Oct 43 26 Jan 44	1 Apr 44	14 Jan 56	Str 1 Jul 60
56	Arneb Moore	20 Apr 43 6 Jul 43	16 Nov 43	12 Aug 71	Ex-*Mischief;* str 13 Aug 71
57	Capricornus Moore	5 May 43 14 Aug 43	31 May 44 12 Oct 50	30 Mar 48 10 Feb 70	Ex-*Spitfire;* str 1 Jan 77
58	Chara Federal-K	6 Dec 43 15 Mar 44	14 Jun 44 25 Jun 66	21 Apr 59	To AE 31 on 1 Nov 65, str
59	Diphda Federal-K	27 Jan 44 11 May 44	8 Jul 44	11 May 56	Str 1 Jul 61
60	Leo Federal-K	17 Mar 44 29 Jun 44	30 Aug 44	11 Feb 55	Str 1 Jul 60
61	Muliphen Federal-K	13 May 44 26 Aug 44	23 Oct 44	31 Aug 70	Str 1 Jan 77
62	Sheliak Federal-K	19 Jun 44 17 Oct 44	1 Dec 44	10 May 46	Str 21 May 46
63	Theenim Federal-K	3 Jul 44 31 Oct 44	22 Dec 44	8 May 46	Sold
64	Tolland NC Ship	22 Apr 44 26 Jun 44	4 Sep 44	1 Jul 46	Str 19 Jul 46
65	Shoshone NC Ship	12 May 44 17 Jul 44	24 Sep 44	28 Jun 46	Str 19 Jul 46
66	Southampton NC Ship	26 May 44 28 Jul 44	8 Oct 44	21 Jun 46	Str 3 Jul 46
67	Starr NC Ship	13 Jun 44 29 Sep 44	21 Oct 44	20 May 46	Ret to owner 1 Jun 46
68	Stokes NC Ship	26 Jun 44 31 Aug 44	4 Nov 44	9 Jul 46	Str 19 Jul 46, to WSA
69	Suffolk NC Ship	11 Jul 44 15 Sep 44	14 Nov 44	27 Jun 46	Ret to owner 28 Jun 46

No.	Name/Builder	LD/Launch	Comm	Decomm	Notes and Fate
70	*Tate* NC Ship	22 Jul 44 26 Sep 44	25 Nov 44	10 Jul 46	Str 19 Jul 46
71	*Todd* NC Ship	10 Aug 44 10 Oct 44	30 Nov 44	25 Jun 46	Ret to owner, 26 Jun 46
72	*Caswell* NC Ship	25 Aug 44 24 Oct 44	13 Dec 44	19 Jun 46	To MarComm 21 Jun 46
73	*New Hanover* NC Ship	31 Aug 44 31 Oct 44	22 Dec 44	30 Jul 46	Str 15 Aug 46
74	*Lenoir* NC Ship	7 Sep 44 6 Nov 44	31 Dec 44	13 Jun 46	To MarComm 14 Jun 46
75	*Alamance* NC Ship	15 Sep 44 11 Nov 44	9 Jan 45	6 Jul 46	To MarComm 30 Jun 46
76	*Torrance* NC Ship	1 Apr 44 6 Jun 44	18 Nov 44	10 Jun 46	Str 3 Jul 46
77	*Towner* NC Ship	8 Apr 44 6 Jun 44	1 Dec 44	8 Jan 46	Ret to owner 13 Jun 46
78	*Trego* NC Ship	14 Apr 44 20 Jun 44	21 Dec 44	21 May 46	Ret to owner 22 May 46
79	*Trousdale* NC Ship	28 Apr 44 3 Jul 44	21 Dec 44	29 Apr 46	Str 8 May 46
80	*Tyrrell* NC Ship	6 May 44 10 Jul 44	4 Dec 44	19 Apr 46	Str 1 May 46
81	*Valencia* NC Ship	20 May 44 22 Jul 44	9 Jan 45	8 May 46	Str 21 May 46
82	*Venango* NC Ship	6 Jun 44 9 Aug 44	2 Jan 45	18 Apr 46	Str 1 May 46
83	*Vinton* NC Ship	20 Jun 44 25 Aug 44	23 Feb 45	16 May 46	Str 5 Jun 46
84	*Waukesha* NC Ship	3 Jul 44 6 Sep 44	20 Sep 44	10 Jul 46	Str 31 Jul 46
85	*Wheatland* NC Ship	17 Jul 44 21 Sep 44	3 Apr 45	23 Apr 46	Ret to owner 26 Apr 46
86	*Woodford* NC Ship	29 Jul 44 5 Oct 44	3 Mar 45	1 May 46	Str 8 May 46
87	*Duplin* NC Ship	18 Aug 44 17 Oct 44	15 May 45	21 May 46	To MarComm 23 May 46
88	*Uvalde* Moore	27 Mar 44 20 May 44	18 Aug 44 18 Nov 61	57 29 Nov 68	Str 1 Jul 60, reacquired 1 Sep 61, str 1 Dec 68
89	*Warrick* Moore	7 Apr 44 29 May 44	30 Aug 44		Tgt 28 May 71
90	*Whiteside* Moore	22 Apr 44 12 Jun 44	11 Sep 44	15 Aug 54	Tgt May 71
91	*Whitley* Moore	2 May 44 22 Jun 44	21 Sep 44	19 Jul 55	To Italy Mar 62
92	*Wyandot* Moore	6 May 44 28 Jun 44	30 Sep 44 Nov 61	10 Jul 59 31 Oct 75	Str 1 Jul 60, reacquired 61, to MSTS Mar 63, to AK 283 on 1 Jan 69, str 31 Mar 86
93	*Yancey* Moore	22 May 44 8 Jul 44	11 Oct 44 17 Nov 61	Mar 58 20 Jan 71	Str 1 Jan 77
94	*Winston* Federal-K	10 Jul 44 30 Nov 44	19 Jan 45 24 Nov 61	1 Feb 57 Nov 69	Str 1 Sep 76
95	*Marquette* Federal-K	30 Aug 44 29 Apr 45	20 Jun 45	19 Jul 55	Sold Mar 76
96	*Matthews* Federal-K	15 Sep 44 29 Apr 45	5 Mar 45 16 Feb 52	4 Apr 47 31 Oct 68	Str 1 Nov 68
97	*Merrick* Federal-K	19 Oct 44 28 Jan 45	31 Mar 45 19 Jan 52	25 Jun 46 17 Sep 69	Str 1 Sep 76
98	*Montague* Federal-K	2 Nov 44 11 Feb 45	13 Apr 45	22 Nov 55	To MarAd 29 Jan 60

No.	Name/Builder	LD/Launch	Comm	Decomm	Notes and Fate
99	Rolette Federal-K	2 Dec 44 11 Mar 45	28 Apr 45 23 Feb 52	Jan 46	Str 23 Apr 47, reacquired 13 Aug 51
100	Oglethorpe Federal-K	26 Dec 44 15 Apr 45	6 Jun 45	Oct 68	Str 1 Nov 68
101	Ottawa NC Ship	5 Oct 44 29 Nov 44	8 Feb 45	10 Jan 46	To MarComm 28 Jan 47, str 14 Mar 47
102	Prentiss NC Ship	10 Oct 44 6 Dec 44	11 Feb 45	31 May 46	To WSA 1 Jun 46, str 19 Jun 46
103	Rankin NC Ship	31 Oct 44 22 Dec 44	25 Feb 45 22 Mar 52	21 May 47 11 May 71	Str 1 Jan 77
104	Seminole NC Ship	7 Nov 44 28 Dec 44	8 Mar 45	23 Dec 70	Str 1 Sep 76
105	Skagit NC Ship	21 Sep 44 18 Nov 44	2 May 45 26 Aug 50	30 Jun 49 Jul 69	Str 1 Jul 69
106	Union NC Ship	27 Sep 44 23 Nov 44	25 Apr 45	5 Jun 70	Str 18 Aug 70
107	Vermillion NC Ship	17 Oct 44 12 Dec 44	23 Jun 45 16 Oct 50	26 Aug 49 14 Apr 71	Str 1 Jan 77
108	Washburn NC Ship	24 Oct 44 18 Dec 44	17 May 45	23 Dec 70	Str 1 Sep 76
109	San Joaquin Federal	17 Aug 45			Canceled 27 Aug 45
110	Sedgwick Federal				Not laid down, canceled 27 Aug 45
111	Whitfield Federal				Not laid down, canceled 27 Aug 45
112	Tulare Beth-SF	16 Feb 53 22 Dec 53	12 Jan 56	15 Feb 80	Str 1 Aug 81, reinstated May 84, str 31 Aug 92
113	Charleston NN Ship	5 Dec 66 2 Dec 67	14 Dec 68	27 Apr 92	
114	Durham NN Ship	10 Jul 67 29 Mar 68	24 May 69	25 Feb 94	
115	Mobile NN Ship	15 Jan 68 19 Oct 68	29 Sep 69	4 Feb 94	
116	Saint Louis NN Ship	3 Apr 68 4 Jan 69	22 Nov 69	2 Nov 92	
117	El Paso NN Ship	22 Oct 68 17 May 69	17 Jan 70	21 Apr 94	

APA/LPA

No.	Name/Builder	LD/Launch	Comm	Decomm	Notes and Fate
1	Doyen Consolidated-W	18 Sep 41 9 Jul 42	22 May 43	22 Mar 46	Ex–AP 2; to MarComm 26 Jun 46, to Massachusetts Maritime Academy 61 as Bay State, ret 73
2	Harris Beth-S	N/A 19 Mar 21	5 Nov 40	16 Apr 46	Ex–SS President Grant, ex–Pine Tree State, ex–AP 8; sold 20 Jul 48
3	Zeilin NN Ship	N/A 11 Dec 20	3 Jan 42	19 Apr 46	Ex–SS President Jackson, ex–Silver State, ex–AP 9; in ordinary commission 6 Aug 40, str 5 Jun 46
4	McCawley Furness	N/A 8 Dec 27	10 Sep 40		Ex–SS Santa Barbara, ex–AP 10; WL 30 Jun 43
5	Barnett Furness	N/A 15 Aug 27	25 Sep 40	21 May 46	Ex–SS Santa Maria; ex–AP 11; to MarComm 3 Jul 46
6	Heywood Beth-A	N/A 24 Dec 18	19 Feb 41	12 Apr 46	Ex–SS City of Baltimore, ex–Steadfast, ex–AP 12; to MarComm 2 Jul 46

No.	Name/Builder	LD/Launch	Comm	Decomm	Notes and Fate
7	Fuller Beth-A	N/A 4 Nov 18	9 Apr 41	20 Mar 46	Ex-SS *City of Newport News*, ex-*Archer*, ex–AP 14; to MarComm 1 Jul 46, BU Apr 57
8	William P. Biddle Beth-A	N/A 20 Oct 18	3 Feb 41	8 Apr 46	Ex-SS *City of San Francisco*, ex–*City of Hamburg*, ex–*Eclipse*, ex–*War Surf*, ex–AP 15; str 5 Jun 46, BU 57
9	Neville Beth-A	N/A 4 Jul 18	14 May 41	30 Apr 46	Ex-SS *City of Norfolk*, ex–*Independence*, ex–*War Harbor*, ex–AP 16; to WSA 16 Jul 46, str 18 Aug 46
10	Harry Lee NY Ship	N/A 18 Oct 30	27 Dec 40	9 May 46	Ex-SS *Exochorda*, ex–AP 17; to Turkey Apr 48
11	Feland Consolidated-W	25 Nov 41 10 Nov 42	21 Jun 43	15 Jan 46	Ex–AP 18; to MarComm 46, BU 64
12	Leonard Wood Beth-S	N/A 19 Sep 21	10 Jun 41	22 Mar 46	Ex-SS *Western World*, ex–*Nutmeg State*, ex–AP 25; to army Mar 46, to WSA, sold 20 Jan 48 for BU
13	Joseph T. Dickman NY Ship	N/A 6 Jul 21	10 Jun 41	7 Mar 46	Ex-SS *President Roosevelt*, ex–*President Pierce*, ex–*Peninsula State*, ex–AP 26; to MarComm 7 Mar 46, sold 48 for BU
14	Hunter Liggett Beth-S	N/A 4 Jun 21	9 Jun 41	18 Mar 46	Ex-SS *Pan America*, ex–*Palmetto State*, ex–AP 27; sold 30 Jan 48
15	Henry T. Allen NY Ship	N/A 24 May 19	22 Apr 42	5 Feb 46	Ex–*President Jefferson*, ex-*Wenatchee*, ex–AP 30; to AG 90 (Seventh Fleet admin flagship) 21 Nov 44, ret to army 5 Feb 46
16	J. Franklin Bell NY Ship	N/A 15 May 20	2 Apr 42	20 Mar 46	Ex–*President McKinley*, ex–*Keystone State*, ex–AP 34; to WSA 3 Apr 48, sold
17	American Legion NY Ship	12 Jan 19 11 Oct 19	26 Aug 41	20 Mar 46	Ex-SS *American Legion*, ex-*Koda*, ex–AP 35; to army 20 Mar 46, sold 48
18	President Jackson NN Ship	2 Oct 39 7 Jun 40	16 Jan 42	6 Jul 55	Ex-SS *President Jackson*, ex–AP 37; to MSTS 22 Oct 49, str 1 Oct 58
19	President Adams NN Ship	10 Jun 40 31 Jan 41	19 Nov 41	14 Jun 50	No merchant name, ex–AP 38; str 1 Oct 58
20	President Hayes NN Ship	26 Dec 39 4 Oct 40	15 Dec 41	30 Jun 49	No merchant name, ex–AP 39; str 1 Oct 58
21	Crescent City Beth-S	8 May 39 17 Feb 40	10 Oct 41	30 Apr 48	Ex-SS *Delorleans*, ex–AP 40; str Oct 58, to California Maritime Academy May 71 as *Golden Bear*, ret 17 Jul 95, to Artship Foundation of Oakland for museum use 6 Aug 99
22	Joseph Hewes NY Ship	N/A 5 Aug 30	1 May 42		Ex-SS *Excalibur*, ex–AP 50; WL(T) 11 Nov 42
23	John Penn NY Ship	N/A 28 May 31	9 May 42		Ex-SS *Excambion*, ex–AP 51; WL 13 Aug 43
24	Edward Rutledge NY Ship	N/A 4 Apr 31	18 Apr 42		Ex-SS *Exeter*, ex–AP 52; WL(T) 12 Nov 42
25	Arthur Middleton Ingalls	1 Jul 40 28 Jun 41	7 Sep 42	21 Oct 46	Ex-SS *African Comet*, ex–AP 55; str 1 Oct 58
26	Samuel Chase Ingalls	31 Aug 40 23 Aug 41	13 Jun 42	17 Apr 47	Ex-SS *African Meteor*, ex–AP 56; str 1 Oct 68
27	George Clymer Ingalls	28 Oct 40 27 Sep 41	15 Jun 42	67	Ex-SS *African Planet*, ex–AP 57; str 1 Nov 67
28	Charles Carroll Beth-S	4 Sep 41 24 Mar 42	13 Aug 42	27 Dec 46	Ex-SS *Deluruguay*, ex–AP 58; str 58, to MarAd 29 Oct 58
29	Thomas Stone NN Ship	12 Aug 40 1 May 41	18 May 42	1 Apr 44	Ex-SS *President Van Buren*, ex–AP 59; CTL 25 Nov 42, salvage failed, str 8 Apr 44
30	Thomas Jefferson NN Ship	5 Feb 40 20 Nov 40	31 Aug 42	18 Jul 55	Ex-SS *President Garfield*, ex–AP 60; to MSTS 1 Oct 49 (T-AP 30), str 1 Oct 58

No.	Name/Builder	LD/Launch	Comm	Decomm	Notes and Fate
31	*Monrovia* Beth-S	26 Mar 42 19 Sep 42	1 Dec 42 30 Nov 50	26 Feb 47 31 Oct 68	Ex–SS *Delargentino,* ex–AP 64; str 1 Nov 68
32	*Calvert* Beth-S	15 Nov 41 22 May 42	1 Oct 42 11 Oct 50	26 Feb 47 30 Jun 66	Ex–SS *Delorleans,* ex–AP 65; cargo handling training hulk at Oakland 30 Jun 66, str 1 Aug 66, sold 11 Mar 77
33	*Bayfield* Western-SF	14 Nov 42 15 Feb 43	22 Nov 43	Jul 68	Ex–SS *Sea Bass,* ex–AP 78; str 1 Oct 68
34	*Bolivar* Western-SF	13 May 42 7 Sep 42	1 Sep 43	29 Apr 46	Ex–SS *Sea Angel,* ex–AP 79; to MarComm 12 Sep 46
35	*Callaway* Western-SF	10 Jun 42 10 Oct 42	11 Sep 43	10 May 46	Ex–SS *Sea Mink,* ex–AP 80; to MarComm 12 Sep 46, sold 49
36	*Cambria* Western-SF	1 Jul 42 10 Nov 42	10 Nov 42 15 Sep 50	30 Jun 49 15 Sep 70	Ex–SS *Sea Swallow,* ex–AP 81; str 15 Sep 70
37	*Cavalier* Western-SF	10 Dec 42 15 Mar 43	15 Jan 44	1 Oct 68	Ex–AP 82; str 1 Oct 68
38	*Chilton* Western-SF	10 Sep 42 23 Jan 43	7 Dec 43	1 Jul 70	Ex–SS *Sea Needle,* ex–AP 83; str 1 Jul 72
39	*Clay* Western-SF	14 Oct 42 23 Jan 43	21 Dec 43	15 May 46	Ex–SS *Sea Carp,* ex–AP 84; sold 12 Sep 46
40	*Custer* Ingalls	19 Feb 42 6 Nov 42	18 Jul 43	24 May 46	Ex–SS *Sea Eagle,* ex–AP 85; to MarComm 11 Sep 46
41	*Du Page* Ingalls	20 Mar 42 19 Dec 42	1 Sep 43	28 Mar 46	Ex–SS *Sea Hound,* ex–AP 86; to WSA 27 Jun 46
42	*Elmore* Ingalls	24 Jun 42 29 Jan 43	25 Aug 43	13 Mar 46	Ex–SS *Sea Panther,* ex–AP 87; to MarComm 15 May 46
43	*Fayette* Ingalls	1 Oct 42 25 Feb 43	14 Oct 43	6 Mar 46	Ex–SS *Sea Hawk,* ex–AP 88; to MarComm 19 Apr 46
44	*Fremont* Ingalls	28 Oct 42 31 Mar 43	23 Nov 43	2 Sep 69	Ex–SS *Sea Corsair,* ex–AP 89; str 1 Jun 73
45	*Henrico* Ingalls	12 Nov 42 31 Mar 43	26 Nov 43	14 Feb 68	Ex–SS *Sea Darter,* ex–AP 90; str 1 Jun 73
46	*Knox* Ingalls	4 Mar 43 17 Jul 43	4 Mar 44	14 Mar 46	Ex–AP 91; str 1 May 46
47	*Lamar* Ingalls	7 Apr 43 28 Aug 43	6 Apr 44	7 Mar 46	Ex–AP 92; to MarComm 3 Jul 46
48	*Leon* Ingalls	6 Feb 43 19 Jun 43	11 Oct 44	7 Mar 46	Ex–SS *Sea Dolphin,* ex–AP 93; to WSA 2 Apr 46
49	*Ormsby* Moore	21 Jul 42 20 Oct 42	28 Jun 43	15 Mar 46	Ex–SS *Twilight,* ex–AP 94; str 17 Apr 46
50	*Pierce* Moore	22 Jul 42 10 Oct 42	30 Jun 43	11 Mar 46	Ex–SS *Northern Light,* ex–AP 95; str 17 Apr 46
51	*Sheridan* Moore	5 Aug 42 11 Nov 42	31 Jul 43	5 Mar 46	Ex–SS *Messenger,* ex–AP 96; str 12 Apr 46
52	*Sumter* Gulf	3 Apr 42 4 Oct 42	1 Sep 43	26 Mar 46	Ex–SS *Iberville,* ex–AP 97; to MarComm 1 Aug 46
53	*Warren* Gulf	19 Apr 42 7 Sep 42	2 Aug 43	11 Mar 46	Ex–SS *Jean Lafitte,* ex–AP 98; str 17 Apr 46
54	*Wayne* Gulf	20 Apr 42 6 Dec 42	30 Apr 43	18 Mar 46	Ex–SS *Afoundria,* ex–AP 99; str 17 Apr 46
55	*Windsor* Beth-S	23 Jul 42 28 Dec 42	17 Jun 43	18 Mar 46	Ex–SS *Excelsior,* ex–AP 100; to MarComm 1 Aug 46
56	*Leedstown* Beth-S	26 Jul 42 13 Feb 43	16 Jul 43	7 Mar 46	Ex–*Wood,* ex–SS *Exchequer,* ex–AP 101; to MarComm 1 Jul 46
57	*Gilliam* Consolidated-W	30 Nov 43 28 Mar 44	1 Aug 44	30 Jun 46	Sunk as tgt at Bikini 1 Jul 46
58	*Appling* Consolidated-W	7 Dec 43 9 Apr 44	22 Aug 44	20 Dec 46	To MarComm 31 Mar 48

No.	Name/Builder	LD/Launch	Comm	Decomm	Notes and Fate
59	*Audrain* Consolidated-W	11 Dec 43 21 Apr 44	1 Sep 44	15 May 46	To MarComm 25 Jul 47, str 1 Aug 47
60	*Banner* Consolidated-W	24 Jan 44 3 May 44	16 Sep 44	27 Aug 46	Bikini tgt, scuttled 16 Feb 48
61	*Barrow* Consolidated-W	28 Jan 44 11 May 44	28 Sep 44	28 Aug 46	Bikini tgt, scuttled 11 May 48
62	*Berrien* Consolidated-W	23 Feb 44 20 May 44	8 Oct 44	17 May 46	To MarComm 12 Aug 47
63	*Bladen* Consolidated-W	8 Mar 44 31 May 44	18 Oct 44	26 Dec 46	Bikini support ship, to MarComm 3 Aug 53
64	*Bracken* Consolidated-W	13 Mar 44 10 Jun 44	4 Oct 44	28 Aug 46	Bikini tgt, scuttled 10 Mar 48
65	*Briscoe* Consolidated-W	29 Mar 44 19 Jun 44	29 Oct 44	28 Aug 46	Bikini tgt, sunk 6 May 48
66	*Brule* Consolidated-W	10 Apr 44 30 Jun 44	31 Oct 44	28 Aug 46	Bikini tgt, sunk 11 May 48
67	*Burleson* Consolidated-W	22 Apr 44 11 Jul 44	8 Nov 44	9 Nov 46	Animal transport for Bikini test, to IX 67 on 5 Oct 56, used for amphibious debarkation training (army and navy), BU Nov 68
68	*Butte* Consolidated-W	4 May 44 20 Jul 44	21 Nov 44	28 Aug 46	Bikini tgt, sunk 12 May 48
69	*Carlisle* Consolidated-W	12 May 44 30 Jul 44	29 Nov 44	11 Jul 46	Sunk at Bikini, 1 Jul 46
70	*Carteret* Consolidated-W	22 May 44 15 Aug 44	3 Dec 44	6 Aug 46	Bikini tgt, sunk 19 Apr 48
71	*Catron* Consolidated-W	1 Jun 44 28 Aug 44	28 Nov 44	29 Aug 46	Bikini tgt, scuttled 6 May 48
72	*Clarendon* Consolidated-W	12 Jun 44 12 Sep 44	14 Dec 44	9 Apr 46	To WSA Jun 46
73	*Cleburne* Consolidated-W	20 Jun 44 27 Sep 44	22 Dec 44	7 Jun 46	Bikini support ship, to MarComm 7 Jul 47
74	*Colusa* Consolidated-W	1 Jul 44 7 Oct 44	20 Dec 44	16 May 46	To MarComm 14 Aug 47
75	*Cortland* Consolidated-W	12 Jul 44 18 Oct 44	1 Jan 45	30 Dec 46	To MarComm 31 Mar 48
76	*Crenshaw* Consolidated-W	21 Jul 44 27 Oct 44	4 Jan 45	19 Apr 46	To WSA 30 Jun 46
77	*Crittenden* Consolidated-W	31 Jul 44 6 Nov 44	17 Jan 45	28 Aug 46	Bikini tgt, tgt 6 Oct 47
78	*Cullman* Consolidated-W	16 Aug 44 18 Nov 44	25 Jan 45	22 May 46	To WSA 30 Jun 46
79	*Dawson* Consolidated-W	29 Aug 44 27 Nov 44	4 Feb 45	20 Apr 48	Bikini tgt, sunk 19 Apr 48
80	*Elkhart* Consolidated-W	13 Sep 44 5 Dec 44	8 Feb 45	12 Apr 46	To MarComm 28 Jun 46
81	*Fallon* Consolidated-W	28 Sep 44 14 Dec 44	14 Feb 45	20 Sep 46	Bikini tgt, sunk 11 Mar 48
82	*Fergus* Consolidated-W	7 Oct 44 24 Dec 44	20 Feb 45	25 Jun 46	To MarComm 4 Sep 47
83	*Fillmore* Consolidated-W	19 Oct 44 4 Jan 45	25 Feb 45	24 Jan 47	Bikini support ship, to MarComm 1 Apr 48
84	*Garrard* Consolidated-W	28 Oct 44 13 Jan 45	3 Mar 45	21 May 46	To MarComm 29 Jun 46
85	*Gasconade* Consolidated-W	7 Nov 44 23 Jan 45	11 Mar 45	28 Aug 46	Bikini tgt, tgt 21 Jul 48
86	*Geneva* Consolidated-W	18 Nov 44 31 Jan 45	22 Mar 45	23 Jan 47	Bikini support ship, str 25 Feb 47, to MarComm 2 Apr 47

No.	Name/Builder	LD/Launch	Comm	Decomm	Notes and Fate
87	*Niagara* Consolidated-W	28 Nov 44 10 Feb 45	29 Mar 45	12 Dec 46	Bikini tgt, used for underwater tests 46–49, BU Feb 50
88	*Presidio* Consolidated-W	6 Dec 44 17 Feb 45	9 Apr 45	20 Jun 46	Str 1 Aug 47, to MarComm 2 Sep 47
89	*Frederick Funston* Sea-Tac	21 Apr 41 27 Sep 41	24 Apr 43	4 Apr 46	Ex–AP 48; served for a time as an army troopship before APA conversion, to army on decomm, reacquired by MSTS as T-AP 178, str Jul 61
90	*James O'Hara* Sea-Tac	16 Jun 41 30 Dec 41	26 Apr 43 28 Apr 50	5 Apr 46 14 Jan 60	Ex–AP 49; served for a time as an army troopship before APA conversion, to army on decomm, reacquired by MSTS as T-AP 179, str 1 Jul 61
91	*Adair* Beth-S	28 Jul 43 29 Feb 44	15 Jul 44	30 Apr 46	Ex–SS *Exchester;* to MarComm 3 May 46
92	*Alpine* Western-SF	12 Apr 43 10 Jul 43	22 Apr 44	5 Apr 46	Ex–SS *Sea Arrow;* to MarComm 3 May 46
93	*Barnstable* Western-SF	6 May 43 5 Aug 43	22 May 44	25 Mar 46	Ex–SS *Sea Snapper;* to MarComm 26 Mar 46
94	*Baxter* Gulf	18 Mar 43 19 Sep 43	15 May 44	22 Mar 46	Ex–SS *Antinous;* to MarComm 22 Mar 46
95	*Burleigh* Ingalls	6 Jul 43 3 Dec 43	30 Oct 44	11 Jun 46	To MarComm 12 Jun 46
96	*Cecil* Western-SF	24 Jun 43 27 Sep 43	15 Sep 44	24 May 46	Ex–SS *Sea Angler;* to MarComm 26 May 46
97	*Dauphin* Beth-S	22 Dec 43 10 Jun 44	23 Sep 44	3 Apr 46	To WSA 4 Apr 46
98	*Dutchess* Beth-S	1 Feb 44 26 Aug 44	4 Nov 44	4 Apr 46	To WSA 5 Apr 46
99	*Dade* Ingalls	2 Sep 43 14 Jan 44	11 Nov 44	25 Feb 46	Ex-*Lorain;* to MarComm 25 Feb 46
100	*Mendocino* Ingalls	20 Sep 43 11 Feb 44	31 Oct 44	25 Feb 46	To MarComm 25 Feb 46
101	*Montour* Ingalls	20 Oct 43 10 Mar 44	9 Dec 44	19 Apr 46	To WSA 23 Apr 46, str 8 May 46
102	*Riverside* Ingalls	11 Nov 43 13 Apr 44	18 Dec 44		Str 8 May 46
103	*Queens* Beth-S	3 Mar 44 12 Sep 44	16 Dec 44	10 Jun 46	to WSA Jun 46, str 19 Jun 46
104	*Westmoreland* Ingalls	8 Dec 43 28 Apr 44	18 Jan 45	5 Jun 46	Str 19 Jun 46
105	*Shelby* Beth-S	13 Jun 44 25 Oct 44	20 Jan 45	18 May 46	Ex–SS *Exeter;* to WSA 16 May 46, str 5 Jun 46
106	*Hansford* Western-SF	10 Dec 43 25 Apr 44	12 Oct 44	14 Jun 46	Ex–SS *Gladwin,* ex–SS *Sea Adder;* to MarComm 20 May 47
107	*Goodhue* Western-SF	7 Jan 44 31 May 44	11 Nov 44	5 Apr 46	Ex–SS *Sea Wren;* to MarComm 5 Apr 46
108	*Goshen* Western-SF	31 Jan 44 29 Jun 44	13 Dec 44	20 Apr 46	Ex–SS *Sea Hare;* to WSA 2 May 46
109	*Grafton* Western-SF	3 Mar 44 10 Aug 44	5 Jan 45	16 May 46	Ex–SS *Sea Sparro;* sold 47
110	*Griggs* Ingalls	1 Dec 43 16 Jun 44	14 Dec 44	27 May 46	To WSA 28 May 46, str 19 Jun 46
111	*Grundy* Ingalls	22 Dec 43 16 Jun 44	3 Jan 45	8 May 46	To MarComm 31 May 46
112	*Guilford* Ingalls	19 Jan 44 14 Jul 44	14 May 45	29 May 46	To MarComm 31 May 46
113	*Sitka* Ingalls	2 Feb 44 23 Jun 44	14 Mar 45	14 May 46	To WSA 15 May 46, str 5 Jun 46

No.	Name/Builder	LD/Launch	Comm	Decomm	Notes and Fate
114	*Hamblen* Ingalls	16 Feb 44 30 Jul 44	12 Jun 45	1 May 46	To MarComm 7 May 46
115	*Hampton* Ingalls	4 Mar 44 25 Aug 44	17 Feb 45	30 Apr 46	To MarComm 7 May 46
116	*Hanover* Ingalls	15 Mar 44 18 Aug 44	31 Mar 45	11 May 46	To MarComm 13 May 46
117	*Haskell* Cal Ship	28 Mar 44 13 Jun 44	11 Sep 44	22 May 46	To MarComm May 46
118	*Hendry* Cal Ship	15 Apr 44 24 Jun 44	28 Sep 44	21 Feb 46	To MarComm 46, str 20 Mar 46
119	*Highlands* Cal Ship	28 Apr 44 8 Jul 44	5 Oct 44	8 Apr 46	To MarComm Apr 46, str 26 Feb 46
120	*Hinsdale* Cal Ship	12 May 44 22 Jul 44	15 Oct 44	8 Apr 46	To MarComm Apr 46, str 1 May 46
121	*Hocking* Cal Ship	30 May 44 6 Aug 44	22 Oct 44	10 May 46	To MarComm May 46, str 21 May 46
122	*Kenton* Cal Ship	13 Jun 44 21 Aug 44	31 Oct 44	28 Mar 46	To MarComm Mar 46, str 12 Apr 46
123	*Kittson* Cal Ship	21 Jun 44 28 Aug 44	5 Nov 44	11 Mar 46	To MarComm Mar 46, str 20 Mar 46
124	*Lagrange* Cal Ship	26 Jun 44 1 Sep 44	11 Nov 44	27 Oct 45	To WSA 27 Oct 45, str 23 Nov 45
125	*Lanier* Cal Ship	25 Jun 44 29 Aug 44	22 Dec 44	5 Mar 46	To WSA 8 Mar 46, str 20 Mar 46
126	*St. Mary's* Cal Ship	29 Jun 44 4 Sep 44	15 Nov 44	15 Feb 46	To MarComm, str 26 Feb 46
127	*Allendale* Cal Ship	1 Jul 44 9 Sep 44	22 Nov 44	14 Mar 46	To MarComm 20 Mar 46, str 28 Mar 46
128	*Arenac* Cal Ship	9 Jul 44 14 Sep 44	8 Jan 45	10 Jul 46	Str 1 Oct 58
129	*Marvin H. McIntyre* Cal Ship	13 Jul 44 21 Sep 44	28 Nov 44	6 Jun 46	To MarComm 12 Jun 46, str 19 Jun 46
130	*Attala* Cal Ship	18 Jul 44 27 Sep 44	30 Nov 44	26 Feb 46	To MarComm 3 Mar 46, str 20 Mar 46
131	*Bandera* Cal Ship	23 Jul 44 6 Oct 44	6 Dec 44	7 May 46	To MarComm 14 May 46, str 21 May 46
132	*Barnwell* Cal Ship	25 Jul 44 30 Sep 44	19 Jan 45	1 Feb 47	Str 1 Oct 48, to MarComm 59
133	*Beckham* Cal Ship	27 Jul 44 14 Oct 44	10 Dec 44	25 Apr 46	To MarComm 29 Apr 46, str 8 May 46
134	*Bland* Cal Ship	2 Aug 44 26 Oct 44	15 Dec 44	27 Apr 46	To MarComm 28 Apr 46, str 8 May 46
135	*Bosque* Cal Ship	7 Aug 44 28 Oct 44	17 Dec 44	15 Mar 46	To MarComm 22 Mar 46, str 28 Mar 46
136	*Botetourt* Cal Ship	22 Aug 44 19 Oct 44	31 Jan 45 23 Sep 50	5 Jun 46 27 Apr 56	Str 2 Jul 61
137	*Bowie* Cal Ship	28 Aug 44 31 Oct 44	23 Dec 44	8 Mar 46	To MarComm 14 Mar 46, str 28 Mar 46
138	*Braxton* Cal Ship	29 Aug 44 3 Nov 44	29 Dec 44	27 Jun 46	To MarComm 29 Jun 46, str 19 Jul 46
139	*Broadwater* Cal Ship	1 Sep 44 5 Nov 44	2 Jan 45	28 Feb 46	To MarComm 1 Mar 46, str 20 Mar 46
140	*Brookings* Cal Ship	5 Sep 44 20 Nov 44	6 Jan 45	25 Jul 46	Str 1 Oct 58
141	*Buckingham* Cal Ship	9 Sep 44 13 Nov 44	23 Jan 45	1 Mar 46	To MarComm 5 Mar 46, str 20 Mar 46
142	*Clearfield* Cal Ship	15 Sep 44 21 Nov 44	12 Jan 45	4 Mar 46	To MarComm 6 Mar 46, str 20 Mar 46

No.	Name/Builder	LD/Launch	Comm	Decomm	Notes and Fate
143	*Clermont* Cal Ship	21 Sep 44 25 Nov 44	28 Jan 45	1 Mar 46	To MarComm 3 Mar 46, str 20 Mar 46
144	*Clinton* Cal Ship	27 Sep 44 29 Nov 44	1 Feb 45	2 May 46	To MarComm 1 Oct 58, str 1 Oct 58
145	*Colbert* Cal Ship	30 Sep 44 1 Dec 44	7 Feb 45	26 Feb 46	To WSA 26 Feb 46, str 12 Mar 46
146	*Collingsworth* Cal Ship	6 Oct 44 2 Dec 44	27 Feb 45	17 Mar 46	To MarComm 20 Mar 46, str 28 Mar 46
147	*Cottle* Kaiser-V	15 Oct 44 25 Nov 44	14 Dec 44	6 Mar 46	To MarComm 11 Mar 46, str 20 Mar 46
148	*Crockett* Kaiser-V	18 Oct 44 28 Nov 44	18 Jan 45	15 Oct 46	To MarComm 1 Oct 58, str 1 Oct 58
149	*Audobon* Kaiser-V	21 Oct 44 3 Dec 44	20 Dec 44	19 Feb 46	To MarComm 28 Feb 46, str 12 Mar 46
150	*Bergen* Kaiser-V	25 Oct 44 5 Dec 44	23 Dec 44	24 Apr 46	To MarComm 26 Apr 46, str 5 May 46
151	*La Porte* Oregon	15 May 44 30 Jun 44	14 Aug 44	25 Mar 46	To WSA 28 Mar 46, str 12 Apr 46
152	*Latimer* Oregon	19 May 44 4 Jul 44	28 Aug 44 23 Sep 50	26 Feb 47 15 May 56	To MarAd Feb 60, str 1 Jul 60
153	*Laurens* Oregon	23 May 44 11 Jul 44	9 Sep 44	10 Apr 46	To WSA 13 Apr 46, str 1 May 46
154	*Lowndes* Oregon	26 May 44 18 Jul 44	14 Sep 44	17 Apr 46	To WSA 17 Apr 46, str 1 May 46
155	*Lycoming* Oregon	30 May 44 25 Jul 44	20 Sep 44	14 Mar 46	To WSA 21 Mar 46, str 28 Mar 46
156	*Mellette* Oregon	3 Jun 44 4 Aug 44	27 Sep 44 18 Oct 50	25 Jun 46 18 Jun 55	To MarAd Jun 60, str 1 Jul 60
157	*Napa* Oregon	7 Jun 44 12 Aug 44	1 Oct 44	24 May 46	To MarComm 30 May 46, str 19 Jun 46
158	*Newberry* Oregon	10 Jun 44 24 Aug 44	7 Oct 44	21 Feb 46	To WSA 3 Mar 46, str 12 Mar 46
159	*Darke* Oregon	14 Jun 44 29 Aug 44	10 Oct 44	17 Apr 46	To MarComm 22 Apr 46, str 1 May 46
160	*Deuel* Oregon	17 Jun 44 4 Sep 44	13 Oct 44 23 Oct 50	17 May 46 27 Jun 56	Str 1 Dec 58
161	*Dickens* Oregon	21 Jun 44 8 Sep 44	18 Oct 44	21 May 46	To MarComm 21 May 46, str 5 Jun 46
162	*Drew* Oregon	30 Jun 44 14 Sep 44	22 Oct 44	10 May 46	To MarComm 19 May 46, str 21 May 46
163	*Eastland* Oregon	4 Jul 44 19 Sep 44	26 Oct 44	15 Apr 46	To WSA 16 Apr 46, str 1 May 46
164	*Edgecombe* Oregon	11 Jul 44 24 Sep 44	30 Oct 44	31 Jan 47	Str 1 Oct 58
165	*Effingham* Oregon	19 Jul 44 29 Sep 44	1 Nov 44	17 May 46	To MarComm 20 May 46, str 5 Jun 46
166	*Fond Du Lac* Oregon	25 Jul 44 5 Oct 44	6 Nov 44	11 Apr 46	Ret to owner 13 Apr 46, str 1 May 46
167	*Freestone* Oregon	4 Aug 44 9 Oct 44	9 Nov 44	17 Apr 46	To WSA 19 Apr 46, str 1 May 46
168	*Gage* Oregon	13 Aug 44 14 Oct 44	12 Nov 44	26 Feb 47	Str 1 Oct 58
169	*Gallatin* Oregon	24 Aug 44 17 Oct 44	15 Nov 44	23 Apr 46	To WSA 24 Apr 46, str 8 May 46
170	*Gosper* Oregon	29 Aug 44 20 Oct 44	18 Nov 44	10 Apr 46	To MarComm 46, str 1 May 46
171	*Granville* Oregon	4 Sep 44 23 Oct 44	21 Nov 44	10 May 46	To MarComm 11 May 46, str 21 May 46

AMPHIBIOUS SHIPS   **509**

No.	Name/Builder	LD/Launch	Comm	Decomm	Notes and Fate
172	*Grimes* Oregon	8 Sep 44 27 Oct 44	23 Nov 44	26 Feb 46	Str 1 Oct 58
173	*Hyde* Oregon	14 Sep 44 30 Oct 44	26 Nov 44	14 May 46	To MarComm 16 May 46, str 5 Jun 46
174	*Jerauld* Oregon	19 Sep 44 3 Nov 44	28 Nov 44	6 May 46	To MarComm May 46, str 21 May 46
175	*Karnes* Oregon	24 Sep 44 7 Nov 44	3 Dec 44	11 Apr 46	To WSA 24 Apr 46, str 1 May 46
176	*Kershaw* Oregon	29 Sep 44 12 Nov 44	2 Dec 44	20 Dec 46	Str 1 Oct 58
177	*Kingsbury* Oregon	5 Oct 44 16 Nov 44	6 Dec 44	19 Apr 46	To MarComm 23 Apr 46, str 1 May 46
178	*Lander* Oregon	9 Oct 44 19 Nov 44	8 Dec 44	29 Mar 46	To MarComm 1 Apr 46, str 17 Apr 46
179	*Lauderdale* Oregon	14 Oct 44 23 Nov 44	12 Dec 44	25 Apr 46	To WSA 25 Apr 46, str 8 May 46
180	*Lavaca* Oregon	17 Oct 44 27 Nov 44	17 Dec 44	31 Jan 47	Str 1 Oct 58
181–186					Canceled 22 May 44
187	*Oconto* Kaiser-V	5 Apr 44 20 Jun 44	2 Sep 44	22 May 46	To MarComm 31 May 46, str 19 Jun 46
188	*Olmsted* Kaiser-V	11 Apr 44 4 Jul 44	5 Sep 44 2 Feb 52	21 Feb 47 27 Feb 59	Str 1 Jul 60
189	*Oxford* Kaiser-V	17 Apr 44 12 Jul 44	11 Sep 44	17 Apr 46	To MarComm 19 Apr 46, str 1 May 46
190	*Pickens* Kaiser-V	22 Apr 44 21 Jul 44	18 Sep 44	12 Apr 46	To MarComm 18 Apr 46, str 1 May 46
191	*Pondera* Kaiser-V	28 Apr 44 27 Jul 44	24 Sep 44	6 Jun 46	To MarComm 6 Jun 46, str 19 Jun 46
192	*Rutland* Kaiser-V	4 May 44 10 Aug 44	29 Sep 44	25 Feb 47	Str 1 Oct 58
193	*Sanborn* Kaiser-V	10 May 44 19 Aug 44	3 Oct 44 6 Jan 51	14 Aug 46 7 Apr 56	Str 1 Jul 60
194	*Sandoval* Kaiser-V	16 May 44 2 Sep 44 20 Nov 61	7 Oct 44 22 Sep 51 3 Mar 70	19 Jul 46 22 Jun 55	Str 1 Jul 60, reinstated 1 Sep 61, str 1 Dec 76
195	*Lenawee* Kaiser-V	22 May 44 11 Sep 44	11 Oct 44 30 Sep 50	3 Aug 46 20 Jun 67	To MarAd 23 Apr 68, str 30 Jun 68
196	*Logan* Kaiser-V	27 May 44 19 Sep 44	14 Oct 44 10 Nov 51	27 Nov 46 14 Jun 55	Str 1 Jul 60
197	*Lubbock* Kaiser-V	2 Jun 44 25 Sep 44	18 Oct 44	14 Dec 46	Str 1 Oct 58
198	*McCracken* Kaiser-V	8 Jun 44 29 Sep 44	21 Oct 44	10 Oct 46	To MarAd Sep 58, str 1 Oct 58
199	*Magoffin* Kaiser-V	20 Jun 44 4 Oct 44	25 Oct 44 4 Oct 50	14 Aug 46 10 Apr 68	Str 1 Dec 76
200	*Marathon* Kaiser-V	4 Jul 44 7 Oct 44	27 Oct 44	8 May 46	Ret to owner 8 May 46
201	*Menard* Kaiser-V	12 Jul 44 11 Oct 44	31 Oct 44 2 Dec 50	14 Jun 48 18 Oct 55	Str 1 Sep 61
202	*Menifee* Kaiser-V	21 Jul 44 15 Oct 44	5 Nov 44 3 Feb 51	31 Jul 46 30 Jun 55	Str 1 Oct 58
203	*Meriwether* Kaiser-V	27 Jul 44 18 Oct 44	8 Nov 44	14 Aug 46	Str 1 Oct 58
204	*Sarasota* Permanente	11 Apr 44 14 Jun 44	16 Aug 44 3 Feb 51	1 Aug 46 30 Jun 55	Str 1 Jul 60
205	*Sherburne* Permanente	18 May 44 10 Jul 44	20 Sep 44	3 Aug 46	Str 1 Oct 58, reinstated 22 Oct 69 after conversion for USAF as Range Sentinel (T-AGM 22), redesignated 16 Apr 69

No.	Name/Builder	LD/Launch	Comm	Decomm	Notes and Fate
206	*Sibley* Permanente	17 May 44 19 Jul 44	2 Oct 44	1 Aug 46	Str 1 Oct 58
207	*Mifflin* Permanente	15 May 44 7 Aug 44	11 Oct 44	5 Jul 46	Str 1 Oct 58
208	*Talladega* Permanente	3 Jun 44 17 Aug 44	31 Oct 44 8 Dec 51	27 Dec 46 Oct 69	Str 1 Dec 76
209	*Tazewell* Permanente	2 Jun 44 22 Aug 44	25 Oct 44	30 Nov 46	Str 1 Oct 58
210	*Telfair* Permanente	30 May 44 30 Aug 44	31 Oct 44 12 Sep 50 22 Nov 61	20 Jul 46 29 Feb 58 31 Oct 68	Str 1 Jul 60, reacquired 24 Aug 60, rem from U.S. Navy custody 14 Jul 69
211	*Missoula* Permanente	20 Jun 44 6 Sep 44	27 Oct 44	13 Sep 46	Str 1 Oct 58
212	*Montrose* Permanente	17 Jun 44 13 Sep 44	2 Nov 44 12 Sep 50	26 Oct 46 28 Oct 69	Str 2 Nov 69
213	*Mountrail* Permanente	3 Jul 44 20 Sep 44	16 Nov 44 9 Sep 50 22 Nov 61	12 Jul 46 1 Oct 55 11 May 69	Str 1 Dec 76
214	*Natrona* Permanente	30 Jun 44 27 Sep 44	8 Nov 44	29 Jul 46	Str 1 Oct 58
215	*Navarro* Permanente	27 Jun 44 3 Oct 44	15 Nov 44	2 Nov 69	Str 1 Dec 76
216	*Neshoba* Permanente	15 Jun 44 7 Oct 44	16 Nov 44	4 Dec 47	Str 1 Oct 58
217	*New Kent* Permanente	11 Jul 44 12 Oct 44	22 Nov 44 10 Oct 51	29 Jul 49 12 Jul 54	Str 1 Oct 58
218	*Noble* Permanente	20 Jul 44 18 Oct 44	27 Nov 44	1 Jul 64	Str Dec 64, to Spain
219	*Okaloosa* Permanente	8 Aug 44 22 Oct 44	28 Nov 44	21 Jul 49	Str 1 Oct 58
220	*Okanogan* Permanente	19 Aug 44 26 Oct 44	3 Dec 44	5 Feb 70	Str 1 Jun 73
221	*Oneida* Permanente	24 Aug 44 5 Nov 44	4 Dec 44	27 Dec 46	Str 1 Oct 58
222	*Pickaway* Permanente	1 Sep 44 5 Nov 44	12 Dec 44	25 Jun 70	FRAM II refit began Aug 63, Str 1 Dec 76
223	*Pitt* Permanente	8 Sep 44 10 Nov 44	11 Dec 44	9 Apr 47	To MarComm Apr 47, str 23 Apr 47
224	*Randall* Permanente	15 Sep 44 15 Nov 44	16 Dec 44	6 Apr 56	Str 1 Jul 60
225	*Bingham* Permanente	22 Sep 44 20 Nov 44	23 Dec 44	17 Jun 46	To MarComm 18 Jun 46, str 3 Jul 46
226	*Rawlins* Kaiser-V	10 Aug 44 21 Oct 44	11 Nov 44	15 Nov 46	Str 1 Oct 58
227	*Renville* Kaiser-V	19 Aug 44 25 Oct 44	15 Nov 44 5 Jan 52	30 Jun 49 23 Apr 68	To MarAd Apr 68, str 30 Jun 68
228	*Rockbridge* Kaiser-V	2 Sep 44 28 Oct 44	18 Nov 44 23 Dec 50	8 Mar 47 29 Nov 68	Str 1 Dec 68
229	*Rockingham* Kaiser-V	11 Sep 44 1 Nov 44	22 Nov 44	17 Mar 47	Str 1 Oct 58
230	*Rockwall* Kaiser-V	19 Sep 44 5 Nov 44	14 Jan 45 3 Mar 51	15 Mar 47 28 Sep 55	Str 1 Dec 58
231	*Saint Croix* Kaiser-V	25 Sep 44 9 Nov 44	1 Dec 44	7 Apr 47	Str 23 Apr 47
232	*San Saba* Kaiser-V	29 Sep 44 12 Nov 44	3 Dec 44	22 Jan 47	Str 1 Oct 58
233	*Sevier* Kaiser-V	4 Oct 44 16 Nov 44	5 Dec 44	30 Apr 47	Str 23 Jan 47

AMPHIBIOUS SHIPS **511**

No.	Name/Builder	LD/Launch	Comm	Decomm	Notes and Fate
234	*Bollinger* Kaiser-V	7 Oct 44 19 Nov 44	9 Dec 44	1 Apr 47	To MarComm 2 Apr 47, str 24 Aug 55
235	*Bottineau* Kaiser-V	11 Oct 44 22 Nov 44	30 Dec 44 24 Mar 51	8 Mar 47 31 Aug 55	Str 1 Jul 61
236	*Bronx* Oregon	22 May 45 14 Jul 45	27 Aug 45	30 Jun 49	Str 1 Oct 58
237	*Bexar* Oregon	2 Jun 45 25 Jul 45	9 Oct 45	Dec 69	Str 1 Dec 76
238	*Dane* Oregon	18 Jun 45 9 Aug 45	29 Oct 45	20 Dec 46	To MarAd 17 Aug 58, str 1 Oct 58
239	*Glynn* Oregon	30 Jun 45 25 Aug 45	25 Aug 45 3 Mar 51	12 Dec 46 9 Sep 55	Str 1 Jul 60
240	*Harnett* Oregon				Canceled 14 Aug 45
241	*Hempstead* Oregon				Canceled 14 Aug 45
242	*Iredell* Oregon				Canceled 14 Aug 45, completed privately
243	*Luzerne* Oregon				Canceled 14 Aug 45
244	*Madeia* Oregon				Canceled 14 Aug 45
245	*Maricopa* Oregon				Canceled 14 Aug 45
246	*McLennan* Oregon				Canceled 14 Aug 45
247	*Mecklenburg* Oregon				Canceled 14 Aug 45
248	*Paul Revere* NY Ship	15 May 52 11 Apr 53	3 Sep 58	1 Jan 80	Ex–SS *Diamond Mariner*; to Spain Jan 80
249	*Francis Marion* NY Ship	30 Mar 53 12 Feb 54	6 Jul 61	1 Jan 80	Ex–SS *Prairie Mariner*; to Spain 11 Jul 80

APH

No.	Name/Builder	LD/Launch	Comm	Decomm	Notes and Fate
1	*Tryon* Moore	26 Mar 41 21 Oct 41	30 Sep 42 1 Mar 50	20 Mar 46 16 Jun 54	Ex–SS *Alcoa Courier*; to army on decomm as APH, renamed *Sgt. Charles E. Mower*, reacquired by MSTS as T-AP 186, str 1 Jul 60
2	*Pinkney* Moore	3 Jun 41 4 Dec 41	27 Nov 42 1 Mar 50	9 Sep 46 27 Dec 57	Ex–SS *Alcoa Corsair*; to army on decomm as APH, renamed *Pvt. Elden H. Johnson*, reacquired by MSTS as T-AP 184, str 27 Dec 57
3	*Rixey* Moore	6 Aug 41 30 Dec 41	30 Dec 42 1 Mar 50	29 Mar 46 27 Dec 57	Ex–SS *Alcoa Cruiser*; to army on decomm as APH, renamed *Pvt. William H. Thomas*, reacquired by MSTS as T-AP 185, str 27 Dec 57

AP

AP 110–119 and 176 were P2-S2-R2 type point-to-point troop carriers built for the U.S. Army. AP 120–129 were similar P2-SE2-R1 type built for the navy. AP 130–159 were engines-aft C4-S-A1–type point-to-point troop carriers. AP 178 and above were MSTS only (T-AP), mainly army transports acquired when MSTS was formed. Many others, as indicated, were taken over by MSTS postwar and were designated T-AP. Missing numbers are for ships redesignated as APAs. AP 50 and 52 were to have become APA 22 and 24, respectively, and are listed with the APAs.

512 APPENDIX C

No.	Name/Builder	LD/Launch	Comm	Decomm	Notes and Fate
1	*Henderson* PHNY	19 Jun 15 17 Jun 16	24 May 17		To *Bountiful* (AH 9), 23 Mar 44, to MarComm 16 Sep 46
3	*Hancock* Glasgow	built 1879			To IX 12 on 24 Apr 22
4	*Argonne* Amer Intl	N/A 24 Feb 20	8 Nov 21		To AS 10 on 1 Jul 24, to AG 31, 25 Sep 40, to MarComm 31 Jul 46
5	*Chaumont* Amer Intl	N/A 31 Mar 20	22 Nov 21		To *Samaritan* (AH 10) 1 Mar 44, to MarComm 29 Aug 46
6	*William Ward Burrows* B&W	N/A 9 Feb 29	25 May 40	14 May 46	Ex–SS *Santa Rita*; sold Jun 46
7	*Wharton* NY Ship	N/A 21 Jun 19	14 Dec 39	26 Mar 47	Ex–SS *Southern Cross*; sold Mar 47
13	*George F. Elliott* Beth-A	N/A 4 Jul 18	19 Oct 18		Ex–SS *City of Los Angeles*; lost before she could be redesignated as an attack transport, WL 8 Aug 42
19	*Catlin* Vulkan	N/A 10 Nov 08	6 Sep 17	26 Sep 41	Ex–SS *George Washington*; served as a World War I transport, sold Mar 21, reacquired 28 Jan 41, to UK 29 Sep 41, ret 17 Apr 42 and to army, sold 13 Feb 51
20	*Munargo* NY Ship	N/A 17 Sep 21	4 Jun 41	18 Oct 43	Str 26 Oct 43, to army as *Thistle*, became army hospital ship
21	*Wakefield* NY Ship	N/A 5 Dec 31	15 Jun 41	15 Jun 46	Ex–SS *Manhattan*; str 59
22	*Mount Vernon* NY Ship	N/A 20 Aug 32	16 Jun 41	18 Jan 46	Ex–SS *Washington*; to Maritime Commission Jan 46 and sold
23	*West Point* NN Ship	22 Aug 38 31 Aug 39	15 Jun 41	14 Feb 46	Ex–SS *America*; str 31 Mar 46, ret to U.S. Lines 31 Oct 46
24	*Orizaba* Cramp	N/A 26 Feb 18	28 May 18 4 Jun 41	4 Sep 19 23 Apr 45	Transferred to army 4 Sep 19, to Brazil 16 Jul 45
28	*Kent* Cramp	N/A 4 Jul 18	18 Nov 18 21 Jul 41	Oct 19 24 Mar 43	Ex–SS *Santa Teresa*; to army Mar 43 (*Ernest Hinds*), str 8 May 43
29	*U. S. Grant* Vulkan	N/A 07	27 Aug 19 16 Jun 41	2 Sep 19 14 Nov 45	Ex–*Madawaska*; ex–*Konig Wilhelm III*, seized during World War I; to army 19, renamed *U. S. Grant* 3 Jun 22, str 28 Nov 45, ret to army Jan 46, sold for BU 24 Feb 48
31	*Chateau Thierry* Amer Intl	N/A 24 Dec 19	6 Aug 41	9 Sep 43	Ex-army; str 20 Sep 43, ret 9 Sep 43
32	*St. Mihiel* Amer Intl	N/A 19 Nov 19	21 Jul 41	24 Nov 43	Ex-army; str 24 Nov 43, ret to army
33	*Republic* H&W	N/A 19 Dec 03	21 Oct 41	27 Jan 45	Ex–*President Grant*, ex-army; ret Feb 45
36	*John L. Clem*				Not acquired
41	*Stratford* Manitowoc	N/A 18	25 Aug 41	16 Apr 46	Ex–AK 45, ex-*Catherine*; sold 47
42	*Tasker H. Bliss* NN Ship	N/A 17 Jul 20	15 Sep 42		Ex–SS *President Cleveland*; lost before she could be redesignated as an attack transport, WL
43	*Hugh L. Scott* Beth-S	N/A 17 Apr 20	7 Sep 42		Ex–SS *President Pierce*; lost before she could be redesignated as an attack transport, WL 12 Nov 42
44	*Willard H. Holbrook*				Not acquired
45	*Thomas H. Barry*				Not acquired
46	*James Parker*				Not acquired
47	*J. W. McAndrew*				Not acquired

No.	Name/Builder	LD/Launch	Comm	Decomm	Notes and Fate
53	*Lafayette* Penhoet	26 Jan 31 29 Oct 32			Ex–SS *Normandie*; completed May 35; largest ship in the world until 1940; taken over by U.S. Navy 24 Dec 41 but CTL 10 Feb 42; redesignated APV 4, 15 Sep 43 but never placed in service; str 11 Oct 45; sold 3 Oct 46 BU
54	*Hermitage* Beardmore	N/A 23 Apr 25	14 Aug 42	20 Aug 46	Ex–SS *Conte Biancamano*; war prize, ret to Italy 14 Aug 47
61	*Monticello* Stab Tecnico	N/A 29 Jun 27	16 Apr 42	22 Mar 46	Ex–SS *Conte Grande*; war prize, ret to Italy Jul 47
62	*Kenmore* NY Ship	N/A 23 Feb 21	5 Aug 42	16 Sep 43	Ex–SS *President Madison*; to *Refuge* (AH 11) Feb 44
63	*Rochambeau* SPC	N/A 14 May 31	28 Sep 42	17 Mar 45	Ex–SS *Marechal Joffre*; seized at Manila as U.S. Army transport 12 Dec 41, renamed 27 Apr 42, str 30 Mar 45, ret to France
66	*Ancon*				Served as an attack transport although not designated as such, became AGC 4 (see AGC entry)
67	*Dorothea L. Dix* Beth-Q	11 Nov 39 22 Jun 40	17 Sep 42	24 Apr 46	Ex–SS *Exemplar*; to MarComm 24 Apr 46
68	*Alameda*				Not acquired
69	*Elizabeth C. Stanton* Moore	21 Sep 39 22 Dec 39	17 Sep 42	3 Apr 46	Ex–SS *Mormacstar*; ret to owner Apr 46
70	*Florence Nightingale* Moore	20 Jun 40 28 Aug 40	17 Sep 42	1 May 46	Ex–SS *Mormacsun*; ret to owner May 46
71	*Lyon* Ingalls	21 Aug 39 12 Oct 40	16 Sep 42	3 May 46	Ex–SS *Mormactide*; ret to owner May 46
72	*Susan B. Anthony* NY Ship	N/A 14 Nov 29	7 Sep 42		Ex–SS *Santa Clara*; WL(M) 7 Jun 44
73	*Leedstown* Federal-K	N/A 3 Oct 32	24 Sep 42		Ex–SS *Santa Lucia*; attack transport, WL 9 Nov 42
74	*Lejeune* B&V	N/A 27 Aug 36	26 Mar 43	9 Feb 48	Ex–SS *Windhuk*; war prize, transport refit completed 12 May 44, to MarComm Jul 57, sold for BU 66
75	*Gemini* Manitowoc	N/A 19	4 Aug 42	6 Apr 46	Ex–AK 52, ex–AG 38, ex–SS *Saginaw*; to MarComm 10 Sep 46
76	*Anne Arundel* Federal-K	18 Jul 40 16 Nov 40	17 Sep 42	2 Mar 46	Ex–SS *Mormacyork*; ret to owner Mar 46
77	*Thurston* Federal-K	12 Sep 41 4 Apr 42	19 Sep 42	1 Aug 46	Ex–SS *Delsantos*, originally commissioned as *Dauphin*; str 28 Aug 46, to WSA
102	*La Salle* Moore	29 Apr 42 2 Aug 42	31 Mar 43	24 Jul 46	Ex–SS *Hotspur*; to MarComm 25 Jul 46
103	*President Polk* NN Ship	7 Oct 40 28 Jun 41	4 Oct 43	26 Jan 46	Ret to owner 26 Jan 46
104	*President Monroe* NN Ship	13 Nov 39 7 Aug 40	20 Aug 43	12 Jan 46	Ret to owner Feb 46
105	*George F. Elliot* NN Ship	10 Apr 39 16 Dec 39	23 Sep 43	10 Jun 46	Ex-*Delbrasil*; str 19 Jun 46
106–109					Became LSV 1–4
110	*Gen. John Pope* Federal-K	15 Jul 42 21 Mar 43	2 Jul 43 1 Aug 50 17 Aug 65	12 Jun 46 14 May 55 1 May 70	To WSA 12 Jun 46, reinstated 20 Jul 50, to MSTS 1 Aug 50, str 30 Jun 70
111	*Gen. A. E. Anderson* Federal-K	7 Sep 42 2 May 43	25 Aug 43	10 Nov 58	To MSTS Oct 49, str 1 Dec 66
112	*Gen. W. A. Mann* Federal-K	28 Dec 42 18 Jul 43	13 Oct 43	66	To MSTS Oct 49, str 1 Dec 66

No.	Name/Builder	LD/Launch	Comm	Decomm	Notes and Fate
113	Gen H. W. Butner Federal-K	23 Mar 43 19 Sep 43	5 Dec 43	28 Jan 60	To MSTS Oct 49, str 1 Jul 60
114	Gen. William Mitchell Federal-K	3 May 43 31 Oct 43	10 Jan 44	66	To MSTS Oct 49, str 1 Dec 66
115	Gen. G. M. Randall Federal-K	20 Jul 43 30 Jan 44	15 Apr 44	2 Jun 61	To MSTS Oct 49, str 1 Sep 62
116	Gen. M. C. Meigs Federal-K	22 Sep 43 13 Mar 44	3 Jun 44 21 Jul 50	4 Mar 46 55	To WSA 4 Mar 46, reacquired for MSTS 21 Jul 50, to MarAd 1 Oct 58, str 9 Jan 72
117	Gen. W. H. Gordon Federal-K	2 Nov 43 7 May 44	29 Jun 44 8 Nov 51 May 61	11 Mar 46 Oct 54 67	To army on decomm, reacquired for MSTS 51, to MarAd 23 Apr 70, str 31 Mar 86
118	Gen. W. P. Richardson Federal-K	2 Feb 44 6 Aug 44	2 Nov 44	14 Feb 46	To army on decomm, to Maritime Commission 10 Mar 48, became liner *La Guardia*
119	Gen. William Weigel Federal-K	15 Mar 44 3 Sep 44	6 Jan 45 1 Aug 50 18 Aug 65	10 May 46 55 67	To army on decomm, reacquired 20 Jul 50, to NDRF 12 Jun 58, reacquired 65, str 30 Jun 70
120	Adm. W. S. Benson Beth-A	10 Dec 42 28 Nov 43	23 Aug 44 1 Mar 50	3 Jun 46 Nov 68	To army on decomm, renamed *Gen. Daniel I. Sultan*, reacquired by MSTS, str 9 Oct 69
121	Adm. W. L. Capps Beth-A	15 Dec 42 20 Feb 44	18 Sep 44 1 Mar 50	8 May 46 4 Nov 68	To army on decomm, renamed *Gen. Hugh J. Gaffney*, reacquired by MSTS, str 9 Oct 69, reacquired 1 Nov 78 as IX 507, str 26 Oct 93
122	Adm. R. E. Coontz Beth-A	15 Jan 43 22 Apr 44	21 Nov 44 1 Mar 50	25 Mar 46 Jan 67	To army on decomm, renamed *Gen. Alexander M. Patch*, reacquired by MSTS, str 20 Aug 90
123	Adm. E. W. Eberle Beth-A	15 Feb 43 14 Jun 44	24 Jan 45 1 Mar 50	8 May 46 69	To army on decomm, renamed *Gen. Simon B. Buckner*, reacquired by MSTS, str 24 Mar 70
124	Adm. C. F. Hughes Beth-A	29 Nov 43 27 Jul 44	11 Jan 45 1 Mar 50	3 May 46 30 Sep 68	To army on decomm, renamed *Gen. Edwin D. Patrick*, reacquired by MSTS, str 9 Oct 69
125	Adm. H. T. Mayo Beth-A	21 Feb 44 26 Nov 44	24 Apr 45 1 Mar 50 14 Aug 65	26 May 46 5 Jun 58 68	To army on decomm, tr 9 Jun 46, renamed *Gen. Nelson M. Walker*, reacquired by MSTS (1965 may be date of second reacquisition rather than of reactivation), str 25 Jan 81
126	Adm. Hugh Rodman Beth-A	24 Apr 44 25 Feb 45	10 Jul 45 1 Mar 50	14 May 46 67	To army on decomm, renamed *Gen. Maurice Rose*, reacquired by MSTS 1 Mar 50, str 30 Apr 70
127	Adm. W. S. Sims Beth-A	15 Jun 44 4 Jun 45	27 Sep 45 1 Mar 50 1 Jul 82	21 Jun 46 68 23 Apr 91	To army on decomm, renamed *Gen. William O. Darby*, reacquired by MSTS 1 Mar 50, to IX 510 (accommodation ship) 27 Oct 81, to MarAd for layup, str 26 Oct 93
128	Adm. D. W. Taylor Beth-A	28 Aug 44 23 Jun 46			Contract canceled 16 Dec 44, completed as liner *President Cleveland*
129	Adm. F. B. Upham Beth-A	27 Nov 44 24 Nov 46			Contract canceled 16 Dec 44, completed as liner *President Wilson*
130	Gen. G. O. Squier Kaiser-R	14 May 42 11 Nov 42	2 Oct 43	10 Jul 46	To WSA 18 Jul 46, sold 7 Apr 64
131	Gen. T. H. Bliss Kaiser-R	22 May 42 19 Dec 42	24 Feb 44	28 Jun 46	To WSA 2 Jul 46, sold Apr 64
132	Gen. J. R. Brooke Kaiser-R	29 Jun 42 21 Feb 43	20 Jan 44	3 Jul 46	To WSA 18 Jul 46, sold Apr 64
133	Gen. O. H. Ernst Kaiser-R	29 Jun 42 14 Apr 43	22 Apr 44	15 Aug 46	To WSA 15 Aug 46, sold Apr 64

No.	Name/Builder	LD/Launch	Comm	Decomm	Notes and Fate
134	*Gen. R. L. Howze* Kaiser-R	22 Jul 42 23 May 43	7 Feb 44 1 Mar 50	1 Apr 46 Jan 56	To WSA 1 Apr 46, to army 58, reacquired by MSTS, to Maritime Administration 17 Jul 58, str 59
135	*Gen. W. L. Black* Kaiser-R	26 Nov 42 23 Jul 43	24 Feb 44 1 Mar 50	28 Feb 46 26 Aug 55	To army on decomm, reacquired by MSTS, to Maritime Administration 26 Aug 55, str 55
136	*Gen. H. L. Scott* Kaiser-R	20 Dec 42 19 Sep 43	3 Apr 44	29 May 46	To WSA 3 Jun 46, sold 31 Jul 64
137	*Gen. S. D. Sturgis* Kaiser-R	15 Apr 43 12 Nov 43	24 Apr 44 1 Mar 50	29 May 46 28 May 55	To army on decomm, reacquired by MSTS, to MarAd 22 Aug 58, str Aug 58
138	*Gen. C. G. Morton* Kaiser-R	20 Sep 43 15 Mar 44	7 Jul 44 1 Mar 50	15 May 46 58	To army on decomm, reacquired by MSTS, to MarAd May 58, str 58
139	*Gen. R. E. Callan* Kaiser-R	16 Dec 43 27 Apr 44	17 Aug 44 28 Apr 50	24 May 46 29 May 58	To army on decomm, reacquired by MSTS, str 29 May 58, reacquired 1 Jul 64, to USAF for conversion to *Gen. H. H. Arnold* (T-AGM 9)
140	*Gen. M. B. Stewart* Kaiser-R	22 Jun 44 15 Oct 44	3 Mar 45 1 Mar 50	24 May 46 29 Apr 55	To army on decomm, reacquired by MSTS, str May 58
141	*Gen. A. W. Greeley* Kaiser-R	18 Jul 44 5 Nov 44	22 Mar 45 1 Mar 50	29 Mar 46 Mar 55	To army on decomm, reacquired by MSTS, to NDRF 29 Aug 59, str Jul 60
142	*Gen. C. H. Muir* Kaiser-R	7 Aug 44 24 Nov 44	12 Apr 45 1 Mar 50	18 Jun 46 7 Feb 55	To army on decomm, reacquired by MSTS, to MarAd 5 Apr 60
143	*Gen. H. B. Freeman* Kaiser-R	28 Aug 44 11 Dec 44	26 Apr 45 1 Mar 50	4 Mar 46 24 Jul 58	To army on decomm, reacquired by MSTS, str 24 Jul 58
144	*Gen. H. F. Hodges* Kaiser-R	21 Sep 44 3 Jan 45	6 Apr 45 1 Mar 50	13 May 46 16 Jun 58	To army on decomm, reacquired by MSTS, to MarAd 16 Jun 58
145	*Gen. Harry Taylor* Kaiser-R	22 Feb 43 10 Oct 43	1 Apr 44 1 Mar 50	13 Jun 46 19 Sep 57	To army on decomm, reacquired by MSTS, to MarAd 10 Jul 58, to USAF 15 Jul 61 for conversion to *Gen. Hoyt S. Vandenber* (T-AGM 10)
146	*Gen. W. F. Hase* Kaiser-R	24 May 43 15 Dec 43	22 Apr 44 1 Mar 50	6 Jun 46 Jul 54	To army on decomm, reacquired by MSTS, to MarAd 8 Jan 60, str Jul 60
147	*Gen. E. T. Collins* Kaiser-R	24 Jul 43 22 Jan 44	20 Jul 44 1 Mar 50	17 Jun 46 Oct 54	To army on decomm, reacquired by MSTS, to MarAd 30 Jun 60, str Jul 60
148	*Gen. M. L. Hersey* Kaiser-R	11 Oct 43 1 Apr 44	29 Jul 44 1 Mar 50	1 Jun 46 11 Jun 54	To army on decomm, reacquired by MSTS, to MarAd 3 Sep 59, str Jul 60
149	*Gen. J. H. McCrae* Kaiser-R	13 Nov 43 26 Apr 44	8 Aug 44 1 Mar 50	27 Feb 46 29 Oct 54	To army on decomm, reacquired by MSTS, to MarAd 30 Jun 60, str Jul 60
150	*Gen. M. M. Patrick* Kaiser-R	24 Jan 44 21 Jun 44	4 Sep 44 1 Mar 50	8 Mar 46 17 Oct 58	To army on decomm, reacquired by MSTS, to MarAd 17 Oct 58, str 58
151	*Gen. W. C. Langfit* Kaiser-R	16 Mar 44 17 Jul 44	30 Sep 44 1 Mar 50	6 Jun 46 30 Sep 57	To army on decomm, reacquired by MSTS, to MarAd 13 May 58, str May 58
152	*Gen. Omar Bundy* Kaiser-R	3 Apr 44 5 Aug 44	6 Jan 45	14 Jun 46	To army on decomm, not reacquired by navy, out of army service 12 Dec 49
153	*Gen. R. M. Blatchford* Kaiser-R	27 Apr 44 27 Aug 44	26 Jan 45 1 Mar 50	12 Jun 46 Jan 67	To army on decomm, reacquired by MSTS, to MarAd Jan 67, str 9 Oct 69
154	*Gen. Leroy Eltinge* Kaiser-R	24 May 44 20 Sep 44	21 Feb 45 20 Jul 50	29 May 46 Jan 67	To army on decomm, reacquired by MSTS, to MarAd Jan 67, str 9 Oct 69
155	*Gen. A. W. Brewster* Kaiser-R	16 Oct 44 21 Jan 45	23 Apr 45 1 Mar 50	10 Apr 46 Dec 54	To army on decomm, reacquired by MSTS, to MarAd 26 Jul 55, str 56
156	*Gen. D. E. Aultman* Kaiser-R	6 Nov 44 18 Feb 45	20 May 45 1 Mar 50	15 Mar 46 4 Jun 58	To army on decomm, reacquired by MSTS, to MarAd 4 Jun 58, str Jun 58
157	*Gen. C. C. Ballou* Kaiser-R	12 Dec 44 7 Mar 45	30 Jun 45 1 Mar 50	17 May 46 Sep 54	To army on decomm, reacquired by MSTS, to NDRF Sep 54, str 1 Jul 60
158	*Gen. W. G. Haan* Kaiser-R	4 Jan 45 20 Mar 45	2 Aug 45 1 Mar 50	7 Jun 46 7 Jan 57	To army on decomm, reacquired by MSTS, to MarAd 22 Oct 58
159	*Gen. Stuart Heintzelman* Kaiser-R	22 Feb 45 21 Apr 45	12 Sep 45 1 Mar 50	12 Jun 46 24 Jun 54	To army on decomm, reacquired by MSTS, to MarAd Jun 60, str Jul 60

No.	Name/Builder	LD/Launch	Comm	Decomm	Notes and Fate
160, 161					Became LSV 5 and LSV 6
162–165					Became AK 221–224, 20 Aug 44
166	Comet Moore	13 Oct 42 21 Dec 42	15 Feb 44	14 Aug 46	To WSA 14 Aug 46
167	John Land Moore	14 Nov 42 22 Jan 43	8 Apr 44	5 Aug 46	To MarComm 6 Aug 46
168	War Hawk Moore	24 Dec 42 3 Apr 43	9 Mar 44	2 Aug 46	To MarComm 46
169	Golden City Moore	16 Aug 43 28 Oct 43	29 May 44	10 Aug 46	To MarComm Aug 46
170	Winged Arrow Moore	26 Jan 43 3 Apr 43	21 Apr 44	2 Aug 46	To MarComm 46
171	Storm King NC Ship	20 Jul 43 17 Sep 43	4 Dec 43	12 Aug 46	To MarComm 46
172	Cape Johnson Consolidated-W	28 Nov 42 20 Feb 43	1 Jun 44	25 Jul 46	Ret to owner 26 Jul 46
173	Herald of the Morning Moore	12 Jun 43 14 Aug 43	22 Apr 44	9 Aug 46	To MarComm 9 Aug 46
174	Arlington Consolidated-W	11 May 42 10 Aug 42	18 Apr 44	20 Mar 46	Ex–SS Fed Morris; served as APA crew training ship, ret to owner Mar 46
175	Starlight NC Ship	9 Oct 43 23 Dec 43	15 Feb 44	14 Aug 46	To MarComm 46
176	Gen J. C. Breckenridge Federal-K	10 May 44 18 Mar 45	30 Jun 45	66	To MSTS 1 Oct 49, to MarComm 1 Dec 66 and str
177	Europa B&V	N/A 15 Aug 28	25 Aug 45	2 May 46	War prize, to France Jun 46, became liner Liberte
178, 179					Ex–APA 89 and ex–APA 90 (see APA entries)
180	David C. Shanks Ingalls	27 Feb 41 21 Oct 42	15 Mar 50	Oct 58	Ex-Gulfport; army transport acquired 1 Mar 50, str 1 Jul 61
181	Fred C. Ainsworth Ingalls	23 Jan 42 20 Nov 42	1 Mar 50	2 Nov 59	Ex–Pass Christian; army transport acquired 1 Mar 50, str 1 Jul 61
182	George W. Goethals Ingalls	7 Jan 41 23 Jan 42	1 Mar 50	29 Sep 59	Ex-Pascagoula; army transport acquired 1 Mar 50, str 1 Jul 61
183	Henry Gibbins Ingalls	23 Aug 41 11 Sep 42	1 Mar 50	2 Dec 59	Ex-Biloxi; army transport acquired 1 Mar 50, to New York State Maritime Academy 2 Dec 59 as Empire State IV
184–186					Ex–APH 1–3 (see APH entries)
187	Pvt. Joe E. Martinez Beth-F	13 Apr 45 29 May 45	1 Mar 50	1 Sep 52	Ex–Stevens Victory; army transport acquired by MSTS, str 8 Nov 52, to Korea
188	Aiken Victory Beth-F	13 Oct 44 30 Nov 44	21 Jul 50	19 Dec 52	Acquired from MarAd, str 12 Feb 53, to Korea
189	Lt. Raymond O. Beaudoin Beth-F	4 Apr 45 21 May 45	1 Aug 50	5 Nov 52	Ex–Marshall Victory; acquired from MarAd 22 Jul 50, str 22 Dec 52, to Korea
190	Pvt. Sadao S. Munemori Beth-F	22 May 45 6 Jul 45	22 Jul 50	9 Oct 52	Ex–Wilson Victory, ex-army transport; str 6 Nov 52, to Korea
191	Sgt. Howard E. Woodford Beth-F	18 Apr 45 2 Jun 45	22 Jul 50	4 Dec 52	Ex–Goucher Victory; acquired 1 Oct 49, str 4 Dec 52, to Korea
192	Sgt. Sylvester Antolak Beth-F	30 Apr 45 11 Jun 45	22 Jul 50	17 Sep 52	Ex–Stetson Victory; str 8 Nov 52, to Korea
193	Marine Adder Kaiser-R	7 Mar 45 16 May 45	1 Aug 50	8 Jun 57	C4-S-B1 type; acquired from MarAd 24 Jul 50, str 6 Jun 58
194	Marine Lynx Kaiser-V	12 Sep 44 17 Jul 45	Sep 50	25 Jul 56	C4-S-B1 type; acquired from MarAd 23 Jul 50, str 1 May 58
195	Marine Phoenix Kaiser-V	16 Dec 44 9 Aug 45	Aug 50	Mar 56	C4-S-B1 type; acquired from MarAd 21 Jul 50, str 3 Nov 58
196	Barrett NY Ship	1 Jun 49 27 Jun 50	15 Dec 51		Ex–SS President Jackson, taken over while under construction; str 1 Jul 73, to New

No.	Name/Builder	LD/Launch	Comm	Decomm	Notes and Fate
197	*Geiger* NY Ship	1 Aug 49 9 Oct 50	13 Sep 52	27 Apr 71	York State Maritime Academy 5 Sep 73 as *Empire State V* Ex–SS *President Adams*, taken over while under construction; str 27 Apr 71, to Massachusetts Maritime Academy 12 Feb 80 as *Bay State*
198	*Upshur* NY Ship	30 Sep 49 9 Jan 51	20 Dec 52	2 Apr 73	Ex–SS *President Hayes*, taken over while under construction; str 2 Apr 73, to Maine Maritime Academy as *State of Maine*
199	*Marine Carp* Kaiser-V	6 Dec 44 5 Jul 45	15 Sep 52	9 Oct 57	Completed in 45 as C4-S-A3 transport for WSA, laid up postwar, acquired from MarAd 17 Mar 52, str 11 Sep 58
200	*Marine Jumper*				C4-type transport not acquired
201	*Marine Marlin*				C4-type transport not acquired
202	*Marine Serpent* Kaiser-V	30 Nov 44 12 Jun 45	Sep 52	Jun 55	Completed in 45 as C4-S-A3 transport for WSA, laid up postwar, acquired from MarAd 8 May 52, str 17 Aug 55

IFS/LFR

No.	Name/Builder	LD/Launch	Comm	Decomm	Notes and Fate
1	*Carronade* Puget Bridge	19 Nov 52 26 May 53	25 May 55 2 Oct 65	31 May 60 24 Jul 70	IFS redesignated to LFR on 14 Aug 68, effective 1 Jan 69, reclassified LFR 1 on 1 Jan 69, str 1 May 73

LHA

No.	Name/Builder	LD/Launch	Comm	Decomm	Notes and Fate
1	*Tarawa* Litton	15 Nov 71 1 Dec 73	29 May 76		
2	*Saipan* Litton	21 Jul 72 18 Jul 74	15 Oct 77		
3	*Belleau Wood* Litton	5 Mar 73 11 Apr 77	23 Sep 78		
4	*Nassau* Litton	13 Aug 73 21 Jan 78	28 Jul 79		
5	*Peleliu* Litton	12 Nov 76 25 Nov 78	3 May 80		

LHD

No.	Name/Builder	LD/Launch	Comm	Decomm	Notes and Fate
1	*Wasp* Litton	30 May 85 4 Aug 87	6 Jul 89		
2	*Essex* Litton	20 Mar 89 7 Jan 91	17 Oct 92		

No.	Name/Builder	LD/Launch	Comm	Decomm	Notes and Fate
3	*Kearsage* Litton	6 Feb 90 26 Mar 92	16 Oct 93		
4	*Boxer* Litton	8 Apr 91 13 Aug 93	11 Feb 95		
5	*Bataan* Litton	22 Jun 94 15 Mar 96	29 Sep 97		
6	*Bonhomme Richard* Litton	18 Apr 95 14 Mar 97	15 Aug 98		
7	*Iwo Jima* Litton	12 Dec 97 4 Feb 00	30 Jun 01		
8					Planned under FY 99–02 programs

LPD

No.	Name/Builder	LD/Launch	Comm	Decomm	Notes and Fate
1	*Raleigh* NYNY	23 Jun 60 17 Mar 62	8 Sep 62	13 Dec 91	Str 25 Jan 92, tgt 4 Dec 94
2	*Vancouver* NYNY	19 Nov 60 15 Sep 62	11 May 63	27 Mar 92	Str 8 Apr 97
3	*La Salle* NYNY	2 Apr 62 3 Aug 63	22 Feb 64		Became AGF 3, 1 Jul 72
4	*Austin* NYNY	4 Feb 63 27 Jun 64	6 Feb 65		
5	*Ogden* NYNY	2 Apr 63 27 Jun 64	19 Jun 65		
6	*Duluth* NYNY	18 Dec 63 7 May 66	21 Apr 67		
7	*Cleveland* Ingalls	30 Nov 64 7 May 66	21 Apr 67		
8	*Dubuque* Ingalls	25 Jan 65 6 Aug 66	1 Sep 67		
9	*Denver* Lockheed	7 Feb 64 23 Jan 65	26 Oct 68		
10	*Juneau* Lockheed	23 Jan 65 12 Feb 66	12 Jul 69		
11	*Coronado* Lockheed	3 May 65 30 Jul 66	23 May 70		Became AGF 11, 1 Oct 80
12	*Shreveport* Lockheed	27 Dec 65 22 Oct 66	12 Dec 70		
13	*Nashville* Lockheed	14 Apr 66 7 Oct 67	14 Feb 70		
14	*Trenton* Lockheed	8 Aug 66 3 Aug 68	6 Mar 71		
15	*Ponce* Lockheed	31 Oct 66 20 May 70	10 Jul 71		
16					Not built
17	*San Antonio* Avondale	11 Dec 00			
18	*New Orleans* Avondale	16 Jul 00			
19	*Mesa Verde*				Not yet laid down
20	*Green Bay*				Not yet laid down

LPH

No.	Name/Builder	LD/Launch	Comm	Decomm	Notes and Fate
1	*Block Island*				Conversion canceled Jun 58
2	*Iwo Jima* PSNY	13 Feb 59 17 Sep 60	26 Aug 61	31 Jul 93	Str 24 Sep 93, island at Texas Air Museum
3	*Okinawa* PHNY	13 Feb 59 19 Aug 61	14 Apr 62	17 Dec 92	Str on decomm, tgt 2000
4	*Boxer* NN Ship	13 Sep 43 14 Dec 44	16 Apr 45	1 Dec 69	To LPH 30 Jan 59, str 1 Dec 69
5	*Princeton* PHNY	1 May 44 8 Jul 45	18 Nov 45	30 Jan 70	To LPH 2 Mar 59, str 30 Jan 70
6	*Thetis Bay* Kaiser-V	22 Dec 43 16 Mar 44	20 Jul 56	1 Mar 64	Ex–CVE 90; str on decomm, comm date is for ship as converted
7	*Guadalcanal* PHNY	1 Sep 61 1 Aug 62	20 Jul 63	31 Aug 94	Str on decomm
8	*Valley Forge* PHNY	7 Sep 44 18 Nov 45	3 Nov 46	10 Dec 69	To LPH 1 Jul 61, str 15 Jan 70
9	*Guam* PHNY	15 Nov 62 22 Aug 64	16 Jan 65	25 Aug 98	
10	*Tripoli* Ingalls	15 Jun 64 31 Jul 65	6 Aug 66	15 Sep 95	To army for trials, 27 Jun 97
11	*New Orleans* PHNY	1 Mar 66 3 Feb 68	16 Nov 68	31 Oct 97	Str 23 Oct 98
12	*Inchon* Ingalls	8 Apr 68 24 May 69	20 Jun 70	30 Sep 94	Decomm for conversion to MCS 12, 6 Mar 95–24 May 96

LSD

No.	Name/Builder	LD/Launch	Comm	Decomm	Notes and Fate
1	*Ashland* Moore	22 Jun 42 21 Dec 42	5 Jun 43 27 Dec 50 29 Nov 61	Mar 46 14 Sep 57 25 Nov 69	To AV 21 on 11 Feb 59 but conversion canceled, str 25 Nov 69
2	*Belle Grove* Moore	27 Oct 42 17 Feb 43	9 Aug 43 27 Dec 50	30 Aug 46 69	Str 12 Nov 69
3	*Carter Hall* Moore	27 Oct 42 4 Mar 43	18 Sep 43 26 Jan 51	12 Feb 47 31 Oct 69	Str 31 Oct 69
4	*Epping Forest* Moore	23 Nov 42 2 Apr 43	11 Oct 43 1 Dec 50	25 Mar 47 68	To MCS 7 on 30 Nov 62, str 1 Nov 68
5	*Gunston Hall* Moore	28 Dec 42 1 May 43	10 Nov 43 5 Mar 49	7 Jul 47 25 May 70	Str 25 May 70, to Argentina 25 May 70
6	*Lindenwald* Moore	22 Feb 43 11 Jun 43	9 Dec 43 18 Feb 49 1 Jul 60	5 Apr 47 1 Feb 58 30 Nov 67	To MSTS 12 Dec 56, str 1 Dec 67
7	*Oak Hill* Moore	9 Mar 43 25 Jun 43	5 Jan 44 26 Jan 51	12 Mar 47 26 Oct 69	Str 31 Oct 69
8	*White Marsh* Moore	7 Apr 43 19 Jul 43	29 Jan 44 8 Nov 50	Mar 46 23 Jan 57	To MSTS 8 Nov 56, to ROC 17 Nov 60
9	NN Ship	23 Nov 42 21 May 43			To UK as HMS *Eastway* (ex-*Battleaxe*) 14 Sep 43, ret May 47, to Greek merch interests May 47, acquired by Greek navy May 53
10	NN Ship	23 Nov 42 19 Jul 43			To UK as HMS *Highway* (ex-*Claymore*) 19 Oct. 43, ret Apr 46, sold 17 Dec 48, merch service until BU Apr 60

No.	Name/Builder	LD/Launch	Comm	Decomm	Notes and Fate
11	NN Ship	24 May 43 18 Nov 43			To UK as HMS *Northway* (ex-*Cutlass*) 15 Feb 44, ret Jan 47, to MarComm 19 Mar 48, merch service 48–62, German navy depot ship, WS 1 in 62, reverted to merch service 66
12	NN Ship	23 Jul 43 29 Dec 43			To UK as HMS *Oceanway* (ex-*Dagger*) 29 Mar 44, ret Jan 47, to Foreign Liquidation Commission 13 Feb 47, merch service 47–52, resold to U.S. Navy, refitted for amphibious service, to France as grant aid May 52
13	*Casa Grande* NN Ship	22 Nov 43 11 Apr 44	5 Jun 44 1 Nov 50	23 Oct 46 6 Feb 70	Scheduled for transfer to UK as HMS *Portway* (ex-*Spear*) but was retained, may have been loaned to Israel for tests, str 1 Nov 76, sold Nov 91
14	*Rushmore* NN Ship	31 Dec 43 10 May 44	3 Jul 44 21 Sep 50	16 Aug 46 30 Sep 70	Scheduled for transfer to UK as HMS *Swashway* (ex-*Sword*) but was retained, str 1 Nov 76; projected sale to Brazil was canceled due to ship's poor condition
15	*Shadwell* NN Ship	16 Dec 43 24 May 44	24 Jul 44 20 Sep 50	10 Jul 47 9 Mar 70	Scheduled for transfer to UK as HMS *Waterway* (ex-*Tomahawk*) but was retained, str 1 Nov 76, damage control training hulk at Mobile, Alabama, on 31 Jan 86
16	*Cabildo* NN Ship	24 Jul 44 28 Dec 44	15 May 45 5 Oct 50	15 Jan 47 70	To MarAd 9 Jul 70, str 15 Oct 76
17	*Catamount* NN Ship	7 Aug 44 27 Jan 45	9 Apr 45	7 Apr 70	FRAM II refit, str 31 Oct 74
18	*Colonial* NN Ship	21 Aug 44 28 Feb 45	15 May 45	30 Jun 70	FRAM II refit, str 15 Oct 76
19	*Comstock* NN Ship	3 Jan 45 28 Apr 45	2 Jul 45	Jan 70	Str 30 Jun 76, to ROC 85, bought for parts but refitted
20	*Donner* BOSNY	16 Dec 44 6 Apr 45	31 Jul 45 15 Sep 49	12 Aug 49 23 Dec 70	FRAM II refit, str 1 Nov 76, to Energy R&D Administration 1 Nov 76, projected sale to Brazil in 82 aborted due to poor material condition
21	*Fort Mandan* BOSNY	2 Jan 45 2 Jun 45	31 Oct 45 25 Oct 50	16 Jan 48 25 Jan 71	FRAM II refit, to Greece 25 Jan 71
22	*Fort Marion* Gulf	15 Sep 44 22 May 45	29 Jan 46	13 Feb 70	FRAM II refit Dec 59–Apr 60, str 31 Oct 74, to ROC 75
23	*Fort Snelling* Gulf	8 Nov 44			Canceled 11 Aug 45, sold and completed as merch *Carib Queen*, taken over by MarAd Mar 58 and by MSTS 15 Jan 59, became *Taurus* (T-AKR 8, then T-AK 273)
24	*Point Defiance* Gulf	28 May 45			Canceled 17 Aug 45
25	*San Marcos* PHNY	1 Sep 44 10 Jan 45	15 Apr 45 26 Jan 51	19 Dec 47 1 Jul 71	To Spain 1 Jul 71
26	*Tortuga* BOSNY	16 Oct 44 21 Jan 45	8 Jun 45 15 Sep 50	18 Aug 47 26 Jan 70	Str 15 Oct 76
27	*Whetstone* BOSNY	7 Apr 45 18 Jul 45	12 Feb 46 2 Dec 50	20 Oct 48 2 Apr 70	Str 2 Jul 70
28	*Thomaston* Ingalls	3 Mar 53 2 Sep 54	17 Sep 54	5 Sep 84	Str 24 Feb 92
29	*Plymouth Rock* Ingalls	5 May 53 7 May 54	29 Nov 54	30 Sep 83	Str 8 Nov 91
30	*Fort Snelling* Ingalls	17 Aug 53 16 Jul 54	24 Jan 55	28 Sep 84	Str 24 Feb 92

No.	Name/Builder	LD/Launch	Comm	Decomm	Notes and Fate
31	*Point Defiance* Ingalls	23 Nov 53 28 Sep 54	31 Mar 55	30 Sep 83	Str 24 Feb 92
32	*Spiegel Grove* Ingalls	7 Sep 54 10 Nov 55	8 Jun 56	2 Oct 89	Str 13 Dec 89, reinstated for stripping, str Feb 94
33	*Alamo* Ingalls	11 Oct 54 20 Jan 56	24 Aug 56	28 Sep 89	To Brazil 2 Nov 90
34	*Hermitage* Ingalls	11 Apr 55 12 Jun 56	14 Dec 56	2 Oct 89	To Brazil on decomm
35	*Monticello* Ingalls	6 Jun 55 10 Aug 56	29 Mar 57	1 Oct 85	Str 2 Aug 91
36	*Anchorage* Ingalls	13 Mar 67 5 May 68	15 Mar 69		
37	*Portland* GD	21 Sep 67 20 Dec 69	3 Oct 70		
38	*Pensacola* GD	15 Mar 69 11 Jul 70	27 Mar 71	30 Sep 99	To ROC 30 Sep 99
39	*Mount Vernon* GD	29 Jan 70 11 Apr 71	13 May 72		
40	*Fort Fisher* GD	15 Jul 70 22 Apr 72	9 Dec 72	27 Feb 98	Planned transfer to ROC aborted due to poor material condition
41	*Whidbey Island* Lockheed	4 Aug 81 10 Jun 83	9 Feb 85		
42	*Germantown* Lockheed	5 Aug 82 29 Jun 84	8 Feb 86		
43	*Fort McHenry* Lockheed	10 Jun 83 1 Feb 86	8 Aug 87		
44	*Gunston Hall* Avondale	26 May 86 27 Jun 87	22 Apr 89		
45	*Comstock* Avondale	27 Oct 86 16 Jan 88	3 Feb 90		
46	*Tortuga* Avondale	23 Mar 87 15 Sep 88	17 Nov 90		
47	*Rushmore* Avondale	9 Nov 87 6 May 89	1 Jun 91		
48	*Ashland* Avondale	4 Apr 88 11 Nov 89	9 May 92		
49	*Harpers Ferry* Avondale	15 Apr 91 16 Jan 93	7 Jan 95		
50	*Carter Hall* Avondale	11 Nov 91 2 Oct 93	30 Sep 95		
51	*Oak Hill* Avondale	21 Sep 92 11 Jun 94	8 Jun 96		
52	*Pearl Harbor* Avondale	27 Jan 95 24 Feb 96	30 May 98		

LSM/LSMR/LFR

LSM 1–200 were ordered as LCT(7) 1501–1700, LSM 233–252 as LCT(7) 1701–1720, LSM 201–230 as LCT(7) 1721–1750, LSM 253–305 as LCT(7) 1751–1803, and LSM 310–336 as LCT(7) 1804–1830. Surviving LSMRs were named after rivers on 1 October 1955. LSMR were redesignated LFR effective 1 January 1969. When no decommissioning date is indicated, but ships are listed as having been transferred, they were transferred upon decommissioning. Israel obtained three LSM from commercial sources in 1972; their original U.S. numbers are not known.

No.	Name/Builder	LD/Launch	Comm	Decomm	Notes and Fate
1	Brown	15 Feb 44 23 Mar 44	5 May 44	9 Apr 46	Str 19 Jul 46
2	Brown	15 Feb 44 23 Mar 44	8 May 44	20 Dec 45	Str 8 Jan 46
3	Brown	1 Mar 44 9 Apr 44	16 May 44	1 May 46	Str 19 Jun 46
4	Brown	1 Mar 44 9 Apr 44	20 May 44	2 May 47	Str 17 Jul 47
5	Brown	1 Mar 44 9 Apr 44	22 May 44	8 Jul 46	Str 15 Aug 46, sunk on beach, Saipan
6	Brown	1 Mar 44 9 Apr 44	23 May 44	21 Mar 46	Str 12 Apr 46
7	Brown	20 Mar 44 24 Apr 44	24 May 44	11 Apr 46	Str 19 Jun 46
8	Brown	14 Mar 44 24 Apr 44	26 May 44	29 Apr 46	Str 19 Jun 46
9	Brown	11 Mar 44 24 Apr 44	27 May 44	20 Dec 45	Str 8 Jan 46
10	Brown	10 Mar 44 24 Apr 44	29 May 44	1 Mar 46	Str 12 Apr 46
11	Brown	20 Mar 44 30 Apr 44	31 May 44	27 Jun 46	Sold 29 Dec 47
12	Brown	20 Mar 44 30 Apr 44	30 May 44		WL 24 Apr 45, broached on beach and broke up
13	Brown	23 Mar 44 30 Apr 44	1 Jun 44		To ROC 31 May 46
14	Brown	23 Mar 44 30 Apr 44	1 Jun 44	6 Jun 46	Sold 13 Jan 48
15	Brown	10 Apr 44 7 May 44	2 Jun 44		Sunk in Okinawa typhoon 9 Oct 45
16	Brown	10 Apr 44 7 May 44	10 Jun 44	15 Mar 46	Str 17 Apr 46
17	Brown	10 Apr 44 7 May 44	14 Jun 44	22 Jul 46	To France for Indo-China Apr 54, ret 14 Oct 55, to Korea 18 Oct 56
18	Brown	10 Apr 44 7 May 44	14 Jun 44	26 Mar 46	Str 17 Apr 46
19	Brown	24 Apr 44 14 May 44	15 Jun 44	1 Jul 46	To Korea 3 Jul 56
20	Brown	24 Apr 44 14 May 44	16 Jun 44		WL 5 Dec 44
21	Brown	24 Apr 44 14 May 44	19 Jun 44	23 Mar 46	Str 12 Apr 46
22	Brown	24 Apr 44 14 May 44	21 Jun 44	27 May 46	Str 19 Jul 46
23	Brown	30 Apr 44 21 May 44	23 Jun 44	1 Apr 46	Str 8 May 46
24	Brown	30 Apr 44 21 May 44	24 Jun 44	29 Mar 46	Str 8 May 46
25	Brown	30 Apr 44 21 May 44	26 Jun 44	18 Dec 45	Str 8 Jan 46
26	Brown	30 Apr 44 21 May 44	27 Jun 44	28 Feb 47	Str 14 Mar 47, scuttled 24 Mar 47
27	Brown	7 May 44 28 May 44	28 Jun 44	3 Apr 46	Str 17 Apr 46
28	Brown	7 May 44 28 May 44	1 Jul 44	27 Mar 46	Str 8 May 46
29	Brown	7 May 44 28 May 44	3 Jul 44	27 Feb 46	Str 12 Apr 46

No.	Name/Builder	LD/Launch	Comm	Decomm	Notes and Fate
30		7 May 44	3 Jul 44	27 Jul 46	To Korea 3 Apr 56
	Brown	28 May 44			
31		14 May 44	7 Jul 44	6 May 46	Str 21 May 46
	Brown	6 Jun 44			
32		14 May 44	8 Jul 44	29 Dec 46	Str 21 Jan 46
	Brown	6 Jun 44			
33		14 May 44	11 Jul 44	19 Apr 46	Str 19 Jun 46
	Brown	6 Jun 44			
34		14 May 44	13 Jul 44	18 Apr 46	Str 5 Jun 46
	Brown	6 Jun 44			
35		21 May 44	14 Jul 44	10 May 46	Str 19 Jun 46
	Brown	14 Jun 44			
36		21 May 44	17 Jul 44	20 May 46	Str 19 Jul 46
	Brown	14 Jun 44			
37		21 May 44	19 Jul 44	26 Jun 46	Str 29 Aug 57
	Brown	14 Jun 44			
38		21 May 44	20 Jul 44	28 Oct 46	Str 22 Jan 48
	Brown	14 Jun 44			
39		28 May 44	21 Jul 44	13 Mar 46	Str 28 Mar 46
	Brown	22 Jun 44			
40		28 May 44	24 Jul 44	5 Apr 46	Str 21 May 46
	Brown	22 Jun 44			
41		28 May 44	3 Aug 44	16 Apr 46	Str 23 Apr 47
	Brown	22 Jun 44			
42		28 May 44	4 Aug 44	4 Jun 46	To ROC merch 4 Jun 46, seized by PRC 50
	Brown	22 Jun 44			
43		6 Jun 44	20 Jul 44	3 Jun 46	To ROC 3 Jun 46, seized by PRC Oct 49
	Brown	30 Jun 44			
44		6 Jun 44	28 Jul 44	17 May 46	Str 19 Jun 46
	Brown	30 Jun 44			
45		6 Jun 44	31 Jul 44	27 Mar 46	Str 3 Nov 58, to Greece, museum ship at Omaha, Nebraska, Nov 98
	Brown	30 Jun 44			
46		6 Jun 44	1 Aug 44	18 Mar 46	Str 12 Apr 46
	Brown	30 Jun 44			
47		14 Jun 44	9 Aug 44	27 Feb 46	Str 12 Mar 46
	Brown	7 Jul 44			
48		14 Jun 44	7 Aug 44	27 May 46	Str 3 Jul 46
	Brown	7 Jul 44			
49		14 Jun 44	14 Aug 44	13 May 46	Str 19 Jun 46
	Brown	7 Jul 44			
50		14 Jun 44	9 Aug 44	19 Apr 46	Str 3 Jul 46
	Brown	7 Jul 44			
51		22 Jun 44	11 Aug 44	8 May 46	Str 19 Jun 46
	Brown	14 Jul 44			
52		22 Jun 44	14 Aug 44	19 Apr 46	Str 23 Apr 47
	Brown	14 Jul 44			
53		22 Jun 44	16 Aug 44	15 Apr 46	Str 19 Jun 46
	Brown	14 Jul 44			
54		22 Jun 44	17 Aug 44	22 May 46	Converted to mine force flag/minelayer, to Korea 3 Jul 56
	Brown	14 Jul 44			
55		30 Jun 44	21 Aug 44	25 May 46	Str 19 Jul 46
	Brown	21 Jul 44			
56		30 Jun 44	31 Aug 44	18 Mar 46	Str 12 Apr 46
	Brown	21 Jul 44			
57		30 Jun 44	18 Aug 44	15 Jul 46	To France for Indo-China 6 Apr 54, ret 29 May 56, to Korea May 56
	Brown	21 Jul 44			
58		30 Jun 44	21 Aug 44	15 Jul 46	To France for Indo-China 6 Apr 54, ret Oct. 55, to Vietnam Apr 56 but ret 29 May 56, sold 8 Feb 57
	Brown	21 Jul 44	8 Sep 50	6 Apr 54	

No.	Name/Builder	LD/Launch	Comm	Decomm	Notes and Fate
59	Brown	7 Jul 44 29 Jul 44	23 Aug 44		WL 21 Jun 45
60	Brown	7 Jul 44 29 Jul 44	25 Aug 44		Bikini tgt ship, destroyed in explosion of bomb suspended from her; replaced LSM 18
61	Brown	7 Jul 44 29 Jul 44	2 Sep 44	21 Mar 46	Str 17 Apr 46
62	Brown	7 Jul 44 29 Jul 44	27 Aug 44		To UNRRA 29 May 46, str 19 Jul 46; also said to have been transferred to ROC
63	Brown	14 Jul 44 5 Aug 44	30 Aug 44	7 May 46	Str 19 Jul 46
64	Brown	14 Jul 44 5 Aug 44	1 Sep 44	21 Jun 46	Str 15 Aug 46
65	Brown	14 Jul 44 5 Aug 44	3 Sep 44	8 Apr 46	Str 9 Jun 46
66	Brown	14 Jul 44 5 Aug 44	6 Sep 44	12 Jun 46	Str 19 Jul 46
67	Brown	21 Jul 44 14 Aug 44	8 Sep 44	12 Jun 46	Sold Dec 46
68	Brown	21 Jul 44 14 Aug 44	10 Sep 44	10 May 46	Str 5 Jun 46
69	Brown	21 Jul 44 14 Aug 44	9 Sep 44	19 Jun 46	Str 23 Jun 47, to ROC 13 Feb 48
70	Brown	21 Jul 44 14 Aug 44	13 Sep 44	1 May 46	Str 19 Jun 46
71	Brown	29 Jul 44 22 Aug 44	15 Sep 44	23 Sep 46	Str 7 Feb 47
72	Brown	29 Jul 44 22 Aug 44	17 Sep 44	5 Apr 46	Str 5 Jun 46
73	Brown	29 Jul 44 22 Aug 44	16 Sep 44	2 Apr 46	Str 5 Jun 46
74	Brown	29 Jul 44 22 Aug 44	20 Sep 44	10 Apr 46	Str 9 Jun 46
75	Brown	5 Aug 44 30 Aug 44	22 Sep 44	19 Apr 46	Str 3 Jul 46
76	Brown	5 Aug 44 30 Aug 44	23 Sep 44	17 Jun 46	To ROC merch on decomm, reacquired by U.S. Navy, to ROC navy May 52
77	Brown	5 Aug 44 30 Aug 44	26 Sep 44	19 Apr 46	Str 3 Jul 46
78	Brown	5 Aug 44 30 Aug 44	27 Sep 44	1 Apr 46	Str 5 Jun 46
79	Brown	14 Aug 44 7 Sep 44	30 Sep 44	29 Apr 46	Sold 6 Jan 47
80	Brown	14 Aug 44 7 Sep 44	2 Oct 44	10 Aug 46	Str 25 Sep 46, to ROC 13 Feb 48
81	Brown	14 Aug 44 7 Sep 44	5 Oct 44	1 Apr 46	Str 23 Apr 47
82	Brown	14 Aug 44 7 Sep 44	7 Oct 44	15 Mar 46	Str 12 Apr 46
83	Brown	22 Aug 44 15 Sep 44	8 Oct 44	22 Mar 46	Str 1 May 46
84	Brown	22 Aug 44 15 Sep 44	10 Oct 44	10 May 46	To Korea 3 Jul 56
85	Brown	22 Aug 44 15 Sep 44	12 Oct 44	30 Jul 46	To France for Indo-China 7 May 54, to Vietnam Oct 56
86	Brown	22 Aug 44 15 Sep 44	13 Oct 44	4 Mar 46	Str 8 May 46, to Argentina Feb 48, to Paraguay 13 Jan 72

No.	Name/Builder	LD/Launch	Comm	Decomm	Notes and Fate
87	Brown	30 Aug 44 23 Sep 44	15 Oct 44	29 Mar 46	Str 8 May 46
88	Brown	30 Aug 44 23 Sep 44	17 Oct 44	6 Jun 46	Str 3 Jul 46, to ROC as merch 13 Feb 48, may have been taken by PRC navy
89	Brown	30 Aug 44 23 Sep 44	18 Oct 44	11 Feb 46	Str 17 Apr 46
90	Brown	30 Aug 44 23 Sep 44	20 Oct 44	1 May 46	Str 5 Jun 46
91	Brown	7 Sep 44 30 Sep 44	22 Oct 44	26 Feb 46	Str 12 Apr 46
92	Brown	7 Sep 44 30 Sep 44	24 Oct 44	16 Apr 46	Str 5 Jun 46
93	Brown	7 Sep 44 30 Sep 44	25 Oct 44	26 Apr 46	Str 5 Jun 46
94	Brown	7 Sep 44 30 Sep 44	27 Oct 44	5 Apr 46	Str 8 May 46
95	Brown	15 Sep 44 7 Oct 44	28 Oct 44	12 Mar 46	Str 28 Mar 46
96	Brown	15 Sep 44 7 Oct 44	31 Oct 44	28 May 46	To Korea Apr 56
97	Brown	15 Sep 44 7 Oct 44	1 Nov 44	20 May 46	Str 5 Jun 46
98	Brown	15 Sep 44 7 Oct 44	2 Nov 44	7 Jun 46	Str 19 Jun 46
99	Brown	23 Sep 44 14 Oct 44	6 Nov 44	14 May 46	Str 14 Jan 59
100	Brown	23 Sep 44 14 Oct 44	6 Nov 44	15 Apr 46	Str 19 Jun 46
101	Brown	23 Sep 44 14 Oct 44	8 Nov 44	6 Nov 46	Str 14 Aug 57
102	Brown	23 Sep 44 14 Oct 44	9 Nov 44	28 Oct 46	To Greece 3 Nov 58
103	Brown	30 Sep 44 21 Oct 44	12 Nov 44	19 Dec 46	Str 8 Feb 47
104	Brown	30 Sep 44 21 Oct 44	13 Nov 44	Mar 47	Str 14 Jan 58
105	Brown	30 Sep 44 21 Oct 44	15 Nov 44	20 Jun 46	Str 27 Aug 57
106	Brown	30 Sep 44 21 Oct 44	17 Nov 44	10 Feb 47	Str 14 Jan 58
107	Brown	7 Oct 44 28 Oct 44	18 Nov 44	16 May 46	Str 14 Jan 58
108	Brown	7 Oct 44 28 Oct 44	19 Nov 44	18 Dec 46	Str 14 Jan 58
109	Brown	7 Oct 44 28 Oct 44	23 Nov 44	18 Dec 46	Str 14 Jan 58, scuttled Jan 76 to form artificial reef in Florida
110	Brown	7 Oct 44 28 Oct 44	25 Nov 44 19 Sep 50	30 Nov 46 Jan 54	To France for Indochina 22 Jan 54, to Vietnam Oct 55
111	Brown	14 Oct 44 4 Nov 44	26 Nov 44	4 May 46	
112	Brown	14 Oct 44 4 Nov 44	29 Nov 44		To ROC merch 23 May 46, taken over by PRC navy
113	Brown	14 Oct 44 4 Nov 44	30 Nov 44	17 Mar 47	To Chile 20 Mar 47
114	Brown	14 Oct 44 4 Nov 44	1 Dec 44	10 Jun 46	Str 19 Jul 46
115	Brown	21 Oct 44 11 Nov 44	2 Dec 44	19 Jun 46	Str 19 Jul 46

No.	Name/Builder	LD/Launch	Comm	Decomm	Notes and Fate
116	Brown	21 Oct 44 11 Nov 44	6 Dec 44	27 Apr 46	Str 14 Jan 58
117	Brown	21 Oct 44 11 Nov 44	7 Dec 44	7 Jun 46	Str 3 Jul 46
118	Brown	21 Oct 44 11 Nov 44	8 Dec 44	25 Apr 47	Str 14 Jan 58
119	Brown	28 Oct 44 18 Nov 44	9 Dec 44	28 May 46	Str 30 Aug 57
120	Brown	28 Oct 44 18 Nov 44	10 Dec 44	14 May 46	Str 30 Aug 57
121	Brown	28 Oct 44 18 Nov 44	13 Dec 44	4 Apr 46	Str 14 Jan 58
122	Brown	28 Oct 44 18 Nov 44	14 Dec 44	29 May 46	Str 3 Jul 46
123	Brown	4 Nov 44 25 Nov 44	16 Dec 44	29 Dec 46	Str 29 Sep 47
124	Brown	4 Nov 44 25 Nov 44	17 Dec 44		To ROC merch 14 Jun 46, taken over by PRC navy
125	Brown	4 Nov 44 25 Nov 44	22 Dec 44 14 Oct 50	18 Dec 46 21 Jan 54	Str 23 Dec 53, to France 21 Jan 54 for Indo-China, to Japan 18 Jul 57
126	CHARNY	29 Jan 44 15 Mar 44	28 Apr 44	17 Jun 46	Str 3 Jul 46
127	CHARNY	29 Jan 44 15 Mar 44	28 Apr 44	22 Mar 46	Str 1 May 46
128	CHARNY	15 Feb 44 1 Apr 44	13 May 44	21 Jun 46	Str 15 Aug 46
129	CHARNY	15 Feb 44 1 Apr 44	13 May 44	7 May 46	Str 19 Jul 46
130	CHARNY	1 Mar 44 12 Apr 44	20 May 44	31 May 46	Str 19 Jun 46
131	CHARNY	1 Mar 44 12 Apr 44	20 May 44	19 Feb 46	Sold 21 Nov 47
132	CHARNY	1 Mar 44 13 Apr 44	22 May 44	16 May 47	Str 23 Jun 47
133	CHARNY	1 Mar 44 13 Apr 44	22 May 44	11 Mar 46	Str 8 May 46
134	CHARNY	14 Mar 44 23 Apr 44	31 May 44	19 Apr 46	Sold 18 Sep 47
135	CHARNY	14 Mar 44 23 Apr 44	31 May 44		WL(K) 25 May 45
136	CHARNY	18 Mar 44 18 Apr 44	30 May 44	26 Apr 46	Str 5 Jun 46
137	CHARNY	18 Mar 44 18 Apr 44	30 May 44	8 Dec 45	Grounded during typhoon at Okinawa, 9 Oct 45, str 21 Jan 46
138	CHARNY	4 Apr 44 1 May 44	10 Jun 44	12 Apr 46	Str 19 Jun 46
139	CHARNY	4 Apr 44 1 May 44	10 Jun 44	23 Apr 46	Str 5 Jun 46
140	CHARNY	17 Apr 44 18 May 44	22 Jun 44	28 Jun 46	Sold 20 Feb 48
141	CHARNY	17 Apr 44 18 May 44	22 Jun 44	8 Mar 46	Str 28 Mar 46
142	CHARNY	17 Apr 44 18 May 44	30 Jun 44	2 Sep 46	Str 25 Sep 46, hulk destroyed
143	CHARNY	17 Apr 44 18 May 44	30 Jun 44	14 Jun 46	Str 19 Jul 46
144	CHARNY	24 Apr 44 10 May 44	12 Jun 44	6 Mar 46	Str 12 Apr 46

No.	Name/Builder	LD/Launch	Comm	Decomm	Notes and Fate
145	CHARNY	24 Apr 44 10 May 44	12 Jun 44	22 May 46	Str 5 Jun 46
146	CHARNY	26 Apr 44 14 May 44	17 Jun 44	24 Sep 46	Str 25 Sep 46, to ROC 13 Feb 48
147	CHARNY	26 Apr 44 14 May 44	17 Jun 44	5 Aug 46	Str 25 Sep 46, to ROC 13 Feb 48
148	CHARNY	4 May 44 27 May 44	8 Jul 44	2 May 46	Str 19 Jul 46
149	CHARNY	4 May 44 27 May 44	8 Jul 44	15 Apr 45	Lost (accident): broached, CTL 5 Dec 44
150	CHARNY	11 May 44 2 Jun 44	15 Jul 44	13 May 46	Str 19 Jun 46
151	CHARNY	11 May 44 2 Jun 44	15 Jul 44	9 Aug 46	Str 28 Aug 46, hulk destroyed 31 Mar 47
152	CHARNY	15 May 44 5 Jun 44	22 Jul 44	17 May 46	Str 19 Jun 46
153	CHARNY	15 May 44 5 Jun 44	22 Jul 44		To ROC merch 4 Jun 46, taken over by PRC navy
154	CHARNY	19 May 44 22 Jun 44	26 Jul 44	13 Jun 46	Str 3 Jul 46
155	CHARNY	19 May 44 19 Jun 44	26 Jul 44		To ROC 12 Jun 46
156	CHARNY	19 May 44 22 Jun 44	8 Aug 44	18 Apr 46	Str 5 Jun 46
157	CHARNY	19 May 44 19 Jun 44	8 Aug 44		To ROC 12 Jun 46
158	CHARNY	29 May 44 18 Jun 44	31 Jul 44	12 Aug 46	Str 22 Jan 48
159	CHARNY	29 May 44 18 Jun 44	31 Jul 44	10 Oct 46	Str 29 Oct 46, to ROC for merch, taken over by PRC navy
160	CHARNY	3 Jun 44 27 Jun 44	16 Aug 44	10 Dec 46	Str 29 Oct 46
161	*Kodiak* CHARNY	3 Jun 44 27 Jun 44	16 Aug 44 1 Jan 51	8 Jun 46 19 Apr 65	Last of this type in U.S. Navy service; named Oct 59, str 21 Apr 65, fire-fighting training hulk at Pearl Harbor, replaced by YAG 80 about Jan 71, sold 14 Aug 72
162	CHARNY	6 Jun 44 26 Jun 44	22 Aug 44	26 Dec 45	Str 21 Jan 46
163	CHARNY	6 Jun 44 26 Jun 44	22 Aug 44	21 Jun 46	Sold 26 Feb 48
164	CHARNY	19 Jun 44 11 Jul 44	26 Aug 44	21 May 46	Str 19 Jun 46
165	CHARNY	19 Jun 44 11 Jul 44	26 Aug 44	10 Feb 47	Str 25 Feb 47
166	CHARNY	20 Jun 44 24 Jul 44	31 Aug 44	2 May 46	Str 3 Jul 46
167	CHARNY	20 Jun 44 24 Jul 44	31 Aug 44	3 Jul 46	Str 14 Jan 58
168	CHARNY	23 Jun 44 28 Jul 44	7 Sep 44	5 Sep 46	Str 30 Aug 57
169	CHARNY	23 Jun 44 28 Jul 44	7 Sep 44	19 Jan 46	CTL (mine) 15 Feb 45, str 7 Feb 46, scuttled 27 Feb 46
170	CHARNY	27 Jun 44 30 Jul 44	14 Sep 44	2 May 46	Str 19 Jul 46
171	CHARNY	27 Jun 44 30 Jul 44	14 Sep 44	19 Jun 46	Str 3 Jul 46, scuttled 26 May 48
172	CHARNY	28 Jun 44 21 Jul 44	20 Sep 44	9 May 47	Str 17 Jul 47

No.	Name/Builder	LD/Launch	Comm	Decomm	Notes and Fate
173	CHARNY	28 Jun 44 21 Jul 44	20 Sep 44	2 May 46	Str 3 Jul 46
174	CHARNY	11 Jul 44 3 Aug 44	25 Sep 44	5 Aug 46	Str 25 Sep 46
175	CHARNY	11 Jul 44 3 Aug 44	25 Sep 44 8 Sep 50	11 Jul 46 30 Oct 55	To Vietnam 1 Aug 61
176	CHARNY	20 Jul 44 12 Aug 44	30 Sep 44	27 Dec 45	Str 21 Jan 46
177	CHARNY	20 Jul 44 12 Aug 44	30 Sep 44	2 May 46	Str 28 Aug 46
178	CHARNY	21 Jul 44 16 Aug 44	5 Oct 44	2 May 46	Str 3 Jul 46
179	CHARNY	21 Jul 44 16 Aug 44	5 Oct 44	21 May 46	Str 5 Jun 46
180	CHARNY	3 Aug 44 26 Aug 44	10 Oct 44	23 Apr 46	Str 19 Jun 46
181	CHARNY	3 Aug 44 26 Aug 44	10 Oct 44	2 May 46	Str 3 Jul 46
182	CHARNY	24 Jul 44 28 Aug 44	14 Oct 44	2 May 46	Str 3 Jul 46
183	CHARNY	24 Jul 44 28 Aug 44	14 Oct 44	2 Aug 46	Str 25 Sep 46
184	CHARNY	13 Aug 44 7 Sep 44	20 Oct 44	22 Jun 46	Sold 26 Feb 48
185	CHARNY	13 Aug 44 7 Sep 44	20 Oct 44	7 Jul 46	Str 19 Jun 46
186	CHARNY	28 Jul 44 5 Sep 44	25 Oct 44	6 May 46	Str 21 May 46
187	CHARNY	28 Jul 44 5 Sep 44	25 Oct 44	19 Dec 45	Str 8 Jan 46
188	CHARNY	17 Aug 44 12 Sep 44	15 Nov 44	23 Jan 46	LSMR, str 7 Feb 46, damaged 29 Mar 45, repaired, converted to ammunition carrier
189	CHARNY	17 Aug 44 12 Sep 44	15 Nov 44	31 Jan 46	LSMR, str 25 Feb 46
190	CHARNY	27 Aug 44 21 Sep 44	21 Nov 44		LSMR, WL 4 May 45
191	CHARNY	27 Aug 44 21 Sep 44	21 Nov 44	12 Feb 46	LSMR, str 26 Feb 46
192	CHARNY	7 Sep 44 5 Oct 44	21 Nov 44	26 Feb 46	LSMR, str 12 Mar 46
193	CHARNY	7 Sep 44 5 Oct 44	21 Nov 44	13 Feb 46	LSMR, str 26 Feb 46
194	CHARNY	29 Aug 44 7 Oct 44	21 Nov 44		LSMR, WL 4 May 45
195	CHARNY	29 Aug 44 7 Oct 44	21 Nov 44		LSMR, WL 3 May 45
196	CHARNY	13 Sep 44 12 Oct 44	12 Oct 44	26 Mar 46	LSMR, str 17 Apr 46
197	CHARNY	13 Sep 44 12 Oct 44	12 Oct 44	13 Feb 46	LSMR, str 26 Feb 46
198	CHARNY	6 Sep 44 14 Oct 44	14 Oct 44	17 Jan 46	LSMR, str 7 Feb 46
199	CHARNY	6 Sep 44 14 Oct 44	14 Oct 44	1 Feb 46	LSMR, str 26 Feb 46
200	CHARNY	21 Sep 44 17 Oct 44	11 Dec 44	10 Jul 46	Str 27 Aug 57
201	Dravo-Wil	24 Dec 43 26 Feb 44	14 Apr 44	16 May 46	Str 3 Jul 46

No.	Name/Builder	LD/Launch	Comm	Decomm	Notes and Fate
202		5 Jan 44	30 Apr 44	19 Nov 45	Str 5 Dec 45
	Dravo-Wil	15 Mar 44			
203		13 Jan 44	21 May 44	14 Jun 46	Sold Jun 46
	Dravo-Wil	10 Apr 44			
204		18 Jan 44	15 May 44	27 Jul 46	Str 14 Jan 58
	Dravo-Wil	29 Mar 44			
205		10 Feb 44	26 May 44	8 May 46	Str 5 Jun 46
	Dravo-Wil	14 Apr 44			
206		22 Feb 44	29 May 44	23 Jul 47	Str 29 Sep 47
	Dravo-Wil	22 Apr 44			
207		2 Mar 44	31 May 44	2 May 46	Str 3 Jul 46
	Dravo-Wil	29 Apr 44			
208		3 Mar 44	12 Jun 44		To ROC 20 May 46, seized by PRC Oct 49
	Dravo-Wil	10 May 44			
209		14 Mar 44	17 Jun 44	29 Dec 45	Str 21 Jan 46
	Dravo-Wil	18 May 44			
210		21 Mar 44	22 Jun 44	2 May 46	Str 3 Jul 46
	Dravo-Wil	23 May 44			
211		30 Mar 44	29 Jun 44	19 Apr 46	Str 19 Jun 46, CTL 19 Feb 45 (mortar hits)
	Dravo-Wil	29 May 44			
212		7 Apr 44	5 Jul 44	19 Jul 46	Str 14 Jan 58
	Dravo-Wil	3 Jun 44			
213		17 Apr 44	11 Jul 44	19 Nov 45	Str 5 Dec 45
	Dravo-Wil	8 Jun 44			
214		27 Apr 44	17 Jul 44	26 Dec 45	Str 21 Jan 46
	Dravo-Wil	13 Jun 44			
215		4 May 44	23 Jul 44	17 Apr 46	Str 8 May 46
	Dravo-Wil	23 Jun 44			
216		13 May 44	29 Jul 44	2 May 46	Str 19 Jul 46, to Dominican Republic Nov 46
	Dravo-Wil	28 Jun 44			
217		19 May 44	4 Aug 44	2 May 46	Str 3 Jul 46
	Dravo-Wil	5 Jul 44			
218		24 May 44	9 Aug 44		To ROC 15 Jun 46, seized by PRC Oct 49
	Dravo-Wil	11 Jul 44			
219		30 May 44	15 Aug 44	25 Mar 47	Str 28 Aug 46
	Dravo-Wil	17 Jul 44			
220		3 Jun 44	20 Aug 44	23 Jul 46	Str 28 Aug 46, later chartered to MSTS as supply ship for U.S. Air Force eastern missile range, wrecked 10 Jan 70
	Dravo-Wil	15 Jul 44			
221		9 Jun 44	26 Aug 44	18 Mar 47	Str 10 Jun 47
	Dravo-Wil	29 Jul 44			
222		15 Jun 44	31 Aug 44	17 Mar 47	Str 10 Jun 47
	Dravo-Wil	5 Aug 44			
223		24 Jun 44	8 Sep 44	29 May 46	Str 19 Jun 46
	Dravo-Wil	15 Aug 44			
224		29 Jun 44	16 Sep 44	24 May 46	Str 3 Jul 46
	Dravo-Wil	22 Aug 44			
225		6 Jul 44	22 Sep 44		To ROC 15 Jun 46, seized by PRC Oct 49
	Dravo-Wil	29 Aug 44			
226		12 Jul 44	30 Sep 44	3 Jul 46	To France for Indo-China 7 Apr 54, to Vietnam Oct 56
	Dravo-Wil	4 Sep 44	6 Sep 50	6 Apr 54	
227		17 Jul 44	5 Oct 44	20 Jun 46	To Greece 3 Nov 58
	Dravo-Wil	9 Sep 44			
228		27 Jul 44	12 Oct 44	27 May 46	Str 19 Jun 46
	Dravo-Wil	16 Sep 44			
229		1 Aug 44	18 Oct 44	6 Nov 46	Str 14 Jan 58
	Dravo-Wil	25 Sep 44			

No.	Name/Builder	LD/Launch	Comm	Decomm	Notes and Fate
230	Dravo-Wil	8 Aug 44 30 Sep 44	26 Oct 44	2 Oct 46	Str 29 Oct 46
231	Dravo-Wil	16 Aug 44 9 Oct 44	31 Oct 44	31 Jul 46	Str 14 Jan 58
232	Dravo-Wil	24 Aug 44 14 Oct 44	6 Nov 44	6 Aug 46	Str 14 Jan 58
233	Western-SP	28 Feb 44 4 Jul 44	3 Aug 44	2 Jan 46	Str 21 Jan 46
234	Western-SP	15 Mar 44 4 Jul 44	17 Aug 44	17 Apr 46	Str 3 Jul 46
235	Western-SP	21 Mar 44 4 Jul 44	24 Aug 44	16 May 46	Str 3 Jul 46, donated to port of Newport, Oregon, on 24 Jun 48, later in PRC navy
236	Western-SP	27 Mar 44 4 Jul 44	2 Sep 44 8 Sep 50	15 Jul 46 17 Oct 55	Str 14 Jan 58, to Philippines 15 Sep 60
237	Western-SP	31 Mar 44 30 Jul 44	7 Sep 44	2 May 46	Str 19 Jul 46, ROC merch, seized by PRC Oct 49
238	Western-SP	5 Apr 44 30 Jul 44	14 Sep 44	5 May 47	Str 17 Jul 47
239	Western-SP	13 Apr 44 31 Jul 44	21 Sep 44	1 Mar 47	Str 14 Mar 47, scuttled 15 Apr 47
240	Western-SP	19 Apr 44 31 Jul 44	28 Sep 44	2 May 46	Str 3 Jul 46
241	Western-SP	16 May 44 20 Aug 44	5 Oct 44	17 May 46	Str 5 Jun 46
242	Western-SP	19 May 44 20 Aug 44	12 Oct 44	27 Dec 46	Sold 17 Feb 48
243	Western-SP	31 May 44 3 Sep 44	19 Oct 44	2 May 46	Str 19 Jul 46
244	Western-SP	6 Jun 44 3 Sep 44	26 Oct 44	19 Apr 46	Str 19 Jun 46
245	Western-SP	31 Jul 44 17 Sep 44	2 Nov 44	3 Jun 46	Str 19 Jul 46
246	Western-SP	1 Aug 44 17 Sep 44	9 Nov 44	20 May 46	Str 29 Sep 47
247	Western-SP	2 Aug 44 1 Oct 44	16 Nov 44	2 May 46	Str 3 Jul 46
248	Western-SP	3 Aug 44 1 Oct 44	23 Nov 44		To ROC merch 23 May 46, taken over by PRC navy
249	Western-SP	21 Aug 44 15 Oct 44	30 Nov 44	17 Jun 46	Str 3 Jul 46
250	Western-SP	23 Aug 44 15 Oct 44	7 Dec 44	3 Jun 49	Str 29 Aug 57
251	Western-SP	4 Sep 44 29 Oct 44	14 Dec 44	22 May 47	Str 17 Jul 47
252	Western-SP	6 Sep 44 29 Oct 44	21 Dec 44	14 Aug 46	Str 29 Oct 46, sunk 47, hulk sold 13 Feb 48
253	Federal-N	24 Feb 44 17 Apr 44	9 May 44	23 Jun 47	Str 10 Jun 47
254	Federal-N	24 Feb 44 17 Apr 44	17 May 44	31 May 46	Sold 20 Feb 48
255	Federal-N	24 Feb 44 22 Apr 44	22 May 44	17 Oct 46	Sold 19 Feb 48
256	Federal-N	24 Feb 44 22 Apr 44	27 May 44		To ROC 3 Sep 46
257	Federal-N	23 Mar 44 12 May 44	30 May 44	7 May 46	Str 5 Jun 46
258	Federal-N	23 Mar 44 12 May 44	2 Jun 44	2 May 46	Str 3 Jul 46

No.	Name/Builder	LD/Launch	Comm	Decomm	Notes and Fate
259	Federal-N	6 Apr 44 26 May 44	16 Jun 44	13 May 46	Str 19 Jun 46
260	Federal-N	18 Apr 44 26 May 44	21 Jun 44	2 May 46	Str 3 Jul 46
261	Federal-N	26 Apr 44 3 Jun 44	28 Jun 44	28 May 46	Str 19 Jun 46
262	Federal-N	26 Apr 44 3 Jun 44	30 Jun 44	20 May 46	Str 14 Jan 58
263	Federal-N	16 May 44 17 Jun 44	3 Jul 44	29 Dec 46	Str 12 Mar 49
264	Federal-N	17 May 44 26 Jun 44	10 Jul 44	2 May 46	Str 3 Jul 46
265	Federal-N	30 May 44 30 Jun 44	18 Jul 44	5 Mar 46	Str 8 May 46
266	Federal-N	31 May 44 3 Jul 44	25 Jul 44		Str 10 Jun 47
267	Federal-N	17 May 44 12 Jul 44	1 Aug 44	19 Apr 46	Str 8 May 46, to Argentina 48
268	Federal-N	17 May 44 17 Jul 44	10 Aug 44 27 Jul 50	13 Jul 46 16 Feb 55	To Korea 16 Feb 55
269	Federal-N	16 Jun 44 17 Jul 44	18 Aug 44	19 Apr 46	Str 8 May 46
270	Federal-N	7 Jun 44 25 Jul 44	28 Aug 44	13 Jul 46	Str 14 Jan 58
271	Federal-N	20 Jun 44 29 Jul 44	2 Sep 44	29 Dec 46	Str 25 Feb 47, destroyed 8 Feb 47
272	Federal-N	27 Jun 44 9 Aug 44	11 Sep 44	16 May 46	Str 5 Jun 46
273	Federal-N	19 Jul 44 26 Aug 44	19 Sep 44	25 Sep 47	Str 31 Oct 47
274	Federal-N	27 Jul 44 4 Sep 44	26 Sep 44	13 Jul 46	Str 14 Jan 58
275	Federal-N	1 Aug 44 11 Sep 44	6 Oct 44 2 Jul 52	21 Apr 47 30 Apr 59	Converted to cable ship *Portunus* (ARC 1) 15 Jan–18 Aug 52, to Portugal 16 Nov 59 as diving tender
276	Federal-N	11 Aug 44 20 Sep 44	16 Oct 44	21 Sep 46	To Vietnam Mar 63
277	Federal-N	17 Jul 44 27 Sep 44	19 Oct 44	18 Jun 46	Sold 10 Mar 48
278	Federal-N	17 Jul 44 6 Oct 44	25 Oct 44	5 Jun 46	Str 29 Sep 47
279	Federal-N	30 Aug 44 14 Oct 44	31 Oct 44	26 Jul 46	Str 28 Aug 46
280	Federal-N	8 Sep 44 21 Oct 44	6 Nov 44	11 Jun 46	Str 3 Jul 46, to ROC 13 Feb 48
281	Federal-N	15 Sep 44 27 Oct 44	13 Nov 44	19 Jul 46	Str 15 Aug 46, scuttled 6 Mar 47
282	Federal-N	23 Sep 44 6 Nov 44	18 Nov 44		To ROC merch 3 Jun 46, taken over by PRC navy
283	Federal-N	2 Oct 44 15 Nov 44	27 Nov 44	7 Aug 46	Str 10 Jun 47
284	Federal-N	10 Oct 44 22 Nov 44	5 Dec 44	5 Jun 46	Str 3 Jul 46
285	Federal-N	18 Oct 44 30 Nov 44	14 Dec 44		To ROC 24 Jun 46
286	Federal-N	25 Oct 44 7 Dec 44	20 Dec 44	17 May 46	Str 3 Jul 46

No.	Name/Builder	LD/Launch	Comm	Decomm	Notes and Fate
287	Federal-N	3 Nov 44 14 Dec 44	30 Dec 44	2 May 46	Str 3 Jul 46
288	Federal-N	10 Nov 44 22 Dec 44	8 Jan 45	18 Jul 46	Str 14 Jan 58
289	Federal-N	18 Nov 44 29 Dec 44	13 Jan 45	25 May 46	Str 14 Jan 58
290	Federal-N	27 Nov 44 8 Jan 45	22 Jan 45	5 Jun 46	Str 3 Jul 46
291	Federal-N	4 Dec 44 15 Jan 45	31 Jan 45	25 May 46	Str 30 Aug 57
292	Federal-N	9 Dec 44 22 Jan 45	7 Feb 45	5 Jun 46	Str 3 Jul 46
293	Federal-N	16 Dec 44 2 Feb 45	16 Feb 45	6 Jul 46	Str 19 Jul 46
294	Federal-N	26 Dec 44 10 Feb 45	26 Feb 45	30 Jul 46	Str 17 Jul 47
295	CHARNY	21 Sep 44 5 Oct 44	11 Dec 44	17 Mar 47	Str 17 Mar 47, to Chile 20 Mar 47
296	CHARNY	5 Oct 44 30 Oct 44	18 Dec 44	23 May 46	Str 14 Jan 58
297	CHARNY	5 Oct 44 30 Oct 44	18 Dec 44	4 Nov 57	Str 14 Jan 58
298	CHARNY	13 Oct 44 13 Nov 44	26 Dec 44	15 Nov 46	Str 14 Jan 58
299	CHARNY	13 Oct 44 13 Nov 44	26 Dec 44	6 Jun 46	Str 14 Jan 58
300	CHARNY	18 Oct 44 19 Nov 44	1 Jan 45	19 Aug 46	Str 14 Jan 58
301	CHARNY	18 Oct 44 19 Nov 44	1 Jan 45	23 May 47	To Greece 22 Dec 52 after minelayer (MMC 6) conversion
302	CHARNY	8 Oct 44 14 Nov 44	6 Jan 45	30 Sep 46	Str 10 Jun 47
303	CHARNY	8 Oct 44 14 Nov 44	6 Jan 45	10 Mar 47	To Greece 22 Dec 52 after minelayer (MMC 7) conversion
304	CHARNY	30 Oct 44 27 Nov 44	11 Jan 45	19 Jul 46	Str 27 Aug 57
305	CHARNY	30 Oct 44 27 Nov 44	11 Jan 45	20 Jun 47	Str 27 Aug 57
306	CHARNY	14 Oct 44 14 Nov 44	16 Jan 45	17 Dec 46	Str 14 Jan 58
307	CHARNY	14 Oct 44 14 Nov 44	16 Jan 45	27 Jun 46	Sold Oct 48
308	CHARNY	19 Nov 44 9 Dec 44	22 Jan 45		To ROC 11 May 46
309	CHARNY	19 Nov 44 9 Dec 44	22 Jan 45	8 Aug 46	Str 29 Aug 57
310	Pullman	15 Feb 44 9 May 44	23 May 44	24 May 46	Str 19 Jun 46
311	Pullman	28 Feb 44 9 May 44	31 May 44	29 May 46	Str 19 Jun 46
312	Pullman	11 Mar 44 17 May 44	17 Jun 44	29 May 46	Str 3 Jul 46
313	Pullman	16 Mar 44 24 May 44	25 Jun 44	6 Jul 46	Str 14 Jan 58, to Vietnam 20 Jun 62
314	Pullman	23 Mar 44 31 May 44	6 Jul 44	22 Jun 46	Sold 9 Feb 48
315	Pullman	29 Mar 44 10 Jun 44	13 Jul 44	14 Jun 46	Str 14 Jan 58, to Australian army 26 Jan 60

No.	Name/Builder	LD/Launch	Comm	Decomm	Notes and Fate
316		6 Apr 44	21 Jul 44	6 Jul 46	To France for Indo-China on decomm, to
	Pullman	18 Jun 44	27 Jul 50	26 Jan 54	Korea 18 Nov 56
317		10 Apr 44	28 Jul 44	25 Mar 47	Str 4 Apr 47
	Pullman	24 Jun 44			
318		17 Apr 44	2 Aug 44		WL 7 Dec 44
	Pullman	6 Jul 44			
319		22 Apr 44	10 Aug 44	14 Jun 46	To Australian army 16 Jul 59
	Pullman	16 Jul 44			
320		1 May 44	19 Aug 44	6 Sep 46	Str 14 Jan 58, to Philippines 27 Apr 62
	Pullman	20 Jul 44			
321		5 May 44	26 Aug 44	10 May 46	Str 3 Jul 46
	Pullman	27 Jul 44			
322		10 May 44	6 Sep 44	19 Nov 46	Str 13 Dec 46
	Pullman	3 Aug 44			
323		18 May 44	16 Sep 44	20 May 46	Sold 12 Aug 47
	Pullman	11 Aug 44			
324		24 May 44	23 Sep 44	20 Jun 46	Str 14 Jan 58
	Pullman	18 Aug 44			
325		29 May 44	30 Sep 44	2 May 46	Str 3 Jul 46
	Pullman	25 Aug 44			
326		30 May 44	7 Oct 44	17 Apr 46	Str 19 Jun 46
	Pullman	1 Sep 44			
327		7 Jun 44	14 Oct 44	2 May 46	Str 19 Jul 46
	Pullman	15 Sep 44			
328		9 Jun 44	21 Oct 44	31 May 46	Str 3 Jul 46
	Pullman	22 Sep 44			
329		16 Jun 44	28 Oct 44	11 Oct 46	Str 14 Jan 58, to Spain 25 Mar 60
	Pullman	29 Sep 44			
330		20 Jun 44	6 Nov 44	28 Oct 46	Str 14 Jan 58
	Pullman	6 Oct 44			
331		27 Jun 44	11 Nov 44	6 Jul 46	Str 14 Jan 58, to Spain 25 Mar 60
	Pullman	13 Oct 44			
332		4 Jul 44	18 Nov 44	23 Jun 47	Str 29 Aug 57
	Pullman	20 Oct 44			
333		13 Jul 44	25 Nov 44	28 Jul 46	To Thailand Oct 46
	Pullman	27 Oct 44			
334		20 Jul 44	2 Dec 44	22 Jun 46	Sold 1 Mar 48
	Pullman	3 Nov 44			
335		26 Jul 44	9 Dec 44	16 May 46	MSTS 27 Jun 51, to AG 335 on 1 Jan 69,
	Pullman	10 Nov 44	15 Aug 51		to Dept of Interior 21 Dec 70
336		3 Aug 44	22 Dec 44		To ROC merch 1 Jun 46, taken over by
	Pullman	17 Nov 44			PRC navy
337		10 Aug 44	30 Dec 44	6 Jul 46	Str 19 Jul 46, to ROC 13 Feb 48
	Pullman	30 Nov 44			
338		17 Aug 44	10 Jan 45	28 Jul 46	To Thailand 14 Oct 46
	Pullman	5 Dec 44			
339		24 Aug 44	23 Jan 45		To ROC 11 Dec 46
	Pullman	8 Dec 44			
340		30 Aug 44	30 Jan 45	20 Nov 47	Str 5 Dec 47
	Pullman	22 Dec 44			
341		14 Sep 44	3 Feb 45	28 Oct 49	Scheduled for France 49 but 478 substi-
	Pullman	29 Dec 44			tuted, str 30 Aug 57
342		22 Sep 44	12 Feb 45	3 Jul 47	To port of Newport, Oregon, 17 Feb 48
	Pullman	12 Jan 45			
343		29 Sep 44	17 Feb 45	29 Oct 46	Str 14 Jan 58, to Spain 25 Mar 60
	Pullman	19 Jan 45			

534 APPENDIX C

No.	Name/Builder	LD/Launch	Comm	Decomm	Notes and Fate
344	Pullman	6 Oct 44 26 Jan 45	24 Feb 45	9 May 47	Str 16 Sep 47
345	Pullman	13 Oct 44 2 Feb 45	3 Mar 45	12 Mar 47	Str 4 Apr 47
346	Pullman	20 Oct 44 9 Feb 45	10 Mar 45	7 Jul 47	Str 16 Sep 47
347	Pullman	26 Oct 44 16 Feb 45	17 Mar 45	13 Mar 47	Str 10 Jun 47
348	Pullman	3 Nov 44 23 Feb 45	24 Mar 45	22 Sep 47	Str 31 Oct 47
349	Pullman	10 Nov 44 2 Mar 45	31 Mar 45		To ROC 26 Aug 46
350	Pullman	17 Nov 44 9 Mar 45	7 Apr 45	23 Jul 47	Str 29 Sep 47
351	Pullman	28 Nov 44 20 Mar 45	14 Apr 45	7 Jul 47	Str 16 Sep 47
352	Pullman	5 Dec 44 23 Mar 45	21 Apr 45	12 Aug 46	Str 22 Jan 48
353	Pullman	13 Dec 44 31 Mar 45	28 Apr 45	12 Aug 46	Str 22 Jan 48
354	Brown	4 Nov 44 25 Nov 44	24 Dec 44	8 Aug 46	Str 29 Aug 57, to Indonesia Mar 59
355	Brown	11 Nov 44 2 Dec 44	24 Dec 44 18 Sep 50	23 Oct 46 18 Jan 54	To France for Indo-China 18 Jan 54, to Vietnam Dec 55
356	Brown	11 Nov 44 2 Dec 44	27 Dec 44	17 Apr 46	Str 21 May 46
357	Brown	11 Nov 44 2 Dec 44	29 Dec 44	10 Aug 46	Str 14 Jan 58
358	Brown	11 Nov 44 2 Dec 44	2 Jan 45	02 Jun 46	Str 14 Jan 58
359	Brown	18 Nov 44 9 Dec 44	31 Dec 44	1 Jun 49	Str 29 Aug 57
360	Brown	18 Nov 44 9 Dec 44	5 Jan 45	31 Jul 46	Str 29 Aug 57
361	Brown	18 Nov 44 9 Dec 44	6 Jan 45		CTL Okinawa typhoon 9 Oct 45
362	Brown	18 Nov 44 9 Dec 44	11 Jan 45 14 Oct 50	18 Dec 46 27 Aug 53	Str 14 Jan 59, to ROC 3 May 62
363	Brown	25 Nov 44 16 Dec 44	12 Jan 45		To ROC 15 May 46
364	Brown	25 Nov 44 16 Dec 44	13 Jan 45	20 Jun 46	Str 14 Jan 58
365	Brown	25 Nov 44 16 Dec 44	14 Jan 45	16 Sep 46	Str 14 Jan 58
366	Brown	25 Nov 44 16 Dec 44	17 Jan 45	20 Jun 47	Str 29 Aug 57
367	Brown	2 Dec 44 23 Dec 44	20 Jan 45	19 Jul 46	Str 14 Jan 58
368	Brown	2 Dec 44 23 Dec 44	21 Jan 45	19 Mar 47	Str 10 Jun 47
369	Brown	2 Dec 44 23 Dec 44	23 Jan 45	30 Sep 46	Str 29 Aug 57
370	Brown	2 Dec 44 23 Dec 44	24 Jan 45	20 Jun 46	Str 15 Jan 58, to Venezuela for parts 15 Jul 59
371	Brown	9 Dec 44 30 Dec 44	25 Jan 45	5 Mar 47	Str 29 Aug 57
372	Brown	9 Dec 44 30 Dec 44	26 Jan 45	26 Aug 46	Str 14 Jan 58

No.	Name/Builder	LD/Launch	Comm	Decomm	Notes and Fate
373	*Lakeland* Brown	9 Dec 44 30 Dec 44	27 Jan 45 24 Feb 58	14 Oct 46 1 Dec 59	Named 14 Oct 59, sold 6 Oct 60
374	Brown	9 Dec 44 30 Dec 44	28 Jan 45	13 May 46	To UNRRA 19 Jun 46; other source claims to ROC
375	Brown	16 Dec 44 6 Jan 45	30 Jan 45	14 Oct 46	Str 23 Apr 47
376	Brown	16 Dec 44 6 Jan 45	31 Jan 45	23 Jul 46	To UNRRA 23 Jun 47; other source claims to ROC
377	Brown	16 Dec 44 6 Jan 45	4 Feb 45	9 Jan 48	Str 22 Jan 48
378	Brown	16 Dec 44 6 Jan 45	9 Feb 45	22 Nov 49	Str 20 Oct 50
379	Brown	23 Dec 44 13 Jan 45	10 Feb 45	17 Jul 47	Str 28 Aug 47
380	Brown	23 Dec 44 13 Jan 45	10 Feb 45	19 Feb 48	Str 12 Mar 48
381	Brown	23 Dec 44 13 Jan 45	12 Feb 45	31 May 47	Str 22 Jan 48
382	Brown	23 Dec 44 13 Jan 45	14 Feb 45	17 Nov 47	Str 5 Dec 47
383	Brown	30 Dec 44 20 Jan 45	15 Feb 45	25 Jan 47	Str 7 Feb 47
384	Brown	30 Dec 44 20 Jan 45	18 Feb 45	12 Feb 46	Str 26 Feb 46
385	Brown	30 Dec 44 20 Jan 45	26 Feb 45	17 Mar 47	Str 10 Jun 47
386	Brown	30 Dec 44 20 Jan 45	26 Feb 45	14 Aug 46	Str 29 Oct 46
387	Brown	6 Jan 45 27 Jan 45	26 Feb 45	16 Aug 46	Str 25 Sep 46
388	Brown	6 Jan 45 27 Jan 45	26 Feb 45	2 May 47	Str 22 May 47
389	CHARNY	13 Nov 44 12 Dec 44	31 Jan 45	20 Jun 46	Str 14 Jan 58
390	CHARNY	13 Nov 44 12 Dec 44	31 Jan 45	30 Dec 46	To Denmark after conversion to minelayer (MMC 8), 18 Jun 54
391	CHARNY	28 Nov 44 17 Dec 44	8 Feb 45		To ROC 13 May 46, seized by PRC Oct 49
392	CHARNY	28 Nov 44 17 Dec 44	8 Feb 45	2 May 47	To Denmark after conversion to minelayer (MMC 9), 22 Jun 54
393	CHARNY	9 Dec 44 29 Dec 44	12 Mar 45		To ROC 1 May 46
394	CHARNY	9 Dec 44 29 Dec 44	12 Mar 45	24 Sep 46	Str 14 Jan 58
395	CHARNY	13 Dec 44 2 Jan 45	24 Mar 45	20 Jun 46	Str 14 Jan 58
396	CHARNY	13 Dec 44 2 Jan 45	24 Mar 45	20 Jun 47	Str 14 Jan 59, to Peru Apr 59
397	CHARNY	18 Dec 44 6 Jan 45	30 Jul 45	24 Feb 58	Logistics support vessel for Naval Facilities Caribbean, May 54–Feb 58, sold Nov 58, str 12 Feb 59
398	*Hunting* CHARNY	18 Dec 44 6 Jan 45	6 Aug 45	23 Nov 62	Converted to sonar test ship Jun 53–Jun 54, reclassified 6 Jan 57 as EAG 398, named 13 Jul 57, sold 26 Jul 63
399	CHARNY	29 Dec 44 18 Jan 45	13 Aug 45	26 Aug 55	To Greece 3 Nov 58
400	CHARNY	29 Dec 44 18 Jan 45	20 Aug 45	17 Mar 47	Str 17 Mar 47, to Chile 20 Mar 47

No.	Name/Builder	LD/Launch	Comm	Decomm	Notes and Fate
401	*Big Black River* CHARNY	2 Jan 45 22 Jan 45	7 Apr 45	15 Feb 54	LSMR, str 1 May 73
402	*Big Horn River* CHARNY	2 Jan 45 22 Jan 45	16 Apr 45	16 Jan 48	LSMR, str 1 Oct 58
403	*Blackstone River* CHARNY	6 Jan 45 26 Jan 45	25 Apr 45	13 May 55	LSMR, str 1 Oct 55
404	*Black Warrior River* CHARNY	6 Jan 45 26 Jan 45	25 Apr 45	10 Nov 54	LSMR, str 59
405	*Broadkill River* CHARNY	18 Jan 45 6 Feb 45	2 May 45 28 Mar 51	10 Feb 47 22 Nov 55	LSMR, str 1 May 73
406	*Canadian River* CHARNY	18 Jan 45 6 Feb 45	2 May 45	10 Feb 47	LSMR, str 1 Feb 60
407	*Chariton River* CHARNY	22 Jan 45 12 Feb 45	9 May 45	10 Feb 47	LSMR, str 1 Oct 58
408	*Charles River* CHARNY	22 Jan 45 12 Feb 45	9 May 45	16 Jan 48	LSMR, str 1 Oct 58
409	*Clarion River* CHARNY	26 Jan 45 18 Feb 45	16 May 45 5 Oct 50	6 Feb 47 26 Oct 55	LSMR, str 26 Oct 55
410	*Clark Fork River* CHARNY	26 Jan 45 18 Feb 45	16 May 45	6 Feb 47	LSMR, str 1 Feb 60
411	*Cumberland River* CHARNY	6 Feb 45 25 Feb 45	23 May 45 26 Jan 51	6 Feb 47 12 Oct 54	LSMR, str 1 Feb 60
412	*Des Plaines River* CHARNY	6 Feb 45 25 Feb 45	23 May 45 30 Aug 50	3 May 50 20 Sep 55	LSMR, str 1 May 73
413	CHARNY	12 Feb 45 3 Mar 45	28 Aug 45	24 May 46	Str 14 Jan 58
414	Dravo-Wil	31 Aug 44 20 Oct 44	14 Nov 44	17 Dec 46	Str 28 Jan 47
415	Dravo-Wil	7 Sep 44 27 Oct 44	18 Nov 44	8 Aug 46	Str 14 Jan 58
416	Dravo-Wil	13 Sep 44 2 Nov 44	25 Nov 44	12 Mar 47	Str 14 Jan 58
417	Dravo-Wil	21 Sep 44 8 Nov 44	30 Nov 44	17 Mar 47	To Chile 20 Mar 47
418	Dravo-Wil	27 Sep 44 14 Nov 44	6 Dec 44	7 Oct 46	Str 29 Aug 57
419	Dravo-Wil	4 Oct 44 18 Nov 44	14 Dec 44	16 Feb 55	To Korea on decomm
420	Dravo-Wil	9 Oct 44 25 Nov 44	31 Dec 44	19 Dec 46	Str 7 Feb 47
421	Dravo-Wil	18 Oct 44 30 Nov 44	24 Dec 44	24 Oct 46	Str 14 Jan 58
422	Dravo-Wil	23 Oct 44 7 Dec 44	31 Dec 44 15 Sep 50	19 Jul 46 2 Feb 54	To France for Indo-China 5 Feb 54, ret Mar 56, to ROC Nov 56
423	Dravo-Wil	28 Oct 44 12 Dec 44	10 Jan 45	10 Aug 46	Str 25 Sep 46, to ROC 13 Feb 48
424	Dravo-Wil	4 Nov 44 23 Dec 44	16 Jan 45	30 Jun 47	Str 29 Aug 57
425	Dravo-Wil	9 Nov 44 23 Dec 44	22 Jan 45	20 May 46	Str 5 Jun 46
426	Dravo-Wil	16 Nov 44 30 Dec 44	27 Jan 45	17 Sep 47	Str 29 Sep 47
427	Dravo-Wil	22 Nov 44 6 Jan 45	2 Feb 45	24 Sep 46	Str 25 Sep 46, to ROC 13 Feb 48
428	Dravo-Wil	25 Nov 44 12 Jan 45	28 Mar 45	19 Jun 46	Sold 13 Jan 48
429	Dravo-Wil	2 Dec 44 18 Jan 45	15 Feb 45	24 Jun 47	To army on decomm, ret 50 for MSTS

No.	Name/Builder	LD/Launch	Comm	Decomm	Notes and Fate
430	Dravo-Wil	7 Dec 44 27 Jan 45	20 Feb 45		To ROC 1 Jun 46, seized by PRC Oct 49
431	Dravo-Wil	14 Dec 44 2 Feb 45	25 Feb 45		To ROC 14 Sep 46
432	Dravo-Wil	18 Dec 44 5 Feb 45	28 Feb 45		CTL 11 Jan 47 (grounded), str 28 Jan 47
433	Dravo-Wil	28 Dec 44 12 Feb 45	6 Mar 45		To ROC 28 Jan 47, WL 49 but salvaged by PRC navy and recommissioned
434	Dravo-Wil	3 Jan 45 17 Feb 45	12 Mar 45	19 Nov 47	Str 22 Dec 48, disposal deferred to complete Sandstone nuclear tests 48, sold May 49
435	Dravo-Wil	10 Jan 45 23 Feb 45	18 Mar 45	20 Nov 47	Str 5 Dec 47
436	Dravo-Wil	18 Jan 45 28 Feb 45	24 Mar 45	23 Dec 47	Str 22 Jan 48
437	Dravo-Wil	23 Jan 45 10 Mar 45	31 Mar 45	11 Dec 47	Str 23 Dec 47
438	Dravo-Wil	31 Jan 45 14 Mar 45	5 Apr 45	8 Dec 47	Str 23 Dec 47
439	Dravo-Wil	3 Feb 45 19 Mar 45	11 Apr 45	14 Aug 46	Str 29 Oct 46, to ROC 13 Feb 48
440	Dravo-Wil	8 Feb 45 24 Mar 45	17 Apr 45	23 Jun 47	Str 1 Aug 47
441	Dravo-Wil	15 Feb 45 30 Mar 45	30 Apr 45	30 Aug 49	Str 29 Aug 57
442	Dravo-Wil	20 Feb 45 6 Apr 45	28 Apr 45		To ROC 19 Mar 47
443	Dravo-Wil	26 Feb 45 11 Apr 45	7 May 45	23 Dec 47	Str 22 Jan 48
444	Dravo-Wil	1 Mar 45 17 Apr 45	12 May 45	Apr 49	Reactivated, transferred to Atomic Energy Commission 13 Nov 57, ret Aug 60, to Chile 2 Sep 60
445	Dravo-Wil	12 Mar 45 24 Apr 45	19 May 45	19 May 60	To radar experimental ship 45, served as radar training ship Mar 50–Mar 54, conversion to drone catapult craft Mar 54 (YV 1), named *Catapult* 1 Jun 57, sold 15 Dec 60
446	Dravo-Wil	16 Mar 45 28 Apr 45	26 May 45 20 Oct 54	20 Dec 51 30 May 60	Conversion to radar training ship Jun–Jul 47, conversion to drone catapult ship Mar 54 (YV 2), named *Launcher* 1 Jun 57, str 1 Apr 60, sold 4 Nov 60
447	Western-SP	18 Sep 44 13 Nov 44	28 Dec 44	3 Jul 47	Str 1 Aug 47
448	Western-SP	20 Sep 44 13 Nov 44	4 Jan 45	25 Jan 57	Str 30 Aug 57
449	Western-SP	2 Oct 44 3 Dec 44	11 Jan 45	2 Jun 47	Str 23 Jun 47
450	Western-SP	4 Oct 44 3 Dec 44	18 Jan 45	17 Jul 47	Str 28 Aug 47
451	Western-SP	16 Oct 44 10 Dec 44	25 Jan 45	13 Mar 47	Str 4 Apr 47
452	Western-SP	18 Oct 44 10 Dec 44	1 Feb 45	24 May 46	Str 10 Jun 47
453	Western-SP	30 Oct 44 22 Dec 44	14 Feb 45	14 Jun 46	Str 19 Jul 46, to ROC merch, probably taken over by PRC navy
454	Western-SP	31 Oct 44 22 Dec 44	15 Feb 45	3 Jul 47	Str 1 Aug 47

No.	Name/Builder	LD/Launch	Comm	Decomm	Notes and Fate
455		14 Nov 44	12 Mar 45	25 Mar 49	Sold 59
	Western-SP	27 Dec 44	13 Nov 51	13 Mar 59	
456		15 Nov 44	13 Mar 45		To ROC 7 Feb 47
	Western-SP	28 Dec 44			
457		4 Dec 44	28 Mar 45		To ROC 12 Jun 46
	Western-SP	28 Jan 45			
458		5 Dec 44	22 Mar 45	24 Dec 47	Str 19 Feb 48, to VFW for use as permanently beached clubhouse, Belleville, New Jersey; used in cycloidal trials
	Western-SP	28 Jan 45			
459		6 Jan 45	28 Feb 45		CTL (aground, broke in half) 9 Dec 46, hulk destroyed 25 Jan 47
	Brown	27 Jan 45			
460		6 Jan 45	28 Feb 45	14 Jan 48	Str 22 Jan 48
	Brown	27 Jan 45			
461		13 Jan 45	1 Mar 45		To ROC 20 Aug 46, seized by PRC Oct 49
	Brown	3 Feb 45			
462		13 Jan 45	4 Mar 45		To Korea 16 Feb 55
	Brown	3 Feb 45			
463		13 Jan 45	7 Mar 45	24 Jun 47	To army 24 Jun 47, ret 19 Jul 55, to MSTS as react, to Philippines 17 Mar 61
	Brown	3 Feb 45	22 Aug 55		
464		13 Jan 45	9 Mar 45	10 Oct 47	Str 31 Oct 47
	Brown	3 Feb 45			
465		20 Jan 45	10 Mar 45	13 Mar 47	Str 10 Oct 47
	Brown	10 Feb 45			
466		20 Jan 45	15 Mar 45	30 Jul 47	Str 28 Aug 47
	Brown	10 Feb 45			
467		20 Jan 45	15 Mar 45	12 Aug 46	Str 25 Sep 47
	Brown	10 Feb 45			
468		20 Jan 45	16 Mar 45	25 Jun 46	Str 3 Jul 46
	Brown	10 Feb 45			
469		27 Jan 45	17 Mar 45	23 Sep 46	Str 14 Jan 59, to Thailand 25 May 62
	Brown	17 Feb 45			
470		27 Jan 45	19 Mar 45		To ROC 25 Jul 46
	Brown	17 Feb 45			
471		27 Jan 45	23 Mar 45	13 May 46	Str 11 Mar 54, to Vietnam, to ROC 15 Nov 56
	Brown	17 Feb 45			
472		27 Jan 45	23 Mar 45	Mar 46	To ROC Feb 59
	Brown	17 Feb 45			
473		3 Feb 45	24 Mar 45	6 May 46	Str 30 Aug 47
	Brown	24 Feb 45			
474		3 Feb 45	25 Mar 45	24 Jul 46	Str 22 May 47, to ROC 6 Feb 59
	Brown	24 Feb 45			
475		10 Feb 45	28 Mar 45		To ROC 24 Jul 46
	Brown	3 Mar 45			
476		10 Feb 45	29 Mar 45	15 Nov 46	To MSTS Apr 50, str 29 Aug 57
	Brown	3 Mar 45	21 Apr 50	21 Oct 53	
477		10 Feb 45	3 Apr 45	15 May 46	Str 14 Jan 58, to Australian army 16 Jul 59
	Brown	3 Mar 45			
478		10 Feb 45	7 Apr 45	28 May 46	To France for Indo-China 1 Apr 54, to ROC Nov 56
	Brown	3 Mar 45			
479		3 Feb 45	3 Apr 45	19 Dec 46	Str 14 Jan 58
	Brown	24 Feb 45			
480		3 Feb 45	5 Apr 45	24 Sep 46	Str 14 Jan 58
	Brown	24 Feb 45			
481		17 Feb 45	8 Apr 45	20 Jun 46	To Turkey 4 Oct 52 after conversion to minelayer (MMC 10), 6 Feb–15 Sep 52
	Brown	10 Mar 45			
482		17 Feb 45	10 Apr 45	8 May 46	To UNRRA 19 Jun 46; other source says to ROC, seized by PRC Oct 49
	Brown	10 Mar 45			

No.	Name/Builder	LD/Launch	Comm	Decomm	Notes and Fate
483	Brown	17 Feb 45 10 Mar 45	13 Apr 45	25 Mar 47	Str 14 Jan 58, to Dominican Republic 10 Mar 58
484	Brown	17 Feb 45 10 Mar 45	15 Apr 45	19 Dec 46	To Turkey 4 Oct 52 after conversion to minelayer (MMC 11) 6 Feb–19 Sep 52
485	Brown	24 Feb 45 17 Mar 45	26 Apr 45	19 Dec 46	Str 10 Jun 47
486	Brown	24 Feb 45 17 Mar 45	20 Apr 45	19 Dec 46	Str 10 Jun 47
487	Brown	24 Feb 45 17 Mar 45	24 Apr 45	19 Dec 46	Str 10 Jun 47
488	Brown	24 Feb 45 17 Mar 45	24 Apr 45	14 Oct 46	Str 29 Aug 57
489	Brown	3 Mar 45 24 Mar 45	27 Apr 45		To ROC 2 May 46, seized by PRC Oct 49
490	Brown	3 Mar 45 24 Mar 45	28 Apr 45	23 May 46	To Turkey 4 Oct 52 after conversion to minelayer (MMC 12), 6 Feb–15 Sep 52
491	Brown	3 Mar 45 24 Mar 45	30 Apr 45	27 Nov 46	To minelayer, to Germany 5 Sep 58
492	Brown	3 Mar 45 24 Mar 45	1 May 45	19 Aug 46	To Norway 4 Oct 52 after conversion to minelayer (MMC 13) 6 Feb–20 Sep 52, to Turkey 1 Nov 60 on ret
493	Brown	10 Mar 45 30 Mar 45	4 May 45	29 Nov 46	To Norway 4 Oct 52 after conversion to minelayer (MMC 14) 6 Feb–20 Sep 52, to Turkey 1 Nov 60 on ret
494	Brown	10 Mar 45 31 Mar 45	8 May 45	22 Nov 46	Str 29 Aug 57
495	Brown	10 Mar 45 31 Mar 45	7 May 45	26 Aug 46	Str 29 Aug 57
496	Brown	10 Mar 45 31 Mar 45	10 May 45	20 Jun 46	Str 14 Jan 58
497	Brown	17 Mar 45 7 Apr 45	11 May 45	16 Sep 46	Str 29 Aug 57
498	Brown	17 Mar 45 7 Apr 45	12 May 45	19 Jun 47	Str 14 Jan 58
499	Brown	17 Mar 45 7 Apr 45	14 May 45	1 Aug 46	Str 14 Jan 58
500	Brown	17 Mar 45 7 Apr 45	17 May 45	1 Aug 46	To Denmark 15 May 53 after conversion to MTB tender 1 Apr 52–15 May 53
501	*Elk River* Brown	24 Mar 45 21 Apr 45	27 May 45	31 Oct 46	LSMR, to IX 501, 1 Apr 67, to support Navy Deep Submergence program at Avondale and then at San Francisco Naval Shipyard, placed "in service, special" Jan 69 and "in service, active" Jan 73, barracks craft Oct 86
502	*Escalante River* Brown	24 Mar 45 21 Apr 45	3 Jun 45	4 May 46	LSMR, str 1 Oct 58
503	*Flambeau River* Brown	24 Mar 45 21 Apr 45	5 Jun 45	3 May 46	LSMR, str 1 Oct 58
504	*Gila River* Brown	24 Mar 45 21 Apr 45	11 Jun 45	11 May 46	LSMR, str 1 Feb 60
505	*Grand River* Brown	31 Mar 45 28 Apr 45	14 Jun 45	20 May 46	LSMR, str 1 Oct 58
506	*Green River* Brown	31 Mar 45 28 Apr 45	19 Jun 45	20 May 47	LSMR, str 1 Oct 58
507	*Greenbrier River* Brown	31 Mar 45 28 Apr 45	22 Jun 45	5 Feb 47	LSMR, str 1 Oct 58
508	*Gunnison River* Brown	31 Mar 45 28 Apr 45	25 Jun 45 7 Apr 61	5 Feb 47 31 Dec 68	LSMR, to YV 3 (*Targeteer*) 9 May 60 Sold Dec 69

APPENDIX C

No.	Name/Builder	LD/Launch	Comm	Decomm	Notes and Fate
509	*Holston River* Brown	7 Apr 45 5 May 45	28 Jun 45	5 Feb 47	LSMR, str 1 Oct 58
510	*James River* Brown	7 Apr 45 5 May 45	1 Jul 45	5 Feb 47	LSMR, str 1 Feb 58
511	*John Day River* Brown	7 Apr 45 5 May 45	3 Jul 45	21 May 47	LSMR, str 1 Feb 60
512	*Lamouille River* Brown	7 Apr 45 5 May 45	5 Jul 45	5 Dec 55	LSMR, str 1 May 73
513	*Laramie River* Brown	21 Apr 45 19 May 45	9 Jul 45	12 Apr 48	LSMR, str 1 May 73
514	*Maurice River* Brown	21 Apr 45 19 May 45	14 Jul 45	30 Jul 54	LSMR, str 1 Feb 60
515	*Owyhee River* Brown	21 Apr 45 19 May 45	16 Jul 45	16 Nov 55	LSMR, str 1 May 73
516	*Pearl River* Brown	21 Apr 45 19 May 45	20 Jul 45	23 Apr 48	LSMR, str 1 Oct 58
517	*Pee Dee River* Brown	28 Apr 45 2 Jun 45	21 Jul 45	13 Apr 55	LSMR, str 1 Feb 60
518	*Pit River* Brown	28 Apr 45 2 Jun 45	24 Jul 45	21 May 47	LSMR, str 1 Oct 58
519	*Powder River* Brown	28 Apr 45 2 Jun 45	28 Jul 45	21 Jan 47	LSMR, str 1 Oct 58
520	*Raccoon River* Brown	28 Apr 45 2 Jun 45	31 Jul 45 2 Feb 52	15 May 46 1 Oct 55	LSMR, str 1 Feb 60
521	*Rainy River* Brown	5 May 45 9 Jun 45	1 Aug 45	20 Jun 46	LSMR, str 1 Oct 58
522	*Red River* Brown	5 May 45 9 Jun 45	6 Aug 45 2 Apr 51	19 Apr 46 15 May 55	LSMR, str 1 May 73
523	*Republican River* Brown	5 May 45 9 Jun 45	10 Aug 45	Mar 46	LSMR, str 1 Feb 60
524	*St. Croix River* Brown	5 May 45 9 Jun 45	14 Aug 45	21 May 46	LSMR, str 1 Oct 58
525	*St. Francis River* Brown	19 May 45 16 Jun 45	14 Aug 45 16 Sep 50 18 Sep 65	28 Mar 46 21 Nov 55 17 Apr 70	LSMR, str 17 Apr 70
526	*St. Johns River* Brown	19 May 45 16 Jun 45	22 Aug 45	28 May 46	LSMR, str 1 Oct 46
527	*St. Joseph River* Brown	19 May 45 16 Jun 45	25 Aug 45 14 Oct 50	28 Mar 46 5 Aug 55	LSMR, to Korea 15 Sep 60
528	*St. Marys River* Brown	19 May 45 16 Jun 45	2 Sep 45	30 Apr 46	LSMR, str 1 Oct 58
529	*St. Regis River* Brown	2 Jun 45 7 Jul 45	5 Sep 45	31 May 46	LSMR, str 1 Feb 60
530	*Salmon Falls River* Brown	2 Jun 45 7 Jul 45	20 Sep 45	25 Jun 46	LSMR, str 1 Oct 58
531	*Smoky Hill River* Brown	2 Jun 45 7 Jul 45	25 Sep 45	29 May 46	LSMR, str 1 May 73, sold to Panama for Panamanian navy 14 Mar 75
532	*Smyrna River* Brown	2 Jun 45 7 Jul 45	12 Oct 45	Mar 46	LSMR, str 29 Aug 58, to Germany 5 Sep 58
533	*Snake River* Brown	9 Jun 45 14 Jul 45	13 Oct 45	10 Jul 46	LSMR, str 1 Oct 58
534	*Thames River* Brown	9 Jun 45 14 Jul 45	18 Oct 45	25 Apr 46	LSMR, to Germany 5 Sep 58
535	*Trinity River* Brown	9 Jun 45 14 Jul 45	1 Nov 45	10 Jun 46	LSMR, str 1 Oct 58
536	*White River* Brown	9 Jun 45 14 Jul 45	28 Nov 45 16 Sep 50 2 Oct 65	31 Jul 46 7 Sep 56 22 May 70	LSMR, str 22 May 70

No.	Name/Builder	LD/Launch	Comm	Decomm	Notes and Fate
537	Brown	16 Jun 45 11 Aug 45	2 Nov 45	24 May 47	To Germany 5 Sep 58
538	Brown	16 Jun 45 11 Aug 45	13 Nov 45	20 Jun 46	Str 14 Jan 58, to Dominican Republic 10 Mar 58
539	Brown	16 Jun 45 11 Aug 45	30 Nov 45	24 May 46	To Ecuador Jan 58
540	*Raritan* Brown	16 Jun 45 11 Aug 45	5 Dec 45 4 Nov 57	29 May 46 1 Dec 59	Str 1 Jan 60
541	Brown	7 Jul 45 18 Aug 45	7 Dec 45	23 May 47	To Greece 30 Jun 58
542	Brown	7 Jul 45 18 Aug 45	19 Dec 45	1 Jul 46	Str 14 Jan 58, to Venezuela 15 Jul 59 for parts
543	Brown	7 Jul 45 18 Aug 45	27 Dec 45	20 May 46	Str 14 Jan 58, to Venezuela 28 Sep 59
544	Brown	7 Jul 45 18 Aug 45	4 Jan 46	20 Jun 46	Str 14 Jan 58, to Venezuela 2 Dec 59
545	Brown	14 Jul 45 25 Aug 45	8 Jan 46	24 May 46	Str 14 Jan 58, to Venezuela 21 Jan 60
546	Brown	14 Jul 45 25 Aug 45	16 Jan 46 22 Sep 50	27 May 46 16 Feb 55	To Korea on decomm
547	Brown	14 Jul 45 25 Aug 45	24 Jan 46 22 Sep 50	11 Mar 47 27 Feb 55	Str 14 Jan 58, to Australian army 26 Jan 60
548	Brown	14 Jul 45 25 Aug 45	1 Feb 46	31 Jul 46	Str 14 Jan 58, to Venezuela 11 Feb 59
549	Brown	25 Aug 45 7 Dec 45	18 Mar 46 8 Aug 51	21 Jan 48 23 Dec 55	To *Gypsy* (ARSD 1) salvage lifting vessel 24 Apr 45, str 1 Jun 73
550	Brown	25 Aug 45 7 Dec 45	8 Mar 46 12 Sep 51	21 Jan 48 20 Dec 55	To *Mender* (ARSD 2) 24 Apr 45, str 1 Jun 73
551	Brown	27 Aug 45 7 Dec 45	22 Mar 46	23 Nov 65	To *Salvager* (ARSD 3) 24 Apr 45, to YMLC 3 on 1 Sep 67, out of U.S. Navy custody 1 Aug 72
552	Brown	27 Aug 45 7 Dec 45	9 Apr 46	23 Nov 65	To *Windlass* (ARSD 4) 24 Apr 45, to YMLC 4 on 1 Sep 67, sold 6 Mar 73
553	CHARNY	12 Feb 45 3 Mar 45	6 Sep 45	19 Apr 46	To Germany 5 Sep 58
554	CHARNY	3 Mar 45 22 Mar 45	14 Sep 45	25 Jun 46	Str 14 Jan 58, to Peru Apr 59
555	CHARNY	3 Mar 45 22 Mar 45	24 Sep 45	3 May 46	Str 14 Jan 58, to Ecuador Nov 58
556	CHARNY	23 Mar 45 10 Apr 45	12 Oct 45	5 Aug 46	Str 14 Jan 58
557	CHARNY	23 Mar 45 10 Apr 45	22 Oct 45	21 May 46	To Greece 30 Oct 58
558	CHARNY	10 Apr 45 28 Apr 45	2 Nov 45	3 Jun 46	To Germany 5 Sep 58

LST

Ships were named 1 July 1955. Those in MSTS service did not receive names.

No.	Name/Builder	LD/Launch	Comm	Decomm	Notes and Fate
1	Dravo-Pitts	20 Jul 42 7 Sep 42	14 Dec 42	21 May 46	Str 19 Jun 46
2	Dravo-Pitts	23 Jun 42 19 Sep 42	9 Feb 43	11 Apr 46	Str 5 Jun 46

No.	Name/Builder	LD/Launch	Comm	Decomm	Notes and Fate
3	Dravo-Pitts	29 Jun 42 19 Sep 42	8 Feb 43	46	Str 19 Jun 46
4	Dravo-Pitts	4 Jul 42 9 Oct 42	14 Feb 43	46	Str 19 Jun 46
5	Dravo-Pitts	12 Jul 42 3 Oct 42	22 Feb 43	46	Str 1 Aug 47
6	Dravo-Wil	20 Jul 42 21 Oct 42	30 Jan 43		WL(M) 17 Nov 44
7	Dravo-Pitts	17 Jul 42 31 Oct 42	2 Mar 43	21 May 46	Str 19 Jun 46
8	Dravo-Pitts	26 Jul 42 29 Oct 42			To UK 22 Mar 43, ret 1 Jun 46, str 3 Jul 46
9	Dravo-Pitts	9 Aug 42 14 Nov 42			To UK 19 Mar 43, ret 1 Jun 46, str 3 Jul 46
10	Dravo-Pitts	15 Aug 42 25 Nov 42	22 Feb 43	6 Sep 46	Renamed *Achelous* (ARL 1) 13 Jan 43, str 1 Jun 73
11	Dravo-Pitts	8 Aug 42 18 Nov 42			To UK 22 Mar 43, ret 13 May 46, str 5 Jun 46
12	Dravo-Pitts	16 Aug 42 7 Dec 42			To UK 25 Mar 43, ret 5 Jan 46, str 20 Mar 46
13	Dravo-Pitts	1 Sep 42 5 Jan 43			To UK 3 Apr 43, ret 27 Feb 46, str 5 Jun 46
14	Dravo-Pitts	23 Aug 42 9 Dec 42	26 Mar 43	6 Apr 46	*Varuna* (AGP 5) 25 Jan 43, sold 47
15	Dravo-Pitts	17 Sep 42 30 Jan 43	5 Aug 43	Jan 47	*Phaon* (ARB 3) 25 Jan 43, str 1 Jul 61
16	Dravo-Wil	1 Sep 42 19 Dec 42	17 Mar 43	8 Mar 46	Str 12 Apr 46
17	Dravo-Pitts	21 Sep 42 8 Jan 43	19 Apr 43	15 Jan 46	SCAJAP 15 Jan 46 (Q015), tgt 15 Aug 56
18	Dravo-Pitts	1 Oct 42 15 Feb 43	26 Apr 43	3 Apr 46	Str 17 Apr 46, merch
19	Dravo-Pitts	22 Oct 42 11 Mar 43	15 May 43	20 Mar 46	Str 1 May 46, LSTH 15 Sep 45
20	Dravo-Pitts	5 Oct 42 15 Feb 43	14 May 43	3 Apr 46	Str 19 Jun 46
21	Dravo-Wil	25 Sep 42 18 Feb 43	14 Apr 43	25 Jan 46	Str 19 Jun 46
22	Dravo-Pitts	5 Nov 42 29 Mar 43	29 May 43	1 Apr 46	Str 17 Apr 46, sold for conversion to merch service (ROC)
23	Dravo-Pitts	27 Oct 42 13 Mar 43	22 May 43	24 May 46	LSTH 15 Sep 45, str 3 Jul 46
24	Dravo-Pitts	19 Nov 42 17 Apr 43	14 Jun 43	26 Feb 46	Str 5 Jun 46, merch
25	Dravo-Wil	12 Oct 42 9 Mar 43	3 May 43	2 Aug 46	Str 8 Oct 46
26	Dravo-Pitts	16 Nov 42 31 Mar 43	7 Jun 43	1 Apr 46	Str 8 May 46, merch
27	Dravo-Pitts	10 Dec 42 27 Apr 43	25 Jun 43	9 Nov 45	Str 28 Nov 45
28	Dravo-Pitts	8 Dec 42 19 Apr 43	19 Jun 43	16 Aug 46	Str 29 Oct 46
29	Dravo-Pitts	8 Jan 43 17 May 43	10 Jul 43	11 Mar 46	Str 8 May 46
30	Dravo-Pitts	12 Jan 43 3 May 43	3 Jul 43	6 Mar 46	Str 8 May 46, merch
31	*Addison County* Dravo-Pitts	2 Feb 43 5 Jun 43	21 Jul 43	8 Jan 46	SCAJAP Jan 46–May 48 (Q005), str 11 Aug 55, tgt

No.	Name/Builder	LD/Launch	Comm	Decomm	Notes and Fate
32	*Alameda County* Dravo-Pitts	17 Feb 43 22 May 43	12 Jul 43 7 Mar 51	Jul 46 25 Jun 62	Flagship of Naval Air Force, Atlantic, 53; redesignated VB 1 on 28 Aug 57; str 25 Jun 62; to Italy 20 Nov 62
33	Dravo-Pitts	23 Feb 43 21 Jun 43	4 Aug 43		To Greece 18 Aug 43
34	Dravo-Pitts	15 Mar 43 15 Jun 43	26 Jul 43	15 Nov 46	To mil gov Ryukyus, str 23 Dec 47, lost Jan 49
35	Dravo-Pitts	20 Mar 43 30 Jun 43			To Greece 18 Aug 43
36	Dravo-Pitts	21 Apr 43 10 July 43			To Greece 23 Aug 43
37	Dravo-Pitts	1 Apr 43 5 Jul 43			To Greece 18 Aug 43, WL
38	Dravo-Pitts	14 Apr 43 27 Jul 43	3 Sep 43	26 Mar 46	LSTH 15 Sep 45, str 1 May 46
39	Dravo-Pitts	23 Apr 43 29 Jul 43	8 Sep 43		Sank summer 44, str 18 Jul 44, to YF 1079, discarded ca Jan 46
40	Dravo-Pitts	3 Jun 43 15 Sep 43	15 Sep 43	18 Feb 46	Str 5 Mar 46, sold to U.S. mil gov Korea, SCAJAP 46 (Q066)
41	Dravo-Pitts	24 May 43 17 Aug 43	24 Sep 43	25 Apr 46	LSTH 15 Sep 45, str 19 Jun 46
42	Dravo-Pitts	17 Jun 43 17 Aug 43	30 Sep 43	26 Jul 46	LSTH 15 Sep 45, str 25 Sep 46
43	Dravo-Phil	19 Jun 43 28 Aug 43	6 Oct 43		WL(A) 21 May 44
44	Dravo-Pitts	7 Jul 43 11 Sep 43	22 Oct 43	20 Feb 46	SCAJAP 46 (Q068), str 28 Aug 47
45	Dravo-Pitts	27 Jun 43 31 Aug 43	15 Oct 43	30 Nov 48	Str 22 Dec 48, to Greece 58
46	Dravo-Pitts	20 Jul 43 16 Sep 43	3 Nov 43	6 Jun 46	Str 19 Jun 46
47	Dravo-Pitts	30 Jul 43 24 Sep 43	8 Nov 43	11 Jan 46	To army, to MSTS 31 Mar 52, to Philippines 13 Sep 76, SCAJAP 46–52 (Q007)
48	Dravo-Pitts	8 Aug 43 2 Oct 43	16 Nov 43	8 Feb 46	Str 5 Dec 47, SCAJAP 46 (Q049)
49	Dravo-Pitts	17 Aug 43 9 Oct 43	20 Nov 43	11 Jun 46	Str 3 Jul 46
50	Dravo-Pitts	29 Aug 43 16 Oct 43	27 Nov 43	6 Feb 46	Str 8 Sep 53, to ARB 13, to Norway, ret 1 July 60, to Greece 16 Sep 60; on decomm had served with SCAJAP (Q046)
51	Dravo-Pitts	29 Aug 43 22 Oct 43	8 Dec 43	6 Mar 46	Str 31 Oct 47
52	Dravo-Pitts	16 Sep 43 20 Oct 43	15 Dec 43	29 Aug 46	Bikini tgt, sunk as tgt 19 Apr 48
53	Dravo-Pitts	24 Sep 43 6 Nov 43	21 Dec 43	46	SCAJAP (Q021), to APL 59 Sep 54, to Korea 29 Apr 55
54	Dravo-Pitts	3 Oct 43 13 Nov 43	24 Dec 43	5 Nov 45	Str 28 Nov 45
55	Dravo-Pitts	10 Oct 43 20 Nov 43	6 Jan 44	11 Dec 45	Str 3 Jan 46
56	Dravo-Pitts	17 Oct 43 27 Nov 43	10 Jan 44	23 May 46	Str 3 Jul 46
57	*Armstrong County* Dravo-Pitts	24 Oct 43 4 Dec 43	15 Jan 44	24 Jan 46	SCAJAP (Q028), str 11 Aug 55, tgt 56
58	Dravo-Pitts	31 Oct 43 11 Dec 43	22 Jan 44	7 Nov 45	Str 28 Nov 45

No.	Name/Builder	LD/Launch	Comm	Decomm	Notes and Fate
59	Dravo-Pitts	7 Nov 43 18 Dec 43	31 Jan 44	21 Jan 46	Str 25 Feb 46
60	*Atchison County* Dravo-Pitts	14 Nov 43 24 Dec 43	7 Feb 44	27 Jun 46	Str 1 Nov 58
61	Jeffersonville	24 Jun 42 8 Nov 42	5 Feb 43	5 Jun 46	Str 19 Jun 46, merch
62	Jeffersonville	5 Aug 42 23 Nov 42			To UK 3 Mar 43, ret 10 Jun 46, str 19 Jul 46
63	Jeffersonville	6 Aug 42 19 Dec 42			To UK 15 Mar 43, ret 17 Dec 45, str 21 Jan 46
64	Jeffersonville	13 Aug 42 8 Jan 43			To UK 2 Apr 43, ret Nov 45, str 5 Dec 45
65	Jeffersonville	14 Aug 42 7 Dec 42			To UK 15 Mar 43, ret 5 Jan 46, str 20 Mar 46
66	Jeffersonville	14 Aug 42 16 Jan 43	12 Apr 43	26 Mar 46	Str 1 May 46
67	Jeffersonville	7 Sep 42 28 Jan 43	20 Apr 43	28 Mar 46	Str 8 May 46
68	Jeffersonville	7 Sep 42 8 Mar 43	4 Jun 43	7 Mar 46	Str 5 Jun 46
69	Jeffersonville	7 Sep 42 20 Feb 43	20 May 43		WL(A) 21 May 44
70	Jeffersonville	13 Nov 42 8 Feb 43	28 May 43	1 Apr 46	Str 1 May 46
71	Jeffersonville	27 Nov 42 27 Feb 43	9 Jun 43	25 Mar 46	Str 8 May 46, merch
72	Jeffersonville	20 Dec 42 17 Mar 43	5 Jun 43	4 Jun 46	Str 19 Jun 46, to Philippines
73	Jeffersonville	10 Dec 42 29 Mar 43	8 Jun 43	13 Jul 46	Str 10 Jun 47
74	Jeffersonville	1 Jan 43 31 Mar 43	15 Jun 43	21 Dec 45	Str 21 Jan 46
75	Jeffersonville	30 Jan 43 7 Apr 43	21 Jun 43	22 Dec 47	To Philippines 30 Dec 47
76	Jeffersonville	19 Jan 43 14 Apr 43	26 Jun 43	24 Dec 44	To UK Dec 44, ret 23 Apr 46, str 19 Jun 46, merch, sank 5 Oct 51
77	Jeffersonville	20 Feb 43 21 Apr 43	3 Jul 43	24 Dec 44	To UK Dec 44, ret 12 May 46, str 19 Jun 46
78	Jeffersonville	9 Feb 43 28 Apr 43	8 Jul 43	8 Mar 46	Str 8 May 46
79	Jeffersonville	28 Feb 43 8 May 43	7 Jul 43		To UK 17 Jul 43, WL 30 Sep 43
80	Jeffersonville	16 Mar 43 18 May 43	12 Jul 43		To UK 19 Jul 43, WL Mar 45
81	Jeffersonville	8 Mar 43 28 May 43	21 Jul 43		ARL 5 on 20 Jul 43, to UK 29 Jul 43 as LSE 1, ret 21 May 46, str 29 Oct 46, to Argentina 20 Aug 47
82	Jeffersonville	25 Mar 43 9 Jun 43	26 Jul 43		ARL 6 on 20 Jul 43, to UK 2 Aug 43 as LSE 2, ret 21 May 46, str 29 Oct 46, to Argentina 20 Aug 47
83	Jeffersonville	31 Mar 43 14 Jun 43	6 Aug 43	10 Oct 46	To *Adonis* (ARL 4) 26 Aug 43, str 1 Jan 60
84	Jeffersonville	13 Apr 43 26 Jun 43	14 Aug 43	2 Mar 46	LSTH 15 Sep 45, str 31 Oct 47
85–116					Canceled 16 Sep 42
117	Jeffersonville	28 Apr 43 10 Jul 43	27 Aug 43	16 Feb 46	LSTH 15 Sep 45, SCAJAP (Q063), MSTS 31 Mar 52, str 10 Jun 73, to Singapore 4 Jun 76, resold for merch service without entering naval service

No.	Name/Builder	LD/Launch	Comm	Decomm	Notes and Fate
118	Jeffersonville	21 Apr 43 21 Jul 43	6 Sep 43	8 Feb 46	SCAJAP (Q048), str 29 Sep 47
119	Jeffersonville	12 May 43 28 Jul 43	1 Sep 43	13 May 46	Str 10 Jun 46, to barge
120	Jeffersonville	5 May 43 7 Aug 43	22 Sep 43	7 Jan 46	SCAJAP (Q004) on decomm, to mil gov Korea Feb 47
121	Jeffersonville	23 May 43 16 Aug 43	29 Sep 43	21 Mar 46	LSTH 15 Sep 45, str 1 May 46
122	Missouri	4 Jun 43 9 Aug 43	3 Sep 43	4 Jun 46	Str 3 Jul 46
123	Missouri	5 Jun 43 14 Aug 43	7 Sep 43	22 Mar 46	LSH 15 Sep 45, str 1 May 46
124	Missouri	7 Jun 43 18 Aug 43	24 Sep 43	26 Jul 46	Str 28 Aug 46
125	Missouri	8 Jun 43 23 Aug 43	29 Sep 43	10 Jun 46	Bikini tgt 14 Aug 46, sunk by gunfire 25 Aug 46
126	Missouri	11 Jun 43 28 Aug 43	2 Oct 43	17 Jun 46	Str 23 Jun 47, merch
127	Missouri	30 Jun 43 31 Aug 43	6 Oct 43	11 Mar 47	Str 10 Jun 47
128	Missouri	20 Jun 43 3 Sep 43	11 Oct 43	23 Mar 46	Str 17 Apr 46
129	Missouri	1 Jul 43 8 Sep 43	23 Oct 43	20 Jan 45	IX 198 on 5 Dec 44, destroyed 16 May 46; may have been intended as ARL 34
130	Missouri	5 Jul 43 18 Sep 43	4 Nov 43	46	To SCAJAP 46 (Q093); to MarComm 27 Mar 48
131	Missouri	7 Jul 43 19 Sep 43	15 Nov 43	20 May 46	Str 10 Jun 47
132	Chicago	27 Jun 43 26 Oct 43	23 Dec 43	30 Aug 46	*Zeus* (ARB 4) 3 Nov 43, str 1 Jun 73
133	Chicago	24 Jun 43 2 Nov 43	20 Nov 43	29 Aug 46	Bikini test Jul 46, tgt 28 May 48
134	Chicago	14 Jun 43 9 Nov 43	7 Dec 43	17 Feb 46	Str 31 Oct 47
135	Chicago	8 Jul 43 16 Nov 43	25 Apr 44	29 Apr 46	To *Orestes* (AGP 10) 3 Nov 43, str 23 Apr 47, to ROC Mar 48
136	Chicago	19 Jun 43 10 Dec 43	9 May 44	Jan 47	To *Egeria* (ARL 8) 3 Nov 43, str 1 Oct 77
137	Amer Bridge	23 Oct 43 19 Dec 43	26 Jan 44	20 Nov 45	Str 5 Dec 45
138	Amer Bridge	27 Oct 43 30 Dec 43	5 Feb 44	20 Nov 45	Str 5 Dec 45, sold for merch Jun 47
139	Amer Bridge	3 Nov 43 12 Jan 44	14 Feb 44	25 Mar 45	Str 8 May 46, sold for merch Apr 47
140	Amer Bridge	10 Nov 43 8 Jan 44	9 Feb 44	5 Jan 46	Str 12 Mar 46
141	Amer Bridge	24 Nov 43 16 Jan 44	16 Feb 44	18 Dec 45	Str 7 Feb 46
142–156					Canceled 16 Sep 42
157	Missouri	25 Jun 42 31 Oct 42	10 Feb 43		To UK 9 Dec 44, ret 11 Apr 46, str 5 Jun 46
158	Missouri	11 Jul 42 16 Nov 42	10 Feb 43		WL 11 Jul 43
159	Missouri	19 Jul 42 21 Nov 42	13 Feb 43		To UK 3 Mar 43, ret 23 Apr 46, str 19 Jun 46, sold Apr 48 for merch
160	Missouri	21 Jul 42 30 Nov 42	18 Feb 43		To UK 4 Mar 43, ret 1 Jun 46, str 3 July 46, sold Dec 47

Appendix C

No.	Name/Builder	LD/Launch	Comm	Decomm	Notes and Fate
161	Missouri	24 Jul 42 7 Dec 42	28 Feb 43		To UK 15 Mar 43, ret 5 Jan 46, str 20 Mar 46, sold May 48
162	Missouri	24 Jul 42 3 Feb 43	15 Mar 43		To UK 22 Mar 43, ret 1 Feb 46, str 19 Jun 46, sold Oct 47
163	Missouri	10 Aug 42 4 Feb 43	24 Mar 43		To UK 29 Mar 43, ret 29 Nov 46, str 1 Aug 47, sold Jul 47 for merch
164	Missouri	13 Aug 42 5 Feb 43	30 Mar 43		To UK 5 Apr 43, ret 29 Nov 46, str 1 Aug 47, sold Oct 47 for merch
165	Missouri	7 Sep 42 2 Feb 43	3 Apr 43		To UK 6 Apr 43, ret 20 Mar 46, str 5 Jun 46
166	Missouri	7 Sep 42 1 Feb 43	22 Apr 43	3 May 46	Str 19 Jun 46
167	Missouri	19 Sep 42 25 Feb 43	27 Apr 43		CTL due to war damage 25 Sep 43, str 6 Dec 43, used as ammo storage hulk until 48
168	Missouri	26 Sep 42 25 Feb 43	3 May 43	14 Mar 46	Str 12 Apr 46
169	Missouri	1 Oct 42 26 Feb 43	22 May 43	12 Apr 46	Str 19 Jun 46
170	Missouri	9 Oct 42 27 Feb 43	31 May 43	6 Apr 46	Str 3 Jul 46
171	Missouri	20 Oct 42 28 Feb 43	5 Jun 43	21 May 46	Str 3 Jul 46
172	Missouri	24 Dec 42 23 May 43	11 Jun 43	8 Jun 46	Str 9 Jun 46, sold 5 Nov 47 for merch, taken into PRC navy
173	Missouri	24 Dec 42 24 Apr 43	18 Jun 43		To UK 24 Dec 44, ret 23 Apr 46, str 19 Jun 46
174	Missouri	1 Jan 43 21 Apr 43	15 Jun 43	21 Dec 45	Str 21 Jan 46, sold 30 Jan 47 for merch
175	Missouri	6 Jan 43 18 Apr 43	19 May 43	1 Mar 46	Str 8 May 46
176	Missouri	18 Jan 43 15 Apr 43	12 May 43	6 Jan 46	SCAJAP (Q002), to MSTS 31 Mar 52, str 1 Nov 73
177	Missouri	5 Feb 43 16 May 43	22 Jun 43	11 Feb 46	Str 12 Apr 46, to France 13 Mar 47
178	Missouri	6 Feb 43 23 May 43	21 Jun 43		To UK 24 Dec 44, ret 12 Dec 46, to Egypt Nov 46, used as blockship in Suez Canal 1 Nov 56
179	Missouri	7 Feb 43 30 May 43	3 Jul 43	18 Jul 44	Accidental loss 21 May 44 but raised, tgt Nov 45
180	Missouri	8 Feb 43 3 Jun 43	29 Jun 43		To UK 10 Jul 43, ret 17 Dec 45, str 21 Jan 46, sold Mar 48 for merch
181	Jeffersonville	7 Apr 43 3 Jul 43	21 Aug 43	4 Mar 46	Str 12 Apr 46
182–196					Canceled 16 Sep 42
197	Chicago	15 Jun 42 13 Dec 42	5 Feb 43	5 Apr 46	Str 5 Jun 46
198	Chicago	22 Jun 42 17 Jan 43	15 Feb 43		To UK 6 Mar 43, ret 23 Jan 46, str 20 Mar 46
199	Chicago	27 Jun 42 7 Feb 43	1 Mar 43		To UK 19 Mar 43, CTL 27 Mar 46 (mined)
200	Chicago	2 Jul 42 20 Feb 43	16 Mar 43		To UK 25 Mar 43, ret 27 Feb 46, str 17 Apr 46, later merch
201	Chicago	13 Jul 42 2 Mar 43	2 Apr 43	2 Apr 46	*Pontus* (AGP 20) 15 Aug 44, str 1 May 46
202	Chicago	15 Jul 42 16 Mar 43	9 Apr 43	11 Apr 46	Str 28 Aug 46
203	Chicago	2 Jul 42 25 Mar 43	22 Apr 43		Grounded, CTL 1 Oct 43

No.	Name/Builder	LD/Launch	Comm	Decomm	Notes and Fate
204	Chicago	24 Jul 42 3 Apr 43	27 Apr 43	23 Feb 46	Str 8 Oct 46
205	Chicago	5 Aug 42 13 Apr 43	15 May 43	2 Apr 46	LSTH 15 Sep 45, str 5 Jun 46
206	Chicago	7 Aug 42 21 Apr 43	7 Jun 43	6 May 46	Str 5 Jun 46
207	Chicago	7 Sep 42 29 Apr 43	9 Jun 43	20 Mar 46	Str 17 Apr 46
208	Chicago	7 Sep 42 11 May 43	8 Jun 43	12 Jun 46	Str 3 Jul 46
209	*Bamberg County* Chicago	7 Sep 42 29 May 43	10 Jun 43	27 Jun 46	To MSTS Jun 46, str 1 Nov 58
210	Chicago	7 Sep 42 1 Jun 43	6 Jul 43	8 Dec 45	Str 3 Jan 46
211	Chicago	7 Sep 42 5 Jun 43	6 Jul 43	20 Nov 45	Str 5 Dec 45
212	Chicago	7 Dec 42 12 Jun 43	6 Jul 43	15 Nov 45	Str 28 Nov 45, sold Jul 47 for merch
213	Chicago	21 Dec 42 16 Jun 43	7 Jul 43	11 Mar 46	LSTH 15 Sep 45, SCAJAP on decomm (Q095), str 5 Mar 47, transferred 26 Jun 47 to U.S. mil gov Korea
214	Chicago	29 Dec 42 22 Jun 43	7 Jul 43		To UK 24 Jul 43, ret 26 Jan 46, str 12 Apr 46
215	Chicago	8 Jan 43 26 Jun 43	12 Jul 43		To UK 19 Jul 43, ret 27 Jul 46, str 29 Oct 46
216	Chicago	23 Jan 43 4 Jul 43	23 Jul 43		To UK 4 Aug 43, WL 7 Jul 44
217	Chicago	2 Feb 43 13 Jul 43	30 Jul 43		To UK 5 Aug 43, ret 12 Feb 46, str 5 Jun 46
218	Chicago	11 Feb 43 20 Jul 43	12 Aug 43	19 Jan 46	SCAJAP Jan 46 (Q020), ret 28 Jan 50, in reserve, to Korea 3 May 55
219	Chicago	18 Feb 43 27 Jul 43	19 Aug 43	29 Nov 48	Str 22 Dec 48
220	Chicago	4 Mar 43 3 Aug 43	26 Aug 43	Mar 46	Bikini tgt, scuttled 12 May 48
221	Chicago	9 Mar 43 7 Aug 43	2 Sep 43	6 May 46	Str 3 Jul 46
222	Chicago	16 Mar 43 17 Aug 43	10 Sep 43	15 Jul 72	LSTH 15 Sep 45, MSTS 31 Mar 52–15 Jul 72, to Philippines 15 Jul 72
223	Chicago	31 Mar 43 24 Aug 43	17 Sep 43		LSTH 15 Sep 45, to State Dept for disposal 13 Mar 47, to France 51
224	Chicago	2 Apr 43 31 Aug 43	23 Sep 43	22 Mar 46	Str 17 Apr 46
225	Chicago	14 Apr 43 4 Sep 43	2 Oct 43	30 Jul 46	Str 28 Aug 46
226	Chicago	16 Apr 43 14 Sep 43	8 Oct 43	8 Jun 46	Str 19 Jun 46
227	Chicago	10 May 43 21 Sep 43	16 Oct 43	22 Jan 46	SCAJAP (Q025) to 6 Jun 50, to Korea 27 Mar 55
228	Chicago	20 May 43 25 Sep 43	25 Oct 43		Grounded, CTL 19 Jan 44
229	Chicago	27 May 43 5 Oct 43	3 Nov 43	12 Feb 46	SCAJAP (Q054), str 31 Oct 47
230	Chicago	10 Jun 43 12 Oct 43	3 Nov 43	4 Mar 46	SCAJAP (Q082), to MSTS 31 Mar 52, to Philippines 13 Sep 76
231	Chicago	3 Jun 43 19 Oct 43	1 Nov 43 1 Jun 51	13 Sep 46 5 Apr 56	To *Atlas* (ARL 7) 3 Nov 43, str 1 Jun 72
232–236					Canceled 16 Sep 42

Appendix C

No.	Name/Builder	LD/Launch	Comm	Decomm	Notes and Fate
237	Missouri	9 Feb 43 8 Jun 43	30 Jun 43		To UK 12 Jul 43, ret 11 Feb 46, str 26 Feb 46, sold Nov 47 for merch, taken into PRC navy
238	Missouri	5 Mar 43 13 Jun 43	9 Jul 43		To UK 16 Jul 43, ret 13 Feb 46, str 12 Mar 46
239	Missouri	6 Mar 43 18 Jun 43	13 Jul 43		To UK 19 Jul 43, ret 5 Feb 46, str 5 Jun 46, sold Apr 48 for merch
240	Missouri	7 Mar 43 25 Jun 43	27 Jul 43	3 May 46	Str 23 Jun 47
241	Missouri	8 Mar 43 29 Jun 43	31 Jul 43	7 Mar 46	Str 5 Jun 46
242	Missouri	8 Mar 43 3 Jul 43	5 Aug 43	9 Feb 46	LSTH 15 Sep 45, SCAJAP (Q051), str 31 Oct 47
243	Missouri	26 Apr 43 9 Jul 43	9 Aug 43	9 Jan 46	LSTH 15 Sep 45, SCAJAP (Q006), str 17 Jul 47
244	Missouri	1 May 43 14 Jul 43	13 Aug 43	28 Mar 46	Str 3 Jul 46
245	Missouri	7 May 43 17 Jul 43	22 Aug 43	1 Apr 46	Str 8 May 46
246	Missouri	12 May 43 22 Jul 43	23 Aug 43	14 Feb 46	SCAJAP (Q061), to army 26 Jun 47, str 12 Mar 48
247	Missouri	17 May 43 30 Jul 43	26 Aug 43	27 Jun 46	LSTH 15 Sep 45, str 15 Aug 46
248–260					Canceled 16 Sep 42
261	Amer Bridge	7 Sep 42 23 Jan 43	22 May 43	22 Feb 46	Str 28 Mar 46
262	Amer Bridge	7 Sep 42 13 Feb 43	15 Jun 43	14 Jan 46	Str 19 Jun 46
263	*Benton County* Amer Bridge	7 Sep 42 27 Feb 43	30 Jun 43	29 May 46	Str 1 Nov 58
264	Amer Bridge	21 Sep 42 13 Mar 43	16 Jul 43	11 Jan 46	Str 19 Jun 46
265	Amer Bridge	31 Oct 42 24 Apr 43	27 Jul 43	11 Dec 45	Str 3 Jan 46
266	*Benzie County* Amer Bridge	1 Nov 42 16 May 43	4 Aug 43	25 Jun 47	Str 1 Nov 58
267	Amer Bridge	21 Nov 42 6 Jun 43	9 Aug 43	25 Jun 46	Str 31 Jul 46
268	Amer Bridge	26 Nov 42 18 Jun 43	19 Aug 43	16 Feb 46	LSTH 15 Sep 45, SCAJAP (Q062), str 31 Oct 47
269	Amer Bridge	28 Dec 42 4 Jul 43	27 Aug 43	7 Feb 46	SCAJAP (Q047), str 23 Dec 47
270	Amer Bridge	13 Jan 43 18 Jul 43	8 Sep 43	N/A	Sold 12 May 50
271	Amer Bridge	21 Jan 43 25 Jul 43	1 Sep 43	22 Apr 46	Str 5 Jun 46
272	Amer Bridge	9 Feb 43 1 Aug 43	17 Sep 43	16 Aug 46	Str 25 Sep 46
273	Amer Bridge	24 Feb 43 8 Aug 43	24 Sep 43	12 Aug 46	Str 8 Oct 46
274	Amer Bridge	11 Mar 43 15 Aug 43	28 Sep 43	6 May 46	Str 23 Jun 47, merch
275	Amer Bridge	22 Apr 43 22 Aug 43	5 Oct 43	16 Aug 46	Str 2 Sep 46
276	Amer Bridge	10 May 43 29 Aug 43	11 Oct 43	Feb 46	LSTH 15 Sep 45, SCAJAP (Q079), to MSTS 31 Mar 52, str 10 Jun 73
277	Amer Bridge	31 May 43 5 Sep 43	24 Oct 43	12 Feb 46	SCAJAP 20 May 49–31 Mar 52 (Q055), to MSTS, str 1 Feb 73, to Chile 2 Feb 73

AMPHIBIOUS SHIPS 549

No.	Name/Builder	LD/Launch	Comm	Decomm	Notes and Fate
278	Amer Bridge	16 Jun 43 12 Sep 43	22 Oct 43	22 Jan 45	To *Seaward* (IX 209), comm 14 Feb 45, surplus, destroyed 16 Oct 46; was barracks and post office at Ulithi
279	*Berkeley County* Amer Bridge	2 Jul 43 19 Sep 43	25 Oct 43	14 Jun 55	To ROC 30 Jun 55
280	Amer Bridge	16 Jul 43 26 Sep 43	2 Nov 43		To UK 26 Oct 44, ret 11 Apr 46, str 5 Jun 46
281	Amer Bridge	25 Jun 43 30 Sep 43	8 Nov 43	9 Mar 46	SCAJAP 20 May 49 (Q092), MSTS 31 Mar 52, str 19 May 54
282	Amer Bridge	12 Jul 43 3 Oct 43	12 Nov 43		WL 15 Aug 44
283	Amer Bridge	2 Aug 43 10 Oct 43	18 Nov 43	13 Jun 46	Str 22 Jan 47, sold Mar 47 for merch, but resold to Peru as LST 21 Dec 51
284	Amer Bridge	9 Aug 43 17 Oct 43	25 Nov 43	13 Mar 46	Str 19 Jun 46
285	Amer Bridge	16 Aug 43 24 Oct 43	13 Dec 43	27 Jun 47	Str 1 Aug 47
286	Amer Bridge	23 Aug 43 27 Oct 43	11 Dec 43	26 Mar 46	Str 8 May 46
287	Amer Bridge	30 Aug 43 31 Oct 43	15 Dec 43		MSTS 29 May 51, to Philippines 13 Sep 76
288	*Berkshire County* Amer Bridge	6 Sep 43 7 Nov 43	20 Dec 43	6 Mar 46	SCAJAP 20 May 49–14 Jun 50 (Q085), to Korea 5 Mar 56
289	Amer Bridge	14 Sep 43 21 Nov 43	31 Dec 43		To UK 9 Dec 44, ret 12 Oct 46, str 15 Oct 46, to Netherlands Jan 47, merch 50
290	Amer Bridge	22 Sep 43 5 Dec 43	10 Jan 44	15 Nov 45	Str 28 Nov 45, sold Dec 46 for merch
291	Amer Bridge	25 Sep 43 14 Nov 43	22 Dec 43	18 Jun 47	Str 19 May 54, tgt Jul 54
292	Amer Bridge	30 Sep 43 28 Nov 43	5 Jan 44	25 Jan 46	Str 12 Apr 46
293	Amer Bridge	5 Oct 43 12 Dec 43	17 Jan 44	3 Dec 45	Str 19 Dec 45
294	Amer Bridge	12 Oct 43 15 Dec 43	20 Jan 44	18 Dec 45	Str 8 Jan 46
295	Amer Bridge	19 Oct 43 24 Dec 43	7 Feb 44	28 Dec 45	Str 12 Apr 46, sold Sep 47 for merch
296–300					Canceled 16 Sep 42
301	BOSNY	26 Jun 42 15 Sep 42	1 Nov 42		To UK 6 Nov 42, ret 20 Mar 46, str Dec 47
302	BOSNY	28 Jun 42 15 Sep 42	10 Nov 42		To UK 14 Nov 42, ret 5 Jan 46, str 20 Mar 46
303	BOSNY	3 Jul 42 21 Sep 42	20 Nov 42		To UK 21 Nov 42, ret 1 Jun 46, str 3 Jul 46
304	BOSNY	3 Jul 42 21 Sep 42	29 Nov 42		To UK 30 Nov 42, ret 29 Nov 46, str 1 Aug 47, sold Oct 47 for merch
305	BOSNY	24 Jul 42 10 Oct 42	6 Dec 42		To UK 7 Dec 42, WL 20 Feb 44
306	*Bernalillo County* BOSNY	24 Jul 42 10 Oct 42	11 Dec 42	13 Jun 46	Str 1 Feb 59
307	BOSNY	15 Sep 42 9 Nov 42	23 Dec 42	13 Jun 46	Str 31 Jul 46
308	BOSNY	15 Sep 42 9 Nov 42	2 Jan 43	17 Oct 46	To State Dept for disposal 5 Dec 47
309	BOSNY	22 Sep 42 23 Nov 42	11 Jan 43	19 Jun 46	Str 23 Jun 47, sold Jun 48 for merch

Appendix C

No.	Name/Builder	LD/Launch	Comm	Decomm	Notes and Fate
310	BOSNY	22 Sep 42 23 Nov 42	20 Jan 43	16 May 45	Conversion to *Aeolus* (ARL 42) canceled 11 Aug 45, str 12 Mar 46, sold Jan 47 for merch
311	NYNY	7 Sep 42 30 Dec 42	11 Jan 43		To UK 20 Nov 44, ret 11 Apr 46, str 5 Jun 46
312	NYNY	7 Sep 42 30 Dec 42	9 Jan 43	12 Jul 46	Str 15 Aug 46
313	NYNY	7 Sep 42 30 Dec 42	13 Jan 43		WL 10 Jul 43
314	NYNY	7 Sep 42 30 Dec 42	15 Jan 43		WL 9 Jun 44
315	NYNY	15 Oct 42 3 Feb 43	3 Feb 43		To UK 9 Dec 44, ret 16 Mar 46, str 26 Feb 46
316	NYNY	15 Oct 42 28 Jan 43	3 Feb 43	24 May 45	Conversion to *Cerberus* (ARL 43) canceled 11 Aug 45, str 12 Mar 46, sold Dec 46 for merch
317	NYNY	15 Oct 42 28 Jan 43	6 Feb 43	18 May 45	Conversion to *Consus* (ARL 44) canceled 11 Aug 45, str 12 Mar 46
318	NYNY	15 Oct 42 28 Jan 43	8 Feb 43		WL 9 Aug 43
319	PHNY	10 Aug 42 5 Nov 42			To UK 15 Dec 42, ret 17 Dec 45, str 21 Jan 46, sold Mar 48 for merch
320	PHNY	10 Aug 42 5 Nov 42			To UK 31 Dec 42, ret 23 Apr 46, str 19 Jun 46
321	PHNY	10 Aug 42 5 Nov 42			To UK 31 Dec 42, ret 11 Apr 46, str 10 Jun 47
322	PHNY	10 Aug 42 5 Nov 42			To UK 9 Jan 43, ret 10 Jul 46, str 29 Oct 46, to Greece 6 Jan 47
323	PHNY	10 Aug 42 5 Nov 42			To UK 18 Jan 43, ret 26 Jan 46, str 19 Jun 46
324	PHNY	10 Aug 42 5 Nov 42			To UK 23 Jan 43, ret 1 Jun 46, str 3 Jul 46
325	PHNY	10 Aug 42 27 Oct 42	1 Feb 43	2 Jul 46	Str 1 Sep 61, to Greece 1 Sep 64; ship was sailed back to the United States winter 2000–2001 by a group of U.S. Navy veterans, an epic undertaking; arrived at Mobile, Alabama, on 10 Jan 2001, having been transferred by the Greek navy on 13 Nov 2000
326	PHNY	12 Nov 42 11 Feb 43	26 Feb 43		To UK 9 Dec 44, ret 25 Feb 46, str 26 Feb, to France 5 Apr 46
327	PHNY	12 Nov 42 11 Feb 43	5 Mar 43	19 Nov 45	CTL mine damage 27 Aug 44, str 5 Dec 45
328	PHNY	12 Nov 42 11 Feb 43	22 May 43	Jan 47	*Oceanus* (ARB 2) 25 Jan 43, str 1 Jul 61
329	PHNY	12 Nov 42 1 Feb 43	18 May 43	15 Jan 47	*Aristaeus* (ARB 1) 25 Jan 43, str 1 Jul 61
330	PHNY	12 Nov 42 11 Feb 43	12 Jun 43	18 Apr 46	*Portunus* (AGP 4) 25 Jan 43, str 13 Nov 46
331	PHNY	12 Nov 42 11 Feb 43	11 Mar 43		To UK 20 Nov 44, ret 16 Mar 46, str 26 Feb 46
332	PHNY	20 Oct 42 24 Dec 42	6 Feb 43	22 May 45	Str 12 Mar 46, sold Oct 46 for merch
333	NORNY	17 July 42 15 Oct 42	20 Nov 42		WL(T) 22 Jun 43
334	NORNY	17 Jul 42 15 Oct 42	29 Nov 42	24 Apr 46	Str 5 Jun 46

No.	Name/Builder	LD/Launch	Comm	Decomm	Notes and Fate
335	NORNY	17 Jul 42 15 Oct 42	6 Dec 42	22 Dec 45	Str 8 Jan 46
336	NORNY	17 Jul 42 15 Oct 42	11 Dec 42		To UK 27 Nov 44, ret 7 Mar 46, str 5 Jun 46
337	NORNY	17 Jul 42 8 Nov 42	16 Dec 42		To UK 2 Dec 44, ret 16 Mar 46, str 17 Apr 46
338	NORNY	17 Jul 42 8 Nov 42	20 Dec 42	6 May 46	Str 23 Jun 47, sold Dec 47 for merch
339	NORNY	17 Jul 42 8 Nov 42	23 Dec 42	13 May 46	Str 23 Jun 47
340	NORNY	17 Jul 42 8 Nov 42	26 Dec 42	24 Oct 44	*Spark* (IX 196) 20 Oct 44, str 1 Sep 45, barracks ship
341	NORNY	21 Aug 42 8 Nov 42	28 Dec 42	14 Mar 46	Str 12 Apr 46, sold for merch
342	NORNY	21 Aug 42 8 Nov 42	31 Dec 42		WL(T) 18 Jul 43
343	NORNY	18 Oct 42 15 Dec 42	9 Jan 43	27 Jan 46	SCAJAP (Q033), to U.S. mil gov Korea 21 Feb 47, str 5 Mar 47
344	*Blanco County* NORNY	18 Oct 42 15 Dec 42	14 Jan 43 66	N/A 3 Oct 69	Str 15 Sep 74
345	NORNY	17 Oct 42 15 Dec 42	21 Jan 43	5 Dec 45	Str 3 Jan 46
346	NORNY	17 Oct 42 15 Dec 42	25 Jan 43		To UK 20 Nov 44, ret 2 May 46, str 19 Jun 46
347	NORNY	10 Nov 42 7 Feb 43	7 Feb 43		To UK 19 Dec 44, ret Jan 48, to France 23 Jan 48
348	NORNY	10 Nov 42 7 Feb 43	9 Feb 43		WL 20 Feb 44
349	NORNY	10 Nov 42 7 Feb 43	11 Feb 43		Grounded, sunk 26 Feb 44
350	NORNY	10 Nov 42 7 Feb 43	13 Feb 43	26 May 45	*Chandra* (ARL 46) 25 May 45, but conversion canceled 11 Aug 45, str 12 Mar 46, sold Dec 46 for merch
351	NORNY	9 Nov 42 7 Feb 43	24 Feb 43		To UK 12 Dec 44, ret 10 Dec 46, to Netherlands between 30 Dec 46 and 17 Jun 47
352	NORNY	9 Nov 42 7 Feb 43	26 Feb 43		To UK 24 Dec 44, ret 2 Aug 46, to Greece between 21 Nov 46 and 6 Jan 47
353	CHARNY	15 Jul 42 12 Oct 42	27 Nov 42		Internal explosion 21 May 44
354	CHARNY	15 Jul 42 13 Oct 42	27 Nov 42	30 Apr 46	Str 16 Dec 47
355	CHARNY	7 Sep 42 16 Nov 42	22 Dec 42	6 Mar 46	SCAJAP (Q084), str 31 Oct 47, sold for BU 48 but later reported in PRC navy
356	*Bledsoe County* CHARNY	7 Sep 42 16 Nov 42	22 Dec 42	21 Sep 45	Str 1 Sep 60
357	CHARNY	24 Oct 42 14 Dec 42	8 Feb 43	8 Jun 46	Str 31 Jul 46
358	CHARNY	24 Oct 42 15 Dec 42	8 Feb 43		To UK 24 Dec 44, ret 27 Feb 46, str 15 Aug 46
359	CHARNY	21 Nov 42 11 Jan 43	9 Feb 43		WL(T) 20 Dec 44
360	CHARNY	21 Nov 42 11 Jan 43	9 Feb 43		To UK 29 Nov 44, ret 10 Jun 46, str 15 Aug 46
361	Beth-Q	10 Aug 42 10 Oct 42			To UK 16 Nov 42, ret 7 Mar 46, str 5 Jun 46
362	Beth-Q	10 Aug 42 10 Oct 42			To UK 23 Nov 42, WL(T) 2 Mar 44

APPENDIX C

No.	Name/Builder	LD/Launch	Comm	Decomm	Notes and Fate
363	Beth-Q	2 Sep 42 26 Oct 42			To UK 30 Nov 42, ret 26 Jan 46, str 12 Apr 46
364	Beth-Q	3 Sep 42 26 Oct 42			To UK 7 Dec 42, WL Feb 45
365	Beth-Q	14 Oct 42 11 Nov 42			To UK 14 Dec 42, str 15 Oct 46, ret 10 Dec 46, merch
366	Beth-Q	1 Oct 42 11 Nov 42			To UK 21 Dec 42, ret 26 Jan 46, str 21 Jan 46
367	Beth-Q	13 Oct 42 24 Nov 42			To UK 29 Dec 42, ret 17 Dec 45, str 21 Jan 46
368	Beth-Q	13 Oct 42 24 Nov 42	16 Mar 43	16 Mar 46	To UK 4 Jan 43 but ret for U.S. service, str 17 Apr 46
369	Beth-Q	13 Oct 42 24 Nov 42	8 Jan 43		To UK 29 Nov 44, ret 29 Nov 46, str 1 Aug 47
370	Beth-Q	31 Oct 42 12 Dec 42	13 Jan 43	7 Jan 46	Str 12 Apr 46, merch
371	Beth-Q	29 Oct 42 12 Dec 42	16 Jan 43		To UK 17 Nov 44, ret and str 26 Feb 46
372	Beth-Q	14 Nov 42 19 Jan 43	23 Jan 43	9 Jul 46	Str 15 Aug 46
373	Beth-Q	14 Nov 42 19 Jan 43	27 Jan 43		To UK 9 Dec 44, ret 16 Mar 46, str 26 Feb 46
374	Beth-Q	12 Nov 42 19 Jan 43	29 Jan 43	29 May 45	Conversion to *Minerva* (ARL 47) canceled 11 Sep 45, str 12 Mar 46
375	Beth-Q	25 Nov 42 28 Jan 43	2 Feb 43	18 Jul 46	Str 10 Jun 47
376	Beth-Q	25 Nov 42 1 Feb 43	5 Feb 43		WL(T) 9 Jun 44
377	Beth-Q	28 Nov 42 1 Feb 43	8 Feb 43	7 Jun 46	Str 31 Jul 46
378	Beth-Q	12 Dec 42 6 Feb 43	10 Feb 43	20 Feb 46	SCAJAP (Q069), str 5 Mar 47, to U.S. mil gov Korea
379	Beth-Q	12 Dec 42 6 Feb 43	12 Feb 43	28 Feb 46	Str 20 Mar 46
380	Beth-Q	10 Dec 42 10 Feb 43	15 Feb 43		To UK 20 Nov 44, ret 11 Apr 46, sold 7 Jun 46 to U.S. mil gov Korea, str 19 Jul 46, ROC merch service, naval service 55
381	Beth-Q	10 Dec 42 10 Feb 43	15 Feb 43		To UK 19 Dec 44, ret 10 Jun 46, str 19 Jul 46
382	Beth-Q	10 Dec 42 3 Feb 43	18 Feb 43		To UK 29 Nov 44, to France 23 Jan 48, to France 21 Mar 49
383	NN Ship	16 Jun 42 28 Sep 42	27 Oct 42		To UK 20 Nov 44, to NEI Maritime Customs 10 Jun 46, str 3 Jul 46
384	NN Ship	16 Jun 42 28 Sep 42	2 Nov 42	22 Apr 46	Str 5 Jun 46
385	NN Ship	19 Jun 42 28 Sep 42	6 Nov 42		To UK 29 Nov 44, str 26 Feb 46
386	NN Ship	9 Jun 42 28 Sep 42	10 Nov 42		To UK 9 Dec 44, str 15 Oct 46, ret 10 Dec 46, merch
387	NN Ship	20 Jun 42 28 Sep 42	17 Nov 42	2 May 46	Saw no action because torpedoed off North Africa 22 Jun 43, later repaired, str 19 Jul 46
388	NN Ship	20 Jun 42 28 Sep 42	20 Nov 42	1 Feb 47	Str 25 Feb 47, to U.S. mil gov Korea Nov 47
389	*Boone County* NN Ship	20 Jun 42 28 Sep 42	24 Nov 42	12 Mar 46	Str 1 Jun 59, to Greece May 60

AMPHIBIOUS SHIPS 553

No.	Name/Builder	LD/Launch	Comm	Decomm	Notes and Fate
390	NN Ship	20 Jun 42 15 Oct 42	28 Nov 42	12 Mar 46	SCAJAP (Q096), str 29 Sep 47
391	*Bowman County* NN Ship	14 Jul 42 28 Oct 42	3 Dec 42		To Greece May 60, to be U.S. museum ship
392	NN Ship	14 Jul 42 28 Oct 42	7 Dec 42	12 Apr 46	Str 19 Jun 46
393	NN Ship	27 Jul 42 11 Nov 42	11 Dec 42	1 Mar 46	Str 14 Mar 47
394	NN Ship	27 Jul 42 11 Nov 42	15 Dec 42		To UK 24 Dec 44, ret 12 May 46, str 19 Jun 46
395	NN Ship	28 Sep 42 23 Nov 42	19 Dec 42	19 Apr 46	Str 1 May 46
396	NN Ship	28 Sep 42 23 Nov 42	28 Dec 42		WL(A) 18 Aug 43
397	NN Ship	28 Sep 42 23 Nov 42	28 Dec 42	26 Apr 46	Str 5 Jun 46
398	NN Ship	28 Sep 42 23 Nov 42	2 Jan 43	27 Feb 46	SCAJAP (Q029), str 28 Aug 46
399	NN Ship	28 Sep 42 23 Nov 42	4 Jan 43 31 Mar 52	3 Dec 45 1 Nov 73	SCAJAP (Q098), MSTS 52–73, str 1 Nov 73,
400	*Bradley County* NN Ship	28 Sep 42 23 Nov 42	7 Jan 43		To Taiwan Sep 58
401	Beth-F	17 Aug 42 16 Oct 42			To UK 30 Nov 42, ret 7 Mar 46, str 5 Jun 46
402	Beth-F	21 Aug 42 9 Oct 42			To UK 9 Dec 42, ret 24 Sep 46, str 10 Jun 47
403	Beth-F	23 Aug 42 24 Oct 42			To UK 8 Dec 42, ret 11 Apr 46, str 5 Jun 46
404	Beth-F	27 Aug 42 28 Oct 42			To UK 16 Dec 42, ret 21 Oct 45, sold
405	Beth-F	30 Aug 42 31 Oct 42			To UK 28 Dec 42, lost in collision 25 Oct 45
406	Beth-F	1 Sep 42 28 Oct 42			To UK 26 Dec 42, ret 11 Apr 46, str 10 Jun 47, to ROC after sale
407	Beth-F	2 Sep 42 5 Nov 42			To UK 31 Dec 42, CTL 24 Apr 44
408	Beth-F	9 Sep 42 31 Oct 42			To UK 23 Dec 42, ret 4 May 46, str 19 Jun 46
409	Beth-F	9 Sep 42 15 Nov 42			To UK 6 Jan 43, ret 2 Jul 46, str 29 Oct 46, to Greece between 21 Nov 46 and 6 Jan 47
410	Beth-F	13 Sep 42 15 Nov 42			To UK 14 Jan 43, ret 16 Apr 47, str 26 Feb 46
411	Beth-F	21 Sep 42 9 Nov 42			To UK 31 Dec 42, WL 1 Jan 44
412	Beth-F	24 Sep 42 16 Nov 42			To UK 26 Jan 43, ret 23 Jan 46, str 20 Mar 46
413	Beth-F	10 Oct 42 10 Nov 42			To UK 5 Jan 43, ret 11 Apr 46, str 10 Jun 47
414	Beth-F	18 Oct 42 21 Nov 42			To UK 19 Jan 43, WL(T) 15 Aug 43, beached and cannibalized; listed in U.S. records as sold to Netherlands navy Jun 46
415	Beth-F	29 Oct 42 21 Nov 42			To UK 19 Jan 43, WL(T) 16 Jan 45
416	Beth-F	25 Oct 42 30 Nov 42			To UK 3 Feb 43, ret 12 Feb 46, str 5 Jun 46, merch

554 APPENDIX C

No.	Name/Builder	LD/Launch	Comm	Decomm	Notes and Fate
417	Beth-F	29 Oct 42 24 Nov 42			To UK 29 Jan 43, ret 31 May 46, str 3 Jul 46
418	Beth-F	2 Nov 42 30 Nov 42			To UK 29 Jan 43, WL 20 Apr 44
419	Beth-F	1 Nov 42 30 Nov 42			To UK 8 Feb 43, ret 4 May 46, str 8 Jul 46
420	Beth-F	6 Nov 42 5 Dec 42			To UK 15 Feb 43, WL 28 Nov 44
421	Beth-F	11 Nov 42 5 Dec 42			To UK 26 Jan 43, ret 29 Nov 46, str 1 Aug 47, merch
422	Beth-F	12 Nov 42 10 Dec 44			To UK 4 Feb 43, WL(M) 26 Jan 44
423	Beth-F	1 Dec 42 14 Jan 43			To UK 24 Feb 43, ret 10 Jun 46, str 19 Jul 46
424	Beth-F	17 Nov 42 12 Dec 42			To UK 1 Feb 43, ret 7 Jan 46, str 21 May 46
425	Beth-F	16 Nov 42 12 Dec 42			To UK 10 Feb 43, ret 30 Aug 46, str 10 Jun 47
426	Beth-F	16 Nov 42 11 Dec 42			To UK 16 Feb 43, ret 23 Apr 46, str 19 Jun 46
427	Beth-F	22 Nov 42 19 Dec 42			To UK 16 Feb 43, ret 11 Apr 46, str 10 Jun 47
428	Beth-F	22 Nov 42 22 Dec 42			To UK 9 Feb 43, ret 10 Jun 46, str 19 Jul 46
429	Beth-F	16 Nov 42 11 Jan 43			To UK 20 Feb 43, WL 3 Jul 43
430	Beth-F	25 Nov 42 31 Dec 42			To UK 19 Feb 43, ret 26 Jan 46, str 8 May 46
431–445					Canceled 16 Sep 42
446	Kaiser-V	15 Jun 42 18 Sep 42	30 Nov 42	13 Jul 46	Str 8 Oct 46
447	Kaiser-V	10 Jul 42 22 Sep 42	13 Dec 42		WL(K) 7 Apr 45
448	Kaiser-V	10 Jul 42 26 Sep 42	23 Dec 42		WL 5 Oct 43
449	Kaiser-V	10 Jul 42 30 Sep 42	31 Dec 42	16 Mar 46	Str 28 Mar 46
450	Kaiser-V	10 Jul 42 4 Oct 42	6 Jan 43	8 Apr 46	Str 17 Apr 46
451	Kaiser-V	20 Jul 42 6 Oct 42	12 Jan 43	22 Jul 46	Str 25 Sep 46
452	Kaiser-V	20 Jul 42 10 Oct 42	16 Jan 43	12 Jun 46	Str 3 Jul 46
453	Kaiser-V	28 Jul 42 10 Oct 42	21 Jan 43	15 Jul 46	To *Remus* (ARL 40) 15 Aug 44, str 15 Aug 46
454	Kaiser-V	10 Jul 42 14 Oct 42	26 Jan 43	25 Mar 46	Str 1 May 46
455	Kaiser-V	3 Aug 42 17 Oct 42	30 Jan 43	19 Jul 46	To *Achilles* (ARL 41) 21 Aug 44, str 28 Aug 46, to ROC 8 Dec 47
456	Kaiser-V	10 Jul 42 20 Oct 42	3 Feb 43 31 Mar 52	46 15 Jun 73	SCAJAP (Q043), MSTS 52–73, str 15 Jun 73
457	Kaiser-V	3 Aug 42 23 Oct 42	6 Feb 43	15 Mar 46	SCAJAP (Q098), str 29 Sep 47
458	Kaiser-V	18 Sep 42 26 Oct 42	10 Feb 43	15 Apr 46	Str 3 Jul 46
459	Kaiser-V	22 Sep 42 29 Oct 42	13 Feb 43	12 Apr 46	Str 19 Jun 46

AMPHIBIOUS SHIPS

No.	Name/Builder	LD/Launch	Comm	Decomm	Notes and Fate
460	Kaiser-V	26 Sep 42 31 Oct 42	15 Feb 43		WL(K) 21 Dec 44
461	Kaiser-V	30 Sep 42 3 Nov 42	18 Feb 43	2 Sep 47	Str 16 Sep 47
462	Kaiser-V	4 Oct 42 6 Nov 42	21 Feb 43	21 Mar 46	Str 1 May 46
463	Kaiser-V	6 Oct 42 9 Nov 42	23 Feb 43	6 Jun 46	Str 19 Jun 46
464	Kaiser-V	10 Oct 42 12 Nov 42	25 Feb 43	16 Apr 46	LSTH 15 Sep 45, str 19 Jun 46
465	Kaiser-V	17 Dec 42 9 Jan 43	27 Feb 43	8 Mar 46	Str 12 Apr 46
466	Kaiser-V	14 Oct 42 18 Nov 42	1 Mar 43	8 Mar 46	Str 12 Apr 46
467	Kaiser-V	17 Oct 42 21 Nov 42	3 Mar 43	28 May 46	Str 5 Jun 46
468	Kaiser-V	20 Oct 42 24 Nov 42	5 Mar 43	12 Apr 46	Str 5 Jun 46
469	Kaiser-V	23 Oct 42 27 Nov 42	8 Mar 43	27 Mar 46	Str 1 May 46
470	Kaiser-V	26 Oct 42 30 Nov 42	9 Mar 43	4 Mar 46	Str 5 Jun 46
471	Kaiser-V	29 Oct 42 3 Dec 42	11 Mar 43	26 Feb 46	Str 12 Apr 46
472	Kaiser-V	31 Oct 42 7 Dec 42	13 Mar 43		WL(K) 21 Dec 44
473	Kaiser-V	10 Jul 42 9 Dec 42	16 Mar 43	18 Mar 46	Str 17 Apr 46
474	Kaiser-V	10 Jul 42 12 Dec 42	19 Mar 43	22 Mar 46	Str 17 Apr 46
475	Kaiser-V	10 Jul 42 16 Nov 42	20 Mar 43	24 Apr 46	Str 5 Jun 46
476	Kaiser-R	5 Aug 42 10 Oct 42	4 Apr 43	12 Feb 46	Str 31 Oct 47
477	Kaiser-R	12 Aug 42 29 Oct 42	19 Feb 43	46	LSTH 15 Sep 45, SCAJAP (Q053), str 28 Aug 47
478	Kaiser-R	17 Aug 42 7 Nov 42	13 Mar 43	23 Mar 46	SCAJAP (Q091), str 28 Aug 47
479	Kaiser-R	25 Aug 42 4 Oct 42	19 Apr 43	28 Feb 46	SCAJAP (Q100), str 28 Mar 46,
480	Kaiser-R	31 Aug 42 29 Oct 42	3 May 43		Explosion 21 May 44
481	Kaiser-R	4 Sep 42 2 Dec 42	15 May 43	28 Feb 46	Str 12 Apr 46
482	*Branch County* Kaiser-R	14 Sep 42 17 Dec 42	20 Mar 43	23 Feb 46	LSTH 15 Sep 45, SCAJAP (Q072), str 11 Aug 55, tgt Mar 56
483	*Brewster County* Kaiser-R	21 Sep 42 30 Dec 42	3 May 43	10 Feb 46	SCAJAP (Q050), str 11 Aug 55, tgt
484	Kaiser-R	28 Sep 42 2 Jan 43	23 Apr 43	27 Jul 46	Str 28 Aug 46
485	Kaiser-R	17 Dec 42 9 Jan 43	19 May 43	30 Jul 46	LSTH 15 Sep 45, str 28 Aug 46
486	Kaiser-R	31 Dec 42 16 Jan 43	29 May 43	13 Jan 46	SCAJAP (Q011) until destroyed 23 Jul 47
487	Kaiser-R	2 Jan 43 23 Jan 43	27 Apr 43	15 Mar 46	Str 1 May 46
488	Kaiser-R	11 Jan 43 5 Mar 43	24 May 43	11 Jan 46	LSTH 15 Sep 45, SCAJAP (Q009), to MSTS 6 Mar 52, to Philippines 15 Jul 72

No.	Name/Builder	LD/Launch	Comm	Decomm	Notes and Fate
489	Kaiser-R	17 Jan 43 2 Apr 43	30 Jul 43	15 Nov 46	To *Amycus* (ARL 2) 13 Jan 43, str 1 Jun 70
490	Kaiser-R	24 Jan 43 3 Apr 43	20 Aug 43	15 Nov 46	To *Agenor* (ARL 3) 13 Jan 43, to France 2 Mar 51, to ROC 15 Sep 57
491	Missouri	29 Jul 43 23 Sep 43	3 Dec 43	12 Jan 46	SCAJAP (Q010), to MSTS 31 Mar 52, str Jun 75, to Philippines 13 Sep 76
492	Missouri	3 Aug 43 30 Sep 43	8 Dec 43	17 Jun 46	Str 23 Jun 47
493	Missouri	9 Aug 43 4 Oct 43	13 Dec 43		Grounded 12 Apr 45, CTL
494	Missouri	10 Aug 43 11 Oct 43	18 Dec 43	29 Jun 46	Str 28 Aug 46
495	Missouri	14 Aug 43 16 Oct 43	23 Dec 43	23 Apr 46	Str 5 Jun 46
496	Missouri	24 Aug 43 22 Oct 43	27 Dec 43		WL(M) 11 Jun 44
497	Missouri	26 Aug 43 27 Oct 43	31 Dec 43	18 Dec 45	Str 8 Jan 46
498	Missouri	31 Aug 43 1 Nov 43	6 Jan 44	8 Nov 45	Str 28 Nov 45
499	Missouri	3 Sep 43 5 Nov 43	10 Jan 44		WL(M) 8 Jun 44
500	Missouri	8 Sep 43 10 Nov 43	13 Jan 44	18 Jul 47	Str 1 Aug 47
501	Jeffersonville	30 Jun 43 22 Sep 43	26 Nov 43	20 Aug 47	SCAJAP (Q097), str 29 Sep 47
502	Jeffersonville	18 Jun 43 25 Sep 43	8 Dec 43	4 Feb 46	SCAJAP (Q041), str 23 Dec 47
503	Jeffersonville	29 Jul 43 8 Oct 43	8 Dec 43	11 Jun 46	To ROC 4 Apr 55
504	*Buchanan County* Jeffersonville	21 Jul 43 19 Oct 43	18 Dec 43	22 Jan 46	SCAJAP (Q016), str 11 Aug 55, tgt Feb 56
505	Jeffersonville	6 Aug 43 27 Oct 43	27 Dec 43	11 Jun 46	Str 16 Sep 47
506	Jeffersonville	19 Aug 43 4 Nov 43	3 Jan 44	24 Jul 47	Str 4 Dec 47
507	Jeffersonville	8 Sep 43 16 Nov 43	10 Jan 44		WL(T) 28 Apr 44
508	Jeffersonville	18 Sep 43 10 Nov 43	14 Jan 44	2 Aug 46	SCAJAP (Q087), str 28 Jan 47, to France for Indo-China Mar 47
509	*Bullock County* Jeffersonville	7 Oct 43 23 Nov 43	20 Jan 44		To Vietnam 8 Apr 70
510	*Buncombe County* Jeffersonville	27 Sep 43 30 Nov 43	31 Jan 44	1 Jul 46	Str 1 Nov 58
511	Chicago	22 Jul 43 30 Nov 43	3 Jan 44	19 Dec 45	Str 8 Jan 46
512	*Burnett County* Chicago	22 Jul 43 10 Dec 43	8 Jan 44	28 Mar 47	Str 18 Feb 57, to Peru 11 Oct 57
513	Chicago	23 Aug 43 17 Dec 43	9 May 44	30 Nov 46	To *Endymion* (ARL 9) 3 Nov 43, str 1 Jun 72
514	Chicago	31 Aug 43 24 Dec 43	23 May 44	Jan 47	To *Midas* (ARB 5) 3 Nov 43, str 15 Apr 76
515	*Caddo Parish* Chicago	3 Sep 43 31 Dec 43	28 Jan 44 2 Aug 63	20 Oct 55 26 Nov 69	To Philippines 26 Nov 69
516	*Calaveras County* Chicago	6 Sep 43 7 Jan 44	31 Jan 44 22 Sep 50	28 Feb 47 1 Oct 58	Str 1 Oct 58
517	Chicago	10 Sep 43 15 Jan 44	7 Feb 44	21 Dec 45	Str 21 Jan 46

No.	Name/Builder	LD/Launch	Comm	Decomm	Notes and Fate
518	Chicago	13 Sep 43 20 Jan 44	24 Jun 44	29 Nov 45	*Nestor* (ARB 6) 3 Nov 43, CTL due to Okinawa typhoon 9 Oct 45
519	*Calhoun County* Chicago	17 Sep 43 25 Jan 44	17 Feb 44	8 Nov 62	Str 8 Nov 62
520	Chicago	24 Sep 43 31 Jan 44	28 Feb 44	13 Jan 46	SCAJAP 13 Jan 46–31 Mar 52 (Q013), to ROC 1 Oct 58
521	*Cape May County* Chicago	4 Oct 43 13 Dec 43	9 Feb 44	21 Oct 45	Str 1 Nov 59
522	Chicago	2 Oct 43 11 Feb 44	1 Mar 44	6 Jun 46	Str 22 Jan 48
523	Jeffersonville	15 Oct 43 6 Dec 43	3 Feb 44		WL(M) 19 Jun 44
524	Jeffersonville	4 Oct 43 13 Dec 43	9 Feb 44	4 Feb 46	SCAJAP (Q040), str 31 Oct 47
525	*Caroline County* Jeffersonville	18 Oct 43 20 Dec 43	14 Feb 44 Oct 50 mid-65	25 Oct 46 15 Sep 54 68	Str 15 Sep 47
526	Jeffersonville	30 Oct 43 27 Dec 43	17 Feb 44	21 Dec 45	Str 21 Jan 46
527	*Cassia County* Jeffersonville	23 Oct 43 3 Jan 44	17 Feb 44 21 Sep 50	28 Feb 45 21 Dec 56	Str 1 Oct 58, tgt
528	*Catahoula Parish* Jeffersonville	13 Nov 43 11 Jan 44	29 Feb 44	Mar 54	Str 21 Nov 60
529	*Cayuga County* Jeffersonville	8 Nov 43 17 Jan 44	29 Feb 44 22 Sep 50	7 Jun 46 17 Dec 63	To Vietnam 17 Dec 63
530	Jeffersonville	23 Nov 43 25 Jan 44	6 Mar 44	46	SCAJAP (Q017), to MSTS 31 Mar 52, str 15 Jun 73
531	Missouri	22 Sep 43 24 Nov 43	17 Jan 44		WL(T) 28 Apr 44
532	*Chase County* Missouri	24 Sep 43 28 Nov 43	20 Jan 44	8 Jun 55	To MSTS 15 Apr 67, str 10 Jun 73, sold to Singapore 4 Jun 76, resold without entering naval service
533	*Cheboygan County* Missouri	29 Sep 43 1 Dec 43	27 Jan 44 18 Nov 61	1 Dec 55 May 69	Str 15 Sep 74
534	Missouri	4 Oct 43 8 Dec 43	31 Jan 44	2 Nov 45	Scuttled because beyond economical repair 9 Dec 45
535	Missouri	19 Oct 43 21 Dec 43	4 Feb 44	14 Jan 46	SCAJAP(Q014), MSTS 31 Mar 52, to ROC 1 Oct 58
536	Missouri	19 Oct 43 27 Dec 43	9 Feb 44	23 Jan 46	SCAJAP (Q024), to Korea 21 Feb 47
537	Missouri	27 Oct 43 31 Dec 43	9 Feb 44	29 May 46	To ROC 29 May 46
538	Missouri	29 Oct 43 5 Jan 44	14 Feb 44	16 Mar 46	Str 26 Feb 46
539	Missouri	9 Nov 43 10 Jan 44	17 Feb 44	22 Jun 46	Str 31 Jul 46
540	Missouri	13 Nov 43 14 Jan 44	22 Feb 44	13 Jan 46	SCAJAP (Q022), lost 20 Aug 47
541	Missouri	22 Nov 43 25 Jan 44	28 Feb 44	9 Nov 45	Str 28 Nov 45
542	*Chelan County* Missouri	29 Nov 43 28 Jan 44	29 Feb 44	56	Str 1 Nov 59
543	Missouri	6 Dec 43 1 Feb 44	6 Mar 44	31 May 46	Str 17 Jul 47
544	Missouri	8 Dec 43 4 Feb 44	16 Mar 44	9 Aug 46	Str 25 Sep 46
545	Missouri	13 Dec 43 12 Feb 44	23 Mar 44	29 Aug 46	Bikini tgt, tgt 12 May 48

No.	Name/Builder	LD/Launch	Comm	Decomm	Notes and Fate
546	Missouri	20 Dec 43 16 Feb 44	27 Mar 44	15 Jul 72	MSTS 31 Mar 52, to Philippines 15 Jul 72
547	Missouri	24 Dec 43 19 Feb 44	30 Mar 44	28 Feb 46	SCAJAP (Q077), str 31 Oct 47
548	Missouri	30 Dec 43 22 Feb 44	3 Apr 44	15 Feb 46	SCAJAP (Q067), MSTS 31 Dec 52, str 1 Jan 60
549	Missouri	4 Jan 44 25 Feb 44	5 Apr 44	28 Feb 46	SCAJAP (Q078), str 5 Dec 47
550	Missouri	13 Nov 43 9 Mar 44	10 Apr 44	13 Jan 46	SCAJAP (Q012), MSTS 31 Mar 52, str 1 Nov 73
551	Chesterfield County Missouri	15 Jan 44 11 Mar 44	14 Apr 44 21 Dec 65	10 Jun 55 70	Str 1 Jun 70
552	Missouri	19 Jan 44 14 Mar 44	19 Apr 44	19 Apr 46	Str 1 May 46
553	Missouri	24 Jan 44 16 Mar 44	22 Apr 44	13 Feb 47	Str 25 Apr 47, to army
554	Missouri	30 Jan 44 18 Mar 44	27 Apr 44	20 Jul 46	Str 25 Sep 46
555	Missouri	5 Feb 44 22 Mar 44	28 Apr 44	6 Jan 46	Str 21 Jan 46, tgt 26 Jan 46
556	Missouri	4 Feb 44 7 Apr 44	1 May 44	14 Mar 46	Str 12 Apr 46
557	Missouri	8 Feb 44 11 Apr 44	5 May 44	29 May 46	To ROC 29 May 46
558	Missouri	11 Feb 44 14 Apr 44	8 May 44	13 Feb 46	SCAJAP (Q056), str 16 Sep 47
559	Missouri	14 Feb 44 18 Apr 44	9 May 44	1 Jun 46	Str 19 Jun 46
560	Missouri	22 Feb 44 21 Apr 44	2 May 44	17 May 46	Str 19 Jun 46
561	Chittenden County Missouri	24 Feb 44 25 Apr 44	15 May 44 18 Sep 50	30 Apr 46 2 Jun 58	Str 27 Jun 58, grounded Mar 58, tgt 21 Oct 58
562	Missouri	28 Feb 44 28 Apr 44	18 May 44	21 May 46	Str 3 Jul 46
563	Missouri	4 Mar 44 1 May 44	20 May 44		Grounded 21 Dec 44, abandoned 9 Feb 45, str 23 Feb 45
564	Missouri	5 Mar 44 4 May 44	25 May 44	8 Mar 46	Str 1 May 46
565	Missouri	16 Mar 44 8 May 44	25 May 44	13 Jun 46	Str 3 Jul 46
566	Missouri	17 Mar 44 11 May 44	29 May 44	11 Mar 46	SCAJAP (Q094), MSTS 31 Mar 52, str 1 Nov 73, to Philippines 13 Sep 76
567	Missouri	20 Mar 44 15 May 44	1 Jun 44	28 Jan 46	SCAJAP (Q032), str 31 Oct 47
568	Missouri	21 Mar 44 18 May 44	3 Jun 44	4 Mar 46	Str 20 Mar 46
569	Missouri	24 Mar 44 20 May 44	5 Jun 44	13 Jun 46	Str 15 Oct 46
570	Missouri	14 Apr 44 22 May 44	9 Jun 44	14 May 46	Str 19 Jun 46
571	Missouri	14 Apr 44 25 May 44	14 Jun 44	12 Mar 46	Str 12 Apr 46
572	Missouri	15 Apr 44 29 May 44	19 Jun 44	8 Mar 46	SCAJAP (Q090), MSTS 31 Mar 52, str 15 Jun 73
573	Missouri	15 Apr 44 31 May 44	21 Jun 44	24 Jan 46	SCAJAP (Q027), str 31 Oct 47
574	Missouri	16 Apr 44 5 Jun 44	26 Jun 44	17 Jun 46	Str 3 Jul 46, to ROC 48

No.	Name/Builder	LD/Launch	Comm	Decomm	Notes and Fate
575	Missouri	3 May 44 9 Jun 44	24 Jun 44	29 May 47	Conversion to LST(M) completed 15 Jan 45, *Wythe* (APB 41) 31 Mar 45, str 1 May 59
576	Missouri	3 May 44 12 Jun 44	8 Jul 44	14 May 46	Str 9 Jun 46
577	Missouri	3 May 44 16 Jun 44	10 Jul 44		WL(T) 11 Feb 45
578	Missouri	4 May 44 19 Jun 44	15 Jul 44	22 Mar 46	SCAJAP (Q099), MSTS 31 Mar 52, to ROC 54
579	Missouri	4 May 44 22 Jun 44	21 Jul 44	24 Feb 46	SCAJAP (Q073), MSTS 31 Mar 52, str 30 Jun 75
580	Missouri	17 May 44 26 Jun 44	25 Jul 44	29 Jan 46	SCAJAP (Q035), str 31 Oct 47
581	Missouri	17 May 44 29 Jun 44	27 Jul 44	28 Jan 46	SCAJAP (Q030), MSTS 31 Mar 52, str 1 Jun 72
582	Missouri	18 May 44 1 Jul 44	31 Jul 44	20 Jan 46	SCAJAP (Q034), str 31 Oct 47
583	*Churchill County* Missouri	18 May 44 5 Jul 44	2 Aug 44 1 Nov 60	Mar 46 11 Dec 68	Str 15 Sep 74
584	Missouri	8 May 44 8 Jul 44	5 Aug 44	12 Apr 46	Str 3 Jul 46
585	Missouri	31 May 44 12 Jul 44	8 Aug 44	31 Jul 46	Str 28 Aug 46, to NEI
586	Missouri	1 Jun 44 15 Jul 44	15 Aug 44	17 Feb 46	SCAJAP (Q065), str 29 Sep 47
587	Missouri	2 Jun 44 19 Jul 44	18 Aug 44		SCAJAP (Q042), MSTS 31 Dec 52, str 1 Jun 72
588	Missouri	6 Jun 44 22 Jul 44	19 Aug 44	8 Jun 46	Str 3 Jul 46
589	Missouri	8 Jun 44 26 Jul 44	24 Aug 44	14 Sep 46	Sold 17 Dec 46
590	Missouri	19 Jun 44 29 Jul 44	26 Aug 44	2 Feb 46	SCAJAP (Q036), MSTS 31 Dec 52, str 15 Jun 73
591	Missouri	21 Jun 44 2 Aug 44	29 Aug 44	5 Feb 46	SCAJAP (Q045), str 29 Sep 47
592	Missouri	24 Jun 44 5 Aug 44	1 Sep 44	11 Jun 46	Str 31 Jul 46
593	Missouri	28 Jun 44 9 Aug 44	5 Sep 44	18 Mar 46	Str 8 May 46
594	Missouri	1 Jul 44 12 Aug 44	6 Sep 44	21 Feb 46	SCAJAP (Q070), str 5 Mar 47, to Korea 4 Jun 47
595	Missouri	7 Jul 44 16 Aug 44	14 Sep 44	3 Jan 46	SCAJAP (Q001), str 5 Mar 47, to Korea 31 May 47
596	Missouri	11 Jul 44 21 Aug 44	14 Sep 44	12 Jun 46	Str 25 Sep 46
597	Missouri	12 Jul 44 28 Aug 44	19 Sep 44	5 Mar 46	Str 29 Sep 47
598	Missouri	14 Jul 44 29 Aug 44	22 Sep 44	10 Jun 46	Str 19 Jul 46
599	Missouri	18 Jul 44 2 Sep 44	27 Sep 44	1 Jun 46	Str 22 Jan 48
600	Chicago	6 Oct 43 28 Feb 44	20 Mar 44	1 Jun 46	SCAJAP (Q074), MSTS 31 Mar 52, str 1 Jun 69 after grounding 23 Dec 68
601	*Clarke County* Chicago	21 Oct 43 4 Mar 44	25 Mar 44 28 Jul 66	23 Nov 55 69	To Indonesia 15 Jul 70
602	*Clearwater County* Chicago	23 Oct 43 9 Mar 44	31 Mar 44 50	46	Operated by U.S. Air Force Sep 57–Sep 69, MarAd custody, str 1 May 72, to Mexico 30 May 72

No.	Name/Builder	LD/Launch	Comm	Decomm	Notes and Fate
603	*Coconino County* Chicago	5 Nov 43 14 Mar 44	5 Apr 44 8 Jun 66	12 May 55 4 Apr 69	To Vietnam 4 Apr 69
604	Chicago	28 Oct 43 20 Mar 44	3 Apr 44	14 Mar 46	*Silenus* (AGP 11) 18 Dec 43, sold 48
605	Chicago	30 Sep 43 29 Mar 44	14 Apr 44	24 May 46	Str 3 Jul 46
606	Chicago	27 Nov 43 3 Apr 44	24 Apr 44	13 May 46	Str 13 Jun 46
607	Chicago	2 Dec 43 7 Apr 44	24 Apr 44	11 Jan 46	SCAJAP (Q008), MSTS 31 Mar 52, to Philippines 13 Sep 76
608	Chicago	4 Dec 43 11 Apr 44	15 Apr 44	1 Jan 46	Str 7 Feb 47, to Korea 31 May 46
609	Chicago	10 Dec 43 15 Apr 44	15 May 44	4 Jan 46	Str 21 Jan 46
610	Chicago	16 Dec 43 19 Apr 44	15 May 44	28 Jun 46	To mil gov Okinawa 18 Sep 46, str 23 Dec 47
611	*Crook County* Chicago	17 Dec 43 28 Apr 44	15 May 44	26 Oct 56	
612	Chicago	18 Dec 43 29 Apr 44	16 May 44	1 Jun 46	Str 3 Jul 46
613	Chicago	21 Jan 44 2 May 44	19 May 44	6 Jan 46	SCAJAP (Q038), MSTS 31 Mar 52, str 30 Jun 75
614	Chicago	28 Jan 44 6 May 44	22 May 44	20 Jun 46	Str 29 Oct 46
615	Chicago	4 Feb 44 9 May 44	26 May 44	14 Mar 46	Str 12 Apr 46
616	Chicago	12 Feb 44 12 May 44	29 May 44	19 Jan 46	SCAJAP (Q019), MSTS 31 Mar 52, str 1 May 61, to Indonesia
617	Chicago	17 Feb 44 15 May 44	1 Jun 44	24 May 46	Str 3 Jul 46
618	Chicago	23 Feb 44 19 May 44	3 Jun 44		To army 24 Oct 46, str 23 Dec 47
619	Chicago	8 Mar 44 22 May 44	5 Jun 44	19 Jun 46	SCAJAP (Q031), str 31 Oct 47
620	Chicago	11 Mar 44 30 May 44	17 Jun 44	7 Jun 46	Str 19 Jun 46
621	Chicago	15 Mar 44 2 Jun 44	21 Jun 44	16 Jun 46	Str 31 Jul 46
622	Chicago	15 Mar 44 8 Jun 44	26 Jun 44	14 Mar 46	Str 12 Apr 46
623	Chicago	17 Mar 44 12 Jun 44	29 Jun 44	46	SCAJAP (Q075), MSTS 31 Mar 52, str 30 Jun 75
624	Chicago	22 Mar 44 16 Jun 44	3 Jul 44	14 Feb 46	SCAJAP (Q057), str 7 Feb 47, to Korea 1 Jun 47
625	Chicago	30 Mar 44 20 Jun 44	10 Jul 44	11 Feb 46	SCAJAP (Q052), MSTS 31 Mar 52, str 19 May 54
626	Chicago	31 Mar 44 27 Jun 44	15 Jul 44	2 Mar 46	SCAJAP (Q081), MSTS 31 Mar 52, str 1 Jun 72
627	Chicago	8 Apr 44 1 Jul 44	20 Jul 44	6 Jun 46	To NEI 15 Jun 46
628	Chicago	10 Apr 44 4 Jul 44	31 Jul 44	3 Apr 46	Str 3 Jul 46
629	Chicago	13 Apr 44 8 Jul 44	28 Jul 44	4 Mar 46	SCAJAP (Q033), MSTS 31 Mar 52, str 15 Jun 73
630	Chicago	14 Apr 44 13 Jul 44	4 Aug 44	13 Feb 46	SCAJAP (Q059), MSTS 31 Mar 52, str 15 Jun 73
631	Chicago	19 Apr 44 18 Jul 44	9 Aug 44	24 May 46	Str 3 Jul 46

No.	Name/Builder	LD/Launch	Comm	Decomm	Notes and Fate
632	Chicago	26 Apr 44 21 Jul 44	12 Aug 44	30 May 44	Str 28 Jan 47
633	Chicago	3 May 44 27 Jul 44	17 Aug 44	15 Feb 46	SCAJAP (Q060), str 29 Sep 47
634	Chicago	13 May 44 1 Aug 44	22 Aug 44	8 Jun 46	Str 19 Jul 46
635	Chicago	17 May 44 7 Aug 44	26 Aug 44	7 Jun 46	To State Dept for disposal 19 Jul 46
636	Chicago	22 May 44 11 Aug 44	31 Aug 44	25 May 46	Str 23 Dec 47
637	Chicago	24 May 44 18 Aug 44	5 Sep 44	29 Mar 46	Str 5 Jun 46
638	Chicago	25 May 44 23 Aug 44	8 Sep 44	8 Jun 46	Str 12 Mar 48, tgt 15 Jun 48
639	Chicago	26 May 44 28 Aug 44	14 Sep 44	1 Jun 46	Str 16 Sep 47
640	Chicago	27 May 44 31 Aug 44	18 Sep 44	30 Apr 46	Str 19 Jul 46, ROC merch service, to ROC naval service 55
641	Chicago	1 Jun 44 4 Sep 44	22 Sep 44	13 Jun 46	Str 19 Jul 46
642	Chicago	5 Jun 44 8 Sep 44	28 Sep 44	30 Jun 47	Abandoned Barter Island, Alaska, 10 Feb 48
643	Chicago	10 Jun 44 12 Sep 44	2 Oct 44		SCAJAP (Q018), MSTS 31 Mar 52, str 15 Jun 73
644	Chicago	16 Jun 44 15 Sep 44	26 Sep 44 22 Sep 50	18 Jun 46	To *Minos* (ARL 14) 14 Aug 44, str 1 Jan 60
645	Chicago	20 Jun 44 20 Sep 44	30 Sep 44 22 Jun 51	26 Feb 47 3 Oct 55	*Minotaur* (ARL 15) 14 Aug 44, to Korea 3 Oct 55
646	Chicago	30 Jun 44 25 Sep 44	13 Oct 44	15 Mar 46	Str 17 Apr 46
647	Chicago	5 Jul 44 28 Sep 44	19 Oct 44	2 Feb 46	SCAJAP (Q047), str 23 Dec 47
648	Chicago	7 Jul 44 3 Oct 44	21 Oct 44	14 Feb 47	Str 25 Feb 47
649	Chicago	19 Jul 44 6 Oct 44	26 Oct 44		SCAJAP (Q058), MSTS 31 Mar 52, str 30 Jun 75
650	Chicago	20 Jul 44 10 Oct 44	21 Oct 44 14 Dec 51	23 Sep 46 30 Sep 68	*Pandemus* (ARL 18) 14 Aug 44, str 1 Oct 68, tgt late 69
651	Chicago	24 Jul 44 16 Oct 44	4 Nov 44	23 Jan 46	SCAJAP (Q023), str 5 Dec 47
652	Chicago	24 Jul 44 19 Oct 44	1 Jan 45	5 Mar 46	LSTH 15 Sep 45, SCAJAP (Q086), MSTS 31 Mar 52, str 1 May 61, to Indonesia Jun 61
653	Amer Bridge	17 Nov 43 23 Jan 44	1 Apr 44	3 Feb 46	SCAJAP (Q039), str 5 Mar 47, to Korea 31 May 47
654	Amer Bridge	9 Dec 43 30 Jan 44	20 Mar 44	12 Jun 46	Str 19 Jul 46
655	Amer Bridge	9 Dec 43 6 Feb 44	28 Mar 44	31 May 46	Str 3 Jul 46
656	Amer Bridge	13 Dec 43 18 Feb 44	7 Apr 44	29 May 46	Str 3 Jul 46
657	Amer Bridge	16 Dec 43 25 Feb 44	10 Apr 44		SCAJAP (Q071), MSTS 31 Mar 52, str 1 May 61, to Indonesia Jun 61
658	Amer Bridge	28 Dec 43 13 Mar 44	17 Apr 44	1 Jun 46	Str 3 Jul 46
659	Amer Bridge	31 Dec 43 20 Mar 44	20 Apr 44	7 Jan 46	SCAJAP (Q003), str 7 Feb 47, to Korea 25 May 47

No.	Name/Builder	LD/Launch	Comm	Decomm	Notes and Fate
660	Amer Bridge	6 Jan 44 24 Mar 44	26 Apr 44	26 Apr 46	Str 5 Jun 46
661	Amer Bridge	9 Jan 44 30 Mar 44	28 Apr 44	29 Aug 46	Bikini tgt, destroyed 25 Jul 48
662	Amer Bridge	14 Jan 44 5 Apr 44	2 May 44	19 Dec 45	Str 8 Jan 46
663	Amer Bridge	22 Jan 44 8 Apr 44	5 May 44	29 May 46	Str 19 Jul 46
664	Amer Bridge	28 Jan 44 13 Apr 44	10 May 44		MSTS 19 Apr 55, str 15 Jun 73
665	Amer Bridge	5 Feb 44 18 Apr 44	12 May 44	11 Jun 46	Str 3 Jul 46
666	Amer Bridge	16 Feb 44 24 Apr 44	16 May 44	20 Jun 46	Str 31 Jul 46
667	Amer Bridge	22 Feb 44 27 Apr 44	20 May 44	5 Jun 46	Str 3 Jul 46
668	Amer Bridge	6 Mar 44 30 Apr 44	23 May 44	24 Jun 46	Str 31 Jul 46
669	Amer Bridge	18 Mar 44 3 May 44	27 May 44	13 Aug 46	Str 25 Sep 46
670	Amer Bridge	22 Mar 44 6 May 44	29 May 44	30 Apr 46	Str 19 Jun 46
671	Amer Bridge	28 Mar 44 11 May 44	2 Jun 44	25 Jun 46	Str 15 Aug 46
672	Amer Bridge	3 Apr 44 14 May 44	5 Jun 44	26 Jun 46	Str 31 Jul 46
673	Amer Bridge	6 Apr 44 22 May 44	9 Jun 44	10 Jul 46	Str 15 Aug 46
674	Amer Bridge	11 Apr 44 26 May 44	19 Jun 44	14 May 46	Str 19 Jun 46
675	Amer Bridge	16 Apr 44 2 Jun 44	24 Jun 44	25 Aug 45	Unsalvageable after battle damage at Okinawa 4 Apr 45
676	Amer Bridge	22 Apr 44 6 Jun 44	20 Jun 44	3 Dec 46	Converted to LST(M) Oct 44–Jan 45, *Yavapai* (APB 42) 31 Mar 45, str 1 May 59
677	Amer Bridge	25 Apr 44 15 Jun 44	30 Jun 44	9 Aug 46	Converted to LST(M) Oct 44–Jan 45, *Yolo* (APB 43) 31 Mar 45, str 1 May 59
678	Amer Bridge	29 Apr 44 16 Jun 44	30 Jun 44	18 Apr 47	Converted to LST(M) Oct 44–Jan 45, *Presque Isle* (APB 44) 31 Mar 45, str 1 May 59
679	Amer Bridge	2 May 44 20 Jun 44	15 Jul 44	24 Jun 46	Str 31 Jul 46
680	Amer Bridge	5 May 44 26 Jun 44	21 Jul 44	5 Jul 46	Str 28 Aug 46
681	Amer Bridge	10 May 44 1 Jul 44	25 Jul 44	6 Sep 46	Str 8 Oct 46
682	Jeffersonville	6 Dec 43 31 Jan 44	18 Mar 44	30 Jul 46	Str 25 Sep 46
683	Jeffersonville	29 Nov 43 7 Feb 44	28 Mar 44	29 May 46	Str 3 Jul 46
684	Jeffersonville	13 Dec 43 12 Feb 44	3 Apr 44	25 Nov 45	Destroyed 22 Mar 46
685	*Curry County* Jeffersonville	21 Dec 43 18 Feb 44	7 Apr 44	22 Jul 46	NRT 13 Jan 47, inactivated 2 Jun 50, str 1 Nov 58
686	Jeffersonville	4 Jan 44 24 Feb 44	14 Apr 44	10 Jul 46	Str 15 Aug 46
687	Jeffersonville	28 Dec 43 28 Feb 44	22 Apr 44	24 May 46	Str 3 Jul 46

No.	Name/Builder	LD/Launch	Comm	Decomm	Notes and Fate
688	Jeffersonville	17 Jan 44 5 Mar 44	27 Apr 44	5 Aug 46	Str 25 Sep 46
689	*Daggett County* Jeffersonville	11 Jan 44 9 Mar 44	2 May 44	Mar 46	Str 1 Oct 59, to Japan Apr 61
690	Jeffersonville	31 Jan 44 14 Mar 44	6 May 44	23 Jul 46	Str 28 Aug 46
691	Jeffersonville	25 Jan 44 23 Mar 44	12 May 44	14 May 46	Str 19 Jun 46
692	*Davies County* Jeffersonville	7 Feb 44 31 Mar 44	10 May 44 51	46	Str 1 Jun 64, to MSTS, to Philippines 13 Sep 76
693	Jeffersonville	18 Feb 44 7 Apr 44	15 May 44	1 May 46	Str 3 Jul 46
694	Jeffersonville	14 Feb 44 16 Apr 44	19 May 44	1 Dec 47	Str 23 Dec 47, to army, to MSTS 1 Mar 50, str 4 Feb 58
695	Jeffersonville	28 Feb 44 24 Apr 44	22 May 44	6 Nov 45	Str 28 Nov 45
696	Jeffersonville	25 Feb 44 27 Apr 44	25 May 44	16 Jul 46	Str 28 Aug 46
697	Jeffersonville	6 Mar 44 1 May 44	30 May 44	12 Jul 46	Str 28 Aug 46
698	Jeffersonville	14 Mar 44 5 May 44	3 Jun 44	26 Nov 45	Str 5 Dec 45
699	Jeffersonville	9 Mar 44 9 May 44	5 Jun 44	24 Jun 46	Str 31 Jul 46
700	Jeffersonville	22 Mar 44 13 May 44	7 Jun 44	27 Jul 46	Str 28 Aug 46
701	Jeffersonville	1 Apr 44 18 May 44	13 Jun 44	13 Jul 46	Str 28 Aug 46
702	Jeffersonville	15 Apr 44 22 May 44	19 Jun 44	5 Jul 46	Str 15 Aug 46
703	Jeffersonville	8 Apr 44 28 May 44	23 Jun 44	10 Jun 46	Str 31 Jul 46
704	Jeffersonville	27 Apr 44 3 Jun 44	27 Jun 44	19 Jun 46	Str 25 Sep 46
705	Jeffersonville	21 Apr 44 7 Jun 44	4 Jul 44	22 Jul 46	Str 25 Sep 46
706	Jeffersonville	4 May 44 12 Jun 44	8 Jul 44	19 Jun 46	Str 31 Jul 46
707	Jeffersonville	1 May 44 16 Jun 44	13 Jul 44	28 May 46	Str 3 Jul 46
708	Jeffersonville	9 May 44 20 Jun 44	17 Jul 44	28 May 46	Str 3 Jul 46
709	Jeffersonville	18 May 44 24 Jun 44	21 Jul 44	3 Jul 46	Str 15 Aug 46
710	Jeffersonville	13 May 44 28 Jun 44	24 Jul 44	9 Aug 46	*Accomac* (APB 49) 1 Aug 45, str 1 May 59
711	Jeffersonville	28 May 44 3 Jul 44	28 Jul 44	11 Aug 46	To army, str 29 Sep 47
712	Jeffersonville	22 May 44 7 Jul 44	2 Aug 44	20 May 46	Str 28 Aug 46
713	Jeffersonville	3 Jun 44 11 Jul 44	7 Aug 44	20 Jun 46	Str 31 Jul 46
714	Jeffersonville	12 Jun 44 15 Jul 44	11 Aug 44	10 May 46	Str 19 Jun 46
715	*De Kalb County* Jeffersonville	7 Jun 44 20 Jul 44	15 Aug 44	17 Apr 46	To army 28 Jun 46, str 29 Sep 46, reacquired 25 Jul 50, MSTS Dec 65, str 1 Nov 73

Appendix C

No.	Name/Builder	LD/Launch	Comm	Decomm	Notes and Fate
716	Jeffersonville	16 Jun 44 25 Jul 44	18 Aug 44	12 Jun 46	To ROC 7 Feb 48
717	Jeffersonville	20 Jun 44 29 Jul 44	23 Aug 44	12 Jun 46	To ROC 17 Feb 48
718	Jeffersonville	28 Jun 44 3 Aug 44	28 Aug 44	25 Jun 46	Str 31 Jul 48
719	Jeffersonville	24 Jun 44 8 Aug 44	31 Aug 44	12 Jul 46	Str 14 Mar 47
720	Jeffersonville	7 Jul 44 12 Aug 44	4 Sep 44	24 Jun 46	Str 31 Jul 46
721	Jeffersonville	3 Jul 44 7 Aug 44	9 Sep 44	24 Jun 46	Str 10 Jun 47
722	*Dodge County* Jeffersonville	15 Jul 44 21 Aug 44	13 Sep 44 16 Nov 51	13 Jul 46 3 Jan 56	Str 15 Sep 74
723	Jeffersonville	11 Jul 44 25 Aug 44	16 Sep 44	20 Jul 46	Str 10 Jun 47
724	Jeffersonville	20 Jul 44 29 Aug 44	22 Sep 44	26 Jun 46	Str 31 Jul 46
725	Jeffersonville	29 Jul 44 2 Sep 44	25 Sep 44	1 May 46	Str 3 Jul 46
726	Jeffersonville	26 Jul 44 6 Sep 44	30 Sep 44	25 Jun 46	Str 31 Jul 46
727	Jeffersonville	8 Aug 44 10 Sep 44	4 Oct 44	26 Jul 46	Str 28 Aug 46
728	Jeffersonville	3 Aug 44 14 Sep 44	10 Oct 44	18 Jun 46	Str 31 Jul 46
729	Jeffersonville	12 Aug 44 18 Sep 44	16 Oct 44	8 Jul 46	Str 28 Aug 46
730	Dravo-Pitts	13 Dec 43 29 Jan 44	30 Mar 44	8 Jun 46	Str 31 Jul 46
731	*Douglas County* Dravo-Pitts	27 Dec 43 12 Feb 44	30 Mar 44	2 Jun 50	LSTH 15 Sep 45, str 1 Nov 58
732	Dravo-Pitts	5 Jan 44 19 Feb 44	10 Apr 44	7 Jun 46	Str 19 Jul 46, to ROC merch service, to ROC navy 55
733	Dravo-Pitts	16 Jan 44 26 Feb 44	15 Apr 44	28 Jun 46	Str 31 Jul 46
734	Dravo-Pitts	25 Jan 44 4 Mar 44	22 Apr 44	7 May 46	Str 5 Jun 46, listed by Argentine navy as transferred, but known to have been BU 48
735	*Dukes County* Dravo-Pitts	30 Jan 44 11 Mar 44	26 Apr 44 3 Nov 50	Mar 46	Supported mine forces in Korea, MinRon flagship postwar, to ROC May 57
736	Dravo-Pitts	2 Feb 44 18 Mar 44	2 May 44	20 Jun 46	Str 31 Jul 46
737	Dravo-Pitts	13 Feb 44 25 Mar 44	6 May 44	2 Nov 46	To army, str 29 Sep 47, to Japan Dec 59
738	Dravo-Pitts	20 Feb 44 1 Apr 44	9 May 44		WL(K) 15 Dec 44
739	Dravo-Pitts	27 Feb 44 8 Apr 44	15 May 44	1 May 46	Str 3 Jul 46
740	Dravo-Pitts	12 Feb 44 8 Apr 44	15 May 44	8 Mar 46	Str 12 Apr 46
741	Dravo-Pitts	5 Mar 44 15 Apr 44	19 May 44	9 Aug 46	Str 25 Sep 46
742	*Dunn County* Dravo-Pitts	12 Mar 44 22 Apr 44	23 May 44 1 Sep 50	26 Apr 46 1 Feb 61	To army 46, ret 50, str 1 Feb 61
743	Dravo-Pitts	20 Feb 44 19 Apr 44	23 May 44	23 Apr 46	Str 19 Jun 46
744	Dravo-Pitts	1 Mar 44 29 Apr 44	29 May 44	28 Jun 46	Str 15 Aug 46

No.	Name/Builder	LD/Launch	Comm	Decomm	Notes and Fate
745	Dravo-Pitts	19 Mar 44 29 Apr 44	31 May 44	9 Jul 46	Str 28 Aug 46
746	Dravo-Pitts	26 Mar 44 6 May 44	3 Jun 44	1 May 46	Str 3 Jul 46
747	Dravo-Pitts	2 Apr 44 20 May 44	15 Jun 44	20 Jun 46	Str 31 Jul 46
748	Dravo-Pitts	2 Apr 44 13 May 44	5 Jun 44	N/A	To MarAd 27 May 48
749	Dravo-Pitts	10 Apr 44 20 May 44	23 Jun 44		WL(K) 21 Dec 44
750	Dravo-Pitts	7 Apr 44 30 May 44	29 Jun 44		WL(T) 28 Dec 44
751	Dravo-Pitts	16 Apr 44 27 May 44	26 Jun 44	21 Aug 46	Str 15 Oct 46
752	Dravo-Pitts	23 Apr 44 3 Jun 44	5 Jul 44	7 Jun 46	Str 19 Jul 46
753	Dravo-Pitts	30 Apr 44 10 Jun 44	10 Jul 44	25 Jun 46	Str 31 Jul 46
754	Amer Bridge	13 May 44 6 Jul 44	29 Jul 44	20 Jun 46	Str 31 Jul 46
755	Amer Bridge	20 May 44 11 Jul 44	3 Aug 44	29 May 46	To ROC on decomm
756	Amer Bridge	25 May 44 15 Jul 44	8 Aug 44	5 Apr 46	Str 17 Apr 46
757	Amer Bridge	1 Jun 44 21 Jul 44	15 Aug 44	28 May 46	Str 3 Jul 46
758	*Duval County* Amer Bridge	5 Jun 44 25 Jul 44	10 Aug 44 3 Nov 50	13 Jul 46 28 Oct 69	Str 15 Sep 74
759	*Eddy County* Amer Bridge	11 Jun 44 29 Jul 44	25 Aug 44	Mar 46	Str 1 Oct 58
760	Amer Bridge	15 Jun 44 3 Aug 44	28 Aug 44	24 May 46	Str 3 Jul 46
761	*Esmeraldo County* Amer Bridge	18 Jun 44 7 Aug 44	2 Sep 44	Mar 46	Tgt use 6 May 59
762	*Floyd County* Amer Bridge	24 Jun 44 11 Aug 44	5 Sep 44 3 Nov 50	46 3 Sep 69	Str 69
763	Amer Bridge	29 Jun 44 16 Aug 44	8 Sep 44	29 Apr 46	Str 15 Aug 46
764	Amer Bridge	4 Jul 44 21 Aug 44	13 Sep 44	30 Apr 46	Str 3 Jul 46
765	Amer Bridge	8 Jul 44 26 Aug 44	18 Sep 44	29 Apr 46	Str 3 Jul 46
766	Amer Bridge	13 Jul 44 30 Aug 44	25 Sep 44	19 Mar 46	Str Jun 46, merch conversion
767	Amer Bridge	19 Jul 44 4 Sep 44	30 Sep 44	7 Mar 46	Str 28 Mar 46 after hurricane damage 9 Mar 46
768	Amer Bridge	22 Jul 44 12 Sep 44	4 Oct 44	15 Apr 46	Str 5 Jun 46
769	Amer Bridge	28 Jul 44 12 Sep 44	9 Oct 44	29 Apr 46	Str 3 Jul 46
770	Amer Bridge	1 Aug 44 17 Sep 44	13 Oct 44	29 Apr 46	Str 31 Jul 46
771	Amer Bridge	5 Aug 44 21 Sep 44	18 Oct 44	14 May 46	Str 5 Jun 46
772	*Ford County* Chicago	3 Aug 44 24 Oct 44	13 Nov 44 3 Nov 50	3 Jul 46	Tgt 19 Mar 58
773	Chicago	15 Aug 44 27 Oct 44	17 Nov 44	27 May 46	*Antigone* (AGP 16) 14 Aug 44, to MarComm 6 Feb 48

No.	Name/Builder	LD/Launch	Comm	Decomm	Notes and Fate
774	Chicago	5 Aug 44 31 Oct 44	20 Nov 44	12 Jul 46	Str 15 Aug 46
775	Dravo-Pitts	22 Apr 44 10 Jun 44	15 Jul 44	15 Jul 46	To MarComm 11 Jun 48
776	Dravo-Pitts	7 May 44 12 Jun 44	20 Jul 44	18 Mar 46	Str 1 May 46
777	Dravo-Pitts	4 May 44 24 Jun 44	25 Jul 44	19 Jul 46	Str 28 Aug 46
778	Dravo-Pitts	14 May 44 24 Jun 44	31 Jul 44	27 May 46	Str 19 Jun 46
779	Dravo-Pitts	21 May 44 1 Jul 44	3 Aug 44	18 May 46	Str 19 Jul 46
780	Dravo-Pitts	28 May 44 10 Jul 44	7 Aug 44	13 Jun 46	Str 31 Jul 46
781	Dravo-Pitts	4 Jun 44 15 Jul 44	18 Aug 44	27 Jun 46	Str 15 Aug 46
782	Dravo-Pitts	11 Jun 44 22 Jul 44	22 Aug 44	14 May 46	Str 5 Jun 46
783	Dravo-Pitts	14 May 44 11 Jul 44	14 Aug 44	22 Aug 46	Str 16 Jun 50
784	*Garfield County* Dravo-Pitts	18 Jun 44 29 Jul 44	1 Sep 44	Mar 46	Str 59, tgt Mar 59
785	Dravo-Pitts	25 Jun 44 5 Aug 44	4 Sep 44	3 May 46	Str 5 Jun 46
786	*Garrett County* Dravo-Pitts	21 May 44 22 Jul 44	28 Aug 44 15 Oct 66	9 Jul 46	To Vietnam 23 Apr 71, reclassified as AGP 786 on 25 Sep 70; had been fitted as PBR/PCF tender for Vietnam War
787	Dravo-Pitts	2 Jul 44 12 Aug 44	13 Sep 44	27 May 46	Str 3 Jul 46
788	Dravo-Pitts	9 Jul 44 19 Aug 44	18 Sep 44	16 Apr 46	Str 5 Jun 46
789	Dravo-Pitts	1 Jun 44 5 Aug 44	11 Sep 44	29 Apr 46	Str 3 Jul 46
790	Dravo-Pitts	11 Jun 44 19 Aug 44	22 Sep 44	28 May 46	LSTH 15 Sep 45, str 3 Jul 46
791	Dravo-Pitts	16 Jul 44 26 Aug 44	28 Sep 44	28 May 46	Str 3 Jul 46
792	Dravo-Pitts	25 Jun 44 2 Sep 44	2 Oct 44	29 Apr 46	Str 19 Jul 46
793	Dravo-Pitts	23 Jul 44 2 Sep 44	5 Oct 44	29 Apr 46	Str 3 Jul 46
794	*Gibson County* Dravo-Pitts	12 Jul 44 16 Sep 44	16 Oct 44	9 Jul 46	Tgt 22 May 59
795	Dravo-Pitts	30 Jul 44 9 Sep 44	9 Oct 44	29 Apr 46	Str 19 Jul 46
796	Dravo-Pitts	6 Aug 44 16 Sep 44	20 Oct 44	17 Apr 46	Str 19 Jun 46
797	Jeffersonville	21 Aug 44 22 Sep 44	20 Oct 44	28 Jun 46	Str 31 Jul 46
798	Jeffersonville	17 Aug 44 26 Sep 44	26 Oct 44	16 Jul 46	Str 15 Aug 46
799	*Greer County* Jeffersonville	25 Aug 44 3 Oct 44	28 Oct 44 26 Aug 50	6 May 46 18 Jan 60	To army, ret 50, str 1 Nov 60, mine squadron flagship 52, to Germany Oct 60
800	Jeffersonville	29 Aug 44 10 Oct 44	2 Nov 44	1 May 46	Str 3 Jul 46
801	Jeffersonville	6 Sep 44 14 Oct 44	8 Nov 44	18 Jul 46	Str 18 Aug 46, to Argentina Dec 47

No.	Name/Builder	LD/Launch	Comm	Decomm	Notes and Fate
802	*Hamilton County* Jeffersonville	2 Sep 44 19 Oct 44	13 Nov 44 30 Aug 50	21 Jul 46 30 Jun 60	SCAJAP 46, ret 50, mine squadron flagship 54, to Japan 20 Apr 60
803	*Hampden County* Jeffersonville	14 Sep 44 23 Oct 44	17 Nov 44 15 Nov 50	26 Mar 49 2 Jan 58	Str 17 Apr 58, tgt 26 Sep 58
804	Jeffersonville	10 Sep 44 27 Oct 44	22 Nov 44	24 May 46	Str 3 Jul 46, to ROC
805	Jeffersonville	22 Sep 44 31 Oct 44	27 Nov 44	25 May 46	Str 19 Jul 46, to Korea Nov 47
806	Missouri	25 Jul 44 7 Sep 44	28 Sep 44	20 May 46	Str 19 Jul 46
807	Missouri	29 Jul 44 11 Sep 44	8 Oct 44	27 May 46	Str 3 Jul 46
808	Missouri	1 Aug 44 15 Sep 44	29 Sep 44		WL(T) 18 May 45
809	Missouri	5 Aug 44 19 Sep 44	10 Oct 44	15 Jul 46	Str 18 Aug 46
810	Missouri	8 Aug 44 21 Sep 44	13 Oct 44	18 Jul 46	Str 28 Aug 46
811	Missouri	12 Aug 44 23 Sep 44	18 Oct 44	26 Jun 46	Str 31 Jul 46
812	Missouri	14 Aug 44 27 Sep 44	19 Oct 44	9 May 46	Str 10 Jun 46
813	Missouri	20 Aug 44 30 Sep 44	24 Oct 44	21 Jun 46	Str 31 Jul 46
814	Missouri	25 Aug 44 4 Oct 44	27 Oct 44	16 Apr 46	Badly damaged 30 Dec 45, str 8 May 46, sunk 14 May
815	Missouri	28 Aug 44 7 Oct 44	30 Oct 44	6 Sep 46	To MarComm 25 May 48, to France May 48
816	Missouri	2 Sep 44 11 Oct 44	2 Nov 44	29 Jun 46	Str 31 Jul 46
817	Missouri	4 Sep 44 14 Oct 44	7 Nov 44	31 Jan 47	Str 7 Feb 47
818	Missouri	8 Sep 44 18 Oct 44	9 Nov 44	16 Jul 46	Str 28 Aug 46
819	*Hampshire County* Missouri	12 Sep 44 21 Oct 44	14 Nov 44 8 Sep 50 9 Jul 66	15 Nov 46 24 Jun 55 19 Dec 70	Str Apr 75
820	Missouri	14 Sep 44 25 Oct 44	16 Nov 44	16 Jan 46	Str 7 Feb 46
821	*Harnett County* Missouri	9 Sep 44 27 Oct 44	14 Nov 44 20 Aug 66	Mar 46	To AGP 821 on 25 Sep 70, to Vietnam 12 Oct 70; had been fitted as PBR/PCF tender for Vietnam War
822	*Harris County* Missouri	20 Sep 44 1 Nov 44	23 Nov 44 23 Nov 50	10 Aug 46 55	MSTS 55, to Philippines 13 Sep 76
823	Missouri	25 Sep 44 4 Nov 44	28 Nov 44	1 Dec 45	Str 3 Jan 46
824	*Henry County* Missouri	28 Sep 44 8 Nov 44	30 Nov 44 5 Sep 59	15 May 46	To Malaysia 1 Aug 74
825	*Hickham County* Missouri	2 Oct 44 11 Nov 44	8 Dec 44 3 Nov 50 22 Mar 63	22 May 46 20 May 56 N/A	To Philippines Nov 69
826	Missouri	6 Oct 44 14 Nov 44	7 Dec 44	N/A	Destroyed 7 May 47
827	*Hillsborough County* Missouri	9 Oct 44 16 Nov 44	12 Dec 44 3 Nov 50	7 Jun 49 22 Jan 58	Str 28 Mar 58, tgt 14 Aug 58
828	Missouri	13 Oct 44 22 Nov 44	13 Dec 44	22 Apr 47	Destroyed 7 May 47, Marianas

No.	Name/Builder	LD/Launch	Comm	Decomm	Notes and Fate
829	Amer Bridge	10 Aug 44 26 Sep 44	23 Oct 44	N/A	To MarComm 19 Mar 48
830	Amer Bridge	15 Aug 44 30 Sep 44	28 Oct 44	29 Apr 46	Str 3 Jul 46
831	Amer Bridge	19 Aug 44 6 Oct 44	8 Nov 44	N/A	To MarComm 19 Dec 47
832	Amer Bridge	25 Aug 44 11 Oct 44	4 Nov 44	30 Apr 46	Str 3 Jul 46
833	Amer Bridge	28 Aug 44 16 Oct 44	10 Nov 44	2 May 46	Str 10 Jun 47
834	Amer Bridge	2 Sep 44 20 Oct 44	10 Nov 44	12 Sep 46	Str 8 Oct 46
835	*Hillsdale County* Amer Bridge	6 Sep 44 25 Oct 44	20 Nov 44	Mar 46	Str Oct 59, to Japan Apr 61
836	*Holmes County* Amer Bridge	11 Sep 44 29 Oct 44	25 Nov 44 3 Nov 50	25 Jul 46 71	FRAM II refit, to Singapore 1 Jul 71
837	Amer Bridge	15 Sep 44 3 Nov 44	29 Nov 44	28 Jun 46	Str 15 Aug 46
838	*Hunterdon County* Amer Bridge	20 Sep 44 8 Nov 44	4 Dec 44 10 Sep 66	7 Aug 46	To AGP 838 on 25 Sep 70, fitted as PBR/PCF tender for Vietnam War, to Malaysia 1 Jul 71
839	*Iredell County* Amer Bridge	25 Sep 44 12 Nov 44	6 Dec 44 18 Jun 66	Mar 46	To Indonesia Jul 70
840	*Iron County* Amer Bridge	28 Sep 44 15 Nov 44	11 Dec 44 3 Nov 50	1 Jun 46 23 Nov 57	To ROC 1 Jul 58
841	Amer Bridge	4 Oct 44 20 Nov 44	18 Dec 44	25 Jun 46	Str 23 Jun 47
842	Amer Bridge	9 Oct 44 24 Nov 44	19 Dec 44	30 Dec 47	To Philippines Dec 47
843	Amer Bridge	13 Oct 44 29 Nov 44	23 Dec 44	18 Dec 47	To Philippines Dec 47
844	Amer Bridge	18 Oct 44 3 Dec 44	30 Dec 44	15 Sep 47	Str 29 Sep 47
845	*Jefferson County* Amer Bridge	23 Oct 44 7 Dec 44	9 Jan 45	61	Str 1 Feb 61
846	*Jennings County* Amer Bridge	27 Oct 44 12 Dec 44	9 Jan 45 3 Nov 50 11 Jun 66	14 Oct 49 7 Dec 55 25 Sep 70	During Vietnam War operated as tender for PBR/PCF (i.e., as AGP) but was not redesignated, str 25 Sep 70 after she caught fire while tending riverine craft, sold 30 Mar 71
847	Amer Bridge	1 Nov 44 17 Dec 44	15 Jan 45	21 Jun 46	Str 31 Jul 46
848	*Jerome County* Amer Bridge	6 Nov 44 21 Dec 44	20 Jan 45 7 Dec 59	10 Aug 46 1 Apr 70	To Vietnam Apr 70
849	*Johnson County* Amer Bridge	10 Nov 44 30 Dec 44	16 Jan 45	13 Jun 46	To Korea Jan 59
850	*Juniata County* Chicago	15 Aug 44 3 Nov 44	27 Nov 44	17 May 46	Str 1 Nov 58, tgt 20 Oct 58
851	Chicago	10 Aug 44 8 Nov 44	30 Nov 44	24 Apr 46	Str 8 May 46, probably to Argentina 1946, misidentified as LST 875
852	Chicago	16 Aug 44 13 Nov 44	24 Nov 44 8 Sep 50 15 Feb 68	1 Aug 47 17 Apr 56 15 Oct 71	*Satyr* (ARL 23) 14 Aug 44, to Vietnam Oct 71, to Philippines 24 Jan 77
853	*Kane County* Chicago	30 Aug 44 17 Nov 44	11 Dec 44	24 Jul 46	To Korea 22 Dec 58
854	*Kemper County* Chicago	30 Aug 44 20 Nov 44	14 Dec 44 20 Nov 50	21 Oct 49 28 May 69	Str 1 Apr 75, to Barbados 24 Jun 75

No.	Name/Builder	LD/Launch	Comm	Decomm	Notes and Fate
855	*Kent County* Chicago	6 Sep 44 27 Nov 44	21 Dec 44 3 Nov 50	15 Feb 50 22 Jan 58	Tgt 22 Jan 58
856	Chicago	16 Sep 44 1 Dec 44	23 Dec 44	29 May 46	Str 3 Jul 46
857	*King County* Chicago	19 Sep 44 6 Dec 44	29 Dec 44	8 July 60	Converted to missile test ship AG 157 beginning Oct 57, reclassified 17 May 58, sold 25 Apr 61
858	Chicago	17 Oct 44 11 Nov 44	22 Dec 44	24 May 48	*Stentor* (ARL 26) 14 Aug 44, str Jul 60
859	*Lafayette County* Chicago	26 Sep 44 15 Dec 44	6 Jan 45	15 Aug 58	To ROC Aug 58
860	Chicago	23 Sep 44 19 Dec 44	13 Jan 45	1 Jun 46	Str 3 Jul 46, to France
861	Jeffersonville	18 Sep 44 4 Nov 44	30 Nov 44	10 Mar 47	Str 4 Apr 47
862	Jeffersonville	26 Sep 44 9 Nov 44	4 Dec 44	46	To MarComm 10 Oct 47
863	Jeffersonville	11 Oct 44 14 Nov 44	9 Dec 44	19 Jun 46	Str 31 Jul 46
864	Jeffersonville	3 Oct 44 18 Nov 44	13 Dec 44	1 May 47	Str 22 May 47
865	Jeffersonville	19 Oct 44 22 Nov 44	16 Dec 44	30 Dec 47	To Philippines 30 Dec 47
866	Jeffersonville	14 Oct 44 27 Nov 44	21 Dec 44	27 Jun 46	Str 31 Jul 46
867	Jeffersonville	23 Oct 44 1 Dec 44	18 Dec 44	2 Jul 46	Str 31 Jul 46
868	Jeffersonville	31 Oct 44 6 Dec 44	30 Dec 44	9 Aug 46	Str 10 Jun 47
869	Jeffersonville	27 Oct 44 11 Dec 44	6 Jan 45	31 Jul 46	Str 28 Aug 46, to Argentina Dec 47
870	Jeffersonville	4 Oct 44 15 Dec 44	10 Jan 45	Jun 46	Str 28 Aug 46
871	Jeffersonville	9 Nov 44 20 Dec 44	18 Jan 45	4 Oct 46	LSTH 15 Sep 45, str 13 Nov 46
872	Jeffersonville	18 Nov 44 28 Dec 44	22 Jan 45	8 Jul 46	Str 15 Aug 46, to Argentina Dec 47
873	Jeffersonville	14 Nov 44 3 Jan 45	27 Jan 45	8 Aug 46	Str 25 Sep 46
874	Missouri	16 Oct 44 25 Nov 44	18 Dec 44	29 May 46	Str 3 Jul 46, to France Jun 48
875	Missouri	18 Oct 44 29 Nov 44	22 Dec 44	22 Apr 46	Str 19 Jul 46, to Philippines 2 Jul 48; this ship also listed as having gone to Argentina as BDT 1, but LST 851 is more likely; for many years this ex–US LST number appeared in the lists of LSTs in both navies
876	Missouri	21 Oct 44 2 Dec 44	27 Dec 44	28 Jun 46	Str 31 Jul 46
877	Missouri	25 Oct 44 6 Dec 44	1 Jan 45	1 May 46	Str 3 Jul 46
878	Missouri	30 Oct 44 9 Dec 44	3 Jan 45	3 May 46	Str 19 Jul 46
879	Missouri	2 Nov 44 13 Dec 44	5 Jan 45	26 Jun 46	Str 25 Sep 46
880	*Lake County* Missouri	6 Nov 44 16 Dec 44	9 Jan 45 20 Aug 51	1 Oct 46 25 Nov 58	Tgt

No.	Name/Builder	LD/Launch	Comm	Decomm	Notes and Fate
881	Missouri	10 Nov 44 20 Dec 44	15 Jan 45	14 Feb 47	Str 5 Mar 47
882	Missouri	14 Nov 44 23 Dec 44	18 Jan 45	5 Jul 46	Str 28 Aug 46
883	La Moure County Missouri	16 Nov 44 30 Dec 44	23 Jan 45 1 Jul 50	20 Apr 46 7 Dec 59	Str 1 Jul 60
884	Dravo-Pitts	23 Jul 44 30 Sep 44	10 Oct 44	16 Feb 46	Extensively damaged by kamikaze 1 Apr 45, hulk sunk 6 May 46
885	Dravo-Pitts	13 Aug 44 23 Sep 44	26 Oct 44	29 Apr 46	Str 3 Jul 46
886	Dravo-Pitts	20 Aug 44 30 Sep 44	2 Nov 44	10 May 46	Str 19 Jun 46
887	Lawrence County Dravo-Pitts	27 Aug 44 7 Oct 44	7 Nov 44 3 Nov 50	23 Jul 46 22 Mar 60	Str 1 Nov 60, to Indonesia
888	Lee County Dravo-Pitts	11 Aug 44 14 Oct 44	13 Nov 44	2 Sep 46	Str 21 Sep 60
889	Dravo-Pitts	3 Sep 44 14 Oct 44	18 Nov 44	28 May 46	Str 19 Jul 46
890	Dravo-Pitts	10 Sep 44 21 Oct 44	24 Nov 44	24 May 46	Str 3 Jul 46
891	Dravo-Pitts	21 Aug 44 28 Oct 44	27 Nov 44	2 Jul 46	Str 31 Jul 46
892	Dravo-Pitts	17 Sep 44 28 Oct 44	30 Nov 44	5 Jul 46	Str 28 Aug 46
893	Dravo-Pitts	24 Sep 44 4 Nov 44	4 Dec 44	8 May 46	Str 19 Jun 46
894	Dravo-Pitts	4 Sep 44 11 Nov 44	12 Dec 44	29 Apr 46	Str 19 Jul 46
895	Dravo-Pitts	1 Oct 44 11 Nov 44	16 Dec 44	17 Aug 46	Str 12 Mar 48
896	Dravo-Pitts	6 Oct 44 18 Nov 44	20 Dec 44	3 Dec 45	CTL in Okinawa typhoon 9 Oct 45, str 3 Jan 46, scuttled 8 Mar 46
897	Dravo-Pitts	19 Sep 44 25 Nov 44	22 Dec 44	23 Jul 46	Str 28 Aug 48
898	Lincoln County Dravo-Pitts	15 Oct 44 25 Nov 44	29 Dec 44 28 Aug 50	9 May 46 24 Mar 61	To army 46, ret 50, to Thailand 31 Aug 62
899	Dravo-Pitts	22 Oct 44 2 Dec 44	1 Jan 45	15 Jul 46	Str 15 Aug 46
900	Linn County Dravo-Pitts	1 Oct 44 9 Dec 44	6 Jan 45	15 May 46	To Korea 2 Dec 58
901	Litchfield County Dravo-Pitts	29 Oct 44 9 Dec 44	11 Jan 45 30 Nov 51	9 Aug 46 6 Dec 69	Sold 14 Jan 77
902	Luzerne County Dravo-Pitts	5 Nov 44 16 Dec 44	15 Jan 45 18 Jan 52 29 Mar 63	3 Aug 46 30 Nov 55 12 Aug 70	Str 12 Aug 70
903	Lyman County Dravo-Pitts	15 Oct 44 23 Dec 44	20 Jan 45	10 Sep 46	Str 1 Nov 58, tgt 28 Mar 59
904	Lyon County Dravo-Pitts	12 Nov 44 23 Dec 44	25 Jan 45	15 Nov 46	Str 1 Nov 58, tgt 13 May 59
905	Madera County Dravo-Pitts	19 Nov 44 30 Dec 44	20 Jan 45 30 Mar 63	11 Sep 46	To Philippines Nov 69
906	Beth-H	24 Jan 44 11 Mar 44	27 Apr 44	20 May 45	Grounded 18 Oct 44, str 22 Jun 45
907	Beth-H	31 Jan 44 18 Mar 44	30 Apr 44	18 Oct 46	To Venezuela 25 Nov 46
908	Beth-H	14 Feb 44 28 Mar 44	8 May 44	30 Jul 46	Str 28 Aug 46

No.	Name/Builder	LD/Launch	Comm	Decomm	Notes and Fate
909		19 Feb 44	11 May 44	21 Jun 46	Str 31 Jul 46
	Beth-H	3 Apr 44			
910		23 Feb 44	24 May 44	27 Jun 46	Str 31 Jul 46
	Beth-H	8 Apr 44			
911		28 Feb 44	14 May 44	24 Jun 46	Str 31 Jul 46
	Beth-H	12 Apr 44			
912	*Mahnomen County*	5 Feb 44	21 May 44	25 Aug 55	Str 31 Jan 67 after grounding 30 Dec 66
	Beth-H	22 Apr 44	27 Mar 63		
913		15 Mar 44	23 May 44	16 Jul 46	Str 14 Mar 47
	Beth-H	26 Apr 44			
914	*Mahoning County*	16 Feb 44	18 May 44	26 Jun 46	To army, ret 50, sold 22 Jun 60
	Beth-H	18 Apr 44	26 Aug 50	5 Sep 59	
915		22 Mar 44	27 May 44	25 Jun 46	Str 31 Jul 46
	Beth-H	3 May 44			
916		22 Mar 44	25 May 44	5 Apr 46	To army, str 29 Sep 47, lost in typhoon 49
	Beth-H	29 Apr 44			
917		31 Mar 44	28 May 44	24 May 46	Str 3 Jul 46
	Beth-H	6 May 44			
918		5 Apr 44	29 May 44	12 Jun 46	Str 31 Jul 46
	Beth-H	7 May 44			
919		11 Apr 44	31 May 44	5 Aug 46	Str 25 Sep 46, to Argentina Jan 48
	Beth-H	17 May 44			
920		26 Apr 44	17 Jun 44	8 Jul 46	Str 14 Mar 47
	Beth-H	29 May 44			
921		1 May 44	23 Jun 44	29 Sep 44	WL(T) 14 Aug 44
	Beth-H	2 Jun 44			
922		26 Apr 44	29 Jun 44	8 Jul 46	Str 28 Aug 46
	Beth-H	7 Jun 44			
923		3 May 44	6 Jul 44	10 Jul 46	Str 15 Aug 46
	Beth-H	11 Jun 44			
924		8 May 44	10 Jul 44	13 Jun 46	Str 3 Jul 46, to Thailand Oct 47
	Beth-H	17 Jun 44			
925		10 May 44	15 Jul 44	26 Nov 45	Str 5 Dec 45
	Beth-H	21 Jun 44			
926		13 May 44	20 Jul 44	14 Jun 46	Str 31 Jul 46
	Beth-H	24 Jun 44			
927		20 May 44	4 Jul 44	20 Jul 46	Str 8 Oct 46
	Beth-H	28 Jun 44			
928		1 Jun 44	30 Jul 44	13 Dec 46	*Cameron* (APB 50) 1 Aug 45, str 59
	Beth-H	5 Jul 44			
929		5 Jun 44	2 Aug 44	24 May 46	To ROC, str 3 Jul 46
	Beth-H	8 Jul 44			
930		9 Jun 44	6 Aug 44	26 Jun 46	Str 31 Jul 46
	Beth-H	12 Jul 44			
931		13 Jun 44	11 Aug 44	26 Jun 46	LSTH 15 Sep 45, str 31 Jul 46
	Beth-H	19 Jul 44			
932		21 Jun 44	15 Aug 44	24 Jun 46	Str 31 Jul 46
	Beth-H	22 Jul 44			
933		23 Jun 44	20 Aug 44	2 Jul 46	Str 15 Aug 46
	Beth-H	26 Jul 44			
934		29 Jun 44	25 Aug 44	13 May 46	Str 19 Jun 46
	Beth-H	29 Jul 44			
935		3 Jul 44	29 Aug 44	2 Jul 46	Str 15 Aug 46
	Beth-H	5 Aug 44			
936		7 Jul 44	1 Sep 44	17 May 46	Str 5 Jun 46
	Beth-H	9 Aug 44			
937		11 Jul 44	6 Sep 44	24 May 46	Str 3 Jul 46
	Beth-H	12 Aug 44			

No.	Name/Builder	LD/Launch	Comm	Decomm	Notes and Fate
938	*Maricopa County* Beth-H	14 Jul 44 15 Aug 44	9 Sep 44 14 Dec 51	Dec 49 29 Feb 56	Str 1 Jun 62, to Vietnam 12 Jul 62
939	Beth-H	21 Jul 44 23 Aug 44	14 Sep 44	22 Jun 46	Str 31 Jul 46
940	Beth-H	25 Jul 44 26 Aug 44	20 Sep 44	13 Jul 46	Str 28 Aug 46
941	Beth-H	28 Jul 44 30 Aug 44	22 Sep 44	1 May 46	Str 3 Jul 46
942	Beth-H	1 Aug 44 6 Sep 44	26 Sep 44	26 Jun 46	Str 31 Jul 46
943	Beth-H	8 Aug 44 9 Sep 44	30 Sep 44	16 Jul 46	Str 25 Sep 46
944	Beth-H	11 Aug 44 13 Sep 44	4 Oct 44	19 Dec 45	Str 8 Jan 46
945	Beth-H	11 Aug 44 16 Sep 44	9 Oct 44	16 Apr 46	Str 19 Jul 46, to ROC merch service, to ROC navy 55
946	Beth-H	15 Aug 44 20 Sep 44	12 Oct 44	25 Jun 46	Str 31 Jul 46
947	Beth-H	18 Aug 44 23 Sep 44	15 Oct 44	16 Aug 46	Str 15 Oct 46
948	Beth-H	25 Aug 44 28 Sep 44	19 Oct 44	7 Jul 47	*Myrmidon* (ARL 16) 11 Aug 44, str 1 Apr 60
949	Beth-H	29 Aug 44 30 Sep 44	23 Oct 44	18 Jul 46	Str 25 Sep 46
950	Beth-H	1 Sep 44 4 Oct 44	27 Oct 44	23 Sep 46	Str 10 Jun 47
951	Beth-H	8 Sep 44 7 Oct 44	31 Oct 44	8 Aug 46	Str 25 Sep 46
952	Beth-H	11 Sep 44 11 Oct 44	3 Nov 44	1 Aug 46	Str 22 Jan 47
953	*Marinette County* Beth-H	15 Sep 44 15 Oct 44	7 Nov 44	12 Nov 46	Str 1 Nov 58
954	Beth-H	19 Sep 44 18 Oct 44	10 Nov 44	1 Jul 47	*Numitor* (ARL 17) 14 Aug 44, str 1 Apr 60
955	Beth-H	22 Sep 44 22 Oct 44	13 Nov 44	2 Oct 46	*Patroclus* (ARL 19) 14 Aug 44, str 22 Aug 52, to Turkey 15 Nov 52
956	Beth-H	11 Jul 44 21 Aug 44	20 Mar 45	29 Jan 47	*Sarpedon* (ARB 7) 14 Aug 44, sold 16 Nov 76
957	Beth-H	30 Sep 44 30 Oct 44	20 Nov 44	20 May 46	Str 22 Jan 48
958	Beth-H	3 Oct 44 31 Oct 44	25 Nov 44	14 Mar 46	Str 28 Mar 46
959	Beth-H	6 Oct 44 4 Nov 44	29 Nov 44	13 Jun 46	Str 3 Jul 46
960	Beth-H	11 Oct 44 8 Nov 44	2 Dec 44	2 Jul 46	Str 15 Aug 46
961	Beth-H	13 Oct 44 11 Nov 44	6 Dec 44	23 Jul 46	Str 28 Aug 46
962	Beth-H	17 Oct 44 15 Nov 44	9 Dec 44 2 Apr 52	3 Feb 47 1 Jun 56	*Romulus* (ARL 22) 14 Aug 44, str Oct 60, to Philippines Oct 61
963	Beth-H	20 Oct 44 18 Nov 44	12 Dec 44 3 Nov 50 16 Dec 67 26 Jul 85	26 May 47 31 Jan 56 30 Sep 71 19 Jun 89	*Sphinx* (ARL 24) 14 Aug 44, used 85–89 as intelligence collection ship, str on decomm
964	Beth-H	24 Oct 44 22 Nov 44	16 Dec 44	27 Jun 46	Str 15 Aug 46
965	Beth-H	27 Oct 44 25 Nov 44	20 Dec 44	3 Jun 46	Str 19 Jul 46

AMPHIBIOUS SHIPS 573

No.	Name/Builder	LD/Launch	Comm	Decomm	Notes and Fate
966	Beth-H	31 Oct 44 29 Nov 44	12 Jun 45	9 May 46	*Callisto* (AGP 15) 14 Aug 44, to MarComm 14 May 48
967	Beth-H	2 Nov 44 2 Dec 44	27 Dec 44	20 Feb 47	*Ulysses* (ARB 7) 14 Aug 44, to Germany Jun 61
968	Beth-H	7 Nov 44 9 Dec 44	3 Jan 45	2 Jul 46	Str 15 Aug 46
969	Beth-H	10 Nov 44 13 Dec 44	9 Jan 45	12 Jul 46	Str 15 Aug 46
970	Beth-H	14 Nov 44 16 Dec 44	13 Jan 45	10 Jul 46	Str 15 Aug 46
971	Beth-H	17 Nov 44 20 Dec 44	15 Jan 45 14 Dec 50	5 Jun 47 5 Sep 55	*Menelaus* (ARL 13) 14 Aug 44, str 1 Jun 60
972	Beth-H	21 Nov 44 22 Dec 44	22 Jan 45	25 Jun 46	Str 15 Aug 46
973	Beth-H	25 Nov 44 27 Dec 44	27 Jan 45 6 Sep 50	24 May 46 7 Nov 51	To army, str 29 Sep 47, but ret for Korea, to France 7 Nov 51
974	Beth-H	28 Nov 44 31 Dec 44	31 Jan 45	14 May 46	Str 19 Jun 46
975	*Marion County* Beth-H	1 Dec 44 6 Jan 45	3 Feb 45 28 Aug 50	16 Apr 46	Navy and MSTS service, to Vietnam 12 Apr 62
976	Beth-H	5 Dec 44 10 Jan 45	5 Feb 45	20 May 47	*Telemon* (ARB 8) 14 Aug 44, str 1 Jun 73
977	Beth-H	12 Dec 44 15 Jan 45	8 Feb 45	28 Jan 46	*Alecto* (AGP 14) 14 Aug 44, to Turkey 10 May 48
978	Beth-H	15 Dec 44 20 Jan 45	15 Feb 45	6 Jun 46	Str 3 Jul 46
979	Beth-H	19 Dec 44 23 Jan 45	20 Feb 45	5 Jul 46	Str 28 Aug 46
980	*Meeker County* BOSNY	9 Dec 43 27 Jan 44	26 Feb 44 23 Sep 66	16 Dec 55 15 Dec 70	Str 1 Apr 75
981	BOSNY	9 Dec 43 27 Jan 44	11 Mar 44	30 Jul 46	Str 28 Aug 46
982	BOSNY	22 Dec 43 10 Feb 44	19 Mar 44	25 Apr 46	Str 19 Jul 46
983	*Middlesex County* BOSNY	22 Dec 43 10 Feb 44	25 Mar 44 27 Sep 61	10 Jan 56 15 Oct 69	Str 15 Sep 74
984	BOSNY	3 Jan 44 25 Feb 44	1 Apr 44	25 Jun 46	Str 31 Jul 46
985	BOSNY	3 Jan 44 25 Feb 44	7 Apr 44	11 Jun 46	Str 3 Jul 46
986	BOSNY	15 Jan 44 5 Mar 44	14 Apr 44	18 Jul 46	Str 28 Aug 46
987	*Millard County* BOSNY	2 Feb 44 5 Mar 44	19 Apr 44	3 Sep 46	Str 1 Jun 60, to Germany Aug 61 after planned ARB conversion canceled, NRT Sep 46–early 50
988	*Mineral County* BOSNY	10 Feb 44 12 Mar 44	25 Apr 44 7 Jun 51	13 Jun 50 11 Oct 57	Str 27 Sep 57, tgt
989	BOSNY	10 Feb 44 12 Mar 44	28 Apr 44	7 Oct 46	Str 13 Nov 46
990	BOSNY	26 Feb 44 27 Mar 44	1 May 44	10 Jul 46	Str 25 Sep 46
991	BOSNY	26 Feb 44 27 Mar 44	6 May 44	3 May 46	To State Dept for disposal 3 May 46
992	BOSNY	5 Mar 44 7 Apr 44	10 May 44	9 Aug 46	Str 25 Sep 46
993	BOSNY	7 Mar 44 7 Apr 44	12 May 44	1 Jun 46	To ROC 7 Feb 48

No.	Name/Builder	LD/Launch	Comm	Decomm	Notes and Fate
994	BOSNY	12 Mar 44 17 Apr 44	17 May 44	31 Jul 46	Str 28 Aug 46, to Argentina Nov 47
995	BOSNY	12 Mar 44 2 May 44	20 May 44	15 Aug 46	Str 25 Sep 46, to Argentina Nov 47
996	BOSNY	27 Mar 44 2 May 44	23 May 44	22 Apr 46	Str 8 May 46
997	BOSNY	27 Mar 44 12 May 44	27 May 44	7 Mar 47	Str 4 Apr 47
998	BOSNY	8 Apr 44 14 May 44	29 May 44	26 Jun 46	Str 31 Jul 46, to Argentina Nov 47
999	BOSNY	8 Apr 44 14 May 44	30 May 44	29 Jul 46	Str 25 Sep 46
1000	BOSNY	18 Apr 44 26 May 44	14 Jun 44	22 Jul 46	Str 28 Aug 46
1001	BOSNY	18 Apr 44 26 May 44	20 Jun 44	26 Feb 46	Str 19 Jun 46
1002	BOSNY	3 May 44 8 Jun 44	25 Jun 44	22 May 46	Str 3 Jul 46
1003	BOSNY	3 May 44 8 Jun 44	29 Jun 44	29 Jul 46	*Coronis* (ARL 10) 14 Aug 44, str 1 Jul 61
1004	Beth-Q	26 Jan 44 3 Mar 44	28 Mar 44	27 Jun 46	Str 7 Feb 47, to Argentina
1005	Beth-Q	2 Feb 44 11 Mar 44	6 Apr 44	6 Apr 46	Str 17 Apr 46 after CTL
1006	Beth-Q	5 Feb 44 11 Mar 44	12 Apr 44	26 Jul 46	Str 28 Aug 46
1007	Beth-Q	8 Feb 44 20 Mar 44	15 Apr 44	2 Mar 46	Str 12 Apr 46
1008	Beth-Q	16 Feb 44 23 Mar 44	18 Apr 44	4 May 46	Str 19 Jun 46, to State Dept for transfer to ROC but transferred to Argentina
1009	Beth-Q	22 Feb 44 23 Mar 44	22 Apr 44		To army 17 Jul 46
1010	Beth-Q	22 Feb 44 29 Mar 44	25 Apr 44		To army 4 Apr 47, ret 1 Mar 50, to Korea 22 Mar 55
1011	Beth-Q	29 Feb 44 29 Mar 44	5 May 44	20 Jun 46	Str 31 Jul 46
1012	Beth-Q	4 Mar 44 8 Apr 44	30 Apr 44	10 Jun 46	Str 19 Jul 46
1013	Beth-Q	13 Mar 44 16 Apr 44	2 May 44	11 Jun 46	Str 19 Jul 46
1014	Beth-Q	15 Mar 44 16 Apr 44	5 May 44	5 Mar 46	Str 17 Apr 46
1015	Beth-Q	22 Mar 44 20 Apr 44	8 May 44	6 May 46	Str 19 Jun 46
1016	Beth-Q	25 Mar 44 25 Apr 44	10 May 44	26 Jun 46	Str 31 Jul 46
1017	Beth-Q	25 Mar 44 25 Apr 44	12 May 44	29 Jun 46	To ROC 14 Dec 46
1018	Beth-Q	31 Mar 44 6 May 44	14 May 44	16 Aug 46	Str 23 Jun 47
1019	Beth-Q	31 Mar 44 6 May 44	17 May 44	30 Jul 46	Str 25 Sep 46
1020	Beth-Q	11 Apr 44 10 May 44	19 May 44	16 Jul 46	Str 28 Aug 46
1021	Beth-Q	18 Apr 44 16 May 44	21 May 44		To UK 24 Dec 44, ret and str 1 Aug 47
1022	Beth-Q	18 Apr 44 16 May 44	24 May 44	31 Dec 47	Str 22 Jan 48

No.	Name/Builder	LD/Launch	Comm	Decomm	Notes and Fate
1023	Beth-Q	20 Apr 44 17 May 44	26 May 44	19 Jul 46	Str 28 Aug 46
1024	Beth-Q	26 Apr 44 22 May 44	28 May 44	27 Jun 46	Str 31 Jul 46
1025	Beth-Q	26 Apr 44 22 May 44	31 May 44	24 May 46	Str 11 Jun 48
1026	Beth-Q	8 May 44 2 Jun 44	7 Jun 44	11 Aug 46	Str 28 Aug 46
1027	Beth-Q	8 May 44 2 Jun 44	7 Jun 44	4 Sep 46	Str 23 Apr 47
1028	BOSNY	15 May 44 18 Jun 44	7 Jul 44	19 Nov 45	Str 5 Dec 45
1029	BOSNY	15 May 44 18 Jun 44	13 Jul 44	1 May 46	Str 10 Jun 47
1030	BOSNY	27 May 44 25 Jun 44	19 Jul 44	29 May 46	To ROC 17 Feb 48
1031	BOSNY	27 May 44 25 Jun 44	25 Jul 44	18 Dec 45	Str 8 Jan 46
1032	Monmouth County BOSNY	9 Jun 44 9 Jul 44	1 Aug 44 28 May 63	14 Nov 55 70	Str 12 Aug 70
1033	BOSNY	9 Jun 44 9 Jul 44	12 Aug 44	1 Aug 46	Str 28 Aug 46
1034	BOSNY	26 Jun 44 4 Aug 44	26 Aug 44	8 Aug 46	To NEI 28 Oct 46
1035	BOSNY	26 Jun 44 4 Aug 44	1 Sep 44	6 Jun 46	Str 3 Jul 46
1036	BOSNY	10 Jul 44 24 Aug 44	16 Sep 44	8 Jun 49	*Creon* (ARL 11) 14 Aug 44, str Jul 60
1037	BOSNY	10 Jul 44 24 Aug 44	22 Sep 44	30 Nov 46	*Poseidon* (ARL 12) 14 Aug 44, str 1 Jul 61
1038	Monroe County Dravo-Pitts	29 Oct 44 6 Jan 45	5 Feb 45	Jun 49	Str 1 Nov 58, NRT May 46–Jun 49
1039	Dravo-Pitts	26 Nov 44 6 Jan 45	9 Feb 45	21 Jun 46	Str 31 Jul 46
1040	Dravo-Pitts	3 Dec 44 13 Jan 45	13 Feb 45	23 Sep 46	To NEI 5 Oct 46
1041	Montgomery County Dravo-Pitts	12 Nov 44 20 Jan 45	19 Feb 45	31 Jan 56	Str 1 Jun 60, to Germany Aug 60 for ARB conversion, conversion aborted
1042	Dravo-Pitts	10 Dec 44 20 Jan 45	22 Feb 45	9 May 46	Str 19 Jun 46
1043	Dravo-Pitts	17 Dec 44 27 Jan 45	24 Feb 45	22 Jul 46	Str 28 Aug 46
1044	Dravo-Pitts	25 Nov 44 3 Feb 45	2 Mar 45	28 Jun 46	Str 31 Jul 46, to Argentina Jan 48
1045	Dravo-Pitts	22 Dec 44 3 Feb 45	27 Mar 45	10 Jul 46	Str 15 Aug 46
1046	Dravo-Pitts	31 Dec 44 10 Feb 45	28 Mar 45	27 Jun 46	Str 31 Jul 46
1047	Dravo-Pitts	9 Dec 44 17 Feb 45	28 Mar 45	6 May 46	To army 25 Jun 46, str 29 Sep 47
1048	Morgan County Dravo-Pitts	7 Jan 45 17 Feb 45	28 Mar 45 26 Aug 50	14 May 46 10 May 56	To army 46–50, MSTS 56, str 1 Aug 59
1049	Dravo-Pitts	14 Jan 45 24 Feb 45	30 Mar 45	18 Jul 46	Str 19 Feb 48
1050	Dravo-Pitts	23 Dec 44 3 Mar 45	3 Apr 45	27 Jan 47	To ROC 27 Jan 47
1051	Dravo-Pitts	21 Jan 45 3 Mar 45	7 Apr 45	11 Jul 46	Str 15 Aug 46

No.	Name/Builder	LD/Launch	Comm	Decomm	Notes and Fate
1052	Dravo-Pitts	29 Jan 45 6 Mar 45	15 Apr 45	11 Jul 46	Str 15 Aug 46
1053	Dravo-Pitts	6 Jan 45 6 Mar 45	23 Apr 45	3 Jun 46	Str 3 Jul 46
1054	Dravo-Pitts	4 Feb 45 17 Mar 45	17 Apr 45	28 Jun 46	Str 31 Jul 46
1055	Dravo-Pitts	10 Feb 45 24 Mar 45	26 Apr 45	13 Feb 47	Str 25 Feb 47
1056	Dravo-Pitts	20 Jan 45 24 Mar 45	2 May 45	12 Jul 46	Str 10 Jun 47, to ROC merch service, to ROC navy 55
1057	Dravo-Pitts	17 Feb 45 31 Mar 45	7 May 45	5 Aug 46	Str 25 Sep 46, to Argentina Jan 48
1058	Dravo-Pitts	24 Feb 45 7 Apr 45	16 May 45	30 Jul 46	Str 25 Sep 46
1059	Dravo-Pitts	3 Mar 45 14 Apr 45	17 May 45	14 Sep 46	Str 23 Apr 47
1060	Beth-H	22 Dec 44 29 Jan 45	24 Feb 45	7 Sep 46	Str 23 Apr 47
1061	Beth-H	26 Dec 44 3 Feb 45	1 Mar 45	1 May 46	Str 3 Jul 46
1062	Beth-H	30 Dec 44 6 Feb 45	5 Mar 45	27 Jun 46	Str 31 Jul 46
1063	Beth-H	3 Jan 45 11 Feb 45	8 Mar 45	13 Jul 46	To MarComm 30 Jun 48
1064	*Nansemond County* Beth-H	9 Jan 45 14 Feb 45	12 Mar 45	21 Aug 46	Str 1 Oct 59, to Japan Apr 61
1065	Beth-H	12 Jan 45 17 Feb 45	16 Mar 45	23 May 46	Str 23 Jun 47
1066	*New London County* Beth-H	18 Jan 45 21 Feb 45	20 Mar 45 21 Dec 65	Mar 46 27 Feb 67	MSTS 27 Feb 67, to Chile 29 Aug 73
1067	*Nye County* Beth-H	24 Jan 45 27 Feb 45	24 Mar 45 21 Dec 65	13 Aug 46 27 Mar 67	MSTS 27 Mar 67, to Chile 29 Aug 73
1068	*Orange County* Beth-H	31 Jan 45 3 Mar 45	27 Mar 45 8 Sep 50	9 Aug 46 27 Sep 57	Str 27 Sep 57, tgt 18 Jun 58
1069	*Orleans County* Beth-H	7 Feb 45 7 Mar 45	31 Mar 45 11 Jan 52	6 Aug 46 20 May 65	Mine warfare flagship 52, redesignated MCS 6 in 59, MSTS May 66, to Philippines 13 Sep 76
1070	Beth-H	8 Feb 45 9 Mar 45	5 Apr 45 6 Oct 46	3 Dec 46 16 Nov 56	*Electron* (AG 146) 27 Jan 49, to AKS 27 on 18 Aug 51, str 1 Apr 60
1071	*Quachita County* Beth-H	13 Feb 45 14 Mar 45	9 Apr 45 3 Jan 51	10 Jun 46 15 Feb 56	Str 1 Nov 59
1072	Beth-H	16 Feb 45 20 Mar 45	12 Apr 45		To MSTS 2 Apr 51, to Philippines 13 Sep 76
1073	*Outagamie County* Beth-H	20 Feb 45 22 Mar 45	17 Apr 45 3 Nov 50	5 Aug 46 21 May 71	To Brazil 24 May 71
1074	*Overton County* Beth-H	24 Feb 45 27 Mar 45	21 Apr 45	4 Sep 46	Str 1 Nov 58
1075	Beth-H	5 Mar 45 3 Apr 45	25 Apr 45	18 Dec 46	To ROC on decomm
1076	*Page County* Beth-H	16 Mar 45 14 Apr 45	1 May 45 28 Nov 60	13 Jun 46 5 Mar 71	To Greece on decomm
1077	*Park County* Beth-H	21 Mar 45 18 Apr 45	8 May 45 6 Sep 50 9 Apr 66	31 Jul 46 12 May 55 Sep 71	To Mexico on decomm
1078	Beth-H	27 Mar 45 25 Apr 45	15 May 45 Feb 51	28 Apr 48 22 Apr 58	*Proton* (AG 147) 27 Jan 49, to AKS 28 on 18 Aug 51, str 1 Jan 59
1079	*Fayette County* Beth-H	30 Mar 45 27 Apr 45	22 May 45 Oct 50	Mar 46 1 Nov 59	Str 1 Nov 59

No.	Name/Builder	LD/Launch	Comm	Decomm	Notes and Fate
1080	*Fender County* Beth-H	5 Apr 45 2 May 45	29 May 45 3 Oct 50	29 Aug 46 2 Jan 58	To Korea Oct 58
1081	*Pima County* Amer Bridge	13 Nov 44 5 Jan 45	30 Jan 45 2 Feb 51	30 Jul 46 12 Dec 56	Str 1 Nov 58, to Germany Aug 61
1082	*Pitkin County* Amer Bridge	18 Nov 44 26 Jan 45	7 Feb 45 6 Sep 50 9 Jul 66	5 Aug 46 1 Sep 55 1 Sep 71	To Brazil 31 Aug 71
1083	*Plumas County* Amer Bridge	22 Nov 44 14 Jan 45	13 Feb 45 8 Sep 50	Aug 46 22 Aug 61	MSTS Dec 65, str 1 Jun 72
1084	*Polk County* Amer Bridge	27 Nov 44 19 Jan 45	19 Feb 45 3 Nov 50	13 Aug 46 3 Oct 69	Str 15 Sep 74, FRAM II refit
1085	Amer Bridge	1 Dec 44 13 Jan 45	21 Feb 45	N/A	*Colington* (AG 148) 27 Jan 49, to AKS 28 on 18 Aug 51, str 1 Apr 60
1086	*Potter County* Amer Bridge	5 Dec 44 28 Jan 45	24 Feb 45	7 Aug 46	To Greece 9 Aug 60
1087	Amer Bridge	11 Dec 44 3 Feb 45	2 Mar 45	11 Aug 47	Str 29 Sep 47
1088	*Pulaski County* Amer Bridge	16 Dec 44 11 Feb 45	27 Mar 45 21 May 63	29 Aug 46	MSTS Jul 67
1089	*Rice County* Amer Bridge	20 Dec 44 17 Feb 45	28 Mar 45 6 Sep 50	16 Aug 46 9 Mar 60	To Germany Oct 60, ret, to Turkey 12 Dec 72, conversion to minelayer in German service
1090	*Russel County* Amer Bridge	28 Dec 44 24 Feb 45	2 Apr 45 3 Nov 50	22 Jul 46 5 Apr 60	Str 1 Nov 60, to Indonesia Dec 60
1091	*Sagadahoc County* Amer Bridge	3 Jan 45 3 Mar 45	6 Apr 45	5 Jul 46	To ROC Oct 58
1092	Amer Bridge	8 Jan 45 24 Mar 45	19 May 45 25 Jul 50	30 Aug 46 4 Apr 52	*Aventinus* (ARVE 3) 8 Dec 44, to Chile Aug 63
1093	Amer Bridge	12 Jan 45 11 Apr 45	31 May 45 28 Jul 50	30 Aug 46 4 Apr 52	*Fabius* (ARVA 5) 8 Dec 44, str 1 Jun 73
1094	Amer Bridge	17 Jan 45 21 Apr 45	19 Jun 45 5 Jan 51 N/A	18 Jun 46 9 Dec 55 9 Dec 66	*Chloris* (ARVE 4) 8 Dec 44, str 1 Jun 73
1095	Amer Bridge	22 Jan 45 25 Apr 45	27 Jun 45 5 Jan 51	3 Jun 46 16 Jan 56	*Megara* (ARVA 6) 8 Dec 44, str 1 Jun 73, to Mexico 1 Oct 73
1096	*St. Clair County* Jeffersonville	27 Nov 44 10 Jan 45	2 Feb 45 3 Oct 50	24 Aug 46 26 Sep 69	Str 1 Apr 75
1097	Jeffersonville	22 Nov 44 16 Jan 45	9 Feb 45 3 Jan 51	19 Dec 46 14 Dec 56	*League Island* (AG 149) 27 Jan 46, str 1 Apr 60
1098	Jeffersonville	6 Dec 44 27 Jan 45	28 May 45	21 Apr 47	*Laysan Island* (ARST 1) 8 Dec 44, str 1 Jun 73
1099	Jeffersonville	1 Dec 44 8 Feb 45	19 Jun 45	5 Aug 46	*Okala* (ARST 2) 8 Dec 44, str 15 Oct 46
1100	Jeffersonville	11 Dec 44 20 Feb 45	28 Jul 45	20 Jun 47	*Palmyra* (ARST 3) 8 Dec 44, str 1 Jun 73
1101	*Saline County* Missouri	22 Nov 44 3 Jan 45	26 Jan 45 3 Nov 50	6 Jun 46 9 Mar 60	Str 1 Nov 60, to Germany, conversion to minelayer in German service, to Turkey 13 Dec 72
1102	Missouri	23 Nov 44 10 Jan 45	29 Jan 45 27 Dec 50	21 Nov 47 22 Apr 58	*Chimon* (AG 150) 1 Feb 49, to AKS 31 on 18 Aug 51, Str 2 Nov 59
1103	Missouri	28 Nov 44 13 Jan 45	31 Jan 45	18 Jun 46	Str 23 Jun 47
1104	Missouri	1 Dec 44 17 Jan 45	8 Feb 45	8 Jul 46	Str 22 May 47, to Argentina
1105	Missouri	5 Dec 44 20 Jan 45	13 Feb 45	29 May 46	Str 19 Jun 46
1106	Missouri	9 Dec 44 24 Jan 45	16 Feb 45	2 Aug 46	Str 8 Oct 46

No.	Name/Builder	LD/Launch	Comm	Decomm	Notes and Fate
1107	Missouri	13 Dec 44 29 Jan 45	21 Feb 45	1 May 46	Str 3 Jul 46
1108	Missouri	16 Dec 44 1 Feb 45	27 Feb 45	15 Aug 46	Str 25 Sep 46, to Argentina 10 Jan 48
1109	Missouri	21 Dec 44 6 Feb 45	28 Feb 45	6 May 46	Str 19 Jun 46
1110	San Bernadino County Missouri	28 Dec 44 9 Feb 45	7 Mar 45	15 Aug 58	To ROC on decomm
1111	Missouri	2 Jan 45 9 Apr 45	9 Jun 45	26 Apr 47	*Blackford* (AKS 16) 8 Dec 44, to APB 45 on 6 Mar 45, str Apr 60
1112	Missouri	5 Jan 45 12 Apr 45	15 Jun 45	16 Oct 46	*Dorchester* (AKS 17) 8 Dec 44, to APB 46 on 6 Mar 45, str 1 Jun 73
1113	Missouri	8 Jan 45 17 Apr 45	27 Jun 45	25 Jan 47	*Kingman* (AKS 18) 8 Dec 44 to APB 47 on 6 Mar 45, str 1 Oct 77
1114	Missouri	12 Jan 45 20 Apr 45	22 Jun 45	8 Jun 46	*Presque Isle* (AKS 18) 8 Dec 44, to *Vandenburgh* 17 Feb 45, to APB 48 on 7 Mar 45, str 1 Apr 72
1115	Chicago	29 Sep 44 22 Dec 44	4 Jan 45	18 Jun 46	Decomm 6 Feb 45, converted, to *Pentheus* (ARL 20) 7 Jun 45, str 1 Jan 60
1116	Chicago	2 Oct 44 28 Dec 44	9 Jan 45 27 Oct 50	18 Jan 47 24 May 56	Decomm 15 Feb 45, converted, to *Proserpine* (ARL 21) 31 May 45, str 1 Jan 60
1117	Chicago	10 Oct 44 2 Jan 45	13 Jan 45	5 Jan 46	Converted, comm as *Tantalus* (ARL 27) 5 Jun 45, sold Jan 47
1118	Chicago	17 Oct 44 5 Jan 45	18 Jan 45	7 Aug 47	Decomm 16 Feb 45, converted, comm as *Typhon* (ARL 28) 18 Jun 45, str Jul 60
1119	Chicago	19 Oct 44 11 Jan 45	23 Jan 45	3 Dec 46	Converted, comm as *Diomedes* (ARB 11) 23 Jun 45, to Germany Jun 61
1120	Chicago	20 Oct 45 16 Jan 45	9 Feb 45	14 Jan 48	Str 19 Feb 48
1121	Chicago	25 Oct 44 19 Jan 45	31 Jan 45	27 May 47	Converted, comm as *Demeter* (ARB 10) 3 July 45, str 1 Mar 59
1122	San Joaquin County Chicago	30 Oct 45 24 Jan 45	14 Feb 45 3 Nov 50	15 Jun 49 26 Sep 69	Str 1 May 72
1123	Sedgwick County Chicago	1 Nov 44 29 Jan 45	19 Feb 45 4 Jun 66	9 Sep 55 6 Dec 69	Str 15 Mar 75, to Malaysia 7 Oct 76
1124	Chicago	6 Nov 44 1 Feb 45	3 Mar 45	1 Jan 47	Converted, comm as *Amphitrite* (ARL 29) 29 Jun 45, str 1 Jul 61
1125	Chicago	15 Nov 44 6 Feb 45	17 Feb 45	14 Mar 46	Converted, comm as *Brontes* (AGP 17) 14 Aug 45, sold 1 Apr 46
1126	Snohomish County Chicago	16 Nov 44 9 Feb 45	28 Feb 45	1 Jul 70	Str 1 Jul 70
1127	Chicago	23 Nov 44 14 Feb 45	26 Feb 45	3 Dec 46	Converted, comm as *Helios* (ARB 12) 23 Jul 45, to Brazil 19 Jan 62
1128	Solano County Chicago	23 Nov 44 19 Feb 45	9 Mar 45	29 Jul 46	Str 1 Nov 58, to Indonesia
1129	Sommerwell County Chicago	29 Nov 44 22 Feb 45	6 Mar 45	31 Jul 46	Str 1 Nov 58
1130	Chicago	5 Dec 44 27 Feb 45	20 Mar 45	23 Mar 48	Abandoned after grounding, str
1131	Chicago	8 Dec 44 2 Mar 45	20 Mar 45 13 Aug 66	21 Mar 56 1 Sep 71	Converted, comm as *Askari* (ARL 30) 23 Jul 45, to Indonesia 1 Sep 71
1132	Chicago	12 Dec 44 7 Mar 45	19 Mar 45	7 Mar 48	Converted, comm as *Bellerophon* (ARL 31) 21 Jul 45, str 1 Oct 77
1133	Chicago	16 Dec 44 10 Mar 45	23 Mar 45	20 Feb 46	Converted, comm as *Chiron* (AGP 18) 18 Sep 45, sold 19 May 47
1134	Stark County Chicago	18 Dec 44 16 Mar 45	7 Apr 45		To Thailand 16 May 66

No.	Name/Builder	LD/Launch	Comm	Decomm	Notes and Fate
1135	Chicago	26 Dec 44 21 Mar 45	12 Apr 45	28 Apr 48	Str 12 Aug 48
1136	Chicago	27 Dec 44 26 Mar 45	6 Apr 45		Decomm 27 Apr 45, converted, comm as *Bellona* (ARL 32) 26 Jul 45, CTL 1 Dec 45
1137	Chicago	3 Jan 45 30 Mar 45	11 Apr 45	8 Mar 48	Decomm 7 May 45, converted, comm as *Chimaera* (ARL 33), 7 Aug 45, str 1 Jul 61
1138	*Steuben County* Chicago	6 Jan 45 5 Apr 45	24 Apr 45		Str 1 Feb 61
1139	Chicago	15 Jan 45 9 Apr 45	27 Apr 45	20 Jul 46	Str 15 Aug 46
1140	Chicago	17 Jan 45 13 Apr 45	4 May 45	3 Jun 49	Str 15 Aug 49
1141	*Stone County* Chicago	22 Jan 45 18 Apr 45	9 May 45 3 Nov 50	24 Aug 49 12 Mar 70	FRAM II refit, to Thailand 15 Aug 73
1142	*Stafford County* Chicago	25 Jan 45 23 Apr 45	12 May 45	15 Nov 46	Str 1 Nov 58
1143	Chicago	31 Jan 45 27 Apr 45	9 May 45	23 Oct 47	Decomm 21 May 45, converted, comm as *Daedalus* (ARL 35) 19 Oct 45, str 1 Jan 60
1144	*Sublette County* Chicago	3 Feb 45 2 May 45	28 May 45	11 Feb 55	To ROC Sep 61
1145	Chicago	5 Feb 45 7 May 45	18 May 45	21 Dec 55	Decomm 11 Jun 45, converted, comm as *Gordius* (ARL 36) 14 Sep 45, str 1 Feb 61
1146	*Summit County* Chicago	10 Feb 45 11 May 45	30 May 45	Dec 69	To Ecuador 1 Nov 76
1147	Chicago	12 Feb 45 21 May 45	28 May 45 Jan 68	6 Oct 47 Apr 70	Converted to *Indra* (ARL 37), comm 2 Oct 45, str 1 Dec 77
1148	*Sumner County* Chicago	15 Feb 45 23 May 45	9 Jun 45 3 Oct 50	11 May 46 9 Oct 69	FRAM II refit, str 16 Mar 70
1149	Chicago	23 Feb 45 25 May 45	3 Dec 45	30 Oct 71	Converted to *Krishna* (ARL 38), comm 3 Dec 45, to Philippines 30 Oct 71
1150	*Sutter County* Chicago	1 Mar 45 30 May 45	20 Jun 45 16 Apr 66	13 Sep 46 1 Dec 70	Str 15 Sep 74
1151	Chicago	3 Mar 45 4 Jun 45	15 Jun 45	27 Jun 47	Converted to *Quirinus* (ARL 39), comm 6 Nov 45, to Venezuela Jun 62
1152	*Sweetwater County* Chicago	5 Mar 45 8 Jun 45	30 Jun 45	1 Jul 46	To ROC 21 Oct 58
1153	*Talbot County* BOSNY	19 Jul 45 24 Apr 47	3 Sep 47	3 Apr 70	Str 1 Jun 73
1154	*Tallahatchie County* BOSNY	4 Aug 45 19 Jul 46	24 May 49	15 Jun 70	To AVB 3 on 3 Feb 62, str after decomm
1155					Canceled 7 Jan 46
1156	*Terrebonne Parish* Bath	2 Jan 52 9 Aug 52	21 Nov 52	29 Oct 71	To Spain on decomm
1157	*Terrell County* Bath	3 Mar 52 6 Dec 53	14 Mar 54	25 Mar 71	To Greece 17 Mar 77
1158	*Tioga County* Bath	16 Jun 52 11 Apr 53	20 Jun 53	25 Nov 70	MSC Jun 72, str 1 Nov 73
1159	*Tom Green County* Bath	2 Sep 52 2 Jul 53	12 Sep 53	5 Jan 72	To Spain 5 Jan 72
1160	*Traverse County* Bath	18 Dec 52 3 Oct 53	19 Dec 53	1 Dec 70	MSC 7 Jun 72, str 1 Nov 73
1161	*Vernon County* Ingalls	14 Apr 52 25 Nov 52	18 May 53	14 Jun 73	To Venezuela 29 Jun 73
1162	*Wahkiakum County* Ingalls	21 Jul 52 23 Jan 53	13 Aug 53	16 Oct 70	MSC 10 Apr 72, str 1 Nov 73
1163	*Waldo County* Ingalls	4 Aug 52 17 Mar 53	17 Sep 53	21 Dec 70	Str 1 Nov 73

No.	Name/Builder	LD/Launch	Comm	Decomm	Notes and Fate
1164	Walworth County Ingalls	22 Sep 52 15 May 53	26 Oct 53	2 Apr 71	MSC 26 May 72, str 1 Nov 73
1165	Washoe County Ingalls	1 Dec 52 14 Jul 53	30 Nov 53	25 Mar 71	Str 1 Nov 73
1166	Washtenaw County Christy	29 Nov 51 22 Nov 52	29 Oct 53	Aug 73	Str 30 Aug 73, to MSS 2 to clear mines in North Vietnam 1 Feb 72
1167	Westchester County Christy	11 Jan 52 18 Apr 53	10 Mar 54	30 Aug 73	To Turkey 27 Aug 74
1168	Wexford County Christy	27 Feb 52 28 Nov 53	15 Jun 54	29 Oct 71	To Spain on decomm
1169	Whitfield County Christy	26 Nov 52 22 Aug 53	14 Sep 54	15 Mar 73	To Greece 17 Mar 77
1170	Windham County Christy	21 Apr 53 22 May 54	15 Dec 54	1 Jun 73	To Turkey on decomm
1171	De Soto County Avondale	Sep 56 28 Feb 57	10 Jun 58	17 Jul 72	To Italy on decomm
1172					Canceled 55
1173	Suffolk County BOSNY	17 Jul 55 5 Sep 56	15 Aug 57	25 Aug 72	Str 16 Feb 89
1174	Grant County Avondale	15 Mar 56 12 Oct 56	17 Dec 57	15 Jan 73	To Brazil on decomm
1175	York County NN Ship	4 Jun 56 5 Mar 57	8 Nov 57	17 Jul 72	To Italy on decomm
1176	Graham County NN Ship	4 Feb 57 9 Sep 57	17 Apr 58	1 Mar 77	AGP 1176 in 71, support ship for *Asheville*s in Mediterranean, str on decomm
1177	Lorain County Amer Ship	9 Aug 56 22 Jun 57	3 Oct 58	1 Sep 72	Str 16 Feb 89
1178	Wood County Amer Ship	1 Oct 56 14 Dec 57	5 Aug 59	1 May 72	Str 1 May 72, reacquired, conversion to AGHS 1 canceled 77, str 16 Feb 89
1179	Newport PHNY	1 Nov 66 3 Feb 68	7 Jun 69	30 Sep 92	
1180	Manitowoc PHNY	1 Feb 67 4 Jan 69	24 Jan 70	30 Jun 93	To ROC 18 Apr 97
1181	Sumter PHNY	14 Nov 67 13 Dec 69	20 Jun 70	30 Sep 93	To ROC 11 Mar 97
1182	Fresno NASSCO	16 Dec 67 28 Sep 68	22 Nov 69	8 Apr 93	
1183	Peoria NASSCO	22 Feb 68 23 Nov 68	21 Feb 70	28 Jan 94	
1184	Frederick NASSCO	13 Apr 68 8 Mar 69	11 Apr 70		
1185	Schenectady NASSCO	2 Aug 68 24 May 69	13 Jun 70	15 Dec 93	To ROC 2000
1186	Cayuga NASSCO	28 Sep 68 12 Jul 69	8 Aug 70	29 Jul 94	To Brazil 26 Aug 94
1187	Tuscaloosa NASSCO	23 Nov 68 6 Sep 69	24 Oct 70	18 Feb 94	
1188	Saginaw NASSCO	24 May 69 7 Feb 70	23 Jan 71	28 Jun 94	To Australia 24 Aug 94
1189	San Bernardino NASSCO	12 Jul 69 28 Mar 70	27 Mar 71	30 Sep 95	To Chile 8 Dec 95
1190	Boulder NASSCO	6 Sep 69 22 Apr 70	4 Jun 71	28 Feb 94	
1191	Racine NASSCO	13 Dec 69 15 Aug 70	9 Jul 71	2 Oct 93	
1192	Spartanburg County NASSCO	7 Feb 70 7 Nov 70	1 Sep 71	16 Feb 92	To Malaysia on decomm

No.	Name/Builder	LD/Launch	Comm	Decomm	Notes and Fate
1193	*Fairfax County* NASSCO	28 Mar 70 19 Dec 70	16 Oct 71	16 Sep 94	To Australia 27 Sep 94
1194	*La Moure County* NASSCO	22 May 70 13 Feb 71	18 Dec 71	17 Nov 00	Str on decomm
1195	*Barbour County* NASSCO	15 Aug 70 15 May 71	12 Feb 72	30 Mar 92	
1196	*Harlan County* NASSCO	7 Nov 70 24 Jul 71	8 Apr 72	14 Apr 94	To Spain on decomm
1197	*Barnstable County* NASSCO	19 Dec 70 2 Oct 71	27 May 72	29 Jun 94	To Spain 26 Aug 94
1198	*Bristol County* NASSCO	13 Feb 71 4 Dec 71	5 Aug 72	9 Jul 94	To Morocco 16 Aug 94

LSV

The LSV classification was revived postwar for ro-ro ships: T-AKV 7 (*Comet*), T-AKV 8 (*Taurus*, ex–T-AK 273, ex–LSD 23), and T-AKV 9 (*Sea Lift*). This classification was superseded by a new AKR classification on 1 January 1969.

No.	Name/Builder	LD/Launch	Comm	Decomm	Notes and Fate
1	*Catskill* Willamette	12 Jul 41 19 May 42	30 Jun 44 6 Oct 67	30 Aug 46 1 Dec 70	Ex–CM 6; to AP 106 on 1 May 43, to LSV 1 on 21 Apr 44, to MCS 1 on 18 Oct 56, str 1 Jul 61, reinstated 1 Jun 64, str 20 Nov 70
2	*Ozark* Willamette	12 Jul 41 15 Jun 42	23 Sep 44 24 Jun 66	N/A 6 Feb 70	Ex–CM 7; to AP 107 on 1 May 43, to LSV 2 on 21 Apr 44, to MCS 2 on 18 Oct 56, str 1 Sep 61, reinstated 1 Jun 64, str 1 Apr 74, to U.S. Air Force 25 Sep 75, tgt 82
3	*Osage* Ingalls	1 Jun 42 30 Jun 43	30 Dec 44	16 May 47	Ex–AN 3; to AP 108 on 1 May 43, converted Tampa Sbldg, to LSV 3 on 21 Apr 44, to MCS 3 on 18 Oct 56, str 1 Sep 61
4	*Saugus* Ingalls	27 Jul 42 4 Sep 43	22 Feb 45	24 Mar 47	Ex–AN 4; to AP 109 on 1 May 43, converted Tampa Sbldg, to LSV 4 on 21 Apr 44, to MCS 4 on 18 Oct 56, str 1 Jul 61
5	*Monitor* Ingalls	21 Oct 41 29 Jan 43	18 Mar 44	22 May 47	Ex–AN 1; to AP 160 on 1 May 43, converted Todd Brooklyn, to LSV 5 on 21 Apr 44, to MCS 5 on 18 Oct 56, str 1 Jul 61
6	*Montauk* Ingalls	14 Apr 42 14 Apr 43	25 May 44	Jul 47	Ex–AN 2, to AP 161 on 2 Aug 43, converted Todd Brooklyn, to LSV 6 on 21 Apr 44, to *Galilea* (AKN 6) 1 Oct 46 but conversion not completed, str 1 Sep 61

LCIL/LSIL/LCFF/LCIG/LCIM/LCIR

Status is as of April 1945 except as noted. All surviving units were redesignated 28 February 1949: LCIL became LSIL, LC(FF) became LSFF. LCIG and LCIM briefly became LSIG and LSIM, but in 1949 they were reclassified as LSIL.

No.	Name/Builder	LD/Launch	Comm	Decomm	Notes and Fate
1	NY Ship	27 Jul 42 1 Sep 42	7 Oct 42		WL 17 Aug 43

No.	Name/Builder	LD/Launch	Comm	Decomm	Notes and Fate
2	NY Ship	27 Jul 42 6 Sep 42	9 Oct 42	23 May 46	LCIG, str 3 Jul 46
3	NY Ship	27 Jul 42 7 Sep 42	7 Oct 42		To UK 19 Oct 44, ret 14 Mar 46, str 17 Apr 46
4	NY Ship	27 Jul 42 9 Sep 42	9 Oct 42		To UK 19 Oct 44, hulked Nhatrung, Indochina, CTL Mar 46, str 17 Apr 46
5	NY Ship	27 Jul 42 11 Sep 42	12 Oct 42		To UK 19 Oct 44, ret May 46, str 5 Jun 46
6	NY Ship	27 Jul 42 13 Sep 42	16 Oct 42		To UK 22 Oct 42, str 5 Oct 46
7	NY Ship	1 Aug 42 15 Sep 42	19 Oct 42		To UK 22 Oct 42, WL Apr 43
8	NY Ship	1 Aug 42 16 Sep 42	21 Oct 42		To UK 19 Oct 44, ret, str 5 Jun 46
9	NY Ship	8 Aug 42 19 Sep 42	23 Oct 42		To UK 20 Oct 44, ret 27 May 46, str 5 Jun 46
10	NY Ship	4 Aug 42 21 Sep 42	24 Oct 42		To UK 31 Oct 44, ret 27 May 46, str 5 Jun 46
11	NY Ship	4 Aug 42 23 Sep 42	28 Oct 42		To UK 3 Nov 44, str 23 Apr 47
12	NY Ship	7 Aug 42 24 Sep 42	29 Oct 42		To UK 20 Oct 44, decomm 25 Mar 47, str 23 Apr 47
13	NY Ship	1 Sep 42 26 Sep 42	31 Oct 42		To UK 20 Oct 44, ret 27 May 46, str 5 Jun 46
14	NY Ship	5 Sep 42 28 Sep 42	3 Nov 42		To UK 20 Oct 44, ret May 46, str 5 Jun 46
15	NY Ship	7 Sep 42 29 Sep 42	5 Nov 42		To UK 18 Oct 44, ret May 46, str 19 Jun 46
16	NY Ship	9 Sep 42 2 Oct 42	7 Nov 42		To UK 18 Oct 44, ret 21 May 46, str 3 Jul 46
17	NY Ship	11 Sep 42 5 Oct 42	9 Nov 42	13 Mar 46	LCIG 15 July 45, selected for Bikini, str 12 Apr 46
18	NY Ship	13 Sep 42 7 Oct 42	11 Dec 42	8 Jan 46	LCIG 15 July 45, str 21 Jan 46
19	NY Ship	15 Sep 42 8 Oct 42	11 Dec 42	9 Apr 46	LCIG 15 July 45, str 21 May 46
20	NY Ship	16 Sep 42 9 Oct 42	16 Nov 42		WL 22 Jan 44
21	NY Ship	19 Sep 42 10 Oct 42	8 Dec 42	10 Apr 46	LCIG, str 19 Jun 46
22	NY Ship	21 Sep 42 13 Oct 42	8 Dec 42	9 May 46	LCIG, str 19 Jul 46
23	NY Ship	23 Sep 42 14 Oct 42	8 Dec 42	23 Apr 46	LCIG, str 28 Aug 46
24	NY Ship	24 Sep 42 15 Oct 42	8 Dec 42	10 May 46	LCIG, str 19 Jun 46
25	NY Ship	26 Sep 42 17 Oct 42	8 Dec 42	N/A	To Maritime Commission 9 Oct 47
26	NY Ship	28 Sep 42 22 Oct 42	8 Dec 42	7 May 46	Str 19 Jun 46
27	NY Ship	29 Sep 42 26 Oct 42	9 Dec 42	9 Apr 46	Str 5 Jun 46
28	NY Ship	4 Oct 42 26 Oct 42	29 Dec 42	20 Feb 46	Str 12 Mar 46
29	NY Ship	5 Oct 42 27 Oct 42	30 Dec 42	9 Apr 46	LCID, str 1 May 46
30	NY Ship	7 Oct 42 29 Oct 42	31 Dec 42	18 Jun 46	Str 31 Jul 46

No.	Name/Builder	LD/Launch	Comm	Decomm	Notes and Fate
31		8 Oct 42	31 Dec 42	15 Jan 46	LCIR, ex-LCIG; str 26 Feb 46
	NY Ship	29 Oct 42			
32		9 Oct 42	3 Dec 42		WL 26 Jan 44
	NY Ship	31 Oct 42			
33		10 Oct 42	3 Dec 42		To UK 1 Nov 44, ret 13 Feb 46, str 26 Feb 46
	NY Ship	2 Nov 42			
34		13 Oct 42	23 Jan 43	15 Jun 46	LCIR, ex-LCIG; str 31 Jul 46
	NY Ship	4 Nov 42			
35		14 Oct 42	7 Dec 42		To UK 14 Nov 44, used for smoke screen tests by RN, str 15 Oct 46
	NY Ship	4 Nov 42			
36		15 Oct 42	11 Feb 43	8 Jan 46	Str 21 Jan 46, conversion to LCIG canceled
	NY Ship	5 Nov 42			
37		17 Oct 42	11 Feb 43	N/A	To Maritime Commission 5 Jun 47
	NY Ship	30 Nov 42			
38		22 Oct 42	9 Feb 43	24 Apr 46	CTL, str 5 Jun 46
	NY Ship	30 Nov 42			
39		26 Oct 42	9 Feb 43	2 May 46	LCIG 15 Jul 45, str 19 Jul 46
	NY Ship	1 Dec 42			
40		26 Oct 42	26 Dec 42	19 Apr 46	Str 19 Jun 46, to Dominican Republic 48
	NY Ship	1 Dec 42			
41		27 Oct 42	26 Dec 42	12 Jul 46	LCIG 15 Jul 45, str 28 Aug 46
	NY Ship	4 Dec 42			
42		29 Oct 42	26 Dec 42	30 May 46	LCIG 15 Jul 45, str 7 Jul 47
	NY Ship	4 Dec 42			
43		29 Oct 42	26 Dec 42	18 Apr 46	LCIG 15 Jul 45, str 5 Jun 46
	NY Ship	7 Dec 42			
44		31 Oct 42	26 Dec 42	18 Jun 46	LCID, str 31 Jul 46
	NY Ship	7 Dec 42			
45		2 Nov 42	26 Dec 42	19 Apr 46	Str 19 Jun 46
	NY Ship	9 Dec 42			
46		4 Nov 42	9 Feb 43	25 Apr 46	LCIG 15 Jul 45, str 13 Dec 46
	NY Ship	8 Dec 42			
47		4 Nov 42	9 Feb 43	24 Apr 46	Str 31 Jul 46
	NY Ship	10 Dec 42			
48		5 Nov 42	9 Feb 43	19 Apr 46	Str 19 Jun 46, to Dominican Republic 48
	NY Ship	10 Dec 42			
49–60					Canceled 6 Nov 42
61		5 Aug 42	12 Nov 42	1 Feb 46	LCIG, str 25 Feb 46
	Consolidated-O	27 Sep 42			
62		4 Aug 42	13 Nov 42	17 Apr 46	Str 15 Aug 46
	Consolidated-O	27 Sep 42			
63		7 Aug 42	16 Nov 42	14 Mar 46	Str 12 Apr 46
	Consolidated-O	27 Sep 42			
64		18 Aug 42	12 Dec 42	25 Mar 46	LCIG, str 17 Apr 46
	Consolidated-O	4 Oct 42			
65		18 Aug 42	14 Dec 42	1 Apr 46	LCIG, str 1 May 46
	Consolidated-O	4 Oct 42			
66		18 Aug 42	14 Dec 42	5 Feb 46	LCIG, str 12 Apr 46
	Consolidated-O	4 Oct 42			
67		1 Sep 42	17 Dec 42	31 Dec 45	LCIG, str 21 Jan 46
	Consolidated-O	11 Oct 42			
68		1 Sep 42	17 Dec 42	N/A	LCIG, sold 5 Dec 46
	Consolidated-O	11 Oct 42			
69		1 Sep 42	24 Dec 42	8 Apr 46	LCIG, str 8 May 46
	Consolidated-O	11 Oct 42			
70		15 Sep 42	24 Dec 42	15 Feb 46	LCIG, str 26 Feb 46
	Consolidated-O	25 Oct 42			

No.	Name/Builder	LD/Launch	Comm	Decomm	Notes and Fate
71	Consolidated-O	15 Sep 42 25 Oct 42	29 Dec 42	31 Jan 46	LCIR, str 25 Feb 46
72	Consolidated-O	15 Sep 42 25 Oct 42	31 Dec 42	5 Feb 46	LCIR, str 26 Feb 46
73	Consolidated-O	28 Sep 42 8 Nov 42	2 Jan 43	23 Apr 46	LCIR, ex-LCIG; str 5 Jun 46
74	Consolidated-O	28 Sep 42 8 Nov 42	2 Jan 43	20 Feb 46	LCIR, str 12 Apr 46
75	Consolidated-O	28 Sep 42 8 Nov 42	5 Jan 43		To UK 15 Nov 44, ret 13 Oct 46, str 10 Jun 47
76	Consolidated-O	29 Sep 42 22 Nov 42	6 Jan 43	21 Jun 50	Str 21 Nov 55, tgt 29 Jun 56
77	Consolidated-O	29 Sep 42 22 Nov 42	11 Jan 43	17 Dec 45	LCIG, str 8 Jan 46
78	Consolidated-O	30 Sep 42 22 Nov 42	11 Jan 43	14 Dec 45	LCIG, str 8 Jan 46
79	Consolidated-O	5 Oct 42 6 Dec 42	15 Jan 43	14 Dec 45	LCIG, str 8 Jan 46
80	Consolidated-O	5 Oct 42 6 Dec 42	15 Jan 43	18 Jan 46	LCIG, str 5 Jun 46
81	Consolidated-O	5 Oct 42 6 Dec 42	18 Jan 43	14 Dec 45	LCIG, str 8 Jan 46
82	Consolidated-O	12 Oct 42 13 Dec 42	20 Jan 43		LCIG, WL 4 Apr 45
83	Consolidated-O	12 Oct 42 13 Dec 42	23 Jan 43	8 Apr 46	Str 19 Jun 46
84	Consolidated-O	12 Oct 42 13 Dec 42	23 Jan 43	13 Apr 46	Str 3 Jul 46
85	Consolidated-O	13 Oct 42 20 Dec 42	25 Jan 43		WL 6 Jun 44
86	Consolidated-O	13 Oct 42 20 Dec 42	29 Jan 43	8 Apr 46	Str 19 Jun 46
87	Consolidated-O	16 Oct 42 20 Dec 42	1 Feb 43	20 Mar 46	Str 1 May 46
88	Consolidated-O	26 Oct 42 20 Dec 42	2 Feb 43	9 Apr 46	Str 8 Jul 46
89	Consolidated-0	26 Oct 42 20 Dec 42	3 Feb 43	6 Mar 46	Str 8 May 46
90	Consolidated-O	26 Oct 42 20 Dec 42	6 Feb 43	8 Apr 46	Str 1 May 46
91	Consolidated-O	23 Nov 42 3 Jan 43	6 Feb 43		WL 6 Jun 44
92	Consolidated-O	23 Nov 42 3 Jan 43	13 Feb 43		WL 6 Jun 44
93	Consolidated-O	23 Nov 42 3 Jan 43	12 Feb 43		WL 6 Jun 44
94	Consolidated-O	21 Dec 42 17 Jan 43	15 Feb 43	19 Apr 46	Str 3 Jul 46
95	Consolidated-O	21 Dec 42 17 Jan 43	15 Feb 43	N/A	To MarComm 8 Mar 48
96	Consolidated-O	21 Dec 42 17 Jan 43	15 Feb 43	2 Apr 46	Str 21 May 46
97	Beth-H	31 Aug 42 24 Sep 42	21 Oct 42		To UK 22 Oct 42, str 15 Oct 46
98	Beth-H	31 Aug 42 9 Oct 42	2 Nov 42		To UK 2 Nov 42, ret 28 Feb 46, str 28 Aug 46
99	Beth-H	31 Aug 42 12 Oct 42	6 Nov 42		To UK 6 Nov 42, WL

No.	Name/Builder	LD/Launch	Comm	Decomm	Notes and Fate
100	Beth-H	31 Aug 42 15 Oct 42	10 Nov 42		To UK 10 Nov 42, ret 18 Dec 46, str 5 Mar 47
101	Beth-H	31 Aug 42 22 Oct 42	17 Nov 42		To UK 17 Nov 42, ret 14 Mar 46, str 17 Apr 46, to France 13 Feb 48
102	Beth-H	31 Aug 42 24 Oct 42	20 Nov 42		To UK 20 Nov 42, CTL Nov 44
103	Beth-H	31 Aug 42 31 Oct 42	30 Nov 42		To UK 2 Dec 42, ret May 46, str 5 Jun 46, to France 13 Feb 48
104	Beth-H	31 Aug 42 30 Oct 42	28 Nov 42		To UK 28 Nov 42, ret 28 May 46, str 19 Jun 46
105	Beth-H	31 Aug 42 7 Nov 42	30 Nov 42		To UK 30 Nov 42, str 13 Nov 44
106	Beth-H	31 Aug 42 9 Nov 42	5 Dec 42		To UK 7 Dec 42, ret Sep 46, str 15 Oct 46
107	Beth-H	31 Aug 42 14 Nov 42	7 Dec 42		To UK 7 Dec 42, WL Sep 43
108	Beth-H	31 Aug 42 18 Nov 42	10 Dec 42		To UK 10 Dec 42, str 15 Oct 46, to France 10 Dec 46
109	Beth-H	31 Aug 42 20 Nov 42	11 Dec 42		To UK 11 Dec 42, ret 14 Mar 46, str 17 Apr 46, to France for Indo-China 19 Apr 46
110	Beth-H	31 Aug 42 20 Nov 42	17 Dec 42		To UK 17 Dec 42, ret, str 15 Oct 46
111	Beth-H	31 Aug 42 27 Nov 42	17 Dec 42		To UK 17 Dec 42, ret, str 5 Jun 46
112	Beth-H	31 Aug 42 27 Nov 42	22 Dec 42		To UK 22 Dec 42, ret Mar 46, str 5 Jun 46
113	Beth-H	7 Sep 42 30 Nov 42	24 Dec 42		To UK 24 Dec 42, ret 24 Sep 46, str 22 Jan 47
114	Beth-H	7 Sep 42 1 Dec 42	30 Dec 42		To UK 30 Dec 42, ret 13 Feb 46, str 26 Feb 46
115	Beth-H	7 Sep 42 5 Dec 42	31 Dec 42		To UK 31 Dec 42, ret 4 Mar 46, str 17 Apr 46
116	Beth-H	7 Sep 42 6 Dec 42	31 Dec 42		To UK 6 Jan 43, ret 13 Feb 46, str 26 Feb 46
117	Beth-H	1 Oct 42 7 Dec 42	5 Jan 43		To UK 5 Jan 43, to France for Indo-China 26 Jan 48
118	Beth-H	2 Oct 42 8 Dec 42	14 Jan 43		To UK 14 Jan 43, ret 27 May 46, str 5 Jun 46
119	Beth-H	5 Oct 42 10 Dec 42	8 Jan 43		To UK 8 Jan 43, ret 9 Jul 46, str 28 Aug 46
120	Beth-H	6 Oct 42 11 Dec 42	13 Jan 43		To UK 13 Jan 43, sold to government of India 24 Dec 45, str 26 Feb 46
121	Beth-H	11 Oct 42 14 Dec 42	13 Jan 43		To UK 13 Jan 43, ret Feb 46, str 26 Feb 46
122	Beth-H	12 Oct 42 14 Dec 42	14 Jan 43		To UK 14 Jan 43, ret 14 Mar 46, str 17 Apr 46
123	Beth-H	17 Oct 42 19 Dec 42	16 Jan 43		To UK 16 Jan 43, ret Apr 47, str 23 Apr 47
124	Beth-H	18 Oct 42 19 Dec 42	16 Jan 43		To UK 18 Jan 43, CTL, str 13 Sep 48
125	Beth-H	19 Oct 42 26 Dec 42	19 Jan 43		To UK 19 Jan 43, ret 14 Mar 46, str 5 Jun 46
126	Beth-H	20 Oct 42 26 Dec 42	20 Jan 43		To UK 20 Jan 43, ret May 46, str 5 Jun 46
127	Beth-H	26 Oct 42 29 Dec 42	22 Jan 43		To UK 22 Jan 43, ret 14 Mar 46, str 17 Apr 46

No.	Name/Builder	LD/Launch	Comm	Decomm	Notes and Fate
128	Beth-H	27 Oct 42 30 Dec 42	23 Jan 43		To UK 23 Jan 43, CTL, str 28 Nov 45
129	Beth-H	6 Dec 42 31 Dec 42	26 Jan 43		To UK 26 Jan 43, ret 9 Jul 46, str 28 Aug 46
130	Beth-H	7 Dec 42 31 Dec 42	26 Jan 43		To UK 26 Jan 43, ret, str 5 Jun 46
131	Beth-H	9 Dec 42 7 Jan 43	28 Jan 43		To UK 28 Jan 43, ret 21 May 46, str 3 Jul 46
132	Beth-H	10 Dec 42 7 Jan 43	29 Jan 43		To UK 29 Jan 43, WL 17 Jun 44
133	Beth-H	12 Dec 42 11 Jan 43	1 Feb 43		To UK 1 Feb 43, to Greece Apr 47
134	Beth-H	12 Dec 42 11 Jan 43	2 Feb 43		To UK 3 Feb 43, ret 4 Mar 46, str 8 May 46
135	Beth-H	22 Dec 42 13 Jan 43	4 Feb 43		To UK 4 Feb 43, ret 14 Mar 46, str 17 Apr 46
136	Beth-H	22 Dec 42 16 Jan 43	4 Feb 43		To UK 4 Feb 43, ret Feb 46, str 26 Feb 46
137–160					Canceled 31 Oct 42
161	Federal-K	10 Aug 42 10 Oct 42	24 Oct 42		To UK 24 Oct 42, str 15 Oct 46
162	Federal-K	10 Aug 42 10 Oct 42	26 Oct 42		To UK 26 Oct 42, WL 7 Oct 43
163	Federal-K	10 Aug 42 12 Oct 42	31 Oct 42		To UK 31 Oct 42, ret 11 Mar 47, str 23 Apr 47
164	Federal-K	10 Aug 42 12 Oct 42	2 Nov 42		To UK 2 Nov 42, ret 20 Mar 46, str 8 May 46
165	Federal-K	26 Aug 42 17 Oct 42	7 Nov 42		To UK 7 Nov 42, ret 20 Mar 46, str 29 Oct 46
166	Federal-K	26 Aug 42 17 Oct 42	9 Nov 42		To UK 9 Nov 42, str 17 Apr 46
167	Federal-K	26 Aug 42 22 Oct 42	11 Nov 42		To UK 12 Nov 42, ret 28 May 46, str 29 Oct 46
168	Federal-K	26 Aug 42 22 Oct 42	13 Nov 42		To UK 13 Nov 42, ret 14 Mar 46, str 17 Apr 46
169	Federal-K	2 Sep 42 2 Nov 42	17 Nov 42		To UK 17 Nov 42, ret 15 Aug 46, str 29 Oct 46
170	Federal-K	2 Sep 42 2 Nov 42	19 Nov 42		To UK 19 Nov 42, ret 24 Sep 46, str 22 Jan 47
171	Federal-K	2 Sep 42 11 Nov 42	21 Nov 42		To UK 21 Nov 42, ret 27 Mar 46, str 5 Jun 46
172	Federal-K	2 Sep 42 11 Nov 42	27 Nov 42		To UK 27 Nov 42, ret 18 Mar 47, str 23 Apr 47
173	Federal-K	7 Sep 42 11 Nov 42	28 Nov 42		To UK 28 Nov 42, to government of India 24 Dec 45 for BU, str 26 Feb 46
174	Federal-K	7 Sep 42 11 Nov 42	30 Nov 42		To UK 30 Nov 42, to France for Indo-China 26 Jan 48
175	Federal-K	7 Sep 42 19 Nov 42	3 Dec 42		To UK 3 Dec 42, ret 16 May 46, str 19 Jun 46
176	Federal-K	7 Sep 42 19 Nov 42	7 Dec 42		To UK 7 Dec 42, ret 27 May 46, str 5 Jun 46
177	Federal-K	10 Sep 42 27 Nov 42	9 Dec 42		To UK 9 Dec 42, ret 14 Mar 46, str 17 Apr 46
178	Federal-K	10 Sep 42 27 Nov 42	12 Dec 42		To UK 12 Dec 42, smoke screen tests, str 15 Oct 46
179	Federal-K	10 Sep 42 7 Dec 42	16 Dec 42		To UK 16 Dec 42, ret 27 Mar 46, str 8 May 46, merch *Vicki*, New York harbor tour boat *Circle Line VIII* in 62

No.	Name/Builder	LD/Launch	Comm	Decomm	Notes and Fate
180	Federal-K	10 Sep 42 7 Dec 42	18 Dec 42		To UK 18 Dec 42, ret 13 Feb 46, str 26 Feb 46
181	Federal-K	17 Sep 42 14 Dec 42	23 Dec 42		To UK 23 Dec 42, ret 20 Mar 46, str 8 May 46
182	Federal-K	17 Sep 42 14 Dec 42	26 Dec 42		To UK 26 Dec 42, ret 21 May 46, str 3 Jul 46
183	Federal-K	17 Sep 42 21 Dec 42	29 Dec 42		To UK 29 Dec 42, ret 13 Feb 46, str 26 Feb 46
184	Federal-K	17 Sep 42 21 Dec 42	31 Dec 42		To UK 31 Dec 42, ret 28 Feb 46, str 28 Aug 46
185	Federal-K	11 Nov 42 4 Jan 43	13 Jan 43		To UK 13 Jan 43, WL 25 Jun 44
186	Federal-K	11 Nov 42 4 Jan 43	16 Jan 43		To UK 16 Jan 43, ret 24 Sep 46, str 22 Jan 47
187	Federal-K	20 Nov 42 11 Jan 43	20 Jan 43		To UK 20 Jan 43, ret 21 May 46, str 3 Jul 46
188	Federal-K	20 Nov 42 11 Jan 43	23 Jan 43	12 Jul 46	Str 25 Sep 46
189	Federal-K	27 Nov 42 18 Jan 43	28 Jan 43	12 Jul 46	LGIG, str 25 Sep 48
190	Federal-K	27 Nov 42 18 Jan 43	30 Jan 43	22 Mar 46	LGIG, str 7 Feb 47
191	Federal-K	7 Dec 42 25 Jan 43	3 Feb 43	19 Apr 46	Str 19 Jun 46, merch, New York tour boat *New Yorker* in 52, *Circle Line VII* in 62
192	Federal-K	7 Dec 42 25 Jan 43	5 Feb 43	10 Jun 46	LCIG, str 3 Jul 46
193	Federal-K	15 Dec 42 30 Jan 43	9 Feb 43		To UK 15 Nov 44, ret 11 Apr 46, str 28 Aug 46
194	Federal-K	15 Dec 42 30 Jan 43	11 Feb 43	19 Apr 46	Str 19 Jun 46
195	Federal-K	21 Dec 42 8 Feb 43	13 Feb 43	15 Jul 46	LGIG, str 28 Aug 46
196	Federal-K	21 Dec 42 8 Feb 43	15 Feb 43	31 May 46	LCIG, str 7 Feb 47
197–208					Canceled 5 Nov 42
209	Lawley	1 Aug 42 7 Sep 42	30 Sep 42		To UK 30 Oct 44, ret May 46, str 5 Jun 46
210	Lawley	14 Aug 42 7 Sep 42	15 Oct 42		To UK 15 Oct 42, ret 28 Feb 46, str 28 Aug 46
211	Lawley	7 Sep 42 23 Sep 42	24 Oct 42		To UK 31 Oct 44, ret 28 May 46, str 19 Jun 46
212	Lawley	30 Aug 42 23 Sep 42	20 Oct 42		To UK 31 Oct 44, ret 9 May 46, str 3 Jul 46
213	Lawley	15 Sep 42 30 Sep 42	28 Oct 42		To UK 31 Oct 44, ret 27 May 46, str 5 Jun 46
214	Lawley	15 Sep 42 8 Oct 42	31 Oct 42		To UK 31 Oct 44, ret May 46, str 5 Jun 46
215	Lawley	23 Sep 42 13 Oct 42	4 Nov 42		To UK 30 Oct 44, ret 14 Mar 46, str 17 Apr 46
216	Lawley	26 Sep 42 19 Oct 42	7 Nov 42		To UK 1 Nov 44, to France for Indo-China 26 Jan 48
217	Lawley	28 Sep 42 25 Oct 42	14 Nov 42		To UK 3 Nov 44, to France for Indo-China 19 Apr 46
218	Lawley	6 Oct 42 31 Oct 42	18 Nov 42		To UK 28 Oct 44, ret 2 Apr 47, str 23 Apr 47
219	Lawley	8 Oct 42 6 Nov 42	20 Nov 42		WL 11 Jun 44

No.	Name/Builder	LD/Launch	Comm	Decomm	Notes and Fate
220	Lawley	11 Oct 42 12 Nov 42	25 Nov 42		LCIG, to army 1 May 46, to ROC 11 Dec 46
221	Lawley	15 Oct 42 14 Nov 42	30 Nov 42	19 Apr 46	Str 19 Jun 46
222	Lawley	17 Oct 42 20 Nov 42	3 Dec 42	26 Apr 46	Str 5 Jun 46
223	Lawley	21 Oct 42 26 Nov 42	8 Dec 42	26 Jun 46	Str 15 Aug 46
224	Lawley	23 Oct 42 30 Nov 42	10 Dec 42	8 Feb 46	LCIR, str 12 Mar 46
225	Lawley	26 Oct 42 3 Dec 42	12 Dec 42	8 Feb 46	LCIR, str 12 Mar 46
226	Lawley	1 Nov 42 6 Dec 42	14 Dec 42	N/A	LCIR, sold 13 Mar 47
227	Lawley	7 Nov 42 4 Dec 42	21 Dec 42	19 Mar 46	LCID, str 12 Apr 46
228	Lawley	13 Nov 42 18 Dec 42	26 Dec 42	21 Mar 46	Str 12 Apr 46
229	Lawley	15 Nov 42 27 Dec 42	2 Jan 43		To UK 15 Nov 44, str 15 Oct 46
230	Lawley	21 Nov 42 31 Dec 42	6 Jan 43	25 Mar 46	LCIR, str 17 Apr 46
231	Lawley	30 Nov 42 5 Jan 43	11 Jan 43		To UK 15 Nov 44, sold to New Delhi, str 5 Mar 47
232	Lawley	8 Dec 42 9 Jan 43	15 Jan 43		WL 6 Jun 44
233	Lawley	13 Dec 42 13 Jan 43	20 Jan 43		LCIG, to ROC 12 Jun 46
234	Lawley	23 Dec 42 16 Jan 43	23 Jan 43	19 Jul 46	LCIG, destroyed 7 Mar 47
235	Lawley	2 Jan 43 21 Jan 43	27 Jan 43	10 Jul 46	LCIG, str 8 Oct 46
236	Lawley	6 Jan 43 26 Jan 43	1 Feb 43	10 Jul 46	LCIG, str 28 Aug 46
237	Lawley	10 Jan 43 30 Jan 43	5 Feb 43	10 Jul 46	LCIG
238	Lawley	13 Jan 43 4 Feb 43	6 Feb 43		To UK 15 Nov 44, ret 21 May 46, str 3 Jul 46
239	NJ Ship	14 Aug 42 12 Sep 42	6 Nov 42		To UK 6 Nov 42, ret 21 May 46, str 3 Jul 46
240	NJ Ship	3 Aug 42 12 Sep 42	5 Dec 42		To UK 5 Dec 42, ret 24 Sep 46, str 10 Jun 47
241	NJ Ship	5 Aug 42 12 Sep 42	19 Nov 42		To UK 19 Nov 42, str 15 Oct 46
242	NJ Ship	10 Aug 42 12 Sep 42	20 Nov 42		To UK 20 Nov 42, ret 27 Mar 46, str 15 Aug 46
243	NJ Ship	14 Aug 42 12 Sep 42	28 Nov 42		To UK 28 Nov 42, ret, str 31 Oct 47
244	NJ Ship	18 Aug 42 12 Sep 42	30 Oct 42		To UK 29 Oct 42, Royal Indian Navy, ret 28 May 46, str 29 Oct 46
245	NJ Ship	9 Sep 42 3 Oct 42	17 Nov 42		To UK 17 Nov 42, ret 21 May 46, str 19 Jul 46
246	NJ Ship	26 Sep 42 19 Oct 42	28 Nov 42		To UK 28 Nov 42, ret 15 Aug 46, str 29 Oct 46
247	NJ Ship	22 Oct 42 20 Nov 42	9 Dec 42		To UK 9 Dec 42, ret 24 Sep 46, str 22 Jan 47
248	NJ Ship	22 Oct 42 21 Nov 42	17 Dec 42		To UK 17 Dec 42, ret 16 May 46, str 19 Jun 46, to France for Indo-China 26 Jan 48

AMPHIBIOUS SHIPS 589

No.	Name/Builder	LD/Launch	Comm	Decomm	Notes and Fate
249	NJ Ship	22 Oct 42 15 Dec 42	24 Dec 42		To UK 24 Dec 42, to France, sold to France on ret from Lend-Lease 21 Mar 49 for Indo-China
250	NJ Ship	22 Oct 42 16 Dec 42	31 Dec 42		To UK 31 Dec 42, ret 27 Mar 46, str 8 May 46
251	NJ Ship	14 Sep 42 18 Oct 42	18 Dec 42		To UK 31 Dec 42, ret 27 Mar 46, str 8 May 46, to France Nov 46
252	NJ Ship	14 Sep 42 18 Oct 42	28 Dec 42		To UK 28 Dec 42, ret 9 Feb 46, str 26 Feb 46
253	NJ Ship	18 Sep 42 18 Oct 42	4 Jan 43		To UK 5 Jan 43, ret, sold to Norway, str 15 Oct 46
254	NJ Ship	18 Sep 42 18 Oct 42	4 Jan 43		To UK 4 Jan 43, to Greece between 1 Jan and 13 Aug 47
255	NJ Ship	14 Sep 42 18 Oct 42	5 Jan 43		To UK 5 Jan 43, str 15 Oct 46
256	NJ Ship	14 Sep 42 18 Oct 42	7 Jan 43		To UK 7 Jan 43, ret 14 Mar 46, str 26 Feb 46
257	NJ Ship	8 Oct 42 30 Dec 42	16 Feb 43		To UK 16 Feb 43, str 15 Oct 46
258	NJ Ship	10 Oct 42 6 Feb 43	27 Feb 43		To UK 27 Feb 43, to Greece Apr 47
259	NJ Ship	10 Oct 42 19 Feb 43	12 Mar 43		To UK 11 Mar 43, to Greece Apr 47
260	NJ Ship	13 Oct 42 13 Mar 43	23 Mar 43		To UK 23 Mar 43, str 15 Oct 46, to France Nov 46
261	NJ Ship	13 Oct 42 19 Mar 43	27 Mar 43		To UK 27 Mar 43, to BU in India, str 26 Feb 46
262	NJ Ship	26 Oct 42 17 Dec 42	4 Jan 43		To UK 4 Jan 43, to France, sold to France when ret from Lend-Lease 21 Mar 49
263	NJ Ship	30 Oct 42 18 Dec 42	7 Jan 43		To UK 7 Jan 43, to France for Indo-China 19 Apr 46
264	NJ Ship	1 Nov 42 22 Dec 42	12 Jan 43		To UK 12 Jan 43, str 15 Oct 46
265	NJ Ship	4 Nov 42 23 Dec 42	19 Jan 43		To UK 19 Jan 43, BU India, str 26 Feb 46
266	NJ Ship	6 Nov 42 6 Jan 43	27 Jan 43		To UK 19 Jan 43, ret 20 Mar 46, str 23 Apr 47
267	NJ Ship	7 Nov 42 16 Jan 43	5 Feb 43		To UK 4 Feb 43, BU India, str 26 Feb 46
268	NJ Ship	9 Nov 42 31 Dec 42	19 Jan 43		To UK 19 Jan 43, ret 20 Mar 46, str 23 Apr 47
269	NJ Ship	16 Nov 42 7 Jan 43	22 Jan 43		To UK 22 Jan 43, ret 20 Mar 46, str 8 May 46
270	NJ Ship	18 Nov 42 12 Jan 43	29 Jan 43		To France for Indo-China 19 Apr 46
271	NJ Ship	19 Nov 42 29 Jan 43	13 Feb 43		To France for Indo-China 19 Apr 46
272	NJ Ship	12 Nov 42 10 Feb 43	26 Feb 43		To UK 26 Feb 43, str 15 Oct 46
273	NJ Ship	25 Nov 42 14 Jan 43	1 Feb 43		To UK 1 Feb 43, WL Mar 44
274	NJ Ship	25 Nov 42 22 Jan 43	9 Feb 43		To UK 9 Feb 43, ret 26 Feb 46, str 8 May 46
275	NJ Ship	21 Dec 42 4 Feb 43	22 Feb 43		To UK 22 Feb 43, ret 27 Mar 46, str 5 Jun 46
276	NJ Ship	31 Dec 42 18 Feb 43	10 Mar 43		To UK 10 Mar 43, str 19 Jun 46

Appendix C

No.	Name/Builder	LD/Launch	Comm	Decomm	Notes and Fate
277	NJ Ship	28 Dec 42 26 Feb 43	22 Mar 43		To UK 20 Mar 43, ret 14 Mar 46, str 17 Apr 46
278	NJ Ship	4 Jan 43 18 Mar 43	26 Mar 43		To UK 26 Mar 43, to Greece Apr 47
279	NJ Ship	28 Dec 42 30 Jan 43	18 Feb 43		To UK 18 Feb 43, BU India, str 26 Feb 46
280	NJ Ship	22 Dec 42 5 Feb 43	24 Feb 43		To UK 24 Feb 43, ret 24 Sep 46, str 22 Jan 47
281	NJ Ship	28 Dec 42 12 Feb 43	5 Mar 43		To UK 5 Mar 43, str 15 Oct 46
282	NJ Ship	4 Jan 43 15 Feb 43	6 Mar 43		To UK 6 Mar 43, ret 20 Sep 46, str 22 Jan 47
283	NJ Ship	8 Jan 43 17 Feb 43	9 Mar 43		To UK 9 Mar 43, str 10 Jun 47
284	NJ Ship	9 Jan 43 23 Feb 43	13 Mar 43		To UK 13 Mar 43, str 15 Oct 46
285	NJ Ship	25 Jan 43 24 Feb 43	13 Mar 43		To UK 13 Mar 43, ret May 46, str 5 Jun 46
286	NJ Ship	8 Jan 43 17 Feb 43	24 Mar 43		To UK 24 Mar 43, BU India, str 26 Feb 46
287	NJ Ship	9 Jan 43 12 Mar 43	24 Mar 43		To UK 24 Mar 43, BU India, str 26 Feb 46
288	NJ Ship	8 Jan 43 27 Mar 43	3 Apr 43		To UK 3 Apr 43, str 15 Oct 46
289	NJ Ship	30 Jan 43 27 Mar 43	20 Mar 43		To UK 19 Mar 43, ret 9 Jul 46, str 28 Aug 46
290	NJ Ship	15 Jan 43 11 Feb 43	2 Mar 43		To UK 2 Mar 43, str 15 Oct 46
291	NJ Ship	20 Feb 43 25 Mar 43	30 Mar 43		To UK 30 Mar 43, ret 28 Feb 46, str 28 Aug 46
292	NJ Ship	10 Feb 43 22 Mar 43	27 Mar 43		To UK 27 Mar 43, ret 25 Aug 46, str 8 Oct 46
293	NJ Ship	4 Feb 43 11 Mar 43	20 Mar 43		To UK 20 Mar 43, to France, sold to France when ret from Lend-Lease 21 Mar 49 for Indo-China
294	NJ Ship	19 Feb 43 25 Mar 43	30 Mar 43		To UK 30 Mar 43, ret 23 Sep 46, str 29 Oct 46
295	NJ Ship	12 Feb 43 27 Mar 43	3 Apr 43		To UK 3 Apr 43, ret 13 Feb 46, str 26 Feb 46
296	NJ Ship	18 Oct 42 21 Nov 42	18 Jan 43		To UK 18 Jan 43, BU India, str 26 Feb 46
297	NJ Ship	21 Oct 42 21 Nov 42	19 Jan 43		To UK 19 Jan 43, ret 24 Sep 46, str 5 Jun 46
298	NJ Ship	25 Oct 42 21 Nov 42	23 Jan 43		To UK 23 Jan 43, ret May 46, str 5 Jun 46
299	NJ Ship	24 Oct 42 21 Nov 42	25 Jan 43		To UK 25 Jan 43, to France for Indo-China 19 Apr 46
300	NJ Ship	18 Oct 42 21 Nov 42	29 Jan 43		To UK 27 Jan 43, ret 17 May 46, str 19 Jun 46
301	NJ Ship	18 Oct 42 21 Nov 42	2 Feb 43		To UK 2 Feb 43, ret 14 Mar 46, str 17 Apr 46
302	NJ Ship	24 Feb 43 31 Mar 43	9 Apr 43		To UK 9 Apr 43, str 5 Jun 46
303	NJ Ship	26 Feb 43 1 Apr 43	9 Apr 43		To UK 9 Apr 43, to Greece Apr 47
304	NJ Ship	1 Feb 43 25 Feb 43	16 Mar 43		To UK 16 Mar 43, ret 9 Feb 46, str 26 Feb 46

No.	Name/Builder	LD/Launch	Comm	Decomm	Notes and Fate
305	NJ Ship	5 Feb 43 9 Mar 43	18 Mar 43		To UK 18 Mar 43, to France for Indo-China 26 Jan 48
306	NJ Ship	13 Feb 43 26 Mar 43	2 Apr 43		To UK 2 Apr 43, ret 27 May 46, str 5 Jun 46
307	NJ Ship	18 Feb 43 23 Mar 43	29 Mar 43		To UK 29 Mar 43, ret May 46, str 5 Jun 46
308	NJ Ship	17 Feb 43 25 Mar 43	1 Apr 43		To UK 1 Apr 43, ret 23 Sep 46, str 29 Oct 46
309	NJ Ship	26 Feb 43 27 Mar 43	5 Apr 43		To UK 5 Apr 43, WL 23 Oct 43
310	NJ Ship	24 Feb 43 31 Mar 43	8 Apr 43		To UK 8 Apr 43, to France for Indo-China, 26 Jan 48
311	NJ Ship	10 Mar 43 1 Apr 43	10 Apr 43		To UK 10 Apr 43, ret 14 Mar 46, str 17 Apr 46
312	NJ Ship	2 Mar 43 30 Mar 43	6 Apr 43		To UK 6 Apr 43, BU India, str 26 Feb 46
313	NJ Ship	21 Nov 42 12 Dec 42	4 Feb 43		To UK 4 Feb 43, ret May 46, str 5 Jun 46
314	NJ Ship	21 Nov 42 12 Dec 42	8 Feb 43		To UK 8 Feb 43, str 15 Oct 46
315	NJ Ship	21 Nov 42 12 Dec 42	10 Feb 43		To UK 19 Feb 43, ret 21 May 46, str 3 Jul 46
316	NJ Ship	21 Nov 42 12 Dec 42	13 Feb 43		To UK 13 Feb 43, str 15 Oct 46
317	NJ Ship	21 Nov 42 12 Dec 42	16 Feb 43		To UK 16 Feb 43, ret May 46
318	NJ Ship	21 Nov 42 12 Dec 42	18 Feb 43		To UK 18 Feb 43, ret 25 Aug 46, str 8 Oct 46
319	Brown	25 Nov 42 21 Dec 42	3 Feb 43	26 Mar 46	Str 17 Apr 46
320	Brown	26 Nov 42 21 Dec 42	8 Feb 43	26 Mar 46	Str 1 May 46
321	Brown	4 Dec 42 21 Jan 43	6 Feb 43	2 Apr 46	Str 19 Jun 46
322	Brown	4 Dec 42 3 Feb 43	15 Feb 43	26 Mar 46	Str 17 Apr 46
323	Brown	21 Dec 42 22 Jan 43	9 Feb 43	15 Apr 46	Str 5 Jun 46
324	Brown	21 Dec 42 28 Jan 43	10 Feb 43	6 Mar 46	Str 20 Mar 46
325	Brown	21 Dec 42 29 Jan 43	12 Feb 43	29 May 46	Str 19 Jun 46
326	Brown	22 Dec 42 2 Feb 43	15 Feb 43	7 May 46	Str 5 Jun 46
327	Brown	27 Jul 42 12 Sep 42	31 Oct 42	29 Aug 46	At Bikini, destroyed 30 Oct 47
328	Brown	27 Jul 42 12 Sep 42	31 Oct 42	19 Apr 46	Str 5 Jun 46
329	Brown	27 Jul 42 12 Sep 42	31 Oct 42		At Bikini, sunk 16 Mar 48
330	Brown	27 Jul 42 22 Sep 42	8 Nov 42	11 Apr 46	Str 8 May 46
331	Brown	27 Jul 42 24 Sep 42	9 Nov 42	8 Feb 46	LCIR, str 12 Mar 46
332	Brown	27 Jul 42 26 Sep 42	16 Nov 42	26 Mar 46	Str 31 Oct 46
333	Brown	27 Jul 42 26 Sep 42	17 Nov 42	2 Jan 46	Str 7 Feb 46

APPENDIX C

No.	Name/Builder	LD/Launch	Comm	Decomm	Notes and Fate
334	Brown	27 Jul 42 3 Oct 42	24 Nov 42	15 Mar 46	Str 12 Apr 46
335	Brown	27 Jul 42 9 Oct 42	27 Nov 42	22 Jan 46	Str 26 Feb 46
336	Brown	27 Jul 42 13 Oct 42	3 Dec 42	21 Jun 46	Str 15 Aug 46
337	Brown	22 Sep 42 21 Oct 42	21 Dec 42	14 Dec 45	LCIR, str 21 Jan 46
338	Brown	22 Sep 42 24 Oct 42	26 Dec 42	19 Apr 46	LCIR, str 22 May 47
339	Brown	30 Sep 42 30 Oct 42	30 Dec 42		WL 4 Sep 43
340	Brown	1 Oct 42 31 Oct 42	12 Dec 42	14 Mar 46	LCIR, str 17 Apr 46
341	Brown	3 Oct 42 7 Nov 42	26 Dec 42	6 Feb 46	LCIR, str 25 Feb 46
342	Brown	13 Oct 42 7 Nov 42	30 Dec 42	25 Jan 46	LCIR, str 12 Apr 46
343	Brown	13 Oct 42 14 Nov 42	8 Jan 43	25 Mar 46	Str 17 Apr 46
344	Brown	26 Oct 42 2 Dec 42	18 Jan 43	15 Mar 46	Str 17 Apr 46
345	Brown	31 Oct 42 2 Dec 42	22 Jan 43	11 Feb 46	LCIG, str 26 Feb 46
346	Brown	31 Oct 42 2 Dec 42	25 Jan 43	22 Mar 46	LCIG, str 5 Jun 46
347	Brown	31 Oct 42 2 Dec 42	28 Jan 43	14 Dec 45	LCIG, str 26 Feb 46
348	Brown	7 Nov 42 21 Dec 42	27 Jan 43	8 Feb 46	LCIG, str 5 Jun 46
349	Brown	7 Nov 42 21 Dec 42	31 Jan 43	2 Apr 46	Str 19 Jun 46
350	Brown	14 Nov 42 21 Dec 42	4 Feb 43	3 May 46	Str 19 Jun 46
351	Lawley	5 Mar 43 8 Apr 43	14 May 43	28 May 46	LCIM, ex-LCIG; str 3 Jul 46
352	Lawley	5 Mar 43 10 Apr 43	18 May 43	19 Jul 46	LCIM, ex-LCIG; str 16 Sep 46
353	Lawley	30 Mar 43 30 Apr 43	26 May 43	25 Jun 46	LCIM, ex-LCIG; str 15 Aug 46
354	Lawley	1 Apr 43 6 May 43	1 Jun 43	19 Jul 46	LCIM, ex-LCIG; str 16 Sep 47
355	Lawley	11 Apr 43 20 May 43	9 Jun 43	3 Jul 46	LCIM, ex-LCIG; str 31 Jul 46
356	Lawley	13 Apr 43 27 May 43	12 Jun 43	4 Jun 46	LCIM, ex-LCIG; str 3 Jul 46
357	Lawley	4 May 43 4 Jun 43	17 Jun 43	23 Mar 46	Str 17 Apr 46
358	Lawley	8 May 43 10 Jun 43	23 Jun 43	14 Mar 46	Str 28 Mar 46
359	Lawley	25 May 43 21 Jun 43	2 Jul 43	31 Jan 46	LCIM, str 25 Feb 46
360	Lawley	1 Jun 43 29 Jun 43	12 Jul 43	29 Jan 46	Str 25 Feb 46
361	Lawley	11 Jun 43 9 Jul 43	23 Jul 43	12 Mar 46	Str 28 Mar 46
362	Lawley	15 Jun 43 16 Jul 43	26 Jul 43	N/A	LCIM, to MarComm 3 May 48

No.	Name/Builder	LD/Launch	Comm	Decomm	Notes and Fate
363	Lawley	26 Jun 43 22 Jul 43	31 Jul 43	7 Jun 46	Str 3 Jul 46
364	Lawley	30 Jun 43 28 Jul 43	6 Aug 43	23 Mar 46	Str 13 Apr 46
365	Lawley	9 Jul 43 3 Aug 43	12 Aug 43		LCIG, hulked due to CTL damage 10 Jan 45 by explosive boat
366	Lawley	18 Jul 43 9 Aug 43	17 Aug 43	23 May 46	LCIG, str 10 Jun 47
367	Lawley	26 Jul 43 17 Aug 43	23 Aug 43	11 Jul 46	LCFF, for disposal Jul 50
368	Lawley	3 Aug 43 23 Aug 43	28 Aug 43	15 Jul 46	LCFF, str 28 Aug 48
369	Lawley	9 Aug 43 27 Aug 43	31 Aug 43	N/A	LCFF, for disposal Jan 46
370	Lawley	11 Aug 43 8 Sep 43	13 Sep 43	20 May 46	LCFF, to MarComm 9 Mar 48
371	Lawley	23 Aug 43 13 Sep 43	18 Sep 43	1 Jul 46	Str 23 Apr 47
372	Lawley	31 Aug 43 19 Sep 43	24 Sep 43	11 Jul 46	LCIG, str 15 Aug 46
373	Lawley	17 Sep 43 23 Sep 43	29 Sep 43	18 Jun 46	LCIG, str 15 Aug 46
374	Lawley	11 Sep 43 27 Sep 43	30 Sep 43	16 May 46	Str 19 Jun 46
375	Lawley	17 Sep 43 3 Oct 43	9 Oct 43	29 Mar 46	Str 1 May 46
376	Lawley	21 Sep 43 8 Oct 43	13 Oct 43		To UK 13 Oct 43, ret 27 Mar 46, str 15 Aug 46
377	Lawley	25 Sep 43 14 Oct 43	18 Oct 43		To UK 18 Oct 43, str 15 Oct 46
378	Lawley	23 Sep 43 18 Oct 43	23 Oct 43		To UK 23 Oct 43, ret 12 Apr 46, str 19 Jun 46
379	Lawley	6 Oct 43 23 Oct 43	28 Oct 43		To UK 28 Oct 43, ret 18 Mar 46, str 3 Jul 46
380	Lawley	11 Oct 43 28 Oct 43	31 Oct 43		To UK 31 Oct 43, ret 27 Mar 46, str 19 Jun 46
381	Lawley	9 Oct 43 29 Oct 43	31 Oct 43		To UK 1 Nov 43, ret 14 Mar 46, str 17 Apr 46
382	Lawley	17 Oct 43 4 Nov 43	8 Nov 43		To UK 8 Nov 43, ret 28 May 46, str 19 Jun 46
383	Lawley	16 Oct 43 7 Nov 43	11 Nov 43		To UK 11 Nov 43, ret 24 Sep 46, str 22 Jan 47
384	Lawley	20 Oct 43 11 Nov 43	15 Nov 43		To UK 15 Nov 43, ret 30 Aug 46, str 22 Jan 47
385	Lawley	24 Oct 43 15 Nov 43	19 Nov 43		To UK 19 Nov 43, ret 19 Jun 46, str 8 Oct 46
386	Lawley	3 Nov 43 17 Nov 43	22 Nov 43		To UK 22 Nov 43, ret 28 May 46, str 19 Jun 46
387	Lawley	1 Nov 43 20 Nov 43	24 Nov 43		To UK 24 Nov 43, ret 20 Mar 46, str 19 Jun 46
388	Lawley	15 Nov 43 25 Nov 43	29 Nov 43		To UK 29 Nov 43, ret 27 Mar 46, str 5 Jun 46
389	Lawley	17 Nov 43 27 Nov 43	30 Nov 43		To UK 30 Nov 43, ret 14 Mar 46, str 17 Apr 46
390	Lawley	21 Nov 43 28 Nov 43	30 Nov 43		To UK 1 Dec 43, ret 19 Jun 46, str 8 Oct 46, New York harbor cruise ship *Knickerbocker*, then *Day Line I*, then *Circle Line IV*

No.	Name/Builder	LD/Launch	Comm	Decomm	Notes and Fate
391	Lawley	25 Nov 43 2 Dec 43	7 Dec 43		To UK 7 Dec 43, ret 28 May 46, str 19 Jun 46
392	Lawley	28 Nov 43 4 Dec 43	8 Dec 43	14 Feb 46	Str 28 Mar 46
393	Lawley	30 Nov 43 7 Dec 43	11 Dec 43	15 Feb 46	Str 20 Mar 46
394	Lawley	3 Dec 43 10 Dec 43	14 Dec 43	15 Feb 46	Str 20 Mar 46
395	Lawley	5 Dec 43 12 Dec 43	20 Dec 43	31 Jan 46	Str 12 Apr 46
396	Lawley	7 Dec 43 15 Dec 43	20 Dec 43		LCIG, to IX 212 on 23 Feb 45, repair ship, gear stowage, receiving barracks, medical center for small craft, str 5 Jan 46, destroyed 21 Mar 46
397	Lawley	10 Dec 43 19 Dec 43	23 Dec 43	12 Apr 46	LCIG, str 7 Feb 47
398	Lawley	13 Dec 43 20 Dec 43	27 Dec 43	11 May 46	LCIG, str 22 Jan 47
399	Lawley	15 Dec 43 23 Dec 43	30 Dec 43	8 Dec 45	LCFF, CTL Oct 45 due to typhoon
400	Lawley	19 Dec 43 29 Dec 43	31 Dec 43		To AMCU 8 on 10 Mar 45, str 14 Mar 47, to MarComm 10 Nov 47
401	Lawley	21 Dec 43 2 Jan 44	8 Jan 44	29 Oct 45	LCIG, str 13 Nov 45
402	Lawley	24 Dec 43 5 Jan 44	11 Jan 44	22 Apr 46	Str 3 Jul 46
403	Lawley	29 Dec 43 7 Jan 44	13 Jan 44	29 Oct 45	LCIG 15 Jul 45, str 13 Nov 45
404	Lawley	2 Jan 44 9 Jan 44	15 Jan 44	25 Jun 46	LCIG, str 16 Sep 47, to Philippines 2 Jul 48
405	Lawley	5 Jan 44 12 Jan 44	18 Jan 44	19 Apr 46	LCIG, str 7 Feb 47
406	Lawley	8 Jan 44 15 Jan 44	20 Jan 44	18 Feb 46	LCIG, str 12 Mar 46
407	Lawley	10 Jan 44 16 Jan 44	22 Jan 44	9 Jul 46	LCIG, str 16 Sep 47
408	Lawley	12 Jan 44 18 Jan 44	24 Jan 44	29 Oct 45	LCIG, str 13 Nov 45
409	Lawley	15 Jan 44 20 Jan 44	25 Jan 44		To AMCU 9 on 10 Mar 45, str 23 Jun 47, to MarComm 8 Oct 47
410	Lawley	17 Jan 44 22 Jan 44	27 Jan 44	2 May 46	Str 23 Apr 47
411	Lawley	18 Jan 44 24 Jan 44	29 Jan 44		To UK 30 Nov 44, ret 25 Sep 46, str 22 Jan 47
412	Lawley	21 Jan 44 26 Jan 44	31 Jan 44	29 Oct 45	LCIG, str 13 Nov 45
413	Lawley	22 Jan 44 29 Jan 44	7 Feb 44	29 Oct 45	LCIG, str 13 Nov 45
414	Lawley	24 Jan 44 1 Feb 44	9 Feb 44	29 Oct 45	LCIG, str 13 Nov 45
415	Lawley	26 Jan 44 4 Feb 44	11 Feb 44	29 Oct 45	LCIG, str 13 Nov 45
416	Lawley	29 Jan 44 7 Feb 44	14 Feb 44		WL 9 Jun 44
417	Lawley	1 Feb 44 9 Feb 44	16 Feb 44	25 Jun 46	LCIG, to ROC 12 Dec 46
418	Lawley	4 Feb 44 11 Feb 44	18 Feb 44		LCIG, to ROC 10 Jul 46

No.	Name/Builder	LD/Launch	Comm	Decomm	Notes and Fate
419	Lawley	7 Feb 44 14 Feb 44	21 Feb 44	1 May 46	LCIG, to army 13 May 46, str 10 Jun 47
420	Lawley	9 Feb 44 16 Feb 44	23 Feb 44	29 May 46	LCIG, to NEI government 29 May 46
421	Lawley	11 Feb 44 19 Feb 44	25 Feb 44	26 Apr 46	LCIG, str 22 Jan 47
422	Lawley	15 Feb 44 21 Feb 44	28 Feb 44	22 Jun 46	LCIG, str 15 Aug 46
423	NJ Ship	13 Mar 43 17 May 43	11 Jun 43	18 Apr 46	LCFF, str 5 Jun 46
424	NJ Ship	27 Mar 43 19 May 43	17 Jun 43	27 Jun 46	LCFF, str 15 Aug 46
425	NJ Ship	16 Mar 43 24 May 43	22 Jun 43	21 Jun 46	LCFF, to MarComm 1 Nov 47
426	NJ Ship	1 Apr 43 2 Jun 43	26 Jun 43	18 Jul 46	LCFF, str 28 Aug 46
427	NJ Ship	6 Apr 43 10 Jun 43	1 Jul 43	19 Jun 46	LCFF, str 31 Jul 46
428	NJ Ship	7 Apr 43 12 Jun 43	5 Jul 43	21 Feb 46	LCIG, str 13 Nov 48
429	NJ Ship	6 Apr 43 16 Jun 43	10 Jul 43	18 May 46	Str 19 Jul 46
430	NJ Ship	9 Apr 43 21 Jun 43	13 Jul 43	23 May 46	Str 19 Jul 46
431	NJ Ship	9 Apr 43 24 Jun 43	15 Jul 43	28 Feb 46	LCIM, str 28 Mar 46
432	NJ Ship	9 Apr 43 29 Jun 43	19 Jul 43	18 Mar 46	Str 17 Apr 46
433	NJ Ship	21 Apr 43 1 Jul 43	21 Jul 43	13 Jun 46	Str 15 Aug 46
434	NJ Ship	22 Apr 43 3 Jul 43	24 Jul 43	30 Apr 46	Str 5 Jun 46
435	NJ Ship	23 Apr 43 8 Jul 43	27 Jul 43	2 May 46	Str 5 Jun 46
436	NJ Ship	23 Apr 43 13 Jul 43	29 Jul 43	28 May 46	Str 3 Jul 46
437	NJ Ship	23 Apr 43 17 Jul 43	31 Jul 43	4 Feb 46	LCIG, str 25 Feb 46
438	NJ Ship	23 Apr 43 17 Jul 43	3 Aug 43	15 Apr 46	LCIG, str 5 Jun 46
439	NJ Ship	29 Apr 43 21 Jul 43	6 Aug 43	18 Jun 46	LCIG, str 15 Aug 46
440	NJ Ship	29 Apr 43 24 Jul 43	6 Aug 43	27 Jun 46	LCIG, str 7 Feb 47
441	NJ Ship	30 Apr 43 20 Jul 43	7 Aug 43	23 May 46	LCIG, str 31 Jul 46
442	NJ Ship	7 May 43 31 Jul 43	11 Aug 43		LCIG, to Korea (U.S. Army) Feb 47, str 29 Feb 47
443	NJ Ship	19 May 43 2 Aug 43	13 Aug 43	15 Mar 46	Str 12 Apr 46
444	NJ Ship	21 May 43 4 Aug 43	15 Aug 43	21 Mar 46	Str 17 Apr 46
445	NJ Ship	27 May 43 5 Aug 43	17 Aug 43	12 Jun 46	Str 23 Dec 47
446	NJ Ship	3 Jun 43 6 Aug 43	19 Aug 43	24 Jun 46	Str 15 Aug 46
447	NJ Ship	12 Jun 43 10 Aug 43	21 Aug 43	5 Apr 46	Str 19 Jun 46

No.	Name/Builder	LD/Launch	Comm	Decomm	Notes and Fate
448		15 Jun 43	22 Aug 43	5 Apr 46	LCID, str 8 May 46
	NJ Ship	12 Aug 43			
449		17 Jun 43	25 Aug 43	8 Feb 46	LCIG, str 8 May 46
	NJ Ship	14 Aug 43			
450		22 Jun 43	26 Aug 43	18 Jun 46	LCIG, str 7 Feb 47
	NJ Ship	18 Aug 43			
451		26 Jun 43	28 Aug 43	19 Jul 46	LCIG, str 28 Aug 46
	NJ Ship	19 Aug 43			
452		30 Jun 43	30 Aug 43	27 Jun 46	LCIG, str 15 Aug 46
	NJ Ship	20 Aug 43			
453		2 Jul 43	31 Aug 43	7 May 46	LCIG, to Korea (U.S. Army) Oct 46, str 29 Sep 47
	NJ Ship	21 Aug 43			
454		5 Jul 43	2 Sep 43	3 Jul 46	LCIG, str 31 Jul 46
	NJ Ship	25 Aug 43			
455		10 Jul 43	4 Sep 43	11 Jul 46	LCIG, str 15 Aug 46
	NJ Ship	26 Aug 43			
456		13 Jul 43	7 Sep 43	25 Apr 46	LCIG, str 7 Feb 47
	NJ Ship	27 Aug 43			
457		15 Jul 43	8 Sep 43	4 Apr 46	LCIG, str 7 Feb 47
	NJ Ship	28 Aug 43			
458		19 Jul 43	10 Sep 43	21 Mar 46	LCIG, str 17 Apr 46
	NJ Ship	1 Sep 43			
459		22 Jul 43	11 Sep 43		LCIG, WL 19 Sep 44
	NJ Ship	2 Sep 43			
460		26 Jul 43	13 Sep 43	9 Apr 46	LCIG, str 7 Feb 47
	NJ Ship	3 Sep 43			
461		29 Jul 43	14 Sep 43	21 Jun 46	LCIG, str 15 Aug 46
	NJ Ship	4 Sep 43			
462		2 Aug 43	16 Sep 43	18 Jul 46	LCIG, str 15 Aug 46
	NJ Ship	8 Sep 43			
463		3 Aug 43	17 Sep 43	21 Mar 46	LCIG, str 17 Apr 46
	NJ Ship	9 Sep 43			
464		5 Aug 43	20 Sep 43	27 Jun 46	LCIG, str 15 Aug 46
	NJ Ship	10 Sep 43			
465		6 Aug 43	21 Sep 43	15 Jul 46	LCIG, str 28 Aug 46
	NJ Ship	11 Sep 43			
466		7 Aug 43	22 Sep 43	8 Feb 46	LCIG, str 5 Jun 46
	NJ Ship	15 Sep 43			
467		11 Aug 43	24 Sep 43	29 May 46	LCIG, str 10 Jun 47, to NEI, later to Indonesia
	NJ Ship	16 Sep 43			
468		13 Aug 43	25 Sep 43		LCIG, WL 18 Jun 44
	NJ Ship	17 Sep 43			
469		15 Aug 43	27 Sep 43	19 Apr 46	LCIG, str 7 Feb 47
	NJ Ship	18 Sep 43			
470		19 Aug 43	28 Sep 43	1 Apr 46	LCIG, str 1 May 47
	NJ Ship	22 Sep 43			
471		20 Aug 43	30 Sep 43	3 Jan 46	LCIG, str 8 May 46
	NJ Ship	23 Sep 43			
472		22 Aug 43	2 Oct 43	28 Jun 46	LCIG, str 15 Aug 46
	NJ Ship	24 Sep 43			
473		23 Aug 43	4 Oct 43	7 Jun 46	LCIG, str 8 Jul 46
	NJ Ship	25 Sep 43			
474		26 Aug 43	5 Oct 43		LCIG, WL 17 Feb 45
	NJ Ship	2 Sep 43			
475		28 Aug 43	7 Oct 43	27 Jun 46	LCIG, str 7 Feb 47
	NJ Ship	30 Sep 43			
476		29 Aug 43	9 Oct 43	18 Feb 46	Str 28 Mar 46
	NJ Ship	2 Oct 43			

No.	Name/Builder	LD/Launch	Comm	Decomm	Notes and Fate
477	NJ Ship	31 Aug 43 4 Oct 43	11 Oct 43	25 Feb 46	Str 28 Mar 46
478	NJ Ship	3 Sep 43 6 Oct 43	12 Oct 43	13 Feb 46	Str 28 Mar 46, to ROC Sep 46
479	NJ Ship	3 Sep 43 7 Oct 43	14 Oct 43	13 Feb 46	Str 28 Mar 46
480	NJ Ship	4 Sep 43 8 Oct 43	15 Oct 43	21 Feb 46	Str 28 Jul 46
481	NJ Ship	6 Sep 43 9 Oct 43	18 Oct 43	14 Feb 46	Str 12 Mar 46
482	NJ Ship	9 Sep 43 13 Oct 43	19 Oct 43	5 Mar 46	Str 12 Apr 46
483	NJ Ship	11 Sep 43 14 Oct 43	20 Oct 43	11 Mar 46	Str 28 Mar 46
484	NJ Ship	12 Sep 43 15 Oct 43	21 Oct 43	29 May 46	LCIG, to MarComm 4 Feb 47
485	NJ Ship	13 Sep 43 16 Oct 43	23 Oct 43	20 Jul 46	LCFF, authorized for use as tgt 28 Jul 56
486	NJ Ship	16 Sep 43 19 Oct 43	25 Oct 43	22 Dec 45	LCFF, str 2 Nov 46
487	NJ Ship	18 Sep 43 20 Oct 43	28 Oct 43	17 May 46	Str 19 Jun 46
488	NJ Ship	19 Sep 43 21 Oct 43	30 Oct 43	17 May 46	Str 5 Jun 46, merch, later PRC
489	NJ Ship	22 Sep 43 22 Oct 43	30 Oct 43		To UK 20 Nov 44, ret 19 Jun 46, str 29 Oct 46
490	NJ Ship	23 Sep 43 23 Oct 43	1 Nov 43		To UK 20 Nov 44, ret 19 Jun 46, str 29 Oct 46
491	NJ Ship	24 Sep 43 27 Oct 43	3 Nov 43		To UK 18 Nov 44, ret 14 Mar 46, str 26 Feb 46
492	NJ Ship	29 Sep 43 28 Oct 43	4 Nov 43		To UK 20 Nov 44, sold in Bermuda, str 3 Jul 46
493	NJ Ship	29 Sep 43 30 Oct 43	6 Nov 43		To UK 30 Nov 44, ret 24 Sep 46, str 13 Dec 46
494	NJ Ship	2 Oct 43 30 Oct 43	8 Nov 43		To UK 30 Nov 44, ret 15 Aug 46, str 29 Oct 46
495	NJ Ship	3 Oct 43 3 Nov 43	10 Nov 43		To UK 30 Nov 44, ret 17 May 46, str 5 Jun 46
496	NJ Ship	5 Oct 43 4 Nov 43	11 Nov 43		To UK 30 Nov 44, ret 25 Sep 46, str 13 Dec 46
497	NJ Ship	7 Oct 43 5 Nov 43	13 Nov 43		CTL 23 Jun 44
498	NJ Ship	9 Oct 43 6 Nov 43	15 Nov 43		To UK 30 Nov 44, ret 25 Sep 46, str 13 Dec 46
499	NJ Ship	10 Oct 43 8 Nov 43	16 Nov 43		To UK 30 Nov 44, ret 14 May 46, str 10 Jun 47
500	NJ Ship	11 Oct 43 10 Nov 43	17 Nov 43		To UK 30 Nov 44, ret 27 Jul 46, str 29 Oct 46
501	NJ Ship	13 Oct 43 11 Nov 43	18 Nov 43		To UK 30 Nov 44, ret 15 Aug 46, str 29 Oct 46
502	NJ Ship	19 Oct 43 13 Nov 43	20 Nov 43		To UK 30 Nov 44, ret 24 Sep 46, str 22 Jan 47
503	NJ Ship	20 Oct 43 17 Nov 43	22 Nov 43	9 Apr 46	LCFF, sold Feb 48
504	NJ Ship	22 Oct 43 18 Nov 43	23 Nov 43	17 Aug 46	LCFF, for disposal 8 Aug 56
505	NJ Ship	18 Oct 43 19 Nov 43	25 Nov 43		To UK 30 Nov 44, ret 24 Sep 46, str 13 Nov 46

No.	Name/Builder	LD/Launch	Comm	Decomm	Notes and Fate
506	NJ Ship	21 Oct 43 20 Nov 43	26 Nov 43	19 Dec 45	LCIG, str 8 Jan 46
507	NJ Ship	27 Oct 43 23 Nov 43	29 Nov 43		To UK 5 Dec 44, ret 21 Sep 46, str 8 Oct 46
508	NJ Ship	1 Nov 43 24 Nov 43	30 Nov 43		To UK 5 Dec 44, str 15 Oct 46, to France for Indo-China 10 Dec 46
509	NJ Ship	2 Nov 43 26 Nov 43	1 Dec 43		To UK 5 Dec 44, ret 9 Aug 46, str 8 Oct 46
510	NJ Ship	3 Nov 43 27 Nov 43	3 Dec 43		To UK 5 Dec 44, ret 27 Jul 46, str 29 Oct 46
511	NJ Ship	4 Nov 43 30 Nov 43	4 Dec 43		To UK 5 Dec 44, ret 30 Aug 46, str 13 Nov 46
512	NJ Ship	3 Nov 43 1 Dec 43	6 Dec 43		To UK 5 Dec 44, ret Sep 46, str 15 Oct 46
513	NJ Ship	5 Nov 43 3 Dec 43	9 Dec 43		To AMCU 10 on 10 Mar 45, str 22 Jan 48, to MarComm 10 Jul 48
514	NJ Ship	5 Nov 43 3 Dec 43	9 Dec 43		LCIG, to ROC 12 Jun 46 on decomm, later PRC
515	NJ Ship	9 Nov 43 4 Dec 43	10 Dec 43		To AMCU 11 on 10 Mar 45, to *Blackbird* (MHC 11) 7 Feb 55, str Jan 60
516	NJ Ship	9 Nov 43 8 Dec 43	11 Dec 43	29 Apr 46	LCIG, to U.S. Army (Korea) 26 Feb 47
517	NJ Ship	14 Nov 43 9 Dec 43	13 Dec 43		LCIG, to ROC 12 Jun 46 on decomm, later PRC
518	NJ Ship	14 Nov 43 10 Dec 43	15 Dec 43		To ROC 8 Jul 46
519	NJ Ship	14 Nov 43 12 Dec 43	17 Dec 43	13 Jun 46	Str 15 Aug 46
520	NJ Ship	17 Nov 43 14 Dec 43	18 Dec 43	27 Mar 46	Str 17 Apr 46, to NEI, later Indonesia
521	NJ Ship	17 Nov 43 15 Dec 43	20 Dec 43		To SU 28 Jul 45, ret Jun 55, str 2 Dec 55
522	NJ Ship	18 Nov 43 17 Dec 43	21 Dec 43		To SU 28 Jul 45, ret Jun 55, str 2 Dec 55
523	NJ Ship	23 Nov 43 17 Dec 43	23 Dec 43		To SU 28 Jul 45, ret Jun 55, str 2 Dec 55
524	NJ Ship	22 Nov 43 21 Dec 43	24 Dec 43		To SU 28 Jul 45, ret Jun 55, str 2 Dec 55
525	NJ Ship	24 Nov 43 22 Dec 43	29 Dec 43		To SU 28 Jul 45, ret Jun 55, str 2 Dec 55
526	NJ Ship	28 Nov 43 22 Dec 43	31 Dec 43		To SU 28 Jul 45, ret Jun 55, str 2 Dec 55
527	NJ Ship	28 Nov 43 23 Dec 43	1 Jan 44		To SU 28 Jul 45, ret Jun 55, str 2 Dec 55
528	NJ Ship	28 Nov 43 23 Dec 43	4 Jan 44	2 Jul 46	LCIG, str 15 Aug 46
529	NJ Ship	29 Nov 43 30 Dec 43	6 Jan 44	1 Apr 46	Str 8 May 46
530	NJ Ship	2 Dec 43 31 Dec 43	7 Jan 44	20 Feb 47	LCIG, str 14 Mar 47
531	NJ Ship	7 Dec 43 1 Jan 44	8 Jan 44	23 May 46	LCFF, delivered to buyer 24 Oct 47
532	NJ Ship	7 Dec 43 5 Jan 44	10 Jan 44	26 Jun 46	LCFF, to MarComm 30 Jan 48
533	NJ Ship	4 Dec 43 6 Jan 44	12 Jan 44		LCFF, to army Jan 46, str 31 Jul 46
534	NJ Ship	4 Dec 43 8 Jan 44	13 Jan 43	14 Feb 47	LCIG, str 14 Mar 47

No.	Name/Builder	LD/Launch	Comm	Decomm	Notes and Fate
535		7 Dec 43	15 Jan 44	23 Sep 46	LCFF, str 8 Aug 56
	NJ Ship	11 Jan 44			
536		10 Dec 43	17 Jan 44	15 Apr 46	LCFF, str 5 Jun 46
	NJ Ship	12 Jan 44			
537		11 Dec 43	20 Jan 44		To UK 30 Nov 44, str 29 Oct 46
	NJ Ship	13 Jan 44			
538		13 Dec 43	20 Jan 44	27 Mar 46	LCIG, to Admiral Farragut Academy on decomm, str 12 Apr 46
	NJ Ship	14 Jan 44			
539		13 Dec 43	22 Jan 44	19 Dec 45	LCIG, str 8 Jan 46
	NJ Ship	18 Jan 44			
540		15 Dec 43	22 Jan 44	19 Dec 45	LCIG, str 8 Jan 46
	NJ Ship	19 Jan 44			
541		16 Dec 43	24 Jan 44	17 Dec 45	LCIG, str 8 Jan 46
	NJ Ship	20 Jan 44			
542		18 Dec 43	26 Jan 44	2 Jul 46	LCIG, str 15 Aug 46
	NJ Ship	21 Jan 44			
543		24 Dec 43	29 Jan 44	16 Mar 49	Str 2 Feb 57
	NJ Ship	21 Jan 44			
544		24 Dec 43	31 Jan 44	16 Mar 49	Str 12 Jul 56
	NJ Ship	25 Jan 44			
545		31 Dec 43	2 Feb 44	10 Feb 49	Str 7 Feb 57
	NJ Ship	26 Jan 44			
546		31 Dec 43	3 Feb 44	22 Oct 46	Str 13 Dec 46
	NJ Ship	28 Jan 44			
547		1 Jan 44	5 Feb 44	1 Nov 46	Str 14 Mar 47
	NJ Ship	28 Jan 44			
548		1 Jan 44	7 Feb 44	1 Nov 46	Str 14 Mar 47
	NJ Ship	2 Feb 44			
549		1 Jan 44	8 Feb 44		At Bikini, str 22 Dec 48 but officially decomm 20 Jan 49
	NJ Ship	4 Feb 44			
550		11 Jan 44	10 Feb 44	13 May 46	Str 19 Jun 46
	NJ Ship	5 Feb 44			
551		12 Jan 44	11 Feb 44		To SU 28 Jul 45, ret Jun 55, str 2 Dec 55
	NJ Ship	7 Feb 44			
552		13 Jan 44	14 Feb 44	16 May 46	Str 3 Jul 46
	NJ Ship	8 Feb 44			
553		14 Jan 44	15 Feb 44		WL 6 Jun 44
	NJ Ship	10 Feb 44			
554		15 Jan 44	16 Feb 44		To SU 28 Jul 45, WL 17 Aug 45
	NJ Ship	11 Feb 44			
555		15 Jan 44	18 Feb 44	18 Apr 46	Str 5 Jun 46
	NJ Ship	14 Feb 44			
556		19 Jan 44	19 Feb 44	19 Dec 46	LCIG, str 8 Jan 46
	NJ Ship	16 Feb 44			
557		18 Jan 44	21 Feb 44		To SU 28 Jul 45, ret Jun 55, str 2 Dec 55
	NJ Ship	17 Feb 44			
558		24 Jan 44	22 Feb 44	22 May 46	LCIG, to NEI government on decomm
	NJ Ship	18 Feb 44			
559		24 Jan 44	24 Feb 44	3 Jun 46	LCIG, to NEI government on decomm
	NJ Ship	19 Feb 44			
560		24 Jan 44	25 Feb 44	25 Jun 46	LCIG, str 31 Jul 46
	NJ Ship	22 Feb 44			
561		26 Jan 44	26 Feb 44	4 Jun 46	LCIG, str 3 Jul 46
	NJ Ship	23 Feb 44			
562		24 Jan 44	28 Feb 44	26 Mar 46	Str 17 Apr 46
	NJ Ship	24 Feb 44			
563		25 Jan 44	1 Mar 44	8 Jun 46	LCIG, str 19 Jun 46
	NJ Ship	25 Feb 44			

No.	Name/Builder	LD/Launch	Comm	Decomm	Notes and Fate
564	NJ Ship	26 Jan 44 26 Feb 44	2 Mar 44	20 Jun 46	Str 15 Aug 46, to Argentina 48
565	NJ Ship	29 Jan 44 28 Feb 44	3 Mar 44	3 Jul 46	LCIG, str 15 Aug 46
566	NJ Ship	29 Jan 44 29 Feb 44	4 Mar 44	14 Feb 46	LCIG, str 5 Jun 46
567	NJ Ship	4 Feb 44 1 Mar 44	6 Mar 44	17 Jun 46	LCIG, str 31 Jul 46
568	NJ Ship	5 Feb 44 2 Mar 44	14 Mar 44	18 Jun 46	LCIG, str 31 Jul 46
569	NJ Ship	8 Feb 44 3 Mar 44	14 Mar 44	12 Jun 46	LCFF, str 31 Jul 46
570	NJ Ship	9 Feb 44 6 Mar 44	15 Mar 44	19 Apr 46	LCIG, to MarComm 15 Aug 47
571	NJ Ship	11 Feb 44 8 Mar 44	16 Mar 44	11 Jun 46	LCFF, str 23 Jun 47
572	NJ Ship	15 Feb 44 9 Mar 44	16 Mar 44	15 Mar 46	LCFF, str 28 Mar 46
573	NJ Ship	15 Feb 44 10 Mar 44	17 Mar 44	16 Mar 49	Str 23 Mar 49
574	NJ Ship	17 Feb 44 11 Mar 44	18 Mar 44	16 Mar 46	LCIG conversion canceled at end of war, to MarComm 12 Nov 47
575	NJ Ship	17 Feb 44 14 Mar 44	20 Mar 44	7 Oct 46	LCFF, str 17 Nov 55, tgt 6 Jun 56
576	NJ Ship	18 Feb 44 15 Mar 44	21 Mar 44	7 Aug 47	LCIG, str 16 Sep 47
577	NJ Ship	21 Feb 44 17 Mar 44	23 Mar 44	28 Oct 46	LCIG 20 Aug 45, str 10 Jun 47
578	NJ Ship	22 Feb 44 18 Mar 44	25 Mar 44	22 Nov 46	Str 10 Jun 47
579	NJ Ship	23 Feb 44 21 Mar 44	27 Mar 44	28 Oct 46	Str 13 Dec 46
580	NJ Ship	24 Feb 44 21 Mar 44	28 Mar 44	25 Jun 46	LCIG, str 15 Aug 46
581	NJ Ship	28 Feb 44 23 Mar 44	30 Mar 44	20 Mar 46	Str 21 May 46, to Argentina 48
582	NJ Ship	29 Feb 44 24 Mar 44	1 Apr 44	14 Dec 45	LCIM, str 8 Jan 46
583	NJ Ship	26 Feb 44 28 Mar 44	3 Apr 44	25 Mar 46	Str 17 Apr 46, to Argentina 48
584	NJ Ship	3 Mar 44 29 Mar 44	4 Apr 44		To SU 9 Jun 45, BU in SU
585	NJ Ship	8 Mar 44 30 Mar 44	5 Apr 44		To SU 9 Jun 45, ret Jun 55, str 2 Dec 55
586	NJ Ship	8 Mar 44 31 Mar 44	7 Apr 44		To SU 14 Jun 45, BU in SU
587	NJ Ship	10 Mar 44 3 Apr 44	8 Apr 44		To SU 14 Jun 45, BU in SU
588	NJ Ship	11 Mar 44 4 Apr 44	10 Apr 44	25 Jun 46	LCIM, str 15 Aug 46, to NEI, later Indonesia
589	NJ Ship	11 Mar 44 5 Apr 44	11 Apr 44		To AMCU 7 on 10 Mar 45, str 14 Mar 57, to MarComm 10 Nov 47
590	NJ Ship	11 Mar 44 6 Apr 44	13 Apr 44		To SU 9 Jun 45, ret Jun 55, str 2 Dec 55
591	NJ Ship	16 Mar 44 10 Apr 44	14 Apr 44		To SU 9 Jun 45, BU in SU
592	NJ Ship	13 Mar 44 11 Apr 44	17 Apr 44		To SU 9 Jun 45, ret Jun 55, str 2 Dec 55

No.	Name/Builder	LD/Launch	Comm	Decomm	Notes and Fate
593	NJ Ship	17 Mar 44 12 Apr 44	18 Apr 44		To SU 9 Jun 45, BU in SU
594	NJ Ship	17 Mar 44 13 Apr 44	20 Apr 44	30 Apr 46	LCIM, to U.S. Army (Korea) 10 Oct 46
595	NJ Ship	23 Mar 44 14 Apr 44	21 Apr 44	3 May 46	LCIM, to army 13 May 46, str 29 Sep 47
596	NJ Ship	22 Mar 44 18 Apr 44	22 Apr 44	18 Sep 46	LCIM, str 22 May 51
597	NJ Ship	22 Mar 44 19 Apr 44	25 Apr 44	17 May 46	Str 15 Aug 46
598	NJ Ship	24 Mar 44 20 Apr 44	26 Apr 44	3 Jun 46	Str 23 Dec 47
599	NJ Ship	25 Mar 44 21 Apr 44	27 Apr 44	29 May 46	Str 3 Jul 46
600	NJ Ship	27 Mar 44 24 Apr 44	29 Apr 44		WL 12 Jan 45
601	NJ Ship	28 Mar 44 25 Apr 44	1 May 44	10 Jun 46	Str 23 Dec 47
602	NJ Ship	30 Mar 44 26 Apr 44	2 May 44	4 Apr 46	Str 21 May 46
603	NJ Ship	1 Apr 44 27 Apr 44	4 May 44	20 May 46	Str 28 Aug 46
604	NJ Ship	3 Apr 44 28 Apr 44	5 May 44	12 Apr 46	Str 8 May 46
605	NJ Ship	3 Apr 44 1 May 44	6 May 44	16 May 46	Str 19 Jun 46
606	NJ Ship	4 Apr 44 2 May 44	8 May 44	13 May 46	Str 10 Jun 47, to Argentina 48
607	NJ Ship	5 Apr 44 3 May 44	10 May 44	18 May 46	Str 19 Jul 46
608	NJ Ship	7 Apr 44 4 May 44	11 May 44	12 Jun 46	Str 15 Aug 46
609	NJ Ship	10 Apr 44 5 May 44	12 May 44	7 May 46	Str 5 Jun 46
610	NJ Ship	11 Apr 44 9 May 44	15 May 44	21 May 46	Str 19 Jun 46
611	NJ Ship	13 Apr 44 10 May 44	17 May 44	30 Apr 46	Str 5 Jun 46
612	NJ Ship	13 Apr 44 11 May 44	17 May 44	N/A	Str 10 Jun 47
613	NJ Ship	17 Apr 44 12 May 44	19 May 44	28 May 46	Str 3 Jul 46
614	NJ Ship	18 Apr 44 15 May 44	20 May 44	14 May 46	Str 19 Feb 48
615	NJ Ship	19 Apr 44 16 May 44	23 May 44		Used for scientific resurvey of Bikini post-explosion then disposed of, str 22 Dec 48 but decomm 15 Feb 49
616	NJ Ship	21 Apr 44 17 May 44	24 May 44		To Chinese Maritime Customs 27 Jun 46
617	NJ Ship	21 Apr 44 18 May 44	25 May 44	17 May 46	Str 19 Jul 46, to ROC Oct 47
618	NJ Ship	24 Apr 44 19 May 44	27 May 44	14 May 46	Str 19 Jun 46
619	NJ Ship	24 Apr 44 23 May 44	29 May 44	18 Apr 46	Str 5 Jun 46
620	NJ Ship	26 Apr 44 24 May 44	31 May 44		At Bikini, sunk 14 Aug 46

No.	Name/Builder	LD/Launch	Comm	Decomm	Notes and Fate
621	NJ Ship	28 Apr 44 25 May 44	31 May 44	13 Apr 46	Str 3 Jul 46
622	NJ Ship	29 Apr 44 26 May 44	2 Jun 44	13 Apr 46	Str 3 Jul 46
623	NJ Ship	1 May 44 29 May 44	5 Jun 44	5 Apr 46	Str 19 Jun 46
624	NJ Ship	2 May 44 30 May 44	5 Jun 44	12 Apr 46	Str 19 Jun 46
625	NJ Ship	4 May 44 31 May 44	6 Jun 44	9 Apr 46	Str 3 Jul 46
626	NJ Ship	5 May 44 1 Jun 44	7 Jun 44	11 Mar 46	Str 19 Jun 46
627	NJ Ship	8 May 44 2 Jun 44	9 Jun 44	10 Jun 46	LCFF, str 7 Aug 56
628	NJ Ship	8 May 44 5 Jun 44	10 Jun 44	26 Jul 46	LCFF, str 4 Aug 56
629	NJ Ship	10 May 44 6 Jun 44	12 Jun 44	25 Jun 46	Str 19 Jul 46
630	NJ Ship	12 May 44 7 Jun 44	15 Jun 44	25 Jun 46	LCIM, ex-LCIG; to ROC 1 Mar 47
631	NJ Ship	15 May 44 8 Jun 44	16 Jun 44	12 Jun 46	LCIM, ex-LCIG; to ROC on decomm
632	NJ Ship	15 May 44 9 Jun 44	17 Jun 44	28 Apr 47	LCIM, ex-LCIG; to ROC on decomm, later PRC
633	NJ Ship	16 May 44 13 Jun 44	19 Jun 44	30 Jan 50	LCIM, ex-LCIG; str 27 Feb 51
634	NJ Ship	18 May 44 14 Jun 44	20 Jun 44	13 Mar 46	Str 28 Mar 46
635	NJ Ship	19 May 44 15 Jun 44	22 Jun 44	8 Apr 46	Str 3 Jul 46
636	NJ Ship	20 May 44 16 Jun 44	23 Jun 44	2 Apr 46	Str 19 Jun 46
637	NJ Ship	27 May 44 19 Jun 44	24 Jun 44	1 Apr 46	Str 17 Apr 46
638	NJ Ship	24 May 44 20 Jun 44	26 Jun 44	30 Apr 54	LCIM, ex-LCIG; NRT postwar, str 27 Aug 57
639	NJ Ship	24 May 44 21 Jun 44	28 Jun 44	2 Jul 47	Str 1 Aug 47
640	NJ Ship	26 May 44 22 Jun 44	30 Jun 44	11 Jul 46	Str 15 Aug 46, to Israel 2 Mar 48
641	NJ Ship	22 May 44 26 Jun 44	1 Jul 44	12 Jul 46	Str 28 Aug 46
642	NJ Ship	27 May 44 27 Jun 44	3 Jul 44	6 Feb 46	LCIR, str 25 Feb 46
643	NJ Ship	30 May 44 28 Jun 44	5 Jul 44	N/A	LCIR, to MarComm 4 Feb 47
644	NJ Ship	31 May 44 29 Jun 44	7 Jul 44	19 Feb 46	LCIR, str 1 May 46
645	NJ Ship	2 Jun 44 30 Jun 44	8 Jul 44	6 Feb 46	LCIR, str 25 Feb 46
646	NJ Ship	2 Jun 44 3 Jul 44	10 Jul 44	7 Feb 46	LCIR, str 25 Feb 46, New York harbor tour boat *Normandy*, then *Knickerbocker*, then *Gotham*, then *Knickerbocker II*, then *Circle Line VI* in 63
647	NJ Ship	5 Jun 44 5 Jul 44	11 Jul 44	5 Apr 46	LCIR, str 19 Jun 46
648	NJ Ship	6 Jun 44 6 Jul 44	13 Jul 44	17 Jun 46	LCIR, str 31 Jul 46

AMPHIBIOUS SHIPS 603

No.	Name/Builder	LD/Launch	Comm	Decomm	Notes and Fate
649	NJ Ship	7 Jun 44 / 7 Jul 44	14 Jul 44	10 May 46	LCIR, str 19 Jul 46
650	NJ Ship	8 Jun 44 / 8 Jul 44	15 Jul 44	19 Feb 46	LCIR, str 12 Mar 46
651	NJ Ship	9 Jun 44 / 11 Jul 44	18 Jul 44	3 Apr 46	LCIR, str 19 Jun 46
652	NJ Ship	10 Jun 44 / 13 Jul 44	19 Jul 44	19 Jul 46	To *Accentor* (AMCU 15) but conversion canceled, str 18 Sep 56, tgt 13 Aug 59
653	NJ Ship	14 Jun 44 / 14 Jul 44	21 Jul 44 9 Dec 53		To *Avocet* (AMCU 16), str 1 Jan 60
654	NJ Ship	15 Jun 44 / 18 Jul 44	22 Jul 44		To *Bluejay* (AMCU 17), not comm when conversion complete, sold 12 May 60
655	NJ Ship	16 Jun 44 / 19 Jul 44	25 Jul 44	14 Aug 46	Str 19 Jul 56, tgt 19 Oct 56
656	NJ Ship	17 Jun 44 / 20 Jul 44	26 Jul 44	24 Mar 47	LCFF, str 17 Nov 55, tgt 6 Apr 56
657	NJ Ship	20 Jun 44 / 21 Jul 44	28 Jul 44	12 Jul 46	LCFF, str 31 Jul 46
658	Lawley	17 Feb 44 / 24 Feb 44	1 Mar 44	24 Jun 46	LCIM, ex-LCIG; str 31 Jul 46
659	Lawley	19 Feb 44 / 26 Feb 44	4 Mar 44	24 Jun 46	LCIM, ex-LCIG; str 25 Sep 48
660	Lawley	21 Feb 44 / 29 Feb 44	7 Mar 44	24 Jun 46	LCIM, ex-LCIG; str 15 Aug 46
661	Lawley	24 Feb 44 / 3 Mar 44	11 Mar 44	8 Nov 46	LCIG, str 10 Jun 47
662	Lawley	26 Feb 44 / 6 Mar 44	13 Mar 44	12 Mar 47	Str 4 Apr 47
663	Lawley	1 Mar 44 / 8 Mar 44	15 Mar 44	8 Nov 46	Str 10 Jun 47
664	Lawley	3 Mar 44 / 11 Mar 44	18 Mar 44 20 Sep 46	May 46 1 Feb 50	LCIM; NRT at Rockport, Illinois, during 46–50, str 22 May 51
665	Lawley	6 Mar 44 / 14 Mar 44	22 Mar 44		To SU 9 Jun 45, BU in SU
666	Lawley	8 Mar 44 / 16 Mar 44	24 Mar 44		To SU 28 Jul 45, BU in SU
667	Lawley	11 Mar 44 / 18 Mar 44	28 Mar 44		To SU 9 Jun 45, ret Jun 55, str 2 Dec 55
668	Lawley	15 Mar 44 / 22 Mar 44	29 Mar 44		To SU 9 Jun 45, BU in SU
669	Lawley	17 Mar 44 / 22 Mar 44	1 Apr 44	1 Mar 46	LCIM, tgt 6 Mar 46
670	Lawley	21 Mar 44 / 28 Mar 44	4 Apr 44	11 Jun 46	LCIM, to Thailand 15 Oct 46
671	Lawley	22 Mar 44 / 30 Mar 44	6 Apr 44		To SU 28 Jul 45, WL 17 Aug 45
672	Lawley	25 Mar 44 / 3 Apr 44	10 Apr 44		To SU 28 Jul 45, WL 17 Aug 45
673	Lawley	28 Mar 44 / 6 Apr 44	13 Apr 44	9 Jul 46	LCIM, str 15 Aug 46, to Israel 2 Mar 48
674	Lawley	31 Mar 44 / 8 Apr 44	15 Apr 44	6 Feb 50	LCIM, NRT, str 20 Dec 50
675	Lawley	3 Apr 44 / 12 Apr 44	19 Apr 44		To SU 9 Jun 45, BU in SU
676	Lawley	6 Apr 44 / 15 Apr 44	22 Apr 44	16 May 46	Str 19 Jun 46
677	Lawley	8 Apr 44 / 18 Apr 44	25 Apr 44	29 Dec 46	Hulk destroyed at Guam 14 Apr 47

No.	Name/Builder	LD/Launch	Comm	Decomm	Notes and Fate
678	Lawley	13 Apr 44 21 Apr 44	28 Apr 44	8 Feb 46	CTL Okinawa typhoon 9 Oct 45, scuttled 13 Feb 46
679	Lawley	17 Apr 44 24 Apr 44	29 Apr 44	25 Sep 46	LCFF, str 17 Nov 55, tgt 17 Mar 56
680	Lawley	18 Apr 44 26 Apr 44	3 May 44	31 May 46	Str 10 Jun 47
681	Lawley	21 Apr 44 29 Apr 44	6 May 44	19 Jun 46	Str 15 Aug 46
682	Lawley	24 Apr 44 2 May 44	9 May 44	20 May 46	Str 19 Jul 46
683	Lawley	26 Apr 44 5 May 44	12 May 44	16 May 46	Str 23 Dec 47
684	Lawley	29 Apr 44 8 May 44	15 May 44		CTL 17 Mar 45, str 30 Mar 45
685	Lawley	2 May 44 11 May 44	17 May 44	26 Apr 46	Str 5 Jun 46
686	Lawley	5 May 44 15 May 44	22 May 44	13 May 46	Str 19 Jun 46
687	Lawley	8 May 44 17 May 44	24 May 44	18 Apr 46	Str 5 Jun 46
688	Lawley	11 May 44 20 May 44	26 May 44	10 Jun 46	Str 23 Dec 47, to Argentina 48
689	Lawley	15 May 44 23 May 44	29 May 44	14 Jun 46	Str 15 Aug 46, to Argentina 48
690	Lawley	17 May 44 25 May 44	31 May 44	11 Apr 46	Str 8 May 46
691	Lawley	20 May 44 27 May 44	3 Jun 44	29 Jul 46	Str 19 Jul 56, tgt 19 Oct 56
692	Lawley	23 May 44 29 May 44	6 Jun 44	20 May 46	Str 26 Aug 46
693	Lawley	25 May 44 3 Jun 44	10 Jun 44	Mar 46	Str 19 Jul 56, tgt
694	Lawley	28 May 44 6 Jun 44	13 Jun 44	Jan 54	To *Chaffinch* (AMCU 18) 7 Mar 52, not comm as converted in 54, sold 28 Apr 60
695	Lawley	31 May 44 9 Jun 44	16 Jun 44	1 Jun 46	Str 19 Jul 56
696	Lawley	3 Jun 44 12 Jun 44	17 Jun 44	23 May 46	Str 5 Jul 56
697	Lawley	6 Jun 44 15 Jun 44	21 Jun 44	N/A	Str 10 Jun 47
698	Lawley	9 Jun 44 17 Jun 44	23 Jun 44	24 Aug 46	Reactivated for transfer to French Indo-China 12 Nov 53, to Vietnam Oct 55
699	Lawley	13 Jun 44 21 Jun 44	25 Jun 44	16 Jul 46	Reactivated for transfer to French Indo-China 12 Nov 53, to Vietnam Oct 55
700	Lawley	22 Jun 44 22 Jun 44	26 Jun 44	27 May 46	Str 20 Jul 56, tgt 6 Aug 57
701	Lawley	17 Jun 44 24 Jun 44	30 Jun 44	N/A	To *Chewink* (AMCU 19) 7 Mar 52, not comm as converted, sold 21 Apr 60
702	Lawley	20 Jun 44 28 Jun 44	6 Jul 44	5 Aug 46	To France for Indo-China 3 Dec 53, to Vietnam navy 56
703	Lawley	22 Jun 44 30 Jun 44	8 Jul 44	N/A	To *Chimango* (AMCU 20) 7 Mar 52, not comm as converted, sold 29 Apr 60
704	Lawley	26 Jun 44 3 Jul 44	12 Jul 44	16 Apr 46	LCIR, str 5 Jun 46
705	Lawley	28 Jun 44 8 Jul 44	15 Jul 44	16 Apr 46	LCIR, str 8 May 46
706	Lawley	1 Jul 44 10 Jul 44	18 Jul 44	16 Apr 46	LCIR, str 6 Jun 46

No.	Name/Builder	LD/Launch	Comm	Decomm	Notes and Fate
707		3 Jul 44	20 Jul 44	15 Feb 46	LCIR, str 20 Mar 46
	Lawley	12 Jul 44			
708		8 Jul 44	24 Jul 44	19 Apr 46	LCIR, str 19 Jun 46
	Lawley	17 Jul 44			
709		11 Jul 44	27 Jul 44	N/A	To *Cockatoo* (AMCU 21) 7 Mar 52, sold 28 Apr 60
	Lawley	20 Jul 44	6 Nov 53		
710		13 Jul 44	29 Jul 44	20 Jun 46	To France for Indo-China 27 Nov 53, to Iran Dec 58
	Lawley	21 Jul 44			
711		20 Jul 44	7 Aug 44	14 Mar 46	Str 19 Jun 46
	Lawley	26 Jul 44			
712		21 Jul 44	15 Aug 44	13 Apr 46	Str 15 Aug 46
	Lawley	31 Jul 44			
713		27 Jul 44	18 Sep 44	7 Oct 46	Str 22 Jan 48
	Lawley	5 Aug 44			
714		7 Aug 44	29 Sep 44	4 Sep 46	Str 22 Jan 47
	Lawley	16 Aug 44			
715		16 Sep 44	30 Oct 44	31 Aug 46	To France for Indo-China 8 Jan 51, ret 56, BU
	Lawley	24 Sep 44			
716		25 Sep 44	31 Oct 44	18 Oct 46	Str 23 Apr 47
	Lawley	3 Oct 44			
717–724					Canceled 19 Aug 44
725		5 Nov 43	24 Dec 43	1 May 46	LCIG, str 10 Jun 47
	Commercial	10 Dec 43			
726		5 Nov 43	31 Dec 43	10 Jun 46	LCIG, str 31 Jul 46
	Commercial	17 Dec 43			
727		12 Nov 43	7 Jan 44	9 Apr 46	LCIG, str 3 Jul 46
	Commercial	24 Dec 43			
728		19 Nov 43	14 Jan 44	9 Apr 46	LCIG, str 3 Jul 46
	Commercial	31 Dec 43			
729		6 Dec 43	21 Jan 44	N/A	LCIG, str 25 Feb 47
	Commercial	7 Jan 44			
730		10 Dec 43	26 Jan 44	12 Mar 46	LCIG, str 17 Apr 46
	Commercial	16 Jan 44			
731		17 Dec 43	30 Jan 44	26 Apr 46	Str 5 Jun 46
	Commercial	16 Jan 44			
732		20 Dec 43	4 Feb 44	10 Jun 46	Str 15 Aug 46
	Commercial	24 Jan 44			
733		24 Dec 43	9 Feb 44	22 Apr 46	Str 5 Jun 46
	Commercial	30 Jan 44			
734		31 Dec 43	14 Feb 44	15 May 46	Str 19 Jun 46, to Argentina 18 Feb 48
	Commercial	30 Jan 44			
735		7 Jan 44	18 Feb 44	23 Apr 46	Str 19 Jun 46
	Commercial	8 Feb 44			
736		16 Jan 44	23 Feb 44	12 Jun 46	Str 23 Dec 47
	Commercial	13 Feb 44			
737		16 Jan 44	28 Feb 44	13 Jun 46	Str 10 Jun 47, to Argentina 18 Feb 48
	Commercial	13 Feb 44			
738		24 Jan 44	3 Mar 44	27 Jun 46	Str 15 Aug 46
	Commercial	19 Feb 44			
739		30 Jan 44	6 Mar 44	2 May 46	LCIM, ex-LCIG; to Thailand 10 May 47
	Commercial	27 Feb 44			
740		30 Jan 44	11 Mar 44	1 Apr 46	LCIM, ex-LCIG; str 1 May 46
	Commercial	27 Feb 44			
741		8 Feb 44	13 Mar 44	1 May 46	LCIM, ex-LCIG; str 19 Jun 46
	Commercial	2 Mar 44			
742		14 Feb 44	18 Mar 44	17 Jun 46	LCIM, ex-LCIG; str 31 Jul 46
	Commercial	5 Mar 44			

Appendix C

No.	Name/Builder	LD/Launch	Comm	Decomm	Notes and Fate
743	Commercial	14 Feb 44 5 Mar 44	22 Mar 44	21 May 46	Str 19 Jun 46
744	Commercial	19 Feb 44 13 Mar 44	25 Mar 44	4 Sep 47	Str 29 Sep 47
745	Commercial	28 Feb 44 19 Mar 44	31 Mar 44	29 Oct 46	Str 10 Jun 47
746	Commercial	28 Feb 44 19 Mar 44	3 Apr 44	4 Sep 47	Str 29 Jul 47, to Argentina 18 Feb 48
747	Commercial	1 Mar 44 25 Mar 44	7 Apr 44	4 Sep 47	Str 29 Jul 47
748	Commercial	6 Mar 44 2 Apr 44	10 Apr 44	4 Sep 47	Str 29 Jul 47, to Argentina 18 Feb 48
749	Commercial	6 Mar 44 2 Apr 44	14 Apr 44	21 Aug 47	Str 29 Sep 47
750	Commercial	13 Mar 44 5 Apr 44	17 Apr 44	2 May 47	Str 22 May 47, to Argentina 48
751	Commercial	20 Mar 44 9 Apr 44	22 Apr 44	17 Jun 46	LCIG, str 15 Aug 46
752	Commercial	20 Mar 44 9 Apr 44	26 Apr 44	16 Apr 46	LCIG, str 5 Jun 46
753	Commercial	25 Mar 44 17 Apr 44	29 Apr 44	21 Aug 47	Str 29 Sep 47, to Argentina 18 Feb 48
754	Commercial	3 Apr 44 23 Apr 44	4 May 44	22 Jun 46	LCIM, ex-LCIG; str 15 Aug 46
755	Commercial	3 Apr 44 23 Apr 44	8 May 44	2 Jul 46	LCIM, ex-LCIG; str 10 Jun 47
756	Commercial	5 Apr 44 28 Apr 44	11 May 44	1 Jul 46	LCIM, ex-LCIG; str 15 Aug 46
757	Commercial	10 Apr 44 7 May 44	15 May 44	N/A	LCIM, ex-LCIG; str 31 Jul 46
758	Commercial	10 Apr 44 7 May 44	20 May 44	7 May 46	Str 19 Feb 48, New York harbor tour boat *Normandy II*, then *Normandy*, then *Circle Line X* in 63
759	Commercial	17 Apr 44 10 May 44	22 May 44	13 May 46	Str 19 Jun 46
760	Commercial	24 Apr 44 14 May 44	15 May 44	17 Jun 46	LCIM, ex-LCIG; str 31 Jul 46
761	Commercial	24 Apr 44 14 May 44	31 May 44	25 Jun 46	Str 19 Jun 46
762	Commercial	28 Apr 44 22 May 44	3 Jun 44	15 May 46	LCIR, str 19 Jun 46
763	Commercial	8 May 44 28 May 44	8 Jun 44	3 Jan 46	LCIR, major hull damage in Okinawa typhoon, str 21 Jan 46
764	Commercial	8 May 44 28 May 44	12 Jun 44	21 May 46	LCIR, str 19 Jul 46
765	Commercial	10 May 44 3 Jun 44	15 Jun 44	22 Apr 46	LCIR, str 19 Jul 46
766	Commercial	15 May 44 11 Jun 44	19 Jun 44	19 Jun 46	LCIR, str 31 Jul 46, merch, by 57 was New York harbor tour boat *Knickerbocker VII*, then *Day Line VII*, then *Circle Line IX* in 62
767	Commercial	15 May 44 11 Jun 44	24 Jun 44	30 Apr 46	LCIR, str 31 Jul 46
768	Commercial	22 May 44 15 Jun 44	28 Jun 44	31 May 46	To France 10 Dec 53 for Indo-China, ret 2 Aug 56, to Iran (MDAP) 25 Oct 56 (delivered 57)
769	Commercial	22 May 44 18 Jun 44	3 Jul 44	26 Apr 46	LCIR, str 5 Jun 46

No.	Name/Builder	LD/Launch	Comm	Decomm	Notes and Fate
770	Commercial	29 May 44 18 Jun 44	8 Jul 44	13 May 46	LCIR, str 19 Jun 46
771	Commercial	3 Jun 44 27 Jun 44	12 Jul 44	22 Dec 45	LCIR, CTL Okinawa 9 Oct 45 due to typhoon, str 21 Jan 46
772	Commercial	12 Jun 44 2 Jul 44	17 Jul 44	8 Mar 46	LCIR, str 12 Apr 46
773	Commercial	12 Jun 44 2 Jul 44	21 Jul 44	9 Nov 46	Used by Fish and Wildlife Survey Party, Philippines; to U.S. Army (Korea) 26 Feb 47
774	Commercial	15 Jun 44 12 Jul 44	25 Jul 44	15 Nov 46	Str 20 Jul 56, tgt 6 Aug 57
775	Commercial	19 Jun 44 9 Jul 44	28 Jul 44	31 Mar 47	LCFF, str 19 Nov 55, tgt 16 Mar 56
776	Commercial	19 Jun 44 9 Jul 44	3 Aug 44	N/A	To *Cotinga* (AMCU 22) 7 Mar 52, not comm as converted, sold 6 May 60
777	Commercial	27 Jun 44 17 Jul 44	7 Aug 44	29 May 46	Scheduled Mar 52 for conversion to *Dunlin* (AMCU 23) but conversion canceled, str 18 Sep 56, tgt 10 Sep 58
778	Commercial	3 Jul 44 23 Jul 44	11 Aug 44	22 May 46	Str 17 Jul 56, to be tgt (no date given)
779	Commercial	3 Jul 44 23 Jul 44	15 Aug 44	13 Mar 46	Str 5 Jun 46
780	Commercial	12 Jul 44 7 Aug 44	19 Aug 44	14 Mar 46	Str 5 Jun 46
781					Canceled 23 Jun 44
782	NJ Ship	21 Jun 44 25 Jul 44	1 Aug 44	4 Oct 46	LCFF, str 17 Nov 55, tgt 56
783	NJ Ship	22 Jun 44 26 Jul 44	31 Jul 44	21 Jun 49	LCFF, str 24 Apr 51
784	NJ Ship	23 Jun 44 27 Jul 44	3 Aug 44	14 Jun 46	Str 19 Jul 46
785	NJ Ship	27 Jun 44 29 Jul 44	4 Aug 44	4 Apr 46	LCIR, str 21 May 46
786	NJ Ship	28 Jun 44 31 Jul 44	5 Aug 44	11 Aug 50	LCFF, str 27 Feb 51
787	NJ Ship	29 Jun 44 2 Aug 44	8 Aug 44	N/A	To MarComm 19 Mar 48
788	NJ Ship	30 Jun 44 3 Aug 44	9 Aug 44	14 Jun 49	LCFF, str 7 Aug 56, tgt 7 Nov 57
789	NJ Ship	1 Jul 44 4 Aug 44	11 Aug 44	20 Nov 46	LCFF, str 17 Feb 47
790	NJ Ship	4 Jul 44 5 Aug 44	12 Aug 44	14 Jun 49	LCFF, str 28 Jul 56
791	NJ Ship	6 Jul 44 8 Aug 44	14 Aug 44	29 Jun 46	LCFF, str 7 Aug 56, tgt 19 Nov 57
792	NJ Ship	7 Jul 44 9 Aug 44	16 Aug 44	29 May 46	LCFF, str 8 Aug 56, tgt 21 Jan 58
793	NJ Ship	8 Jul 44 10 Aug 44	17 Aug 44	14 May 46	LCFF, str 18 Sep 56, tgt Jun 58
794	NJ Ship	10 Jul 44 11 Aug 44	19 Aug 44	13 Mar 46	Str 28 Mar 46
795	NJ Ship	12 Jul 44 15 Aug 44	21 Aug 44	8 Feb 46	Str 13 Nov 46
796	NJ Ship	14 Jul 44 16 Aug 44	22 Aug 44		CTL in Okinawa typhoon Oct 45, str 19 Jun 46
797	NJ Ship	15 Jul 44 17 Aug 44	23 Aug 44	8 Mar 46	Str 12 Apr 46

No.	Name/Builder	LD/Launch	Comm	Decomm	Notes and Fate
798		19 Jul 44	25 Aug 44	7 Feb 46	Str 4 Apr 47
	NJ Ship	19 Aug 44			
799		20 Jul 44	26 Aug 44	26 Sep 46	NRT 46–50 and 50–54, str 12 Jul 56
	NJ Ship	21 Aug 44	46	1 Jul 50	
			22 Nov 50	7 May 54	
800		21 Jul 44	28 Aug 44	18 Oct 46	Str 22 Jan 48
	NJ Ship	22 Aug 44			
801		22 Jul 44	30 Aug 44	9 May 46	LCIM, ex-LCIG; str 23 Jun 47
	NJ Ship	23 Aug 44			
802		26 Jul 44	31 Aug 44	4 Apr 46	LCIM, ex-LCIG; str 8 May 46
	NJ Ship	25 Aug 44			
803		27 Jul 44	2 Sep 44	N/A	LCIM, ex-LCIG; str 10 Jun 47
	NJ Ship	26 Aug 44			
804		28 Jul 44	4 Sep 44	14 May 46	LCIM, ex-LCIG; str 19 Jun 46
	NJ Ship	29 Aug 44			
805		31 Jul 44	6 Sep 44	14 May 46	LCIM, ex-LCIG; str 19 Jun 46
	NJ Ship	30 Aug 44			
806		1 Aug 44	7 Sep 44	11 Apr 46	LCIM, ex-LCIG; str 5 Jun 46
	NJ Ship	31 Aug 44			
807		3 Aug 44	8 Sep 44	N/A	LCIM, ex-LCIG; to MarComm 24 Mar 47
	NJ Ship	2 Sep 44			
808		4 Aug 44	11 Sep 44	18 Apr 46	LCIM, ex-LCIG; str 5 Jun 46
	NJ Ship	5 Sep 44			
809		7 Aug 44	12 Sep 44	1 Apr 46	LCIM, ex-LCIG; str 3 Jul 46
	NJ Ship	6 Sep 44			
810		7 Aug 44	15 Sep 44	N/A	LCIM, ex-LCIG; to MarComm 31 Mar 47
	NJ Ship	7 Sep 44			
811		9 Aug 44	15 Sep 44	29 May 46	Str 19 Jun 46
	NJ Ship	8 Sep 44			
812		10 Aug 44	16 Sep 44	12 Jul 46	Str 28 Aug 48
	NJ Ship	9 Sep 44			
813		11 Aug 44	19 Sep 44	14 May 46	Str 3 Jul 46
	NJ Ship	13 Sep 44			
814		12 Aug 44	21 Sep 44	18 Oct 46	Str 22 Jan 48
	NJ Ship	14 Sep 44			
815		16 Aug 44	21 Sep 44	17 Oct 46	Str 19 Feb 48
	NJ Ship	15 Sep 44			
816		17 Aug 44	23 Sep 44	11 Oct 46	Str 19 Feb 48
	NJ Ship	16 Sep 44			
817		18 Aug 44	25 Sep 44	18 Oct 46	Str 22 Jan 48
	NJ Ship	19 Sep 44			
818		21 Aug 44	26 Sep 44	N/A	To France for Indo-China 8 Jan 51, to ROC 54
	NJ Ship	21 Sep 44			
819		22 Aug 44	27 Sep 44	15 Oct 46	Str 22 Jan 48
	NJ Ship	21 Sep 44			
820		23 Aug 44	29 Sep 44	17 Oct 46	Str 22 Jan 48
	NJ Ship	22 Sep 44			
821		24 Aug 44	30 Sep 44	15 Oct 46	Str 22 Jan 48
	NJ Ship	23 Sep 44			
822–837					Canceled Aug 44
838–844					Canceled 5 Jun 44
845–859					Canceled 23 Jun 44
860–865					Canceled 5 Jun 44
866		26 Aug 44	3 Oct 44	18 Mar 46	Str 23 Dec 47
	NJ Ship	26 Sep 44			
867		28 Aug 44	4 Oct 44	25 Jan 47	NRT 50–54, str 7 Aug 56
	NJ Ship	27 Sep 44	13 May 48	26 Mar 54	

No.	Name/Builder	LD/Launch	Comm	Decomm	Notes and Fate
868		30 Aug 44	5 Oct 44	27 Mar 47	Str 27 Feb 51
	NJ Ship	28 Sep 44	13 May 48	10 Feb 50	
869		31 Aug 44	7 Oct 44	N/A	To *Goldcrest* (AMCU 24) 7 Mar 52, str 1
	NJ Ship	29 Sep 44	53	Mar 55	Jan 60
870		1 Sep 44	9 Oct 44	N/A	To *Jackamar* (AMCU 25) 7 Mar 52, sold
	NJ Ship	2 Oct 44	12 Jan 54		21 Jul 60
871		4 Sep 44	10 Oct 44	20 Aug 46	To France for Indo-China 4 Sep 52, to
	NJ Ship	3 Oct 44			Vietnam navy 56
872		6 Sep 44	11 Oct 44	N/A	To France for Indo-China Dec 53, to
	NJ Ship	4 Oct 44			Vietnam navy 56
873		6 Sep 44	12 Oct 44	27 Jul 46	Str 28 Jul 56, tgt 23 Nov 57
	NJ Ship	5 Oct 44			
874		7 Sep 44	13 Oct 44	N/A	To *Kestrel* (AMCU 26) 7 Mar 52, sold 28
	NJ Ship	6 Oct 44	8 Feb 54	2 Dec 57	Jun 60
875		8 Sep 44	16 Oct 44	19 Jul 46	To France for Indo-China 24 Feb 53, to
	NJ Ship	9 Oct 44			Cambodia 10 May 56
876		9 Sep 44	17 Oct 44	N/A	To MarComm 7 Mar 47
	NJ Ship	10 Oct 44			
877		14 Sep 44	18 Oct 44	9 May 46	Str 27 Jan 47, to Chile 12 Feb 47
	NJ Ship	11 Oct 44			
878		16 Sep 44	19 Oct 44	7 Feb 46	Str 22 Jan 47, to Chile 12 Feb 47
	NJ Ship	12 Oct 44			
879		16 Sep 44	20 Oct 44	2 May 46	Str 19 Feb 48
	NJ Ship	13 Oct 44			
880		18 Sep 44	23 Oct 44	22 Oct 46	Str 13 Dec 47
	NJ Ship	16 Oct 44			
881		19 Sep 44	24 Oct 44	25 Apr 46	Str 15 Aug 46
	NJ Ship	17 Oct 44			
882		22 Sep 44	26 Oct 44	15 May 47	To *Shearwater* (AMCU 40) in place of
	NJ Ship	18 Oct 44			LCIL 1053 on 7 Mar 52 but conversion canceled, str 12 Jul 56
883		22 Sep 44	26 Oct 44	24 Mar 47	To *Kildeer* (AMCU 27) 7 Mar 52 but conversion canceled, str 17 Nov 55, to be
	NJ Ship	19 Oct 44			used as tgt
884		22 Sep 44	27 Oct 44	N/A	To *Longspur* (AMCU 28) 7 Mar 52, not
	NJ Ship	20 Oct 44			comm on conversion, sold 18 May 60
885–901					Canceled Aug 44
902–910					Canceled 5 Jun 44
911–928					Canceled Aug 44
929–942					Canceled 23 Jun 44
943		15 Jan 44	18 Mar 44		To SU 9 Jun 45, WL 17 Aug 45
	Consolidated-O	10 Feb 44			
944		15 Jan 44	15 Apr 44	N/A	To *Magpie* (AMCU 29) 7 Mar 52, not
	Consolidated-O	10 Feb 44			comm on conversion, sold 18 May 60
945		17 Jan 44	22 Mar 44		To SU 28 Jul 45, ret Jun 55, str 2 Dec 55
	Consolidated-O	10 Feb 44			
946		20 Jan 44	30 Apr 44		To SU 28 Jul 45, ret Jun 55, str 2 Dec 55
	Consolidated-O	16 Feb 44			
947		18 Jan 44	30 Apr 44	7 Aug 46	Str 22 May 51
	Consolidated-O	16 Feb 44			
948		21 Jan 44	24 Mar 44	30 May 46	LCIG, purchased by NEI 6 Jun 46, to
	Consolidated-O	16 Feb 44			Indonesia on independence
949		31 Jan 44	27 Mar 44		To SU 9 Jun 45, ret Jun 55, str 2 Dec 55
	Consolidated-O	3 Mar 44			
950		31 Jan 44	27 Mar 44		To SU 9 Jun 45, BU in SU
	Consolidated-O	3 Mar 44			
951		31 Jan 44	29 Mar 44	N/A	LCIM, str 22 Jan 47
	Consolidated-O	3 Mar 44			

Appendix C

No.	Name/Builder	LD/Launch	Comm	Decomm	Notes and Fate
952	Consolidated-O	10 Feb 44 7 Mar 44	1 Apr 44	20 Aug 46	LCIM, str 20 Dec 50
953	Consolidated-O	10 Feb 44 7 Mar 44	1 Apr 44	6 Oct 45	Str 28 Aug 47
954	Consolidated-O	10 Feb 44 7 Mar 44	1 Apr 44	1 Feb 49	Uncomm service craft Newport 1 Feb 49, str 29 Jan 57
955	Consolidated-O	14 Feb 44 10 Mar 44	4 Apr 44	13 Feb 47	Str 14 Mar 47
956	Consolidated-O	14 Feb 44 10 Mar 44	5 Apr 44	28 Oct 46	Str 13 Dec 46
957	Consolidated-O	14 Feb 44 10 Mar 44	5 Apr 44	13 Feb 47	Str 14 Mar 47
958	Consolidated-O	18 Feb 44 14 Mar 44	10 Apr 44	4 Sep 47	Str 29 Sep 47
959	Consolidated-O	18 Feb 44 14 Mar 44	10 Apr 44	28 Oct 46	Str 13 Dec 46
960	Consolidated-O	21 Feb 44 14 Mar 44	10 Apr 44	4 Sep 47	Str 29 Sep 47
961	Consolidated-O	23 Feb 44 20 Mar 44	12 Apr 44	16 Jul 47	Str 28 Aug 47
962	Consolidated-O	24 Feb 44 20 Mar 44	12 Apr 44	8 Aug 46	Str 20 Dec 50
963	Consolidated-O	25 Feb 44 20 Mar 44	16 May 44	N/A	To *Mallard* (AMCU 30) 7 Mar 52, not comm on conversion, str 1 Jan 60
964	Consolidated-O	3 Mar 44 23 Mar 44	18 Apr 44	5 Jun 46	Str 22 Jan 47
965	Consolidated-O	4 Mar 44 23 Mar 44	19 Apr 44	24 Sep 46	Str 29 Oct 46
966	Consolidated-O	6 Mar 44 23 Mar 44	19 Apr 44	8 Jul 46	To *Medrick* (AMCU 31) 7 Mar 52 but conversion canceled, str 27 Oct 55, tgt Oct 56
967	Consolidated-O	29 Feb 44 27 Mar 44	21 Apr 44	7 May 46	Str 21 May 46
968	Consolidated-O	29 Feb 44 27 Mar 44	21 Apr 44	7 May 46	Str 5 Jun 46
969	Consolidated-O	29 Feb 44 27 Mar 44	21 Apr 44	3 Oct 46	To *Minivet* (AMCU 32) 7 Mar 52 but conversion canceled, str 10 Sep 56
970	Consolidated-O	8 Mar 44 31 Mar 44	27 Apr 44	26 Aug 46	Str 22 May 51
971	Consolidated-O	8 Mar 44 31 Mar 44	27 Apr 44	25 Jul 46	Str 28 Aug 46
972	Consolidated-O	9 Mar 44 31 Mar 44	27 Apr 44	26 Sep 46	Str 29 Oct 46
973	Consolidated-O	11 Mar 44 5 Apr 44	30 May 44 20 Feb 54	N/A 7 Jul 55	To *Oriole* (AMCU 33) 7 Mar 52, str 1 Jan 60
974	Consolidated-O	12 Mar 44 5 Apr 44	30 Apr 44		WL 10 Jan 45
975	Consolidated-O	12 Mar 44 5 Apr 44	30 Apr 44	21 Oct 46	LCIM, ex-LCIG; str 13 Dec 46
976	Consolidated-O	15 Mar 44 10 Apr 44	5 May 44 21 Nov 53	N/A 23 Jun 55	To *Ortolan* (AMCU 34) Feb 53, str 1 Jan 60
977	Consolidated-O	15 Mar 44 10 Apr 44	5 May 44	24 Jan 49	At Bikini, str 10 Feb 49
978	Consolidated-O	17 Mar 44 10 Apr 44	5 May 44	15 Jan 51	NRT; str 18 Apr 52; to U.S. Naval Salvage School Bayonne to replace ex–PC 469 on 31 Jan 52, stripped of all machinery, spares, equipage, provisions; to school 14 May 52

Amphibious Ships 611

No.	Name/Builder	LD/Launch	Comm	Decomm	Notes and Fate
979	Consolidated-O	21 Mar 44 14 Apr 44	10 May 44	17 May 46	Str 19 Jun 46
980	Consolidated-O	21 Mar 44 14 Apr 44	15 May 44	6 May 46	Str 5 Jun 46
981	Consolidated-O	21 Mar 44 14 Apr 44	15 May 44	N/A	Sold 13 Nov 46
982	Consolidated-O	23 Mar 44 18 Apr 44	16 May 44 19 Dec 53	24 Jun 46 1 Nov 54	To *Owl* (AMCU 35) 7 Mar 52, str 17 Oct 57
983	Consolidated-O	24 Mar 44 18 Apr 44	15 May 44	22 Nov 46	Str 10 Jun 47
984	Consolidated-O	26 Mar 44 18 Apr 44	20 May 44	13 Jun 46	Str 15 Aug 46
985	Consolidated-O	28 Mar 44 22 Apr 44	18 May 44	2 May 47	Str 12 Jul 56, had been nominated for AMCU conversion
986	Consolidated-O	28 Mar 44 22 Apr 44	18 May 44	N/A	Str 15 Aug 46
987	Consolidated-O	30 Mar 44 22 Apr 44	18 May 44	15 Jun 46	Str 3 Jul 46
988	Consolidated-O	31 Mar 44 25 Apr 44	22 May 44	N/A	LCFF, CTL, str 19 Jun 46
989	Consolidated-O	31 Mar 44 25 Apr 44	22 May 44	30 Nov 49	Str 17 Jul 56, tgt
990	Consolidated-O	3 Apr 44 25 Apr 44	22 May 44	19 Apr 46	Str 5 Jun 46
991	Consolidated-O	5 Apr 44 29 Apr 44	25 May 44	19 Apr 46	Str 5 Jun 46
992	Consolidated-O	6 Apr 44 29 Apr 44	27 May 44	18 Jun 46	Str 23 Dec 47
993	Consolidated-O	6 Apr 44 29 Apr 44	27 May 44	4 Apr 46	Str 1 May 46
994	Consolidated-O	12 Apr 44 4 May 44	27 May 44	1 Oct 46	LCFF, str 8 Aug 56, tgt 6 Dec 57
995	Consolidated-O	11 Apr 44 4 May 44	27 May 44	27 Aug 47	LCFF, str 8 Aug 56
996	Consolidated-O	12 Apr 44 4 May 44	30 May 44	7 Jun 46	Str 5 Jul 56, tgt 14 Aug 56
997	Consolidated-O	14 Apr 44 11 May 44	31 May 44	17 May 46	Str 19 Jul 46
998	Consolidated-O	14 Apr 44 11 May 44	31 May 44	9 Aug 46	LCFF, str 28 Jul 56, tgt 28 Mar 58
999	Consolidated-O	15 Apr 44 11 May 44	31 May 44	17 May 46	Str 23 Dec 47
1000	Consolidated-O	19 Apr 44 13 May 44	3 Jun 44	24 Jun 46	Str 12 Jul 56, had been nominated for AMCU conversion
1001	Consolidated-O	18 Apr 44 13 May 44	3 Jun 44	30 Oct 46	To *Partridge* (AMCU 36) 7 Mar 52 but conversion canceled, str 7 Aug 56
1002	Consolidated-O	19 Apr 44 13 May 44	10 Jun 44	8 Feb 46	Str 12 Mar 46
1003	Consolidated-O	24 Apr 44 20 May 44	8 Jun 44	3 Jun 46	Str 23 Dec 47
1004	Consolidated-O	24 Apr 44 20 May 44	8 Jun 44	13 Jun 46	Str 15 Aug 46
1005	Consolidated-O	26 Apr 44 20 May 44	8 Jun 44	28 May 46	Str 23 Dec 47
1006	Consolidated-O	25 Apr 44 6 May 44	19 Jun 44	6 Mar 47	Str 10 Jun 47, merch, to Dominican Republic navy 52
1007	Consolidated-O	27 Apr 44 27 May 44	19 Jun 44	4 Jun 46	Str 19 Jun 46

APPENDIX C

No.	Name/Builder	LD/Launch	Comm	Decomm	Notes and Fate
1008	Consolidated-O	27 Apr 44 27 May 44	19 Jun 44 7 Oct 53	14 May 46 N/A	To *Sandpiper* (AMCU 38), str 1 Jan 60
1009	Consolidated-O	6 May 44 27 May 44	19 Jun 44	19 Jun 46	Str 15 Aug 46
1010	Consolidated-O	5 May 44 1 Jun 44	24 Jun 44	N/A	LCIM, ex-LCIG; str 19 Feb 48
1011	Consolidated-O	8 May 44 1 Jun 44	24 Jun 44	23 Oct 46	LCIM, ex-LCIG; str 13 Dec 46
1012	Consolidated-O	9 May 44 1 Jun 44	28 Jun 44	11 Jul 46	LCIM, ex-LCIG; str 15 Aug 46
1013	Albina	16 Dec 43 21 Feb 44	21 Mar 44	N/A	Str 19 Feb 48
1014	Albina	5 Jan 44 25 Feb 44	25 Mar 44	21 Aug 47	Str 29 Sep 47
1015	Albina	11 Jan 44 2 Mar 44	28 Mar 44	4 Sep 47	Str 29 Sep 47
1016	Albina	24 Jan 44 7 Mar 44	5 Apr 44	13 Feb 47	Str 14 Mar 47
1017	Albina	31 Jan 44 14 Mar 44	12 Apr 44		Retained as towing vessel Sep 47, reclassified as noncommissioned service craft Dec 52, out of service (Subic Bay) 30 Nov 56, str 27 Aug 57
1018	Albina	7 Feb 44 21 Mar 44	18 Apr 44	14 Aug 47	Str 16 Sep 47, to Argentina 18 Feb 48
1019	Albina	14 Feb 44 28 Mar 44	25 Apr 44	24 Oct 46	Str 13 Dec 46
1020	Albina	22 Feb 44 5 Apr 44	4 May 44	26 Apr 46	Str 5 Jun 46
1021	Albina	26 Feb 44 12 Apr 44	11 May 44	26 Aug 46	Str 22 May 51
1022	Albina	3 Mar 44 17 Apr 44	18 May 44 13 Sep 52	N/A 13 Oct 57	To *Rail* (AMCU 37) 7 Mar 52, str 1 Jan 60
1023	Albina	8 Mar 44 24 Apr 44	25 May 44	2 Jul 46	LCIM, ex-LCIG; str 15 Aug 46
1024	Albina	15 Mar 44 29 Apr 44	5 Jun 44	18 Apr 46	LCIR, str 5 Jun 46
1025	Albina	22 Mar 44 6 May 44	12 Jun 44	18 Sep 46	To Chile 20 Mar 47
1026	Albina	6 Apr 44 12 May 44	19 Jun 44	2 Apr 46	LCIR, str 1 May 46
1027	Albina	13 Apr 44 18 May 44	26 Jun 44	27 Nov 46	To Chile 20 Mar 47
1028	Albina	18 Apr 44 24 May 44	4 Jul 44	8 Apr 46	LCIR, str 15 Aug 46
1029	Albina	24 Apr 44 2 Jun 44	11 Jul 44	8 Apr 46	LCIR, str 19 Jun 46
1030	Albina	1 May 44 8 Jun 44	18 Jul 44	12 Apr 46	LCIR, str 15 Aug 46
1031	Albina	8 May 44 15 Jun 44	25 Jul 44	14 Mar 47	LCFF, str 23 Apr 47
1032	Albina	12 May 44 21 Jun 44	4 Aug 44	11 Oct 46	To U.S. Air Force 10 Feb 51
1033	Albina	19 May 44 27 Jun 44	11 Aug 44	20 May 46	Str 19 Jun 46
1034–1051					Canceled 16 Jun 44
1052	Defoe	8 Feb 44 9 Mar 44	21 Mar 44	N/A	To *Sentinel* (AMCU 39) 7 Mar 52, not comm on conversion, str 1 Jan 60

No.	Name/Builder	LD/Launch	Comm	Decomm	Notes and Fate
1053	Defoe	17 Feb 44 15 Mar 44	27 Mar 44	15 Jan 51	NRT 46–51, to *Shearwater* (AMCU 40) 7 Mar 52, out of service 3 Jul 53, str 10 Sep 56 but retained at Navy Salvage School, Bayonne
1054	Defoe	24 Feb 44 21 Mar 44	31 Mar 44		At Bikini, str 28 Apr 49 and sold
1055	Defoe	2 Mar 44 25 Mar 44	4 Apr 44	21 Aug 46	LCIM, ex-LCIG; str 13 Dec 46
1056	Defoe	10 Mar 44 28 Mar 44	10 Apr 44		LCIM, ex-LCIG; to U.S. Army (Korea) Feb 47, to Korean navy
1057	Defoe	17 Mar 44 2 Apr 44	15 Apr 44	15 Jul 46	LCIM, ex-LCIG; str 28 Aug 46
1058	Defoe	23 Mar 44 7 Apr 44	18 Apr 44		LCIM, ex-LCIG; to Philippine naval patrol 3 Oct 47, to Philippine navy 2 Jul 48
1059	Defoe	29 Mar 44 12 Apr 44	21 Apr 44	18 Jun 46	LCIM, ex-LCIG; str 16 Sep 47, to Philippines 2 Jul 48
1060	Defoe	3 Apr 44 17 Apr 44	26 Apr 44	4 Sep 47	Str 29 Sep 47, to Argentina 48
1061	Defoe	8 Apr 44 22 Apr 44	4 May 44	21 Aug 47	Str 29 Sep 47
1062	Defoe	13 Apr 44 28 Apr 44	5 May 44	N/A	To MarComm 9 Mar 48
1063	Defoe	18 Apr 44 2 May 44	11 May 44	13 Feb 47	Str 14 Mar 47
1064	Defoe	23 Apr 44 7 May 44	15 May 44	6 Jun 46	Str 15 Aug 46
1065	Defoe	28 Apr 44 11 May 44	16 May 44		WL 24 Oct 44
1066	Defoe	3 May 44 16 May 44	21 May 44	N/A	Str 10 Jun 47
1067	Defoe	8 May 44 20 May 44	25 May 44	N/A	To Argentina 18 Feb 48
1068	Defoe	12 May 44 24 May 44	31 May 44	19 Mar 46	LCIR, str 12 Apr 46
1069	Defoe	17 May 44 30 May 44	5 Jun 44	1 Apr 46	LCIR, str 31 Jul 46
1070	Defoe	21 May 44 5 Jun 44	10 Jun 44	9 May 46	LCIR, str 19 Jul 46
1071	Defoe	25 May 44 10 Jun 44	16 Jun 44 17 Jan 47	22 Nov 46 1 Jun 54	NRT 47–54, str 27 Aug 57
1072	Defoe	31 May 44 16 Jun 44	21 Jun 44	18 Sep 46	To Chile 20 Mar 47
1073	Defoe	6 Jun 44 21 Jun 44	26 Jun 44	27 Nov 46	To Chile 20 Mar 47
1074	Defoe	12 Jun 44 26 Jun 44	30 Jun 44	21 Jun 46	Str 15 Aug 46
1075	Defoe	17 Jun 44 30 Jun 44	7 Jul 44	13 Jun 46	Str 15 Aug 46
1076	Defoe	22 Jun 44 6 Jul 44	12 Jul 44	26 Jun 46	Str 23 Dec 47
1077	Defoe	27 Jun 44 11 Jul 44	15 Jul 44	15 Feb 46	LCIR, str 20 Mar 46, to Chile 47
1078	Defoe	1 Jul 44 15 Jul 44	20 Jul 44	12 Apr 46	LCIR, str 19 Jun 46
1079	Defoe	7 Jul 44 20 Jul 44	25 Jul 44	13 Feb 47	LCFF, str 17 Jul 47
1080	Defoe	12 Jul 44 25 Jul 44	16 Aug 44	6 Jun 46	LCFF, str 4 Aug 56, tgt 14 Aug 58

No.	Name/Builder	LD/Launch	Comm	Decomm	Notes and Fate
1081	Defoe	17 Jul 44 29 Jul 44	22 Aug 44	1 Oct 46	LCFF, str 13 Aug 56, tgt 7 Nov 57
1082	Defoe	21 Jul 44 3 Aug 44	23 Aug 44	9 Jul 46	LCFF, str 13 Aug 56, tgt 2 Nov 57
1083	Defoe	26 Jul 44 8 Aug 44	24 Aug 44	20 May 46	LCFF, str 7 Aug 56, tgt Jul 58
1084	Defoe	31 Jul 44 12 Aug 44	25 Aug 44	7 Sep 46	Str 22 May 51
1085	Defoe	4 Aug 44 17 Aug 44	26 Aug 44	16 Jul 47	Str 28 Aug 47
1086	Defoe	9 Aug 44 22 Aug 44	28 Aug 44	24 Jun 46	Str 27 Feb 51
1087	Defoe	14 Aug 44 26 Aug 44	31 Aug 44	7 Feb 46	Str 28 Mar 46
1088	Defoe	18 Aug 44 31 Aug 44	6 Sep 44	16 Apr 46	LCIM, ex-LCIG; str 15 Aug 46
1089	Defoe	23 Aug 44 5 Sep 44	9 Sep 44	2 Apr 46	LCIM, ex-LCIG; str 1 May 46
1090	Defoe	28 Aug 44 9 Sep 44	14 Sep 44	14 Mar 50	At Bikini, str 20 Dec 50
1091	Defoe	31 Aug 44 14 Sep 44	21 Sep 44	N/A	Sold 1 Feb 61; at the LCI reunion in 95
1092	Defoe	6 Sep 44 19 Sep 44	23 Sep 44	15 Sep 49	To France for Indo-China 8 Jan 51, to ROC Mar 58
1093	Defoe	11 Sep 44 23 Sep 44	28 Sep 44 12 Jan 54	N/A 1 Jul 55	To *Skimmer* (AMCU 41) 7 Mar 52, str 1 Jan 60
1094	Defoe	15 Sep 44 28 Sep 44	3 Oct 44	28 May 46	Str 3 Jul 46
1095	Defoe	20 Sep 44 3 Oct 44	7 Oct 44	N/A	To Miami Power Squadron of U.S. Power Squadrons 5 Feb 48
1096	Defoe	25 Sep 44 7 Oct 44	13 Oct 44	22 May 46	Str 19 Feb 48
1097	Defoe	29 Sep 44 12 Oct 44	17 Oct 44		NRT, in comm in reserve 10 Jun 50, out of comm in reserve 26 Mar 54, str 8 Aug 56
1098	Defoe	4 Oct 44 17 Oct 44	23 Oct 44 23 Oct 53	5 Jun 46 12 Apr 55	To *Sparrow* (AMCU 42), str 1 Jan 60
1099–1139					Canceled Aug 44

LCS(L)/LSSL

No.	Name/Builder	LD/Launch	Comm	Decomm	Notes and Fate
1	Lawley	28 Apr 44 15 May 44	20 Jun 44 13 Jan 47	12 Sep 46 7 Jul 50	NRT 47–50, str 20 Dec 50
2	Lawley	10 May 44 10 Jun 44	19 Jul 44	25 May 46	To France for Indo-China 15 Aug 50
3	Lawley	14 Jun 44 5 Jul 44	31 Jul 44	14 May 46	Str 20 Dec 50
4	Lawley	5 Jul 44 15 Jul 44	11 Aug 44	N/A	To France 15 Aug 50, to Vietnam 1 Sep 55
5	Lawley	17 Jul 44 27 Jul 44	21 Aug 44	31 Oct 46	Str 10 Jun 47
6	Lawley	17 Jul 44 3 Aug 44	27 Aug 44 13 Nov 47	1 Oct 46 6 Feb 50	NRT 47–50, str 27 Feb 51
7	Lawley	31 Jul 44 9 Aug 44	31 Aug 44		WL 14 Feb 45

No.	Name/Builder	LD/Launch	Comm	Decomm	Notes and Fate
8	Lawley	28 Jul 44 11 Aug 44	31 Aug 44	12 Jun 46	Str 19 Jul 46
9	Lawley	7 Aug 44 17 Aug 44	6 Sep 44	1 Jun 46	To France for Indo-China 15 Aug 50
10	Lawley	10 Aug 44 19 Aug 44	10 Sep 44	1 May 46	To France for Indo-China 15 Aug 50
11	Lawley	12 Aug 44 22 Aug 44	13 Sep 44	30 Jul 46	Str 20 Dec 50
12	Lawley	14 Aug 44 24 Aug 44	17 Sep 44	1 Jun 46	To Japan 30 Jun 53, ret 30 Jun 58, tgt 12 Mar 59
13	Lawley	17 Aug 44 26 Aug 44	21 Sep 44	5 Jun 46	To Japan 30 Jun 53, ret 30 Jun 58, tgt 10 Feb 59
14	Lawley	19 Aug 44 28 Aug 44	23 Sep 44	31 May 46	To Japan 16 Feb 53
15	Lawley	23 Aug 44 3 Sep 44	26 Sep 44		WL 22 Apr 45
16	Lawley	25 Aug 44 4 Sep 44	28 Sep 44	8 Jun 46	Str 20 Dec 50
17	Lawley	28 Aug 44 4 Sep 44	30 Sep 44	14 Sep 46	Str 20 Dec 50
18	Lawley	29 Aug 44 6 Sep 44	30 Sep 44	3 Jun 46	To Japan 30 Jun 53, ret 30 Jun 58, str 1 Jan 59
19	Lawley	4 Sep 44 11 Sep 44	7 Oct 44	8 Jun 46	Str 20 Dec 50
20	Lawley	4 Sep 44 13 Sep 44	10 Oct 44	27 Jun 46	To Japan 29 Aug 53, ret 29 Aug 58, str 1 Jan 59
21	Lawley	12 Sep 44 20 Sep 44	14 Oct 44	14 Sep 46	Str 27 Feb 51
22	Lawley	13 Sep 44 22 Sep 44	16 Oct 44	23 May 46	To Japan 11 Mar 53
23	Lawley	20 Sep 44 29 Sep 44	18 Oct 44	8 May 46	Str 27 Feb 51
24	Lawley	23 Sep 44 1 Oct 44	20 Oct 44	N/A	To Japan 30 May 53, ret 31 May 59
25	Lawley	29 Sep 44 8 Oct 44	24 Oct 44	6 Jun 46	To Japan 30 Jun 53, ret 30 Jun 58, tgt 12 Mar 59
26	Commercial	10 Jul 44 13 Aug 44	26 Aug 44		WL 16 Feb 45
27	Commercial	10 Jul 44 13 Aug 44	31 Aug 44	7 Jun 46	To Japan 30 Apr 53
28	Commercial	17 Jul 44 19 Aug 44	8 Sep 44	5 Jun 46	To France for Indo-China 15 Aug 50
29	Commercial	24 Jul 44 27 Aug 44	12 Sep 44	22 May 46	Str 20 Dec 50
30	Commercial	24 Jul 44 27 Aug 44	16 Sep 44	3 Jun 46	Str 20 Dec 50
31	Commercial	7 Aug 44 2 Sep 44	20 Sep 44	23 Oct 46	Str 13 Nov 46
32	Commercial	14 Aug 44 10 Sep 44	23 Sep 44	10 Oct 46	NRT, inactivated Mar 50, str 20 Dec 50
33	Commercial	14 Aug 44 10 Sep 44	27 Sep 44		WL 12 Apr 45
34	Commercial	19 Aug 44 16 Sep 44	30 Sep 44	24 Jul 46	To Italy 25 Jul 51
35	Commercial	28 Aug 44 17 Sep 44	3 Oct 44	13 Aug 46	NRT 1 Jul 47, to France for Indo-China 12 Mar 53, ret Jul 56, to Greece 25 Oct 56
36	Commercial	28 Aug 44 17 Sep 44	4 Oct 44	18 Sep 46	Str 27 Feb 51

APPENDIX C

No.	Name/Builder	LD/Launch	Comm	Decomm	Notes and Fate
37	Commercial	2 Sep 44 23 Sep 44	10 Oct 44	19 Jan 46	Str 7 Feb 46, destroyed 7 Mar 46
38	Commercial	11 Sep 44 1 Oct 44	13 Oct 44	26 Jun 46	To Italy 25 Jul 51
39	Commercial	11 Sep 44 1 Oct 44	16 Oct 44	28 Oct 46	Str 23 Jun 47
40	Commercial	16 Sep 44 7 Oct 44	20 Oct 44	28 Oct 46	Str 23 Jun 47
41	Commercial	18 Sep 44 8 Oct 44	24 Oct 44	28 Oct 46	Str 23 Jun 47
42	Commercial	18 Sep 44 8 Oct 44	26 Oct 44	18 Oct 46	Str 23 Jun 47
43	Commercial	23 Sep 44 14 Oct 44	30 Oct 44	28 Oct 46	Str 23 Jun 47
44	Commercial	2 Oct 44 22 Oct 44	2 Nov 44	28 Oct 46	Str 23 Jun 47
45	Commercial	2 Oct 44 22 Oct 44	6 Nov 44	28 Oct 46	Str 23 Jun 47
46	Commercial	7 Oct 44 27 Oct 44	9 Nov 44	28 Oct 46	Str 23 Jun 47
47	Commercial	9 Oct 44 29 Oct 44	13 Nov 44	28 Oct 46	Str 23 Jun 47
48	Albina	25 May 44 14 Jul 44	26 Aug 44	18 Oct 46	Str 23 Jun 47
49	Albina	2 Jun 44 20 Jul 44	31 Aug 44		WL 14 Feb 45
50	Albina	8 Jun 44 29 Jul 44	11 Sep 44	28 Oct 46	Str 23 Jun 47
51	Albina	15 Jun 44 5 Aug 44	16 Sep 44	18 Oct 46	Str 23 Jun 47
52	Albina	21 Jun 44 14 Aug 44	23 Sep 44	22 May 46	To Japan 29 Aug 53, ret 29 Aug 58, tgt
53	Albina	27 Jun 44 22 Aug 44	30 Sep 44	15 Nov 46	Str 20 Dec 50
54	Albina	19 Jul 44 5 Sep 44	9 Oct 44	31 May 46	Reactivated for Ryukyu Coast Guard 1 Feb 52, to Korea Sep 52, on loan as of 20 Oct 52
55	Albina	31 Jul 44 2 Sep 44	16 Oct 44	27 Jun 46	Str 20 Dec 50
56	Albina	5 Aug 44 7 Sep 44	23 Oct 44	12 Jun 46	Reactivated for Ryuku Coast Guard 1 Feb 52 but inactivated 10 May 52, str 23 Dec 52, to ROC 19 Feb 54
57	Albina	15 Aug 44 14 Sep 44	30 Oct 44	20 Aug 46	In service 30 Nov 51 for MDAP, then to Japan 14 Jan 53, ret 27 Mar 53, str 1 Oct 58, tgt
58	Albina	22 Aug 44 22 Sep 44	6 Nov 44	N/A	To Japan 31 Mar 53, ret 31 Jul 59
59	Albina	2 Sep 44 2 Oct 44	14 Nov 44	4 Apr 46	NRT, inactivated May 50, str 20 Dec 50
60	Albina	7 Sep 44 7 Oct 44	22 Nov 44	2 Oct 46	To Japan 30 Sep 53, ret 30 Sep 58, str 1 Apr 59
61	Albina	15 Sep 44 14 Oct 44	29 Nov 44	26 Jun 46	Str 22 May 51
62	Albina	22 Sep 44 23 Oct 44	4 Dec 44	N/A	To Italy Jul 51
63	Albina	2 Oct 44 2 Nov 44	11 Dec 44	N/A	To Italy Jul 51

AMPHIBIOUS SHIPS **617**

No.	Name/Builder	LD/Launch	Comm	Decomm	Notes and Fate
64	Albina	9 Oct 44 7 Nov 44	18 Dec 44	N/A	To Italy Jul 51
65	Albina	16 Oct 44 14 Nov 44	20 Dec 44	N/A	To France 51 for Indo-China, to Greece Jun 58
66	Albina	24 Oct 44 23 Nov 44	2 Jan 45	30 Sep 46	Str 29 Oct 46
67	Albina	2 Nov 44 2 Dec 44	8 Jan 45	20 Aug 46	To Japan 29 Aug 53, ret 29 Aug 58, tgt May 59
68	Albina	9 Nov 44 7 Dec 44	15 Jan 45	16 Sep 46	To Japan 30 Jun 53, str 1 Feb 59
69	Albina	15 Nov 44 14 Dec 44	22 Jan 45	30 Sep 46	Str 22 May 51
70	Albina	24 Nov 44 22 Dec 44	29 Jan 45	22 Aug 46	Str 20 Dec 50
71	Albina	2 Dec 44 2 Jan 45	5 Feb 45	30 Aug 46	Str 20 Dec 50
72	Albina	8 Dec 44 8 Jan 45	14 Feb 45	N/A	To Japan 30 Apr 53, ret 15 Feb 72
73	Albina	15 Dec 44 16 Jan 45	19 Feb 45	1 Aug 46	Str 20 Dec 50
74	Albina	23 Dec 44 30 Jan 45	26 Feb 45	N/A	To Japan 16 Feb 53, ret 6 Aug 71
75	Albina	3 Jan 45 9 Feb 45	6 Mar 45	N/A	To Japan 30 Sep 53, ret 30 Sep 58
76	Albina	9 Jan 45 15 Feb 45	12 Mar 45	N/A	To Japan 30 Apr 53, ret 30 Apr 69
77	Albina	17 Jan 45 26 Feb 45	19 Mar 45	N/A	To Korea Jan 52
78	Albina	31 Jan 45 28 Feb 45	26 Mar 45	N/A	To Japan 16 Feb 53, ret 8 Mar 76
79	Commercial	9 Oct 44 29 Oct 44	20 Nov 44	N/A	To Japan 30 Apr 53, ret 23 Apr 76
80	Commercial	14 Oct 44 7 Nov 44	21 Nov 44	16 Apr 46	To France for Indo-China 15 Aug 50, ret, to Japan
81	Commercial	23 Oct 44 12 Nov 44	24 Nov 44	15 Nov 46	In service for Ryukyu Coast Guard 1 Feb 52, out of comm in reserve 28 May 52, to ROC 19 Feb 54
82	Commercial	23 Oct 44 12 Nov 44	27 Nov 44	N/A	To Japan 30 Mar 53, ret 6 Aug 71
83	Commercial	27 Oct 44 17 Nov 44	30 Nov 44	N/A	To Japan 30 May 53, ret 30 Mar 59
84	Commercial	30 Oct 44 19 Nov 44	9 Dec 44	N/A	To Japan 30 May 53, ret 30 Mar 59
85	Commercial	30 Oct 44 19 Nov 44	14 Dec 44	N/A	To Japan 30 May 53, ret 30 Mar 59
86	Commercial	7 Nov 44 30 Nov 44	18 Dec 44	N/A	To Korea 20 Oct 52
87	Commercial	13 Nov 44 3 Dec 44	22 Dec 44	N/A	To Japan 30 Mar 53, became drone carrier 58, str 67, to Philippines for parts 17 Nov 75
88	Commercial	13 Nov 44 3 Dec 44	27 Dec 44	N/A	To Japan 30 Apr 53, ret 64, to Philippines 17 Nov 75, not comm
89	Commercial	17 Nov 44 7 Dec 44	30 Dec 44	N/A	To Japan 30 May 53, ret 31 Mar 59
90	Commercial	20 Nov 44 17 Dec 44	4 Jan 45	N/A	To Japan 30 May 53, ret 31 Mar 59
91	Commercial	20 Nov 44 17 Dec 44	8 Jan 45	N/A	To Japan 30 May 53, ret 31 Mar 59

No.	Name/Builder	LD/Launch	Comm	Decomm	Notes and Fate
92	Commercial	30 Nov 44 22 Dec 44	13 Jan 45	15 Nov 46	Nominated for Indo-China but str instead, sold 27 Feb 51
93	Commercial	4 Dec 44 23 Dec 44	16 Jan 45	13 Jul 46	Nominated for Indo-China but str instead, sold 27 Feb 51
94	Commercial	4 Dec 44 23 Dec 44	20 Jan 45	N/A	To Japan 30 May 53, ret 24 Oct 69
95	Commercial	7 Dec 44 3 Jan 45	24 Jan 45	27 Sep 46	For Ryukyu Coast Guard activated and then deactivated, to ROC 19 Feb 54
96	Commercial	18 Dec 44 6 Jan 45	29 Jan 45	N/A	To Japan 30 Jun 53, ret 28 Apr 65, to Vietnam 8 Dec 65, escaped 75 to Philippines
97	Commercial	18 Dec 44 6 Jan 45	2 Feb 45	12 Sep 46	Str 27 Feb 51
98	Commercial	23 Dec 44 13 Jan 45	5 Feb 45	29 May 46	To Japan 16 Feb 53, ret 27 Mar 58, tgt
99	Commercial	23 Dec 44 13 Jan 45	9 Feb 45	9 Apr 46	Str 20 Dec 50
100	Commercial	6 Jan 45 27 Jan 45	13 Feb 45	24 Aug 46	To Japan 11 Mar 53, ret 27 Mar 58, tgt
101	Commercial	6 Jan 45 27 Jan 45	17 Feb 45	N/A	To Japan 30 Apr 53, ret 28 Apr 65, to Vietnam 2 Oct 65, escaped to Philippines 75, used for parts
102	Commercial	13 Jan 45 3 Feb 45	24 Feb 45	N/A	To Japan 30 Apr 53, ret 18 Apr 66, to Thailand 66
103	Commercial	13 Jan 45 3 Feb 45	28 Feb 45	22 Jul 46	To Japan 30 Sep 53, ret 30 Sep 58, str 1 Apr 59
104	Commercial	27 Jan 45 17 Feb 45	5 Mar 45	N/A	To Japan 14 Jan 53, ret 31 Mar 59
105	Commercial	27 Jan 45 17 Feb 45	9 Mar 45	9 Aug 46	For Ryukyu Coast Guard, in service 12 Mar 52, to France for Indo-China 28 Dec 53
106	Commercial	24 Feb 45 24 Feb 45	9 Mar 45	N/A	To Japan 30 Mar 53, ret 27 Jun 75
107	Commercial	24 Feb 45 10 Mar 45	13 Mar 45	N/A	To Japan 14 Jan 53, ret 27 Mar 58
108	Commercial	17 Feb 45 10 Mar 45	22 Mar 45	13 Aug 46	Str 27 Feb 51
109	Lawley	1 Oct 44 10 Oct 44	26 Oct 44	7 Jun 46	To Japan 30 Jun 53, ret 30 Jun 58, str 1 Dec 58
110	Lawley	3 Oct 44 12 Oct 44	29 Oct 44	N/A	To Japan 30 Mar 53, ret 6 Aug 71
111	Lawley	9 Oct 44 17 Oct 44	3 Nov 44	N/A	To Japan 16 Feb 53, ret 21 Jul 70
112	Lawley	11 Oct 44 19 Oct 44	6 Nov 44	22 May 46	Str 27 Feb 51
113	Lawley	12 Oct 44 22 Oct 44	9 Nov 44	6 May 46	Str 27 Feb 51
114	Lawley	17 Oct 44 26 Oct 44	12 Nov 44	N/A	To Japan 30 Apr 53, ret 14 Feb 66
115	Lawley	19 Oct 44 28 Oct 44	14 Nov 44	N/A	To Japan 16 Feb 53, ret 11 Feb 66
116	Lawley	22 Oct 44 30 Oct 44	17 Nov 44	N/A	To Japan 30 May 53, ret 31 Mar 59
117	Lawley	26 Oct 44 4 Nov 44	19 Nov 44	16 Jul 46	Str 27 Feb 51
118	Lawley	28 Oct 44 6 Nov 44	23 Nov 44	6 Jun 46	To Italy 25 Jul 51

No.	Name/Builder	LD/Launch	Comm	Decomm	Notes and Fate
119		31 Oct 44	27 Nov 44	5 May 47	To Japan 29 Aug 53, ret 29 Aug 58, str 1 Jan 59
	Lawley	8 Nov 44			
120		4 Nov 44	29 Nov 44	N/A	To Japan 30 Apr 53, ret 12 Jul 71
	Lawley	14 Nov 44			
121		7 Nov 44	1 Dec 44	6 Mar 47	NRT 47–50, str 20 Dec 50
	Lawley	16 Nov 44	18 Mar 47	Mar 50	
122		9 Nov 44	8 Dec 44	29 May 47	Str 27 Feb 51
	Lawley	18 Nov 44			
123		14 Nov 44	11 Dec 44	27 Jan 47	Str 22 May 47
	Lawley	24 Nov 44			
124		16 Nov 44	14 Dec 44	27 Jan 47	NRT 47–50, str 20 Dec 50
	Lawley	27 Nov 44	18 Mar 47	Mar 50	
125		19 Nov 44	17 Dec 44	21 Mar 47	Str 27 Feb 51
	Lawley	1 Dec 44			
126		25 Nov 44	20 Dec 44	N/A	To Japan 30 May 53, ret 30 May 58
	Lawley	5 Dec 44			
127		27 Nov 44	26 Dec 44		WL(A) 5 Mar 45
	Lawley	6 Dec 44			
128		1 Dec 44	29 Dec 44	26 Nov 46	NRT 47–50, str 20 Dec 50
	Lawley	9 Dec 44	29 Jan 47	Mar 50	
129		5 Dec 44	31 Dec 44	N/A	To Japan 30 Jun 53, ret 28 Apr 65, to Vietnam 19 Feb 66, escaped 75, to Philippine navy
	Lawley	13 Dec 44			
130		7 Dec 44	3 Jan 45	31 Jan 47	To Japan 14 Jan 53, ret 27 Mar 58, tgt 14 Nov 58
	Lawley	15 Dec 44			

Notes

Chapter 1. Introduction

1. The development of a multi-divisional Marine Corps designed to assault and hold bases was not inevitable, even at this late date. In 1940 President Franklin D. Roosevelt was much impressed by British practice. At the outbreak of war, the Royal Marines had asked to form a Marine Division, as in World War I, but that was rejected. Instead, they reshaped themselves into a commando force. In 1940 the president suggested that the U.S. Marines follow a similar course. They resisted successfully, but formed a raider force ("Edson's Raiders") within the Corps as a way of following the president's directive.

2. In 1948 the Fleet Marine Force was organized into two understrength divisions (11 battalions, compared to 18 at full strength) and two corresponding air wings (23 squadrons), a total of 35,086 men. Strength in 1950 was only 8 battalions and 12 squadrons, and at the time of the North Korean invasion of South Korea in June 1950 plans called for a further cut to 6 battalions and 12 squadrons with a total strength of 23,952 marines. Existing units were badly understrength; companies in the battalions fielded two rather than three infantry platoons, and light artillery batteries had four rather than six howitzers. Thus a Marine division was cut to about the strength of a World War II regimental combat team, which was something less than a third of a World War II division. The cuts were not entirely due to prejudice against the Marines; army units were also badly understrength on the eve of the Korean War.

3. The General Board had already lost its primacy over ship characteristics. In 1945 OpNav formed a new Ship Characteristics Board (SCB), consisting of representatives of the bureaus; SCB members decided characteristics by majority vote. The SCB was little concerned with policy issues, which were decided at a higher level within OpNav.

4. These numbers may be somewhat illusory. In his December 1963 planning memorandum McNamara claimed that a shortage of combat vehicle lift (mainly LSTs) limited the navy to only 60 percent of the two-division goal (i.e., 1.2 divisions' worth). The main difference between his plan and that proposed by the U.S. Navy for FY 65–69 was almost a doubling in the number of LSTs, from 18 to 30. LSDs would be increased from 9 to 12. Amphibious cargo ships (AKA) would be more than doubled, from 3 to 7. More expensive ships would be cut: amphibious carriers (LPH) from 9 to 4, and amphibious transports (LPD) from 12 to 7. The U.S. Navy program had been designed to create two divisions' worth of 20-kt lift by FY 76.

5. Incidentally, prior to 1943 the APA designation was reserved for animal transports. None was ever built or converted.

Chapter 2. Developing the Techniques

1. Prior to Normandy, the army did not have any consistent concept of how to assault a defended beach; in particular it did not appear to have much faith in naval fire support. That lack of faith may have been a reaction to the failure of British naval fire support at Gallipoli. For example, plans for the Salerno landing in September 1943 called for a parachute assault to neutralize a coast defense position. Given the army's strong preference for a night attack, the paratroopers still needed some light, so the assault was scheduled for a bright moonlit night that ruined the element of tactical surprise. The naval commander, Adm. H. Kent Hewitt, who was responsible for all the major European landings, argued that surprise was impossible anyway, given the sheer mass of shipping involved. According to Adrian R. Lewis, *Omaha Beach: A Flawed Victory* (Chapel Hill: University of North Carolina Press, 2001), the U.S. Army practices can be traced to undue reliance on

British-developed amphibious doctrine. The British thought in terms of small forces that could land anywhere along a lengthy coastline; indeed, this capability could cause an enemy to stretch its defenses to the breaking point along that coast. The Germans did find themselves fortifying the entire European coast against possible attack. It seemed to follow that the main threat would be troops quickly moved to the invasion beach, hence any delay in perceiving which beach was to be hit would buy invaluable time for the attackers. That in turn made a preattack bombardment extremely dangerous. Thus the British doctrine developed just before and early in the war emphasized speed of assault and night landings. Lewis argues that this doctrine was ill-adapted to the sort of massive attacks mounted in North Africa, Sicily, Italy, and France, and that it should have been rethought. He also points out that, as overall commander in Normandy, Field Marshal Sir Bernard Montgomery imagined that the area was only lightly held, in accord with earlier British thinking, and he was unable to take into account developing intelligence information indicating that the Germans were reinforcing the Normandy beach defenses.

Chapter 3. Transports

1. In 1908 the Marines first considered a formal organization into regiments, battalions, and companies suitable as a fleet striking force or advanced base force. At this time considerable pressure was being applied to eliminate the old detachments aboard ship in favor of a new concentrated Marine force. The Marines managed to stay aboard ship, but the concept of larger organized units took hold. Prior to this the largest units afloat were battalions, which were little larger than current companies. Regiments were formed on a provisional basis. For example, in 1903 a provisional regiment comprising two battalions of Marines (infantry) plus an artillery company landed at Culebra. The early twentieth century experiments in maintaining floating Marine forces concerned battalions. Battalions had first been organized during the Civil War. For example, a battalion landed in Panama in 1885 had 10 officers and 212 enlisted men. A Marine battalion fought at Guantanamo Bay in 1898. Similarly, a 250-marine battalion went to the Philippines to fight the rebels after the United States annexed the islands. The first true Marine brigade (1,678 men) was formed to fight in the Boxer Rebellion in China in 1900. It comprised two regiments (two battalions each). By 1904, the battalion had grown to 12 officers and 486 enlisted men, so a Marine brigade was about 2,000 men plus supporting units. The brigade that fought at Vera Cruz in 1914 comprised two regiments of two battalions each: 96 officers and 2,373 enlisted men. By this time battalions consisted of varying numbers of companies. Nine more regiments were raised to fight in World War I. They mirrored contemporary army practice; they were specialized (as infantry or artillery) and each comprised three rather than two battalions (plus a machine gun company and headquarters units). Brigades formed specially to fight in France comprised two such regiments (about 3,600 marines each) plus a machine gun battalion. They slotted into four-brigade army divisions. The nominal three-battalion organization became standard postwar, but authorized manning levels were too low to fill them out. For example, the 4th Marines (regiment) at Shanghai typically had two rather than three battalions, each of which had two rather than the rated three rifle companies. These regiments were single purpose; the reinforced regimental landing team (RLT) added supporting arms (logistical support was added at the brigade level). Marine brigades (an infantry regiment plus supporting arms or two infantry regiments) functioned throughout the interwar period and were employed during World War II. In 1940 the Marines formed the next higher size unit, a division built around three regiments.

2. Although the navy took *Republic* over (as AP 33), she never served as an attack transport because she was far too unstable, with zero metacentric height in light condition. The other ships listed did serve as navy combat loaders.

Chapter 4. Landing Craft

1. Standard 30-ft whaleboats were tested. They were suitable for light surf, but could not negotiate seas heavy enough to break over the motor. The boats had neither rudders nor backing power (for retracting from a beach). When under power, they steered by turning their outboard motors. They relied on oars to back off a beach and to steer when coming alongside.

2. According to the British official staff history of World War II amphibious ships and craft, the LCP(L) was preferred to the LCP(R) because the latter's flat low ramp made for poor seakeeping. According to the report, the high-ramped bow of the LCVP was still "much less popular than the spoonbill bow of the original LCP(L) for trips in choppy seas." Nor did the British like the LCV; according to their official report, disembarkation often led to flooding over the low ramp sill, with disastrous effects on the craft's stability.

3. As of May 1941, the standard Marine division included a tank battalion of 54 army-type light tanks

(each 11½–13½ tons), plus 17 of the earlier very light tanks (6–9 tons).

4. In July 1942 landing craft designations were standardized and U.S. and British designations combined in a common L (for landing craft or ship) series. Craft were distinguished by function and by Mark number. Thus the British MLC became LCM Mk 1 or LCM(1) and the bureau 45-footer became LCM Mk 2 or LCM(2). Similarly, the early British tank landing craft (TLCs) became LCTs of various types (Mks 1 through 4), and the U.S.-built tank ferry become LCT Mk 5 or LCT(5). Documents sometimes use Roman numerals for versions of the LCT, and sometimes the number is rendered with a dash, as in LCM-3. For simplicity, the parenthetical version is always used in this book, for example, LCM(2) or LST(2). These common designations did not, for some reason, embrace the converted U.S. merchant ships, which were designated as naval auxiliaries (AKA for cargo ship, attack; APA for personnel ship, attack). The equivalent British designations were in an LSI (landing ship, infantry) series. To further confuse matters, after World War II the U.S. Navy redesignated its LCI(L)s (landing craft, infantry, large) as LSILs (landing ships, infantry, large)—a designation previously applied to some much larger British ships.

5. Accounts of detailed timing disagree. All sources agree that Higgins was officially asked for a tank lighter on 27 May. Higgins' biographer (Jerry E. Strahan, *Andrew Jackson Higgins and the Boats that Won World War II* [Baton Rouge: Louisiana State University Press, 1994]) dates the test to 29 or 30 May; other sources claim that the boat was ready for testing on 5 June.

6. In their official history of amphibious ships and craft, the British took credit; they demanded a test due to their preference for the Higgins lighter. On the basis of Senate committee documents, Higgins' biographer, Jerry E. Strahan, credits Senator Harry Truman and his war production investigation committee, which was looking into the landing craft program.

7. Ironically, the British, who were largely responsible for the existence of the Higgins LCM(3), had complaints. According to their official report on the Normandy landing, opinions of these craft differed. The U.S. Navy clearly liked them, but the British considered them inefficient due to their low serviceability and their unsuitability for unloading vehicle-carrying ships, and more generally for open-anchorage work (an Admiralty comment written into the draft of this report stated that this opinion was quite contrary to reports of the North African and Sicilian invasions). On the whole, the British felt that their LCM(1) gave less trouble than the U.S. LCM(3). It had problems with its "A" bracket (to the propeller shaft) and with its skeg. The after fuel tank was integral with the hull, hence it was leaky and inaccessible for repair. Yet the use of bulk fuel to cool the diesels necessitated its use (and its position).

8. According to 1942 and 1943 editions of the standard U.S. classified handbook of amphibious ships and craft, the bureau type was also designated LCM(3), but that usage was soon dropped.

9. Different manufacturers did vary the design. Some LCM(3)s had vertical ramps instead of the usual slightly sloped ones. In most, the bulwark sloped down and then angled to the horizontal; in others, it sloped all the way to the stern. The angled bulwark can be mistaken for the broken bulwark of the slightly lengthened LCM(6).

10. The evocative DUKW designation was derived from the army's alphabetical symbols for 4-wheel drive and the year the vehicle entered service, 1942. D indicated 1942, U indicated amphibious, K indicated front-wheel drive, and W indicated rear-wheel drive.

11. DUKWs were usually used to unload cargo ships after the assault had been completed. At Anzio, however, assault boats (LCVPs) were in short supply, because there were no attack transports, and not enough of the six-davit LSTs (see Chapter 5) that acted as surrogate attack transports. To free the remaining LCVPs to carry personnel, their role carrying 105-mm howitzers ashore as part of the assault was assigned to DUKWs, which the LSTs would launch before they beached. Each of 13 LSTs releasing LCVPs carried 11 DUKWs, some carrying howitzers and some carrying their ammunition. One advantage of this technique was that the LSTs could be loaded below their beaching draft, because they would be lightened when the DUKWs were launched. To further complicate the situation, the beaches off Anzio sloped very gently, so the DUKWs had to be launched well offshore. The landing was conducted at night, and the DUKW drivers were ordinary army truck drivers without any naval experience. In his after-action report, Rear Adm. Frank J. Lowry, commanding the naval force (Task Force 81), argued that, when released at night 3–6 miles offshore, DUKWs "present a very serious problem." He found "a growing tendency in the Army to use DUKWs in this manner. Training films have done a great deal to stretch their capabilities." If they were to carry vital equipment in future assaults, it would be necessary to prove that they were really all-weather craft, and also to have qualified seamen in charge of them. Because the DUKW's speed of 4 kts was so much slower than LCVP speed, they had to be

launched before the first wave of boats, and as a result they were in danger of being run down by the boats. It would have been better to launch them last, and to assign them to the final wave. Lowry also noted that, although the first DUKWs reached the beach roughly on schedule, it took about 25 minutes for them to complete their wave, far longer than conventional boats would have taken. In a supplemental final report, Lowry reported that DUKWs were poor seaboats. In a rehearsal, with wind at Force 5 and the sea at Force 2, many were lost at sea, but LCVPs encountered no problems and suffered no losses. In the actual assault, with a smooth sea and light airs, two DUKWs were lost at sea, one being swamped at launch and one due to leaks (which would have been detected had the army inspected it before launch).

Chapter 5. The British Connection

1. The British had actually considered converting merchant ships to carry tanks for a projected 1918 assault on Zeebrugge in Belgium. According to Lt. Comdr. The Hon. J. M. Kenworthy, *Sailors, Statesmen—And Others: An Autobiography* (London: Rich and Cowan, 1933), 109, the Admiralty war staff proposed "to fit out old merchant ships to carry a dozen tanks each in their holds. The bows were to be reconstituted in such a way that they would run up on the beach in the known state of the tide. . . . Slung up in the forepart of each vessel was to be a kind of drawbridge which would be lowered, and over which the tanks would trundle into the shallow water and so ashore." This proposal was ultimately rejected, but it was surely widely known within the war staff, and to the commander of the Zeebrugge attack, Adm. Sir Roger Keyes—who in 1940 was Churchill's director of combined operations. Kenworthy was on the Admiralty war staff when the proposal, for what amounted to an LST, was raised.

2. The C4/Seatrain saga illustrates just how complicated wartime shipbuilding could be. The army was given permission to build 50 Seatrains, mainly as tank transports, during a 19 February 1942 conference in President Roosevelt's bedroom. Its main outcome was an emergency directive to build many more ships in 1942 and 1943 to overcome losses due to German U-boats and to transport a large U.S. Army overseas. Sun Shipbuilding received the Seatrain contract. Within a few weeks, however, the navy had pointed out that a Seatrain could not survive a single torpedo hit, and that even a minor-caliber shell hitting at the waterline might sink one. The army therefore shifted to the Maritime Commission C4 design, conceived for the American-Hawaiian Line, but intended to be adaptable to troop transport duty. Because, like a Seatrain, a C4's engines were aft, it offered large uninterrupted cargo spaces amidships, and the contract was rewritten for 50 C4s in August 1942. It called for C4-S-B1 tank carriers, with special tank ramps; in effect, they would have been ro-ros. Only one ship was finished to this design, because in September 1943 the Joint Chiefs ordered the ships completed as point-to-point troopships (C4-S-B2) to support the planned invasion of Europe, just approved at the Quebec Conference. Ultimately six became U.S. Navy hospital ships. The last 5 of the 50 were completed as cargo ships (C4-S-B5 design). The decision that these be *army* transports was significant, because the army used civilian crews, and required much smaller crews than the navy. The navy view was that the ships should be built as close to naval standards as possible, because they might ultimately be navy operated. In February 1943, for example, it was decided that crew spaces should be made as capacious as possible, specifically for naval use. In March the army agreed that the navy would man its transports. Then it turned out that the Naval Transportation Service could not in fact man the ships, and they had to be reconverted to merchant standards. However, many of the C4s actually did serve as navy-manned APs.

3. The first 36 conversions were LST 1, 3, 4, 6, 197, 309, 310, 314–318, 326, 327, 332, 347–352, 357–360, 372–382. Formally ordered in March 1943, they were carried out at Pier 45 North River in New York; conversion was a shipyard job because the davits were powered. At the same time 40-mm guns were installed and the forward 20-mm guns relocated. Two further series of six-davit conversions were ordered. On 18 September 1943 the VCNO ordered 62 LSTs converted, mostly while under construction; on 28 September another 72 (of the 1944 program) were ordered completed with six davits each. The 62 ships were described as "for special employment." They were to have 150 bunks on the tank deck, 150 ft of hose suitable for delivering potable water, barrage balloon equipment, and cradles for LCTs. In October 1943 plans were developed to arrange their tank decks to transport 300 wounded from a beach for up to 6 hrs (a "morphine ride") and 300 litters ordered for each ship in place of the 150 bunks. Another 23 special ships were designated in November 1943. They retained the usual pair of davits and lacked hospital facilities, but they had the water hose, balloon fittings, and the LCT cradle. The 62 six-davit ships were LST 46–60, 133, 134, 230, 280–294, 491–512, 515–517, 523, and 531–533. The other 72 six-davit ships were LST 141,

528–530, 539–545, 548–551, 554–557, 561–564, 573–576, 600–602, 653–662, 664–667, 669–672, 675–678, 683–686, 689–692, 695–698, 701–704, and 707–710. That ended the program. However, in answer to a 20 September 1944 request for hull numbers of British and U.S. six-davit LSTs, BuShips listed the following, which do not quite match any of the lists above: LST 46–60, 133, 134, 230, 280–294, 491–495, 497, 498, 500–506, 508–512, 532, 533, 548–551, 554–557, 561–564, 573–576, 601–603, 655, 656, 659–662, 664–667, 670–672, 675–678, 689–692, 695–698, 701–704, 707–710, all of which were U.S. ships, although some were later transferred to the Royal Navy. On 3 October the Pacific amphibious force complained that the six-davit ships were more difficult to load with LCTs, because the forward davits limited access to the cargo hatch or ramp, and the ships had less upper deck space for vehicles. They also had more boat weight and suffered from more interference when going alongside. It would be preferable to limit new construction LSTs to four davits, all on the deckhouse aft. As a consequence, on 4 January 1945 BuShips ordered that all six-davit LSTs coming home for refit have their forward davits removed. Two-davit LSTs, including those being completed, would have two sets of davits added. Since the davits themselves were in short supply, those removed from six-davit ships would be installed on two-davit ships. Among those modified were LSTs withdrawn from the Atlantic Amphibious Training Command with the conclusion of amphibious operations in Europe. At least one of the three LSTs transferred to Japan postwar (ex-LST 689, 835, and 1064) had six davits when it was transferred.

4. In March 1944 LST Flotilla Five, operating in the Southwest Pacific, reported to BuShips that its LSTs mounted eighteen 20-mm guns in addition to their seven single 40-mm. BuShips objected, and the route sheet on its copy of the report asked where the flotilla got the personnel to man the additional guns. Ships converted as light plane carriers had their forward 40-mm guns blanked off by the impromptu take-off deck. As compensation, the army provided a pair of 20-mm guns on portable mounts.

5. Rearmament was hardly complete by the spring of 1945, when the advent of kamikazes made it more urgent. Similarly, LSMs were all wired to take a twin 40-mm gun forward, although they were generally completed with a single mount. Many LCS(L), wired for the twin 40-mm forward, were completed with either a single 3-in/50 or with a single 40-mm. Wiring was significant, because the single 40-mm gun was not power operated, and it did not have a director. The twin mount was power operated and also had to be wired to a director, generally an Mk 51. In April 1945 three LSTs (LST 308, 355, and 392) were assigned to install twin 40-mm mounts and directors on amphibious units wired for these weapons. Each could carry about 55 twin mounts and was equipped with a boom and provided with necessary personnel.

6. Ships were completed to U.S. standard and then modified for the British, the U.S.-supplied 3-in/50 being removed and a 12-pounder fitted. PAC was a modified Royal Air Force weapon intended to be equivalent, for a short time, to a barrage balloon. It was a cable, supported at either end by a parachute, carried to 500 ft by a rocket. A bomber trying to attack at low altitude would, in theory at least, snag the cables. At the least masthead attacks would be discouraged. In at least one case such a barrage entirely stopped a bomber. Many British coastal craft were fitted with PAC from the summer of 1940 onward. The U.S. Navy preferred barrage balloons, for which both LSTs and attack transports were fitted. FAM was a PAC with an explosive attached to its cable. The standard wartime U.S. Navy handbook of allied landing craft lists FAM as the British LST weapon, but a recent unofficial account of British LSTs, *Ships Without Names*, lists only the PAC.

7. For the Sicily invasion (Operation Husky), Eighth Fleet converted numerous LSTs, in addition to the six-davit units described above, the aircraft-carrying units described below, and the fighter-control units described in Chapter 9. Following experiments in North Africa, ten were fitted to side-carry enough pontoons (175 ft length on each side) to form a complete causeway for each ship. These ships were given an enlarged hole on the underside of their ramps so that they could pivot on the kingpost welded to one of the pontoon causeways. Another twenty had cross-connections fitted between their fire and bilge pumps and freshwater tanks to permit these pumps to be used to discharge potable water to shore tanks through 2½-in hoses. They could pump over a distance of as much as 750 ft. This type of operation was practical because the LST had so much tankage, thanks to its dual-displacement design. In its after-action report, Eighth Fleet recommended that all LSTs be fitted with the necessary cross-connections, and that all LSTs fitted to side-carry pontoons be fitted with the enlarged kingpost holes. For Salerno, nine U.S. LSTs side-carried pontoons, and three British LSTs towed them. For Normandy, 15 U.S. and 15 British LSTs were fitted with rails (three tracks abreast) on their tank decks to carry rolling stock. For Southern France (August 1944), 20 U.S. LSTs, 3 Greek LSTs, and 1 British LST side-carried pontoons. For Iwo Jima in

February 1945, 6 LSTs carried pontoon causeways and 14 more carried pontoon barges. Pontoon causeways and bridges proved difficult to handle in the rough water around Iwo Jima. Of 60 LSTs in the Iwo Jima operation, 38 were preloaded with troop supplies, and 24 with U.S. Navy ammunition.

8. In mid-1944 Admiral Nimitz, CinCPac, asked for a total of eight LSTs converted to full hospital ships (i.e., disarmed, with special markings, painted white). Four conversions (LST 575, 676, 677, and 678) were approved in September 1944, each ship to accommodate 196 patients plus 6 officers and 50 enlisted medical personnel. With design work in full swing, on 11 October BuShips stopped it, and on 23 October Cominch formally killed the project. The four ships earmarked for conversion became small craft mother ships instead. Some LSTs, however, had already been converted to hospital ships. For example, as of November 1944 the BuShips amphibious ship directorate believed that LST 476 was serving as a 100-bed hospital. For the Iwo Jima operation, four LSTs were converted to medical supply and casualty evacuation ships; they handled 6,136 casualties, and were credited with saving many lives. More were converted for Okinawa. At Iwo Jima, the landing ship *Ozark* was also specially outfitted for casualty handling, and, according to the after-action report, she proved her worth.

9. In May 1945 Cominch approved conversion of several LSTs to ammunition carriers; this involved cutting an additional hatch in the upper deck at frames 19–22. There was some concern that this break in the ship's strength deck would reduce her torsional strength excessively, but the hatch could not be foregone. The first conversions were LST 126 and 761, and on 1 July 1945 more were added: LST 127, 224, 270, 271, and 272. Work on these conversions was stopped on 17 August 1945.

10. The flight deck was proposed by Capt. Brenton A. Devol Jr., artillery air officer of the 3rd Infantry Division, who had flown off a carrier during the North African invasion. A detachment of the division's 10th Engineer Combat Battalion installed the 200 × 10 ft runway, built of steel matting on timbers, in 36 hrs. Captain Devol made the first take-off on 4 July 1943.

11. For many years *Jane's Fighting Ships* carried the statement that LST 525 was the sole U.S. LST fitted with a flight deck, but that several British LSTs were so fitted. No documentary evidence of such conversions has emerged. The British did convert three ships (LST 13, 216, and 217) to fighter-direction ships specifically for the Normandy invasion, redesignating them as FDTs. They were fitted with standard RAF (rather than Royal Navy or U.S. Navy) radars, and ballasted with pig iron.

12. The device was conceived by 1st Lt. James Brodie of the U.S. Army Ordnance Corps, specifically to allow a light airplane to take off from and land onto a ship, without using a flight deck. He built a test device and had a pilot awaiting movement overseas demonstrate it. Although the army showed no interest, both the U.S. Navy and the Office of Strategic Services liked it. After seeing it demonstrated in February 1944, Gen. Leslie McNair, chief of army ground forces, wanted to buy a few, but the army's air staff blocked any procurement; McNair died in action in July 1944, ending army interest. That left the navy, which installed the Brodie device on board LST 776, which the army unofficially called USS Brodie. Although the LST also had a launching ramp, according to the army history airplanes could take off from the wire. A winch attached to the wire helped slow the airplane on landing and boosted it for take-off. The original Brodie installation was modified in November 1944. After that tests showed that average time between launch and recovery was 9 minutes (3 minutes minimum). Other average times (minimum times in parentheses) were recovery to launch, 15 (12) minutes; launch to launch, 13 (10) minutes; and recovery to recovery, 12 (10) minutes. Installation of a carrier-type gasoline tank (7,500 gals) was proposed but rejected on the theory that such a tank would make beaching impossible; instead, the ship carried her aviation fuel in jettisonable drums topside. For the Okinawa assault she carried nine uncrated and two crated L-4 observation aircraft, as well as a complete Bailey bridge and three DUKWs. In December 1945 her Brodie gear was ordered removed and stored aboard a reserve fleet LST. In May 1945 four more LSTs were ordered fitted with Brodie gear: LSTs 325, 338, 388, 393. They would retain their original two davits. All these conversions, as well as the original one, were done in New Orleans. By July 1945 all were completed, and gear for three other ships remained in New Orleans (no further LSTs seem to have been designated). None but LST 776 saw service because only that ship was available for the last major assault of the war, at Okinawa. Installations clearly survived the end of the war. In October 1945 LST 393 had hers disabled in a collision with the cargo ship *Amherst Victory*. In addition to the Brodie craft, on 27 June 1945 conversion of one LST to launch JB-2 missiles (the U.S. Army version of the German V-1) was ordered. By this time BuShips had prepared plans. Surviving records do not show that any particular LST was selected for conversion before the end of the war, but apparently a large-scale program was envisaged.

13. The ARB project seems to have begun with a 3 December 1942 request by BuShips for three LSTs to support big floating drydocks in forward areas. OpNav approved the three, but argued that the floating drydocks should be served by unpowered barges, because mobile repair facilities (ARBs) were too valuable to tie down.

14. Conversions generally entailed extending the superstructure well forward, to provide additional officers' quarters and workshops. The tank deck housed electrical, machine, sheet metal, shipfitter, blacksmith, and pipe shops and storage. The elevator was changed to a hatch, the ramp removed, and the bow sealed. An A-frame and winch atop the extended deckhouse could lift 50 tons, and 10-ton booms were slung from two kingposts forward. Characteristics for the ARL dated 23 December 1942 required berthing for 22 officers, 30 petty officers, and 200 personnel, for which two 4,000 gal per day distilling plants were provided. However, in-service complement was only 7 officers and 104 enlisted men. The characteristics showed a pair of 36-ft motor launches instead of the usual landing craft, although ships often carried LCVPs instead. Armament was increased to one 3-in/50 (on the fantail), two quadruple 40-mm Bofors (fore and aft), and twelve single 20-mm guns. The first conversions were ordered on 6 December 1942. Three more landing craft repair ships (ARL 4–6) were apparently ordered early in 1943, two of them going to the Royal Navy under Lend-Lease with the designations LSE 81 and 82; the U.S. Navy retained the former LST 83 as ARL 4. Plans for this particular conversion were dated 20 July 1943. ARL 7–9 were ordered converted on 1 December 1943. Later more ARLs were ordered. ARL 10 and later units had heavier booms (60- and 25-ton capacities); they carried an LCM as well as the motor launches. They retained the bow door and ramp, and had increased storage space at the expense of shop size. This version lacked the 3-in gun of the earlier ARLs. In addition to these ships, LST 387, beached at Bizerte, was fitted out as an immobile base ship for the invasion of Southern France. On 26 January 1943 BuShips designated five ships for two new types of conversion: three to ARB 1–3 and two to AGP 4 and 5. Gibbs & Cox acted as design agents. The ARB was similar in configuration to an ARL but had two 5-ton booms and one heavy lift boom (about 50 tons). More conversions were ordered on 1 December 1943: ARB 4–6 and AGP 10 and 11. The initial conversions were not enough to serve a rapidly expanding fleet. In June 1943 the Auxiliary Vessels Board proposed further units. Beside merchant hulls, it wanted to convert 8 LST hulls in FY 44 (beginning 1 July 1943) and another 8 in FY 45 to provide 3 battle damage repair ships (ARBs), 3 landing craft repair ships (ARLs), and 2 torpedo boat tenders (AGPs) each year: a total of 9 ARB, 12 ARL (including the British ones), and 6 AGP, including the 1942 ships. Thus by March 1944 the U.S. Navy was planning, for example, ARL 10 and 11, of the FY 45 group. This group also included ARB 7–9 and AGP 14 and 15. Huge expansions in the number of landing craft and motor torpedo boats demanded many more tenders. On 13 March 1944 the CNO proposed, and the secretary of the navy approved, a massive expansion of the landing craft program, with supporting units: 12 ARL and 12 unpowered barracks ships (APLs). That brought the total of ARLs to 24. On 2 May 1944 the Auxiliary Vessels Board called for another 15 ARLs to service landing craft (bringing the total to 39) plus 3 more ARBs and 4 more AGPs (AGP 16, 17, 18, and 19), all of them converted LSTs. Conversion of the other 13 was formally authorized on 9 August 1944. ARL 42–47, the final six, were added in May 1945, but canceled in August when the war ended.

Four aircraft repair ship conversions of LSTs, requested by the Pacific Fleet naval air arm as early as 7 March 1944, were approved in November 1944: ARV(E) 3 and 5 and ARV(A) 4 and 6 for, respectively, engine and airframe repairs (ARV 1 and 2 were converted Liberty ships). They were similar in outline to the ARLs, ARBs, and AGPs, with deckhouses extended forward.

On 14 August 1944 the Pacific Fleet service force complained that it was increasingly difficult to service and supply the numerous salvage vessels and tugs engaged in each major amphibious operation. Moreover, operations were now so far from bases that these small ships, which had so little onboard space for bulky items, could not get back to replenish. In the Marianas and in earlier operations salvage, tug, and fire-fighting services had been seriously handicapped due to the lack of a salvage tender from which supplies and relief personnel could have come. A salvage tender would carry salvage pumps, air compressors, blowers, anchors, winches, wire, blocks, rigging, and other heavy salvage machinery; plates, shapes, pipe, timbers, etc., for repairs; fire pumps, hose, foam, and other fire-fighting gear; personnel, such as divers, to relieve those on board salvage ships; repair facilities to maintain and overhaul salvage equipment, fire-fighting gear, boat engines, etc.; and additional boats, lifting gear, work floats, etc., that smaller ships could not accommodate. The tender would also function as command ship for the salvage task group, and would be able to distribute limited amounts of diesel oil, fresh water, and gasoline to tugs and other salvage craft. She might be roughly similar to a PT boat tender (AGP). Three were proposed: two for the Central Pacific, one for the Southwest Pacific. CinCPac approved the plan on 14 August 1944. BuShips noted that OpNav had already received a 31 January 1944 report outlining the need for such tenders. In European waters coasters or similar craft had sufficed, but distances were far greater in the Pacific. Three salvage tenders, ARST 1–3, were approved in November 1944. Unlike the ARLs, ARBs, and AGPs, they had their bridges relocated forward.

Despite the massive conversion program, units in forward areas needed more tenders. Thus on 4 June 1944 the Seventh Fleet (Southwest Pacific) PT boat command recommended characteristics to convert an LST into a mobile torpedo boat base. The idea was to install parts of the standard EII PT operating base below decks, including all shop equipment, the torpedo unit, the sick bay equipment, and the flake ice machines. Topside would be one 10-ton and one 5-ton crane, 14,150 cu ft refrigerated lockers, two 75 kW generators, two 250 gal per hr distiller units, a communications unit, and a laundry expanded to 600-man capacity. Pontoons and fittings ("jewelry") for two PT drydocks would be carried on board, together with one LCM, two LCV, four jeeps, and two 2½-ton trucks. The ship would also be equipped with two A-frame derricks. She would tow a drydock and a 5,000- or 7,000-barrel aviation gas (PT fuel) barge from station to station. Upon arrival at a new station, she would beach, lower her ramp, and establish communications with air and army units by jeep and telephone. Living quarters and floored tent shops would be erected ashore. The unit would move as required, from point to point. The PT command considered her 95 percent as effective as a more sophisticated tender. The proposal reflected the command's experience with the forward-converted LST *Pontus*. Seventh Fleet wanted the ship converted in the United States, and loaded with all available PT engines and spares before departure. BuShips was initially quite favorable, but the conversion was canceled on 1 July because several adequate tenders were due in the area.

The Seventh Fleet proposal was for a sort of semi-tender, which would help set up a series of bases. On 11 August 1945 Cominch ordered six LSTs designated for a related role, to carry and set up a forward repair base, called an E-3. They would load one such base component, which would arrive at Pearl Harbor by 1 September 1945, in the form of various tray- and skid-mounted machine tools and equipment, plus material and extra personnel. These LSTs would become a repair and maintenance group. They would unload their E-3 to become a shore-based repair activity. Another six LSTs would carry not only a second E-3 component, but also material for a 1,000-man camp to man it. This program ended with the end of the war.

15. The six purpose-built APBs were intended specifically to provide personnel for forward repair activities, such as floating drydocks; they were initially designated in the same series as unpowered barracks (labor) barges (APLs). They were not conceived as receiving ships. BuShips submitted a design on 12 August 1944, to fulfill requirements stated the previous July. Two surviving units, *Mercer* (IX 502, ex–APB 39) and *Nueces* (IX 503, ex–APB 40), disarmed and with propulsion deactivated, reverted to APL rather than APB status in 2001. Unlike the converted LSTs, four of the specially built APBs (the two above plus *Benewah* and *Echols*) had substantial postwar careers, including service in Vietnam as riverine mother/headquarters ships.

Even though several of the converted LSTs ultimately were designated as APBs, they had a very different role: they were landing craft tenders or mother ships. This distinction is quite clear in wartime correspondence files. The first were LSTs 453 and 455 of the Seventh Fleet (Southwest Pacific). LST 453 was converted at Brisbane in May 1943. She became flagship of the landing craft control officer for the Southwest Pacific in September 1943, and beginning in January 1944 she towed a 400-ton floating drydock with her as she advanced from base to base. Later a 250-ton pontoon drydock was added. LST 455 was converted by USS *Rigel* after suffering battle damage. LST 453 became ARL 40 and LST 455 became ARL 41 in August 1944. They replaced ARL 25 and 34, planned conversions of which were canceled (they reverted to LST status). Each of the Brisbane ARL conversions had a pair of Quonset huts welded down on the upper deck instead of the extended superstructure typical of ships converted in the United States. They accommodated 40 officers; bunks on the tank deck accommodated 196 enlisted men. The ship's ballast tanks were converted to store fresh water, and 4 distilling plants were added, plus a bake shop and 16 refrigeration boxes. Similarly, LST 201 was converted into a motor torpedo boat tender (AGP 20, *Pontus*). She replaced AGP 19.

Similar efforts proceeded independently in the European theater. Experience at Sicily and at Salerno showed that some sort of landing craft "parent" ship was needed, and an LST was duly modified for Anzio as a supply and minor repair tender. The final after-action report, dated May 1944, described the necessary attributes of such a ship, although it is not clear if she met all the requirements: she would accommodate an LCT repair and maintenance crew and plenty of spares; she would have a crawler crane of 3-ton capacity; she would carry fuel and water (after the first few days, demands for water exceeded those for fuel by a ratio of 3 to 1); she would have manifolds and pumps to fuel and water craft rapidly; she would carry a large supply of smoke oil, which was used heavily during an assault; she would carry medical material and personnel; she would carry clothing and provide berths for survivors of landing craft sunk during the operation; she would have extra refrigerator capacity; and she would carry depth charges for the small craft. During Operation Dragoon, the invasion of Southern France, one LST was stationed as mother ship in each of the three assault areas. For this occasion the conversion was temporary. Each such mother ship (LST 74 was one) was a 6-davit elevator (not ramp) LST fitted out with 150 portable bunks, two 625 cu ft refrigerators, and one portable Cleaverbrook distilling unit. Off the Southern France beaches, LST 74 watered and fueled 225 ships and craft between D-day (15 August 1944) and 8 September 1944. She repaired 109 ships and craft, issuing 552 tons of water, 952 tons of fuel oil, and 200 drums of fog oil. On average, she had 256 men berthed on board each day, issuing rations for 307. The total number of ships and craft (excluding LCVPs) alongside was 416. The three U.S. LSTs were supplemented by four British LCIL mother ships.

The Pacific Fleet mounted a parallel effort, reportedly inspired by Capt. Stanley Leith, operations officer to Vice Adm. Richmond Kelly Turner, commander of amphibious forces in the Pacific. On 23 August 1944 Pacific Fleet's amphibious force ordered its administrative command to prepare and submit to CinCPac a plan to produce four tenders for small craft used in amphibious operations: YMS, PC, PCS, SC, LCI, and LCT. Recent operations had shown an urgent need for such units. A special board was convened, and on 1 September the administrative command proposed that four LSTs be converted, to be ready by 1 January 1945. They would be six-davit ships equipped with evaporators. Requirements included accommodation for 40 officers and 400 enlisted men, more fresh water, evaporators, refrigerated stores, lubricating oil, fog oil, a laundry, a galley, a bakery, boats, hospital spaces, and cargo handling facilities. The bow ramp and LST doors would be retained. A sketch was submitted. Conversion was likely to take about six months, so the board recommended that four old LSTs be converted on an interim basis, with Quonset huts on the upper decks and reefers on the tank decks. A more detailed plan showed that such a craft could accommodate 30 officer and 300 enlisted transients (by covering the upper deck with berthing areas), with about 115,000 gals of fresh water (deliverable at 10,000–15,000 gal per hr with evaporators for 8,000 gals per day), diesel oil (120,000 gals, deliverable at 25,000 gal per hr to fuel diesel-powered destroyer escorts), and refrigerated spaces filling most of the tank deck. A more detailed plan for the ultimate conversion, submitted by the board on 30 August showed 80,000 gals of fresh water in converted ballast tanks (with two 500 gpm diesel centrifugal pumps) and two 4,000 gal per day distillers as on an ARL. Internal stowage would be provided by extending the ship's second deck forward and building new dry storerooms and refrigerated spaces above and below it. Diesel tankage would be adequate. Two 5-ton booms would be installed, one at each corner of the main deckhouse, to plumb the existing ramp opening. At this stage the ship would have resembled an ARL, with her deckhouse extended forward but with her bridge aft. A CinCPac review, dated 20 October 1944, recommended raising the bridge a level by adding a new deckhouse below it. Ultimately ships had their bridges moved forward.

The four interim conversions were approved in September 1944; they were to be done at the Amphibious Repair Base, Waipio, Hawaii, prior to 1 January 1945 (in fact they were slightly late). These four ships (LSTs 575, 676, 677, and 678) were redesignated LST(M), initially meaning simply modified, but later "mother ship." Each was intended to service small craft, up to and including the subchasers used in amphibious operations. Two Quonset huts (20 × 48 ft) were welded to the upper deck, one for officers and the other for a bakery, and four shore-type salt water evaporators (capacity 12,000 gals per day) were fitted forward to service small craft. Large heads and showers were installed on the tank deck for transient personnel, together with 196 bunks at the after end of the tank deck. More bunks were added to the crew spaces. In the fore end of the tank deck were 16 shore-type refrigerators. Dry stores were also carried on the tank deck. Fresh water and other dry stores were carried in the hold. Two 75 kW diesel generators were fitted forward of the refrigerators on the tank deck, to power the new equipment. Two auxiliary boilers, each with a capacity of 4,000 lbs of steam per hr, were installed. Capacities on APB 44 (ex–LST 678) were 235,000 gals of fuel oil, which could be pumped at a rate of 6,000 gal per hr; 119,000 gals of water, distilled at about 20,000 gal per day; provisions (200 tons dry, 100 tons frozen, 150 tons fresh); berths for 40 officer and 300 enlisted transients and 14 bed patients. Such a ship could mess 750 at a meal, food being served continuously, and she could fuel and provision 25 ships each day. She could fuel six ships simultaneously. Two of these LST(M)s performed very well at Iwo Jima. At Iwo Jima and Okinawa APB 44 serviced a wide variety of ships, from destroyers down through LCTs and rescue tugs. She reprovisioned and refueled from APAs, AKAs, and gasoline tankers (AOG), some of which carried diesel fuel. All four ships were later redesignated APBs.

Conversion of four more sophisticated ships, initially called simply mother ships for small craft, was approved by Cominch on 4 November 1944. At the same time the three salvage tenders and four ARVs were approved. After considerable wrangling over proper designation, they were officially described as general stores issue ships and type-designated AKS 16–19; they were soon redesignated APBs. Visually, they were quite different from the *Benewah*-class APBs: they had one rather than two superstructure decks, and they had one pair of triple-Welin davits amidships (carrying LCPRs), rather than three pairs of single davits running down the side. Unlike the purpose-built APBs, they had a pair of cargo booms forward of the bridge. Both types had a prominent funnel aft. As converted, these ships displaced 1,995 tons light and 3,900 tons fully loaded, compared to 1,780 tons light and 3,640 tons fully loaded for a typical LST. Complement was 10 officers and 107 enlisted, and transient capacity was 30 officers and 292 enlisted, close to the stated requirement. Capacities were 18,300 cu ft of refrigerated stores, 13,900 of dry provisions, 14,353 of supplies, 10,950 of ships' stores, 1,428 of medical stores, 5,415 of clothing and small stores, and 2,300 of paints and inflammable liquids. Liquid capacities were 180,000 gals of diesel oil, 188,000 gals of fresh water, 11,400 gals of reserve feed water, 14,700 gals of lubricating oil, and 39,600 gals of fog oil. Distilling capacity was 12,000 gals per day. Armament was two quadruple 40-mm guns and eight 20-mm guns (six twin 20-mm were planned), compared to a more typical LST armament of seven single 40-mm and twelve 20-mm.

Two more ships, APB 49 and 50, were converted in the United States in the summer of 1945 to match the interim units (250 bunks each, however, and mess tables and benches for 206 men). They had the planned ultimate LST battery of two twin and four single 40-mm and eight twin 20-mm.

Aside from these tenders, under a 3 December 1943 policy designated LSTs carried spare LCVP engines. On 19 January 1945 the following ships were designated for this role: LST 848, 872, 873, 905, 974, 1043, 1062, 1087, 1089, 1096, 1106, 1107, 1120, and 1126.

While the small craft tenders were being designed and converted, on 28 September 1944 the commanding officer of YMS 140 proposed a tender for small minesweepers that, like salvage craft, certainly had no space for bulky spares. The Pacific Fleet mine force passed on his request, and OpNav suggested that, like other small craft, the YMS might use the new mother ships. BuShips demurred: mine craft needed something more. At this time three former army mine planters, designated ACM, were serving partly as temporary YMS tenders. This requirement explains the early, ultimately abortive, postwar attempt to modify an LSV as a minecraft tender, designated AKN.

16. A 19 December 1958 summary of the projected program showed a total of 14 LSTs: 3 in FY 60, 4 in FY 61, 4 in FY 62, and 3 in FY 63. The FY 60 program modernized LST 836 and 1148, and LST 1084 and 1141 in FY 61, LST 1073 in FY 62, and LST 1076 in FY 63. Reactivation or operations and maintenance funds paid for the modernization of other ships. Other LSTs scheduled for FRAM, and possibly modernized using other funds, were LST 762, 848, 854, 1073, 1096, 1126, and 1146.

17. In December 1941 the U.S. Navy added 8 ships to the Lend-Lease order for 7 ex-Winettes, making a total of 15. Of these only four (LSD 9–12) served with the Royal Navy. In January 1943 the VCNO rejected any further LSD orders until the first ships had been tested. However, 12 more were approved on 15 June 1943, to be delivered at the rate of 3 per quarter during FY 45 (beginning 1 July 1944). Two were canceled in August 1945: LSD 23 (*Fort Snelling*) and LSD 24 (*Hilton Head*, later *Point Defiance*). Laid up incomplete, *Fort Snelling* was completed for a commercial operator in 1956 as *Carib Queen*. When the operator went bankrupt, she was taken over by the Military Sea Transportation Service as *Taurus* (AK 273, later T-LSV 8).

18. For example, according to the after-action report on the invasion of Southern France, an operation of that size needed six LSDs; in fact, only two were available. They were needed to transport warping barges (to handle pontoon barges), pontoons, LCM(3)s, and other small craft. By this time, LSDs were needed to transport a wide variety of special craft, such as LCCs (command boats), Apex (explosive) boats, the LCM(R) "Woofus" obstacle destruction boat, smoke-producing boats, and LCVPs converted to shallow-water minesweepers. For Southern France, priority went to the LCM "smokers."

19. Design work on a crane ship conversion of *Chiwawa* (AO 68)–class tankers was completed about January 1944; *Enoree* (AO 69) and *Niobrara* (AO 72) received their cranes at Norfolk. Two other tankers, *Pasig* (AO 91) and *Abatan* (AO 92), apparently received theirs later, after the Marianas operation. They were probably fitted with cranes at the

same time as they were modified as distilling ships and water tankers (AW 3 and 4, respectively).

20. Postwar British evaluations of the LCT were more complementary. According to the British staff history of the Salerno landing, the LCT(5) was well liked because it could get over bars as well as LCMs. It could marry up to LSTs day or night, helping them unload. Its main defect was a very noisy engine. According to the British staff history of the Normandy landing, the curved ramp of the LCT(5) caused problems when unloading onto a causeway. The U.S. Navy fitted completely new ramps to its LCT(5)s, but they could not be supplied in time to the Royal Navy. According to the U.S. after-action report for Normandy, all LCTs were fitted with a "Mulock ramp" devised by Lt. Comdr. Mulock, RNVR, to make it easier to handle "awkward vehicles." It was an extension to the main ramp, to reduce the angle to the beach, and was particularly important for LCT(5)s. The extension was hinged so that it did not interfere with the line of sight from the craft's pilothouse.

21. For Operation Husky, the Sicilian invasion, Eighth Fleet made portions of LCT(5) bulwarks detachable. For the invasion of Southern France, bulwarks of 50 LCTs were made removable.

22. Four Southwest Pacific LCI(L)s supported UDT units; they were unofficially designated LCI(D). Each carried two six-man (one officer, five enlisted) naval combat demolition units (NCDUs) and 30–40 tons of explosives, mainly in the form of the hand-emplaced flexible line (hose) Mk 8 demolition charges, which remained in service as late as the Vietnam War, when they were used to blast a canal in the Mekong Delta. LCI(D) 227 was apparently unique, being rearmed with a bow 40-mm gun and two twin 0.50 caliber machine guns to supplement her 20-mm guns. She was briefly designated a gunboat. The demolition support craft were LCI(D) 29 for NCDUs 20 and 24; LCI(D) 44; LCI(D) 227 for NCDUs 19 and 21; and LCI(D) 448 for NCDUs 2 and 3. LCI(D) 44 acted as support vessel for others, and had no NCDU assigned.

23. LCI(L)s were adapted to several other roles. At Normandy, even though they were not specially fitted, several were used as combat tugs. LCI(L) 494 and 495 towed several large ammunition barges to the beach. LCI(L) 490 towed so many craft that at one point she was asked to tow a Liberty ship back to England; fortunately, the request was withdrawn. LCI(L)s were apparently first used for salvage at Salerno. They could pull LCM(3)s and LCVPs off the beach by securing at a 45 degree angle on the seaward side and running their engines ahead. Their propeller wash scoured sand from alongside and beneath the stranded craft, which could then be towed off. They were also easily adapted to fire fighting by fitting high-pressure pumps and several were so fitted. These craft were later used in the invasion of Southern France. For the invasion of Southern France, four British LCI(L)s served as mother craft for smaller amphibious craft. Nine U.S. LCI(L)s were converted for salvage operations. In December 1945 CinCPac made 12 LCI(L)s available, in advance of any BuShips approval, for urgent conversion to express and passenger vessels, to meet a 3 December 1944 request by the Pacific Fleet service force. Not only BuShips, but also OpNav had not been consulted, and a testy OpNav ordered that no future alterations be made without its express approval. OpNav had also recently discovered (from an after-action report) that Pacific Fleet had converted far more LCIs to gunboats than it had authorized. Side ramps and catheads and the after anchor and its winch were removed. Troop holds 1, 2, and 4 were converted into cargo holds 1, 2, and 3 by removing all pipe berths and lockers. Existing booby hatches and inclined ladders providing troop access from holds 1 and 2 were removed, and vertical ladders mounted to provide access from holds to forecastle deck and to the top of the deckhouse. Cargo hatches were cut in the main deck to the two forward holds. They were served by a new 3-ton boom on a new mast. The existing troop access hatchway to No. 4 troop hold (now No. 3 cargo hold) was adapted as a cargo hatch, and a portable davit and winch (capacity half a ton) was provided. The original No. 3 troop hold and troop officers' quarters became passenger accommodations. Estimated capacity, based on the BuShips limiting displacement of 387 tons (draft 5 ft 8 in), was 72 passengers and 78 tons of freight. Material removed during conversion would be retained for later reconversion. The BuShips route sheet for correspondence on this project calls it "a new 'temporary' conversion," which the bureau could not reject. It confined itself to pointing out that the planned hatches (8 × 10½ ft) would cut away too much of the ship's strength deck; no hatch should be wider than 6 ft, with well-rounded corners to minimize stress, and with deep coamings to carry the deck's strength around the cuts. Clearly BuShips doubted that any of the ships would ever be reconverted. The bureau's correspondence files make it clear that the conversions were carried out, but do not specify the ships involved, nor is the specific service involved mentioned. In addition, on 24 December 1944 the commander of Task Force 73 in the South Pacific asked that six LCIs be modified as spare parts issuing ships, on the ground that such units in forward areas would make it possible for many more combatant ships to be maintained there. The spares problem affected all LCIs and their derivatives. For example, according to the after-action report of RCM and Rocket Division Two, an LCI(R)

unit at Okinawa, "if promises were spare parts, we could rebuild every LCI in the Pacific and have enough left to stock a flotilla of ARLs [landing craft repair ships]." As it was, ships never got the required 24 hrs per month each for overhaul. They suffered badly because they were used as independent units even though they were poorly equipped to operate independently. For example, they lacked centrifuges to purify contaminated oil. Several LCI(R) at Okinawa lost all power and damaged the injectors in their engines due to water in oil taken on board from a diesel oil tanker. Similarly, LCIs had insufficient radio operators, and received little help from larger units in, for example, monitoring the fleet broadcasts (Fox schedules). They did not even receive paint routinely. The parts issuing ships were all approved, but BuShips files do not show which LCI(L)s were involved. In the Mindanao invasion in January 1945, LCI Flotilla 24 was part of a deception force, Task Group 77.11, simulating an amphibious force. Hoisting 3 × 5 ft corner reflectors on their masts, the LCI(L)s simulated transports. Accompanying PT boats with corner reflectors simulated escorting destroyers, and fired rockets to simulate naval support gunfire. The unit was built around Beach Jumper Units (BJU) 6 and 7, which reached the Pacific in October 1944; in the Mediterranean beach jumpers used 63-ft air-sea rescue boats (AVRs, which they called ARBs). In the final beach jumper campaign of the war, begun the first week of January 1945, LCI(L) 24 acted as flagship. Task Unit (TU) 77.11.1 comprised eight LCI(L) with BJU 6 aboard, plus six PT boats; TU 77.11.2 (flagship LCI[L] 702) comprised seven LCI(L) (BJU 7) and six more PT boats. Radar and radio countermeasures operators were on board LCI(L) 625 and 1001. They were equipped with APT-1 and APT-3 jammers, with ARR-1 intercept receivers, RBK S7 intercept receivers, SPA-1 pulse analyzers, RBK S36 VHF receivers, and RDK panoramic adapters. LCI(L) 702 produced notional communications traffic. This particular operation featured a dummy paratrooper attack by C-47 aircraft. In a second operation, TU 77.11.1 simulated a bombardment group, TU 77.11.2 (4 LCI[L], 8 PT boats) simulated a "hydro-smoke group," and TU 77.11.3 (12 LCI[L] with BJU 7, plus 8 PT boats) simulated a transport group. This time the LCI(L)s had 6 × 8 ft trihedral reflectors on their masts, the PT boats the earlier 3 × 5 ft reflectors. LCI(L)s 712 and 1076 flew reflecting box kites. The same LCI(L)s acted as jammers and deception transmitters. When the "transport group" reached the notional transport area, LCI(L)s 699, 1025, and 1072 released corner reflectors supported by balloons, the cables of which had skip anchors to keep them from moving. Thus enemy radars showed large numbers of "transports" in place. LCI(L) sent voice and code signals on the CinCPac circuit. The PT boats produced a smoke screen, from behind which the deception force fired its rockets, thus they could simulate 5-in shells without needing a large ship to support them. In December 1944 the British decided to convert six early LCI(L)s to smoke makers specifically to shield assaults. They replaced adapted LCM(3)s; at the invasion of Southern France the British had six LCM "smokers" supported by three LCT(1) converted as smoker tenders. Smoker LCI(L)s would carry sufficient acid and gasoline to form smoke continuously for 12 hrs. Acid was stowed in the troop spaces and gasoline in the double bottom. Acid was pumped by compressed air through pipes laid along the upper deck to nozzles in the stern. A Merlin aircraft engine was installed on the stern to disperse the discharged smoke. The first was completed at Chatham in March 1946. By this time the war was over, but the U.S. Navy allowed the British to retain the craft for trials. They were satisfactory, but in the end only two of the six craft were converted.

Chapter 6. Attack Transports and Conversions

1. Of the ships retained as combat loaders, AP 67 and 69–71 were all C3-Cargo ships. AP 76 was also a C3-Cargo conversion, but AP 77 was a C2-F. Note that APA 55/56 (initially AP 100/101) were originally sisters of AP 67 and also of AKA 10 (initially AK 27). All of these ships had only two pairs of triple Welin davits, on each side of the deckhouse; when functioning as attack transports they carried most of their boats on deck. Ships of similar types designated as APAs generally had two or three sets of Welin davits on each side (two on the superstructure, often a third forward). Designed boat outfits were 2 LCM and 24 smaller craft AP 67, while all others had 2 LCM and 20 smaller craft. In June 1945 *Thurston* (AP 77) had on board 2 LCM(3), 1 LCPL, 4 LCPR, and 18 LCVP. For the invasion of Iwo Jima, she replaced USS *Briscoe* (APA 65), which had to remain at Pearl Harbor due to engine problems. Thus *Thurston* (and AP 67 and AP 76) functioned as a full attack transport, operating alongside ships designated as attack transports. In March 1944 AP 67, 76, and 77 were all ordered modified for the Normandy landing, with extra potable water tanks (two 600-gal tanks each), barrage protection (four barrage balloons and three Sauls kites each), additional hoisting slings for equipment, and salvage equipment sufficient to equip one LCM(3) each.

2. During Operation Torch, *Susan B. Anthony* carried 2 LCM(3) plus 12 LCP(R), 12 LCV, and 1 LCPL; her designed outfit was 2 LCM and 16 smaller landing

craft. However, as of January 1943 BuShips said that adding even four 20-mm guns to the existing eight (as then planned) would jeopardize her stability unless landing craft were removed to compensate. Inclined in May 1943, she showed negative stability in the light condition. Her cargo capacity was limited to 450 tons to preserve two-compartment survivability; during Operation Torch, she carried 1,195 tons of cargo. BuShips recommended that she revert to permanent transport (AP) status, but even so she served as an attack transport at Sicily, carrying 2 LCM(3) and 18 smaller boats (LCPL, LCVP). On 31 July 1943 the VCNO ordered all APA features removed and the ship limited to duty as a convoy transport.

3. The assault force for Southern France also included, unusually, four army transports (non-combat loaders): *James Parker*, *Marine Robin*, *General G. O. Squier*, and *Santa Rosa*.

4. These ships were troop transport variants of the C3 design, built to U.S. Army account under September 1940 contracts, let before the army agreed to transfer its transports to the navy.

5. The Auxiliary Vessels Board report recommending acquisition referred to a 19 September 1942 proposal to acquire three C2s—*Dashing Wave*, *Typhoon*, *Young America*—on a bareboat charter basis. SS *Hotspur*, which became USS *La Salle*, was a sistership. The War Shipping Administration was willing only to provide them on a single-voyage basis, which the navy found unacceptable. AP 103–105 were faster and larger. Of the three proposed ships, *Dashing Wave* was allocated to the army; the other two were operated by the War Shipping Administration.

6. These ships were obtained by shifting around authorizations. First a navy request for ten Liberty hulls for freighters (AK) was approved; the Joint Military Transportation Committee then approved a navy bid to replace five of them with C-type hulls as troop transports. Because the transports were to be under the Naval Transportation Service, they did not form the sort of pool of dedicated ships to which the War Shipping Administration objected. Once the ships were in service, they were temporarily allocated to the Pacific Fleet. As of November 1943 two C2s, *Comet* and *John Land*, had already been nominated as part of this group. A second group of five met a CinCPac request for ten C-type ships (half AK, half AP), with the proviso that their conversion to naval service not interfere with the conversions of Liberty ships (AK 114–123 and 129). The War Shipping Administration member of the Joint Military Transportation Committee agreed that C-type ships could be provided in place of Liberties; the navy wanted C types for the transports but would accept Liberties as the five cargo ships.

7. The prospective CO of USS *Golden City* (AP 169) complained to OpNav that "BuShips has insisted that the vessel will not be an APA. No effort has been made to make her adaptable to amphibious operations because the PCO is the only one who believes that the assignment of 14 landing craft has any significance." Kingposts, which would have simplified landing craft handling, had been cut down (they were retained in AP 166–168). Booms that could easily have been modified to handle LCVPs (as on AP 168) were left with insufficient (4 ton) capacity. Because all the landing craft were stowed on the hatches, they could not be hoisted aboard while the ship unloaded. LCVPs had to nest in each other, further complicating boat handling. The ship had too little distilling and fresh water capacity, and no troop magazine or packaged gasoline stowage. The CO wanted a second 20,000 gal per day distilling plant and a third turbo-generator, the latter so that all of his booms could operate simultaneously to unload the ship. Even accommodation was insufficient.

8. Ships were generally designed to carry two LCM and four LCVP, but they were altered in service. Thus at Mindoro in February 1945, *John Land* (AP 167) and *Herald of the Morning* (AP 173) each carried four LCM and six LCVP. By April 1944 *Winged Arrow* (AP 170) was arranged to stow four LCM and ten LCVP, although she would only carry two LCM and six LCVP. By the end of the war she had her full complement on board, however. At Okinawa *Starlight* (AP 175) carried four LCM(3), nine LCVP, and one LCPR. *Cape Johnson* (AP 172), a C1-B, was initially fitted for two LCM and four LCVP, but her CO asked for more boats. A BuShips route sheet asked "is this an AP or APA?" The answer was that she was an AP, but that she would carry four LCM and ten LCVP, similar to AP 171. She fought at Iwo Jima, all her boats being launched from booms. Initial plans in December 1943 for the C1-B conversion described the ship as an XAK, modified for quick conversion to a convoy transport to carry 70 officers and 1,350 enlisted men (roughly an APA load), with the usual APA 20,000 gal per day evaporator for water. She would stow two LCM(3) and four LCVP on deck or on hatches, handled by one 30-ton boom on No. 2 hatch and two 10-ton booms on No. 4 hatch.

9. The first 11 conversions were ordered in October 1943, roughly coincident with the decision to make *La Salle* and 10 other transports into quasi-APAs. Ships were fitted with 1,000 temporary berths, 50 of them for officers, and with stowage and booms to service 2 LCM and 6 LCVP. No distilling plants were added, but for the short hauls envisaged freshwater tankage sufficed. Ultimately 18 ships

were converted: AK 76, 94, 105, 108, 114–119, 121, 123, 127, 129, 221–224. They could not economically be brought to AP standards, but they had to be kept in service due to the chronic shortage of true transports, which were used as APAs. One Liberty ship, *Prince Georges* (AP 165), was converted into a full-fledged transport, but she was so unsatisfactory in that role that she and three other Liberty APs were reclassified as freighters (AK); *Prince Georges* became AK 224.

10. *Oglethorpe* (AKA 100), *Skagit* (AKA 105), and *Washburn* (AKA 108) were modernized under the FRAM program in the early 1960s.

11. *Hercules* (AK 41, ex–SS *Exporter*) was a C3-E completed in September 1939 and taken over in July 1941. She was a sister ship to *Exemplar*, which became the XAP *Dorothea L. Dix*. *Mercury* (AK 42, ex–SS *Mormactern*), one of the first C2-S-B1s, was completed in September 1939 (as SS *Lightning*) and taken over in June 1941. *Jupiter* (AK 43, ex–SS *Santa Catalina*) was a sister to *Mercury*, completed in November 1939 (as SS *Flying Cloud*) and taken over in June 1941. By June 1944 *Hercules* was being used as an AKA, serving at Saipan, Peleliu, and Iwo Jima, with eight LCVP assigned to her. At the end of the war she was being converted into an ammunition ship, although not redesignated as such. Despite being described as one of the best naval cargo ships, earmarked for retention, after the war she was returned to her civilian owners. *Mercury* was passed to the 5th Amphibious Force in January 1944 in lieu of *Jupiter*. Beside normal cargo runs, she was part of the invasion forces at Kwajalein in February 1944 and at Leyte and Lingayen Gulf in the Philippines. She was then modified for underway replenishment, and ultimately redesignated a stores issue ship (AKS 20). *Jupiter* entered combat at Angaur in September 1944, and fought in the Philippines invasion. Despite operating as an AKA, she had serious inadequacies. Troop living space on board was quite limited (10 officers and 110 enlisted men, including mess cooks), so she could carry either a cargo-handling platoon or drivers for trucks on board, but not both; at Lingayen Gulf she carried drivers. Part of the ship's platoon had to be placed on board another ship, an APA or an AP. Without diesel oil tanks, she had to carry fuel for any landing craft assigned to her in 55-gal drums on deck and in the No. 5 hold. At Lingayen Gulf in January 1945, the ship had 4 army LCM, 1 DUKW, and 12 LCVP on board. Because she was not nominally an AKA, all were assigned to other ships when the force unloaded. The LCMs and the DUKW never came back. The LCVPs proved useless except during daylight one day. After the battle, her allowance was 5 LCM(3) and 13 LCVP, which she carried at Iwo Jima. Because the ship had never been modified to carry boats, all had to ride atop hatch covers. With a rough sea and a moderate swell, it was difficult to launch, load, and hoist boats by boom. Once the ship began to unload, they had to remain in the water. They suffered structurally due to frequent hoisting and lowering. As specially converted AKAs appeared, the need for such ships declined, and by July 1945 *Jupiter* was being converted to an aviation supply ship, AVS 8.

12. According to the British report on Normandy, the LST's ability to dry out on a beach, discharging cargo dryshod, ensured the success of the operation. Prior to the operation drying out had been considered but rejected on technical grounds. LST commanders were ordered specifically not to allow their ships to dry out "except in case of dire operational necessity."

13. Except for LSM 231, 232, and 306–309, LSM 1–336 were ordered as LCT(7) 1501–1830, within the LCT series. Ships were all completed as LSMs, and 1944 orders certainly described them as LSMs rather than as LCT(7)s. The precise date of the changeover is not clear.

Chapter 7. Increasing Demand for Landing Craft

1. The section was further broken down into teams. An assault section comprised a five-man rifle team, a four-man wire-cutting team (with bangalore torpedoes), a four-man BAR (automatic rifle) team, a four-man bazooka team, a four-man mortar team, a two-man flamethrower team, and a five-man demolition team using satchel charges. The support section comprised a rifle team, a six-man machine gun team, a wire-cutting team, an eight-man 81-mm mortar team, and a demolition team. Each boat was commanded by an officer with a NCO. Boat teams were heavily loaded with weapons, which it was hoped could counter German fortifications including flamethrowers, mortars, and bazookas. However, bazookas and flamethrowers were in short supply, and some boats lacked them. Many senior army officers objected to this reorganization, which broke unit integrity. One argument favoring the team concept was that each team was so balanced that men from each boat could fight as a unit. Once past the fortifications, the boat teams would reorganize into platoons (two rifle squads and a weapons squad) as a step toward re-creating the original company organization.

2. LCVPs used as sweepers had a special Size 5 sweep gear (Size 5 Type O auxiliary minesweeping) to sweep moored or snag-line mines in water as shallow as 4½ ft, at speeds up to 7 kts. The LCVP was

adapted by placing the Otter and float on the port side and the hand reels in a former gun tub on the starboard side. LCVPs were used as assault sweepers both during World War II and postwar, for example, in Korea in 1950. During World War II, LCPRs were also used as shallow-water sweepers.

3. According to the U.S. Eighth Fleet after-action report for the invasion of Southern France, that operation showed that a boat similar to the British LCA would be desirable for a stealthy landing on a rocky shore. The reference was to a surprise landing in advance of the main force. The British official history of the Southern France landing notes that the LCA proved easier to maneuver. Better silencing had always been a goal in British landing craft design, the hope being that a force could land stealthily (the U.S. view was that beaches would be so limited that stealth would be pointless).

4. The OSS-developed "surfboard" was demonstrated to the U.S. Navy staff developing UDT capabilities at their training school at Fort Pierce, Florida, on 11 September 1943. OSS contracted with the Ohio Rubber Company of Willoughby, Ohio for production models. The device, which could carry two men and their equipment (total 900 lbs), was a rubber boat in the form of a surfboard, with a round bow and a tapered after section, 10 ft 6 in long and 3 ft 7 in wide, weighing about 310 lbs. It came with a compressed air cylinder to inflate it. Power was provided by a ¾ HP battery-powered motor, driving the craft at 5 kts for 10 nm. Unlike those riding a rubber boat, the riders lay atop the "board." The OSS ordered 200 on 11 January 1944, and another 400 on 7 March 1944; unit cost was $1,200. The OSS figures reflect direct procurement (OSS purchase orders), but at least some of those bought by the navy were probably for OSS use. A 23 December 1943 Cominch letter referred to "repeated and urgent requirements" for 100 such craft from "the theaters," i.e., from deployed commands. Navy records show three shipped on 7 January 1944 to FRAY-82 (Eighth Fleet, based in the Mediterranean) and another three shipped 11 January 1944. At that time plans called for increasing production from one per day (12–16 January) to two per day (17–23 January), then three per day (24–30 January), and finally to four per day beginning 31 January 1944. Meanwhile, CNO requests for 50, then 60, and then another 50 (made 2, 17, and 26 January 1944, respectively) were approved, and on 21 March 1944 the first 50 were ordered delivered to the commander of the Eighth Fleet. Another 50 were approved on 12 April 1944, another 6 on 25 April, and another 120 on 17 July 1944, making a total of at least 336 approved during 1944. On 13 February 1945 the secretary of the navy approved a request for another 200 surfboards, specifically for UDTs, and another 600 were approved on 13 April 1945. In UDT service, one man operated the craft, while the other rode it or observed. UDTs were bought for the UDT team leaders. Each "mattress" would carry an enlisted men armed with a carbine, a radio, and the officer directing a team. No other platform would have afforded the officer radio contact with supporting units. The UDTs found that the mattresses formed excellent targets; at Saipan, the first time they were used, four were hit, and one of the enlisted operators was killed. UDT accounts of Saipan indicate disillusionment with the "mattress," but procurement continued until the end of the war.

5. The LCM(6) designation was applied as early as December 1943, so it was probably intended to indicate a "super" or "double" LCM(3). The LCM(4) and LCM(5) designations were apparently not applied, at least at the time. They do not appear in the standard handbook of Allied Landing Craft (ONI 226) issued by the Office of Naval Intelligence in April 1944, which does list LCM(6). The 1945 supplement to ONI 226 listed several Australian equivalents to the LCM but does not associate any of them with LCM designations. According to the British official history of wartime amphibious ships and craft, the LCM(4) and LCM(5) designations were retroactively applied to some of the Australian craft, but the 1945 ONI supplement does not identify them as such; it merely says that Australian craft do not fit the U.S.-British joint designation system.

6. The U.S. Army tried to build another 76-mm amphibious vehicle, based on its M18 "Hellcat." The Army Ordnance Department asked for this vehicle late in 1943, and the preliminary study was done by the New York naval architectural firm of Sparkman and Stevens. The project having proved feasible, detailed design began in February 1944 and construction of the new vehicle, designated T86, was approved in September 1944. Two were built, one propelled by tracks in water, the other by propellers. A third (T87) had a 105-mm gun. The vehicle gained sufficient buoyancy by having a greatly expanded hull. The M18 turret was retained. No version proved nearly as fast in the water as had been expected, and at the end of the war the army was trying LVT-type grousers welded to their tracks. Work ended with the war.

Chapter 8. Fire Support

1. The U.S. Navy gunfire support unit for Omaha Beach was organized in September 1943 and sailed for Scotland in November, commanded by Capt. L.

S. Sabin. It comprised 1 LCH, 7 LCR, 5 LCG, 28 LCPL, 9 LCT(A), and 10 LCT(HE). Sabin was also responsible for the Utah Beach support units. One Utah Beach support unit, which was part of the escort group, was commanded by Lt. Comdr. L. H. Hart. It comprised LCH 209, 4 LCG, 13 LCT, 12 LCS, and 20 LCPL. In addition, Unit 7 of the Utah Beach minesweeper group, under Lt. (j.g.) Irving Kramer, comprised 8 LCT, 4 LCG, and 4 LCF. In these lists, taken from Rear Adm. Samuel Eliot Morrison's *History of U.S. Naval Operations in World War II*, Vol. XI, *The Invasion of France and Germany* (Boston: Little, Brown, 1957), LCH are LCI(L) converted to headquarters (H) craft. The LCPL were "smokers." Adrian R. Lewis, *Omaha Beach: A Flawed Victory* (Chapel Hill: University of North Carolina Press, 2001), lists the Omaha Beach force as 9 LCR, 7 LCF, 5 LCG, 18 LCT(A), and 32 LCPL and the Utah Beach force as 5 LCR, 4 LCF, 4 LCG, 8 LCT(A), and 16 LCPL. Because many of these craft were hurriedly produced, there was little or no time for training. For example, two of the rocket craft crews had never handled their craft except on passage from base, and had never fired rockets except at the Assault Gunnery School. Sabin considered that only the LCF crews had been adequately trained, but that these craft were useless (there was no air opposition at Normandy). LCG and LCPL training was fair. The training of other crews was poor; some of the LCT(HE) beached for the first time at Normandy. Station-keeping on the passage to the beaches was poor, partly because some of the larger craft were towing LCMs to be used by the demolition teams. Sabin considered rocket craft the most useful of those he had. LCGs were useful, but their fire control systems were far too crude. The smokers were not used.

2. The early message to BuShips describing these ships referred to them as anti–landing craft gunboat, but they were first used in action as shore bombardment gunboats, in the invasion of the Treasury Islands. Service Squadron South Pacific reported the conversion to BuShips in October 1943, enclosing a drawing that has apparently been lost. It also provided weight data, dated 11 October 1943, showing a displacement of about 351 tons in gunboat configuration, and the loss of about half a foot in metacentric height.

3. These craft are described on page 45 of the ONI supplement, as reprinted in U.S. Navy Office of Naval Intelligence, *Allied Landing Craft of World War II* (Annapolis, Md.: Naval Institute Press, 1985). Early LCI(G) conversions are described in the non-paginated main manual, with some hull numbers. The type designations seem to correspond to the arrival of information in the United States, because the South Pacific version, which was done first, is Type E (the early rocket-firing version is Type F). Types A through C were described in a 24 March 1944 memo to OpNav, based on information from the Fifth Fleet (Central Pacific) amphibious force and the Pacific amphibious training command.

4. These craft were covered by a 17 August 1944 request to the Pacific Fleet service force for 18 LCI(G) with 30 more Mk 7 launchers on sponsons (total of 40 per ship), with rockets stowed in new magazines in troop spaces 1 and 4 (total 960 rockets). Ramps would be removed. This was a priority project, to be completed by 10 September 1944. On 19 August the desired battery was increased to 42 Mk 7 launchers (1,056 rockets), to be installed on LCI(G) 365, a Type B unit then in the yard. Of the Type B boats, LCI(G) 365, 366, 437, 439, 340, and 442 were converted. Of the Type D series, 24 units were converted by early February 1945: LCI(G) 407, 422, 450, 464–467, 469, 471–475, 558–561, 565–568, 580, 751, 752.

5. The first four units, LCI(L) 739–742, were converted at Pearl Harbor in August 1944, becoming LCI Group 40; LCI(L) 739 was the flagship. They retained troop facilities, and had an army detachment (22 personnel) on board to fire the army mortars. These units retained their original five 20-mm guns. LCI(L) 739 retained her troop ramps until 5 January 1945, and she did not take a navy mortar team on board until 6 February. On 18 October 1944 CinCPac requested 30 west coast conversions of side-ramp LCI(L)s, to be armed with three mortars and one 40-mm gun, the latter replacing the bow 20-mm. Ammunition amounted to 1,500 rounds of 40-mm in No. 1 hold and 1,200 rounds of mortar ammunition in No. 2 hold. Cominch approved this order on 28 October. Boats converted in the San Pedro area were LCI(L) 351–356, 630–633, 638, and 1010–1012. The other conversions in this program were LCI(L) 756, 757, 760, 801–810, 1023, 1088, and 1089, of which 801–810, 1088, and 1089 were ordered converted in December 1944, to be ready by 5 January 1945. A second series retained their bow 20-mm guns; as of November 1944 they amounted to LCI(L) 658–660, 754, 1056–1059, and the original 739–742. By 4 February 1945 additional units had been converted: LCI(L) 755, 974, 975, and 1055. A simpler conversion, in which mortar ammunition was simply stacked 4 ft high in No. 2 hold, applied to LCI(L) 550, 731, 732, and 734. This was a temporary conversion, and ships so fitted were not redesignated LCI(M). Planned further conversions, LCI(L) 730, 1032, and 1038, apparently were not carried out. In all these ships, the mortars were fixed on

deck. BuOrd developed the 4.2-in mortar Mk 1, on a 3-in gun mount, which could be trained and elevated. On 20 April 1945 the Pacific Fleet service force announced plans to convert LCI(L) 570, 582, 588, 594–596, 664, 669, 673, 674, 951, and 952. Each would have a 40-mm gun forward and two of the new mortars in the waist. In these ships the bow was cut down to allow the mortars to fire forward at reduced elevation. In at least one ship, LCI(M) 596, converted at San Diego in March–April 1945, the bow was cut down and the crew added a flying bridge. Of this group, LCI(L) 570 apparently was not converted, or at least completed, by the end of the war. Other units were converted in forward areas. For example, LCI(L) 359, 362, and 431 were converted in Mios Woendi, near Biak, in November–December 1944. LCI(L) 670 was also apparently converted in a forward area. LCI(L) 755 was lost under an LCI(G) designation in March 1945, the LCI(M) designation not yet being official.

6. This tally is taken from wartime *BuOrd Armament Summaries;* it tallies with the hull numbers given in the 1945 edition of *Ships Data U.S. Naval Vessels* (Vol. II, pp. 214–15). Of the units listed, only the following had *not* previously been designated LCI(G): LCI(M) 359, 362, 431, 582, 588, 594–596, 664, 669, 670, 673, 674, 951, 952. The ONI supplement claims 42 conversions, but does not list them by hull number; presumably it is based on information current as of the spring of 1945, and omits new conversions that had never been designated LCI(G).

7. The first of these rocket ships, LCI(L) 31, was reportedly converted on the personal orders of Adm. Daniel E. Barbey, Seventh Fleet amphibious force commander. He felt that a bombardment beginning ten minutes before troops were to come ashore would be quite effective. Rocket launchers were mounted on her troop gangways, and the 20-mm gun in the bows replaced by a 40-mm Bofors. LCI(L) 31 first fired her rockets at Saidor and Hollandia. Two more LCI(L)s, including LCI(L) 34, were then converted, and the three were first used at Cape Gloucester on 26 December 1943.

8. The division, Task Group 52.20, was organized as six divisions (divisional flagships are starred): TG 52.20.1, LCI(R) 642,* 650, 707, 771, 772, 1077; TG 52.20.2, LCI(R) 651,* 708, 1028, 1029, 1030, 1078; TG 52.20.3, LCI(R) 643, 644, 645, 646,* 769, 770; TG 52.20.4, LCI(R) 647, 648,* 649, 762, 763, 764; TG 52.20.5, LCI(R) 785, 1024, 1026, 1068,* 1069, 1070; TG 52.20.6, LCI(R) 704, 705, 706, 765,* 766, 767. It is not clear whether all these craft had RCM gear on board. On 9 December 1944 RCM gear was ordered installed on board LCI(L) 642, 651, 707, 708, 771, 772, 1029, 1030, and 1077.

9. There were two separate types of LCT(A). According to the British DNC history of World War II ship development, LCT(A) trials were undertaken in August 1943, and 69 were produced. In this account, LCT(A) carried a pair of Centaur tanks mounting 95-mm guns, with their engines removed; it was a permanent if improvised gunboat, with 2-in side and 1-in deck armor over engine room, fuel tanks, and wheelhouse. Craft were not successful, and they were converted back into LCT(5)s after the Normandy landing. However, the units described as LCT(A) in the U.S. (western) sector were in fact LCT(5) conversions carrying three Shermans at a time, not the British version. The U.S. LCT(A)s were intended to cover combat engineers and naval demolition units trying to destroy German beach obstacles. At one point in planning it was proposed that the LCT(A)s simply ram their way through the obstacles, but that idea was abandoned because the Germans had placed mines on the seaward side of the obstacles. Any LCT(A) ramming them would have been destroyed, and its hulk would have blocked further access to the beach. Although the official British report lists only 16 LCT(A) in the U.S. sectors at Normandy (8 on each beach), a recent account by Lewis, in *Omaha Beach,* based on U.S. Army records, lists a total of 26: 18 on Omaha Beach and 8 on Utah Beach. Lewis does not give the number of LCTs carrying self-propelled guns.

10. Of a total of 24 "Woofus" boats, 21 were used in the invasion of Southern France. They performed extremely well, firing at H-2 minutes. According to the Eighth Fleet after-action report, the only desirable improvement would have been 100 yds more range. In the invasion, Woofus shots were preceded by LCS(S) firing at H-5 minutes, following their initial barrage of explosive rockets with smoke, which provided a very effective cloak for the assault. Naval demolition units following up the Woofus boats had their own LCVPs. In tests early in 1944, single Woofus rockets detonated German Teller mines over a 13–14 ft radius. The rockets destroyed barbed wire over a similar radius. Besides Woofus, several special obstacle-breaching devices were developed. They were demonstrated before the Joint Army-Navy Engineer Task (JANET) board, sponsored by the Supreme Commander Allied Expeditionary Force (SHAEF) Engineer Division, at Fort Pierce, Florida, on 8–12 February 1944, but were not used at Normandy. There, obstacles on the U.S. beaches were all attacked by hand, by demolition personnel landed from 16 LCMs. The British, however, used small landing craft (LCAs called LCA(HR) or Hedgerows), roughly equivalent to the U.S. Woofus, to blast paths through wire and antipersonnel

mines, followed up by special AVRE (armored vehicle Royal Engineers) combat engineer tanks landed from their LCTs, to blast a path on the beach. Nine were assigned to each British (not U.S.) assault force. They were towed from England, and some foundered or had to be cut adrift. In Southern France, the new U.S. technology was used. Besides Woofus, the force used explosive-laden remote-controlled Apex boats (converted LCVPs) and a line charge called Reddy Fox (Demolition Outfit Mk 119, also known as Demolition Charge Mk 13 or, to the army, Aqueous Snake). Each Apex boat carried 8,000 lbs of high explosive. Apex was tested in the Bay of Salerno prior to the invasion. It was clear that, properly placed, it could indeed clear obstacles. However, even when tracked visually, it was very difficult to place correctly. Guidance problems led to the development of a parasite technique, in which one such boat could tow two or four unguided LCVPs, with their engines running slightly slower, so that they lagged slightly; all were connected by fuse cord so that they would explode together. The resulting wide breach would include the desired sector of beach. This proved too complex and had to be abandoned. For the Southern France operation there were 54 Apex boats and 27 control boats; the Apex boats were carried on LST davits or in LSD well decks. Some proved effective, but one turned 180 degrees while out of sight of its control boat. Estimating that it had reached the beach, the control boat detonated it—unfortunately, 15–20 ft from SC 1029, which was acting as a beach control boat. The subchaser was badly damaged. Some of the Apex boats also failed to explode on reaching the beach. Apex was also tried, unsuccessfully, at Kwajalein (Roi-Namur) in January 1944; control was lost, and the boat blew up. Apex was carried to, but not used at, Saipan. Reddy Fox, also used in Southern France, was a buoyant line charge, 100 ft long in 50 ft units, normally deployed from an LCM. It carried 25–45 lbs of explosive per foot. Normally it was towed to an obstacle. An electric circuit cut the canvas float surrounding the charge and also set a 5-minute delay fuse. The resulting crater was about 30 ft wide and 4 ft deep. Parallel to a beach, it could blast a gap about 120 ft wide in a single line of obstacles. Perpendicular to the beach line, Reddy Fox could blast a gap 60–75 ft deep. By February 1944 there was interest in using a torpedo to propel Reddy Fox into place, but those used in Southern France were hand-emplaced. Reddy Fox remained in service postwar, at least through the late 1940s.

Other devices shown at the February demonstrations were the Bucket Boat, Porcupine, and Bobsled. The Bucket Boat was an LCM(3) loaded with 50,000 lbs of explosives. Unlike Apex, it was driven to a designated point, then scuttled to explode, a time fuse being set by the coxswain. Porcupine was an LCM(3) with 32 rockets, all mounted outboard of the well at elevation angles of 21 degrees 20 minutes to 43 degrees 40 minutes (stern). It could fire two ranging shots, then three salvoes (10, 10, and 8 rockets), then a final two rockets to mark the area hit. The attack would be mounted from about 250 yds offshore. Bobsled was a set of 42 launchers in the well of a DUKW, all at 45 degree elevation. Two Bobsleds would fire together from 300 yds offshore. Neither the Bucket Boat, Porcupine, nor Bobsled was used in action. At about the same time BuOrd had several projects of its own, including Hellion, using the torpedo-size (but unpowered) Mk 12 demolition charge projected from two or three Mk 18 or Mk 19 torpedo tubes in the well of an LCM. The mechanical and standoff devices were largely abandoned after the Southern France invasion in favor of charges placed manually by swimmers. As a consequence, the underwater demolition team (UDT) organization grew enormously as the invasion of Japan approached. This growth in turn may account for many of the numerous APDs converted from destroyer escorts; others supported beach reconnaissance and, presumably, beach jumper (deception) units.

11. Units assigned to Charleston, and subcontracted to Texas Gulf yards, included LCI(L) 18, 36, 37, 39, 40, 45, 76, 189, 191, 194, 530, 953, and 954. Flotilla 8 included LCI(L) 543–546, 548, 597, 661–663, 744, 745, and 753. The training LCI(L)s were 428, 534, 573, 574, and 576; a sixth, LCI(L) 570, was scheduled for conversion into a mortar boat.

12. As of 16 August 1945, LCI 539 and 540 were 65 percent complete; LCI 506, 541, and 556 were 55 percent complete. Cominch was interested in reconverting them to LCI(L)s so that they could carry troops along the U.S. coast, because there was difficulty with other modes of transportation; one such craft could carry 150 troops. Reconversion would cost about $125,000. Although nothing seems to have been done at Jacksonville, the 42 LCI(G) and 45 LCI(M) at Pearl Harbor were considered for quick conversions to troop carriers as part of the larger Magic Carpet program. The maximum number of bunks, four high, would be installed in Nos. 1, 2, and 3 holds; all rockets, mortars, and 40-mm guns would be removed.

Chapter 9. Command and Control

1. According to the official British report of the Sicily landing, fitting out the four U.S. command

ships was a constant source of anxiety. The attack transport *Monrovia* was chosen as force flagship. She was to have been fitted out in the United States, but arrived in North Africa, the jumping-off place for the invasion, unequipped. She was fitted out sufficiently in two weeks by local units. The British described her as the greatest disappointment, the other ships being either satisfactory or quite successful. *Ancon*, serving as flagship of the central attack force, was considered practically ideal. The attack transport *Samuel Chase,* flagship of "Dime" force, was fitted out at Oran under the orders of the commander of the amphibious squadron there; she was described as reasonably satisfactory. The locally converted small seaplane tender *Biscayne,* flagship of "Joss" force, was described as never very suitable, but she was successful.

2. LSTs 13, 216, and 217 became FDTs with the same numbers. From a design point of view, conversion presented two major problems, which must have afflicted extemporized GCI LSTs. First, an LST, particularly an unladen one, was far too lively in a seaway; its motion had to be reduced. That meant drastically reducing its stability. Second, at least in the British case, the air search (GCI) radar antenna could not be more than 35 ft above the surface of the water, presumably to limit multipath reflection off the water. The two conflicted, because to increase draft would be to load the hull, which in turn would make the ship stiffer. The British solution was twofold. First, the ship's center of gravity was raised by placing 720 tons of pig iron on the upper (tank) deck. This ballast also helped protect the radar and control spaces installed on the tank deck. Second, the longitudinal bulkheads, the ballast tanks in the sides of the ship, and the dry compartments amidships were pierced and all tanks kept only partly full to create free surfaces that would further reduce the ship's stability (i.e., her stiffness). Some tanks had to be filled, at least partly, to keep the aerials close enough to the water. Stability reduction had to be finely calculated, because it was vital that the ship still be able to float with three compartments open to the sea. In fact FDT 216 sank after being torpedoed, the shifting of her ballast contributing to her loss. Bow doors were sealed. GCI involved not only antennas and radios but, equally important, the operations room (equivalent to a CIC) where radar data were translated into a plot and the latter used to direct fighters. On the tank (third) deck a 118 × 60 ft space was built to house the filter room (the plot), a plot radar display room (from which operators phoned information to the plotters), a fighter direction room, a communications room, a radio receiving room, an ELINT (Y-service) room, a coding room, and a small workshop. Protection included 1.5-in armor on the main deck (about 70 × 30 ft). Also on the tank deck were VHF direction finding spaces, a transmitting room, a radar countermeasures space, and three generator rooms (for four generators in all). On the upper deck was the GCI office, and on the bridge was the beacon office. Four extra masts were fitted. GCI beacon and fighter direction spaces were built of 0.75-in D1HT steel, presumably equivalent to U.S. STS. Standard ground-based radars were installed permanently on the upper decks. To operate the fighter direction facilities, the ships carried 18 RAF officers and 150 RAF enlisted men. Armament was six twin 20-mm, four FAM, and one 12-pounder. The first conversion took ten weeks, the other two being carried out (by the same yard) within three months. The British seem to have conducted their first GCI trials with LST 301 early in 1943, off Portland. LST 407 carried a GCI unit in the Sicilian invasion, and LSTs 305 and 430 were used for GCI at Anzio. Besides the converted LSTs, the British also used merchant ships as FDTs.

3. BuShips observed that in July 1943 the Auxiliary Vessels Board had called for construction of three more *Currituck* (AV 7)–class tenders, two of which would be completed in 1945, and one in 1946: AV 18–20. Four such ships were already under construction for completion in 1944–45, and ten other seaplane tenders were either in service or under construction. Yet seaplanes were being replaced by land-based naval aircraft, which the tenders could not easily support. More AGCs were clearly needed, but shipbuilding manpower was in short supply. A quick study of converting AV 11–13, already under construction, to AGCs suggested that they were unusually well adapted to this role, given their high speed and inherent survivability. They were considered much better than converted merchant ships, due to excellent subdivision and to the use of two widely separated machinery plants. The AVs already accommodated the considerable staff required for seaplane operations. The big hangar offered just the sort of volume the AGC role demanded. A parallel study suggested that a seaplane tender would make an excellent tender for destroyers or smaller craft. Ultimately only one ship of the projected group, *Townsend* (AV 18), was ordered, but she was not built.

4. In addition to the ships converted into AGCs, the presidential yacht *Williamsburg* (AG 369) was redesignated AGC 369 in November 1945. She had no amphibious function.

5. A 17 June 1943 meeting of the Landing Force Development Board decided that LCCs should have

SO-series radars, larger generators, and QBGs modified so that they could be installed below decks in the navigator's space. It also decided that there should be no change in armament or in the fitting of smoke generators to accommodate this equipment.

6. Reference vessels (PC or SC) were introduced for the invasion of Sicily. They lay as close as one mile from the beach. In the North African landings, scout boats approaching the beach had to navigate by stars, in some cases with the assistance of beacon submarines near the beach. One lesson learned in Sicily was that the PC, though valuable as a reference vessel, had too large a silhouette. PT boats were used in the Southern France landing.

7. According to the after-action report, the LCC was too wet and too uncomfortable. It seemed impossible to work in the below-decks control room. The boats were criticized as too cramped, and they had inadequate electrical power. It would be better to transfer the equipment into a larger seagoing hull. At Normandy LCCs were used as scout boats or survey boats rather than as navigational leaders.

8. In 1969 most amphibious ships previously listed as auxiliaries (A) were redesignated in the L category; thus APAs became LPAs. However, the APD could not become the LPD because there was already an LPD (amphibious transport [L], personnel [P], dock [D]). Because the APD was used for raiding and reconnaissance, it became an LPR (amphibious transport [L], personnel [P], raiding and reconnaissance [R]).

9. The British took the flagship idea more seriously at this time. In July 1943 their chief of combined operations, Lord Louis Mountbatten, proposed that several LCI(L), including LCI(L) 98 and 168, be converted into headquarters ships, which were designated LCH. They received foremasts and a new aerial yard on their mainmasts; accommodations for 6 officers and 27 enlisted men were added; a generator was added; and a store room was added. They were fitted with echo sounders and bottom logs for accurate navigation. The first conversion was completed at Chatham Dockyard in October 1943, and a total of 16 were converted. Their perceived value is indicated by British plans to retain eight in their postwar assault fleet, although all were returned to the United States. A later version was the administrative flagship, which the British designated an LCQ. It was an LCI(L) 351–class hull modified to accommodate landing craft squadron commanders and flotilla staffs. This craft was broadly equivalent to the U.S. LCFF. Additional accommodation was provided for 14 officers, 8 petty officers, and 48 enlisted men, plus squadron and flotilla offices, and a sick bay (to service the squadron or flotilla). The first was completed in October 1944; 21 were ultimately converted, but the British never planned to retain any in their postwar fleet.

10. The list of units converted comes from the official *Ships' Data U.S. Naval Vessels* produced by BuShips (1945 and 1953 editions). It does not agree, however, with hull lists derived from BuShips correspondence files. The *Ships' Data* lists add LCFF 503, 504, 531–533, 535, 536, 569, 571, 572, 575, 627, 628, 679, 988, and 1079, a total of 15 units, to those listed in correspondence files. They omit LCFF 582, 588, 594, 629, 666, 672, 951, 952, 985, 993, and 997, a total of 11 units, that are mentioned in wartime BuShips files. It is not clear whether the official publications were misprints. The wartime files do not indicate any substitutions for the units listed. The list of the first units converted comes from correspondence files. According to the correspondence files, the first 13 units converted on the mainland, all slightly delayed, were LCI(L) 367–370, 399, 423–427, and 484–486. After the first wave of conversions, LCI(L)s returning from Europe were ordered converted on 2 December 1944. They were one flotilla flagship (LCI 666), three group flagships (LCI 582, 594, 672), and three division flagships (LCI 588, 951, 952). On 18 December 1944 Cominch approved a November CinCPac request that 18 side-ramp units (divisions) be converted on the mainland: LCI 656, 657, 775, 1031, and 1080 for Division 134; LCI 782, 783, 786, 788, 1081, and 1082 for Division 135; LCI 789, 790, 791, 792, 793, and 1083 for Division 136. This exhausted available side-ramp units. Another 10 units of the LCI(L) 1–350 class were to have been converted during the summer of 1945, but that was not done.

Chapter 10. Perfecting the Technique

1. At least a few LCC survived; two were among the initial equipment of the revived postwar Atlantic Fleet beach jumper unit (BJU-2).

2. The wartime designation system differentiated between armored and unarmored LVTs. The highest-numbered unarmored (i.e., personnel-carrying) version was LVT(4). The postwar personnel carrier naturally became LVTP5. The "P" for personnel was added to distinguish this vehicle from other versions. However, the howitzer carrier was not designated LVTH5 because the turreted versions were LVT(A)4 and LVT(A)5; thus the howitzer version became LVTH6. This system was then further corrupted, the tacit assumption being that LVTH6 was a follow-on to LVTP5. Hence the next amphibian was LVTP7, which was later designated AAV7.

Chapter 11. The Fast Submarine and the Fast Amphibious Force

1. Bright's proposal influenced the SCB 77A project, evidenced by the fact that photostats of his remarks and his proposal survive in the BuShips SCB 77A design history.

2. Like its predecessors, the *Newport* class required massive ballast tankage to bring it to seagoing displacement. LSD tankage could not be similarly used because reserving it for valued liquids would preclude the use of the wet well for small craft.

3. Six ships were modernized in the early 1960s under the FRAM II program: APD 60, 89, 90, 119, 123, 130, and 135. APD 60, 89, 107, and 135 were activated under the FY 62 program for the Berlin crisis. APD 90 was recommissioned on 15 January 1965 to replace APD 107. APD 119 was recommissioned on 17 March 1967 to replace APD 60. A FRAM II refit planned for APD 107 apparently was not carried out. Modernization details varied. Thus *Beverly W. Reid* (APD 119) had the enlarged CIC, but a pole mast. *Ruchamkin* (APD 89) and *Weiss* (APD 135) had tripod masts.

Chapter 12. The Bomb and Vertical Envelopment

1. In the vocabulary of U.S. Navy–type abbreviations, V stood for fixed-wing aircraft, as in CV, an aircraft carrier. Thus adopting the designator LPV instead of LPH meant replacing the helicopters of the LPH with fixed-wing aircraft.

2. The *American Challenger* (C4-S-57A) class figured in several amphibious projects of this period. For example, it was proposed as the basis of a command ship (the LPH hull was used instead). It was part of a major Maritime Administration initiative to revive the U.S. merchant fleet in the 1960s by funding 46 replacement ships. All these ships were designed for high speed. Each had three engine ratings: normal, ABS (American Bureau of Shipping) maximum, and national defense, the last two not being at full load displacement. This class was designed by Gibbs & Cox for U.S. Lines, which wanted fast freighters for the North Atlantic run. Contracts were let for 16 ships: 5 by Newport News, 6 by Bethlehem-Quincy, and 5 by Sun Shipbuilding. Subsidies were 50.2 percent for Newport News ships and 46.9 percent for Bethlehem-Quincy ships. *American Challenger* herself was built at Newport News, and launched 15 June 1962. Her potential military role is suggested by the fact that she was christened by Mrs. Fred Korth, the wife of the secretary of the navy. Dimensions were 560 ft 6 in overall/529 ft between perpendiculars × 75 ft 7 in (moulded) × 42 ft 6 in (depth) × 28 ft 6 in, with a displacement of 21,053 tons and a gross deadweight tonnage of 11,400. Capacity was 625,775 cu ft plus 1,191 tons of liquids in deep tanks. She had six holds with 10- and 15-ton booms, plus a 70-ton boom, the largest in any American merchant ship to that date. National defense features, which justified the subsidies, included oversize machinery (23,500 SHP for 27 kts), strengthening, and heavy-lift booms. The ships were unusually fast for their time. For merchant service, they were rated at 16,500 SHP (maximum 18,150 SHP). Rated speed was given as 21 or 22 kts. However, *Pioneer Moon* of this class averaged 24.11 kts on an Atlantic round trip. According to a contemporary description in the October 1962 issue of *Marine Engineer and Naval Architect* magazine, their exceptionally fine lines fore and aft (for speed) limited internal volume available on lower levels and aft. The 70-ton Stuelken derrick served Nos. 3 and 4 hatches. Unlike the derricks in the later *Charleston*-class attack cargo ships, and in contemporary Hansa Line ships, it had vertical kingposts. As break-bulk freighters, these ships became obsolescent with the advent of freight containers. Several were chartered to the Military Sealift Command, including *American Challenger*, *American Charger*, and *American Chieftain* (ex–*Pioneer Moon*) in 1971. *American Champion* was part of the original maritime prepositioning force; she was laid up in 1983. Two ships of this class, the former *American Commander* and *American Constructor*, were still in the Ready Reserve Fleet as of 2000; a third, the former *American Crusader*, reverted to the Maritime Administration National Defense Reserve Fleet in 1998.

3. JPTDS, the Junior Participating Tactical Data System, was a far more compact equivalent to the earlier NTDS (naval tactical data system). It was made possible by the advent of the powerful UYK-7 computer. The "J" was reportedly used because it was well known that installation of NTDS in large warships had been painful and expensive. It would have been embarrassing to admit, a few years later, that equivalent capacity could now be installed on board a destroyer (the *Charles F. Adams* class was involved). The "P" meant that the system could participate in a Link 11 net. *Perry*-class frigates initially had JTDS, a nonparticipating version.

4. VSS was the VSTOL support ship; it was not a potential type designation. There was talk at the time of a VSTOL-capable carrier designated CVV, and VSS was intended to indicate that this ship was not a substitute carrier. See Norman Friedman, *U.S. Aircraft Carriers: An Illustrated Design History*

(Annapolis, Md.: Naval Institute Press, 1983), 354–57, for the VSS story.

5. At this time and later it was common to use X in a notional designation to indicate an undefined or future design. By calling the new amphibious ship LX instead of, for example, LPD(X) or LPH(X), programmers were leaving themselves free to choose among many possible types. LPH(X), on the other hand, indicated a successor to the existing LPH.

6. According to A. D. Baker III, *The Naval Institute Guide to Combat Fleets of the World, 2000–2001* (Annapolis, Md.: Naval Institute Press, 2000), 950, the ships are aging more quickly than expected due to thin flight decks, cramped command and berthing spaces, electric power distribution problems, control deficiencies, and obsolete piping. The first ship was scheduled for SLEP under the FY 03 program. Instead, they are to be retired after about 35 years' service, in 2010–13. Limited improvements have been made. New combat direction systems and SPS-48E radars from discarded missile cruisers were installed on board all five ships. Sea Sparrow missiles and 5-in guns were removed (the port aft mount early in the 1990s, the remaining two mounts in 1997), and the corresponding radars of the Mk 86 fire control system were removed. Two rolling airframe missile (RAM) missile launchers were installed.

Chapter 13. New Landing Craft

1. See Norman Friedman, *U.S. Small Combatants: An Illustrated Design History* (Annapolis, Md.: Naval Institute Press, 1987), for details.

Chapter 14. Fire Support Revisited

1. The third ship of the class, *Newport News*, was decommissioned after suffering a serious accident in her No. 2 turret. Although stricken, she remained in the Philadelphia Navy Yard for many years, presumably to provide spares for the other two cruisers if they were reactivated.

Chapter 15. A New-Generation Amphibious Flagship

1. See Norman Friedman, *U.S. Battleships: An Illustrated Design History* (Annapolis, Md.: Naval Institute Press, 1985), 401–2.

2. See Norman Friedman, *U.S. Cruisers: An Illustrated Design History* (Annapolis, Md.: Naval Institute Press, 1984), 432–45.

3. Here J stands simply for joint. The background of such designations is not always obvious. AGC began as a general-purpose auxiliary (AG) with a command function (C). To some extent this was probably a cover designation, the AGC concept being highly classified. In 1969 AGCs were redesignated LCC because there were already command ships designated CC (not for command and control but rather for command-type cruisers, the first C meaning cruiser). There was clearly some hope that the LCCs would perform CC functions. In this system the initial L became a prefix modifying CC. J then replaces L to indicate that the ship will house a joint staff. Note that command ships converted from LPDs are designated AGF—miscellaneous auxiliaries (AG) with flagship role (F).

Chapter 16. Over the Horizon Assault

1. R&D funds for a new LCU(X) program were provided in FY 01, and by the spring of 2001 several contractors were offering designs.

2. The implication was that the LCU was not quite finished, but there was little money for new units. Two new LCUs, LCU 1680 and 1681, were built in 1983–89 for the naval reserves, based on the army's LCU 1646 class. More than 25 design improvements were determined to merit incorporation in the design of the craft. The major change was to replace the previous twin General Motors 6-71 diesel (rated 500 BHP at 2,100 rpm) with Detroit Diesel 12V-71 engines (rated 425 BHP at 2,300 rpm). The keel cooling system was redesigned. Air conditioning capacity was increased, so the earlier 40 kW diesel generators were replaced by 60 kW units. The conning tower was made portable. The forward part of the beaching plate was thickened from ½ in to 1 in. Displacement is 221.2 tons light and 406.24 tons fully loaded. Experience gained in the LCU 1680 design was applied to a program to rehabilitate existing LCU 1646s. Plans initially called for 12 rehabilitated craft by 1996, but only one (LCU 1650) was completed, on the east coast. Work on LCU 1651, the west coast craft, was not completed due to budget limits and cost overruns.

3. By this time the V prefix was already being used for VSTOL aircraft of various types. As in earlier projects, X stood for whatever VSTOL (V) was chosen. J indicated the hope that the product would be used by all the services (i.e., jointly).

4. LCM(6) HPI was an improved version of the 56-ft LCM(6) Mod 2. Improvements included improved controls, Kort nozzles (with a new hull tunnel), and an inexpensive remote magnetic heading system (RMHS). The new powerplant was the 600 BHP Detroit Diesel 8V71T1 engine (the 225 BHP Detroit Diesel 6-71 was rejected). The heavier

engines trimmed the craft down by the stern, the ramp then blocking the operator's forward view. The problem was solved by placing permanent wedges in each of the hull tunnels aft of the Kort nozzles at the transom. The heavier engines and gearing added enough weight to preclude attainment of the desired 13 kts; the boat gained about a knot at full load.

5. The 1974–88 SLEP applied to 20 aluminum hull and 64 steel hull LCM(8)s, half of which were modified on the east coast (at Newport, Rhode Island; Charleston, South Carolina; and New Bern, North Carolina) and half on the west coast (at San Diego). The existing Detroit Diesel twin 6-71 engines were replaced with Detroit Diesel 12V-71 engines with the twin disc 514C marine gear. Other improvements were new steering gear; the helm was replaced by levers with automatic backup and the pneumatic system was replaced by hydraulics. Bow ramp systems were overhauled or replaced.

6. An unofficial designation, LKD-X recalled the AKD designation the Military Sealift Command was then using for well-deck ships; the L rather than A prefix indicated an amphibious combatant rather than an auxiliary. As before, the X suffix merely indicated that this was an undefined design.

Appendix A. Maritime Prepositioning and Sealift

1. The three liners, the American President Lines' P2-S1-DN1 round-the-world type, became T-AP 196-198. That line between a point-to-point ship and an attack transport was thin; note that on 26 July 1950 OpNav asked for a sketch design of a transport squadran flagship (APA) version to carry 80 officers and 1,500 enlisted men plus a staff of 20 officers and 50 enlisted men. The boat complement would be 4 LCM, 23 LCVP, 2 LCPL, and 1 LCPR, and the ship would carry 1,500 tons of combat-loaded cargo. Armament would be six twin 3-in/50 with five directors (presumably two of the mounts would be controlled by one director.) No sketch has been found, and no SCB project number was ever assigned. Only in September 1950, well after the OpNav study had been ordered, did the Maritime Administration announce that the Navy was taking over the three ships. The big liner *United States*, then under construction, was also to have been taken over, but that did not happen.

2. In June 1961 Secretary of Defense Robert S. McNamara called for reactivating 15 troop transports, but by 1962 he said he did not need even the ones already in service. In April 1962 he rejected calls for operating subsidies for two liners that would have been convertible into wartime troopships. By this time he stated sealift was important only for logistical support of forces that had already been delivered by air. Sealift concepts were revived due to a series of studies of the requirements of rapid deployment. In July 1964 the Special Studies Group of the JCS concluded that strategic mobility was necessary, because forces could not be prepositioned to meet crises around the world. This report led to a series of studies of the requirements of strategically mobile forces, which in turn led to the call for the FDL ships. The original navy concept was to move the troops themselves by fast sealift ships.

3. In fact only three ships, designated AG 172–174 (*Phoenix*, *Provo*, and *Cheyenne*), were acquired. Authorization to acquire 12 more Victory ships as AG 179–190 was canceled 31 October 1966. The army wanted a total of 16 ships, presumably to support a single division.

4. This was not a concept for deploying marines. The objective was rapid deployment of the U.S. Army's central reserve, the three-division STRAC (the Strategic Command), which had been formed in 1958—but for whose deployment no real provision had ever been made. Originally plans called for using airlift, but STRAC had no organic airlift. In March 1961 McNamara asked the Joint Chiefs to plan to integrate STRAC with an air component, and an air force–army Strike Command was formed. Its army components were the two airborne divisions (82nd and 101st) and the 4th Infantry Division.

5. One of the school ships, *Empire State*, was taken over to return U.S. troops from Somalia in 1994.

Bibliography

This book is based mainly on official material held in various archives, supported and amplified by published works. As in earlier volumes in this series, the most important material came from the papers of the U.S. Navy's Bureau of Construction and Repair (C&R) and Bureau of Ships (BuShips) and their successors, particularly their preliminary design section. General Board files, which are invaluable for the period prior to World War II and for some of the postwar period, were formerly held at the U.S. Navy Operational Archives of the Naval Historical Center at the Washington Navy Yard, but they are now at the National Archives downtown (Washington, D.C.) branch. This branch also holds pre-1940 files of C&R and the Office of the Chief of Naval Operations (OpNav). The latter includes pre–World War II building programs. Marine Corps material for the pre-1940 period, including the typed reports of the prewar fleet landing exercises (FLEXes), is also at this branch. Its microfilm room holds the confidential correspondence file of the secretary of the navy/OpNav for 1917–27. This file includes the key papers on the "beetle boat" and initial artillery lighter concepts. U.S. Navy and U.S. Marine Corps World War II files are now available at the National Archives at College Park, Maryland. Also at College Park are the wartime Joint Chiefs of Staff records, relevant because the Joint Chiefs controlled allocation of merchant shipping that could be converted into assault transports. Many of the navy's preliminary design files, for the period into the early 1960s, have been released to the National Archives at College Park. In 2001 the Maritime Commission and U.S. Shipping Board files, formerly at the National Archives, Washington, D.C., were moved to the National Archives at College Park. Some World War II material remains in the Command File of the U.S. Navy Operational Archives at the Washington Navy Yard. This archive also contains most postwar U.S. Navy files. Of particular value to this study were the "double-zero" files, those of the immediate office of the Chief of Naval Operations (CNO), which have been declassified through 1974.

As in earlier volumes, the records of the General Board were used extensively. They were particularly important for the period before 1940. General Board correspondence files, which included a few hearings, were the primary source for the U.S. Marines' initial interest in transports and for the development of the *Henderson* and *Heywood* designs. They were also the main source for discussions of the "ideal transport" (1938–39) and of the role and characteristics of the *Doyen*-class attack transports (in General Board hearings for 1940–41). The hearings in particular make it obvious that the *Doyen* design was not a U.S. Navy initiative. These hearings are also a rich source for more general amphibious practice. The General Board had no responsibility for the smaller landing craft or for wartime amphibious ships. Its files do include early postwar Marine Corps discussions (through 1949), including some material on landing craft and on the Marines' early postwar LVT program, as well as the conclusions of the Amphibious Type Conference of 1947. These records are to be found among those collected for the General Board's last major study, a projection of naval needs for 1951–60. Also useful is the General Board's 1948 study of the reserve fleet, which includes evaluations of existing warship types. General Board files and the related Command File include Ship Characteristics Board memoranda for 1945–50.

C&R and BuShips files for the period through the 1960s are in National Archives Record Group 19. Box 12 of the prewar confidential correspondence includes the 1938 parametric study of a future transport. Secretary of the navy/CNO files for the World War II period are in Record Group 80. The files used for this book were the classified correspondence for 1940–45, arranged by filing manual symbol. Some of the files of the Landing Boat Board are coded

AP/KK. The reports of the Auxiliary Vessels Board are coded AA/QB. Record Group 38 contains the U.S. Navy after-action reports. The World War II records of the Joint Chiefs of Staff (Record Group 218) include conclusions on merchant ship allocations, particularly in 1943.

Marine Corps headquarters files (Record Group 127) were also useful, particularly the historical files at the National Archives at College Park. They include a detailed and illustrated account of the 1924 Culebra maneuvers. A copy of the Mason manuscript on World War II amphibious operations is included in this file, with its source materials.

Also important, particularly for the prewar period, were the archives of the U.S. Naval War College at Newport, Rhode Island. They include both lectures and source documents. Because the War College was used extensively to test navy tactical concepts, the students needed current information, and the lectures were given by senior practitioners.

Because the Maritime Commission was also intimately involved in prewar and wartime construction, the files of its wartime chairman, Rear Adm. (Retired) Emory S. Land, were also important. They are held by the National Archives in Washington, D.C., in Record Group 127. Box 2 of Land's personal papers includes his report to President Roosevelt on the new Marine transport (1940), including the note that Roosevelt had personally requested the design.

The Ships' Histories Branch of the U.S. Navy Historical Center at the Washington Navy Yard holds card files on small craft, including descriptions of contracts and cards for all craft of the pre-1935 period. In addition, British files on wartime landing craft were used.

Primary Sources

British Official Files

British files are located at the Public Record Office, Kew, near London. The bulk of the relevant files are DEFE (Defense) rather than ADM (Admiralty), because to the British amphibious operations were combined, i.e., interservice.

ADM 239/242, "The Development of Landing Ships and Craft" (Technical Staff Monographs 1939–1945, dated 1948; marked CB 3206 and also TSD 32/48; approved 22 November 1948). A draft of this report, somewhat different from the final version, is in the Naval Historical Branch in London.

DEFE 2/1, Combined Operations War Diary, Vol. 1 (includes ships taken up from trade in 1940–41).

DEFE 2/733–735, Combined Operations Summaries, including reports on US developments.

DEFE 2/829, Prime Minister's Directives etc. for Tank Landing Craft and Tank Landing Ships, 1940–42; this file includes the November 1941 decision to requisition the Higgins LCMs.

DEFE 2/835, Conversion of train ferries to tank transports, 1940–41; this file includes the Popper patent and memoranda about using this design.

DEFE 2/1097, LST(3), LST(Q), including remarks on use of six-davit LSTs in Sicily.

DEFE 2/1287, Visit of Captain Hussey to United States to survey Joint R&D (JANET), May 1945.

DEFE 2/1327, Notes on various landing craft for Admiralty monograph (more complete than the printed version referenced above).

Royal Navy Battle Summaries

The following Royal Navy staff histories of battle summaries were based on both British and U.S. reports; sometimes they included details missing from U.S. lessons-learned reports. They are located in the British Naval Historical Branch in London. The Normandy and Southern France reports were published by H. M. Stationery Office in 1994 as part of a boxed collection on *Invasion Europe*.

Anzio (Report of Proceedings, not a full Battle Summary).

Normandy (No. 39).

North Africa (No. 31).

Salerno (not numbered).

Sicily (No. 35).

Southern France (not numbered).

U.S. Government Publications

Clifford, Lt. Col. Kenneth J., USMCR. *Progress and Purpose: A Developmental History of the United States Marine Corps, 1900–1970.* Washington, D.C.: History and Museums Division, HQMC, 1973.

Raines, Edgar F. *Eyes of Artillery: The Origins of Modern U.S. Army Aviation in World War II.* Washington, D.C.: Center of Military History, 2000.

U.S. Army. *Report of Army Small Boat Construction 1 July 1940–31 May 1945.* Washington, D.C.: 18 December 1945; in files of NAVSEA small boat office.

U.S. Navy, Administrative Command, Amphibious Forces, U.S. Pacific Fleet. *Ship Characteristics,* 7 November 1944 and 20 July 1945 editions. Gives full data for amphibious forces except displacement. Some armament data were not up to date.

U.S. Navy, Bureau of Ordnance. *Armament Summary,* various issues, 1941–August 1945.

———. *Missile Launchers and Related Equipment Catalog* (OP 1855, 1 June 1953); includes LCI(G) rocket launchers.

U.S. Navy, Bureau of Ships. *Boats of the United States Navy* (NAVSHIPS 250-452, May 1967).

———. *LST Operational Information* (NAVSHIPS 250-420, revised edition, January 1945).

———. *Naval Vessel Register*, 27 July 1946 edition. Gives light displacements for current and past auxiliary ships.

———. *Ships and Aircraft Types: Lessons Learned in Pacific War* (Comstock Board), 13 December 1945 (Naval Historical Center, microfilm AR-174-75).

———. *Ships' Data U.S. Navy Vessels*, 1945 and 1953 issues.

———. *Ships Laid Down Since the Washington Treaty by Authorizing Acts* (NAVSHIPS 250-282, 1 January 1949).

U.S. Navy, CinCPac. *History* (World War II, dated 27 December 1945).

U.S. Navy, Cominch. *Amphibious Operations* lessons-learned publications: volumes for *The Period August to December 1943* (P-001), *The Marshall Islands January–February 1944* (P-002), *Excluding Marshall Islands Operations January–March 1944* (P-004), *Invasion of Northern France Western Task Force June 1944* (P-006), *Invasion of the Marianas June–August 1944* (P-007), *Invasion of the Philippines October 1944 to January 1945* (P-008), *Capture of Iwo Jima 16 February to 16 March 1945* (P-0012), *Capture of Okinawa (Ryukyus Operation) 27 March to 21 June 1945* (OpNav 34-P-0700).

———. *Description of Amphibious Force Flagship (AGC)* (P-02, 1944).

———. *Description of APAs Equipped as Relief AGCs* (OpNav 20-18).

U.S. Navy, Office of Naval Intelligence. ONI 226, *Allied Landing Ships and Craft*. Washington, D.C.: 1944, with 1945 supplement; reprinted together by Naval Institute Press as *Allied Landing Craft of World War II* in 1985.

U.S. Navy, OpNav. Naval History Division, *Dictionary of American Naval Fighting Ships*, 9 vols. Washington, D.C.: 1959–1991.

———. USF 66, *Amphibious Tactics and Doctrine* (1947). Available at Naval Historical Center, microfilm AR 56-78, Reel 4.

———. WPL 10, *Transportation Service Readiness Plan* (29 April 1924). This is the mobilization plan. The microfilm of this plan in the U.S. Navy Operational Archives includes its 1925, 1929, and 1930 versions. Microfilm NRS 286 is the September 1941 version, describing actual mobilization.

U.S. Marine Corps Files, 1940–55

Marine Corps war plans and policy files are located in Division of Plans and Policy—War Plans Division, 1940–1955, U.S. Marine Corps File, Record Group 127, National Archives at College Park, Maryland.

Box 1, Amphibious Operations and Doctrine, general.

Box 2, Amphibious Operations and Training, Atlantic, includes a detailed illustrated account of the 1924 exercises.

Box 3, Mason monograph with source materials.

Box 17, Landing Craft, General.

Box 18, Equipment—DUKW, LCVP, LSMR, LST, LVT, etc. This box includes the 1936 small boat estimates (program) and also doctrine for LCI(R) and LCI(M) as of March 1945.

U.S. Marine Corps, HQ Marine Corps Files

HQ Marine Corps: Division of Operations and Training, U.S. Marines Corps File, Record Group 127, National Archives, Washington, D.C. Box 4 contains detailed reports of FLEXes 4–6 of 1938–40, including some landing craft evaluations.

U.S. Maritime Commission Reports

Permanent Report of Completed Ship Construction Contracts (Report B-1: no date [1946]).

Gerald J. Fisher. *A Statistical Summary of Shipbuilding Under the U.S. Maritime Commission during World War II*. Historical Reports of War Administration, United States Maritime Commission, No. 2, 1949.

U.S. Naval War College Archives

The following lectures and other documents are located at the Naval War College Archives in Newport, Rhode Island.

Lectures Given at the Naval War College

Coolidge, J. W. *Landing Craft and Landing Ships* (27 March 1945).

Earle, Ralph. *Landing Operations of the Control Force, Nov. 1921–May 1922* (18 December 1922).

Miller, Col. E. B., USMC. *The Marine Corps: Its Mission, Organization, Powers, and Limitations, with Special Reference to Advanced Bases in Support of the Fleet* (16 January 1922).

Shertz, R. H. *Landing Craft and Landing Ships* (25 September 1944).

Utley, Lt. Col. Harold H. *The Marine Corps in Support of the Fleet, Part II: Landing Operations* (December 1935).

Other Documents

Commander, Control Force (Adm. M. M. Taylor), *Suggested Doctrine for Joint Army-Navy Operations* (22 July 1925).

Commanding General, Fleet Marine Force (L. McCarty Little), *Surf Landing* (22 April 1938).

Control Force, *Lighters for Landing Expeditionary Forces: Transport of Artillery and Motor Transport with Expeditionary Detachments* (24 April 1922).

U.S. Navy After-Action Reports

NRS numbers, where indicated, designate microfilms at the U.S. Navy Operational Archives of the Naval Historical Center at the Washington Navy Yard.

Anzio Landing (Shingle), CTF 81 initial and final reports, February and May 1944, OpNav Files, Record Group 38, Boxes 295 and 296, National Archives at College Park, Maryland.

Iwo Jima Invasion, NRS 1984-27. Amphibious Group Two report.

Normandy Invasion (Overlord), NRS 1969-38; this microfilm includes both CTF 122 report and reports by some subordinate units.

Salerno Invasion (Avalanche). Eighth Fleet Report, OpNav Files, Record Group 38, Box 52, National Archives at College Park, Maryland.

Sicilian Invasion (Husky), NRS 1974-11. Eighth Fleet Report.

Southern France Invasion (Anvil-Dragoon), NRS 1979-19. Eighth Fleet Report.

U.S. Navy Bureau of Construction and Repair Files

Pre–World War II correspondence files are located in Pre–World War II Correspondence Files (Bureau of Construction and Repair), Record Group 19, Confidential Series, National Archives, Washington, D.C.

Box 12, AP/S1-1 to AR5/S1-1, includes parametric study of ideal transport, 1939.

U.S. Navy Bureau of Ships Files

World War II correspondence files are located in Record Group 19, National Archives at College Park, Maryland. When ships were redesignated, correspondence was collected under their final designation. For example, all USS *Ancon* files are located under AGC 4, rather than *Ancon*'s earlier AP 66 designation.

Box 255, LC/A1-3 through LC/A4-5 programs, including possible changes in LCI(L) and LST programs, 1943.

Boxes 312–314, LCC construction records.

Box 327, C-LCI/A16-3 Vol. 5 to LCI/L8-3.

Box 328, C-LCI/L9 to C-LCI/L9-3 Vol. 4, modifications of LCI(L)s 1944–45 including fire support craft. LCI(G), etc., files include only after-action reports.

Box 329, C-LCI/L9-3 Vol. 5 to C-LCI/S82, includes termination of LCI(G) program, 1945.

Box 504, LCI(R), including an after-action report from Okinawa.

Box 658, AGC 4 Vol. 7 to C-AGC 4 Vol. 1.

Box 790, C-AK/A1-3 to AK/L7-1; includes a list of ships in the Naval Transportation Service at various times in 1945, showing those assigned to Seventh Fleet and to CinCPac, including AP acting as APA.

Box 792, AK/L9-3 Vols. 2–4; includes origin of quadrupod mast for U.S. attack cargo ships.

Box 793, C-AK/L9-3 to C-AK/P16-3.

Box 794, AK/P16-4 to AK/S11-3.

Box 815, AK17 Vol. 6 to AK21.

Box 821, C-AK41 Vol. 2 to AK45.

Box 912, C-AKA/L9-3 to C-AKA/S1-3.

Box 981, AKS 20 Vol. 4 to AKV/S38-1.

Box 1013, LST/L9-3 Vol. 3 to C-LST/L9-3 Vol. 2.

Box 1014, C-LST/L9-3 Vol. 3 to C-LST/L9-3 Vol 7.

Box 1058, LST/S82-1 Vol. 3 to C-LST/S82-(19).

Box 1182, LST 317 to C-LST 326.

Box 1195, LST 391 to C-LST 394.

Box 1313, LST 771 to C-LST 776 (Brodie LST).

Box 1486, QS1/S1-1, including a May 1943 discussion of the P2-S2-R2 design and the effect of reallocating it from the army to the navy; also contains material on proposed XAPH conversions of 1941. This box includes correspondence calling for construction of 50 Seatrains and the Maritime Commission's description of the new Victory ship (24 March 1943).

Box 1788, C-AP/A1-3 to AP/L6-2.

Box 1791, C-AP/L9-3 Vol. 1. This box includes the 1941 assessment of the adequacy of transports for the amphibious units.

Box 1792, C-AP/L9-3: wartime conversions, including C1s.

Box 1793, AP/P16-1 to C-AP/S1-3; includes design material on the *Doyen* class.

Box 1856, C-AP67 to AP69 Vol. 2.

Box 1861, AP72 Vol. 2 to AP73.

Box 1862, C-AP73 to AP74 Vol. 3.

Box 1865, C-AP74 to AP75 Vol. 2.

Box 1868, C-AP76 to AP77 Vol. 2.

Box 1869, AP77 Vol. 3 to AP94.

Box 1871, C-AP102 to AP103 Vol. 2.

Box 1872, AP103 Vol 3 to AP104 Vol. 1.

Box 1874, C-AP104 to AP105 Vol. 2.

Box 1875, AP105 Vol. 3 to AP109.
Box 1926, AP166 to C-AP166.
Box 1928, C-AP168 to C-AP169.
Box 1929, AP170 to AP171-5/P16.
Box 1932, C-AP171 to AP172 Vol. 2.
Box 1933, AP172 Vol. 3 to AP173 Vol 2.
Box 1934, C-AP173 to C-AP174.
Box 1935, AP175 to C-AP175.
Box 1945, APA/L8-2 to C-APA/L9-3 Vol. 1. This includes conversion of C1-Bs to APAs for the British.
Box 1947, APA/P11-1 to C-APA/S1-3 including combat characteristics of new APA and AKA, 18 October 1944.
Box 1951, AP171 Vols. 1–3.
Box 1965, APA1 *Doyen* alterations.
Box 2241, APA239 to APB/S48-2.
Box 2242, APB/S48-2 to APB35-40/S1-3.
Box 2247, C-APB39 to APB45-8/L7.
Box 2248, C-APB45-8/L8 to APB45-8/S11-5.
Box 2658, ARL39 to C-ARL41.
Box 3324, AVS8 Vol. 1 to C-AVS8.

U.S. Navy CNO/SECNAV Files, 1917–27

Formerly classified correspondence files are located in Formerly Classified Correspondence Files (CNO/SECNAV) for 1917–1927, Record Group 80, Microfilm Publication M1140, National Archives, Washington, D.C.

Roll 31/ SC 120-11, Control Force discussion of lighters to land heavy Marine Corps artillery. Adjacent documents lay out the requirements for other landing craft.

U.S. Navy CNO/SECNAV Files, 1927–39

Formerly classified pre–World War II correspondence files are located in Pre–World War II Formerly Classified Correspondence Files (CNO/SECNAV) for 1927–1939, Record Group 80, National Archives, Washington, D.C.

Box 1, A1-1 to A1-3/S72, includes projected building program as of late 1939.
Box 2, A1-3/AA to A2-8, includes more building program material.
Box 146, AO2 to ARV, includes AP data.

U.S. Navy COMINCH/SECNAV Files

Formerly classified World War II correspondence files are located in World War II Formerly Classified Correspondence Files (COMINCH/SECNAV), Record Group 80, National Archives, Washington, D.C. There are separate series by year.

1940–41

Box 166, AO24/L9-3 to AP/L9-3 includes prewar conversion plans for *Harris, Zeilin,* and *McCawley.*
Box 167, AP/N33-2 to AR3/L9-3, including orders to divert "truck transports" from British account (November 1941) and an 11 May 1940 War Department description of a 550-ft transport then in the design stage (which presumably became the P2). This box also includes material on the revised APD mission.
Box 170, AVP/A6-1 to BB48/S3-1, includes the 1941 Seatrain conversion drawing.

1942

Box 439, L9-3/AM to L9-3/AP38, includes conversions to AP 78-96 (all later designated APA).
Box 525, AC/L6-3 to CV/S1-1, includes landing craft armoring discussion.
Box 530, LC/L6-3 to LC/P19-3, includes initial LST conversions.

1943

Box 718, LC/L9-3 to LC/S41-5.
Box 890, L9-3/AM341 to L9-3/AP 1943.
Box 891, L9-3/AP, includes the AKN conversion to what became the LSV.
Box 965, P19-1/Stamps to S1-1/55, includes LST and APA conversion material.
Box 1022, AA/QB to DE/S1-1, includes District Craft Development Board and Auxiliary Vessels Board reports.
Box 1024, LC to LC/A4-3, including a December 1943 memo on possible postwar civilian use of landing craft, and also the army tests of the 62-ft LC(X).
Box 1025, LC/A4-3 to LC/L21-3.
Box 1027, LC/L21-4 to LC/37(S7). This box includes the 1943 landing craft characteristics book and plans for the LCS(L) and for LCT conversions to fire support craft.
Box 1028, LC/S8-2 to LC/S41-5, includes camouflage photographs.

1944

Box 1111, A1-3/AA to A1-3/NH.
Box 1265, L9-3/APA to L9-3/ATR57, includes conversion of Victory ships to attack transports and also conversion characteristics for APA 106–116. This file also includes calls for heavier armament on board existing APAs, and BuShips replies that they are overweight.
Box 1468, JJ14 to LC, includes Tucker-Higgins LCM test.
Box 1470, LC/A4-1 to LC/L21, includes orders for the electrically-powered surfboard.

Box 1471, LC/L21-3 to NA92/P16-1, includes LCI(G) armament.
Box 1746, S1-1, includes comments on limitations of Victory APA conversion.
Box 1859, P17-1(6) to QB(6).
Box 2005, P20-1/QRS to S1-1/LVT, contains two LVT Development Board reports.

1945
Box 2419, S82-3/EF30 to JJ7 (26 July).

U.S. Navy Documents

Barrett. *Landing Operations: Quantico.* Available at Naval Historical Center, microfilm NRS 1972-35. This is actually a set of drafts of sections of a landing operations manual, probably written about 1931–32 (the file is officially dated 1931–37, but it shows no signs of later material).

Mason, Col. A. J., USMC. *Amphibious Warfare Monograph.* Washington, D.C.: Navy Department, 1950. Available at Naval Historical Center, microfilm NRS 15 M 1973-3. Mason covered only about half of World War II, delivering his final chapters in June 1950. He did outline the arguments about a landing in France, and he described the proposals to use small craft and river steamers for transportation. Mason refers to Lt. Col. B. W. Gally, USMC, "A History of U.S. Fleet Landing Exercises, Atlantic Squadron, 1939," a mimeographed confidential pamphlet that contains a "General Record of Landing Boat Development" by Lt. William F. Royall, USN. Although the latter has apparently been lost, a set of photographs in the Marine Corps photo section of the National Archives at College Park appears to be Royall's set of illustrations, with his comments.

University of Pennsylvania. *History of U.S. Navy Research and Development in World War II* (no date [1950s]; paper written for ONR).

U.S. Navy "Double-Zero" Files

Files are located in Command File, U.S. Navy Operational Archives of the Naval Historical Center at the Washington Navy Yard. Each year has its own series.

1964
Box 19, on characteristics of new LST and other amphibious ships including LFS and LPD.
Box 30, on 20-kt LST characteristics.

1965
Box 25, FDL operational concepts, LHA concept development.
Box 44, includes account of successful LST ramp test.

1966
Box 30, includes rationale for size of LHA force.

1967
Box 69, FDL material.
Box 70, LHA material, goes back to 1966.

1968
Box 34, includes LHA characteristics and program as of 1968.

1969
Boxes 93 and 94, include FDL material.

1970
Box 107, includes material on the LCA program and some on the landing force support weapon (LFSW). This box also includes the rationale for cutting the LHA program short.

1973
Box 112, includes LST 1179 characteristics.

U.S. Navy Preliminary Design Files

Declassified design history files are located in Record Group 19, National Archives at College Park, Maryland.

Entry 13419, Preliminary design files for auxiliaries, including wartime 520-ft APA (A-7), small wartime APA (A-8), small wartime AKA (A-9), AP converted from AKN (A-11), 20-kt AKA (A-15), and SCB 77A ro-ro postwar APA (A-25); A numbers are in a general auxiliary series.

Entry 13455, Wartime and postwar landing craft, including plastic LCVP, Mk 4 LCPL. This series includes material on landing boats for 1937–41 and on LCM orders for 1942 (including the consequences of substituting Higgins for BuShips craft). Box 2 contains the postwar LCM file, through 1951 and wartime Higgins boat files.

Entry 13941, Landing craft general file, 1939–43. This box includes early correspondence with Higgins as well as some LVT files. It also includes characteristics for the 56-ft LCC.

Entry 15411, Design data including LST 1153 class as well as LCI(L) and LCI(G) material.

Entry 60A1716, Includes SCB 15 postwar AKA, also LCT design work 1941–48 (including calculations for the abortive fire support conversions) and the project to convert an LSV into an AKN postwar.

Entry 61A2306, LST material, including calculations for the original 280-ft design (in Box 13, coded L-3), also IFS design material. This series includes miscellaneous landing craft material. Box 13 includes British September 1940 specifications for the LCT(1), the British February 1941 plans for LST(1), the LST 1153 (L-8), 1156 (L-15), and 1171 (L-18) designs, and some material on the IFS design (coded X-37). Box 12 includes material on the LSD 28 design (X-36). Box 14 includes miscellaneous landing craft material, plus the George G. Sharp "Carport" LST, an LVT historical file (L-9), the BuShips LCM(3) (L-2), a sketch of a 50-ft lighter for LVTs (L-10), an LSM file (L-7/X-51), a file on LCI(L) history (L-5), and a file on the 26-ft outboard landing craft of 1944 (L-1).

Entry 61A3115, LCP(L), LCVP wood and plastic versions, LCMs; general data on landing craft, 1943–52 are in Box 1. This box includes the 56-ft LCC and its heavy support craft predecessor, plus the 70-ft LCM and its LCM(8) postwar successor.

Entry 62A3200, Box 2 includes material on the Mariner APA conversion.

Entry 67A6018, Exotic LCVP versions (hydrofoils, etc.) and material from the Boat Design Panel charged with developing new types of small craft.

Secondary Sources

Alexander, Joseph H., and Merrill L. Bartlett. *Sea Soldiers in the Cold War: Amphibious Warfare 1945–1991*. Annapolis, Md.: Naval Institute Press, 1995.

Brown, D. K., ed. *The Design and Construction of British Warships 1939–45, Vol 3: Amphibious Warfare Vessels and Auxiliaries*. London: Conway Maritime Press, 1996. This is the edited version of the official Director of Naval Construction (DNC) department reports on different warship classes, written soon after World War II.

Brunner, John W. *OSS Weapons*. Williamstown, N.J.: Phillips Publications, 1994.

Cawne, Jonathan. *Spearheading D-Day: American Special Units in Normandy*. Paris: Histoire et Collections, 1998.

Clifford, Kenneth J. *Amphibious Warfare Development in Britain and America from 1920–1940*. Laurens, N.Y.: Edgewood, 1983.

Croizat, Col. Victor A., USMC (Ret.). *Across The Reef: The Amphibious Tracked Vehicle at War*. London: Arms and Armour, 1989.

Dockery, Kevin. *Navy SEALS: A History of the Early Years*. New York: Berkeley Books, 2001.

Dwyer, John B. *Commandos From The Sea: The History of Amphibious Special Warfare in World War II and the Korean War*. Boulder, Colo.: Paladin, 1998.

———. *Scouts and Raiders: The Navy's First Special Warfare Commandos*. Westport, Conn., Praeger, 1993.

———. *Seaborne Deception: The History of U.S. Navy Beach Jumpers*. Westport, Conn.: Praeger, 1992.

Fergusson, Bernard. *The Watery Maze: The Story of Combined Operations*. London: Collins, 1961.

Friedman, Norman. *U.S. Aircraft Carriers: An Illustrated Design History*. Annapolis, Md.: Naval Institute Press, 1983.

———. *U.S. Battleships: An Illustrated Design History*. Annapolis, Md.: Naval Institute Press, 1985.

———. *U.S. Cruisers: An Illustrated Design History*. Annapolis, Md.: Naval Institute Press, 1984.

———. *U.S. Destroyers: An Illustrated Design History*. Annapolis, Md.: Naval Institute Press, 1982.

———. *U.S. Small Combatants: An Illustrated Design History*. Annapolis, Md.: Naval Institute Press, 1987.

Goldberg, Mark H. *"Caviar and Cargo": The C3 Passenger Ships*. Kings Point, N.Y.: American Merchant Marine Museum, 1992.

———. *The "Hog Islanders": The Story of 122 American Ships*. Kings Point, N.Y.: American Merchant Marine Museum, 1991.

———. *The "Stately President" Liners: American Passenger Liners of the Interwar Years, Part I: The "502"s*. Kings Point, N.Y.: American Merchant Marine Museum, 1996.

Grover, David H. *U.S. Army Ships and Watercraft of World War II*. Annapolis, Md.: Naval Institute Press, 1987.

Haffa, Robert P., Jr. "Rapid Deployment Strategies for the Half War: Planning U.S. General-Purpose Forces to Meet a Limited Contingency, 1960–1980." Ph.D. dissertation, MIT, 1982.

Heinl, Robert D., Jr. *Soldiers of the Sea: The United States Marine Corps, 1775–1962*. Annapolis, Md.: Naval Institute Press, 1962.

Hunnicutt, R. P. *Bradley: A History of American Fighting and Support Vehicles*. Novato, Calif.: Presidio, 1999.

———. *Stuart: A History of the American Light Tank*. Novato, Calif.: Presidio, 1992.

Jordan, Roger. *The World's Merchant Fleets 1939: The Particulars and Wartime Fates of 6000 Ships*. Annapolis, Md.: Naval Institute Press, 1999.

Kenworthy, Lt. Comdr. The Hon. J. M. *Sailors, Statesmen—And Others: An Autobiography*. London: Rich and Cowan, 1933.

Ladd, J. D. *Assault from the Sea, 1939–1945*. Newton Abbott, England: David and Charles, 1976.

Lane, Frederic C., with Blanche D. Coll, Gerald J. Fisher, and David B. Tyler. *Ships for Victory: A*

History of Shipbuilding under the U.S. Maritime Commission in World War II. Baltimore: Johns Hopkins, 1951.

Lewis, Adrian R. *Omaha Beach: A Flawed Victory.* Chapel Hill: University of North Carolina Press, 2001.

Lorelli, John A. *To Foreign Shores: U.S. Amphibious Operations in World War II.* Annapolis, Md.: Naval Institute Press, 1995.

LSM-LSMR: WW II Amphibious Forces, vols, 1 and 2. Paducah, Ky.: Turner Publishing, 1994 and 1997. Unedited collections of LSM-LSMR veterans' recollections.

Macdermott, Brian. *Ships without Names: The Story of the Royal Navy's Tank Landing Ships of World War II.* London: Arms and Armour, 1992.

Maund, Rear Adm. L.E.H., RN. *Assault from the Sea.* London: Methuen and Co., 1949.

Metcalf, Col. Clyde H. *A History of the United States Marine Corps.* New York: Putnam, 1939.

Morrison, Rear Adm. Samuel Eliot. *History of U.S. Naval Operations in World War II,* Vol. II, *Operations in North African Waters.* Boston: Little, Brown, 1950.

———. *History of U.S. Naval Operations in World War II,* Vol. XI, *The Invasion of France and Germany.* Boston: Little, Brown, 1957.

O'Dell, James D. *The Water Is Never Cold: The Origins of the U.S. Navy's Combat Demolition Units, UDTs, and SEALs.* Washington: Brassey's, 2000.

Pawle, Gerald. *The Secret War, 1939–45.* London: Harrap, 1956.

Rielly, Robin L. *Mighty Midgets at War: The Saga of the LCS(L) Ships from Iwo Jima to Vietnam.* Central Point, Ore.: Hellgate Press, 2000.

Sawyer, L.A., and W. H. Mitchell. *From America to United States: The History of the Long-Range Merchant Shipbuilding Programme of the United States Maritime Commission,* 4 vols. Kendal, England: World Ship Society, 1979–86.

———. *Victory Ships and Tankers: The History of the 'Victory' Type Cargo Ships and of the Tankers Built in the United States of America During World War II.* Newton Abbott, England: David and Charles, 1974.

Scheina, Robert L. *U.S. Coast Guard Cutters and Craft of World War II.* Annapolis, Md.: Naval Institute Press, 1982.

Strahan, Jerry E. *Andrew Jackson Higgins and the Boats that Won World War II.* Baton Rouge: Louisiana State University Press, 1994.

Terzibaschitsch, Stefan. *50 Jahre Amphibishe Schiffe der U.S. Navy* (50 Years of U.S. Amphibious Ships). N.p., n.d. [1998].

———. *70 Jahre Flottenhilfsschiffe der U.S. Navy* (70 Years of U.S. Auxiliaries). N.p., n.d. [1998].

Triplet, William S. *A Colonel in the Armored Divisions: A Memoir, 1941–1945.* Columbia: University of Missouri, 2001.

USS LCI, vols. 1 and 2. Paducah, Ky.: Turner Publishing, 1993 and 1995. Unedited collections of LCI veterans' recollections.

Wakefield, Ken. *Lightplanes at War: U.S. Liaison Aircraft in Europe, 1942–1947.* Gloucestershire, England: Stroud, 1999.

Index

AAAV (advanced amphibious assault vehicle), 461–64, *464*
AAFSF (amphibious assault fuel supply facility), 478
AALC (advanced assault landing craft), 435
AAV (amphibious assault vehicle), 491–93
AAV7, 301, 461
ABFS (amphibious bulk fuel system), 478
ACEs (air component elements), 476
ACV (air cushion vehicle), 395–96
Adirondack (AGC 15), 273, 275
Adirondack-class, 267–68
Admiral William B. Callaghan, 470
ADO 4-27X, 408
Advance Base School, 5, 6
AE (assault element), 447
AEW (airborne early warning), 263
AFOE (assault follow-on element), 447
African Comet, 151
African Meteor, 151
African Planet, 151
AFS (combat stores ship), 427
AGC (command ship), 261–78; air support, 266, 272; communications on, 262–65; conversions to, 188–89, 425–31; criticism of, 268–74; functions, 425–27; relief, 163, 265–66, 274, 278; refitting, 275–78
AGF 3, 365, 431
AGF 11, 431
AGS (survey ship), 174
aircraft, light artillery-spotting, 126
aircraft carriers. *See* flight decks; helicopter carriers
Aisne, 30
AK (cargo ship), 15, 46, 62, 64
AKA (attack cargo ship): boom capacity, 165, 314–17, 327; cargo-handling ability, 320, 327; construction, 188, 191–95; conversions, 49, 62–64, 188–89, 314–20, 320–25; designation, 15, 46; smaller version, 165–77; speed, 311, 312–20; wartime, 176
AKR, *322, 323*
AKV (aircraft transport), 113
ALC (assault landing craft), 103
Alchiba (AK 23), 60, 64, *95*
alfalfa assaults, 20
Alhena (AK 26/AKA 9), 57, 60, 64
Alligator, 99–100, 214–15
Almaack (AK 27/AKA 10), 57, 60, 64
ALPS (amphibious landing platform slipway), 370–72
AMC(U) (coastal minehunter), 148
America, 55

American Challenger (C4-6-57A), 33, 373
American Champion, 472
American Courier, 472
American Legion (AP 17), 187
AMFIST (amphibious fire support study), 450, 452
AMFOR (Amphibious Forces), 443
Amphibious Corps, 8
amphibious development, 3, 17–21; post–WW II, 205, 287–88; wartime, 218
Amphibious Type Conference, 311
Amphibious Warfare Conference, 351
Anchorage (LSD 36), 313
Anchorage-class, 331
Ancon (AP 66/AGC 4), *263, 264*; air support, 272; characteristics, 273; communications on, 264–65; conversion, 159, 263–64
Andromeda-class attack cargo ships, 163, 165, *171, 183, 192*
Anne Arundel (AP 76), 160
AN (netlayer), 178–79, *179*, 181
Antares, 36
antiaircraft craft, 246
AP (transport ship): civilian operation of, 468; in combat zones, 177–78; conversions, 49, 54–57, 151–63, *179, 179–81*; designation, 15, 151
APA (attack transport ship): armament, 161, 188; capacity, 186–88, 315; in combat zones, 177–78; conversions, 49, 54–57, 151–63, 186–94, 323–25; designation, 15, 151; as flagship, 274–75, 430; shortage, 182–86; small, 167, 169, 171–75; speed, 311–13; wartime camouflage plan, *43*
APA-M, *349*, 349–50
APD (transport destroyer), *16*, 33, 195–97, 283, 344–45
APF (administrative flagship), 261
APH (casualty evacuation ship), 64–65
Appalachian (AGC 1), 265, 268, 273
Appalachian-class, 261
APP designation, 73
APSRON 4 (Afloat Prepositioning Ship Squadron Four), 479
APSS (transport submarine), 348–49
APY (special troop transport (short-radius)), 142
ARC (cable layer ship), 174
Arctic operations, 288
Arcturus, 49, 57, *59*, 62, 64
ARG (amphibious ready group), 11, 378, 458
Argonne, 30
Arlington (AP 174), *185*, 186

Arneb (AKA 56), *171*, 315
Artemis-class attack transport (AKA 21), *177*
Arthur Middleton (AP 55/APA 25), *152, 153, 154*, 275
artillery lighters, *71, 72, 72, 73*; designation, 77; program, 484
Ashland (LSD 1), 131
Ashtabula, 474
Askari (ARL 30), *102*
assault tactics, 369–70
ASW (antisubmarine warfare), 281, 354, 356
ATACMS (army tactical missile system), 453
Atlantic amphibious assault lift, 11–13
ATL (Atlantic TLC), 117
atomic bomb, 347
attack transports: acquisitions in 1940s, 45–46; cargo ships for, 49–54; characteristics, 26, 51; conversions, 39–44, 57; davits on, 48–49; and landing craft, 38; large war-built, 168–69; priority of, 157; small wartime, 175. *See also* AKA (attack cargo ship); APA (attack transport ship)
ATU (amphibious task unit), 11
Auburn (AGC 10), 266
Augusta, 261
Auriga (AK 98), 178
Austin (LPD 4, SCB 187B), 365
Austin-class, *366, 367*
Auxiliary Vessels Board, 9
AVP 52 (small seaplane tender), 283
AWR-3 (army warfare readiness), 479

Bachaquero, 112
Baker, Chauncey B., 24
Baker, Rowland: on LCT, 133, 134; on LCT(5) alternatives, 138; on LCVP, 208; LST design, 118, 121; on TLC carriers, 111, 117, 127; on Winette design, 115
Baltimore-class cruisers, 425, 426
Baltimore Mail liners, 30, 45
BARC (barge, amphibious resupply, cargo), 303
Barge A, *69*, 70, 71
barges, *69*; as alternatives to LSD and LHD, 444–46; BARC, 303; development, 70–74, 468; transporters, 115; troop, 483
Barnett (AP 11/APA 5), 43, 45, *45, 47*, 188
Barnstable (APA 93), *163*, 275
Bayfield (APA 33), as relief AGC, 274
Bayfield-class, 163, 183, 190, 274, 325
Bay Head boats, 77–78
beachjumpers, 75
beetles, 67, *68–69*, 70
Bellatrix (AK 20/AKA 3), 51, *58*, 151

654 INDEX

Benewah (APB 35), *12*
Berg, J. V., 120
Bikini atomic tests, 174, 347
Biscayne (AVP 11/AGC 18), 268, 273, 274
Blessman (APD 48), *66*
Block Island (CVE 106), 355
BLTs (battalion landing teams), 6–7, 11, 37, 369, 458
Blue Ridge (LCC 19/AGC 2), *262*, *428*, *429*, *430*, *431*; characteristics, 273
Blue Ridge–class, 429; designation as LCC, 429
Board of Inspection and Survey (INSURV), 441
boats, 70, 72–78, 484, 491
Boxer (LHD 4), 359, *379*
Boxer-class, 378
Bradley, M2, 439
break bulk ships, 468
Briareus (AR 12), 57
Bright, C. B., 320
broadband sleeve antennas, 262–63
Brodie system, 126
Brown, Harold, 443
Brunettes, 112–13
BuAer (Bureau of Aeronautics), 3
Buckley-class destroyer escort, *195*, 195
BuEng (Bureau of Engineering), 75
Bullpup missile, 407
Bulolo, 261
Bunker Hill, USS, 358
Bureau of Aeronautics (BuAer), 3
Bureau of Construction and Repair (C&R), 3
Bureau of Engineering (BuEng), 75
Bureau of Medicine and Surgery, 64
Bureau of Ordnance (BuOrd), 3, 4, 13. See also Bureau of Weapons (BuWeps); Naval Ordnance Systems Command (NAVORD)
Bureau of Ships (BuShips), 12–13. See also Naval Sea Systems Command (NAVSEA)
Bureau of Steam Engineering, 3
Bureau of Weapons (BuWeps), 426. See also Bureau of Ordnance (BuOrd)
Bureau of Yards and Docks, 3
Burke, Arleigh, 12, 358
Burke (APD 65), *195*
Burleigh (APA 95), 275, *275*
Burleson (APA 67), 174
Bush, George, 448
BuShips. See Bureau of Ships (BuShips)
Butte (APA 68), *173*

C1-B, 184
C2 (cargo ships): conversions, 49–57, 151–55, 161, 183, 188–89, 194; designations, 32–33
C2-Fs, 151
C2-S-A1s, 54, 65
C2-S-AJ1s, 184, 266, 267
C2-S-B1s, 163, 184, 191, 261–64
C2-S-E1s, 163
C3 (combat loaders): capacity, 188; conversions to AKA, 53–57, 163, 188–89; conversions to APA, 54–55, 64, 151–57, *155*, 161–63, 183, 188–90; designations, 32–33; hulls, 188–89, 192; speed, 314; triple davits on, 49
C3-S-A2s, 163
C3-S-A3s, 163, 184
C3-ST-14a designation, 468
C4, 182
C4-S-1As. *See* Mariner (C4-S-1A)
C-1981, *71*
Callaway (APA 35), 275

Calvert (AP 65/APA 32), 157, 275
Cambrai, 30
Cambria (APA 36), 274, 278
Cantigny, 30
Capella (AK 13), 36, 90
cargo ships, 157, 468, 472–74. See also C1-B; C2 (cargo ships); C3 (combat loaders); C4
Carney, Robert, 317
Carronade, USS, *257*, 257–58, 405
Carter, Jimmy, 13, 446, 472
Casablanca-class escort carrier, 349, 350
Casa Grande (LSD 13), *150*
casualty evacuation, 124
Catoctin (AGC 5), 263–64, 272
Catskill, 180
causeways, 197–99, 337, *340*, 478–79
Cavalier (APA 37), 274
Cecil (APA 96), 275
Center for Naval Analyses (CNA), 13
CF/CD (concept formulation/contract definition), 407–17
Charles Carroll (AP 58/APA 28), *156*, 275
Charleston (AKA 113), *328*, *329*; characteristics, 312
Charleston-class, 295, 327
chartering, 475–76
Chateau Thierry (AP 31), *29*, 30, 54
Chaumont, 30, 31, 37
Cheboygan County (LST 533), *124*
Chemold LCP(N) (landing craft, personnel, nested), *211*, 211
Chief of Naval Operations (CNO), 4
Chilton (APA 38), *10*, 274, 325
Christie amphibious tanks, 18
Churchill, Winston, 103, 111, 120
Clarion River (LSMR 409), *259*
Clay (APA 39), 275
Clemson (APD 31), *38*
Cochrane, Edward: on design of attack ships, 167; on LCIs, 141, 142; on LCT(5) alternatives, 138; on LSTs, 117–18, 120; on tank lighter design, 98; on TLC carriers, 127
Cole, Eli K., 24
Colwell, John B., 340
Combined Operations Headquarters (COHQ), 111, 117, 118, 261
Comet (SCB 236), 373, 468–70
Comfort (T-AH 20), 477
command ships, 261
Commencement Bay–class (CVE), 355
CONFORM (concept formulation), 12
Conolly, Richard, 122, 232
Constitution, 314
container ships, 468. See also cargo ships
control craft, 103, 261
control fleet, U.S., 467
control organization, 281–83
control ships, 265
Convair design study, 348
Copenhaven, G. H., 165
Coronado (LPD 11), 365, *367*, 431
COTS (container offloading and transfer system), 478
Courageous, 30
C&R (Bureau of Construction and Repair), 3
Crescent City (AP 40/APA 21), 49, *63*, 64
Crosby, 85
cube, 37
Curtiss, 90
Custer (APA 40), 275
Custer-class, 163, *164*, 274
CVE (escort carrier), 350, 351, 355–56
CVHA (helicopter assault carrier), 350
CVHE (helicopter escort carrier), 350
CVL (light carrier), 348, 350, 351

Dade (APA 99), 275
Daniels, Josephus, 5
davits, 46, 48–49
DEC (destroyer escort, control), 283
deck space, 37–38
DE (destroyer escort), 195–97, *198*, 233, *233*
Defense Department Posture Statement, 429–30
Delargentino (AP 64/APA 31), 55, 157
Delbrasil (APA 105), 178
Delbrasil–class, 54–55
Deloreans, 157
Delsantos (C2), 160
Delta (AK 29/AR 9), 57, 59, 62
Del Uruguay, 151
Department of the Navy, 3
Design 1024, Hog Island Type B ships, 30
Design 1095, 29–30
designations, 15
Des Moines–class cruisers, 403, 406, 426
De Soto County (LST 1171), 305, 308, 335
destroyers, 33–36
destroyer tenders, 57, 59–61
Development of Landing Boats for Training Operations, 77. See also Landing Boat Board
Dewey, George, 3, 4
Dewey-class missile destroyer, 403
Diamond Mariner, SS, *317*
Dickman, USS, 224
diesel-electric powerplants, 169–71
District Craft Board, 74–75
Donner (LSD 20), *127*, *129*
Dorling, J.W.S., 119, 140–41
Dorthea L. Dix (AP 67), *32*, 160
Doyen (AP 2/APA 1), *106*, *107*, *108*, *109*, 109–10
Doyen-class, 45, *45*, 103, 105–10, *106*, 172
dual-drive tanks, 213
Duane (W 33/AGC 6), 265, *266*, 273
DUKWs, 101, *101*, *181*, 188, *494*
Du Page (APA 41), 55, 275
Durham (LKA 114); *329*
DX destroyer, 408, 412
DXG destroyer, 408, 412

Earle, Ralph, 67, 72
Eastway (LSD 9), HMS, *116*
Edward Rutledge (AP 52), 49, 151
Effective U.S. Control Fleet (EUSC), 477
Eisenhower, Dwight D., 11, 13, 403
ELCAS (elevated causeway), 478–79
Eldorado (AGC 11), *267*, *424*; configuration, 266; functions, 274, 275; refitting, 278
Electra (AK 21/AKA 4), 51
electric surfboat/surfboard program, 491
Elizabeth C. Stanton (AP 69), 160
Ellis, Earl ("Pete"), 5
Elmore (APA 42), 274, 275
El Paso (LKA 117), *328*
Empire Weapons class, 183, 184
engines, Snadecki, 292
enhancements, general, 478
Enright (APD 66), *197*
ENTPS (enhanced near term prepositioning squadron ship), 472
Essex (LHD 2), *451*, *455*
Essex-class, 358–63, *360*, *362*
Estes (AGC 12), 274, 275, *277*
Eurekas, 75–86, 79
Evergreen Mariner, 323
Excalibur, 151
Excambion, 151
Exceller (C2-S-A1), 54

Exemplar (AP 67), 159
Exeter, 151

Fayette (APA 43), 275
FDL (fast deployment logistics), 408, 470–71
FDT (fighter direction tender), 265
Feland (AP 18/APA 11), 109, *110*
FFD (forward-floating depots), 470
FIE (fly-in element), 447
fire support: craft, 103, *222*, 227; for infiltration, 457–58; in 1980s, 452–54; ships, 13, 227, *409*, *411*, 454. *See also* guns; missiles; rocket launchers; rockets
1st Lt. Harry L. Martin, USNS, 479
fixed defense regiments, 23
flag merchant fleet, U.S., 467, 477
flagships: British role in, 103; characteristics, 273, 426; LCI(L) as, 283–84; LPD as, 430, 431; LST as, 264, 283; need for, 261, 265, 425. *See also* AGC (command ship)
flatracks, 478
Fleet Employment Plan for fiscal year (FY) 49–50, 10–11
fleet landing exercises (FLEXs). *See entries beginning with FLEX*
Fleet Marine Force (FMF), 6, 11, 19
Fleet Rehabilitation and Modernization (FRAM) programs. *See* FRAM programs
FLEX 1, 19
FLEX 2, 19
FLEX 3, 19, 73, 77
FLEX 4, 19, 75, 77, 87
FLEX 5, 19, 35, 100
FLEX 6, 19
FLEX 7, 19, 20
flight decks, 31, *125*, 126
flo-flo, 115
Florence Nightingale (AP 70), *158*, 159
flotillas, 283–84
flycatchers, 251
Flying Mattress, 211
Folta, George W. Jr., 410
Force X, 184
foreign ships, U.S.-owned, 467
Formalhaut (AK 22), 49, 61, 62
Forrestal, 359
Forrest Sherman–class destroyer, 403, 411, 453
Fort Snelling (LSD 30), *310*
four aces, 49, 151
FRAM programs, 127, 190, 278, 325, 360
Francis Marion (APA 249), 325
Frederick Funston (APA 89), 177
Freeport boats, 77, 78
freighters. *See* cargo ships
Fremont (APA 44), *162*, *276*, 278

gantry ships, 111
GEM (ground effect machine), 395–96
General Board, 3–4; dissolution of, 12; and ship characteristics, 9; and transports, 23–27, 31, 32
George Clymer (APA 27), *151*, *154*, 275
George F. Elliott (AP 13), 64, 178
Germantown (LSD 42), *444*
Giant Y designation, 142
Gilliam (APA 57), *177*
Gilliland, (T-AKR 298), *466*
Glen liners, 103
Global Positioning Satellite (GPS), 458
Good, Roscoe F., 323
Goodall, Stanley, 115
Good Neighbor Policy, 6
Grand Joint Exercise No. 4, 75

Grant County (LST 1174), *307*
GRC (giant raiding craft), 142
Guadalcanal (LPH 7), 354, *357*
Guam, USS, 380
Gun Fire Board, 18
gunfire support ships, 126
guns, 18, 223, 403–4; 16-in, *409*, 409–10; 155-mm, 453, *456*
Gunston Hall, 442

Hall, John L., 230, 282
Halsey, William F., 148
Hamblen (APA 114), *22*, 164
Hamul (AK 30/AD 20), 57, 59, 62
Harpers Ferry (LSD 49), 313, *448*, 448
Harris (AP 8/APA 2), 39, 43, 274
Harry Lee Exochorda (AP 17/APA 10), 45–46, 49, *54*
Haskell (APA 117), *191*
Haskell-class, 190, *191*, 325
Haynes, Fred E. Jr., 14
Hayward, Thomas, 449–50, 472
Helena (CA 75), 425
helicopter carriers, 351, 356. *See also* LHA (landing ships, helicopter, assault); LHD (landing ship, helicopter, dock); LPH (landing ship, personnel, helicopter)
Henderson (AP1), 24–27, *25*, 67
Henrico (APA 45), 274, *276*, 278
Henry T. Allen, 51
Hercules (AK 41), 191
Hewitt, Kent, 261
Heywood (APA 6), 27–28, *28*, *52*, 188
Heywood-class, 30, 45, *51*
Higgins, Andrew Jackson, 75, 77, 78, 80–83, 292–94
Higgins boats, *20*, *35*, *79*, *83*, *86*; conversion of LCS(S), 232; Eureka, 75–86, *79*; LCM(3), 212; personnel, 99; ramped, 84; support, 223; tank landing craft, *94*, 94–98, *95*
High Pockets hydrofoil, 390
HMX-1, 348
Hogaboom, Robert E., 355
Hog Island transports, 30
Holcomb, Thomas, 100
Hollis, James E., 231
Holloway, R. E., 115
Hooper Committee, 406
Hopkins, Harry, 120
Horne, Frederick J., 9
hovercraft, 395–96
Howard, Herbert S., 120
HR2S helicopter, 353
HSC (heavy support craft), 223
Hugh L. Scott (AP 43), 157, 159
Hull (destroyer), 404, 422
hull specifications, Maritime, 33
Hunter Liggett, 51, *60*
Hussey, Thomas A., 115, 119, 138
hydrofoil, 383, 390–94, *391*, 394
hydrokeel program, 397

IFS (inshore fire support), 257–58, 411–12
Illinois, 472
ILS (infantry landing ship), 103
Inchon (LPH 12), 370, *371*, 380
Independence, 314
Independence-class, 428
infiltration tactics, 457–58
Inter-Service Committee on Communications, 261
Interservice Training and Development Centre (ISTDC), 103
Iowa-class battleships, 403, 406, 454

iron horse, 75
Iwo Jima (LPH 2), 363, *363*
Iwo Jima-class, 362–63

J. Franklin Bell, 54
Jacob Jones, 33–34
James O'Hara (APA 90), *21*, 177
JCC (joint craft, control), 431
jeeps, amphibious, *101*
Jeff (air cushion vehicle), 434–35
Jersey Sea Skiff, 289
Jim (air cushion vehicle or planing craft), 434–35
Joe (planing craft), 434
John Penn (AP 51), 49, *55*, 57
Johnson, Hubert S., 77
Johnson, Louis, 10
Johnson, Lyndon, 470
Joint Action of the Army and Navy, 6
Joseph Hewes (AP 50), 49
Joseph Hewes–class, 35
Joseph T. Dickman, 51, *59*
JPTDS (Junior Participating Tactical Data System), 375
Jupiter (AK 43), 191
JVX (troop carrier), 439

Kalbfus, Edward C., 100
Kane (APD 18), *36*
Keating, Frank A., 19–20
Kenmore (AP 62), 30, 65
Kennedy, John F., 13, 403, 469, 470
Kent (AP 28), 57
Keyes, Roger, 115
Keystone State (T-ACS 1), 477
Kidd-class destroyers, 415
King, Ernest J.: and AGC, 266, 267; appointment as Chief of Naval Operations, 8–9; and attack ship conversions, 157, 160, 183, 188; and control of bureaus, 4; on 16 in guns on LFS, 410; and LCT, 202; and LST, 120; on small transports, 105, 171–72; on troop accommodations, 38
Kinzer (APD 91), *196*
Kirk, Alan G., 186, 230
Kirsten, Frederick, 384
Kirsten cycloidal propellers, 384–90, *388*
Knox (APA 46), 275
Knox-class, 403, 411, 452, 454
Krulak, Victor H., 36, 82, 85

Lacrosse surface-to-surface weapon, 404
Lake Champlain, 359
Lamar (APA 47), 275
Lance missile, 404–5
Land, Emory S., 7, 45, 104–5
landing boat allowance, 43
Landing Boat Board: on amphibians, 100; creation of, 9; and Eurekas, 81; on LCC, 278–79; on LCS designs, 223; on LCVs, 84–85; on tank landing craft, 87, 94, 95; on transport conversions, 33–35
landing craft: designations, 77, 98; ideal design, 77; prototype shallow-draft, 75, 77; shortage, 157, 219–21; small, 80; speed, 383; for tanks, 86–87; 30-ft program, 484
Landing Craft Coordinating Group, 434
Landing Operations Doctrine (FTP 167), 6
LARC (lighter, amphibious resupply, cargo), 302–3
Largs, HMS, 261
La Salle (AP 102), 178
La Salle (LPD 3), 365, 431
LASH (lighter aboard ship), 444

Laycock, John W., 197
LBP (landing boat, personnel), 82
LBS (landing boat, support), 223
LBV (landing boat, vehicles), 84
LCAC (landing craft, air cushion), *436, 437, 438;* ancestor of, 213; on barge carriers, 444–46; future plans for, 479; inability of LHA to support, 381; and LSD design, 441–43, 449, 452; and LX design, 443–44; OTH capabilities of, 433; post–WW II construction, 303, 390; program, 484; and Reagan administration, 14; requirements, 435–39; SLEP prototype, *459*
LCA (landing craft, assault), 302
LCA (landing craft, assault (British)), 138, 208
LCA(X)1, *302*
LCAX2, 302
LCC (landing craft, control), *278;* and AGC conversions, 425–26, 429–31; designation, 223–24; Mk 1, 279, *280;* Mk 2, 279, *281;* program, 484; uses, 278–83
LCF (landing craft, flak), 226, 230
LC(FF) (landing craft, flotilla flagship), 284
LCG (landing craft, gun), 226, 230–31
LCI (landing craft, infantry), 138, 140–42, *144, 222*
LCI(D), 148
LCI(G) (landing craft, infantry, gunboat), *234, 235, 255;* conversion from LCI(L), 251–53; development, 233–38; Type A, *241;* Type C, *241, 242;* Type D, *243, 244*
LCI(L) (landing craft, infantry, large), *9, 143;* adaptation for fire support, 230–33, 236–39, 251–53; alternative configurations, 142–43; British role in, 103; cancellation of, 220; in combat, 145; development, 140–43; as flagship, 283–84; hull formation, *142;* internal arrangement, *145;* modifications, *144,* 145–48, *146;* program, 481; ramps, *147;* redesignation as LSIL, 288; in salvage efforts, 208, 209; seaworthiness, 143–44; studies of, 136
LCI(M) (landing craft, infantry, mortar), 237–38
LCI(R), *235,* 239–42, *245,* 246
LCI(S) (landing craft, infantry, small), 140, 142, 226
LCM (landing craft, mechanized): armament, 246; carriers, 113, 186; complaints about, 212–13; designation, 91, 94; development, 86; forerunners of, 18, 25; inferiority to LCT(5)s, 136; post–WW II stock, 288; program, 487–90; replacement, 433–34; in salvage efforts, 208; stowage on APAs, 186–88
LCM(1), 91, 98
LCM(2), *92, 93,* 98, 226, 487–88
LCM(3), *97;* armament, 259; carriers, 186–88; conversion to LCM(6), 212; designation, 98–99; Higgins version, *98, 99,* 212; post–WW II development, 292; program, 488–90; replacement of *Doyen*-class transports, 109; in salvage efforts, 208–9
LCM(6), *12;* development, 290–95; Mod 1, *291, 293;* Mod 2, 439; need for, 212–13; program, 488–90; replacement, 434; speed, 439
LCM(8), *294, 296;* development, 213, 290–95; and LCACs, 438; replacement, 434; SLEP program, 439
LCM(9), *434,* 434–36
LCM(K) (landing craft, mechanized, hydrokeel), 397
LCM(R) "Woofus," *247*
LCM(X), 212

LCP(L) (landing craft, personnel, large), *81;* basis for Mk 2 reconnaissance boat, 398; comparison to Eureka, 80; comparison to LCC, 278; comparison to LCV, 85, 86; designation, 82; Mk 1, *86;* Mk 2, 295; Mk 3, *290,* 296; Mk 4, 296–98, *297;* Mk 11, *297,* 298, *298;* post–WW II, 288, 295–98; program, 484–86
LCP(M) (landing craft, personnel, medium), 82
LCP(N) (landing craft, personnel, nested), 80, 211, *211,* 487
LCP(R) (landing craft, personnel, ramped), *16, 36, 37, 76, 85, 87, 89;* characteristics, 80; post–WW II stock, 288; program, 484–85; as support for UDTs, 86
LCP (Sy) (landing craft survey), 278
LCPV(H) (landing craft, vehicles and personnel, hydrojet), 289
LCR(L) (landing craft, rubber, large), 101, 490
LCR(S) (landing craft, rubber, small), 101, 490
LCS(L) (landing craft, support, large), *239, 480;* adaptation for fire support, 231–32, *240;* designation, 223; Mk 3, *231, 238, 402*
LCS(L)1, 226
LCS(L)2, 226
LCS(L)3 (landing craft, support, large, Mk 3), 236, 242, 481
LCS(M) (landing craft, support, medium), 223, 226
LCSR (landing craft, swimmer reconnaissance), 397–98, 490–91
LCSR(K) (landing craft, swimmer reconnaissance, hydrokeel), 397, 398
LCSR(L) (landing craft, swimmer reconnaissance, large), 398
LCS(S) (landing craft, support, small): armament, *224, 225, 226,* 232; program, 491; uses, 223–26, 279
LCS(S)2, 232–33
LCT (landing craft, tank): assembly in water, *132;* British role in, 103; BuShips design, *132;* cancellation of, 220; development, 131–34; and dual-drive tanks, 213; for fire support, *229;* on LSTs, 122–23, *136;* modification, 201–5; post–WW II uses, 288; program, 481–83; redesignation, 383; in salvage efforts, 208
LCT(2), 227
LCT(3), 113
LCT(4), 114
LCT(5), *135, 384;* alternatives to, 136, 138; designation, 117, 134; preference for, 136; in salvage efforts, 209
LCT(6), *133, 138;* configuration, *140;* detail poster, *139;* and dual-drive tanks, 213; enlargement, 201–3; evaluations of, 136
LCT(7), 203
LCT(8), 205
LCT 501, 136
LCT(A) (landing craft, tank armor), 242, 246
LCT(CB) "concrete buster," 246
LCT(HE), 246
LCT(R) (landing craft, tank, rocket), 227
LCT(SP) (landing craft, tank, self-propelled [gun]), 242, 246
LCU (landing craft, utility), *385, 386, 389, 460–61;* air cushion design, *387;* designation, 303, 383; Kirsten cycloidal propellers on, *384–90;* and LCACs, 438; maneuverability, 320; program, 481–83; replacement, 434
LCV (landing craft, vehicle), *83, 84, 88;* comparison to Eureka, 80; program, 486–87; uses, 84–85, 86

LCVP (landing craft, vehicles and personnel), *206, 209, 210;* armament, *228,* 259; hydrofoil, 390–93, *391;* hydrokeel, *395;* internal arrangement, *208;* plastic, *286,* 289–90; post–WW II stock, 288; program, 486–87; prototypes, 289, *391;* relationship to Eureka, 80; replacement, 433; speed, 219; stowage of, 186–88; uses, 85, 207–11, 279–82
LCVP(H) (landing craft, vehicles and personnel, hydrojet), *396,* 397
LCVP(K) (landing craft, vehicles and personnel, hydrokeel), 397
LCVP(T) (landing craft, vehicles and personnel, planing), 397
Leedstown (AP 73), *158, 159, 167*
Lejeune, John A., 5
Lemnitzer, Lyman L., 320
Leon (APA 48), 275
Leonard Wood (AP 25/APA 12), *41,* 51, 274
Leopold, Reuven, 376
LFS (landing ship, fire support), *405, 406;* armament, 452–53; arrangement sketches, *417–22;* delays in production, 13, 407–11; development, 405–6; maximum- and minimum-cost designs, 417; as multimission ship, 413–16; *Spruance*-class version, *413,* 417–22; studies of, 408–17
LFSC designation, 412
LFSR (landing ship, fire support, rocket), 408, 411–12
LFSW (landing ship fire support weapon), 404, *405*
LHA (landing ships, helicopter, assault): design alternatives, 372–76, 430–31; designation, 374; plans, *371;* production, 377–78, 408; redesign as LHD, 448–49; Scheme B-3, *372;* Scheme C, *373;* Scheme D, *374;* uses, 13, 380–81
LHA-X, 430
LHD (landing ship, helicopter, dock), 448–49
LHDX, 448–49
Liberty ships, 7, 157, 186, 188
Libra (AK 53/AKA 12), 151, 165
Lighter B (APB), 72
Lighter C, Experimental, 72
Linsert, Ernest E., 82–83, 94
Lipscomb, 472
Little, Louis M., 100
LKA (cargo ship, attack), 15
LKD-X, 446
Lloyd's of London, 111–12
loading, 17, 28
LOGLAND study, 470
Logsden, E. H., 141, 142
Long, John D., 3
long-range attacks, 17
Lothian, HMS, 184, 272–74
LPA (transport ship, attack), 15
LPD (landing ship, personnel, dock): characteristics, 364–69; development, 331, 350; and flagship design, 430, 431; SLEP program, 448
LPD 20, 458–61
LPDX, 448
LPH (landing ship, personnel, helicopter): CVEs as, 355–56; development, 350–63; *Essex*-class carriers as, 358–63; modifications, 353–59, 370; as modified AGC flagship, 427–31; preliminary characteristics, 350–53; replacement, 440
LPH(X), 440
LPR (amphibious transport, personnel, raiding and reconnaissance), 283
LPV (landing ship, personnel, fixed-wing aircraft), 363

LRGS (long-range gun system), 404
LSC (light support craft), 223
LSD (landing ship, dock), *116;* armament, 130; British role in, 103, 115; capacity, 327–29; final design, 128–31; LCACs on, 435–37; modifications, 131; plan, *128;* problems, 130; replacement, 440; SLEP program, 441; speed, 311–13, 327–36; superdeck plan, *131*
LSD 41, 441–46, 448
LSFF (landing ship, flotilla flagship), 288
LSFH (landing ship, fire support, heavy), 408
LSI (landing ship infantry), 103
LSI(L) (landing ship, infantry, large), 183, 184
LSIL (landing ship infantry, large), 142, 288
LSM (landing ship, medium), *202, 204, 388, 480;* armament, 246; carriers, *150;* development, 201–5; internal arrangement, *203;* origin, 134; post–WW II uses, 288; program, 483
LSM(R) (landing ship, medium, rocket), *248, 249, 250, 252, 253, 254, 480;* development, 246–49; modifications, 256–58; program, 483
LSMR (large sealift ship, medium speed, ro-ro), 479
LSSL (landing ship, support large), 242, *402*
LST (landing ship, tank), *102;* armament, 123, 242, 411–12, 423; beaching, 337–41; bow and bilges, *121;* British role in, 103, 114–27; capacity, 122–26, 341–44; carport, 335, *339;* catamaran, 337, *341;* causeway-regurgitating, 337, *340;* compartmentation, *120;* as control ship, 265; conversion to LFS, 452–53; davits on, *122;* elevators, 121, 124; first use of, 120–21; 542 class, 124; as flagship, 264, 283; flight decks on, *125, 126;* flying, 348; hull design, 122; LCTs on, 134, *136;* modifications, 123–27, *126,* 199–201; *Newport*-class, *2;* original configuration, *119;* pontoons on, *137,* 197–99, 338; post–WW II, 288, 303–8; production, 20–21, 120, 122; program, 483; ramps, 121, 124, 199–200, 339–40; refitted, *123;* as ro-ro, 468; speed, 127, 199, 303, 311–13, 334–41, *336–37, 338;* turntables on, 124; ventilation system, 121; 1942 version, *117*
LST(1), 115, 117, 127
LST(2), 117
LST(3), 200–201
LST(H) (landing ship, tank, hospital), 124
LSTS (landing ship, tank, small), 383
LSU (landing ship, utility), 303, 383, *385*
LSV (landing ship, vehicle), *179,* 181–82
LVA (landing vehicle, assault), 439
LVHX1 hydrofoil, *394,* 395
LVHX2 hydrofoil, 395
LVT (landing vehicle, tracked), *100, 101;* demand for, 213–19; development, 6, 99–100; forerunners of, 46; on LSTs, 124; post–WW II, 298–302; program, 491–93; in salvage efforts, 208; speed, 398–99; transportation of, 181, 348, 349, 435–36; use of, 207, 279–82
LVT(1), 214, 217
LVT(2), 217
LVT(3), 219, 300
LVT(3) Bushmaster, 218
LVT3C (covered), 300
LVT(4), *216,* 217–19
LVT(A), 300
LVT(A)1, *215,* 216–18, *217*
LVT(A)4, *217,* 218
LVT(A)5, 218, 300
LVTAA-1, 301

LVT(A)2 (landing vehicle, tracked, armored), 214
LVTC7, 400
LVTCR-1, 301
LVTE-1, 301
LVTH6, *300,* 301
LVTHX4, 302
LVTP5, *299,* 300–301, 398–400
LVTP6, 302
LVTP7, 301, *399,* 400, 440
LVTP7A1, 440
LVTPX1, 299
LVTPX2, 301–2
LVTPX3, 299
LVTPX12, 302, 399–400
LVTR-1, 301
LVTR7, 400
LVT(U) (landing vehicle, tracked, unarmored), 299, 300, 302
LVT(U)X1, 302
LVT(U)X2, 302
LVT(2) Water Buffalo, 213–14, *214*
LVT(X), 439–40
LVW (landing vehicle, wheeled), 101, 395
LX (landing ship), 380, 440–41
LXA (assault amphibious ship), 443
LXS (support amphibious ship), 443–44
Lykes, 472
Lynde, USS, 414
Lyon (AP 71), 160

MAB (marine amphibious brigade), 11, 446, 472, 473
MacArthur, Douglas, 9
MAF (marine amphibious force), 10, 11, 380, 440–43, 446
MAG (marine air group), 11, 369
MAGTF (marine air-ground task force), 11
Mahnomen County (LST 912), 339
Maine-class ro-ros, 472, 474–75
Manhattan, 55
Manley, USS, *34,* 35, *35*
Marine Brigade, 23
Mariner (C4-S-1A), 33, 318, 321–23, 325, 430–31
Maritime Commission, 7, 31, 32
Markab (AK 31/AD 21), 57, 59, 62
Marmon-Herrington tanks, 87
Marne, 30
Mars (AFS 1), 427
Marshall, George C., 8, 119, 120, 219
MARS (multiple army rocket system), 450
Martin design study, 348
Massachusetts Institute of Technology (MIT), 416
Matthews (AKA 96), *170, 192*
MAU (marine amphibious unit), 11
MAW (marine air wing), 11
McCawley (AP 10/APA 4), *7, 20,* 43, 45, *46*
McCormick (DDG 8), 414
MCDEC (Marine Corps Development Center) project 30-64-01, 406
McDonald, David L., 413
McDowell, Ralph S., 77
MCLWG (major caliber lightweight gun), 403–4, 419
MCMOBE (Marine Corps mobility enhancement), 472–74
McNamara, Robert S., 13, 370, 376, 407–8, 471
MCS (mine countermeasures mother ship), 408
M-day, 51
measurements of ships and craft, 15
MEB (marine expeditionary brigade), 11, 378

MEF (marine expeditionary force), 369, 370, 378, 406
Mendenhall, William K. Jr., 258
Mendocino (APA 100), 275
merchant ships: and AKA design, 316–18; conversion to transports, 31, 41, *42,* 61, 73–75, 111–12, 188; production, 30–31; as sealift assets, 467–68, 477
Mercury, USNS, *473, 474*
Mercury (AK 42), 191
Mercury (T-AH 19), 477
Meteor (T-LSV 9), 469, 472
MEU (marine expeditionary unit), 11, 458
Mexafloat, 199
Military Sealift Command (MSC), 468, 470
Military Sea Transportation Service (MSTS), 468–69
Miller, Ellis B., 5
missiles, 353, 404–5, 407, 426
MLC (mechanized landing craft): designation, 91, 95, 96, 98; development, 103; launching, 111
MLC MkII, 97
MLRS (multiple long-range rocket system), 416, 449–50
Mobile Base Plan, 28
mobile defense regiments, 23
MOBs (mobile offshore bases), 458
Monitor (LSV 5), *181*
Monrovia (AP 64/APA 31), 157, *157,* 274, 275
Montauk (LSV 6), 182, *182,* 183
Monterey (AP 68), 159
Monticello (LSD 35), *332,* 382
Montour (APA 101), 275
Montrose (APA 212), *12*
Mormacstar (AP 69), 159
Mormacsun (AP 70), 159
Mormactide (AP 71), 159
Mormacyork (C3), 160
mortars, 237–38, 249–51
Moses, E. P., 100
motor launches, *18,* 77
Mountbatten, Lord Louis, 138
Mount McKinley (AGC 7), *260, 268, 269–71;* accommodations, 268; characteristics, 273; refitting, 278
Mount Olympus (AGC 8), 275
Mount Vernon (LSD 39), *333*
Mount Whitney (LCC 20), 429
Mousetrap projectors, 236–37
MPS (maritime prepositioning squadron), 458, 473–76, 477
Mulberry (artificial port), 21, 199
"must" program, 188–95, 266
Myers, Charles, 412–13

Nassau (LHA 4), *379,* 449
National Defense Reserve Fleet (NDRF), 467, 477–78
NATO (North Atlantic Treaty Organization), 14, 477
"Naval Gunfire Requirements for the Long Range Period Through 1975 for a Division/Wing Team (MEF)", 406
Naval Landing Force Equipment Depot (NLFED), 232
Naval Ordnance Systems Command (NAVORD), 13, 403, 410
Naval Sea Systems Command (NAVSEA), 12, 13, 407, 478. *See also* Bureau of Ships (BuShips)
Naval Ship Engineering Center (NAVSEC), 409, 410, 415–16
Naval Ship Systems Command (NAVSHIPS), 13, 410–17

"Naval Surface Fire Support [NSFS] Improvement Study," 453
Naval Transportation Readiness Plan (WPL-10), 28, 223
Naval War College, 3
NAVWAG 4 ("Capabilities and Requirements for the Fire Support of Amphibious Assault"), 405
NAVWAG 36, 407, 410
NAVWAG 44 ("Amphibious Assault Shipping in the Mid-Range Period"), 370, 374–75
Navy-Marine Continuing Board for Tracked Landing Vehicle Development, 217
Nevada, USS, 245
Neville (APA 9), 282
Newberry, Truman H., 4
New Jersey (BB 62), 412–13, 452
New Orleans (LPH 11), 370
Newport (LST 1179), 313, 341
Newport-class, 2, 115, 342, 344, 403
New River exercises, 100, 114
Niedermair, John C., 117–18
Nifty Nugget, 472
Nitze, Paul, 374
Nixon, Richard, 13
NL (Naval Landing Force Equipment), 197
NTDS (naval tactical data system), 425
NTPF (near-term prepositioning force), 472–73
NTPS (near-term prepositioning squadron), 472–76
nuclear weapons, 347, 352, 403

Oberon (AK 56/AKA 14), 151
Ocean, HMS, 446
OEG (operational evaluation group), 13
Office of Strategic Services (OSS), 211
Office of the Chief of Naval Operations (OpNav), 4; Ad Hoc Committee to Study the Long-Range Shipbuilding Program, 12–13, 350, 353; authority in 1960s, 13; post–WW I role, 8–9
Office of the Secretary of Defense (OSD), 13
Ogden (LPD 5), 368, 398
Okanogan (APA/LPA 220), 191
Okinawa, 352
OMFTS (operational maneuver from the sea), 457, 479
Op-36, 406, 419
Op-37, 13
Op-93 (long-range objectives group), 12, 363–64, 405, 457
OPDS (offshore petroleum discharge system), 478
OpNav. See Office of the Chief of Naval Operations (OpNav)
Oriente, 55
Orizaba (AP 24), 182
Ormsby-class transports, 163, 165
Osprey, M-22A, 458
Osprey, MV-22, 439, 479
OTH (over the horizon) craft, 433–34
Ourcq, 30
Outboard Landing Craft, 211–12, 487

P2, 314, 428
P2-SE2-R1 class troopships, 428
P5Y, 348
Pacific amphibious assault lift, 11–13
Palau (CVE 122), 348
palletization, 213
Panama, 55
Panamint (AGC 13), 266

P&C (passenger and cargo), 54–55, 64, 155
passenger ships, 54–55, 64, 155, 467
patrol boats (PB Mk IIIs), 15
Paul Revere (APA 248), 312, 325, 326, 327
PC (patrol craft), 282–83
PCC (patrol craft, control), 283
PCE (patrol craft, escort), 233
PCEC (patrol craft, escort, control), 283
PCS (patrol craft, small), 282–83
PD-214 "Security class" ships, 474–75
Peleliu (LHA 5), 376
Pensacola (LSD 38), 333, 437, 438
personnel boats, 67–70, 99
PGM(K) (gunboat, hydrokeel), 397
Piasecki HRP-1, 348
planing-hull amphibian, 393, 395
platform improvements, 478
Plymouth Rock (LSD 2), 335
Pocono (AGC 16), 272, 275
Point Barrow (T-AKD 1), 469
Polana (AKA 35), 174
Pollux, 57
pontoons, 75, 137, 197–99, 338
Pontoon T-6, 197
Pontoon T-7, 197
Popper, Otto, 115
Portland (LSD 37), 15, 244
Prairie, 24
prepositioning, 13, 467, 470–75, 479
President Adams (AP 38/APA 19), 49, 61, 62, 64
President Cleveland, 55
President Coolidge, 55
President Garfield, 151
President Hayes (AP 39/APA 20), 49, 64
President Jackson (AP 37/APA 18), 49, 64
President Jackson-class, 54–55
President Madison, 65
President Monroe (APA 104), 151, 178
President Pierce, 55
President Polk (APA 103), 178
President Taft, 55
President Van Buren, 151
Prince Baudouin–class ferry steamers, 111
Princeton, 359, 360, 362
Procyon, 49, 58, 62
"Project Flattop," 457
Protector, HMS, 178–79
P-X-L designation, 103

"Quadrant" conference, 188
quadrupod masts, 64, 165, 315

radars: SK air search, 262; SPS-6B, 263
RA designation, 101
Raleigh (LPD 1, SCB 187), 364, 365
ramjet, 454
ramps: on boats, 83–84, 94; on LST, 121, 124, 199–200, 339–40
RAP (rocket-assisted projectile), 403, 454
RB designation, 101
RCT radios, 194
RDF (rapid deployment force), 472, 478
Ready Reserve Fleet (RRF), 467, 475–79
Reagan, Ronald, 14, 422–23, 446, 467, 477
reconnaissance boat Mk 2, 397
Red Bank boats, 77, 78
Refuge (AH 11), 30
repair ships, 126–27, 131
Republic, 54
Rhino ferries, 197, 199
rigid floating causeways, 197–99
River Clyde, 111
riverine troop carriers, 12
Riverside (APA 102), 275

RLTs (regimental landing teams), 6–7, 11, 369
Robinson, S. M., 117, 119
rocket launchers: Mk 20, 239; Mk 51, 239; Mk 102, 246–47; Mk 105, 247, 254, 256
rockets, shore bombardment, 405. See also rocket launchers
Rocky Mount (AGC 3), 265, 265
Roebling, Donald, 99–100, 215
Roebling, John, 99–100
Roosevelt, Franklin D.: and *Doyen*-class ships, 103, 105; and landing craft shortage, 157, 219–20; on LSTs, 120; and need for auxiliaries, 30; and transports, 41, 45, 51, 64
Roosevelt, Theodore, 4–5
Roper (APD 20), 37
ro-ro (roll-on roll-off), 320, 324, 468, 473, 476
RPVs (remotely piloted vehicles), 454
Ruchampkin (APD 89), 198
Rudderow-class destroyer escort, 195, 196
Russell, Richard B., 413
R3Y Tradewind, 348

SACEUR (Supreme Allied Commander, Europe) Order of Battle Report (30 September 1951), 11
SADARM (sense and destroy armor munition), 454
Saipan (LHA 2), 346
salvage landing craft, 208–9
Samuel Chase, 274
San Antonio (LPD 17), 365, 462, 463
San Antonio-class, 448, 458–61
Sangay (AE 10), 49
Santa Barbara, 43, 48. See also *McCawley* (AP 10/APA 4)
Santa Clara (AP 72), 8, 159
Santa Lucia (AP 73), 158, 159
Santa Maria, 43
Santa Rita, 39, 41
Saratoga (CV 60), 331
SC (subchaser), 281, 283
SCB 9, 303
SCB 9A, 303
SCB 14, 311
SCB 15, 311
SCB 15 AKA (Scheme B), 316
SCB 16, 311
SCB 17, 311, 327–29, 330
SCB 25, 383
SCB 32, 311
SCB 36, 254
SCB 37, 254, 256
SCB 75, 329–30
SCB 77A, 319, 320, 321–22
SCB 94, 292
SCB 119, 306, 307
SCB 122, 350
SCB 141, 292
SCB 149, 384, 386
SCB 152, 337
SCB 157, 353
SCB 187A, 365
SCB 221, 398
SCB 247, 340–41
SCB 404.65, 331–34, 334
SCB 409.68, 374
SCB project 60A, 300
SCB project 60B, 300
SCC (subchaser, control), 283
Schippersgracht, 422
Schlesinger, James, 14
scout boats, 279
scows (unpowered artillery lighters), 67

"Sea-Based Expeditionary Force–1980" (SEF-80), 457
Sea Bass, 163
Seabee (naval construction battalion), 197
Seabee (sea barge), 178, 444
Sea Knight, CH-46, 362, 439
sealift, 467–70, 477–79
Sea Lift (T-LSV 9/T-AK 278), *469*
Sea Otter, 136, 138
Sea Sheds, 478
sea sleds, 75, 77, 78
Seatrain New Jersey, SS, 113
Seatrains, *112–13*, 113–14
Secretary of the Navy, 3
Seminole (AKA 104), *193*, *286*
SES (surface effect ship), 397
Sheffer, A. T., 112
Shelby (APA 105), *170*
Shepherd, Lemuel C. Jr., 347
Sherburne (APA 205), *189*
Sheridan (APA 51), *165*
Sherman tanks, 88, 212
Sherwood, M. B., 143
Shinshu Maru, 36
Ship Acquisition and Improvement Panel (SAIP), 440, 442–43
Ship Characteristics Board (SCB), 13, 429
Short, Norman S., 320
SHP (shaft horsepower) steam plants, 199
Sikorsky CH-53, 362
Sikorsky HR2S (S-56), 362
skiffs, 78, 289
SLC (support landing craft), 223
SLFs (special landing forces), 11
SL-7s, 475, 476
Small Boat Development Board, 224–25
Smith, Holland M., 114, 216
smokers, 251
Snyder, Russell, 320
Somers, 77
Somme, 30
Sommervel, Brehon, 142
Southern Cross, 29, 31, 39, 41
Spica (AK 16), 36
Spruance-class destroyers, 412, *413*, 415, 417–22, 454
square, 37
S2-ST-23a designation, 469
St. Mihiel (AP 32), 30, 54, 182
Standing Committee on the Long Range Shipbuilding Program, 351
Standing Committee on the Shipbuilding and Conversion program, 359
Stark, Harold R., 39, 114, 117, 119
Starlight (AP 175), *185*
STOM (ship-to-objective maneuver), 457
submarines, 311, 348–49
Sumter (APA 52), *166*
Sumter-class, 163, *166*, 188
surf boats, 74, 76, 77, 78, 491
Susan B. Anthony (AP 72), *8*, 159
S-X-DY, 316, 317–18
Systems Commands, 13

Taconic (AGC 17), 275
T-ACS (MSTS/MSC crane ship), 477
TAC (tank assault carrier), 112
T-AH (MSTS/MSC hospital ship), 477
T-AKR designation, 470
TAKX (MSC multi-purpose mobilization ship), 446, 474–75
Talbot County (LST 1153), *201*, 305
Talladega (APA 208), 189, *190*
tank battalions, 128
tankers, 478

tank landing craft, *90*, *96*; 1938–42, 92–93; British, 111–18, 127–29; designations, 77, 90–91; development, 86–98; prototypes, 90
tanks: Christie amphibious, 18; Marmon-Herrington, 87; medium, 95; Sherman, 88, 212
Tarago, 479
Tarawa (LHA 1), 359, *375*, *377*, 377
Tartar antiaircraft missile, 353, 426
Tasker H. Bliss (AP 42), *42*, 157, 159
Taurus, USNS, 469
Taurus missile (ZRGM-59), 404
TAVB (MSC aviation deployment ship), 477
T-boats, 82
tenders, 126–27
Tentative Landing Force Manual, 75
Tentative Landing Operations Manual, 6
Terrebonne Parish (LST 1156), 305
Terror, 108, 110
Thetis Bay (CVHA 1), 350
Thomas Stone (AP 59), 151, *155*
Thomaston (LSD 28), 312, *331*
Thomaston-class, 331, *333*
Thurston (AP 77), 160, *160*, 178
Titania (AK 55/AKA 13), 151
TLC (tank landing craft), 111–18, 127–29
TLC-C (tank landing craft carrier), 115
TLL designation, 91
T-LSV, 468–70
T-LSV 8, 469
Todhunter, R. C., 127, 134
Tolland-class attack cargo ships, 191, *193*, 194
Tours, 30
tractor groups, 124
train ship, 412
Transdivs (transport divisions), 46
transports, 23–66 ; Army, 30; British inspired, 103–7; characteristics, 30–31, 37–39; convoy, 26; destroyers converted to, 33–36; expeditionary, 26; marine battalion, 105; merchant ships converted to, 31, 41, *42*, 61, 73–75, 111–12, 188; Navy, 7, 30; point-to-point, 26; quasi-attack, 187; special purpose, 35; wartime attack, 29. See also AK (cargo ship); AKA (attack cargo ship); AP (transport ship); APA (attack transport ship); APD (transport destroyer)
Treasury-class Coast Guard cutters, 265
Trenton (LPD 14), *366*, 368
Triplet, William S., 218
Tripoli (LPH 10), *354*, 359
TR (landing boat, vehicles), 84
troop barges program, 483
troopships, 29–30, 183–84, 478
trucks, amphibious, 100–101
Truman, Harry, 10, 314, 316
Tryon (APH 1), *64*, 65
Tryon-class, 65
TS (Higgins support boat), 223
Tulare (AKA 112), 312, 323, *324*, 325
Turner, Richmond Kelly, 216

UDTs (underwater demolition teams), 86, 195, 197, 211
Union (AKA 106), *194*
United States, 467
United States Shipping Board, 30–31
universal ship, 412
"USMC in 1985: A Long-Range Study," 433, 457
USNS (U.S. Naval Ship) designation, 468
Utah, 37

Valley Forge (LPH 8), 359, 360–62
Van Keuren, Alexander H., 140
vee-drives, 83, 84
vertical envelopment. See helicopter carriers
Vestal, 24
Vickery, Howard L., 119–20
Victory ship (VC 2): conversions to attack ships, 33, 188–90, *190*, 192–95, 315; conversion to FFDs, 470; reactivation for reserve, 467–68
VLC (vehicle landing craft), 84
VSS (VSTOL support ship), 380
VTOL (vertical take-off and landing) aircraft, 457–58

Wahkiakum County (LST 1162), *304*
Ward (APD 16), *196*
War Hawk (AP 168), *184*
War Shipping Administration (WSA), 161, 163, 186
Washington, 55
Washington Treaty (1922), 6
Washoe County (LST 1165), *306*
Wasp (LHD 1), *450*, 452
Watt, Richard M., 24, 25
W-boats, 90
Wedemeyer, Albert C., 140
Weiss (APD 135), *198*
Wells, Forrest H., 292
Weschler, Thomas R., 416
Westmoreland (APA 104), 275
Weymss, Henry C. B., 119
whaleboats, modified, 77–78
Wharton (AP 7), 29, 41
Whidbey Island (LSD 41), 313, *432*, *445*, 446
Whidbey Island-class, 444
White Marsh (LSD 8), 320
White River (LSMR/LFR 536), *259*
William P. Biddle (AP 15/APA 8), *47*, *50*
William Ward Burrows (AP 6), 39, 41
Wilson, Louis A., 14
Windsor-class transports, 163, *167*, *170*, 183
Winettes, 112–15, 117
Winged Arrow (AP 170), *184*
Winstons, 112
Winterhalter, Albert G., 25
WL-boats (light tank capacity), 90–91
WM-boats (medium tank capacity), 90–91
Woofus boats, 228, 246, *247*
Worcester (CL 144), 426
Wright (AZ 1), 30
Wyoming, 36

XAP (expeditionary transport ship), 28, 30, 159–60
XAPB barges, 74
XAPP barges, 73
XAPT barges, 73
XAS (submarine tender), 30
X-boats, 75
XCA (auxiliary cruiser), 30
XCM (minelayer), 30, 178–81
XCVs (aircraft carriers), 30

Y-boats, 75, 82
York County (LST 1175), *308*, *309*
YTL (yard [harbor] tank lighter), 134

ZA boats, 100
Z-boats, 100
Zeilin (AP 9/APA 3), 39, *39*, *40*, 274
Zumwalt-class land attack destroyers, 458

The Naval Institute Press is the book-publishing arm of the U.S. Naval Institute, a private, nonprofit, membership society for sea service professionals and others who share an interest in naval and maritime affairs. Established in 1873 at the U.S. Naval Academy in Annapolis, Maryland, where its offices remain today, the Naval Institute has members worldwide.

Members of the Naval Institute support the education programs of the society and receive the influential monthly magazine *Proceedings* and discounts on fine nautical prints and on ship and aircraft photos. They also have access to the transcripts of the Institute's Oral History Program and get discounted admission to any of the Institute-sponsored seminars offered around the country.

The Naval Institute also publishes *Naval History* magazine. This colorful bimonthly is filled with entertaining and thought-provoking articles, first-person reminiscences, and dramatic art and photography. Members receive a discount on *Naval History* subscriptions.

The Naval Institute's book-publishing program, begun in 1898 with basic guides to naval practices, has broadened its scope to include books of more general interest. Now the Naval Institute Press publishes about one hundred titles each year, ranging from how-to books on boating and navigation to battle histories, biographies, ship and aircraft guides, and novels. Institute members receive significant discounts on the Press's more than eight hundred books in print.

Full-time students are eligible for special half-price membership rates. Life memberships are also available.

For a free catalog describing Naval Institute Press books currently available, and for further information about subscribing to *Naval History* magazine or about joining the U.S. Naval Institute, please write to:

Membership Department
U.S. Naval Institute
291 Wood Road
Annapolis, MD 21402-5034
Telephone: (800) 233-8764
Fax: (410) 269-7940
Web address: www.navalinstitute.org